Contemporary Canadian

Business Law

PRINCIPLES AND CASES

TENTH EDITION

Contemporary Canadian Business Law

PRINCIPLES AND CASES

TENTH EDITION

John A. Willes, QC

BA, LLB, MBA, LLM

Barrister-at-Law

Barrister and Solicitor, Notary

Professor Emeritus, Queen's University

John H. Willes

BComm, LLB, MBA, LLM, CIM, FSALS

Barrister and Solicitor, Notary

Contemporary Canadian Business Law
Tenth Edition

The Internet addresses listed in the text were accurate at the time of publication. The inclusion of a website does not indicate an endorsement by the authors or McGraw-Hill Ryerson. The authors and McGraw-Hill Ryerson do not guarantee the accuracy of information presented at these sites.

ISBN-13: 978-0-07-040185-3
ISBN-10: 0-07-040185-3

1 2 3 4 5 6 7 8 9 0 QDB 1 9 8 7 6 5 4 3 2

Printed and bound in the United States of America

Care has been taken to trace ownership of copyright material contained in this text; however, the publisher will welcome any information that enables it to rectify any reference or credit for subsequent editions.

DISCLAIMER
The wide range of topics in a text of reasonable size limits the treatment of the law to only the most general statements of what are often complex, specialized, and constantly changing areas of the law; consequently, the text content is not intended nor should it in any way be considered or used as a substitute for the prompt and timely advice of competent legal counsel in the jurisdictions concerned. No professional relationship is created, nor legal opinion rendered, between the authors and any user of this text. The names and facts used in discussion cases and examples are entirely fictitious, and any similarity to persons or corporations is entirely coincidental.

Sponsoring Editor: *James Booty*
Marketing Manager: *Cathie Lefebvre*
Developmental Editor: *Kamilah Reid-Burrell*
Supervising Editor: *Cathy Biribauer*
Photo/Permissions Researcher: *Allison McDonald*
Editorial Associate: *Erin Catto*
Copy Editor: *Tracey Haggert*
Production Coordinator: *Tammy Mavroudi*
Cover and Inside Design: *Liz Harasymczuk*
Composition: *Bookman Typesetting Co. Ltd.*
Cover Image: *Glen E. Farrelly (Pillars of Justice by Edwina Sandys)*
Printer: *Quad/Graphics*

Library and Archives Canada Cataloguing in Publication Data

Willes, John A.
Contemporary Canadian business law : principles and cases / John A. Willes, John H. Willes. — 10th ed.

Includes bibliographical references and index.
ISBN 978-0-07-040185-3

1. Commercial law — Canada — Textbooks.
I. Willes, John H. II. Title.

KE919.W54 2012 346.7107 C2011-903995-8
KF889.W54 2011

John A. Willes

John A. Willes is an Ontario Barrister and Solicitor, and Emeritus Professor of Labour Relations and Business Law at the School of Business, Queen's University. He was called to the Bar of Ontario in 1960, and joined the faculty of Queen's University on a full-time basis in 1969, where he assumed responsibility for the business law program. Throughout his teaching career, he carried on an extensive commercial and labour arbitration practice, practised law as Counsel to a Kingston, Ontario, law firm, and from 1986 to 2000, acted as a Vice-Chair of the Ontario Public Service Grievance Board. During his long career as a lawyer, he has provided legal advice to many clients with extensive business interests in Canada, the United States, and abroad. He was appointed as a Queen's Counsel by the Lieutenant Governor of the Province of Ontario in 1984.

He holds a Bachelor of Arts degree from Queen's University, Bachelor of Laws and Master of Laws degrees from Osgoode Hall Law School, York University, and a Master of Business Administration degree from the University of Toronto.

In addition to *Contemporary Canadian Business Law* (10th ed.) (McGraw-Hill Ryerson), he has authored numerous academic monographs and cases, and is the author of *Contemporary Canadian Labour Relations* (McGraw-Hill Ryerson), *Canadian Labour Relations* (Prentice Hall), and *Out of the Clouds*, the official military history of the First Canadian Parachute Battalion in World War II. Professor Willes is the co-author with John H. Willes of *International Business Law* (McGraw-Hill, U.S.A., 2004), and *Fundamentals of Canadian Business Law* (2nd ed.) (McGraw Hill Ryerson, 2008).

John H. Willes

John H. Willes is an Ontario Barrister and Solicitor. For many years, he taught at Queen's University in both the Faculty of Law and the School of Business, in the LLB, BComm, and MBA programs. He is the former co-ordinator and principal instructor of the International Business and International Law Programme at Herstmonceux Castle International Study Centre in the United Kingdom.

He holds Bachelor of Commerce, Bachelor of Laws, and Master of Business Administration degrees from Queen's University, and a Master of Laws from Vrije University, Brussels, Belgium. He also holds a Canadian Investment Manager designation from the Canadian Securities Institute. He was elected as a Fellow of the Society for Advanced Legal Studies, in London, England, and was a Visiting Fellow at the University of London, England, in 2002.

In addition to serving clients in North America and Western Europe, his business activities included advising on legislative transition in the republics of the former Soviet Union and enterprise restructuring and business management in the People's Republic of China. He is a past member of the editorial advisory boards of *Financial Crime Review* and the *European Financial Law Review*. He is the co-author, with John A. Willes, of *International Business Law* (McGraw-Hill, U.S.A., 2004), and *Fundamentals of Canadian Business Law* (2nd ed.) (McGraw Hill Ryerson, 2008).

BRIEF CONTENTS

TABLE OF CONTENTS

PREFACE

This 10th Edition of *Contemporary Canadian Business Law* also marks its 30th anniversary providing students with a clear understanding of the legal environment for business professionals, owners and managers in Canada.

This text provides students with the legal knowledge and edge they require, matched to the course offerings of Canada's universities and colleges. This text covers all the core topics of business law, as well as many emerging topics, allowing instructors to design courses appropriate for the particular needs and interests of their students and business program.

Covering the full span of business law in the everyday commercial world, the text is divided into six parts. Part 1 introduces the law and the legal system, establishing the nature of law and its system of administration. Part 2 delves into torts, one of the oldest and most interesting areas of the law, and one that rapidly comes to the fore when business ventures cause injury to others. Part 3 looks at the heavy-lifting of commercial relationships, the law of contract, while Part 4 examines the various forms of business organizations as well as common commercial relationships. Part 5 explores the rights and responsibilities associated with property, including intellectual property. Part 6 treats a variety of special legal rights and relationships, from consumer protection and bankruptcy to environmental law and international legal issues and obligations.

Contemporary Canadian Business Law adopts a learning goals approach to the law, with clear learning goals leading the way as each chapter opens. In addition to clarity in development and explanation, scope, depth, interest and debate are further fuelled through use of special features. These range from management advisories and checklists, to ethical questions, case summaries and topical media items that will be familiar to many students. Tying it all together, review questions, mini-cases and discussion cases provide not only review, but drive home the application of legal principles and rules to business problems.

JAW
JHW

A Student's Guide

Contemporary Canadian Business Law, 10th edition, offers pedagogical features to help students learn and apply the concepts found throughout the text. Each chapter is organized to enhance the learning process, and includes a concise outline of the important business and legal principles relating to each topic. The following will help to guide you through the text.

Part Openers

The text is organized into six parts and each part opens with a list of the chapters that are included within the part.

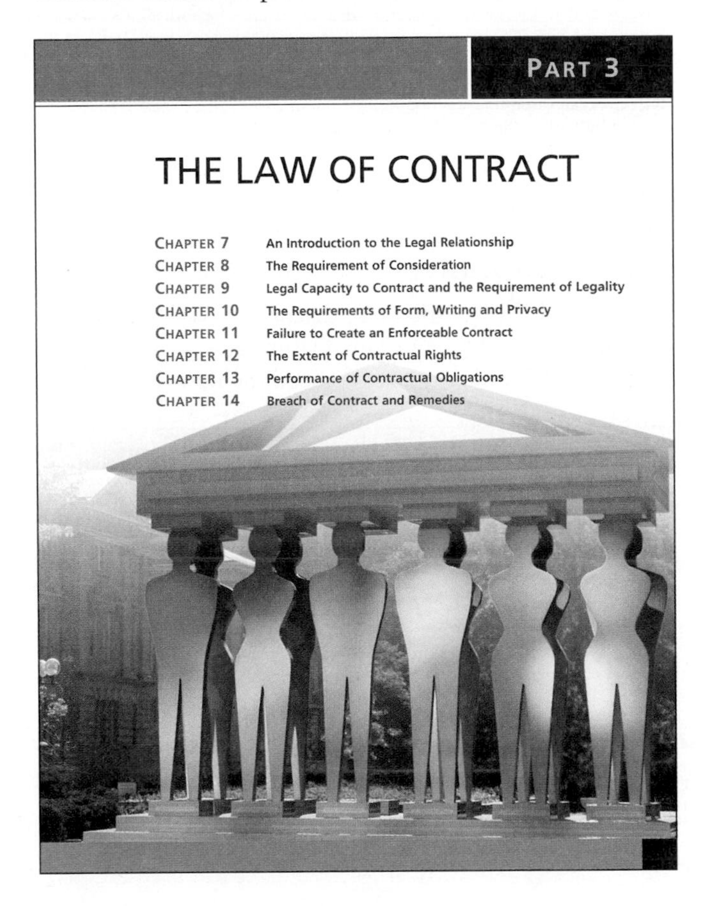

Chapter Objectives

Each chapter begins with a list of its primary learning objectives. This useful tool enables students to see what they should expect to learn in each chapter.

Chapter Objectives

After study of this chapter, students should be able to:

- Explain how the legal doctrine of mistake may result in the failure to create an enforceable contract.
- Distinguish between different kinds of misrepresentation, and the results of each.
- Explain the difference between undue influence and duress, and the effect of these factors on enforceability of a contract.

Case In Point

Case In Point features appear in most chapters. Case In Point boxes give a brief description of a recent legal decision that affects the business world, reinforcing the lessons learned in the text and the Court Decisions.

CASE IN POINT

A dentist moved his dental practice to a city mall, and announced the move in the local newspaper. As a result, the mall placed a "welcome" advertisement in the newspaper, and a reporter interviewed him. The reporter's article described the new office, and included a photograph of the dentist. When the provincial dental board became aware of this, it began disciplinary action against the dentist on the basis that the publicity violated the board's rule which prohibited this form of advertising. The dentist made an application to the provincial Supreme Court, which stayed the board's actions against him as he had neither solicited nor paid for the article but was only exercising his freedom of expression.

Carmichael v. Provincial Dental Board of Nova Scotia, 1998 CanLII 1773 (N.S.S.C).

A Question of Ethics

This box raises ethical questions for discussion with respect to particular business activities and compliance with the law. This feature acts as a springboard for exploration of legal issues in business that are in today's headline news and are timeless.

A QUESTION OF ETHICS

Exemption clauses can allow one party to escape performance or liability while compelling the other party to perform. In interpreting them, judges exercise a certain degree of social conscience. If you were a judge, what persuasive factors would enter your mind in deciding to uphold such a clause, or to strike it down?

Court Decisions

Court Decisions convey to students the overall importance of "classic" cases in the development of the law and their continuing relevance to business today.

COURT DECISION: Practical Application of the Law

Contract—Tender in Error—Right to Withdraw—Reliance on Tender
***Gloge Heating and Plumbing Ltd. v. Northern Construction Co. Ltd.*, 1986 ABCA 3 (CanLII).**

Gloge submitted a tender to do certain mechanical work for Northern Construction. Northern Construction entered into a construction contract to do work on the strength of Gloge's bid. Gloge later discovered that the bid was too low and refused to do the work. Northern Construction was obliged to have the work done by another company but the cost was much higher. Northern Construction then brought an action against Gloge for the amount of its loss, and was successful. Gloge appealed.

THE APPEAL COURT: ... The owner [of the construction project] duly accepted Northern's tender and awarded it the work. Gloge refused to perform the mechanical subcontract, so that Northern was required to make alternative arrangements by employing a subsidiary at an increase of $341,299 in the mechanical subcontract price. Was Gloge's tender revocable after close of tender? Gloge knew that Northern would select a mechanical tender and rely on it, and Gloge also knew that the tenders of the general contractors to the owner would be irrevocable for the time set out in the contract documents. Perhaps these facts of themselves might justify holding the Gloge tender to be irrevocable. But, in addition, the trial judge accepted certain expert evidence given at the trial by two witnesses who had been in the construction industry for many years. That evidence demonstrated that it was normal and standard practice for general contractors to accept last minute telephone tenders from subcontractors, and that it was understood and accepted by those in the industry that while such tenders could be withdrawn, that such tenders must remain irrevocable for the same term that the general contractors' tenders to the owner are irrevocable ....

Front-Page Law

This feature takes its inspiration from headlines in the media. The authors have identified the relevant topic that relates to the news article, explained the legal/business context, and included a question for discussion.

FRONT-PAGE LAW | **Yes, our chimneys kiss. We like it that way. Life on a tilt: Faulty foundations on street in Beaches root of the problem**

Chris Goddard places a ball on his living room floor and watches as it rolls to the south side of his Beaches home.

This is no poltergeist at work but rather a structural problem that has left his home, and a handful of others along the southeast side of Glen Manor Drive, slightly tilted to one side.

"It was one of the features and characteristics that drew us to it," he said.

The crookedness also had little effect on his neighbours, who were unable to resist the posh two-storey duplexes with a beach for a backyard. Unfortunately, they did not put in proper foundations to counter the effects of a stream that ran below.

The tilt on some homes is more obvious than on others. For example, Derek Ferris's home and his neighbour's home lean toward each other, causing his eavestrough to fit snugly underneath his neighbour's and their chimneys to kiss.

Because both homes are effectively trespassing on each other's property, owners have "encroachment agreements."

He did have some initial concerns when he first saw the place, but those worries were soon dismissed after the house was checked out by a structural engineer and given a clean bill of health.

"Any settlement that will take place will occur in the first few years of life," said John Zimnoch, a sales associate with Re/Max.

Mr. Zimnoch said the crookedness does slightly affect the value of the homes and limits the number of prospective buyers.

"A lot of buyers will come along and see the houses and think there is probably a structural problem," he

Your Business At Risk

Most chapters open with a feature example that relates the chapter content to the management of risk in a business organization.

YOUR BUSINESS AT RISK

Negligent acts and accidents happen. Businesses must be aware of when, how, and to what degree they will be responsible for unintentional or careless acts, and to the extent that they may recover for careless acts done to them. A single liability suit for an accident that could have been avoided can run to millions of dollars and may bankrupt the dreams of the owner or investors.

Management Alert: Best Practice

These features illustrate risky situations in which business law principles can be used to avoid serious hazards or seize business opportunities.

MANAGEMENT ALERT: BEST PRACTICE

Business managers can take concrete steps toward reducing the risk of negligence and liability.

- Make employee safety a genuine corporate priority.
- Mark products with serial numbers to allow tracking and recall. Take proactive steps for safety's sake.
- Document business processes into manuals for employee guidance.
- Document and discipline employee misbehaviour.
- Use appropriate warning labels.
- Archive customer correspondence and complaints.
- Participate actively in industry and professional associations to keep abreast of best practices.
- Devote meaningful resources to employee training.
- Instill pride in a job well done. Refuse to reward poor performance.
- Keep complete production records.
- Record sources and dates of raw materials and supplies.
- Conduct in-house product testing.
- Record and learn from feedback and problems.
- Issue complete instruction manuals, documentation and registration cards to purchasers.

Clients, Suppliers or Operations in Quebec

Businesses in the Common Law provinces that deal with Quebec firms will be subject to the rules imposed by Quebec's *Civil Code*. To emphasize this reality, the Clients, Suppliers or Operations in Quebec feature illustrates how common transactions may be treated differently in Quebec.

CLIENTS, SUPPLIERS *or* OPERATIONS *in* QUEBEC

Most of the Common Law jurisdictions in Canada have or are moving toward a basic two-year limitation period for commencement of most civil actions (from the date at which the claimant became aware or should have become aware of the wrong). The *Civil Code of Quebec*, however, provides a wide array of limitation periods ranging between one year and ten years for civil actions, for action commenced in that province. See CCQ, Book Eight, Prescription.

Checklists

These numbered lists substantiate the key points in the chapter.

CHECKLIST FOR INTENTIONAL TORTS

1. Was an act committed that is prohibited by Common Law?
2. Was it intentional?
3. Was it outside of a contract (or trust)?

If the answer is "yes" to all three, an intentional tort exists. Regardless, other liability may still exist under criminal law, contract law, or the law of negligence (unintentional torts).

Shaded Examples

Examples are shaded throughout the text to reinforce the chapter material.

Alport and Bush enter into an agreement that will provide a benefit for Cooke, who is not a party to the agreement. Under the privity of contract rule, Cooke could not enforce the promise to confer the benefit. Only the person who gave consideration for the promise could enforce the agreement. However, if that person refuses to take court action, Cooke would be entitled to do so. The reasoning here is that if the trustee of the benefit refused or was unable to act to enforce the promise, this in turn would affect Cooke's rights as beneficiary of the trust. This would also be the case if the trustee, once in possession of the funds or benefit, refused to confer the benefit on the beneficiary.

Charts and Diagrams

Charts and diagrams have been included throughout the text serving to illustrate and clarify important concepts.

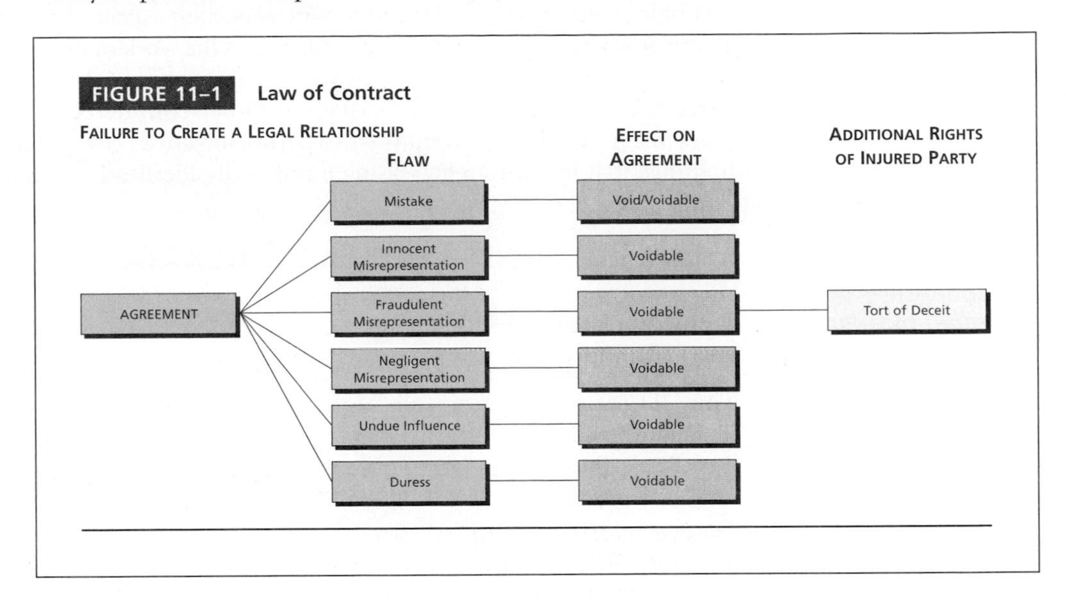

End of Chapter Material

Chapter Summary

A summary at the end of each chapter reviews the important concepts of the chapter.

Key Terms

Within each chapter, key terms are highlighted in boldface type when they first appear, with a running glossary in the margins of the text for quick reminders. For reference, there is a comprehensive list of key terms at the end of each chapter with page references for each term, and a full glossary at the back of the text.

Review Questions

These questions assist students with their review of the chapter material.

Mini-Case Problems

The brief mini-case problems allow the students to apply the concepts learned in the chapter.

Case Problems for Discussion

Each chapter concludes with extensive case material that offers students the opportunity to apply the law to specific fact situations and arrive at their own conclusions as to how a case should be decided.

Weblinks

Weblinks are now found online at the ***Online Learning Centre***, mcgrawhill.ca/olc/willes. Canadian courts and legislatures have recognized that public electronic access to the full text of legal judgments and statutes is an important aspect of access to justice, and virtually all are now available via the Internet. Of note is the work of the Canadian Legal Information Institute (CanLII) whose Web site, www.canlii.org, provides free comprehensive access to current Canadian cases and legislation. Other commercial data banks cross-reference cases in a number of different ways to permit advanced computer-accessible searches, with holdings that include an increasing number of older leading cases.

Comprehensive Learning and Teaching Package

For the Instructor:

Instructor's Online Learning Centre (OLC)
The OLC at www.mcgrawhill.ca/olc/willes includes a password protected Web site for instructors. The site offers downloadable supplements, newsletters, the computerized Test Bank, PowerPoint presentations, Instructor's Manual, and Case Citator.

Course Management
McGraw-Hill Ryerson offers a range of flexible integration solutions for Blackboard, WebCT, Desire2Learn, Moodle and other leading learning management platforms. Please contact your local McGraw-Hill Ryerson *i*Learning Sales Specialist for details.

Integrated Learning
Your Integrated Learning Sales Specialist is a McGraw-Hill Ryerson representative who has the experience, product knowledge, and training to provide the support to help you assess and integrate any of our products, technology, and services into your course for optimum teaching and learning performance. The Integrated Learning Sales Specialist can show you tools that will help your students improve their grades, or help you put your entire course online — your *i*Learning Sales Specialist is there to help. Contact your local *i*Learning Sales Specialist today to learn how to maximize all of McGraw-Hill Ryerson's resources.

iLearning Services Program
At McGraw-Hill Ryerson, we take great pride in developing high-quality learning resources while working hard to provide you with the tools necessary to utilize them. We want to help bring your teaching to life, and we do this by integrating technology, events, conferences, training, and other services. We call it *i*Services. For more information, visit **www.mcgrawhill.ca/olc/iservices**.

Teaching, Learning and Technology Conference Series
The educational environment has changed tremendously in recent years, and McGraw-Hill Ryerson continues to be committed to helping you acquire the skills you need to succeed in this new milieu. Our innovative Teaching, Technology and Learning Conference Series brings faculty from across Canada together with 3M Teaching Excellence award winners to share teaching and learning best practices in a collaborative and stimulating environment. Pre-conference workshops on general topics, such as teaching large classes and technology integration, will also be offered. We will also work with you at your own institution to customize workshops that best suit the needs of your faculty.

McGraw-Hill's Create Online gives you access to the most abundant resource at your fingertips—literally. With a few mouse clicks, you can create customized learning tools simply and affordably. McGraw-Hill Ryerson has included many of our market-leading textbooks within Create Online for eBook and print customization as well as many licensed readings and cases. For more information, go to **www.mcgrawhillcreate.com**.

Instructor's Manual

Prepared by the authors, each chapter presents the chapter objectives, chapter commentaries to assist in lecture presentation, reproductions of the figures from the text, and answers to *Review Questions* and *Mini-Case Problems*. Comments are also provided for each of the *Case Problems for Discussion*.

EZ Test

The test bank, which contains over 1,000 multiple-choice, true or false, and essay questions, is provided in McGraw-Hill Ryerson's EZ Test. This electronic testing program is flexible, easy to use, and allows instructors to create tests from book-specific items. It accommodates a wide range of question types and instructors may add their own questions. Multiple versions of the test can be created and any test can be exported to use with course management systems such as WebCT and Blackboard. The program is available for Windows and Macintosh environments.

Microsoft® PowerPoint® Presentations

This presentation system offers material that can be edited and manipulated to fit a particular course format. Figures included in the text have been included in this package.

Case Citator

The updated Case Citator presents a reference list of the most current case law divided by Canadian jurisdictions and points of law. The Case Citator is organized on a regional basis and citations from recent cases relate to chapter topics. It is available as a downloadable supplement from the Instructor's Centre of the OLC.

For the Student:

The Online Learning Centre at www.mcgrawhill.ca/olc/willes offers aids for each chapter including: Learning Objectives, Quick Quizzes, Internet Application Questions, Newsletters, Key Terms, and a Searchable Glossary.

Acknowledgements

Text revisions require constructive comments from users as well as from others in the academic community, and we, the authors, are grateful to many reviewers who have provided suggestions and advice to assist us in the preparation of this new edition. Their feedback and comments during the review process were most appreciated, and to them we give credit for making this edition a more useful and responsive text for the business law student. In particular, our special thanks go to the following reviewers who took the time to review our manuscript and provide us with their kind advice:

Douglas Beatty, *Lambton College of Applied Art & Technology*
Mark D. Bridge, *University of Victoria*
John Cavaliere, *Sault College*
George M. Cummins, *Memorial University of Newfoundland*
John Lane, *Holland College*
Hugh Laurence, *University of Toronto*
Ralph Lembcke, *Fanshawe College*
Sandra Malach, *University of Calgary*
Gerald E. Palmer, *University of the Fraser Valley*
Jim Silvos, *Mount Royal University*

As always, our very special thanks go to Fran Willes. For all ten editions she has not only managed the many administrative details associated with the preparation of the manuscript for publication with unflagging dedication, but also used her unique management skills to organize the authors and their work to ensure that production deadlines were met. Our special thanks also go to Tracey Haggert, our copy editor, for her skill, enthusiasm and professionalism in her careful editing work on the text. Special thanks also go to the McGraw-Hill Ryerson staff: James Booty, Sponsoring Editor, Kamilah Reid-Burrell, Developmental Editor, and Cathy Biribauer, Supervising Editor for their kind support and encouragement in the preparation of the text for publication. Our thanks as well to Liz Harasymczuk, cover and interior designer, and Chris Hudson at Bookman Typesetting for their professional and thorough work on the layout and design.

This text provides a general outline of the current law. However, the constraints imposed by the wide range of subject matter permit only a very limited treatment of what are often rapidly changing, highly specialized and complex areas of the law. The text is in no way intended, and should not be used as a substitute for the advice of legal counsel in the jurisdictions concerned.

JAW / JHW

Abbreviations of Law Reports

A. & E.	Adolphus and Ellis Reports	(UK)
A.C.	Law Reports Appeal Cases	(UK)
A.P.R.	Atlantic Provinces Reports	
A.R.	Atlantic Reports	
All E.R.	All England Reports	(UK)
Alta. L.R.	Alberta Law Reports	
App. Cas.	Law Reports Appeal Cases	(UK)
Atk.	Atkyn's Chancery Reports	(UK)
Atl.	Atlantic Reports	
B. & C.	Barnewall & Cresswell Reports	(UK)
B. & S.	Best & Smith Reports	(UK)
B.C.L.R.	British Columbia Law Reports	
B.C.W.L.D.	British Columbia Weekly Law Digest	
B.L.R.	Business Law Reports	
Black.W. (Bl.W.)	William Blackstone Reports	(UK)
C. & P. /Car. & P.	Carrington & Payne's Nisi Prius Reports	(UK)
C.B.R.	Canadian Bankruptcy Reports	
C.C.C.	Canadian Criminal Cases	
C.C.E.I.	Canadian Cases in Employment and Immigration	
C.C.E.L.	Canadian Cases in Employment Law	
C.C.L.	Canadian Current Law	
C.C.L.I.	Canadian Cases on the Law of Insurance	
C.C.L.S.	Canadian Cases on the Law of Securities	
C.C.L.T.	Canadian Cases on the Law of Torts	
C.C.P.B.	Canadian Cases on Pensions and Benefits	
C.E.L.R.	Canadian Environmental Law Reports	
C.E.R.	Canadian Customs and Excise Reports	
C.H.R.R.	Canadian Human Rights Reporter	
C.I.P.R.	Canadian Intellectual Property Reports	
C.L.L.C.	Canadian Labour Law Cases	
C.L.R.	Construction Law Reports	
C.L.R.B.R.	Canadian Labour Relations Board Reports	
C.O.H.S.C.	Canadian Occupational Health and Safety Cases	
C.P.	Law Reports Common Pleas	(UK)
C.P.C.	Carswell's Practice Cases	
C.P.D.	Law Reports Common Pleas Division	(UK)
C.P.R.	Canadian Patent Reporter	
Ch. App.	Law Reports Chancery Appeal	(UK)
Ch.	Law Reports Chancery	(UK)
Ch.D.	Law Reports Chancery Division	(UK)
Co. Rep.	Coke Reports	(UK)
Cranch	Cranch's United States Supreme Court Reports	
D.L.R.	Dominion Law Reports	
De G.F. & J.	De Gex Fisher & Jones Chancery Reports	(UK)
De G.M. & G.	De Gex Macnaughten & Gordon's Bankruptcy Reports	(UK)
E.R.	English Reports	(UK)
E.T.R.	Estates and Trusts Reports	
Exch.	Exchequer Reports	(UK)
F.	Federal Reporter	(USA)
F.C.	Federal Cases	
F.T.R.	Federal Trial Reports	
Godb.	Godbolt Reports	(UK)

H. & C.	Hurlstone & Coltman Reports	(UK)
H.L. Cas.	Clark's House of Lords Cases	(UK)
Hare	Hare Reports	(UK)
Ill. App.	Illinois Appellate Court Reports	(USA)
J. & H.	Johnson & Hemming Reports	(UK)
Jur. N.S.	Jurist Reports, New Series	(UK)
K.B.	Law Reports King's Bench	(UK)
L.A.C.	Labour Arbitration Cases	
L.C.R.	Land Compensation Reports	
L.R.B.R.	Labour Relations Board Reports	
L.R.C.P	Law Reports Common Pleas	(UK)
L.R. Exch.	Law Reports Exchequer	(UK)
L.R.H.L.	Law Reports House of Lords	(UK)
M. & W.	Meeson & Welsby's Exchequer Reports	(UK)
M.P.L.R.	Municipal And Planning Law Reports	
M.P.R.	Maritime Provinces Reports	
M.V.R.	Motor Vehicles Reports	
Man. R.	Manitoba Reports	
Mod.	Modern Reports	(UK)
My. & Cr.	Mylne & Craig Reports	(UK)
N.B.R.	New Brunswick Reports	
N.E.	North Eastern Reporter	(USA)
N.R.	National Reporter	
N.S.R.	Nova Scotia Reports	
N.W.T.R.	North West Territory Reports	
N.Y.	New York Reports	(USA)
Nfld. and P.E.I.R.	Newfoundland and Prince Edward Island Reports	
Nfld. R.	Newfoundland Reports	
O.A.C.	Ontario Appeal Cases	
O.L.R.	Ontario Law Reports	
O.L.R.B. Rep.	Ontario Labour Relations Board Reports	
O.R.	Ontario Reports	
O.T.C.	Ontario Trial Cases	
O.W.N.	Ontario Weekly Notes	
P. (Pac.)	Pacific Reporter	(USA)
P.E.I.R.	Prince Edward Island Reports	
P.P.S.A.C.	Personal Property Security Act Cases	
Popham (Pop.)	Popham Reports	(UK)
Q.B.	Law Reports Queen's Bench	(UK)
Q.B.D.	Law Reports Queen's Bench Division	(UK)
R.J.Q.	Quebec Law Reports	
R.P.A.	Quebec Practice Reports	
R.P.R.	Real Property Reports	
S.C.R.	Supreme Court Reports	
Sask. R.	Saskatchewan Reports	
Sty.	Style Reports	(UK)
T.C.T.	Trade Commission Tribunal	
T.L.L.R.	Tenant and Landlord Law Reports	
T.L.R.	Times Law Reports	(UK)
T.T.R.	Trade and Tariff Reports	
W.C.A.T.R.	Workers' Compensation Appeals Tribunal Reporter	
W.W.R.	Western Weekly Reports	
Y.B.	Year Book	(UK)

Abbreviations of Legislation

C.C.C.	Criminal Code of Canada
C.C.S.M.	Continuing Consolidation of the Statutes of Manitoba
C.S. <province>	Consolidated Statutes of <province>
Edw.	Edward, King of England, followed by Roman numeral
Eliz.	Elizabeth II, Queen of England
Geo.	George, King of England, followed by Roman numeral
Hen. IV	Henry IV, King of England
R.S. <province>	Revised Statutes of <province>, followed by year of revision
R.S.C.	Revised Statutes of Canada, followed by year of revision
S. <province>	Statutes of <province>, followed by year of enactment
S.C.	Statutes of Canada, followed by year of enactment
Vic.	Victoria, Queen of England
Wm. IV	William IV, King of England

Abbreviations of Courts of Canada or a Province, Appearing in Neutral Citation

SCC	Supreme Court of Canada
FC	Federal Court of Canada
FCA	Federal Court of Appeal
TCC	Tax Court of Canada
CACT	Competition Tribunal
ABCA	Alberta Court of Appeal
ABQB	Alberta Court of Queen's Bench
BCCA	British Columbia Court of Appeal
BCSC	British Columbia Supreme Court
MBCA	Manitoba Court of Appeal
MBQB	Manitoba Court of Queen's Bench
NBCA	Court of Appeal of New Brunswick
NBQB	Court of Queen's Bench of new Brunswick, Trial Division
NLCA	Supreme Court of Newfoundland and Labrador (Court of Appeal)
NLTD	Supreme Court of Newfoundland and Labrador (Trial Division)
NSCA	Nova Scotia Court of Appeal
NSSC	Nova Scotia Supreme Court
NUCA	Nunavut Court of Appeal
NUCJ	Nunavut Court of Justice
NWTCA	Court of Appeal for the Northwest Territories
NWTSC	Supreme Court of the Northwest Territories
ONCA	Court of Appeal for Ontario
ONSC	Ontario Superior Court of Justice
ONSCDC	Ontario Superior Court of Justice, Divisional Court
PESCAD	Prince Edward Island Supreme Court, Appeal Division
PESCTD	Prince Edward Island Supreme Court, Trial Division
QCCS	Quebec Superior Court
QCCA	Quebec Court of Appeal
SKCA	Court of Appeal of Saskatchewan
SKQB	Court of Queen's Bench of Saskatchewan
YKCA	Court of Appeal of the Yukon Territory
YKSC	Supreme Court of the Yukon Territory

Table of Cases

TABLE OF STATUTES

THE LEGAL ENVIRONMENT FOR BUSINESS

CHAPTER 1

The Law and the Legal System

Chapter Objectives

The rights and obligations of businesses and business persons stem from the law and our legal system. After study of this chapter, students should be able to:

- Describe the sources, role and development of Canadian law.
- Distinguish between statute and Common Law, and describe the significance of *stare decisis*.
- Recognize matters of federal versus provincial jurisdiction.
- Describe the fundamental rights and freedoms set out in the *Charter of Rights and Freedoms*.

LEARNING THE LAW

You are a player in the game of life, and Canadian society has its own set of rules for playing the game. Many of these rules — laws — you know already through education, experience or the media. However, you probably know there are gaps in your education and experience in this field, and the media tends to dwell on only the sensational stories.

Together we will bridge this gap, for business law is important to business persons. To begin with, this knowledge keeps your business out of trouble. Better yet, you can use the law and your rights to advance your business interests. You can be sure you will encounter people who are well aware of *their* rights — your customers, employers and investors, to name three — and they will be unforgiving if you and your business do not know the business law applicable to your enterprise.

There are some peculiarities, however, about learning the law. For instance, you will notice that some of the cases in this book are centuries old, while others are only a few months old. This is not a random choice. Old cases making clear statements of the law are the "fine wine" of our legal system. They have withstood the test of time, and are used daily in our courts. New cases become important when the legal principles of our society change. Thus you will find a case in this book from the year 1348 — when it was established that it was a bad thing to throw an axe at someone. It is still a bad thing to chuck nasty objects at people, so the case remains a valid statement of the law. You will also find cases from 2011, for example, an appeal before the Supreme Court of Canada by a British Columbia resident, regarding cell phone airtime charges. Old laws or new developments, we include cases that display the current law as clearly as possible.

Clarity is important and it comes from understanding the big picture as well as necessary details. It is nice to learn that there is a good restaurant "on the east side of town," but it is even more useful to get the address. Rest assured that we will provide you with the big picture, and we will give you the details where they are important. In fact, when we plunge into some serious detail, that is your cue to watch out for things that can have a serious impact on your business.

One final thought on the big picture: many of our chapters have an historical section telling you the roots of today's law. We are not history buffs, but these sections are included because they are sometimes the best explanation of why rather strange aspects of the law exist today. In addition, knowing where the law has come from is a big help in understanding where it is going — which makes this book useful long after the course is over.

The Legal Environment for Business

In Canada, one would expect that any individual or group of individuals would be free to establish and carry on a business activity in any manner that they see fit. This is not the case, however. Modern business operates in an exceedingly complex legal environment. Laws govern the formation of many types of business organizations, the products they may manufacture or sell, the conditions under which the employees of a business may work, the relationships between customers and competitors and, indeed, between the business owners themselves. These laws present a complicated web of rules and regulations for the business person, and they may ensnare the unwary as well as those who deliberately ignore them.

Very briefly, business law may be divided into a number of general areas. These include tort law, which represents an area of the law largely concerned with injury to others. These injuries may arise as a result of the negligent production of goods causing injury to the consumer, careless professional service causing physical or economic loss, unsafe operating premises, or injury to others in a myriad of other ways either directly or through the actions of employees. A second major area of the law is concerned with the basis of most business activity. This is contract law, and represents the law that has, perhaps, the greatest application in the day-to-day operations of a business organization. Contract law facilitates the purchase and sale of goods, the employment of staff, the reduction of risk (through insurance contracts), and some forms of organization of the business itself. Special contractual relationships and laws that control these and other business relationships include bailment, labour law and employment, negotiable instruments, consumer protection, and the law relating to restrictive trade practices.

A knowledge of the law relating to the formation and operation of business organizations is also important in order to determine the appropriate vehicle by which to conduct business activities. For this reason, an understanding of partnerships and corporations is essential. Also important is a knowledge of how premises are acquired for business operations. Land law, which covers the purchase or leasing of premises or the financing of the purchase of land and buildings, sets out the rules for these business activities. The final areas of the law that are of interest to business include the securing of debt in credit transactions, bankruptcy law, international trade, environmental law, and the protection of intellectual and industrial property. Persons engaged in business must have at least a rudimentary knowledge of these areas of the law in order to function effectively in a business environment.

Many of these laws simply reflect behavioural norms — how fair-minded people would expect to be treated by others — but some laws go much further in that they are designed to enforce or control policy. Legislation that requires the licensing of the various professions or laws that control the possession or use of certain goods, such as drugs or explosives, are examples of the latter policies.

It is important for all business persons to be aware of the areas of the law that affect them in the conduct of their business activities, not only from the point of view of knowing their rights at law, but to ensure their own compliance with all relevant legislation.

The Nature of Law

The law holds a special fascination for most people. This is perhaps partly due to the fact that the law has become so all-pervasive that we are constantly reminded of its presence. This cannot be the only reason, however. The law is also of interest because it reflects the society in which we live. It determines the rights and freedoms of the individual, and the extent to which privileges may be enjoyed. At any given time the law represents the values and concerns of the people in the jurisdiction from which it arose, and examining the historical development of the laws of a society is much like examining the society itself, for the two are inextricably intertwined. The law touches so many facets of human endeavour that it would be difficult to imagine how a modern society might exist without it. It has become, in fact, the very essence of society. This in itself is reason enough to justify the study of its principles and application.

Definitions of "Law"

The word "law" has been applied to so many rules, principles, and statements that it is probably incapable of exact definition. Legal philosophers have agonized over the meaning of the term and have wrestled with its sources and nature since the earliest of times. Part of the difficulty in reaching a precise definition is the nature of law itself. It is very much a concept rather than an object or thing that has clearly defined limits or parameters. Simple definitions may be attempted, however, bearing in mind that the definition may not be precise or all-encompassing.

Irwin Dorfman, a prominent Canadian lawyer, explained the term very simply in a speech to the Canadian Bar Association in 1975. He defined the law as "merely a set of rules that enable people to live together and respect each other's rights."[1] This definition covers much of the law that affects interpersonal relationships and, in particular, the Common Law (which is simply the recorded judgments of the courts). Modern society has prompted others to offer definitions of the law, each in an attempt to explain the nature and purpose of the law as succinctly and precisely as possible. Salmond, for example, described the law as the "body of principles recognized and applied by the state in the administration of justice."[2]

Oliver Wendell Holmes, the distinguished American jurist, once described the law as "a statement of circumstances in which the public force will be brought to bear through the Courts."[3] Blackstone, in his famous *Commentaries on the Law of England*, defined the law as "a rule of civil conduct, prescribed by the supreme power in a state, commanding what is right, and prohibiting what is wrong."[4]

Rights versus Privileges

Each of these definitions implies that something will happen if an individual does not obey the rules. Irwin Dorfman mentions respect for the "rights" of others. To understand the operation or application of the law, then, it is necessary to know what constitutes a right, and to distinguish this right from that which constitutes a privilege. When we say we have the "right" to do something, we are saying, in essence, that we may do the particular act with impunity, or with the force of the state behind us. Because rights are closely associated with duties, our right to do an act usually imposes a duty on others not to interfere with our actions. What the law does is set out our rights and duties so that everyone will know what they are, and to whom they apply. In a similar fashion, the law sets out actions that are not rights and duties, but privileges.

Privileges are actions that may be taken by an individual under specific circumstances and that may be withdrawn or curtailed by the state. Rights enjoyed by individuals often become privileges as a result of social pressure or public policy. Occasionally, rights may also become privileges, out of a desire on the part of legislators to increase the flow of funds to the public coffers through licence fees. Statutes relating to the ownership and operation of automobiles are examples of laws of this nature.

1. Statement by Irwin Dorfman, President, Canadian Bar Association, at the annual meeting of the Association, August 1975, Halifax, Nova Scotia.
2. Williams, G., *Salmond on Jurisprudence*, 11th ed. (London: Sweet & Maxwell Ltd., 1954), p. 41.
3. Corley, R.N., and Black, R.L., *The Legal Environment of Business*, 2nd ed. (New York: McGraw-Hill Book Company, 1968), p. 4.
4. Lewis, W.D., *Blackstone's Commentaries on the Law of England* (Philadelphia: Rees, Welsh & Company, 1897), book 1, s. 2, para. 44.

THE ROLE OF LAW

The law

The body of rules of conduct that are obligatory in the sense that sanctions are normally imposed if a rule is violated.

The law represents a means of social control, and a law in its most basic form is simply an obligatory rule of conduct.[5] **The law**, in contrast to a single law, consists of the body of rules of conduct laid down by a sovereign or governing body to control the actions of individuals in its jurisdiction. It is normally enforced by sanctions. The law develops to meet the needs of the people in a free society and changes with their changing needs. For this reason, the law tends to respond to the demands of a free society, rather than shaping its nature. However, laws may arise in other ways as well.

Social Control

Laws established and enforced by legislators that are not in response to the demands of the majority of citizens of the state may be introduced to shape or redirect society in ways that legislators perceive as desirable. Laws of this type represent a form of social engineering that frequently restricts individual rights and freedoms, and very often transfers individual rights and powers to the governing body. Laws of this nature represent a growing proportion of Canadian law, but this form of legislation is not a recent phenomenon.

Laws that legislatures have attempted to impose on society to alter the behaviour of the majority normally prove to be ineffective unless enforced by oppressive penalties or complete government control of the activity. In a business context, compulsory, provincially operated automobile-insurance schemes in a number of provinces in Canada represent examples of legislation of the latter kind. It tends to be of a confiscatory nature: the government, by decree, transfers the right to engage in the activity to itself and virtually excludes all private-sector insurers. Provinces with this type of legislation have decided that the government should engage in the activity for the public good as a matter of public policy and without interference or competition from the private sector. In a free society, the only way a governing body may implement such a policy is through laws that establish an insurance system that, in a variety of ways, excludes all others from engaging in the activity.

The desirability of this type of legislation, which impinges on the freedom of the individual and alters the rights and behaviour patterns of those living in the jurisdiction, has been the subject of debate amongst legal philosophers for centuries. The basis of much of this debate is rooted in political philosophy, rather than in the law as such. It does, however, illustrate how the law may be used to implement public policies that may not represent the particular desires or wishes of the people, but rather the desires of those in a position of political power at a given point in time.

Settlement, Rules, and Protection

While the principal role of law in a society is, broadly speaking, social control, the law may be subdivided into three functions: settling disputes, establishing rules of conduct, and providing protection for individuals. Laws must, first of all, be the vehicles by which disputes between individuals are settled. Some of the earliest laws were simply procedural rules outlining the manner in which parties would deal with each other when they settled their dispute by combat. The sole purpose of these laws was to ensure some degree of fairness in the combat and a minimum of disruption in the lives of others not directly involved in the dispute. In this sense, the law served a second function: it established rules of conduct for individuals living in close association with others. Such rules direct the energies of individuals in an orderly fashion and minimize conflict between persons engaged in similar activities, or in activities that have the potential for conflict. In the distant past these were rules that determined which hunter was entitled to the game if more than one party was responsible for bringing the animal down. Classic examples of modern applications of this type of law are the "rules of the road" under the various

5. Osborn, P.G., *The Concise Law Dictionary*, 7th ed. (London: Sweet & Maxwell Ltd., 1983), p. 194.

Highway Traffic Acts. These rules dictate where and how a motor vehicle may be driven, to facilitate the orderly flow of traffic and to minimize the risk of accident or injury to those who drive motor vehicles.

The third major function of the law is to provide protection for individuals in a society. Self-preservation is of paramount importance to any person, and the law represents a response to this desire or need. Individuals in the earliest societies realized that they were dependent upon one another for much of their security. Some of the first laws, therefore, related to the protection of individuals from intentional or careless assaults by others. These laws soon expanded to include the protection of possessions as well, and established the concepts of individual freedom and property rights. The development of these laws can best be understood by an examination of law in its historical context.

THE EARLY DEVELOPMENT OF LAW

The origins of some of our most basic laws and principles are lost in antiquity. Records of early civilizations make reference to a great many laws similar to those that stand today on our statute books or in our Common Law as the law of the land. These early writings no doubt simply recorded what had then been long-standing rules or customs. By using logic and an understanding of human nature, one may speculate on how the law evolved in early times.

Within the Family

The earliest laws were handed down by word of mouth from generation to generation long before the first words were recorded on parchment or clay tablets. These early laws, or rules of behaviour, probably had their origins in rules of conduct established to maintain orderly relationships in the families of early humankind. Later on, they were perhaps expanded to govern conduct between families in a tribe as those families began living in close proximity to one another.

The concentration of families in a relatively small area was not without its disadvantages, however. Disruptive behaviour in a village affected a far greater number of individuals than it would in an isolated family setting, and the need to curb activities of this nature took on a new importance. During this period, early attempts to control disruptive behaviour took the form of the selection of individuals, usually family elders (or the strongest members of the community), to hear disputes and recommend non-disruptive, or at least controlled, methods of settlement. The disputants were required to accept their decision as the binding resolution of the conflict. Initially, this group decided each dispute on its merits, but gradually some consistency emerged as similar disputes were decided on the basis of the previous decisions. At this point, rules of behaviour for the entire community began to take on meaning. In the beginning, these decisions were imposed by the community if they were considered just and equitable. However, as the decision-makers assumed more responsibility for the orderly operation of the affairs of the village, so, too, did they acquire the power to impose their decisions on disputants. With the establishment of a governing body or a sovereign with the power to impose a decision on an individual in the interests of the common good, the law, as an instrument of social control, took form.

Within the City-State

These changes did not happen in any significant way until the village form of living gave way to the establishment of the city and the city-state. The city arose with the development of trade and the production of goods and services by individuals for others. This concentration of activity in a relatively small area gave the individual an identification with the community, and a desire to act in a concerted way to solve common problems. The answer was usually to direct the efforts of the community as a whole to form some sort of organization or government to deal with these matters. These formal bodies — with the authority to decide how individuals in the community should conduct themselves in their dealings with others, and with the power to enforce their decisions — created the first state-legislated laws.

Needless to say, these laws developed gradually over a relatively long period of time, and at different rates among different peoples. The earliest known written laws were undoubtedly accepted as the officially sanctioned rules of behaviour long before they were formally recorded. In some cases, when a tribe or race reached the stage of the city-state and became rich and weak from "easy living," it would then be plundered by "less-civilized" tribes. The invaders would then adopt the new city life and much of its form of organization. The laws of the city-state, necessary for order and tranquility, would often be adopted by the invaders as well. As a tribe or a race, they would then move directly from a relatively lawless behavioural system directed by custom to a society governed by law. In other cases, the law was imposed upon less-developed tribes or races by the invading armies of city-states. The Roman Empire was a notable example of a state that spread the rule of law over much of Western Europe and the Middle East, and established the first legal institutions in many of those parts of the world.

THE RISE OF THE COURTS AND THE COMMON LAW

Customary Law

Without some system of determining its application or imposing its sanctions, the law has little more than a persuasive effect on human behaviour. The rise of the city-state brought with it the establishment of the mechanism for law enforcement. The inhabitants of large communities (and, later, the city-states) were quick to realize that disputes between individuals would increase as the population's density increased. At first, these communities established tribunals or other bodies officially authorized to hear disputes. With the aid of the society, these tribunals could force restitution or undertake vengeance; however, they lacked sufficient power to compel the use of their system. Eventually, by requiring certain formalities that had to be undertaken before vengeance, and by providing the inducement of monetary compensation as a remedy, the state ultimately became powerful enough to force everyone to use the tribunal or court. At this stage, because of its negative effect on the community, vengeance itself became a crime.[6]

Of particular importance in the development of the courts was the nature of their decisions. For the most part, the law relating to relationships between individuals, and the rights and responsibilities of one to another, were gradually established in the form of consistent decisions, first by tribunals, and, later, by the courts. As a result, the judgments of the courts preceded the law. The correctness of these early decisions, or judgments, was usually based upon either a religious foundation (if the head of the state held his or her position by some divine right), or the approval of the assembled citizens of the community. Consistency in the decisions handed down by these courts in similar cases eventually created the body of rules known as the law.

Pre-Norman England

In England, the early courts and the law were imposed on the inhabitants by successive invaders. Following the Roman conquest of England in 43 B.C., the country was subject to *lex romana* (Roman law) and its administrative machinery. In those areas under Roman control, the law was uniform in nature and application, in theory at least. This lasted throughout the period of Roman occupation.

The disintegration of the Roman Empire and the invasion of England by the Germanic tribes produced a decentralized system of government under a king. This system divided the land into shires and counties, each with its own government and, to varying degrees, freedom from royal interference. The shires were further divided into

6. *Hemmings and Wife et al. v. The Stoke Poges Golf Club, Ltd. et al.*, [1920] 1 K.B. 720.

smaller communities called hundreds, administrative units about the size of a township. Within these units were the boroughs or local communities.

In the pre-Norman period, the law in each of these shires developed according to local custom or need. It was not until after 1066 that the trend was reversed. At that time, the only laws that were common throughout the land were the small number of written laws, relating to general crimes, that several of the kings had pronounced as law. These laws frequently set out the penalty in monetary terms, with a part of the money to be paid to the king by the perpetrator of the crime, and the balance to be paid to the injured party or the party's next of kin. The laws were enforced by a relatively weak central government, called a *witenagemot*, that governed the country along with the king.

Norman England and the Common Law

The Norman Conquest in 1066 brought with it a more centralized system of administration and, shortly thereafter, the establishment of a central judicial system. The Conquest had very little effect on English custom, but it did have a considerable effect on the administration of the country. The power of the shires was brought under the control of the king, and the right of the shire court or county court to hear cases concerning land and certain criminal cases (known as pleas of the Crown) was transferred to the king's justices.

The establishment of a central judiciary under King Henry II to hear the more serious cases was an important factor in the development of the Common Law in England. After 1180, justices of the royal court travelled regularly throughout the country to hear cases. On their return to London, they discussed their cases with one another or exchanged notes on their decisions. Amongst themselves they gradually developed what later became a body of law common throughout the land. The "Common Law" had been born.

During the twelfth and thirteenth centuries, the administration of justice became more centralized, and after 1234 the king's justices began keeping their own records. Early decisions had been largely based on local custom. This gradually changed, however, as the judges based their decisions more upon the written records of the court, and less upon what was alleged to be local custom. After 1272, written records of decisions were maintained to assist the fledgling legal profession. Initially sparse, the records improved over the next three centuries to the point where they were useful as statements of the Common Law. Better case reports, and their consolidation in the years that followed, developed the body of law known as the Common Law of England. The settlement of English colonies in North America established the Common Law in Canada and the United States. Today, these principles represent a large and flexible body of law that is adaptable and responsive to the needs of our society.

THE SOURCES AND COMPONENTS OF MODERN CANADIAN LAW

The Common Law

Common Law

The law as found in the recorded judgments of the courts.

Statute law

A law passed by a properly constituted legislative body.

The **Common Law** is an important source of law in Canada. It is sometimes referred to as "case law," because that is where statements of the Common Law may be found. Another reason it is referred to this way is to distinguish it from the second major source of law: **statute law**.

In Common Law provinces, and in those countries with Common Law systems, the law (except statute law) is not found in a code but in the recorded judgments of the courts. These judgments were not always recorded, but in 1290, during the reign of Edward I, the Year Books were commenced. These books provided reports of the cases, but in the early years reasons for the decisions were seldom included. With the introduction of printing in England in 1477, the Year Books were improved, and printed copies became available to the legal profession. The reporting of cases in the form of law reports took place in the sixteenth century, and from that point on the decisions of judges were reported without a break by various law reporters. Judicial reasoning and the principles applied by the judges were readily available by way of these reports, and the Common Law could be determined from them through the doctrine of *stare decisis*.

Stare Decisis

Stare decisis ("to let a decision stand," or "to stand by a previous decision") is the theory of precedent in Common Law. The doctrine means that a judge must apply the previous decision of a case similar to the one before the court if the facts of the two cases are the same, providing such a decision was (1) from the judge's own court, (2) from a court of equal rank, or (3) from a higher (or superior) court.

In cases of identical facts, only the Supreme Court of Canada has the unrestricted ability to overrule its previous decisions, and even then it does so only with caution. Decisions of the Supreme Court of Canada are binding on all lesser courts. All other lesser courts, again in cases of identical facts, are bound to follow the decisions taken by higher courts in their own jurisdiction. Where a previous decision of a court on identical facts comes before that court a second time, unless there are compelling reasons, the expectation is that the earlier precedent will be followed. Decisions of higher courts or courts of equal rank from other jurisdictions are persuasive, but not binding.

The need for certainty in the law (in the sense that it must be clear in its meaning and predictable in its application) was quickly realized by judges. The adoption of the theory of precedent provided a degree of stability to the Common Law without sacrificing its flexibility, although at times the courts became so reluctant to move from previous decisions that their application of the law made no sense at all to the case at hand. Fortunately, judges have been adaptable in their formulation of the law. Over the years they have maintained the Common Law as a blend of predictable yet flexible principles, capable of conforming to the changing needs of society. This has been due, in part, to a reluctance on the part of the judiciary to accept precedent as a hard-and-fast rule. The facts of any two cases are seldom precisely the same. Differences in the facts or circumstances are sufficient to permit a judge to decide that a particular obsolete precedent should not apply to the case before the court if the application of such a precedent would produce an unsatisfactory result. In this fashion, the courts have gradually adapted the Common Law to changing times.

The adaptability of the Common Law has enabled it to absorb, over a long period of time, many legal principles, customs, and laws from other legal systems and sources. The law of England before the Norman Conquest of 1066 was, for the most part, local in both form and application. It consisted of a mixture of early customs, a few remnants of early Roman law, and the laws and customs brought to England by the Anglo-Saxon invaders. Decisions were handed down by judges based upon local custom, which prior to the Conquest was the only precedent available. The Norman Conquest brought with it a central system for the administration of justice and, through this centralized system, the incorporation of the customs and laws from all parts of the country into the Common Law.

Other rules of law were incorporated into the Common Law by more direct means. The law relating to land tenure, which had its roots in feudal law, was introduced by the Normans, and the courts thereafter were obliged to apply these rules in dealing with land disputes.

Even customs or practices that have developed over time have found their way into the Common Law. The courts have often recognized long-standing practices in determining the rights of parties at law, and, in this fashion, have established the custom or practice as a part of the Common Law.

Canon Law

Other laws were also incorporated into the Common Law as the courts in England expanded their jurisdiction. Originally, the church had jurisdiction over religion, family and marriage, morals, and matters relating to the descent of personal property of deceased persons. The law relating to these matters was initially administered by ecclesiastic courts, but cases concerning some of these church-administered areas of the law gradually found their way before judges of the civil courts. After the Reformation (during the years 1534–38) much of the ecclesiastic courts' jurisdiction passed to the royal courts. In dealing with cases that had previously fallen within the province of the ecclesiastic courts, the judges naturally looked at the decisions of those courts in reaching their own decisions. As a result, many of the rules of **Canon Law** or church law became a part of the Common Law.

Canon Law

The law developed by the church courts to deal with matters that fell within their jurisdiction.

Law Merchant

In much the same fashion, a substantial part of the law relating to commerce and trade was incorporated into the Common Law. Early merchants belonged to guilds, as did the artisans. Customs of the various trades gradually developed into a body of rules that were similar throughout much of Western Europe, and disputes that arose between merchants were frequently settled by the application of these rules.

Initially, most of the merchants sold their wares at fairs and markets, and any disputes that arose were settled by the senior merchants, whose decisions were final and binding. Later, decision-making became somewhat more formal, and the decisions more uniform. Gradually, rules of law relating to commercial transactions began to emerge, as the decisions of the guild courts became firmly established and consistent in their application by the merchant guilds. These courts had jurisdiction only over their members, so for a long period of time the body of law known as the **Law Merchant** was within the exclusive domain of the merchant guilds. Eventually, merchants who were not guild members began to trade, and when disputes arose, they appealed to the courts of the land for relief. In dealing with these disputes, the judges applied the Law Merchant. By way of their decisions the large body of law relating to commerce gradually became a part of the Common Law.

Law Merchant

The customs or rules established by merchants to resolve disputes that arose between them, and that were later applied by Common Law judges in cases that came before their courts.

Equity

The last important source of law administered by the Common Law courts was the body of law called **equity**. The rules of equity are not, strictly speaking, a part of the Common Law but, rather, a body of legal principles that takes precedence over the Common Law when the Common Law and rules of equity conflict. The rules of equity developed largely because the Common Law in England had become rigid in its application by the fifteenth century, and litigants often could not obtain a satisfactory remedy from the courts. To obtain the kind of relief desired, they would frequently petition the king. The king, and, later, his chancellor, heard these cases and in each case made what was called an equitable decision: one not necessarily based upon the law, but one that the king considered to be fair. The ideas of fairness that the king expressed as the basis for his decisions gradually took on the form of principles or rules, which he applied in other cases that came before him. Over time, these rules became known as principles of equity. These principles were later followed by the chancellor, and later still by the Court of Chancery (the Chancellor's Court). Eventually, they took on the form of rules of law. In the late nineteenth century, the Court of Chancery and the Common Law courts merged, and the rules of equity became a part of the body of law that the courts could apply in any civil case coming before them. As a result, a judge may apply either the Common Law rules or the principles of equity to a case before the court and, if the Common Law is inappropriate, the equitable remedy is usually available to ensure a fair and just result. How these laws are administered is the subject matter of the next chapter.

Equity

Rules originally based on decisions of the king rather than on the law, and intended to be fair.

Statute Law

Statutes are laws that are established by the governing bodies of particular jurisdictions and have their root in the Latin word *statutum*, meaning "it is decided." Governments are vested with the power to make laws, either under the terms of a written constitution, such as that of Canada or the United States, or as a result of long-standing tradition, such as in England.

Statutes are the product or end result of a legislative process. Under this process, the wishes of the people, as interpreted by the members of a provincial legislature or the Parliament of Canada, are brought forward for debate in the legislative assembly. They then finally become law if the majority of the legislators believe that the law is necessary. The process provides time for study and amendment of the proposals and, in the case of the Parliament of Canada, a thorough examination not only by the House of Commons but by the Senate as well. It is by this democratic process that statute laws are created.

Bill

A proposed law presented to a legislative body.

Motion

The decision to read a bill a first time.

A statute law has its beginnings in a **bill**, which is a proposed law presented to a legislative body (such as the House of Commons or a provincial legislature). The bill then requires a **motion** (or decision) to have the bill read a first time and printed for circulation. Members of the legislature are then given a period of time to read the bill and prepare to debate its contents before it is brought forward again. Some time later, the bill comes before the legislature (or House) for a second reading. The bill is then debated in principle.

If the bill passes the second reading, it is then sent to a committee of the House or legislature for study on a clause-by-clause basis. The committee is made up of a number of members of the legislature appointed to examine the proposed legislation and amend it if necessary. Any member of the committee may propose amendments, as the bill must be passed one clause at a time. Once the bill (perhaps in amended form) has been passed by the committee, the Chair of the Committee reports the bill in final form to the legislature.

The report of the bill may then be subject to further debate and amendment before the bill is given a third reading. If passed (in the sense that a motion to have the bill read a third time is carried by a majority vote) the bill is then sent to the Senate at the federal level. It must then be approved by a similar process at the Senate level. The Senate may also initiate bills, but if it does, it must send the bill to the House of Commons, where the process is repeated.

Royal assent

Needed in order for a bill to become law.

Once a bill has been passed by the House of Commons and Senate or the provincial legislature, it must receive **royal assent** by the Governor General (federally) or the Lieutenant-Governor (provincially). Royal assent is largely automatic, as it has never been refused at the federal level. This has not been the case with provincial bills, however, as royal assent has been refused on numerous occasions in the past.

Proclaimed

When a law becomes effective.

A bill does not become law until it receives royal assent, but, in some cases, the government may not wish to implement the law (or parts of it) until some time in the future. In these cases, the bill will not become law until it is **proclaimed**, or becomes effective.

Revised statutes

Updated or amended statutes.

Properly passed, and within the jurisdiction of a legislature, a statute affects the residents of that province or, in the case of a federal statute, all residents of Canada. Prior to recent digitization, it was hard to keep track of the creation, amendment, and repeal of legislation. A consolidation of laws and their changes (known as **revised statutes**) was usually only printed by governments once in a decade (for example, the *Revised Statutes of Canada, 1985*), which was out-of-date even before being delivered to law libraries. Legal research to determine the valid text of a statute involved starting from the last revision, consulting an annual volume for each intervening year thereafter, then monthly and weekly reporter series up to the current date. Today, authoritative online electronic consolidation of changes has made printed statute books obsolete, with governments across Canada providing continuous full-text updates of their current law.

As noted previously, when a statute is properly enacted within the legislative jurisdiction of a province or the federal government it will, when declared to be law, apply to all those persons within that jurisdiction. Statute law can be used to create laws to cover new activities or matters not covered by the Common Law, or to change or abolish a Common Law rule or right. Statutes may also be used to codify the Common Law by collecting together in one written law the Common Law rules or principles relating to a specific matter.

The particular advantage of statute law over the Common Law is the relative ease by which the law may be changed. The Common Law is generally very slow to respond to changing societal needs. It follows a gradual, evolutionary pattern of change, rather than a quick response. Statute law, on the other hand (in theory at least), may be quickly changed in response to the demands of the public. The disadvantage of statute law is that it will be strictly interpreted by the courts. Unless it is very carefully drafted, it may not achieve its intended purpose. Occasionally, a badly drafted statute only serves to compound the problems that it was intended to solve, and may require additional laws to respond to the problems it created. In spite of the potential problems inherent in statute law, the general direction of the law appears to be toward more, rather than less, statute law to deal with social change.

Quebec's *Civil Code*

Civil Code

A body of written law that sets out the private rights of the citizens of a state.

In contrast to the rest of Canada, the province of Quebec has codified much of the law that is normally found in the Common Law of other provinces. As a result, this body of law, known as the ***Civil Code***, may be consulted in the determination of rights and duties. These same rights and duties would ordinarily be found in the Common Law in other jurisdictions. This particular method of establishing the law of a jurisdiction is not a new approach. It is simply an alternate method of setting out the law that has a long history. The first codification of law of major significance took place under the direction of the Roman emperor Justinian, who ordered a compilation of all the laws of Rome dating back to the time of Cicero. The collection of the laws was an enormous task that took seven years to complete, and on completion in 534 A.D. became the famous Corpus Juris Civilis. This body of law formed the basis of the law in a large part of continental Europe for the next 1,200 years. It was not until the eighteenth century that major revisions were made.

Frederick the Great of Prussia directed the preparation of a new code during his reign, but it was not adopted until 1794, some eight years after his death. Shortly thereafter, France (under Napoleon), began a codification of French law, in 1804. It was this code that influenced the codification of the law in Spain, Italy, Belgium, and, by way of colonization, much of South America, the state of Louisiana, and the province of Quebec. The 1900 codification of the law in Germany, which replaced Frederick the Great's Prussian code, found favour as a model for many other countries, notably Japan, Switzerland, and Greece. Quebec, however, continued to follow the 1804 French code. It made its own *Civil Code of Lower Canada* in 1866, a right preserved to it by the *Quebec Act* of 1774, more than a decade after the colony had been ceded by France to Britain. A complete review of the *Code* was not undertaken for over a century, before the new *Civil Code of Quebec* came into force in 1994.

The *Civil Code of Quebec* is more than just an act of a legislature setting down rules. It was a short evolution from its original philosophical roots shortly after the French Revolution. Its makers intended it to be a complete legal pathway for life: birth, family, business relationships, death, inheritance, and management of their chief obligations and assets along the way.

The modern *Code* of 3,168 articles preserves this philosophical journey, set into 10 books as follows:

Book 1: Persons
Book 2: The Family
Book 3: Successions (inheritances)
Book 4: Property
Book 5: Obligations (forms of contracts)
Book 6: Prior Claims and Hypothecs (mortgages)
Book 7: Evidence
Book 8: Prescription
Book 9: Publication of Rights
Book 10: Private International Law

Quebec does, however, have other laws beyond its *Civil Code*. It creates statute law for specific matters just as the other provinces do, for example, labour or environmental laws. Even the code itself is a statute of the Quebec legislature. Still, as it contains all the general principles of law that apply to behaviour and relationships, it serves as a starting point and framework for more specific laws to build upon or create exceptions.

Codification of Common Law

The codification of the Common Law in England was never seriously considered, even though it was urged by such respected English writers as Sir Francis Bacon. Nor did the idea find much favour in the United States or the Common Law provinces of Canada. Codification of some parts of the Common Law in all three countries has taken place, however. England codified the Common Law as it stood in 1882 with respect to bills of exchange and negotiable instruments and, in 1890, codified the laws of partnership. The Common Law relating to the sale of goods was codified in 1893, but after that the process lost its impetus. Since then, no major effort has been made to codify other branches of mercantile law except to modify or settle matters of difficulty relating to particular issues.

Proponents of codification have long argued that the advantage of this method over the Common Law is *certainty*. According to them, if the law is written down, it is there for all to see and know. In theory, the judge decides a dispute by reference to the appropriate part of the code. However, if no specific article covers the dispute, then the decision is based upon general principles of law set out in the code. The particular difficulty with the code is that it might be interpreted differently in some cases by different judges. Unless some uniformity exists between judges in deciding similar cases, one of the important advantages is lost. To avoid this, judges in Quebec (and other places with civil codes) consider the decisions of other judges who have decided similar cases.

The American Bar Association, during the late nineteenth and early twentieth centuries, proposed a number of uniform statutes relating to commercial practices in an effort to eliminate the differences that existed in state legislation, but its efforts were unsuccessful. It was not until after World War II that there was sufficient interest in codification to produce the *United States Uniform Commercial Code*. This code, which governs commercial-law practices, was first drafted in 1952 and, following a number of amendments, eventually was adopted by all states by 1975. Unfortunately, the goal of true conformity of legislation, as envisaged by the Bar Association, was not realized. Not all of the states adopted the entire code, and some altered the code to suit their own particular needs. Substantial conformity was achieved in the United States, however, at least with respect to a number of areas of commercial law.

Administrative Law

Administrative law

A body of rules governing the application of statutes to activities regulated by administrative tribunals or boards.

Regulations

Procedural rules made under a statute.

Administrative tribunals

Agencies created by legislation to regulate activities or do specific things.

A growing part of statute law is an area called **administrative law**. The primary focus of this body of law is directed toward the **regulations** made under statute law and enforced by administrative bodies. While legislation usually creates laws or repeals old laws, it may also create agencies or **administrative tribunals** to regulate activities or do specific things. The activities of these tribunals and agencies are said to be administrative acts, and the body of law that relates to their activities is administrative law.

Administrative law is not a new area of the law, but rather an area that has increased substantially in size and importance since World War II. In early times in England, Parliament would authorize the king and his officials to carry out such activities as the collection of taxes, the maintenance of the armed forces, and the operation of the courts. Gradually, a public service was established to perform these tasks. In this fashion, Parliament did not directly supervise the activity, but merely authorized it and set out guidelines for the officials to follow in the performance of their duties. Today, the Canadian Parliament and the provincial legislatures use this method to regulate many activities that fall within their legislative jurisdiction. Examples of some of the activities under the control of regulatory agencies include the sale of securities by public companies, labour relations, employment standards, aeronautics, broadcasting, the sale and consumption of alcoholic beverages, land use, and a wide variety of commercial activities. How these regulatory bodies operate and how business persons and corporations must respond to them are examined in Chapter 3.

THE CONSTITUTIONAL FOUNDATIONS OF CANADIAN LAW

The decisions of the courts required some authoritative source to ensure compliance. In England, this source of authority was for many centuries the king, who could also establish laws by virtue of his supreme authority as head of state. As the parliamentary form of government developed, the supreme authority of the king was gradually eroded and the authority to make laws shifted to the elected legislative body. Over the last three centuries, most of Western Europe and the Americas have shifted to some form of democratically elected government with the authority to make laws governing their peoples. With this change also came the need for some means by which the powers of the government could be circumscribed, in order to protect the fundamental rights of citizens from the abuse of power by governments.

Constitution

The basis upon which a state is organized, and the powers of its government defined.

Most countries have adopted some form of written authority that sets out both the fundamental rights and freedoms of their citizens and the law-making powers of the various legislative bodies of the state. This document is generally referred to as a **constitution**. It represents a source of law in the sense that it not only establishes certain legal rights, but also law-making authority. A constitution may not always be contained in a single document, however. The "constitution" of Great Britain, for example, consists of a large number of proclamations, statutes, legal decisions, and traditions. Over many centuries, these have come to establish the fundamental rights of individuals and to represent the source of a number of limits on the legislative authority of the British Parliament. These sources of law or rights are nevertheless very fragile, as they might conceivably be overridden by Parliament (the supreme law-making authority) or ignored by the courts. Strong public support for these traditions requires both the government and the courts to give them greater force and effect than they might otherwise possess.

In contrast to this type of constitution, the United States of America has a formal, written document setting out the rights and freedoms of the individual (the Bill of Rights) and the powers that may be exercised by the executive, legislative, and judicial branches of government. The written constitution is the supreme law of the United States, and every law enacted must be in conformity with the constitution, or it is invalid.

The Canadian Constitution

The Canadian constitution is also a formal written document that sets out the rights and freedoms of Canadians (the *Canadian Charter of Rights and Freedoms*) and the powers of the federal government and the provinces. However, unlike the constitution of the United States, which imposes an elaborate system of checks and balances on the different branches of government, the Canadian constitution recognizes the supremacy of the legislative bodies for passing certain laws within their sphere of jurisdiction. This is notwithstanding the possibility that they might thereby violate certain rights and freedoms set out in the constitution. This particular distinction is examined in greater detail in the review of the specific sections of the *Charter of Rights and Freedoms* set out in this chapter.

The constitution of Canada is divided into two major parts: the first contains a *Canadian Charter of Rights and Freedoms*; the second contains an amending formula, some additional changes in the powers of government, and what was previously the contents of the *British North America Act*, 1867, as amended. This new constitution not only entrenches the basic rights and freedoms of the citizens of Canada, but also establishes the organization and jurisdiction of the federal and provincial governments.

Part of Canada's constitution was originally in the form of an English statute, the *British North America Act*, 1867.[7] This statute created the governing bodies at the federal and provincial levels, and also divided and assigned the legislative jurisdiction or powers between the two levels of government. English statute law also gave Canadian citizens the right to enjoy the basic rights and freedoms traditionally possessed by their English

7. The *British North America Act*, 1867, and amendments, when patriated by the *Canada Act*, 1982, cited together became the *Constitution Acts*, 1867 to 1981.

counterparts. In 1982, the division of powers in the *British North America Act*, 1867, was embodied in the constitution by the *Canada Act*, 1982.[8]

Under the Canadian constitution, the jurisdiction of the provinces to make laws over non-renewable natural resources was confirmed. The provinces were also given certain rights to levy indirect taxes on these resources. Section 92 specifies the exclusive authority of the provinces to make laws pertaining to such matters as property and civil rights (heading 13); matters of a local or private nature in a province (heading 16); the incorporation of companies (heading 11); the licensing of certain businesses and activities (heading 9); the solemnization of marriage (heading 12); and local works and activities (heading 10). Section 91 gives the federal government exclusive jurisdiction over the regulation of trade and commerce (heading 2); criminal law (heading 27); bankruptcy and insolvency (heading 21); navigation and shipping (heading 10); bills of exchange and promissory notes (heading 18); and a wide variety of other activities, totalling 31 in all. Most importantly, s. 91 gives the federal government residual powers over all matters not expressly given to the provinces. As well, it is under this particular power that the federal government has assumed authority over relatively recent technological developments, such as communications, radio, television, aeronautics, and nuclear energy.

While ss. 91 and 92 appear to clearly divide the authority to make laws, the nature of the wording of the two sections has raised problems. For example, would the regulation of a particular provincial business practice constitute regulation of trade and commerce under s. 91, or would it be property and civil rights, under s. 92? More particularly, would the construction of a pipeline from an oil well in the bed of a river be navigation and shipping and, hence, a federal matter, or would it be a local work or undertaking and therefore within the jurisdiction of the provinces? These are the kinds of problems associated with the interpretation of this statute. Because the legislative jurisdictions of the federal government and the provincial legislatures are not always clear and precise, overlaps tend to occur. When legislation passed by one legislative body appears to encroach on the jurisdiction of another level of government, the matter of constitutionality must be answered by the courts.

THE *CANADIAN CHARTER OF RIGHTS AND FREEDOMS*

The *Canadian Charter of Rights and Freedoms* sets out the basic rights and freedoms of all Canadians. For the most part, many of these have been the same basic freedoms, democratic rights, and legal rights Canadians have enjoyed for many years. Now they are entrenched in the constitution, which may only be repealed or amended by an Act of Parliament consented to by at least two-thirds of the provinces that together contain at least 50 percent of the country's population. This restriction makes the constitution very difficult to change and, in a sense, provides a level of protection to these rights and freedoms. This does not mean that the rights set out in the *Charter* are clear and absolute, however, because the *Charter* itself states in Section 1 that:

> The *Canadian Charter of Rights and Freedoms* guarantees the rights and freedoms set out in it, subject only to such reasonable limits prescribed by law as can be demonstrably justified in a free and democratic society.

As a consequence of this provision in the *Charter*, the rights set out may be limited or proscribed wherever Parliament or a provincial legislature can justify legislation placing a limit or restriction on the exercise of a right or freedom. This limitation may be necessary, in some cases, so that the rights or freedoms of a single individual will not be exercised to the detriment of the public generally. For example, the *Charter* permits freedom of expression, but should an individual be allowed the unbridled freedom to defame others with impunity? Surely not. Some restriction on freedom of expression is necessary if a person's reputation is to be protected as well. For this reason, the law must attempt to balance these rights.

8. *The Canada Act*, 1982 (U.K.), c. 11.

Apart from this general qualification on the rights and freedoms, the *Charter* specifically permits Parliament or provincial legislatures to pass legislation that conflicts with or overrides fundamental freedoms, legal rights, and certain equality rights, by way of a "notwithstanding clause."[9] This allows statutes to be enacted to meet special situations without the necessity of a constitutional amendment. In each case, the legislation would automatically expire at the end of a five-year period, unless specifically renewed.

CLIENTS, SUPPLIERS *or* OPERATIONS *in* QUEBEC

Quebec's language law (Bill 101) was held by the Supreme Court of Canada to offend the *Charter of Rights and Freedoms* with respect to the languages that businesses in that province might use on their signs and establishments. With Bill 101 unconstitutional with respect to this provision, Quebec passed new legislation that would require French language only on exterior business signs. In doing so, the province invoked s. 33 of the *Charter*, which allowed the legislature to have the new statute override the *Charter* rights of the business persons for a period of five years.

The fundamental freedoms enshrined in the *Charter* are those that in the past were established under British tradition, custom, and law. The *Charter* describes them this way:

Fundamental Freedoms

2. Everyone has the following fundamental freedoms:
   (a) freedom of conscience and religion;
   (b) freedom of thought, belief, opinion, and expression, including freedom of the press and other media of communication;
   (c) freedom of peaceful assembly; and
   (d) freedom of association.

COURT DECISION: Practical Application of the Law

Charter of Rights—Freedom of Expression

***RJR MacDonald Inc. and Imperial Tobacco Ltd. v. Canada* (1995), 187 N.R. 1, 1995 CanLII 64.**

In September 1995, the Supreme Court of Canada settled a manufacturer's challenge to the Tobacco Products Control Act, which had imposed a virtual ban on tobacco advertising, sales displays, and displays of brand names at sporting and cultural events. The Supreme Court concluded that the statute violated s.2(b) expression rights of the Charter, and were not demonstrably justified in a free and democratic society. The test for justification was set out by the Madam Justice McLachlin:

There is merit in reminding ourselves of the words chosen by those who framed and agreed upon s. 1 of the *Charter*. First, to be saved under s. 1 the party defending the law (here the Attorney General of Canada) must show that the law which violates the right or freedom guaranteed by the *Charter* is "reasonable." In other words, the infringing measure must be justifiable by the processes of reason and rationality. The question is not whether the measure is popular or accords with the current public opinion polls. The question is rather whether it can be justified by application of the processes of reason. In the legal context, reason imports the notion of inference from evidence or established truths. This is not to deny intuition its role, or to require proof to the standards required by science in every case, but it is to insist on a rational, reasoned defensibility.

Second, to meet its burden under s. 1 of the *Charter*, the state must show that the violative law is "demonstrably justified." The choice of the word "demonstrably" is critical. The process is not one of mere intuition, nor is it one of deference to Parliament's choice. It is a process of demonstration. This reinforces the notion inherent in the word "reasonable" of rational inference from evidence or established truths.

9. See the *Canadian Charter of Rights and Freedoms*, s. 33.

The bottom line is this. While remaining sensitive to the social and political context of the impugned law and allowing for difficulties of proof inherent in that context, the courts must nevertheless insist that before the state can override constitutional rights, there be a reasoned demonstration of the good which the law may achieve in relation to the seriousness of the infringement. It is the task of the courts to maintain this bottom line if the rights conferred by our Constitution are to have force and meaning. The task is not easily discharged, and may require the courts to confront the tide of popular public opinion. But that has always been the price of maintaining constitutional rights. No matter how important Parliament's goal may seem, if the state has not demonstrated that the means by which it seeks to achieve its goal are reasonable and proportionate to the infringement of rights, then the law must perforce fail.

In determining whether the objective of the law is sufficiently important to be capable of overriding a guaranteed right, the court must examine the actual objective of the law. In determining proportionality, it must determine the actual connection between the objective and what the law will in fact achieve; the actual degree to which it impairs the right; and whether the actual benefit which the law is calculated to achieve outweighs the actual seriousness of the limitation of the right. In short, s. 1 is an exercise based on the facts of the law at issue and the proof offered of its justification, not on abstractions... Context is essential in determining legislative objective and proportionality, but it cannot be carried to the extreme of treating the challenged law as a unique socio-economic phenomenon, of which Parliament is deemed the best judge. This would be to undercut the obligation on Parliament to justify limitations which it places on *Charter* rights.

In spite of this judgment, today there are no tobacco advertisements. How do you think government(s) accomplished this, faced with a Charter protection of expression?

Similarly, democratic rights, which include the right to vote and the right to stand for election, are contained in the *Charter*. For the protection of the public, a limit is placed on the duration that a legislative assembly may continue without an election. This time period is limited to five years, except in times of war, invasion, or insurrection.

Mobility and Personal Liberty

An important right enshrined in the *Charter* is found in s. 6, which provides that Canadian citizens are free to remain in, enter, or leave Canada, and to move freely within the country. Canadians have always assumed that they enjoyed these rights, but in the past, some Canadian citizens have had these rights restricted in times of national emergency. For example, during World War II, Canadian citizens of Japanese ancestry were stripped of their citizenship and interned in special camps during hostilities with Japan. The purpose of this provision in the *Charter* is to prevent a recurrence of this type of treatment of Canadian citizens and to permit free movement throughout Canada. Mobility rights essentially grant to every permanent resident the right to take up residence anywhere in the country and to move for the sake of earning a livelihood in any province, as well.

Of equal importance to the right to move freely about the country is the right to enjoy life without interference by the state. This broad right, which includes the right to life, liberty, and the security of the person, is supported by the right to be free from unreasonable search and seizure, and free from arbitrary detention or imprisonment.

Right to Due Process

Accompanying this latter right is the right to be informed, on arrest or detention, of the reasons for the arrest, and the right to retain and instruct a lawyer promptly after the arrest or detention has been made. The arrest itself must be for an alleged act or omission that at the time would constitute an offence under Canadian or international law, or if the person is considered to be a criminal, according to the general principles of law recognized by the community of nations.

In cases where the state has reasonable grounds for the arrest, the accused person is entitled to be brought to trial within a reasonable time. In addition, the accused is presumed innocent until proven guilty at a fair and impartial public hearing of the matter. To ensure fairness at trial, accused persons are entitled to have the services of an interpreter if they do not speak the language in which the trial is conducted, or if they are deaf. The trial itself must be conducted in a fair manner, and an accused person cannot be compelled to testify against himself or herself. In addition, whether found innocent or guilty, no one can be tried a second time for the same offence. If a person is found guilty, the punishment for the offence must not constitute cruel or unusual treatment.

The legal rights in the *Charter* also provide that, pending trial, a person should not be denied reasonable bail unless the denial can be justified. Moreover, the *Charter* provides that persons who give evidence at a trial that would incriminate them may not have that evidence used against them in any other proceedings, except where they have committed perjury or given contradictory evidence.

Most of these rights are not new, as they existed in the past in the form of precedent or practice followed by the courts, or have their roots in statute law, some dating back as far as the English *Magna Carta* (in 1215).[10] The federal government had included many of these rights in the *Canadian Bill of Rights* in 1960. By enshrining these rights in the constitution, however, not only must the state and its law enforcement agencies respect them, but they may not be taken away except through the difficult amendment process, which must be followed to effect constitutional change.

Equality Rights

Equality rights in the *Charter* complement the legal rights of Canadians by providing that every individual is equal before the law. He or she has the right to equal protection and benefit without discrimination on the basis of race, creed, colour, religion, sex, age, national or ethnic origin, or any mental or physical disability. This section of the *Charter* also permits "affirmative-action" programs or laws designed to improve the conditions of disadvantaged individuals or groups, and laws which might otherwise be precluded by the *Charter* as providing special rights or benefits for one group and not others.

CASE IN POINT

The federal government permitted three aboriginal bands to communally fish for salmon in the mouth of the Fraser River for a period of 24 hours and to sell their catch. Another group of commercial fishers, mainly non-aboriginal, were excluded from the fishery during this 24-hour period. The second group staged a protest fishery and were charged and convicted with fishing at a prohibited time. Their defence was that the communal fishing licence discriminated against them on the basis of race. The Supreme Court of Canada ruled that the communal fishing licence was constitutional as an affirmative action program, and upheld the conviction.

R. v. Kapp, 2008 SCC 41.

The *Canadian Charter of Rights and Freedoms* confirms that the two official languages in Canada are English and French. It also confirms the right of the individual to communicate with the federal government in either official language and receive federal government services in either language, where there is a significant demand for services. The *Charter* requires neither the public nor the public service to be bilingual, but, instead, ensures that the federal government will provide its services in both languages. New Brunswick is the only officially bilingual province, but Quebec and Manitoba residents are entitled to use either language in their courts or their legislatures (under the *British North America Act*, 1867, in the case of Quebec, and *The Manitoba Act*, 1870, in the case of Manitoba).[11] The remaining provinces are free to offer their services in a language other than English if they so desire, as they do not specifically fall under the official-languages provisions of the *Charter*.

10. The English *Magna Charta* (or *Magna Carta*), which was proclaimed by King John of England in 1215, became the foundation of the fundamental rights and freedoms of the individual.

11. These statutes have now become the *Constitution Act*, 1867, and the *Manitoba Act*, 1870 [33 Vict., c. 3 (Can.)]. They both represent a part of the Canadian constitution.

Minority English- or French-language groups in a province where the official language is other than their own language are granted special rights or guarantees under the *Charter*, so that their children can be educated in their own language.

Enforcement of Rights

A charter of rights and freedoms is of no value to anyone unless some means are provided to enforce the rights granted under it. To provide for enforcement, the *Charter* states that persons who believe their rights or freedoms under the *Charter* have been infringed upon may apply to a court of law for a remedy that would be appropriate in the circumstances. For example, unless good cause exists for denial of bail, a person accused of a crime and denied bail would be entitled to a bail order on application to a proper court. Similarly, any restriction placed upon a person's religious activity by a public servant would, perhaps, constitute a denial of religious freedom, and the person so affected could apply to a competent court for an order preventing such a restriction. With both of these examples, it is important to note that the rights are not absolute, and, if the actions of the public servant were found to be justified, the courts might not provide the relief requested. To carry the bail example further, if an accused person represents a clear danger to the public if released pending trial, the denial of bail in the first instance may be justified, and the court may very well agree that the denial was a reasonable decision in order to protect the public.

With respect to enforcement of rights under the *Charter*, it is also important to realize that the *Charter* would appear to apply only to governments, and not to individuals. This does not mean, however, that individuals may infringe on the freedoms of others; action can be taken against them under other laws. What the *Charter* does is protect individual rights and freedoms from unreasonable restriction or interference by governments or persons acting on their behalf. It also protects the rights of these same individuals to carry on business without undue government interference or restriction.

If a government passes legislation that encroaches on the individual rights or freedoms set out in the *Charter*, then the individual affected must take steps to have a court decide if the statute has affected a particular right guaranteed under the *Charter*. This is necessary because all statutes passed by a legislature or Parliament are presumed to be valid until determined otherwise by the courts. This means that the individual must bring the matter before the court for a decision. It is then up to the government to show that the statute or law in question is not invalid but a reasonable limitation on the right or freedom, which "can be demonstrably justified in a free and democratic society," as provided in s. 1 of the *Charter*. If the government cannot satisfy the court that the limits imposed by the statute are reasonable and necessary, the court may very well decide that the statute is invalid.

CASE IN POINT

Subjected to ten years of Canada Customs seizures of its imported gay and lesbian erotica, a Vancouver bookstore challenged the actions under its *Charter* Rights of Equality and Expression. The Supreme Court of Canada preserved the government's right to prohibit obscenity, but a community standard cannot suppress minority expression and, moreover, must not discriminate, as was the case with more lenient treatment of heterosexual erotica.

Little Sisters Book and Art Emporium v. Canada, [2000] 2 S.C.R. 1120.

Since the establishment of the *Charter* in 1982, the Supreme Court of Canada has ruled on many other statutes in addition to those that affect business activity. Many decisions were related to the *Charter* provisions of equality rights and the fundamental freedoms of religion, expression, media rights, and assembly, and a significant number dealt with the application of *Charter* rights to the *Criminal Code.* While it is difficult to clearly define the attitude of the Supreme Court toward the *Charter*, the Court has tended to stress the rights of the individual over collective rights in many of its decisions. It has also, in some cases, restricted or delineated the legislative and other activities of governments and its agencies.[12]

COURT DECISION: Practical Application of the Law

Charter of Rights—Freedom of Commercial Expression

***The Attorney General of Quebec v. La Chaussure Brown's Inc., et al.*, [1988] 2 S.C.R 712, 1988 CanLII 19.**

This 1988 decision by the Supreme Court of Canada is one of the most important cases upholding the freedom of expression as guaranteed under s. 2(b) of the Charter of Rights and Freedoms. Business firms in Quebec were charged for violating provincial law that required public signs, commercial advertising and firm names on business premises to be in the French language only. Firms challenged the statute on the basis that it violated their s. 2(b) Charter right.

THE COURT — It is apparent to this Court that the guarantee of freedom of expression in s. 2(b) of the Canadian *Charter* and s. 3 of the Quebec *Charter* cannot be confined to political expression, important as that form of expression is in a free and democratic society. The pre-*Charter* jurisprudence emphasized the importance of political expression because it was a challenge to that form of expression that most often arose under the division of powers and the "implied bill of rights," where freedom of political expression could be related to the maintenance and operation of the institutions of democratic government. But political expression is only one form of the great range of expression that is deserving of constitutional protection because it serves individual and societal values in a free and democratic society.

The post-*Charter* jurisprudence of this Court has indicated that the guarantee of freedom of expression in s. 2(b) of the *Charter* is not to be confined to political expression. In holding, in *RWDSU v. Dolphin Delivery Ltd.*, [1986] 2 S.C.R. 573, that secondary picketing was a form of expression within the meaning of s. 2(b) the Court recognized that the constitutional guarantee of freedom of expression extended to expression that could not be characterized as political expression in the traditional sense but, if anything, was in the nature of expression having an economic purpose. Although the authority canvassed by McIntyre J. on the importance of freedom of expression tended to emphasize political expression, his own statement of the importance of this freedom clearly included expression that could be characterized as having other than political significance, where he said of freedom of expression at p. 583: "It is one of the fundamental concepts that has formed the basis for the historical development of the political, social and educational institutions of western society."

In order to address the issues presented by this case it is not necessary for the Court to delineate the boundaries of the broad range of expression deserving of protection under s. 2(b) of the Canadian *Charter* or s. 3 of the Quebec *Charter.* It is necessary only to decide if the respondents have a constitutionally protected right to use the English language in the signs they display, or more precisely, whether the fact that such signs have a commercial purpose removes the expression contained therein from the scope of protected freedom.

In our view, the commercial element does not have this effect. Given the earlier pronouncements of this Court to the effect that the rights and freedoms guaranteed in the Canadian *Charter* should be given a large and liberal interpretation, there is not sound basis on which commercial expression can be excluded from the protection of s. 2(b) of the *Charter.* It is worth noting that the courts below applied a similar generous and broad interpretation to include commercial expression within the protection of freedom of expression contained in s. 3 of the Quebec *Charter.* Over and above its intrinsic value as expression, commercial expression which, as has been pointed out, protects listeners as well as speakers plays a significant role in enabling individuals to make informed economic choices, an important aspect of individual self-fulfillment and personal autonomy. The Court accordingly rejects the view that commercial expression serves no individual or societal value in a free and democratic society and for this reason is undeserving of any constitutional protection.

What are the wider implications of this decision for business persons across Canada?

12. *McKinney v. University of Guelph*, [1990] 3 S.C.R. 229.

While many of the decisions have required the Court to decide if legislation (or a part thereof) offends the *Charter*, the Court has been much more willing of late to "read into" legislation words or provisions that the legislature did not include, either deliberately or by error. This raises the question of the propriety of an unelected judiciary assuming the functions of an elected government, which has become a matter of considerable debate.

Overall, the *Charter of Rights and Freedoms* has set certain parameters for governments and their boards and agencies in their dealings with business activities. Today, these governments and bodies must be more aware of the rights of the individual in both the legislative and administrative activities that they conduct.

Occasionally, the government may also exercise its right to place *proposed* legislation before the Supreme Court of Canada (a reference case) to determine its constitutionality in advance. This feature of the Canadian Constitution does not exist in the United States, and was most recently exercised by our federal government in the 2004 Reference to Same Sex Marriage.[13]

By placing its proposed same sex marriage legislation before the Supreme Court of Canada, the federal government received the assurance that it had the power to define civil marriage as a lawful union between the two persons, and the reminder that freedom of religion could not compel religious groups to change their practices.

Business persons must recognize the significance of this case and make accommodation for same sex married couples in the administration of their employee benefit programs (pensions, insurance, and the like).

A QUESTION OF ETHICS

Recent proposals have brought the notion of including religious-based legal systems, such as Islamic sharia law, or social systems of aboriginal law, into the structure of the Canadian legal system. At one end of the spectrum this is purely a matter of interest between persons who consent to the application of alternative legal systems, and at the other end, critics who believe that only one system should exist, open and applicable to all. Is there room for this kind of choice in multi-cultural Canada, or is this a slippery slope toward an unworkable system?

MANAGEMENT ALERT: BEST PRACTICE

Constitutional cases, when they do affect business, do so in a significant way. These cases are, by definition, fundamental. Often they arise from a new law prohibiting old practices as a result of social change — for example, tobacco advertising. Businesses large and small must monitor the news of constitutional cases, for they have little influence over the outcome and they must immediately absorb the changes. It's also important to be aware of similarities or common themes that are applicable across a variety of industries — once a decision opens the floodgates, there can be many such court actions.

Protection of Other Special and General Rights and Freedoms

The rights and freedoms set out in the *Charter* are intended to be the basic or minimum rights entrenched or guaranteed for all citizens. The *Charter* also recognizes that other rights and freedoms exist that all individuals, or certain of them, may enjoy. Among the groups assigned special rights are Canada's aboriginal peoples. These groups were granted

13. 2004 SCC 79.

a number of freedoms and rights under treaties they entered into with the government in the past. The *Charter* recognizes these rights, in a sense, by stating that it will not interfere with or limit special rights previously acquired, or that may be acquired in future through land-settlement claims.

The framers of the *Charter* also recognized that a charter of rights could not include all of the rights Canadians possessed at the time the *Charter* came into effect. They realized that while many of these rights or freedoms were not included in the *Charter*, they were, nevertheless, important rights and freedoms. In their view, the *Charter* should ensure that its specific guarantee of certain rights and freedoms would not deny Canadians these other freedoms and rights. Consequently, the *Charter* simply entrenches the fundamental or basic rights and freedoms.

Those rights not specifically entrenched may be infringed upon by governments, as they would not have the special protection of the *Charter*, however. For example, unlike the U.S. constitution, the Canadian *Charter of Rights and Freedoms* does not enshrine the right to own property. Property rights remain outside the *Charter*. Consequently, governments may at any time impose limits on this right, or could, theoretically, abolish the right entirely if they so desired.

The remainder of the *Charter* deals with governments and their powers. It specifically provides in s. 30 that the *Charter* applies to the territories as well as the provinces, and stipulates (in s. 31) that the legislative powers of the various bodies or authorities are not changed by the *Charter*. This latter section states that the powers divided between the federal government and the provincial legislatures under the *British North America Act* remain unchanged. Moreover, the legislative authority of each level of government is subject to the limitation imposed upon its powers by the *Charter*. In effect, the *Charter* defines those rights and freedoms that both levels of government must respect in their legislative jurisdictions.

CLASSIFICATION OF LAWS

Statute law and the Common Law may be classified in two broad, general categories. The first is **substantive law**, and includes all laws that set out the rights and duties of individuals. The second broad classification is **procedural law**. This area of the law includes all laws that set out the procedures by which individuals may enforce their substantive law rights or duties.

Substantive law

All laws that set out the rights and duties of individuals.

Procedural law

The law or procedures that a plaintiff must follow to enforce a substantive law right.

To illustrate these two classifications of law we might note an old observation of the law relating to assault, which says that "your right to swing your arm stops just short of your neighbour's nose." Put another way, you owe a duty to your friends not to injure them if you swing your arm in their presence. If you should strike one of your friends through your carelessness, your friend has a right to redress for the injury you cause. This right is a **substantive right**, and it represents a part of the substantive law. To enforce the right, your friend would institute legal proceedings to obtain redress for the injury, and the steps taken would be part of the procedural law.

Substantive right

An individual right enforceable at law.

Substantive law may be further subdivided into two other types of law: **public law** and **private law**. Public law deals with the law relating to the relationship between the individual and the government (or its agencies). The *Criminal Code* and the *Income Tax Act* are two examples of this kind of law at the federal level, and the various Highway Traffic Acts are examples of similar public laws at the provincial level. Under these statutes, if an individual fails to comply with the duties imposed, it is the Crown that institutes proceedings to enforce the law.

Public law

The law relating to the relationship between the individual and the government.

Private law concerns the relationship between individuals, and includes all laws relating to the rights and duties that the parties may have, or may create between themselves. Much of the Common Law is private law, but many statutes also represent private law. The law of contract is private law. Such statutes as the *Partnerships Act*[14] and the *Sale of*

Private law

The law relating to the relationship between individuals.

14. R.S.O. 1990, c. P.5.

Goods Act[15] are examples of private law. Legal rights, if private law in nature, must be enforced by the injured party.

Anderson, a construction worker, is digging a trench alongside a sidewalk and is placing the excavated soil on the walkway. While walking along the sidewalk, Brown complains to Anderson that the soil is blocking his way. Anderson is angered by Brown's complaint and strikes him with his shovel. In this situation, Anderson has violated s. 265 of the *Criminal Code*[16] by striking Brown with the intention of causing him injury. The *Criminal Code* is a public law, and it is the Crown that will take action against Anderson for his violation of the law.

Anderson, in this example, also owes a duty to Brown not to injure him (a private law duty). If Anderson injures Brown, as he did in this case, Brown has a Common Law right to recover from Anderson the loss that he has suffered as a result of Anderson's actions. Brown's right at Common Law is a private law matter, and Brown must take steps himself to initiate legal proceedings to enforce his right against Anderson.

In this case, both the Crown in the enforcement of the *Criminal Code*, and Brown in taking legal action against Anderson, use procedural law to enforce their rights.

Private law is sometimes referred to as civil law, to distinguish private laws of a non-criminal nature from public laws (principally the *Criminal Code*). Unfortunately, this has caused some confusion in Canada, because most of the private law in the province of Quebec has been codified, and the law there is referred to as the *Civil Code*. While both the Common Law and the *Code* deal with private law, care must be taken to note the distinction between the two bodies of law when reference is made to civil law.

SUMMARY

The law is the principal means by which the state maintains social control, and the system of courts is the vehicle used for its enforcement. The first laws were not laws as such, but family behavioural rules, and, later, religious and non-religious taboos. These were enforced, first, by the family elders, and, later by the community. As the community became stronger, it gradually assumed more and more of the duties of law-making and, with the development of the early city-states, law-making and enforcement took on an organized character. As the power of the state increased, so too did the areas of human endeavour that the state brought under its control.

Virtually every modern state has a constitution which sets out the powers of government and the rights of its citizens. In Canada, this is found in our *Constitution Act* and *Charter of Rights and Freedoms*.

The law that we have today has evolved over a long period of time, and has become a very complex system comprising the Common Law, equity, constitution-based statute law, and a subordinate type of law called administrative law. These laws may be classified as either substantive law or procedural law, with the former setting out rights and duties and the latter, as the name indicates, setting out the procedure for the enforcement of the substantive law rights.

15. R.S.O. 1990, c. S.1.

16. R.S.C. 1985, c. C-46, s. 265, as amended.

KEY TERMS

The law, 5
Common Law, 8
Statute law, 8
Canon Law, 10
Law Merchant, 10
Equity, 10
Bill, 11
Motion, 11
Royal assent, 11
Proclaimed, 11
Revised statutes, 11
Civil Code, 12
Administrative law, 13
Regulations, 13
Administrative tribunals, 13
Constitution, 14
Substantive law, 22
Procedural law, 22
Substantive right, 22
Public law, 22
Private law, 22

REVIEW QUESTIONS

1. What impact does the *Canadian Charter of Rights and Freedoms* have on rights and freedoms not mentioned specifically in the *Charter*? Could these "other rights and freedoms" be curtailed or extinguished by governments?
2. What is the difference between a "right" and a "privilege"?
3. Why are "rights" and "duties" often considered together when one thinks of laws?
4. Could a society exist without laws? If not, why not?
5. "Advanced civilizations are generally characterized by having a great many laws or statutes to control the activities of the citizenry." Comment on the validity of this statement.
6. On what basis are *Charter* fundamental rights and freedoms open to restriction by Parliament or the provincial legislatures?
7. Why is the doctrine of *stare decisis* an important part of the Common Law system?
8. How does the Common Law differ from the principles of equity? From statute law?
9. How does a legislature establish a new law? Explain the procedure.
10. Define substantive law, and explain how it differs from procedural law.
11. Describe the difference between the Common Law and the *Civil Code* of the province of Quebec. What are the relative merits of each system?
12. "The supremacy of the state was reached when it managed to exercise a sufficient degree of control over the individual to compel him or her to use the state judicial system rather than vengeance to settle differences with others." Why was it necessary for the state to require this of the individual?
13. How does a "regulation" made under a statute differ from other "laws"?
14. Explain how the enforcement of a public law differs from the enforcement of rights under private law.
15. The *Canadian Charter of Rights and Freedoms* has been described as being "supreme" law, or law that is "entrenched." Why, or in what sense, is this the case?
16. Explain the Common Law system, and how it relates to the function of the courts.
17. Does the *Canadian Charter of Rights and Freedoms* permit the Supreme Court of Canada to override the will of Parliament or a provincial legislature? If so, in what way?
18. If changing social attitudes or values were to dictate a change in the *Canadian Charter of Rights and Freedoms*, how would this be accomplished?
19. Why is the word "law" so difficult to define in a precise manner?
20. In what way does the *Constitution Act*, 1982, affect the legislative jurisdiction of the Parliament of Canada and the provincial legislatures? How are questions of jurisdiction decided?

MINI-CASE PROBLEMS

1. A freight train derailed, dumping dangerous chemicals into a small stream in British Columbia, causing significant environmental damage. What jurisdictional issues are raised by this scenario?
2. A coastal province passed a law prohibiting boats and ships from dumping waste along its shoreline. A ship's captain was later charged with commission of such an offence. What defence may exist to such a charge?
3. Simone believes that genetically-modified foods are extremely dangerous for people to consume. What paths can she take in fighting (legally) for her belief? Which one would be the most efficient? Why?

Case Problems for Discussion

CASE 1

Mary applied for a job at the Millstone Restaurant. She was told her uniform would be a white blouse and black skirt, with a hem an inch above the knee. She agreed, but when she started work, she realized that female staff was dressed accordingly, but the men wore white shirts and black pants. At a later date, Mary appeared at work also dressed in a white shirt and black pants. The manager of the restaurant told her she was "out of uniform," words were exchanged, and Mary was fired. Is there a *Charter* issue here?

Discuss.

CASE 2

FM 96 Tiger Radio wanted to set up a three-day live-remote event in a provincial park for the Labour Day weekend. The provincial parks commission approved the plan, subject to a payment of $4,100, comprised of a park event permit ($1,000), a sanitation charge ($1,000), a broadcasting permit ($500), a beer/wine premises special occasion permit ($1,500), and a fire inspection fee for the beer/wine consumption premises ($100). How should FM 96 respond to the provincial commission?

CASE 3

In the year 1555, Maxwell was drunk, lost control of his horse and killed a child. He was executed for his crime. In the year 1955, Sharon was sober, lost control of her car and killed a child. She was given a suspended sentence and ordered to pay a $100 fine. In the year 2005, Karl was drunk, lost control of his car and killed a child. He was given one year imprisonment, a $1,000 fine and is subject to a lifetime licence suspension.

In what way is the principle of *stare decisis* at work here, if at all?

The Online Learning Centre offers more ways to check what you have learned so far. Find quizzes, Weblinks, and many other resources at www.mcgrawhill.ca/olc/willes.

CHAPTER 2

The Judicial System and Alternative Dispute Resolution

Chapter Objectives

Knowing where and how to enforce rights and obligations is a key business survival skill. After study of this chapter, students should be able to:

- Describe the development, content and structure of the judicial system.
- Explain the sequence of steps in court procedure, particularly civil court procedure.
- Identify how and why alternative dispute resolution may be the best option for the settlement of business disputes.
- Explain the role of barristers and solicitors, the range of services provided by the legal profession, and the concept and limitations of court costs.

INTRODUCTION

In a complex, modern society, such as in Canada or the United States, the courts play an important role in the lives of citizens. First and foremost, the courts decide disputes between individuals (including corporations), and between individuals and the state. This is the role that generally comes to mind when the function of the courts in society is considered. The courts have other very important functions to perform as well, however. They are the chief interpreters of the constitution, and in this capacity decide if legislation passed by either the federal government or the provincial legislatures exceeds the governments' respective powers or violates the rights and freedoms of individuals. In this sense, the courts are the guardians of these rights and freedoms, and through their interpretation of the constitution may enlarge or restrict its provisions.

For example, s. 2(b) of the *Canadian Charter of Rights and Freedoms* provides for freedom of expression, including freedom of the press. Section 1 permits a government to establish a limit on the exercise of this freedom if the limit is "demonstrably justified in a free and democratic society." If the government passed a law prohibiting the publication in a newspaper of any literary review of certain kinds of literature, would this violate freedom of the press under s. 2(b) of the *Charter*? This would be a matter for the court to decide, as it would be necessary to establish the validity of the legislation in terms of whether the law represented a reasonable limit on the freedom of the press, as provided in s. 1 of the *Charter*. If the court decided that the law represented a reasonable limit, then the fundamental freedom would be diminished by that interpretation. If the court ruled that the law was an unreasonable limitation, then it would be invalid, and the court would have preserved the freedom in its broader sense. How the courts interpret the provisions of the *Charter* has a profound effect on the rights and freedoms that an individual will, in reality, enjoy. Consequently, the Supreme Court of Canada, which has the ultimate say in how the provisions of the *Charter* may be interpreted, is expected to take great care and deliberation in reviewing challenged legislation that touches upon these rights and freedoms.

The courts are also responsible for deciding the legislative jurisdiction of the federal and provincial governments under the *Constitution Act*. This is not a new role for the courts, as the court has been charged with this duty since the original *British North America Act* was passed in 1867. Again, the Supreme Court of Canada has the last word in deciding the question of whether a government has exceeded its legislative powers under the constitution. If the Supreme Court rules that a law is beyond the powers of the particular legislative body, the law is a nullity and unenforceable.

Much of the law relating to business concerns only disputes between corporations or individuals. However, as the regulatory powers of governments at both the federal and provincial levels intrude to a greater degree into areas of business activity previously untouched by government, the need for judicial review of legislation in light of the *Charter* becomes more important. For this reason, the courts may, in a sense, also be considered the guardian of business rights and freedoms.

In a business context, the courts are generally called upon to interpret contracts between individuals and business firms in order to determine the rights and duties of each party. As well, when a person or business is injured through the negligence of others, the courts will play an important role in establishing the responsibility for the loss and the compensation payable. For example, if a construction firm carelessly constructs a wall of a building and it falls upon a passerby, the injured person may take legal action against the construction firm for its negligence in the construction of the wall. The court will hear the dispute and decide if the firm is liable for the injuries to the person. If liability is found, the court will also establish the amount of the loss payable as compensation.

DEVELOPMENT OF THE LAW COURTS

The legal system as we know it today evolved over a long period of time, developing gradually as the law developed. It was not until the state was strong enough to require resort to the courts and the rule of law that the courts assumed an important place in the social system.

The Canadian and American judicial systems had their beginnings in England. While the two North American jurisdictions have since evolved in slightly different ways, they are both an outgrowth of the English court system, and consequently share the same heritage. In both Canada and the United States, courts were established in the colonies as part of the British colonial administrative structure. These courts applied British law to the cases that came before them during the colonial period.

A significant change took place in court procedure after the War of American Independence. The war severed both political and legal ties with England, but the law, and much of the legal system, remained in the former colonies. Pennsylvania, for example, continued to call its civil trial court the Court of Common Pleas, and the law that the court dispensed was the Common Law. Most states at the time of the War of American Independence adopted English Common Law as it stood in 1776, with appropriate modifications to meet local needs and conditions.

Canada did not experience the wrenching changes brought about by the War of American Independence. Instead, the Canadian system of courts changed gradually as the country moved from a colonial status to that of a dominion. After 1663, the law in what is now Quebec, Ontario, Prince Edward Island, and part of Nova Scotia consisted of the Customs of Paris, which had been declared the law of New France by the Conseil Superieur of France. The Seven Years War resulted in the capture by England of the first colonies in 1758 and 1759, however, and the imposition of English law in New France. France, by the Treaty of Paris in 1763, formally surrendered what is now Quebec, Cape Breton, and Prince Edward Island to England and the law administered by the new civil courts established in 1764 was English law. The *Quebec Act*[1] of 1774 made a further change in the legal system by introducing a Court of King's Bench and a significant change in the law — the criminal law in the new British province of Quebec was declared to be English law, but the civil law was to be based upon "Canadian" law, i.e., the Customs of Paris.

1. 14 Geo. III, c. 83.

Quebec, as a result, continued to follow this law until 1866, when the first *Quebec Civil Code* was compiled and put into effect.

When the War of American Independence officially ended in 1783, more than 10,000 United Empire Loyalists left the United States and moved into what was then the western part of Quebec. The new settlers did not understand, nor did they like, French civil law. Initially, in response to their demands, the western part of Quebec (which is now Ontario) was divided into four districts, where English law would apply. Each new district had a Court of General Sessions to hear minor criminal cases, a civil Court of Common Pleas to hear Common Law matters, and a Prerogative Court to deal with wills and intestacy. In 1791, the *Canada Act*[2] created two separate provinces: the province of Upper Canada (now Ontario) with its own governor, executive council, and elected House of Assembly; and the province of Lower Canada (now Quebec), which continued as it had before.

The first Act passed by the new House of Assembly for the Province of Upper Canada was a *Property and Civil Rights Act.* This Act provided that English law (as it existed on October 15, 1792), would be the law of the province in all matters concerning property and civil rights. The law in amended form is still in force.[3] The following year, the courts were reorganized by replacing the Prerogative Courts in each district with a Surrogate Court. A Probate Court was also established to hear appeals in cases concerning wills and intestacy on a province-wide basis.

In 1794, the *Judicature Act*[4] was passed. The effect of this Act was to abolish the former Court of Common Pleas, and replace it with the Court of King's Bench. The new court had jurisdiction over both civil and criminal matters for the entire province. The judges of the new court were also different from the old (all of the new judges having been members of the Bar of either Upper Canada, England, or some other English jurisdiction). A later act provided for the appointment of judges "during the period of their good behaviour," and required an address from both the council and the Assembly to remove a judge from office.[5]

The Court of Chancery was established in 1837 for the province of Upper Canada, and was composed of the chancellor (who was also the governor) and a vice-chancellor. All judicial powers were exercised by the vice-chancellor, who handled all of the legal work. In 1849, the office of chancellor became a judicial office, and a second vice-chancellor was appointed.

The courts were again changed in 1849, this time by the creation of a Court of Common Pleas. Its jurisdiction was concurrent with that of the Court of Queen's Bench, and in 1856 the procedure of the two courts was modernized and unified. In the same year, a new Appeal Court for Upper Canada was established — the Court of Error and Appeal. This court consisted of all of the judges of the Courts of Queen's Bench, Common Pleas, and Chancery sitting together as an appeal body at Toronto. This particular court later became the Court of Appeal. The *British North America Act,* 1867, did not change the composition of these courts, but simply changed the name from Upper Canada to Ontario.[6] The names of the courts in the other provinces also remained unchanged upon joining Confederation. The courts in most of the older provinces had developed along lines similar to those of Ontario, although legislation that brought about the change had been passed at different times, and with some variation.

Early Law Reform

In the latter part of the nineteenth century, changes that took place in the structure of the courts in England soon appeared in Canada. The many Law Reform Acts that were passed in England between 1873 and 1925 were reflected in a number of changes in Canadian law and its judicial system. The Prairie provinces were influenced not only by law reform in

2. 31 Geo. III, c. 31.
3. *Property and Civil Rights Act,* R.S.O. 1990, c. P-29.
4. 34 Geo. III, c. 2.
5. *An Act to Render the Judges of the Court of King's Bench in This Province Independent of the Crown,* 4 Wm. IV, c. 2.
6. *British North America Act,* 1867, 30-31 Vict., c. 3, as amended.

England, but also by other Canadian provinces in the establishment of their court systems. When Ontario introduced its own law-reform legislation (in 1881, 1909, and 1937), which reorganized the Supreme Court into two parts (a High Court of Justice as a trial court, and a Court of Appeal to hear appeals), a number of other provinces followed suit.[7]

Law reform in Canada has been an evolutionary process. The general trend has been toward a streamlining of the judicial process, with a gradual reduction in formal procedure. Although much variation exists from province to province, a provincial judicial system generally consists of a Small Claims or Small Debts Court, a Superior or Supreme Court, and a Court of Appeal. In those provinces where a Court of Appeal as a separate court does not exist, the Supreme Court is empowered to hear appeals from courts of original jurisdiction. Criminal cases are usually heard by the Supreme Court. In addition, each province has a Provincial or Magistrate's Court (presided over by either a provincial judge or magistrate), which hears minor criminal matters and cases concerning violations of non-criminal provincial statutes.

In 1982, the introduction of the *Charter of Rights and Freedoms* created a new role for the courts — that of chief interpreter of the rights and freedoms of the individual. Prior to this time, the role of the courts in constitutional matters was for the most part limited to the determination of the validity of provincial or federal statutes under ss. 91 and 92 of the *British North America Act.* However, with the advent of the new constitution, this role was expanded. The court is now the body charged not only with the responsibility to determine the jurisdiction of the law-making bodies, but also with the responsibility to determine if statutes offend the *Charter of Rights and Freedoms* as well. In a very real sense, the Supreme Court of Canada has become the ultimate interpreter of the constitution.

THE STRUCTURE OF THE JUDICIAL SYSTEM

In Canada, there are many different courts, each with a different jurisdiction. In this context, jurisdiction means the right or authority of a court to hear and decide a dispute. Jurisdiction may take a number of different forms. Usually the court must have the authority to deal with cases of the particular type brought before it, in addition to authority over either the parties or the property in dispute. With respect to the first type of jurisdiction, the authority of the court may be monetary (in the sense that the court has been authorized to hear cases concerning money up to a set amount), or geographic (it hears cases concerning land within the particular province or area where the land is situated). In the case of jurisdiction over the parties to a dispute, the court must have the authority or power to compel the parties' attendance or to impose its decision on them.

Courts of law may be placed in two rather general classifications. The first group is called courts of original jurisdiction. These are courts before which a dispute or case is heard for the first time by a judge, and where all the facts are presented so that the judge can render a decision. Courts of original jurisdiction are sometimes referred to as **trial courts**, where both civil and criminal cases are first heard.

Trial court

The court in which a legal action is first brought before a judge for a decision.

Courts that fall into the second group are called courts of appeal. These courts, as their name implies, hear appeals from the decisions of courts of original jurisdiction. Courts of appeal are superior or "higher" courts in that their decisions may overrule or vary the decisions of the "lower" or trial courts. Their principal function is to review the decisions of trial courts if one of the parties to the action in the lower court believes that the trial judge made an erroneous decision. They do not normally hear evidence, but instead hear argument by counsel for the parties concerning the decision of the trial court. Usually an appeal alleges that the judge hearing the case at trial erred in the application of the law to the facts of the case. Sometimes the appeal is limited to the amount of damages awarded, or, in a criminal case, to the severity of the penalty imposed. On occasion, an appeal court may find that the judge at trial failed to consider important evidence in reaching his or her decision, in which case the appeal court may send the case back to the lower court for a new trial.

7. See the *Judicature Act*, 9 Edw. VII, c. 5, as an example of the New Brunswick Acts of the General Assembly.

Federal Courts

In Canada, each province has a number of courts of original jurisdiction and at least one appeal court. In addition, the federal court system also exists to deal with matters that fall within the exclusive jurisdiction of the federal government. The Federal Court Trial Division hears disputes between provincial governments and the federal government; actions against the federal government; admiralty, patent, trademark, copyright, and taxation matters; and appeals from federal boards, tribunals, and commissions. In some cases, the court has exclusive jurisdiction to hear the dispute, in others, the jurisdiction is concurrent with that of the superior provincial courts, so that a person may sue in either the Federal Court Trial Division or the appropriate provincial court. A trial decision of the Federal Court may be appealed to the Federal Court of Appeal, and the appeal with leave to the Supreme Court of Canada.

A special Tax Court is a part of the federal court system that hears disputes between taxpayers and the Canada Customs and Revenue Agency. The jurisdiction of the court is limited to appeals from tax assessments, and, if either party wishes, the decision of the Tax Court may be "appealed" to the Federal Court of Appeal.

At the federal level, the Supreme Court of Canada is maintained to hear important appeals from the appeal courts of the various provinces, as well as those from the Federal Court of Appeal.

Provincial Courts

There is no uniform system of courts in the provinces, as each province has the authority to establish its own system and to assign to each court a specific jurisdiction. Fortunately, however, most of the provinces have established courts somewhat similar in jurisdiction. Bearing this in mind, and the fact that variation in names and powers do exist, the following list represents what might be considered to be typical of the court systems found in a province.

Criminal Courts

Magistrate's or Provincial Court

A Magistrate's Court or Provincial Court is a court of original jurisdiction that is presided over by a provincially appointed magistrate or judge. This court generally deals with criminal matters relating to accused individuals or corporations, although many provinces have empowered the court to hear cases involving the violation of provincial statutes and municipal by-laws where some sort of penalty is imposed.

The Provincial or Magistrate's Court initially deals with all criminal cases, either as a court with jurisdiction to dispose of the matter (as in cases involving less serious offences, where the court has been given absolute jurisdiction), or where the accused has the right to elect to be tried by a magistrate or Provincial Court judge. It will also hold a preliminary hearing of the more serious criminal cases to determine if sufficient evidence exists to have the accused tried by a higher court. All provinces except Quebec have a Magistrate's Court or Provincial Court. The Quebec counterpart of the Magistrate's Court is the Court of Sessions of the Peace.

Provincial Supreme Court

Each province has a Supreme Court or Superior Court empowered to hear the most serious criminal cases. Ontario calls this court the Superior Court of Justice. In many provinces, justices of the court periodically travel throughout the province to hear these cases (usually at the County court house) at sessions of the court called ***assizes.*** In some provinces, the court follows the English custom of presenting the judge with a pair of white gloves at the beginning of the session, if no criminal cases are scheduled to be heard.

Assizes

Sittings of the court held in different places throughout the province.

Youth Courts

Youth Courts are particular courts designated in each province to deal with cases where young persons are accused of committing criminal offences. Under the *Youth Criminal*

Justice Act,[8] a young person is defined as a person 12 years of age or older and under 18 years of age. Persons over this age limit are treated as adults and would be tried in the ordinary courts if accused of a criminal offence.

Youth Courts are presided over by judges who have the powers of a justice or magistrate of a summary conviction court. They may, if the young person is found guilty, impose a fine, order compensation or restitution, direct the youth to perform community-service work, commit the youth to custody, or provide an absolute discharge, depending upon what is believed to be in the best interest of the young person. In some instances, the case may be transferred to ordinary courts. Unlike other courts, the names of young persons and the offences they have committed may not be published by the press or any other media and, under certain circumstances, the public may be excluded from the hearing.

Family Courts

Family Courts, while not criminal courts in the ordinary sense, have jurisdiction to deal with domestic problems and the enforcement of federal and provincial legislation that relates to family problems. Most of the cases in Family Court deal with non-support of family members, or family relationships that have deteriorated to the point where the actions of one or more members of a family have become a serious threat to others. Family Courts are usually presided over by a magistrate or provincial court judge in those provinces where they have been established; in others, where separate courts do not exist, family disputes or legislation relating to family matters fall under the jurisdiction of the Superior or Supreme Court of the province.

Criminal Courts of Appeal

Each province has a Court of Appeal to review the convictions of accused persons by the Youth Court, Supreme Court, or Magistrate's (or Provincial) Court. A panel of judges presides over the Appeal Court, and the decision of the majority of the judges hearing the appeal decides the case. The final Court of Appeal in criminal matters is the Supreme Court of Canada. It will hear appeals from the decisions of provincial Courts of Appeal; however, neither the accused nor the Crown generally may appeal to the Supreme Court of Canada as a matter of right. Instead, they must obtain leave to appeal. A right to appeal, however, does exist in the case of an indictable offence if the decision of the provincial Court of Appeal on a question of law was not unanimous. Figure 2–1 outlines a typical appeal process.

FIGURE 2–1 **Criminal Appeals***

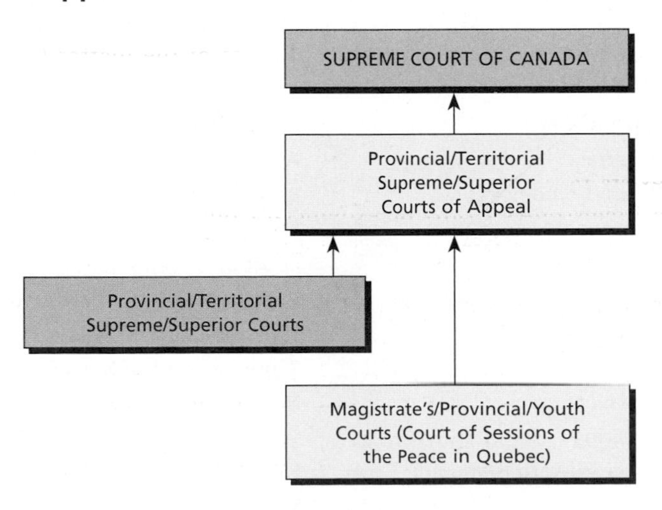

* Court names and, in some cases, appeal routes, differ for some provinces.

8. *Youth Criminal Justice Act*, S.C. 2002, c. 1.

Civil Courts

Most provinces have a number of civil courts to deal with disputes that arise between corporations or individuals, or between corporations or individuals and the government. Some courts have limited jurisdiction and hear only special kinds of disputes; others hear only appeals from inferior courts. Although most civil courts of original jurisdiction permit cases to be heard by both a judge and jury, in some courts, such as Small Claims Courts, cases are heard by a judge sitting alone. Courts of Appeal are always non-jury courts.

Small Claims Court

Small Claims Courts have jurisdiction to hear cases where the amount of money involved is relatively small. In those provinces that have a Small Claims Court, the court's jurisdiction is usually limited to hearing only those cases where the amount of money involved is less than $10,000. In some provinces, the Small Claims Court is limited to a much smaller monetary amount, although at the time of writing a number of jurisdictions are seriously considering raising their monetary limits, with British Columbia having raised its limit to $25,000.

Small Claims Courts are usually informal courts that may be presided over by a superior court judge in a number of provinces, and by a magistrate (as a part of the Provincial Court) in others. The cases the courts hear are usually small debt or contract disputes, and damage cases, such as claims arising (in some provinces) out of minor automobile accidents. Litigants in Small Claims Court frequently present their own cases, and court costs are usually low. The right to appeal a Small Claims Court decision is sometimes restricted to judgments over a specific amount, and, where an appeal exists, it is usually to a single judge of the Court of Appeal of the province.

Provincial Supreme Court

Each province has a Supreme (or Superior) Court to hear civil disputes in matters that are beyond or outside of the jurisdiction of the lower courts. Alberta, Manitoba, and Saskatchewan call their Supreme Court the Court of Queen's Bench. The provincial Supreme or Superior Court has unlimited jurisdiction in monetary matters, and is presided over at trial by a federally appointed judge. As shown in Table 2–1, although the Supreme Court of each province (or territory) is similar in jurisdiction, no one designation applies to all of these courts.

TABLE 2–1 Trial Level

Jurisdiction	Designation of Court
Alberta	Alberta Court of Queen's Bench
British Columbia	British Columbia Supreme Court
Manitoba	Manitoba Court of Queen's Bench
New Brunswick	Court of Queen's Bench of New Brunswick, Trial Division
Newfoundland & Labrador	Supreme Court of Newfoundland & Labrador (Trial Division)
Northwest Territories	Supreme Court of the Northwest Territories
Nova Scotia	Supreme Court of Nova Scotia
Nunavut	Nunavut Court of Justice
Ontario	Superior Court of Justice
Prince Edward Island	Supreme Court of Prince Edward Island, Trial Division
Quebec	Superior Court
Saskatchewan	Court of Queen's Bench of Saskatchewan
Yukon Territory	Supreme Court of Yukon Territory

Appeal Level	Designation of Court
Alberta	Alberta Court of Appeal
British Columbia	British Columbia Court of Appeal
Manitoba	Manitoba Court of Appeal
New Brunswick	Court of Appeal of New Brunswick
Newfoundland & Labrador	Supreme Court of Newfoundland & Labrador (Court of Appeal)
Northwest Territories	Court of Appeal for the Northwest Territories
Nova Scotia	Nova Scotia Court of Appeal
Nunavut	Nunavut Court of Appeal
Ontario	Court of Appeal for Ontario
Prince Edward Island	Supreme Court of Prince Edward Island, Appeal Division
Quebec	Court of Appeal
Saskatchewan	Court of Appeal of Saskatchewan
Yukon Territory	Court of Appeal of the Yukon Territory

Cases in the Supreme (or Superior) Court (both jury and non-jury) may be heard by judges who travel throughout the province to the various County court houses (the assizes), or in specified cities where the court sits without a jury on a regular basis. An appeal from a decision of the Supreme Court is to the Appeal Court of the province.

Surrogate or Probate Court

The Surrogate Court (or Probate Court, as it is called in New Brunswick and Nova Scotia) is a special court established to hear and deal with wills and the administration of the estates of deceased persons. The provinces of Newfoundland, Quebec, British Columbia, Ontario, Manitoba, and Prince Edward Island do not have special courts to deal with these matters, but, instead, have placed them under the jurisdiction of their Supreme Courts. In provinces that do have Surrogate or Probate Courts, the presiding judge is usually the same judge appointed as a local superior court judge.

Civil Courts of Appeal

Provincial Court of Appeal

Each province (or territory) has an Appeal Court, although (as indicated in Table 2–1) no one designation applies to all of these courts. The lines of appeal in civil cases are not always as clear-cut as those in criminal matters. Sometimes the right of appeal from an inferior court does not go directly to the Appeal Court of the province or the territory. In Ontario, for example, an appeal from a decision of the Small Claims Court (where the amount is above a certain minimum) would be appealed to a single judge of the Divisional Court, which is a part of the Superior Court of Justice. Supreme Court trial judgments, however, would be appealed to the provincial Court of Appeal or the Appeal Division of the provincial Supreme Court, as the case may be.

Supreme Court of Canada

The Supreme Court of Canada is the final and highest Appeal Court in Canada. It hears appeals from the provincial Appeal Courts, but the right to appeal is restricted. In civil cases, leave (or permission) to appeal must be obtained before a case may be heard by the Supreme Court of Canada, and normally the issue or legal point must be of some national importance before leave will be granted. The Court also hears appeals from the Federal Court, and it is the body that finally determines the constitutionality of statutes passed by both the federal and provincial governments. The civil appeal process is outlined in Figure 2–2.

FIGURE 2–2 Civil Appeals*

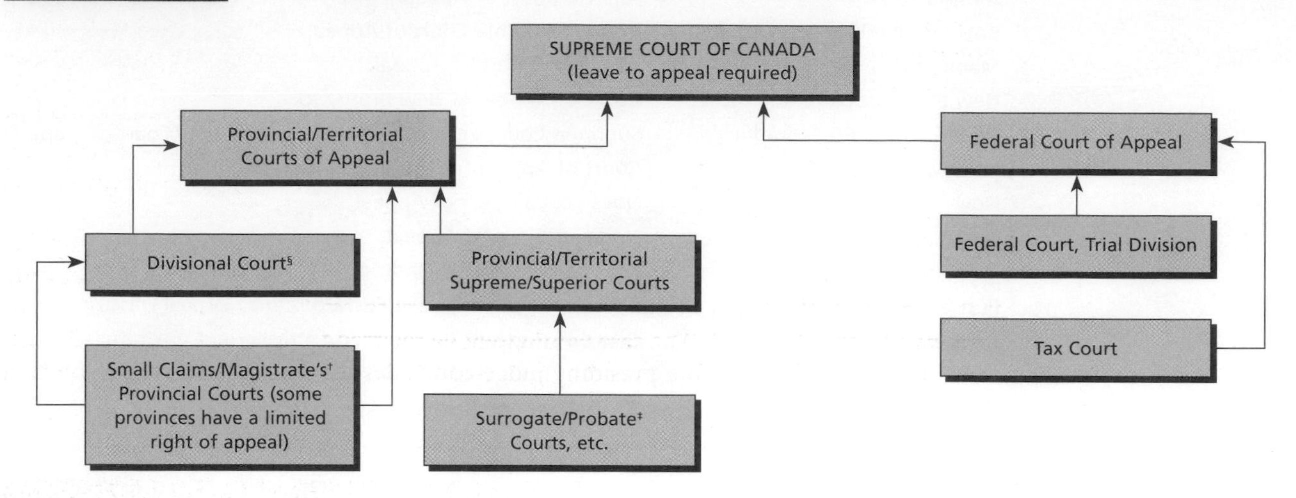

* Some provinces do not have all of the courts shown on this chart.

† Appeal routes vary from province to province with respect to Small Claims Courts.

‡ Special courts, such as Probate or Surrogate, usually have disputes litigated in the Supreme Court of the province.

§ Ontario only. The Divisional Court is a unique court in that it can conduct both trials and certain types of appeals.

THE JUDICIAL SYSTEM IN ACTION

Criminal Court Procedure

As the name implies, the Criminal Court is concerned with the enforcement of the criminal law. A Criminal Court is often the same court that deals with civil law matters, or it may be a court organized to deal exclusively with law of a criminal or quasi-criminal nature. The Provincial Court (Criminal Division) in Ontario, or the Magistrate's Court in most other provinces, is frequently the court with jurisdiction to deal with criminal matters of a minor nature, or to act as a court where a preliminary hearing of a more serious criminal offence would be held.

In a criminal case involving a minor or less serious offence, the Crown brings the case before the court by way of the summary conviction rules of procedure. In a serious case, it will bring the case by way of indictment. In either situation, the case is first heard in the Provincial Court (Criminal Division), or in the Magistrate's Court. These courts have absolute or elective jurisdiction to dispose of the case if the matter is minor in nature. When the offence is of a more serious nature, the court will conduct a preliminary hearing to determine whether the Crown has sufficient evidence to warrant a full hearing of the case by a superior court.

The procedure of the Magistrate's or Provincial Court tends to be very informal. The normal procedure at the hearing is to have the charge which the Crown has placed before the court read to the accused. The accused is then asked how he or she pleads. If the accused admits to the commission of the offence, a plea of guilty is entered, and the court will then hear evidence from the Crown to confirm the act and the circumstances surrounding it. A conviction will then be lodged against the accused, and a penalty imposed.

If a plea of not guilty is entered by the accused, the Crown is then obliged to proceed with its evidence to show that, in fact, the criminal act was committed (*actus reus*) by the accused, and that the accused had intended to commit (*mens rea*) the crime. Witnesses are normally called by the Crown to identify the accused as the person who committed the act and to establish this evidence. Counsel for the accused (the defence) then has the opportunity to cross-examine the witnesses.

On completion of the Crown's evidence, the defence counsel may ask the judge or magistrate to dismiss the case if the Crown has failed to prove beyond any reasonable doubt that the accused committed the crime. If the judge does not accept the defence counsel's motion, then the defence may proceed to introduce evidence to refute the Crown's case. This, too, is usually done by calling witnesses, although, in this instance, they testify on the accused's behalf. If defence witnesses are called, they are open to cross-examination by the Crown counsel.

Once all of the evidence has been presented to the court, both parties are entitled to sum up their respective cases and argue any legal points that may apply to the case. The judge or magistrate then determines the accused either not guilty or guilty of the crime, and his or her decision, with reasons, is recorded as the **judgment**.

Judgment

A decision of the court.

In the case of a preliminary hearing, the proceedings will normally end at the conclusion of the Crown's evidence (which will not necessarily be all of the evidence, but only that part which the Crown believes will be necessary to establish sufficient evidence to warrant a further hearing). The case would then be referred to the court with jurisdiction to try the matter in full if the presiding judge concludes that the case should go on to a full trial.

LATEST NEWS

FRONT-PAGE LAW

Two Arrested on Money Laundering Charges

RCMP News Release, Toronto, June 14, 2004—Earlier today, members of the Toronto Integrated Proceeds of Crime Section arrested one lawyer and his business associate on money laundering and related charges. The charges stem from four alleged money laundering transactions totaling approximately $750,000 Canadian currency.

"Investigations into alleged criminal activity involving a lawyer are uniquely difficult and complex," stated Chief Superintendent Ben Soave, head of the Combined Forces Special Enforcement Unit. "Investigative tools that police could normally employ are extremely restricted when the subject of an investigation is a lawyer. The unfortunate consequence of these restrictions is that the legal profession can be exploited to cloak criminal activity."

"For example, 'solicitor/client privilege' is a principle of law that protects the disclosure of all communication between a lawyer and his or her client. However, criminals or organized crime takes advantage of this needed, crucial and fundamental cornerstone of our justice system," commented Chief Superintendent Soave. "They go out of their way to engage, recruit, compromise, and corrupt lawyers and other members of the judicial system in order to further their criminal enterprises. The result is that criminals, with the assistance of a complicit lawyer, can hide behind the law, enabling them to shield their criminal activity from the legitimate scrutiny of law enforcement."

Are you comfortable with the Chief Superintendent's view? Should there be no concept of privilege? Where and how would you draw the line?

Source: Excerpted from Royal Canadian Mounted Police News Release "Two Arrested on Money Laundering Charges," June 14, 2004, www.rcmp-grc.gc.ca.

Civil Court Procedure

Civil cases follow a much different procedure. Before a civil case may proceed to trial, the parties must exchange a number of documents called **pleadings** that set out the issues in dispute and the facts surrounding the conflict.

Pleadings

Written statements prepared by the plaintiff and defendant that set out the facts and claims of the parties in a legal action, and are exchanged prior to the hearing of the case by the court.

Claim

Civil cases may begin in a number of ways, depending upon the court and the relief sought. In some provinces, the usual procedure in a simple dispute is for the plaintiff (the injured party) to issue a writ of summons against the defendant alleging the particular injury suffered by the plaintiff, and notifying the defendant that the plaintiff intends to hold the defendant responsible for the injury set out in the claim. The writ of summons is usually prepared by the plaintiff's lawyer and taken to the court's office, where the writ is issued by the court. It is then served personally on the defendant, usually by the sheriff or someone from that office.

Once the defendant receives the writ, he or she must notify the court's office that a defence will follow. This is done by filing a document called an appearance.

The next step in the proceedings is for the plaintiff to provide the defendant (and the court) with details of the claim and the facts that the plaintiff intends to prove when the case comes to trial. This document is called a statement of claim. In some provinces (Ontario and Nova Scotia, for example), a civil action is usually commenced by the issue of this pleading.

Defence

The defendant, on receipt of the statement of claim, must prepare a statement of defence setting out the particular defence that the defendant has to the plaintiff's claim and, if necessary, the facts that he or she intends to prove at trial to support the defence. The statement of defence is filed with the court and served upon the plaintiff. If the defendant also has a claim against the plaintiff, the defendant will file a pleading called a counterclaim, which is essentially a statement of claim. When a counterclaim is filed, the roles of the parties change. The defendant on his or her counterclaim becomes the plaintiff by counterclaim, the plaintiff, the defendant by counterclaim.

On receipt of the defendant's statement of defence, the plaintiff may wish to respond. In this case the response to the statement of defence will be set out in a document called a reply, which is filed and served on the defendant. If the defendant has served the plaintiff with a counterclaim, then the plaintiff (who at this point is also the defendant by counterclaim) will usually file a statement of defence to counterclaim, which will set out his or her defence. This usually ends the exchange of documents, and the pleadings are then noted closed.

Occasionally, a pleading may not contain sufficient information to enable the opposing party to properly prepare a response. If this should be the case, a further document called a demand for particulars may be served to obtain the necessary information.

Close of Pleadings

Once the pleadings have been closed, either party may set the action down on the list for trial by filing and serving a notice of trial on the other party. In some instances, where a jury may be appropriate, a jury notice may also be served. This indicates that the party serving the notice intends to have the case heard by a judge and jury.

Since the mid-1990s, most provinces have introduced a change in procedure for certain civil trials whereby the parties are expected to explore the possibilities of resolution of the dispute before trial with the assistance of a mediator. The mediator meets with the parties and attempts to resolve the issues in dispute through negotiation, but if the mediator fails to effect a settlement, the case then proceeds to the discovery and trial stage.

Discovery

To clarify points in the statement of claim or statement of defence, the parties may hold examinations under oath, called **examinations for discovery**. The right to an examination for discovery of the other party is conditional upon delivering an affidavit of documents, which is a list of all the documents that the party intends to rely on at trial that relate in any manner to the issue in the action. This affidavit must include all documents that the party controls and will produce, those that are privileged and will not be produced, and those that are relevant but are no longer in possession of the party. The opposing party may request a copy of the documents listed in an affidavit of documents, in order to better understand and respond to the case against them.

Examination for discovery

A pretrial oral or written examination under oath, of a person or documents.

Oral or written examinations for discovery in civil commercial actions are reserved to matters above the level of Small Claims Courts, with claims in the order of $50,000 and above. Only parties to the action can be examined, unless the court has approved the examination of a non-party, and only one representative from a corporation can be examined, again unless permission of the court has been obtained. At an examination for discovery, any question of fact relating to the issue in the action may be asked. The question and the answer are recorded and may be introduced as evidence at trial.

CHECKLIST FOR EXAMINATIONS FOR DISCOVERY

1. Attend the discovery with legal counsel, and keep calm.
2. Say nothing in the presence of others before or after the examination.
3. When questioned, be truthful, and tell the whole truth. Your forthrightness or deception will be evident in the transcript or at trial.
4. Listen to questions and do not rush to respond.
5. Request clarification of questions you do not really understand, and request simplification of overly complex questions.
6. Think, form your answer, and speak clearly. Your answer will be transcribed from a tape recording, so do not answer with an expression or gesture; your oral answer must stand on its own.
7. Speak only of facts within your personal knowledge.
8. If you don't know, say so. Don't guess, give opinions, exaggerate or estimate.
9. Reject attempts of questioners to summarize or re-characterize your answers for your agreement.
10. Once you have answered the question, stop speaking.
11. Look to your counsel for guidance when you need it.

Trial

At trial, the case follows a procedure that differs from that of a criminal action. In a civil matter, the counsel for the plaintiff usually begins the case with an opening statement that briefly sets out the issues and the facts that the plaintiff intends to prove. Witnesses are called, and evidence is presented to prove the facts in the claim. All witnesses may be subject to cross-examination by defence counsel.

On the completion of the plaintiff's case, counsel for the defendant may ask the judge to dismiss the plaintiff's case if the evidence fails to establish liability on the defendant's part. Again, if the judge does not agree with the defendant, the action will proceed, and the defendant must enter evidence by way of witnesses to prove that the plaintiff's claim is unfounded. Defence witnesses, like the plaintiff's witnesses, may be subject to cross-examination.

Witnesses may be of two kinds: ordinary witnesses who testify as to what they saw, heard, or did (direct evidence); and expert witnesses who are recognized experts on a particular subject and give opinion evidence on matters that fall within their area of special knowledge. A medical expert testifying as to the likelihood of a plaintiff suffering permanent physical damage as a result of an injury would be an example of this type of expert witness. An accountant testifying about the financial accounts or transactions of a corporation would be another example of an expert witness.

Courts will generally insist that only the "best evidence" available be presented to the court, so, for this reason, a court will not normally allow hearsay evidence. Hearsay evidence is evidence based upon what a person has heard someone else say, but that is not within that person's own direct knowledge. Because the statements would not be open to challenge on cross-examination, the courts will not usually admit such evidence. Consequently, a party wishing to have the particular evidence placed before the court would be obliged to bring the person with the direct knowledge before the court to testify about it.

When all of the evidence has been entered, counsel argue the relevant points of law and sum up their respective cases for the judge. The judge will then render a decision, which, with his or her reasons, represents the judgment.

CLIENTS, SUPPLIERS *or* OPERATIONS *in* QUEBEC

Be aware that civil law legal proceedings are conducted very differently. In addition to significant procedural differences, civil law judges play a much less passive role in the trial than their Common Law counterparts. For example, the civil law judge is active in the conduct of investigation of the facts at issue, questioning witnesses from the bench in both criminal and civil matters.

Appeal

If either of the parties believe that the trial judge erred in some manner (such as in the application of the law, or the admission of certain evidence), an appeal may be lodged with the appropriate Appeal Court. A notice of appeal must be served within a relatively short time after the trial judgment is handed down. Then an appeal book containing all material concerning the appeal is prepared by counsel for the Appeal Court. The Appeal Court will review the case and, if it finds no errors, will affirm the decision of the Trial Court and dismiss the appeal. On the other hand, if it should find that the Trial Court erred in reaching its decision, it may admit the appeal and reverse the decision of the Trial Court, vary the decision, or send the case back for a new trial.

A Small Claims Court judge refused to allow a defendant to enter certain evidence concerning her defence. The defendant appealed the resulting judgment against her to the provincial Court of Appeal. The Court of Appeal dismissed the appeal. The defendant then appealed the appeal court decision to the Supreme Court of Canada. The Supreme Court of Canada gave leave to appeal, heard the appeal, and sent the case back to the Small Claims Court for a new trial, directing the court to hear the defendant's evidence in the course of the trial.

Very recently, a number of provinces have made efforts to streamline the litigation process. Most have attempted to remove some of the unnecessary steps in the pleadings process, eliminated archaic terms and the use of Latin terminology, and have a greater onus on legal counsel to expedite trial matters.

Court Costs

Court costs represent a part of the expense that litigants incur when they bring their dispute before the court for resolution. Most of these costs are imposed to help defray the expense of maintaining court offices and the services that they provide, such as the acceptance and recording of pleadings, the issue of certificates, and the preparation of copies of court documents. The service of certain pleadings by court officers, such as the sheriff, are also court costs. The fees of counsel according to a fixed schedule or tariff are also considered to be a part of the court costs. These fees usually represent only a portion of the actual fee that the client must pay for the legal services rendered. While these costs must be paid initially by the plaintiff (and defendant), the plaintiff usually asks the court to order the defendant to pay these costs along with the damages or relief requested in the action. Conversely, the defendant may ask the court to dismiss the plaintiff's claim and order the plaintiff to pay the defendant's cost of defending the action.

Court costs and counsel fees are awarded at the discretion of the court. However, judges will usually award the successful party to the litigation the costs that that party has incurred, plus a counsel fee that the court may fix in amount or that may be calculated according to a tariff or schedule. Costs awarded on this basis are frequently referred to as "costs on a party-and-party basis." In some cases, such as when the plaintiff had an indefensible claim and the defendant insisted on proceeding with the case even though no valid defence was put forward, the court may award costs on a "solicitor-and-client" basis. In this case, the court orders the unsuccessful party to pay the plaintiff's entire legal expenses associated with the court action. In cases where the plaintiff's claim was entirely unfounded, the court might make a similar award to the defendant who was obliged to defend the unfounded claim. In both of these situations the court is, in effect, compensating the party who was obliged to undertake court action in a case where the matter should have been resolved outside the court, and, in a sense, punishing the party who caused the unnecessary litigation. In most cases, however, the courts will only award costs on a party-and-party basis, because the action merits a judgment or determination of issues that the parties are unable to resolve themselves.

Contingency Fees

Contingency fee

A lawyer's fee payable on the condition of winning the case.

Canada's system of awarding costs to the victor, even if often only a fraction of the victor's full costs, was long contrasted with the practice of **contingency fees** in the United States. Under a contingency fees system, the lawyer's collection of a fee from his or her client is contingent on the lawyer winning the case. The lawyer will then receive a previously-agreed amount (and frequently a significant percentage) of the award to be received from the losing party.

Throughout Canada's judicial history, contingency fees were prohibited on the basis that they encouraged frivolous litigation and compromised a lawyer's sense of justice in the search for profit. Over the past decade however, all Canadian jurisdictions have come to permit contingency fee arrangements. While some lawyers exclusively use contingency fees, others never agree to the method. To its supporters, this system allows for greater access to justice, as potential plaintiffs who could never have afforded to hire a lawyer (despite having a good case) now have a route to being heard. Those opposed say it brings the worst of U.S. justice to Canada: frivolous cases brought for all manner of slight injuries on slim hopes of success. Regardless of choices between lawyers and their clients, it must be remembered that losing a case under a contingency fee relationship does not rule out a court order of costs payable to the victor.

Class Actions

Class action

An action where a single person represents the interests of a group, who will share in any award.

Some circumstances lend themselves to class proceedings (commonly known as **class actions**) in which a single person could stand as the representative for an entire class of similar individuals. This person usually represents a group of plaintiffs, but the law does also provide for a representative of a class of defendants.

Consider a case of a single act that derails a passenger train, injuring a hundred persons. There is nothing to be gained through a series of a hundred trials on the same facts, so judicial economy would be achieved by combining all the actions into one. Secondly, the cost of proving a complex case (for instance, faulty research behind a dangerous medicine) could possibly be greater than the possible damages and costs awarded to a single person. Thus the large collective (and then shared) damage award makes the case worth mounting, where otherwise no one would have stepped forward. This increases the access to justice for all claimants. Finally, class actions remind society that a large profit from a tiny wrong done to many people is not a technique that will escape justice, and thus hopefully prevents such opportunism in the first place.

Class proceedings first require identification and legal recognition that a class of two or more people really exists, with issues in common that could shelter under the representative claim. The representative plaintiff must be someone who can fairly represent everyone with enough time, money, and understanding of the issues, and not have a conflict of interest with the other claimants.

Once certified as a class, the claim moves forward much like any other litigation. It can however, be difficult to assess the correct amount of damages in class actions. Accommodation exists as a result for different degrees of injury among claimants, uncertainty as to how many persons were affected by the wrong done, how many claimants actually participate, and different remedies for different claimants. Often a settlement is reached between the class and (usually) a defendant. The representative of the class may accept the settlement, but first it must be approved by the court as being in the interest of all the class.

CASE IN POINT

A term in a cell phone contract excluded court proceedings as a dispute remedy, substituting binding arbitration in its place. Regardless, a subscriber sued the carrier with the intention of recovering certain airtime charges on behalf of a class, representing all subscribers. To shut down the class action, the carrier relied on the agreement to go to arbitration. The Supreme Court of Canada disagreed, and allowed the class action to proceed, as justice would not be well served by a series of isolated low-profile private and confidential arbitrations.

See: *Seidel v. TELUS Communications Inc.*, 2011 SCC 15 (CanLII).

MANAGEMENT ALERT: BEST PRACTICE

While simple actions can be undertaken by a paralegal in small claims court, the complexities of cases and senior courts rapidly require the services of a lawyer. Many lawyers charge hundreds of dollars per hour for services that range from an hour's work to weeks of preparation for trial. A business person must coldly analyze whether a lawsuit is cost-effective in the dispute at hand, or whether settlement or alternative dispute resolution is more appropriate.

The Law Reports

The Common Law, as noted in Chapter 1, consists of the recorded judgments of the courts. Each time a judge hands down a decision, the decision constitutes a part of the body of Common Law. Most of these decisions simply confirm or apply existing Common Law principles. However, when a Common Law principle is applied to a new or different situation, the decision of the court is usually published and circulated to the legal profession. These published decisions are called law reports.

Judicial decisions have been reported in England for hundreds of years. The first series were the Year Books, which were very sketchy reports of the judges' decisions. The reporting of cases was variable for many years, and the accuracy of the reports was dependent to no small degree on the skill of the reporter. The early reports varied from an account of the cases heard during a part of a single year, to many volumes spanning several decades. It was not until the nineteenth century that the reporting of English cases was rationalized, and a council set up to oversee the reporting and publishing of the cases. The council also collected many of the older, important cases, and reproduced them in a series called the *English Reports* (E.R.).

In Canada, judicial decisions have been reported for many years in a number of different series covering different parts of the country. Cases decided in the Maritime provinces, for example, may be found in the *Maritime Reports.* These represent a series separate from those of Quebec, Ontario, and Western Canada. Quebec cases, due to the different nature of Quebec law, have a distinct and separate use from those in the Common Law provinces and are seldom referred to as precedent outside of that province. Ontario has maintained its own law reporting in a number of different series of reports, the most important today being the *Ontario Reports.* Western Canadian cases are found in the *Western Weekly Reports* and its predecessor series, the *Western Law Reports.*

In addition to these various regional series, a comprehensive national series is also available in the form of the *Dominion Law Reports* (D.L.R.), which document cases from all parts of Canada. A second national series, which reports only Supreme Court of Canada decisions, is the *Supreme Court Reports* (S.C.R.). The *Supreme Court Reports* are limited to the decisions of a single court, but, nevertheless, a most important one.

Specialized series of reports are also available. These usually deal with particular types of cases or courts. For example, criminal matters are covered in the *Canada Criminal*

Cases (C.C.C.), bankruptcy in the *Canadian Bankruptcy Reports* (C.B.R.), and patent matters in the *Canadian Patent Reporter* (C.P.R.).

Since 2000, Canadian courts and legislatures have recognized that public electronic access to the full text of legal judgments and statutes is an important aspect of access to justice, and virtually all are now available via the Internet. Of note is the work of the Canadian Legal Information Institute (CANLII) whose website, www.canlii.org, provides free comprehensive access to current Canadian cases and legislation. Other commercial data banks cross-reference cases in a number of different ways to permit advanced computer-accessible searches, with holdings that include an increasing number of older leading cases.

Case Citations

The cases are cited as authority for the statements of the law contained therein. In order that a legal researcher may readily find the report and read the statement of the law, a concise method of case identification has been developed by the legal profession. This involves the writing of a case reference in a particular manner.

Cases are usually cited in terms of the names of the parties to the action, followed by the year in which the case was decided, the volume and name of the report, and, finally, the page in the report where the case may be found. For example, the case of *Wilson v. Taylor*, which was decided by an Ontario court in November 1980, was reported in volume 31 of the second series of the *Ontario Reports*, at page 9. The case would be cited as follows: *Wilson v. Taylor* (1980), 31 O.R. (2d) 9.

The names of all law reports are referred to by their initials, or some other short form. In the above example, the second series of the *Ontario Reports* is written as shown. If the series is produced on an annual basis, rather than by volume number, the citation is written with the year in square brackets followed by the abbreviation of the report series. Viz: [1937] 1 D.L.R. 21, where the volume of the law report is the first of the 1937 *Dominion Law Reports* for that year, and the case reported at page 21.

If a particular case is reported in several different reports, or subsequently appealed to a higher court, the case citation may list all of the reports in which the case may be found, together with information as to the disposition of the case to show if the decision was affirmed, reversed, or appealed. Thus, in the case of *Read v. J. Lyons and Co.*, the trial was reported in [1944] 2 All E.R. 98, appealed to the Court of Appeal, where the judgment was reversed, and reported in [1945] 1 All E.R. 106. The case was then appealed to the House of Lords, where it was reported in [1947] A.C. 156. The appeal was dismissed by the House of Lords, affirming the Court of Appeal decision. The case, as a result, might be reported in the following manner: *Read v. J. Lyons and Co.*, [1944] 2 All E.R. 98; reversed, [1945] 1 All E.R. 106; C.A. affirmed, [1947] A.C. 156.

With the digital release of decisions by the courts themselves and services such as CANLII, a manner of citation has grown that is independent of the title of a reporter series and page number. The electronic citations usually follow the name of the case, taking the form of <year><province, court><judgment number>. For example, a 2011 British Columbia Supreme Court judgment is neutrally cited by CANLII as *BRZ Holdings v. JER Envirotech International Corp.*, 2011 BCSC 356.

Administrative Tribunals

Administrative tribunals are often boards or commissions charged with the responsibility of regulating certain business activities. These tribunals are established under specific legislation, and their duties and responsibilities are set out in the statute, along with the power to enforce those provisions in the law that affect persons or businesses that fall under the legislation. For example, administrative tribunals regulate the radio and television industry, telephone companies, and the securities market. They also regulate certain aspects of business practices, such as labour relations, when employees wish to be represented by a trade union.

Administrative tribunals have only the powers granted to them under the legislation that they are directed to enforce. In most cases, when their decisions affect the rights of parties, the tribunal is expected to hold a hearing, at which the affected parties may attend and present their case. The hearing is usually less formal than a court hearing. Nevertheless, the tribunal must conduct the hearing in such a way that the parties are treated fairly and have a full opportunity to present their case to the tribunal before a decision is made. If a tribunal fails to treat a party in a fair manner, then recourse may be had to the courts, where the decision of the tribunal may be "quashed" or rendered a nullity.

Ad hoc tribunal

A tribunal established to deal with a particular dispute between parties.

While many administrative tribunals are permanent bodies, legislation may provide for the establishment of ***ad hoc* tribunals** to deal with disputes that fall under the legislation. Most *ad hoc* administrative tribunals tend to be boards of arbitration, established to deal with specific disputes between parties, and are frequently found in the area of labour relations. Such boards usually comprise one nominee selected by one interested party, and a second by the other party to the dispute. A neutral third party who acts as chairperson is then selected by the two nominees to complete the tribunal. In some cases, only a single, neutral arbitrator may be appointed. For example, under the Ontario *Labour Relations Act*,[9] the Minister of Labour may appoint a single arbitrator to hear and decide a labour dispute arising out of a collective agreement if either the employer or the union request the Minister to do so.

ALTERNATIVE DISPUTE RESOLUTION (ADR)

In addition to regulatory tribunals, tribunals are sometimes used as alternatives to the courts to deal with a wide variety of disputes between individuals. The advantages associated with these bodies are the speed at which hearings may be held, the informality and confidentiality of the proceedings, and the lower cost of obtaining a decision. Given the high cost of resolving disputes in the courts, business contracts often contain clauses that provide for alternative methods of resolving disputes between the parties. The usual method is by **arbitration**, whereby the parties agree that if any dispute arises between them, they will refer the dispute to an arbitrator and that they will be bound by the arbitrator's award or ruling on the dispute.

Arbitration

A process for the settlement of disputes whereby an impartial third party or board hears the dispute, then makes a decision that is binding on the parties. Most commonly used to determine grievances arising out of a collective agreement, or in contract disputes.

Mediation

The parties may also use mediation as a preliminary step to the arbitration process in an effort to resolve the dispute without a formal arbitration decision. The mediation process usually involves a third party who is skilled at the process to meet with the parties, usually jointly at first, then individually. The mediator will then move back and forth between the parties with suggestions or proposals for settlement in an attempt to resolve the issues in dispute. The mediator will often clarify the issues and establish a framework for discussion that may lead to a resolution of the dispute, and often the settlement of the case itself. Where mediation fails to settle the dispute, the next step in the process would usually be binding arbitration.

Arbitration

An arbitration is conducted much like a court, but the process tends to be less formal. It normally begins with the selection of an arbitrator or the establishment of an arbitration board. Unlike a court, the parties in most cases may choose their own arbitrator. If a board of arbitration is to be established, the parties each nominate a member of the board, and the nominees select an impartial chairperson, who is usually knowledgeable in that area of business.

Once the arbitrator is selected or the board of arbitration established, an informal hearing is held, where each of the parties present their side of the dispute, and their evidence. Often, the parties will also make written submissions to the arbitrator. Depending

9. S.O. 1995, c. 1, s. 49.

upon the nature of the dispute, the hearing may be conducted in the same manner as a court, with witnesses called to give evidence, and legal counsel conducting the case for each of the parties. When all of the evidence has been submitted to the arbitrator or board of arbitration, the arbitrator or board will then make a decision that will be binding on the parties. In the case of a board of arbitration, the decision of the majority is generally the decision of the board.

Most provinces[10] have passed legislation to provide for the establishment of arbitration boards, and to give arbitrators or arbitration boards the power to require witnesses to give evidence under oath, to conduct their own procedure, and to provide for the enforcement of their awards. The parties themselves are expected to bear the cost of the arbitration.

In Canada, recent private-sector developments patterned after similar systems in the United States provide various kinds of civil litigants with an alternative to the courts as a means of resolving their disputes. Numerous organizations have now been established by specialists in mediation and arbitration, including retired judges. These organizations offer arbitration by experts in a number of fields as a means of dispute resolution that avoids the slow, costly, and formal court system. The organization usually provides a panel of experienced persons (lawyers for the most part) in various fields of law who will hear disputes in much the same manner as an administrative tribunal, then render a decision that the parties agree will be binding upon them. The dispute-resolution procedure frequently involves a two-step process, with the first step being a settlement conference that attempts to resolve the dispute under the auspices of a mediator. If the conference does not result in a settlement, then the parties move to a second step — arbitration — consisting of a streamlined discovery and hearing, and, finally, an award.

Advantages of Arbitration

The arbitration process as a means of dispute resolution is used frequently to resolve disputes in contract matters. When a dispute arises concerning the interpretation of a term in a contract, or a determination as to whether one party has properly performed the contract, business persons and corporations usually do not wish to undertake the lengthy process of civil litigation to resolve their differences, and will frequently agree to have their differences resolved by an arbitrator. This is particularly the case when the parties have had a long-standing business relationship and wish to continue to do so. The arbitration process allows the parties to take their dispute to a third party for a prompt decision, then move on with their business.

The confidentiality of the proceedings is often a part of the dispute-resolution process, and many agreements specifically provide that the arbitration proceedings and the decision of the arbitrator be kept confidential. In this way corporations may avoid the release of important (and sometimes confidential) business data to the public and competitors that might occur if the dispute was taken to court.

CASE IN POINT

A telecommunications manufacturer and one of its component suppliers have a dispute over ownership in a design modification done by the telecom to a part provided by the supplier. The modification is very valuable, increasing the efficiency of the part by 1,000 times. A public trial (which a competitor's employees may attend) would require documentation and discussion of the nature of the modification, while arbitration would be limited to the participants in the case. If heard in open court, the competitive advantages regarding the design for both parties would be lost.

10. See, for example, *Commercial Arbitration Act*, R.S.B.C. 1996, c. 55 (B.C.); *Arbitration Act*, S.N.B. 1992, c. A-10.1 (N.B.); *Arbitration Act*, S.O. 1991, c. 17 (Ont.).

Commercial Arbitration

Business parties will often include a clause in complex contracts that provides for arbitration as a means of resolving any differences that may arise out of the performance of the agreements. This approach is commonly taken with contracts of an international nature, where the businesses are located in different countries, and where the use of the courts would raise jurisdictional issues or questions as to the enforceability of the courts' decision.

Arbitration between international businesses is often handled by arbitrators associated with international bodies, such as the London Court of International Arbitration in London, England. These organizations arrange arbitrations under an internationally recognized set of procedural rules, and are frequently used by businesses that have disputes arising out of their international business activities.

Domestically, arbitration is frequently used in many business situations. Parties to a long-term commercial lease will often include a clause in the lease that provides for arbitration of disputes concerning the establishment of periodic rental changes over the term of the lease. For example, the owner of a commercial building may enter into a ten-year lease of 30,000 square feet of floor space to a retailer. The lease may provide for a further ten-year renewal of the lease at a rental rate to be agreed upon, or, if the parties are unable to agree, to have the rental rate fixed by an arbitrator.

If, at the end of the ten-year term, the parties wish to continue with their lease agreement, but are unable to agree on a rental rate, they could request the services of an arbitrator to resolve their dispute. The arbitrator would then hold an informal hearing, at which time the parties would each present their evidence as to what they consider an appropriate rental rate should be. This might take the form of evidence given by professional appraisers or rental agents, and their opinions as to an appropriate rental rate based upon the rental of comparable space in other nearby buildings. The parties would also provide any other information that would assist the arbitrator in reaching a proper conclusion as to what an appropriate rental rate should be. After the hearing is concluded, the arbitrator would then provide the parties with a decision that would fix the rental rate for the renewal period.

Arbitration may also be used as a means of resolving disputes arising out of many other contract-based agreements. Franchise agreements, equipment leasing, and service contracts often may contain arbitration clauses. Motor-vehicle manufacturers have also turned to arbitration as a means of resolving disputes with the purchasers of their vehicles. The Canadian Motor Vehicle Arbitration Plan, supported by all automobile manufacturers in Canada, provides an arbitration service to settle consumer complaints against the manufacturers when new vehicle defects are not properly corrected by the manufacturers' dealers.

Labour Arbitration

Most provinces have turned to arbitration to resolve labour disputes by mandating arbitration as a means of resolving disputes arising out of collective agreements between employers and labour unions. When an employer and a union have not included an arbitration clause in their collective agreement, the provincial labour legislation provides for an arbitration process that is deemed to be included in the collective agreement. In effect, the province requires employers and unions to use arbitration to resolve all disputes that may arise out of their collective agreements. This method of settling disputes is examined in greater detail in Chapter 19, "Employment and Labour Relations."

While arbitration and mediation as alternative dispute-resolution processes do not fit every situation, they are becoming increasingly common in contracts and agreements established between parties with more or less equal bargaining power, and where the parties wish or are required to have the relationship continue. As noted earlier in this chapter, in some provinces the process is also incorporated into the court process for certain kinds of cases.

THE LEGAL PROFESSION

Attorney

A lawyer.

The legal profession has a long and noble history, dating back to early Roman times. In England, **attorneys** were not necessary to plead cases before the Anglo-Saxon courts, as the parties to a dispute presented their own cases. By Norman times, however, a person who was old or infirm was permitted to appoint an attorney to act for him, if he agreed to be bound by the attorney's acts. This gradually changed over the next century to the point where the king's justices permitted any person to bring an action in the king's court by an attorney. The attorney, at that time, required no formal training in the law.

The first legislation regulating persons who might practise as attorneys was passed in 1402.[11] This statute required anyone who wished to practise before the courts to be first examined by the justices of the king's court, and, if found to be fit, to have his name put on a Roll. Only those persons whose names appeared on the Roll were entitled to practise law as attorneys. If an attorney failed to carry out his duties properly, his name could be struck from the Roll and he would be barred forever from appearing before the king's court.

Solicitor

A lawyer whose practice consists of the preparation of legal documents, wills, etc., and other forms of non-litigious legal work.

Persons who became familiar with proceedings in the chancellor's court were called **solicitors** because the special proceedings in the chancellor's court were begun by a petition soliciting some form of equitable relief from the court.

Attorneys and solicitors continued to be examined by the judges and the chancellor (or his senior master) for almost three centuries. By 1729, however, the examination process had become so perfunctory that complaints were made to the courts that some solicitors and attorneys were not "learned in the law." To remedy this situation, legislation was passed that required both solicitors and attorneys to serve a five-year apprenticeship with a practising attorney or solicitor before taking the examination for admission to practise alone.

As early as the fourteenth century, practitioners in England banded together in societies for the purpose of jointly purchasing and owning law manuscripts and law books, and to establish premises where they could maintain their libraries and live during the time the courts were in session. These establishments were known as inns. The inns were governed by senior members who, by tradition, dined together at a high table on a bench. Consequently, these senior members acquired the name benchers.

Barrister

A lawyer who acts for clients in litigation or criminal court proceedings.

As part of their training at the inns, junior members were expected to attend after-dinner lectures and engage in "practice" moots. A barra, or bar similar to the bar found in the courts, was usually set up in the hall. When an apprentice was considered to be sufficiently experienced, the benchers would allow the apprentice to plead his case from the bar. The apprentice's call to the bar was recognition by the profession that he had acquired the necessary skill to act as a **barrister**. Once the apprentice had been called to the bar he would then be permitted to plead cases in the courts (although until 1846 the Court of Common Pleas would only allow practitioners who held the degree of serjeant to plead before it). The training of apprentices unfortunately declined during the late eighteenth century and early nineteenth century to the point where the four Inns of Court decided (in 1852) to establish a law school to provide instruction and to conduct examinations of apprentices before the students could be called to the bar. Formal training has been required since that time for the legal profession in England.

When the *Law Society Act* was passed in Upper Canada in 1797,[12] the tradition of the Inns of Court, including the titles, was followed by the then fledgling legal profession. Similar legislation in the other colonies adopted the same traditions. The small number of lawyers in North America did not, however, permit the division of the profession (barristers and solicitors) as it existed in England. From the beginning, lawyers in the North American colonies combined both avenues of practice into one. Formal training continued to be patterned after the English model, and, for many years, formal legal training was possible in some provinces by apprenticeship and examination by the provincial

11. *Punishment of an Attorney Found in Default*, 4 Hen. IV, c. 18.

12. *Law Society Act*, 37 Geo. III, c. 13.

bars. Over time, however, legal training gradually shifted to the universities, where courses in law were offered. At first, these courses were part of general university programs; later, they became part of the curriculum of university law schools.

McGill University in Montreal, Quebec, established a law school in 1848, and claims to be the oldest university in Canada with a law faculty. Many other universities established law schools shortly thereafter and, by the late nineteenth century, law schools existed in New Brunswick, Quebec, Ontario, Nova Scotia, and Manitoba. One of the most famous, and one of the oldest law schools, Osgoode Hall Law School, dates from 1797. However, it did not grant an academic law degree until the benchers of the Law Society of Upper Canada transferred the formal teaching function to degree-granting institutions (including its own school) in 1957. Formal training of the legal profession in most Common Law provinces now requires a law degree as well as some form of apprenticeship, to be served in the office of a practising solicitor or barrister for a specified period of time.

CLIENTS, SUPPLIERS *or* OPERATIONS *in* QUEBEC

Not only is civil law quite different in terms of approach, but members of the legal profession also differ from those in the Common Law provinces. Much of the work normally considered a part of a solicitor's practice in other provinces is performed in Quebec by notaries, and the "trial work" is conducted by avocats.

THE ROLE OF THE LEGAL PROFESSION

The legal profession in each province is governed by legislation that limits the right to practise law to those persons admitted to practise pursuant to the legislation. This legislation is designed to ensure that persons who offer legal services to the public are properly trained to do so. The legislation also provides for control of the profession, usually through a provincial law society or association that is charged with the responsibility of administering the governing Act. The law society or association enforces rules of conduct, and usually has the power to discipline or disbar members who fail to comply with the rules or the standards of competence set for the profession.

The legal profession provides a wide variety of services to the public by advising and assisting them with their legal problems. Those members of the profession who undertake to provide service to business persons practise their profession in the relatively broad and loosely defined area of "business law." This area of the law includes legal work associated with the formation and financing of business organizations, the broad range of activities engaged in by business organizations generally, and in the many specialized areas of law such as intellectual and industrial property, real property, labour relations, taxation, and security for debt. Large law firms who offer their services to business firms tend to have members of the firm who specialize in particular areas of the law that affect business, while the smaller law firms may either specialize in one or a few areas of business law. Many small firms engage in a wide variety of legal work, and call upon the services of other lawyers who specialize in a particular area of business law when a client has a particularly difficult problem in that area of the law.

The usual role of the legal profession in a business transaction is to advise a client of the legal implications of the course of action proposed by the client and, if the client decides to undertake the matter, to act on behalf of the client to protect his or her interests and give effect to the action undertaken. For example, if a client wishes to enter into a contract with another business person to sell certain assets, the lawyer will advise the client of the nature of the sale agreement required, the tax implications of the sale, and perhaps the need for any special licence or permit required to sell the assets if the buyer resides abroad and the assets are goods subject to export restrictions. The lawyer will also either prepare the agreement of sale, or review the sale agreement if it is prepared by the other party's lawyer, to make certain that it protects the rights of the client and gives effect to the client's wishes in the sale. In the event that the other party fails to complete the transaction, a lawyer will advise the client of his or her rights under the agreement and, if retained to do so, will take the necessary legal action on behalf of the client to enforce the agreement.

Apart from advising clients on the legal implications of business transactions, firms frequently engage lawyers to assist in the negotiation of collective agreements with the labour unions that represent the firm's employees. Collective agreements are enforceable contracts that set out the terms and conditions of employment under which the employees will work, and often require much discussion before the agreement is reached. Lawyers who specialize in labour relations are frequently called upon to assist organizations of employees (unions) in their negotiations as well. Both employers and unions call upon members of the legal profession to process disputes arising out of collective agreements at the stage where the dispute is brought before an arbitrator or arbitration board for a determination.

Some law firms specialize in the area of patents, trademarks, and copyright law, and these firms assist inventors and firms that develop new products to establish patent protection for their products or processes. They also assist businesses with trade names or trademarks by attending to the necessary legal work associated with the protection of the name or mark.

In large financial centres, many law firms specialize in providing advice and assistance in the incorporation of firms, the mergers of firms, and legal work associated with the financing of take-overs. This is often complex work, as it frequently involves not only expertise in the area of securities (bonds, debentures, and shares) but also taxation and, to some extent, public policy related to restrictive trade practices.

A QUESTION OF ETHICS

Eighty-eight-year-old Edgar O'Malley, a 70-year cigarette smoker diagnosed with lung cancer, commenced legal action against the cigarette maker of the brand he had smoked for over 40 years. Over the next three years, the case was adjourned six times at the request of lawyers for the manufacturer, and 26 motions were made before the court for O'Malley to produce various bits of evidence. At age 91 O'Malley died, leaving only a possible claim in the hands of his widow, and extinguishing any possibility of O'Malley giving evidence toward the claim. The suit was settled on undisclosed terms. What do you make of this?

SUMMARY

The legal system is the vehicle by which the law is enforced. This was not always the case, however. The development of the legal system has been evolutionary in nature. The earliest courts were not courts as we know them today, but simply meetings of the community. Community pressure to support a decision was common until the Norman Conquest of England in 1066. After the Conquest, the gradual centralization of power began, and, with it, the courts' increased authority to demand that the parties comply with their decisions. Courts established special procedures that the parties were required to follow if they wished to be heard, and these special procedures in turn gave rise to the legal profession. Modern courts in Canada and the United States still retain many of the traditions and names of the English courts, even though the kinds of cases that they hear are quite different today.

While the judicial process appears to be cumbersome and complex, much of the procedure is designed to ensure that justice is done. Safeguards in the form of rules of evidence and judicial review form an important part of the mechanism designed to eliminate arbitrary or unfair decision-making. The appeal process in the court system

www.mcgrawhill.ca/olc/willes

is designed to ensure that lower court decisions are both fair and impartially made. As well, the election process for judges has produced an experienced and well qualified judiciary.

Business persons and corporations may also use alternative dispute resolution (ADR) as an expedient and cost-effective method of resolving disputes that would otherwise require a lengthy and expensive judicial process.

KEY TERMS

Trial court, 29
Assizes, 30
Judgment, 35
Pleadings, 35
Examination for discovery, 36
Contingency fee, 39
Class action, 39
Ad hoc tribunal, 42
Arbitration, 42
Attorney, 45
Solicitor, 45
Barrister, 45

REVIEW QUESTIONS

1. Why is arbitration sometimes a more attractive means of settling contract disputes between business persons?
2. Discuss the importance of an independent judiciary.
3. If a provincial government passed a law prohibiting any person from expressing any criticism of any elected government official, on penalty of imprisonment, how might the law itself be challenged?
4. Explain the differences between a Small Claims Court and a Magistrate's Court.
5. On what basis is it possible to justify the right of the court to declare unconstitutional a law enacted by a legislature?
6. How does a criminal case differ from a civil action?
7. In criminal proceedings, what obligation rests on the Crown in order to obtain a conviction?
8. What is the purpose of "pleadings" in a civil case?
9. How does "direct" evidence differ from "opinion" evidence? How do these kinds of evidence differ from "hearsay" evidence?
10. Describe the role performed by legal counsel in the administration of justice.
11. Distinguish between a "barrister" and a "solicitor."
12. On what basis might an appeal be heard from a judgment of the court of original jurisdiction?
13. When a judgment is reviewed by a Court of Appeal, what type of decisions might the court make?
14. What is a "court of original jurisdiction"? How does it differ from a court of appeal?
15. How does arbitration differ from a court action?
16. Explain the nature of mediation, and how it is used in the resolution of disputes.
17. "In a free and democratic society, the courts perform the important role of guardians of the rights and freedoms of the individual. While important, this is far from being the only part they play in society." How do the courts perform this important role? What other functions do they have in society?

MINI-CASE PROBLEMS

1. Michael, an otherwise well-behaved boy of 11, fell in with Gavin, 13, a repeat offender as a delinquent. The boys were arrested by police while spray painting a school and breaking into a portable outside classroom. Under the provisions of the *Young Offenders Act*, there was no public disclosure of their names, and the Youth Court proceedings were sealed from public scrutiny now and future. Are the provisions of the law appropriate for both boys? Whether yes or no, explain why.
2. Anne is unsuccessful in her suit at trial and on appeal to the provincial appeal court. If she wishes to proceed, where must she go, and what burden will she face before her case will be heard again? What factors may prevent her case from being heard again?
3. Carlson invented a new valve for natural gas pipelines, and sold his drawings to Pipeco, a manufacturer of such items. A dispute arose before production began, and both parties immediately agreed to ADR. What would lie behind this mutual aversion to court proceedings?

Case Problems for Discussion

CASE 1

A backhoe owned by Digger Ltd. was involved in the trenching for a water pipeline located along a city sidewalk. In moving a bucketful of rock into a waiting truck, the backhoe operator accidentally struck and cut an overhead power line. The cut power line fell to the sidewalk, injuring and electrocuting a pedestrian who had been observing the trenching. The pedestrian required extensive hospitalization and reconstructive surgery, and later commenced legal proceedings against Digger for the injuries suffered, including six month's lost wages while hospitalized.

Outline the various steps the parties to the action would take to bring the case to trial, and briefly describe the trial process.

CASE 2

As the Head of Information Systems at Equity Brokerage, you maintain the computer systems which support all aspects of stock trading accounts at Equity, including access for trading purposes. Your firm is being sued by a customer who alleges trading irregularities occurred on her account. When you testify at trial as to the identity of the person who conducted the transactions and the actions that affected the status of the account, what kind of witness are you, and what kind of evidence are you providing?

CASE 3

Farhan is personally injured and his home is badly burned by a toaster oven that, on first use, burst into flame. With Farhan successful in an action for damages, what factors should a judge take into consideration in determining the issue of costs?

The Online Learning Centre offers more ways to check what you have learned so far. Find quizzes, Weblinks, and many other resources at www.mcgrawhill.ca/olc/willes.

www.mcgrawhill.ca/olc/willes

CHAPTER 3

Business Regulation

Chapter Objectives

Regulation can have more immediate impact on how we do business than either legislation or case law. After study of this chapter, students should be able to:

- Describe the effect of regulation and distinguish it from legislation.
- Identify and describe the activities of administrative tribunals.
- Recognize matters of federal versus provincial jurisdiction.
- Describe the elements of natural justice.

YOUR BUSINESS AT RISK

Many businesses are tempted to gloss over or ignore seemingly burdensome administrative rules and procedures. However, government regulation and administrative tribunals have more power than most business persons realize. Failing to comply with administrative law provisions can result in immediate and drastic consequences for businesses and their owners.

INTRODUCTION

Administrative law, broadly speaking, encompasses much more than what we normally consider to be "laws," because it includes not only legislative acts of all levels of governments, but the rules, decisions, and other directives of public officials, agencies, boards, and commissions created by these statutes to carry out the policies set out in the legislation. Unlike legislation that clearly establishes laws that are enforceable through the courts, administrative laws generally set out broad policy objectives of the government, then delegate the enforcement of the policy to an entity that is created to administer the policy and ensure compliance with it by affected parties.

The process is usually quite uniform. In most cases, a statute is passed to create a board, agency, or commission to supervise an activity. The statute will also set out broad policy guidelines for the regulation of the activity by the particular agency. These boards or commissions are sometimes referred to as **administrative tribunals**, and to enable the tribunals to carry out their public policy goals they are generally allowed to set their own procedures and rules. These may either be approved by the government as an Order-in-Council, or approved by the minister in charge of the tribunal, depending upon the importance attached to the regulations.

Administrative tribunals

Agencies created by legislation to regulate activities or do specific things.

The regulations or procedures of administrative agencies, boards, and commissions represent a body of subordinate legislation that governs activities subject to supervision by an agency or board. The general trend of governments at all levels to exercise greater control over the activities of citizens within their respective jurisdictions has resulted in a proliferation of boards and agencies, at times acting in conflict with one another. The net result of this trend has been the creation of a large bureaucracy at provincial and federal levels of government, and a substantial restriction of the freedom of the individual through many and varied regulations. In recent years, complaints from the business community concerning the actions of many boards and tribunals have prompted governments to study why so many administrative agencies are necessary. While many changes have taken place in the public service, little has changed in the overall burden of business regulation for enterprises. Given the propensity of governments at all levels to create more, rather than fewer, laws, it is unlikely that this form of government regulation of business will diminish significantly in the foreseeable future.

A QUESTION OF ETHICS

The election of any government brings with it speculation about who will be named to the many executive positions of the various administrative law agencies of the nation, province, or even municipality. These jobs are appointments, thus lacking direct voter approval, but carry with them considerable influence in the creation and enforcement of regulations, and social and business policy. Critics call this pork-barrel politics; others see it as the best mechanism to get interested and competent people into positions where their expertise can make a real difference. Others see this in terms of efficiency, getting on seamlessly with the job of regulating the economy. What is your opinion, and why? Is reform necessary to have the situation square with your opinion? If so, what changes would you make?

GOVERNMENT REGULATION OF BUSINESS

A large part of the body of administrative law is directed at the control or regulation of business-related activity, and business persons and corporations alike must be aware of the many administrative bodies or agencies that can have a serious impact on their operations. These regulatory bodies usually have regulations and procedures that must be complied with either before a business may lawfully be commenced or during its operation; a failure to comply with these rules or regulations may result in serious penalties or a closure of the business. Examples of some of the activities that fall under the control of regulatory agencies, commissions, boards, or non-governmental bodies include the regulation of professions, trades, travel agents, and real-estate salespersons; the incorporation of business; the sale of securities by public companies; labour relations; human rights; employment standards; workers' compensation; employment insurance; land use; environmental protection; aeronautics; broadcasting telecommunications; and many other forms of business or commercial activity.

In most cases, a regulatory body is established because of a perceived need by either a federal, provincial, or municipal government to protect the public or to ensure that services provided to the public meet certain standards. Enabling legislation will then be passed to establish the particular administrative body that regulates the activity or the persons engaged in the activity. These administrative bodies may take the form of a government agency, commission, or board, or a non-government body established for a particular purpose, such as a law society, college of physicians and surgeons, or a college of pharmacists. For example, most professions (e.g., law, medicine, accounting, engineering) are controlled by a self-governing body created by statute to establish and maintain the standards of the profession. These bodies are granted the right to determine the entrance requirements that permit a person to enter the profession and to lawfully perform the work of the profession. Membership and adherence to the professional standards must also be maintained if the person wishes to continue as a professional. Because the non-government entity (professional body) controls entry to the profession and the maintenance of professional standards on an ongoing basis, the public is assured of the delivery of skilled service by the members of the profession.

Many other business persons and groups that offer services to the public have also been brought under the regulatory control of governments during the last half century. These include real-estate salespersons and brokers, travel agents, securities dealers and salespersons, and taxi operators, as well as electricians, plumbers, beauticians, housing contractors, mechanics, and others. Overall, there are perhaps very few service providers who are not in some way and to some degree subject to some licensing or regulatory control by government.

Regulatory Control of Professions and Skilled Service Providers

The regulatory control of the professions and other service providers is relatively similar in most provinces. As noted, a statute is passed by the legislature, which creates an entity to regulate or control the particular profession or group of individuals. For the professions, this usually takes the form of a self-governing college, society, or association; for business persons or trades, the legislation may designate a branch of a particular ministry of the government to handle the regulation of the individuals. The Ministry of Labour may be the designated ministry for service groups such as trades or semi-professional skills; the ministry responsible for consumer or business relations might oversee, for example, travel or real-estate agencies.

These statutes take on the names of the particular professions, businesses, trades, or services that the legislatures wish to regulate. For example, the various professions may fall under statutes called the *Law Society Act*, *College of Physicians and Surgeons Act*, *Society of Professional Engineers Act*, *Accounting Profession Act*, *Pharmacists Act*, or *Architects Act*.

The statute will generally set out the broad policy guidelines or goals for the governing entity, and authorize the entity to establish its own procedures to achieve these policy goals. The procedures established by the entity will usually first determine the skill requirements for the particular profession, trade, or service, then arrange an educational or training process to ensure that those persons wishing entry to the profession, trade, or service possess the minimum required skills. This may take the form of specialized education or training as well as some form of formal examination or testing to satisfy the requirement that the person possess the necessary skills. The process will then provide for registration or a licensing procedure that will entitle the person to practice the profession or engage in the trade, business, or service to the public. Most of these regulatory statutes will also provide that no person may engage in the particular service or profession unless licensed or registered, and this in turn enables the governing agency to exercise control over those persons who practise the profession or engage in the trade or service.

This control over the profession, trade, or service also requires the governing agency or policy to provide a process for the revocation or termination of licences of those who fail to maintain standards of skill. Because the revocation of a license can have devastating consequences on a person's status in society and their economic well-being, in most cases a process must be established either by the legislation or by a regulation of the body or agency for the review of any proposed decision to revoke a licence or registration of the individual.

MANAGEMENT ALERT: BEST PRACTICE

Compliance with the regulations of administrative agencies is a fact of life in business today. Individuals with special skills sometimes fail to realize the importance of licences (or registration) before the start-up of their business and, as a result, may undergo the embarrassment of being closed down and penalized for their ignorance or oversight.

CLIENTS, SUPPLIERS *or* OPERATIONS *in* QUEBEC

While all provinces have created administrative law agencies that differ somewhat from one another in either composition or activity, Quebec is no different than the rest of the provinces in creating administrative law tribunals to oversee many major aspects of commercial activity. Quebec counterparts exist (as they do in all provinces) for those agencies.

THE HEARING PROCESS

The courts have clearly stated that any agency or organization that has the power to deny or revoke the rights of a person to practise a profession, or engage in a lawful trade or business, is acting in a quasi-judicial manner, and may not revoke or deny these rights until and unless the affected person has been given the opportunity to present his or her case to the decision-makers. What the courts have said, essentially, is that the rules of natural justice must be adhered to in any decision-making process whereby a person's rights to their professional practice or trade practice may be adversely affected or denied.

Compliance with the rules of natural justice can take a number of forms, but the most common approach requires the society or agency to give the affected persons notice of the charges made against them, and provide them with the opportunity to prepare and present their side of the issue to the decision-makers. This will usually take the form of a hearing before the decision-makers, where the affected persons may appear and present their evidence or argument, and challenge the evidence raised against them. While the court rules of evidence may not apply in their entirety, the hearing procedure in many cases will represent that of an informal court: lawyers may represent the parties, evidence may be presented, witnesses may be heard and cross-examined, and arguments may be presented to the decision-makers. Finally, the decision-makers will be expected to provide a decision (with reasons) after a full hearing has been held.

An appeal process may occasionally be permitted under the legislation or regulations, but, more often than not, no provision will be made for appeal of the decision. However, if the decision is flawed or natural justice is denied, a judicial review may be available to the affected party. If the decision is found to be unreasonable by the courts, it may be quashed, and the organization or agency may be directed to hold a new hearing in accordance with the court's directives.

In some cases, a member of a professional organization may feel that the board that oversees the profession has exceeded its authority or rules in commencing disciplinary proceedings against the professional. In such a case, the professional may apply to the courts to have the disciplinary proceedings stayed or terminated.

CASE IN POINT

A dentist moved his dental practice to a city mall, and announced the move in the local newspaper. As a result, the mall placed a "welcome" advertisement in the newspaper, and a reporter interviewed him. The reporter's article described the new office, and included a photograph of the dentist. When the provincial dental board became aware of this, it began disciplinary action against the dentist on the basis that the publicity violated the board's rule which prohibited this form of advertising. The dentist made an application to the provincial Supreme Court, which stayed the board's actions against him as he had neither solicited nor paid for the article but was only exercising his freedom of expression.

Carmichael v. Provincial Dental Board of Nova Scotia, 1998 CanLll 1773 (N.S.S.C).

BROAD-POLICY ADMINISTRATIVE LAW

Employment

In addition to administrative laws regulating the professions, trades, and business services, provincial legislatures and the federal government have passed broad-policy administrative laws that affect business practices throughout the province or the country. For example, each province and the federal government have in place labour-relations legislation designed to regulate the relationship between labour unions, the employees they represent, and employers. This legislation, broadly speaking, provides an administrative structure and process to determine if a union is entitled to represent the employees of an employer, the process for the negotiation of the terms and conditions of employment of employees represented by a union, and a procedure for the resolution of disputes that may arise out of collective-employment agreements. These boards are usually called labour relations boards, and may include members selected from employers' organizations and labour unions.

In addition to a Labour Relations Board that is responsible for matters concerning the relationship between employers and unions, provincial (and federal) labour relations legislation also provides for dispute resolution processes to resolve workplace disputes between employers and their employees. Dispute resolution processes are discussed in detail in Chapter 19 of the text, but it is worth noting here that the dispute resolution processes usually include mediation and arbitration. The selection of these processes may vary depending upon the type of union–employer relationship, as in some provinces some essential service employees are not permitted to engage in strikes. In most cases, however, these employees are not engaged in business, but in government service.

Provinces have also established other administrative agencies or boards to implement broad policies of the government with respect to human rights, pay equity, and employment. As well, Workers' Compensation Boards exist in all provinces to provide procedures to address the compensation and rehabilitation of injured workers. These boards not only administer the compensation and rehabilitation programs, but have hearing processes to determine claims and address complaints by injured workers. The regulations and policies of these boards and agencies must be carefully followed by business firms, and are examined in greater detail in later chapters.

Securities Regulation

In addition to employment-related administrative bodies, there are numerous other boards or agencies that have been established by the provincial legislatures and the federal government that affect businesses and their operations. Provincial securities legislation sets out requirements for the issue of securities to the public; in addition, persons who engage in securities trading must be registered in order to sell securities to the public as investment dealers or investment salespersons. The Ontario Securities Commission (OSC), for example, administers the provincial *Securities Act*; it protects investors from unfair or fraudulent practices and fosters confidence in the capital markets. The Commission administers the Act and deals with the licensing of investment dealers and salespersons. It also supervises the issue of corporate securities by imposing certain disclosure requirements on the issuers. For example, if a corporation wishes to issue shares to the public, it must first prepare a prospectus setting out details of the security and its purpose, as well as financial information about the corporation and those who manage and direct it. This document must be filed with and accepted by the regulator before shares may be sold. Prospective purchasers of the shares must also be given a copy of the prospectus before a sale may be made, to ensure that they have the necessary knowledge about the corporation to make an informed purchase. The Securities Commission also requires all persons with "inside" information about a corporation (such as directors and senior management) to inform the Commission of any dealings in their corporation's securities on a timely basis, so that investors may be made aware of these persons' actions. This regulation ensures that these "insiders" do not take advantage of special information about the corporation's fortunes that is not available to other investors.

Provincial securities commissions also have broad investigative powers to ensure that investors' confidence in the securities market is maintained, and for this purpose the commissions are given the power to revoke the licences of investment dealers or securities salespersons if they act in violation of the Securities Act or securities regulations.

FRONT-PAGE LAW

OSC approves Lee settlement

The Ontario Securities Commission has approved a settlement with Lee Simpson, the former president, CEO and CFO of defunct brokerage Thomson Kernaghan & Co. Ltd. that will see him banned permanently and pay costs of $50,000. The OSC said Simpson agreed in the settlement that while employed at Thomson Kernaghan he failed to ensure that terms of a loan guaranteed by the company were properly disclosed to the Investment Dealers Association of Canada. He also agreed that he failed to properly supervise the actions of Mark Valentine, the firm's maverick ex-chairman, in a number of transactions.

Valentine has been banned for life from being an officer or director of any Canadian company and must pay $100,000 in a settlement with regulators and was convicted of securities fraud in the United States earlier this year. Simpson was permanently prohibited from being registered under Ontario securities law and from acting as an officer or director of any registrant in the province. He may not act as a director or chief financial officer of a reporting issuer for five years and has agreed to never re-apply for membership in or approval from the IDA anywhere in Canada.

Is it fair or reasonable that a penalty should include a lifetime ban on acting as a director of any Canadian corporation?

Source: Excerpted from *Canadian Press*, "OSC approves Lee settlement," August 17, 2005.

Control of Supplies and Services

Control of specific industries or groups of industries in each province is also managed through administrative law. Energy boards in some provinces may fix or establish rates that electrical-power corporations may charge consumers; other boards or commissions may fix rates for natural gas supplied to consumers. Many agricultural products also are regulated by controlling the number of suppliers that may produce and sell particular agricultural products. Milk- and egg-marketing boards or commissions are examples of these government agencies.

Federal boards and commissions regulate telecommunications, airlines, nuclear-power generation, radio and television broadcasting, interprovincial transportation, banking, and many other commercial activities that offer services to the public. As well, the federal government has established boards and commissions to regulate labour relations and other activities in those industries and businesses that fall within federal jurisdiction.

Many of these boards and commissions have established a hearing process or procedure whereby interested parties may present evidence or arguments concerning decisions that the board or commission is required to make about the duties or responsibilities delegated to them.

> The Canadian Radio-television and Telecommunications Commission (CRTC) holds hearings concerning the issue of a television-station licence to ensure that interested parties have an opportunity to be heard with respect to any conditions that may be imposed on the station and its operation. The Commission is acting quasi-judicially in much the same fashion as any other administrative body established under statute.

Municipal Regulation

Administrative law reaches to the local-government level as well. Most provinces have delegated the authority to regulate many business activities to municipalities when the business activity or property is local. For example, municipalities in most provinces may regulate taxi businesses, tradespeople, street vendors, restaurants, and many other types of local businesses. Regulation usually takes the form of a licensing commission (for taxis), and a licensing department for most other regulated activities.

One very important business regulation that is, for the most part, delegated to local levels of government is that of land development. Because land development can have a major impact on municipalities and the essential services they provide, it is important for most municipalities to have some control over the activities of land developers. The construction of large housing subdivisions, apartment buildings, or condominiums may require the construction of additional public services, such as schools, water-treatment plants, sewage-disposal facilities, and roads, as well as additional police, fire, and hospital facilities. Consequently, some control is necessary at the local level to ensure that the development is appropriate for the municipality and that the costs of development are properly apportioned.

While land-development policies vary from province to province, most provinces have passed planning legislation, which sets out broad policy guidelines and procedures. The authority to implement these policies is then delegated to the municipalities or lower-tier governments (such as counties or regional governments), often with some authority retained by the province to ensure that the lower levels of government adhere to the policies set out in the legislation.

At the municipal level, development is usually controlled by way of an official plan, which designates suitable development areas, and by zoning by-laws. Development is normally expected to take place in accordance with the plan and the by-law. These by-laws are administered by the municipal council, a designated planning body, or a committee of adjustment, which hear requests for land severance or development. (Committees of adjustment will usually deal with small development proposals, such as the creation of a few lots for housing construction.) After an informal hearing, consent to the proposed development may be granted if it meets the criteria set out in the official plan and zoning by-law. Larger developments, and changes to zoning by-laws or official plans usually require approval by the municipal government and, in many cases, by the provincial authority. In all cases, however, development is subject to a hearing process, where interested parties may address the decision-makers before consent or approval is issued.

In some provinces, if objection is raised to a planning decision at the local level, decisions may be subject to appeal to a provincial body. For example, decisions of a municipal committee of adjustment may be appealed to the provincial Municipal Board, which has the authority to hold a hearing on the matter and review the decision.

THE APPEAL PROCESS FOR ADMINISTRATIVE DECISIONS

Administrative boards, commissions, and agencies are generally considered to be entities created for the purpose of carrying out government policies of a regulatory nature, and while most of their work follows set procedures of a routine nature, some aspects of their work involves decision-making that may frequently affect the rights of business persons, corporations, or individuals. In this respect, their decisions are quasi-judicial in nature, and so it is essential that the decisions are reached in a fair and just manner. In recognition of this fact, some form of appeal process within the system is usually necessary to ensure that the regulatory process is conducted fairly, or that access to the courts is possible when no appeal process is provided.

In many cases, the decisions of administrative agencies, boards, and commissions are deemed to be final and binding, and usually no provision is made for an appeal to the courts or other appeal body. In some cases, however, an appeal is provided, such as when land-development decisions are appealed to the provincial Municipal Board, but this procedure is less common. When the decision-making is internal, in the sense that the regulatory work is performed in a government ministry, an appeal of the decision may sometimes be provided to the deputy minister.

Judicial review

The judicial process whereby the decision of an administrative tribunal is examined by the court to determine if the decision-making process and decision was unfair or flawed.

When no appeal process is provided for by the legislation or the regulations, a decision by a regulatory body acting in a quasi-judicial manner will normally be final and binding, unless the decision or the process is flawed. Should this be the case, the aggrieved party may apply to the courts for a **judicial review** of the decision. The courts, however,

Natural justice

Procedural fairness in decision-making.

will generally not disturb a decision of a regulatory agency, board, or commission, unless its decision is patently unreasonable, the decision-maker is biased, or the rules of **natural justice** were not followed in the process of reaching a decision. If the court should find any of these flaws, it may quash the decision, revise it, or send it back to be properly heard again, in accordance with the court's directives.

CHECKLIST FOR NATURAL JUSTICE

1. Prejudicial rules cannot be made retroactive.
2. Notice of hearing to be given to all parties.
3. Time provided to prepare for hearing.
4. Permission to have counsel or representation.
5. Fair hearing of evidence, and the chance to test adverse evidence.
6. Impartial judge.
7. Timely decision with reasons.

CASE IN POINT

A Quebec City FM radio station, CHOI, employed an outspoken and controversial on-air host. In response to complaints from the public about racist and sexually explicit statements made by the host, the CRTC withdrew CHOI's broadcast licence on July 13, 2004, silencing the station. CHOI's parent firm appealed the CRTC administrative decision to the Federal Court of Appeal. In August 2004, the court granted a stay of the suspension, allowing CHOI to continue broadcasting until such time as a trial of the issues of law (primarily Charter rights of expression) and CRTC jurisdiction could be held. In August 2005, the Court of Appeal upheld the right of the CRTC to revoke the licence and to regulate the content of programming.

An attempt by the radio station to overturn the decision by an appeal to the Supreme Court of Canada was dismissed in 2007.

Genex Communications v. CRTC, 2007 CanLII 22312 (S.C.C.).

SUMMARY

Administrative law is concerned with the actions of the many regulatory boards, agencies, and commissions created by statute to oversee, regulate, or control an extensive array of business entities and activities. These regulatory bodies usually establish a registration or licensing system whereby persons, businesses, or organizations wishing to engage in a particular activity must be registered or licensed. In some cases, the responsibility for regulation and control is delegated by statute to a non-governmental organization, such as a professional society, but apart from these special bodies, regulation is usually delegated to a specific government body.

Boards, commissions, and agencies not only regulate by registration or licences, but use the power of licence revocation to control the regulated activities. Because the licensing and revocation process may affect the rights of individuals or businesses, regulatory bodies are said to be acting in a quasi-judicial manner in the conduct of their duties. Consequently, if they act unfairly, unreasonably, or deny natural justice in their decisions, the courts may intervene to correct their actions.

www.mcgrawhill.ca/olc/willes

Key Terms

Administrative tribunals, 50
Judicial review, 56
Natural justice, 57

Review Questions

1. What are the "rules of natural justice" and how do they apply to a board that acts quasi-judicially?
2. Explain how an administrative law differs from other laws.
3. Explain how governments regulate the professions. What type of body is usually created for this purpose?
4. What administrative controls are used to control or regulate business activities that offer services to the public?
5. Outline the process that an administrative body must follow if it wishes to revoke an individual's licence.
6. What is an administrative law?
7. Under what circumstances may a decision of a board or commission be "appealed" to the courts?

Mini-Case Problems

1. Norman seeks a licence to operate a taxi in a major city, and applies in writing to the City Taxi Commission. Two days later, he receives a letter in reply: "Dear Norman: Upon careful review of your application, your request for a licence is denied. Signed: Chair of Taxi Commission." Advise Norman of his rights.
2. Janine is a herbalist, preparing certain natural tonics she believes to be "beneficial to the human spirit." Her activities are located in a province which does not regulate such matters. Carla is similarly occupied, as a registered herbalist, in a province which does regulate the trade. Aside from federal regulation of drugs, in what way would these two persons be differently affected if one of their preparations turns out to have adverse effects?
3. A provincial Liquor Control Agency discovers that a licensee tavern has sold liquor to minors on two occasions and orders suspension of the tavern's licence for a period of seven days. A month later, a different tavern is found in violation on essentially identical facts, and the Agency suspends its licence for a period of fourteen days. Is there a ground for appeal of the harsher suspension?

Case Problems for Discussion

CASE 1

A restaurant licensed to sell liquor served several bottles of wine to a couple who had visited the restaurant for dinner. Both of the patrons had after dinner drinks, and when they were ready to leave, the restaurant owner offered to send them home in a taxi, since they were, in his opinion, unfit to drive. The patrons accepted the offer, and took a taxi home. Upon reaching home, they decided to return to the restaurant and pick up their car. They called another taxi and were taken to the restaurant parking lot. They picked up the car, and on their way home, the driver lost control of the car, and crashed through the side of a wood-frame house, killing one occupant and injuring another. Investigating police charged the driver with criminal negligence causing death (among other charges), and the provincial liquor authority revoked the licence of the restaurant.

Discuss the issues raised in this case. What recourse might the restaurant have to restore its licence?

CASE 2

A radio station employed a program host that repeatedly made disparaging and often sexist remarks about Canadian celebrities to the point where complaints to the Broadcast Regulator resulted in a cancellation of the station's license to broadcast.

The station objected to the Regulator's decision on the basis that it had violated the station's right to freedom of speech and expression.

Discuss the issues in this case. Would your answer be any different if the station had been located off-shore on a ship beyond the 12 mile limit? Using the Internet from an off-shore location?

The Online Learning Centre offers more ways to check what you have learned so far. Find quizzes, Weblinks, and many other resources at www.mcgrawhill.ca/olc/willes.

THE LAW OF TORTS

CHAPTER 4

Intentional Torts

Chapter Objectives

After study of this chapter, students should be able to:

- Identify intentional interference with a person that constitutes a tort.
- Explain the concept of vicarious liability.
- Explain how land and chattels may be the subject of intentional torts.
- Recognize business-related intentional torts and those that are crimes.

YOUR BUSINESS AT RISK

This chapter is about civil wrongs: people and events that cause injury, and the rights of injured parties. You and your business are capable of causing serious injury or you or your business may be injured by others. When this happens, it is vital to know what sort of injury is recognized by the law. You must know where responsibility for that injury begins and ends, and what kind of compensation is available.

TORT LAW DEFINED

The term "tort" is a legal term derived from the Latin word *tortus*, meaning a "wrong." Its use in law is to describe a great many activities that result in damage to others, with the exception of a breach of trust, a breach of duty that is entirely contractual in nature, or a breach of a merely equitable obligation. The term has been used in English law for many hundreds of years to characterize a wrong committed by one person against another, or against the person's property or reputation, either intentionally or unintentionally. It also generally covers cases where a person causing an injury has no lawful right to do so. Unfortunately, the term is not capable of precise definition, because the area of the law that it encompasses is so broad that to determine its limits with any degree of precision would be an impossible task. We can, however, identify many of the more important areas of tort law and examine those that have a direct bearing on ordinary personal and business activities.

THE DEVELOPMENT OF TORT LAW

Some of the earliest laws made by the community or the early judges pertained to actions that are now the subject of tort law. Indeed, most of our more familiar criminal law was once tort law, and many criminal actions today still have a concurrent tort liability attached to them. Assault causing bodily harm, for example, is an act covered by the *Criminal Code*, under which the Crown would proceed against the person who committed the assault. Under tort law, the victim of the assault would also have the right to seek redress from the accused for the injury by way of civil proceedings for assault and battery.

In the past, no distinction was made between acts that were criminal and acts that were civil in nature. Both were treated as torts in the sense that compensation would be due to the victim or the victim's family. This is not so today. Because so many torts in the past have now become crimes, modern legal writers, particularly those in the United States, distinguish the two classes by referring to crimes as "public wrongs" or "wrongs against society," and the remainder of tort law as "private wrongs" or "wrongs against the individual."

The earliest tort cases tended to reflect the injuries that a plaintiff of that era might suffer: violent assault and battery, the seizure of one's goods, or the slander of one's name. Today, the scope of the law of torts has broadened to the point where it encompasses a great many areas of human endeavour. Yet, for all of its applications to modern-day activity, many of the principles and concepts date back to much simpler, earlier times when the courts first saw the need to remedy a wrong.

Tort law, by and large, is an attempt by the courts to cope with the changes that take place in society, and to balance as best they can individual freedom of action with protection or compensation for the inevitable injury to others that the exercise of such freedom occasionally produces. The development of the law of torts is essentially the development of the principles and legal fictions used by the courts to effectively maintain some sort of equilibrium between these two individual desires in an increasingly complex society.

A tort situation might arise in the course of a number of business activities. It would, for example, arise when a business person negligently produces a product that injures the user or consumer. It would also arise when a surgeon carelessly performs an operation on a patient, causing the patient further unnecessary pain and suffering or permanent injury.

While many areas of the law affect business, the law of torts and the law of contract are two major areas of importance in terms of the rights of the parties arising out of business transactions or business activity. As a very general rule, contract law applies to business activities where the parties have voluntarily agreed to their rights and responsibilities and which the courts will enforce through civil action. In the case of tort law, a party affected by a business activity that causes injury need not necessarily be associated with the business transaction or activity, and in many cases may be a complete stranger to the transaction, but nevertheless injured by it. In this instance, tort law may provide a remedy, and, in a sense, has a much broader application than contract law.

Many common business torts relate to carelessness resulting in injury or loss to a client or customer, but some torts may arise out of deliberate acts. Because business persons are often responsible for the actions of their employees in the course of carrying out their duties, liability for any tort committed by an employee in the course of business may fall on the employer as well. In Chapters 4, 5 and 6, the law of torts and its impact on business activity is explored under the general headings of intentional torts, unintentional torts, and professional liability.

MANAGEMENT ALERT: BEST PRACTICE

Business is risky business, but we do not often think of it as risk resulting from intentional acts done to us or by us. Sadly, a reality check is needed. Sometimes a business is a victim, be it of fraud or the malicious actions of other competitors. Other times, it is our own business doing wrong to others, hopefully innocently; we may be infringing on the trademarks of another firm, impairing a neighbour's use and enjoyment of their land, or any one of a galaxy of other civil wrongs. These risks demand thoughtful general business management, particular care in supervision of employees, and insurance to cover the unforeseen actions that may arise in us and others.

INTENTIONAL INTERFERENCE WITH THE PERSON

Interference with the person in tort law includes both willful and unintentional interference (negligence), but for ease in distinguishing between the two, we shall look at each in separate chapters. The principal forms of willful or intentional interference are the torts of assault and battery, and false imprisonment. These torts are ancient, and represent breaches of the "king's peace" as well as intentional injury to the person. Today, assault and battery constitute criminal offences under the *Criminal Code*[1] in both serious and less serious forms, a distinction that has been carried forward for these particular torts since the thirteenth century. Similarly, false imprisonment is an offence under the *Criminal Code* when it takes the form of kidnapping or abduction.[2] It remains a tort in those instances where persons are restrained without their consent. Needless to say, the Crown is concerned with the criminal aspects of these torts, but the individual must normally look to the civil courts for compensation for the injury caused to the person.

Most victims of violent tort do not seek redress for their injuries, however, because the persons who cause the injuries seldom have sufficient financial assets to make civil action worthwhile to recover the damages. In view of this fact, some governments have recognized that victims of crime are for all practical purposes left without compensation for their injuries. Consequently, in some provinces, boards (usually called Criminal Compensation Boards) have been established to review cases of criminal injury. They have the power to award compensation to the victims of these and other types of physical violence. Compensation is paid from a fund established by the province, and in this sense is a quite different procedure from court action.

Assault and Battery

Assault

A threat of violence or injury to a person.

Battery

The unlawful touching or striking of another person.

Assault and **battery** are torts that occasionally arise in ordinary business relationships, generally in instances where employees act improperly in dealing with unruly patrons of food and drink establishments, or in professions where dealing with people in a physical manner occurs.

Assault and battery are frequently considered to be a single tort, but in fact each term refers to a separate tort. Assault originally referred to a threat of violence, and battery to the application of force to the person. The distinction between the two torts has become blurred by the passage of time, and today, judges occasionally refer to the application of force as an assault. This evolutionary change is understandable, since an assault usually occurs before the application of force, or accompanies it. The distinction is still important, however, because not every application of force by one person on another is a battery within the meaning of the law. For example, two people passing in a dark and narrow hall may accidentally collide with one another, yet neither are injured or bruised in any way from the contact. Technically, each would have applied force to the other, but neither act was intended to cause harm, and this would distinguish the unintentional act from a battery. For the battery in this case to be actionable, it must be something more than the mere application of force. The force must be applied with the intention of causing harm. Where it does not cause harm, it must be done without consent, in anger, or accompanied by a threat of injury or violence in order to constitute a tort.[3]

An assault need not be accompanied by the application of force to be actionable, however.

In an early case, a man hammered on a tavern door with a hatchet late one night and demanded entry. When the plaintiff told him to stop, he struck at her with the hatchet but missed. He caused her no harm, but he was held to have made an actionable assault.[4] In a more modern case involving a plea of self-defence, the defendant was driving his car close to the rear of the plaintiff's vehicle. After the two cars stopped, the plaintiff got out of his car and shook his fist at the defendant. The plaintiff's actions were held to be an assault.[5]

1. R.S.C. 1985, c. C-46, ss. 265–278, as amended.
2. ibid., ss. 279–286.
3. *Cole v. Turner* (1705), 6 Mod. 149, 87 E.R. 907.
4. *De S and M v. W De S* (1348), Year-Book. Liber Assisarum, folios 99, pl. 60.
5. *Bruce v. Dyer,* [1966] 2 O.R. 705; affirmed, [1970] 1 O.R. 482 (ONCA).

In some cases, a battery need not be violent. It is sufficient for it to be any situation that involves the touching of a person without consent in such a way that the recipient of such action is injured. In a number of recent cases, surgeons have been found liable for battery when they failed to inform their patients of the risks involved in operations,[6] or when they used experimental techniques without fully explaining the risks to the patients.[7]

The damages that courts may award in assault and battery cases are designed to compensate the plaintiffs for the injuries suffered. But in many cases, particularly when the attack on the plaintiff is vicious and unprovoked, the court may award punitive or exemplary damages as well. The principal thrust of these awards is to deter the defendant from any similar actions in the future, and to act as a general deterrent for the public at large.

In some cases of assault and battery, a defendant may raise the defences of provocation or self-defence, but each defence is subject to particular limitations imposed by the courts. Generally, the defence of provocation will only be taken into consideration in determining the amount of punitive damages that may be awarded to the plaintiff. As a defence, it would not absolve the defendant from liability. Self-defence, on the other hand, may be a complete defence if the defendant can satisfy the court that he or she had a genuine fear of injury at the hands of the plaintiff, and that he or she only struck the plaintiff as self-protection from a threatened battery. Normally, the courts also require the defendant to establish that the amount of force used was reasonable and necessary under the circumstances.[8]

COURT DECISION: Practical Application of the Law

Assault—Use of Force

***Bruce v. Dyer*, [1966] 2 O.R. 705, affirmed ONCA.**

The defendant attempted to pass the plaintiff's car on the highway on a number of occasions, but on each attempt, the plaintiff increased his vehicle's speed to prevent the defendant from re-entering the traffic lane. After the defendant had followed the plaintiff some 16 km, the plaintiff stopped his car on the paved portion of the road, blocking passage of the defendant. The plaintiff got out of his car, gesturing with his fist and walked back to the defendant's car. The defendant got out, and when the parties met, the defendant struck the plaintiff with his fist on his diseased jaw, causing a serious fracture. The plaintiff brought this action against the defendant for damages for assault.

THE COURT — The question for decision, therefore, is whether Dr. Dyer is liable in damages for the assault suffered by the plaintiff. The law concerning assault goes back to earliest times. The striking of a person against his will has been, broadly speaking, always regarded as an assault. It has been defined in the 8th American edition of *Russell on Crime* as "an attempt or offer with force and violence to do a corporeal hurt to another." So an attempted assault is itself an assault; so an attempt to strike another is an assault even though no contact has been made.

Usually, where there is no actual intention to use violence there can be no assault. When there is no power to use violence to the knowledge of the plaintiff there can be no assault. There need not be in fact any actual intention or power to use violence, for it is enough if the plaintiff on reasonable grounds believes that he is in danger. When the plaintiff emerged from his vehicle waving his fist, I think the defendant had reasonable grounds for believing that he was about to be attacked and that it was necessary for him to take some action to ward it off. Bruce had not only emerged from his vehicle shaking his fist, in addition he blocked the defendant's passage on the road. In my opinion, that blocking action on his part was an assault...

The right to strike back in self-defence proceeds from necessity. A person assaulted has a right to hit back in defence of himself, in defence of his property or in defence of his way. He has, of course, no right to use excessive force and so

6. *Reibl v. Hughes* (1977), 16 O.R. (2d) 306.
7. *Zimmer et al. v. Ringrose* (1978), 89 D.L.R. (3d) 646.
8. *Veinot v. Veinot* (1978), 81 D.L.R. (3d) 549.

cannot strike back in defence of his way if there is a way around. Here, however, the evidence is that...the defendant was effectively blocked for the time being at least.

The law requires that the violence of the defence be not disproportionate to the severity of the assault. It is, of course, a fact that severe damage was done to the plaintiff. In my opinion, the plea of self-defence is still valid. The defendant struck one blow only. The law does not require him to measure with nicety the degree of force necessary to ward off the attack even where he inflicts serious injury. This is not a case of "beating up." The defendant was highly provoked by the plaintiff's conduct which was quite unjustified in my view. The plaintiff knew the condition of his own physical state and one would have thought that he would have, for that reason alone, refrained from such highly provocative conduct. He invited the treatment he received.

So self-defence is a legitimate defence to assault; however, take particular note that if a "way out" is available, it should be taken, and excessive use of force in self-defence is not permitted.

In the case of *MacDonald v. Hees* (1974), 46 D.L.R. (3d) 720, the judge described the manner in which the parties to an alleged assault must present their cases before the court, and the onus on the defendant who raises "self-defence" as justification for the assault (battery):

> In an action for assault, it has been, in my view, established that it is for the plaintiff to prove that he was assaulted and that he has sustained an injury thereby. The onus is upon the plaintiff to establish those facts before the jury. Then it is upon the defendant to establish the defences, first, that the assault was justified and, secondly, that the assault even if justified was not made with any unreasonable force and on those issues the onus is on the defence.

Assault and battery are, obviously, torts that have a criminal side as well. The state has an interest in preserving the peace and protecting its citizens from injury. As a consequence, the Crown may take steps to charge a person who commits an assault or battery. The nature of the criminal charge, however, depends to some extent on the type of assault and the severity of the injury inflicted.

Assault and battery under the *Criminal Code* usually occur outside the ordinary business relationship, but occasionally an employee of a business may assault a customer. The employee may be personally liable for the tort, but, at Common Law, the employer may also be held liable for the actions of his or her employees in the ordinary course of business. This is known as **employer vicarious liability**. However, in an instance where an employee assaults a customer in the course of business, it is important to note the difference between the criminal and civil consequences. While the employer may be liable for the actions of the employee in the case of a tort committed by the employee in the ordinary course of business, only the employee will be liable for the criminal consequences of the act, unless the employer had in some way directed or authorized the commission of the offence by the employee, or the employer was aware of the propensity of the employee to commit violent acts.

Employer vicarious liability

The liability of an employer for acts of his or her employees in the course of business.

MANAGEMENT ALERT: BEST PRACTICE

Torts happen. An employee in a company uniform or truck may get involved in a fight on the way home, and the employer may learn of it only through receiving a Statement of Claim. The combination of employer responsibility for acts of employees while on duty and often richer business finances makes including an employer in a claim, if at all possible, attractive to potential plaintiffs. Business operators should clearly define when employees are "on-duty" and "off-duty," and understand the risks in allowing company vehicles to be used as personal transportation. Failure to do this leads to increasing chances of employer liability in tort.

FRONT-PAGE LAW

"Feisty" passenger bites and punches attendants on trans-atlantic flight

TORONTO — A woman is facing charges of assault and aggravated assault after she went on a two-hour rampage aboard a flight from Paris to Toronto, attacking passengers and punching and biting flight attendants.

Constable Harry Tam of Peel Regional Police said the woman began drinking early in the eight-hour flight Monday afternoon and became progressively more intoxicated.

The woman, one of 186 passengers on the crowded Air Canada Boeing 767, became more disruptive as the flight wore on, staggering up and down the aisles and accosting fellow passengers.

Police say flight attendants asked her several times to return to her seat, until the woman finally turned on them and began swinging.

Two attendants were punched during a melee in the aisle and another was bitten badly enough that his finger required medical attention.

"They tried to restrain her and she got loose," said Const. Tam.

"She certainly wasn't a large woman, but she was feisty .... She wasn't going down without a fight."

She is charged with four counts of assault, aggravated assault and mischief.

How will this business (Air Canada) become involved in this legal affair? How do the civil and criminal justice systems connect with this single set of facts?

Source: Excerpted from Chris Wattie, *National Post*, "'Feisty' passenger bites and punches attendants on transatlantic flight," September 13, 2000, p. A6.

False Imprisonment

False imprisonment is another type of intentional interference with the person that Canadian courts recognize as a tort. As an actionable civil wrong, it represents any restraint or confinement of the individual by a person who has no lawful right to restrict the freedom of another. It most often arises when store security personnel seize and hold a person suspected of taking goods from a place of business, only to discover later that the person was innocent. In such cases, the imprisonment need not involve actual physical restraint if the shopkeeper makes it clear to the suspect that any attempt to leave the premises will result in the embarrassment of seizure, or pursuit accompanied by calls for help.

The law, as a matter of public policy, usually views the restraint of one individual by another with disfavour, and the defences available to the defendant are meagre. The *Criminal Code* permits citizens to seize persons who have committed a criminal offence (such as shoplifting) and hold them without warrant until a police officer can take the offender into custody, but care must be taken to ensure that an innocent person is not apprehended.

As a general rule, a person may restrain another when the person apprehended is in the process of committing a crime, or when a person attempting to seize a criminal mistakenly apprehends the wrong person. In the latter case, however, the person falsely seizing the innocent person must have reasonable and probable grounds for believing that the innocent person had committed an offence and was escaping custody; otherwise, it would be no defence to a claim of false imprisonment.[9] Peace officers, of course, may mistakenly restrain innocent persons without committing the tort of false imprisonment or false arrest, provided they had reasonable grounds for believing the person was an offender at the time of the restraint.[10]

Forcible confinement

Confinement against a person's will.

False imprisonment is also a criminal offence in the form of **forcible confinement**. Under the *Criminal Code*, anyone who without lawful authority confines, imprisons, or forcibly seizes another person may be prosecuted for the offence. If found guilty, that person would be liable to imprisonment for a term of up to ten years.[11] If forcible confinement is alleged, the lack of resistance is not a defence unless the accused can successfully prove that the failure to resist was not caused by threats, duress, the use of force, or the exhibition of force. In an ordinary business situation, if staff members apprehend a person suspected of shoplifting or theft, they would probably not be subject to criminal charges if they immediately called the police to take charge of the suspect, and the suspect was later found not guilty of the alleged offence that led to the confinement.

9. *Criminal Code*, R.S.C. 1985, c. C-46, s. 494.

10. ibid., s. 495.

11. ibid., s. 279.

Defamation

Defamation

False statements that injure a person's reputation.

Libel

Defamation in some permanent form, such as in writing, a cartoon, etc.

Slander

Defamatory statements or gestures.

The law of tort relating to the interference with a person's reputation is called **defamation.** Defamation may take the form of e ither **libel** or **slander**. Slander generally consists of false statements or gestures that injure a person's reputation. Libel takes the form of printed or published slander. Defamatory statements that slandered a person's good name or reputation were originally dealt with as moral matters by the old ecclesiastic courts, but with the passage of time, fell under the jurisdiction of the Common Law courts. Before the introduction of printing, defamation took the form of slander. This was largely a "localized" injury, but with the invention of printing the extent of the injury changed, both in terms of geographical limits and permanency. The printed word was capable of widespread circulation, and thus disseminated the scandal over a larger area. In addition, it provided a permanent record of the defamation that would remain long after slanderous statements normally would be forgotten. The criminal aspects of the printed defamation (libel) originally fell within the jurisdiction of the English Court of the Star Chamber, where the person who published a libel was punished both criminally as well as civilly for the tort. When the Star Chamber Court was abolished in 1641, the jurisdiction fell to the Common Law courts.

Five centuries of Common Law must now contend with the development of the tort of Internet defamation. Cases such as the Barrick Gold case (next page), few as they are, but sure to grow in number, indicate that the Common Law copes quite well. The principles of defamation remain unchanged: published slander (libel), whose untruth is given wide circulation as fact, to the injury to the victim's reputation. Since the Internet casts everywhere, the notion of a limited geographic area of harm is gone, and one can expect to see damage awards to victims rise to reflect this.

Generally, in a defamation action the plaintiff must establish that the defendant's statements have seriously injured his or her reputation; otherwise, the court will award only nominal damages. If the defendant's statements are true, the plaintiff will not succeed, as the truth of the statements will constitute a good defence to the plaintiff's claim.

CASE IN POINT

The CBC investigative TV program "The Fifth Estate," used a "cut and paste" technique portraying a research doctor as biased, incompetent, and a puppet of drug manufacturers. The court concluded that highly selective interview footage was used to create this impression, including a repeated segment showing the doctor fumbling with his eyeglasses. As to the CBC defence of fair comment, the court concluded that "no comment can be fair if it is based upon facts which are invented or misstated," adding that a portrayal of ineptitude handling eyeglasses "goes to the depth to which the defendants tried to destroy the plaintiff not just professionally but as a human being."

Damages awarded: $950,000.

Leenen v. Canadian Broadcasting Corp., 2001 CanLII 4997 (ONCA). Leave to appeal refused by S.C.C.

Qualified privilege and absolute privilege are also recognized by the court as defences to a claim for defamation. Absolute privilege, as the name implies, protects the speaker of the words absolutely, regardless of the words' truth or falsity, and even if they are made with malicious intent. This defence is limited, however, to those cases where it is in the public's interest to allow defamatory statements to be made. Consequently, statements made in Parliament, before a Royal Commission, in court, at coroners' inquests, and in any proceeding of a quasi-judicial nature[12] are not subject to an action for defamation by a person injured by the statements.

In some instances, a qualified privilege may apply if the defendant can show that the statements were made in good faith and without malicious intent, even though the facts

12. *Byrne v. Maas*, 2007 CanLII 49483 (ONSC).

that he or she believed to be true at the time were subsequently proven to be false. The most common example of this situation would be where an employer provides a letter of reference containing derogatory statements (which the employer believes to be true, and a fair assessment) about an employee. The justification for these exceptions is based upon the importance of allowing free speech on matters of public importance, and balancing this intent with the protection of the individual's reputation.[13] Some provinces now have legislation dealing with the action of libel and slander; to some extent, this legislation has modified the Common Law.[14]

CASE IN POINT

An aggrieved BC claimant to a Chilean gold property made highly critical Internet bulletin board postings, threatening a further range of disclosures about Barrick Gold Corp. if he was not paid $3 million. Unpaid, he did just that. Barrick appealed the $15,000 for defamation it was awarded at trial. The appeal judge determined that the Internet "makes instantaneous global communication available cheaply to anyone with a computer and an Internet connection. It enables individuals, institutions, and companies to communicate with a potentially vast global audience...[It heralds] a new and global age of free speech and democracy, [but] the Internet is also potentially a medium of virtually limitless international defamation." The damage award was increased to $125,000 including punitive damages.

Barrick Gold Corporation v. Lopehandia, 2004 CanLII 12938 (ONCA).

The use of the Internet to make and disseminate defamatory statements about persons (or corporations) raises a special problem for the victims in that the maker must be identified and located. However, once the person who made and published the defamatory statements is identified, action through the courts is possible, and where the defamation is proven, the damages for the defamation is often substantial.

Note that the full range of remedies for defamation are available to the aggrieved party where the Internet is used to publish the defamation. In a 2008 case[15] where the Internet was used to damage a person's reputation, the court awarded the injured plaintiff not only substantial general damages, damages for breach of privacy, special damages, and aggravated damages, but an injunction as well.

Defamation has a criminal element as well. Under the *Criminal Code*, defamatory libel arises when a matter is published, without lawful excuse or justification, that is likely to injure the reputation of any person by exposing the person to hatred, contempt, or ridicule, or that is designed to insult the person concerned. Publishing includes not just newspapers but any exhibition in public of the material, which causes it to be read or seen, and includes books, pamphlets, or other printed matter. Persons who publish a defamatory libel that they know to be false may be charged with the offence and, if found guilty, would be subject to imprisonment for a term of up to five years.

Publishers of newspapers as a group are most concerned with this aspect of the law, and usually take care to avoid defamatory libel in material that they include in their newspapers. Defences to an allegation of defamatory libel may include the publication for the public benefit of matters that in themselves are true, and fair comment of fair reporting in good faith of lawful public meetings. As a general rule, the publisher of a newspaper is in the same position as the writer of the material, and in order to establish the defence of fair comment, the newspaper must be able to say that the comment represents the honest opinion of the newspaper.[16]

13. *Stopforth v. Goyer* (1979), 23 O.R. (2d) 696.
14. See, for example, *Libel and Slander Act*, R.S.O. 1990, c. L-12.
15. *Griffin v. Sullivan* 2008 BCSC 827 (CanLII).
16. *Cherneskey v. Armadale Publishers Ltd.*, [1979] 1 S.C.R. 1067.

COURT DECISION: Practical Application of the Law

Tort—Defamation—Test for "Fair Comment" Defence

***WIC Radio Ltd. and Rafe Mair v. Simpson,* 2008 SCC 40 (CanLII).**

The appellant Mair is a well-known and sometimes controversial radio talk show host. The target of one of his editorials was Simpson, a widely known social activist opposed to any positive portrayal of a gay lifestyle. The two took opposing sides and Mair compared Simpson in her public persona to Hitler, the Ku Klux Klan and skinheads. Simpson brought an action for defamation; at trial the appellant testified that he intended to convey that Simpson was an intolerant bigot. The trial judge ruled that the statements were defamatory, but the defence of fair comment applied. The Court of Appeal reversed the trial judgment.

THE SUPREME COURT of Canada: The trial judgment dismissing the action should be restored. M's expression of opinion, however exaggerated, was protected by the law. M's editorial was defamatory, but the trial judge was correct to allow the defence of fair comment. Although this is a private law case that is not governed directly by the *Canadian Charter of Rights and Freedoms*, the evolution of the common law is to be informed and guided by *Charter* values. The law of fair comment must therefore be developed in a manner consistent not only with the values underlying freedom of expression, including freedom of the media, but also with those underlying the worth and dignity of each individual, including reputation.

It is therefore appropriate to modify the "honest belief" element of the fair comment defence so that the test, as modified, consists of the following elements: (a) the comment must be on a matter of public interest; (b) the comment must be based on fact; (c) the comment, though it can include inferences of fact, must be recognizable as comment; (d) the comment must satisfy the following objective test: could any person honestly express that opinion on the proved facts? Even though the comment satisfies the objective test of honest belief, the defence can be defeated if the plaintiff proves that the defendant was subjectively actuated by express malice. The defendant must prove the four elements of the defence before the onus switches back to the plaintiff to establish malice.

The public debate…clearly engages the public interest, and the facts giving rise to the dispute between M and S were well known to M's listening audience, and referred to in part in the editorial itself. The third element of the defence is also satisfied since the sting of the libel was a comment and it would have been understood as such by M's listeners. M was a radio personality with opinions on everything, not a reporter of the facts.

Does this mean that so-called "shock-jocks" can say anything they wish, as long as they remain "understood by their listeners"?

CASE IN POINT

A human-rights inquiry concluded that a teacher was anti-Semitic, and the teacher's employment was terminated by the school board. Shortly thereafter, a cartoonist gave a presentation at a workshop using a number of cartoons he had drawn of the teacher, which depicted the teacher as one who held Fascist beliefs.

The teacher brought an action for defamation against the cartoonist on the basis that the cartoons were malicious.

The New Brunswick Court of Appeal concluded that while the cartoons were erroneous, they were based on information that the cartoonist had relied upon, and were made without malice. The court concluded that the cartoons represented fair comment, and dismissed the case.

Ross v. Beutel, 2001 NBCA 62 (CanLII).

Cartoons that often ridicule or satirize political figures are also subject to defamation law, but most newspapers may successfully defend against libel claims in these cases on the basis of fair comment, as a person's political life may be open to a much greater degree of criticism and comment than his or her private life.[17]

Figure 4–1 describes defamation and its defences.

FIGURE 4–1 **Defamation**

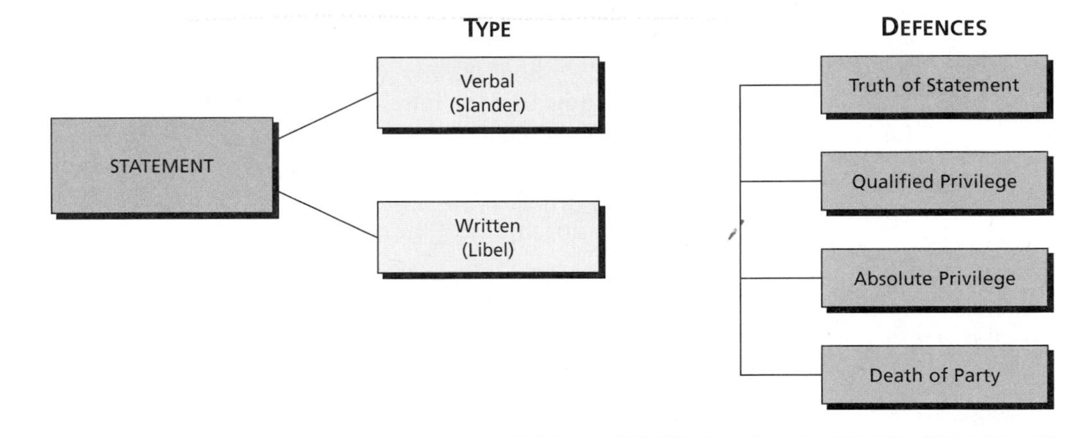

FRONT-PAGE LAW

Clark guilty of slandering critic of BC Ferries

VICTORIA — Glen Clark, the former B.C. premier, has been ordered to pay $150,000 for slandering a critic of his fast-ferry project.

In a judgment released yesterday, Judge Dermod Owen-Flood found that Mr. Clark had slandered Bob Ward, a marine engineer who had raised early warnings that the ferries would not perform as promised.

Mr. Ward sued after the former premier called him "a disgruntled bidder on this project who is constantly feeding misinformation on this issue," in a report in *The Vancouver Sun*.

Judge Owen-Flood rejected Mr. Clark's argument that even if his comments were not true, they were defensible because he was speaking as the minister responsible for BC Ferries.

The judge ruled that if he had merely repeated what he had been told by his officials, he would have been covered by the defence of qualified privilege, but said he had exceeded that advice with his statement to reporters.

Mr. Ward, the most prominent early critic of the troubled catamaran project, told the B.C. Supreme Court that Mr. Clark's comments caused him to lose years of work. Once the comments were published in 1996, clients started avoiding him, Mr. Ward testified.

He testified he was not a "disgruntled bidder" because there was no bidding process.

Virtually all of his criticisms proved accurate. He warned the ships were underpowered, predicting both substandard performance and engine problems. He also warned they would be late and over budget. The first ship was delivered more than two years late and costs rose to $150-million per vessel. They are now for sale for about $40-million each.

From these facts, and the text material on defamation, prepare Bob Ward's claim in a concise, logical, and persuasive manner.

Source: Excerpted from *National Post*, "Clark guilty of slandering critic of BC Ferries," June 23, 2000.

INTENTIONAL INTERFERENCE WITH LAND AND CHATTELS

Trespass
A tort consisting of the injury of a person, the entry on the lands of another without permission, or the seizure or damage of goods of another without consent.

Intentional interference with land and with chattels are matters that relate to property law, but also contain an element of tort liability. The two principal classes of torts that relate to property law are **trespass** to land and conversion of goods. In both of these cases

17. *Vander Zalm v. Times Publishers et al.* (1980), 18 B.C.L.R. 210 (C.A.).

(as with assault and battery, and false imprisonment), there is an element of intention associated with the act of interference.

Trespass to Land

The law relating to trespass to land represents one of the oldest actionable torts. It is the act of entering the land of another without the express or implied consent of the person in lawful possession. It is also trespass if a person, once given permission to enter the lands, refuses to leave when requested to do so.[18] This tort also is relatively broad in its application to interference with the land. For example, the acts of tunnelling under another's land without permission or lawful right, erecting a wall or fence on another's land, or stringing wires or lines over another's land, all constitute trespass. Even things can trespass. Trespassing trees, overhanging eaves, and the like are common sources of dispute among urban neighbours. Involuntary entry on the lands of another, however, would not constitute trespass, as the act of entry would be unintentional.[19]

In one significant case,[20] the court described the law of trespass in the following terms:

> *Prima facie*, every invasion of property, be it ever so minute and negligible, is a trespass and therefore unlawful. Such trespass may not amount to crime since "breaking the close" simpliciter is not a criminal offence. Nonetheless, entry into the premises of another without either consent or specific legal authorization has always been tortious and unlawful. It is an interference with the common law right to peaceful enjoyment of one's property that has been recognized at least since *Semayne's Case* (1604), 5 Co. Rep. 91a, 77 E.R. 194 [at 195] (K.B.), where it was said: "that the house of everyone is to him as his castle and fortress, as well as for his defence against injury and violence, as for his repose."

In a British Columbia case, a building contractor, during the course of construction of a large building, inserted a large number of steel rods under the adjacent property to shore up the walls of the new structure. The eight-metre-long rods caused no physical damage to the neighbouring property, but were installed without the neighbour's consent, and without attempting to contact the property owner to obtain consent. The court found the contractor liable for trespass and awarded the neighbouring property owner general damages in the amount of $500 and exemplary damages in the amount of $47,500, because the contractor had made no effort to obtain the consent of the landowner.[21]

Much of the law relating to trespass is well settled due to its long period of existence, but occasionally new forms of technology give rise to new claims in trespass. In England, for example, the invention and use of aircraft resulted in actions for trespass by property owners,[22] but the control of this activity by statute and government regulatory bodies has, for all intents and purposes, removed it from the realm of actionable trespass.

The principle of trespass remains important however, for any business whose operations invite the public onto its property for specific purposes. It is the principle that allows the business to eject the customer if the customer departs from the rules of the establishment.

18. *The Six Carpenters Case* (1610), 8 Co. Rep. 146a, 77 E.R. 695.
19. *Smith v. Stone* (1647), Style 65, 82 E.R. 533.
20. Reference pursuant to s. 27(1) of the *Judicature Act*, [1985] 2 W.W.R. 193.
21. *Austin v. Rescon Construction (1984) Ltd.* (1987), 18 B.C.L.R. (2d) 328, 45 D.L.R. (4th) 559, on damages: 57 D.L.R. (4th) 591 (C.A.).
22. *Pickering v. Rudd* (1815), 4 Camp. 219, but see *Bernstein v. Skyviews & Gen.*, [1977] 2 All E.R. 902, for the current view that trespass is limited to the landowner's useable air space above the property.

A patron of a bar becomes obnoxious and loud. The owner has the right to require the patron to leave the premises. If the patron refuses to leave, he or she becomes a trespasser. The bar owner may then take steps to have the patron removed.

Conversion and Willful Damage to Goods

Conversion

The refusal to deliver up a chattel to its rightful owner by a bailee.

The intentional interference with the goods of another person constitutes the torts of conversion or trespass to goods. **Conversion** is the wrongful taking of the goods of another or, where the goods lawfully come into the possession of the person, the willful refusal to deliver up the goods to the lawful owner. For the tort of conversion to exist, the lawful owner must be denied possession and enjoyment of the goods, and the defendant must retain the goods without colour of right. The remedy granted by the courts is usually monetary damages equal to the value of the goods.

The second form of trespass to goods involves the willful damage to the goods while they are in the possession of the owner. (For example, the deliberate smashing of the windshield of an automobile by vandals while it is parked in an automobile dealer's parking lot, or the deliberate act of killing a farmer's livestock.) Both of these torts normally have a criminal element attached to them, and while a great many of these cases reach the criminal courts, the widespread use of insurance to protect against loss of goods through conversion or willful damage has reduced the number of tort actions that might otherwise come before the courts. The two torts, nevertheless, remain as actionable wrongs.

CLIENTS, SUPPLIERS *or* OPERATIONS *in* QUEBEC

Governments (and the employees for whom they are vicariously liable) can also commit wrongful acts, for instance trespass, or as we shall see in the next chapter, acts of negligence. In the Common Law provinces the limitation period for commencing legal action against governments is set by specific statutes and is usually very short (days or months); but in Quebec it is longer (three years). This is because the limitation provisions are contained in the *Civil Code*, and the *Code* applies to everyone, even public bodies serving the public interest.

Business-Related Torts and Crimes

The conduct of business activities in a free-enterprise economy involves competition between business firms for customers. While the vast majority of business firms compete with each other on a fair and equitable basis, occasionally some will engage in improper practices that cause injury to others. These practices often take the form of untrue statements about competitors or the goods or services they provide to the public. In other cases, they may involve improper attempts to acquire the trade secrets of others. As well, they may involve agreements to restrict trade between businesses to the detriment of the public. Most of these improper practices were originally (and have remained) actionable torts at Common Law, but others have become the subject matter of legislation such as the *Competition Act*, and, in some cases, been made offences under the *Criminal Code*. It should be noted, however, that not all business-related torts and crimes are committed by persons engaged in business. Consumers may attempt to obtain goods under false pretenses from business firms, or may attempt to injure the reputation of a firm by making untrue or false statements about the firm or its dealings. These consumer-related activities would also be actionable torts and, in some cases, criminal offences as well.

Slander of goods

A statement alleging that the goods of a competitor are defective, shoddy or injurious to the health of a consumer.

Slander of Goods and Injurious Falsehood

Perhaps the most common business-related tort an unscrupulous business person might commit is the **slander of goods**. Slander of goods involves making a statement alleging that the goods of a competitor are in some way defective or shoddy, or are injurious to the health of the consumer. Statements of this nature, if untrue, could cause injury to the competitor, and would be actionable at law. This tort is not limited to business persons: consumers who make untrue and unfounded statements of this nature would also commit the tort of slander of goods.

Injurious falsehood

False statements about a firm, its products or business practices intended to dissuade others from doing business with the firm.

Injurious falsehood is closely related to slander of goods, but its scope is wider. Not only may the goods be the subject of the slander, but the disparagement can be aimed at the business enterprise as a whole, or its owners in particular. The nature of injurious falsehood encompasses slander of title (see below), the quality of goods offered, or the nature and conduct of business by the enterprise. The essential element is that the disparaging remarks are calculated to dissuade other parties from conducting business with the target. This tort has arrived in Canada largely from the United States,[23] though it is known throughout the British Commonwealth as well.

Slander of Title

Slander of title

An untrue statement about the right of another to the ownership of goods.

A similar tort is **slander of title**. This arises when a person makes an untrue statement about the right of another to the ownership of goods. It may take the form of statements that the competitor or seller has improperly acquired the goods put up for sale. For example, the allegation might be that the goods were stolen, improperly imported, or produced in violation of the copyright or patent rights of the rightful owner. The allegation might also be that the seller is passing off his or her goods as those of another well-known manufacturer in violation of that manufacturer's trademark or trade name.

Customer: Are these iPads for sale cheaper at Boombox WarehouseDepot?

Salesperson:

- Maybe, but they deal in cash only, no refunds or exchanges, if you get my drift.
- Lots of things fall off the back of a truck, you know.
- Oh sure, cheap price there, but never any in stock.
- Sure, just check the box to make sure the charger and everything else is actually there.

Are any of these slander of goods, injurious falsehood, or slander of title?

Untrue allegations of any of these aspects of the title of the seller to the goods would constitute the tort of slander of title, and would be actionable at law. However, if, in fact, the goods were produced and sold in violation of the copyright or patent of the true owner, their production would in itself constitute an actionable right by the copyright or patent owner. Patent and copyright law in Canada give the inventor or creator the exclusive right to the production of the work for the life of the patent or copyright, and any unauthorized production of the work would entitle the true owner to bring an action against the unauthorized producer. Similarly, the production of goods and their presentation to the market as being the goods of another well-known producer (such as packaging the goods in the same distinguishing guise as the well-known producer or marking the goods with the same trademark) is also actionable, as it constitutes plagiarism or passing-off, and, in effect, trading on the good will that belongs to another. Patents, trademarks, and copyright are controlled by statute, and under each of these acts the penalty for improperly dealing with the property rights of the owner is usually an award of money damages and an accounting for profits lost as a result of the violation. These topics are dealt with in detail in Chapter 26 of the text, but, for the purpose of studying business torts and crimes, it is sufficient here to note that such improper activities are actionable at law.

23. *Restatement (Second) of Torts* § 623a (1977).

MANAGEMENT ALERT: BEST PRACTICE

While management cannot control every act of their employees, sound risk management practices can reduce the possibility of intentional tort liability. Businesses should establish a wide range of internal protocols such as:

- An official spokesperson to deal with the press.
- Job descriptions and training for employees so that their obligations are understood.
- A clearly communicated and understood code of conduct for all employees.
- A cultivated culture of ethics and respect for employees, customers and competitors.
- Dedicated management effort in effective supervision of people, products and processes.

Breach of Confidence

Improper dealings with a competitor's employees may also constitute an actionable tort. For example, if a business person were to attempt to acquire trade secrets by offering money or some inducement to a competitor's employee, the competitor would have a right of action for the loss suffered against both the employee and the firm or person who offered the inducement. The remedies in these cases are usually money damages against the employee for breach of confidence (along with the right to dismiss the employee for the breach) and an injunction against the other business to prevent it from using the improperly acquired trade secret.

In a broader context, there are some signs that the courts are attempting to enforce a form of corporate or commercial morality by applying the confidentiality rules to corporate dealings. For example, in one case where two companies were exploring the possibility of a joint venture to develop a mining property, one of the participants revealed information of a confidential nature about an adjacent property it was considering purchasing. The second company then used the information for its own benefit after the discussions ceased, and purchased the adjacent property. In the litigation that followed, the court held that the use of the confidential information constituted a breach of confidentiality, and ordered the company to transfer the property to the party who had discussed the purchase in confidence.[24]

Restaint of trade

Agreement between firms to fix prices, injure competition, or prevent others from entering a market.

Agreements in **restraint of trade**, such as combinations or conspiracies to eliminate competition, to fix prices, to restrict the output of goods in order to enhance the price of the goods or services to the public, or to prevent the entry of others into the market, are treated as "business crimes" under the *Competition Act* and, if proven, are subject to penalty. As with criminal law, in these cases the Crown enforces the law, rather than the injured party, because these activities are treated as contrary to the public interest generally. The *Competition Act*, and the various offences that fall under it, are examined at some length in Chapter 32, "Restrictive Trade Practices."

If one business firm induces another to break or sever a business relationship, it may have committed an actionable tort. For example, if a firm induces a manufacturer to sever its contract to supply a competitor, on the threat of ceasing to deal with the manufacturer itself, the action would constitute an attempt to interfere with the business contract between the manufacturer and the competitor. The competitor could take action against the firm for its interference. This type of interference, if proven, would also be in violation of the *Competition Act*, which establishes the activity as a punishable offence.

24. *Lac Minerals Ltd. v. International Corona Resources Ltd.*, [1989] 2 S.C.R. 574.

CASE IN POINT

A company sold seat covers to a large automotive retail chain for over 30 years. Another manufacturer of seat covers wished to obtain a contract to supply seat covers to the retailer, and hired a marketing company to determine if a contract could be obtained. If it was successful, the marketing company would receive a commission on all seat covers sold to the retailer. The marketing company successfully obtained a contract to supply seat covers for the retailer by offering the retailer's purchasing manager a kickback of two percent on sales. The manufacturer was unaware of the kickback arrangement.

The retailer later discovered the kickback arrangement, fired the manager, and cancelled the contract.

When the former supplier discovered the facts, it sued the marketing company and the other manufacturer for its lost sales to the retailer.

The Supreme Court of Canada found the marketing company liable for the loss, but held that the manufacturer was not liable, because it was unaware of the illegal actions of the marketing company.

671122 Ontario Ltd. v. Sagaz Industries Canada Inc. et al. (2001), 274 N.R. 366.

Deceit

A tort that arises when a party suffers damage by acting upon a false representation made by a party with the intention of deceiving the other.

Two final torts that fall within the realm of business activity are the tort of **deceit** (arising from fraudulent misrepresentation) and the tort of fraudulent conversion of goods. Deceit as a tort may arise when one person induces another to enter into a contract by way of false statements. To constitute fraudulent misrepresentation, the statements made must be of a material nature, and must be made with the intention of deceiving the other party. In addition, the plaintiff must have relied on the misrepresentation. The statements themselves must be known to be false or made recklessly, without caring as to their truth or falsity, and must be relied upon by the other party. If proven, fraudulent misrepresentation constitutes the tort of deceit, and would permit the injured party to rescind the contract made as a result of misrepresentation (provided that the party does so promptly on discovery of the fraud). For the tort of deceit, the injured party would also be entitled to damages for any loss suffered, and perhaps to punitive damages as well.

Fraudulent conversion of goods is also a tort, and usually arises when the person has obtained goods under false pretenses. It differs from the theft of goods (which is the taking of goods without the owner's consent) in the sense that the goods are voluntarily delivered by the owner to the person who obtains them through the fraud. For example, a person may obtain goods on credit by posing as the agent or employee of a well-known customer of the seller, or by the presentation of a cheque as payment for the goods, if the cheque is drawn on a non-existent bank account. Fraudulent conversion of goods is also a criminal offence.

Unfair Business Practices

Unfair business practices

Business practices designed to take advantage of consumer inexperience or ignorance.

Some provincial governments, in response to business and consumer-group pressure, have introduced consumer-protection legislation meant to control a number of **"unfair" business practices** that sometimes take advantage of consumer ignorance or inexperience. For example, the provinces of Saskatchewan and Ontario have established by statute a list of business practices recognized as "unfair." Any business engaging in them would be subject to penalty. The injured customer would also be free to rescind any agreement made as a result of the unfair practice. Many of the unfair practices listed also relate to torts (such as fraudulent misrepresentation), but some, such as contracts imposing onerous payment terms on the buyer, become "actionable" by way of the complaint procedure set out in the statute. Although these laws are concerned with consumer protection, they are also designed to protect honest and ethical business persons from unfair competition by the few firms that take advantage of consumer inexperience or ignorance. Consumer-protection legislation is examined in greater detail in Chapter 27, "Consumer-Protection Legislation."

A QUESTION OF ETHICS

Montecristo is an old, cavernous, inner-city warehouse that has been converted into a high-energy, late-night dance club. Hundreds pack into the place, from 10 p.m. Saturday until 5 a.m. Sunday. The entrance is separated from the exit by a cloakroom marked with three signs: "All coats here"; "One-night membership: $10 — pay here"; and "All illicit substances will be reported to police." The entrance and exit doors are each manned by a pair of impossibly large bouncers. The only other staff are those taking coats, the DJ, and waitstaff behind the bar (which closes precisely at the provincially required hour). Seeking their coats at the exit side of the cloakroom, patrons note another sign: "Coat-check $10, pay here." In the past 12 months, three drug-related deaths of patrons have occurred in local hospitals in the hours after the 5-a.m. closing.

This situation raises a number of fundamental issues. What, in your view, is "demonstrably justified" in regulating our "free and democratic" society? Your viewpoint could include the interests of the owners of the Montecristo Club, patrons, and police, or local, provincial, and federal officials.

Many business torts also may be considered business crimes, and are subject to prosecution under the *Criminal Code*. These offences include fraudulent property transfers, fraud, obtaining goods under false pretenses, importing prohibited articles, falsifying books and records, theft, extortion, charging criminal rates of interest (over 60 percent), and theft. Ethical business persons do not engage in these criminal activities in the conduct of their business. Charges against business persons under these provisions of the *Criminal Code*, while not uncommon, are relatively rare, given the large amount of business activity that takes place in Canada.

CHECKLIST FOR INTENTIONAL TORTS

1. Was an act committed that is prohibited by Common Law?
2. Was it intentional?
3. Was it outside of a contract (or trust)?

If the answer is "yes" to all three, an intentional tort exists. Regardless, other liability may still exist under criminal law, contract law, or the law of negligence (unintentional torts).

SUMMARY

Tort law is concerned with the injury that one person causes to another, or to his or her property. As a result, tort law is not limited to a specific type of injury or activity. It includes the intentional injury to another in the form of assault and battery, the intentional restraint of a person in the form of false imprisonment, and injury to a person's reputation in the form of libel and slander.

Injury to a person's reputation is called defamation, and may take the form of slander (verbal defamatory statements) or libel (published statements of a defamatory nature). Both are torts, but libel is the more serious of the two due to the permanency of its nature.

Trespass to land is a form of intentional injury involving either the unlawful entrance upon the land of another or willful damage to the land in some fashion (such as tunnelling under it). Trespass may also include the willful damage to goods if the act is done deliberately. If the goods are simply retained and the owner is denied possession, the tort is conversion. Any denial of the true owner's title, or any dealing with the goods without colour of right that would deny the true owner possession, also constitutes conversion.

Business torts, which involve unfair practices or statements by unethical business persons, are generally actionable at law, but many are now subject to special statutes designed to deal specifically with the improper acts.

KEY TERMS

Assault, 62
Battery, 62
Employer vicarious liability, 64
Forcible confinement, 65
Defamation, 66
Libel, 66
Slander, 66
Trespass, 69
Conversion, 71
Slander of goods, 71
Injurious falsehood, 72
Slander of title, 72
Restraint of trade, 73
Deceit, 74
Unfair business practices, 74

REVIEW QUESTIONS

1. Distinguish between the civil and criminal aspects of intentional torts, such as assault and battery or false imprisonment.
2. How does an assault differ from a battery? Must both have a violent element?
3. Under what circumstances might a person accused of assault and battery raise self-defence as justification?
4. Explain the circumstances under which the tort of false imprisonment might arise.
5. Distinguish between slander and libel. Why is libel generally considered to be more serious in the eyes of the law?
6. Define qualified and absolute privilege, and explain the circumstances in which each might be claimed.
7. How does the tort "trespass to land" arise? Must damage occur for the tort to be actionable?
8. Explain the difference between conversion and theft of goods.
9. Explain "slander of title." How does this differ from ordinary tort law related to defamation?
10. Explain the rationale behind the law that condemns as a tort any third-party interference with the performance of contracts made between other persons.
11. Define: "passing-off," "plagiarism,"and "slander of goods."
12. Give an example of "involuntary entry on the lands of another."
13. Does false imprisonment always have a criminal aspect to it? Explain.
14. To what extent, if any, does libel differ from the publication of a slanderous statement made by a person other than the publisher?

MINI-CASE PROBLEMS

1. A lawyer, who was also a professional boxer, was travelling by train from Winnipeg to Regina. He was sitting next to a woman who, during the course of a conversation, said: "Professional boxers should be charged with assault and battery each time they engage in a prize fight." As a lawyer, how should he respond to her statement?
2. The house in which X resided was located on a busy street. From time to time, vandals had broken into his garage and stolen tools and equipment he had stored there. One evening, just at dusk, he saw someone enter his garage. He quickly rushed out to the garage, closed the door, locked it, then called the police. When the police arrived, X discovered that he had locked the municipal building inspector in the garage. Discuss the legal position of X and the building inspector.
3. At a packed public meeting of town council, an irate taxpayer named Kilmer questioned the council repeatedly, at length and in detail, about a proposed new sewer system. Exasperated with Kilmer's domination of the questioner's microphone (but despite the audience's apparent contentment to let Kilmer ask questions), one of the councillors blurted: "Kilmer, you're an idiot. In fact, you come from a long line of idiots. Can't you see we don't have all the sewer answers yet? It's just the early stage of the planning process for this." Embarrassed, Kilmer left the room. If Kilmer approached you for advice, what would you offer?
4. You have decided to open a new nightclub in a university town and have decided to hire part-time employees as "security staff" for the dance hours of 7 p.m. – 1 a.m. Write a set of notes regarding the policies you want your security staff to observe and the legal liability that may result if they fail to adhere to those policies.
5. Joe hits Bob. Discuss the legal issues raised as a result.

www.mcgrawhill.ca/olc/willes

Case Problems for Discussion

CASE 1

Lukas owned a small factory of long-standing reputation, and bid successfully on a contract from Atlas Aircraft. The aerospace manufacturer's contract was for Lukas to supply custom-made hydraulic cylinders for landing-gear assemblies. To execute the job, Lukas required a heavy metal milling machine capable of handling large pieces with high degrees of precision. He had heard of the impressive capability of the Stormsen 1500 mill and placed an order for one, which in time was delivered and installed. Over a period of months, Lukas found the machine was constantly working out of adjustment, causing much of its output to fail quality control. Six times, Lukas caught as many as 2 faults in a job lot of 20 cylinders, and on one early occasion Atlas Aircraft returned a cylinder as substandard. After consulting all the Stormsen manuals, looking for an adjustment solution, Lukas called in the area distributor for consultation. No long term solution to the wandering adjustment seemed apparent, and Lukas began placing calls to the Stormsen Company itself. In the meantime more cylinders had been rejected by Atlas, whose chief engineer (a personal friend of Lukas) wrote Lukas a letter wondering why "suddenly [it] was receiving crap." Lukas' calls to Stormsen were fielded by an engineer/manager, Lewis Cranston, but Lukas remained far from satisfied. Within another month, Lukas visited a tradeshow where Stormsen machines were displayed, and in the display area was Cranston. The two, meeting for the first time, exchanged words before the shouting began. In the end, they had drawn quite a crowd of factory owners and two writers from a trade journal. The exchange ended with Lukas brandishing the Atlas letter before the onlookers, shouting "The 1500 produces crap, Atlas calls the 1500 crap, and I call the 1500 crap!"

Discuss the tort issues that may arise from this situation.

CASE 2

The plaintiff, a nurse, was injured in a motor-vehicle accident and was taken to a local hospital. She was examined by the defendant, who could find no physical injury other than a few minor bruises. She was discharged from the hospital the next day when she admitted that she "felt fine." Within 24 hours after her release, she returned to the hospital. She complained to the defendant of painful headaches and remained in hospital for a month. During her second stay in hospital she was examined by three neurosurgeons who could find nothing wrong with her.

On her release from the hospital, she instituted legal proceedings against the parties responsible for her automobile accident, and her solicitor requested a medical opinion from the defendant to support her case.

In response to the solicitor's request, the defendant wrote two letters that were uncomplimentary and suggested that the plaintiff had not suffered any real physical injury. In addition, the defendant had indicated on the plaintiff's medical records that the plaintiff was suffering from hypochondriasis.

After her discovery of the uncomplimentary letters and medical reports, the plaintiff brought an action against the defendant for libel.

Examine the arguments that might be raised in this case and identify the defences (if any) to the plaintiff's claim.

Render a decision.

CASE 3

Gretel was shopping at a large shopping centre and, while walking through the crowded mall area, she saw a youth pushing his way through the crowd in what appeared to be an attempt to escape from a man in a dark-blue uniform, who was following him.

At that time, Gretel was standing near the exit from the building. When the youth finally broke through the crowd and attempted to leave the building, she stepped in front of him to block his path. The youth collided with Gretel, and the two parties fell to the floor.

Gretel seized the fallen youth by the arm as he attempted to stand up and tried to pull him back down to the floor. The youth then struck Gretel a blow on the side of the head with his fist, causing her to lose consciousness.

The youth, as it turned out, was hurrying through the crowd in an attempt to catch a bus, and the older man, who was following him through the crowd, was his father. The youth's father was employed as a security guard at the shopping centre and was leaving work for the day.

Explain this incident in terms of tort law and tort liability.

CASE 4

A university operated a tavern on its premises for the benefit of its students. One student, who attended the tavern with some friends for the purpose of celebrating the end of the fall semester, became quite drunk. The tavern bartenders realized that the student was drunk around 11:00 p.m. and refused to serve him any additional alcoholic beverages. They also asked him to leave the premises. The student, however, remained and drank two additional beers that were purchased for him by his friends.

Some time later, around 12 a.m., one of the bartenders noticed the student drinking and instructed the tavern bouncer to ask the student to leave. The bouncer did so, but the student refused, and the bouncer took the student by the arm and escorted him to the door. Along the hallway to the door the student was abusive and resisted leaving, but the bouncer managed to eject him from the building.

A few minutes later, the student returned to the tavern and slipped by the doorman for the alleged purpose of obtaining an explanation as to why he had been ejected. About eight feet from the door, he was apprehended by the bouncer and once again expelled from the tavern, but not without some resistance in the form of pushing and shoving and abusive language on the part of the student. In the course of ejection, the student fell against the door and smashed a glass pane in the door, which caused severe lacerations to his hand. The injury to the student's hand required medical treatment and took several months to heal.

The student brought an action against the university and the bouncer, claiming damages and claiming as well that the injury he received caused him to fail his mathematics course in the semester that followed the accident.

Discuss the issues raised in this case and the various arguments that each party might raise.

Render a decision.

CASE 5

The Silver Sports and Recreation League was a women's hockey league that operated under Canadian Amateur Hockey Association rules. Under the rules, no bodily contact was permitted by the players.

During the course of a semi-final playoff game between the Silver Lake Lions and the Calabogie Cats in the Silver Lake Municipal Arena, April, the star centre of the Silver Lake Lions, was attempting to regain control of the puck in her own end of the ice when Carol, a defence player of the Calabogie Cats, collided with her from behind, driving her into the boards. As a result of the collision, April suffered a serious injury to her neck and spine.

Immediately after the collision, the referee, who had witnessed the collision, stopped the game and awarded a "match penalty" against Carol on the basis that she had deliberately attempted to injure April. According to the referee, Carol and April were both skating towards the puck, with April ahead of Carol. Upon reaching the puck, April stopped abruptly. Carol, in the process of stopping, raised her hockey stick to a horizontal position in front of her body just before she collided with April from behind. The blow from the horizontally held hockey stick prevented Carol's body from striking April, but the impact of the hockey stick propelled April into the boards.

The linesman, who also witnessed the incident, reported that from his point of view, Carol had pushed April from behind either to move her away from the puck or to avoid a more violent collision. The impact, however, in his opinion, only caused April to lose her balance and it was her loss of balance that resulted in her fall against the boards.

April subsequently instituted legal proceedings against Carol for her injuries.

Discuss the various legal arguments that each party might raise, and render, with reasons, a decision.

CASE 6

Jonas purchased a picnic basket at a hardware store in a nearby shopping mall. The basket was not wrapped by the sales clerk at the conclusion of the transaction. Jonas carried his new basket with him to a supermarket located in the same mall, where he intended to purchase a quantity of grapefruit.

At the produce counter he could not find grapefruit on display, and asked the clerk if the store had any in stock. The clerk offered to check in the storage room for him. While he waited for the clerk to return, Jonas picked a quantity of grapes from a display case and ate them. A few moments later, the clerk returned to inform him that all the grapefruit had been sold.

As Jonas left the store, he was seized by the store owner and requested to return to the owner's office. Jonas obediently followed him back inside the store. Once inside the owner's office, the owner accused Jonas of theft; then, without further explanation, telephoned the police.

When the police officer arrived, the store owner informed him that Jonas was a thief and that he had apprehended him just outside the store. Jonas admitted eating the grapes, then to his surprise, he discovered that the owner had apprehended him because he (the owner) thought Jonas had stolen the picnic basket.

Both the supermarket and the hardware store sold similar baskets; even on close examination, the products appeared identical. With the aid of the sales clerk at the hardware store, Jonas was able to convince the police officer that he had purchased the basket which he had in his possession.

He later decided to bring an action against the owner of the supermarket for false imprisonment.

Discuss the issues raised in this case and determine the respective arguments of the parties.

Render a decision.

CASE 7

Justin and Therese met during his business studies and her electrical engineering studies. They lived together and eventually both worked for Cosmic Star, a company producing Global Positioning System navigation receivers. Justin worked in the marketing department and Therese worked in the design department. With considerable in-house research, Therese developed an algorithm that reduced fuzziness in the GPS signal, making the receiver ten times more sensitive and accurate. It was a fundamentally simple equation, and with considerable pride she had explained it to Justin more than once. Not all that long later, Justin left Cosmic Star to work for its competitor, NavDirect Industries. About a year behind Cosmic Star and six months after Justin joined the company, NavDirect began producing a GPS that contained an algorithm remarkably similar to the Cosmic Star model. Some months later, NavDirect was prospering while Cosmic Star began showing signs of financial problems, and Therese joined Justin at NavDirect. Her departure soon caused Cosmic Star to look closely at the technology in NavDirect products. Upon finding the algorithm that Therese developed, Cosmic Star headed to its lawyers for assistance. Discuss the advice Cosmic Star would likely receive. What advice would you give to each of Justin, Therese, and NavDirect?

CASE 8

Nico owned a strip mall development along a busy suburban city avenue. One of his tenants operated a small shop selling new and used sewing machines. Over a period of six months, it was apparent to Nico that his tenant was in trouble. Originally, the rent was paid in full and on time. Then the tenant paid in full but late; then it was on time but short; then finally short and late, then for two months nothing at all was paid. Realizing that this would only go from bad to worse, Nico called a locksmith who drilled out and replaced the door lock. Nico moved the inventory of twenty sewing machines into the back storage room, clearing away the stock from the shelves visible from the front window. He placed a "For rent" sign in the window, set the alarm, and left. When he returned the next morning, he found the door lock smashed, the alarm disarmed, and the inventory gone.

Discuss the rights, responsibilities, and defences the parties may raise in this situation.

The Online Learning Centre offers more ways to check what you have learned so far. Find quizzes, Weblinks, and many other resources at www.mcgrawhill.ca/olc/willes.

CHAPTER 5

Negligence and Unintentional Torts

Chapter Objectives

After study of this chapter, students should be able to:

- Explain the chief elements of negligence: duty, breach, damage and causation.
- Explain the relevance of foreseeability and the "Reasonable Person."
- Explain the concepts of strict and vicarious liability.
- Describe instances of tort liability arising from product defects.
- Describe the chief defences and remedies in unintentional tort cases.

YOUR BUSINESS AT RISK

Negligent acts and accidents happen. Businesses must be aware of when, how, and to what degree they will be responsible for unintentional or careless acts, and to the extent that they may recover for careless acts done to them. A single liability suit for an accident that could have been avoided can run to millions of dollars and may bankrupt the dreams of the owner or investors.

NEGLIGENCE

There is no end to the number of ways in which individuals and businesses may unintentionally interfere with the person or property of another. Careless driving of a highway transport may cause injury or death to others on the road, the careless manufacture of a food product may poison or injure a consumer, or the careless performance of an audit by an accountant may lead to financial losses; all are examples that come to mind. These cases have something in common:

1. Someone owes a duty not to injure,
2. There is an act or omission in breach of that duty, and,
3. An injury is suffered as a direct result of that breach.

The Duty Not to Injure

The basic premise upon which tort liability is founded is that individuals and corporations living in a civilized society should not intentionally cause injury to one another, or to the property of others. It is not much of a stretch to extend this duty to acts or omissions which, despite a lack of intention, have fallen below some expected standard of behaviour and have caused harm to another person. The question is: in what situations can it be said that a person owes such a **duty of care** to another person?

Duty of care
The duty not to injure another person.

That duty exists where a legal right exists. At the moment you hail a taxi, you have no legal right against the taxi driver, just as the driver has no duty toward you. However, once you have engaged the driver and have been accepted as a passenger, things change. A duty of safe driving is matched with a right to safe transportation. Those rights and duties end at the end of the journey.

Standard of Care and Breach

Reasonable person

A standard of care used to measure acts of negligence.

The standard of care to be met in the execution of that duty is not absolute perfection, for the expectation is that of the mythical "**reasonable person**," someone of average intelligence who will prudently exercise reasonable care, considering all of the circumstances. The expected standard of care may be increased when persons owing the duty hold themselves out to have special or professional skill. Breach of the standard of care occurs when the performance rendered falls below that which the injured party had a right to expect. Those expectations and the standard of care may be high or low as circumstances change. For example, we expect greater care and prudence from drivers when the roads are slick with rain than we might demand of them when the weather is clear. All along it is really only one standard, which is what a reasonably prudent person would do when faced with differing situations. This standard offers great flexibility to the courts to assess a wide range of behaviour in many situations with a single test. It is judgmental, but then again, that is what judges must routinely do.

Damage and Foreseeability

It is axiomatic that if there is no damage, there can be no liability. However, there can be damage for which there is no liability, when that damage was not reasonably capable of being foreseen by our "reasonable person." This tort concept is not easy to apply, as parties in tort actions have widely divergent views on what is, or is not, reasonably foreseeable. It should be noted that children may be held liable for the torts they commit, provided they have the degree of maturity to appreciate the nature of their acts.

CASE IN POINT

While replacing a water dispenser bottle, the plaintiff saw a dead fly in the unopened replacement bottle. Obsessed with this, he developed a major depressive disorder, phobia and anxiety. He successfully sued the supplier for psychiatric injury. The Supreme Court of Canada overturned this; the plaintiff failed to show it as foreseeable that a person of ordinary fortitude would suffer serious injury from seeing a fly in the bottle of water he was about to install. Unusual or extreme reactions to events caused by negligence are imaginable but not reasonably foreseeable, and thus do not give rise to a cause of action.

Mustapha v. Culligan of Canada Ltd., 2008 SCC 27 (CanLII).

Causation

Causation or "proximate cause"

A cause of injury directly related to an act of a defendant.

Linked to both duty and reasonableness, the plaintiff is also required to show that the acts or omissions of the defendant represent **causation or "proximate cause"** of the injury suffered. This does not mean that the defendant should be expected to answer for every careless act, no matter how remotely connected or traceable between the injury and the defendant. In fact, just the opposite, the acts or omission of the defendant must be related in a relatively direct way to the injury, without intervening events.

The "But For" Test in Causation

The test is general in nature, and asks the question: "but for" the defendant's actions, would the injury or damage not have occurred? An affirmative answer to the question would identify the defendant's actions as a cause of the injury or damage. This is however, a less helpful test (from the plaintiff's view) where there are other intervening incidents that also result in injury or damage. Any break in the chain of events running from the defendant's act to the plaintiff's injury will normally defeat the plaintiff's claim that the proximate cause was the defendant's act. For example, if a person slips while walking in the wet foyer of an office building and breaks a leg, the building owner may be liable if its failure to remove the water was the proximate cause of the injury. However, if the injured person is being transported to hospital to have the broken leg treated and along the way is severely injured when the ambulance is involved in a traffic accident, the building owner's negligence cannot be said to be the proximate cause of the injuries suffered in the traffic accident. The intervening event of driving the person to the hospital is sufficient to break the chain of causation between the building owner and the traffic accident injuries.

CASE IN POINT

An employee of the municipal electrical utility marked an area of a construction site as "safe to dig." Later, an employee of one of the construction firms working on-site was preparing to dig a trench with a mechanical digger. A site foreman, hired to manage the project by the site owner, asked him if he wanted the job-site power to be cut for a period of time. The worker replied that his work area was safe, and proceeded to dig the trench. Severing a live cable, the worker was seriously injured. Discuss the issues of proximate cause raised by these facts.

McCain v. Florida Power Corporation, 593 So.2d 500 (Fla. 1992).

CASE IN POINT

A newspaper subscriber suspended home delivery of daily papers during her two-week vacation, but the newspaper kept arriving on her porch in any event. She returned to find her home broken into and robbed. Is the publisher the proximate cause of her losses? What facts and legal reasoning would you use to determine that the newspaper publishers were negligent and liable for her losses?

Crotin v. National Post, 2003 CanLII 64305 (ONSC).

CASE IN POINT

The Supreme Court of Canada has acknowledged that exceptional cases do exist. If it is *truly* impossible to apply the "but for" test, and the injury is within a duty of care owed by the defendant, then the defendant's material contribution to the plaintiff's injury can be recognized. It is important to realize that the "but for" test remains as the basic test.

Resurface v. Hanke, 2007 SCC 7 (CanLII).

FIGURE 5–1 Negligence

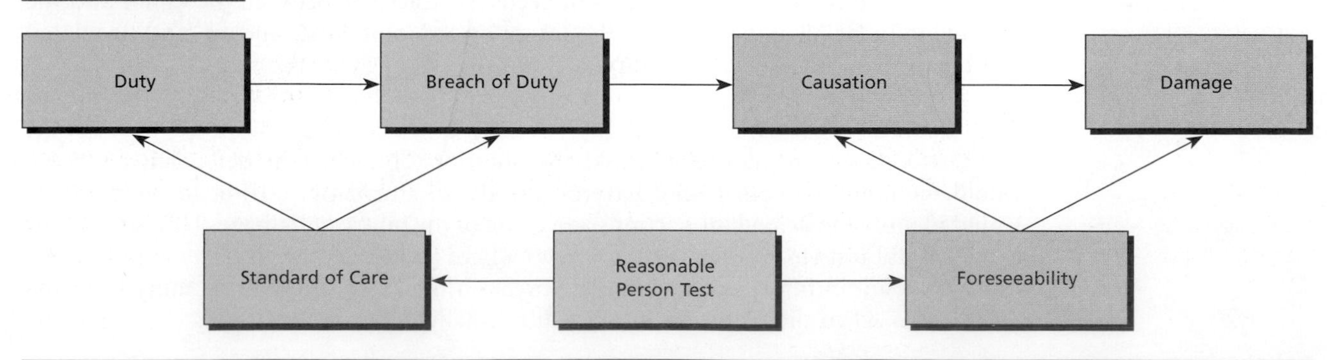

FOR NEGLIGENCE

1. Was there a duty not to injure?
2. What was the standard of care?
3. Did the defendant breach that standard?
4. Was there damage?
5. Were the defendant's actions the proximate cause (causation) of that damage?
6. Was the damage foreseeable by a reasonable person to result from such actions?

If the six elements are proved against a defendant, an action in negligence would likely succeed, resulting in an award of damages.

Strict Liability

Strict liability

Responsibility for loss regardless of the circumstances.

Some activities are so inherently dangerous that the public has a right to recovery regardless of how remote or unforeseeable the causation of an injury may be. Today, any person or business who maintains a potentially dangerous thing (such as hazardous chemicals, toxic contaminants, or hazardous waste) on its land may be held strictly liable for any damage it may occasion should it escape. This may be the case even if the landowner takes every precaution to protect others from injury, and if the escape of the dangerous thing is not due to any act or omission on the landowner's part. Much attention has been paid recently to the issue of dog attacks, which remain largely subject to a presumption of strict liability for breeds such as pit bulls. Even "safe" things fall within the law when they behave dangerously.

> In order to have a supply of water for the purpose of operating his mill, Rylands built a water reservoir on land that he occupied. He employed engineers and contractors to construct the reservoir and, when the reservoir was completed, began filling it with water. Unknown to Rylands, the land under the reservoir was a part of a coal mine that was being worked by Fletcher. Water from the reservoir found its way into the old shafts and passageways in large quantities and flooded areas of the mine some distance away, where Fletcher was working. Fletcher sued Rylands for damages. The court held that Rylands was liable for the damage caused to Fletcher even though Rylands did nothing to deliberately harm Fletcher. The court, in reaching its decision, stated that anyone who accumulates anything non-natural on his land or uses it in a non-natural way that might injure his neighbour does so at his peril, and if it should escape through no fault of his own, he should nevertheless be liable.[1]

This form of liability, which is based neither upon intent nor negligence, remains, largely because the conduct, while not wrongful or improper, is so inherently dangerous or so unusual that any risk associated with it should be borne entirely by the individual wishing to proceed with it upon his or her property.

Vicarious Liability

The same principle is sometimes applied when one person controls the activity of another to such an extent that the act of one may be attributed to the other. In these cases, the courts carry the liability for the tort back to the person who (in a sense) initiated the act, and in this manner impose liability on a person not directly associated with the tortious act.

Vicarious liability

The liability at law of one person for the acts of another.

At Common Law, an employer is usually considered to be **vicariously liable** for the torts of the firm's employees, provided that the torts are committed by the employees in the course of the employer's business. The reasoning of the courts on this rule is that employees may not have the financial means to compensate for the damage they might cause,

1. See *Rylands v. Fletcher* (1868), L.R. 3 H.L. 330.

but their employers probably would, if only through insurance coverage. Additionally, because the employee's tort is committed during the course of the employer's business, the employer should have some responsibility for the loss as he or she is presumably directing or controlling the employee at the time. Similarly, all partners in a partnership are vicariously liable for the torts committed by a partner in the conduct of partnership business. In a similar vein, provincial statute law has imposed a vicarious liability on the owner of a motor vehicle, when the driver of the vehicle is negligent in its operation.

The imposition of liability on persons not directly associated with a tort reflects a departure from the tort theory that the individual who injures another should bear the loss. It also reflects a move away from the former *laissez-faire* policy to a new social policy that recognizes that some satisfactory method must exist for the distribution of what is essentially a social loss among those best able to bear it. The modern concept of insurance is, for the most part, responsible for this shift in tort liability and represents a further extension of this philosophy. Similar examples may be found in other rights and statutes that deal with specific areas of tort, such as the various Workers' Compensation Acts. These Acts essentially spread the risk of loss to workers injured in employment-related accidents to all employers in an effort to reduce the burden that would otherwise fall on the individual or the injured worker's employer.

Res Ipsa Loquitur

One of the particular difficulties that sometimes faces a person injured by the negligence of another is proving the negligent act. In some cases, the injured party is unaware of how the injury occurred, or what the act of the other party was that resulted in the loss. For example, a freight forwarder sending cargo on an aircraft has a right to expect a safe flight and delivery from the departure point to the destination. However, if the aircraft should crash on landing and the cargo is seriously damaged, can the forwarder, as a plaintiff, reasonably be expected to satisfy the courts that certain specific acts of the pilot constituted negligence? Only the pilot or the flight crew would know the circumstances that led to the crash, causing damage to the goods. This dilemma of the injured plaintiff first came before the courts in an English case[2] in the mid-nineteenth century, when a person standing in the street was injured by a barrel of flour that fell from the upper level of a shopkeeper's building. The person sued the shopkeeper for his injuries, but could do nothing more than relate the facts of the case available to him and plead ***res ipsa loquitur*** ("the thing speaks for itself"). The court accepted this line of reasoning and ruled that, since the plaintiff had established that he was injured by the defendant's barrel of flour, it was now up to the defendant to satisfy the court that he was not negligent in the handling of the barrel.

Res ipsa loquitur

"The thing speaks for itself."

Since this decision, the rule or principle of *res ipsa loquitur* has been applied in a wide variety of negligence cases where the plaintiff has been unable to ascertain the particular circumstances surrounding the injury inflicted. For the rule to apply, however, the cause of the injury must be something that is exclusively in the care and control of the defendant at the time of the injury. As well, the circumstances surrounding the accident must be unusual in the sense that they constitute events that do not ordinarily occur if proper care has been taken by the defendant. If the plaintiff is in a position to satisfy the courts on these two points, the burden of proof then shifts to the defendant to show no negligence; otherwise, the defendant will be held liable for his or her apparent actions.

A QUESTION OF ETHICS

How far should foreseeability extend? After a serious late-night accident, a car is towed to the curbside of the locked yard of an auto-repair shop. Hours later, after the slowly leaking fuel tank pools gasoline in the gutter, a passing pedestrian tosses away a match after lighting a cigarette. The result is as might be expected. Are there some things that simply must be described as accidents? Should we split up liabilities between the car driver, the tow-truck driver, and the pedestrian, or just conclude that some things are simply inevitable?

2. *Byrne v. Boadle* (1863), 2 H. & C. 722, 159 E.R. 299.

OCCUPIER'S LIABILITY

The first business concern with respect to occupier's liability is that it applies to acts and omissions of occupiers, not owners. There are many business owners renting properties who do not realize that regardless of their arrangements with their landlord, the first obligation lies upon them to protect persons who enter on the property.

Formerly, different classes of persons ranging from trespassers to those entering for the benefit of the occupier were owed increasing degrees of duty of care. While trespassers rarely find favour in the courts, all persons are owed a duty of common humanity and a warning of eminent dangers, and thus it is not permissible to set traps for vandals or graffiti artists. As a result of codification of this area of the law under provincial occupier's liability acts, visitors of any kind to a property are owed the general duties and standard of care expected of a reasonable person.

CASE IN POINT

A farm family lived 300 feet from the main road and did not take care of their driveway. After a particularly severe storm left ice beneath a dusting of snow, a visitor slipped on their parking area and was badly injured. Despite that it was not a local practice to spread salt or ice in such conditions, and despite that the visitor was aware of the slick state of the driveway, the Supreme Court of Canada agreed with lower courts that this did not absolve the farm family of the requirement to take greater precautions against such injuries.

Waldick v. Malcolm, [1991] 2 SCR 456.

CASE IN POINT

During a special sale at a crowded supermarket grocery store, a container of cheese spread fell from a customer's grocery cart and smashed on the floor. An elderly customer, who did not notice the broken jar, slipped on the cheese spread and fell, breaking her hip. She sued the grocery store for damages. At trial, the grocery store owner and staff testified that they had a regular floor-cleaning and inspection system in place and the accident had happened shortly after an inspection had been made. They were unaware of the broken jar until after the accident occurred.

The court found that the accident was not foreseeable, and nothing could have been done by the store owner to prevent the injury. The claim was dismissed.

Kosteroski v. Westfair Properties Ltd. (1997), 155 Sask. R. 37.

MANUFACTURER'S LIABILITY

Before the Industrial Revolution, goods were normally made by craftsmen serving the local population. The Industrial Revolution ushered in the mass production of goods and their distribution over a wide area. This in turn established a remoteness of the manufacturer from the ultimate consumer in the distribution system. Under the law of contract, the seller of goods (as an implied condition in the sale agreement) provides that the goods sold are of merchantable quality and reasonably fit for the use intended. If a purchaser buys such goods by description and the goods contain a defect, the law allows the purchaser to recover any loss from the vendor. While this aspect of the law of contract will be dealt with in greater detail in subsequent chapters, it is sufficient to say at this point that the law of contract provides a remedy to the purchaser or the person for whom the goods were intended if the goods prove to be defective or cause some injury.

The limitation imposed by the law of contract is that it does not provide a remedy for the user or the consumer of the goods if that person is not the purchaser. This particular shortfall of the law was eventually remedied by the law of torts in the United States. The case involved the sale of a motor vehicle that contained a defect in its construction, resulting in an injury to the purchaser. The purchaser sued the manufacturer, rather than the seller of the car, and was met by the defence that the manufacturer owed a duty of care only to the immediate purchaser, i.e., the retailer. The court, however, ruled that the duty of care extended beyond the immediate purchaser to the plaintiff, who was the ultimate user of the goods.[3]

English and Canadian courts discussed the duty of care of manufacturers in a number of cases,[4] but it was not until the case of *M'Alister (or Donoghue) v. Stevenson* came before the courts that the issue of the responsibility of manufacturers was established.

COURT DECISION: Practical Application of the Law

Negligence—Duty of Care—Manufacturer's Liability

***M'Alister (or Donoghue) v. Stevenson*, [1932] A.C. 562.**

This famous case of humble facts is the foundation of all modern product liability cases and consumer protection legislation. The plaintiff, a Scottish woman (and under Scots law entitled to sue in both her maiden and married names — thus the odd citation) entered a shop in Paisley, Scotland with a friend. The friend purchased a bottle of ginger beer manufactured by the defendant, and gave it to her. The glass bottle was opaque and the contents could not be inspected before opening. The plaintiff consumed part of the drink, and when she emptied the remainder into a glass, discovered the remains of a decomposed snail. She became violently ill following the incident and brought an action for damages.

Here is an example of monumental and sudden change in the law: until this case, if a person purchased goods under a contract, then the contract implicitly contained a warranty that the goods were fit for the use intended, i.e., consumption. But the friend had the contract with the shop, which was similarly connected to the manufacturer. To M'Alister, the bottle resulted from a gift, not a contract; no contract meant no warranty; no warranty meant no damages. Tort liability for products was unheard of. And so, to Court:

Lord MacMillan: The law concerns itself with carelessness only where there is a duty to take care and where failure in that duty has caused damage. In such circumstances carelessness assumes the legal quality of negligence and entails the consequences in law of negligence. What, then, are the circumstances which give rise to this duty to take care? In the daily contacts of social and business life human beings are thrown into, or place themselves in, an infinite variety of relations with their fellows; and the law can only refer to the standards of the reasonable man in order to determine whether any particular relation gives rises to a duty to take care as between those who stand in that relation to each other. [T]o descend from these generalities to the circumstances of the present case I do not think that any reasonable man or any twelve reasonable men would hesitate to hold that, if the appellant establishes her allegations, the respondent has exhibited carelessness in the conduct of his business. For a manufacturer of aerated water to store his empty bottles in a place where snails can get access to them, and to fill his bottles without taking any adequate precautions by inspection or otherwise to ensure that they contain no deleterious foreign matter, may reasonably be characterized as carelessness without applying too exacting a standard. But, as I have pointed out, it is not enough to prove the respondent to be careless in his process of manufacture. The question is: Does he owe a duty to take care, and to whom does he owe that duty? Now I have no hesitation in affirming that a person for gain engages in the business of manufacturing articles of food and drink intended for consumption by members of the public in the form in which he issues them is under a duty to take care in the manufacture of these articles. That duty, in my opinion, he owes to those he intends to consume his products. He manufactures his commodities for human consumption; he intends and contemplates that they shall be consumed. By reason of that very fact he places himself in a relationship with all of the potential consumers of his commodities, and that relationship which he assumes and desires for his own needs imposes on him a duty to take care to avoid injuring them. He owes them a duty not to convert by his own carelessness an article that he issues to them as wholesome and innocent into an article which is dangerous to life and health. It is said that liability can only arise where a reasonable man would have foreseen and could have avoided the consequences of his act or omission. In the present case the respondent, when he manufactured his ginger-beer, had directly in contemplation that it would be consumed by members of the public. Can it be said that he could not be expected as a reasonable man to foresee that if he conducted his process of manufacture carelessly he might injure those whom he expected and desired to consume his ginger-beer? The possibility of injury so arising seems to me in no sense so remote as to excuse him from foreseeing it.

3. *MacPherson v. Buick Motor Co.* (1916), 217 N.Y. 382.
4. See, for example, *Heaven v. Pender* (1883), 11 Q.B.D. 503; *Le Lievre v. Gould*, [1893] 1 Q.B. 491.

In the cases that have followed the *M'Alister (or Donoghue) v. Stevenson* decision, the courts have held that when there is physical injury to persons or damage to property, the manufacturer will be liable to the ultimate consumer if it can be shown that the manufacturer was negligent in the manufacture of the goods,[5] or where the goods had some danger associated with them, and that the producer failed to adequately warn the consumer of the danger.[6] In the United States, the liability of the manufacturer is not limited to cases where negligence may be proven, but in many cases has been held to be strict. Accordingly, for certain types of goods, the manufacturer may be held liable for any injury, regardless of the efforts made by the manufacturer to prevent faulty products from reaching the consumer.[7]

FRONT-PAGE LAW

Child injuries spur toy recalls

Fisher-Price voluntarily recalled selected models of two products for preschoolers and infants yesterday after two Canadian children required stitches and another nearly strangled himself.

In Canada, about 32,000 Big Action Construction toys and 75,000 Fisher-Price Hop, Skip, Jumpers for babies have been recalled.

Two models of the jumper, an activity seat for babies to sit in while suspended from a doorway and sold between 1987 and 1994, were recalled after the company discovered a spring attached to a suspension strap had broken.

Fisher-Price has received 13 reports in Canada of the string breaking and the security cord failing to keep the jumper suspended. Eighteen children were injured in the United States.

Two Canadian children suffered serious injures while nine others suffered minor injuries.

The company will replace the defective spring with a suspension strap.

Fisher-Price stopped making the product in 1993. Both products are being recalled across North America.

The Big Action Construction set, modelling a work site with a crane on top of a dirt hill, is being recalled for products manufactured prior to January of 1998.

In the United States, the toy company received 16 reports of children wrapping the strap around their necks, seven of whom suffered minor injuries.

A recall of jumpers produced two decades ago and intended for consumers who were infants at that time seems like a case of closing the stable door after the horse is gone. Some jumpers may, however, still be in use, making the recall sensible and prudent. Aside from the goal of making the jumper safe, what added legal protection does the recall give to the manufacturer?

Source: Excerpted from Natalie Armstrong, *National Post,* "Child injuries spur toy recalls," July 22, 2000, p. A4.

Nuisance

Nuisance

Interference with the enjoyment of real property or, in some cases, material interference with a person's physical comfort.

The tort of **nuisance** has been applied to a great many activities that cause injury to landowners or occupiers, and, because of its wide use, it is now, to some extent, incapable of precise definition. In practice, it generally refers to any interference with a person's enjoyment of his or her property, and includes such forms of interference as noise, vibration, smoke, fumes, and contaminants of all sorts that may affect the use of land. Unlike some torts, nuisance is very much dependent upon the circumstances surrounding the interference, or the degree of interference, rather than the fact that the interference occurred.

The courts have long recognized that in the case of nuisance they must essentially balance the reasonable use of land by one person with the decrease in enjoyment that the reasonable use of the property produces for that person's neighbour and, in some cases, the community as a whole. To be actionable, then, the interference must be such that it results in a serious decrease in the enjoyment of the neighbour's property, or that it causes specific damage to the land. What is a reasonable use of land is usually determined by an examination of the uses of land in the immediate vicinity. For example, if individuals insist on making their residence in an area of a city where heavy manufacturing activity is carried on, they must expect the reasonable use of adjoining property to include the emission of noise and, perhaps, odours, smoke, and dust.

5. *Arendale v. Canada Bread Co. Ltd.*, [1941] 2 D.L.R. 41; *McMorran v. Dominion Stores Ltd. et al.* (1977), 14 O.R. (2d) 559.
6. *Austin v. 3M Canada Ltd.* (1975), 7 O.R. (2d) 200, but see also *Lem v. Barotto Sports Ltd. et al.* (1977), 69 D.L.R. (3d) 276.
7. See, for example, *Escola v. Coca-Cola Bottling Co. of Fresno* (1944), 150 P. (2d) 436.

Injunction

An equitable remedy of the court that orders the person or persons named therein to refrain from doing certain acts.

When a nuisance is determined, the remedies available to the party subjected to the nuisance would be damages and, at the court's discretion, an **injunction** ordering the defendant to cease the activity causing the interference. If the nuisance is such that its restraint would be detrimental to the community as a whole, the court, in balancing the two interests, will normally place the interests of the community before those of the individual, and will limit the landowner's remedy to damages. For example, when the issuance of an injunction would have the effect of closing down a mine or smelter upon which the community depends for its very existence, public interest would dictate that the remedy take only the form of monetary compensation.[8] On the other hand, if the interference is localized, the courts have often considered an injunction to be appropriate as well.[9]

In the case of *Segal et al v. Derrick Golf & Winter Club* (1977), 76 D.L.R. (3d) 746, the court dealt with a complaint by the owners of property located adjacent to a golf course that members of the golf club were interfering with their enjoyment of their property by driving golf balls into their yards. The property owners complained to the court that the careless golfing was a nuisance, and sought court assistance to have it stopped. The judge in reviewing the law of nuisance stated that the nuisance was essentially interference with the occupier's beneficial interest and enjoyment of the property. He described the law as it applied to this case in the following manner:

> That there has been interference with the plaintiff's pleasure, comfort and enjoyment of their land is clearly established. They are unable to use and enjoy their backyard during the golfing season. Their fears for the safety of their children and for their own safety are well-founded and their home will continue to receive damage from hard-driven golf balls. Their inconvenience is serious and substantial.
>
> As I have previously stated, this interference was entirely foreseeable from the layout of the 14th hole. The defendant as owner and operator of the golf-course has known of this interference at least since 1972, and has permitted it to continue although it was and is within its power to prevent it by control and supervision which it has over the players using the course.
>
> In the circumstances, the defendant as occupier of the golf course is liable for the private nuisance to the plaintiffs.

The current concern for the environment, and the difficulty in determining the sources of pollution in a precise manner, have prompted all levels of government to establish legislation for the purpose of controlling the sources of many of the more common environmental nuisances that interfere with the use and enjoyment of property. The statutes have for the most part replaced actions for nuisance as a means of controlling pollution, but on an individual basis, when the cause of the pollution is isolated and identified by a landowner, an action for nuisance may still be used to stop the offending activity.

It should be borne in mind that, as a tort, nuisance applies to many activities that are not environmental in nature. For example, picketers unlawfully carrying picket signs at the entrances to a hotel have been declared a nuisance by the courts.[10] Similarly, while awaiting the opening of a theatre, a queue of patrons who habitually blocked the entrance to an adjoining shopkeeper's premises were held to be a nuisance.[11] These are simply two of the many kinds of nuisances that the courts may recognize as interference with the enjoyment of property.

Because nuisance generally involves a public-policy decision, much of the legislation relating to property use at the municipal level is directed to this particular problem. Local zoning laws attempt to place industrial uses and residential uses of land some distance apart. This is to permit the users maximum freedom in the respective uses of their properties, and to minimize the interference of one with the other that might take the form of nuisance. The larger problems of industrial nuisances and the environment are also

8. *Black v. Canada Copper Co.* (1917), 12 O.W.N. 243.

9. *Russell Transport Ltd. v. Ontario Malleable Iron Co. Ltd.*, [1952] 4 D.L.R. 719.

10. *Nipissing Hotel Ltd. v. Hotel & Restaurant Employees and Bartenders International Union* (1962), 36 D.L.R. (2d) 81.

11. *Lyons, Sons & Co. v. Gulliver*, [1914] 1 Ch. 631.

the subject of increasing control by senior levels of government, rather than the courts. The right of the individual to redress for interference with the enjoyment of property nevertheless still remains for many nuisances that are not subject to statutory regulation.

General Tort Defences

Liability does not always rest on a person who causes injury or damage to another person or to his or her property. Certain persons have the lawful right in the course of carrying out the duties of their work or office to interfere with others or, in some cases, to cause them injury or inconvenience. For example, a police officer, in pursuit of a person believed to have committed a criminal offence, may trespass on the property of a third party in the course of the pursuit. A customs officer may search the luggage of a Canadian citizen returning to Canada from abroad and may prohibit the entry of goods that may not be lawfully brought into the country. Similarly, goods smuggled into Canada illegally are forfeited to the Crown and may be seized by customs officers or the police. All of these acts would normally be actionable at law if done by an ordinary citizen under different circumstances, but persons such as police and customs officers possess these special rights or powers by nature of their office. Ordinary citizens nevertheless may have defences that would excuse them from liability in cases where they are accused of committing a tort against another. Some of the more common defences follow.

Contributory Negligence and *Volenti Non Fit Injuria*

A further development in tort law relating to negligence occurred when the courts expanded their examination of the circumstances surrounding a tort to include the actions of the injured party. One of the earliest defences raised by a defendant in a negligence case was the argument that the injuries suffered by the plaintiff were due in some measure to the plaintiff's own carelessness, or to the plaintiff's voluntary assumption of the risk of injury by undertaking the activity resulting in the injury. Initially, both of these defences, if accepted by the courts, would allow the defendant to escape liability even in those cases where the defendant was, for the most part, responsible for the loss or injury suffered by the plaintiff.

Volenti non fit injuria
Voluntary assumption of the risk of injury.

The voluntary assumption of risk (***volenti non fit injuria***) has remained as a valid defence when a defendant is able to satisfy the court that the plaintiff voluntarily assumed the risk of the injury that occurred. To some extent this was a reflection of the early *laissez-faire* individualism prevalent in the nineteenth century, which was based upon the premise that the law should not protect those who were capable of protecting their own interests and, in particular, should not protect those who were prepared to assume the risk of loss or injury on a voluntary basis.

COURT DECISION: Practical Application of the Law

Negligence—Duty—Standard of Care

***More v. Bauer Nike Hockey Inc.*, 2010 BCSC 1395 (British Columbia Supreme Court).**

Darren More, age 17, suffered a serious brain injury while playing AAA Midget level hockey. He sued Bauer, the designer and manufacturer of his helmet, and the Canadian Standards Association (CSA) which sets the minimum standards for hockey helmets, which are required to be worn. More claimed Bauer had a duty to consumers of ice hockey helmets to design, manufacture and distribute for sale only helmets that were adequate to provide protection to such consumers from risk of serious head injury caused by foreseeable impacts by wearers. Further he claimed the CSA knew, or ought to have known, that the standard set by them was inadequate.

THE COURT: Players of all ages and sizes skate fast, get hit, or fall, and hit the ice or the boards, or both. There is nothing startling or unusual in the manner in which Darren hit the boards in this case. Unfortunately, what is very unusual and possibly unprecedented in Canadian organized amateur hockey is the injury Darren sustained that left him with such devastating effects, despite his wearing a CSA certified helmet. According to one of the CSA biomechanical experts, Darren's case is the only reported instance of a helmeted player sustaining a subdermal hematoma (SDH) while playing organized hockey.

The CSA facilitates the development of standards and tests products for certification against those standards. [Standard M90 for helmets] states that its objective is to: [R]educe the risk of head injury to ice hockey participants, when the helmet is used as intended and in accordance with the manufacturer's instructions, without compromising the form and appeal of the game. It also states that ice hockey is a sport with "intrinsic hazards" and that participation "implies the acceptance of some risk of injury." It further states that, while the use of a helmet certified under the standard "will not prevent all injuries" it "is intended to reduce the frequency and severity of head injuries."

There is no issue that Bauer, as a manufacturer of consumer products, had a duty to take reasonable steps to ensure that its hockey helmets were safe for their intended use. Bauer must design products to minimize the risks arising from their intended use and to minimize the loss that may result from reasonably foreseeable mishaps involving the product. Counsel for CSA further contends that there is insufficient proximity between the plaintiffs and the CSA for a duty of care to exist. I am satisfied that it was reasonably foreseeable that a hockey player and wearer of a mandatory certified hockey helmet might suffer harm if the CSA set the certification standard unreasonably low in the circumstances.... Similarly, the claim against the CSA is not like a free insurance scheme. The specific allegation is that M90 did not set the bar high enough to reduce the risk of the particular injury that Darren sustained.

The basic test for causation in negligence is the "but for" test: the plaintiff has the burden to show that but for the negligent act or omission of the defendant, the injury would not have occurred.

In all the circumstances, I am satisfied that the Bauer helmet offered a reasonable level of safety for rear impacts having regard to the risk of the wearer sustaining a serious head injury like an SDH while playing hockey. There was no substantial likelihood of the alleged harm associated with its ordinary use. Nor did the evidence demonstrate that it was feasible to design the helmet in a safer manner to protect against such risk... I point out that Bauer was not obligated to design an "accident-free" helmet. It is not an insurer nor is it to be held to a standard of perfection. I dismiss the claims in negligence against Bauer and the CSA.

The above is extracted from a 50 page judgment of medical evidence of Darren's injury and an engineering review of the helmet design and manufacture. What key evidence would have been required to convince this judge that either Bauer or the CSA should be liable (in whole or part) for Darren's injury?

CASE IN POINT

A manufacturer of fairgrounds equipment sold and assembled bleacher seating along the end zones of a small university stadium, such that it could be temporarily used, removed, and used in future for special events seating. At the first event, a homecoming football game, an unlimited number of tickets had been sold as "general admission and standing room." The regular bleachers had fixed seats limiting their capacity, but the temporary bleachers were packed with people, and collapsed mid-game. Discuss the issues of manufacturer's liability, occupier's liability, and voluntary assumption of risk in this scenario.

Contributory negligence, on the other hand, underwent a number of largely unsuccessful efforts by the courts to relieve the hardship in cases where the plaintiff was responsible in only a very minor way for the injury or loss sustained. Eventually, it became clear that the problem could only be solved by legislation. In the 1920s, the provinces passed legislation[12] that required the courts to determine the degree of responsibility of each of the parties in a tort action, and to apportion the damages accordingly. These statutes provided a general framework for the apportionment of loss in contributory negligence cases in all jurisdictions, as all say that where harm is contributed by two or more parties, loss is apportioned according to the degree of fault.

12. For example, the *Contributory Negligence Act*, 1924, S.O. 1924, c. 32.

Negligence law remains a major area of tort law, notwithstanding its relatively short history, and is an important part of the law of torts dealing with unintentional interference with individuals and property. There are, however, a number of other principles and doctrines that relate to the law of torts. The development of these particular principles and concepts and the defences that may apply to them may be best illustrated by an examination of the specific areas of tort law to which they generally apply.

Act of God

Act of God

An unanticipated event that prevents the performance of a contract or causes damage to property.

A good defence in the instance of property damage might arise when the defendant can establish that the resulting loss was caused by an **act of God**. For example, assume lightning struck a person's house and set it on fire, and that the fire spread to the neighbour's house. If the neighbour claimed compensation for the damage, the person whose house was struck by lightning might argue that the cause of the fire was an act of God, and therefore beyond personal control. In this instance, assuming there were no unusual circumstances (such as the storage of large quantities of explosive or highly inflammable substances in the residence), the defence of act of God may prevent the neighbour from recovering for the property loss resulting from the spread of the fire.

Waiver

Waiver

An express or implied renunciation of a right or claim.

The defence of **waiver** may be raised by a defendant if the defendant can satisfy the court that the injured party specifically renounced or waived his or her right to claim against the defendant for the particular injury that did occur. Waivers are usually in writing and, to be effective, must cover the injury contemplated by the parties at the time. However, because the circumstances surrounding the injury are often different from those contemplated at the time the waiver is given, a waiver is not always an effective way to avoid liability.

Release

Release

A promise not to sue or press a claim, or a discharge of a person from any further responsibility to act.

Sometimes, when damage or injury in a tortious situation has already occurred, the injured party may **release** the person who caused the injury or damage. They may do so either because the person has paid compensation for the injury or damage, or because the injured party has decided not to claim for the tort. It is usually required by the person who compensates the injured party for loss to avoid future claims arising out of the same incident. When a proper release is given under circumstances where the injured party is fully aware of the nature and extent of the loss or damage suffered, such a person may not normally take action against the released party at a later time. If the injured party should do so, the released party may raise the release as a defence to the tort action.

Statute of Limitations

Persons who are injured or who suffer a loss as a result of the tortious acts of another must take steps to bring a claim before the court within a reasonable time after the loss or injury occurs. To delay for a long period of time places the defendant at a disadvantage, as records may be destroyed, witnesses may die or move elsewhere, and memories of the event may fade. Consequently, a person with a right of action is expected to pursue the claim without undue delay. At Common Law, judges may refuse to hear a case when the plaintiff fails to bring a claim before the court within a reasonable length of time, if the defendant can satisfy the court that he or she has been prejudiced by the unreasonable delay. The barring of this relief is known as the **doctrine of laches**.

Doctrine of laches

An equitable doctrine of the court which provides that no relief will be granted when a person delays bringing an action for an unreasonably long period of time.

While the Common Law doctrine may still be applied, the limitation period for tort actions is now largely covered by statute law. In most provinces, specific laws have been passed stating the time within which different claims must be brought before the courts; otherwise, claimants lose their right to do so. Limitation periods are calculated from the time when a victim could reasonably have been expected to discover the loss.[13] The time

13. *City of Kamloops v. Nielsen et al.*, [1984] 2 S.C.R. 2.

limits vary from relatively short periods for certain specified torts (two years, in some provinces, for motor-vehicle accidents and claims of defamation) to a general maximum of six years for tort claims without specified time limits. Subject to certain exceptions, if a person fails to institute a tort action within the period fixed by the statute, the defendant may raise as a successful defence the fact that the claim is statute-barred by the legislation, and thereby avoid liability.

MANAGEMENT ALERT: BEST PRACTICE

Business managers can take concrete steps toward reducing the risk of negligence and liability.

- Make employee safety a genuine corporate priority.
- Mark products with serial numbers to allow tracking and recall. Take proactive steps for safety's sake.
- Document business processes into manuals for employee guidance.
- Document and discipline employee misbehaviour.
- Use appropriate warning labels.
- Archive customer correspondence and complaints.
- Participate actively in industry and professional associations to keep abreast of best practices.
- Devote meaningful resources to employee training.
- Instill pride in a job well done. Refuse to reward poor performance.
- Keep complete production records.
- Record sources and dates of raw materials and supplies.
- Conduct in-house product testing.
- Record and learn from feedback and problems.
- Issue complete instruction manuals, documentation and registration cards to purchasers.

TORT REMEDIES

Once liability for a tort has been established, the court will provide the injured party with a suitable remedy for the damage or injury suffered. The general approach is to attempt to place the injured party in the same position that the party would have been in had the tort not been committed. For this reason, the courts have fashioned a variety of different remedies, each designed to compensate for specific losses.

Compensatory Damages

The most common remedy is money damages. Often, the loss suffered by a person in a negligence case is the loss of or damage to property. In these instances, an award of money damages may be sufficient to undo the harm done by the tort. For example, if a person negligently operated construction equipment in a manner that caused $10,000 damage to another person's automobile parked nearby, a judgment of the court that obliged the negligent person to pay the injured party $10,000 (plus court costs) would cover the cost of having the damaged automobile repaired to its previous condition. In this way, the injured party would be returned to the same position as before the tort incident.

Torts that cause physical injury to persons are often compensated for by money damages, as the injury often results in medical expenses, the loss of wages, and many other identifiable costs that usually flow from the sudden incapacity of the injured party. An award of money damages to cover these specific costs or losses suffered by the injured party is referred to as special damages. These may be determined from invoices or calculated (for example, in the case of lost wages). In a typical motor-vehicle accident, where an innocent party is seriously injured, special damages might include ambulance expenses, hospital charges, the fees of medical practitioners attending to the injuries, the cost of drugs and therapy, lost wages, rehabilitation costs, and the cost of repairs to the injured party's automobile. To this would be added a sum intended to compensate the injured party for pain and suffering, for future health problems that might arise from the injuries suffered, and for any permanent incapacity resulting from the accident.[14] These losses are

14. See, for example, *Andrews v. Grand & Toy Alberta Ltd. and Anderson*, [1978] 2 S.C.R. 229, where the various components were noted and a cap placed on the amount awarded for pain and suffering.

referred to as general damages; the monetary amount to cover them may be estimated by the court, based upon the evidence of expert witnesses at the trial.

Nominal Damages

If a tort does not result in a monetary loss to the person whose rights have been infringed by the act, such as when a person trespasses on the land of another without inflicting physical damage to the property, a court may sometimes award nominal damages in the amount of $1. This is done by the court to emphasize to the trespasser that the property owner's right to exclusive possession had been affected by the trespass. In most of these cases, however, the cost to the injured party of bringing the case before the court is significantly higher than $1, as court costs and legal fees must also be paid to have the matter decided by the court. In keeping with the concept of compensation for the loss suffered, courts will usually order the trespasser to pay not only the nominal damages of one dollar, but the plaintiff's court costs as well. On this basis, the cost to the defendant might very well be substantial, even though the tort committed caused no actual damage to the "injured" party.

Punitive Damages

Exemplary/ punitive damages

Damages awarded to "set an example" or discourage repetition of the act.

When a tort is intentionally committed for the purpose of injuring another, or with reckless disregard for the other party, a court may award the injured party **punitive** or **exemplary damages**. For example, if one person assaults another, or spreads malicious lies about an individual, punitive damages may be awarded by the court to punish the wrongdoer. The award of punitive damages runs contrary to the concept of compensation for the injury suffered. However, because one person may sometimes cause injury to another in a vindictive or malicious manner, punitive damages are awarded as a deterrent to further similar behaviour, as well as to compensate for the injury suffered.

Court Orders

Contempt of court

Refusal to obey a judge's order.

When the tort committed is of an ongoing nature, such as trespass to land, pollution of a person's water supply, or interference in any way with an individual's exercise of lawful rights, the court may issue an injunction as a remedy, to prevent further occurrences of the tort. An injunction is a court order directed to a person (or persons) ordering the person named to cease doing the act described in the order. Failure to comply with the injunction, or continuation of the prohibited act, is considered to be **contempt of court**. It may be punished by a fine or jail sentence, if the court finds the person ignored the court order.

Order of replevin

Court action that permits a person to recover goods unlawfully taken by another.

In cases where a person has wrongfully retained goods that belong to another, the retention of the goods may constitute the tort of conversion. When this occurs, the owner of the goods usually wishes to regain possession of the property, rather than receive an award of money damages equal to their value. Common Law courts in such cases may issue an order of replevin that directs the sheriff or bailiff of the county or district to take possession of the goods until their ownership can be determined by the court. An **order of replevin** is a remedy often granted in addition to money damages, where the owner has requested court assistance to recover goods and to obtain compensation for the loss of use of the goods while they were wrongfully retained.

CLIENTS, SUPPLIERS *or* OPERATIONS *in* QUEBEC

Where a negligence harms a business (for example, a factory set on fire), Common Law courts indeed award damages to replace it, but have been reluctant to go further and award damages for economic losses (e.g., lost profits from downtime). In Quebec however, there is little distinction, and full restitution is required. On the other hand, Quebec courts are much less willing than Common Law courts to award punitive damages, seeing them as belonging more properly to the realm of criminal law and punishment.

SUMMARY

The law of torts is one of the oldest areas of law. It is concerned with injury caused by one person to another person, or to property, when the courts have determined that a duty exists not to injure. Unintentional injury to a person, when a duty not to injure is owed, is a tort. Most torts of this nature fall under the general classification of negligence, which includes not only unintentional injury to the person, but injury to property as well.

Injury to the property of another person through carelessness is actionable in tort if a duty is owed not to damage the property. While trespass to land normally is a willful act and an actionable tort, unintentional interference with the enjoyment of the lands of another constitutes the tort of nuisance. Landowners or occupiers of land, however, owe a duty not to injure persons who enter on their land, although the extent of the duty differs between trespassers, licensees, and invitees. The manufacturers of goods also owe a duty not to injure the users of their products, and are subject to a very high duty of care.

The courts have developed a number of principles and doctrines applicable to both old and new forms of tort. Two important principles, the concept of a duty of care and the standard of the "reasonable person," have made the law flexible in its application. In addition, the development of "foreseeability" as a test for tort liability has been an important advancement in the law. In tort cases, the courts usually attempt to compensate the injured party for the loss suffered insofar as monetary damages permit. However, in cases where injury is caused by the deliberate acts of one party, the courts may award punitive damages as well.

KEY TERMS

Duty of care, 80
Reasonable person, 81
Causation or "proximate cause", 81
Strict liability, 83
Vicarious liability, 83
Res ipsa loquitur, 84
Nuisance, 87
Injunction, 88
Volenti non fit injuria, 89
Act of God, 91
Waiver, 91
Release, 91
Doctrine of laches, 91
Exemplary/punitive damages, 93
Contempt of court, 93
Order of replevin, 93

REVIEW QUESTIONS

1. Distinguish between a moral obligation not to injure and a duty not to injure. Why is this distinction important?
2. Why do the courts impose strict liability for damage in certain instances?
3. Explain the concept of duty of care as it relates to liability in a tort action.
4. Why are the concepts of duty of care and the reasonable person important in a case where negligence is alleged?
5. Explain the concept or doctrine of proximate cause in tort law.
6. How are the concepts of the reasonable person and foreseeability related?
7. Identify and explain the essential ingredients of unintentional tort liability.
8. Does strict liability apply to manufacturers of goods when the goods are defective and cause injury or damage?
9. What is the duty of care of an occupier of land to a trespasser?
10. In some cases, the courts consider the public interest as an important aspect of a nuisance action. Why is this so? Give an example.
11. Is "nuisance" simply another name for negligence? If not, why not?
12. Define "*res ipsa loquitur.*"
13. Why was it necessary for the Common Law provinces to introduce contributory-negligence legislation?
14. Is the defence of *volenti non fit injuria*, in effect, the defence of waiver?

MINI-CASE PROBLEMS

1. X, a trained and licensed plumber, carelessly installed a steam heater in Y's restaurant, and, as a result, Y was seriously burned when a heater pipe exploded. Z, a qualified medical practitioner who treated Y at the hospital, prescribed an improper medication as treatment for the burn, which aggravated the injury. This obliged Y to undergo an expensive skin-graft operation to correct the condition. Discuss the responsibility of X and Z to Y.
2. At dusk one evening, as X was hauling a load of rock in his truck, he saw a large piece of rock fall from the truck onto the travelled portion of the roadway. X continued on his way without stopping. Y, who was travelling in her automobile along the same road some time later, collided with the rock and damaged her automobile. Discuss the liability of X.
3. Would your decision in Question 2 be any different if:
   (a) X was unaware that the piece of rock had fallen from his truck?
   (b) Y was travelling along the road at a high rate of speed?
   (c) Y noticed the rock in her driving lane and swerved to the other side of the road to avoid it, thereby colliding with Z, who was travelling in the opposite direction?
4. Roger, a delivery driver with Golden Bakery, is driving a company van across town when he hits a pedestrian who stepped from the curb in order to cross the street. Describe how Golden Bakery might respond to a claim of vicarious liability.
5. A factory stood near a self-storage facility, separated by a gravel parking lot owned entirely by the factory. In the summer of one year, the factory paved its parking lot. Late the following spring, heavy rains which formerly soaked into the gravel ran directly from the parking lot into the self-storage facility, causing hundreds of thousands of dollars in damage to goods of the unit tenants. Discuss the tort issues raised as result.
6. Iron Engineering was engaged to design a bridge to be built by Gorge Construction. Upon internal review, one of Iron's design team caught a fatal error in the design, and drafted a plan amendment accordingly. The engineer placed the amended plan as "version 2.0" in a second file, in front of the first in the file cabinet, as was the practice. A young file clerk, trying to be helpful, reorganized the cabinet, reversing the file order. Gorge Construction ultimately received version 1.0 of the plan and built the bridge, which collapsed soon after completion. Discuss the tort issues raised as a result.

CASE PROBLEMS FOR DISCUSSION

CASE 1

Angus, a bicycle courier, picked up an envelope at a law office. It was addressed to "City Works, Office 1212, For attention of: Janet Bari, Re: #2650/06." Angus put the letter in his bag, making it the sixth identical letter that he received for the same address while doing his afternoon rounds. He dropped the letters off at City Hall before the office closed, and headed home. The next morning he realized that one of the letters was still in his bag, stuck inside a fold in the nylon lining. Angus returned to City Hall, but delivery of the letter was refused. The reference 2650/06 was a call for bids on city construction work, and the competition for lowest bid closed when City Hall closed the day before.

Identify the legal issues raised between Angus, the law office, and the presumably irate client on whose behalf the law firm had created the bid, and render a decision. Would it matter if the bid, in any event, would not have been the lowest?

CASE 2

Basil, aged 14, lived in a large metropolitan city, but spent his summer vacations with his parents at a cottage in a remote wilderness area of the province. On his 14th birthday, his father presented him with a pellet rifle and provided him with instruction on the safe handling of the weapon. The father specified that the gun was only to be used at the cottage. Basil used the pellet rifle to rid the cottage of area rodents during his vacation.

On their return to the city, Basil's father stored the weapon and the supply of pellets in his workshop closet. He warned Basil that he must not touch the rifle until the next summer, but did not lock the cabinet in which the weapon was stored.

One day, when Basil was entertaining a few friends at his home, he mentioned his summer hunting activities. At the urging of his friends, he brought out the pellet gun for examination. Basil demonstrated the ease with which the magazine could be filled, and how the weapon operated. He then emptied the magazine and allowed his friends to handle it. The gun was returned to him and, as he was replacing it in the cabinet, the weapon discharged. He was unaware that a pellet had remained in the weapon, and that he had accidentally charged the gun when he handled it. The pellet struck one of his friends in the eye, and an action was brought against Basil and his parents for negligence.

Discuss the liability (if any) of Basil and his parents, and the defences (if any) that might be raised.

Render a decision.

CASE 3

Khalid lived in a residential area some distance from where he worked. One morning, he found himself late for work because his alarm clock had failed to wake him at the usual time. In his rush to leave his home, Khalid backed his automobile out of his garage after only a cursory backward glance to make certain the way was clear. He did not see a small child riding a tricycle along the sidewalk behind his car, and the two came into collision. The child was knocked from the tricycle by the impact and was injured.

The child's mother, at the front door of her home (some 70 metres away), heard the child scream and saw the car back over the tricycle. She ran to the scene of the accident, picked up the child, and carried him home. Khalid called an ambulance, and the child was taken to the hospital and treated for a crushed leg.

The mother brought a legal action against Khalid for damages resulting from the shock of seeing her child struck by Khalid's car. An action was also brought on behalf of the child for the injuries suffered.

Discuss the validity of the claims in this case, and identify the issues and points of law that are raised by the actions of Khalid. How would you decide the case?

CASE 4

Thompson operated an ice-cream truck owned by Smith Foods Ltd. During the summer months Thompson travelled throughout the residential areas of a large city selling ice-cream products. Thompson's principal customers were children, and Thompson would drive along the streets ringing a series of bells attached to his truck to signal his arrival in the area.

Alberta, a five-year-old child, and her brother were regular customers of Thompson. On the day in question the two children heard the bells that signalled the approach of Thompson's ice-cream truck. Martha, Alberta's mother, was talking to her husband on the telephone at the moment that the ice-cream truck arrived. In response to the cries of her two small children for money to buy ice-cream, she gave them enough money to buy an ice-cream bar each. The children ran across the street to where the truck was parked, and each ordered a different ice-cream product. Thompson served Alberta first, and then turned to serve her brother. At that instant, Alberta ran into the street, with the intention of returning home, and was struck by a car driven by Donaldson.

Alberta was seriously injured as a result of the accident and an action for damages was brought against Thompson, the operator of the ice-cream truck, Smith Foods Ltd., the owner of the truck, and Donaldson, the owner and driver of the automobile.

Discuss the basis of the action on Alberta's behalf against the owners and drivers of the vehicles, and determine the basis of the liability of each party under the law of torts.

Render a decision.

CASE 5

Marie-Claude operated a bowling alley in a commercial area that was adjacent to a residential area. Many small children used the parking lot near the bowling alley as a playground, and Marie-Claude was constantly ordering the children off the premises for fear that they might be injured by motor vehicles.

One young boy, about six years old, was a particular nuisance in that he would climb onto the flat roof of the bowling alley by way of a fence at the back of the building. Marie-Claude ordered the child off the roof on several occasions, but to no avail. The child continued to climb on the roof at every opportunity in spite of Marie-Claude's instructions to the contrary.

On one occasion, when Marie-Claude was away from the premises, the child climbed to the roof and, while running about, tripped and fell to the ground. The fall seriously injured the child, and an action was brought on his behalf against Marie-Claude.

Discuss the liability of Marie-Claude and her defences, if any.

Render a decision.

CASE 6

On a clear September day, Mario walked across a parking lot adjacent to his home to buy a newspaper at a convenience store in a shopping plaza. He purchased the newspaper and left the store, but before he had walked more than a couple of feet, his foot struck a raised portion of a sidewalk slab, and he fell heavily to the concrete.

Mario was unable to work for a month as a result of his injuries, and he brought an action for damages against the tenant who operated the convenience store and the owner of the shopping plaza.

When the case came to trial, the evidence established that the maintenance of the sidewalk outside the store was the sole responsibility of the owner of the shopping plaza, but the owner had not inspected the sidewalk for some months. The sidewalk, while not in disrepair, had been slightly heaved by the previous spring frost. It presented an uneven surface upon which the patrons of the plaza were forced to walk. Many of the concrete slabs were raised from 1/2 to 1 1/4 centimetres, but some protruded above abutting slabs by as much as 4 centimetres. The slab that caused Mario's fall protruded by approximately 3 centimetres.

The tenant who occupied the convenience store was aware of a "certain unevenness of the slabs," but did not realize that it was hazardous.

Discuss the liability (if any) of the defendant and render, with reasons, a decision.

CASE 7

Kamikazi Ski Lodge Limited operated a ski resort in a mountainous ski area. While the company offered a number of runs suitable for skiers of all levels of competence, it maintained one particularly fast and adventurous run on a slope that required a relatively high degree of proficiency to successfully ski its course. In its advertising brochure the company warned potential patrons of the lodge that they should not attempt the run unless they had considerable experience. The company also required all guests to sign a guest-registration form at the time of check-in at the lodge. The form provided as follows:

> "Guests assume all risks associated with their use of the owner/operator's lodge and facilities. The owner/operator, its employees, and agents shall not be liable for any loss or injury incurred by any guest, however caused. Registration as a guest shall constitute a waiver of all claims that the guest may have against the owner/operator or its employees for any injury or loss suffered while on the premises."

Each year the lodge held a special challenge competition on its most difficult slope that required the contestants to travel down the slope in large rubber inner tubes rather than on skis. The competition was appropriately described as the "$500 Kamikazi Challenge Race," and was open to all contestants on the payment of a $15 race-registration fee. Contestants were also required to sign an entry form at the time of payment, which provided: "Contestant recognizes that the Kamikazi Challenge is a dangerous undertaking that could result in injury. The contestant assumes all risk and responsibility for any or all injuries suffered while engaged in the competition."

Two days before the competition, Crocker, a novice skier who was a guest at the lodge, paid his entry fee and signed the entry form without bothering to read it. He had been drinking prior to the point in time that he entered the contest, but did not appear to be drunk or unaware of his actions.

Later that day, the contestants (including Crocker) were shown a videotape of the previous year's race. The tape described the race and provided footage of incidents that involved contestants being thrown from their inner tubes, as well as those who managed to remain in their inner tubes to the end of the race.

On the day of the contest, Crocker appeared at the starting point for the race in an obviously drunken state. His speech was slurred, and he had difficulty standing on the slope. Nevertheless, he managed to obtain his inner tube from the lodge employees in charge of the race and prepared for the start of the race. At the starting gun, he threw himself into the inner tube and managed thereafter to more or less successfully navigate the slope. He placed among the five finalists, suffering only a cut above one eye when he was thrown from his inner tube on one occasion.

Before the final run, he took a large drink of brandy that had been offered to him by a spectator. The race operator noticed Crocker drinking brandy, and asked him if he was in any condition to compete in the final race. Crocker replied that he was, and the final race began.

Partway down the slope, Crocker failed to negotiate a sharp turn, and was thrown from his inner tube and into a rock gully. His injuries rendered him a quadriplegic, and he brought an action against the resort for damages.

Discuss the liability of the lodge, and speculate, with reasons, as to how the case might be resolved.

CASE 8

Kevin and a companion spent most of a Saturday afternoon and evening drinking alcoholic beverages at Kevin's home. Around 10:00 p.m., they decided to visit a local tavern, where they had arranged to meet some friends. The two drove in Kevin's automobile to the tavern, which was located on a highway some 5 kilometres away. Kevin was obviously quite drunk when he arrived at the tavern, and for the next two hours he continued to drink beer.

At closing time, the bartender informed the group (who were all quite drunk by then) that they must leave the premises. Kevin and his companion did so, and drove from the tavern parking lot, with Kevin behind the wheel and his companion attempting to direct him into the traffic. When the way appeared to be clear for Kevin to enter his traffic lane, his friend said: "Okay! Clear to go!" Kevin accelerated into the traffic, but was unable to control the automobile, and it careened across the centre line of the highway and into the path of an oncoming vehicle.

The collision destroyed the two automobiles, killed the driver of the other vehicle, and permanently crippled both Kevin's friend and the passenger in the other automobile. The estimated loss suffered by those injured in the accident was $1,900,000. Apart from some cuts, bruises, and a broken arm, Kevin suffered no serious effects from the accident. Blood tests taken shortly after the accident indicated that Kevin's blood-alcohol level was almost three times the legal limit.

Advise the parties of their rights in this case and indicate how the liability, if any, would be determined.

CASE 9

Sayeeda operated a small food stand at an open-air market where farmers in close proximity to one another had set up makeshift stalls or tables to display their produce. To protect themselves from the hot sun, most had erected canvas or cloth canopies over their tables. The stand operated by Sayeeda was located in the midst of the covered stalls.

For the purpose of cooking a particular delicacy that she sold, Sayeeda had placed a tiny propane-fuelled stove on her tabletop and used it to boil a pot of oil. While she was serving a customer, a youth about 16 years old turned up the flame on the burner, and the oil immediately caught fire. In a frantic attempt to put out the fire, Sayeeda seized the flaming pot and flung it from her food stand. The youth was splashed with flaming oil and was seriously burned when his clothes caught fire.

The pot of burning oil landed in a nearby farmer's stall and set fire to the canopy that he had placed over it. The farmer kicked the flaming pot out of his stall and into a neighbouring stall, where it set fire to a quantity of paper that was used for food packaging.

Before the fire was finally extinguished, three stalls had been destroyed, and both the youth and Sayeeda hospitalized with serious burns.

Discuss the rights and liabilities of the parties involved in this incident.

CASE 10

Zhao and Henry set out for a day of fun at "Fearsome," a white-water rafting attraction. They both signed a waiver stating they "release the operator and its employees from all liability however caused." Halfway through the course, their raft (piloted by one of the attraction staff) flipped over, an event considered as all part of the "extreme" factor. Both Zhao and Henry were wearing life-jackets. Henry was thrown to one side from the raft, and broke his hip in collision with a submerged rock. Zhao was initially unhurt, and both men were carried on the current to the lagoon at the end of the course. A Fearsome boat was waiting for them, and its crew used a hooked pole to snare the loops of their lifejackets. In thrusting forward with the pole, the crewman seriously sliced Zhao's hand. Henry spent three months in a cast before he could work again, and Zhao missed six weeks of work.

Discuss the issues arising from this which will affect a legal action brought by Henry and Zhao against Fearsome, and render a decision.

CASE 11

Sandra purchased a timer for her living-room lamp, so that it would come on in the evening while she was out working night-shifts. One afternoon Sandra carried out her energy conservation routine of shutting off lights and appliances, except for the living room lamp in the northwest corner. She set the lamp timer and left for work. She returned home after midnight to find the fire department packing away its hoses, and her home gutted by fire. The fire marshal told her the fire was electrical in origin, and that its spread located the origin along the north wall of the living room.

Describe the case in tort that Sandra is now facing in order to receive compensation for her losses.

CASE 12

A 45-year-old patient was under doctor's care for the past twenty years for chronic arthritis, trying many different drugs without success. A new drug appeared, having been certified by Health Canada and the doctor explained that like all the rest, the drug would have a range of side effects, among which was possible depression. The patient took the drug, and within a week suffered a mood swing and committed suicide. Concerned by this news and two other cases, Health Canada suspended the drug from Canada, pending further review. The family of the deceased sued the pharmaceutical maker and Health Canada.

Aside from issues of the doctor's possible professional liability, discuss the issues of manufacturer's liability and negligence raised in this scenario.

The Online Learning Centre offers more ways to check what you have learned so far. Find quizzes, Weblinks, and many other resources at www.mcgrawhill.ca/olc/willes.

CHAPTER 6

Special Tort Liabilities of Business Professionals

Chapter Objectives

After study of this chapter, students should be able to:

- Describe how and why the liability of professionals differs from the norm.
- Explain the nature and responsibilities of a fiduciary relationship.
- Explain the significance of informed consent.
- Describe the key aspects of professional liability in specific cases.

YOUR BUSINESS AT RISK

Professionals are not only held to a higher standard of care, but they tend to be even more financially desirable targets of litigation. Professionals must understand the scope of their additional responsibility and consider how they will conduct their professional practice to reduce their risks. Wise professionals will obviously consider the advice of insurance professionals as part of their risk reduction strategy.

THE PROFESSIONAL

Professional

A person with special skills not possessed by most individuals.

A **professional** is a person who possesses special knowledge or exercises special skills not normally possessed by most individuals. The possession of these skills distinguish the professional from other business persons or service providers in the sense that the person who engages the services of the professional will normally rely on the professional's expertise to perform services that they cannot or do not wish to perform themselves. When we think of a professional person, the traditional professions usually come to mind: lawyers, doctors, accountants, engineers and architects. However, there are many other professional service providers and experts, such as pharmacists, nurses, actuaries, chiropractors, physiotherapists and radiologists, to name only a few. There are also many emerging groups of service providers that have developed special business-related skills that may be relied upon for their expertise by clients, for example, financial planners.

One of the difficulties in the establishment of the professional status of these emerging professions is the recognition and determination of professional standards for their skills. In most cases, professional recognition requires the creation of a professional body to establish these skill standards with the authority to enforce adherence by those who wish to acquire and use the professional designation.

An important number of characteristics distinguish the professional from other service providers. These include a structured educational and training program, a system for recognition and accreditation of the person, a standard for the performance of the skills, and an oversight body (such as Law Society, Institute of Chartered Accountants, or College of Physicians and Surgeons) to perform these duties and enforce the standards for performance. The oversight body must also have the authority to revoke the professional designation from persons who fail to maintain the standards of performance for the profession. These duties are usually performed by a professional association created in accordance with or empowered by government legislation.

PROFESSIONAL STANDARDS AND PROFESSIONAL ASSOCIATIONS

Provincial governments have recognized the unique roles that professionals play in the delivery of specialized services to business and individuals. In order to ensure that the public is protected (as well as the members of the various professions), legislatures have passed legislation for the establishment of professional associations to control and regulate their members. These duties include the oversight of training and education of potential members to ensure that they meet a competence level to practice the profession, to ensure that competence levels are maintained, and to discipline members who fail to maintain the standards of competence or the standards of the association.

Professional associations are normally given the authority to limit the use of the professional designation to individuals who are members in good standing in the association, and to take action under the legislation to prevent anyone not a member in good standing from holding himself or herself out to the public as a particular professional, or offering the professional services to the public.

The right of the association to bar a member from practicing the profession also acts as a powerful incentive for a member to maintain the standards of the profession in his or her dealings with the public, as expulsion from the association would probably have a devastating effect on the professional's career.

Most of the professional associations have established standards of competence for their members, and these are usually set out as a set of rules of conduct. These standards are often referred to by the courts in cases where a professional is accused of negligence in the performance of professional services. The Law Society of Upper Canada, typical of all provinces, sets out the professional standard for lawyers practicing law in Ontario in its *Rules of Professional Conduct* manual in the following manner:[1]

> "competent lawyer" means a lawyer who has and applies relevant skills, attributes, and values in a manner appropriate to each matter undertaken on behalf of a client including
>
> (a) knowing general legal principles and procedures and the substantive law and procedure for the areas of law in which the lawyer practises,
>
> (b) investigating facts, identifying issues, ascertaining client objectives, considering possible options, and developing and advising the client on appropriate courses of action.

The definition then sets out the particular steps the lawyer should follow in acting for a client. These steps include legal research, and an obligation to perform all functions in a conscientious and diligent manner.

It also states that a lawyer must recognize the limitations on his or her ability to handle a legal matter, and to take such steps as may be required to ensure that the client has appropriate advice, such as directing the client, if necessary, to another lawyer with greater expertise in the particular legal matter.

THE PROFESSIONAL–CLIENT RELATIONSHIP

Contract

The professional–client relationship is usually created by a contract between the professional and the client. Under the terms of the contract, the professional person agrees to

1. Law Society of Upper Canada, *Rules of Professional Conduct* Rule 2.01 — Relationship to Clients (Extract only).

perform certain specified services for the client in return for a fee. It is important to note that the professional under a contract impliedly promises to perform the services required in accordance with the standard for his or her profession, and a failure to do so would constitute a breach of contract. In a sense, the contract establishes a *contractual duty* on the part of the professional to perform the agreed professional services. Contract law, and the implications of breach of contract are covered in the next section of the text, but it is sufficient to say here that in the case of a breach of contract by the professional, the client may be able to hold the professional liable for any foreseeable loss that the client suffered as a result of the professional's breach of the contract.

Breach of contract cases usually are a result of careless or negligent performance of the professional service, but a breach of contract may also apply where a professional person enters into a contract to perform a service for a client, then refuses to perform the service. In this case, the professional would be in breach of the contract for a failure to perform the promised service. For example, an architect enters into a contract with a client to design a new commercial complex. Shortly thereafter, and before commencing work, the architect advises the client that she does not wish to design the complex. In this case, the architect would be in breach of contract by refusing to perform the service, but would not be liable for careless performance of the professional service.

A QUESTION OF ETHICS

The principal distinction of the professional relationship, as opposed to the ordinary business relationship, is the extent of the knowledge imbalance between the parties. While the Internet has made strides in addressing this imbalance, accountants still know more of financial affairs than the majority of their clients. Doctors know more of the intricacies of surgery, therapy and drugs than do patients. Engineers are more keenly aware of forces, movement and materials. As with all relationships, superior knowledge translates to superior power and issues of dependence and dominance. The professional has the moral obligation (aside from the legal obligation) to use that superior knowledge for the benefit of his or her client, and not for self-benefit at the client's expense.

Fiduciary Duty of Care

Fiduciary duty

A duty to place a client's interest above the professional's own interests.

The law imposes a special duty of care on professionals where the client, as a result of the relationship, places his or her trust in the professional to act in the best interests of the client. This is known as a **fiduciary duty**.

The fiduciary duty on the part of the professional is to place the interests of the client above his or her own, and to avoid any conflict of interest with those of the client. For example, if a client reveals confidential business information to his accountant about an intended business development, the accountant may not use this confidential information for his or her own benefit, or reveal it to any other business client.

A fiduciary duty of care may also arise where a professional agrees to perform a service without a fee. By way of illustration, if a professional accountant agrees to audit the financial records of a religious or charitable organization without charging a fee, no contract would exist, but the accountant would nevertheless have a duty to perform the audit in accordance with the standards of the profession. If the audit was performed carelessly, and the organization suffered a loss, the accountant may be liable for breach of the fiduciary duty of care arising out of the relationship between the two parties.

A professional engaged to provide advice or information in a business activity where the professional advice would be relied upon by one or both parties to the transaction must ensure that the professional advice or information provided is not only carefully prepared, but that the advice is open and unbiased. The professional must also ensure that his or her role or interest in the transaction is fully explained and brought to the attention of both parties. This is particularly important if the professional has a financial interest in the transaction that is not readily apparent. The following case illustrates this issue.

CASE IN POINT

An inexperienced investor sought advice from a chartered accountant about tax shelters that might be suitable investments. The accountant recommended a MURB (Multiple Unit Residential Building) project as a suitable investment, and the inexperienced investor made an investment in the project. Unbeknownst to the investor, the accountant had a financial relationship with the project developer, whereby he received a fee for each investment he sold on the development.

The investor suffered a serious financial loss a year later when the real-estate market crashed. When the investor discovered the involvement of the accountant with the developer, he sued the accountant for his loss.

The court held that the accountant was in breach of his fiduciary duty to his client by failing to reveal his relationship with the developer, and found the accountant liable for the investor's loss.

Hodgkinson v. Simms, [1994] 9 W.W.R. 69 (S.C.C.).

Tort Duty of Care

In general, the person who claims to be a professional must maintain the standard of proficiency or exercise the degree of care in the conduct of his or her duties that the profession normally imposes on its members. This does not mean that the professional person will not make mistakes, nor does it mean that the results will always be perfect. In many professions, the work of the professional requires much skill and judgment, often more closely resembling an "art" than an exact "science." Consequently, a successful negligence suit against a member of a profession must measure the performance or duty of the particular practitioner with that prescribed by the profession in general, and must show in the evidence that the practitioner failed to meet that standard.[2] This is not always an easy thing to do. However, in many cases the professional's performance is so far below the standard set for the profession that the determination of the duty of care and the breach thereof are not difficult.[3]

In the course of a tonsillectomy a surgeon used a number of sponges that did not have strings or tapes attached for easy retrieval from the child's throat. The attending nurse did not do a sponge count before and after the operation. On completion of the surgery, the anesthetist present suggested that all of the sponges may not have been removed. The surgeon made a cursory search in the child's throat with his forceps and, when he found none, did nothing more to determine the number used. The child suffocated and died as a result of a sponge left in his throat.

At trial, the evidence submitted indicated that the surgeon had performed the operation carefully by using the proper techniques and suture. The evidence also indicated that it was not a common practice to use sponges with tapes attached, although some hospitals followed this practice. The surgeon's search at the end of the operation was also a normal practice.

The action in negligence brought against the surgeon was dismissed by the trial judge on the basis that the surgeon had followed the same practices as any other careful practitioner. On appeal, however, the court viewed the matter in a slightly different light. It concluded that the surgeon had a duty to make a thorough, rather than routine, search when a fellow professional informed him that he might not have removed of all the sponges. Because he did not employ all of the safeguards available to him, his failure to do so constituted negligence.[4]

2. *Karderas v. Clow*, [1973] 1 O.R. 730.
3. *McCormick v. Marcotte*, [1972] S.C.R. 18, 20 D.L.R. (3d) 345.
4. *Anderson v. Chasney*, [1950] 4 D.L.R. 223; affirming [1949] 4 D.L.R. 71; reversing in part [1948] 4 D.L.R. 458.

In the case of *Kangas v. Parker and Asquith,* [1976] 5 W.W.R. 25, the court was called upon to consider the standard of care imposed upon professionals. The judge in that case described the standard by reference to other Common Law decisions in the following manner:

> The medical man must possess and use that reasonable degree of learning and skill ordinarily possessed by practitioners in similar communities in similar cases and it is the duty of the specialist who holds himself out as possessing special skill and knowledge to have and exercise the degree of skill of an average specialist in the field. Vide: *McCormick v. Marcotte,* [1972] S.C.R. 18, 20 D.L.R. (3d) 345.

In *Rann v. Twitchell* (1909), 82 Vt. 79, at 84 appears the following comment:

> He is not to be judged by the result nor is he to be held liable for an error of judgment. His negligence is to be determined by reference to the pertinent facts existing at the time of his examination and treatment, of which he knew, or in the exercise of due care, should have known. It may consist in a failure to apply the proper remedy upon a correct determination of existing physical conditions, or it may precede that and result from a failure to properly inform himself of these conditions. If the latter, then it must appear that he had a reasonable opportunity for examination and that the true physical conditions were so apparent that they could have been ascertained by the exercise of the required degree of care and skill. For, if a determination of these physical facts resolves itself into a question of judgment merely, he cannot be held liable for his error.

Lord Hewat C.J., in *Rex v. Bateman* (1925), 41 T.L.R. 557 at 559, 19 Cr.App. R.8, clearly and succinctly stated the desired standard of care as follows:

> If a person holds himself out as possessing special skill and knowledge and he is consulted as possessing such skill and knowledge by or on behalf of a patient, he owes a duty to the patient to use due caution in undertaking the treatment. If he accepts the responsibility and undertakes the treatment and the patient submits to his direction and treatment accordingly, he owes a duty to the patient to use diligence, care, knowledge, skill and caution in administering the treatment ... The law requires a fair and reasonable standard of care and competence.

In each of these case comments to which the judge referred in his judgment, the common thread is that the professional has a duty to maintain the standard of skill that the profession itself has fixed for its members. The standard generally is that of the competent and careful "reasonable person" trained to exercise the skills of the profession. Here, the level of competence to be maintained is that of the average member of the profession, rather than the skill level of the most highly skilled member of the professional group.

Informed Consent

One of the obligations upon professionals acting in fields which can create tort liability is the obligation to explain procedures and obtain consent to action from the party who will be affected by the professional service. It is a moral obligation (a "right thing to do") that rapidly translates into a defence to tort liability (if done), and a cause of action for a plaintiff (if not done). The explanation of proposed action and the consent following must be meaningful. If it is to be meaningful, the professional must provide a full and understandable explanation of possible risks resulting from the action, and the client must provide a corresponding clear statement that he or she understands those possible results and still consents to the action. If this happens, **informed consent** exists, and some of the professional liability for foreseeable adverse results is lifted. If something less than informed consent is given, the professional is exposed to full liability for foreseeable harm. He or she is dangerously relying on luck in order to avoid harm to the client, which is neither professional nor ethical.

Informed consent

The full and understandable explanation of the risks associated with a course of action, and the clear understanding by the client or patient.

FRONT-PAGE LAW

Doctors negligent in diagnoses, court told

WHITEHORSE — A doctor from White Rock, B.C., yesterday testified that doctors, including Allan Reddoch, the former president of the Canadian Medical Association, failed to take proper steps to diagnose and treat Mary-Anne Grennan, a 16-year-old girl suffering from botulism in 1995.

In a report filed as evidence in Dr. Reddoch's malpractice suit, Paul Assad, a general practitioner and specialist in emergency medicine, said doctors failed to do a neurological examination on the girl that might have revealed she was suffering from double vision or other eye problems, which when considered with some of her other symptoms, might have made a diagnosis of botulism more likely.

Dr. Reddoch and the Whitehorse General Hospital are being sued for malpractice by the girl's father, Ed Grennan, for the potential lost income of the girl, had she not died from brain damage suffered under respiratory arrest caused by the disease. Both defendants are denying negligence.

In cross-examination, however, Chris Hinkson, Dr. Reddoch's lawyer, noted there is no evidence on the charts that Ms. Grennan ever complained about her vision to doctors or nurses, and that under treatment, some aspects of her condition were improving.

Dr. Assad also conceded doctors do not do differential diagnoses in many cases, and doctors are taught to look for common diseases over rare ones. There had never been a reported case of botulism in the Yukon before Ms. Grennan's, court heard yesterday. There are usually fewer than 10 cases in Canada per year.

Do you think these doctors were negligent? Which fact is most critical in shaping your opinion and your expected standard of care?

Source: Excerpted from Adam Killick, *National Post*, "Doctors negligent in diagnoses, court told," September 9, 2002, p. A9.

In most cases, the duty of care of the professional is owed only to the client or patient who engages the services of the professional. The relationship is usually one of contract, and the contract will normally specify the services to be performed by the professional. The professional person has an obligation to perform the services set out in the contract in accordance with the standards of the profession. Even if the contract does not specifically set out the standard, or if a lower standard is agreed upon, the contract normally must specifically refer to the lower standard, otherwise the standard of the profession would probably apply.

For many professions, the standard of care extends beyond the single performance of the service, and the professional may also be required to consider third parties. This is particularly the case where third parties may be affected.

Architects must consider the safety of the occupants when designing buildings. Similarly, engineers who design and construct buildings, bridges, roads, and equipment must also consider the loads, stresses, and use of the structures or equipment in the performance of their services. For example, if an engineer designs a bridge without consideration for the weight of traffic that would travel over the bridge, and the bridge collapses, the engineer may be negligent in the performance of his or her professional duty, and liable for the loss or injury that results.

Negligent misrepresentation

Negligent misstatements made by a professional to a client.

Professional negligence extends beyond carelessness in carrying out duties, and extends as well to statements made and information provided to clients. Lawyers, accountants, and others (such as brokers) who provide advice or information must ensure that they are not negligent in this regard, as negligent misstatements or **negligent misrepresentation** may constitute a breach of duty of care on the part of the professional. The courts in the past were reluctant to impose this type of liability on those who provided advice or information except in certain kinds of cases, such as where a fiduciary relationship existed, or where the contract between the parties provided for the provision of the information. The restriction on the scope of liability was deliberately limited because the courts had no wish to impose virtually unlimited liability on professionals for words negligently expressed. The scope of liability was extended significantly by the case of *Hedley Byrne & Co. Ltd. v. Heller & Partners Ltd.*[5] which is the root of all development in this area of the law since the mid-1960s.

In that case, an advertising agency asked its own bank to determine that a client was in sound financial condition. The bank made enquiries of the client's bank and replied in the

5. [1964] A.C. 465.

affirmative, advising that the client's bank had issued a disclaimer of responsibility when it provided the information. The information proved to be false, and, as a result, the agency suffered a financial loss. The agency then brought an action against the client's bank claiming negligence. While the bank successfully avoided liability on the basis of its disclaimer of responsibility, the court held that an action could lie for negligent misstatements or negligent misrepresentation when a plaintiff could prove that it relied upon the skill or expertise of the person or party that made the negligent statement or representation.

The courts subsequent to the *Hedley Byrne* case have delineated the limits of liability for negligent misrepresentation. In the case of *Haig v. Bamford* (1976), 72 D.L.R. (3d) 68, the Supreme Court of Canada decided that the test for liability to third parties for negligent misrepresentation by auditors would be actual knowledge of the limited class that would use or rely on the information.

Normally, a professional is only responsible in tort to the patient or client. However, in recent cases, the courts have held some professionals liable to third parties when the professional's expertise or skill was intended to be relied upon by the third party and the professional was aware of this fact. Accountants, in particular, have been subject to this extended tort liability. This is partly because they are generally aware that their financial statements will be relied upon by third parties, and partly because securities legislation in most jurisdictions imposes a liability on the accountant to third parties if the third party purchases securities based upon negligently prepared financial statements published in a prospectus. Liability to third parties is not unlimited, however, as the professional accountant is generally responsible only to those third parties that the accountant can reasonably expect to rely on the information provided. Liability for negligently prepared financial information is not necessarily limited to the professional–accountant third-party relationship. It appears to be applicable to all persons and institutions engaged in providing financial advice.[6]

A later case in Great Britain[7] has suggested that a threefold test be used to determine liability for economic loss caused by negligent misrepresentation. The test would consider (1) whether the harm was foreseeable, (2) whether there was a relationship between the parties of sufficient proximity, and (3) that in terms of public policy and the circumstances it would be just and reasonable to impose the duty on the party making the statement.

MANAGEMENT ALERT: BEST PRACTICE

With increased liability for statements and opinions, professionals need solid management control systems over the release and distribution of official (and sometimes worse, unofficial) information. This requires foresight by professionals as to potential end-users (likely and unlikely) to ensure that sufficient warnings and limitations appear on the face of documents. Additional effort is also required in the training and awareness of support staff to prevent inadvertent and inappropriate disclosures.

ACCOUNTANTS

For many professionals, the negligent performance of professional work is often limited in impact to the party who contracted for the professional service. For example, a surgeon who negligently performs an operation on a patient may be liable in tort to the patient, and only the patient. The impact of negligence by a professional accountant, however, in many cases may fall on others who did not contract for the accountant's services, but who relied upon the accountant's advice or information to their detriment. For these 'third parties,' accountants and auditors have often been the target of their complaints, but with only limited success.

6. See, for example, *Hedley, supra.; Goad v. Canadian Imperial Bank of Commerce*, [1968] 1 O.R. 579 (bankers); *Haig v. Bamford*, [1974] 6 W.W.R. 236 (accountants); *Cari-Van Hotel Ltd. v. Globe Estates Ltd.*, [1974] 6 W.W.R. 707 (real-estate appraisers); *Surrey Credit Union v. Wilson et al.* (1990), 73 D.L.R. (4th) 207 (auditors).

7. *Caparo Industries v. Dickman*, [1990] 2 A.C. 605, 1 All E.R. 568. See also *Surrey Credit Union v. Wilson* (1990), 49 B.C.L.R. (2d) 102 for a Canadian case.

In 1997, the Supreme Court of Canada considered the responsibility of auditors to shareholders in the Hercules Managements case — much to the relief of auditors, but not without establishing important principles. The case illustrates how judges refer to earlier precedent cases; the court makes reference to *Foss v. Harbottle* in which the court ruled that an individual shareholder cannot take action for a wrong committed against the corporation that results in a loss. Only the corporation can bring such an action. Note how the court makes use of this precedent in reaching its decision to reject the individual shareholder's claims.

COURT DECISION: Practical Application of the Law

Professional Liability—Professional Accountants—Negligent Preparation of Financial Statements for a Corporation.

***Hercules Managements Ltd. v. Ernst and Young* (1997), 146 D.L.R. (4th) 577 (S.C.C.), 1997 CanLII 345.**

A firm of accountants negligently prepared the financial statements for a corporation. On the basis of the financial statements, additional shares in the corporation were purchased by Hercules and other shareholders. The corporation a short time later ran into financial difficulties, and the accountant's negligence in the preparation of the financial statements was discovered. Hercules and another shareholder sued the accountants for the loss they suffered on the shares. Their claim was based upon their use of the financial statements when they made an additional investment in the corporation. The Supreme Court considered the issue of whether an individual shareholder or the corporation has the right to sue the auditors for their negligence.

THE COURT: ... audited reports are provided to the shareholders as a group in order to allow them to take collective (as opposed to individual) decisions. Let me explain.

The rule in *Foss v. Harbottle* provides that individual shareholders have no cause of action in law for any wrongs done to the corporation and that if an action is to be brought in respect of such losses, it must be brought either by the corporation itself (through management) or by way of a derivative action. The legal rationale behind the rule was eloquently set out by the English Court of Appeal in *Prudential Assurance Co. v. Newman Industries Ltd. (No. 2)*, [1982] 1 All E.R. 354, at p. 367, as follows:

> The rule [in *Foss v. Harbottle*] is the consequence of the fact that a corporation is a separate legal entity. Other consequences are limited liability and limited rights. The company is liable for its contracts and torts; the shareholder has no such liability. The company acquires causes of action for breaches of contract and for torts which damage the company. No cause of action vests in the shareholder. When the shareholder acquires a share he accepts the fact that the value of his investment follows the fortunes of the company and that he can only exercise his influence over the fortunes of the company by the exercise of his voting rights in general meeting. The law confers on him the right to ensure that the company observes the limitations of its memorandum of association and the right to ensure that other shareholders observe the rule, imposed on them by the articles of association. If it is right that the law has conferred or should in certain restricted circumstances confer further rights on a shareholder the scope and consequences of such further rights require careful consideration.

To these lucid comments, I would respectfully add that the rule is also sound from a policy perspective, inasmuch as it avoids the procedural hassle of a multiplicity of actions. The manner in which the rule in *Foss v. Harbottle, supra,* operates with respect to the appellants' claims can thus be demonstrated. As I have already explained, the appellants allege that they were prevented from properly overseeing the management of the audited corporations because the respondents' audit reports painted a misleading picture of their financial state. They allege further that had they known the true situation, they would have intervened to avoid the eventuality of the corporations' going into receivership and the consequent loss of their equity. The difficulty with this submission, I have suggested, is that it fails to recognize that in supervising management, the shareholders must be seen to be acting *as a body* in respect of the corporation's interests rather than as individuals in respect of their own ends. In a manner of speaking, the shareholders assume what may be seen to be a "managerial role" when, as a collectivity, they oversee the activities of the directors and officers through resolutions adopted at shareholder meetings. In this capacity, they cannot properly be understood to be acting simply as individual holders of equity. Rather, their collective decisions are made in respect of the corporation itself. Any duty owed by auditors in respect of this aspect of the shareholders' functions, then, would be owed not to the shareholders *qua* individuals, but rather to all shareholders as a group, acting in the interests of the corporation. And if the decisions taken by the collectivity of the shareholders are in respect of the corporation's affairs, then the shareholders' reliance on negligently prepared audit reports in taking such decisions will result in a wrong to the corporation for which the shareholders cannot, as individuals, recover.

Is this absolute? Under what circumstances would you think that accountants or auditors would be liable to individuals?

As a result of the rash of corporate scandals in the United States and the misleading financial information which these corporations provided to their shareholders, (and prospective shareholders) efforts have been made through legislation to impose a greater duty on auditors to accurately prepare the financial statements of publicly traded corporations.

In Canada, the Province of Ontario in 2002[8] introduced legislation that would hold liable not only the directors of publicly traded corporations, but also their accountants and auditors, and other 'experts' such as appraisers, lawyers, or engineers for public statements that are inaccurate. The legislation is intended to encourage compliance by providing a statutory remedy to persons who relied upon the misstatements to their detriment.

The legislation provides that where an investor has purchased or disposed of securities of the corporation that has issued a misleading statement, the investor is deemed to have relied upon the misleading statement,[9] and it is up to the directors or auditors to establish that they used reasonable care in the preparation of the statement. This change in the Common Law shifts the burden of proof to the accountant or auditor of a public corporation to prove that he or she was not careless in the preparation of financial statements that would be made available to the public.

While this legislation is limited to public corporations in Ontario, other provinces may also follow this approach. It is important to note, however, that the legislation is applicable only to the misstatements issued by public corporations, and the 'deemed reliance' provision would not apply to investors in other types of business transactions where accountants or auditors prepare financial statements.

LAWYERS

The legal profession is engaged in providing legal services to clients. This often takes the form of advice and the enforcement of a client's legal rights, but it also includes the preparation of legal documents and the negotiation of business transactions on behalf of clients.

Clients are entitled to expect that the advice that they receive from their lawyers would not only be carefully considered, but would also be accurate in terms of the client's rights or duties under the law. In terms of the preparation of legal documents, clients are also entitled to expect that the documents will legally and accurately carry out the client's wishes. For example, a client is entitled to expect that a deed prepared by his or her lawyer will convey the title to a property that the client wishes to sell to a purchaser, or that a mortgage will accurately set out the rights and duties of the parties.

The standard of care imposed on a lawyer is that of a skilled and careful member of the profession, and the failure to maintain this level of competence in the performance of the lawyer's duties to the client may constitute negligence.

8. *Keeping the Promise for a Strong Economy Act (Budget Measures)* 2002, S.O. 2002, c. 22 (Part XXVI — Securities Act).

9. ibid. Part XXVI s.138.3(1).

A bank agreed to make a loan to a lumber mill on the condition that the mill obtained business interruption insurance prior to advancement of the loan. The lawyer for the bank was so informed, but was under the impression that the bank would discuss the insurance issue with the mill's insurance broker. The lawyer proceeded with the loan transaction, and eventually advanced the loan to the lumber mill without confirming with the bank that the mill had obtained the insurance coverage. The mill burned down a short time later. It was then discovered that no insurance had been obtained by the mill. The bank sued the lawyer. The court held that the lawyer had an obligation to check with the bank to confirm that insurance coverage had been obtained before advancing the loan.[10]

Solicitor–client privilege

The duty of a lawyer to keep confidential information provided by a client.

Lawyers are unique in a business sense, in that they are expected to hold confidential any information concerning legal matters that is related to them in the course of their lawyer–client relationship. This is known as **solicitor–client confidence** or **privilege**, and the lawyer normally has no legal obligation to reveal any of this confidential information to legal authorities or to anyone, except with the client's authorization. Recent federal legislation concerning money laundering, however, may represent an exception to this long-standing confidentiality right attached to the solicitor–client relationship.

ENGINEERS

Professional engineers may be held liable for their negligence to those who rely on their expertise. The most common relationship between a professional engineer and a client is the relationship between a building contractor and the civil engineer, but there are many different engineering specialties, each providing engineering expertise to a particular clientele. Mechanical engineers, aeronautical engineers, and electrical engineers are examples that immediately come to mind, but there are many recognized areas of engineering expertise available in the business world.

The standard of care of the professional engineer, in general, is that of a competent and careful engineer trained in that particular area of engineering specialty. For example, an aeronautical engineer would be expected to maintain and exercise the skills related to the design and construction of aircraft that would normally be possessed by a careful and experienced engineer trained in that branch of the engineering specialty.

Because engineering is a field where engineering work is divided into many special areas, engineers generally limit their practice to the particular area of expertise for which they are trained. This is sometimes a problem where skills may overlap. For example, where a construction project involves civil engineering, electrical engineering and mechanical engineering skills, a civil engineer may well have the skill to do simple mechanical or electrical engineering design work, but at what point should the engineer defer to the other particular skill? A failure to recognize this dividing line may result in design error and a loss to the client.

It is important to note that the liability of an engineer is not limited to the client that engages the services of the engineer, but also to the subsequent purchasers of the construction project.

Faulty design in the construction of a condominium building resulted in defects that could conceivably affect the health and safety of the occupants. The corporation sued the engineer and the architect for negligence, and when the case reached the Supreme Court of Canada, the court held that the engineers (and architects) would be liable in tort to subsequent purchasers of the building if their construction design made future defects foreseeable.[11]

10. See: *ABN Amro Bank Canada v. Gowling Strathy & Henderson* (1995), 20 O.R. (3d) 779.

11. *Winnipeg Condominium Corporation No. 36 v. Bird Construction Co.*, [1995] 1 S.C.R. 85.

ARCHITECTS

Architects are professionals whose area of expertise is generally limited to the design and supervision of the construction of buildings and structures, and the surrounding grounds. Architects are required to complete an extensive program of education and experience in order to attain the professional designation, and once they become architects they are expected to maintain that level of professional skill in the performance of their duties to their clients. The failure to maintain this level of skill may result in a breach of their duty of care, and if loss results to the client, it may result in tort liability for negligence.

The normal approach for the engagement of the services of an architect is through a contract. Under the terms of the contract, the architect is usually expected to design the project, and to supervise its construction. In this instance, the architect is expected to not only use his or her skill to properly design the project, but also to ensure that the construction is carried out in accordance with the design. If the architect is negligent in designing the project, or fails to properly supervise the construction, liability in tort for negligence may arise in either or both instances. Note, however, that the relationship between the architect and the client is contractual, and the architect may limit his or her liability for loss to a specific amount in the contract, or the architect may by contract be only required to prepare the design, and not supervise the construction of the project.

Architects and engineers often work together on building projects, and when problems arise, the client is often faced with the question of where liability should lie for the resulting loss or damage. If the client acts as the general contractor for the project, the liability of the architect or engineer is often limited to faulty design, and liability will not extend to the execution of the design by the contractor.

An owner-contractor wished to construct a large parking garage, and an architect and engineer were engaged to provide the design for the building. The client corporation agreed to engage its own subcontractors to perform the construction work. The completed building was found to contain a roof top water 'ponding' problem. This was due to an inadequate slope of the surface that did not allow the water to drain from the roof. The owner-contractor sued the architect for negligence in the design of the roof. At trial, the court found that the slope determined by the architect was barely adequate, but was within accepted slope factors used by the profession. The water ponding was due to changes by the contractor in other aspects of the roof design. Because the architect was not engaged to supervise the construction, he was not liable for the work done.[12]

CLIENTS, SUPPLIERS *or* OPERATIONS *in* QUEBEC

Engineers and architects under the Quebec *Civil Code* are subject to a strict liability for negligence in the performance of their professional duties, and can only avoid liability if they can establish that the client's loss was not due to their design work or their failure to supervise the construction of the project. They may also be relieved of liability if they can establish that the fault was due to decisions made by their client, but they are obliged to advise their client of the best course of action if the client suggests or requests alternatives. See: *Lac St. Charles v. Construction Choiniere Inc.* (2000), R.R.A. 639 (C.A.).

The *Civil Code of Quebec* also establishes a fixed time limit of five years from the date of completion of the project for any claim of negligence against an engineer or architect in Quebec.

12. See: *792132 Ontario Inc. v. Ernest A. Cromarty Inc. et al.*, 2002, Ont. Superior Court file Kingston CV 95-8700.

SEMI-PROFESSIONAL AND OTHER SKILLED PERSONS

While the courts have traditionally held professional persons responsible for their acts of negligence, persons engaged in business activities where some expertise, special knowledge, or skill is required may also be held responsible for losses others suffer as a result of their careless acts or omissions. For example, insurance agents, even though they are agents of companies, may be held liable for clients' losses if, when requested to do so, they fail to provide proper coverage for the contemplated loss. In one instance, a husband and wife contacted an insurance agent for coverage for the wife's business, financed by the husband. The agent negligently arranged for insurance in the husband's name only. When the wife suffered a loss she thought would be covered by the policy, she discovered that her business was not covered by the insurance. In the court action that followed, the agent was held liable for the loss the wife suffered, because he had failed in his duty to make proper arrangements for coverage of the wife's interest.[13]

CHECKLIST FOR PROFESSIONAL LIABILITY

1. Is the professional, in fact, a professional in the eyes of the law?
2. Does a relationship, contractual or otherwise, link the injured party and the professional?
3. Did the injured party rely on the professional to his or her detriment?
4. Do any damages arise on the *Hedley Byrne* principle?
5. Do any damages arise on the basis of a fiduciary duty?
6. Does any statutory extension of liability apply?

The answers to these questions may provide an indication of professional liability.

SUMMARY

A professional is a person who possesses specialized knowledge or skills, and who must meet and maintain a particular standard in the performance of his or her services to the public. Most professionals are members of professional associations that are responsible for the establishment of training, education and the standards of performance of the members, as well as the enforcement of the standards by an accreditation or licensing process.

Professional liability may arise under contract or tort, or as a breach of fiduciary duty by the professional. A failure to perform a service in accordance with the standard of care established for the profession may constitute negligence.

Professionals in a sense have not only a duty to their clients, but to their professional organizations as well, as a failure to maintain professional levels of performance may result in the revocation of their right to practice their profession.

KEY TERMS

13. *Knowles v. General Accident Assurance Co. of Canada* (1984), 49 O.R. (2d) 52.

Review Questions

1. Does a professional person have any responsibility to a person other than the one who engaged the professional services?
2. How is the standard of care determined for most professionals?
3. Define "negligent misrepresentation."
4. Explain the role of a professional organization and the issue of professional responsibility to clients.
5. How would the duty of care be determined for persons who are not members of a professional association, but who hold themselves out as persons in possession of special expertise? Use an insurance agent as an example.
6. What is the "threefold test" and how is it applied in cases of negligent misrepresentation? In what way has the Supreme Court of Canada changed the application of this test?
7. In what way or ways would a professional engineer have a responsibility to third parties?
8. Why are the courts reluctant to make auditors responsible to every person who obtained a copy of a financial statement negligently prepared by the accountant? What limits do the courts apply to an accountant's responsibility?
9. Outline the responsibility of a surgeon to a patient. Is it limited only to the actual performance of the medical procedure?
10. Outline the steps that an emerging profession must take in order to establish a recognized professional status.
11. Explain "fiduciary duty."
12. Should professional associations be responsible for the negligent acts of their members? What should be the limits of their responsibility to the public-at-large?

Mini-Case Problems

1. A lawyer negligently prepared a deed of land for a client who was selling his property, and in doing so described the wrong parcel of land in the deed. The lawyer acting for the purchaser failed to notice the error, and registered the deed. The purchaser of the land later discovered that her deed was for the wrong parcel of land. Discuss the responsibility of the lawyers.
2. An engineer designed an outdoor elevated patio that was expected to accommodate conference meetings of up to 100 persons. The engineer considered the load bearing structure on the basis of the weight of 200 persons as a safety factor. When the structure was completed, the owner decided to use the patio for a rock concert, and sold 400 tickets. In actual fact over 400 patrons attended, and packed the patio for a 'standing room only' concert. The enthusiastic crowd at one point began jumping to the music, and the additional load pressure caused the structure to collapse, injuring many of the patrons. How would liability be determined in this case?

Case Problems for Discussion

CASE 1

Central Land Development Ltd. sold three commercial building lots to Commercial Builders Ltd. for the sum of $400,000. As a part of the purchase price, Perros and Masson, the two principal shareholders of Commercial Builders Ltd., and the corporation gave Central Land Development Ltd. a mortgage for $250,000. The balance of the $400,000 selling price, in the amount of $150,000 was paid at the time of the purchase.

A year and six months later, Commercial Builders Ltd., and its two shareholders wished to pay off the mortgage, and requested Central Land Development Ltd. to provide them with a pay-out amount. Central Land Development Ltd. requested its accountant, Hamilton, to calculate the balance owing on the mortgage. Hamilton calculated the balance owing to be $172,459. Commercial Builders Ltd. paid the amount and received a discharge of the mortgage.

Several months later, Central Land Development Ltd. discovered that the amount calculated by Hamilton was in error, and the correct balance was in fact $202,459. Central Land Development requested payment from Commercial Builders Ltd. of the $30,000 difference, but Commercial Builders Ltd. refused to pay the amount.

Discuss the issues raised in this case, and the arguments that might be raised by each of the parties.

CASE 2

Road Builders Ltd. was the successful bidder for a contract with a provincial government to construct a 23 km. highway by-pass of a large city. As a part of its preparation for the contract bid, Road Builders Ltd. engaged the services of Highway Engineering Corporation Ltd. to prepare engineering design and specifications for the construction project. This was done, and Road Builders Ltd. submitted the engineering design and specifications as a part of its bid for the contract. The bid was successful, and the engineering design was incorporated in the contract by the government.

During construction of the roadway, errors were discovered in the design, and Road Builders Ltd. was obliged to correct the design errors at considerable expense. As a result, Road Builders Ltd. suffered a loss on the construction project, and demanded compensation for the loss from Highway Engineering Corporation Ltd.

Highway Engineering Corporation Ltd. refused to pay for the loss on the basis that it was not involved in the construction, and also that the province had accepted the design by including it in the contract.

Discuss the arguments of Highway Engineering Corporation Ltd., and the arguments that Road Builders Ltd. might raise if it took legal action against the engineering firm for its loss. Render a decision.

CASE 3

Alex carried on a relatively successful business as a manufacturer of a cleaning product that was well received by the users. After several years of slow but steady growth in sales, his accountant suggested that he expand the business by the incorporation of a company. This would permit the sale of shares to acquire the capital necessary for a new plant and equipment.

A company was eventually incorporated, but instead of selling shares to acquire the necessary capital to expand the business, Alex decided to arrange a $200,000 loan from his banker. To do so, however, Alex needed the accountant to prepare financial statements for the company to deliver to the bank.

Preparing the financial statements, the accountant failed to notice that the existing land and plant building were not acquired by the corporation, but retained by Alex, and simply leased to the company on an annual basis. The accountant had included the land and building (which had a value of $250,000) as an asset of the corporation on the financial statements, without checking with Alex to determine if the property had been transferred. When the error was discovered later during negotiations with the bank, the bank insisted that Alex guarantee the loan as a principal debtor, and use the land and building as additional security for the loan.

A few weeks later, Alex decided that it would be necessary to acquire additional capital to complete the expansion program. He contacted a private investor with a view to selling her a block of shares in the corporation. The investor was interested in the financial status of the corporation, and Alex informed her that the corporation had borrowed $200,000 from the bank for the purpose of expanding the business. He also suggested that the investor contact either his accountant or the bank for information on the corporation's assets and financial position. The investor contacted the bank and requested copies of the financial statements that the accountant had prepared. A bank employee, who was unaware that the statements were in error, gave them to the investor without comment.

On the strength of the financial statements, the investor invested $50,000 in shares of the corporation. Some months later, she discovered that the corporation did not own the land or buildings, and that the financial statements were in error.

Advise the investor of her legal position, and her rights (if any) against Alex, the accountant, and the bank.

CASE 4

Community Sportsplex Ltd. purchased a block of land from a small municipality for the development of a sports facility consisting of an outdoor sports field, and an indoor arena with a standard ice surface with a seating capacity for 3,000 spectators.

The municipality had zoned the area for 'recreational use' and prohibited the construction of residential housing on the land. This was so because a part of the land had been a former landfill site, and the surface soil subject to some settling as the garbage buried below gradually decomposed.

Community Sportsplex Ltd. was unaware of the former use, but nevertheless, engaged the services of Geo Engineering Corporation Ltd., a soil testing business, to determine the suitability of the surface and subsurface to support the proposed arena building. The proposed location of the building was at the edge of the landfill area, and soil testing indicated that the land was suitable for the construction of the building.

Shortly after the soil had been tested, the architect that designed the arena building in consultation with Community Sportsplex Ltd. decided to move the location of the proposed building some 10 metres closer to the playing field. As a result, the new location placed one wall of the proposed building on the landfill area.

A contract for construction of the building was given to a contractor who proceeded to excavate for the footings of the building. In the area over the former landfill the contractor discovered the poor soil conditions, and extensive work was required to stabilize the ground for the building footings. The contractor determined that this added cost was approximately $50,000. Community Sportsplex Ltd. refused to pay the additional cost.

Discuss the rights of the parties and the possible outcome, if the dispute should be taken to court for a decision.

The Online Learning Centre offers more ways to check what you have learned so far. Find quizzes, Weblinks, and many other resources at www.mcgrawhill.ca/olc/willes.

THE LAW OF CONTRACT

CHAPTER 7

An Introduction to the Legal Relationship

Chapter Objectives

After study of this chapter, students should be able to:

- Describe and explain the first three elements of a valid contract: intention, offer, and acceptance.

YOUR BUSINESS AT RISK

Almost all of your business activity will be governed by contracts, and you must learn how to create contracts correctly. If a contract is not created correctly, it does not come into existence, and this is a dangerous circumstance for your business. If the other party to your transaction abandons the deal or falls short in performance, you will have incurred time and expense for nothing. You will be left with little or no means to enforce your "rights," as no contract ever existed.

INTRODUCTION

A contract may be defined as an agreement made between two or more persons that is enforceable at law. It is not something that is tangible (although in some cases evidence of its existence may take that form): it is a legal concept. It comes into existence, in a legal sense, when the parties have established all of the elements that make the contract enforceable. Until they have done so, no enforceable agreement exists. This part of the text, consisting of eight chapters, will examine the nature of each of these elements.

Contract law differs from the law of torts and many other areas of law in a rather remarkable way: if the parties comply with the principles laid down for the creation of an enforceable contract, they are free to create specific rights and duties of their own that the courts will enforce. In some respects, they create their own "law," which they are obliged to follow. How these concepts were developed is largely a matter of history, although, as a body of law, the law of contract is not old. Its development parallels the rise of the mercantile class in England and North America, but it did not reach a position of importance until the nineteenth century. The next 100 years, however, saw the growth of contract law to the point where it became one of the most important areas of the Common Law. Today, contract law forms the basis of most commercial activity.

HISTORICAL DEVELOPMENT OF THE LAW OF CONTRACT

The law of contract is essentially an area of law relating to business transactions. Until the rise of the merchant class in England, business transactions as we know them today were not common. Under the feudal system, each manor was relatively self-sufficient, and such trade as did exist was frequently by barter or by purchase at a local fair or market. Transactions were usually instantaneous, in the sense that goods changed hands as soon as the exchange was agreed upon. Disputes that arose between merchants were promptly settled by the merchants themselves, and later by way of the rules set down by their guilds. Disputes arising out of the informal contracts or agreements that did reach the courts most often fell within the jurisdiction of the ecclesiastic courts on the basis that a breach of a solemn promise was a moral issue to be dealt with by the church. Minor cases (the equivalent of our breach of contract), however, were occasionally handled by manor courts, where damages were sometimes awarded for breach of the promise.

The establishment of England as a trading nation in the seventeenth century brought with it the need for a legal response to trade disputes. As a consequence, the seventeenth century saw the development of the bargain theory of a contract, whereby each party to the agreement derived some benefit from the agreement in return for a promise to do or give something in return. This benefit–detriment approach formed the basis of modern contract law. From that point on, the law developed rapidly as it responded to the changes taking place in society. The decline of the feudal system and the rise of the merchant class precipitated a change in the economic order that required new methods of dealing with the changing activities of the people. The development of the law of contract was essentially a response to those needs, rather than a gradual evolution of older Common Law.

In the twenty-first century, our contracts and commerce are governed by case law and statutes that, in some instances, are hundreds of years old. Certainly, there has been legal evolution, but as you study cases dating from the 1800s and 1900s it is important to remember that they reflect the timeless precepts of fairness, promise, and obligation. These rules govern commerce today just as much as they did when they were new, and are a reflection of respected business practices in the twenty-first century.

THE ELEMENTS OF A VALID CONTRACT

Intention to be bound

The assumption at law that strangers intend to be bound by their promises.

The creation of a binding contract that the courts will enforce requires the contracting parties to meet a number of requirements prescribed by the law of contract. While these requirements are not numerous, they must nevertheless be met before the agreement creates rights and duties that may be enforceable at law. These requirements are referred to as the elements of a valid contract, and consist of:

(1) an intention to create a legal relationship,
(2) offer,
(3) acceptance,
(4) consideration,
(5) capacity to contract, and
(6) legality.

In addition to these six basic elements, to be enforceable, certain types of contracts must be in writing, in an electronic substitute, or take on a special form. In general, however, all contracts must have these six elements present to be valid and binding. Contracts must also be free from any vitiating elements, such as mistake, misrepresentation, or undue influence. These elements are examined in the text, following an examination of the basic elements. In this chapter, the first three elements are examined, to identify the rules applicable to the establishment of these requirements for a contract.

THE INTENTION TO CREATE A LEGAL RELATIONSHIP

The concept of a contract as a bargain or agreement struck by two parties is based upon the premise that the end results will be a meeting of the parties' minds on the terms and conditions that will form their agreement with each other. Each will normally agree to do, or perhaps refrain from doing, certain things in return for the promise of the other to do certain things of a particular nature. In the process of reaching this meeting of the minds, the parties must establish particular elements of the contract itself.

Consensus ad idem

Agreement as to the subject or object of the contract.

Closely related to the intention of the parties to be bound by their promises is the notion of ***consensus ad idem*** or agreement as to the subject or object of the contract. In short, the parties must be of one mind and their promises must relate to that subject or object. A promise begins this process. "Do you want my car?" which is a rather vague statement that *might* contain a promise, begs the question: "What do you mean?" Does the speaker intend to lend the car, gift the car, lease the car, or sell the car? Until this is established — consensus reached — no sensible acceptance can follow, and no contract can result.

CASE IN POINT

Klemke Mining Corporation and Shell Canada conducted extensive discussions with respect to mining work on Shell's oil sands lease. After almost six months, KMC prepared a June meeting agenda outlining its future participation, which was verbally confirmed by Shell. Further discussions resulted in KMC making a December investment in a joint venture operation, and Shell provided a set of operating terms for KMC, to which KMC agreed. Some months later, Shell advised KMC that a different mining firm had been selected to carry out the mining work. Shell took the position that no formal contract had been created. The trial judge found that sufficient substance existed to show a meeting of the minds and that the June discussions constituted an oral contract, which was further memorialized in December. A formally titled contract was not necessary in order to bind the parties. The judgment of over $21 million (plus interest) was upheld on appeal.

Klemke Mining Corporation v. Shell Canada Limited, 2008 ABCA 257 (CanLII).

Negotiations relating to the agreement must, of necessity, have a beginning. If the agreement, by definition, consists of promises made by the parties, then one of the essential elements of an agreement must be a promise. Obviously, not all promises can be taken as binding on the party making them. Some may be made by persons who have no intention of becoming legally obligated to fulfill them. This type of promise cannot be taken as the basis for a contract. The first requirement, then, for a valid contract, must be the intention on the part of the promisor to be bound by the promise made. This intention to create a legal relationship is an essential element of a valid contract. It is generally presumed to exist at law in any commercial transaction where the parties are dealing with one another at arm's length.

The intention to create a legal relationship is a presumption at law, because the creation of the intention would otherwise be difficult to prove. Presuming that the party intended to be bound by the promise shifts the onus to prove otherwise, if the intention did not exist. If the intention is denied, the courts will usually use the conduct of the party at the time that the statements were made as a test, and assess such conduct and statements from the point of view of the "reasonable person."

The reason for the presumption — that strangers who make promises to one another intend to be bound by them — is essentially an approach that permits the courts to assume that the promises are binding, unless one or both of the parties can satisfy the courts that they were not intended to be so.

CASE IN POINT

Bell Aliant made an agreement to allow Rogers Communications to use its utility poles. The agreement stated that it "shall continue in force for a period of five years from the date it is made, and thereafter for successive five year terms, unless and until terminated by one year prior notice in writing by either party." With one year's notice, Aliant terminated the contract. Rogers felt that the position of the commas meant it should get the benefit of a full five year locked-in term, before anything could be terminated. Professors of English concluded that the second comma allowed the notice period to modify the whole sentence. The governing tribunal, the Canadian Radio-television and Telecommunications Commission (CRTC), found that the contract had been made in English and in French. While the English version could be read two ways, the French version could only be read one way, which prevailed: notice could only be effective a year before the end of each five year term. What do you think of that resolution? Where do you (or do you not) find the meeting of the minds in this situation?

Telecom Decision CRTC 2007-75, available at www.crtc.gc.ca.

The law, nevertheless, recognizes certain kinds of promises or statements as ones that are normally not binding, unless established as such by the evidence. For example, promises made between members of a family would not normally be considered to be an enforceable contract. As well, generally speaking, advertisements in newspapers, magazines, and other written media are not normally taken as enforceable promises that are binding on the advertiser.

The basis for these two exceptions is, for the most part, obvious. Members of a family frequently make promises to one another that they would normally not make to strangers. And advertisers, in the presentation of their goods to the public, are permitted to describe their products with some latitude and enthusiasm, provided, of course, that they do not mislead the prospective purchaser. While these two groups are not normally subject to the presumption that their promises represent an intention to create a legal relationship, they may, nevertheless, be bound by their promises if the party accepting their promises can convince a court that the promisor intended to be bound by the promise.

COURT DECISION: Practical Application of the Law

Offer to Public at Large—Acceptance—Notice to Offeror

***Carlill v. Carbolic Smoke Ball Co.*, [1893] 1 Q.B. 256.**

This English case created the principle that persons who make promises will be bound to come good on their promises. The defendants were manufacturers of a medical preparation called the "carbolic smoke ball." To sell their product they inserted an advertisement promoting a £100 reward in a number of newspapers. In 1893, one pound sterling (£) was valued at $5.00, thus £100 amounted to two years of a worker's wages. The plaintiff purchased and used the preparation according to the instructions and was then attacked by influenza. She brought an action against the defendant for the £100 reward. The advertisement read as follows:

"£100 reward will be paid by the Carbolic Smoke Ball Company to any person who contracts the increasing epidemic influenza, colds, or any disease caused by taking cold, after having used the ball three times daily for two weeks according to the printed directions supplied with each ball. £1,000 is deposited with the Alliance Bank, Regent Street, shewing our sincerity in the matter. During the last epidemic of influenza many thousand carbolic smoke balls were sold as preventives against this disease, and in no ascertained case was the disease contracted by those using the carbolic smoke ball. One carbolic smoke ball will last a family several months, making it the cheapest remedy in the world at the price, 10s., post free. The ball can be refilled at a cost of 5s. Address, Carbolic Smoke Ball Company, 27, Princes Street, Hanover Square, London."

THE COURT: The first observation I will make is that we are not dealing with any inference of fact. We are dealing with an express promise to pay £100 in certain events. Read the advertisement how you will, and twist it about as you will, here is a distinct promise expressed in language which is perfectly unmistakeable — "£100 reward will be paid by the Carbolic Smoke Ball Company to any person who contracts the influenza after having used the ball three times daily for two weeks according to the printed directions supplied with each ball."

We must first consider whether this was intended to be a promise at all, or whether it was a mere puff which meant nothing. Was it a mere puff? My answer to that question is No, and I base my answer upon this passage: "1,000 pounds is deposited with the Alliance Bank, shewing our sincerity in the matter." Now, for what was that money deposited or that statement made except to negative the suggestion that this was a mere puff and meant nothing at all? The deposit is called in aid by the advertiser as proof of his sincerity in the matter — that is, the sincerity of his promise to pay this

£100 in the event which he has specified. I say this for the purpose of giving point to the observation that we are not inferring a promise; there is the promise, as plain as words can make it. It appears to me, therefore, that the defendants must perform their promise....

This case is over 100 years old. It established principles of law then that continue to apply as law today. What message should business persons of today take away from this?

The rule that can be drawn from this case is that, while an advertiser is not normally bound by the claims set out in an advertisement, if a clear intention to be bound by them is expressed, then the courts will treat the promise as one made with an intention to create a legal relationship.

As a general rule, the courts view an advertisement (or for that matter, any display of goods) as a mere *invitation to do business*, rather than an intention to enter into a contract with the public at large. The purpose of the advertisement or display is merely to invite offers that the seller may accept or reject. This particular point becomes important when determining if a contract is made where goods are displayed for sale in a self-serve establishment. The issue was decided in an English case where the court held that the display of goods in a self-serve shop was not an offer to sell the goods to the patron of the shop, but merely an invitation to the public to examine the goods and, if the patron desired, to offer to purchase the goods. The possession of the goods by the prospective purchaser was of no consequence, as the offer to purchase and the acceptance of the offer by the seller did not take place until the seller dealt with the goods at the check-out counter. It was at this point in time that the contract was made — not before.[1] Figure 7–1 outlines the effects of a statement.

FIGURE 7–1 **Intention of the Parties**

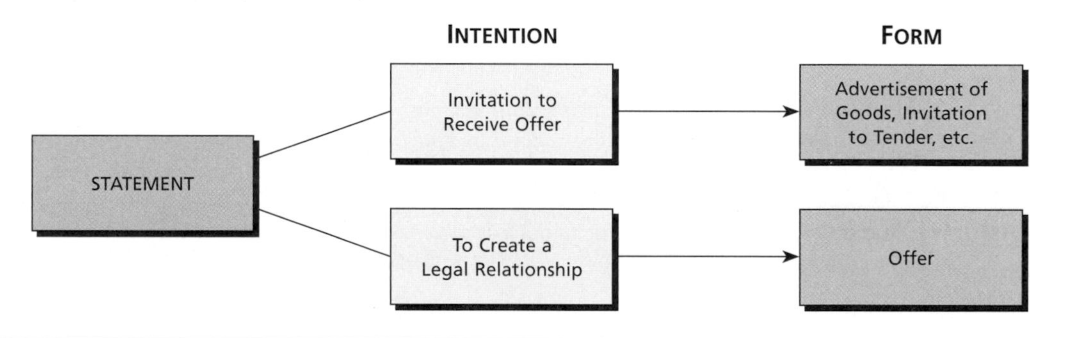

COURT DECISION: Practical Application of the Law

Offer—Acceptance—Time of an Offer—Time of Acceptance

***Pharmaceutical Society of Great Britain v. Boots Cash Chemists (Southern) Ltd.*, [1952] 2 Q.B. 795.**

The defendants operated a self-serve drug store. Some drugs, which were classed as "poison" under government legislation, were displayed on self-serve shelves, but the sales were recorded by a qualified pharmacist at the check-out counter. The defendant was charged with selling poisons contrary to the Act by failing to have a registered pharmacist supervise the sale of the product. The case turned on when and where the sale took place, and who was doing the offering and accepting.

1. *Pharmaceutical Society of Great Britain v. Boots Cash Chemists (Southern) Ltd.*, [1952] 2 Q.B. 795.

THE COURT: The question which I have to decide is whether the sale is completed before or after the intending purchaser has passed the scrutiny of the pharmacist and paid his money, or, to put it in another way, whether the offer which initiates the negotiations is an offer by the shopkeeper or an offer by the buyer.

I think that it is a well-established principle that the mere exposure of goods for sale by a shopkeeper indicates to the public that he is willing to treat but does not amount to an offer to sell. I do not think I ought to hold that that principle is completely reversed merely because there is a self-service scheme, such as this, in operation. In my opinion it comes to no more than that the customer is informed that he may himself pick up an article and if, but only if, the shopkeeper then expresses his willingness to sell, the contract for sale is completed. In fact, the offer is an offer to buy, and there is no offer to sell; the customer brings the goods to the shopkeeper to see whether he will sell or not. In 99 cases out of a 100 he will sell and, if so, he accepts the customer's offer, but he need not do so. The very fact that the supervising pharmacist is at the place where the money has to be paid is an indication to the purchaser that the shopkeeper may not be willing to complete a contract with anybody who may bring the goods to him.

... Therefore, in my opinion, the mere fact that a customer picks up a bottle of medicine from the shelves in this case does not amount to an acceptance of an offer to sell. It is an offer by the customer to buy and there is no sale effected until the buyer's offer is accepted by the acceptance of the price. The offer, the acceptance of the price, and therefore the sale, take place under the supervision of the pharmacist.

This case establishes the nature of the relationship between buyer and seller, under normal retail shop circumstances, as offeror and acceptor. Aside from the specific consequences for Boots, why would this arrangement be preferable to the reverse in retail transactions?

OFFER AND ACCEPTANCE

The Nature of an Offer

The second element of a binding contract deals with promises made by the parties. Only a promise made with the intention of creating a legal relationship may be enforced. But in the normal course of negotiations a person seldom makes such a promise unless some condition is attached to it, requiring the other party (the promisee) to do some act or give a promise in exchange. Consequently, such a promise is merely tentative, until the other party expresses a willingness to comply with the condition. The tentative promise made subject to a condition is therefore not binding on the offering party (the promisor or offeror) until the proposal is accepted. It is only when a valid acceptance takes place that the parties may be bound by the agreement. These two additional requirements constitute the second and third elements of a valid contract: offer and acceptance.

Communication of an Offer

Offer

A tentative promise subject to a condition.

If the analysis of the negotiations is carried further, an obvious observation can be made. An **offer** must be communicated by the offeror to the other party (the offeree) before the offer is capable of being accepted. From this observation flows the first rule for offer and acceptance: *An offer must be communicated by the offeror to the offeree before acceptance may take place.*

This rule may appear to be self-evident, but an offer is not always made directly to the offeree by the offeror. In some cases the parties may deal with each other by letter, facsimile, e-mail, or other means of communication. It is important for the offeror to know when the offeree becomes aware of the offer. This is so because an offer is not valid until it is received by the offeree, and the offeror is not bound by the offer until such time as it is accepted. This means that identical offers that cross in the mail or electronic transmission do not constitute a contract, even though there is an obvious meeting of the minds of the parties, as evidenced by their offers. The essential point to make here is that no person can agree to an offer of which he or she is unaware.

If, for example, the acceptance takes place before the offer is made, the offeror is not bound by the promise. This is particularly true in the case of offers of reward.

Jones reaches her construction site on a Monday morning to find her firm's front-end loader missing from where it had been left the previous Friday. Advising police, she further places an advertisement in the newspaper offering a $2,000 reward for information leading to the return of the loader. Later in the week, Smith discovers the loader abandoned on his farm property, and from the name and telephone number painted on the loader, calls Jones to find out how it got there. Jones recovers the loader and says nothing of the reward. Smith later reads the newspaper, becomes aware of the reward, and attempts to claim it from Jones.

In this case, Jones's offer was not communicated to Smith until after Smith had fully performed what was required of him under the terms of the offer of reward. Smith, therefore, cannot accept the offer, because he returned the loader without the intention of creating a contract. His act was gratuitous, and he cannot later claim the right to payment. This concept will be examined more closely with respect to another element of a contract, but for the present, it may be taken as an example of the communication rule.

If the negotiation process is examined closely, another rule for offer and acceptance can be drawn from it. A person who makes an offer frequently directs it to a specific person, rather than to the public at large. A seller of goods may wish to deal with a specific person for a variety of sound reasons. For example, a seller of a business may wish to sell it only to a person who would maintain its level of product quality and service, or a seller of a specific type of goods may wish to sell the goods only to those persons trained in the use of the goods if some danger is attached to their use. Hence, we have the general rule that *only the person to whom an offer is made may accept the offer.*[2] If an offer is made to the public at large, this rule naturally does not apply; for the offeror is, by either words or conduct, implying in such an offer that the identity of the offeree is not important in the contract.

Acceptance of an Offer

Acceptance

A statement or act given in response to and in accordance with an offer.

While both an offer and its **acceptance** may be made or inferred from the words or the conduct of the parties, the words or conduct must conform to certain established rules before the acceptance will be valid. These rules have been formulated by the courts over the years as a result of the many contract disputes that came before them. At present, the major rules for acceptance are now well settled.

The first general rule for acceptance when a response is necessary is simply the reverse of the rule for offers. It states that *the acceptance of the offer must be communicated to the offeror in the manner requested or implied by the offeror in the offer.* The acceptance must take the form of certain words or acts in accordance with the offer that will indicate to the offeror that the offeree has accepted the offer. These words or conduct need not normally be precise, but they must convey the offeree's intentions to the offeror in the manner contemplated for acceptance. The rationale behind this general rule is that the offeror is entitled to some indication from the offeree of the offeree's intention to enter into the agreement, failing which the offeror may then take steps to revoke the offer and make it to someone else.

If Maple Leaf Electronics sends a fax to Wholesale Notebooks ordering 10 notebook computers at a particular catalogue price, requesting delivery by courier (or some other means), the fax would constitute an offer to purchase. The acceptance would take place when the seller acted in accordance with the instructions for acceptance set out in the fax. It would not be necessary for the seller to write a reply conveying acceptance of the offer, because the offer contemplates acceptance by the act of sending the goods to the offeror. The acceptance would be complete when the seller did everything required by the terms of the offer contained in the fax.

2. *Cundy v. Lindsay* (1878), 3 App. Cas. 459.

This particular issue was raised in the case mentioned previously, where a pharmaceutical product was offered to the public at large as a cure and prevention for influenza. One of the arguments raised by the manufacturer was the fact that the plaintiff had not communicated her acceptance of the offer before using the product. Hence, no contract existed. The court disposed of this argument by saying that if the terms of the offer indicate a particular mode of acceptance to make the promise binding, it is sufficient for the offeree to comply with the indicated mode of acceptance. Notification to the offeror of the acceptance would then be unnecessary.[3]

In the case of an offer requiring some expression of acceptance by written or spoken words, a number of specific rules for acceptance have been set down. If acceptance is specified to be by oral means, the acceptance would be complete when it is communicated by the offeree either by telephone or when the offeree meets with the offeror and speaks the words of acceptance directly to the offeror. With this form of acceptance there is no question about the communication of the words of acceptance. It takes place when the words are spoken.

The time of acceptance, however, is sometimes not as clear-cut with other modes of acceptance, and the courts have been called upon to decide the issues as the particular modes of acceptance came before them. In the case of an offer that invites acceptance by post, the rule that has been established is that *the acceptance of the offer takes place when the letter of acceptance, properly addressed and the postage paid, is placed in the postbox or post office.*[4] The reasoning behind this decision is sensible. The offeree, in preparing a letter of acceptance and delivering it to the post office, has done everything possible to accept the offer when the letter moves into the custody of the postal system. The postal system, as the agent of the addressee, is responsible for delivery from that point on. If the acceptance should be lost while in the hands of the post office, the contract would still be binding, as it was formed when the letter was posted. The offeror, by not specifying that acceptance would not be complete until the letter is received, assumes the risk of loss by the post office and any uncertainty that might accompany this specified mode of acceptance.

The courts have also held that when an offer does not specifically state that the mail should be used for acceptance, but when it is the usual or contemplated mode of acceptance, then the posting of the letter of acceptance will constitute acceptance of the offer.[5]

For all other modes of communication, the acceptance would not be complete until the offeror was made aware of the acceptance.

Electronic Offer and Acceptance

Electronic documents are governed by federal law, the *Personal Information Privacy and Electronic Documents Act*, unless a province has moved to enact equivalent legislation of its own. An electronic document is deemed to be sent when it enters an information system outside the sender's control or, if the sender and the addressee use the same information system, when it becomes capable of being retrieved and processed by the addressee. It is presumed to be received by the addressee:

(a) if the addressee has designated or uses an information system for the purpose of receiving information or documents of the type sent, when it enters that information system and becomes capable of being retrieved and processed by the addressee; **or,**

(b) if the addressee has not designated or does not use an information system for the purpose of receiving information or documents of the type sent, when the addressee becomes aware of the information or document in the addressee's information system and it becomes capable of being retrieved and processed by the addressee.

It is also important to understand that validating electronic contracts is not the same as forcing their use, and thus parties are free to carry on using traditional paper. Just as

3. *Carlill v. Carbolic Smoke Ball Co.*, [1893] 1 Q.B. 256.
4. *Household Fire and Carriage Accident Ins. Co. v. Grant*, [1878] 4 Ex.D. 216.
5. *Henthorn v. Fraser*, [1892] 2 Ch. 27.

with traditional contracts, electronic contractors may opt out of these statutory provisions (such as what constitutes the time of acceptance) and create their own terms of operation for electronic contracts and relationships.

A QUESTION OF ETHICS

Faint and/or fine print in many application forms often says "Check this box ❑ if you would prefer not to receive offers from other carefully screened alliance partners." This note is, of course, easily overlooked. How would a court logically deal with something designed to deceive? Are there any parallels to offer-and-acceptance rules?

In addition to the rules relating to the time and place, a number of other rules also apply to acceptance. Of particular importance is the nature of the acceptance itself. When an offer is made, the only binding acceptance would be one that clearly and unconditionally accepted the offeror's promise, and complied with any accompanying condition imposed by the offeror. Anything less than this would constitute either a counteroffer, or an inquiry. If an acceptance is not unconditional, but changes the terms, then it would have the effect of rejecting the original offer. It would represent, in itself, an offer that the original offeror could then either accept or reject.

Smith, a retailer, writes a letter to another retailer, in which he offers to sell his inventory for $30,000 cash. The other retailer writes a letter of reply, in which she "accepts" Smith's offer, but states that she will buy the inventory on the payment of $10,000 cash and give Smith a promissory note for $20,000, to be payable at $2,000 per month over a ten-month period.

In this example, Smith offers to sell his inventory for a cash payment of $30,000. The other retailer has expressed her willingness to purchase the inventory, but has changed the offer by altering the payment provision from $30,000 cash to $10,000 cash and a promissory note for $20,000. The change in terms represents for Smith (who now becomes the offeree) a counteroffer that he must accept or reject. The counteroffer submitted by the other retailer has the effect of terminating the original offer that was made by Smith. If Smith should reject the counteroffer, the other retailer may not accept the original offer unless Smith wishes to revive it.

The desirable approach for the other retailer to follow in a situation where some aspect of the offer is unacceptable to her would be to inquire if Smith would be willing to modify the terms of payment before a response is made to the offer in a definite manner. In this fashion, she might still retain the opportunity to accept the original offer if Smith should be unwilling to modify his terms of payment.

CASE IN POINT

A contractor wished to lease a crane from an equipment-rental agency for a construction project. The rental agency proposed the terms of the rental and the contractor, in response, made a counter-proposal for the rental. The rental agency did not respond to the counter-proposal, but delivered the crane to the contractor's work site. The rental agency later attempted to fix new terms for the rental, but the contractor refused to agree to the terms.

When the case came before the court, the court held that delivery of the crane by the rental agency constituted acceptance of the contractor's counter-proposal for the rental agreement.

J. Lowe (1980) Ltd. v. Upper Clements Family Theme Park Limited, 1990 CanLII 4194 (NSSC).

A somewhat different matter is a rule stating that silence cannot be considered to be acceptance unless a pre-existing agreement to this effect has been established between the parties. The rationale for this rule is obvious. The offeree should not be obligated to refuse an offer, nor should he or she be obliged to comply with an offer made to him simply because he or she has failed to reject it. The only exception to this rule would be where the offeree has clearly consented to be bound by this type of arrangement.

The question of whether a pre-existing agreement exists that would make silence acceptance is not always answerable in a definitive way. In some cases, persons may conduct themselves in such a way that, even though they remain silent in terms of acceptance, they lead the offeror to believe that acceptance has been made of the offer. This is particularly true if a person offers to perform a service with the intention of receiving payment for the service, and the offeree stands by in silence while the service is performed, with full knowledge that the offeror expects payment.

For example, in the case of *Saint John Tug Boat Co. Ltd. v. Irving Refinery Ltd.* (1964), 49 M.P.R. 284, the court quoted authorities for the Common Law rules, which stated:

1. Liabilities are not to be forced upon people behind their backs any more than you can confer a benefit upon a man against his will.
2. But if a person knows that the consideration is being rendered for his benefit with an expectation that he will pay for it, then if he acquiesces in its being done, taking the benefit of it when done, he will be taken impliedly to have requested its being done: and that will import a promise to pay for it.

What the court was saying, in essence, is that a person cannot be forced to refuse an offer presented on the basis that silence will constitute acceptance. However, if offerees so conduct themselves that their actions would lead a reasonable person to believe that they had assented to terms of the offer, then the offerees may not stand by and watch the offeror perform a benefit for them if it is clear that the offeror expects to be compensated for the work performed. Under these circumstances, offerees would have an obligation to immediately stop the offeror from performing the work, or must reject the offer.

It is important to note that this exception to the silence rule would normally apply to those situations where the offeree's actions constitute some form of acquiescence. It would not apply, for example, when sellers send unsolicited goods to householders, because no acquiescence could take place before the delivery of the goods, at least in terms of communication of the interest to the seller.

Recent consumer-protection legislation in a number of provinces has reinforced this Common Law rule. The legislation generally provides that members of the public shall not be obliged to pay for unsolicited goods delivered to them, nor should they be liable for the goods in any way if they are lost or damaged while in their possession. Neither the Common Law nor the legislation, however, affect a pre-existing arrangement whereby silence may constitute acceptance of a subsequent offer. This is a common characteristic of most book, CD, and DVD "clubs." These clubs operate on the basis that a contract will be formed and that the book, CD, or DVD will be delivered to the offeree if the offeree fails to respond to the offeror within a specific period of time after the offer has been made. Contracts of this nature are generally binding, because silence is considered acceptance due to the pre-existing agreement governing the future contractual relationship of the parties.

Acceptance, while it must be unconditional and made in accordance with the terms of the offer, may take many forms. The normal method of accepting an offer is to state or write, "I accept your offer," but acceptance may take other forms as well. For example, it may take the form of an affirmative nod of the head and a handshake. At an auction sale, it may take the form of the auctioneer dropping the hammer and saying "Sold" to the person making the final offer.

When a particular method of accepting the offer is specified, the offeree must, of course, comply with the requirements. If the offeror has stated in an offer that acceptance must only be made by facsimile, then the offeree, if he or she wishes to accept, must use this form of communication to make a valid acceptance. Offerors usually do not impose such rigid requirements for acceptance, but often suggest that a particular method of communication would be preferred. In these cases, *if a method other than the method mentioned in the offer is selected, the acceptance would only be effective when it was received by the offeror.*[6]

Offers that require offerees to complete their part of the contract as a mode of acceptance form a special class of contracts called unilateral agreements. These agreements usually do not call for the communication of acceptance before the contract is to be performed, but rather signify that the offer may be accepted by the offeree completing his or her part of the agreement. Once completed, the offeror would then perform his or her part. On the surface, there would appear to be a danger with this mode of acceptance. If the offeror should withdraw the offer before the offeree has fully performed the acceptance, then no contract would exist, and any expense or inconvenience incurred by the offeree would not be recoverable. To remedy this situation, the courts have held that when an offeree is obliged to perform his or her part of the contract in order to accept the offer, then the offeror will not be permitted to withdraw the offer so long as the offeree is in the course of performing his or her part. This approach, however, assumes that the offeror has not expressly reserved the right to withdraw the offer at any time during the offeree's act of acceptance. The offer of a discounted price upon presentation of a newspaper coupon would be an example of a contract where the offer would be accepted by the act of the offeree. Figure 7–2 illustrates the different forms of acceptance.

FIGURE 7–2 Forms of Acceptance

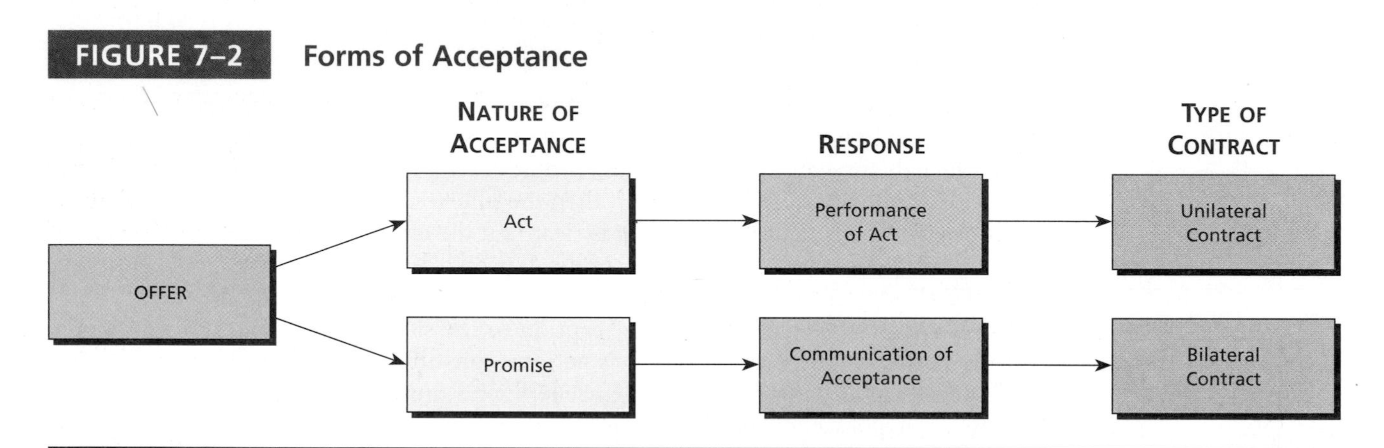

MANAGEMENT ALERT: BEST PRACTICE

Not understanding or ignoring the binding nature of offer and acceptance has resulted in many lawsuits. Businesses sometimes sign contracts with little intention of performing a duty, but merely to "tie up" a buyer or seller. Others believe that they can create a "short list" of qualified suppliers in this way. And some business persons think there is a "cooling-off" period available, during which they can cancel their contracts. Others accept the verbal promise of the other party to "tear up the contract if it doesn't work out." All of these are recipes for a lawsuit.

Location of the Transaction

The place where an offer is made and the place where acceptance occurs has always been a significant (but by no means complete) determinant of the location of the contract and the jurisdiction that governs it.

Place of offer and acceptance are dealt with under UNCITRAL/ULCC-based electronic business legislation. A data message is deemed to be dispatched at the place where the originator's business lies, and it is deemed to be received at the place where the

6. *Henthorn v. Fraser,* [1892] 2 Ch. 27.

addressee has its place of business. If either has more than one place of business, it is the place with the closest relationship to the underlying transaction that is used. Failing that (if no transaction is involved), the parties' principal place of business is the default, followed by their habitual residence.

These provisions mark geography for offer and acceptance, but that of location of the contract is still a factual question, and a large one in the Internet world. Location of the transaction helps determine the jurisdiction whose laws apply to the contract and what, if any, sales or income taxes will be applicable.

Online buyers and sellers may be half a world apart, but this was a problem faced in telegraph and telex commerce as well. Historically, the solution has been to take into account the otherwise express intention of the parties, their previous contracts, their physical residence, and the place where the seller conducts his or her business (including its degree of permanence, the specific place where the goods are stored or delivered, etc.). To all of this, our courts have added a dash of good sense in either asserting or rejecting their jurisdiction over a particular contract. In so doing, courts have largely ignored the electronic aspect by treating the contract as though it had been formed face-to-face by instant means.

The Internet cannot be summarily ignored, however. Instead of just being a conduit, like a telegraph or facsimile, it has dispensed with the need for permanent establishments for business. A Web catalogue of goods can be stored in a server sitting on a tiny Pacific island, entertaining browsing customers from anywhere on the globe, redirecting orders to manufacturers anywhere for shipment to addresses around the planet. In turn, a Canadian purchaser may make a credit-card payment to a merchant account on a Caribbean island, under the name of a Hong Kong-registered company that is wholly owned by an Australian.

An emerging distinction at law is this difference between active and passive Web sites. If a site can be merely accessed from a foreign jurisdiction, this does not provide a foreign court with sufficient authority to exercise its own jurisdiction. American law has recognized this distinction, and cases involving foreign Web sites have been dismissed on the grounds that accessibility, and, most importantly, passivity, does not create jurisdiction in the U.S. courts.

In all probability, Canadian courts will take the same route, and the door has been opened for them to do so, as in the British Columbia case below.

COURT DECISION: Practical Application of the Law

Electronic Commerce—Jurisdiction

***Braintech v. Kostiuk* 1999 B.C.C.A. 169 (CanLII).**

Braintech, a Canadian company in British Columbia, sued Kostiuk, also a Canadian in British Columbia, in a Texas court on the grounds that the company's investors, or potential investors, in Texas could read allegedly defamatory statements about the company that had been posted by Kostiuk. The British Columbia Supreme Court did not recognize the Texas judgment, worth some $300,000 (US), largely because Kostiuk was not properly served notice of the Texas trial. More importantly, on appeal it was evident that the British Columbia Court of Appeal was unwilling to recognize such a tangential connection with Texas simply because the alleged defamatory statements could be read by Texans. Leave to appeal to the Supreme Court of Canada was denied in 2000.

THE COURT: From what is alleged in the case at bar it is clear Kostiuk is not the operator of Silicon Investor. It is equally clear the bulletin board is "passive" as posting information volunteered by people like Kostiuk, accessible only to users who have the means of gaining access and who exercise that means. In these circumstances the complainant must offer better proof that the defendant has entered Texas than the mere possibility that someone in that jurisdiction might have reached out to cyberspace to bring the defamatory material to a screen in Texas. There is no allegation or evidence Kostiuk had a commercial purpose that utilized the highway provided by Internet to enter any particular jurisdiction. It would create a crippling effect on freedom of expression if, in every jurisdiction the world over in which access to Internet could be achieved, a person who posts fair comment on a bulletin board could be hauled before the courts of each of those countries where access to this bulletin could be obtained.

In my opinion the trial judge erred in failing to consider whether there were any contacts between the Texas court and the parties which could, with the due process clause of the 14th Amendment to the Constitution of the United States, amount to a real and substantial presence. In the circumstances revealed by record before this Court, British Columbia is the only natural forum and Texas is not an appropriate forum. That being so, comity does not require the courts of this province to recognize the default judgment in question. I would allow the appeal, set aside the judgment below and dismiss the action.

The "passive" approach is, thankfully, being adopted in most jurisdictions. Why did the plaintiff sue in Texas when both parties were residents of British Columbia?

Until such time as more Canadian cases are heard, and specialized statute laws are developed, only three solutions can prevail: either the parties to the contract agree in their contract as to applicable jurisdiction (for governing law between the parties, subject to a possible dim view of that choice in the eyes of taxing authorities), Canadian courts declare jurisdiction over a particular transaction (as a result of a breach or harm occurring in Canada), or Canadian government agencies (e.g., the Canada Customs and Revenue Agency) impose jurisdiction and taxation on one or both of the parties.

As the example suggests, international transactions are the most problematic, though inter-provincial transactions (by virtue of sales tax issues and jurisdiction for trials) are a close second. Historically, Canada has imposed income tax on foreigners who do business in Canada, but not on those who merely do business at a distance with Canadians. The difference is subtle, and it is one that, in the eyes of the Canadian government, may have to change lest great amounts of tax revenue are lost.

In the main, we can expect that our government's primary, near-term focus will be on proper and fair taxation. Issues related to jurisdiction for the purposes of governing the rights and responsibilities of parties to a contract will be secondary. As a result, prudent parties will make their own provisions for applicable law and private arbitration of Web-based disputes.

Effect and Timing of the Click as Offer and Acceptance

Like any other valid contract, electronic business contracts must include valid offer and acceptance, and the current boundary of technology in signifying selected options is the radio button and the click box. These are probably quite sufficient tools for signifying offer and acceptance, and making adjustments to accommodate new technology under old rules of contract more likely means changing ourselves, rather than changing the rules.

Click-wrap agreement
An Internet click box of "I agree," which constitutes valid acceptance of enumerated contractual responsibilities.

The Ontario Superior Court of Justice has recognized the **click-wrap agreement** — the click box of "I Agree" — as valid acceptance of contractual responsibilities. While scrolling is required to see the entire content of an agreement, the court has rendered this analogous to turning the pages of a paper contract.

COURT DECISION: Practical Application of the Law

Electronic Commerce—Clickwrap Agreements
***Rudder v. Microsoft Corp.*, 1999 CanLII 14923 (ONSC).**

Microsoft used a clickwrap agreement with a clause requiring all disputes to be heard before a court in Washington State, U.S.A. As this presented a hardship to clients intending to dispute their accounts, one such client, Rudder, attempted to challenge the jurisdiction clause.

THE COURT: This is a motion by the defendant Microsoft for a permanent stay of this intended class proceeding. The Microsoft Network ("MSN"), is an online service, providing, inter alia, information and services including Internet access to its members. The plaintiffs allege that Microsoft has charged members of MSN and taken payment from their credit cards in breach of contract and that Microsoft has failed to provide reasonable or accurate information concerning accounts.

The contract which the plaintiffs allege to have been breached is identified by MSN as a "Member Agreement." Potential members of MSN are required to electronically execute this agreement prior to receiving the services provided by the company. Each Member Agreement contains the following provision:

> 15.1 This Agreement is governed by the laws of the State of Washington, U.S.A., and you consent to the exclusive jurisdiction and venue of courts in King County, Washington, in all disputes arising out of or relating to your use of MSN or your MSN membership.

The defendant relies on this clause in support of its assertion that the intended class proceeding should be permanently stayed.

The plaintiffs contend, first, that regardless of the deference to be shown to forum selection clauses, no effect should be given to the particular clause at issue in this case because it does not represent the true agreement of the parties. It is the plaintiffs' submission that the form in which the Member Agreement is provided to potential members of MSN is such that it obscures the forum selection clause. Therefore, the plaintiffs argue, the clause should be treated as if it were the fine print in a contract which must be brought specifically to the attention of the party accepting the terms. Since there was no specific notice given, in the plaintiffs' view, the forum selection clause should be severed from the Agreement which they otherwise seek to enforce.

The argument advanced by the plaintiffs relies heavily on the alleged deficiencies in the technological aspects of electronic formats for presenting the terms of agreements. In other words, the plaintiffs contend that because only a portion of the Agreement was presented on the screen at one time, the terms of the Agreement which were not on the screen are essentially "fine print."

I disagree. The Member Agreement is provided to potential members of MSN in a computer readable form through either individual computer disks or via the Internet at the MSN website. In this case, the plaintiff Rudder, whose affidavit was filed on the motion, received a computer disk as part of a promotion by MSN. As part of the sign-up routine, potential members of MSN were required to acknowledge their acceptance of the terms of the Member Agreement by clicking on an "I Agree" button presented on the computer screen at the same time as the terms of the Member Agreement were displayed.

Rudder admitted in cross-examination on his affidavit that the entire agreement was readily viewable by using the scrolling function on the portion of the computer screen where the Membership Agreement was presented. Moreover, Rudder acknowledged that he "scanned" through part of the Agreement looking for "costs" that would be charged by MSN. He further admitted that once he had found the provisions relating to costs, he did not read the rest of the Agreement. It is plain and obvious that there is no factual foundation for the plaintiffs' assertion that any term of the Membership Agreement was analogous to "fine print" in a written contract. What is equally clear is that the plaintiffs seek to avoid the consequences of specific terms of their agreement while at the same time seeking to have others enforced. Neither the form of this contract nor its manner of presentation to potential members are so aberrant as to lead to such an anomalous result.

To give effect to the plaintiffs' argument would, rather than advancing the goal of "commercial certainty," to adopt the words of Huddart J.A. in *Sarabia*, move this type of electronic transaction into the realm of commercial absurdity. It would lead to chaos in the marketplace, render ineffectual electronic commerce and undermine the integrity of any agreement entered into through this medium. The defendant shall have the relief requested. The action brought by the plaintiffs in Ontario is permanently stayed.

To do its part in avoiding "commercial absurdity," any business controlling an e-commerce website should ensure that its click-wrap agreement includes a clause establishing the jurisdiction and court which is to hear claims of unsatisfied customers. The wisdom that suggests people should read contracts before signing them applies equally to electronic contracts.

Lapse of an Offer

Until an offer is accepted, no legal rights or obligations arise. And offers are not always accepted. Even in cases where the offeree may wish to accept, events may occur or conditions may change that will prevent the formation of the agreement. The death of either party, for example, will prevent the formation of the contract, because the personal representative of the deceased normally may not complete the formalities for offer and acceptance on behalf of the deceased. When an offeree dies before accepting an offer, the offer lapses, because the deceased's personal representative cannot accept an offer on behalf of a deceased person. By the same token, acceptance cannot be communicated to a deceased offeror, as the personal representative of the offeror would not be bound by the acceptance, except under special circumstances where the offeror has bound them to the offer. The same rule would hold true in the case of the bankruptcy of either of the parties, or if a party should be declared insane before acceptance is made.

Lapse

The termination of an unaccepted offer by the passage of time, a counteroffer, or the death of a party.

An offer will also **lapse** as a result of a direct or indirect response that does not accept the offer unconditionally and in accordance with its terms. If the offeree rejects the offer outright, it lapses and cannot be revived except by the offeror. Similarly, any change in the terms of the offer in a purported acceptance will cause the original offer to lapse, as the modified acceptance would constitute a counteroffer.

Offers may also lapse by the passage of time, or the occurrence of a specified event. Obviously, an offer that must be accepted within a specified period of time or by a stipulated date will lapse if acceptance is not made within the period of time or by the particular date. An offer may also lapse within a reasonable time if no time of acceptance has been specified. What constitutes a reasonable time, needless to say, will depend upon the circumstances of the transaction and its subject matter. An offer to sell a truckload of perishable goods would have a much shorter "reasonable" time for acceptance than an offer to sell a truckload of non-perishable goods that are not subject to price or market fluctuation.

As a general rule, when an offer is made by one person in the company of another and where no time limit for acceptance is expressed, the offer is presumed to lapse when the other party departs without accepting the offer, unless the circumstances surrounding the offer would indicate otherwise.

FRONT-PAGE LAW

Death of F1 deal leads to legal action
No franchise for Canada

Montreal-based Vector Motorsports Group Inc. has ended a frantic attempt to salvage a multi-million-dollar agreement that would have brought a Formula One franchise to Canada. Instead it has announced plans to take legal action against the Canadian investors who killed the deal.

Marc Bourdeau, Vector chairman and CEO, yesterday officially withdrew his company's offer to acquire 100% of the Formula One team operated by racing legend Alain Prost.

"I met with Alain several times over the weekend of the Canadian Grand Prix, and we very quickly came to an agreement on a transaction, which simply required a guarantee of funds," Mr. Bourdeau said, adding that Vector had assurances of funding by a group that claimed control of a US$100-million instrument. But "deliberately or accidentally, they did not deliver the financial instrument and we could not get them to do it," Mr. Bourdeau said.

With the deadline past, yesterday morning Mr. Bourdeau waited to see whether the Quebec government, which he says had offered a multi-million-dollar package of tax credits to support the deal, would save the sale.

But Vector said Quebec's offer was too low and had too many unacceptable conditions.

Mr. Bourdeau said he did not rule out a marketing partnership with the Prost team.

Former major-general Lew MacKenzie, a company director, said Vector had about $250-million in long-term sponsorships in place before running into the short-term financing problem. He said the five-year-old marketing company wanted the Prost team as a way to promote new light-weight technology it believes will revolutionize the automotive industry.

Why did Bourdeau "officially" withdraw his firm's offer? Why not just let it lapse?

Source: Excerpted from Thomas Watson, *National Post*, "Death of F1 deal leads to legal action," September 19, 2000, p. C3.

Revocation of an Offer

Revocation

The termination of an offer by notice communicated to the offeree before acceptance.

Revocation, as opposed to lapse, requires an act on the part of the offeror in order to be effective. Normally, the offeror must communicate the revocation to the offeree before the offer is accepted; otherwise, the notice of revocation will be ineffective. With an ordinary contract, as a general rule, an offeror may revoke the offer at any time prior to acceptance, even when the offeror has gratuitously agreed to keep the offer open for a specified period of time. If the offeree wishes to make certain that the offeror will not revoke the offer, the

Option

A separate promise to keep an offer open for a period of time.

method generally used is called an option. An **option** is a separate promise that obliges the offeror to keep the offer open for a specified period of time, either in return for some compensation, or because the promise is made in a formal document under seal. (The effect of the seal on a document will be examined in the next chapter, as will the effect of compensation paid to an offeror in return for the promise. However, for the present it is sufficient to note that either of these two things will have the effect of rendering the promise to keep the offer open for a specified period of time irrevocable.)

A second aspect of revocation of an offer is that it need not be communicated in any special way to be effective. The only requirement is that the notice of revocation be brought to the attention of the offeree before the offer is accepted. This does not mean, however, that the same rules apply to revocation as do to acceptance when some form of communication other than a direct form is used. Because the offeree must be aware of the revocation before it is effective, the courts have held that the posting of a letter revoking an offer previously made does not have the effect of revoking the offer. The notice of revocation is only effective when it is finally received by the offeree.[7]

The question of whether indirect notice of revocation will have the effect of revoking an offer is less clear.

Anderson Auto Sales offers to sell a car to Burton and promises to keep the offer open for three days. On the second day, Anderson Auto sells the car to Coulson. The sale of the car would clearly be evidence of Anderson Auto's intention to revoke the offer to sell to Burton. If a mutual friend of Burton and Coulson told Burton of the sale of the car to Coulson, would this indirect notice prevent Burton from accepting the offer?

This very question arose in an English case in which an offer had been made to sell certain property and the offeree was given a number of days to accept. Before the time had expired, the offeror sold the property to another party, and a person not acting under the direction of the offeror informed the offeree of the sale. The offeree then accepted the offer (within the time period for acceptance) and demanded conveyance of the property. The court held in this case that the offeree was informed of the sale by a reliable source, and this knowledge precluded acceptance of the offer by him.[8]

The judge in the case described the situation in the following manner:

> If a man makes an offer to sell a particular horse in his stable, and says, "I will give you until the day after tomorrow to accept the offer," and the next day goes and sells the horse to somebody else, and receives the purchase-money from him, can the person to whom the offer was originally made then come and say, "I accept," so as to make a binding contract, and so as to be entitled to recover damages for the non-deliver of the horse? If the rule of law is that a mere offer to sell property, which can be withdrawn at any time, and which is made dependent on the acceptance of the person to whom it is made, is a mere *nudum pactum*, how is it possible that the person to whom the offer has been made can by acceptance make a binding contract after he knows that the person who has made the offer has sold the property to some one else? It is admitted law that, if a man who makes an offer dies, the offer cannot be accepted after he is dead, and parting with the property has very much the same effect as the death of the owner, for it makes the performance of the offer impossible. I am clearly of the opinion that, just as when a man who has made an offer dies before it is accepted it is impossible that it can then be accepted, so when once the person to whom the offer was made knows that the property has been sold to some one else, it is too late for him to accept the offer, and on that ground I am clearly of opinion that there was no binding contract for the sale of this property ....[9]

7. *Byrne & Co. v. Leon Van Tienhoven & Co.* (1880), 5 C.P.D. 344; *Henthorn v. Fraser*, [1892] 2 Ch. 27.
8. *Dickinson v. Dodds* (1876), 2 Ch.D. 463.
9. ibid.

The essential point to note, in cases where notice of revocation is brought to the attention of the offeree by someone other than the offeror or the offeror's agent, is the reliability of the source. The offeror must, of course, prove that the offeree had notice of the revocation before the offer was accepted. The onus would be on the offeror to satisfy the court that the reliability of the source of the knowledge was such that a reasonable person would accept the information as definite evidence that the offeror had withdrawn the offer. Cases of indirect notice, consequently, turn very much on the reliability of the source when the notice comes from some source other than the offeror or the offeror's agent. The case cited represents only an example of how a court may deal with the problem of indirect notice, rather than a statement of the Common Law on this issue. Figure 7–3 summarizes offer and acceptance.

FIGURE 7–3 **Offer and Acceptance**

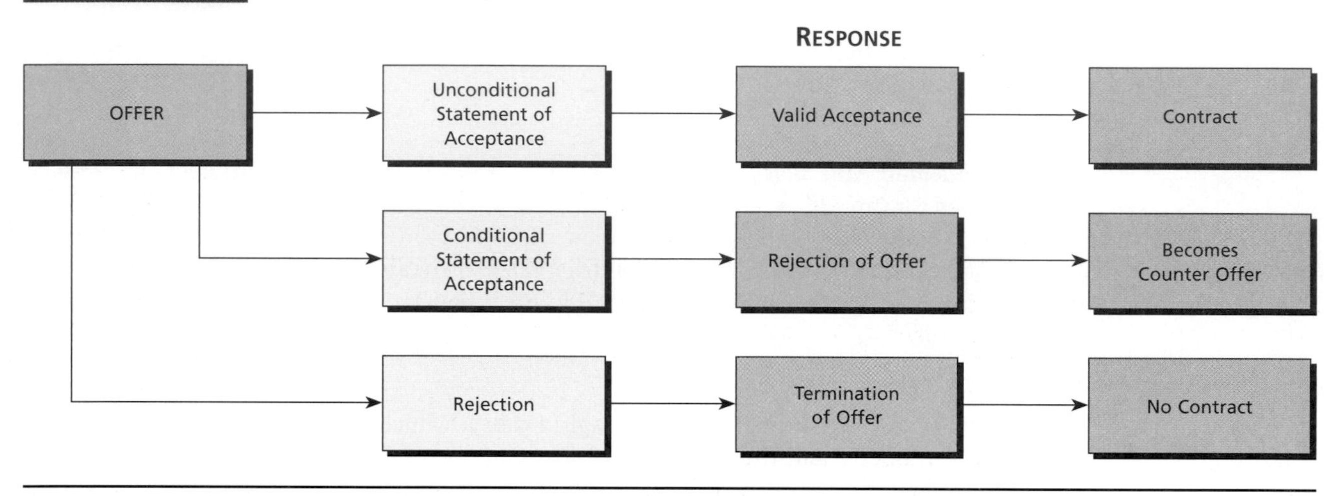

CLIENTS, SUPPLIERS *or* OPERATIONS *in* QUEBEC

When compared to the Common Law, there are number of differences (some slight, others considerable) in offer and acceptance in the *Civil Code of Quebec*, Articles 1385 to 1397. A sampling from among them are:

1387. A contract is formed when and where acceptance is received by the offeror, regardless of the method of communication used, and even though the parties have agreed to reserve agreement as to secondary terms.

1393. Acceptance which does not correspond substantially to the offer or which is received by the offeror after the offer has lapsed does not constitute acceptance. It may, however, constitute a new offer.

1394. Silence does not imply acceptance of an offer, subject only to the will of the parties, the law or special circumstances, such as usage or a prior business relationship.

S.Q., 1991, c. 64, Book Five, Obligations.

MANAGEMENT ALERT: BEST PRACTICE

Good document management and control are essential to preserving your rights and limiting your responsibilities. Court cases usually hinge on which party can show a sensible interpretation of a comprehensive offering of documents related to the transaction in issue. Make sure your requests for quotations, your offers, your acceptances are clear corporate communications, and insist on timely delivery of relevant documents from other parties. Do not accept promises to "paper up the transaction later" — it often never gets done, and more often memories of "who does what" become very short and hazy.

CHECKLIST FOR LEGAL RELATIONSHIP

1. Intention to create legal relationship and *consensus ad idem*?
2. Offer?
3. Acceptance?
4. Consideration?
5. Capacity?
6. Legality?

If the answer is "yes" to all six, a legal relationship in contract will probably exist. We will investigate consideration, capacity and legality in subsequent chapters.

SUMMARY

The law of contract, unlike many other laws, does not set out the rights and duties of the parties. Instead, it permits the parties to establish their own rights and duties by following a series of principles and rules for the formation of a contract. A contract is essentially an agreement that is enforceable at law. As such, it must contain all of the elements of a true agreement. It must, first of all, be a "meeting of the minds of the parties," and an intention on the part of both parties to create a legal relationship must be present. There must also be an offer of a promise by one party subject to some condition, and the offer must be properly accepted by the other party. If the offer is not properly accepted, no contract will exist, and the offer may be terminated or replaced by a counteroffer. If nothing is done to accept the offer it will lapse, or the offeror may withdraw it. In either case, if validly done, no contract will exist. If an unconditional acceptance has been properly made in accordance with the terms of the offer, and before the offer lapses or is withdrawn, then the parties will be bound by the agreement they have made, and their promises will be enforced by the courts. Offer, acceptance, and an intention to create a legal relationship represent three very important elements of a contract. There are other elements that the parties must establish to have a valid contract. These are the subject-matter of the following chapters.

KEY TERMS

Intention to be bound, 115
Consensus ad idem, 116
Offer, 119
Acceptance, 120
Click-wrap agreement, 126
Lapse, 128
Revocation, 128
Option, 129

REVIEW QUESTIONS

1. Describe the burden placed upon the offeror by the courts when the offeror has alleged that indirect notice of revocation was received by the offeree.
2. Why is an intention to create a legal relationship an important element of a valid contract?
3. Explain why communication of an offer must take place before the offer may be accepted. Why is this important?
4. Describe the "rules" for acceptance, and explain why such rules are necessary.
5. Under what circumstances would "silence" be acceptance?
6. Explain the term "counteroffer," and describe how it might arise.
7. Is an advertisement containing an offer of a reward for a lost pet a valid offer? How does it differ from an advertisement of goods for sale?

8. Describe four instances where an offer might lapse before acceptance.
9. Explain the rationale behind the rule that states that an offer by mail invites acceptance by mail, and that the acceptance is complete when a properly addressed letter of acceptance is dropped in a mailbox.
10. What condition must be met before revocation of an offer is effective?
11. "The parties to a contract, if they do things right, create their own rights and duties." Is this a valid observation?
12. How does acceptance of a unilateral agreement differ from that of an ordinary agreement?
13. Written forms often contain "YES" or "NO" check boxes. Why should online forms go one beyond "OK" and "CANCEL" boxes to establish "CONFIRM" checks?
14. What is a "click-wrap agreement"?

MINI-CASE PROBLEMS

1. Potter, the owner of Industrial Castings Ltd., placed a for-sale ad in a trade paper. The ad read: "High-Speed Caster 750 for Sale, $25,000." Yates and Zabel both arrive at the shop and wish to acquire the machine. Yates is the first to meet Potter and says: "I accept your offer. Here is my $25,000 cheque." Zabel then states: "Here is a $26,000 cheque. I will take your machine." Must Potter sell the machine to Yates?
2. D offered to sell his automobile to E for $5,000 cash. E responded to D's offer by saying: "I will buy your automobile for $5,000. I will pay you $2,500 now, and $2,500 in one week's time." Have D and E formed a binding agreement?
3. The president of A Co. wrote a letter to B Co. offering to sell B Co. a large quantity of steel at a specific price. B Co. did not respond to the letter, but A Co. sent a "sample load" of one tonne that B Co. used in its manufacturing process. A month later, A Co. sent the quantity of steel specified in its letter along with an invoice at the price specified. Is B Co. bound to accept and pay for the steel?
4. Joe met Larry on the main street of their town. Both owned used car lots in the town, and were casual acquaintances. Parked at the curb beside them was a vintage muscle-car Joe had long admired. Pointing to it, Joe said "I love that beauty. I'd pay $40,000 in a heartbeat for that one." "Yeah?" said Larry, "Well, I accept your offer. That's a $5,000 profit for me. I bought it this morning for $35,000. It's yours now, buddy." Joe protested that he had not been serious about the car. Larry intended to hold him to his word. Advise the parties.
5. Alexandre attended an auction sale of the property of a bankrupt restaurant. Among the items for auction was a stainless steel food preparation table. The auctioneer commenced by asking for bids: "Do I hear seven hundred dollars?" Alexandre raised his hand, and the auctioneer cried "SOLD." Is this a binding contract?
6. Ann's electric clothes dryer was unreliable, and she sought a replacement. She noticed an advertisement in the newspaper for a used dryer, in like-new condition, at a price of $200. She visited the residential address given, and was shown the dryer, sitting against the wall inside a garage. The elderly lady who owned it had died, and her son was selling her possessions. Ann was delighted, paid $200 and took the dryer home. When she tried to install it, she discovered it was designed to run on natural gas rather than electricity, a fact that had not even occurred to her to check. Does a contract exist in this situation?
7. Alphatech, an Ontario firm, employs Jane. She sends an offer by e-mail on behalf of her company to Fayeed, to purchase equipment. Fayeed replies by e-mail three days later, accepting the offer, but in the interval Jane leaves her job at Alphatech. Her e-mail account is closed, and Fayeed's reply is bounced back as undeliverable. After a cumbersome effort of some days to find out who replaced Jane, Fayeed is informed that the offer has lapsed as Jane's replacement found an alternate supplier for the equipment. Was there an enforceable contract between Fayeed and Alphatech?

CASE PROBLEMS FOR DISCUSSION

CASE 1

Ming's home was located on a rural residential lot overlooking a pretty ravine, and he thought that he might enjoy a small terrace behind his home for a patio. He placed a sign at the road in front of his home with the words "Clean Fill Wanted" and his telephone number. While Ming was at work one morning, a caller rang about the sign, and Ming's teenage son answered. He told the caller that his father wanted fill in the ravine, at which the caller was delighted. Dump trucks began arriving non-stop through the afternoon, bearing the logo of Rock-kut Highway Excavators Ltd. When Ming returned from work, he was aghast at the sight before him.

Discuss the issues raised in this situation.

CASE 2

McKay operated a large farm on which he grew a variety of vegetables for commercial canners. He also grew a smaller quantity for sale to local retailers and wholesalers as fresh produce. On August 5th, Susan Daigle of Daigle Wholesale Foods Ltd. approached McKay and offered to purchase 5,000 5-kilogram bags of carrots from him at a fixed price per bag. McKay stated that the price was acceptable to him, but he was uncertain as to whether his crop would be sufficient to make up the 5,000 bags. He told Daigle he could definitely supply 4,000 bags, and that he would be in a position to tell Daigle Wholesale Foods Ltd. by the next week if the additional 1,000 bags would be available. Daigle nodded approval and left.

A few days later, McKay discovered that crop failures in other parts of the province had pushed carrot prices substantially above the price offered by Daigle Wholesale Foods Ltd. McKay's crop, however, was abundant, and he discovered that he had 6,000 bags when the crop was harvested.

At the end of the week, Susan Daigle called to determine if McKay could supply her with 5,000 bags, or only 4,000. McKay refused to supply Daigle Wholesale Foods Ltd. with any carrots, and informed her that it was his intention to sell the crop elsewhere.

Discuss the negotiations between the parties, and determine the rights (if any) and liabilities (if any) of the parties. Assume that Daigle Wholesale Foods Ltd. brought an action against McKay.

Discuss the nature of the action and render, with reasons, a decision.

CASE 3

Armstrong Aggregates Co. wrote a letter to Bishop on May 2nd offering to sell him 200 tonnes of scrap mica at $180 per tonne. Bishop received the letter on May 3rd. A few weeks later, Bishop checked the price of mica, and discovered that the market price had risen to $185 per tonne. On May 22nd, Bishop wrote Armstrong Aggregates Co., accepting the offer. Armstrong did not receive Bishop's letter until May 30th. Armstrong refused to sell the mica to Bishop at $180 per tonne, but expressed a willingness to sell at the current market price of $187 per tonne.

Bishop instituted legal proceedings against Armstrong for breach of the contract that he alleged existed between them.

Discuss the rights (if any) and the liabilities (if any) of the parties, and render a decision.

CASE 4

The Clear Fruit Juice Company purchased a quantity of a special variety of apple that Ely grew in his orchards. The apple was of a type that kept well in winter storage, and the company would market a portion of the apples purchased as fresh fruit and process the remainder into apple juice that it sold under a private brand name. The company purchase represented approximately 50 percent of Ely's total harvest each year.

Initially, the Clear Fruit Juice Company and Ely entered into a formal, written purchase agreement for each year's crop. However, over time the agreement became less formal. Eventually it consisted at first of a telephone order for the "usual supply," and later to an arrangement whereby Ely would deliver the normal quantity to the company each year, and in due course would receive payment at the going market price for the crop. This latter arrangement carried on for a period of about ten years.

In the last year, the president of the Clear Fruit Juice Company fell seriously ill and retired. He had been responsible for the original contracts with Ely, and had made the later informal arrangements for the supply of apples. For many years, the two men had been good friends. The new president of the company moved to the area from a subsidiary corporation and was not aware of the arrangement between Ely and the company. He decided that for the current year he would purchase the company's apple requirements from another orchard.

Over the course of the summer, Ely heard rumours that an orchard some kilometres distant had acquired a contract to supply his variety of apple to the company, but he did nothing to investigate the matter further. In the fall of that year, he delivered his usual supply in large pallet boxes to the company and placed them in the storage yard. No employees were in the yard at the time, but Ely did not find the fact unusual, as that was typically the case when he made his deliveries in the past. He was not concerned about identification of the crop as each pallet box bore his name and address as well as the variety and quantity.

The yard foreman noticed the apples in the supply yard some time later on the day of delivery, and informed the plant manager. The plant manager did nothing about the apples until the next day, when he informed the company president. The company president decided to write a letter to Ely requesting him to take back his apples, but it was Friday, so he left the letter until Monday of the next week. Ely received the letter on the Wednesday, some six days after delivery of the fruit to the company.

During the six-day period the apples had remained in the hot sun and had deteriorated from the exposure. Ely refused to take back the apples, and the company refused to pay for them.

Advise the parties of their rights in this case, and determine the probable outcome if Ely should bring an action against the company for the value of the goods.

CASE 5

The Garden Book Company advertised in a local newspaper the publication of its latest book on flower growing. The advertisement indicated that orders would be taken by mail at a price of $30 per copy, but the book would be available at all book stores as well.

In response to the advertisement, Laurel Bush sent her cheque for $30 to the Garden Book Company, and requested in her letter that the company send her a copy of the book by return mail. Laurel mailed her letter on February 19th. On February 21st, Laurel noticed the same book on sale at a local book store at a price of $9.95. She purchased a copy, then went home and immediately wrote a letter to the Garden Book Company revoking her offer to purchase the book, and requested a return of her $30 cheque. Laurel mailed her letter at 4:20 p.m. on February 21st.

The Garden Book Company received Laurel's letter of February 19th on February 22nd, and mailed her a letter the same day that acknowledged her order and advised her that the book would be sent to her by courier within a few days. The Garden Book Company received Laurel's letter of February 21st on February 23rd. The company ignored her second letter, and delivered the book to the courier on February 25th.

Discuss the rights of the parties in this case. Identify and explain the legal principles and rules applicable, and indicate in your answer how the case would be decided if the matter came before the court.

CASE 6

The Golden Lake Mining Company decided to dispose of two of its undeveloped mining properties (A and B) and authorized the company president to find a buyer. On September 10th, the president wrote a letter to the East Country Exploration Corporation offering to sell the properties "en bloc" for $500,000.

On September 16th, East Country Exploration Corporation replied by mail to the letter in which it expressed an interest in the purchase of property A at a price of $275,000 if Golden Lake Mining was prepared to sell the properties on an individual basis.

On September 22nd, the president of Golden Lake Mining replied by fax that he would prefer to sell both properties, but if he could not find a buyer for the two parcels within the next few weeks, the Company might consider selling the properties on an individual basis.

On September 30th, the East Country Exploration Corporation made an inquiry by fax to determine if Golden Lake Mining had decided to sell the properties on an individual basis. The president of Golden Lake responded with a fax that stated: "Still looking for a buyer for both properties."

Following this response, East Country Exploration decided to examine property "B," and sent out its two geologists to do a brief exploratory evaluation. On October 6th they reported back to say that they had examined property B and found some surface evidence of what might be a potentially economic ore body worth between $5 million and $15 million. East Country Exploration then prepared a letter accepting the offer of Golden Lake Mining to sell properties A and B for $500,000. The letter was mailed on October 9th.

On October 10th the president of Golden Lake Mining Company found a buyer for the B property at a price of $300,000, and signed a sale agreement the same day. He then wrote a letter to East Country Exploration in which he accepted their offer to buy the A property for $275,000. The president of Golden Lake Mining received the October 9th letter of East Country Exploration on October 11th.

Discuss the issues raised in this case, and indicate in your answer how the case might be decided if it was brought before the court.

CASE 7

Tamiko lived in Calgary, Alberta, and owned a cottage on Vancouver Island. Percy, who lived in Victoria, British Columbia, was interested in purchasing the cottage owned by Tamiko. On September 10th, he wrote a letter to Tamiko offering to purchase the cottage and lot for $275,000. Tamiko received the letter on September 15th, and sent Percy a fax offering to sell the cottage to him for $298,000.

Percy did not respond to the fax immediately, but on a business trip to Calgary on September 22nd, he spoke to Tamiko about the cottage in an effort to determine if Tamiko might be willing to reduce the price. Tamiko replied that the price was "firm" at $298,000.

When Percy returned to Victoria, he sent a letter to Tamiko accepting her offer to sell the cottage at $298,000. The letter was posted at 11:40 a.m. on September 23rd, but, through a delay in the mail, was not delivered to Tamiko in Calgary until 4:20 p.m. on September 28th.

In the meantime, when Tamiko had not heard from Percy by September 26th she offered to sell the cottage to Johnson, who had expressed an interest in purchasing the cottage some time before. Johnson accepted the offer, and the two parties executed a written purchase agreement in Johnson's office on the morning of September 27th.

Identify the various rights and liabilities that developed from the negotiations set out above.

CASE 8

Tom was tired of running his business and was looking for a buyer so he could retire before the winter. Bill wanted to buy the business, but needed another month to raise the necessary finances. Tom said to Bill: "Look, for $5,000, I will give you an option to purchase in 30 days, but if you are not ready to buy at that time for fifty thousand, I am going to sell it to Dan." Bill paid Tom the $5,000. Thirty days later, Bill appeared at Tom's door with $45,000. Tom said: "Sorry, I am going to sell it to Dan. I said the price was $50,000." Bill was confused, and sought legal advice.

What advice would the lawyer provide to Bill?

CASE 9

Janine owned a shop selling fine bone china. After a customer dropped an expensive bowl and shattered it, Janine made a new policy by posting a sign reading "Lovely to look at, lovely to hold, if you break it, consider it sold." In time, another customer broke a vase she was examining but refused to pay for it.

Discuss the legal position of both Janine and her customer.

The Online Learning Centre offers more ways to check what you have learned so far. Find quizzes, Weblinks, and many other resources at www.mcgrawhill.ca/olc/willes.

www.mcgrawhill.ca/olc/willes

CHAPTER 8

The Requirement of Consideration

Chapter Objectives

After study of this chapter, students should be able to:

- Illustrate how contracts are intended to exchange a mutual benefit.
- Explain how "consideration" represents this mutuality.
- Identify exceptions to the rule requiring "consideration."
- Explain how estoppel prevents injury from reliance on a gratuitous promise.

YOUR BUSINESS AT RISK

To earn a profit, a business cannot operate as a charity. It expects something in return for its actions. As simple as that sounds, there are times when one-way relationships exist, and these create problems in enforcing contractual rights and obligations. You must understand the rules of "consideration" to avoid inadvertently creating these difficult business law traps.

CONSIDERATION

Nature of Consideration

The bargain theory of contract suggests that a contract is essentially an agreement between parties wherein each gets something in return for his or her promise. If this is the case, then every promise by an offeror to do something must be conditional. The promise must include a provision that the offeree, by conveying acceptance, will promise something to the offeror. The "something" that the promisor receives in return for the promisor's promise is called **consideration** — an essential element of every simple contract.

Consideration

Something that has value in the eyes of the law, and which a promisor receives in return for a promise.

Consideration can take many forms. It may be a payment of money, the performance of a particular service, a promise not to do something by a promisor, the relinquishment of a right, the delivery of property, or a myriad of other things, including a promise in return for a promise. In every case, however, the consideration must be something done with respect to the promise offered by the promisor. Unless a promisor gets something in return for his or her promise, the promise is merely gratuitous. Generally, consideration for a promise must exist for the contract to be legally binding.

There are certain exceptions to this rule, but they are few in number, most dating back before the concept of a modern contract was developed. It has long been a rule of English law that a gratuitous offer of a service, if accepted, must be performed with care and skill; otherwise, the promisor will be liable for any loss suffered as a result of careless performance or negligence. This liability, however, would not flow from any breach of contract but, rather, from the tort committed.

A second major exception, dating back to the days of the law merchant, remains today as a part of our legislation dealing with negotiable instruments. A person may be liable on a promissory note, or other negotiable instrument, to a subsequent endorser, even though no consideration exists between them. Under the same body of law, a party who endorses a bill of exchange (e.g., a cheque) to enable another to negotiate it may be held liable on the bill, even though no consideration was given as a result of the endorsement.

A modern-day exception to the rule concerns the promise of a donation to a charitable organization. If the rule relating to consideration is strictly applied to a promise of a donation to charity, the agreement would not be enforceable, as the promise would be gratuitous. This is because the donor would receive nothing in return for the promise made. While this is the usual situation when a donation is made to a charity, the courts have made exceptions in some cases. If the charity can show that it undertook a specific project on the strength of the donor's pledge, then the promise may be enforced. This, of course, would only be applicable if the donor's promised donation was such that it represented a substantial part of the funds necessary for the project.

A Manitoba YMCA branch, while soliciting donations for the construction of a new building, obtained a substantial pledge from one person. That pledge prompted the YMCA to commit itself to build the new building. After construction was underway, the pledgor refused to honour his promise of payment, and the YMCA sued for the amount. The pledgor's defence was that the promise was unenforceable due to a lack of consideration, but the court ruled otherwise. The charity satisfied the court that it would not have begun the project without the particular pledge, and had incurred liability on the strength of the promise of funds. The court decided that this was sufficient consideration for the promised donation.[1]

If the promised donation is not significant, the courts will not enforce the gratuitous promise. In the case of *Governors of Dalhousie College v. Boutilier*,[2] the university solicited funds for the maintenance and construction of new facilities. Mr. Boutilier promised a donation, but died before the payment was made. When his estate refused to pay the pledge, the university sued to obtain the amount promised. The amount pledged, however, was small in comparison with the total funds contributed. As a result, the court refused to enforce the pledge.

Even where the dollar value of the pledge has been significant, there is a Canadian reluctance to enforce such promises, taking a more English than American approach in this regard.

CASE IN POINT

In 1998 a wealthy widow made a pledge to a hospital of $1 million over five years. A little over a year later she paid a $200,000 instalment, then died. She and her late husband had donated millions in the past to the hospital. Rather than providing another $800,000 to honour the pledge, her will gave a significant share of her estate to the hospital. Receiving this, the hospital sued her estate in addition for the balance of the pledge. The court determined that the hospital's idea to name a wing after her was a matter of indifference to the woman, and was not consideration for the pledge. The hospital had not specifically relied on her pledge, thus there was no estoppel, and its fundraising plans had merely faced a setback. For lack of consideration or estoppel, the pledge was not enforced.

Brantford General Hospital Foundation v. Marquis Estate (2003), 67 O.R. (3d) 432 (S.C.J.).

The question of estoppel is discussed later in this chapter. In this interesting case, one wonders about the range of other factors which may have weighed heavily on the mind of the court, but did not form part of the written judgment.

1. *Sargent v. Nicholson* (1915), 25 D.L.R. 638.
2. [1934] 3 D.L.R. 593.

Seal as Consideration

Seal

A formal mode of expressing the intention to be bound by a written promise or agreement. This expression usually takes the form of signing or affixing a wax or gummed paper wafer beside the signature, or making an engraved impression on the document itself.

A final, major exception to the requirement for consideration in a contract is a device that was used by the courts to enforce promises long before modern contract law emerged. This particular device is the use of a **seal** on a written contract. In the past (as early as the thirteenth century), a written agreement would be enforced by the court if the promisor had placed his seal on the document. The original purpose of the seal was to prove the authenticity of the agreement, leaving the promisor free to prove the document fraudulent if the seal was not present. As time passed, the document became a formal agreement that the courts would enforce if the seal was present. The general thinking of the judges of the day was that any person who affixed a seal to a document containing a promise to do something had given the matter considerable thought, and the act of affixing the seal symbolized the intention to be bound by the agreement. In a sense, the promisor, by affixing the seal to the written promise, was formalizing his intention to be bound. This particular method of establishing an agreement, and this ritual, distinguished the formal contract from the ordinary or simple contract that may or may not be in writing.

Originally, the person entering into the agreement would affix a seal (a wax impression of his or her family crest or coat of arms) to the document, then place a finger on the impression and express the intention to be bound by the promise. This formal act gave the document its validity. Over time, the act became less formal. Today, most formal legal documents that require a seal either have the seal printed on the form, or have a small gummed wafer attached to the form by the party who prepares the document before it is signed by the promisor. The binding effect of a formal contract under seal, however, persists today. The courts will not normally look behind a contract under seal to determine if consideration exists, because the agreement derives its validity from its form (i.e., the signature plus the seal).

In spite of its ancient roots, the contract under seal is a useful form of contract today. For example, when parties wish to enforce a gratuitous promise, the expression of the promise in writing with the signature of the promisor and a seal affixed is the usual method used.

Many formal agreements still require a special form and execution under seal to be valid. For example, in some provinces, where the Registry System applies to land transfers, the various Short Forms of Conveyances Acts require a conveyance of land to be in accordance with a particular form, as well as signed, sealed, and delivered, to effect the transfer of the property interest to the grantee. A power of attorney in some provinces must also be executed under seal in order to authorize an attorney to deal with a grantor's land.

Some legal entities execute formal documents by way of a seal. A corporation, for example, can only act through its agents to carry out its objects. For many kinds of simple contracts it may be bound by its properly authorized agent without the use of the corporate seal. To be bound by a formal contract, in some provinces, those corporations that have corporate seals must have the corporate seal affixed by the agent. It should be noted, however, that not all corporations are required to have seals. Under some business corporations' statutes (such as the *Canada Business Corporations Act*), formal documents need only be signed by the proper officers of the corporation. The general trend has been to eliminate the need for business corporations to use seals to execute documents such as contracts.

CASE IN POINT

A corporation was given certain lands by its true owners to hold on their behalf. The corporation mortgaged the land, and the proper signing officers of the corporation affixed the corporate seal to the mortgage. Later, the corporation defaulted on the mortgage, and the mortgage company sued the owners for payment when it discovered that they were the true owners of the property. The true owners claimed they were not party to the sealed mortgage and not liable on the mortgage. The Supreme Court of Canada, on appeal, stated that only the party who seals a document has obligations under it, and so in this case only the corporation was liable on the mortgage.

Friedmann Equity Developments Inc. v. Final Note Ltd., [2000] 1 S.C.R. 842.

A QUESTION OF ETHICS

What is the use of a seal today? No one wears a signet ring anymore, and almost everyone can sign his or her name. Should we refuse to block rights and responsibilities on the basis of an ancient practice that most people are unlikely to understand? What is your view?

Tenders

The tender in contract law, as it relates to the formation of a legal relationship, differs from the ordinary offer. The tender process usually involves the advertisement of the particular needs of the firm to potential suppliers of the goods or services, either by way of newspapers or by direct mail contact. This step in the process is known as calling for tenders, and has no binding effect on the firm that makes the call. It is merely an offer to negotiate a contract. In most cases it represents an invitation to persons or business firms to submit offers that the firm calling for the tenders may, at its option, accept or reject. The firm making the call is not bound to accept the lowest offer nor, for that matter, any of the offers.

Tenders are frequently used by business firms, government organizations, and others that wish to establish a contractual relationship for the supply of goods or services or the construction of buildings, machinery, or equipment. Municipalities commonly use the tender method in the acquisition of supplies or services as a means of fairly opening competition to all firms in the municipality (and elsewhere). The tendering process generally uses the seal to render an offer irrevocable, and often uses the payment of a money deposit as a special type of consideration.

As a general rule, unless provided to the contrary in the call for tenders, an offer made in response to the call may be revoked at any time before acceptance. To avoid this, the call for tenders frequently requires offerors to submit their offers as irrevocable offers under seal. In this manner the offer may not be revoked, and will stand until such time as it is either accepted or it expires. Businesses and organizations calling for tenders may also require the offerors to provide a money deposit as well to ensure that the successful offeror will execute the subsequent contract that is usually required to formalize the agreement between the parties. When a deposit has been submitted with the tender under seal, a failure or refusal on the part of the successful bidder to enter into the formal contract and perform it according to its terms would result in forfeiture of the deposit, as well as entitle the party who made the call to take legal action.

A government call for tenders was made and the defendant submitted a tender along with the required deposit. Later, the defendant was notified that it was the successful bidder. The defendant, however, discovered an error in the tender that would result in a loss for the company if the contract was performed. The defendant then refused to enter into a formal contract to perform the work. In the litigation that followed, the court held that the call for tenders stipulated that the deposit would be forfeited if the successful bidder refused to execute the formal contract. The defendant had agreed to these terms, and, as a result, the defendant was not entitled to a return of the deposit.[3]

Adequacy of Consideration

In general, the courts are not concerned about the adequacy of consideration because they are reluctant to become involved as arbiters of the price or value that a person receives for

3. *The Queen in Right of Ontario v. Ron Engineering and Construction Eastern Ltd.* (1981), 119 D.L.R. (3d) 267 (S.C.C.).

COURT DECISION: Practical Application of the Law

Contract—Tender in Error—Right to Withdraw—Reliance on Tender

***Gloge Heating and Plumbing Ltd. v. Northern Construction Co. Ltd.*, 1986 ABCA 3 (CanLII).**

Gloge submitted a tender to do certain mechanical work for Northern Construction. Northern Construction entered into a construction contract to do work on the strength of Gloge's bid. Gloge later discovered that the bid was too low and refused to do the work. Northern Construction was obliged to have the work done by another company but the cost was much higher. Northern Construction then brought an action against Gloge for the amount of its loss, and was successful. Gloge appealed.

THE APPEAL COURT: ... The owner [of the construction project] duly accepted Northern's tender and awarded it the work. Gloge refused to perform the mechanical subcontract, so that Northern was required to make alternative arrangements by employing a subsidiary at an increase of $341,299 in the mechanical subcontract price. Was Gloge's tender revocable after close of tender? Gloge knew that Northern would select a mechanical tender and rely on it, and Gloge also knew that the tenders of the general contractors to the owner would be irrevocable for the time set out in the contract documents. Perhaps these facts of themselves might justify holding the Gloge tender to be irrevocable. But, in addition, the trial judge accepted certain expert evidence given at the trial by two witnesses who had been in the construction industry for many years. That evidence demonstrated that it was normal and standard practice for general contractors to accept last minute telephone tenders from subcontractors, and that it was understood and accepted by those in the industry that while such tenders could be withdrawn, that such tenders must remain irrevocable for the same term that the general contractors' tenders to the owner are irrevocable ....

Applying the industry practice to this case, Gloge could not withdraw its tender to Northern after tenders had closed, and its tender was irrevocable for the same period as Northern's. Accordingly, Gloge was obliged to perform the work when Northern awarded it the subcontract, and is liable for failing to do so. We observe that Gloge's conduct after its tender error was discovered is consistent with the industry practice already described. Gloge did not take the position with Northern that it had the right to withdraw its bid. Instead, it sought Northern's assistance in an attempt to persuade the owner to consent to the substitution of a higher mechanical tender to replace Gloge's. In the result, the judgement at trial is affirmed. The appeal is dismissed with costs to be calculated on the same basis as ordered by the trial judge.

The lesson of the case is clear: tender only for things you can perform, ensure the accuracy of your tender, and be prepared to execute it.

a promise. Apart from the requirement that the consideration be legal, their main concern is with the presence or absence of consideration, rather than with whether the promisor received proper compensation for his or her promise. In some cases, however, the courts will look more closely at the adequacy of the consideration. If the promisor can satisfy the court that the promise was made under unusual circumstances (such as when an error occurred that rendered the consideration totally inadequate in relation to the promise made), the courts may intervene.

Able writes a letter to Baker offering to sell his machine shop equipment to Baker for $35,000. Baker refuses the offer, but makes a counteroffer to Able to purchase the equipment for $30,000. Able sends an e-mail in return, rejecting Baker's offer to purchase and offering to sell his equipment for $32,000. In sending the e-mail, Able mistakenly types in a price of $3.20, instead of $32,000. If Baker should "snap up" the offer, he could not enforce Able's promise to sell, because the courts would reject his claim on the basis of the obvious error in the offer. Able, after offering his equipment for sale for $35,000 and rejecting an offer for $30,000, would not then offer to sell the equipment for a nominal consideration of $3.20.

If the error, however, was in Able's original letter, in which he intended to sell the equipment for $35,000, but inadvertently offered it to Baker for $30,000 and Baker accepted the offer, he would probably be bound by the contract. In this case, the courts would have no way of determining Able's intention at the time the offer was made, and would not inquire into the adequacy of the consideration.

The Click as Consideration

While historically the Common Law recognizes "even a peppercorn" as being sufficient valuable consideration to bind a contract, and the seal has been the substitute, we are now faced with the question of whether a mouse click is an acceptable equivalent. Indeed, some Web sites attract hits by offering a benefit to consumers (a prize, discounted merchandise, etc.) in return for the willingness of the browsing party to click through the site and at least be exposed to banner advertising. The sale of these banners to advertisers is the prime source of the site's revenue, balanced against the cost of prizes or discounts given to consumers.

It seems sensible that a click should at least be equivalent to the application of a seal, but it hardly retains the gravity of a deliberate act that a wax seal and signet ring once represented. Who can remember how many times they clicked on a computer screen yesterday, or in the past week? The click will become the new seal when it is tested in court by the first party who attempts to enforce upon a promised, but undelivered, Internet-based benefit.

Past Consideration

To be valid, consideration must also be something of value that the person receives in return for the promise made. It cannot be something that the person has received before the promise is made, nor can it be something that a person is already entitled to receive by law, or under another enforceable agreement. In the first case, if a person has already received the benefit for which the promise is offered, nothing is received in return for the promise. The consideration is essentially past consideration, which is no consideration at all, and the promise is gratuitous. The consideration offered must be something that the promisee will give, pay, do, or provide, either at the instant the promise is made (present consideration), or at a later date (future consideration).

Aird looks after Brett's shop for a weekend, and, on Brett's return, Brett promises to give Aird $250 for his kind act. In this case, Aird's act of looking after Brett's shop was gratuitous, and past consideration for Brett's later promise of payment. If Brett fails to pay the $250, no contract exists for Aird to enforce because no consideration can be shown except the past gratuitous act.

However, if Brett requested Aird to look after his shop while he was on vacation, and promised Aird $250 if he did so, when Brett returned he would be obliged to pay the $250, because the consideration would be future consideration, and the contract enforceable.

In the second case, the consideration that the promisee agrees to provide in return for the promise must not be something that the promisee is already bound to do at law or under another agreement. In both instances it would not constitute a benefit that the promisor would receive in return for the promise, and it would not constitute valuable consideration. If a person already has a duty to do some act, provide some service, or pay something to the promisor, then the promisor receives nothing from the promisee in return for the promise, other than that which he or she is already entitled to receive. There is no consideration given in return for the promise; again, the promise would be gratuitous.

Smith enters into a contract with Jones to construct a house for her on her property for $90,000. Smith underestimates the cost of constructing the house and, when the house is partly finished, he refuses to proceed with the construction unless Jones agrees to pay him an additional $10,000. If Jones agrees to pay the additional amount, and Smith completes the construction, Jones is not bound by the promise if she should later decide to withhold the $10,000. Smith is already under a duty to construct the house for Jones, and Jones receives nothing in return for her promise to pay the additional amount of money to Smith.[4] Her promise is gratuitous and unenforceable by Smith.

FRONT-PAGE LAW | **Details of the deal**

Some details in the six-year tentative deal announced Wednesday by the NHL and NHL Players' Association:

- A 24 percent salary rollback on all existing contracts.
- The upper limit on the salary cap for 2005-06 will be $39 million U.S.
- Players salaries cannot — on a league-wide basis — take up more than 54 percent of revenues; in ensuing years, the cap levels will be decided by the previous year's revenues.
- No player can earn more than 20 per cent of the team cap, which for 2005-06 means no player can earn more than $7.8 million.
- Teams will not be allowed to re-structure existing player contracts in an attempt to fit a big salary under the cap.

Why is this not an example of past consideration?

Source: Excerpted from *Canadian Press,* "Details of the deal," July 14, 2005.

Legality of Consideration

Quite apart from the aspect of lack of consideration, the enforcement of a contract of the kind illustrated in the Smith/Jones example would be contrary to public policy. If this type of contract were enforceable, it would open the door to extortion — an unethical contractor could threaten to cease operations at a critical stage in the construction, unless additional funds were made available to complete the remainder of the contract.

As noted earlier, the consideration in a contract must be *legal.* If Able should promise Baker $1,000 on the provision that Baker sets fire to the lumberyard of Charlie, and later Able fails to pay Baker the $1,000 when the arson is committed, the courts would not enforce Able's promise. Public policy dictates that the contract must be lawful in the sense that the promises do not violate any law or public policy. For this reason, an ordinary business contract containing a clause requiring the buyer to resell goods at a fixed or minimum price would be unlawful under the *Competition Act.* It would be illegal, as well as unenforceable.

QUANTUM MERUIT

Occasionally a person may request goods or services of another person, and the goods will be provided or the services rendered without mention of the price. In effect, no mention of consideration is made. No remedy was available for the unfortunate seller or

4. The courts in the United States have, on occasion, taken a slightly different approach in cases of this type. Because the contractor is free to abandon the contract at any time and compensate the owner for any damages that he or she might suffer, some American courts have held that the promise to pay an additional amount constitutes part of a new contract between the parties. This is because the original contract is executory and there is consideration on the part of both parties to terminate it and replace it with a new agreement. See, for example, *Pittsburgh Testing Laboratory v. Farnsworth & Chambers Co. Inc.* (1958), 251 F. (2d) 77. See also *Owens v. City of Bartlett* (1974), 528 P. (2d) 1235, for a discussion of the exception. Most American courts, however, would probably follow the consideration rule.

tradesperson in this kind of situation until the seventeenth century. Prior to that date, an unpaid seller could not recover for the goods by way of an action of debt because the sum certain had not been agreed to for the debt. Nor could the tradesperson recover by way of assumpsit, because the recipient of the goods or service had not made an express promise to pay. The law simply did not provide a remedy. By the seventeenth century, the courts had come to recognize this problem. They determined that by requesting the goods or the service the parties had made an agreement, whereby the goods would be supplied or the service rendered in return for the implied promise of payment of a reasonable price for the goods (*quantum valebant*) or a reasonable price for the services rendered (***quantum meruit***). Because contracts of this type frequently involve the supply of goods as well as services (such as when an electrician is called to replace a faulty light switch or a plumber replaces a worn faucet) the term *quantum meruit* has come to be used to cover both situations.

Quantum meruit

"As much as he has earned." A quasi-contractual remedy that permits a person to recover a reasonable price for services and/or materials requested, where no price is established when the request is made.

If the parties agree upon a price for the services or goods at any time after the services have been performed or the goods supplied, the agreed price will prevail, and the contract will become an agreement for a fixed price. If the parties cannot agree upon a reasonable price, then the courts will decide what is reasonable, based upon the price of the goods or the service in the area where the contract was made. As a general rule, the rate charged for similar goods or services by other suppliers of goods and services in the immediate area will be treated by the courts as a reasonable price. If the price charged for the service or goods is comparable to the "going rate" charged by similar suppliers, then the court will fix the contract price accordingly.

For example, a homeowner may call a plumber to fix a leak in a water pipe. No price for the service call is mentioned, but the plumber responds to the call and repairs the broken pipe. The plumber later submits an account for $200 for the work done. If the homeowner considers the account excessive, then recourse may be had to the courts to have a reasonable price fixed for the work done. However, if the parties agreed upon $200 as the price at any time after the request for the service was made by the homeowner, the price of $200 would stand.

THE DEBTOR–CREDITOR RELATIONSHIP

Under the law of contract, a debt paid when due ends the debtor–creditor relationship, as the debtor has fully satisfied his or her obligations under the contract. Similarly, the creditor has no rights under the contract once the creditor has received payment in full. The law relating to consideration applies when the debtor and creditor agree that the amount payable on the due date should be less than the full amount actually due. At first glance, it would appear that this common business practice is perfectly proper. The creditor, if he agrees to accept a lesser sum than the amount due, should be free to do so, and his promise should be binding upon him. Unfortunately, this practice runs counter to the doctrine of consideration. Unless the parties bring themselves within an exception to this rule, the creditor's promise is simply gratuitous and unenforceable.

Under the doctrine of consideration, when a creditor agrees to accept a lesser sum than the full amount of the debt on the due date, there is no consideration for the creditor's promise to waive payment of the balance of the debt owed. To recover this amount the creditor can, if desired, sue for payment of the balance immediately after receiving payment of the lesser sum.

The difficulties that the application of this principle raise for the business community are many, and the courts, as a result, have attempted to lessen the impact or harshness of the law in a number of ways. The most obvious method of avoiding the problem of lack of consideration would be for the parties to include the promise to take a lesser sum in a written document that would be under seal and signed by the creditor. The formal document under seal would eliminate the problem of lack of consideration entirely. A second method would be for the creditor to accept something other than money in full satisfaction of the debt. For example, the debtor could give the creditor his delivery truck as payment in full, and the courts would not inquire into the adequacy of the consideration.[5] Payment of the lesser sum in full satisfaction of the debt before the due date would also be consideration for the creditor's promise to forgo the balance, since the payment before the time required for payment would represent a benefit received by the creditor. A final exception to the consideration rule arises when the lesser sum is accepted as payment in full by the creditor under circumstances where a third party makes the payment in settlement of the creditor's claim against the debtor.

Mario's small business is indebted to Equity Finance for $4,000. Tony, Mario's father, offers Equity $3,500 as payment in full of his son's indebtedness. If Equity accepts Tony's payment as settlement of the $4,000, it cannot later sue Mario for the remaining $500, as it would be a fraud on the stranger (Tony) to do so.[6]

The difficulties that this particular rule for consideration raises in cases where indebtedness has been gratuitously reduced have been resolved in part by legislation in some jurisdictions.[7] In all provinces west of Quebec, statute law (such as the *Mercantile Law Amendment Act* in Ontario) provides that a creditor who accepts a lesser sum in full satisfaction of a debt will not be permitted to later claim the balance once the lesser sum has been paid. The eastern Common Law provinces remain subject, however, to the requirement of consideration, and the parties must follow one of the previously mentioned methods of establishing or avoiding consideration if the debtor is to avoid a later claim by the creditor for the balance of the debt.

The relationship between an individual creditor and debtor must be distinguished, however, from an arrangement or a *bona fide* scheme of consolidation of debts between a debtor and his or her creditors. This differs from the isolated transaction in that the creditors agree with each other to accept less than the full amount due them. Each of the creditors promise the other creditors to forgo a portion of the claim against the debtor (and forbear from taking legal action against the debtor to collect the outstanding amount) as consideration for their promise to do likewise.

It is also important to note that an agreement between a creditor and debtor, whereby the creditor agrees to accept a lesser sum when the amount owed is in dispute, does not run afoul of the consideration rule. If there is a genuine dispute concerning the amount owed, and the creditor accepts a sum less than the full amount claimed, the consideration that the creditor receives for relinquishing the right to take action for the balance is the debtor's payment of a sum that the debtor honestly believes he owes the creditor.

MANAGEMENT ALERT: BEST PRACTICE

In the course of longer-term business arrangements, business persons often give or receive "sweeteners" — an extension of time to pay or extra services — that are not part of the original deal. (Reread the concepts of past consideration and gratuitous promises.) Often these sweeteners become an undocumented part of the relationship. Later, one of the parties may seek to avoid or enforce them in court. Success or failure in enforcement hinges on finding consideration for them or finding that they are agreed, formal amendments to the original contract. Sweeteners may wind up being either temporary or permanent, depending on how they are documented or treated by the parties.

5. *Pinnels' Case* (1602), 5 Co. Rep. 117a, 77 E.R. 237.
6. For a discussion of this point see the dictum of Willes, J., in *Cook v. Lister* (1863), 13 C.B. (N.S.) 543 at p. 595, 143 E.R. 215.
7. See, for example, Ontario: *Mercantile Law Amendment Act*, R.S.O. 1990, c. M.10, s. 16.

EQUITABLE OR PROMISSORY ESTOPPEL

In the previous section debtors and creditors had to act carefully, with either a legislative remedy or the use of a seal to turn a gratuitous reduction of a debt into an enforceable promise to reduce it. Recall too from early in the chapter, that lacking a seal, a promised donation of a gift is a gratuitous promise, and is also unenforceable except where someone actually *relies* on the promise of that donation.

The courts remain reluctant to enforce gratuitous promises but a class of such promises has emerged that can be enforced, based on this aspect of reliance. These are cases where it would be fair (equitable) that the person making the promise is prevented (estopped) from later denying the promise or withdrawing it. These cases are known as cases of equitable **estoppel**, or promissory estoppel. The essential element is that the promisee relies on the promise to their later detriment. For many years it was considered that estoppel should only be used as defence to a claim, however the door to its use as a means to advance a claim has opened wider, particularly in the United States.

Estoppel

A rule whereby a person may not deny the truth of a statement of fact made by him or her when another person has relied and acted upon the statement.

Evans wanted to build a cottage on a small island in a large lake. Conrad, who lived in a distant city, owned a cottage on the mainland as well as a flat barge. In June, Evans called Conrad and asked if he could borrow the barge occasionally to haul building supplies to the island. Conrad replied, "I don't have a problem with that." In August, Conrad visited the lake, sought out Evans and asked him how many times he had used the barge. Evans told him he had used it ten times. Conrad multiplied this by $250 and demanded $2,500 rent from Evans for use of the barge. Depending on the court's interpretation of what Conrad said in June, Evans may raise equitable estoppel as a defence to the claim.

CASE IN POINT

The British judge Lord Denning modernized the concept of equitable estoppel in a case where a landlord had reduced the rent on a long-term lease because the tenants could not occupy the apartments in the building during the Second World War. At the end of the war, creditors of the landlord and the premises claimed against the tenants for the full amount of the rent. There was no consideration for the promise to reduce the rent, but Lord Denning was of the opinion that that the tenants relied on the promise to their detriment, and that the full amount of the rent should not be payable for the entire period in which the difficult conditions existed.

Central London Property Trust, Ltd., v. High Trees House, Ltd., [1947] K.B. 130.

As a result of Denning's influential opinion, equitable estoppel has continued as an effective defence against the enforcement of contractual rights when the promisee has relied upon a gratuitous promise to their detriment.

CLIENTS, SUPPLIERS *or* OPERATIONS *in* QUEBEC

Under the *Civil Code of Quebec*, a contract does not require consideration to be enforceable.

1381. A contract is onerous when each party obtains an advantage in return for his obligation. When one party obligates himself to the other for the benefit of the latter without obtaining any advantage in return, the contract is gratuitous.

1434. A contract validly formed binds the parties who have entered into it not only as to what they have expressed in it but also as to what is incident to it according to its nature and in conformity with usage, equity or law.

S.Q. 1991, c. 64, Book Five, Obligations.

CHECKLIST FOR CONSIDERATION

1. The consideration is not past consideration?
2. Is the proposed consideration legal?
3. Is there consideration flowing in both directions?
4. Does it have value in each direction?
5. If not, is there an exception (a seal, donations, negotiable instruments, third-party settlement of debt, *quantum meruit*, injurious reliance leading to estoppel)?

Numbers 1 and 2 are vital, but a failure in 3 or 4 can be saved by 5. If not, there is no contract, but rather a gratuitous promise which (almost invariably) cannot be enforced in court.

SUMMARY

Consideration is an essential requirement for any contract not under seal. If the promise is gratuitous, the promise must be made in writing and under seal to be enforceable. Consideration must be legal and must have some value in the eyes of the law. The courts, however, will not consider the adequacy of the consideration unless it is grossly inadequate given the circumstances surrounding the transaction. Since consideration is the price that a person receives for his or her promise, the consideration must move from the promisee to the promisor. In cases where the consideration is not specified, and a request is made for goods or services, the consideration will be determined by the courts at a reasonable price.

The courts have permitted the injurious reliance on the part of a party to be used as a defence in cases where a plaintiff attempts to enforce an original agreement that had been altered by a gratuitous promise. In cases involving promises of donations to charitable organizations, gratuitous promises have been enforced when the promise was such that it induced the charity to act to its detriment on the strength of the promise.

KEY TERMS

REVIEW QUESTIONS

1. In what way does an agreement on the price for services made after the services have been performed affect the right of the person who performed the services to claim *quantum meruit*?
2. Must consideration always be present to render a promise enforceable?
3. Does a contract under seal require consideration? If not, why not?
4. Consideration may be "present" or "future." How do these forms of consideration differ?
5. A creditor's promise to accept the lesser sum as payment in full for a larger debt is considered gratuitous at Common Law. Why is this so?
6. Explain promissory estoppel. What are its uses in a contract setting?
7. In what way does the Mercantile Law Amendment legislation in some provinces alter the Common Law rules for gratuitous reduction of debt?
8. What conditions must be established to put forward a claim of *quantum meruit*?
9. Explain the nature of "consideration" as it applies to a contract.
10. How have the Common Law courts, in some cases, permitted a debtor to enforce a creditor's gratuitous promise to reduce a debt?

Mini-Case Problems

1. X offered to deliver a parcel for Y to a downtown shop, as he would pass the shop on his way to work. Y agreed and gave X the parcel. The next day, Y told X he would give him $5 for delivering the parcel. X accepted Y's offer, but, later, Y refused to pay the $5. What are X's rights?
2. A, a wealthy widow, was approached by a women's group for a donation that the group intended to use to construct a women's centre in the community. A offered to donate an amount that would cover 25 percent of the cost of the proposed building. If the group commenced construction based upon A's promise, could A later refuse to pay?
3. Markus owed Landers $5,000, and since Markus had fallen on hard times, Landers cut the debt in half. Some months later, Markus was successful in getting a good paying job, and wrote a cheque payable to Landers for $2,500 with his thanks. Landers wrote back suggesting that Markus now pay the remaining $2,500. What is Markus' obligation?
4. A municipality called for tenders for the construction of a recreation centre. A construction firm submitted a bid of $8 million, under seal. The municipality accepted the bid. Shortly thereafter, the construction firm realized it had made a simple but serious error in calculation and refused to complete the contract, unless the award was increased by $500,000. The municipality refused. Discuss the rights of the parties.

Case Problems for Discussion

CASE 1

Jane and Henry were married and Jane's father, a farmer, told her and Henry that he had no use for the old pasture north of the meadow. He suggested that they could build a house on the property, saying, "It will be all yours anyhow when I'm gone." Jane and Henry did so, and five years later, Jane and her father had a terrible falling out. Her father changed his will to give all of his property to the Humane Society, and within the year he died. When the executor of the will read its terms, he began proceedings to evict Jane and Henry from the property and to sell their house.

Discuss the legal issues raised in this scenario.

CASE 2

Hansen and Dhafir owned cottage lots that abutted each other along a line at right angles to a lake front. Nothing marked the boundary between the two lots, and neither landowner could recall with any degree of accuracy where the lot line actually lay. The parties had mentioned a survey of the lots on several occasions, but nothing had been done to fix the boundaries.

One summer, when both Hansen and Dhafir were vacationing at their respective cottages, Dhafir noticed a large diseased limb on a tall tree that was growing at a point approximately equidistant from each cottage. As he was concerned that the limb might fall on his cottage, Dhafir suggested to Hansen that he do something about the situation. To this suggestion Hansen replied: "That tree is growing on your lot, and it is up to you to cut the limb. If it should fall my way and damage the roof of my cottage, I would look to you for the repairs."

Dhafir decided to cut down the entire tree instead of just the diseased limb. He did so with some reluctance because the elimination of the tree would remove the only protection Hansen had from the hot afternoon sun.

Hansen did not object to the cutting of the tree at the time. Some time later, however, when a survey that he had requested revealed that the tree had been entirely on his side of the property line between the two cottage lots, he brought an action against Dhafir for damages.

Identify the defences that might be raised by Dhafir and explain how the courts might decide this case.

CASE 3

On a cold evening in January, the furnace in the Bisrams' home failed to operate properly, causing the water pipes to freeze and later burst, flooding their basement. Mr. Bisram made a hurried telephone call to Brown's Plumbing when he discovered the water in his basement, and asked Brown if he could come over immediately and fix the leaking pipes.

Brown agreed to do so, and arrived about an hour later, only to find that Mr. Bisram had discovered the leak and fixed it himself. Brown left immediately, and the next day he submitted an account to the Bisrams for $95 for services rendered. The Bisrams refused to pay Brown's account.

If Brown should sue the Bisrams on the account, explain the arguments that might be raised by each of the parties and indicate how the case might be decided.

CASE 4

Nuptial Creations carried on business as a custom manufacturer of wedding dresses and accessory products. The company offered a standard line of dresses that it sold through its own retail establishments. The prices for the basic dress line were available to prospective purchasers, but the final price was dependent upon the amount of lace, beading, and changes that a customer requested in the product.

For her wedding, Marie selected a particular dress design that carried a basic price of $1,200. She requested a number of changes in the design that included a great deal of beadwork, requiring time-consuming hand application. No price was quoted by the company for the custom dress at the time that Marie requested that it be made for her.

Some time later, when the dress was finished and ready for delivery, Marie went to the shop and enquired as to the price. She was informed by a salesperson that the completed dress would probably cost "around $2,000." Nevertheless she accepted the dress, with the complaint that she was surprised at the cost.

Two weeks later, the company submitted an account to Marie for $2,368 for the dress. Marie refused to pay it.

Advise the parties of their rights and, if legal action should be taken, discuss the way in which the case might be resolved.

CASE 5

A service club in a small community decided to raise funds for the purchase of a special wheelchair that it intended to donate to a disabled child. The wheelchair was of a special design and made only on special order. The estimated purchase price was $5,600.

Donations were solicited throughout the community, and a total of $2,100 was raised toward the purchase price. The club, unfortunately, found it difficult to solicit further donations, as the $2,100 represented the contributions of nearly every family in the community.

A meeting of the club members was held to discuss ways and means of raising the further $3,500. After some discussion, one member stood up and promised to "match dollar for dollar every additional contribution received, if the club can raise a further $1,750."

The next day, the local newspaper reported the club member's pledge, and the publicity produced a large number of donations to the fund. By the end of the week, the donations totalled $3,700, of which $1,600 had been donated that week. The club immediately placed an order for the special wheelchair, confident that the additional $150 could be raised by the club members.

A few weeks later, the club was advised that the wheelchair was ready for delivery, and that the price would be $5,300 instead of the $5,600 originally quoted. The club was delighted to receive the news that the wheelchair would only cost $5,300, and it immediately notified the club member who had promised to match the contributions.

When the member was advised that the $1,600 donated, together with his matching pledge, would be sufficient (along with funds previously collected) to pay for the wheelchair, he responded: "I promised to match dollar for dollar only if the club could raise a further $1,750. Since it did not do so, I have no intention of matching the donations."

Discuss the position at law of the parties in this case and explain how this matter might be determined if each party exercised his (or its) legal rights (if any).

CASE 6

Great Adventures Ltd. offered white-water rafting trips involving a relatively short 50-kilometre journey down a swift river. The price of the trip, including overnight hotel and meals, was advertised at $450. Jack and Jill, in response to the advertisement, entered into a verbal arrangement with the operator of the tour to join in on a journey, and they agreed to appear at the designated hotel the evening before the date of the excursion.

At the hotel, they met the president of the tour company and paid him the tour price. The next morning, as Jack and Jill assembled their gear with the nine other passengers, a representative of the company spoke to the participants and instructed each of them to sign a form entitled "Standard Release." The form stated that the operator of the tour was "not responsible for any loss or damage suffered by any passenger for any reason, including any negligence on the part of the company, its employees, or agents."

Jack and Jill were reluctant to sign the release, but when they were informed by the tour representative that they would not be allowed on the raft unless they signed it, they did so. When the release was signed, the representative gave each of them a life jacket with a 10-kilogram buoyancy rating. After they donned the life jackets, they were allowed to climb aboard the raft.

During the course of the journey, the raft overturned in very rough water, and Jack and two other persons drowned.

An investigation of the accident by provincial authorities indicated that the life jacket Jack had been wearing was too small to support the weight of a person his size. The investigation also revealed that, due to the swiftness of the river at the place where the accident occurred, a more suitable life jacket would probably not have saved Jack's life.

Jill survived the accident and brought an action against the tour company under the provincial legislation that permitted her to institute legal proceedings on behalf of her deceased husband.

Discuss the issues raised in this case and indicate the arguments that might be raised by each party. Render a decision.

CASE 7

Speedy Delivery Service Ltd. had been engaged by the Commercial Bank as its regular courier for documents and other bank correspondence for both local and long-distance deliveries. The parties had an ongoing agreement for services under which Speedy's maximum liability to the bank for any matter arising out of their relationship was limited to $100,000. Speedy also held business insurance in the amount of $10,000,000, from which it could claim indemnification should any claim be made against Speedy by the bank or by any other party with which Speedy did business. A term of its insurance policy required Speedy to make an immediate report to the insurer if it suspected that there may be a claim made against it.

A few days after Speedy had made a delivery of documents to one of Commercial Bank's branches, a letter arrived from the bank stating that the documents had been lost. Apparently the Speedy driver had placed the documents in the usual outside receptacle at the branch, which was used for deliveries after the branch has closed, but he could not remember whether he had properly secured the receptacle afterwards. The bank's letter stated that bank employees were trying to locate the missing documents and that they related to an important mortgage transaction that was to have been finalized that week. The letter further stated that there would be losses to the bank if the documents could not be found, but, in any event, the bank would further advise Speedy at a later date of its progress.

Speedy heard nothing more from the bank about the documents and assumed that they had been located. Almost three years later the bank wrote to Speedy stating that it was suing for its damages for the lost documents in the amount of $100,000. When Speedy refused to pay, the bank threatened to bring legal action.

Discuss the legal issues raised in this case and the arguments that all of the parties, including the insurance company, will rely on.

CASE 8

Levine's car stopped running on the highway, and a passing motorist called a service station for him on a mobile phone. The service station called an independent tow truck operator, who soon towed the car to the service station, some 5 km distant. A mechanic examined the car, and realized that a wire had come loose from the distributor cap. He snapped the wire back into place, and wrote up a bill. Levine was charged $5 for the repair and $175 for the tow. Levine was outraged and refused to pay.

What would be the nature of the claim, and what factors would the court consider in deciding the case?

CASE 9

Able and Laryssa had been good friends for many years. Able wished to make a gift to Laryssa of a house he owned that was subject to two mortgages held by a third party. In order to avoid the tax legislation in force at the time, the parties agreed that Able would convey the property to Laryssa and take back an interest-free mortgage equal to the difference between the two existing mortgages and the value of the property. The third mortgage would be forgiven each year in an amount equal to the permissible tax-free gift allowed under the legislation. The gift transaction at the time was a lawful method of disposing of the property, and Laryssa accepted the gift of the property. Laryssa moved into the house and made the payments on the first two mortgages for a number of years; then, after a difference of opinion on another matter, Able demanded payment of the third mortgage. Laryssa refused to pay, and Able sued Laryssa for the amount owed.

Discuss the respective positions of the parties to the transaction and render a decision.

www.mcgrawhill.ca/olc/willes

CASE 10

Devon, a contractor, checked out the yard of Adams Equipment Sales and noticed a backhoe that looked to be in good shape. Beside it was a set of buckets and a rock breaking point. Adams noticed his customer and yelled across the yard "Seventeen-five takes it away." Devon nodded and waved, and returned later with a certified cheque for $17,500. Adams accepted the cheque, but when Devon started to load the other buckets on his flatbed, Adams stopped him. "They're not in the deal," he said, "They are worth $2,500 on their own." Devon agreed on the value, but said the backhoe wasn't worth $17,500 on its own, because the going rate for similar models was $15,000.

What position will the court support, if Devon and Adams proceed to litigation over the agreement?

The Online Learning Centre offers more ways to check what you have learned so far. Find quizzes, Weblinks, and many other resources at www.mcgrawhill.ca/olc/willes.

CHAPTER 9

Legal Capacity to Contract and the Requirement of Legality

Chapter Objectives

After study of this chapter, students should be able to:

- Explain and illustrate the various situations where capacity to contract is absent or diminished, and its effect on the contractual relationship.
- Explain and illustrate what types of contracts would be illegal contracts.
- Demonstrate situations where business activities may be illegal as contracts in restraint of trade.

YOUR BUSINESS AT RISK

Public policy makes some transactions voidable and others illegal. The consequences of failing to appreciate these risks are serious, unnecessary and avoidable.

THE MINOR OR INFANT

Infant

A person who has not reached the age of majority.

Not everyone is permitted to enter into contracts that would bind them at law. Certain classes of promisors must be protected as a matter of public policy, either for reasons of their inexperience and immaturity, or because of their inability to appreciate the nature of their acts in making enforceable promises. The most obvious class to be protected is the group of persons of tender age called minors or **infants**. An infant at Common Law is a person under the age of 21 years, but in most provinces this has been lowered to 18 or 19 years of age by legislation. Nevertheless, "infants" are a highly desirable segment of the commercial marketplace, with a buying power that is not reflected in the employment income of their age group. Consequently, business seeks their commercial activity, but not without unique dangers in contract.

Public policy dictates that minors should not be bound by their promises; consequently, they are not liable on most contracts that they might negotiate. The rule is not absolute, however, because in many cases, a hard and fast rule on the liability of a minor would not be in his or her best interest. For example, if a minor could not incur liability on a contract for food or clothing, the hardship would fall on the minor, rather than the party with full power to contract, as no one would be willing to supply an infant with food, shelter, or clothing on credit.[1] The law, therefore, attempts to balance the protection of the minor with the need to contract by making only those contracts for necessary items enforceable against a minor.

Enforceability and the Right of Repudiation

The enforceability of any contract for non-necessary goods will depend to some extent on whether the contract has been fully executed by the minor, or whether it has yet to

1. *Zouch v. Parsons* (1765), 3 Burr. 1794, 97 E.R. 1103.

be performed. If the contract made by the minor has been fully performed (and, consequently, fully executed) then the minor may very well be bound by the agreement, unless he or she can show that he or she had been taken advantage of by the merchant, or can return all the goods purchased to the other party. For example, a minor purchases a gift for a friend and pays the merchant the full purchase price. The contract at this point has been fully performed (executed) by both parties. If the minor some time later wishes to repudiate the contract, the court would not allow the minor to do so unless the minor could convince the court that the merchant had taken advantage of the minor by charging the minor an excessively high price. The minor would also be obliged to return the goods in order to succeed in the court action. If the contract has not been fully performed, then the agreement (if for a non-necessary item) may be voidable at the minor's option. This rule would probably apply as well to a necessary item, if the minor has not taken delivery of the goods. For example, if a minor orders a clothing item from a mail-order house, and then repudiates the contract before delivery is made, the minor would probably not be bound by the agreement, even though the item is a necessary.

The adult with full capacity to contract is bound in every case by the contract negotiated with an infant, since the adult person has no obligation to do business with an infant unless the adult wishes to do so. Any business firm or merchant that decides to enter into a contract with a minor assumes the risk (in the case of a contract for a non-necessary) that the minor might repudiate the agreement.

A minor or infant of tender age is normally under the supervision of a parent or guardian, and the need to contract in the infant's own name is limited. The older minor, however, is in a slightly different position, with a need in some cases to enter into contracts for food, clothing, shelter, and other necessaries. For this latter group, the law provides that an infant will be bound by contracts for necessaries, and will be liable for a reasonable price for the goods received or the services supplied. The effect of this rule is to permit a merchant to provide necessaries to an infant or minor, yet limit the infant's liability to a reasonable price. This is eminently fair to both contracting parties: the merchant is protected because the minor is liable on the contract; the infant is protected in that the merchant may only charge the infant a reasonable price for the goods.

Criteria for Necessary

The unusual aspect of the law relating to minors is the criteria used by the courts to determine what is a necessary for a minor. The courts will examine the social position of the infant in deciding the question. Such an approach smacks of different standards for different minors and has its roots in a number of older English cases.[2] Nevertheless, it remains as the law today. Other requirements, which are perhaps of more concern to a modern-day merchant, are that the goods supplied to the minor are actually necessary, and that the minor is not already well supplied with similar goods. Since the merchant in each case has the onus of proving these facts, some care is obviously essential on the part of merchants supplying goods of this nature to infants. The entire issue becomes rather obscure when the merchant must also distinguish between what constitutes a necessary, and what might be a luxury, bearing in mind the minor's station in life.

In many cases, contracts of employment or apprenticeship are contracts considered to be beneficial to minors, and are enforceable against them. Although some educational ventures that involve minors may not be considered by the courts as being beneficial, many are held enforceable, even when the educational aspect is unusual.

A minor entered into a contract to take part in a world tour with a professional billiard player as his playing opponent. He later repudiated the contract before the tour was scheduled to take place, and the adult contracting party took action

2. *Ryder v. Wombwell* (1868), L.R. 3 Ex. 90; affirmed (1869), L.R. 4 Ex. 32; *Nash v. Inman*, [1908] 2 K.B. 1.

against him for breach of contract. The court held that the minor was bound by the contract because the experience of playing billiards with a professional would be valuable instruction for a minor who wished to make billiards his career.[3]

The Effect of Repudiation

The general rule relating to contracts that have not been fully performed (executory contracts) for non-necessary goods or services is that the minor or infant may repudiate the contract at any time, at his or her option. This rule applies even when the terms of the contract are very fair to the infant. Once the contract has been repudiated, the minor is entitled to a return of any deposit paid to the adult contractor. However, if the minor has purchased the goods on credit and taken delivery, the minor must return the goods before the merchant is obliged to return any monies paid. Any damage to the goods that is not a direct result of the minor's deliberate act is not recoverable by the merchant: the merchant may not deduct the "wear and tear" to the goods from the funds repayable to the infant. The reasoning of the courts in establishing this rule is that the merchant should not be permitted to recover under the law of torts what he or she cannot recover by the law of contract.[4] However, if the minor deliberately misrepresents the use intended for the goods, and the goods are damaged, then the merchant may be entitled to recover the loss by way of an action for tort.[5]

In a New Brunswick case,[6] a merchant sold an automobile to a minor. The minor acquired possession of the automobile and agreed to pay for the vehicle on an installment basis over a period of time. Shortly after obtaining possession of the car, the infant was involved in an accident, and the vehicle was seriously damaged. The infant repudiated the contract and the merchant took legal action against the infant in tort for the damage to the automobile in the amount of the unpaid balance of the purchase price. The judge dismissed the tort action against the infant, in the following words:

> While an infant may be liable in tort generally, he is not answerable for a tort committed in the course of doing an act directly connected with, or contemplated by, a contract which, as an infant, he is entitled to avoid. An infant cannot, through a change in the form of action to one ex delicto, be made liable for the breach of a voidable contract.
>
> The plaintiff, when selling the car, clearly contemplated the defendant would drive it. The conditional sale contract provided the car was to be at the defendant's risk and placed no restriction on the manner in which it should be driven. That the plaintiff had in mind the possibility of physical damage resulting from the driving of the car is evidenced by the stipulation in the contract that the defendant would procure and maintain insurance against all physical damage risks.
>
> The amount of the damages which the plaintiff claimed was the equivalent of the deferred balance of the purchase-price less the financing charge. The action was framed in tort, but the real purpose was to recover under the contract. As at the time of its destruction the car was being used in a manner contemplated by the parties to the conditional sale agreement, any claim against the plaintiff in tort, founded on his negligent driving, must fail.

Fraudulent Misrepresentation as to Age

The protection extended to a minor under the rules of contract may not be used by a minor to perpetrate a fraud on an unsuspecting merchant. On the other hand, an adult entering into a contract with a minor where the minor has represented himself as having

3. *Roberts v. Gray*, [1913] 1 K.B. 520.
4. *Jennings v. Rundall* (1799), 8 T.R. 335, 101 E.R. 1419; *Dickson Bros. Garage & U Drive Ltd. v. Woo* (1957), 10 D.L.R. (2d) 652, affirmed 11 D.L.R. (2d) 477.
5. *Burnhard v. Haggis* (1863), 3 L.J.C.P. 189; *Ballett v. Mingay*, [1943] K.B. 281.
6. *Noble's Ltd. v. Bellefleur* (1963), 37 D.L.R. (2d) 519.

attained the age of majority will not be permitted to hold the minor to the contract. The mere fact that a minor misrepresents his or her age does not generally alter the fact that the minor cannot be bound in a contract for non-necessaries.[7]

In contracts for non-necessary goods where the minor has falsely represented that he or she is of full age when, in fact, he or she is not, the merchant may be entitled to recover the goods on the basis of the minor's fraud. Additionally, where a minor attempts to use age incapacity to take advantage of merchants, the criminal law relating to obtaining goods under false pretenses may also be applicable. What the law attempts to do in providing protection to the minor is to impose only a limited liability toward others, based upon what is perceived as being in the best interests of the minor. The treatment of minors is, in a sense, a matter of equity. When a minor has attempted to use age minority in a manner that is contrary to public policy, the courts will either provide the other contracting party with a remedy for the loss caused by the infant, or prevent the infant from avoiding liability under the contract.

COURT DECISION: Practical Application of the Law

Capacity to Contract—Minor—Fraud on Part of Minor

Gregson v. Law and Barry (1913), 5 W.W.R. 1017.

The minor made a transfer of land to a person who was unaware of her minority in return for the market value of the property. In the affidavits accompanying the deed, the minor swore that she was of full age. She later repudiated the conveyance and brought an action for a return of the property.

THE COURT: In this action I am forced to hold on the evidence that the plaintiff well knew when she executed the final deed to Law that, being a minor, she could not legally do so, and that, with such knowledge, she proceeded to complete and execute the same, including the making of the acknowledgment and representing herself to be of full age. No hint of the true condition of things was given to Law, and I hold this was done knowingly, and that therefore the plaintiff is now coming into court to take advantage of her own fraud. Whilst, apparently, it is true to say that being an infant she could not be made liable on a contract thus brought about, it is, I think, an altogether different proposition to say the court will actually assist her to obtain advantages based entirely on her own fraudulent act.The authorities cited in argument show in fact, I consider, that infants are no more entitled than adults to gain benefits to themselves by fraud, or at any rate establish the proposition that the courts will not become active agents to bring about such a result.

The minor's action was dismissed. The final sentence carries the full message, applicable to legal actions of every kind.

Ratification and Repudiation

When the minor has entered into a contract of a continuing or permanent nature under which the minor receives benefits and incurs obligations (such as engaging in a partnership, or purchasing non-necessary goods on a long-term credit contract), the contract must be **repudiated** by the minor within a reasonable time after attaining the age of majority. Otherwise, the contract will become binding on the minor for the balance of its term.

Repudiation

The refusal to perform an agreement or promise.

Ratification

The adoption of a contract or act of another by a party who was not originally bound by the contract or act.

The reverse is true for contracts for non-necessary items purchased by a minor, when the contract is not of a continuing nature. The infant must expressly **ratify** (acknowledge and agree to perform) such a contract on attaining the age of majority in order to be bound by it. For example, a minor enters into a contract to purchase a sailboat in the fall of the year. The minor gives the merchant a deposit to hold the boat until spring and promises to pay the balance at that time. The contract would be voidable at the infant's option, and the infant would be free to ignore the contract or repudiate it at any time before reaching the age of majority. If the minor wished to be bound by the transaction, he or she would be obliged to ratify it after attaining the age of majority.

7. *Jewell v. Broad* (1909), 20 O.L.R. 176.

Statutory Protection of Minors

Statute law has modified the Common Law on the question of ratification to some extent. The provinces of New Brunswick, Newfoundland, Nova Scotia, Ontario, and Prince Edward Island have all passed legislation requiring the ratification to be in writing before it will be binding on the infant. British Columbia has carried the matter one step further in its protection of minors: an infant cannot ratify a contract of this nature in any fashion that would render it enforceable by the adult contracting party. This legislation also has the effect of rendering contracts for non-necessaries and debt contracts "absolutely void." This particular term is open to question, however, as to whether it will only be applied to the liability of the minor, or to both the minor and the other contracting party if the issue should be raised. If the adult contracting party fails to perform the contract, the question arises: Is the contract absolutely void and is the infant deprived of a remedy, or would the words "absolutely void" be construed to mean void as against the minor? The question remains unresolved at the present time, but given the intent of the legislation, an interpretation that deprives the minor of a right of action would appear to be counter to the intention of the statute.

Minors Engaged in Business

Contracts of employment, if they are lawful and contain terms that are not onerous, are generally binding on minors. Since agreements of this type are generally contracts of indefinite hiring, the minor need only give reasonable notice to terminate, and the infant would be free of all obligations imposed by the agreement. A minor engaging in business as a sole proprietor, or in a partnership, is in quite a different situation. That is because the law generally does not support the thesis that an infant must, of necessity, engage in business activity as a principal.

The rules relating to contracts engaged in by an infant merchant are, for the most part, consistent with those for minors in general. Since it is not necessary for an infant to engage in business, any attempt to purchase business equipment, even if the equipment is necessary for the business, will probably be treated by the courts as a contract for non-necessaries. This renders the contract voidable at the option of the minor, and, if the minor has not taken delivery of the goods, it may permit the minor to repudiate the contract and obtain a refund of any deposit paid. Similarly, if a minor has taken delivery of goods on credit, the minor may return the goods and cancel the obligation. If the goods are accidentally damaged while in the minor's possession, the minor will not normally be liable for the damage. However, if the minor has sold the goods, he or she will be required to deliver up any monies received.

In the case of a sale of goods by an infant merchant, the infant merchant cannot be obliged to perform the contract if he or she does not wish to do so, as this type of contract is also voidable at the infant's option. If the infant merchant received a deposit or part-payment with regard to a sale of goods or services, he or she would not be required to deliver the goods, but would, of course, be required to return the deposit.

These general rules for contracts engaged in by infant merchants are also consistent with the general rules for the enforcement of infant contracts. Even though they tend to place hardship on adults dealing with minors, public policy, nevertheless, dictates protection of the minor over the rights of persons of full contracting age. With freedom to contract, an adult is under no obligation to deal with an infant merchant any more than there is an obligation on the adult merchant to deal with an infant customer. Since the opportunity to take advantage of the infant's inexperience in business matters exists in both cases, unfair treatment is avoided by the extension of infant protection rules to the infant merchant as well.

In the case of a minor joining a partnership, the protection afforded the minor is, again, consistent with the general public policy concerning infants' contracts. A partnership agreement involving a minor in a contract for non-necessaries is voidable by the minor, even though the adult parties remain bound. Since the contract is a continuing contract, the infant must repudiate it promptly on or before reaching the age of majority if the infant wishes to avoid liability under it.[8] If a minor continues to accept benefits under the contract after reaching the age of majority, the minor will be bound by the contract, even though he or she did not expressly ratify it. Ratification will be implied from the action of taking benefits under the contract.

Should the minor repudiate the partnership agreement, the minor would not be liable for any debts incurred by the partnership during minority. It would also appear that a minor would not be entitled to withdraw any contribution to the partnership until the debts of the partnership had been settled.[9]

The Parent–Infant Relationship

Parents are not normally liable at Common Law for the debts incurred by their infant children. This rule, however, has been modified to some extent by family-law legislation in some provinces. This legislation obligates parents to support a child under the age of 16 years, and renders parents jointly and severally liable with the child for any necessaries supplied to the child by merchants. The minor may, as the parents' "agent of necessity," pledge the parents' credit to obtain the necessaries of life, and the parents will be bound by the minor's actions. Apart from this statutory requirement, there are a number of other circumstances under which parents may become liable for the debts of their infant children. In some cases, a parent may have appointed the child to act as his or her agent to purchase goods on credit. In these situations, a true agency situation would exist in which the parent, as principal, would be liable for payment. However, if the child were to later purchase goods on credit for his or her own use, and the parents continued to pay the merchant, the parents would be bound to pay the debts of the infant in the future. This is because the parents implied by their conduct that they would continue to honour the debts incurred. In such cases, the parents must specifically state to the merchant that they do not intend to be bound by any subsequent purchases by their infant child if they wish to avoid liability for future purchases negotiated by the minor.

A QUESTION OF ETHICS

Kids aren't kids anymore, it seems. Fourteen-year-olds today know more about life and responsibilities than 21-year-olds did 50 years ago: television and the Internet ensure that knowledgeable 14-year-olds should not be able to escape contracts when they know full well the consequences of signing their name. What do you think?

DRUNKEN AND INSANE PERSONS

The courts treat drunken and insane persons in much the same way as infants with respect to the capacity to contract. Those persons who have been committed to a mental institution cannot normally incur any liability in contract. Persons who suffer mental impairment from time to time are distinguished from the insane and are subject to a number of special contract rules.

In general, persons who suffer from some mental impairment caused either as a result of some physical or mental damage, or as a result of drugs or alcohol, will be liable on any contract for necessaries negotiated by them, and they will be obliged to pay a reasonable price for the goods or services. In this respect, the law makes no distinction between infants and persons suffering from some mental disability. The merchant involved would be entitled to payment even if he or she knew of the insane or drunken state of the purchaser. Again, public policy dictates that it is in the best interests of the drunken or insane

8. *Hilliard v. Dillon*, [1955] O.W.N. 621. Should the minor repudiate the partnership agreement, he would not be liable for any debts incurred by the partnership during his minority. It would also appear that he would not be entitled to withdraw his contribution to the partnership until the debts of the partnership had been settled.

9. See, for example, the *Partnerships Act*, R.S.O. 1990, c. P.5, s. 44, as amended by S.O. 1998, c. 2, s. 5.

person to be entitled to obtain the necessaries of life from merchants and to be bound by such contracts of purchase.

Contracts for non-necessary items, however, are treated in a different manner from contracts for necessaries. If a person is intoxicated or insane when entering into a contract for what might be considered a non-necessary item or service, and the person's mental state renders him or her incapable of knowing or appreciating the nature of his or her actions (and if he or she can establish by evidence that he or she was in such a condition, and the other contracting party knew he or she was in that condition), then the contract may be voidable by that person when he or she becomes aware of the contract on his or her return to a sane or sober state.

It is important, in the case of an intoxicated or insane person, that the contract be repudiated as soon as it is brought to the person's attention after his or her return to sanity or sobriety. If the contract is not repudiated promptly, and all of the purchased goods returned, the opportunity to avoid liability will be lost. Similarly, any act that would imply acceptance of the contract while sane or sober would render the contract binding.

Able attended an auction sale while in an intoxicated state. Everyone at the sale, including the auctioneer, was aware of his condition. When a house and land came up for auction, Able bid vigorously on the property and was the successful bidder. Later, when in a sober state, he was informed of his purchase, and he affirmed the contract. Immediately thereafter he changed his mind. He repudiated the contract on the basis that he was drunk at the time, and the auctioneer was aware of his condition.

At trial, the court held that Able had had the opportunity to avoid the contract when he became sober, but, instead, he affirmed it. Having done so, he was bound by his acceptance, and he could not later repudiate the contract.[10]

The effect of the affirmation renders the contract binding, and the insane or intoxicated person would be liable for breach of contract if the agreement was later rejected. If the contract concerned goods or services, the injured party would be entitled to monetary damages for the loss suffered. In the case of a contract concerning land, the equitable remedy of specific performance would be available to force the person to complete the transaction.

CORPORATIONS

A corporation is a creature of statute, and as such may possess only those powers that the statute may grant it. Corporations formed under Royal Charter or letters patent are generally considered to have all the powers to contract that a natural person may have. The statute that provides for incorporation may specifically give the corporation these rights as well. The legislature need not give a corporation broad powers of contract if it does not wish to do so; indeed, many special-purpose corporations do have their powers strictly controlled or limited. Many of these corporations are created under a special Act of a legislature or Parliament for specific purposes. If they should enter into a contract that is beyond their limitations the contract will be void, and their action ***ultra vires***. While this may appear to be harsh treatment for an unsuspecting person who enters into a contract with a "special-Act" corporation that goes beyond the limits of its powers, everyone is deemed to know the law, and the statute creating the corporation and its contents, including the limitations on its contractual powers, are considered to be public knowledge and familiar to everyone.

Ultra vires

An act that is beyond the legal authority or power of a legislature or corporate body.

Business corporations in most provinces are usually incorporated under legislation that gives the corporations very wide powers to contract, and in many cases, they have all the powers of a mature, natural person. This is not always the case, however. A corporation, in its articles of incorporation (the legal documents creating it), may limit its own powers

10. See *Matthews v. Baxter* (1873), L.R. 8 Ex. 132.

for specific reasons and, depending upon the legislation under which it was incorporated, the limitation may bind third parties. A full discussion of the effect of these limitations on the capacity of a corporation to contract is reserved for Chapter 17, "Corporation Law."

First Nations Bands

The question of capacity of First Nations bands is a window on their difficult relationship with paternalistic government and Canada's *Indian Act*. As far as bands are concerned, their capacity is an inherent aspect of sovereignty, just as the Government of Canada's capacity to contract goes unquestioned. On the other hand, the *Indian Act* does not recognize First Nations bands as either legal entities or as persons.

Courts in Canada are obligated to interpret legislation as it stands, and thus have usually considered First Nations, out of necessity, to at least be unincorporated associations with a right to sue. This is only a partial solution, for different provinces give differing powers to unincorporated associations, rather than the universality that comes with being a "person." The practical effect is that many First Nations bands choose to take title to assets in trust through the Chief or band councilors.

The *Indian Act* provides that a band exercises its powers through a majority of electors, and its council can exercise powers through a majority resolution at a meeting of band councilors.[11] This would be essential in establishing that the band had entered into a contract, but there are serious limitations on a non-Indian who wants to enforce rights under that contract. If the band breaches the contract, the *Indian Act* still prevents anyone except an Indian from gaining an interest in reserve lands. Physical assets located on a reserve are also exempt from seizure by creditors, as are monies paid by the Crown directly pursuant to treaty obligations.

CASE IN POINT

A lumber company supplied building materials to a Manitoba First Nations band. When the account went unpaid, the company obtained a judgment and attempted to seize funds held by the band in a bank in Winnipeg. The Supreme Court of Canada ruled that the funds were not situated on the reserve, nor were they funds paid by the Crown under its treaty with the band, and thus were available to the creditor.

McDiarmid Lumber Ltd. v. God's Lake First Nation, 2006 SCC 58 (CanLII).

Labour Unions

A labour union is an organization with powers that vary considerably from province to province. Originally, a union was an illegal organization, since its object was restraint of trade, but legislation in the nineteenth century changed the status of the union to a legal one. An agreement that a labour union negotiates with an employer would not normally be enforceable were it not for specific legislation governing its negotiation and enforcement. Apart from a brief period in Ontario when the courts had the authority to enforce collective agreements,[12] the courts have not normally been concerned with labour "contracts." The reason for this may be found in the law itself. The legislation in most provinces, and at the federal level, provides for the interpretation and enforcement of collective agreements by binding arbitration rather than through the courts. In addition, the legislation in all provinces specifically provides that a labour union certified by the Labour Relations Board has the exclusive authority to negotiate a collective agreement for

11. *Indian Act*, R.S.C. 1985, c. I-5, s.2(3)a and s.2(3)b.

12. *Collective Bargaining Act, 1943*, S.O. 1943 (Ont.), c. 4.

the employees it represents. The capacity of a labour union in this regard is examined in detail in Chapter 19, "Employment and Labour Relations."

Bankrupt Persons

A person who has been declared bankrupt has a limited capacity to contract. Until a bankrupt person receives a discharge, he or she may not enter into any contract except for necessaries. All business contracts entered into before bankruptcy become the responsibility of the trustee in bankruptcy, and the bankrupt, on discharge, is released from the responsibility of the contracts and all related debts, except those relating to breach of trust, fraud, and certain orders of the court. To protect persons who may not realize that they are dealing with an undischarged bankrupt, the *Bankruptcy and Insolvency Act*[13] requires the undischarged bankrupt to reveal the fact that he or she is an undischarged bankrupt before entering into any business transaction whatsoever or credit arrangement involving more than $500.

The Requirement of Legality

Agreements that offend the public good are not enforceable. If parties enter into an agreement that has an illegal purpose, it may not only be unenforceable, but illegal as well. Under these circumstances, the parties may be liable to a penalty or fine for either making the agreement, or attempting to carry it out. However, certain contracts are only rendered void by public policy in general, or by specific statutes. In these cases, the law has simply identified certain contractual activities that it will not enforce if the parties fail to comply with the statute or the policy. In other cases, the law declares certain activities to be not only illegal, but the contract pertaining thereto void, or absolutely void. As a result of these various combinations, the absence of legality does not always neatly classify contracts as those that are unlawful and those that are void, since an overlap exists.

Legality Under Statute Law

Generally

An illegal contract, if considered in a narrow sense, includes any agreement to commit a crime, such as an agreement to rob, assault, abduct, murder, obtain goods under false pretences, or commit any other act prohibited under the *Criminal Code*.[14] For example, an agreement by two parties to commit the robbery of a bank would be an illegal contract and subject to criminal penalties as a conspiracy to commit a crime, even if the robbery was not carried out. If one party refused to go through with the agreement, the other party would not be entitled to take the matter to the courts for redress because the contract would be absolutely void and unenforceable.

Another type of agreement that would be unenforceable is an agreement relating to the embezzlement of funds by an employee if the employee, when the crime is discovered, promises the employer restitution in return for a promise not to report the crime to the police. The victim of the theft is often not aware that the formation of an agreement to accept repayment of the funds in return for a promise to not report the matter is improper. The contract would, accordingly, be unenforceable.

Competition Act

A statute that affects certain kinds of contracts and that, in part, is criminal law, is the *Competition Act*.[15] This statute renders illegal any contract or agreement between business firms that represents a restraint of competition. The *Act* covers a number of business

13. R.S.C. 1985, c. B-3, as amended by S.C. 1992, c. 27.

14. R.S.C. 1985, c. C-46, as amended.

15. Formerly the *Combines Investigation Act*, R.S.C. 1985, c. C-34. Renamed the *Competition Act* by S.C. 1986, c. 26.

practices that are contrary to the public interest, the most important being contracts or agreements that tend to fix prices, eliminate or reduce the number of competitors, allocate markets, or reduce output in such a way that competition is unduly restricted. The Act applies to contracts concerning both goods and services, and it attempts to provide a balance between freedom of trade and the protection of the consumer. The formation of mergers or monopolies that would be against the public interest may also be prohibited under the Act, and all contracts relating to the formation of such a new entity would also be illegal. Similarly, any agreement between existing competitors that would prevent new competition from entering the market would be prohibited by the Act, and the agreement would be illegal.

FRONT-PAGE LAW | **Heinz agrees to open up to competition**

OTTAWA — In a decision that shows the baby-food business is as grown up as any other, the federal Competition Bureau said yesterday that H.J. Heinz Co. has agreed to change several "anti-competitive" practices, including paying retailers not to stock competitors' products.

The voluntary agreement saves both sides from a costly and likely time-consuming battle before the quasi-judicial Competition Tribunal, but the bureau made it clear it felt it had a good case.

An investigation revealed several "anti-competitive practices," the bureau said in a release. "A major and dominant supplier had engaged in activities which created a significant additional barrier to entry for competitors, preventing or lessening competition substantially."

The company has agreed to stop a raft of acts including paying large lump-sum, up front amounts to retailers to refrain from stocking non-Heinz products; signing multi-year exclusive arrangements with retailers; and giving discounts based on being the only baby food on the shelves.

A spokeswoman for Heinz, the sole supplier of jarred baby food in Canada and the major supplier of the infant cereal under the Pablum and Heinz brand names, said the deal should have no impact on consumers. "Although the bureau's allegations weren't proved in a court of law, Heinz chose to negotiate a settlement to avoid a costly court battle," said Anna Relyea.

The bureau said it is happy with the result. "The undertaking is designed to open the market to competition for existing or new suppliers," said André Lafond, deputy commissioner of competition.

What do you think of the statement that "the deal should have no impact on consumers"? This would suggest that the law is ineffective and pointless. What has been achieved in this case?

Source: Excerpted from Ian Jack, *Financial Post,* "Heinz agrees to open up to competition," August 2, 2000, p. C3.

Administrative Acts

Statute law, other than criminal law, may also render certain types of contracts illegal and unenforceable. Some statutes (such as workers' compensation legislation, land-use planning legislation in some provinces, and wagering laws) render any agreement made in violation of the Act void and unenforceable. In contrast to illegal contracts, void contracts usually carry no criminal penalties with them. The contract is simply considered to be one that does not create rights that may be enforced. For example, under most land-use control legislation any deed purporting to convey a part of a lot that a landowner owns and requiring consent of the planning authority for the severance is void unless consent to sever the parcel is obtained from the planning authority and endorsed on the deed. With respect to gambling, the courts have long frowned upon gamblers using the courts to collect wagers and, as a matter of policy, have treated wagering contracts as unenforceable unless specific legislation has made them enforceable.

Insurance Policies

One type of contract that contains an element of wager, which is not treated as being illegal or void, is the contract of insurance. An insurance transaction, while it bears a superficial resemblance to a wager, is essentially an attempt to provide protection from a financial loss that the insured hopes will not occur. This is particularly true in the case of life insurance. In this sense, it is quite different from placing a wager on the outcome of a football game or a horse race, where the gambler hopes to be the winner.

What distinguishes the contract of insurance from a simple wager is the insurable interest of the party taking out the insurance. Since the insured is presumed to have an interest in the insured event not happening (particularly if the insurance is on his or her life) this type of wager assumes an air of legitimacy. The provincial legislatures have recognized this difference and passed legislation pertaining to insurance contracts that render them valid and enforceable, provided that an insurable interest exists, or that the provisions of the Act permit the particular interest to be insured. The importance of this type of agreement as a risk-spreading or risk-reducing device for business and the public in general far outweighs the wager element in the agreement. Public policy has, on this basis, generally legitimated insurance contracts by legislative enforcement.

(Un)Licensed Persons

One type of contract that the courts treat as illegal is a contract between an unlicensed tradesperson or professional, and a contracting party. If the jurisdiction in which the tradesperson or professional operates requires the person to be licensed in order to perform services for the public at large, then an unlicensed tradesperson or professional may not enforce a contract for payment for the services if the other party refuses to pay the account. In most provinces, the licensing of professionals occurs on a province-wide basis, and penalties are provided if an unlicensed person engages in a recognized professional activity. The medical, legal, dental, land-surveying, architectural, and engineering professions, for example, are subject to such licensing requirements in an effort to protect the public from unqualified practitioners. The same holds true for many trades, although, in some provinces, these are licensed at the local level.

When a licence to practise is required, and the tradesperson is unlicensed, it would appear to be a good defence to a claim for payment for the defendant to argue that the contract is unenforceable on the tradesperson's part because he or she is unlicensed.[16] However, if the unlicensed tradesperson supplies materials as well as services, the defence may well be limited only to the services supplied and not to the value of the goods. In a 1979 case,[17] the Supreme Court of Ontario held that an unlicensed tradesperson may recover for the value of the goods supplied, because the particular by-law licensing the contractor did not contain a prohibition on the sale of material by an unlicensed contractor. It should be noted, however, that the reverse does not apply. If the tradesperson fails to perform the contract properly, and the injured party brings an action against the tradesperson for breach of contract, the tradesperson cannot claim that the contract is unenforceable because he or she does not possess a licence. The courts will hold the tradesperson liable for the damages that the other party suffered.[18]

These various forms of illegality and their impact on a contract were described by the court in the following manner:

> The effect of illegality upon a contract may be threefold. If at the time of making the contract there is an intent to perform it in an unlawful way, the contract, although it remains alive, is unenforceable at the suit of the party having that intent; if the intent is held in common, it is not enforceable at all. Another effect of illegality is to prevent a plaintiff from recovering under a contract if in order to prove his rights under it he has to rely upon his own illegal act; he may not do that even though he can show that at the time of making the contract he had no intent to break the law and that at the time of performance he did not know that what he was doing was illegal. The third effect of illegality is to avoid the contract ab initio and that arises if the making of the contract is expressly or impliedly prohibited by statute or is otherwise contrary to public policy.[19]

16. *Kocotis v. D'Angelo* (1958), 13 D.L.R. (2d) 69.
17. *Monticchio v. Torcema Construction Ltd. et al.* (1979), 26 O.R. (2d) 305.
18. *Aconley and Aconley v. Willart Holdings Ltd.* (1964), 49 W.W.R. 46.
19. *Archbolds (Freightage) Ltd. v. S. Spanglett Ltd.*, [1961] 1 Q.B. 374.

LEGALITY AT COMMON LAW: PUBLIC POLICY

Public policy

The unwillingness of the courts to enforce rights that are contrary to the gereral interests of the public.

There are a number of different circumstances at Common Law under which a contract will not be enforceable. Historically, these activities were contrary to **public policy**, and remain so today. Contracts designed to obstruct justice, injure the public service, or injure the state are clearly not in the best interests of the public: they are illegal as well as unenforceable. An agreement that is designed to stifle prosecution, for example, or influence the evidence presented in a court of law, is contrary to public policy.[20]

Contracts prejudicial to Canada in its relations with other countries would also be void on the grounds of public policy. An example of this type of contract would be an agreement between a resident of Canada and an enemy alien during a period when hostilities had been declared between the two countries.

Public policy and the *Criminal Code* also dictate that any contract that tends to interfere with or injure the public service would be void and illegal. For example, an agreement with a public official, whereby the official would use his or her position to obtain a benefit for the other party in return for payment, would be both illegal and unenforceable.

Another class of contract that would be contrary to public policy is a contract involving the commission of a tort, or a dishonest or immoral act. In general, agreements of this nature, which encourage or induce others to engage in an act of dishonesty or immoral conduct, will be unenforceable.

Contracts for debts where the interest rate charged by the lender is unconscionable are contrary also to public policy. If a lender attempts to recover the exorbitant interest from a defaulting debtor, the courts will not enforce the contract according to its terms but may set aside the interest payable or, in some cases, order the creditor to repay a portion of the excessive interest to the debtor.[21]

The law with respect to contracts of this nature is not clear as to what constitutes an unconscionably high rate, as there is often no fixed statutory limit for contracts where this issue may arise. Interest rates fall within the jurisdiction of the federal government, and, while it has passed a number of laws controlling interest rates for different types of loans, the parties in many cases are free to set their own rates. To prevent the lender from hiding the actual interest rate in the form of extra charges, consumer-protection legislation now requires disclosure of true interest rates and the cost of borrowing for many kinds of consumer-loan transactions. For others, the courts generally use, as a test, the rate of interest that a borrower in similar circumstances (and with a similar risk facing the lender) might obtain elsewhere.[22] To charge an interest rate in excess of 60 percent would also violate the *Criminal Code*, and render the creditor liable to criminal action.

CONTRACTS IN RESTRAINT OF TRADE

Contracts in restraint of trade fall into three categories:

(1) agreements contrary to the *Competition Act* (which were briefly explained earlier in this chapter),
(2) agreements between the vendor and purchasers of a business that may contain an undue or unreasonable restriction on the right of the vendor to engage in a similar business in competition with the purchaser, and
(3) agreements between an employer and an employee that unduly or unreasonably restrict the right of the employee to compete with the employer after the employment relationship is terminated.

Prima facie

"On first appearance."

Of these three, the last two are subject to Common Law public-policy rules that determine their enforceability. The general rule in this respect states that all contracts in restraint of trade are ***prima facie*** void and unenforceable. The courts will, however,

20. *Symington v. Vancouver Breweries and Riefel*, [1931] 1 D.L.R. 935.
21. *Morehouse v. Income Investments Ltd.* (1966), 53 D.L.R. (2d) 106.
22. *Miller v. Lavoie* (1966), 60 D.L.R. (2d) 495; *Scott v. Manor Investments*, [1961] O.W.N. 210.

enforce some contracts of this nature, if it can be shown that the restraint is both reasonable and necessary and does not offend the public interest.

Restrictive Agreements Concerning the Sale of a Business

When a vendor sells a business that has been in existence for some time, the goodwill that the vendor has developed is a part of what the purchaser acquires and pays for. Since goodwill is something that is associated with the good name of the vendor, and represents the propensity of customers to return to the same location to do business, its value will depend in no small part on the vendor's intentions when the sale is completed. If the vendor intends to set up a similar business in the immediate vicinity of the old business, the "goodwill" that the vendor has developed will probably move with the vendor to the new location. The purchaser, in such a case, would acquire little more in the sale than a location and some goods. The purchaser would not acquire many of the vendor's old customers and, consequently, may not value the business at more than the cost of stock and the premises. On the other hand, if the vendor is prepared to promise the purchaser that he or she will not establish a new business in the vicinity, nor engage in any business in direct competition with the purchaser of the business, the goodwill of the business will have some value to both parties. The value, however, is in the enforceability of the promise (a **restrictive covenant**) of the vendor not to compete or do anything to induce the customers to move their business dealings from the purchaser after the business is sold.

Restrictive covenant

A contractual clause limiting future behaviour.

The difficulty with the promise of the vendor is that it is *prima facie* void as a restraint of trade. The courts recognize, however, the mutual advantage of such a promise with respect to the sale of a business. If the purchaser can convince the courts that the restriction is reasonable and does not adversely affect the public interest, then the restriction will be enforced. It is important to note, though, that the court will not rewrite an unreasonable restriction to render it reasonable. It will only enforce a reasonable restriction, and nothing less. Such agreements are regularly encountered in the sale of almost all businesses, and thus for any given situation there are plenty of precedents (good and bad) which may be used for guidance.

COURT DECISION: Practical Application of the Law

Contract in Restraint of Trade—Reasonableness of Covenant in Restraint of Trade Where Vendor Becomes Employee of Purchaser of Business

***J.G. Collins Insurance Agencies Ltd. v. Elsley*, 1976 CanLII 48 (ONCA), appeal dismissed 1978 CanLII 7 (SCC).**

The defendant sold an insurance business and agreed to work for the purchaser as his sales manager. At that time, the defendant signed a covenant whereby he would not compete with the plaintiff as a general insurance agent for a period of five years within a given area. The defendant left the employ of the plaintiff and set up his own business within the five-year period and within the defined area.

THE APPEAL COURT: The general rule is that clauses restraining the scope of a man's future business activities, whether contained in agreements of employment or sale of a business, must be reasonable both as between the parties and with reference to the public interest. Otherwise such a clause is unenforceable as being in restraint of trade and contrary to public policy. Public policy is not a fixed and immutable standard but one which changes to remain compatible with changing economic and social conditions. The old doctrine that any restraint on trade was void as against public policy must be balanced against the principle that the honouring of contractual obligations, freely entered into by parties bargaining on equal footing, is also in the public interest. These competing principles of public policy are frequently in conflict in the commercial world and the question whether a particular non-competition agreement is void and unenforceable is one of law to be determined on a consideration of the character and nature of the business, the relationship of the parties, and the relevant circumstances existing at the time the agreement was entered into.

Courts recognize that some restraints must be imposed, otherwise the purchaser of a business could not with safety buy the goodwill of the business unless the vendor could be enjoined from setting up next door in competition. A similar problem would arise in certain employer and employee situations where, because of the confidential nature of the relationship, the employee has access to computer lists, trade secrets or other matters in which the purchaser or the employer has a proprietary interest.

.... There can be no doubt in the present case that the plaintiff had a substantial proprietary interest which he was entitled to have protected. At the time of purchase of the business, both parties recognized that goodwill represented by the customer lists had an economic value and that this business connection was a substantial asset. The defendant moved from owner-vendor to manager and ostensible owner of the business that had been his originally without any break in time or in his relationship with the customers. I am satisfied, as was the trial Judge that the covenant against competition is not invalid as an unreasonable restraint of trade and any challenges to the covenant on that ground cannot succeed.

Accepting this judgment, what business activities could the defendant conduct without offending the covenant? Moreover what sort of covenant in your opinion would serve as the threshold of unreasonableness, so it could be attacked?

The danger that exists with restrictions of this nature is the temptation on the part of the purchaser to make the restriction much broader than necessary to protect the goodwill. If care is not taken in the drafting of the restriction, it may prove to be unreasonable, and will then be struck down by the courts. If this should occur, the result will be a complete loss of protection for the purchaser, as the courts will not modify the restriction to make it enforceable. For example, Victoria operates a drugstore in a small town and enters into a contract with Paul, whereby Paul will purchase Victoria's business if Victoria will promise not to carry on the operation of a drugstore within a radius of 160 kilometres of the existing store for a period of 30 years.

In this case, the geographical restrictions would be unreasonable. The customers of the store, due to the nature of the business, would be persons living within the limits of the town, and perhaps within a few kilometres' radius. No substantial number of customers would likely live beyond a ten-kilometre radius. Similarly, the time limitation would be unreasonable, as a few years would probably be adequate for the purchaser to establish a relationship with the customers of the vendor. The courts, in this instance, would probably declare the restriction unenforceable; with the restriction removed, the vendor would be free to set up a similar business in the immediate area if she wished to do so.

Agreements of this nature might be severed if a part of the restriction is reasonable, and a part overly restrictive.

A manufacturer of arms and ammunition transferred his business to a limited company. As a part of the transaction, he promised that he would not work in or carry on any other business manufacturing guns and ammunition (subject to certain exceptions), nor would he engage in any other business that might compete in any way with the purchaser's business for a period of 25 years. The restrictions applied on a worldwide basis. The court, in this case, recognized the global nature of the business and held that the restriction preventing the vendor from competing in the arms and ammunition business anywhere in the world was reasonable, and the restriction enforceable. However, it viewed the second part of the restriction (which prevented the vendor from engaging in any other business) as overly restrictive. The court severed it from the contract on the basis that the two promises were separate.[23]

An unenforceable restriction may only be severed if the agreement's meaning will remain the same after the severance. This rule, the "blue-pencil rule," was once applied by English courts in a number of cases concerning restraint of trade where the unreasonable restriction was reduced to a reasonable covenant by the elimination of a few words.[24] This

23. *Nordenfelt v. Maxim Nordenfelt Guns & Ammunition Co. Ltd.*, [1894] A.C. 535.

24. *Goldsoll v. Goldman*, [1915] 1 Ch. 292.

approach was rejected in later cases,[25] however, and is now limited to "blue pencilling" an entire or severable unreasonable restriction, leaving the remainder of the restrictions intact. Even this practice would appear to be limited to the restraint of trade restrictions contained in agreements between the vendors and purchasers of a business.[26]

Balancing Business Freedom with Restraint of Trade

In the case of *Stephens v. Gulf Oil Canada Ltd. et al.* (1974), 3 O.R. (2d) 241, the judge discussed the freedoms that the courts attempt to balance in dealing with public policy as it relates to restraint of trade. The judge in the case outlined the policies in terms of "freedom," and the way in which the courts have reconciled the differences. He described the process in the following words:

> As usual in all such cases it is necessary to choose between or to reconcile, according to principle, important public policies that may be or appear to be in conflict with one another. Here, these policies are:
>
> Freedom under the law — Freedom of a person to conduct himself as he wishes subject to any restraint imposed upon him by the law.
>
> Freedom of commerce — The right of a person or firm to establish the business of his choice assuming that he has the capital and ability to do so, to enter a market freely, to make contracts freely with others in the advancement and operation of his business and to gain such rewards as the market will bestow on his initiative.
>
> Free competition — The right of business to compete for supplies and customers on the basis of quality, service and price, and the right of the public at large, whether industrial consumers or end users, to a choice in the market of competing goods and services of competitive quality and at prices established by competition.
>
> It will at once be apparent that these public policy principles are inherent in the private enterprise economy which depends for its proper working on a free competitive market. The market under this concept will determine what is produced, who produces it, who obtains the product or service and at what price.
>
> It must be recognized, however, that the Canadian economy is not universally a free market economy; it is a mixed economy, substantial areas of which are subject to regulation by Governments or public agencies in accordance with federal or provincial statutes. In these areas there may exist to a greater or lesser degree, economic or social controls which replace in whole or in part the free market operations in those areas. There, market forces as the main influence in economic decision-making may be replaced by state agencies or officials who in accordance with valid statutes may determine questions of entry, production, distribution and price. Inroads are thus made on the three public policy principles mentioned. But in the areas of the economy not so regulated, the private enterprise competitive market system prevails. This is the fundamental mechanism whereby economic resources are allocated to their most productive uses and economic efficiency is encouraged, inefficiency penalized and waste controlled.
>
> It will also be apparent that the public policy in relation to freedom of commerce, particularly freedom of contract, may and frequently does, come into conflict with the public policy in relation to free competition. For example, an agreement between two or more suppliers to deny supplies of an essential commodity or service to a new entrant to a market may foreclose that market to the newcomer and so destroy his right to establish or expand the business of his choice.
>
> Such conflicts are nowadays ordinarily adjusted, if public policy requires it, by appropriate legislation. But it is also open to the Common Law Courts to do so by the development of jurisprudence ....

25. *Putsnam v. Taylor*, [1927] 1 K.B. 637.
26. *Mason v. Provident Clothing Co.*, [1913] A.C. 724; *E.P. Chester Ltd. v. Mastorkis* (1968), 70 D.L.R. (2d) 133.

CASE IN POINT

A large oil corporation entered into an agreement with a small, local corporation to sell its oil products in a particular community only. The agreement contained a restrictive covenant whereby the local corporation agreed not to solicit customers of the oil supplier for 12 months after the agreement was terminated. The agreement was later terminated, and the local corporation permitted its principal shareholders to use its trucks for the purpose of selling oil obtained from another supplier. The large oil corporation brought an action for breach of the restrictive covenant.

The court held that the restrictive covenant was reasonable and enforceable against the local corporation, but not against the particular individuals who were not party to the covenant.

Imperial Oil Ltd. v. Westlake Fuels Ltd., 2001 MBCA 130 (CanLII).

Restrictive Agreements Between Employees and Employers

The law distinguishes restrictive agreements concerning the sale of a business from restrictive agreements made between an employer and an employee. In the latter case, an employer, in an attempt to protect business practices and business secrets, may require an employee to promise not to compete with the employer upon termination of employment. The legality of this type of restriction is generally subject to close scrutiny by the courts, however, and the criteria applied differ from those that the law has established for contracts where a sale of a business is concerned.

Balancing Private Interests

The justification for the different criteria is based upon the serious consequences that may flow from a restriction of the employee's opportunities to obtain other employment and to exercise acquired skills or knowledge. In general, the courts are reluctant to place any impediment in the way of a person seeking employment. As a consequence, a restrictive covenant in a contract of employment will not be enforced unless serious injury to the employer can be clearly demonstrated. This reluctance of the courts stems from the nature of the bargaining relationship at the time the agreement is negotiated between the employer and the employee. The employee is seldom in a strong bargaining position vis-à-vis the employer when the employment relationship is established. As well, the employment contract is often an agreement on the employer's standard form that the employee must accept or reject at the time. Public policy recognizes the unequal bargaining power of the parties by placing the economic freedom of the employee above that of the special interests of the employer.

In some cases, however, the special interests of the employer may be protected by a restrictive covenant in a contract of employment. The courts have held, for example, that when an employee has access to secret production processes of the employer, the employee may be restrained from revealing this information to others after the employment relationship is terminated.[27] The same view is taken when the employee has acted on behalf of the employer in his or her dealings with customers, and then later uses the employer's customer lists to solicit business for a new employer.[28] The courts will not, however, prevent an employee from soliciting business from a previous employer's customers under ordinary circumstances, nor will the courts enforce a restriction that would prevent a person from exercising existing skills and ordinary production practices acquired while in the employment relationship after the relationship is terminated.[29]

In contrast, contracts of employment, containing restrictions on the right of employees to engage in activities or business in competition with their employer while the employment relationship exists, are usually enforceable. This is provided that they do not

27. *Reliable Toy & Reliable Plastics Co. Ltd. v. Collins,* [1950] 4 D.L.R. 499.

28. *Fitch v. Dewes,* [1921] 2 A.C. 158; *Western Inventory Service Ltd. v. Flatt and Island Inventory Service Ltd.* (1979), 9 B.C.L.R. 282.

29. *Herbert Morris Ltd. v. Saxelby,* [1916] 1 A.C. 688.

unnecessarily encroach on the employees' personal freedom and that they are reasonable and necessary. The usual type of clause of this nature is a "devotion to business" clause, in which the employees promise to devote their time and energy to the promotion of the employer's business interests, and to refrain from engaging in any business activity that might conflict with it.

Over a lengthy period of time, two employees of a log scaling company prepared bids on log scaling contracts for their own profit. Eventually they succeeded in competing against their own employer, and secured a substantial portion of their employer's business. At trial, the court found them to be in breach of their general duty of good faith to their employer, and for misusing confidential information.[30]

A second type of restriction sometimes imposed by an employer is one requiring the employee to keep confidential any information of a confidential nature concerning the employer's business that should come into the employee's possession as a result of the employment. An employee subject to such a covenant would conceivably be liable for breach of the employment contract (and damages) if he or she should reveal confidential information to a competitor that results in injury or damage to the employer. Restrictions of this type are frequently framed to extend beyond the termination of the employment relationship, and, if reasonable and necessary, they may be enforced by the courts.[31] The particular reasoning behind the enforcement of these clauses is not based upon restraint of trade, but rather upon the duties of the employee in the employment relationship. The employer has a right to expect some degree of loyalty and devotion on the employee's part in return for the compensation paid to the employee. Actions on the part of the employee that cause injury to the employer represent a breach of the employment relationship, rather than a restraint of trade. It is usually only when the actual employment relationship ceases that the public policy concerns of the court come into play with respect to restrictive covenants.[32]

Balancing Public Interests

With some types of employment, where the service offered to the public by the employer is essential for the public good, the courts will generally take into consideration the potential injury to the public at large if a restrictive covenant in an employment contract is enforced. For example, if a medical clinic employs a medical specialist under a contract of employment prohibiting the specialist from practising medicine within a specified geographic area if the specialist should leave the employ of the employer, the courts might refuse to enforce the restriction even if it is reasonable, if the court concluded that enforcement would deprive the community of an essential medical service.

In the case[33] upon which this example was based, the court considered the criteria it would use to decide the question of enforceability of a restrictive covenant. The words of the judge are as follows:

> I find that the defendant signed the agreement, ex. 3, being well aware of the restrictive covenant therein and that he was under no duress or at any disadvantage in so doing. The agreement was supported by the mutual promises of the plaintiffs and in any event, it was under seal. Were it not for other aspects of this case, I would hold that it was binding on the defendant and the plaintiffs were entitled to the injunction which they seek.
>
> The first other aspect of the matter which, in my view, renders the covenant unenforceable is the public interest ....

30. *Woodrow Log Scaling Ltd. v. Halls*, [1997] B.C.J. No. 140 (Q.L.) (B.C.S.C.).
31. *Reliable Toy & Reliable Plastics Co. Ltd. v. Collins*, supra.
32. An interesting restrictive covenant concerning pension rights may be found in the case of *Taylor v. McQuilkin et al.* (1968), 2 D.L.R. (3d) 463.
33. *Sherk et al. v. Horowitz*, [1972] 2 O.R. 451.

In my view, the public interest is the same as public policy.

In the *Harvard Law Review*, vol. 42 (1929), p. 76, Professor Winfield, then of St. John's College, Cambridge, England, at p. 92, defines public policy as: "a principle of judicial legislation or interpretation founded on the current needs of the community." He goes on to say that public policy may change not only from century to century but from generation to generation and even in the same generation.

This author goes on to say at p. 97 that in ascertaining what is public policy at any time, one guide that Judges are certain to employ whenever it is available is statutory legislation in *pari materia*.

It will, therefore, be apposite to consider how statute law in the Province of Ontario affects medical care for the residents of this Province ....

Now the beneficial purpose of this legislation is to provide the widest medical care for the residents of the Province. Clearly this is in the public good. And I think it follows that the public are entitled to the widest choice in the selection of their medical practitioners.

In the light of this modern development, ex. 26 at the trial may have some special significance where in May, 1971, a resolution of the council of the Ontario Medical Association was passed as follows:

"RESOLVED that the Ontario Medical Association disapproves the concept of restrictive covenants in the contracts of one physician with another."

A further feature to be considered is whether a restrictive covenant between medical people tends to further limit the right of the public to deal with a profession which has a strong monopoly position. I believe that it does and I think that to widen that monopoly would be injurious to the public.

MANAGEMENT ALERT: BEST PRACTICE

"Illegality" conjures up images of things you can go to jail for, but this is a serious shortcoming in business thinking. It means acting contrary to any law, including administrative ones. Business operators must ask their lawyers what compliance responsibilities they face, for there is an often surprising array for any business. When a business is not in compliance, it may not be able to enforce its own rights, or defend against the claims of others. Competitors, suppliers, customers, and others may seek to exploit this vulnerability.

CLIENTS, SUPPLIERS *or* OPERATIONS *in* QUEBEC

When business contracts are concluded in Quebec, firms across Canada should be aware of its *Civil Code* provisions.

Capacity and Minority

Article 153. Full age or the age of majority is eighteen years. On attaining full age, a person ceases to be a minor and has the full exercise of all his civil rights.

Article 156. A minor fourteen years of age or over is deemed to be of full age for all acts pertaining to his employment or to the practice of his craft or profession.

Article 157. A minor may, within the limits imposed by his age and power of discernment, enter into contracts alone to meet his ordinary and usual needs.

S.Q. 1991, c. 64, Book One, Persons.

Legality

In cases where the *Competition Act* applies, the Act prevails over the *Civil Code* as a matter of federal jurisdiction. That aside, the rich term "ordre public" (literally translated as "public order," and somewhat more appropriately as "matters of public policy") acts as an umbrella requirement in the *Civil Code of Quebec*. Among others, note the following Articles:

1411. A contract whose cause is prohibited by law or contrary to public order is null.

1413. A contract whose object is prohibited by law or contrary to public order is null.

CHECKLIST FOR CAPACITY AND LEGALITY

Capacity

1. Assuming you possess capacity: Is the other party to your transaction of full age and is neither drunk nor insane?
2. Is the contract for a good or service which is a necessary?

If the answer to both is "no," the contract is in danger of repudiation by the other party.

Legality

1. Does the subject or object of your proposed contract contravene any law of Canada or a province, or a by-law of a municipality?
2. Does the subject or object of your proposed contract contravene "public policy?"
3. Where you may require a license or qualification to perform an act or service, do you possess such a license or qualification?

A "no" in the first and second case would render the contract illegal and unenforceable, and in the third case, "no" would prevent you from launching or defending a suit for enforcement.

SUMMARY

Not everyone has the capacity at law to enter into a contract that the courts will enforce. Minors or infants, as a general rule, are not liable under any contract they may make; but in the interests of the minor, the courts will hold the minor liable in contracts "for necessaries." Necessaries generally include food, shelter, clothing, employment, and education, but do not include contracts negotiated by an infant or minor engaged in business as a proprietor. All contracts, other than contracts for necessaries, are voidable at the infant's option.

Other persons may also lack the capacity to contract. Intoxicated persons and insane persons are bound to pay a reasonable price for necessaries contracted for while in a drunken or insane state. However, they will not generally be liable for other contracts if they can show that they were drunk or insane at the time of making the contract, and the other party was aware of it. They must, however, repudiate the contract promptly on becoming sane or sober. Bankrupt persons are prohibited from entering into contracts (except for necessaries) until they are discharged by the courts.

Corporations, First Nations and labour unions, because of their nature, may be subject to certain limitations on their capacity to contract. The limitations are based, in part, on their activities, with the rights and powers in respect to their contracts or collective agreements clearly delineated by statute. Apart from the restrictions placed upon these groups, and certain limitations on a few others, the courts recognize all persons that have reached the age of majority as persons with full capacity to contract.

The legality of the subject matter of a contract determines its validity and enforceability. Legality is generally determined on the basis of public policy or public interest; any agreement contrary to these considerations may be illegal or void. Contracts or agreements to commit a criminal offence are clearly illegal, and as such are unenforceable. So, too, are agreements between businesses that restrain competition contrary to the *Competition Act*. Agreements that violate specific laws are generally treated in this legislation as being either void or illegal. Other contracts in restraint of trade at Common Law may be simply unenforceable in a court of law. Some contracts, such as loan agreements that require the borrower to pay an unconscionable rate of interest, may only be enforceable in part; others that tend to offend the public good, such as agreements that injure the public service or obstruct

www.mcgrawhill.ca/olc/willes

justice, will not be enforceable at all. Some restraint of trade agreements that are beneficial to the parties and do not offend public policy may be enforced if the restrictions can be shown to be reasonable and necessary.

KEY TERMS

Infant, 151
Repudiation, 154
Ratification, 154
Ultra vires, 157
Public policy, 162
Prima facie, 162
Restrictive covenant, 163

REVIEW QUESTIONS

1. Why are courts reluctant to enforce restrictive covenants in contracts of employment? Under what circumstances would such a covenant be enforceable?
2. Discuss the application of public policy or the "public interest" to contracts of employment containing a covenant that would limit the ability of an employee to compete with the employer in a given area for a period of time.
3. If an adult entered into a contract without realizing that he was dealing with an infant, would the adult be in a position to enforce the agreement? Would your answer to this question apply under all circumstances?
4. If a minor or infant is engaged in business, how are the courts likely to view business contracts entered into by the minor?
5. Identify the three major classes of contracts considered to be in restraint of trade.
6. How does the capacity to contract of a minor differ from the capacity to contract of an insane or drunken person?
7. What are the limits on the powers of a "special-Act" corporation to bind itself in contract?
8. How does the law limit the capacity of a bankrupt person to enter into a binding contract?
9. Distinguish between "illegal" and "void" with respect to contract law.
10. What is the basis upon which the requirement of legality of a contract is determined?
11. Explain the risk assumed by an unlicensed tradesperson when entering into a contract to perform a service that may only be performed by a person possessing a licence.
12. Under what conditions or circumstances would an "agency of necessity" arise?
13. Explain the rationale behind the passage of the *Competition Act.*
14. Under what circumstances would a restrictive covenant in the contract for the sale of a business be enforceable? Why is this so, when contracts in restraint of trade are contrary to public policy?
15. Explain the reasoning behind the Common Law rules that limit the capacity of certain persons to bind themselves in contract.
16. Indicate the purpose of a devotion-to-business clause in a contract of employment, and explain the conditions under which this type of clause would be enforceable.
17. Why do the courts make certain exceptions to the general rule concerning the capacity of infants to bind themselves in contract?
18. An employee is caught stealing money from her employer, confesses to the theft, and agrees to repay the money taken. The employer, in response, promises not to report the incident to the police. Discuss the validity or enforceability of the employer's promise.

MINI-CASE PROBLEMS

1. X, aged 17, purchased a bicycle on credit for the purpose of transportation to and from her place of employment. She made no payments on the bicycle, and the seller brought an action against her for the debt. Discuss the issues raised in this case and render a decision.
2. A and B agree to carry on a business in partnership as hardware merchants. A and B agree that if either party wishes to end the partnership he must not carry on a similar business within 80 kilometres for a period of ten years. A leaves the business a year later and sets up a competing hardware business across the street from their old shop, which B continues to operate. Advise B and A.
3. A and B are rival salespersons in the same firm and each have earned 2 tickets to the Grey Cup based on last quarter's sales. Each wants to take two friends to the game, and thus A and B agree to compete for the top sales position in the next quarter, winner taking the Cup tickets of the other. If B takes top spot in sales, and A refuses to hand over the tickets, can B sue successfully?

4. Monica, at age 16, is an Olympic silver-medal swimmer. A swimwear company wishes to engage her to endorse a line of new racing suits. The endorsement package will require her to make public appearances, and appear for photo-shoots and in advertising copy. Discuss the difficulties that the firm may face in holding Monica to her commitments in the future.
5. On the above facts, if it is later found that Monica won her medal using banned performance-enhancing drugs, what challenges will the company face in terminating the endorsement arrangement?
6. Fred purchases a vibrant nightclub from Amin. Amin is a dynamic impresario with a track record of success in being able to gauge the pulse and tempo of entertainment tastes of young people, and being able to build clubs and bars suited perfectly for university students. Fred's club is located in a university town with a population of 350,000. What terms might be appropriate for a non-compete agreement between Fred and Amin? How dependent on population is your answer — if at all?

Case Problems for Discussion

CASE 1

A major manufacturer of advanced electronic game play units agreed with an equally large international retailer on certain terms for the production and distribution of the newest such game units. It was agreed that the manufacturer would provide the retailer with the first 1 million units to the exclusion of all other retailers. The retailer would receive its units for an exclusive product launch week, before any other units would be sold wholesale to other retailers. In turn, the retailer agreed to sell the units at precisely $299.99, this price being ideal, according to the manufacturer's market research. Every purchaser would also receive a T-shirt with the manufacturer's logo and $20 in gift coupons for the retail store.

If the business relationship later fell apart, which party could enforce which parts of this agreement? Could it be attacked by others?

CASE 2

Ilsa Sharp had operated Sharp's Wholesale Grocery for a number of years with limited success. Eventually, she found herself seriously in debt as a result of a number of unfortunate purchases of goods that spoiled before she could find a market for them, and she made a voluntary assignment in bankruptcy.

A few weeks later, while still an undischarged bankrupt, she purchased $600 worth of farm produce on credit from a farmer who was unaware of the fact that she was an undischarged bankrupt. Sharp sold part of the goods at a profit to a friend, and kept the remainder of the food for her own use.

When Sharp failed to pay the farmer, the farmer instituted legal proceedings to collect the $600.

Explain the rights of the farmer in this case, and render a decision.

CASE 3

Tuma entered into a rental agreement with Cross-Moto-Cycle for a one-year lease of a motorcycle by misrepresenting his age as being 20 when, in reality, he was only 17. The agreement that Tuma signed prohibited the use of the motorcycle in any race or contest, and required Tuma to assume responsibility for any damage to the machine while it was in his possession.

A week after Tuma acquired the machine, he made arrangements to enter a motorcycle race that was to be held in a nearby city. On his way to the race, he lost control of the motorcycle on a sharp turn in the road, and the machine was badly damaged in the ensuing accident.

Tuma refused to pay for the rental and the damage on the basis that he had not attained the age of majority and he was not liable on the contract.

Discuss the rights of Cross-Moto-Cycle, and comment on its likelihood of success if it should take legal proceedings against Tuma.

CASE 4

Linda and John, aged 17 and 18, respectively, entered into a partnership agreement to carry on business as a local parcel-delivery service. The business was to be operated under the name "L & J Parcel Delivery." In order to conduct the business, the two partners purchased a small truck on credit from a local truck dealer. The purchase agreement for the truck was signed "L & J Parcel Delivery" by John, who negotiated the purchase.

Linda purchased a motorcycle on credit from a local dealer for the twofold purpose of (1) delivering parcels, and (2) transportation to and from her home to the place of business of L & J Parcel Delivery, a distance of some eight kilometres. She had informed the seller that the motorcycle would be used by L & J Parcel Delivery and for personal transportation, but signed the purchase agreement in her own name only.

A few days before Linda's 18th birthday, John and Linda decided to cease their business operations. A substantial part of the purchase price remained owing to the sellers of both the truck and the motorcycle, and, with the intention of avoiding liability on the two purchase agreements, Linda repudiated the contracts and the partnership agreement. Over the next few months, John and Linda retained possession of the truck and motorcycle, while they argued between themselves and with the two sellers as to responsibility for the payment of the balance of the purchase price on each vehicle. Finally, after three months of fruitless discussion and argument, the sellers each brought an action against John and Linda for payment of the debts.

Discuss the rights of the parties and the issues that might be raised in the case. Render a decision.

CASE 5

A company owned a parcel of land upon which it wished to have a commercial building constructed. An architect was engaged to design the building, and a contractor was contacted to carry out the construction. Contracts were signed with both.

Before the construction was completed, it was discovered that the building violated a municipal by-law that required certain safety features to be included in the building. Neither the architect nor the contractor were aware of the by-law at the time they entered into their respective agreements with the company.

The safety features required by the by-law could be incorporated in the building at a cost of approximately $10,000, but the contractor refused to do so unless he was paid for the work as an "extra" to the contract price. The company refused to do so, and it withheld all payment to the contractor on the basis that the construction contract was illegal. The contractor then instituted legal proceedings against the company.

Explain the nature of the contractor's claim, and explain the defence raised by the company. Discuss the issue of responsibility in the case. Render a decision.

CASE 6

The Suburban Medical Centre was founded in 1981 as a medical clinic by eight physicians and surgeons. In 2008, the clinic advertised in the medical press for an obstetrician. Umesh, a medical specialist, answered the advertisement. Following an interview, Umesh was employed by the clinic and signed an employment contract that contained the following clause:

> Should the employment of the Party of the Second Part by the Parties of the First Part terminate for any reason whatsoever, the Party of the Second Part COVENANTS AND AGREES that he will not carry on the practice of medicine or surgery in any of its branches on his own account, or in association with any other person or persons, or corporation or in the employ of any such person or persons or corporations within the said City of Suburbia or within ten kilometres of the limits thereof for a period of five (5) years thereafter.

Umesh proved to be a difficult, but hard-working employee, and after some years an argument arose between Umesh and one of the founders of the clinic. As a result of the argument, Umesh resigned. He immediately set up a practice in the same city. The clinic continued to operate without the services of Umesh and later brought an action for damages and an injunction against him.

Discuss the factors the courts should consider in deciding this case. Render a decision.

CASE 7

In 2008, Herbert entered into the employ of TOPE Limited as an electrical engineer. He was employed to design electronic testing equipment, which the company manufactured. At the time he was hired, he signed a written contract of indefinite hiring as a salaried employee. The contract contained a clause whereby he agreed not to disclose any confidential company information. The contract also required him to agree not to seek employment with any competitor of the company if he left the employ of TOPE Limited.

Some years later Herbert was asked to develop a dwell tachometer suitable for sale to home mechanics through a particular hardware store chain under the chain's brand name. He produced a prototype in less than a week, and then he went to the president's office to discuss the development and production of the equipment.

During the course of the discussion, Herbert and the company president became involved in a heated argument over manufacturing methods. At the end of the meeting, the president suggested that Herbert might begin a search for employment elsewhere, as his job would be terminated in three months' time.

The next morning, Herbert went to the president's office once more, ostensibly to discuss the dwell tachometer. Instead, Herbert informed the president as soon as he entered the room that he no longer intended to work for the firm. He complained that the company had never given him more than a two-week vacation in any year, and that he often worked as much as 50 hours per week, with no overtime pay for the extra hours worked. In a rage, he smashed the dwell tachometer prototype on the president's desk, breaking it into a dozen small pieces. He then left the room.

The following week, Herbert accepted employment with a competitor of TOPE Limited to do a type of work similar to that which he had done at his old firm. He immediately developed a dwell tachometer similar in design to the previous model; he then suggested to the management of his new employer that they consider the sale of the equipment through the same hardware chain that TOPE Limited had contemplated for its product. The competitor was successful in obtaining a large order for dwell tachometers from the hardware chain a short time later.

TOPE Limited presented its new product to the hardware chain a week after the order had been given to the competitor, and only then discovered that Herbert had designed the equipment for that firm. The hardware chain had adopted the competitor's product as its own brand and was not interested in purchasing the product of TOPE Limited, in view of its apparent similarity in design.

TOPE Limited had expected a first year's profit of $31,000 on the dwell tachometer if they obtained the contract from the hardware chain.

Discuss the nature of the legal action (if any) that TOPE Limited might take against Herbert, and indicate the defences (if any) that Herbert might raise if TOPE Limited should do so.

CASE 8

Lui was a qualified journeyman electrician who was employed by a municipal public-utilities commission on a full-time basis. On weekends and evenings, he occasionally assisted friends who were constructing their own homes by installing their electrical wiring for them. In most cases he did the installation work gratuitously, but from time to time he would be given a sum of money in appreciation of his services.

One day, while on vacation at his summer cottage, a neighbouring cottage owner who was renovating his cottage approached Lui and inquired if he might be interested in taking on the job of rewiring the cottage. Lui thought about the offer, then agreed to do so, and a price was agreed upon. Lui would do the work and supply the materials (estimated at $950) for $1,400.

A few days later, Lui purchased the necessary materials and proceeded to rewire the cottage. Upon completion of the job, Lui presented his account for $1,550, which represented the cost of materials at $1,100 and his labour at $450.

Lui's neighbour refused to pay the account, insisting that the agreed price was $1,400. Lui's argument was that the $950 price quoted was only an estimate, and subject to change. The only firm part of his quote, he maintained, was his labour charge of $450.

The two parties continued to argue over the price for several months, and eventually Lui instituted legal proceedings to collect the account.

Identify and discuss the legal issues that might arise in this case that could affect Lui's right to recover payment. If you were called upon to act as counsel for the defendant, what inquiries would you make?

CASE 9

Mala Anand was a computer scientist with over 25 years of experience in the computer field. For the last five years of her 12 years with a large computer manufacturer, her mandate was to develop "next-generation" computer hardware. She was, in essence, responsible for most of the "high-tech" research in the area of data-reading technology in the company. She was also a recognized international authority in this highly specialized area of research.

A competitor offered Mala a position in its firm to carry out the same type of research, as it, too, was interested in producing next-generation data-reading equipment. Mala accepted the position and began working for her new employer.

Mala's previous employer then sought an injunction to prevent Mala from engaging in any work for the competitor that was similar to the work she had carried on at her previous place of employment.

The above employment scenario raises a number of significant legal and public policy issues. Identify and discuss these issues. In your answer, indicate how the courts attempt to deal with them.

CASE 10

Alice, a young woman who was 17 years of age, saw an advertisement by Silver Flatware Ltd. in a magazine that offered a 24-piece set of silver flatware for sale on the following terms: "$100 payable with order, and monthly payments of $50 each, payable over a three-year term."

The advertisement was accompanied by a coupon setting out the terms of payment and requiring the purchaser to provide his or her name and address and signature in the space provided.

Alice completed the coupon and mailed it, together with a cheque for $100, to the company. A few weeks later, the 24-piece set of silverware arrived by post.

Alice made a number of payments according to the terms of the agreement, then decided that she did not wish to continue with the agreement. A week before her eighteenth birthday, she wrote to the seller and repudiated the contract, but did not offer to return the silverware because she had lost several of the teaspoons.

The company and Alice then engaged in a protracted round of correspondence in which the company demanded a return of the silverware and the retention of all money paid as the price of her release from the agreement. Alice refused to return the silverware. She maintained that she was entitled to a return of the payments as she was a minor at the time she entered into the agreement.

The company, some 10 months later, brought an action against Alice for the balance of the purchase price.

Discuss the defences (if any) that Alice might raise in this case and render a decision.

CASE 11

A company produced a line of sophisticated toys using a secret special plastic and process. The company hired a chemist to work on the special plastic, but before she was allowed out into the lab, she was obliged to sign a confidentiality agreement with respect to all aspects of her work. Some years later, she left the firm and joined a competitor, and in time, that competitor began producing toys with plastics of very similar properties. The first company commenced legal action against her and her employer on the basis that she had revealed confidential information to her new employer.

Discuss the issues raised in this case, and how it may be decided if it reached the courts.

The Online Learning Centre offers more ways to check what you have learned so far. Find quizzes, Weblinks, and many other resources at www.mcgrawhill.ca/olc/willes.

www.mcgrawhill.ca/olc/willes

CHAPTER 10

The Requirements of Form, Writing and Privacy

Chapter Objectives

After study of this chapter, students should be able to:

- Identify situations in which the *Statute of Frauds* (or similar legislation) applies.
- Explain the importance of a written memorandum of a contract at Common Law and under the *Sale of Goods Act*.
- Recognize the chief obligations imposed by privacy legislation.

YOUR BUSINESS AT RISK

Some contracts may be verbal, but others must be in writing, or in a certain form to be enforceable. Business persons must respond to these differences to secure their rights and to limit their obligations. Business operators are also accountable to their customers and regulators for properly using and safeguarding the private information provided by customers, or generated in the course of business relationships.

FORMAL AND SIMPLE CONTRACTS

Under the law of contract, there are two general classes of contracts. A contract deriving its validity from the form that it takes is referred to as a formal contract, or, sometimes (under English law), a covenant. The second class of contract is the informal or simple contract, which may be implied, oral, or written. These two classes of contracts evolved in different ways under early English law, with the formal contract being the older of the two. It is important to bear in mind that both forms of contract were not normally matters that the early king's justices felt should fall under their jurisdiction. As a consequence, agreements in the nature of mutual promises were either under the jurisdiction of the ecclesiastic courts or the early local, or communal, courts. A breach of an agreement, then, took on the character of a breach of a promise that, if solemnly made, was considered to be a breach of faith, and, hence, a religious matter. Promises that bore no solemn or ritualistic aspect presumably fell under the jurisdiction of local or communal court for consideration. In any event, the king's courts did not concern themselves with the forerunners of modern contracts until the thirteenth century, when the royal court began to expand its jurisdiction.

Contractual disputes between merchants of a business-related nature were normally resolved by the merchants themselves, in accordance with the rules established by the various merchant guilds. Later, they were resolved by the courts, which largely adopted the merchant rules in rendering a decision. By the end of the thirteenth century, covenants in writing (except for debt), if under the seal of the promisor, were enforced by the king's courts on the basis that the impression of the seal was an expression of the promisor's intention to be bound by the promise made. While these early agreements were not the same as modern contracts, the use of a seal has continued to the present day.

Power of attorney

A legal document usually signed under seal in which a person appoints another to act as his or her attorney to carry out the contractual or legal acts specified in the document.

In most provinces, many kinds of agreements must still be made under seal to be enforceable. An example of a modern formal "covenant" would be a **power of attorney**. This is a formal document frequently used to empower a person to deal with the land of another. At Common Law, the grant of the power must be made under seal to be valid. Another formal covenant is a deed of land under the Registry System in a number of provinces in Eastern Canada. To be valid, the deed must be in writing, and signed, sealed, and delivered in order to convey the property interest in the land to the grantee. Apart from these, and a number of other special types of agreements that must be in a specific form and under seal, the formal agreement has been largely replaced by the second type of agreement, the informal written contract.

Informal contracts developed along a distinctly different route. In English law, the informal agreement, like the formal agreement, was initially enforced by the church or communal courts if some formality was attached to the agreement to render it morally binding. In this respect, the actions of the parties assumed immense proportions in determining the question of enforceability. The handshake, for example, rendered a promise binding; in this respect, the informal agreement and the early formal agreement were similar. At that time, the ceremonial aspects surrounding the agreement were important determinants of enforceability.

In the early cases, the courts would enforce the duties promised by persons in particular trades or professions if they improperly carried out their duties. No action would lie, however, if they simply did nothing to fulfill their promises, because there could be no trespass if a person did nothing. This early deficiency was remedied in part by the application of the action of deceit (also a tort) in the early sixteenth century. This provided a remedy if one party had fully performed, but if the other refused to do so. If neither party had performed there was still no remedy. It was not until the seventeenth century that the courts were finally prepared to enforce executory promises by way of a writ of assumpsit. From that point on, the theory of consideration and the modern concept of contract developed rapidly, and with them the enforceability of the informal contract by Common Law, rather than the law of tort. Today, the informal or simple contract does not depend upon a prescribed form for its enforceability. Had it not been for a statute passed in 1677,[1] no simple contract would have been required to be evidenced in writing under any circumstances to be enforceable at law.

The requirement is steadily evolving. Canada and many provinces have specifically provided for electronic memoranda (e-mail/Internet and computer archives) as evidence of a contract.

THE *STATUTE OF FRAUDS*

The particular statute that imposed the requirement of writing for certain informal contracts was the *Statute of Frauds* — an Act that was passed by the English Parliament and introduced to Canada as a colony. The law still remains as a statute in Nova Scotia, New Brunswick, Prince Edward Island and Ontario, even though it has been repealed in the balance of the Common Law Canadian jurisdictions.[2] The *Statute of Frauds* was originally passed following a period of political upheaval in England. It was ostensibly designed to prevent perjury and fraud with respect to leases and agreements concerning land. The statute went further than perhaps was intended at the time, and encompassed, as well, a number of agreements that today are simple contracts in nature.

Those provinces that have repealed[3] the *Statute of Frauds* have incorporated the requirement of writing into statutes applicable to particular transactions where a written record is considered necessary: usually land, long-term contracts, wills, and the like. In

1. *An Act for Prevention of Frauds and Perjures*, 29 Car. II, c. 3.
2. See, for example, *Statute of Frauds*, R.S.N.S. 1989, c. 442.
3. British Columbia, Alberta, Saskatchewan, Manitoba, Newfoundland and Labrador, Yukon, Northwest Territories, Nunavut.

the provinces retaining the Statute, contracts that cannot be fully performed within one year also must be in writing, but usually the Statute provides:

> No action shall be brought whereby to charge any executor or administrator upon any special promise to answer damages out of his own estate, or whereby to charge any person upon any special promise to answer for the debt, default or miscarriage of any other person, or upon any contract or sale of lands, tenements or hereditaments, or any interest in or concerning them, unless the agreement upon which the action is brought, or some memorandum or note thereof, is in writing and signed by the party to be charged therewith or some person thereunto by him lawfully authorized.

The effect of the *Statute of Frauds* (or any other statutory provision in provinces that have repealed it) is that none of the following may be brought in a court of law unless they are in writing and signed by the party to be charged: a contract concerning an interest in land, a promise by an executor or administrator to settle a claim out of his or her own personal estate, a guarantee agreement, or, in some provinces, a contract not to be fully performed in a year.

CASE IN POINT

Yvette entered into an agreement to produce graphic arts designs for advertising materials promoting the Metropolitan Opera Company in the upcoming year. The year would feature a series of productions in which the lead role would be played by an internationally famous opera singer, whose image would be featured in posters and other materials. The verbal agreement was made on August 14, with work to begin in October, to run through performances in December of the following year. Her work would not be completed until she prepared the art work in November for the final performances in December. Before the contract was reduced to writing, the opera star retired, and the company notified Yvette that her services were not required. Since the contract between Yvette and the company could not have been completed within one year, it had to be in writing to be enforceable against the opera company.

The law does not prohibit or render void these particular agreements if they do not comply with the statute or equivalent provision — it simply renders them unenforceable by way of the courts. The agreement continues to exist and, while rights cannot be exercised to enforce the agreement, it may be possible to appeal to the courts in the event of breach under certain circumstances. For example, if a party had paid a deposit to the vendor in an unwritten agreement to buy land, the vendor's refusal to convey the land would entitle the prospective purchaser to treat the agreement as at an end and recover his or her deposit. The courts would not enforce the agreement, however, since it would not be evidenced in writing and signed by the vendor. The agreement was caught by the statute, but once it was repudiated the purchaser could bring an action to recover the deposit.

The justification for the statutory requirement is obvious. Each of the four particular kinds of contracts at the time were agreements that were either important enough to warrant evidence in writing to clearly establish the intention of the particular promisors to be bound by the agreement, or the nature of the agreement was such that some permanent form of evidence of the terms of the agreement would be desirable for further reference.

The application of the statute to each of these contractual relationships produced a number of responses by the courts to avoid the hardships imposed by the law. Each response was an attempt to assist innocent parties who were unaware of the implications of the lack of evidence in writing of their agreement.

Contracts by Executors and Administrators

The protection that the statute or equivalent provision provides to the executor or administrator of an estate from a claim that the executor promised to answer for a debt or default out of his or her own estate is perhaps the most justifiable reason for the continued existence of the statute. An executor or administrator undertakes to collect, care for, and distribute the assets of a deceased person, and essentially to keep the assets of the deceased's estate separate from his or her own personal funds. However, an executor might be tempted to personally pay outstanding debts of the deceased's estate should the state of affairs of the estate render prompt payment inopportune. This temptation might stem from a concern to protect the good name of the deceased, or it might simply provide the executor with immediate relief from persistent creditors. In any event, it is important to note that an executor has no obligation to pay the debts of the estate out of personal funds. However, should an executor decide to do so, such an intention must be clearly indicated in writing, as the statute requires.

Assumed Liability: The Guarantee

Guarantee

A collateral promise (in writing) to answer for the debt of another (the principal debtor) if the debtor should default in payment.

The second legal agreement that the statute embraces is an agreement whereby a person agrees to answer for the debt, default, or tort of another. One particular type of agreement of this nature, which requires a memorandum to be in writing and bear the signature of the party to be charged, is the **guarantee**. This relationship always involves at least three parties: a principal debtor, a creditor, and a third party, the guarantor. The guarantor's role in a guarantee agreement is to provide a promise of payment in the form of a contingent liability. If the principal debtor does not make payment when the debt falls due, the creditor may then look to the guarantor for payment. The guarantor is never the party who is primarily liable. The guarantor's obligation to pay is always one that arises if and when the principal debtor defaults. The consideration for the guarantor's promise is usually based upon the creditor's act or promise to provide to the principal debtor goods on credit or funds, in circumstances where the creditor would not ordinarily do so. Because of the unique relationship between the parties, the guarantee must be in writing to be enforceable. The province of Alberta has added an additional procedural step, which a guarantor in certain circumstances must follow in order to be bound by his or her promise. Under the Alberta *Guarantees Acknowledgement Act*,[4] the guarantee must not only be in writing, but it must be made before a notary public, who must signify in writing that the guarantor understands the obligation. The statute does not apply to corporations that act as guarantors, nor does the statute apply to guarantees given in the sale of land or interests in chattels.

The legal nature of the guarantee was succinctly described by the court in the case of *Western Dominion Investment Co. Ltd. v. MacMillan*,[5] where the judge said:

> Reduced to its simplest terms, a guaranty is the promise of one man to pay the debt of another if that other default. In every case of guaranty there are at least two obligations, a primary and a secondary. The secondary — the guaranty — is based upon the primary, and is enforceable only if the primary default. It is so completely dependent upon the unchanged continuance of that primary, that if any, even the slightest, unauthorized changes are made in the primary, as, e.g., by extension of time for payment, or by reducing the chances of enforcing payment, as, e.g., by releasing any part of the securities, the secondary thereby falls to the ground. In other words, the secondary is not only collateral to, but is exactly co-extensive with the primary, as the primary existed when the secondary came into existence. Lastly, if the secondary obligor pays the debt he is entitled, as of right, to step into the creditor's shoes.

If a principal debtor fails to make payment when required to do so, the creditor may call upon the guarantor to pay. The guarantor is then liable for payment of the principal debtor's indebtedness. If the guarantor pays the obligation, the guarantor may demand an assignment of the debt. Once the debt is paid, the guarantor possesses the rights of the creditor, and may demand payment from the debtor if he or she should choose to do so. (See Figure 10–1.)

4. R.S.A. 2000, c. G-11.
5. [1925] 1 W.W.R. 852.

FIGURE 10–1 **Guarantee Relationship**

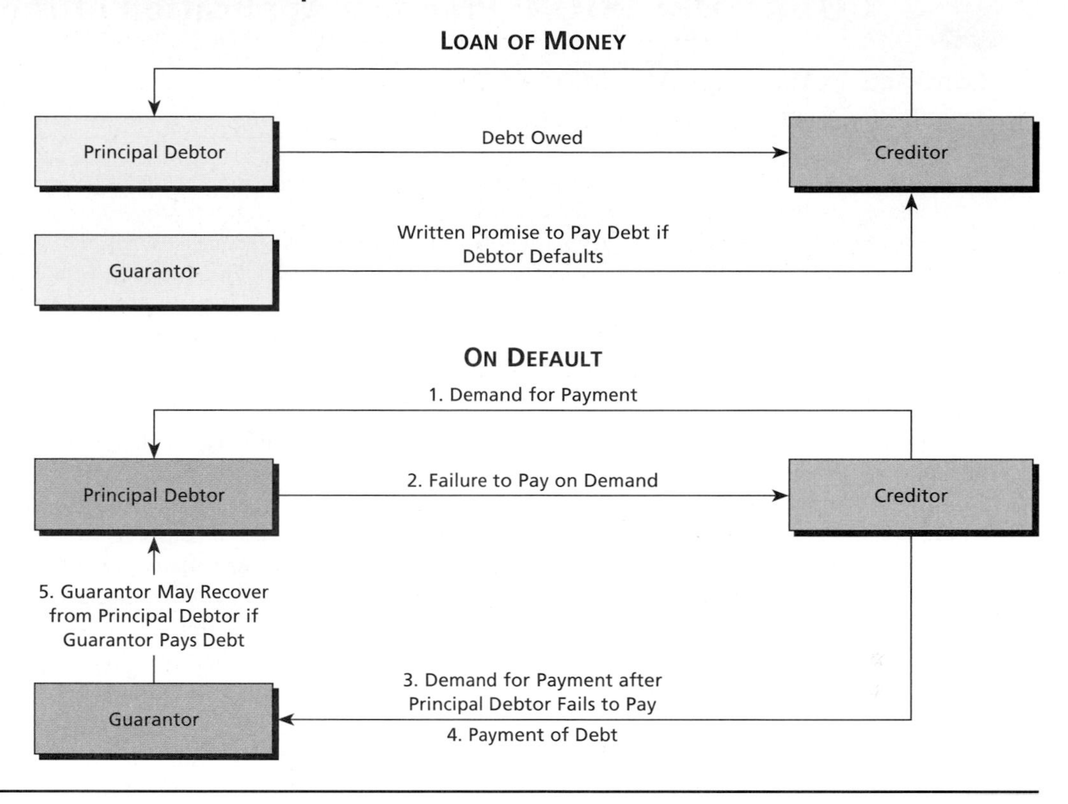

The distinction between a guarantee and a situation in which a person becomes a principal debtor by a direct promise of payment is important. If the promise to pay is not conditional upon the default of the principal debtor, but a situation where both parties become principal debtors, then the agreement need not be in writing or signed to be enforceable. By the same token, the third party can ask the creditor to release the principal debtor from the debt and promise to assume payment of the indebtedness personally. This transaction would also be outside the statute, because the agreement would simply be to substitute principal debtors. (See also Figure 10–1.)

A guarantee agreement between parties is not a simple arrangement, because the guarantor's potential liability is of a continuous nature. Consequently, the requirement that the guarantee be reduced to writing and signed by the guarantor is not unreasonable. As with any agreement extending over a long period of time, memories become hazy, facts may be forgotten, and interpretations may change. Far from being onerous, the requirement of evidence of the agreement in writing makes good sense. As a result, the courts have not attempted to circumvent the statute with respect to guarantees to avoid injustice. One form of relief that the courts have employed in guarantee cases, however, relates to agreements made between the creditor and principal debtor subsequent to the guarantee agreement. If these two parties alter the security that the guarantor may look to in the event of default, or alter the debt agreement without the consent of the guarantor, the alteration may release the guarantor. If the change in the agreement is detrimental to the guarantor, the courts will normally not enforce the guarantee if the principal debtor should later default.

Addy Finance Co. loans Bristo Construction Co. $100,000 on a promissory note guaranteed by Columbus Construction Co. If Addy Finance Co. and Bristo Construction Co. later agree to a higher interest rate on the promissory note without Columbus Construction Co.'s consent, Columbus Construction Co. may be released from its guarantee if Bristo Construction Co. should later default on payment of the loan.

COURT DECISION: Practical Application of the Law

Contracts in Writing—Alteration of Terms

***Manulife Bank of Canada v. Conlin et al.*, 1994 CanLII 1357 (ONCA), appeal dismissed 1996 CanLII 182 (SCC).**

In 1987, the defendant Dina Conlin gave a mortgage to the bank in the amount of $275,000 on an apartment building that she owned. As additional security, the bank required Ms. Conlin's spouse, John Conlin, and his company to guarantee the debt. The mortgage was renewed in 1990 at a higher interest rate without the knowledge of the guarantors, as Dina and John had separated and were living apart by this time. In 1992 the mortgage went into default, and the bank obtained summary judgment against the mortgagor and the guarantors. The guarantors then appealed the decision on the basis that the renewal of the mortgage at the higher interest rate without the guarantors' knowledge or consent released the guarantors from their guarantee.

THE COURT OF APPEAL: John Joseph Conlin appeals to this court from the decision of Killeen J. I would allow his appeal. While the two guarantors could have contracted out of their equitable rights to the extent of agreeing to this renewal agreement, it is not clear to me that they have done so. If the contention of the respondent mortgagee is accurate, the mortgagor and mortgagee could extend indefinitely the mortgagor's liability on the principal debt by successive renewal agreements at any interest rate, no matter how onerous, and the surety would have no right to notice of these extensions to his contingent liability nor any ability to terminate the liability. The portion of cl. 7 that authorizes the renewal agreement appears to me to contemplate something more than the normal accommodation of a debtor by a creditor in order to facilitate the debtor meeting his obligation. These accommodations often benefit the guarantor as well and are clearly contemplated by the language used in this guarantee. The renewal agreement, however, brings about a change that is not plainly unsubstantial or necessarily beneficial to the guarantor.... These latter changes must clearly be contemplated by the language of the guarantee to bind the guarantors. ... On balance, and keeping in mind that these documents were all drawn and presented by the mortgagee, I conclude that the renewal agreement was a material change to the original mortgage debt not contemplated by the language of the guarantee and has the effect of releasing the guarantors from their obligations as sureties.

The only person to benefit from an alteration in terms under guarantee is the guarantor itself. This is perilous to the underlying business because if at that later point the guarantor refuses to accept the alteration, the only choices are for the transaction to proceed without the guarantee, continue under the former terms, or be terminated (if the agreement provides for it).

A QUESTION OF ETHICS

A mother provides a bank a guarantee of the business debts of her daughter's company "up to $100,000." If the daughter's business receives the funds and repays them properly, is it fair that the guarantee survives the end of the loan? Consider this in light of the daughter returning to the bank for a second $100,000, for a more risky business venture. The guarantee arrangement has not changed, but the mother's perception of risk may be different. Should there be a positive duty on the bank to inform the mother before advancing the second $100,000?

Assumed Liability: Tort

A second promise of a somewhat similar nature to the guarantee is also covered by this particular section of the *Statute of Frauds* or equivalent statutory provision. Any agreement whereby a third party promises to answer for the tort of another must be in writing and be signed by the party to be charged, otherwise the promise will not be enforceable. This is not unlike the guarantee, but it applies when a third party promises to compensate a person who is injured by the tortious act of another, rather than by the person's failure to pay a debt. For example, Thompson, Jr., a young man aged 16 years, carelessly rode his bicycle on the sidewalk and collided with Varley. The collision caused injuries to Varley and placed her in the hospital. Thompson, Sr., Thompson, Jr.'s father, promised to compensate Varley for her injuries, if Varley would promise not to sue Thompson, Jr. If Varley wishes to enforce the promise of Thompson, Sr., she must insist that Thompson,

Sr. put his promise in writing and sign it. Otherwise it would be caught by the *Statute of Frauds*, and it would be unenforceable against Thompson, Sr.

Contracts Concerning Interests in Land

Of the remaining two kinds of agreements subject to the statute, the requirement of writing for contracts concerning the sale of land (or other dealing) has given the courts the most concern. The vagueness of the wording initially gave rise to much litigation. This forced the courts to struggle with an interpretation that would limit the application of the statute to those cases concerned specifically with the sale or other disposition of interests in land. The courts gradually excluded agreements that did not deal specifically with the land itself, agreements concerned with the repair of buildings, and contracts for "room and board." A great many other agreements that were remotely concerned with the disposition of land were also held to be outside the statute. For those cases encompassed by the statute, it was necessary to devise ways and means to prevent the law itself from being used to perpetrate a fraud on an unsuspecting party by way of an unwritten agreement.

Part performance

A doctrine that permits the courts to enforce an unwritten contract concerning land where certain conditions have been met.

The most important relief developed by the courts to avoid the effect of the statute was the doctrine of **part performance**. This concept allowed the courts, on the basis of equity, to enforce an unwritten agreement concerning land. The doctrine, unfortunately, is quite limited in its application. A party adversely affected by a failure to place the agreement in writing must be in a position to meet four criteria to successfully avoid the statute:

(1) The acts performed by the party alleging part performance must be demonstrated to be acts that refer only to the agreement of the lands in question, and to no other.
(2) It must be shown that to enforce the statute against the party who partly performed, for the lack of a written memorandum would perpetrate a fraud and a hardship on the person.
(3) The agreement must relate to an interest in land.
(4) The agreement itself must be valid and enforceable apart from the requirement of writing, and verbal evidence must be available to establish the existence of the agreement.[6]

To meet these four criteria is seldom an easy task.

> Anderson enters into a verbal agreement with Baxter to purchase Baxter's farm for $300,000. Anderson gives Baxter $100 in cash to "bind the bargain," and takes possession of the buildings and property. Anderson removes an old barn on the premises and makes extensive repairs to the house. After Anderson has completed the repairs, Baxter refuses to proceed with the transaction. He raises the absence of a written agreement as a defence, which will probably fail.

To meet the first criterion, the payment of $100 cash will not qualify, as it was not an act that would solely relate to this particular transaction (it could represent payment of rent). The acts of removing the old barn and repairing the house, however, might meet this requirement. A person would not normally undertake activities of this nature unless the person believed that he or she had some interest in the land. Therefore, the purchaser's acts would refer to such a contract, and to no other, under the circumstances.

The second criterion would also be met by Anderson's expenditure of time and expense in making renovations and removing the barn. These actions would represent acts that a person would only perform in reliance on the completion of the unwritten agreement. They would constitute a detriment or loss if the agreement was not fulfilled. To allow the landowner to refuse to complete the transaction at that point would constitute a fraud on the purchaser and represent unjust enrichment of the vendor.

6. See *Rawlinson v. Ames*, [1925] Ch. 96 at p. 114; *Brownscombe v. Public Trustee of the Province of Alberta* (1969), 5 D.L.R. (3d) 673.

The third criterion would be met by the nature of the agreement itself: it constitutes a contract for an interest in land, and one that equity would enforce by way of an action for specific performance.

The last criterion would be one that the purchaser might be able to prove by showing the court that the agreement, apart from the requirement of writing, contained all of the essential components of a valid agreement. This might be done by way of the evidence of witnesses who were present at the time of the making of the agreement, and who might be in a position to establish the terms.

An example of how the courts regard the doctrine of part performance was illustrated in the case of *Brownscombe v. Public Trustee of the Province of Alberta.*[7] The judge in that case described the event and the law as follows:

> In 1932 when Canada and the world in general were in a severe business depression, the plaintiff, whose home was in Prince George, B.C., and who was then 16 years of age, applied to the late Robert Marcel Vercamert at the latter's home, not far from Rockyford in Alberta, for work. The said Vercamert, a bachelor, somewhat severely crippled by heart trouble and able to do but little work on the farm where he lived and which he conducted, took the plaintiff into his home. On the evidence I find that plaintiff worked faithfully for his employer with but little financial reward for a considerable number of years. I find that on a number of occasions when the plaintiff thought of leaving Vercamert's employ he was dissuaded by the latter's promised assurance that on his demise the farm would go to plaintiff by will. In January 1961, Vercamert died intestate and this action is the result.
>
> The contract relating to land is within s. 4 of the *Statute of Frauds*, and there is no memorandum in writing. Therefore, part performance is necessary for the plaintiff to succeed on his claim for specific performance. Per Cranworth, L.C., in *Caton v. Caton* (1866), L.R. 1 Ch. App. 137 at p. 147: Part performance will afford relief from the operation of the statute "... in many cases ... when to insist upon it would be to make it the means of effecting instead of preventing fraud". However, not all acts done in pursuance of the unenforceable contract will constitute part performance in law. They may be found to relate only to a contract of service as in *Maddison v. Alderson* (1883), 8 App. Cas. 467, and *Deglman v. Guaranty Trust Co. of Canada and Constantineau*, [1954] 3 D.L.R. 785, [1954] S.C.R. 725, except where such acts are "unequivocally referable in their own nature to some dealing with the land which is alleged to have been the subject of the agreement sued upon ...": Per Duff, J., in *McNeil v. Corbett* (1907), 39 S.C.R. 608 at p. 611, approved by the Supreme Court of Canada in *Deglman*. The issue for decision by this Court is whether the acts relied upon by the appellant over the period 1932 to 1961 are acts which are "unequivocally referable in their own nature to some dealing with the land which is alleged to have been the subject of the agreement sued on", as stated by Duff, J. (as he then was), in *McNeil v. Corbett* (1907), 39 S.C.R. 608, and approved by this Court in *Deglman v. Guaranty Trust Co. of Canada and Constantineau*, [1954] 3 D.L.R. 785, [1954] S.C.R. 725.
>
> It is clear that not all the acts relied on as testified to by the appellant and his wife can be regarded as "unequivocally referable in their own nature to some dealing with the land", but in my view the building of the house on the lands in question in the years 1946 and 1947 at the suggestion of Vercamert almost, if not wholly, at the appellant's expense was, as the learned trial Judge found "unequivocally referable" to the agreement which the appellant alleged had been made and inconsistent with the ordinary relationship of employee or tenant.

Requirements for the Written Memorandum

To comply with either the *Statute of Frauds* or most statutory writing requirements, evidence of the contract in writing need not be embodied in a formal document. It is essential, however, to include in the written document all of the terms of the contract.

7. ibid.

The first requirement is that the parties to the agreement be identified either by name or description, and that the terms of the agreement be set out in sufficient detail that the contract may be enforced. For example, an agreement may consist of an exchange of letters that identify the parties, contain the offer made, describe the property as well as the consideration paid, or to be paid, and include a letter of acceptance. The two documents taken together would constitute the written memorandum. The final requirement is that the written memorandum be signed by the party to be charged. It is important to note that only the party to be charged need sign the memorandum. The party who wishes to enforce the agreement need not be a signatory, since the statute requires only that it be signed by the party to be charged.[8]

Parol Evidence Rule

Parol evidence rule

A rule that prevents a party from introducing evidence that would add to or contradict terms of a contract.

Of importance, where written agreements are concerned, is the **parol evidence rule**, which limits the kind of evidence that may be introduced to prove the terms of a contract. By this rule, no evidence may be adduced by a party that would add new terms to the contract, or change or contradict the terms of a clear and unambiguous written agreement. Evidence may only be admitted to rectify or explain the terms agreed upon, or to prove some fact such as fraud or illegality that may affect the enforceability of the agreement.

The application of the rule is not arbitrary, however, and the courts have accepted a number of different arguments that allow parties to circumvent the effect of the rule.

Condition Precedent

Condition precedent

A condition that must be satisfied before a contract may come into effect.

A **condition precedent**, as the name implies, is an event that must occur before the contract becomes operative and is the first exception to the parol evidence rule. The parties frequently place this term in the written agreement, but they need not do so. If the condition is agreed to by the parties, or, in some cases, if it can be implied, then the written agreement will remain in a state of suspension until the condition is satisfied. If the condition cannot be met, then the contract does not come into existence, and any money paid under it may usually be recovered.

For example, Allan and Brewster discuss the purchase of Brewster's excavation equipment by Allan. Allan agrees to purchase the equipment for $85,000 if she can successfully negotiate a loan from her banker. Allan and Brewster put the agreement in writing. However, they do not include in the agreement the term that the purchase is conditional upon Allan obtaining a loan for the purchase. While the parol evidence rule does not permit evidence to be admitted to add to the contract, the court will admit evidence to show that the agreement would not come into effect until the condition was met. The distinction here is that the evidence relating to the condition precedent does not relate to the contract terms but, rather, to the circumstances under which the written agreement would become enforceable.

Doctrine of Implied Term

Implied term

The insertion by the court of a standard or customary term omitted by the parties when the contract was prepared.

A second exception to the parol evidence rule is the application of the doctrine of **implied term**. Occasionally, in the writing of an agreement, the parties may leave out a term that is usually found in contracts of the type the parties negotiated. If the evidence can establish that the parties had intended to put the term in, and that it is a term normally included in such a contract by custom of the trade or by normal business practice, the courts may conclude that the term is an implied term. They could then enforce the contract as if it contained the term. Generally, the type of term that will be implied is one that the parties require in the contract in order to implement the agreement. It must be noted, however, that if the term conflicts in any way with the express terms of the agreement, the parol evidence rule will exclude it. Similarly, an express term may be incorporated in a written agreement by reference if (a) the agreement is a "standard form" type of contract, and (b) the term is expressed before the agreement is concluded. For example, in a parking

8. *Daniels v. Trefusis*, [1914] 1 Ch. 788; *McLean v. Little*, [1943] O.R. 202.

lot, a large sign, which states that the owner will not be responsible for any damages to a patron's vehicle, may be binding upon the patrons. This may be the case even though the limitation is not expressly stated on the front of the ticket, but is referred to in small print on the back.

Collateral Agreement

Collateral agreement

An agreement that has its own consideration, but supports another agreement.

A third important exception to the parol evidence rule is the **collateral agreement**. A collateral agreement is a separate agreement that the parties may make that has some effect on the written agreement, but that is not referred to in it. One of the difficulties with the collateral agreement is that it usually adds to, or alters, the written contract. If it were allowed at all times, it would effectively circumvent the parol evidence rule. For this reason, the courts are reluctant to accept the argument that a collateral agreement exists, unless the parties can demonstrate that it does, in fact, exist as a separate and complete contract with its own consideration. The application of this criteria usually defeats the collateral agreement argument because the collateral agreement seldom contains separate consideration from that of the written agreement. However, in those cases where a separate agreement does exist, the courts will enforce the collateral agreement even though it may conflict to some extent with the written one.

COURT DECISION: Practical Application of the Law

Contracts—Parol Evidence Rule—Collateral Warranty

***Gallen et al. v. Allstate Grain Co. Ltd. et al.* (1984), 9 D.L.R. 496 (B.C.).**

Allstate Grain Co. Ltd. was engaged in the purchase and sale of seed grains and sold the plaintiff, Gallen, buckwheat seeds. The seller assured the buyer that weeds wouldn't grow amongst the grain. The written purchase agreement subsequently entered into stated that the seller gave no warranty as to the germination of the seeds "or any other matter pertaining to the seed." The agreement also stated that the seller was in no way responsible for the crop. The seeds were planted, but the crop was destroyed by weeds. The purchaser sued the seller on the basis that the defendant's verbal statement was a collateral warranty. The trial judge found in favour of the plaintiff. The defendant then appealed the trial decision.

THE COURT: The parol evidence rule is not only a rule about the admissibility of evidence. It reaches into questions of substantive law. But it is a rule of evidence, as well as a body of principles of substantive law, and if the evidence of the oral representation in this case was improperly admitted, the appeal should be allowed.

The rule of evidence may be stated in this way: Subject to certain exceptions, when the parties to an agreement have apparently set down all its terms in a document, extrinsic evidence is not admissible to add to, subtract from, vary or contradict those terms.

So the rule does not extend to the cases where the document may not embody all the terms of the agreement. And even in cases where the document seems to embody all the terms of the agreement, there is a myriad of exceptions to the rule .... So, if it is said that an oral representation, that was made before the contract document was signed, contains a warranty giving rise to a claim for damages, evidence can be given of the representation, even if the representation adds to, subtracts from, varies or contradicts the document, if the pleadings are appropriate, and if the party on whose behalf the evidence is tendered asserts that from the factual matrix it can be shown that the document does not contain the whole agreement....

Is the oral representation a warranty? A warranty is one of the terms that may form a part of a contractual relationship and affect the scope of the relationship. It may be either a representation as to the existence of a present fact ("This car has traveled only 10,000 kilometres."); or it may be a promise to bear the risk of the loss that will flow from a failure of a fact to occur in the future ("This car is guaranteed rust-proof.") ... Mr. Justice Paris [the trial judge] concluded that Mr. Nunweiler's statement regarding weed control constituted a warranty. There is ample evidence to support that conclusion, much of it referred to by Mr. Justice Paris in his reasons. I do not think that it is open to me to consider that matter afresh, or, if I were to reach a different conclusion on the facts than Mr. Justice Paris, to substitute my view of the facts for his.

Since, in my opinion, there is no contradiction in this case between the specific oral warranty and the signed standard form Buckwheat Marketing Agreement, 1980, I have concluded that the warranty has contractual effect and that the defendant, Allstate Grain Co. Ltd., is liable to the plaintiffs for breach of that warranty. But if it were correct, in this case, to conclude that the oral representation and the Buckwheat Marketing Agreement, 1980 contradicted each other, then, on the basis of the facts found by the trial judge and his conclusion that the oral representation was intended to affect the contractual relationship of the parties, as a warranty, I would have concluded that, in spite of the strong presumption in favour of the document, the oral warranty should prevail.

I would dismiss the appeal.

With all of these exceptions to the parol evidence rule, one element is common. In each case, the modifying term precedes, or is concurrent with, the formation of the written agreement.

Subsequent Agreement

Subsequent agreement

An agreement made after a written agreement that alters or cancels the written agreement.

Any verbal agreement made by the parties after the written agreement is effected may alter the terms of the written contract[9] or cancel it.[10] The parol evidence rule will not exclude evidence of the **subsequent agreement** from the court. The reason for this distinction is that the subsequent agreement represents a new agreement made by the parties that has as its subject matter the existing agreement.

REDUCTION TO WRITING

It is not uncommon for business persons to enter into either verbal or written negotiations with a view to making a formal contract. During these negotiations the parties may reach agreement upon a sufficient number of key issues to agree in principle to proceed with a formal contract. This agreement would embody the issues agreed upon in principle but not include the details yet to be agreed upon. If a final agreement is reached and reduced to writing in its formal form, the negotiations would be complete. The parties sometimes do not proceed beyond the agreement in principle stage, however, and one party or the other may attempt to enforce the agreement in principle on the basis that an enforceable contract had been reached. The formal written agreement, in their view, would be merely the fine-tuning of the existing agreement.

If one party alleges that an enforceable agreement exists, the courts are generally obliged to determine the stage at which the parties intended to be bound by their negotiations. In the case of *MacLean v. Kennedy* (1965), 53 D.L.R. (2d) 254, the court noted that two principles of law must be kept in mind in dealing with this type of case:

> It appears to be well settled by the authorities that if the documents or letters relied on as constituting a contract contemplate the execution of a further contract between the parties, it is a question of construction whether the execution of the further contract is a condition or term of the bargain or whether it is a mere expression of the desire of the parties as to the manner in which the transaction already agreed to will in fact go through.
>
> The second principle of law to be kept in mind is that the material terms of the contract must not be vague, indefinite or uncertain. This principle has been variously stated.
>
> If an oral agreement is vague, indefinite or uncertain, it would appear that this fact may be taken into account in deciding whether the execution of a formal agreement is a condition or term of the oral agreement. At least this would seem to follow from certain observations of Meredith, C.J. in *Stow v. Currie* (1910), 21 O.L.R. 486 at pp. 493 and 494, and of Clute, J., at p. 496. In *Stow v. Currie*, the uncertain agreement was written, not oral, but it was held that it was intended to be subject to a new and formal agreement, the terms of which were not expressed in detail, and one reason for so holding was its uncertainty in certain respects.

9. *Johnson Investments v. Pagritide*, [1923] 2 D.L.R. 985.
10. *Morris v. Baron and Co.*, [1918] A.C. 1.

FRONT-PAGE LAW

Arrington settles contract spat with Redskins

Washington Redskins linebacker LaVar Arrington won't be getting the $6.5 million he claims was removed from the final draft of his contract. Instead, Arrington settled his long-standing dispute with the team by agreeing to a deal that could make him a free agent or net him a $3.25 million bonus if he makes two Pro Bowls in the next four years. The disagreement stemmed from an eight-year, $68 million contract Arrington signed near the end of the 2003 season. Arrington and his agent, Carl Poston, then claimed that the Redskins bilked the linebacker out of a promised $6.5 million bonus that was agreed upon but did not appear in the final contract. The Redskins denied the allegations, and Arrington filed a grievance through the players' union. Arrington's case did not appear strong because he and Poston signed the final contract — with Poston initialing every page — without the bonus included. Under the settlement, the Redskins have the option of giving Arrington a $3.25 million bonus if he makes the Pro Bowl twice in the next four years. If they decline to give him the bonus, Arrington can void the contract and become a free agent.

What rule of law is probably applied here?

Source: Excerpted from *Associated Press*, "Arrington settles contract spat with Redskins," August 26, 2005. Used with permission of the Associated Press © 2005.

MANAGEMENT ALERT: BEST PRACTICE

Every business is different, and preprinted legal documents purchased in stationery stores (including sales books, blank contracts, and leases) may not cover the unique requirements of a particular business. A lawyer should review these forms prior to their first use, because they can easily contain promises you do not wish to make and omit protections you may wish to have. Forms printed in the United States or in another province may conflict with your own provincial laws and therefore be worthless. It is better to avoid resort to the parol evidence rule than be forced to rely on it later.

SALE OF GOODS ACT

A second important statute that contains a requirement of writing is the *Sale of Goods Act*.[11] The particular requirement of writing was originally a part of the *Statute of Frauds*, and had remained there for several hundred years. After 1893, when separate legislation concerning the sale of goods was passed in England, the requirement of writing was removed from the *Statute of Frauds* and embodied in the new Act. The provincial legislatures in Canada copied the English legislation and varied the value of the goods to which the requirement of evidence of the agreement in writing applied. The legislation fortunately provided, as well, a number of activities on the part of the parties that would permit them to enforce the agreement, even though the contract was not evidenced by a written memorandum. These activities include the payment of a deposit, acceptance of delivery of part of the goods, or the giving of "something in earnest" (such as a trade-in) to bind the bargain. Because the parties normally comply with one of the exceptions, if the contract of sale is not in writing, the requirement does not pose a hazard for most buyers and sellers. Of more importance today is the consumer-protection legislation applicable to many kinds of contracts. This legislation often requires certain types of sales contracts to be in writing. It also imposes penalties for the failure to provide consumers with a written purchase agreement, disclosing information concerning the sale and any credit terms. This Act and its application are discussed in detail in Chapter 22, "The Sale of Goods."

11. All provinces (except Quebec) have legislation pertaining to the sale of goods.

CLIENTS, SUPPLIERS *or* OPERATIONS *in* QUEBEC

With respect to legal agreements, loans, or deeds ("juridical acts"), the Quebec *Civil Code* imposes the following requirements of writing on business transactions within its jurisdiction:

2861. Where a party has been unable, for a valid reason, to produce written proof of a juridical act, such an act may be proved by any other means.

2862. Proof of a juridical act may not be made, between the parties, by testimony where the value in dispute exceeds $1,500. However, failing proof in writing and regardless of the value in dispute, proof may be made by testimony of any juridical act where there is a commencement of proof; proof may also be made by testimony, against a person, of a juridical act carried out by him in the ordinary course of business of an enterprise.

2863. The parties to a juridical act set forth in a writing may not contradict or vary the terms of the writing by testimony unless there is a commencement of proof.

Willes: "commencement of proof" is other established evidence which would directly support such oral (or, at least, other than written) testimony.

CHECKLIST FOR FORM AND WRITING

1. Is it simply prudent business for the agreement be placed in writing?
2. Does a *Statute of Frauds*, a *Sale of Goods Act* or other equivalent statutory provision require the agreement to be in writing to be enforceable in court?
   - If the answer to either is yes, put the agreement in writing.
3. Is the written agreement subject to:
   a. Fulfilment of a condition precedent?
   b. Any possible implied term?
   c. Collateral agreement?
   d. Subsequent agreement?
   - If not, then the written agreement should stand on its own, protected by the parol evidence rule.

PRIVACY LEGISLATION

While not strictly a contract formation issue, business persons will gather considerable quantities of personal information about the person with whom they are creating a contract. This information must be gathered and maintained (at least) according to *PIPEDA*, the federal government's *Personal Information Protection and Electronic Documents Act.*[12] Provincial Acts may require even higher standards of accountability.

One goal of the *PIPEDA* was to ensure effective alternatives to paper documents for all manner of modern government operations — information, filings, payments, secure signatures, and submissions of evidence. Secondly, on the personal information side, the Act requires essentially all private sector enterprises and health care providers to obtain the consent of individuals to collect, use or disclose personal information for commercial activity or health care. Further, such information must only be used for the pre-identified purposes for which it was collected, and organizations are legally liable for maintaining privacy and control over that personal data.

The *PIPEDA* applies across Canada as a federal Act, where a province does not enact federally-approved substantially similar legislation of its own. The provincial electronic business Acts have tended to separate[13] out the twin goals of privacy and

12. S.C. 2000, c. 5

13. With some jurisdictions using two separate Acts.

commercial certainty in new technological media. It is essential for business to comply with the requirements of *PIPEDA* (or its provincial counterparts), not only just to obey the law, but to meet the expectations of clients. With the ever-present threat of identity theft, businesses who do not secure their client's data will find themselves liable for their failures.

Application of the Act and Personal Information

Personal information

Information about an identifiable individual.

As of 2012, the privacy legislation of British Columbia, Quebec and Alberta meets *PIPEDA* standards, and for all other jurisdictions, the *PIPEDA* has direct application to all commercial activity. The Act covers all **personal information** collected, used and retained by an organization engaged in commercial activity, with the very limited exceptions of journalistic, artistic or literary purposes.

Personal information includes, but is not limited to: name, date of birth, age, medical facts, ethnicity, personal description, employee records, earnings, credit and loan files, survey responses, beliefs, opinions or intentions. Name, title and address as business contact information are exempt from *PIPEDA* provisions.

The central obligation of *PIPEDA* is that all organizations conducting commercial activity must comply with the 10 principles of privacy, as set out below:

CHECKLIST OF PIPEDA COMPLIANCE REQUIREMENTS FOR BUSINESS

1. **Accountability** — An organization is responsible for personal information (PI) under its control and shall designate an individual(s) accountable for the compliance.
2. **Identifying Purposes** — The purposes for which PI is collected shall be identified and documented at or before the time the information is collected.
3. **Consent** — The knowledge and consent of the individual are required for the collection, use, or disclosure of PI, except where inappropriate (with high standards for disclosure) and subject to withdrawal of consent.
4. **Limiting Collection** — The collection of PI shall be limited to that which is necessary for the purposes identified by the organization. Information shall be collected by fair and lawful means.
5. **Limiting Use, Disclosure, and Retention** — PI shall not be used or disclosed for purposes other than those for which it was collected, except with the consent of the individual or as required by law. PI shall be retained only as long as necessary for the fulfilment of those purposes.
6. **Accuracy** — PI shall be as accurate, complete, and up-to-date as is necessary for the purposes for which it is to be used.
7. **Safeguards** — PI shall be protected by security safeguards appropriate to the sensitivity of the information.
8. **Openness** — An organization shall make readily available to individuals specific information about its policies and practices relating to the management of PI.
9. **Individual Access** — Upon request, an individual shall be informed of the existence, use, and disclosure of his or her PI and shall be given access to that information. An individual shall be able to challenge the accuracy and completeness of the information and have it amended as appropriate.
10. **Challenging Compliance** — An individual shall be able to address a challenge concerning compliance with the above principles to the designated individual(s) responsible for compliance.

The objective of *PIPEDA* has been to establish and encourage best practices, rather than to entrench an enforcement regime. The Office of the Privacy Commissioner of Canada has this ombudsman or conciliation role, but it also has the capacity to investigate complaints. After investigation, a complaint pursuant to *PIPEDA* can also be heard in Federal Court, which can award damages to a complainant.

Proposed amendments to *PIPEDA* in 2011 will create a positive obligation to notify the Commissioner in cases of material breach of security around personal information holdings. Further, the individuals concerned must be notified where the breach of security around their information creates a real risk of significant harm. This feature already exists in Alberta's privacy legislation, and as it exceeds current federal requirements, is the governing law in the case below.

In 2011, a contractor managing a customer loyalty program for Best Buy Canada Ltd., and Air Miles Reward Program suffered a major data security breach. The breach compromised personal information of their customers, and once notified by the contractor, Best Buy and Air Miles notified their customers, pursuant to the Alberta *PIPA*[14], which since 2010 has required mandatory notification of a security breach. The Alberta Privacy Commissioner subsequently confirmed[15] this as the correct course of action, even where the breach occurs with a third party. The expected federal *PIPEDA* amendments will likely mirror this expectation.

Significant harm is not limited to bodily harm, but also includes humiliation, damage to credit records, reputation and relationships, business opportunities, financial loss and identity theft. Whether the risk of that harm is real requires consideration of both the sensitivity of the information and the probability of its misuse. Where the risk of significant harm is real, the notification must be conspicuous, prompt and meaningful, with sufficient detail for it to be understood and to permit mitigating steps to be taken.

SUMMARY

Formal and informal contracts developed along distinctive lines, and each has a different legal history. Formal contracts generally derive their validity from the form they take. They may be required to effect particular transactions, for example. All formal contracts are evidenced by writing, and most are subject to the requirement that they be signed, sealed, and delivered before they become operative.

Informal contracts may be written, oral, or, in some cases, implied. However, certain informal contracts (those subject to the *Statute of Frauds*) must be evidenced by a memorandum in writing setting out their terms, and must be signed by the party to be charged before they are enforceable by a court of law.

Written contracts are subject to a number of rules, principles, and doctrines that have developed to mitigate the hardship that is sometimes imposed by the statute. A notable example of one of these special rules is the doctrine of part performance, which may be applied in some cases where land is sold without written evidence of the transaction.

Not all of the rules, however, are designed to prevent hardship. Written agreements are also subject to the parol evidence rule, which excludes evidence of any prior or concurrent agreement that might add to, or contradict, the terms of the written agreement in question. Exceptions to this rule, nevertheless, exist in the form of

14. *Personal Information Protection Act*, SA 2003, c.P-6.5, as amended.

15. Alberta Privacy Commissioner, Breach Notification Decisions P2011-ND-011 and -012.

conditions precedent, implied terms, and genuine collateral agreements. All of these may take either a written or oral form. Additionally, agreements made subsequent to a written agreement may alter or terminate the contract, even though the subsequent agreement is verbal in nature.

Special requirements for consumer contracts have been established by legislation in many provinces in recent years. These statutes usually require certain kinds of transactions to be in writing. While ostensibly designed to require the seller to disclose information concerning the sale to the buyer, the statutes also require the written memorandum to contain specific information, otherwise fines or penalties may be imposed upon the seller. The contract of sale and the legislation pertaining thereto are examined at length in Chapter 21, "The Sale of Goods" and Chapter 27, "Consumer-Protection Legislation."

Privacy legislation creates the responsibility of businesses to be accountable for the personal information they collect, hold, and use in the course of commercial activity. The fundamental concepts of privacy are based on consent of the individual, minimal use, and a commitment by the business enterprise to safeguard the information. The limitations and required safeguards represent national minimum standards, and continue to evolve.

KEY TERMS

Power of attorney, 176
Guarantee, 178
Part performance, 181
Parol evidence rule, 183
Condition precedent, 183
Implied term, 183
Collateral agreement, 184
Subsequent agreement, 185
Personal information, 188

REVIEW QUESTIONS

1. Explain the rationale behind the general Common Law rule stating that an agreement in writing may be terminated by a subsequent verbal agreement. Is this always the case, or is it subject to exception?
2. What are the legal implications of failing to comply with the requirements of writing under the Statute of Frauds or other equivalent provisions?
3. Distinguish a guarantee from an indemnity. How does the *Statute of Frauds* affect these two relationships?
4. Explain the doctrine of part performance and the rationale behind the establishment of the doctrine as a means of avoiding the *Statute of Frauds.*
5. Describe the minimum requirements for a written memorandum under the *Statute of Frauds* or other equivalent provisions.
6. What exceptions from the *Statute of Frauds* have the courts established for contracts that do not require immediate performance?
7. How does the parol evidence rule affect evidence related to a contract in writing?
8. Explain the effect of a collateral agreement and the doctrine of implied term on a written agreement subject to the parol evidence rule.
9. Explain the effect of the *Statute of Frauds* or other equivalent provisions on the law of contract.
10. Why have requirements of privacy in consumer information been created now, when they have not been required in the past 500 years of commercial relations?

MINI-CASE PROBLEMS

1. X enters into a verbal agreement with Y to purchase Y's farm for $150,000. X pays Y a deposit of $1,000 cash. What are X's rights if Y later refuses to go through with the agreement?
2. How would your answer to the problem in question 1 differ if X refused to fulfill his part of the agreement?
3. A offers to buy B's sailboat for $10,000, provided that he can obtain a loan of $5,000 from his banker. A and B put the agreement in writing, but the agreement does not mention the loan. If A cannot borrow the $5,000, can B sue A for failing to comply with the agreement of purchase?
4. Alex agreed to purchase a used light aircraft for $180,000 (as is) from Aeroventure Aviation Company. On the day agreed for transfer, Alex appeared with his certified cheque for $180,000

and was met with an invoice for $207,000 being the price plus provincial sales tax and GST. Discuss the issues raised as a result.

5. Tom agreed to sell his car to George for $11,000 and the two exchanged e-mails to that effect. Later meeting at a party, George asked if Tom would throw in his boat trailer along with the car. Tom agreed. A week later, George brought an $11,000 certified cheque to Tom's home and received the keys to the car. The boat trailer remained locked in Tom's garage. In response to George's protest, Tom said: "We had no agreement about a trailer." Advise George.
6. Zhao, the owner of a thriving drug store, discusses the trade with Victor, the son of a friend. Victor decides to open a drug store, and Zhao loans him the money. As matters turn out, Victor purchases a franchise that is as much a general store as it is a drug store. Zhao considers this venture much more risky and demands that Victor find another investor willing to buy out Zhao. What argument in law can Zhao raise to suggest that Victor must do so?
7. Betacorp holds personal information about its customers. What limitations exist as to how Betacorp may use that information?
8. The All-Prov Insurance Company holds personal information about its clients, including health questionnaires and personal profiles. What rights does a client of All-Prov have regarding this information?

Case Problems for Discussion

CASE 1

Maria intended to open a store in a major mall. She invited Yasmin to join her through investment and management of the operation. Yasmin agreed on the condition that she could raise the money and work it into her schedule as an assistant in an accounting firm. She did raise the money, and sent it to Maria, and on the strength of that, Maria signed the shopping centre lease agreement. When Yasmin's employer discovered her plans, it informed her that "no moonlighting" was a condition of her employment contract.

Discuss the implications of this for Maria and Yasmin.

CASE 2

Habitation Apartments Ltd. borrowed $500,000 from the Good Times Bank and secured the loan by way of a three-year mortgage on its apartment building. The bank demanded additional security for the loan, and Simple, the president of the corporation, personally guaranteed repayment of the loan.

Several years later, as a result of a dispute between shareholders, Simple was voted out of office as president along with most of the Board of Directors, and a new president and Board of Directors were selected by the shareholders. During the months that followed, the new president and Board of Directors reorganized the corporation's operations. As a part of the reorganization, it was necessary for the corporation to rearrange its mortgage loan with the bank. The bank agreed to extend the loan for a further three-year term but at a higher interest rate. Simple, who was still a shareholder of the corporation, was unaware of the new refinancing arrangement the corporation had made with the bank.

A year later, as a result of tenant problems and a high vacancy rate, the corporation was unable to meet its mortgage payments, and the mortgage went into default. When the corporation failed to pay the mortgage, the bank turned to Simple and demanded payment under the guarantee.

Discuss the rights of the parties in this case and explain the possible outcome if the bank should take legal action against the corporation and the guarantor.

CASE 3

Clement entered into a verbal agreement with Calhoun to purchase Calhoun's farm for $140,000. In the presence of his friend Saunders, Clement gave Calhoun $500 in cash "to bind the bargain." The farm adjoined the farm that Clement already owned. Immediately after the deal was made, both he and Saunders proceeded to remove an old fence that separated the two farms.

A few days later, Clement plowed a large field on his "new" farm, and Saunders cut down a few trees. Later that day, he prepared a cheque in the amount of $139,500 and took it to the farmhouse where Calhoun was still living.

Calhoun met Clement at the door and said that he had changed his mind. He did not wish to move off the land and had decided not to sell the farm.

Discuss Clement's rights (if any) in this case. Explain the possible outcome if Clement should decide to take legal action against Calhoun.

www.mcgrawhill.ca/olc/willes

CASE 4

Slippery Silica Mining Corporation entered into a contract with Highgrade Transport Company to haul its ore from its mine to a railway terminal, a distance of some 50 kilometres. The contract called for the hauling of approximately 60 tonnes a week for a one-year period. Because Slippery Silica Mining Corporation was a very small company, Highgrade Transport requested that Wilson and Rose, its two principal shareholders, personally guarantee the payment required under the terms of the contract.

In due course, Wilson and Rose provided a written guarantee of payment. It bore their signatures and that of their witness, Sheila Drew, a young woman who worked in their office as a receptionist and typist. The guarantee was not under seal, so the owner of the transport company immediately drove down to the mine office to have the two owners affix seals to the document.

At the mine office, the owner of the transport company met Sheila, who informed him that both Wilson and Rose were away for the day. When he told her the purpose of his visit, she took a box of red legal seals from her desk and offered them to him with the comment, "I don't think they would mind if you put on the seals yourself."

The owner of the transport company took two red seals and affixed them next to the signatures of Wilson and Rose on the guarantee. Then he left the office with the document.

Some time later, the mining company fell into arrears in its payments under the contract. The transport company notified Wilson and Rose that it intended to look to them for payment under their guarantee.

Discuss the issues raised in this case and determine the legal position of the parties.

CASE 5

Karl and Wilbur were older men who each owned small apartment buildings in the same city. Karl renovated his buildings in his spare time, and he was owed $2,000 by Wilbur for a job. Karl's health was poor and he told Wilbur he wanted to retire. He told Wilbur that he had a mortgage on his building for $300,000, and that he was prepared to sell it to Wilbur for $500,000. Wilbur did not have access to such money, but by the end of the conversation, Karl and Wilbur agreed that Wilbur would make Karl's mortgage payments, and in four-and-a-half years, when $250,000 in GICs that Wilbur owned came due, Wilbur would pay Karl $245,000 (the $200,000 balance plus some interest) in cash.

Wilbur made the next three monthly trips to the bank, at the end of December, January, and February, paying the appropriate $1,750 on each trip. During that time, the men had discussed Karl's impending retirement. In the first week of March, Karl died, and Karl's executor told Wilbur he knew of the deal, but that it was "off." He offered Wilbur a cheque for $3,250, which represented a refund of Wilbur's payments to the bank, less the money Wilbur owed to Karl for renovations. Wilbur refused the cheque, wrote his own cheque to the executor for $2,000 in payment of his repair bill, and told him that he would sue to enforce the deal as "the deal was really good for me, and you want to hold out for more cash." The property, independently appraised, was worth $670,000.

Assess the likelihood of Wilbur succeeding in obtaining an order of specific performance, compelling the sale on the agreed terms.

CASE 6

Reid owned a car and travel trailer that he wished to sell. The trailer was outfitted with a stove and refrigerator as built-in equipment. Reid had added a small plasma television set as a part of the equipment, but the television set was not built into the trailer.

Calder expressed an interest in the car and trailer, and also examined the equipment. Reid advised Calder that the price was $21,000 and that the television set would be $1,000 extra if Calder wished to buy it as well. Calder indicated that he wished to do so.

Reid prepared a written purchase agreement that itemized the car and trailer, but simply referred to the appliances as "equipment." The contract price was $21,000, and the agreement called for a deposit of $1,000. Both parties signed the agreement, and Calder gave Reid a deposit cheque in the amount of $2,000.

Reid changed his mind about the sale of the television set. Shortly before Calder was due to return for the car and trailer, Reid telephoned to say that he was selling only what was specified in the written agreement.

Explain Calder's rights (if any) in this case.

CASE 7

Marcus and his father Louis visited Megabank, where Marcus borrowed $25,000 to purchase a car. The manager was satisfied with Marcus' ability to repay the loan, but in closing added to Marcus: "if you start getting into financial trouble, be sure your father knows, because I don't want to be forced to look to him to pay your debts." Louis spoke up and said, "Not to worry, I'll keep things on track and keep my boy out of trouble." In time, Marcus did face financial difficulties, after having a falling out with his father. Louis refused to assist Marcus, and ignored the Megabank demand that he pay the debt of his son.

Discuss the resolution of this case.

CASE 8

When approached by Norman, Ivan decided to sell him a 250-acre parcel of farm land. Ivan intended to retire and move into the city, leaving his home located on a lake within the parcel. After the paperwork went to their lawyers, Ivan discovered that Norman intended to use the house as a summer home only, and would be absent for the rest of the year. Ivan, an avid hunter, called Norman and asked if it would be OK if he continued to hunt on the land for one week in each November. Norman agreed, and in time the lawyers completed the deal and Ivan received his payment. That November, Ivan found the fencing posted with "No Hunting" signs, and called Norman again to assure himself that their deal still stood. Norman told Ivan that his wife did not want hunting on the land, and when Ivan protested, Norman pointed out that the paperwork showed no mention of their telephone discussion.

Advise Ivan.

CASE 9

Evelyn is delighted at the engagement of her niece Anne. She writes to her niece, "Kevin is such a nice fellow. I want you to have a nest-egg for your future, so on your wedding day I will give you $50,000." Anne breaks her engagement to Kevin, and marries Philip.

Is the letter enforceable?

CASE 10

At an Agricultural Exhibition and Fair, Calvin sets up a dozen small tables at the exhibition, each with a free raffle for a small prize, for guessing the number of jelly beans in different size jars. Calvin was paid by the County Fair Board and the Municipal Tourism Office to use the raffle to build a database for a tourism promotion campaign. Over the nine days of the fair, Calvin obtained 25,000 sets of names, addresses and email addresses. Some weeks after Calvin compiles the database, his laptop containing the data is stolen. Discuss the range of privacy issues that arise in this scenario.

The Online Learning Centre offers more ways to check what you have learned so far. Find quizzes, Weblinks, and many other resources at www.mcgrawhill.ca/olc/willes.

CHAPTER 11

Failure to Create an Enforceable Contract

Chapter Objectives

After study of this chapter, students should be able to:

- Explain how the legal doctrine of mistake may result in the failure to create an enforceable contract.
- Distinguish between different kinds of misrepresentation, and the results of each.
- Explain the difference between undue influence and duress, and the effect of these factors on enforceability of a contract.

YOUR BUSINESS AT RISK

Having brought a contract into existence, some events (by accident, neglect or malice) can easily create a contract that a court will not enforce. You must learn to identify these dangers so that you do not create a situation where you cannot enforce your rights. Equally, you must learn how these enforcement rules protect you from being a victim, if someone attempts to take advantage of you via a flawed contract.

INTRODUCTION

In their negotiations, the parties may meet all of the essentials for the creation of a binding agreement but nevertheless, may occasionally fail to create an enforceable contract. Offer and acceptance, capacity, consideration, legality of object, and an intention to create a legal relationship all must be present, together with the requirements of form and writing, under certain circumstances. But even when these elements are present, the parties may not have an agreement that both may enforce until they also show that they both meant precisely the same thing in their agreement. There are essentially four situations of this general nature that could arise and render the agreement unenforceable. They are of critical importance in determining who bears the brunt of losses, particularly if the subject matter of the contract has dramatically changed in value or has been destroyed.

MISTAKE

Mistake

A state of affairs in which a party (or both parties) has formed an erroneous opinion as to the identity or existence of the subject matter, or of some other important term.

If, in their negotiations, the parties are mistaken as to some essential term in the agreement, they may have failed to create a contract. **Mistake** at law does not mean the same thing to both the layperson and the legal practitioner, however. Mistake from a legal point of view has a relatively narrow meaning. It generally refers to a situation where the parties have entered into an agreement in such a way that the contract does not express their true intentions. This may occur when the parties have formed an untrue impression concerning an essential element, or when they have failed to reach a true meeting of the minds as to a fundamental term in the agreement.

Ann offers to sell her share in a business to Burt for $75,000, then realizes that her share is worth $80,000. The courts would probably not allow Ann to avoid the agreement on the basis of mistake. Because Ann made the offer to Burt, and then later alleged that she had made a mistake as to the value of the subject matter, the courts would have no real way of knowing Ann's true state of mind at the time the offer was made.

On the other hand, if the consideration is clearly out of line and the mistake is obvious, the courts may not allow the other party to "snap up" the bargain.[1] When the mistake is due to the party's own negligence, however, the contract, under certain circumstances, may be binding.[2]

Mistake of Law

At one time, a mistake could take the form of a mistake of law or a mistake of fact. Recovery of money paid under a mistake of law was often difficult, because everyone was (and is) presumed to know the law. As a consequence, recovery was only possible if the statute provided for recovery of the money paid, or there were some other conditions related to the mistake that permitted the court to direct the repayment of the money. For example, a rent-restriction law prohibited the collection of a rental premium by the landlord for leasing premises to a tenant. In a case where the tenant paid the premium, then later applied to the courts to recover it on the basis of a mistake of law, the court held that the collection of the premium was illegal but the restriction applied only to the landlord. The court allowed the tenant to recover the premium.[3]

In 1989, the Supreme Court of Canada decided that money paid under a mistake of law should not be distinguished from mistake of fact. In essence, the difference should be abolished. In the case of *Air Canada et al. v. British Columbia*[4] the court stated:

> Where an otherwise constitutional or intra vires statute or regulation is applied in error to a person to whom in its true construction it does not apply, the general principles of restitution for money paid under mistake should be applied, and subject to available defences and equitable considerations, the general rule should favour recovery. No distinction should be made between mistakes of fact and mistakes of law.

Mistake of Fact

Mistake of fact

Mistake as to the existence of the subject matter of a contract or the identity of a party.

Mistake of fact may take many forms, and, for many of these, the courts do provide relief. As a general rule, if the parties are mistaken as to the existence of the subject matter of the contract, then the contract will be void.[5] For example, Sunset Sails Co. offers to sell Beverley a yacht moored in the Caribbean, and Beverley accepts the offer. Unknown to both Sunset Sails Co. and Beverley, the previous night a fire completely destroyed the marina where the yacht was. The subject matter did not exist at the time that Sunset Sails Co. and Beverley made the contract. The contract is void due to a mistake as to the existence of the subject matter. In essence, there was no yacht to sell at the time the parties made their agreement. The same rule might well apply if the yacht had been badly damaged in the fire and was no longer usable. Under the Common Law, the courts would not require the purchaser to accept something different from what she had contracted to buy.

A second type of mistake of fact applies when there is a mistake as to the identity of one of the contracting parties. This is essentially an extension of the rule for offer and acceptance, which states that only the person to whom an offer is made may accept the offer. With a mistake of fact of this nature, the courts will generally look at the offer to

1. *Hartog v. Colin & Shields*, [1939] 3 All E.R. 566; *Imperial Glass Ltd. v. Consolidated Supplies Ltd.* (1960), 22 D.L.R. (2d) 759.
2. *Timmins v. Kuzyk* (1962), 32 D.L.R. (2d) 207; *Hydro Electric Comm. of Township of Nepean v. Ontario Hydro* (1980), 27 O.R. (2d) 321.
3. *Kiriri Cotton Co. Ltd. v. Dewani*, [1960] 1 All E.R. 177.
4. [1989] 1 S.C.R. 1161 at 1167.
5. *Barrow, Lane, & Ballard Ltd. v. Phillips & Co.*, [1929] 1 K.B. 574.

determine if the identity of the person in question is an essential element of the contract. If the identity of the party is not an essential element of the agreement, then the agreement may be enforceable.[6] However, if one party to the contract does not wish to be bound in an agreement with a particular contracting party, and is misled into believing that he or she is contracting with someone else, the contract may be voidable when the true facts are discovered.[7]

Able Engineering may wish to engage the services of a soil-testing company to do a site inspection for it. Able Engineering used the services of soil-testing company B in the past and found their services to be unsatisfactory. On this occasion they request soil-testing company C to do the work.

Unknown to Able Engineering, soil-testing company B has purchased company C, and all the work of company C is directed to company B. Company B accepts the offer. When Able Engineering becomes aware of the acceptance by company B it may successfully avoid the contract on the basis of mistake as to the identity of the contracting party, if the identity of the party is an important element in the contract.

Non Est Factum

Mistake may also occur when one of the parties may be mistaken as to the true nature of a written contract. However, this is a very narrow form of mistake that represents an exception to the general rule that a person will be bound by any written agreement that he or she signs. The important distinction here is that the circumstances surrounding the signing of the written document must be such that the person signing the document was led to believe that the document was of a completely different nature from what it actually was. Had the person known what the agreement really was, he or she would not have signed it. This exception is subject to a number of constraints. It has a very limited application, because a person signing a written agreement is presumed to be bound by it. A failure to examine the written agreement does not absolve a person from any liability assumed under it. Nor is a person absolved from liability if the party is aware of the nature of the agreement as a whole, but remains ignorant of a specific term within it.[8] To avoid liability, a person must be in a position to establish that the document was completely different in nature from the document described, and that due to some infirmity or circumstances he or she was obliged to rely entirely on another person to explain the contents. The person must also establish that it was not possible to obtain an independent opinion or assistance before signing the written form and that he or she was not in any way careless. This particular exception, which represents a form of mistake, is a defence known as ***non est factum*** ("it is not my doing").

Non est factum

A defence that may allow illiterate or infirm persons to avoid liability on a written agreement if they can establish that they were not aware of the true nature of the document, and were not careless in its execution.

It is important to note, however, that the Supreme Court of Canada has essentially limited this defence to a very narrow group of contracting parties. In *Marvco Color Research Limited v. Harris*,[9] the Supreme Court held that if a person was careless in signing a document, the defence of *non est factum* would not be available to the person. This was the case even if the person had some infirmity, such as a reading difficulty or partial blindness.

The narrowness of this defence may be illustrated by the following example. An elderly person with failing eyesight and no opportunity to get legal or other advice on a document was induced to sign that document. The person believed it to be a letter of reference, when in fact it was a guarantee. Under these conditions, the person may be able to avoid liability. However, first she must show that she was not careless, but obliged to rely upon the person presenting it for her signature. She also must prove that it was

6. *Ellyatt v. Little*, [1947] O.W.N. 123.
7. *Said v. Butt*, [1920] 3 K.B. 497; *Boulton v. Jones* (1857), 2 H & N 564, 157 E.R. 232; *Cundy v. Lindsay* (1878), 3 App. Cas. 459.
8. *Sumner v. Sapkos* (1955), 17 W.W.R. 21.
9. [1982] 2 S.C.R. 774.

described to her as being a completely different document. The infirmity that made a personal examination and understanding of the document impossible must, of course, also be established to the satisfaction of the court, which will require evidence to prove that the party was not otherwise careless in signing the document. Once this is done, the court may decide that the party would not be bound by the document.[10]

An additional point to note here is the true nature of the document. The signed document must be completely different in nature from the document that the party believed he or she was signing, for a plea of *non est factum* to succeed. If, however, the document is not of a different nature, but, rather, the same type of document as described, differing only in degree, then a defence of *non est factum* would be unsuccessful. The party would have been aware of the true nature of the agreement at the time of signing, and no mistake as to the nature of the document would have existed.[11] The justification for this rule of law is obvious: public policy dictates that a person should be bound by any agreement signed; the excuse that it was not read before signing is essentially an admission of carelessness or negligence on the part of the signor. The courts are not prepared to offer relief to those persons who are so careless in the management of their affairs that they are unwilling to take the time to read the terms and conditions that are contained in an agreement. There are, however, persons who, as a result of advanced years, some infirmity, or simply a lack of knowledge, are unable to read the written agreement. It is this group that the courts are prepared to assist if, through their reliance on another, they have been misled as to the true nature of the agreement that they have signed. Even here, the disadvantaged persons are expected to assume some responsibility for their own protection. If the opportunity for independent advice is available, and they refuse to avail themselves of it, the courts will probably treat their actions as careless and not permit them to avoid the contract. For example, when a person heard the contract read aloud and then later pleaded *non est factum*, the claim was rejected and the contract enforced.[12]

Unilateral and Mutual Mistake

Unilateral mistake

A mistake by one party to the agreement.

Mutual mistake

A mistake where both parties have made mistaken assumptions as to the subject matter of the agreement.

Mistake may take one of two forms insofar as the parties are concerned. The mistake may be made by only one party to the agreement, in which case it is called **unilateral mistake**. Or it may be that both parties are unaware of the mistake; in this case, the mistake is a **mutual mistake**. In a unilateral mistake, usually one of the parties is mistaken as to some element of the contract, and the other is aware of the mistake. Cases of this nature closely resemble misrepresentation — one of the parties is aware of the mistake, and either allows it to exist or actively encourages the false assumption by words or conduct. The major difficulty with this form of mistake is establishing a general rule for its application. The best that might be said in this instance is that the courts tend to treat contracts as being unenforceable when a party makes or accepts an offer that he or she knows the other party thinks or understands to be materially different from what he or she makes or accepts.

COURT DECISION: Practical Application of the Law

Errors in Written Agreement—Failure to Inform Other Party at the Time

***Vukomanovic v. Cook Bros. Transport Ltd. and Mayflower Transit Co. Ltd.* (1987), 65 Nfld. R. 181 (NLTD).**

The plaintiff entered into a contract with the defendant to have his household goods moved from Calgary, Alberta, to St. John's Newfoundland. The contract was entered into by the parties in January of 1985. At the time of loading, January 25th, the plaintiff was presented with a bill of lading with the loading date marked as February 25th and delivery date marked as March 22nd. The plaintiff noticed the errors in the dates on the written agreement. He had been advised that delivery would take "approximately four weeks." He did not inform the defendant of the error in

10. *W.T. Rawleigh Co. v. Alex Dumoulin*, [1926] S.C.R. 551; *Commercial Credit Corp. v. Carroll Bros. Ltd.* (1971), 16 D.L.R. (3d) 201.
11. *Dorsch v. Freeholders Oil Co.*, [1965] S.C.R. 670.
12. *Prudential Trust Co. Ltd. v. Forseth*, [1960] S.C.R. 210.

the contract at the time. The goods, however, were delayed in transit, and were not delivered until March 14th. The plaintiff was obliged to stay at a motel from February 22nd to March 14th, due to the late delivery. He then brought an action for damages against the moving companies.

THE COURT: The question posed by counsel for the plaintiff is whether the dates appearing on the bill of lading are the correct dates and the bill of lading representing the true contract between the parties or, whether an error was made by the defendants whereby the delivery date ought to have been February 22nd as per an oral agreement that preceded the signing of the bill of lading.

[20] In the case at hand there is no allegation or suggestion that the defendants or their agents in Calgary were guilty of misrepresentation, fraud, deceit or for that matter any improper or irregular conduct. It is possible that somehow, perhaps by inadvertence, the incorrect dates were put on the bill of lading but, certainly, at most the charge can be only that of inadvertence. It was never established how the defendants got the dates that appeared on the bill of lading. Cook attempted to explain how the dates could only have come from the plaintiff. However, in view of what subsequently occurred I hardly think it matters. The plaintiff was presented with the bill of lading at the time of loading and by his own evidence he examined it carefully. As I have already indicated, he was fully aware of the dates already filled in on the bill of lading showing the loading date as February 25th and a delivery date of March 22nd. The plaintiff says that he was not concerned by the delivery date shown as both he and the driver signed the bill of lading on January 25th. In my view it was the plaintiff who carelessly or neglectfully failed to bring the matter to the attention of the driver. He did nothing. By signing the bill of lading that indicated a delivery date of March 22nd the plaintiff quite unintentionally misled the defendants into believing that March 22nd was the proper delivery date. For that carelessness or neglect he can hardly now blame the defendants. The defendants or their agents in Calgary were innocent of any negligence, carelessness or wrong doing and it was the failure of the plaintiff to bring the alleged irregularity to the attention of the driver that ultimately resulted in the goods not reaching St. John's until March 13th. The plaintiff's error was in assuming that if the goods were picked up on January 25th, they would be delivered within four weeks and that, therefore, the delivery date of March 22nd was meaningless. That was the crucial mistake and for that error assume his loss.

[21] Without again reviewing the evidence, I should also state I am satisfied the defendants acted reasonably once they became aware of the plaintiff's predicament. Cook made every effort to cooperate and cannot be faulted. Unfortunately for the plaintiff it was the time of year when the carriage of household goods are at its lowest level and it simply was not possible to expedite the move. The plaintiff's claim is dismissed.

On countless occasions and as simple as this is, people have and will continue to sign documents with errors, sign documents without reading them, or sign without understanding them. The results will reflect their carelessness.

Unilateral mistake may arise, for example, when a seller, offering to sell a particular product to a buyer, knows that the buyer believes the offered product is something different from what it is. In this case, if the court is satisfied that the seller was aware of the buyer's mistake but allowed it to go uncorrected, the court may permit the buyer to rescind the agreement.

Mutual mistake, on the other hand, is generally the easiest to deal with. It encompasses common forms of mistake, such as mistake as to the existence of the subject matter or mistake as to its identity. Only the latter sometimes presents problems. When it does so, the courts frequently decide that a mistake has occurred and the contract is therefore unenforceable.[13] Cases of this sort tend to place a hardship on the plaintiff, because the courts, in effect, reject the plaintiff's interpretation of the contract. Nevertheless, if a reasonable interpretation is possible, it may be accepted by the court in an effort to maintain the agreement.

The nature of mistake and the differences between its various forms was discussed by the court in the case of *McMaster University v. Wilchar Construction Ltd. et al.*[14] The judge described mistake in the following terms:

> The distinction between cases of common or mutual mistake and, on the other hand, unilateral mistake, must be kept in mind. In mutual or common mistake, the error or mistake, in order to avoid the contract at law, must have been based either upon a fundamental mistaken assumption as to the subject matter of the contract or upon

13. See *Raffles v. Wichelhaus* (1864), 2 H.&C. 906, 159 E.R. 375, for an example of a case of mutual mistake that the court found insoluble insofar as an interpretation of the interest was concerned.

14. [1971] 3 O.R. 801.

> a mistake relating to a fundamental term of the contract. There, the law applies the objective test as to the validity of the contract. Its rigour in this aspect has been designed to protect innocent third parties who have acquired rights under the contract.
>
> Normally a man is bound by an agreement to which he has expressed assent. If he exhibits all the outward signs of agreement, at law it will be held that he has agreed. The exception to this is in the case where there has been fundamental mistake or error in the sense above stated. In such case, the contract is void *ab initio.* At law, in unilateral mistake, that is when a mistake of one party relating to the contract is known to the other party, the Courts will apply the subjective test and permit evidence of the intention of the mistaken party to be adduced. In such case, even if one party knows that the other is contracting under a misapprehension, there is, generally speaking, no duty cast upon him to disclose to the other circumstances which might affect the bargain known to him alone or to disillusion that other, unless the failure to do so under the circumstances would amount to fraud. This situation, of course, must be distinguished from the case in which the mistake is known to or realized by both parties prior to the acceptance of the offer.
>
> The law also draws a distinction between mistake simply nullifying consent and mistake negativing consent. Error or mistake which negatives consent is really not mistake technically speaking in law at all, as it prevents the formation of contract due to the lack of consensus and the parties are never *ad idem.* It is rather an illustration of the fundamental principle that there can be no contract without consensus of all parties as to the terms intended. This is but another way of saying that the offer and the acceptance must be coincident or must exactly correspond before a valid contract results.
>
> A promisor is not bound to fulfil a promise in a sense in which the promisee knew at the time that the promisor did not intend it. In considering this question, it matters not in what way the knowledge of the meaning is brought to the mind of the promisee, whether by express words, by conduct, previous dealings or other circumstances. If by any means he knows there was no real agreement between him and the promisee, he is not entitled to insist that the promise be fulfilled in a sense to which the mind of the promisor did not assent.

Rectification

The correction of a mistake in an agreement that would have rendered the agreement impossible to perform.

A special form of relief is available in the case of mistake in a written agreement that renders performance impossible. This is known as **rectification**. It is sometimes used to correct mistakes or errors that have crept into a written contract, either when a verbal agreement has been reduced to writing, or when a written agreement has been changed to a formal agreement under seal. In each of these cases, if the written agreement does not conform with the original agreement established by the parties, the courts may change the written words to meet the terms of the original agreement. The purpose of this relief is to "save" the agreement that the parties have made. It is not intended to permit alteration of an agreement at a later date to suit the wishes or interpretation of one of the parties. It is, essentially, a method of correcting typographical errors or errors that have crept into the writing through the omission of a word or the insertion of the wrong word in the agreement.

To obtain rectification, however, it is necessary to convince the court through evidence that the original agreement was clear and unequivocal with regard to the term that was later changed when reduced to writing. The court must also be convinced that there were no intervening negotiations or changes in the interval between the establishment of the verbal agreement and the preparation of the written document. It would also be necessary to establish that neither party was aware of the error in the agreement at the time of signing.[15]

A Co. and B Co. enter into an agreement by which A Co. agrees to supply a large quantity of fuel oil to B Co.'s office building at a fixed price. The building is known municipally as 100 Main Street. When the agreement is reduced to writing, the address is set out in error as 1000 Main Street, an address that does not exist.

15. *Paget v. Marshall* (1884), 54 L.J. Ch. 575.

After the contract is signed, B Co. discovers that it could obtain fuel oil at a lower price elsewhere. It attempts to avoid liability on the basis that A Co. cannot perform the agreement according to its terms. In this case, A Co. may apply for rectification to have the written agreement corrected to read 100 Main Street, the address that the parties had originally agreed would be the place of delivery.

MISREPRESENTATION

Misrepresentation is a statement or conduct that may be either innocent or fraudulent and that induces a person to enter into a contract. Normally, a person is under no obligation to make any statement that may affect the decision of the other party to enter into the agreement. Any such statement made, however, must be true. Otherwise, it may constitute misrepresentation if it is material (important) to the contract. Additionally, the law recognizes a small group of contractual relationships when the failure to disclose all material facts may also amount to misrepresentation. Misrepresentation does not, however, render a contract void *ab initio*. Misrepresentation, whether innocent, fraudulent, or by means of non-disclosure, will only render the agreement voidable at the option of the party misled by the misrepresentation. In every instance, it is important that the injured party cease accepting benefits under the agreement once the misrepresentation is discovered. Otherwise the continued acceptance of benefits may be interpreted as a waiver of the right to rescind the contract. Exceptions have been made to this general rule by both statute law[16] and recent cases concerning fraudulent misrepresentation.[17] However, the behaviour of the injured party, once the misrepresentation is discovered, is still of paramount importance.

The false statement must be a statement of fact and not a mere expression of opinion. Whether the fact is material or not is determined on the basis of whether the innocent party to the negotiations would have entered into the agreement had he or she known the true fact at the time. If the innocent party did not rely on the particular fact, or was aware of the falsity of the statement made, then he or she cannot avoid the contract on the basis of misrepresentation by the other party. **Rescission** is only possible if the innocent or injured party relied on the false statement of fact made by the other party.

Rescission
The revocation of a contract or agreement.

Misrepresentation seldom arises out of a term in a contract. It is generally something that takes place before the contract is signed, and that induces a party to enter into the agreement. Misrepresentation must be of some material fact, and not simply a misstatement of a minor matter that does not go to the root of the contract. If the parties include the false statement as a term of the contract (such as a statement as to quality or performance), then the proper action, if the statement proves to be untrue, is an action for a breach of contract, rather than misrepresentation.

Innocent Misrepresentation

Innocent misrepresentation
A false statement of a material fact made by a party that honestly believed the fact to be true.

Innocent misrepresentation is the misrepresentation of a material fact that the party making the statement honestly believes to be true, but is discovered to be false after the parties enter into the contract. If the statement can be shown by the injured party to be a statement of a material fact that induced him or her to enter into the agreement, then he or she may treat the contract as voidable and bring an action for rescission. If the injured party acts promptly, the courts will normally make every effort to put the parties back in the same position they were in before the contract was made. For example, Lakeside Land Development Ltd. and High Rise Construction Ltd. enter into negotiations for the purchase of a building lot that Lakeside Land Development Ltd. owns. The president of High Rise Construction Ltd. asks the president of Lakeside Land Development Ltd. if the land is suitable for the construction of a small apartment building. The president of

16. See, for example, insurance legislation in each province that does not permit the insurer to avoid liability on a policy when the insured failed to disclose a material fact many years before a claim is made on the policy.

17. See, for example, *Siametis et al. v. Trojan Horse (Burlington) Inc. et al.* (1979), 25 O.R. (2d) 120, affirmed, ONCA.

Lakeside Land Development Ltd. (who had inquired from the municipality some months before and determined that the land was indeed suitable and approved for the proposed use) answers "Yes." Unknown to Lakeside Land Development Ltd., the lands had subsequently been rezoned for single-family dwellings, and the construction of apartment buildings was prohibited. High Rise Construction Ltd., on the strength of Lakeside Land Development Ltd.'s statement, enters into an agreement to purchase the lot. A short time later, before the deed is delivered, High Rise Construction Ltd. discovers that the land is not zoned for multiple-family dwellings and refuses to proceed with the contract. In this case, High Rise Construction Ltd. would be entitled to rescission of the agreement on the basis of Lakeside Land Development Ltd.'s innocent misrepresentation. At the time that Lakeside Land Development Ltd.'s statement was made, the land was not zoned for the use intended by High Rise Construction Ltd. Even though the president of Lakeside Land Development Ltd. honestly believed the land to be properly zoned at the time that he made the statement, it was untrue. Since High Rise Construction Ltd. had relied on Lakeside Land Development Ltd.'s statement, and it was material to the contract, the courts would probably provide the relief requested by High Rise Construction Ltd. and rescind the contract. The courts would probably order Lakeside Land Development Ltd. to return any deposit paid by High Rise Construction Ltd., but would not award punitive damages.[18]

A classic statement on the Common Law related to innocent misrepresentation was pronounced in the case of *Newbigging v. Adam.*[19] The court described innocent misrepresentation and the remedies available in the following words:

> If we turn to the question of misrepresentation, damages cannot be obtained at law for misrepresentation which is not fraudulent, and you cannot, as it seems to me, give in equity any indemnity which corresponds with damages. If the mass of authority there is upon the subject were gone through, I think it would be found that there is not so much difference as is generally supposed between the view taken at common law and the view taken in equity as to misrepresentation. At common law it has always been considered that misrepresentations which strike at the root of the contract are sufficient to avoid the contract on the ground explained in *Kennedy v. Panama, New Zealand, and Australian Royal Mail Company*; but when you come to consider what is the exact relief to which a person is entitled in a case of misrepresentation it seems to me to be this, and nothing more, that he is entitled to have the contract rescinded, and is entitled accordingly to all the incidents and consequences of such rescission. It is said that the injured party is entitled to be replaced *in statu quo*. It seems to me that when you are dealing with innocent misrepresentation you must understand that proposition that he is to be replaced *in statu quo* with this limitation — that he is not to be replaced in exactly the same position in all respects, otherwise he would be entitled to recover damages, but is to be replaced in his position so far as regards the rights and obligations which have been created by the contract into which he has been induced to enter.

CASE IN POINT

A landowner and farmer entered into a share-crop agreement whereby the farmer would farm the land provided it was free of weeds. The landowner, although not a farmer, represented the land to be free of weeds, in the presence of witnesses. The farmer seeded the land, but later discovered the land to be so weedy that he had to abandon his crop. The landowner sued for the first year's rent and damages. The farmer counterclaimed that the landowner had misrepresented the land and claimed for the expenses of seeding the land. The court held that the landowner had innocently misrepresented the land as weed-free, and dismissed the landowner's claim.

Kooiman v. Nichols (1991), 75 Man. R. (2d) 298.

18. *Derry v. Peek* (1889), 14 App. Cas. 337; *Alessio v. Jovica* (1973), 42 D.L.R. (3d) 242.
19. (1887), 34 Ch.D. 582.

Fraudulent Misrepresentation

Fraudulent misrepresentation

A false statement of fact made by a person who knows, or should know, that it is false, and made with the intention of deceiving another.

Deceit

A tort that arises when a party suffers damage by acting upon a false representation made by a party with the intention of deceiving the other.

Unlike innocent misrepresentation, whereby a party honestly believes a fact to be true when the fact is stated, **fraudulent misrepresentation** is a statement of fact that, when made, is known to be false. It is made with the intention of deceiving the innocent party. If a party makes a false statement recklessly, without caring if it is true or false, it may also constitute fraudulent misrepresentation. In each case, however, the statement must be of a material fact and must be made for the purpose of inducing the other party to enter into the agreement.[20]

In the case of fraudulent misrepresentation, the innocent party must prove fraud on the part of the party making the false statement. This is because the action is based upon the tort of **deceit**, as well as a request for the equitable remedy of rescission. Rescission is limited to those cases where the courts may restore the parties to the position they were in before entering into the contract. However, this is not the case with tort. If the innocent party is able to prove fraud on the part of the party making the statement, then the courts may award punitive damages against that party committing the tort as punishment for the act. This remedy would be available in all cases where fraud may be proven, even if it would not be possible to restore the injured party to the same position that he or she was in before the contract was established. As with innocent misrepresentation, the injured party must refrain from taking any benefits under the agreement once the fraud is discovered. The continued acceptance of benefits may prevent a future action for rescission. Usually the parties must act promptly to have the agreement rescinded, because the remedy would not be available if a third party should acquire the title to any property that may have been the subject matter of the agreement.

Insofar as the tortious aspect of the misrepresentation is concerned, prompt action by the innocent party is usually also important, in order to avoid any suggestion that that party had accepted the agreement notwithstanding the fraud. Delay does not always preclude relief, however, as the courts have awarded damages under certain circumstances even after the passage of a lengthy period of time.[21]

COURT DECISION: Practical Application of the Law

Contracts—Fraudulent Misrepresentation—Damages

***Siametis et al. v. Trojan Horse (Burlington) Inc. et al.* (1979), 25 O.R. (2d) 120, affirmed, ONCA.**

The plaintiffs purchased a restaurant business based upon fraudulent statements and information provided by the defendants. The fraud was not discovered until some months after the plaintiffs had operated the business and expended funds of their own in an attempt to make the business profitable. The plaintiffs brought an action against the defendants for damages as well as rescission of the contract.

THE COURT: The plaintiff could not have discovered the fraud by examining the statements, including the Trojan Horse statement, for the first year's business. A cheat may not escape even if the fraud could be uncovered by a more sophisticated purchaser. That Siametis may have been a foolish and unwise purchaser did not absolve the deceivers; ... If the misrepresentation is fraudulent, the case for relief is overwhelming. The person deceived is entitled to damages and rescission: Waddams, *The Law of Contracts* (1977), p. 248. The measure of damages in a case such as this was stated simply in the inimitable style of Lord Denning., *Doyle v. Olby*, supra, at p. 167, where he stated:

> The person who has been defrauded is entitled to say: 'I would not have entered into this bargain at all but for your representation. Owing to your fraud, I have not only lost all the money I paid you, what is more, I have been put to a large amount of extra expense as well as suffered this or that extra damages.'

20. *Derry v. Peek*, supra.
21. *Siametis et al. v. Trojan Horse (Burlington) Inc. et al.* (1979), 25 O.R. (2d) 120, affirmed, ONCA.

All such damages can be recovered and the defrauders cannot plead that the damages could not reasonably have been foreseen. Siametis is entitled to damages for all of his loss "subject to allowances for any benefit that he has received."

So Siametis receives rescission and damages to compensate for fraudulent misrepresentation; what would have been the result had the misrepresentation been innocent?

The requirement for a tort action for deceit may not be an easy matter to establish. In the case of *Charpentier v. Slauenwhite*[22] the defendant, Mrs. Slauenwhite, made certain false statements concerning water supply to a property. The plaintiff brought an action in tort for the alleged deceit, and the court in its judgment described the requirements for maintaining a successful deceit action as follows:

> First, in order to sustain an action of deceit, there must be proof of fraud, and nothing short of that will suffice. Secondly, fraud is proved when it is shown that a false representation has been made, (1) knowingly, or (2) without belief in its truth, or (3) recklessly, careless whether it be true or false. Although I have treated the second and third as distinct cases, I think the third is but an instance of the second, for one who makes a statement under such circumstances can have no real belief in the truth of what he states.

Occasionally, a party injured by the false statements of the other contracting party may not be able to satisfy the court that the statements constituted fraudulent misrepresentation. In these instances the court may provide only the contract remedy of rescission if the statements represent an innocent misrepresentation.

A purchaser, in response to an advertisement of a model 733 BMW automobile for sale, contacted the seller (who had bought the car in Germany) and was assured that it was a model 733. In fact, the vehicle was a 728 European model that did not meet Canadian safety standards, nor were parts readily available. While visually identical to the exported 733, the 728 was only manufactured for the European market. The purchaser brought an action for fraudulent misrepresentation against the seller, but the court found the statements of the seller to be innocent misrepresentation and awarded only rescission of the contract.[23]

Misrepresentation by Non-Disclosure

Generally, a contracting party is under no duty to disclose material facts to the other contracting party. However, the law does impose a duty of disclosure in certain circumstances where one party to the contract possesses information that, if undisclosed, might materially affect the position of the other party to the agreement. This duty applies to a relatively narrow range of contracts, called contracts of *utmost good faith*. It also applies to cases where there is an active concealment of facts, or where partial disclosure of the facts has the effect of rendering the part disclosed as false. With respect to this latter group, the courts will normally treat the act of non-disclosure, or of partial disclosure, as a fraud or an intention to deceive. In contracts of "utmost good faith," the failure to disclose, whether innocent or deliberate, may render the resulting contract voidable.

Fortunately, contracts of utmost good faith constitute a rather small group of contracts. The most important are contracts of insurance, partnership, and those where a relationship of special trust or confidence exists between the contracting parties. The courts have indicated that the class of contracts that may be identified as being of utmost good faith are not limited, but they have generally been reluctant to expand the class.[24]

22. (1971), 3 N.S.R. (2d) 42.

23. *Ennis v. Klassen* (1990), 66 M.R. (2d) 117 (C.A.).

24. See, for example, *Hogar Estates Ltd. in Trust v. Shelbron Holdings Ltd. et al.* (1979), 25 O.R. (2d) 543; *Laskin v. Bache & Co. Inc.*, [1972] 1 O.R. 465.

This is perhaps due in part to the fact that the duty of disclosure, in many cases, has been dealt with by statute, rather than by Common Law.[25]

Contracts of insurance, in particular, require full disclosure by the insurance applicant, who knows essentially everything about the risk that he or she wishes to have insured, while the insurer knows very little. The reasoning behind the law under these circumstances is that an obligation rests on the prospective insured to reveal all material facts. This is, first, to enable the insurer to determine if it wishes to assume the risk, and second, to have some basis upon which to fix the premium payable for the risk assumed. This is particularly important in the case of life insurance, where the insurer relies heavily on the insured's statement as to his or her health record in determining insurability and the premium payable. Even here, limits have been imposed on innocent non-disclosure. Most provinces have legislation for particular kinds of insurance that limit the insurer's right to avoid liability on a contract of insurance for non-disclosure beyond a fixed period of time.

Partnership agreements, and all other contracts representing a fiduciary relationship, are similarly subject to the rules requiring full disclosure of all material facts in any dealings that the parties may have with each other. In all of these circumstances, withholding information of a material nature by one party would entitle the innocent party to avoid liability under the agreement affected by the non-disclosure.

The question of non-disclosure of material facts arose in the case of *Re Gabriel and Hamilton Tiger-Cat Football Club Ltd.*[26] The issue was whether an employment contract was one that fell within the definition of a contract of this special type because the club failed to reveal the length of the playing season. The judge explained the nature of a contract of utmost good faith by saying:

> ... there is a limited class of contract in which one of the parties is presumed to have means of knowledge which are not accessible to the other and is, therefore, bound to tell him everything which may be supposed likely to affect his judgment. They are known as contracts *uberrimae fidei*, and may be voided on the ground of non-disclosure of material facts. Contracts of insurance of every kind are in this class. There are other contracts, though not contracts *uberrimae fidei*, in the same sense, which impose a duty of full disclosure of all material facts by the parties entering into them. Contracts for family settlements and arrangements fall into this category. I am dealing here with a contract of personal service. The House of Lords in *Bell et al. v. Lever Bros. Ltd. et al.*, [1932] A.C. 161, refused to extend the duty of disclosing material facts to contracts for service.

MANAGEMENT ALERT: BEST PRACTICE

Misrepresentation is an area of special vulnerability for business. Business operators must take special steps in their advertising and employee training to address this area. Ignoring outright fraud, business persons must review advertising copy through the eyes of their customers and ask themselves how their customers might respond. Courts will hold ambiguity and misrepresentation against the business. Employees, particularly a commissioned sales force, must be cautioned against making unreasonable promises about products, especially anything in excess of the manufacturer's claims. And, financial advisors may be the authors or targets of claims of misrepresentation of an investment. Tape-recording of telemarketing calls is one management response to this. Failure in these responsibilities means incurring an unnecessary risk of future liability.

Negligent Misrepresentation

In the last half of the 20th century, Common Law courts began testing the proposition that some misrepresentations were neither innocent nor fraudulent, and opposite to non-disclosure. The proposition was that these representations were statements that were sufficiently reckless that they should be considered as negligent misrepresentations. The

25. See, for example, consumer-protection legislation and recent securities legislation in most provinces.

26. (1975), 57 D.L.R. (3d) 669.

body of law gained ground with an English case, *Esso Petroleum v Mardon*,[27] in which a fuel company induced a prospective gas station owner to enter into a franchise contract by forecasting sales which were three times higher than were likely (or in fact) achieved.

A famous English judge, Lord Denning, summed up the situation in these terms:

> ... it was a forecast made by a party, Esso, who had special knowledge and skill. It was the yardstick (the "e a c") by which they measured the worth of a filling station. They knew the facts. They knew the traffic in the town. They knew the throughput of comparable stations. They had much experience and expertise at their disposal. They were in a much better position than Mr. Mardon to make a forecast. It seems to me that if such a person makes a forecast — intending that the other should act on it and he does act on it — it can well be interpreted as a warranty that the forecast is sound and reliable in this sense that they made it with reasonable care and skill. ... If the forecast turned out to be an unsound forecast, such as no person of skill or experience should have made, there is a breach of warranty.

In 1993, the Supreme Court of Canada affirmed that the concept exists in Canada as well, in *Queen v. Cognos Inc.*[28] An Ottawa software firm had hired an accountant away from a secure job in Calgary, to work on a "major two-year project." The employment agreement did, however, provide for termination on one month's notice. Five months after taking up employment the accountant received notice that he would be reassigned elsewhere in the company for lack of funding for his work. The internal approval of corporate funding of the project had always been speculative, a fact known to the hiring officer at the time of hiring the accountant. The Court found a duty of care existed to ensure that the representations made at that time were accurate and not misleading, and those regarding the "major project" were negligent. Damages in excess of $67,000 were awarded.

FRONT-PAGE LAW

SEC goes after ex-Kmart executives; says disclosures 'materially false and misleading'

The U.S. Securities and Exchange Commission yesterday accused two former Kmart Corp. executives of misleading investors about the company's financial condition before the retailer's bankruptcy filing in early 2002. The civil charges allege former chairman and chief executive officer Charles Conaway and former chief financial officer John McDonald were responsible for disclosures that were "materially false and misleading." According to the SEC, Kmart's filings failed to disclose the reasons for a massive inventory buildup in 2001...[which the company attributed to]..."seasonal inventory fluctuations and actions taken to improve our overall in-stock position," the SEC said. The commission alleges that was misleading because [the buildup] was caused by a Kmart officer's "reckless and unilateral purchase of $850-million (U.S.) of excess inventory," the statement said. The SEC alleges Mr. Conaway and Mr. McDonald lied about why vendors were not being paid on time and misrepresented the impact that the problem had on Kmart's relationship with its vendors.

Should bad buying decisions be treated as 'false and misleading' disclosures?

Source: Excerpted from Joseph Altman, AP and Reuters, *The Globe and Mail,* "SEC goes after ex-Kmart executives; says disclosures 'materially false and misleading,'" August 24, 2005, p. B11. Used with permission of The Associated Press © 2005.

UNDUE INFLUENCE

Undue influence

A state of affairs whereby a person is so influenced by another that the person's judgment is not his or her own.

The law of contract assumes that the parties to a contract have freely assumed their respective duties under the agreement. Such is not always the case, however. Occasionally, a party entering into an agreement may be so dominated by the power or influence of another that he or she is unable to make a free and deliberate decision to be bound by his or her own act. In essence, **undue influence** occurs when one party is so dominated by another that the decision is not his or her own. A contract obtained under these circumstances would be voidable, if the dominated party acts to avoid the contract as soon as he or she is free of the dominating influence.

27. [1976] Q.B. 801.

28. [1993] 1 S.C.R. 87, 1993 CanLII 146 (S.C.C.).

Undue influence must be established before the courts will allow a contracting party to avoid the agreement. If no special relationship exists between the parties, the party alleging undue influence must prove the existence of such influence. In certain cases, however, when a special relationship exists between the parties, a rebuttable presumption of undue influence is deemed to exist. These cases are limited to those relationships of trust or good faith and frequently have a confidential aspect to them. These special relationships include solicitor–client, medical doctor–patient, trustee–beneficiary, parent–child, and spiritual advisor–parishioner relationships. In all of these relationships, if undue influence is alleged the onus shifts to the dominant party to prove that no undue influence affected the formation of the contract. The onus is usually satisfied by showing the courts that the fairness of the bargain or the price (if any) paid for the goods or service was adequate; that a full disclosure was made prior to the formation of the agreement; and that the weaker party was free to seek out the advice of others, and to seek out independent legal advice, if appropriate. If the presumption cannot be rebutted by evidence, then the contract is voidable by the weaker party, and the courts will grant rescission. Again, prompt action is necessary to obtain relief from the courts. If the weaker party fails to take steps promptly on being free of the undue influence, or ratifies the agreement either expressly or by inaction for a long period of time, the right to avoid the agreement may be lost, and the agreement will be binding.

The presumption of undue influence does not apply to the husband–wife relationship. Consequently, undue influence must be proven by the party raising the allegation. The relationship, however, is treated in a slightly different manner from one where it is deemed to exist and where the presumption applies. In the husband–wife relationship, the courts normally look at the degree of domination of the subordinate party by the dominant party, and the fairness or unfairness of the bargain struck between them, in deciding the question of enforceability.

> Armstrong convinces his wife (who is inexperienced in business matters) to convey to him a valuable property that she owns in exchange for some worthless company shares that he holds. If Armstrong's wife does so, and later discovers that the shares are worthless, she may be able to convince a court that the contract should be rescinded. To do so, however, she must establish undue influence on the part of Armstrong. If she can satisfy the court that the transaction was unfair, and that Armstrong so dominated her judgment that the decision was not her own, the court may provide relief.

Historically the presumption has been present in husband and wife cases, and inferred in Common Law marital relationships. The test of any presumption of undue influence in same-sex marriages has not yet arisen; one can only assume that the same test of dominance and bargain fairness will also prevail.

A common business situation that frequently gives rise to an allegation of undue influence is related to the requirement made by banks for a married person to guarantee his or her spouse's indebtedness. No presumption of undue influence exists in these cases. However, banks often require an assurance that the spouse has had independent legal advice before signing a guarantee of the married partner's loan to avoid any later claim that the guarantee is unenforceable on the basis of undue influence.[29]

MANAGEMENT ALERT: BEST PRACTICE

Business persons must recognize and react to situations where the possibility of undue influence exists — a trusted person assisting the elderly or persons with language difficulty, or any other special case. Where substantial amounts are concerned with a party who may be unduly influenced, you should demand evidence that the person has in fact had an opportunity to privately receive independent legal advice regarding the transaction. You are well within your rights to demand written proof that this opportunity was available to the other party.

29. See, for example, *Bank of Montreal v. Stuart*, [1911] A.C. 120.

A QUESTION OF ETHICS

Despite advances in gender equality, women remain vulnerable to pressure from their spouses. There should be a presumption of undue influence between husbands and wives, just as there is in other "special" relationships. Do your agree with this statement?

Duress

Duress

The threat of injury or imprisonment for the purpose of requiring another to enter into a contract or carry out some act.

The last basis for avoiding a contract is, fortunately, a rare business occurrence. Nevertheless, it is grounds for rescission. If a person enters into a contract under a threat of violence, or as a result of actual violence to his or her person or to a family member (or a close relative),[30] the contract may be avoided on the basis of **duress**. The threat of violence must be made to the person, however, and not simply directed toward the person's goods or chattels. Again, it is important that the victim of the violence take steps immediately on being free of the duress to avoid the contract. Otherwise, the courts are unlikely to accept duress as a basis for avoiding the agreement.

CLIENTS, SUPPLIERS *or* OPERATIONS *in* QUEBEC

The *Civil Code of Quebec* provides the following Articles, which chiefly govern failure to create enforceable contracts within the province:

1375. The parties shall conduct themselves in good faith both at the time the obligation is created and at the time it is performed or extinguished.

1399. Consent may be given only in a free and enlightened manner. It may be vitiated by error, fear or lesion. *(Willes: Lesion is roughly equivalent to "exploitation")*

1407. A person whose consent is vitiated has the right to apply for annulment of the contract; in the case of error occasioned by fraud, of fear or of lesion, he may, in addition to annulment, also claim damages or, where he prefers that the contract be maintained, apply for a reduction of his obligation equivalent to the damages he would be justified in claiming.

Figure 11–1 illustrates the effect of flaws on a contract.

FIGURE 11–1 Law of Contract

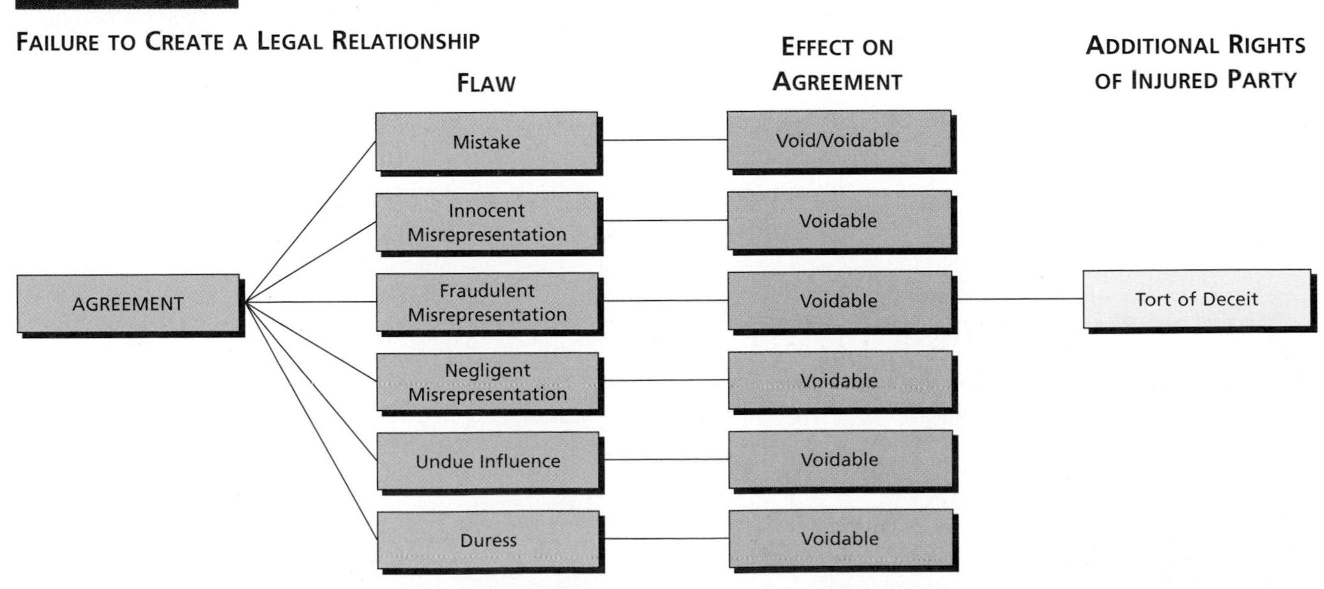

30. *Kaufman v. Gerson*, [1904] 1 K.B. 591.

CHECKLIST FOR AN ENFORCEABLE CONTRACT

Given an otherwise valid contract, **as per previous chapters including form and writing**, are there any grounds to believe that any of the following exist?

1. Mistake of fact (or law), be it unilateral or mutual.
2. Innocent or fraudulent misrepresentation.
3. Misrepresentation by non-disclosure or by negligence.
4. Undue influence.
5. Duress.

If not, the contract should be enforceable.

SUMMARY

The parties may comply with all of the essentials for the creation of a valid contract, but if there is not a true meeting of the minds, or if some mistake occurs that affects the agreement in an essential way, the contract may be either void or voidable. If a mistake should occur as to the existence of the subject matter, the identity of a contracting party, or, in some cases, the nature of the agreement, then the agreement may be unenforceable. If a party is induced to enter into a contractual relationship as a result of innocent misrepresentation of a material fact by a party, then the contract may be voidable at the option of the injured party when the misrepresentation is discovered. Non-disclosure in some instances may also constitute misrepresentation. If the misrepresentation is made deliberately, and with the intention of deceiving the other party, then the party making the fraudulent misstatement may also be liable in tort for deceit. A contract may also be voidable if a party enters into the agreement as a result of undue influence or duress. In each case, the victim must take steps to avoid the agreement as soon as he or she is free of the influence or threat. In all cases where the contract is voidable, if the party continues to take benefits under the agreement, or affirms the agreement after becoming aware of the defect, misrepresentation, undue influence, or duress, the right to rescind the contract may be lost.

KEY TERMS

REVIEW QUESTIONS

1. In what way(s) would mistake in its legal context differ from what one would ordinarily consider to be mistake?
2. How does unilateral mistake differ from mutual mistake?
3. What effect does mistake, if established, have on an agreement that the parties have made?
4. Explain the difference between mistake and misrepresentation.
5. Distinguish innocent misrepresentation from fraudulent misrepresentation.
6. What obligation rests upon a person who, having made an innocent misrepresentation, discovers the error?
7. Why do the courts consider non-disclosure to be misrepresentation under certain circumstances? Identify the circumstances where this rule would apply.

8. Explain the rationale behind the rule that a person who applies for insurance from an insurer must disclose all material facts concerning the subject matter of the insurance.
9. The presumption of undue influence applies to contracts made by individuals in specific types of relationships. Why is this so?
10. What obligation rests on the "dominant party" in a contract where the presumption of undue influence applies, if the other party alleges undue influence?
11. How does undue influence differ from duress?
12. Identify the situations where duress may be raised as a legal means for avoiding a contract.

MINI-CASE PROBLEMS

1. A and B entered into a verbal agreement whereby A would sell an automobile to B for $6,500. B drew up a written agreement (which A and B signed) that set out the price as $6,000. A failed to notice the change at the time of signing the agreement. What advice would you give A?
2. How would your answer to question 1 differ if A was illiterate and B read the contract aloud to him, stating the price to be $6,500 instead of the $6,000 price set out in the written agreement?
3. At a sports bar, Martin offers to sell an authentic Wayne Gretzky Edmonton Oilers hockey jersey to his friend Alfred for $400. Unknown to both, Martin's dog at home has just finished shredding the jersey. Carl arrives at the bar, and hearing of the deal, tells Alfred he has previously seen the jersey, and that it is not "game-worn" by Gretzky, but rather a current authorized production official replica jersey, made by an NHL-licensed supplier. What legal principles apply to this deal?
4. An industrial bakery agrees to supply one hundred thousand loaves of bread monthly to a major Canadian grocery chain at a price of $1.00 per loaf. Before deliveries begin, the baker rethinks the deal, feeling it should have charged more. On review, it discovers that the contract recites "one hundred" loaves of bread monthly at $1.00 per loaf. What recourse will the grocery chain have if the bakery refuses to honour its price beyond 100 loaves?
5. Eric, at age 80, owns a large property of about 100 forested acres, half of which is covered with walnut and butternut trees he had planted as a boy. Eric is approached by a real-estate agent who tells him he has a client who is looking for land with trees where a new house can be built. Eric severs the treed half of his property and sells it, and the new owner builds a family home. In talking with his new neighbour, Eric discovers to his horror that the buyer is the owner of a custom sawmill, and intends to cut down all of the hardwood trees for furniture and veneer. Does Eric have any recourse?

CASE PROBLEMS FOR DISCUSSION

CASE 1

An industrial firm purchased a rivet-making machine from a distributor whose advertising materials claim that the machine will produce up to 300,000 rivets per day. After purchase and installation the firm discovers that the machine, at maximum settings, will produce only 100,000 rivets per day. After an exchange of e-mails, the distributor asked if the firm was running three 8-hour shifts, or just one. The firm was running only one production shift, and could not afford to consider 24-hour operations.

Discuss the legal issues that apply here, and what factors might influence a judge to decide the case in favour of one party over the other.

CASE 2

Chamberlain of Chamberlain's Antiques carried on business as a used-furniture dealer. He would occasionally purchase all of the furnishings in a person's home if the person was leaving the country or moving a long distance.

Mrs. Lyndstrom, an elderly widow, indicated to Chamberlain that she intended to sell her home and furniture, as she was planning to move into her daughter's home to live with her. Chamberlain expressed an interest in purchasing her furniture, so an appointment was arranged for Chamberlain to examine the contents of her home.

When Chamberlain arrived, Mrs. Lyndstrom took him to each room and indicated the furniture that she intended to sell. Chamberlain listed the different items by groups, such as "all chairs and table in kitchen," or "bedroom suite in front bedroom." When they reached the living room, Mrs. Lyndstrom said, "all the furniture in here," and Chamberlain recorded, "all furniture in living room." Chamberlain offered $3,500 for the lot when the list was complete. Mrs. Lyndstrom agreed, and Chamberlain added to the bottom of the list: "Agreed price of all above furniture, $3,500." Both signed the document.

www.mcgrawhill.ca/olc/willes

The next day, Chamberlain arrived with a truck to pick up the furniture and a cheque for $3,500. As he was about to have the grand piano moved from the living room, Mrs. Lyndstrom informed him that he was to take only the furniture, not the piano. Chamberlain protested that the piano was included. However, Mrs. Lyndstrom argued that a piano was not furniture, but a musical instrument.

Explain how this matter might be resolved.

CASE 3

Corgi was the breeder of prize-winning pedigree dogs that often sold for very high prices. Reynolds, a wealthy businessman who had recently retired, decided to purchase one of these dogs. His intention was to enter the animal in the various dog shows that were held from time to time across the country.

Reynolds knew very little about dogs. He explained to Corgi that he wished to purchase a young dog that was already a prize-winning specimen of the breed. Corgi took Reynolds to a fenced run where several young dogs were caged. He pointed to one dog that he said, in his opinion, had the greatest potential, and that it had already won a prize at a local dog show. Corgi pointed to a red ribbon pinned to the opposite wall of the kennel building and explained that it was a first-prize ribbon that the dog had won. Reynolds did not bother to examine the ribbon.

Reynolds purchased the dog for $1,000 and took it home. His neighbour later saw the dog in Reynolds' backyard. He instantly recognized it as the dog that had recently won the first-prize ribbon in the children's pet show at the neighbourhood park. When he told Reynolds where he had last seen the dog, Reynolds telephoned Corgi immediately and demanded his money back.

Corgi refused to return Reynolds' money or take back the dog, and Reynolds threatened to take legal proceedings against him. Reynolds was unable to do so immediately, however, as he was called out of town on a family matter the next day. He was obliged to leave the dog with his neighbour during his absence. Reynolds advised the neighbour to take care of the animal as if it were his own.

Reynolds was out of town for several weeks. During that time, his neighbour entered the dog in a dog show sponsored by a kennel club. The dog won first prize in its class for its breed. On Reynolds' return, the neighbour advised him of his success. The two men decided to enter the dog in another dog show that was scheduled to be held in a nearby city.

At this second show, the dog placed only third in its class, and Reynolds was disappointed. He returned home and immediately took legal action against Corgi.

Discuss the basis of Reynolds' claim and the defences (if any) of Corgi. Render, with reasons, a decision.

CASE 4

Silica Mining Ltd. acquired a lease from the province that would allow it to mine mineral rights under a large block of land. The surface rights were owned by McCarthy. Silica Mining Ltd. assigned the mining rights for $25,000 and certain tonnage royalties to Granite Exploration Ltd., which intended to mine the silica deposits lying beneath the surface of the land. Both parties assumed that silica was a mineral that was reserved by the province and that the lease of mineral rights would allow the silica to be mined.

When Granite Exploration Ltd. entered on McCarthy's land to begin mining operations, McCarthy ordered Granite Exploration Ltd. employees off the land. He informed them that silica was not a mineral reserved by the province and that he owned the silica under the surface. Granite Exploration Ltd. contacted the provincial ministry and was advised that McCarthy was correct. Silica was not reserved by the province, and McCarthy owned the deposits.

Granite Exploration Ltd. demanded a return of its $25,000 payment from Silica Mining Ltd., and when Silica Mining Ltd. refused to do so, Granite Exploration Ltd. took legal action.

Explain the nature of the Granite Exploration Ltd. claim and indicate the probable outcome of the action.

CASE 5

A general contractor invited subcontractors to submit bids for mechanical work in the construction of a large office building. Several written bids were received by the contractor, as well as a telephone call from a subcontractor who responded with a bid about 12 percent below the lowest bid price of the written submissions. All written bids were within one percent of each other.

The subcontractor who submitted the telephone bid checked his estimates the day following the telephone call. He discovered that he had made an error in his calculations, and immediately called the contractor to withdraw his bid. Unfortunately, the contractor was out of town, and the subcontractor could not reach him at his office. However, he left a message with the contractor's secretary to the effect that his bid was in error, and that his bid price for the contract would be 13 percent above the figure quoted on the telephone.

The contractor, while out of town, prepared his contract price for the construction of the building, using the original telephone bid price for the mechanical work. He was awarded the construction contract. When he returned to the office, his secretary informed him of the subcontractor's error.

The subcontractor refused to enter into a written contract or to perform the mechanical work at the original price. The contractor was obliged to get the work done by the subcontractor who had submitted the next lowest bid. When the contract was fully performed, the contractor brought an action against the telephone bidder for the difference between the contract price quoted on the telephone and the actual cost incurred in having the work done.

Discuss the arguments that might be raised by the parties in this case. Render a decision.

CASE 6

Fatima attended an auction sale of race horses at a well-known and respectable racing stable. Prior to the sale, Fatima, who was interested in acquiring a particular horse, examined the animal's lineage, breeding history, and veterinary records, as well as the horse itself. At the time of the sale, the horse appeared to be ill, and Fatima asked the vendor's veterinarian about its health. The veterinarian replied that she thought the horse was suffering from a cold. On the basis of this information, Fatima entered the bidding on the sale of the horse, and she was the successful bidder at a price of $9,000. The conditions of the sale were that the purchaser assumed all risks and responsibility for the horse from the instant of the sale, when the title to the animal was deemed to pass.

Fatima took the horse to her own stable and informed her veterinarian that the horse was ill with a cold. Without examining the horse, the veterinarian prescribed some medicine for it. The medicine did not appear to help the horse. It remained ill for two weeks, becoming increasingly more feeble, as the time passed. Sixteen days after the horse had been purchased, it died. An autopsy revealed that the cause of death was rabies.

Fatima brought an action for a return of the purchase money paid for the horse. At the trial, expert witnesses for the plaintiff and defendant disagreed as to when the horse might have contracted the disease, but both agreed that it could possibly have had the disease at the time of the sale. Both also agreed that rabies was a disease most difficult to detect in its early stages of development. Neither the purchaser nor the vendor had any suspicion that the horse had contracted rabies, because of the rarity of the disease among horses vaccinated against it. For some reason, the horse Fatima had purchased was not vaccinated against rabies.

Identify the issues raised in this case. Indicate how it might be decided.

CASE 7

Mary McDonald was an 86-year-old widow who lived in her own home. She complained frequently about the high cost of maintenance of her house and the high property taxes she paid. Mrs. McDonald eventually felt it was necessary to cancel her fire insurance to reduce her household expenses. However, she handled all her own business affairs, and maintained herself in her home.

Mary's daughter, a real-estate agent, was aware of her mother's concern about her property. On a summer's day, she drove to her mother's home and informed her that she had some papers at the lawyer's office, which would protect her. The two women went down to the lawyer's office, where he advised Mary McDonald that the document was the deed to the property and required her signature. Mary briefly glanced at the document and said: "Where do I sign?"

Some months later, the municipal tax bill arrived and Mary McDonald was surprised to see that the property was in her daughter's name. She immediately brought a legal action against her daughter to have the property transferred back into her own name.

Discuss the arguments that the parties might raise in this case and render a decision.

CASE 8

Tower Installation Co., by way of a newspaper advertisement, learned of an invitation to bid on the engineering work for an electric-power transmission-tower line that a hydro-electric corporation wished to have erected in a remote area of the province. The advertisement indicated that the right-of-way had been laid out, and preliminary site preparation was in the process of completion. Engineers from Tower Installation Co. visited the site before making a bid. They noticed that many downed trees and branches cluttered the right-of-way throughout the entire length of the proposed line. The engineers assumed that the downed trees and branches were part of the preliminary site preparation.

The proposed contract stipulated that the successful bidder would be "responsible for the removal of standing trees and brush in those areas where they had not been cleared, and the final site preparation."

The proposal also stated that the bidders must inform themselves of all aspects of the site and the work to be done. Tower Installation Co. submitted its bid to perform the required work based upon its examination of the site and the proposed contract terms.

Some three months later, the contract was awarded to Tower Installation Co. It immediately sent its engineers and work crews to the construction site to begin work. To the surprise of the engineers, the downed trees and branches still remained on the site, as no further work had been done since the site was visited before. The company engineers complained to the hydro-electric corporation, but were told that they should have enquired about the logs and downed trees before making their bid on the work.

Tower Installation Co. proceeded with the contract and completed it in accordance with its terms. The company, however, lost $2 million as a result of the extra work required to clear the downed trees and branches from the right-of-way.

Tower Installation Co. took legal action against the hydro-electric corporation for the amount of its loss on the contract.

Discuss the basis of the claim in this case and the possible defences of the corporation. Render a decision.

CASE 9

Efficient Management Inc. offered to purchase the shares of Clean Management Ltd. from its shareholders, Andrews and Beatty. Efficient Management Inc. also agreed to lease premises from shareholder Andrews under a long-term lease.

Under the share-purchase agreement, Efficient Management Inc. agreed to purchase all of the outstanding shares, which were owned 50 percent by Andrews and 50 percent by Beatty. The total price payable for the shares was $2, $1 for all of the shares of Andrews and $1 for all of the shares of Beatty. Efficient Management Inc. also agreed to repay to Andrews and Beatty loans that they had made to their corporation, Clean Management Ltd., in the amount of $450,000. The purchase agreement provided that Clean Management Ltd. had accounts receivable outstanding in the amount of $350,000, which Andrews and Beatty warranted were all in good standing and could be collected by Clean Management Inc. from its customers within 60 days.

Under the terms of the agreement, Efficient Management Inc. agreed to pay $100,000 as a deposit on the signing of the agreement and the balance of the money in 90 days. The shares of Clean Management Ltd. were to be delivered at the time of signing on payment of the $1 to each shareholder. The agreement was signed on May 1 and the $100,000 paid. Andrews and Beatty signed over their shares on the same day and each received $1 as payment in full.

By June 30, Efficient Management Inc. had collected only $225,000 of the accounts receivable, and the balance had to be considered uncollectible. At that time they informed Andrews and Beatty that they expected them to provide the company with the remaining $125,000 that Andrews and Beatty had warranted were collectible accounts. Andrews and Beatty did not pay the balance owing.

On July 31 the accountant at Efficient Management Inc. inadvertently sent Andrews and Beatty a cheque for the $350,000 balance owing under the agreement. When the error was discovered, the company claimed repayment from Andrews and Beatty of the $125,000. Andrews and Beatty refused to do so, and Efficient Management Inc. instituted legal proceedings against Andrews and Beatty to recover the $125,000, or, alternatively, to set off rent owing to Andrews against the amount unpaid.

Outline the nature and basis of the claim against the shareholders, and any defences that Andrews and Beatty might raise. Render a decision.

CASE 10

Schuster & Co. owned two volumes of a rare edition of Geoffrey Chaucer's *Canterbury Tales*. One volume was in excellent condition. The second volume was in poor shape, but nevertheless intact. Schuster & Co. sold both volumes to MacPherson, a rare book merchant.

MacPherson loaned the two volumes to a local library for a rare-book display. Unknown to MacPherson, only the volume in excellent condition was put on display with a collection of other rare books. The second copy was placed in a display designed to show how rare books might be repaired, but the book was placed in such a position that neither its title nor its contents could be determined.

A week after the books had been returned to their owner, Holt, a collector of rare books, telephoned MacPherson to determine if he had a copy of the *Canterbury Tales* for sale. MacPherson replied that he did, but it was "not in top shape." Holt then asked if the copy had been on display at the library, and MacPherson said, "Yes."

Holt informed MacPherson that she had seen the display of books at the library and would be interested in purchasing the volume. A price was agreed upon, and Holt sent a cheque to MacPherson for the agreed amount.

MacPherson sent the volume that was in poor condition to Holt by courier. On its receipt, Holt complained that the volume was not the same one that had been on display at the library. MacPherson maintained that it was and refused to return Holt's money.

Holt brought an action against MacPherson for a return of the money that she had paid MacPherson.

Indicate the nature of Holt's claim and express an opinion as to the outcome of the case.

CASE 11

Bob and Jane are looking for an older home with "character." Their real-estate agent shows them a number of properties, and the one most suitable is described by the agent as a "real heritage home," an old stone rural farmhouse. Bob and Jane purchase the property and complete extensive modern interior renovations. With the interior complete, they embark on restoring the exterior, but are stopped by a township building inspector who informs them that the property is a "designated heritage site," and that no alterations are permitted without town council approval. Moreover, when the interior modern improvements are discovered, Bob and Jane are charged a significant fine for their ignorance.

Discuss the aspects of misrepresentation that apply in this case.

CASE 12

Gavin is a real-estate agent, and suggests to his dentist, Ravi, that he should invest $250,000 in commercial condominium real estate in a distant city. Gavin provides Ravi with legal papers to sign, but in the end the property address is not a commercial development, but rather a residential condominium. Ravi claims "non est factum."

Discuss the nature of Ravi's claim against Gavin.

The Online Learning Centre offers more ways to check what you have learned so far. Find quizzes, Weblinks, and many other resources at www.mcgrawhill.ca/olc/willes.

CHAPTER 12

The Extent of Contractual Rights

Chapter Objectives

After study of this chapter, students should be able to:

- Explain how the rule of privity establishes who may have rights or obligations under a contract.
- Illustrate the exceptions to the rule of privity.
- Explain how rights and obligations in a contract may be transferred to others who were not a party to the original contract.
- Illustrate the advantages and flexibility such transfers offer to business operators.

> **YOUR BUSINESS AT RISK**
>
> While a simple contract may be between two persons, the law often allows third parties into the picture. Business persons must know when and how their rights and obligations can be extended to others. Without understanding this process you may retain responsibilities you may have wished (or paid!) to transfer to someone else, or you may be expecting benefits under a contract which you no longer have a right to receive.

PRIVITY OF CONTRACT

Privity

A person cannot incur liability under a contract to which he or she is not a party.

Once a valid contract has been negotiated, each party is entitled to performance of the agreement according to its terms. In most cases, the contract calls for performance by the parties personally. However, the parties may either attempt to confer a benefit on the third party by way of contract, or impose a liability on a third party to perform a part of the agreement.

At Common Law, the rule relating to third-party liability is relatively clear. Apart from any statutory obligation or obligation imposed by law, a person cannot incur liability under a contract to which he or she is not a party. By this rule, parties to an agreement may not impose a liability on another who is not a party to the contract except in those circumstances where the law imposes liability. For example, under the law of partnership, a partner (in the ordinary course of partnership business) may bind the partnership in contract with a third party. The remaining partners, although not parties to the agreement, will be liable under it and obligated to perform. The law provides in such a case that the partner entering the contract acts as the agent of all of the partners and negotiates the agreement on their behalf as well.

Under certain circumstances, a person may acquire liability under a contract negotiated by others if the person accepts land or goods that have conditions attached to them as a result of a previous contract. In this case, the person would not be a party to the contract but, nevertheless, would be subject to liability under it. This situation may be distinguished, however, from the general rule on the basis that the person receiving the goods or the property accepts them subject to the conditions negotiated by the other parties. This person must also be fully aware of the liability imposed at the time that the goods or property are received. The acceptance of the liability, along with the goods or property, resembles, in a sense, a subsidiary agreement relating to the original agreement under which one of the original contracting parties retains rights in the property transferred.

> Alton Ship Leasing Ltd. owns a large charter power boat that it wishes to sell, but that it presently leases to Chambers for the summer. Burrows enters into a contract with Alton to purchase the boat, aware of the lease that runs for several months. Burrows intends to make a gift of the boat to his son. As a part of the contract he requires delivery by Alton to his son of the ownership papers pertaining to the vessel. If Burrows's son accepts the ownership of the boat, aware of the delay in delivery, he would probably be required to respect the contract and accept the goods under the conditions imposed by the contract between Alton and his father.[1]

Prior Interests in Land

An important exception to the privity of contract rule concerns contracts that deal with the sale, lease, or transfer of land. In general, the purchaser of land takes the land subject to the rights of others who have acquired prior interests in the property or rights of way over the land. The purchaser is usually aware of the restrictions before the purchase is made, however, with the exception of some tenancies, all restrictions running with land and all rights of third parties must be registered against the land in most jurisdictions. Consequently, the person acquiring the land has notice of the prior agreements at the time of the transfer of title. Actual notice of an unregistered contract concerning the land may also bind the party, depending upon the jurisdiction and the legislation relating to the transfer of interests or land. Apart from this limited form of restriction or liability imposed by law, a person who is not a party to a land contract may not normally incur liability under it.

Constructive Trusts

The second part of the rule relating to third parties concerns the acquisition of rights by a person who is not a party to a contract, but upon whom the parties agree to confer a benefit. The principle of consideration comes into play, however, if the third party attempts to enforce the promise to which he or she was not a contracting party. When the principle of consideration is strictly applied, it acts as a bar to a third party claiming rights under the contract. This occurs because the beneficiary gives no consideration for the promise of the benefit. Only the person who is a party to the agreement and gives consideration for the promise would have the right to insist on performance.

The strict enforcement of the rule had the potential for abuse at Common Law whenever a party to a contract who had the right to enforce the promise was unable or unwilling to do so. To protect the third party, the courts of equity provided a remedy in the form of the *doctrine of constructive trust.* Under this equitable doctrine, the contract is treated as conferring a benefit on the third party. As well, the promisor is obliged to perform as the trustee of the benefit to be conferred on the third-party beneficiary. Under the rules of **trust**, the trustee, as a party to the contract, has the right to sue the contracting party required to perform. However, if the trustee refuses or is unable to take action, the third-party beneficiary may do so by simply joining the trustee as a party defendant.

Trust

An agreement or arrangement whereby a party (called a trustee) holds property for the benefit of another (called a beneficiary or *cestui que* trust).

> Alport and Bush enter into an agreement that will provide a benefit for Cooke, who is not a party to the agreement. Under the privity of contract rule, Cooke could not enforce the promise to confer the benefit. Only the person who gave consideration for the promise could enforce the agreement. However, if that person refuses to

1. See, for example, *Lord Strathcona Steamship Co. v. Dominion Coal Co.*, [1926] A.C. 108.

take court action, Cooke would be entitled to do so. The reasoning here is that if the trustee of the benefit refused or was unable to act to enforce the promise, this in turn would affect Cooke's rights as beneficiary of the trust. This would also be the case if the trustee, once in possession of the funds or benefit, refused to confer the benefit on the beneficiary.

Promises Under Seal

A formal contract under seal also represents an exception to the general rule concerning privity of contract. If the formal agreement is addressed to the third party and contains covenants that benefit the third party, the delivery of the agreement to the third party would enable that person to maintain an action against the promisor to enforce the rights granted under the agreement, should the promisor fail or refuse to perform.

Statutory Rights and Liabilities

Since rights are frequently conferred on third parties in certain types of contracts, the legislation governing these types of contracts generally provides the third party with the statutory right to demand performance directly from the contracting party. This is without regard for consideration or the Common Law rule concerning privity of contract.[2] The right of a beneficiary under a contract of life insurance is a notable example of the legislative establishment of third-party rights to a benefit under such a contract. Without statutory assistance (and assuming that a trust cannot be ascertained) the doctrine of privity of contract would apply and the beneficiary would be unable to collect from the insurance company. This is so because under the privity of contract rule a person not a party to a contract would not acquire rights under the agreement. This rule also applies to the liability of outside parties. Unless the person can be shown to be liable by way of some statutory or Common Law rule, a person who is not a party to a contract cannot be liable under it.

The acquisition of rights and the assumption of liabilities by third parties may, nevertheless, be expanded at Common Law to include parties who are closely related to the agreement and aware of the terms. In a number of cases, the courts have held that a party who receives a benefit under a contract may take advantage of implied terms within it,[3] or be subject to the liabilities that were negotiated.[4]

Other Privity Exceptions

From a practical point of view, another route is available whereby a third party may acquire a right against another. By the law of contract, only the purchaser under a contract of sale would have a right of action if the goods purchased proved to be unfit for the use intended. While under the law of torts, if the user or consumer can establish a duty on the part of the manufacturer not to injure the consumer, and if injury ensues as a result of use by the third party, a right of action would lie against the manufacturer. This would be the case even though no contractual relationship existed. The availability of these alternate remedies to third parties has eased the pressure for changes in the privity of contract rule, and perhaps slowed the move toward broadening the exceptions to it. Apart from the use of the law of torts, the general trend seems to be to provide for specific cases in legislation or by alternate remedies, rather than to alter the basic concept of privity of contract.

A similar agreement that attempts to confirm a right on a party occurs when a third party attempts to enforce rights or benefits under an agreement to which the third party is not a party but a beneficiary. This issue arose in the case of *Shanklin Pier Ltd. v. Detel Products Ltd.*,[5] where a warranty had been given that certain paint substituted for the type specified was suitable for the purpose intended. The contract was negotiated by

2. See, for example, the *Insurance Act*, R.S.O. 1990, c. I.8, s. 127.
3. *Shanklin Pier Ltd. v. Detel Products Ltd.*, [1951] 2 K.B. 854.
4. *Pyrene Co. Ltd. v. Scindia Navigation Co. Ltd.*, [1954] 2 Q.B. 402; *Anticosti Shipping Co. v. Viateur St.-Armand*, [1959] S.C.R. 372.
5. Supra.

the painting contractor and the paint supplier. However, when the paint proved to be unsuitable, the third-party owner of the painted pier attempted to enforce the warranty.

The judge discussed the issue in the following manner:

> This case raises an interesting and comparatively novel question whether or not an enforceable warranty can arise as between parties other than parties to the main contract for the sale of the article in respect of which the warranty is alleged to have been given.
>
> The defence, stated broadly, is that no warranty such as is alleged in the statement of claim was ever given and that, if given, it would give rise to no cause of action between these parties. Accordingly, the first question which I have to determine is whether any such warranty was ever given.

His Lordship reviewed the evidence about the negotiations which led to the acceptance by the plaintiffs of two coats of D.M.U. in substitution for the paint originally specified, and continued:

> In the result, I am satisfied that, if a direct contract of purchase and sale of the D.M.U. had then been made between the plaintiffs and the defendants, the correct conclusion on the facts would have been that the defendants gave to the plaintiffs the warranties substantially in the form alleged in the statement of claim. In reaching this conclusion, I adopt the principles stated by Holt, C.J. in *Crosse v. Gardner* and *Medina v. Stoughton* that an affirmation at the time of sale is a warranty, provided it appear on evidence to have been so intended.
>
> Counsel for the defendants submitted that in law a warranty could give rise to no enforceable cause of action except between the same parties as the parties to the main contract in relation to which the warranty was given. In principle this submission seems to me to be unsound. If, as is elementary, the consideration for the warranty in the usual case is the entering into of the main contract in relation to which the warranty is given, I see no reason why there may not be an enforceable warranty between A and B supported by the consideration that B should cause C to enter into a contract with A or that B should do some other act for the benefit of A.

When employees negligently perform their duties under a contract made between their employer and a customer of the firm, the customer may have the opportunity to bring an action in tort against both the employer and the employees. If a duty of care was owed by the employees to the customer, it would appear that the court may consider the employees personally liable to the customer for their tort. In the Supreme Court of Canada case of *London Drugs v. Kuehne & Nagel International Ltd.*,[6] a customer entered into a storage contract with a warehouse operator for the storage of a transformer. Employees of the warehouse operator negligently damaged the transformer in the course of moving it. The customer brought an action for damages in contract and tort for negligence against both the warehouse operator and the employees. The court decided that both the employees and the warehouse operator were liable for the loss, but allowed the employees who were not party to the contract between the warehouse operator and the customer to limit their liability to the amount agreed upon in the contract. On the basis of this case, it would appear that employees who are not parties to their employer's contract may nevertheless use the contract's provisions to limit their liability in tort.

CLIENTS, SUPPLIERS *or* OPERATIONS *in* QUEBEC

The *Civil Code of Quebec*, Articles 1440-1450, provides counterparts to the chief Common Law principles of third party involvement in contract. These are:

1440. A contract has effect only between the contracting parties; it does not affect third persons, except where provided by law. (Willes: privity)

6. 1992 CanLII 41 (SCC).

1443. No person may bind anyone but himself and his heirs by a contract made in his own name, but he may promise in his own name that a third person will undertake to perform an obligation, and in that case he is liable to reparation for injury to the other contracting party if the third person does not undertake to perform the obligation as promised. (Willes: vicarious performance)

1444. A person may make a stipulation in a contract for the benefit of a third person. The stipulation gives the third person beneficiary the right to exact performance of the promised obligation directly from the promisor. (Willes: third party beneficiaries)

Assignment of Contractual Rights and Obligations

Novation

Novation

The substitution of parties to an agreement, or the replacement of one agreement by another agreement.

A third party may, of course, wish to acquire rights or liability under a contract by direct negotiation with the contracting parties. Should this be the case, the third party may replace one of the parties to the contract by way of a process called **novation**. This process does not conflict with the privity of contract rule because the parties, by mutual consent, agree to terminate the original contract and establish a new agreement. In this agreement the third party (who was outside the original agreement) becomes a contracting party in the new contract and subject to its terms. The old agreement terminates, and the original contracting party, now replaced by the third party, becomes free of any liability under the new agreement. By the same token, the original contracting party, being no longer a contracting party, is subject to the privity of contract rule.

The legal nature of novation involves a number of elements that must be present to establish a complete novation. These were set out by the British Columbia Court of Appeal in the case *of Re Abernethy-Lougheed Logging Co.*; *Attorney-General for British Columbia v. Salter*,[7] where the requirements were noted as follows:

> ... three things must be established: First, the new debtor must assume the complete liability; second, the creditor must accept the new debtor as a principal debtor, and not merely as an agent or guarantor; third, the creditor must accept the new contract in full satisfaction and substitution for the old contract; one consequence of which is that the original debtor is discharged, there being no longer any contract to which he is a party, or by which he can be bound.
>
> All these matters are in our law capable of being established by external circumstances; by letters, receipts, and payments and the course of trade or business.
>
> In other words, in the absence of an express agreement the intention of the parties may be inferred from external circumstances including conduct.

The process of novation, as a means of transferring contractual rights to a third party, is a useful method of avoiding the privity of contract rule. However, in modern business practice, it is not only cumbersome but inappropriate for certain kinds of transactions, where the third party may only be interested in a particular aspect of the transaction.

Textile Imports Ltd. sells goods to Best Clothing Co. on credit. Textile Imports Ltd. may not wish to have its funds tied up in a large number of credit transactions — it may wish to have access to its money to finance its own business. Credit Finance Co. is prepared to buy Best Clothing Co.'s promise to pay from Textile Imports Ltd., but if novation is the only method of transfer of these rights, all parties would be obliged to consent. Assuming Best Clothing Co. is willing to surrender its rights against Textile Imports Ltd. in the contract of sale, a new agreement between Best Clothing Co. and Credit Finance Co. would have to be formed. In this contract, Credit Finance Co. would be obliged to give some consideration for Best Clothing

7. [1940] 1 W.W.R. 319, (BCCA).

Co.'s promise to pay, or the contract would have to be under seal. Novation would clearly be an unwieldy tool for such a simple business transaction.

Equitable Assignments

There was a difficulty at Common Law that prevented the development of a simpler, more streamlined method of bringing a third party into a contractual relationship. This was the fact that, originally, the Common Law courts would only recognize rights in contracts between parties as personal rights that were not subject to transfer. These particular rights, called **choses in action** (in contrast to choses in possession, which are goods and things of a physical nature), were treated differently by the court of equity. Equity recognized the need for flexibility in the transfer of rights under contracts and business agreements. It would enforce rights that had been transferred to a third party if all of the parties could be brought before the court. If the court could be satisfied that a clear intention to assign the rights had been intended by the parties to the agreement, the contract would be enforced. While this process, too, was cumbersome, it nevertheless permitted the assignee a right to enforce it against the promisor.

Choses in action

A paper document that represents a right or interest that has value (e.g., a share certificate).

Equity did not normally permit an assignee to bring an action on an assignment in the assignee's own name. Rather, it imposed a duty on the assignor to attach his or her name to the action. This step was normally necessary in order to have all parties before the court, and to prevent a further action in the Common Law courts by the assignor against the debtor at a later date. The presence of the assignor, while sometimes inconvenient, was necessary since **equitable assignments** need not take any particular form. They may be either oral or written. The only essential part of the assignment was that the debtor be made aware of the fact that the assignment had been made to the third party. If this was done, any other evidence surrounding the assignment or the original agreement would be available to the court by way of evidence at trial.

Equitable assignment

An assignment that could be enforced if all parties could be brought before the court.

Under the rules of equity, an assignment did not bind the debtor or promisor until notice had been given to him of the assignment. From that point on, the assignment was effective, and he was obliged to comply with it. It was, however, subject to any rights that existed between the debtor and the assignor, for the assignee took the assignment subject to any defences to payment that existed between the original parties to the contract.

Vicarious performance

Certain contracts were recognized as unassignable in both the Common Law courts and the Courts of Equity. Any contract that required the personal service or personal performance by a party to the contract could not be performed by a third party to the agreement. For example, if a person engaged an artist to paint her portrait, the artist who was engaged would be required to do the painting. The only procedure enabling a third party to perform would be novation, which would require all parties to consent to the change. In some circumstances the courts did permit a modified form of personal performance to take place if the contract did not specifically state that only the contracting party could perform. In these contracts, the party to the contract remains liable for the performance according to the terms of the agreement. However, the actual work done, or performance, is carried out by another person under a separate agreement with the contractor. This type of performance, known as **vicarious performance**, involves two or more contracts. The first contract is the contract between the parties in which the contractor agrees to perform certain work or services. The contractor, in turn, enters into a second contract. This may be a contract of employment with one of the contractor's employees, or it may be a contract with an independent contractor to have the actual work done. In all cases, the primary liability rests with the contractor if the work is done improperly. The unsatisfied party to the contract would not sue the person who actually performed the work, but would sue the contractor. The contractor, in turn, would have the right under the second contract to take action against the party who actually performed the work, if the work was done negligently.

Vicarious performance

A performance of a contract by a third party, where the contracting party remains liable for the performance.

CASE IN POINT

Cast-tek Metal Works receives an unexpectedly large order for cast aluminium wheel rims from one of its regular clients, a major Canadian auto parts retailer. In order to meet the order delivery date, Cast-tek farms out part of the production to another metalworking firm, Allied Metal. Allied, in attempt to raise its own profit margin on the deal, uses lower grade aluminium and cuts corners in production. The faulty rims show up as returns made to the retailer from dissatisfied final customers. The retailer takes action against Cast-tek for damages. It will succeed against Cast-tek, and Cast-tek will have a similar right of recovery against Allied Metal.

These contracts conflict neither with the privity of contract rule nor with the rules relating to novation and the assignment of contractual rights. Both contracts remain intact, and the third party does not acquire rights under the second contract. The only difference is that the actual performance of the work in one contract is done by a party to the second contract.

As a general rule, most contracts for the performance of work or service may be vicariously performed if there is no clear understanding that only the parties to the contract must perform personally. Most parties to business transactions do not contemplate that the other party to the agreement will personally carry out the work, nor in many cases would the parties consider it desirable. Consequently, by customs of the trade in most business fields, the contracts may be vicariously performed. Only in the case of professionals, entertainers, and certain other specialized activities where special skills or talents are important would personal service be contemplated.

MANAGEMENT ALERT: BEST PRACTICE

If your business is sourcing inputs from a particular supplier, and you wish to ensure those inputs originate only from that source directly (perhaps because the source always exceeds your quality specifications, uses only best-quality sub-components, etc.), you may wish to include a clause in supply contracts prohibiting vicarious performance, unless with your specific consent.

Statutory Assignments

Statutory assignment

An assignment of rights that an assignee may enforce if certain conditions are met by the assignment.

By the middle of the nineteenth century, the need for a more streamlined method of transferring contractual rights (other than novation, vicarious performance, and equitable assignments) became apparent. Businesses frequently assigned contractual rights, but when difficulties arose, the practice of the courts of equity to require all parties to be present often proved inconvenient. This was particularly so if the assignor had made a complete assignment of all rights to the assignee, and the dispute concerned only those events that transpired after the assignment had been made. To eliminate these difficulties, the law was altered by an English statute in 1873[8] to give the assignee of a chose in action a right to institute legal proceedings in the assignee's own name if the assignee could satisfy four conditions:

(1) the assignment was in writing and signed by the assignor,
(2) the assignment was absolute,
(3) express notice of the assignment was given in writing to the party charged, the title of the assignee taking effect from the date of the notice, and
(4) the title of the assignee was taken subject to any equities between the original parties to the contract.[9]

8. *Supreme Court of Judicature Act*, 36 & 37 Vict., c. 66, s. 25(6). Similar legislation was passed in most common law provinces of Canada.

9. See, for example, *Conveyancing and Law of Property Act*, R.S.O. 1990, c. C.34, s. 54(1).

Essentially, the change in the law did nothing more than permit the assignee to bring an action in his or her own name to enforce a contractual right that had been assigned absolutely, and to provide the form in which notice of the assignment should take. In effect, however, it enormously increased the efficiency by which assignments could be made. In all other respects, the rights and duties of the parties remained much the same as those for equitable assignments that the courts of equity had enforced before the statutory change.

Mall Developments Inc. sells Maple Leaf Mall and assigns absolutely the tenant leases to Realty Management Corporation. MDI delivered notices of the assignment of the leases to RMC.This statutory assignment is now complete, and tenants will thereafter make their regular lease payments to RMC.

The statutory requirement of written notice of the assignment was a beneficial change in procedure. In the past, the notice for equitable assignments could be either written or oral. Indeed, in many cases where the debtor was illiterate, the creditor saw no need to prepare a written notice, only to be obliged to explain it to the debtor on delivery. The alteration reflected, however, the changes that had taken place in society and the relative rarity of illiteracy, particularly in the business community, by the end of the century. It also had the added advantage of fixing the time at which the title in the assignee is established as far as enforcement of the debt is concerned. Until the written notice is received by the debtor, any payment could properly be made to the creditor. If the assignee was tardy in delivering the notice of the assignment, the payment of the debt to the original creditor would discharge the debtor. The assignee would then be obliged to recover the money from the assignor. Conversely, any payment made to the original creditor after the debtor received notice of the assignment would be at the debtor's risk, for the assignee would be entitled to payment of the full amount owing from the time the notice was given. If the debtor failed to heed the notice, he or she could conceivably be obliged to pay the amount over again to the assignee, if he or she was unable to recover it from the original creditor.

In the event that a creditor has assigned the same debt to two different assignees, by either accident or design, the assignee first giving notice to the debtor would be entitled to payment, provided that he or she had no notice of any prior assignment. Thus, if the first assignee delays giving notice to the debtor, and the second assignee of the same debt gives notice to the debtor without knowledge of the prior assignment, and is paid by the debtor, the debtor is discharged from any obligation to pay the first assignee.

From the assignee's point of view, some risk is involved when an assignment takes place, because the assignee takes the contract as it stands between the parties at the time of the assignment. While the assignee can usually obtain some assurance as to the amount owing on the debt, the risk that the debtor-promisor may have some defence to payment, or some **set-off**, is always present. The assignee gets the same title that the assignor had. If the assignor obtained the title or rights by fraud, undue influence, duress, or some other improper means, the debtor may raise this as a defence to any claim for payment. While the assignee would not be liable in tort for any deceit, the defence would allow the debtor to avoid payment, as the contract would be voidable against both the assignor and assignee. The same rule would also apply if the assignor became indebted to the debtor on a related or unrelated matter before the notice of the assignment was made. The debtor, in such circumstances, would be entitled to deduct the assignor's debt from the amount owing by way of set-off. He or she would be obliged to pay the assignee only the difference between the two debts. If the assignor's obligation was greater than the amount of the debt assigned, then the assignee would be entitled to no payment at all. He or she would not, however, be liable to the debtor for the assignor's indebtedness. Figure 12–1 illustrates statutory assignments.

Set-off

When two parties owe debts to each other, the payment of one may be deducted from the other, and only the balance paid to extinguish the indebtedness.

FIGURE 12–1 Statutory Assignment

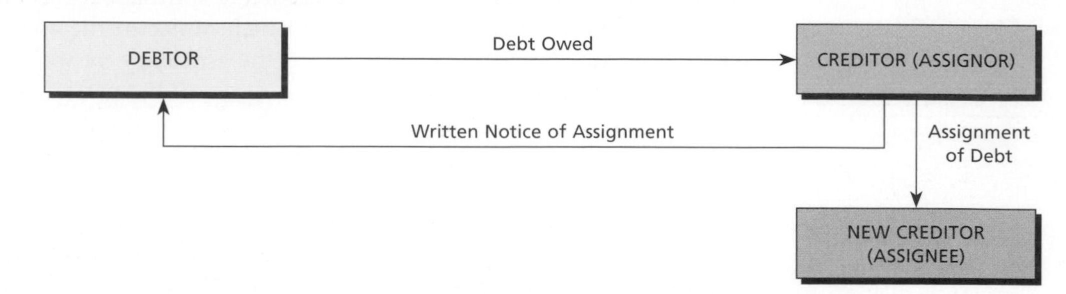

MANAGEMENT ALERT: BEST PRACTICE

"Factoring receivables" is a common business tool for raising funds quickly. A business assigns its existing accounts receivable (due now or in the future) to a "factor" (a collections company or bank) in return for an immediate lump-sum percentage payment (often 75 percent to 95 percent of the face value). The factor then collects the debts as they normally become due. Such assignments can create problems, as the debtors may have significant defences to payment against the original creditor, which can be raised against the factor. When considering factoring, a business should winnow out these potentially problematic accounts. Too often, businesses are excessively optimistic on the quality of their accounts. Difficulties then encountered by the factor result in claims back against the original creditor and lower percentages offered in future. They may also result in the loss altogether of factoring as a future financing option if factors believe that the business cannot be trusted in its assessment of the probability of collection of its accounts.

A QUESTION OF ETHICS

The assignment of contractual rights raises the potential of a clash of values within the ongoing relationship between the parties. The sale contract is a closed environment (recall: privity) where buyer and seller have found a set of common values sufficient to support a single transaction. In the case of ongoing contracts capable of assignment, the environment is not closed, and any new participant need only find common values with one of the two original parties. This raises the possibility that one original party may work well with the other, but the third (assignee) may not. Picking an odd example illustrates all those which are more frequently encountered: an animal rights activist leases a shop to an agreeable tenant. The tenant then makes a valid sub-lease to a butcher shop. The result is relationship tension. Think carefully before providing for the assignment of contracts before knowing who you might get as your new counterparty!

Assignments by Law

In addition to the provision made for the assignment of ordinary contractual rights by which the parties must prepare a document in writing and give notice, there are a number of other statutory assignments that come into effect on the death or bankruptcy of an individual. Certain other statutory rights also come into play in some cases where a person is incapable of managing his or her own affairs. Under all of these circumstances, some other person assumes all of the contractual rights and obligations of the individual, except for those requiring personal performance. For example, when a person dies, all of his or her assets, and all contractual rights and obligations by operation of law, are assigned to the executor named in the deceased's will or to the administrator appointed if the person should die intestate. Similarly, when a person makes a voluntary assignment in bankruptcy or is adjudged bankrupt, a trustee is appointed. The trustee acquires an assignment of all contractual rights of the bankrupt for the purpose of preservation and distribution of the assets to the creditors. The rights and duties of both the executor and trustee are governed by statute, as are the rights and duties of persons similarly appointed under other legislation to handle the affairs of incapacitated persons or corporations.[10]

Negotiable Instruments

Negotiable instrument

An instrument in writing that, when transferred in good faith and for value without notice of defects, passes a good title to the instrument to the transferee.

The assignment of contractual rights must be distinguished from the assignment of **negotiable instruments**, such as promissory notes and cheques. Negotiable instruments are subject to special legislation called the *Bills of Exchange Act*, which governs the rights of the parties and assignees. Assignments under the *Bills of Exchange Act* are examined in some detail in Chapter 28, "The Law of Negotiable Instruments." However, for the purpose of assignments generally, it is important to note that these instruments are subject to a different set of assignment rules.

CHECKLIST FOR EXTENT OF CONTRACTUAL RIGHTS

Privity prevails against third parties except for:

1. Interests running with land.
2. Constructive trusts.
3. Promises under seal.
4. Statutory exceptions.

Transfer of rights can be accomplished through:

1. Novation.
2. Equitable and statutory assignment.
3. Vicarious performance.
4. By operation of law.

Summary

The general Common Law rule with respect to contractual rights and liabilities states that no person may acquire rights or liability under a contract to which he or she is not a party. By novation, a party may replace another in a contract and become bound if all parties consent. The contract is a new agreement, however, and the replaced party no longer has rights or liability under it. In some cases, parties may acquire rights if they are in a position to establish that a constructive trust for their benefit was created by a contract. Apart from these exceptions, the Common Law rule applies to all contracts unless, of course, a statute specifically provides a right to a third party.

Rights under a contract may also be performed by another person under vicarious performance, unless the contract calls for personal performance. At law, rights may be assigned either by an equitable assignment or by a statutory assignment. In each case, notice of the assignment must be given to the promisor before the assignment is effective. Even then, the assignee receives only as good a title to the right as the assignor had. Any defence that the promisor or debtor is entitled to raise to resist or avoid a payment demand by the assignor may be raised against the assignee.

When parties are no longer capable of dealing with their own affairs, due to death, incapacity, or bankruptcy, statute laws assign their contract rights to others to manage.

Key Terms

10. For example, the public trustee in the case of persons committed to a mental institution.

www.mcgrawhill.ca/olc/willes

Review Questions

1. What risk does an assignee take when a contract assignment is made?
2. What are the major exceptions to the privity of contract rule? Why are these exceptions necessary?
3. Define novation and explain its role in the assignment of contractual rights.
4. Under what circumstances would a debtor be entitled to set off a debt against a claim for payment that the debtor has against another debtor?
5. Discuss vicarious performance as an exception to the assignment of contractual rights. Why is this the case?
6. Explain the requirements that must be met in order to make a proper statutory assignment of contractual rights.
7. Explain the importance of the privity of contract rule in contract law.
8. Indicate the rights and duties of a debtor when notice of assignment of the debt is received.
9. How is an equitable assignment of a contract made?
10. Must a person always consent to a statutory assignment before it is effective?

Mini-Case Problems

1. C, a contractor, enters into a contract with D to construct a house for him. C engages E, a plumber, to install the plumbing in the house. If E installs the plumbing negligently, what are D's rights?
2. X owes Y $500. Some time later, X sells Y her canoe for $200, but Y makes no payment for it. Y assigns the $500 debt that X owes him to Z. Notice is given by Z to X that Z is now the assignee of the debt, and Z is to be paid the $500. What are X's rights? What are Z's rights?
3. Each year since he was born, Aunt Mabel has sent Dan $1,000 toward his future education. The money is always in the form of a cheque payable to Dan's father. Explain the legal principle which will allow Dan to claim these funds when he goes on to higher education.
4. Thompson ran an auto service shop and bought certain parts directly from the automobile manufacturers. He bought such a transmission to repair a customer's vehicle, installed it, and soon after went out of business. Should that transmission fail, the customer may be able to claim under a warranty given to Thompson. Explain how that would be possible under the Common Law.
5. Angelo sells his flower shop business to Carla for $60,000 including the inventory, equipment and the balance of his lease. When the landlord hears of this, she draws up a new lease for Carla, and gives Angelo a release. What has happened regarding this lease arrangement, at Common Law?

Case Problems for Discussion

CASE 1

On January 20th, the Morrison Manufacturing Company assigned a block of its receivables to Acme Finance Company. On January 21st, it assigned a second block to Zenith Finance Company. One of Morrison's accounts assigned to Zenith — that of Jenkin — had already been assigned to Acme. Jenkin received notice of the assignment to Zenith on January 27th, and that of Acme on January 28th.

To whom should Jenkin's next payment be made, and why?

CASE 2

Personalized Performance Garage advertised "personalized service" for its customers. A large sign at the front of the garage depicted the owner of the establishment placing a check mark in a box on a work order that read "personally inspected and repaired by a certified A-1 Mechanic."

Rita Morales was impressed with the advertisement and the prospect of obtaining careful repair of her expensive automobile. She delivered the automobile for a minor engine adjustment and inquired if the owner did provide personal service. In response to her question, the owner replied, "I look at every one."

Rita left the car at the garage and returned home. About an hour later, she received a telephone call from the garage owner who informed her that he had fixed the engine, but the automobile had a leak in its radiator. Rita requested that the leak be repaired as well.

Some time later, Rita went to the garage and was informed that the car was "ready." She paid her account and drove away in the automobile. A few blocks away the vehicle broke down, because the radiator had not been filled with coolant after it had been repaired.

The garage refused to repair the damage. They informed Rita that the radiator had been repaired by another repair shop that specialized in radiator repairs. It was apparently not a custom of the trade for ordinary mechanics to repair radiators in view of the special skill and equipment involved.

Discuss the rights of the parties in this case.

www.mcgrawhill.ca/olc/willes

CASE 3

Sly operated a used-car business. He sold Fox an automobile that he indicated was in good condition and suitable for use by Fox as a taxi. In the course of this discussion, Sly stated that, in his opinion, the vehicle was "hardly broken in," as the odometer registered only 26,000 kilometres.

Sly suggested that Fox test drive the car to satisfy himself as to its condition. Fox did so and, on his return from a short drive, agreed to purchase the automobile for $12,000. Fox signed a purchase agreement whereby he would pay for the car by monthly installments over a three-year period.

On the completion of the transaction, Sly immediately sold the purchase agreement to the Neighbourhood Finance Company for $10,000. Fox was duly notified in writing of the assignment.

A week later, the automobile broke down. When the vehicle was examined by a mechanic, Fox was informed that most of the running gear and the engine were virtually worn out. Unknown to both Sly and Fox, the previous owner had driven the automobile 226,000 kilometres and the odometer, which registered only six digits (including tenths of kilometres), was now counting the third time over.

When Fox discovered the condition of the automobile, he refused to make payments to the finance company.

Discuss the rights of the parties in this transaction.

CASE 4

Carlos sold his business inventory to Avi for $10,000. Under their agreement, Avi was to pay Carlos $1,000 per month over a 10-month period, commencing on March 1st of that year. Before the first payment was due, Carlos became indebted to Avi for $5,000. Carlos then assigned the $10,000 debt agreement to Malcolm. Notice of this assignment was given to Avi. When he discovered that he was to make all payments under the agreement to Malcolm, he informed Malcolm that he would not make a payment until August. He would then make only the five remaining payments under the agreement and consider the debt fully satisfied.

Discuss the rights of Avi and Malcolm and the reasoning behind Avi's response to the notice.

CASE 5

Yolanda Adams was a naturalist who made a wildlife film that she hoped to use in conjunction with lectures to be given to conservation and hiking groups throughout the country. She engaged Basso, a well-known musician in her community, to prepare the musical background for the film. Adams paid Basso $5,000 for his work. However, unknown to Adams, Basso had given the work to another musician, Smith, and paid him $1,000 for his efforts.

Adams used the film on one of her lecture tours, and the newspaper reviews without exception declared the film to be the highlight of her performance. Many of the reporters commented favourably about the beautiful musical background to the film.

Some months later, while on a second lecture tour, a musician in the audience recognized Smith's musical style. He mentioned to Adams how much he enjoyed listening to the musical accompaniment that the musician had provided. Adams was annoyed when she discovered that the background music had not been played by Basso, but by another. On her return home, she confronted Basso with the evidence. Basso admitted that he had been too busy to do the musical background.

Discuss the outcome of the case if Adams should decide to take legal action against Basso.

CASE 6

On August 4, Anchor Service Company entered into an agreement with Rolston Products Company. By this agreement, Rolston Products Company would assign to Anchor Service Company its right to payment under a contract it had with Werner Supply Company. This would settle its indebtedness for certain services rendered by Anchor Service Company.

Under the terms of the contract between Werner Supply Company and Rolston Products Company, Werner Supply Company was obliged to make monthly payments in the amount of $5,000 over a three-year term. Johnson, a trustee, would collect the payments and forward them to Rolston Products Company. Rolston Products Company assigned the right to these payments to Anchor Service Company on August 4th. On the same day they gave written notice to Johnson that, as trustee, he was to make all future payments to Anchor Service Company.

On August 10th, Rolston Products Company gave a general assignment of its accounts receivable to its bank as security for a loan. The bank notified Werner Supply Company in writing on December 27th that it held a general assignment of the accounts receivable of Rolston Products Company, and that all future payments of the company should be made to the bank. Included with the notice was a written request from Rolston Products Company for payment in full to the bank if Werner Supply Company could possibly manage it, even though the contract only called for monthly payments.

When Werner Supply Company received the notice, the president of the company contacted Rolston Products Company by telephone, and asked

www.mcgrawhill.ca/olc/willes

if payment should not be made to Johnson, rather than to the bank. The response of Rolston Products Company was: "Pay the bank." On this advice, Werner Supply Company paid the balance of the money to the bank.

A month later, when no payment had been received, Johnson contacted Werner Supply Company and was informed that the money had been paid to the bank. He immediately notified Anchor Service Company of the fact. Anchor Service Company then instituted legal proceedings against Werner Supply Company for payment under the contract.

Discuss the legal issues raised in this case and the respective arguments of the parties. Render a decision.

CASE 7

Maya Black wished to purchase Seamus Green's farm. After lengthy negotiation, the two parties drew up and signed an agreement of purchase and sale for the farm with the assistance of and in the presence of Green's real-estate agent, Simms. The agreement that the parties signed was a standard, preprinted form used by all local real-estate agents that contained a clause which was an irrevocable direction by the vendor of the land, Green, to the purchaser, Black, to deduct from the sale price and pay the real-estate agent's commission directly to Simms, on the closing day of the transaction. The preprinted form had printed on it a black circle that looked like a seal and had the word "Seal" written under it. Also, above the line for Green's signature were printed the words "In witness whereof I have hereunto set my hand and seal." Beside Green's signature line was a place for a witness to sign the agreement. Immediately above this place appeared the preprinted words "Signed, sealed and delivered in the presence of." Green and Black each had lawyers looking after the details of the transaction and each gave a copy of the agreement to their respective lawyers. Shortly before the closing of the sale, Green's lawyer prepared some final paperwork and had Green sign several important documents. Among them was a direction to Black and Black's lawyer to make the cheque for the full amount of the purchase price payable to Green. On closing, Black's lawyer presented to Green's lawyer a cheque made payable to Green for the full purchase price of the farm. Green's lawyer later turned over the cheque to Green.

After the closing, Simms contacted Green to request delivery of his commission cheque. Green replied that neither he nor his lawyer had received a cheque for Simms from Black's lawyer on closing and suggested that Simms contact Black himself since that was the arrangement. When Simms called Black, Black stated that as far as she was concerned she had paid for the farm and didn't owe anyone any more money.

Discuss the legal issues raised here and the respective arguments, rights, and liabilities of the parties.

Render a decision.

CASE 8

Vivian carried on business as a plumber for many years; she used the Home Bank for her business-banking needs. In 1997, Vivian incorporated the business as Vivian's Plumbing Inc., and she asked the bank to change the bank accounts to the new corporation. The bank did so and accepted all deposits and payments made in the corporation's name. Some time later, the corporation requested a line of credit from the bank in the amount of $50,000. This was granted, provided that Vivian personally guarantee the credit extended. Vivian agreed and signed a guarantee.

The line of credit was never used, but the bank did lend $75,000 to the corporation on the security of a large contract that the corporation had obtained. Unfortunately, the contract proved to be virtually worthless for the corporation when the other contracting party became bankrupt, and the corporation was unable to repay the loan. The bank then took legal action against Vivian personally for payment.

Explain the nature of the bank's claim and indicate the defences (if any) that Vivian might raise.

Render a decision.

CASE 9

Morgan was visited by a travelling salesperson who talked him into buying a machine that produced pin-on buttons suitable for selling in bulk as advertising gimmicks, novelty slogans, and political party campaigns. A number of good contracts would produce enough money to cover the five instalments of $200 owed for the machine. Morgan agreed to the deal, and the machine was delivered. It took much more skill to operate than Morgan could muster. The salesperson had effortlessly demonstrated the machine, but clearly had skills borne of long practice. The guide to operation was sketchy at best, and the machine consistently produced buttons with off-centre or crumpled graphics. Morgan tried to find the salesperson without success. Morgan was, however, soon found by the finance company to which the salesperson had assigned the debt. Advise Morgan.

The Online Learning Centre offers more ways to check what you have learned so far. Find quizzes, Weblinks, and many other resources at www.mcgrawhill.ca/olc/willes.

www.mcgrawhill.ca/olc/willes

CHAPTER 13

Performance of Contractual Obligations

Chapter Objectives

After study of this chapter, students should be able to:

- Explain the significance and effect of tender.
- Recognize how a contract may be discharged other than by performance.
- Explain the significance and effect of a condition precedent.
- Recognize and explain the effect of a frustrated contract.

YOUR BUSINESS AT RISK

Contract disputes often centre on whether the parties have performed as they have promised and, if not, whether they can be excused from their obligations. This becomes a risk management decision for a business faced with a lawsuit. If you are certain your actions or rights fall within the legal interpretation of your contract, you may want to go ahead with a lawsuit. Any uncertainty should be a consideration in favour of settlement, knowing that the party on the other side will be weighing the same issues.

THE NATURE AND EXTENT OF PERFORMANCE

A contract that contains all of the essential requirements for a binding agreement, and that does not contain an element enabling a party to avoid the agreement, must be performed by the parties in accordance with its terms. Performance must always be exact and precise in order to constitute a discharge of a contractual obligation. Anything less than complete compliance with the promise would render the party in default liable for breach of the contract. For example, a party entered into a contract to supply a quantity of canned fruit, packed 30 cans to the case, and, on delivery, supplied some of the goods in cases containing 24 cans. The failure to supply the goods in the correct size of case entitled the buyer to reject the goods, even though the total number of cans was correct.[1]

If the performance of the promises of the parties is complete, the contract is said to be discharged. If, however, one of the parties does not fully perform the promise made, then the agreement remains in effect until the promise is fulfilled or the agreement is discharged in some other way. Whether the performance is complete or not must be determined by comparing the promise made with the act performed. The act of offering to perform the promise is called **tender** of performance, and may take one of two general forms: tender of payment or tender of performance of an act.

Tender
The act of performing a contract or the offer of payment of money due under a contract.

1. *Re An Arbitration Between Moore & Co. Ltd. and Landauer & Co.*, [1921] 2 K.B. 519.

COURT DECISION: Practical Application of the Law

Contracts—Performance

***Farnel v. Main Outboard Centre Ltd.* (1987), 50 Man. R. (2d) 13.**

The plaintiff, who owned a boat with an inboard/outboard engine, requested the defendant company to "winterize the engine" before the onset of cold weather. The defendant allegedly did so, but water left in the cooling system froze in the cold weather and cracked the engine block and exhaust manifold. The boat owner took legal action against the company for breach of contract and negligent performance of the work. The trial judge dismissed the action, but the boat owner appealed the decision.

THE COURT: This appeal involves the question: What is the duty of a boat marina owner who agrees to winterize a boat engine? A promise to winterize a boat engine is more than a promise to exercise care in doing so. It is a promise to put the engine in a condition in which it will survive the winter without damage due to water left in it. The promise is not, of course, a promise to remove water retained within the engine as a result of a defect unknown to the person engaged in the winterization. The promisor may raise the defect as a defence. If the water remained in the engine because of a defect, the existence of which could not have been detected by the promisor in the ordinary course of winterization, he will not be liable. It may be that, once facts consistent with such a defence are proved, the onus of showing that the damage resulted from the promisor's fault will revert to the owner as a shift in the evidentiary burden, but that does not arise in this case.

In the case at bar, the defendant raised only two defences. It said that it had drained all of the water from the engine and that, in any event, it had not been negligent. No explanation was offered as to how there could have been water in the engine if the defendant had performed its contract. The defendant did suggest that mud, from the river water which cooled the engine, might have accumulated within the engine causing a blockage. That may be so, but the defendant was aware of that possibility and cranked the engine over to eliminate water which might be trapped behind blockage. The finding of the learned trial judge that the cause of the engine cracking was the freezing of water within the engine can mean only one thing in the circumstances: the defendant did not drain all the water from the engine. As the defendant did not fulfill its contractual obligation, the fact that it followed the normal and proper winterizing procedure does not relieve it of liability. The absence of negligence is irrelevant ....In the result, I would set aside the judgment appealed from, and substitute judgment for the plaintiff in the amount of $2,552.27.

Tender of Payment

If a promisor simply agrees to purchase goods from a seller, performance is made when payment is offered to the seller at the required time and place fixed for delivery under the contract. The sum of money offered in payment at that time must be in accordance with the terms of the agreement. If the form of payment is not specified, then currency, or legal tender, must be offered to the seller. Legal tender may not be refused when offered in payment, providing that it is the exact amount required. Unless specified in the agreement, a personal cheque, credit card, bill of exchange, or other form of payment may be rejected by the seller. This would constitute a failure to perform by the buyer. For this reason, buyers will often include in a purchase agreement that payment may be made by personal cheque or some other form of payment in lieu of legal tender.

The importance of the rule for tender was underscored by the court in the case of *Blanco v. Nugent*,[2] where the plaintiff made a tender conditional upon the defendant executing a transfer of certain property to the plaintiff. The court rejected the plaintiff's action and noted the requirements for a valid tender in the following words:

> In order to support a plea of tender, there must be evidence of an offer of the specific sum, unqualified by any circumstance whatever: *Brady v. Jones* (1823), 2 Dow & Ry KB 305. The condition here sought to be imposed — that the defendant execute a transfer to the plaintiff — is entirely unwarranted and vitiates the tender.

2. [1949] 1 W.W.R. 721.

And at p. 387:

> The tender was bad for the further reason that Mr. McFadden did not tender the exact amount payable, but tendered a larger amount and asked for change. The amount tendered ought to be the precise amount that is due. If the debtor tenders a larger amount and ... requires change it is not a good tender.

In the case of a debt owing, once the debtor tenders payment to the creditor in the proper amount of legal tender at the required time and place, the tender of payment is complete. If the creditor is unwilling to accept payment, the debtor need not attempt payment again. Once a proper tender of payment is made, interest ceases to run on the debt. While the debtor is not free of the obligation to pay, the debtor need only hold the amount of the debt until the creditor later demands payment, then pay over the money. If he or she should be sued by the creditor, or if the creditor attempts to seize an asset of the debtor, the debtor may prove the prior tender and pay the money into court. The courts, in such circumstances, will normally penalize the creditor with costs for causing the unnecessary litigation or action.

When the contract concerns the purchase and sale of land, the purchaser, on the date fixed for closing the transaction, has an obligation to seek out the seller and offer payment of the full amount in accordance with the terms of the contract. Once this is done, any refusal to deliver up the deed to the land would probably entitle the purchaser to bring an action in court for specific performance. By this action, if the purchaser can satisfy the court that he or she was ready and willing to close the transaction and was prepared to pay the required funds, the purchaser may obtain an order from the court ordering the seller to deliver up the land.

Tender of Performance of an Act

The seller's performance is not by tender of money, but by the tender of an act. For the sale of goods, the seller must be prepared to deliver the goods to the buyer at the appointed time and place, and in accordance with the specifications set out in the agreement. If the buyer refuses to accept the goods when the tender is made, the seller need not tender the goods again. He or she may simply institute legal proceedings against the buyer for breach of the contract. If the contract concerns land, the equitable remedy of specific performance may be available to the seller.

The remedy of specific performance is a discretionary remedy. To obtain this relief, the seller must show that he or she was prepared to deliver the title documents for the property to the purchaser as required under the agreement. It must also be shown that on the closing date the seller attempted to transfer the deed, but the purchaser was unwilling to accept it. Unless the purchaser had a lawful or legitimate reason to refuse the tender of performance by the seller, the courts may order the payment of the funds by the purchaser and require the purchaser to accept the property. In this respect, the tender of performance of a seller of land differs from the seller of goods. Apart from this difference, the tender itself remains the same. The seller must do everything required in accordance with the promise made, if he or she wishes to succeed against the purchaser for breach of contract.

A variation of this situation arose in a case[3] where a company agreed to sell to the defendant the entire quantity of a specific type of scrap metal that the company produced at its manufacturing plant during a one-year period. The defendant later refused to perform the agreement. The plaintiff argued that the agreement constituted a contract, but the defendant submitted that the agreement was only an agreement to sell, and not an agreement on the defendant's part to buy. The court dismissed the defendant's argument with the words:

> The case appears to me clearly to fall within the class referred to by Cockburn, C.J., in *Churchward v. The Queen* (1865), L.R. 1 Q.B. 173, at p. 195: "Although a contract may appear on the face of it to bind and be obligatory only upon one party,

3. *Canada Cycle & Motor Co. Ltd. v. Mehr* (1919), 48 D.L.R. 579.

> yet there are occasions on which you must imply — although the contract may be silent — corresponding and correlative obligations on the part of the other party in whose favour alone the contract may appear to be drawn up. Where the act done by the party binding himself can only be done upon something of a corresponding character being done by the opposite party, you would there imply a corresponding obligation to do the things necessary for the completion of the contract. If A covenants or engages by contract to buy an estate of B, at a given price, although that contract may be silent as to any obligation on the part of B to sell, yet as A cannot buy without B selling, the law will imply a corresponding obligation on the part of B to sell: *Pordage v. Cole* (1607), 1 Wm. Saund. 319 i (85 E.R. 449).

A contract may, of course, involve performance in the form of something other than the delivery of goods or the delivery of possession of land. It might, for example, require a party to carry out some work or service. In this case, the other contracting party must permit the party tendering performance to do the required work. Any interference with the party tendering performance of the act might entitle that party to treat the interference as a breach of contract. The two forms of tender are shown in Figure 13–1.

FIGURE 13–1 **Tender**

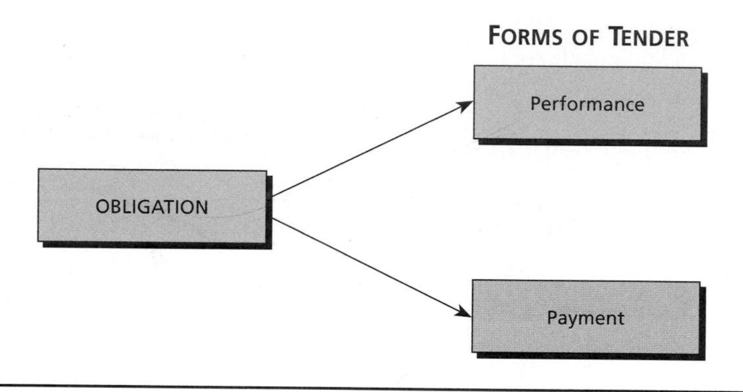

Breach of Contract

Breach of contract
The failure to perform a contract in accordance with its terms.

Imperfect tender, and hence imperfect performance, technically represents a **breach of contract**. This gives rise to a right of action by the party affected by the breach. It is the court's disposal of the action that serves as discharge of the agreement and its replacement with a judgment. In addition, the breach of contract by one party in certain circumstances will have the effect of discharging the injured party from any further performance under the agreement. Discharge of a contract by breach covers a wide range of activities and remedies. It is examined in detail in the next chapter, while the special case of substantial performance and the lien remedy is discussed in our chapter entitled "Security for Debt."

Discharge by Means Other Than Performance

Contracts may be discharged in a number of other ways in addition to performance. The law may, under certain circumstances, operate to terminate a contract, or the parties themselves may agree to end the contract before it is fully performed. Often, the parties may specifically provide in the agreement that the contract may be terminated at any time on notice, or on the happening of a subsequent event. The contract may also provide that the contract does not operate unless some condition precedent occurs. In addition, the parties may, by mutual agreement, decide to terminate the contract and replace it with a revised agreement, or simply decide before either has fully performed that the contract should be discharged.

Many other methods of discharge may also be possible. The parties could replace the existing agreement with a substituted agreement. This can be done either by a material alteration of the terms, or by accord and satisfaction. (The latter is a means by which one party offers different goods in satisfaction of a promise made under the agreement, and the buyer is prepared to accept the different goods as a substitute for the goods originally requested.) Each of these, in turn, has a different effect on the obligation of the parties to perform their specific promises in the agreement.

Termination as a Right

An option to terminate is a method of discharging an agreement that is usually effected by either party giving notice to the other. The option frequently has a time limit attached to the notice. At the expiry of the notice period, the agreement comes to an end. Agreements that contain a notice or option to terminate often provide for some means of compensating the party who has partly performed at the time the notice is given, but this is not always the case.

The right to terminate, if exercised in accordance with the specific terms of the agreement, may entitle a party to terminate the agreement without liability for any loss suffered by the other.

Brown's Motors agrees to sell Arnold a new car. As part of the contract, Arnold reserves the right to cancel the agreement, without incurring liability, on notice to Brown's Motors at any time before Brown's Motors has the car ready for delivery. After Brown's Motors receives the automobile from the factory, but before it services the car in preparation for delivery, Arnold notifies Brown's Motors of his intention to cancel the order. In this instance, Arnold would not be obliged to purchase the car. Under the terms of the contract, the notice would have the effect of terminating the agreement, and Arnold would be free of any liability under it. Brown's Motors, which had assumed the risk of Arnold's cancellation, would have no rights against Arnold, but it would also no longer be liable to deliver an automobile to Arnold.

Express term

Discharge by the occurrence of an event specified in the contract.

Condition subsequent

A condition that alters the rights or duties of the parties to a contract, or that may have the effect of terminating the contract if it should occur.

Force majeure

A major, unforeseen, or unanticipated event that occurs and prevents the performance of a contract.

Act of God

An unanticipated event that prevents the performance of a contract or causes damage to property.

Implied term

Discharge by the occurrence of an event that by custom of the trade would normally result in exemption from liability.

External Events

Express Terms

If a contract is discharged upon the occurrence of a particular event, the circumstance that gives rise to the termination is called a **condition subsequent**. It is not uncommon to make provision in contracts of a long-term or important nature for events that might arise to prevent the performance of the agreement by one party or both. These are sometimes referred to as ***force majeure*** clauses. They may either specifically or generally set out the circumstances under which the contract may be terminated. (In Roman law, the term is interpreted to mean a major force in the nature of an **Act of God**.) *Force majeure* usually indicates an unforeseen and overpowering force affecting the ability of a party to perform the contract (such as war, insurrection, or natural disasters), although the parties may indicate in their contract that the interference need not be that serious to constitute discharge. Since the parties making the contract are free to insert whatever clauses they are prepared to mutually agree upon, either or both of the parties may set out a large number of circumstances that, if any one should occur, would discharge the agreement.

Implied Terms

In certain fields, conditions subsequent are sometimes implied in contracts by the courts from customs of the trade. For example, common carriers, who are normally liable for any ordinary loss or damage to goods carried, may be exempted from liability if the loss is due to an Act of God or some other event that could not have been prevented by the carrier.[4] Partial destruction of the goods would not discharge the carrier from its obligation to deliver, however, but the carrier would not be liable for the damage caused by the Act of God. If the goods are inherently dangerous and self-destruct while in the possession of the carrier, the carrier would also be absolved from any liability. Indeed, if dangerous goods were deliberately mislabelled by the shipper, the carrier might have a right of action

4. See, for example, *Nugent v. Smith* (1875), 1 C.P.D. 423.

against the shipper for any damage caused by the goods. Rather than relying on implied terms, most carriers take the added precaution of making the terms express by including them on their bills of lading.

Implied Terms and the Doctrine of Frustration

Frustrated contract

A contract under which performance by a party is rendered impossible due to an unexpected or unforeseen change in circumstances affecting the agreement.

If performance is rendered impossible due to circumstances not contemplated by the parties at the time the agreement was entered into, and through no fault of their own, the agreement may be said to be **frustrated** and thereby discharged. This particular doctrine has been applied to a number of different situations, the simplest being one where the agreement could not be performed because of the destruction of something essential to the performance of the contract. For example, Allen, a theatre owner, enters into a contract with Black to provide the premises for a concert that Black wishes to perform. Before the date of the concert, the theatre burns to the ground. Allen, as a result of the fire, is unable to perform her part of the contract through no fault of her own. In this case, the courts would assume that the contract was subject to an implied term that the parties would be excused from performance if an essential part of the subject matter should be destroyed without fault on the part of either party.[5]

This case was decided in 1863, and yet it is a matter for discussion in the 21st century. Such is the application of Common Law precedent, for this case was the basis upon which all similar cases have been decided. Two hundred years later, the case underscores the need to consider impossibility of performance, as in the following case, where frustration was not proved.

A contractor engaged to set pilings and footings for Vancouver's Second Narrows bridge discovered a wholly expected layer of rock and material deep underground. The bid contract was straight-forward and required performance, but the contractor claimed that the unforeseen sub-layers frustrated the contract. The Supreme Court of Canada disagreed: the fact that the job was made more difficult and more expensive still did not render the performance impossible.[6]

When the contract involves the sale of goods, the *Sale of Goods Act* in most provinces provides that the destruction of specific goods (through no fault of either the buyer or seller) before the title to the goods passes to the buyer will void the contract. This particular section of the Act represents a codification of the doctrine of implied term with respect to the sale of goods. It would apply in most cases where specific goods are destroyed before the title to the goods passes to the buyer.

The doctrine of frustration is also applicable in cases where the personal services of one of the parties is required under the terms of the agreement but, through death or illness, the party required to provide the personal service is unable to do so. For example, Albert enters into a contract with Roberts, whereby Roberts agrees to perform at Albert's theatre on a particular date. On the day before the performance, Roberts falls ill with a severe case of influenza and is unable to perform. In this case, the courts would include as an implied term the continued good health of the party required to personally perform. The occurrence of the illness has the effect of discharging both of the parties from the contract.[7]

A third and somewhat different application of the doctrine may apply in cases where the event that occurs so alters the circumstances under which the agreement is to be performed that performance will be virtually impossible for a promisor. Many of these cases arose during World War I and World War II, when goods were diverted for war purposes. In these cases the hostilities were not contemplated by the parties. Therefore, to impose the contract after hostilities ceased would, in effect, be imposing an entirely different agreement on the parties.[8]

5. *Taylor v. Caldwell* (1863), 3 B. & S. 826, 122 E.R. 309.
6. *Peter Kiewit Sons' Co. v. Eakins Construction Ltd.*, [1960] S.C.R. 361.
7. *Robinson v. Davidson* (1871), L.R. 6 Ex. 269.
8. *Morgan v. Manser*, [1948] 1 K.B. 184.

Although less common, the doctrine may also apply to cases where the performance of the agreement is based upon the continued existence of a particular state of affairs. The state of affairs changes to prevent the performance of the agreement, however.

> Canada Safeway Ltd. sold land to KBK Ventures who planned on building a mixed commercial and residential condominium. The project required certain permits from the city to go ahead, but the zoning did allow for these. KBK paid Safeway an installment of $150,000 on the purchase price of $8.8 million. Thereafter the city rezoned the property making the proposed development impossible. KBK believed that the contract was frustrated, and demanded Safeway return the installment paid. Safeway refused, but at trial the court found the contract had been frustrated, and ordered repayment to KBK.[9]

The courts will not release a party from performing simply because the performance turned out to be more difficult or expensive than expected at the time the agreement was made. Inability to obtain credit or a shortage of funds does not constitute legal frustration of a contact, which is a reason why it is often provided for explicitly as a term of the contract. Courts will also not allow a person to claim frustration where their own performance has been made impossible by their own deliberate act in an effort to avoid the agreement. For example, Baxter Co. enters into an agreement with the local municipality whereby it agrees to spray the road allowances alongside all rural roads with a particular herbicide to control the growth of weeds. The company later discovers that the herbicide costs twice the price that it contemplated at the time that it made the agreement. As a result, it can only perform its part of the agreement at a loss. The company then sells its spraying equipment and claims that it cannot perform the contract. Under the circumstances, Baxter Co. would be liable to the municipality for breach of contract, as the courts would not allow the company to avoid the contract on the basis of self-induced frustration.

MANAGEMENT ALERT: BEST PRACTICE

In an employment situation, an employment contract may become frustrated should matters turn out that a person hired for particular work simply cannot perform the work, not because of any self-induced factor (disobeying instructions, laziness, inattention), but rather of inate factors, such as insufficient skill or lack of legally-required certification (class of driver's licence, trade certificate) or the like. Dismissal of this nature, due to frustration, is termed **non-culpable dismissal**. In such cases the employer must ensure that it is not, itself, self-frustrating the contract by failing to provide training or supervision, lest it be open to a claim of unjust dismissal.

Non-culpable dismissal

Dismissal of an employee where the inability to perform is not self-induced but due to frustrating factors.

When an event occurs that renders performance impossible, or changes the conditions under which the contract was to be performed to such an extent that the parties would have provided in the contract for its discharge in such circumstances, the courts will treat the agreement as frustrated and relieve both parties from any further performance after the frustrating event occurs. The frustrating event in the eyes of the courts would have the effect of bringing the contract to an end automatically. This was a reasonable conclusion for the courts to reach, but in some of the early cases the rule worked a hardship on one of the parties.

Initially, the courts let the loss fall on the parties as at the time the event occurred. If rights had accrued to a party at the time of the event, they could still be enforced.[10] This was later modified to provide that if the contract was wholly executory by one party, and if the other party had paid money under the agreement, but received no benefit for it, the money could be recovered on the basis that no consideration had been received for the payment.[11]

9. *KBK v. Safeway*, 2000 B.C.C.A. 295.
10. *Chandler v. Webster*, [1904] 1 K.B. 493.
11. *Fibrosa Spolka Akcyjna v. Fairbairn Lawson Combe Barbour, Ltd.*, [1943] A.C. 32.

In the case of *Cahan v. Fraser*[12] the court described the impact of frustration at Common Law on a contract:

> The rights of the parties fall to be determined at the moment when impossibility of further performance supervened, that is, at the moment of dissolution ... The effect of frustration is that while the parties are relieved from any further performance under it, it remains a perfectly good contract up to that point, and everything previously done in pursuance of it must be treated as rightly done.
>
> The same event which automatically renders performance of the consideration for the payment impossible not only terminates the contract as to the future, but terminates the right of the payee to retain the money which he has received only on the terms of the contract performance ....
>
> The payment for the option in *Goulding's Case*, [1927] 3 D.L.R. 820, was an "out and out" payment, that is, there was no provision for its return or for its application on the purchase-price if the option were exercised. An "out and out" payment cannot be recovered in the event of frustration ...
>
> In the case at bar the sums paid were not "out and out" but were to be applied on to the purchase-price; the payment was originally conditional. The condition of retaining it was eventual performance. Accordingly, when that condition failed, the right to retain the money must simultaneously fail ....

Frustrated Contracts Acts

The unsatisfactory state of the Common Law prompted the English Parliament to pass a Frustrated Contracts Act in 1943.[13] This legislation permits a court to apportion the loss somewhat more equitably. This is done by providing for the recovery of deposits and/or advances and the retention of part of the funds to cover expenses, when a party has only partly performed the contract at the time the frustrating event occurs. The legislation also permitted a claim for compensation when one party, by partly performing the contract, had conferred a benefit on the other party.

However, a party who has received no benefit and paid no deposit under the contract will not be obliged to compensate the other party to the contract for any work done prior to the frustrating event. Under these circumstances the Act does not protect the party who undertakes to perform or must perform a contract without the benefit of a deposit.

CASE IN POINT

A transportation company leased a special type of flat-bed trailer to haul an 80-ton transformer from Montreal to St. John's, Newfoundland. The leasing company agreed to supply the trailer and personnel to supervise the loading of the trailer on board a ship in Montreal. At the ship, the parties discovered that the clearance of the trailer would not allow the trailer to pass over a hump in the loading ramp, and it could not be loaded on the ship. The transportation company demanded return of the rental price paid.

The court held that the contract had been frustrated and ordered the return of most of the rental paid.

Summers Transport Ltd. v. G.M. Smith Ltd. (1990), 82 Nfld. & P.E.I.R. 1.

12. *Cahan v. Fraser*, [1951] 4 D.L.R. 112.
13. *Law Reform (Frustrated Contracts) Act*, 1943, 6 & 7 Geo. VI, c. 40.

FRONT-PAGE LAW

Yashin is back, but not happy to be a senator; "Can't play anywhere else"

If it's true that nice guys finish last, the Ottawa Senators would seem immediate Stanley Cup favourites now that Alexei Yashin is back in camp.

The unrepentant 26-year-old Russian superstar centre formally ended his one-year holdout with the Ottawa Senators in a chilly press conference Saturday.

"I'm here because I can't play hockey anywhere else in the world," said Yashin, stunning a group of reporters assembled in the Corel Centre. Yashin said the only reason he was returning to Ottawa was because an arbitrator's decision forced him to do so.

The one-time Senators captain was unapologetic for sitting out after the 1998–99 playoffs. He was in line to make US$3.6-million last year, a sum he felt wasn't enough of a reward for his six years of service in Ottawa.

On June 28, Boston-based arbitrator Lawrence Holden decided that Yashin still had to honour his contract with the Senators, and that he could not catch on with any other NHL team until he honoured the deal. The ruling was upheld last week in an Ontario court.

Yashin will now join his teammates on the ice for training camp, as well as in an icy dressing room.

"We don't need an apology, but certainly an explanation would be good," Senators defenceman Chris Phillips said.

After this season, Yashin will be a restricted free agent, meaning any team that wants to sign him may have to provide the Senators with up to five first-round draft picks in compensation. Of course, he could also be traded.

How does the doctrine of frustration apply to contracts of personal services such as Alexei Yashin's?

Source: Excerpted from *National Post*, with files from *The Canadian Press*, "Yashin is back, but not happy to be a Senator," September 11, 2000.

Seven provinces — Alberta,[14] Manitoba,[15] New Brunswick,[16] Newfoundland,[17] Ontario,[18] British Columbia,[19] and Prince Edward Island[20] — subsequently passed legislation somewhat similar to the English Act. The remainder of the provinces remain subject to the Common Law. This legislation does not apply, however, to an agreement for the sale of specific goods under the *Sale of Goods Act*, where the goods have perished without fault on the part of the seller (or buyer) and before the risk passes to the purchaser. Nor does it apply to certain types of contracts, such as insurance contracts or those contracts that are expressly excluded by the Act.[21]

CHECKLIST FOR FRUSTRATED CONTRACTS

1. Is performance impossible?
2. The impossibility was not self-induced?
3. The contract is not excepted (sale of specific goods perished, maritime carriage, insurance).
4. Has a benefit been received or deposit been paid?

If all the answers are true, the contract is frustrated, is discharged, and such benefit and/or deposit must be returned.

14. *Frustrated Contracts Act*, R.S.A. 2000, c. F-27.
15. *Frustrated Contracts Act*, C.C.S.M., c. F190.
16. *Frustrated Contracts Act*, R.S.N.B. 1973, c. F-24.
17. *Frustrated Contracts Act*, R.S.N.L. 1990, c. F-26.
18. *Frustrated Contracts Act*, R.S.O. 1990, c. F.34.
19. *Frustrated Contract Act*, R.S.B.C. 1996, c. 166.
20. *Frustrated Contracts Act*, R.S.P.E.I. 1988, c. F-16.
21. See, for example, the *Frustrated Contracts Act*, R.S.A. 2000, c. F-27, ss. 3 and 4.

Condition Precedent

Condition precedent

A condition that must be satisfied before a contract may come into effect.

The parties may also provide in their agreement that the contract does not come into effect until certain conditions are met or events occur. These conditions, if they must occur before the contract is enforceable, are called **conditions precedent**.

Often, when a condition precedent is agreed upon, the agreement is prepared and signed; only the performance is postponed pending the fulfillment of the condition. Once fulfilled, performance is necessary to effect discharge. If the condition is not met, it then has the effect of discharging both parties from performance. It may be argued that an agreement cannot exist until the condition is satisfied, in which case the agreement only then comes into effect. Regardless of the position adopted, however, the condition is the determining factor with respect to the termination of the agreement or the establishment of contractual rights between the parties.

The case of *Turney et al. v. Zhilka*[22] provides an example of a condition precedent in a contract. It supplies a description of how the courts apply this legal principle in a case involving the sale of land, where the sale was subject to a condition that the property would be annexed to a village and the village council would approve a plan of subdivision proposed for the parcel of land in question. The agreement provided that the transaction was to be performed (i.e., the deed given) 60 days after the plan of subdivision was approved. However, the agreement did not specify which party was to satisfy the condition. The purchaser in the case then attempted to hold the vendor liable for the non-performance. The court discussed the nature of the condition in the following terms:

> The date for the completion of the sale is fixed with reference to the performance of this condition — "60 days after plans are approved." Neither party to the contract undertakes to fulfil this condition, and neither party reserves a power of waiver. The obligations under the contract, on both sides, depend upon a future uncertain event, the happening of which depends entirely on the will of a third party — the Village Council. This is a true condition precedent — an external condition upon which the existence of the obligation depends. Until the event occurs there is no right to performance on either side. The parties have not promised that it will occur. In the absence of such a promise there can be no breach of contract until the event does occur. The purchaser now seeks to make the vendor liable on his promise to convey in spite of the non-performance of the condition and this to suit his own convenience only. This is not a case of renunciation or relinquishment of a right but rather an attempt by one party, without the consent of the other, to write a new contract. Waiver has often been referred to as a troublesome and uncertain term in the law but it does at least presuppose the existence of a right to be relinquished.

Operation of Law

A contract may be discharged by the operation of law. For example, Importco Ltd., a Canadian corporation, and Axelrod, a resident and citizen of a foreign country, enter into a partnership agreement. Shortly thereafter, formal hostilities break out between the two countries. The contract between Importco Ltd. and Axelrod would be dissolved, as it would be unlawful for Importco Ltd. to have any contractual relationship with an enemy. Similarly, if Importco Ltd. and Axelrod entered into a partnership to carry on a type of business in Canada that was subsequently declared unlawful, the agreement between them would be discharged.

Specific legislation also discharges certain contracting parties from contracts of indebtedness. The *Bankruptcy and Insolvency Act*,[23] for example, provides that an honest but unfortunate bankrupt debtor is entitled to a discharge from all debts owed to his or her creditors when the bankruptcy process is completed. The *Bills of Exchange Act* provides that a bill of exchange that is altered in a material way without the consent of all of the parties liable to it has the effect of discharging all parties, except the person who made

22. [1959] S.C.R. 578.

23. S.C. 1992, c. 27.

the unauthorized alteration and any subsequent endorsers.[24] In due course, however, a holder would still be entitled to enforce the bill according to its original tenor if the alteration is not apparent.[25]

The law also comes into play when a person allows a lengthy period of time to pass before attempting to enforce a breach of contract. At Common Law, in cases where a party fails to take action until many years later, the courts will sometimes refuse to hear the case. The reasoning here is that the undue and unnecessary delay would often render it impossible for the defendant to properly defend against the claim. Undue delay in bringing an action against a party for failure to perform at Common Law is known as the *doctrine of laches*, under which a court may refuse to hear a case not brought before it until many years after the right of action arose. It is important to note, however, that the doctrine only bars a right of action; it does not void the agreement. In effect, it denies a tardy plaintiff a remedy when a defendant fails to perform.

While the doctrine of laches still remains, all of the provinces have passed legislation stating the time limits for bringing an action before the courts following a breach of an agreement. These statutes, or Limitations Acts,[26] provide that actions not brought within the specified time limits will be statute-barred, and the courts will not enforce the claim or provide a remedy. As with laches, the statutes do not render the contracts void — they simply deny the injured party a judicial remedy. The contract still exists, and, if liability should be acknowledged (such as by part-payment of a debt or part-performance), the contract and a right of action may be revived.

CLIENTS, SUPPLIERS *or* OPERATIONS *in* QUEBEC

Most of the Common Law jurisdictions in Canada have or are moving toward a basic two-year limitation period for commencement of most civil actions (from the date at which the claimant became aware or should have become aware of the wrong). The *Civil Code of Quebec*, however, provides a wide array of limitation periods ranging between one year and ten years for civil actions, for action commenced in that province. See CCQ, Book Eight, Prescription.

Merger may also discharge an agreement. For example, Amber Land Development Co. and Brown enter into an informal written agreement, whereby Amber Land Development Co. agrees to sell Brown a parcel of land. If the informal written agreement is later put into a formal agreement under seal and is identical to the first except as to form, then a merger of the two takes place, and the informal agreement is discharged. The delivery of a deed on the closing of a real-estate transaction normally has the same effect on an agreement of purchase and sale (relating to the same parcel of land), although there are a number of exceptions to this general rule.

Agreement

Waiver

Waiver

An express or implied renunciation of a right or claim.

Often, the parties to an agreement may wish to voluntarily end their contractual relationship. If neither party has fully performed his or her duties, the parties may mutually agree to discharge each other by **waiver**. In the case of a waiver, each party agrees to abandon his or her right to insist on performance by the other. As a result, there is consideration for the promises made by each party. However, if one of the parties has fully performed the agreement, it would be necessary to have the termination agreement in writing and under seal in order for it to be enforceable. In a classic, illustrative example, Alford and Brown enter into an agreement whereby Brown agrees to drive Alford to a nearby town. Upon

24. *Bills of Exchange Act*, R.S.C. 1985, c. B-4, s. 144(1).

25. Supra, s. 144(2).

26. See, for example, R.S.O. 1990, c. L.15.

arrival at their destination, Alford will pay Brown $10. Alford and Brown may mutually consent to terminate their agreement at any time before they reach their destination, and the mutual waivers will be binding. However, as soon as they reach the destination, Brown would have fully performed his part of the contract. If Brown chose to waive his rights to payment under the agreement after they reached their destination (after he had fully performed his driving under the contract), his promise to do so would be gratuitous. Alford must obtain a signed and sealed written promise to that effect before Alford would be protected from a later claim by Brown.

Novation

Novation

A mutual agreement to amend the terms or parties to an existing agreement.

The parties may also discharge an existing agreement by mutually agreeing to a change in the terms of the agreement or to a change in the parties to the agreement. Both of these changes require the consent of all parties and have the effect of replacing the original agreement with a new contract. A substituted agreement differs from merger in several ways. In the case of merger, the terms and the parties to the agreement remain the same — only the form of the agreement changes. The parties are simply replacing a simple agreement with a written one, or replacing a written agreement with a particular type of formal agreement dealing with the same subject matter (e.g., replacing an agreement for the sale of land under seal with a deed for the same land). On the other hand, a substituted agreement may involve a change in the parties to the agreement, or a change of a material nature in the terms of the contract. For example, Appleby, Ballard, and Crawford enter into an agreement. Appleby later wishes to be free of the contract, and Donaldson wishes to enter the agreement and replace Appleby. This may be accomplished only with the consent of all parties. The arrangement would be a novation situation, where Appleby would be discharged from her duties under the agreement with Ballard and Crawford, and the parties would establish a new contract between Ballard, Crawford, and Donaldson.

Material Alteration of Terms

Material alteration

The major alteration of an agreement that has the effect of discharging the contract and replacing it with another.

A **material alteration** of the terms of an existing agreement has the effect of discharging the agreement and replacing it with a new one containing the material alteration. The alteration of the terms of the existing agreement must be of a significant nature before the contract will be discharged by the change. As a general rule, the change must go to the root of the agreement before it constitutes a material alteration. A minor alteration, or a number of minor alterations, would not normally be sufficient to create a new contract unless the overall effect of the changes completely altered the character of the agreement.

> A highway transportation company places an order to purchase a truck of a standard type with a truck sales dealer. Later it decides to have the vehicle equipped with a radio and a special brand of tires. The changes would constitute only a variation of terms of the agreement. If, on consent, the order was changed after acceptance to a special-bodied truck of a different size and with different equipment, the changes would probably be sufficient to constitute a discharge of the first contract, and the substitution of a new one.

The nature of the agreement (i.e., the purchase of a truck) would still be the same, but the subject matter would be altered to such an extent by the changes that it would represent a new contract.

A QUESTION OF ETHICS

When a company purchases equipment, often it buys consumable supplies as well. These supplies make the equipment perform — photocopiers need toner, and casting equipment requires raw metal of a particular alloy, for example. Often, too, equipment warranties hinge on the buyer using only "original-brand" supplies, and an initial stock is provided at a given price in the sales contract. Later, when more supplies are required, they are sold under a separate contract at much higher prices. The buyer has committed itself to the equipment and faces the blackmail choice of voiding the warranty or paying the higher price for "brand-name" supplies. Should a price for supplies in one contract be required to be reflected or limited in later contracts (in cases where a buyer is "hooked"), or is it enough to let the buyer ask ahead of time what follow-on prices can be expected?

Substitute Agreement

If it is the intention of the parties to discharge an existing contract by a substitute agreement, the substitution may effect the discharge, even if it is unenforceable in itself. This situation is likely to arise when the parties enter into a written contract to comply with the *Statute of Frauds*, and then later agree to discharge the written agreement by a subsequent one. The statute simply requires that agreements be in writing to be enforceable, but does not require compliance with the statute to dissolve such a contract. Consequently, if the parties, by way of a subsequent mutual agreement, agree to discharge an existing contract and replace it with a verbal agreement that is rendered unenforceable by its non-compliance with the *Statute of Frauds*, the subsequent agreement will discharge the prior one, but will be unenforceable with respect to the remainder of its terms.[27]

MANAGEMENT ALERT: BEST PRACTICE

Once a contract is made in writing, business operators regularly find reasons to modify it in practice and often do so over the telephone. If litigation occurs, there is no record of these changes, which often tend to be critical. Prudent business persons make a point of actually substituting a new written agreement, or at minimum ensure some written record of the change exists — for example, a confirming fax or e-mail. These provisions catch misunderstandings before they start, and provide evidence of the change. At a later date, verbal amendments are extremely hard to prove on their own.

FIGURE 13–2 **Discharge by Means Other than Performance**

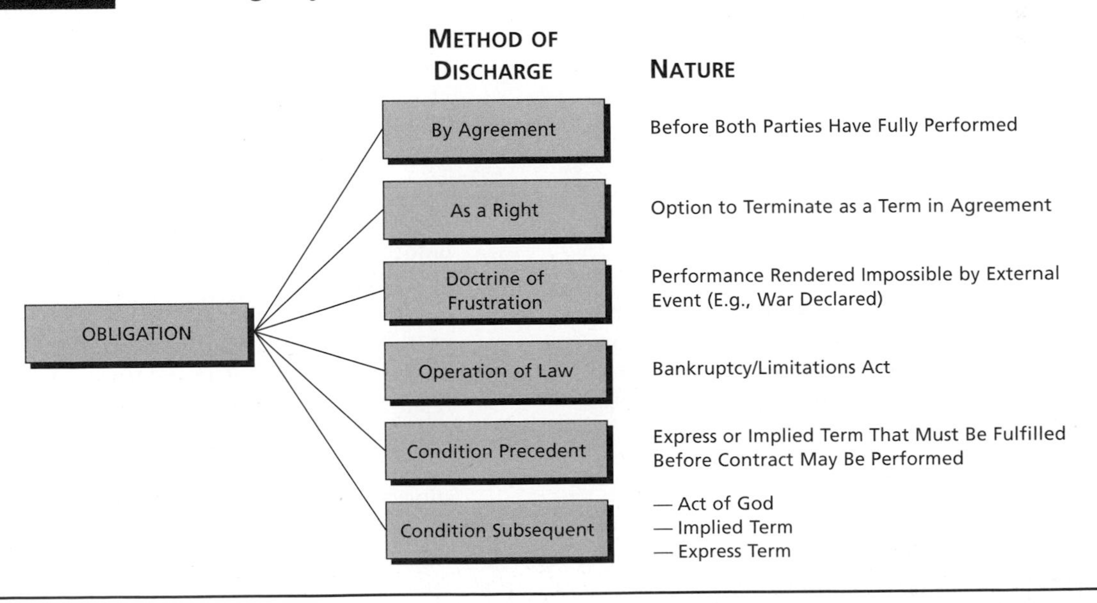

27. See, for example, *Morris v. Baron and Co.*, [1918] A.C. 1.

SUMMARY

The usual method of discharging a contract is by performance. To constitute a discharge the performance must, however, exactly match the one required under the terms of the agreement. Anything short of full and complete performance will not discharge the contract. The parties may also provide in the agreement that it may be terminated at the option of one or both of the parties, or that the agreement will automatically terminate on the occurrence of a particular event.

When events occur that were not contemplated by the parties and that render the promises of one or both of the parties impossible to perform, the courts may treat the contract as frustrated and exempt both parties from further performance. Specific legislation in the form of a *Frustrated Contracts Act* (in some provinces) attempts to equitably distribute the loss when performance is rendered impossible. In the remainder of the provinces, the Common Law rules apply.

Some contracts are discharged by the operation of law upon the happening of particular events that are set out in the statutes or provided for at Common Law. In other cases, the failure to enforce rights under a contract may result in what amounts to a discharge of the agreement by the extinguishment of the right to enforce performance.

The parties may also, by mutual agreement, decide to discharge a contract either by waiver or by the substitution of a new agreement for the existing contract.

A final method of discharge arising out of a breach of the contract has special consequences that are dealt with in the next chapter.

KEY TERMS

REVIEW QUESTIONS

1. What are the implications at Common Law when an unanticipated event renders performance of a contract impossible? Has this been altered by frustrated contracts legislation in some provinces?
2. Does performance by one party terminate a contract?
3. Describe the effect of a valid tender of payment of a debt.
4. Distinguish performance from specific performance.
5. What are the usual consequences that flow from a failure or refusal to perform a valid contract?
6. How might custom affect the performance of a valid contract?
7. Explain the nature and purpose of a *force majeure* clause in a contract. Illustrate your answer with an example.
8. Other than by performance, in what ways might a contract be discharged?
9. Describe the effect of a material alteration on the enforceability of a contract.
10. What is meant by "performance" with respect to contract law?
11. Distinguish between a void and a frustrated contract.
12. Outline the nature of a valid tender. How does it relate to performance?
13. Explain the effect of a condition precedent on the obligation of the parties to perform a contract subject to the condition.

MINI-CASE PROBLEMS

1. X agreed to lease her automobile to Y for the month of July in return for the payment of $300. Before the end of June, X was involved in an automobile accident and the automobile was severely damaged. What are the rights and obligations of X and Y?
2. A agreed to purchase B's snowshoes from him for $100 on the Saturday of the next week. On the

specified date, A went to B's home with 10,000 pennies and demanded the snowshoes. B refused to deliver them. Discuss the rights and obligations of the two parties.

3. A Canadian exporter sends goods to an importer in Phantasia. While the goods are at sea, the United Nations declares a trade embargo on Phantasia, and the ship is prohibited from entering Phantasian waters. What has happened to the contract between the Canadian and Phantasian business persons?
4. Through their lawyers, Wilson agreed to buy Jackson's business for $500,000. On the date fixed for closing the transaction, Wilson's lawyer gave Jackson's lawyer a certified cheque for $550,000, drawn in error. Describe whether Jackson can keep the $50,000 extra, is obligated to return the $50,000, or can avoid the transaction altogether.

CASE PROBLEMS FOR DISCUSSION

CASE 1

The Ebert Electrical Company contracted and paid for 80 air-conditioners at a special discount price made by the Kool-Air Equipment Company. The units were to be installed by Ebert in a new condominium development planned by Carson Developments Ltd. Delivery of the air-conditioners was to take place within 7 days of Carson's architects confirming that final electrical inspections were complete. Carson eventually went bankrupt and the confirmation was never issued.

What is the legal position of Ebert and Kool-Air?

CASE 2

Hamish, an experienced painting contractor, entered into an agreement with Mr. McPhail to paint the McPhail residence both inside and out for $3,200. Mrs. McPhail selected the colours, and Hamish proceeded with the work. During the time Hamish was painting the interior of the house, Mrs. McPhail constantly complained that he was either painting too slowly and interfering with her housecleaning, or painting too fast and splattering paint on the wood trim. Hamish never responded to her remarks.

By the fifth day, Hamish had painted all of the house except the eavestroughs and down-spouts. As he was climbing the ladder to begin painting the eavestroughs, Mrs. McPhail appeared and warned him not to drop paint on her prize azaleas.

At that point, Hamish turned and, without a word, climbed down from the ladder and left the premises.

The next day, he presented his account for $3,200 to Mr. McPhail. When Mr. McPhail refused to make payment, Hamish instituted legal proceedings to collect the amount owing.

Discuss the nature of the claim and indicate the defence (if any) which Mr. McPhail might raise. Render a decision.

CASE 3

Gina Potter owned a farm that had a rather odd-shaped land formation at its centre. Gill, a road contractor, suspected that the land formation might contain a quantity of gravel. He entered into a purchase agreement with Gina to purchase the farm for $280,000. On the offer to purchase Gina deleted the words "or cheque" and insisted that she be paid only in cash.

Before the date fixed for Gill and Gina to exchange the deed and money, Gina heard rumours that Gill wished to buy the farm because it contained a "fortune in gravel." Gina accused Gill of trying to steal her land by offering her only a fraction of its true worth. She instructed her lawyer not to prepare the deed. In actual fact, the $280,000 was slightly more than the value of most similar farms in the area.

On the date fixed for closing, Gill arranged with the bank for $280,000 in cash. He drove to Gina's farm with the money in a brief case. At the gate, he met Gina who refused to allow him entry. Pointing to the brief case, Gill said that he had the money and wanted the deed. Gina refused, so Gill returned to the city.

Discuss the actions of Gill and Gina in this case. Determine the rights and obligations of each if Gill should institute legal proceedings to enforce the contract. Render a decision.

CASE 4

Hansen admired a sports car that Sports Motor Sales Ltd. wished to sell. Hansen informed the company salesman that he would buy the automobile if he could obtain a loan from the bank to cover part of the $17,000 asking price. The salesman agreed to hold the car until Hansen could check with his bank.

Hansen discussed a loan with his bank manager. The manager stated that he would be prepared to

www.mcgrawhill.ca/olc/willes

make a $5,000 loan, but, due to the nature of the purchase, he must first get approval from the regional office. He indicated that this was usually just a formality, and he did not anticipate any difficulty in obtaining approval for the loan.

Hansen then entered into a written agreement with Sports Motor Sales Ltd. to purchase the sports car, with payment to be made in 10 days' time. Both parties signed the agreement, and Hansen paid a $100 deposit. The company retained the sports car pending payment of the balance.

A few days later, the bank manager telephoned Hansen to say that he had encountered a problem with the loan approval. The most he could lend would be $4,000. As a result of the reduction in the loan amount, Hansen found himself $1,000 short.

Advise Hansen of this position at law. Indicate how the case might be decided if Sports Motor Sales Ltd. wished to enforce the agreement.

CASE 5

The Metro Mallards Baseball Club was concerned about the practice of "reselling" tickets to its games by persons who would appear at the entrances of the baseball stadium just before game time and offer tickets for sale to ticketless patrons at premium prices. These persons, known as "scalpers," would purchase a quantity of tickets in advance of the games with the intention of selling them at a profit, a practice that the club wished to curtail. The practice was also prohibited under provincial laws.

In an effort to stop the resale of tickets (except through authorized agents) the club printed the following terms on the back of each ticket:

> This ticket is a personal, revocable license to enter the stadium and view the game described on the front side hereof, and shall not be resold to any other person without the express consent in writing of the issuer or its authorized agents. Management shall have the right to refuse admission to the stadium to anyone in possession of this ticket on return of the stated purchase price.
>
> The issuer, on notice to the person in possession of this ticket, may revoke at any time the license represented by this ticket and recover the same without compensation if the person in possession attempts to resell the ticket.

The notice was included in very fine print on the back of each ticket, along with the usual clauses related to liability for injury and cancellation-of-game matters.

John and Jane Doe purchased a number of tickets with the intention of reselling them at the stadium before a game. They were not aware of the change in the printed information on the back of each ticket.

At the stadium, an employee of the club noticed John and Jane attempting to sell their tickets. She immediately confronted them and demanded that they turn over their tickets to her. John and Jane refused to do so, and a nearby police officer was called over to the scene. John and Jane denied that they were attempting to sell at a profit the tickets they had in their possession. The club employee then pointed out the terms on the back of the tickets to the police officer; after reading the notice, the police officer suggested to John and Jane Doe that it might be advisable for them to surrender their tickets. John and Jane did so, but only on the condition that the police officer hold the tickets until they could obtain legal advice. The officer agreed to do so.

Advise John and Jane Doe. In your answer clearly identify the legal principles that would be applied in this case to determine the rights of the parties.

CASE 6

Community Clipper Airlines Ltd. was a small airline that operated a scheduled service to a number of remote northern communities in the province. In addition, it operated a charter service that consisted for the most part of flying hunting and fishing enthusiasts to tourist camps located on otherwise inaccessible northern lakes.

On a day in late September, the dispatcher received a radio telephone request from the operator of a remote tourist camp, requesting an aircraft to fly in and pick up two late-season fishermen who had to return to their offices for work the next day. The dispatcher passed the flight instructions to one of the charter pilots, "Red Baron."

The pilot checked the aircraft, a four-passenger, single-engine, wheel- and float-equipped craft, then called the weather office for a weather briefing. The weather office reported marginal flying weather until 4:00 p.m. that day. The forecast for the period after 4:00 p.m. indicated rain and fog conditions in the area of the airport, and in the area generally. The weather was expected to close all of the neighbouring airports as well, but remain clear in the area of the tourist camp.

The tourist camp was approximately two hours flying time from the airport, and given that the time was then 11:00 a.m., Red Baron decided to make the flight. She arrived at the tourist camp at approximately 1:00 p.m., but the fishermen had not yet packed their bags and fishing tackle for the trip. Red Baron warned the two fishermen that the weather was deteriorating, and that they must leave immediately, otherwise they would be unable to land at the airport. One of the fishermen was particularly slow at getting his equipment and clothes packed, and in spite of the pilot's urging, the fisherman was not ready to leave until almost 2:00 p.m. At 2:05 p.m., the plane flew from the camp with the two fishermen and their gear.

The weather, which was still appropriate for flying, held until the aircraft was within approximately 20 minutes' flying time of the airport. At that point the aircraft encountered rain and deteriorating visibility. The pilot advised the two passengers that she might not be able to proceed to the airport if the weather became worse. Undaunted, the passengers urged her to proceed on rather than fly back to the camp, as they were obliged to make flight connections at the airport in order to be at their offices the next morning.

With some misgivings and warnings about the weather, Red Baron continued on. She contacted the airport and was informed that rain and fog had closed the runways, but that the lake approach might still be possible since a very slight on-shore breeze was keeping the lake from being totally closed in.

Red Baron elected to try the lake approach rather than turn back. Visibility was poor in the rain, and fog patches obscured parts of the lake. Due to the weather, she decided to land some distance out in the lake, then taxi to the dock area. She made a careful approach, but after successfully landing on the water, collided with a floating log. The log damaged the aircraft and caused it to flip forward and begin to sink.

A rescue boat was dispatched to the aircraft immediately, and the pilot and passengers were rescued from the cold water. While not seriously injured, the passengers suffered bruises, some minor lacerations, and exposure. After being kept in hospital overnight, they were released the next day to return home.

If the passengers instituted legal proceedings against the airline, discuss the possible arguments and legal principles each side would raise.

Render, with reasons, a decision.

CASE 7

Taylor owned and operated a large gravel truck. He entered into a contract with Road Construction Contractors to haul gravel for them at a fixed price per load for a period of six months, commencing May 1st. On April 24th, he appeared at the construction site with his truck. When he examined the distance he would be required to haul the gravel, he realized that he had made a contract that he could only perform at a substantial loss.

He approached one of the partners of the firm with which he had contracted, to inform him that he could not perform the agreement. The partner persuaded Taylor not to be hasty in his decision, but to wait until the next day, when he could discuss the matter with the other partners. Taylor agreed to wait and left his truck in the contractor's garage overnight.

During the night, a fire at the garage destroyed Taylor's truck and some of the contractor's equipment that had also been stored in the garage.

Analyze the events that occurred in this case and discuss the legal position of both parties.

CASE 8

A shipment of liquid milk in refrigerated tank cars was sent by rail from Toronto to Edmonton. At Winnipeg, an electrical connection from the locomotive's generator to the tank car's cooling equipment was found to be faulty. Three of six tank cars of milk had spoiled, and the remainder was sold by the railway company to a creamery in Edmonton.

Discuss the legal positions of the Toronto seller, the Edmonton buyer, and the railway company.

CASE 9

Francis Building Co. contracted in January with Lomas Warehousing Inc. to install a new steel roof in April on the Lomas structure. As matters turned out, the price of steel quadrupled within two months, and Francis advised that it would not proceed on the existing terms. Lomas answered that it would commence legal action against Francis.

Discuss the law upon which each of Francis and Lomas base its argument.

The Online Learning Centre offers more ways to check what you have learned so far. Find quizzes, Weblinks, and many other resources at www.mcgrawhill.ca/olc/willes.

www.mcgrawhill.ca/olc/willes

CHAPTER 14

Breach of Contract and Remedies

Chapter Objectives

After study of this chapter, students should be able to:

- Distinguish between the types of breach of contract.
- Explain the principles of compensation for loss.
- Describe the remedies available at law to an injured party.

YOUR BUSINESS AT RISK

Not every contract will have a happy ending. Business persons must understand the elements of breach and the extent to which they may recover or be liable, as the case may be. This is essential in order to be able to know whether you can walk away from your remaining responsibilities. It is also essential to know these elements in order to assess the cost of pursuing or settling a lawsuit. If you do not understand the remedies the court has available you may overestimate or underestimate the power of your opponent, or you may miss out on options if your lawyer fails to set them out clearly.

THE NATURE OF BREACH OF CONTRACT

The express or implied refusal to carry out a promise made under a contract is a form of discharge. When the refusal occurs, it creates new rights for the injured party that entitle that party to bring an action for the damages suffered as a result of the breach. Under certain circumstances, a breach of contract may also permit the injured party to treat the agreement as being at an end, and to be free from any further duties under it. The courts may either grant compensation for the injury suffered as a result of the non-performance or, in some cases, issue an order requiring performance according to the terms of the contract by the party who committed the breach.

Express Repudiation

Breach of contract may be either express or implied. When a party to a contract expressly repudiates a promise to perform, either by conduct or by a form of communication, the **repudiation** is said to be an express breach.

Repudiation

A refusal to perform a contract.

A Co. and Baxter enter into a contract under which A Co. agrees to sell Baxter a tankerload of fuel. They agree to make delivery at Baxter's service station on September 1st, but A Co. later refuses to deliver the fuel on that date. In this case, A Co. has committed an express breach of contract by its refusal to deliver the goods on the date fixed in the agreement. The company's breach of the agreement would give Baxter a right of action against it for damages for breach of the contract.

Anticipatory breach

An advance determination that a party will not perform his or her part of a contract when the time for performance arrives.

Condition

An essential term of a contract.

Repudiation of a promise before the time fixed for performance is known as **anticipatory breach.** If the repudiated promise represents an important **condition** in the agreement, then the repudiation of the promise would entitle the other party to treat the agreement as at an end. The injured party has an alternate remedy available, however: he or she may also treat the contract as a continuing agreement. The injured party may wait until the date fixed for performance by the other party, notwithstanding the express repudiation, and then bring an action for non-performance at that time. If the injured party should elect to follow the latter course, presumably with the hope that the party who repudiated the agreement might experience a change of mind, the injured party must assume the risk that the agreement may be discharged by other means in the interval.

Maxwell and Fuller Co. enter into a contract for the purchase of heavy machine tools that Fuller Co. has on display in its showroom. Maxwell is to take delivery of the machinery at the end of the month. Before the end of the month, Fuller Co.'s sales manager advises Maxwell that the company does not intend to sell the machinery, but plans to keep it as a display model. Maxwell does nothing to treat the contract as at an end. He continues to urge Fuller Co. management to change its mind. The day before the date fixed in the agreement for delivery, the machinery is destroyed in a fire at Fuller Co.'s showroom. The destruction of the specific goods would release both Maxwell and Fuller Co. from their obligations under the agreement. As a consequence, Maxwell would lose his right of action for breach of contract.

Generally, a breach of contract that takes the form of express repudiation would entitle the injured party to a release from his or her promise of performance under the contract. But if the promises are such that each party must perform independently of the other, the injured party may not be entitled to treat the contract as at an end. For example, Russell and Hall, two farmers, enter into an agreement. Russell agrees to cut Hall's hay, and Hall consents to harvest Russell's wheat crop for him in return. The parties agree that the value of each service is approximately equal, and if both services are performed they will cancel each other out in terms of payment. If Russell should later refuse to cut Hall's hay crop, Hall is not necessarily released from his agreement to harvest Russell's wheat crop. However, he would be entitled to bring an action against Russell for damages arising out of the breach.

Similarly, if the repudiated promise has been partly fulfilled, the party injured by the repudiation may not be entitled to avoid the contract unless the repudiation goes to the very root of the agreement. If the repudiated promise is one that has been substantially performed before repudiation, the injured party is usually bound to perform the agreement in accordance with its terms, subject only to a deduction for the damages suffered as a result of the breach by the other party.

The particular rule of law that may be applied in cases where a contract has been substantially performed before the breach occurs is known as the *doctrine of substantial performance.* It is frequently employed by the courts to prevent the injured party from taking unfair advantage of the party who commits a breach after his promise has been largely fulfilled.

Smith enters into an agreement with Bradley Construction Co. to have them erect a garage on her premises. Payment is to be made in full by Smith upon completion of the construction. Bradley Construction Co. purchases the materials and erects the garage. When the garage has been completed, save and except for the installation of some small trim boards, Bradley Construction Co. leaves the job to work on another, more important project. Because the agreement had been substantially completed by Bradley Construction Co. before it repudiated the contract, Smith could not treat the contract as at an end. She would be required to perform her part of the agreement, but would be entitled to deduct from the contract price the cost of having the construction completed by some other contractor.

The doctrine of substantial performance would prevent Smith from taking unfair advantage of Bradley Construction Co., and from obtaining benefit from that company's breach that would be disproportionate to the injury that she suffered as a result of the breach.

Similarly, in an English case[1] where the court was required to consider defects that would cost approximately £56 to complete a £750 contract, the court held that the work done, excluding the defects (which amounted to 92 percent completion), was sufficient to constitute substantial performance. In the later case of *Bolton v. Mahadeva*,[2] the judge accepted the reasoning in the previous case and made clear in the judgment that the court not only considers the nature of the defects but also the percentage of the work completed in arriving at a conclusion as to the application of the doctrine of substantial performance. In that case, the judge stated:

> In considering whether there was substantial performance I am of opinion that it is relevant to take into account both the nature of the defects and the proportion between the cost of rectifying them and the contract price. It would be wrong to say that the contractor is only entitled to payment if the defects are so trifling as to be covered by the *de minimis* rule.

A rule somewhat similar to the doctrine of substantial performance is also applicable in cases of express repudiation, where the repudiation is of a subsidiary promise rather than an essential part of the agreement. These subsidiary promises are referred to as warranties where a sale of goods is concerned. They do not permit a party to avoid the agreement as a result of the repudiation or non-performance by the other party. The general thrust of this rule is similar to that of the doctrine of substantial performance. If the repudiated promise does not go to the root of the agreement, or is not a condition (an essential term), then the parties should both be required to fulfill their obligations under the agreement. Appropriate compensation would go to the injured party for the incomplete performance of the other.

This approach is consistent with the general policy of the courts to uphold the contract, whenever it is just and reasonable to do so.

Implied Repudiation

The most difficult form of anticipatory breach to determine is implied repudiation of a contract. This occurs when the repudiation must be ascertained from the actions of a party, or implied from statements made before the time fixed for performance. For example, if a party acts in a manner indicating that he or she might not perform on the specified date, the other party to the agreement is faced with a dilemma. To assume from the actions of a party that performance will not be forthcoming in the future is hazardous, yet to wait until the date fixed for performance may only exacerbate the problem if the performance does not take place.

The same problem exists when a party is required to perform over a period of time, or when a seller promises to deliver goods to a buyer from time to time in accordance with the terms of a contract. In each of these cases, a failure to perform in accordance with the contract initially may not permit the other party to treat the substandard performance as a breach. Continued failure to meet the requirements, however, may permit the injured party to be free of the agreement. This would occur if the failure on the part of the other party falls so short of the performance required in the agreement that performance of the promise as a whole becomes impossible.

When a party infers from the circumstances that performance will be below standard in the future, or if the party decides, on the basis of incomplete information, that performance may not take place as required, it becomes risky indeed to treat the contract as at an end. For example, A Co. enters into a contract with B Co. to clear snow from B Co.'s premises during the winter months. After the first snowfall is cleared, A Co. complains

1. *Hoenig v. Isaacs*, [1952] 2 All E.R. 176.
2. [1972] 1 W.L.R. 1009.

bitterly about the poor contract that it made with B Co. A week later, and before the next snowfall, B Co.'s president is informed by Carter, a business acquaintance, that he has just purchased A Co.'s snow-removal equipment. B Co. assumes that A Co. has no intention of performing the snow-removal contract, so it enters into a contract with D Co. for snow removal for the balance of the winter.

The next evening, a snowstorm strikes the area, and B Co.'s president discovers that both A Co. and D Co. drivers are at his door arguing over who has the right to remove the snow. A Co. had sold its old snow-removal equipment and purchased a new and larger snowblower. It had no intention of repudiating the contract.

The dilemma of B Co. in this example illustrates the hazard associated with the determination of repudiation when the intention of a party must be inferred from the party's conduct. In this case, B Co.'s own actions placed it in a position where it was now in breach of the contract.

CASE IN POINT

A corporation owned an auto-body repair business. It offered to sell the business and lease the equipment and building to another corporation, with an option to buy the building and equipment at a later time. The buyer moved into the building and operated the business for several months, but the transaction was never finalized because a third party interfered with the transaction by urging the selling corporation to try to get a better deal. When no agreement could be reached, the buyer sued the selling corporation and third party for breach of contract and conspiracy to induce a breach of contract.

The court held that the selling corporation was in breach of contract and that the third party was liable for inducing a breach of the contract.

Mount Baker Enterprises Ltd. v. Big Rig Collision Inc. (1990), 64 Man. R. (2d) 180.

Fundamental Breach

Fundamental breach

A breach of the contract that goes to the root of the agreement.

Occasionally, when the performance by a party is so far below that required by the terms of the contract, it may be treated as a **fundamental breach** of the agreement. Fundamental breach permits the party injured by the breach to be exonerated from performance, even though the contract may specifically require performance by the party in the face of a breach. This particular doctrine was developed by the courts in a line of English and Canadian cases that dealt with contracts containing exemption clauses. In part, the doctrine was a response by the courts to the problems that have arisen as a result of the unequal bargaining power between buyers and sellers in the marketplace. In the past, contract law was based upon the premise that the agreement was freely made between two parties with equal bargaining power, or at least equal knowledge of the terms of the agreement and their implications. The shift of marketing power in favour of sellers permitted them to insert exemption clauses in standard form contracts. These clauses protect sellers from the risks of liability for defects, price changes, and the obligation to comply with implied warranties and other terms designed to protect the buyer.

Exemption clauses usually require the buyer to perform, even though the seller may avoid performance or substitute performance of a different nature by way of the exemption clause. While the courts will normally enforce exemption clauses (except where the clauses are excluded by legislation), they construe them strictly, and against the party who inserts them in the agreement. Even so, if the breach on the part of the party who seeks to hide behind an exemption clause is so serious as to constitute non-performance of a fundamental term of the agreement, the courts may not allow that party to use the clause to avoid liability.

A person purchased a brand-new truck from a dealer under a contract containing a broadly worded exemption clause. The buyer discovered that the truck had many defects and was very difficult to drive. The court held that the seller had delivered a truck that was totally different from that which the buyer had contracted for, and the buyer was entitled to rescind the contract.[3]

The court, in this particular case, decided that the buyer was entitled to a vehicle that was relatively free from defects and reasonably fit for the use intended. The truck turned out to be wholly unsatisfactory, and the court decided that the delivery of such a vehicle by the dealer constituted a repudiation of the agreement. The failure to deliver a vehicle that the parties had contracted for constituted a fundamental breach of the agreement. Notwithstanding the exemption clause, the buyer was entitled to treat the contract as at an end.

The fact that fundamental breach works to relieve the aggrieved party from performance almost ensured that parties on the other side would draft broad exclusion clauses that would exempt themselves from liability. The Supreme Court of Canada has laboured toward a balance. Its golden objective: aggrieved parties receiving virtually nothing should escape their obligation to perform, but freedom of contract should allow their honest counterparties to demand exclusion of their liability when freely negotiated.

COURT DECISION: Practical Application of the Law

Contracts—Fundamental Breach

***Tercon Contractors Ltd. v. British Columbia (Transportation and Highways),* 2010 SCC 4 (CanLII).**

If the "golden objective" has been achieved by the Supreme Court of Canada, it has required the hearing of four landmark cases over the last 100 years (1918, 1961, 1989, and now 2010), at which the language of the Supreme Court of Canada seems blunt, to the point of exasperation.

THE COURT: With respect to the appropriate framework of analysis, the doctrine of fundamental breach should be "laid to rest." The following analysis should be applied when a plaintiff seeks to escape the effect of an exclusion clause... The first issue is whether, as a matter of interpretation, the exclusion clause even applies to the circumstances established in evidence... If the exclusion clause applies... (is it) unconscionable and thus invalid at the time the contract was made? If the exclusion clause is ... valid at the time of contract formation and applicable to the facts (then the court may) refuse to enforce the exclusion clause because of an overriding public policy. The burden of persuasion lies on the party seeking to avoid enforcement of the clause to demonstrate an abuse of the freedom of contract that outweighs the very strong public interest in their enforcement. Conduct approaching serious criminality or egregious fraud are but examples of well-accepted considerations of public policy that are substantially incontestable and may override the public policy of freedom to contract and disable the defendant from relying upon the exclusion clause.

The burden upon the party seeking to avoid enforcement of an exclusion clause: is it inapplicable, is it unconscionable, or does it offend public policy?

A QUESTION OF ETHICS

Exemption clauses can allow one party to escape performance or liability while compelling the other party to perform. In interpreting them, judges exercise a certain degree of social conscience. If you were a judge, what persuasive factors would enter your mind in deciding to uphold such a clause, or to strike it down?

3. *Cain et al. v. Bird Chevrolet-Oldsmobile Ltd. et al.* (1976), 69 D.L.R. (3d) 484.

A problem for an injured party may arise in cases where that party continues to hold the agreement in effect after the other contracting party has failed to properly perform an important part of the promise. The effect of the delay may change a condition into a mere warranty. This has important implications for the parties to a contract because of the nature and effect of these terms. Generally, the essential terms of the contract constitute conditions. If they are not performed, conditions may entitle the other party to treat the contract as being at an end. **Warranties**, on the other hand, are generally minor promises or terms that may be express or implied and are collateral to the object of the agreement.

Warranty

In the sale of goods, a minor term in a contract. The breach of the term would allow the injured party damages, but not rescission of the agreement.

Anil purchases a used camera on eBay for a few hundred dollars. He discovers the camera is broken and cannot be repaired. He may claim a refund from the seller on the basis of breach of condition. He buys another camera on eBay and discovers that while the camera functions perfectly, its fitted case is badly soiled. In this instance he would be only entitled to damages for breach of warranty.

A breach of a warranty usually does not permit the injured party to treat the contract as being at an end; it only entitles the injured party to sue for damages. However, if a party should refuse to perform a condition or important term that would entitle the other party to avoid performance, and that party does not act at once to do so, the condition may become a mere warranty. The same holds true if the party injured by the breach of the condition continues on with the contract and accepts benefits under it. Then the condition becomes a warranty *ex post facto*. In effect, the injured party's actions constitute a waiver of the right to avoid the agreement. The injured party will be obliged to perform with only the right to damages as compensation for the breach by the other party.[4]

A purchaser of a parcel of land wishes to avoid the purchase of the land and recover the deposit paid, because the vendor of the land did not have the deed available on the date fixed for closing the transaction. Unless the purchaser can satisfy the court that the purchase monies had been paid or tendered, the court may treat the purchaser's failure to show the ability to complete the contract as the purchaser's default or his rejection of the contract. The court may refuse to assist the purchaser in the recovery of the deposit paid.

In the case of *Zender et al. v. Ball et al.*,[5] similar to the example above, the judge described the purchaser's position as follows:

> ... ordinarily a want of title would entitle the purchaser to commence an action for rescission and the return of all moneys paid. However, the purchaser is only entitled to rescind on the neglect or refusal of his vendor to deliver a registrable conveyance on the date fixed by the contract if he has tendered or paid all his purchase money or otherwise performed his part of the contract. The purchaser cannot recover, at law, his deposit where he has constructively or expressly abandoned the contract, wrongfully repudiated, or unequivocally manifested his inability to carry out his part of the contract and the vendor is willing to complete. The deposit, under the terms of the contract, is in the nature of a guarantee or security for the performance of a contract and is not merely a part payment of the purchase price. To permit a purchaser to recover his deposit on his default, where he has abandoned or wrongfully repudiated the contract, would be to permit him to take advantage of his own wrong. Ordinarily, a demand by the purchaser for the return of his deposit from the defaulting vendor is an election to rescind and specific performance is no longer available.

4. *Couchman v. Hill*, [1947] K.B. 554.
5. (1974), 5 O.R. (2d) 747.

FIGURE 14–1 Breach of Contract

FRONT-PAGE LAW

L.A. TV producer sues Vancouver company

VANCOUVER — A Los Angeles-based television producer is suing Peace Arch Entertainment of Vancouver for breach of contract after he was fired from the new television series The Immortal.

James Margellos, who was fired as line producer June 15, was hired Feb. 14 to produce 22 episodes of the one-hour action science-fiction series.

He was to be paid US$279,000, about US$12,700 an episode.

His contract also stipulated an under-budget bonus of US$55,000, a car allowance of Cdn$350 a week and living expenses of Cdn$2,800 a month, plus a cellphone and 11 round-trip airline tickets between Vancouver and Los Angeles.

Named as defendants are Peace Arch Entertainment, company president Timothy Gamble, chief financial officer Juliet Jones, Immortal Productions, Hill Top Entertainment of California, the international distributor and executive producer of the The Immortal.

Margellos says in his suit he was fired "to deflect responsibility ... for delays and budgeting overruns which were largely beyond the control of the plaintiff and occurred largely as a result of the action of the individual plaintiffs."

It says he had no prior warning of dissatisfaction with his performance.

Peace Arch is expected to file a statement of defence at a later date.

If Margellos appeared before you, and as a judge you found in his favour, what remedy would you give?

Source: Excerpted from *National Post*, "L.A. TV Producer sues Vancouver company," July 24, 2000.

REMEDIES

The Concept of Compensation for Loss

A breach of contract gives the party injured by the breach the right to sue for compensation for the loss suffered. Loss or injury as a result of the breach must be proven. If this is done, the courts will attempt to place the injured party in the same position he or she would have been in had the contract been properly performed. Compensation may take the form of monetary damages or it may, in some circumstances, include the right to have the contract promise, or a part of it, performed by the defaulting promisor. It may also take the form of *quantum meruit*, a quasi-contract remedy.

The usual remedy for a breach of contract is monetary damages. The reason that the courts usually award compensation in this form is that most contracts have as their object something that can be readily translated into a monetary amount in the event of non-performance.

Fuller offers to sell Brown 600 crates of apples at $50 a crate. On the date fixed for delivery, Fuller delivers the apples to Brown, but Brown refuses to take delivery. Fuller later sells the apples to Caplan, but the price by then has fallen to $40 a crate. Fuller has suffered a loss of $10 a crate, or $6,000 in total, as a result of Brown's breach of the contract. If Fuller should sue Brown for breach of contract, the courts would probably award Fuller damages in the amount of $6,000 to place Fuller in the same position that he would have been in had Brown carried out his part of the agreement.

Restitutio in integrum

To restore or return a party to an original position.

This basic principle of damages is sometimes referred to as the principle of ***restitutio in integrum***, which originally meant "a restoration to the original position." However, this is not what the Common Law courts attempt to do. In the case of a breach of contract, they attempt to place the injured party in the same position as if the contract had been performed. *Restitutio in integrum* was originally a principle in the old courts of equity. It was applied in cases where it was desirable to place the parties in the position they were in before the agreement had been formed. Today, the term is usually used by the courts to mean "to make the party whole," or to compensate for the loss suffered.

Types of Damages

General damages

Restitution for losses naturally expected from a breach of contract.

The principal forms of damages are general and special damages. **General damages** are monies awarded by the court, intended to place the injured party in the position they would have been had the contract been performed in accordance with its terms, and represent compensation for losses that would naturally (generally) flow from the breach. For example, where a seller was in breach of a contract to sell a business, general damages would be those that restored the purchaser's investment, plus interest. General damages are not subject to particular proof, other than proof of the breach, as they will be assessed by the court as those that are expected to occur from such a breach.

Lawyers (and their clients) have had a particular interest in identifying certain general damages as "consequential damages," if only to be able to avoid and exclude liability for them when drafting agreements. For example, it is difficult for a courier company to exclude liability to repay its freight charge to a client when it fails to deliver a package on time, but the courier will be interested in excluding liability for the *consequences* to the *client* of late delivery. The courier has no knowledge of the contents of packages; a day late in delivery may mean nothing to one client, but may spell disaster for another. This topic is addressed further in the next section of the text.[6]

Special damages

Specific damages that would flow from a breach of contract.

Special damages are monies awarded by the courts for damages that do not arise naturally or generally from the normal course of events following such a breach; they are in fact, special. In the same example on the previous page, if the business purchaser plaintiff had to sell her home in Halifax and move to Vancouver to operate the business, and this was known to the defendant, the breach would result in special damages (for the cost of the move, capital loss on a house), in addition to the general damages. These damages must be carefully itemized in any claim before the courts, and be proven.

In addition to general and special damages are punitive damages. These are not compensation of the plaintiff, but rather punishment of the defendant for a breach of contract which is deceitful, malicious, or offensive to ordinary standards of morality. One should be able to imagine cases involving fraudulent misrepresentation, vulnerable parties, or some cases of fundamental breach that would qualify for punitive damages. Recall, however, that civil courts are not criminal courts, and thus their primary goal is restitution, not punishment. Punitive damages are therefore rare, for cases when compensation is clearly not enough. Where a breach of contract also amounts to a criminal act, the criminal courts will be the forum where the Crown will exact its punishment, in addition to restitution to the plaintiff in civil courts.

6. The curious will also find an interesting example cited by a notable judge within [1861-73] All E.R. Rep. 340 (C.P.), involving a blacksmith, a horse, a fiancé, a bride at the altar, and heartbreak.

The Extent of Liability for Loss—Reasonable Foreseeability

While damages may be readily determined in the event of a breach of a simple contract, some contracts may be such that a breach or failure to perform may have far-reaching effects. This is particularly true when a contract is only a part of a series of contracts between a number of different parties, and the breach of any one may adversely affect the performance of another. A manufacturer of automobiles, for example, depends heavily upon the supply of components from many subcontractors, while the manufacturer's assembly plant performs the function of merging the various parts into the finished product. The failure of any one supplier to provide critical parts could bring the entire assembly process to a standstill and produce losses of staggering proportions. Fortunately, automobile manufacturers usually take precautions to prevent the occurrence of such a state of affairs. This example illustrates how a breach of contract may have ramifications that extend beyond the limits of the simple contract.

Since a party may generally be held liable for the consequences of his or her actions in the case of a breach of contract, it is necessary to determine the extent of the liability that might flow from the breach. At law, it is necessary to draw a line at some point that will end the liability of a party in the event of a breach of contract. Beyond this line, the courts will treat the damages as being too remote. An early English case[7] involved a contract between a milling firm and a common carrier to deliver a broken piece of machinery to the manufacturer to have a replacement made. The mill was left idle for a lengthy period of time because the carrier was tardy in the delivery of the broken mill part to the manufacturer. The miller sued the carrier for damages resulting from the undue delay. In determining the liability of the carrier, the court formulated a principle of remoteness that identified the damages that may be recovered as *those that the parties may reasonably contemplate as flowing from such a breach.* The case, in effect, established two rules to apply in cases where a breach of contract occurs. The first identifies the damages that might obviously be expected to result from a breach of the particular contract as contemplated by a *reasonable person.* The second "rule" carries the responsibility one step further, and includes *any loss that might occur from special circumstances relating to the contract that both parties might reasonably be expected to contemplate at the time the contract is made.*

These two "rules" for the determination of remoteness in the case of a breach of contract were enunciated in 1854. With very little modification they were used as a basis for establishing liability for over a century. More recently, however, the two rules were rolled into a single one that states: "*...any damages actually caused by a breach of any kind of contract is recoverable, providing that when the contract was made such damage was reasonably foreseeable as liable to result from the breach.*"[8]

This particular rule would hold a person contemplating a breach liable for any damages that would reasonably have been foreseen at the time that the contract was formed. However, the person would only be liable for those damages that would be related to the knowledge available to the party that might indicate the likely consequences of the contemplated breach.

COURT DECISION: Practical Application of the Law

Contracts—Damages for Breach—Reasonable Foreseeability

***Hadley et al. v. Baxendale et al.* (1854), 156 E.R. 145.**

The plaintiffs were millers who operated a steam-powered mill. The crankshaft of the engine cracked, and it was necessary to obtain a new part from the engine manufacturer. A common carrier contracted to deliver the old part to the manufacturer the next day if he received the part before noon. The old part was accordingly delivered to the

7. *Hadley et al. v. Baxendale et al.* (1854), 9 Ex. 341, 156 E.R. 145.

8. *C. Czarnikow Ltd. v. Koufos*, [1966] 2 W.L.R. 1397 at 1415.

carrier, but through neglect was not delivered to the engine manufacturer for some time. The plaintiffs instituted legal proceedings against the carrier for the loss suffered as a result of the delay.

THE COURT: Now we think the proper rule in such a case as the present is this: Where two parties have made a contract which one of them has broken, the damages which the other party ought to receive in respect of such breach of contract should be such as may fairly and reasonably be considered either arising naturally, i.e., according to the usual course of things, from such breach of contract itself, or such as may reasonably be supposed to have been in the contemplation of both parties, at the time they made the contract, as the probable result of the breach of it. Now, if the special circumstances under which the contract was actually made were communicated by the plaintiffs to the defendants, and thus known to both parties, the damages resulting from the breach of such a contract, which they would reasonably contemplate, would be the amount of injury which would ordinarily follow from a breach of contract under these special circumstances so known and communicated. But, on the other hand, if these special circumstances were wholly unknown to the party breaking the contract, he, at the most, could only be supposed to have had in his contemplation the amount of injury which would arise generally, and in the great multitude of cases not affected by any special circumstances, from such a breach of contract. For, had the special circumstances been known, the parties might have specially provided for the breach of contract by special terms as to the damages in that case; and of this advantage it would be very unjust to deprive them. Now the above principles are those by which we think the jury ought to be guided in estimating the damages arising out of any breach of contract.

Lawyers (and judges) have argued over the meaning of this case for over 150 years. As a business person, do you really need to know anything more than what the judge is telling you here?

In rare instances, the compensation may also be extended to cover damage in the nature of mental stress when it is associated with the transaction and where the actions of the party in breach compound the problems of the injured party. In the case of *Vorvis v. Insurance Corporation of British Columbia*,[9] the Supreme Court of Canada indicated that under certain circumstances in cases of breach of contract, aggravated damages for mental suffering may be appropriate. The court characterized such damages, however, as compensatory rather than punitive, and stated that the award would hinge upon whether the party in breach should have reasonably expected that mental suffering would result from the breach. Nevertheless, the court emphasized that an injured plaintiff is normally entitled only to have what the contract provided for, or the equivalent in compensation for the loss.

The Duty to Mitigate Loss

Mitigation

The obligation of an injured party to reduce the loss flowing from a breach of contract.

In the case of breach, the injured party is not entitled to remain inactive. The prospective plaintiff in an action for damages must take steps to mitigate the loss suffered. Otherwise the courts may not compensate the injured party for the full loss. If the party fails to take steps to reduce the loss that flows from a breach, then the defendant, if he or she can prove that the plaintiff failed to mitigate, may successfully reduce the liability by the amount that the plaintiff might otherwise have recovered, had it not been for the neglect. For example, Ashley enters into a contract with Bentley for the purchase of a truckload of California grapes. The purchase price is fixed at $10,000, but when Bentley delivers the grapes, Ashley refuses to accept delivery. If Bentley immediately seeks out another buyer for the grapes and sells them for $5,000, Bentley would be entitled to claim the actual loss of $5,000 from Ashley. On the other hand, Bentley may do nothing after Ashley refuses to accept delivery of the grapes, and, as a result, the grapes become worthless. Then a claim against Ashley for the $10,000 loss suffered by Bentley may be reduced substantially, if Ashley can successfully prove that Bentley did nothing to mitigate the loss.

It should also be noted that if Ashley refused to accept the grapes, and Bentley sold them to Carter for $10,000, Bentley would still have a right to action against Ashley for breach of contract. Bentley, however, would only be entitled to nominal damages under the circumstances, because he suffered no actual loss.

9. [1989] 1 S.C.R. 1085.

The question arises: to what lengths must a person go in order to mitigate a loss? In the case of *Asamera Oil Corp. Ltd. v. Sea Oil & General Corp. et al.*,[10] the Supreme Court of Canada reviewed the requirements:

> We start of course with the fundamental principle of mitigation authoritatively stated by Viscount Haldane, L.C., in *British Westinghouse Electric & Mfg. Co., Ltd. v. Underground Electric R. Co. of London, Ltd.*, [1912] A.C. 673 at p. 689:
>
> > The fundamental basis is thus compensation for pecuniary loss naturally flowing from the breach; but this first principle is qualified by a second, which imposes on a plaintiff the duty of taking all reasonable steps to mitigate the loss consequent on the breach, and debars him from claiming any part of the damage which is due to his neglect to take such steps. In the words of James L.J. in *Dunkirk Colliery Co. v. Lever* (1898), 9 Ch.D. 20, at p. 25. "The person who has broken the contract is not to be exposed to additional cost by reason of the plaintiffs not doing what they ought to have done as reasonable men, and the plaintiffs not being under any obligation to do anything otherwise than in the ordinary course of business."
> >
> > As James L.J. indicates, this second principle does not impose on the plaintiff an obligation to take any step which a reasonable and prudent man would not ordinarily take in the course of his business. But when in the course of his business he has taken action arising out of the transaction, which action has diminished his loss, the effect in actual diminution of the loss he has suffered may be taken into account even though there was no duty on him to act.

Liquidated Damages

Liquidated damages

A *bona fide* estimate of the monetary damages that would flow from the breach of a contract.

Punitive damages

Damages awarded by a court to punish a wrongdoer. Not normally awarded for ordinary breach of contract.

At the time the contract is entered into, the parties may attempt to estimate the damages that might reasonably be expected to flow from a breach of contract, and they may insert the estimate as a term. The courts will generally respect the agreement, provided that the estimate is a genuine attempt to estimate the loss. Usually the clause takes the form of a right in the seller to retain a deposit as **liquidated damages** in the event that the buyer refuses to complete the contract. However, the parties may occasionally insert a clause that requires a party in default to pay a fixed sum. If the amount is unreasonable in relation to the damage suffered, the sum may be treated as a penalty rather than liquidated damages, and the courts will not enforce the clause. Similarly, if a party has paid a substantial portion of the purchase price at the time the contract is entered into, and the contract contains a clause that entitles the seller to retain any payments made as liquidated damages, a failure to perform by the buyer would not entitle the seller to retain the entire part-payment. The seller, instead, would only be entitled to deduct the actual loss suffered from the partial payment and would be obliged to return the balance to the purchaser. The reasoning of the courts behind this rule is that **punitive damages** will not be awarded for an ordinary breach of contract. Only in cases where the actions of a party are reprehensible will a party be penalized. In the case of contract, the circumstances would probably be limited to those relating to contracts negotiated under fraud or duress.

The distinction between a part-payment and a deposit arose in the case of *Stevenson v. Colonial Homes Ltd.*[11] A purchaser of cottage building materials had paid a "deposit" of $1,000 under a contract that provided that if the purchaser failed to complete the contract the deposit would be forfeited. The $1,000 deposit in this case represented a substantial part of the purchase price, and the purchaser attempted to recover a part of it when default did occur. The purchaser argued that the payment was a part-payment and not a deposit.

In the case, the court reviewed the law as it related to liquidated damages by saying:

> Whether or not the appellant is entitled to the return of the $1,000, in the view I take of the case, depends upon whether the $1,000 was paid as a deposit or whether it was part payment of the purchase price.

10. (1978), 89 D.L.R. (3d) 1.
11. [1961] O.R. 407.

A useful summary of the law upon this point is to be found in the judgment of Finnemore, J., in *Gallagher v. Shilcock*, [1949] 2 K.B. 765 at pp. 768–9:

> The first question is whether the [money] which the plaintiff buyer paid on May 17 was a deposit or merely a pre-payment of part of the purchase price .... When money is paid in advance, it may be a deposit strictly so called, that is something which binds the contract and guarantees its performance; or it may be a part payment — merely money pre-paid on account of the purchase price; or, again it may be both: in the latter case, as was said by Lord Macnaghten in *Soper v. Arnold* (1889), 14 App. Cas. 429, 435: "The deposit serves two purposes — if the purchase is carried out it goes against the purchase-money — but its primary purpose is this, it is a guarantee that the purchaser means business." If it is a deposit, or both a deposit and prepayment, and the contract is rescinded, it is not returnable to the person who pre-paid it if the rescission was due to his default. If, on the other hand, it is part-payment only, and not a deposit in the strict sense at all, then it is recoverable even if the person who paid it is himself in default. That, I think, follows from *Howe v. Smith*, 27 Ch.D. 89, and from *Mayson v. Clouet*, [1924] A.C. 980, a case in the Privy Council. As I understand the position, in each case the question is whether the payment was in fact intended by the parties to be a deposit in the strict sense or no more than a part-payment: and, in deciding this question, regard may be had to the circumstances of the case, to the actual words of the contract, and to the evidence of what was said.

As was stated by Lord Dunedin in *Mayson v. Clouet*, at p. 985: "Their Lordships think that the solution of a question of this sort must always depend on the terms of the particular contract." The contract between the appellant and the respondent should be critically examined to see if from it can be drawn the intention of the parties as to whether the $1,000 was to be a deposit or a part payment of purchase price only.

Remedies for Particular Situations

Specific Performance

Specific performance

An equitable remedy of the court that may be granted for breach of contract where money damages would be inadequate, and that requires the defendant to carry out the agreement according to its terms.

In rare cases, where monetary damages would be an inadequate compensation for breach of contract, the courts may decree specific performance of the contract. The decree of **specific performance** is a discretionary remedy that has its origins in the English Court of Chancery. The remedy requires the party subject to it to perform the agreement as specified in the decree; a failure to comply with the decree would constitute contempt of court. Unlike an ordinary monetary judgment, the decree of specific performance carries with it the power of the courts to fine or imprison the wrongdoer for failure to comply with the order.

Specific performance is generally available as a remedy when the contract concerns the sale of land. The unique nature of land is the reason why the courts will enforce the contract, as no two parcels of land are exactly the same. Even then, the courts expect the injured party to show that the fault rests entirely on the party in breach before the remedy will be granted. The plaintiff (the injured party) must satisfy the court that he or she was willing and able at all times to complete the contract, and did nothing to prompt the refusal to perform by the party in breach. To satisfy this particular onus, the plaintiff must usually make a tender of either the money or the title documents as required under the contract. This must be done strictly in accordance with the terms of the contract on the day, and at the time and place fixed for performance. The plaintiff must also satisfy the court that the other party refused to perform at that time. If the court is satisfied on the evidence presented that the plaintiff did everything necessary to perform, and that the other party was entirely at fault for the breach, it may issue a decree of specific performance that would require performance of the contract by the party in breach.

Specific performance may apply to either a vendor or a purchaser in a land transaction. The courts may order performance by either a defaulting seller or buyer in the contract. The remedy of specific performance may also be available in a case where the contract has a "commercial uniqueness" or has as its subject matter a chattel that is rare and unique.[12] But for most contracts that involve the sale of goods, monetary damages would normally be the appropriate remedy. Moreover, the courts will not grant specific performance of a contract of employment or any contract that involves the performance of personal services by an individual. The principal reason for not doing so is that it will not enforce promises that it would be obliged to continually supervise.

Injunction

A remedy similar to specific performance may also be available in the case of a breach of contract where the promise that the party refuses to perform is a promise to forbear from doing something. The difference between this remedy, known as an **injunction**, and a decree of specific performance is that the injunction usually orders the party to refrain from doing something that the party promised that he or she would not do. On the other hand, a decree of specific performance usually requires the party to do a positive act.

Injunction

An equitable remedy of the court that orders the person or persons named therein to refrain from doing certain acts.

Like a decree of specific performance, an injunction is an equitable remedy, and may be issued only at the discretion of the court. Its use is generally limited to the enforcement of "promises to forbear" contained in contracts. However, the courts are sometimes reluctant to grant the remedy in contracts of employment if the effect of the remedy would be to compel the promisor to perform the contract to his or her detriment. For example, Maxwell and Dixon enter into an agreement. Maxwell agrees to work exclusively for Dixon for a fixed period of time, and to work for no one else during that time. If Maxwell should repudiate her promise and work for someone else, Dixon may apply for an injunction to enforce Maxwell's promise not to work for anyone else. If the injunction should be granted, it would enforce only the negative covenant, and not Maxwell's promise to work exclusively for Dixon. In other words, Maxwell need not remain in the employ of Dixon, but because of the injunction, she would not be permitted to work for anyone else. It should be noted, however, that if circumstances were such that Maxwell did not have independent means, and was obliged to work for Dixon in order to support herself, the courts may not issue an injunction. The reasoning here is that the injunction, in effect, would constitute an order of specific performance of the entire contract. Usually contracts containing a negative promise limit the party to the acceptance of similar employment, rather than employment of any kind. By placing only a limited restriction on the employee's ability to accept other employment, the plaintiff may argue that the defendant is not restricted from other employment, but only employment of a similar nature. Therefore, the employee would not be restricted to working only for the plaintiff.

In other types of contracts, an injunction may be issued to enforce a negative covenant if the covenant is not contrary to public policy. It may be granted, for example, in the case of a contract for the sale of a business to enforce a covenant made by the vendor, where the vendor agrees not to compete with the purchaser within a specific geographic area for a specified period of time. It may also be available to enforce a negative covenant with respect to the use of premises or equipment. For example, Dawson may enter into an agreement with Ballard to allow Ballard the use of certain premises for business purposes. In turn, Ballard promises that he will not operate the business after a certain hour in the evening. If Ballard should continue to operate the business past the stipulated hour, Dawson may be entitled to an injunction to enforce Ballard's negative covenant. It is important to note, however, that an injunction, like a decree of specific performance, is discretionary; the courts will not issue an injunction unless it is fair and just to do so.

Quantum meruit

"As much as he has earned." A quasi-contractual remedy that permits a person to recover a reasonable price for services and/or materials requested, where no price is established when the request is made.

Quantum Meruit

In some cases, when a contract is repudiated by a party and the contract is for services, or goods and services, the remedy of ***quantum meruit*** may be available as an alternative for the party injured by the repudiation. *Quantum meruit* is not a remedy arising out of the contract; rather, it is a remedy based upon quasi-contract. In the case of *quantum*

12. *Re Wait,* [1927] 1 Ch. 606.

meruit, the courts will imply an agreement from a request for goods and services. They will also require the party who requested the service to pay a reasonable price for the benefit obtained.

Quantum meruit may be available as a remedy if the contract has only been partly performed by the injured party at the time the breach occurred. To succeed, however, the injured party must show that the other party to the contract repudiated the contract, or did some act to make performance impossible. The breach by the party cannot be of a minor term, but must be of such a serious nature that it would entitle the party injured by the breach to treat the contract as at an end. *Quantum meruit* is not normally available to the party responsible for the breach. Under the doctrine of substantial performance, however, the party may be entitled to recover for the value of the work done. Similarly, *quantum meruit* would not apply if a party had fully performed his or her part of the contract at the time the breach occurred. The appropriate remedy in that case would be an action for the price if the party in breach refused or failed to pay. *Quantum meruit* would also be inapplicable if the contract itself required complete performance as a condition before payment might be demanded.

The distinction between the two remedies is also apparent in the approach the courts may take to each. In the case of an ordinary breach of contract, the remedy of monetary damages is designed to place the injured party in the position that the party would have been in had the contract been completed. This is not so with *quantum meruit*. When a claim of *quantum meruit* is made, the courts will only be concerned with compensation to the party for work actually done. The compensation will be the equivalent of a reasonable price for the service rendered. This may differ substantially from the price fixed in the repudiated agreement. It is not designed to place the injured party in the same position that he or she would have been in had the other party not broken the agreement.

CLIENTS, SUPPLIERS *or* OPERATIONS *in* QUEBEC

In Quebec, contracts of enterprise (such as building projects and the like) are subject to a statutory right of unilateral repudiation by the party receiving a service (Article 2125 — essentially cancelling the work and ejecting the service provider), and an equivalent statutory right (Article 2129) of *quantum meruit*, for the ejected service provider. The *quantum meruit* given is a share of the price proportionate to the amount of work performed. Thus, the service provider, depending on the case, could be overpaid but will likely be underpaid if there were large upfront costs in the project and the cancellation takes places early on. Thus, it is common to find waivers of the Article 2125 right as part of standard service provider contracts.

FIGURE 14–2 Remedies for Breach

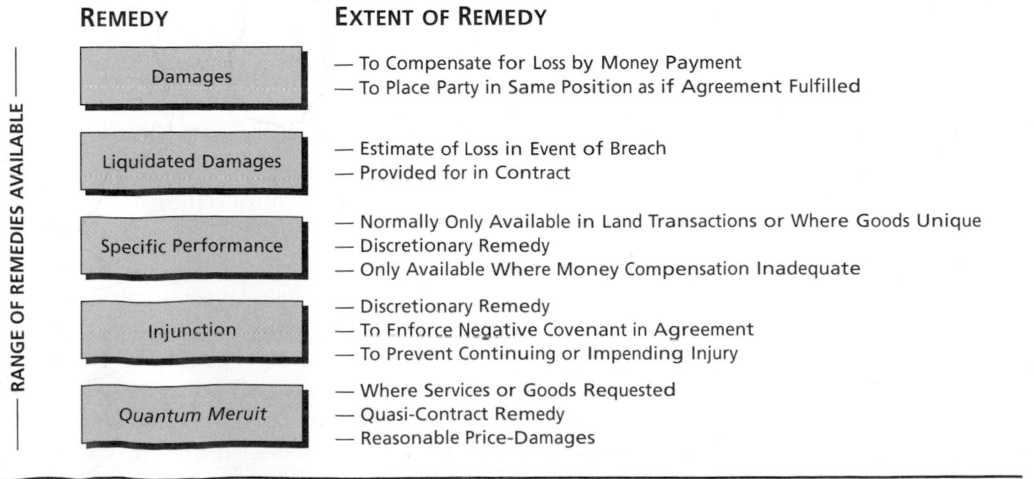

MANAGEMENT ALERT: BEST PRACTICE

A breach of contract gives rise to a remedy, but at what cost? Business persons must obviously consider the time and expense of litigation, and whether the potential defendant has the ability to pay a judgment. They should consider as well the distraction from business that a lawsuit creates, the emotional cost, public disclosure of the facts, and damage to any remaining longer-term business relationship with the defendant. Often, a negotiated settlement, or the choice to "live and learn" is preferable. If a claim gives rise to a counterclaim, it is difficult to just walk away from the litigation. These factors should be discussed with a lawyer; pure emotion should never rule one's decisions.

CHECKLIST FOR BREACH OF CONTRACT CASES

Breach

1. Is the act or omission, in fact, a repudiation of a term of the contract?
2. Is it of an important condition, rather than merely of a warranty?
   - If so, the aggrieved party is relieved of performance.
3. Is the breach sufficiently grave to be fundamental breach?
   - Exemption clauses will not protect the party in breach.

Remedies — General Considerations

1. Has the aggrieved party failed to mitigate its losses?
2. Are any special remedies available?
3. Does a valid term exist for liquidated damages?
4. *Restitutio in integrum* may be available for damages caused by breach that were reasonably foreseeable as liable to result from that breach.
5. Is there further entitlement to special or punitive damages?

SUMMARY

Breach of contract is the express or implied refusal by one party to carry out a promise made to another in a binding contract. Express or implied repudiation before the date fixed for performance is called anticipatory breach. If the refusal to perform is such that it goes to the root of the agreement, and is made before the agreement has been fully performed by the other party, the injured party may be released from any further performance and may sue for the damages suffered. However, if the party who refused to complete the contract has substantially performed the agreement, the doctrine of substantial performance may apply, and the injured party may only obtain damages for the deficient performance.

The remedies available in the case of breach are: (1) monetary damages; (2) specific performance; (3) an injunction; and (4) the quasi-contract remedy of *quantum meruit*.

Specific performance and the injunction are equitable remedies that may be awarded only at the discretion of the court. They are normally only awarded to enforce contract clauses if it is fair and just to do so. These circumstances are usually limited to contracts concerning land, commercial uniqueness, and rare chattels (in the case of specific performance), and to the enforcement of a negative promise (in the case of an injunction). In most other cases, monetary damages are adequate compensation. As an alternative to damages, if the contract has not been fully performed by a party, the remedy of *quantum meruit* may be available. *Quantum meruit*, however, only entitles the party to a reasonable price for the service or the work done. It is not designed to put the party in the position the party would have been in had the contract been performed.

Key Terms

Repudiation, 244
Anticipatory breach, 245
Condition, 245
Fundamental breach, 247
Warranty, 249
Restitutio in integrum, 251
General damages, 251
Special damages, 251
Mitigation, 253
Liquidated damages, 254
Punitive damages, 254
Specific performance, 255
Injunction, 256
Quantum meruit, 256

Review Questions

1. What tests are applied by the courts to determine the remoteness of a damage claim for breach of contract?
2. What are the rights of one party to a contract when informed by the other party that performance will not be made?
3. In a situation where a contract is expressly repudiated, what are the dangers associated with waiting until the time for performance to determine if breach will actually occur?
4. Explain why mitigation of loss by the injured party is important when a breach of contract occurs.
5. Explain the doctrine of fundamental breach as it applies to a contract situation.
6. Does repudiation of a subordinate promise permit the party affected by the repudiation to avoid the obligation to perform the contract itself?
7. Describe the concept of damages as it applies to Common Law contracts.
8. What does "*restitutio in integrum*" mean?
9. Explain the difference between "express" and "implied" repudiation of a contract. Give an example of each.
10. Apart from money damages, what other remedies are available from the courts in cases involving a breach of contract? Under what circumstances would these remedies be granted?
11. How does the doctrine of substantial performance affect the rights of a party injured by repudiation when the contract is not fully performed?
12. Under what circumstances would a contractor be entitled to partial payment if the work was not fully performed?

Mini-Case Problems

1. X engaged Y to repair his lawn mower. Y dismantled the machine. However, before Y had time to carry out the repairs, X informed him that he had purchased a new mower and did not require the repairs on the old machine. Discuss the rights and duties of the parties.
2. D agreed to make certain alterations to an expensive dress that E had purchased. A violent storm caused the roof of D's shop to leak, and the dress was stained by the rain entering the building. D offered to have the dress professionally cleaned to remove the water stains, but E refused and attempted to remove the stains herself. Her attempt was unsuccessful, and the dress was ruined. Advise D and E.
3. Eduard purchases an expensive coffee-maker from a store, advertised as an "end-of line," "all sales final" special. The coffee-maker leaks constantly, sending most of the coffee to the countertop. What kind of breach of contract does this represent?
4. James is contracted to build a cedar deck and fencing around Albert's newly installed pool. James completes the carpentry but fails to finish applying a coat of weather-protecting varnish as per the agreement. How would a dispute of payment owing to James be resolved by the court?

Case Problems for Discussion

CASE 1

Megamalls Inc. is known for its ability to create "retail experiences" for customers visiting any one of its six malls in major Canadian cities, with each mall being in the 80-120 store range. To further its reputation, Megamalls specially selects its qualifying tenant stores for a very particular mix of goods and services it feels will create the greatest customer draw. Moreover, it draws up tenant contracts to ensure that retail offerings are in accordance with its wishes. Recently one chain of stores with tenancies in each of the six Megamalls began changing its own retail image by altering its line of goods and marketing tactics. Megamalls feels that this is in contravention of the tenant contract.

What remedy or remedies should Megamalls seek and why?

CASE 2

Mrs. Field listed her home for sale with a local real-estate agent. The agent introduced Mr. Smith to Mrs. Field as a prospective purchaser. After Mr. Smith had inspected the house, the agent obtained a written offer to purchase from him. The offer provided that he would purchase the house for $260,000 if Mrs. Field could give him vacant possession of the premises on September 1st, some three weeks hence. The offer was accompanied by a deposit in the amount of $1,000.

Mrs. Field accepted the offer in writing, then proceeded to lease an apartment under a two-year lease. She moved her furniture to the new premises immediately and vacated her home in preparation for closing. A few days before the date fixed for delivery of the deed, Mrs. Field was informed by one of her new neighbours (who was a friend of Mr. Smith) that Mr. Smith's employer intended to transfer him to a new position in another city some distance away.

Discuss the rights and obligations (if any) of the parties in this case. Suggest a course of action that Mrs. Field might follow.

CASE 3

Trebic was a skilled cabinetmaker of European ancestry. Moldeva, who had emigrated to Canada from the same country, requested him to build a set of kitchen cupboards "in the old-country style." The two men discussed the general appearance desired, then Trebic drew up a list of materials that he required to construct the cupboards. Moldeva obtained the necessary lumber and supplies for Trebic, then took his family on a vacation.

On his return, Moldeva found the work completed, and admired the craftsmanship and design that Trebic had exhibited in the making of the cabinets. Trebic had carefully carved the "old-country designs" on the trim boards. He had skillfully constructed the drawers and cabinets using wooden dowels, rather than nails, again in accordance with "old-country" tradition. In the execution of this skill he had used only hand tools, and then only the tools used by "old-country" craftsmen in the cabinet-making trade. In every detail, the cabinets were "old-country style."

When Moldeva indicated that he was completely satisfied with the cabinets, Trebic submitted his account in the amount of $4,800. The sum represented 120 hours work at $40 per hour, the normal rate charged by skilled cabinetmakers in the area.

Moldeva, who was a building contractor himself, objected to the amount of Trebic's account. He stated that carpenters in his shop could manufacture kitchen cabinets of the general size and shape of those made by Trebic in only a few days' time. He offered Trebic $800 as payment in full.

Trebic refused to accept the $800 offer and brought an action against Moldeva on the $4,800 account.

Discuss the possible arguments of the parties. Render a decision.

CASE 4

Awwad, a skilled carpenter, agreed to construct a garage for Henderson for a contract price of $3,000. Henderson was to supply the plans, foundation, and materials.

Awwad constructed the garage according to the plans. When the building had been framed, he discovered that the siding boards that Henderson had purchased were of poor-grade lumber. The boards could only be made to fit with a great deal of hand labour and cutting.

Awwad complained to Henderson and demanded that he provide siding boards that were of "construction-grade" lumber. Henderson refused to do so. An argument followed in which Awwad refused to complete the work until Henderson provided suitable materials.

At the time of the argument, the foundation, the roof, and the walls had been erected. The work that remained included the installation of the wall siding, the doors and windows, and the trim.

Discuss the rights of the parties, and the nature of the claims and defences of each. Indicate the possible outcome, if the case should come before the courts.

CASE 5

The 18 Wheel Trucking Co. purchased a new truck from C.K. Motors Ltd., a firm that specialized in the sale of large tractor-trailer vehicles. In the six months following delivery, 18 Wheel Trucking found the new truck to be the subject of numerous breakdowns, both major and minor. The truck was out of service on 18 occasions, usually for minor problems such as headlight or brake light failures. On five occasions, the breakdown was serious and required major parts replacement, causing the vehicle to be out of service for several days. All of the repair work was done under warranty by the manufacturer's dealer, at no cost to 18 Wheel Trucking Co.

At the end of the six months, on the 19th breakdown, 18 Wheel Trucking Co. left the vehicle with the dealer and demanded that the purchase price be returned. C.K. Motors Ltd. refused, and 18 Wheel Trucking Co. decided to take legal action against C.K. Motors Ltd. for a return of the purchase price.

Indicate the nature of the claim of 18 Wheel Trucking Co. and the defences (if any) of C.K. Motors Ltd.

Render a decision.

CASE 6

Valentino entered into a contract with TV Production Co. to perform the leading role in a television play the company wished to produce. The contract called for the actor to devote his time exclusively to the play until taping was complete, a period of some four weeks. His compensation was to be $20,000. Three days after the contract was signed, Valentino notified the company that he did not intend to perform the role, and that the company should find a new leading actor for the production.

The company attempted to find a substitute for Valentino, but after an exhaustive search could find no one suitable. As a consequence, they were obliged to abandon their plans for the production. During the three-day period after signing Valentino, the company incurred liability of $36,500, under contracts they had entered into for services and commitments made in anticipation of his starring in the production. They also incurred the sum of $2,500 in expenses paid to find a substitute, when Valentino refused to perform.

The company instituted legal proceedings against Valentino to recover the total expenses incurred as a result of his repudiation. In response, Valentino offered a settlement of $2,500 to cover expenses incurred in their search for a substitute actor.

Discuss the arguments of the parties and render a decision.

CASE 7

Kenja Marshall wished to purchase an automobile that would not only provide her with basic transportation to and from her home to her place of employment, but would have a sporty appearance. She also wanted a vehicle with sufficient power to enable her to engage in rally racing.

She visited a local dealer in imported automobiles and enquired about a sports sedan displayed on the premises. The salesman on duty informed her that the vehicle was a used car that had been driven only 12,000 kilometres. It had been purchased the year before as a new car by a customer who then traded it in for a similar model of the current year's production. To the salesman's knowledge, it was a "good car." Since it had a turbo-charged engine, the policy of the company was to sell it "as is," without a warranty of any kind.

After a careful inspection of the car, Kenja asked the salesman to start the engine, in order for her to determine the condition of the turbo-charger. The salesman did so, and, after running the engine for a few moments at various engine speeds, he shut it down. He offered Kenja a test drive, but at that point she noticed a coolant leak at the engine water pump. The salesman examined the pump. He then stated that the water pump would be repaired or replaced if necessary, if she wished to buy the car. Kenja said she would think about the purchase overnight and contact the salesman the next day if she was still interested.

Kenja returned the next day. She informed the salesman that she was prepared to purchase the automobile if the dealer would repair the water pump and take $500 less than the advertised price of $24,000. With some reluctance, the dealer agreed to sell the car to her, and a written agreement of sale was prepared that contained the following terms:

> 9. It is expressly agreed that used goods are not warranted by the dealer as to year, make, model, or otherwise, unless so stated in writing.
>
> 10. The dealer agrees to make the following repairs to the vehicle as a part of this sale:
>
> (1) repair or replace water pump, as necessary.

Kenja was anxious to use the car in a local car rally the following day. She enquired if she might take the car immediately, then return it early the next week to have the water pump dealt with at that time. The dealer agreed, but cautioned her not to drive the car too hard until the pump was fixed. He also told her not to worry if she heard a slight "popping" noise from the pump.

Kenja paid for the car, then drove it home, a distance of some 16 kilometres. Along the way she heard what she described as a "clangy" or "tinny" noise from the engine. However, she was not concerned about it, believing it to be the noise the dealer had described to her.

The next morning, when she attempted to start the car, the engine made a number of "clangy" sounds, then stopped. A mechanic who came to her home in response to her call examined the engine and informed her that the noise came from the engine bearings. He indicated that the engine had been seriously damaged and could cost up to $2,000 to repair.

Kenja immediately informed the automobile dealer that she wished to have her money back. When the dealer refused, she brought an action for rescission of the agreement. At the trial, an expert testified that the problem was indeed a breakdown of the engine bearings, something that could occur in only a few minutes if insufficient oil was supplied to them. There was no evidence to indicate that the damage was in any way related to the water pump.

Discuss the nature of Kenja's claim, as well as the possible defences to it.

Render a decision.

www.mcgrawhill.ca/olc/willes

CASE 8

On September 1st, Rothwell entered into an agreement to purchase Andrea's Restaurant. The purchase price was $375,000, with a down payment of $75,000. The balance was payable December 1st, when Rothwell was to take possession of the business. In anticipation of his start in the restaurant business, Rothwell quit his job and enrolled in a three-month community-college course on restaurant management.

On November 1st, the owner of the restaurant notified Rothwell that she had received another offer to purchase the restaurant for $450,000, and she intended to sell the business to the offeror. Rothwell objected to the restaurant owner's actions and threatened to take legal action against her if she proceeded with the proposed sale.

A few days later, the restaurant owner did in fact enter into an agreement to sell the business to Polonek, the new purchaser, for the purchase price of $450,000. The closing date of the transaction was to be December 1st. She then mailed a cheque to Rothwell for the $75,000 she had received from him previously as his deposit.

Rothwell immediately returned the cheque and insisted that the restaurant owner proceed with the sale of the restaurant to him in accordance with their agreement.

On November 28th, the local newspaper contained an announcement of the opening of a new restaurant in a large office building across the street from Andrea's Restaurant. The office building housed most of the customers of Andrea's Restaurant, and the new restaurant could be expected to take about 70 percent of the lunch customers and 40 percent of the dinner customers from Andrea's Restaurant.

The announcement came as a surprise to all parties. Polonek immediately wrote a letter to the owner of Andrea's Restaurant, in which he indicated that he did not intend to proceed with the transaction unless the owner reduced the purchase price to $150,000. Rothwell was out of town on other business on November 28th, and he did not become aware of the new competitor until December 1st, the proposed closing date for his purchase of the restaurant.

Advise each of the parties of their legal position in this case. Assuming that each party exercised their rights at law, indicate how the issues raised in the case would be resolved by a court.

CASE 9

A commercial vegetable grower decided to grow a variety of open-pollinated cabbage as a market garden crop, based upon the success that his relative (who was also a commercial grower) had had with the variety several years before. He purchased seeds for the cabbage variety from the catalogue of a commercial seed supplier.

The seeds were planted according to proper planting instructions and cultivated in accordance with accepted agricultural practices. Weather conditions were "normal" throughout the growing season, but, in spite of this, the seeds produced a very poor crop.

The grower informed the seed supplier that the crop had failed, even though he had used proper growing techniques. He demanded that the seed company compensate him for his loss. The seed company rejected his complaint and noted the seed purchase contract term which stated:

> The vendor warrants seeds only as to variety named and makes no warranty express or implied as to quality or quantity of crop produced from the seed supplied. Any responsibility of the vendor is limited to the price paid for the seed by the purchaser.

When the seed company refused to entertain his complaint, the grower decided to take legal action to recover his loss.

Discuss the basis for the grower's action and the defence, if any, of the seed company. Render a decision.

CASE 10

Complex Software Corporation produced sophisticated software programs for computer-assisted product development. Complex Software was engaged by Turbine Engines Ltd. to develop software that would enable its engineers to develop the most efficient blade design and angles for its large turbines, which were used in hydro-electric power generators. Turbine Engines provided the engineering data necessary to develop the program, and Complex Software prepared the software.

The software was tested by both Complex Software and Turbine Engines using a simple turbine with known design and performance characteristics as a model. The software appeared to work properly, and Turbine Engines used the program to design a new multi-stage turbine engine.

Unknown to Turbine Engines, the striking of certain computer keys in a particular sequence had the effect of cancelling out the safety factor to be built into the turbine blades. The key sequence was not the sequence used in the test, but a technician used the particular key sequence in designing the new model turbine blades. As a result, when the new turbine was tested in a high-speed operational mode, the blades disintegrated and destroyed the engine.

www.mcgrawhill.ca/olc/willes

Turbine Engines brought a legal action against Complex Software claiming $850,000 in damages as its loss in the construction of the faulty prototype turbine.

Discuss the various arguments that may be raised by the parties in this case, and prepare a decision as if you were the judge. Outline your reasoning in reaching your decision.

CASE 11

The Canada Tea Co. entered into a contract to purchase 400 wooden chests of tea from an off-shore tea merchant. On receipt, it was discovered that some of the tea chests had insects in the wood, but not in the tea, which was enclosed in sealed plastic bags inside the chests. Canada Tea Co. warned the supplier that it would refuse acceptance of any future shipments that were infested with insects.

A second order for 400 wooden chests of tea was placed. On arrival, the tea chests were examined. Evidence that insects had been in the wood was revealed, but no insects were found in the shipment. Canada Tea Co. rejected the shipment. The off-shore tea merchant then took legal action against Canada Tea Co. for the value of the shipment.

Discuss the nature of defence (if any) that Canada Tea Co. might raise. Indicate the possible outcome of the case.

CASE 12

Hatfield owned a large farm on which he grew grain. His combine was inadequate in relation to the acreage of grain that he harvested annually. As a result, on several occasions his crops had been adversely affected by rain and poor weather conditions. He reasoned that a larger machine could reduce the time spent harvesting by as much as two-thirds and, thereby, reduce the chances of bad weather affecting his harvest.

At an agricultural exhibition, he examined a new self-propelled combine that was advertised as capable of harvesting grain at three times the speed of his old equipment. The machine was much larger and more powerful than his old combine and appeared to be of the correct size for his farm.

On his return home, he contacted the local dealer for the combine. After explaining his needs, he was assured by the dealer that the size he was considering would be capable of harvesting his crop in one-third of the time taken by his older model. He placed an order for the combine, with delivery to be made in early July, well before he would require the equipment.

The machine did not arrive until the beginning of the harvest, and Hatfield immediately put the machine into service. Unfortunately, the machine was out of adjustment, and Hatfield was obliged to call the dealer to put it in order. The equipment continued to break down each time Hatfield operated it at the recommended speed. In spite of numerous attempts by the dealer to correct the problem, the equipment could not be operated at anything more than a very slow speed without a breakdown. Hatfield found that despite the large size of the equipment, his harvest time was no faster. When the harvest was completed, he returned the machine and demanded his money back.

The equipment dealer refused to return his money. He pointed to a clause in the purchase agreement that Hatfield had signed, which read: "No warranty or condition, express or implied, shall apply to this agreement with respect of fitness for the use intended or as to performance, except those specifically stated herein."

The only reference in the agreement to the equipment stated that it was to be a "new model XVX self-propelled combine."

Advise Hatfield of his rights (if any).

CASE 13

One summer, Gianni visits an All-Terrain-Vehicle dealer, and, thinking about the autumn hunting season, signs a contract to purchase a $7,500 ATV, not putting the slightest effort to reading or understanding the terms of the deal. He puts down the requested 10% deposit of $750 and sets about to finding the balance, which is due a week prior to hunting season. As the date approaches, Gianni admits financial failure to the dealer, who tells him, "You had better keep trying to find the money. Your deposit will be forfeit as liquidated damages if you do not come up with the balance. Did you not read your copy of the contract?"

Advise Gianni, and explain the likely outcome of the situation. What facts would have to change to reverse your opinion of the outcome?

www.mcgrawhill.ca/olc/willes

CASE 14

Roland Exploration and Drilling Co. purchased mobile two-way radios for use by its crews at outlying natural gas wells. One day a small fire broke out in a storage shed near one such well, and the crew called for help on its radio. The radio failed internally, and made no transmission. As a result, what was a small fire burned continuously and grew until it enveloped the wellhead, while a member of the crew drove forty kilometres to the nearest telephone. Once the call got through to an aerial fire crew, it was only ten minutes before a firebomber dropped chemical smothering powder onto the well and extinguished the flames. Damage to the shed amounted to $2,000. Damage to the wellhead amounted to $1.5 million. The purchase agreement between Roland and the radio dealer stated: "Damages limited to cost of replacement radio. Not responsible for consequential damages of any kind, regardless whether caused by defect in materials or manufacturing." The same terms appeared on a card inside the packing box in which the radio was originally delivered.

Advise Roland.

The Online Learning Centre offers more ways to check what you have learned so far. Find quizzes, Weblinks, and many other resources at www.mcgrawhill.ca/olc/willes.

PART 4

THE LAW OF BUSINESS RELATIONSHIPS

CHAPTER 15

Law of Agency

Chapter Objectives

After study of this chapter, students should be able to:

- Understand the nature of the agency relationship.
- Know the duties of an agent.
- Explain the relationship between agents and third parties.
- Recognize the liability of all parties in tort.

YOUR BUSINESS AT RISK

An agent may be useful and even essential for certain jobs, but when an agent acts on your behalf, you become responsible for the actions of that agent. Aside from just good outcomes, an agent can bind you or your business to impossible promises, create huge financial liabilities, or tarnish your reputation. Proceed with caution in using agents!

THE ROLE OF AN AGENT

Principal

A person on whose behalf an agent acts.

Agent

A person appointed to act for another, usually in contract matters.

The law of agency is concerned with the relationship that arises when one individual (a **principal**) either expressly or impliedly uses the services of another (an **agent**) to carry out a specific task on his or her behalf. The relationship may arise through an express agreement, conduct, or necessity, but in every case the relationship involves three parties: the principal, the agent, and a third party. The purpose of the agency relationship is to enable the principal to accomplish some particular purpose, usually the formation of a contract with the third party. If the law has been properly complied with, the end result will be the accomplishment of the task without direct dealings between the principal and the third party.

Agents are generally engaged in business activities that involve the negotiation of contracts, but agents may also be used for many other purposes. For example, a lawyer may be engaged as an agent to perform certain legal services on behalf of a client, or a real-estate agent may be engaged to bring a buyer and seller together, but with no authority to bind the principal (the seller) in contract. Still, the most common use of an agent is to bind the principal in contract with a third party.

HISTORICAL DEVELOPMENT OF THE LAW OF AGENCY

The law of agency has its roots in the law relating to tort, contract, quasi-contract, and equity. Some evidence indicates that agency existed in the early medieval period. However, agents were not widely used during that period because it was seldom necessary for a person to engage another to perform a service in a community that was small and mostly self-sufficient. It was only with the growth of trade and the rise of the mercantile class that the use of agents became important and widespread.

During the late eighteenth century, and throughout the nineteenth century, the rapid growth of industry and commerce created the need for a Common Law response to the problems that rapid industrial change had brought about. The law of contract was a part of this response, and, with it, the law of agency was refined. In their respective spheres of influence, both the Common Law courts and equity moulded the concepts and ideas that now form the basis of modern agency law. The law of agency itself was distinguished from the law of employment by means of a subtle blend of maritime and mercantile law, tort, contract, and trust concepts that the courts modified and merged into the present-day body of law. These rules of law relating to agency may be divided into a number of distinct categories, where each refers to a particular aspect of the agency relationship.

THE NATURE OF THE RELATIONSHIP

By definition, an agent is a person who is employed to act on behalf of another. If the act of an agent is done within the scope of the agent's authority, the act will bind the principal.[1] A great many agents are independent business persons who may be employed by a number of principals at any one time to act for each in a variety of different business transactions. The employment of an agent should not be viewed merely as the ordinary relationship of employer–employee.

The general rules of contract normally apply to an agency relationship. An infant, for example, may be a principal, but any contract negotiated on the infant's behalf (except a contract for necessaries) would remain voidable at the infant's option, even if the agent was of full age. Since the agent is simply a conduit by which a contract may be effected between the principal and a third party, the capacity of the agent is, to some extent, unimportant. An agent must not, however, be insane. An agent may still be a minor and, despite this minority, negotiate a binding agreement on behalf of a principal who possesses full capacity to contract. The agency agreement between the principal and agent is subject, nevertheless, to the ordinary rules of contract. If one of the parties is an infant, it may be voidable at the option of the infant, even though the other party may be bound.

FIGURE 15-1 **Simple Agency Relationship**

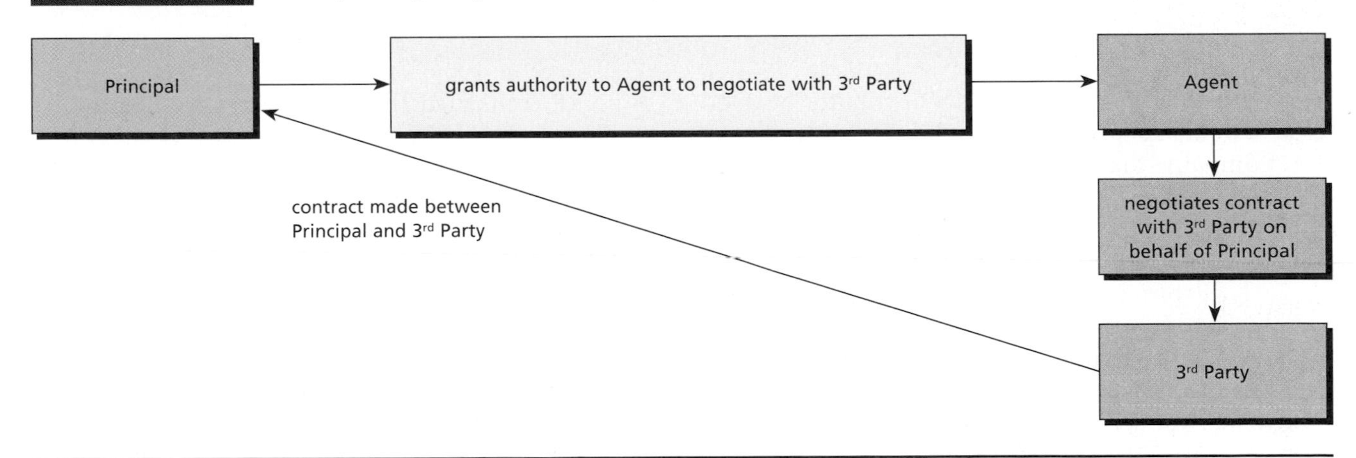

Agency by express agreement
An agency relationship established by an express oral or written agreement.

Agency by conduct
An agency relationship inferred from the actions of a principal.

Agency by Express Agreement

The principal–agent relationship may arise as a result of an express agreement, which may be either oral or in writing. The relationship may also be inferred from the actions of the principal, in which case it is sometimes called **agency by conduct** or agency by estoppel.

1. Osborn, P.G., *The Concise Law Dictionary*, 4th ed. (London: Sweet & Maxwell, 1954), p. 22.

A third type of agency that may arise in certain circumstances is referred to as an agency of necessity.

Agency that arises out of an express agreement is contractual in nature, and is subject to the ordinary rules of contract with respect to its formation and performance. While it may be oral and binding under most circumstances, it must comply with the *Statute of Frauds* if the agreement cannot be fully performed within a year. The agreement must also comply with any special requirements for formal contracts, if the agent, as a part of his or her duty, is expected to execute such a document on behalf of the principal. For example, suppose that Anderson engages Baker as her agent to sell a parcel of land for her and to execute a conveyance of the land to the purchaser. Baker must be empowered to do so by a grant of the power. The particular document under seal that would authorize the execution of the conveyance would be a power of attorney — a legal document that would appoint the agent as the principal's attorney for the purpose of conveying the land.

The advantage of a written agency agreement is that the terms and conditions of the agency, and in particular the duties of the principal and the agent, are set out in a document for future reference. Not every agency agreement requires this formality, since a verbal agreement is perfectly adequate in many cases. For example, if Martin asks Thomas to purchase an item for him from the local hardware store, an agency relationship would be created whereby Thomas would be authorized to act as Martin's agent in the purchase from the shopkeeper. A written agreement would not normally be required for this simple agency task.

The ordinary agency relationship extends beyond the agreement between the principal and the agent. It usually involves a second contract with a third party. The two contracts are separate, but in the case of the contract negotiated by the agent and the third party, if the agent acts within the scope of his or her authority, the rights and duties under the agreement become those of the principal and the third party. The agent, in effect, drops from the transaction once the agreement is executed, and has neither rights nor liabilities under it.

CASE IN POINT

NRF Distributors Inc. was a distributor of wood flooring products purchased from Korca, a company that purchased logs from Starwood Manufacturing Inc. and turned the wood into finished flooring products. NRF was aware that Korca had a business relationship with Starwood.

NRF purchased finished flooring for several years from Korca, but eventually the quality of product could not (or did not) meet NRF's specifications. Korca then suggested to NRF that NRF deal directly with Starwood, as it had the ability to produce flooring that would meet NRF's specifications. Korca at this time had on hand three purchase orders from NRF. Korca then sent the purchase orders that it had received from NRF to Starwood, and advised Starwood that in future NRF would be dealing directly with Starwood. Starwood notified NRF that it was in receipt of the purchase orders sent to it by Korca.

NRF confirmed the three orders for flooring with Starwood and paid Starwood $112,056.00 in advance of delivery. Starwood did not deliver the wood, but instead used the money to set off Korca's indebtedness to it.

NRF sued Starwood for non-delivery. In its defence, Starwood argued that Korca was NRF's agent, and it was entitled to set-off Korca's debt with the funds. The court held that Korca was not an agent of NRF at any time, as all transactions were on the basis of buyer and seller, and at no time did NRF ask Korca to negotiate a contract with Starwood on its behalf. No agency was involved. The court ordered Starwood to pay NRF for its breach of contract.

NRF Distributors Inc. v. Starwood Manufacturing Inc., 2009 CanLII 3786 (ONSC).

FRONT-PAGE LAW

The real Wayne's world

Player agent Michael Barnett and Wayne Gretzky, his most famous client, go as far back as a meeting in an Edmonton bar in 1978. Since then, they have forged a professional bond and an even stronger friendship.

In retrospect, it seems almost poetic that the most powerful partnership in National Hockey League history was formed over a cold beer.

But in the waning months of 1978, an underaged, undersized, overachieving Edmonton Oilers centre named Wayne Gretzky walked into a local sports bar owned in part by a man named Mike Barnett.

"They became friends then," said Vancouver-based columnist Jim Taylor. After a few years, Barnett became Gretzky's agent.

Barnett wheeled and dealed Gretzky into money-making ventures with Coca Cola, Toronto-based Zurich Insurance and Domino's Pizza, among others. One of the more notable deals was the one Barnett helped him ink with Easton in 1989 to use their brand of aluminum shaft hockey sticks. He signed a fee-plus-royalty agreement with the company, which, in turn, saw its sales jump to more than $40-million from $14-million in the first five years.

"I think Mike had a fair bit of imagination that got Wayne into some things he might not have thought of," said longtime *Edmonton Journal* hockey writer Jim Matheson.

Through the years, success had made them both fantastically more wealthy than they were in that bar that autumn day in 1978. "It is a relationship absolutely based on friendship," Barnett said. "With years of satisfactory business results compounding the relationship."

Even though Gretzky no longer needs contract re-negotiations, or hockey stick contracts, he and Barnett are still very close. "They are joined at the hip," Matheson said. "You have to go through Mike to get Wayne for just about anything right now."

The tale of Barnett and Gretzky is a success story. Where and what sort of problems could have emerged, and what protective steps would you recommend to others in a similar situation?

Source: Excerpted from Sean Fitz-Gerald, with files from Alan Adams, *National Post*, "The real Wayne's world," July 6, 2000.

Duties of the Parties

A point to note with respect to agency is the nature of the relationship between the principal and the agent. Unlike many contractual relationships, the parties in an agency relationship must act in good faith in their dealings with each other. Under agency law, the principal has a duty to pay the agent either the fixed fee or a reasonable fee for the services rendered. The principal must also indemnify the agent for any reasonable expenses that the agent properly incurs in carrying out the agency agreement. At Common Law, the agent is entitled to payment immediately on the completion of the service. However, it is customary for the parties to fix the time for payment at some later date, usually when the accounts have been settled.

The agent has a number of obligations toward the principal. First, the agent must obey all lawful instructions of the principal and keep confidential any information given to him by the principal. Secondly, the agent must keep in constant contact with the principal and inform the principal of any important developments that might occur as negotiations progress. This is a particularly important duty on the part of the agent, for the law holds that any notice to an agent is notice to the principal, and the principal is therefore deemed to know everything that is communicated to the agent by the third party.

MANAGEMENT ALERT: BEST PRACTICE

When business operators seeking premises approach agents who hold property listings, they often forget where the agent's duties lie. The agent is not neutral; his or her duty is to the seller or lessor. Mistakenly, buyers and lessees often disclose too much information, too early, about their finances and willingness to pay. This is a negotiation, and an agent should be told nothing more than what the buyer or lessee would be prepared to say directly to the seller or lessor at first meeting.

If an agent possesses special skills or the competence to perform the act required under the agency agreement, then the agent must maintain the standard required for that skill in the performance of his or her duties. The agent may be liable to the principal if the agent fails to maintain that standard, and if the failure results in a loss to the principal. The agent normally may not, except with the express permission of the principal, delegate agency duties to a subagent. The principal is entitled to rely on the special skills and judgment of the agent alone. While there are certain exceptions to this rule, generally a principal is entitled to personal service by the agent.

If an agent is authorized to receive funds or goods on behalf of the principal, the agent has a duty to account for the goods or the money. To fulfill this obligation, the agent must keep records of the money received and keep the funds separate from his or her own. The usual practice of agents in this regard is to place all funds in a trust account, or a special account identified as the principal's, and to remit money to the principal at regular intervals. If the agent is entitled to deduct an earned commission from the funds received under the terms of the agency agreement or by custom of the trade, then the deduction is usually made at the time that the balance of the account is remitted to the principal.

The agency relationship is a relationship of the utmost good faith. The agent is obliged to always place the principal's interest above his or her own. To fulfill this duty, the agent must bring to the principal's attention any information that the agent receives that might affect the principal. And, when the agent engages in any activity on behalf of the principal, the agent must act only in the best interest of the employer. For example, an agent must endeavour to obtain the best price possible for goods that he or she sells for the principal, or, if engaged to purchase goods, the agent must seek out the lowest price that can be found in the marketplace. In both cases, the agent must act in the best interests of the principal, and seek the most favourable price, rather than the quickest commission.

An agent has a duty only to the principal and not to any third party with whom the agent negotiates. The agent may not act for both parties without the express consent of the principal and the third party. If the agent should obtain a commission or benefit from the third party without disclosing that fact to the principal, the agent would not be entitled to claim a commission from the principal.[2]

Newsome contracts with agent Wilson to sell Newsome's business, and agrees to pay Wilson a commission if he is successful. Wilson, without Newsome's knowledge or permission, enters into an agreement with Bray to find Bray a business and to be paid a commission if he is successful. Wilson negotiates the sale of Newsome's business to Bray and collects a commission from both Newsome and Bray. In this case, there would be no enforceable commission contract, as Wilson would be liable to Newsome for the return of any commission paid for his actions. Similarly, if Newsome engaged Wilson to sell a quantity of goods for him, and Wilson sold the goods to Bray as if they were his own and at a higher price than that reported to Newsome, Newsome would be entitled to recover the secret profit made by Wilson in the transaction.[3]

COURT DECISION: Practical Application of the Law

Dishonest Agent—Entitlement to Commission

***Andrews v. Ramsay & Co.*, [1903] 2 K.B. 635.**

A dishonest agent colluded with a purchaser to acquire land that the agent was authorized to sell — at a price lower than what a purchaser might otherwise pay for it. When the collusion was discovered, the vendor (the principal) refused to pay a commission to the agent on the sale.

2. *S.E. Lyons Ltd. v. Arthur J. Lennox Contractors Ltd.*, [1956] O.W.N. 624; *Andrews v. Ramsay & Co.*, [1903] 2 K.B. 635; *McPherson v. Watt* (1877), 3 App. Cas. 254.
3. *Fullwood v. Hurley*, [1928] 1 K.B. 498; *Industries & General Mortgage Co. v. Lewis*, [1949] 2 All E.R. 573.

THE COURT: It seems to me that this case is only an instance of an agent who has acted improperly being unable to recover his commission from his principal. It is impossible to say what the result might have been if the agent in this case had acted honestly. It is clear that the purchaser was willing to give £20 more [ed. — *a substantial sum at the time*] than the price which the plaintiff received, and it may well be that he would have given more than that. It is impossible to gauge in any way what the plaintiff has lost by the improper conduct of the defendants. I think, therefore that the interest of the agents here was adverse to that of the principal. A principal is entitled to have an honest agent, and it is only the honest agent that is entitled to any commission. In my opinion, if an agent directly or indirectly colludes with the other side, and so acts in opposition to the interest of his principal, he is not entitled to any commission. That is, I think, supported both by authority and on principle; but if, as is suggested, there is no authority directly bearing on the question, I think that the sooner such an authority is made the better.

The courts generally take a very strong position against agents who place their own interests above those of their principal. Does this court decision stand the "test of time" for you?

A QUESTION OF ETHICS

An "appraiser and sales agent" is engaged by a business to sell off its surplus equipment. Asked "What is it worth?" the appraiser/agent says, "I know I can get $25,000 for it." An agency and commission agreement is then executed, and within two days, a buyer is found. A month of searching might have brought a $30,000 buyer for the equipment. In your opinion, is a duty of utmost good faith offended in this situation, or are you comfortable with the situation? Are there other considerations you would like to weigh before forming an opinion?

CLIENTS, SUPPLIERS *or* OPERATIONS *in* QUEBEC

A solicitor in Common Law Canada is a lawyer handling issues other than litigation, acting as an agent for his or her client. In Quebec, a solicitor's work is performed by a notary, who can be hired to give confidential advice, but the Quebec notary performs a far larger and more critical role as a public officer. When a Quebec notary prepares a contract, agreement or land title deed, he or she is guided by a requirement of impartiality and the mandate to advise all parties. In many instances documents cannot take on an "official" character until they are authenticated by a notary, and the notary's fireproof file serves as the public depository for these official documents.

Agency by Conduct or Estoppel

Agency may arise in ways other than by express agreement. A person may, by actions, convey the impression to another that he or she has conferred authority on a particular person to act as an agent in specific matters. In this case, an agency relationship may be created by conduct. If a person permits this state of affairs to occur, and the agent enters into a contract with a third party on the person's behalf, the person may not be permitted to later deny it. In this instance, the person may be said to have created an agency relationship by **estoppel**. The authority of the agent under these circumstances would not be real, but apparent. The binding effect of the agent's actions, however, would be real if the principal led the third party to believe that the agent had the authority to act on his or her behalf.[4]

Agency by estoppel

A representation by words or conduct that a person is an agent cannot be later denied if a third party relies on the representation.

4. *Reid and Keast v. McKenzie Co.* (1921), 61 D.L.R. 95; *Agnew v. Davis* (1911), 17 W.L.R. 570.

COURT DECISION: Practical Application of the Law

Apparent Authority of Agent—Liability of Principal

***Freeman & Lockyer v. Buckhurst Park Properties (Mangal) Ltd. et al.*, [1964] 2 Q.B. 480.**

K, a director of the defendant company, engaged the plaintiffs to do certain professional work on behalf of the company. The plaintiffs completed the work and submitted an account for their fees. When the company failed to pay, legal action was instituted to collect the account. The defendants argued that the director K was not a managing director (the firm did not have one), and therefore he did not have authority to bind the company.

THE COURT: It is necessary at the outset to distinguish between an "actual" authority of an agent on the one hand, and "apparent" or "ostensible" authority on the other. Actual authority and apparent authority are quite independent of one another...

An "actual" authority is a legal relationship between principal and agent created by a consensual agreement to which they alone are parties. Its scope is to be ascertained by applying ordinary principles of construction of contracts, including any proper implications from the express words used, the usages of the trade, or the course of business between the parties. To this agreement the contractor is a stranger; he may be totally ignorant of the existence of any authority on the part of the agent. Nevertheless, if the agent does enter into a contract pursuant to the "actual" authority, it does create contractual rights and liabilities between the principal and the contractor....

An "apparent" or "ostensible" authority, on the other hand, is a legal relationship between the principal and the contractor created by a representation, made by the principal to the contractor, intended to be and in fact acted upon by the contractor, that the agent has authority to enter on behalf of the principal into a contract of a kind within the scope of the "apparent" authority, so as to render the principal liable to perform any obligations imposed upon him by such contract. To the relationship so created the agent is a stranger. He need not be (although he generally is) aware of the existence of the representation but he must not purport to make the agreement as principal himself. The representation, when acted upon by the contractor by entering into a contract with the agent, operates as an estoppel, preventing the principal from asserting that he is not bound by the contract. It is irrelevant whether the agent had actual authority to enter into the contract.

The representation which creates "apparent" authority may take a variety of forms of which the commonest is representation by conduct, that is, by permitting the agent to act in some way in the conduct of the principal's business with other persons. By so doing, the principal represents to anyone who becomes aware that the agent is so acting, that the agent has authority to enter on behalf of the principal into contracts with other persons of the kind which an agent so acting into the conduct of his principal's business has usually "actual" authority to enter into.

What sort of behaviour by a potential agent and potential principal would create "apparent" authority in the "agent"? Are you in favour of reading "apparent" authority into situations, or would you take a narrow approach?

Agency by estoppel arises most often from a contractual relationship wherein the principal has adopted a contract negotiated by another. As a result, the principal has given the third party the impression that the contract was one of agency. For example, a person may engage another to make cash purchases of goods on his or her behalf on a number of occasions. The same person may be engaged at some later time to purchase goods on credit without the authority of the person for whom they were intended. If the principal adopts the contract by paying the account, the third-party seller would be led to believe that the agent had the authority to pledge the principal's credit. This would be inferred from the principal's conduct of settling the account. Unless the principal makes it clear to the seller that the agent does not have authority to pledge his or her credit, the principal may be estopped from denying the agent's authority to pledge credit in the future.[5] The failure to act would in effect "clothe the agent with apparent authority" to act on behalf of the principal.There are other instances where apparent authority may exist apart from the situation where the conduct of the principal creates the agency relationship. A spouse, for example, is presumed at law to have the authority to purchase goods and services for household use as agent of the spouse.[6] This particular presumption at law arises out of

5. ibid.

6. Under Common Law see: *Miss Gray Ltd. v. Earl Cathgart* (1922), 38 T.L.R. 56; as modified by provincial family law acts.

cohabitation as a rebuttable presumption only, rather than a right at law. The presumption may be rebutted by the spouse in several ways. He or she may establish one of the following: that notice was given to the shopkeepers that they were not to supply goods on credit, or that the goods were not necessaries of life. However, if a spouse had pledged the credit of the other previously, and if the accounts had been paid by the other spouse without giving notice to the shopkeepers that future purchases were not to be on credit, the other spouse may be bound by his or her conduct on future credit purchases.

Agency by conduct may result in liability for the principal if the principal fails to notify third parties that the agency relationship has terminated. Until such time as the third party becomes aware of the termination of the agency relationship, the third party is entitled to assume that the agency continues to exist and that the agent has authority to bind the principal. Again, the authority of the agent would only be apparent, because the termination of the agency relationship would have the effect of ending the agent's real authority. The third party may, nevertheless, hold the principal liable on a contract negotiated by the agent on the basis of the agent's **apparent authority**.[7]

Apparent authority

The ability of an agent to bind a principal where the principal has not notified third parties of the restricted or terminated authority of the agent.

An agent may also possess the implied authority to bind the principal in circumstances where the agency agreement expressly withholds real authority to do so from the agent. When an agent is engaged to perform a particular service, the agent customarily has the authority to engage in other forms of contract on behalf of the principal. If a principal engages an agent to perform a particular service, any restriction on the agent's authority must be brought to the attention of the third party, otherwise the principal may be bound if the agent should exceed his or her actual authority and negotiate a contract within what may be described as the agent's implied authority.

A retailer has the implied authority to sell goods placed in his or her possession. If a principal sends goods to a retailer on the express agreement that they must not be sold before a particular time, a sale by the retailer before that time would be binding on the principal. This is so because a retailer is normally clothed with the authority to sell goods in the retailer's possession without restriction as to time of the sale. The agent in this instance would be liable to the principal for breach of the agency agreement by selling the goods, but this would not affect the validity of the sale to the third party.

MANAGEMENT ALERT: BEST PRACTICE

Many local business firms act as dealers, authorized dealers or distributors of manufacturers' products. In most cases, these local firms are independent business entities with no authority to bind the manufacturer in contract, but for many products, they are authorized to perform warranty repairs under the manufacturer's warranty agreement with the purchasers of the product. To avoid holding out the local firm as an agent with authority to bind the manufacturer, most manufacturers will require the sales contracts of the local firm to clearly indicate to the public that it is an independent business with no authority to bind the manufacturer.

An example of a case where agents bound their principal on the basis of apparent authority arose in the case of *Reid & Keast v. McKenzie Co. Ltd.*[8] In this instance, a buyer's agents were authorized to enter into contracts on the principal's contract form, subject to a proviso that the agents inform the third parties that the contracts would not come into effect until the principal had approved samples of the product. An agent failed to so inform the third parties, and the outcome of the case turned on the agent's apparent authority. The judge described the case as follows:

7. *Trueman v. Loder* (1840), 11 Ad. & E. 589, 113 E.R. 539.
8. [1921] 3 W.W.R. 72.

> The defendants admit that they sent [the agent] out armed with the forms of the contracts actually entered into. They admit also that he had authority to get the plaintiffs to sign these contracts, and that he had authority to sign them on behalf of the defendants, subject to this: that he must stipulate that the contracts were not to come into effect until the company had approved of the samples. He represented to the plaintiffs that he had authority to enter into these contracts on behalf of the defendants. As evidence of that authority he produced the contract forms. He did not mention, as the trial Judge has found, the limitations on his authority which had been verbally given to him, but which did not appear on the forms.
>
> The arming of the agent with the defendants' contract forms and the sending him out to have these forms executed by the plaintiffs was, in my opinion, a clear holding out by the defendants that he had their authority to make the contracts. The defendants were, therefore, bound by the contract entered into by Thompson, as the plaintiffs had no notice of the limitations which had been placed upon his authority.

Agency by Operation of Law

Agency by operation of law

Agency that may arise in certain circumstances out of necessity where it is not possible to obtain the authority of the principal to act.

Agency may also arise by operation of law. In certain circumstances it may be necessary for a person to act as an agent (such as in an emergency), where it is impossible to obtain authority to perform the particular acts from the property owner. With modern-day communications, the circumstances under which an agency of this nature may arise are limited. Nevertheless, in an emergency, certain persons at Common Law may act as agents of necessity on behalf of others and bind them in contract. For example, a shipmaster at Common Law is presumed to have the power to bind the ship owner in contract in the event of an emergency if it is necessary to do so to preserve the ship and its cargo. However, this right would only arise if the shipmaster was unable to communicate with the owner, and if it was necessary to do so to secure the safety of the ship. The same right at law would permit the shipmaster to sell or otherwise deal with the cargo in an emergency, even though no express authorization would be given by the owner of the cargo to do so.

The Common Law limits the circumstances under which a person may act as an agent of necessity. The courts have recognized a number of instances where an agency of necessity may arise, such as in a true emergency. However, the relationship is generally limited to those cases where a pre-existing legal relationship exists between the principal and the agent of necessity. For example, in one case, a railway carrying perishable goods sold the goods to avoid total loss, because a labour strike prevented the railway from making delivery. The court in that instance ruled that the railway acted as an agent of necessity, since a true emergency existed, and the owner could not be reached for authority to act.[9]

However, the court will not normally find an agency of necessity where no pre-existing relationship can be shown. In an eighteenth-century case, a man found a dog and maintained it until its master came to retrieve it. The court would not hold that the man was an agent of necessity and entitled to compensation for the expense of caring for the animal, because no pre-existing legal relationship could be shown between the agent and the owner.[10] The rationale of the court in reaching this conclusion was that no person should be entitled to force an obligation upon another unless express or implied consent has been given.

Ratification of Contracts by a Principal

A principal may in certain circumstances wish to take advantage of a contract that an agent negotiated on his or her behalf, and that the agent clearly had no authority to make. The process of acceptance of a contract of this type is called ratification. If properly done, it has the effect of binding the principal in contract with the third party as of the date it was negotiated by the agent.

9 *Sims & Co. v. Midland Railway Co.*, [1913] 1 K.B. 103. See also *Hastings v. Village of Semans*, [1946] 3 W.W.R. 449; *Campbell v. Newbolt* (1914), 20 D.L.R. 897.

10. *Binstead v. Buck* (1776), 2 Black., W. 1117, 96 E.R. 660.

A principal may ratify a contract if the principal was in existence at the time the agent made the contract and was identified in the agreement as the principal. This particular rule generally applies to corporations rather than to individuals, since corporations have a particular time at which they come into existence. Unless the legislation under which a corporation is created permits the adoption of pre-incorporation contracts,[11] the corporation may not ratify a contract made before it was created. In addition, the subject matter of the contract must be something that the principal would have been capable of doing at the time the contract was made by the agent, and at the time of ratification. Again, this rule would have direct application to corporations, rather than to individuals. Corporations in some jurisdictions are limited to the objects for which they are incorporated. If the subject matter of the contract should be something that the corporation is not permitted to do, the subsequent change of the objects clause (to permit the corporation to undertake a particular activity) would still not permit the corporation to ratify the contract, unless statutory authority permitted the ratification.[12]

Ratification in every case must be made within a reasonable time after the agent enters into the contract and the principal becomes aware of its existence. Effective ratification must be of the whole agreement, and not simply of the favourable parts. The ratification need not be expressly stated, however. It may be implied from the conduct of the principal. The principal, for example, may accept benefits under the contract, or perform the promises made on the principal's behalf to signify ratification. The principal's silence would not normally constitute acceptance. However, in circumstances where the agent has exceeded his or her authority in negotiating the contract, a refusal to promptly repudiate the agreement on becoming aware of it may imply acceptance.

In each case, ratification of the contract by the principal dates the time of acceptance back to the date on which the contract was made. It does not run from the date of ratification. As a result of this rule, the actions of the third party may be altered by the subsequent ratification of the contract by the principal. The law on this particular point is unclear at the present time. In an English case, the court held that repudiation by the third party, before ratification, was ineffective if the principal subsequently ratified.[13] However, this decision has been questioned in Canada and reversed in the United States. The law in the United States presently seems to state that withdrawal from the agreement by the third party, or the institution of legal proceedings against the agent for breach of warranty of authority, would prevent later ratification of the agreement by the principal.[14] The position in Canada appears to fall somewhere between these two extremes. Some case law suggests that the principal may not ratify if the ratification would adversely affect parties other than the third party,[15] and others follow the English law.[16]

The English Common Law on this point has been described by the courts in the following manner:

> The rule as to ratification by a principal of acts done by an assumed agent is that the ratification is thrown back to the date of the act done, and that the agent is put in the same position as if he had had authority to do the act at the time the act was done by him. Various cases have been referred to as laying down this principle, but there is no case exactly like the present one .... The rule as to ratification is of course subject to some exceptions. An estate once vested cannot be divested, nor can an act lawful at the time of its performance be rendered unlawful, by the application of the doctrine of ratification.[17]

11. See, for example, the *Canada Business Corporations Act*, R.S.C. 1985, c. C-44, s. 14(2).
12. *McKay v. Tudhope Anderson Co., Ltd.* (1919), 44 D.L.R. 100.
13. *Bolton Partners v. Lambert* (1889), 41 Ch.D. 295; *Pickles and Mills v. Western Assurance Co.* (1902), 40 N.S.R. 327.
14. See, for example, *LaSalle National Bank v. Brodsky*, 51 Ill. App. (2d) 260 (1964). See also American Law Institute, *Restatement of the Law Second Agency*, 2d., American Law Institute Publishers, St. Paul, Minn., 1958, Vol. 1, p. 226, S 88.
15. *Goodison Thresher Co. v. Doyle* (1925), 57 O.L.R. 300; *Peterson v. Dominion Tobacco Co.* (1922), 52 O.L.R. 598.
16. *Farrell & Sons v. Poupore Lumber Co.*, [1935] 4 D.L.R. 783.
17. *Bolton Partners v. Lambert* (1889), 41 Ch.D. 295.

THIRD PARTIES AND THE AGENCY RELATIONSHIP

Disclosed Agency

In an ordinary agency relationship between principal, agent, and third party, the agent (if negotiating a contract within the scope of his or her authority) will bind the principal. The performance of the contract will be by the principal and the third party. The agent must clearly indicate to the third party that he or she is acting only as an agent, and will usually identify the principal for whom he or she acts. This is normally done by the agent signing the principal's name on the agreement and adding his or her own, together with words to indicate that the signature is that of the agent only. For example, an agent may sign as follows: "Mary Smith per Jane Doe" where Mary Smith is the principal, and Jane Doe the agent. The use of the term per is a short form of *per procurationem*, which means "on behalf of another," or, in agency law, "by his agent." It is also possible to specifically state in the agreement that a party is only acting as agent for another; that is, John Doe may sign the contract as follows: "John Doe as agent for John Smith."

Where the agent has revealed to the third party that he or she is acting as an agent only, and acts in accordance with the given authority, the principal alone is liable to the third party. The agent has no rights and duties under the contract with respect to the third party, nor may the agent claim any of the benefits that flow to the principal. If the principal does not wish to have his or her identity revealed, and instructs the agent to enter into an agreement without revealing the principal's identity, the agent may proceed in either of two ways. The agent may enter into the agreement in the agent's own name, without revealing that he or she is acting as an agent, or the agent may enter into the agreement as agent for an unnamed principal.

Undisclosed Agency

If the agent enters into an agreement without disclosing the fact that he or she is an agent, the third party may assume that the agent is acting as a principal. If the agreement is reduced to writing, and if the agent in the negotiations holds himself or herself out as a principal, describing himself and signing as principal, then the agreement from the third party's point of view will be one of direct contractual relations with a principal, rather than with an agent. The agent alone in this case would be liable.[18] Under these circumstances, the third party may look to the agent for damages if the contract is not performed, because the agent under the contract would be personally responsible for performance. The agent, by the same rule of law, would be entitled to enforce the agreement against the third party if the third party should fail to perform the agreement in accordance with its terms.

Fictitious Agency

A different liability would fall upon the agent if the agent contracts on behalf of a fictitious or non-existent principal, and the third party discovers the non-existence of the principal. The third party may sue the agent for breach of warranty of authority.[19] The same right would also be available to the third party if the agent entered into the agreement on behalf of a principal for whom the agent did not have authority to act.[20] In each of these cases, the agent would not be liable on the contract, but would be liable to the third party for damages arising from the agent's warranty that he or she had authority to act for the named principal.[21] In the first instance, where the principal was fictitious or non-existent, if the intention of the agent was to deceive the third party, the agent's actions would amount to fraud. An action for deceit would be available to the third party as well.

---

18. *Lawson v. Kenny* (1957), 9 D.L.R. (2d) 714; *Collins v. Associated Greyhound Racecourses, Ltd.*, [1930] 1 Ch. 1.
19. *Gardiner v. Martin and Blue Water Conference Inc.*, [1953] O.W.N. 881.
20. *Wickberg v. Shatsky* (1969), 4 D.L.R. (3d) 540.
21. *Austin v. Real Estate Exchange* (1912), 2 D.L.R. 324.

In the case of an agreement negotiated by an agent (where the agent describes himself or herself neither as a principal or agent), if the principal should decide to come forward and reveal his or her identity, the principal may enforce the contract. However, if the principal should do so, the third party may then bring an action against the principal instead of the agent if a breach should occur. The third party, however, is restricted in this regard. The third party may sue either the principal or the agent, but not both.[22] A particular exception to this rule does exist with respect to a contract under seal in which the agent signs as a contracting party. In this case, only the agent may enforce the agreement, and only the agent would be liable under it.[23]

Undisclosed Principal

If the agent enters into a contract with a third party in which the agent expressly describes himself or herself as an agent, but without disclosing the identity of the principal, the agent will not be personally liable. The fact that the agent describes himself or herself as an agent, and that the third party is willing to enter into a contract with the agent on that basis, protects the agent from personal liability. However, if the agent simply does not disclose that he or she is acting as an agent, and enters into a contract with the third party, the third party may elect to hold either the principal (if the third party discovers the principal's identity) or the agent liable. Of importance, in the case of an undisclosed principal, is the position of the principal. If the principal makes his or her existence known after the contract has been made by an agent who did not disclose the principal's existence, the principal may be in a position somewhat analogous to that of an assignee of a contract. The principal may take the agent's place, but, in doing so, he or she would also be obliged to accept the relationship as it stands between the agent and the third party. If the third party had contracted in the belief that the agent was in fact a principal, then any defence that the third party might have had against the agent may be raised against the principal.

Smith entered into a contract with Jones without disclosing that she was acting as an agent for Brown. If Brown should come forward after the contract is made and sue on it when Jones defaults, any defence that Jones might have against Smith could be raised against Brown. If Smith owed Jones a sum of money, Jones might legitimately claim the right to deduct Smith's debt with him from any sum owing under the contract.[24]

LIABILITY OF PRINCIPAL AND AGENT TO THIRD PARTIES IN TORT

The general rule in agency law is that a principal may be held liable for a tort committed by the principal's agent, if the tort is committed by the agent in the ordinary course of carrying out the agency agreement. A tort that an agent might commit is sometimes based upon fraudulent misrepresentation. This constitutes the tort of deceit. If a third party should be induced to enter into a contract by an agent as a result of a fraud on the part of the agent, then both principal and agent will be liable if the tort was committed in the ordinary course of the agent's employment. However, if the tort is committed outside the scope of the agent's employment, only the agent will be liable. Also, the principal will not be liable for damages for a failure to perform such an agreement, unless the principal adopts the contract or accepts benefits under it.

If a third party is induced to enter into a contract on the basis of a false statement that the agent innocently makes, the third party may repudiate the contract on the basis of innocent misrepresentation. Similarly, if the third party is in a position to prove that

22. *M & M Insulation Ltd. v. Brown* (1967), 60 W.W.R. 115; *Phillips v. Lawson*, [1913] 4 O.W.N. 1364.

23. *Pielstricker and Draper Dobie & Co. v. Gray*, [1947] O.W.N. 625.

24. *Campbellville Gravel Supply Ltd. v. Cook Paving Co.*, [1968] 2 O.R. 679.

the principal knew the statement was false, but allowed the agent to innocently convey it to the third party, the principal may be liable for fraud.[25]

TERMINATION OF THE PRINCIPAL–AGENT RELATIONSHIP

In many cases, agency relationships created by express agreement also provide for termination. If the agreement specifies that either party may terminate the agency relationship by giving the other party a particular period of notice and, if such notice is given, the agency will then terminate on the expiry of the notice period. If no specific time for termination is fixed in the agreement, the right to terminate may be implied, and either party may give notice to end the relationship. An agency may also terminate in other ways. Agency agreements may be made for the purpose of accomplishing a particular task. When the task is complete, the agency relationship will automatically terminate. For example, B Co. may own a quantity of wood and engage D Co. to sell it as its agent. Once the wood is sold, the task is complete, and the agency ends.

CASE IN POINT

A manufacturer of musical instruments entered into an exclusive distributorship agreement with the plaintiff for the Vancouver and Burnaby British Columbia area. The agreement could be terminated by either party on reasonable notice. Five years later, the manufacturer appointed another distributor for the area without notice to the plaintiff.

The court held that the manufacturer was in breach of the agreement for failing to give reasonable notice, and concluded that six months would be reasonable notice. Damages were awarded for the loss the plaintiff incurred during the six-month notice period.

See *Yamaha Canada Music Ltd. v. MacDonald and Oryall Ltd.* (1990), 46 B.C.L.R. (2d) 363.

The incapacity of the principal or the agent (by either death or insanity) has the effect of terminating the relationship, but the general legal incapacity of an agent or principal, if the agent or principal is a minor, may not. As with infants' contracts in general, in some jurisdictions the agency contract, in which one party is a minor, may be voidable only at the option of the infant or minor.

The bankruptcy of the principal terminates the agency agreement. The principal would not be bound by any agreement negotiated by the agent after the point at which the bankruptcy took place. The agent is expected to be in constant touch with the principal, and therefore aware of the principal's financial state. Consequently, a contract negotiated by an agent after the principal becomes bankrupt may render the agent liable to the third party for damages for breach of warranty of authority.

In all cases where the agency relationship is for more than a specific task, it is of the utmost importance that the principal inform all third parties that had dealings with the agent that the agency has been terminated. If the principal fails to notify the third parties of the termination of the agency, the agent may still bind the principal in contract on the basis of the agent's apparent authority.

For many years Joliet, as Alford's agent, purchased goods on Alford's credit from Clayton. Alford terminated the agency relationship, but did not notify Clayton. Joliet later purchased goods from Clayton on Alford's credit and sold them. Alford would be liable to Clayton for the payment, as Joliet had the apparent authority to purchase goods as Alford's agent in the absence of notice to the contrary.[26]

25. *Campbell Motors Ltd. v. Watson (Crescent Motors)*, [1949] 2 W.W.R. 478.

26. *Consolidated Motors Ltd. v. Wagner* (1967), 63 D.L.R. (2d) 266; *Watson v. Powell* (1921), 58 D.L.R. 615.

CHECKLIST FOR AGENCY

1. Does a non-fictitious apparent principal exist?
   - If yes, is there any issue of an incapable principal?
2. If capacity is not an issue, then is there at least one of:
   a) Express agreement with agent?
   b) Agency by conduct or estoppel?
   c) Agency by necessity or operation of law?
3. Has the fact of agency been disclosed to the third party contractor?
   - If yes, and the contract is within the scope of the agent's apparent authority, the principal will be bound.

SUMMARY

The law of agency is concerned with the relationship that exists when one person, either expressly or impliedly, uses the services of another to carry out a specific task on his or her behalf. This relationship may be created by express agreement, through conduct, or by necessity. If the agent has negotiated a contract on behalf of the principal within the scope of the agent's authority, only the principal and the third party will be bound by the contract. An infant may be an agent and still bind the principal if the principal is of full age, but for a contract negotiated on behalf of an infant, the infant principal would not be bound except for necessaries. If an agent should negotiate a contract on behalf of a principal without the authority to do so, it may be possible for the principal to ratify the agreement. By ratification, the contract would be effective from the date that it was entered into by the agent.

Ordinarily, if a contract is negotiated by an agent within the scope of the agent's authority, only the principal will be liable on the agreement and entitled to enforce the rights under it. However, if the agent negotiated a contract on behalf of a fictitious or non-existent principal, the agent would be liable to the third party on the basis of breach of warranty of authority. Similarly, if the agent negotiated an agreement on behalf of a principal and had no express or apparent authority to do so, the agent, and not the principal, would be liable. In cases where the agent did not disclose the identity or existence of the principal to the third party, depending upon the circumstances, the third party may elect to hold one or the other liable. If a fraud is involved in the contract negotiated, both the principal and agent may be liable, provided that the agent acted within the scope of the agency agreement.

Agency may be terminated on the death or insanity of either the principal or the agent. It may also be terminated in the case of bankruptcy of the principal. Agency may automatically end on the completion of the task for which it was formed or by notice, either as specified in the agreement, or at any time, if no notice period is required. When an agency agreement is terminated (other than by death, insanity, or bankruptcy), notice to third parties is important. Otherwise, the agent may still bind the principal in contract with an unsuspecting third party.

KEY TERMS

Principal, 266
Agent, 266
Agency by express agreement, 267
Agency by conduct, 267
Agency by estoppel, 271
Apparent authority, 273
Agency by operation of law, 274

www.mcgrawhill.ca/olc/willes

Review Questions

1. List the various ways in which an agency relationship may be terminated.
2. What types of agency relationships may arise or be formed?
3. If an agent exceeds his or her authority in negotiating a contract, under what circumstances would the agent alone be liable?
4. Why do the courts sometimes recognize the existence of agency relationships as based upon the conduct of parties?
5. How does the implied authority of an agent differ from express authority? Give examples of each.
6. Define an "agent of necessity." Explain how this agency might arise.
7. Describe briefly the duties of an agent to his or her principal.
8. Explain "agency by estoppel."
9. Under what circumstances would a principal be entitled to ratify a contract made by an agent?
10. Is a written document setting out the terms of an agency relationship always necessary? If not, why not?
11. When would a principal be liable for the acts of an agent who exceeded his or her authority?
12. What is the role of an agent?
13. Is a principal liable for the torts of his or her agent? Under what circumstances would the principal not be liable?

Mini-Case Problems

1. Alice engaged Kelly as her agent to negotiate a contract for the purchase of a large quantity of china for her shop. Before Kelly completed his negotiations, Alice became insolvent and was declared bankrupt. Kelly completed the transaction for the purchase of the supplies some days later. What are the supplier's rights?
2. The Acme Co. frequently sold goods through B Co., as its agent. The Acme Co. terminated the agency agreement, but failed to pick up goods at B Co.'s warehouse. B Co. later sold the goods to D Co. However, before D Co. took delivery, Acme Co. demanded return of the goods from B Co. Discuss.
3. Realty Limited employed Louis as a salesman. Martine listed a building lot with Realty Limited for sale at $178,000. Louis found a buyer, Jeanne, but Jeanne could not raise the full purchase price. Louis and Jeanne agreed that Louis would purchase the lot personally and to sell it to Jeanne once Jeanne had finally raised the necessary $178,000. Louis then purchased the lot from Martine for $170,000 and six months later resold it to Jeanne for $178,000. Discuss.
4. Chantal held herself out as a principal while negotiating the purchase of a large stock of cosmetics and perfume from Radiance, a retailer that had run into financial difficulty. In fact, Chantal was making the purchase on behalf of her employer, Salon. Radiance was unaware that Chantal was employed by its competitor Salon; however, Chantal gave the Salon address as the delivery address. Upon delivery, Salon changed its mind about buying the goods and refused to pay and rejected the goods. Advise Radiance.

Case Problems for Discussion

CASE 1

Custom Conveyor Ltd. was a manufacturer of conveyor systems for use in warehouses and shipping facilities. Most systems were custom designed to customer specifications, but used standard components or units manufactured by the company. Conveyor systems were sold locally by authorized dealers, who assisted the customer in the design of the system for the customer's needs.

Custom Conveyor Ltd. advertising brochures stated:

> "The design of the conveyor system is a team effort between us and our local dealer to ensure that the conveyor system will perform efficiently and trouble-free in your facility."

Max Warehouse Ltd. contacted Ace Conveyor Installations Ltd., a local dealer for Custom Conveyor Ltd. systems with a view to replacing its existing conveyor system with a new system. Ace Conveyor examined the warehouse facility, and designed a new conveyor system for the warehouse. Ace Conveyor advised the president of Max Warehouse Ltd. that the proposed system would be 50% more efficient than its existing system and would be trouble-free. The cost was $175,000, and included installation.

Max Warehouse Ltd. signed an Ace Conveyor Ltd. purchase order form for the conveyor system which contained a banner across the top of the order form that read: "We work for you to ensure your success." The conveyor system was eventually delivered and installed in the warehouse by Ace Conveyor Ltd., but failed to work in a satisfactory manner. Break-downs were frequent, and the system required constant servicing to operate as was intended. Discuss the nature of the relationships established by the parties. If Max Warehouse Ltd. should decide to take legal action, speculate as to the arguments of the parties, and the possible outcome of the case.

CASE 2

Alex was the agent for a number of professional dancers. As agent for the dancers, he would seek out work in local musical stage, television and movie productions. For his efforts, Alex received a percentage of the earnings that the dancers received for their performances. During a particularly slow season, a local theatre owner approached Alex, and suggested that they organize a special 'theatre night' to promote local talent in the municipality. Alex and the theatre owner agreed that admission would be charged to the public, but the performers would donate their services, since it would be a mixed amateur/professional event to promote local theatre and local talent. Alex and the theatre owner would share the proceeds from the ticket sales as payment for their efforts in the promotion of the event. Alex arranged for his clients to perform at the event, but did not tell them that he would be sharing in the ticket sales. Alex and the theatre owner worked very hard to promote the special theatre night, and it was a success. Alex and the theatre owner shared ticket sales of $20,000. Some time later, one of Alex's dancers discovered that Alex had shared in the proceeds from the special theatre night. Advise the dancers.

CASE 3

Amin was interested in the purchase of the shares of Holt Manufacturing Ltd., which was in financial difficulties due to a high debt load. He contacted Jones, a business consultant, to provide him with an assessment of the firm. Jones was to negotiate the purchase on Amin's behalf if his investigation indicated that the purchase of the shares represented a good investment.

Jones suggested that Brown, a consulting engineer, be engaged to assess the condition and value of the manufacturing equipment. Brown was also to provide some advice on what might be done to improve the profitability of the operation. Amin agreed, so Jones and Brown proceeded with their assessment of the firm.

During their examination, Jones and Brown realized that the firm represented a good investment if the equity-to-debt ratio could be altered and some manufacturing processes changed to improve efficiency. The two then established a corporation. They indicated to the present owners of the manufacturing firm (whom they had met through Amin) that they also represented a corporation that might be interested in the purchase if Amin should decide against the investment.

Jones and Brown completed their assessment of the business that in their written opinion to Amin was worth approximately $1.1 million. They submitted accounts of $3,000 and $3,500 respectively, which Amin promptly paid.

A few days later, Amin presented the owners of the manufacturing firm with an offer to purchase the shares for $1 million. The offer was promptly rejected. Before Amin could submit a new offer, the corporation that Jones and Brown had incorporated made an offer of $1.1 million for the business. The second offer was accepted, and the shares transferred to the corporation for the $1.1 million.

When Amin discovered that Jones and Brown were the principal shareholders of the corporation that had made the $1.1 million offer, he brought an action against them for damages.

Describe the nature of Amin's action. Discuss the possible arguments that might be raised by both the plaintiff and the defendants. Identify the main issues and render a decision.

CASE 4

Chelsea Pets Ltd. engaged Kent to act as its agent in the sale of 3,000 live rabbits. Without disclosing the fact that he was only an agent, Kent offered the rabbits to Somerset Pet Shops Ltd. at $2 each. Somerset agreed to purchase the lot, but at the conclusion of the discussion, reminded Kent that he owed them $1,500. Kent responded, "When this sale is completed you will get your $1,500."

Kent advised Chelsea Pets Ltd. of the sale. Chelsea Pets Ltd. delivered the rabbits to Somerset's Pet Shops Ltd. and informed them that it was delivering the rabbits in accordance with the sale that Kent had negotiated on its behalf.

Somerset Pet Shops Ltd. took delivery and mailed Chelsea Pets Ltd. its cheque in the amount of $4,500. Chelsea Pets Ltd. demanded the full $6,000. Chelsea Pets Ltd. threatened to sue Somerset Pet Shops Ltd. if it did not pay the purchase price in full.

Advise Chelsea Pets Ltd. and Somerset Pet Shops Ltd. in this case. Offer a possible outcome if Chelsea Pets Ltd. should carry out its intention to institute legal proceedings against Somerset Pet Shops Ltd.

CASE 5

Johnson used Birkett as his stockbroker for most of his investments. Occasionally, when Johnson had spare funds, he would seek the advice of Birkett as to investments he should make. On one such occasion, Birkett recommended two companies as investments with potential for profit. At the time,

Birkett indicated that he personally preferred stock B over stock A and intended to purchase some shares on his own account. Johnson ignored his advice, however, and purchased stock A.

During the next few weeks, stock A dropped in value as a result of unexpected political upheaval in an unstable country where the company had extensive holdings. Stock B, on the other hand, gradually increased in value during the same period. It had reached a price approximately 20 percent above what its value had been when Birkett recommended it as a possible investment.

Johnson discussed his investment with Birkett, and Birkett suggested he sell stock A. Johnson did so and requested Birkett to invest the proceeds in stock B. Birkett cautioned Johnson that the stock had already climbed in price and might not be as attractive an investment as it was some weeks earlier. Johnson, nevertheless, insisted that he buy the shares. Birkett then transferred some of his own shares to Johnson at the current market price, without disclosing the fact to Johnson.

The shares almost immediately declined in value for reasons unknown to both Johnson and Birkett.

A month later, Johnson discovered that the shares that he had acquired had been transferred to him by Birkett. He immediately brought an action against Birkett for the amount of his loss.

Explain the nature of the claim that Johnson might make against Birkett. Indicate how the case might be decided.

CASE 6

John Rowe had been a successful businessman during his lifetime. When he died, he left his business to his son and daughter and, under the terms of his will, left his widow, Florence, a life annuity that paid her $40,000 per year. Florence was quite elderly at the time of her husband's death. Her son and daughter concluded that the $40,000 annuity might not be sufficient for her to maintain her home and cover her living expenses, since she required a housekeeper to look after the premises. To provide her with additional income, the two children placed $300,000 in the hands of the family stockbroker in their mother's name.

The funds were placed with Margaret Lawson, an investment advisor with the brokerage firm. She was instructed to invest the money in the shares of Canadian corporations only, in order to provide Florence with income and dividend tax credits. No part of the funds was to be placed in bonds or the securities of foreign corporations.

Florence had no investment experience, and so advised Margaret. She informed Margaret that she intended to leave the choice of investment with her, as she did not wish to have "the worry of making investment decisions."

During the next two years, Margaret invested the funds in the shares of Canadian corporations. The investment income was approximately $40,000 per year. Florence was pleased with the results and, at the end of the second year, wrote a note to Margaret that read: "Many thanks for your hard work and success to date. Invest as you see fit, until further notice from me."

During the third year, Margaret switched most of the Canadian shares to bonds, foreign currency holdings, and speculative issues, at a very high investment turnover rate. The high trading activity resulted in very high sales commissions for Margaret, but very little in earnings for Florence. By the end of the third year, Florence's income was down to $3,000, and the net worth of her investment fund had diminished to $180,000.

At the end of the third year, Florence notified her son that "something seemed to be wrong" with her investment income. Her son immediately contacted the investment firm. At that point, Margaret's trading practices were discovered, and the value of the investment fund was determined. On the advice of her son, Florence brought an action against the stockbroker and Margaret, an employee of the firm.

Discuss the nature of the action that Florence might bring, and the issues involved. Render a decision.

CASE 7

The Rockland Corporation carried on business as a representative for investor groups that wished to acquire active business firms. Peter Rockland, the President of Rockland Corporation, met with a group of business people who had recently established a small brewery, and who wished to purchase some of the assets of another brewery business, the Best Brewery Ltd.

At the direction of the investor group, Peter Rockland met with the president of the Best Brewery Ltd. and negotiated a purchase price of $4.5 million for the assets. Peter Rockland then prepared a "letter of intent," which set out the purchase price of the assets and the terms of payment. The letter called for the payment of a deposit of $450,000 within 48 hours of the signed acceptance of the offer by Best Brewery Ltd. The "letter of intent" was signed: "Rockland Corporation on behalf of an investor group. per Peter Rockland."

Peter Rockland delivered the letter to the president of Best Brewery Ltd. On receipt of the letter, the president inquired as to the identity of the group of investors, and Peter Rockland arranged for Graham, a member of the investor group, to telephone the president and assure him that Rockland Corporation had authority to sign the letter on the investor group's behalf. Graham telephoned the president of Best Brewery Ltd. to assure him, but did not reveal the fact that he was a principal in the group that had established the competitor brewery. The president then wrote to Peter Rockland advising him that the company would accept the offer.

The investor group was unable to arrange financing and decided to abandon the purchase. When the investor group did not pay the deposit within the 48-hour period, Best Brewery Ltd. delivered a written notice of termination of the agreement to Rockland Corporation. Peter Rockland felt, however, that he could assemble another group of investors to acquire the assets, so he decided to enforce the agreement. Rockland Corporation then instituted legal proceedings against Best Brewery Ltd., claiming for breach of the agreement and damages.

Explain the basis upon which Rockland Corporation might bring the claim; indicate the defences, if any, that the Best Brewery Ltd. might raise. Render a decision.

CASE 8

The Clear Cut Lumber Company engaged the services of Timber Harvesters Ltd. to selectively cut certain hardwood species of trees on an extensive acreage that it owned. The contract provided that Timber Harvesters Ltd. was free to cut any access roads required, and it was required to remove and stack the cut timber at certain collection points in the area. Timber Harvesters Ltd. was provided with survey maps of the lands owned by Clear Cut Lumber Company, and the boundaries of the property were shown to Timber Harvesters Ltd. supervisors.

Timber Harvesters Ltd. began cutting logging roads immediately thereafter, in order to gain access to the particular stands of hardwood required by Clear Cut Lumber. In doing so, Timber Harvesters Ltd. inadvertently entered into neighbouring lands with its road, and cut a stand of trees that belonged to the property owner.

The property owner immediately took legal action against Clear Cut Lumber Company for trespass and the cutting of the trees. Clear Cut Lumber denied liability on the basis that it had informed the contractor of the boundaries of the property, and it was not responsible for the actions of the contractor.

Discuss the arguments that may be raised in this case, and discuss the possible outcome if the court is required to determine the issue.

CASE 9

A property owner listed her property for sale. She provided the agent with authority to sell the property on her behalf if the terms of any offer received met the terms set out in the listing agreement. A prospective buyer inspected the property during the period of time that the property was listed for sale, but did not make an offer to purchase.

After the agency agreement had expired, the prospective buyer made an offer to purchase the property that corresponded with the terms of the listing agreement. The agent accepted the offer on behalf of the property owner.

When the buyer discovered that the agency agreement had expired, he brought an action against the agent.

Explain the nature of the buyer's action and indicate how the case may be decided. Could the property owner ratify the agreement? What factors would affect the ratification?

CASE 10

The Acme Company, which frequently acted as agents for sea food processors and buyers, was contacted by the Gourmet Fish Company to find a supply of a particular fish for its new product line. Under the terms of the agreement, the Acme Company was entitled to a flat commission rate based upon the quantity of fish purchased. The Acme Company contacted several fish-processing plants and arranged for each to supply a quantity of the fish species required by the Gourmet Fish Company. In each case, the Acme Company charged the processing plant a $3,000 fee for arranging the supply contract. In due course, the fish were delivered to Gourmet Fish Company, and the agreed-upon commission was paid to the Acme Company, based upon the quantity of fish supplied.

Some time later, when the Gourmet Fish Company discovered that the Acme Company also charged a fee to the processing plant for arranging the supply contracts, it took legal action against the Acme Company.

Discuss the nature of the claim that would be made by the Gourmet Fish Company and the defences, if any, of the Acme Company. Render a decision.

The Online Learning Centre offers more ways to check what you have learned so far. Find quizzes, Weblinks, and many other resources at www.mcgrawhill.ca/olc/willes.

CHAPTER 16

Law of Sole Proprietorship and Partnership

Chapter Objectives

After study of this chapter, students should be able to:

- Recognize a sole proprietorship.
- Explain the nature of a partnership.
- Understand the rights and duties of partners to one another.
- Know how a partnership is dissolved.
- Recognize special types of partnerships such as LLPs.
- Understand the importance of registration of partnerships.

YOUR BUSINESS AT RISK

Sole proprietors go it alone with absolute control but limited access to resources. Two partners have double the personal assets and skills, but the human factor of sharing responsibilities and rewards can introduce a minefield of potential problems. Be careful in choosing the partner you select in your commercial marriage!

FORMS OF BUSINESS ORGANIZATION

A business organization may be established to carry out a particular project or venture, produce a certain product or product line, or provide a service to the public or a part thereof. In the classic view, a business begins when an entrepreneur identifies a particular need in a community for a product or service and proceeds to satisfy the need. The business is often established in a small way in the beginning. Then it grows, if the entrepreneur has accurately assessed the market, until the business, through investment and the development of new or better products or services, becomes a large entity. In doing so, the business entity may begin as a sole proprietorship, expand to include others and become a partnership, then continue to develop and take on the form of a corporation. This developmental sequence of a business organization, needless to say, is not the only way that large firms may develop. Large organizations are often created to carry on business activities that have been established by larger organizations. Not infrequently, businesses that require a large amount of capital for their establishment generally begin their existence as large organizations. In these cases, the organizations do not pass through different forms of organization. Nevertheless, the classic type of business growth pattern remains as one of the most common patterns of development. It illustrates the number of forms of organization that a business may take during the period of its progression from a small entrepreneurial venture to a large corporate organization. The purpose of this part of the text is to examine some of the more general laws that affect the nature of the firm, as it passes through the various stages of development.

Sole Proprietorship

Sole proprietorship

A business where the sole owner is responsible for the management and the debts of the business.

The **sole proprietorship** represents the simplest form of business organization as far as the law is concerned. The sole proprietor, as the owner of the business, owns all of the assets, is entitled to all of the profits, and is responsible for all of the debts. The sole proprietor also makes all of the decisions in the operation of the business and is directly responsible for its success or failure. The major disadvantages of the sole proprietorship are usually the limited ability to raise capital and the management-skill limitations of the proprietor, who must assume responsibility for all aspects of the business.

One of the important requirements for a sole proprietorship is the registration or licensing of the business, which can be completed online in a few minutes in most jurisdictions. Persons who offer to the public services of a professional nature must generally be licensed to practise the particular skill in a province before they may carry on a professional practice. The legislation governing professions such as medicine, dentistry, law, architecture, and others is usually provincial. It must be complied with before the practice may be established. Many semi-professional and skilled trade activities are also subject to provincial licensing or registration. Municipalities often impose their own registration or licensing of certain businesses in an effort to protect or control the activity. Skilled trades, the operation of taxis, and many other service-oriented businesses frequently require municipal licences to operate within the confines of the municipality.

In spite of the organizational disadvantages of the sole proprietorship, the freedom that it allows the individual in the operation of the business frequently makes it the most attractive form of business organization for a new, small enterprise. The flexibility of the operation during the formative period is often as important as the speed by which decisions may be made. However, as the business becomes larger, the need for new and varied management skills may require a change in the form of the organization.

Partnerships

From a progressive point of view, the second phase that a small business may enter is that of the partnership. A partnership differs in a number of ways from a sole proprietorship, but it, too, is essentially a simple form of business organization. The increase in the number of proprietors to two or more provides additional management expertise in the operation of the firm, and usually additional capital as well. Lending institutions, if they are in a position to look to the assets of several proprietors, are generally more willing to extend credit in larger amounts to a partnership than to a sole proprietorship. In addition, partners usually possess a greater interest in the success of the business than do employees. An overall advantage of the partnership is that the introduction of new proprietors may substantially increase the chances for growth and development of the enterprise. The partnership form is not without risk, however, as the partnership, and each partner individually, is responsible for the carelessness or poor judgment of each other partner in his or her conduct of partnership business.

HISTORICAL DEVELOPMENT OF PARTNERSHIP

Partnership

A legal relationship between two or more persons for the purpose of carrying on a business with a view to profit.

A **partnership** is a relationship that subsists between two or more persons carrying on business in common with a view to profit.[1] This definition represents a narrower delineation of the relationship than most people might realize, as it excludes all associations and organizations that are not carried on for profit. Social clubs, charitable organizations, and amateur sports groups are not partnerships within the meaning of the law, and many business relationships are excluded as well. For example, the simple debtor–creditor relationship and the ordinary joint ownership of property do not fall within the definition of a partnership.

1. *The Partnerships Act*, R.S.O. 1990, c. P.5, s. 2. All Common Law provinces have similar legislation relating to partnership. Quebec partnership law is found in the *Civil Code*.

The partnership is an ancient form of organization, and undoubtedly one that existed long before the advent of the written word and recorded history. The first associations had their roots in early trading ventures, when merchants banded together for mutual protection from thieves and vandals. The natural extension of these loose associations was for the merchants to join together in the trading itself and to share the fortunes of the ventures. The partnership form of organization has been used by merchants for thousands of years. References to early partnerships are found both in biblical writings[2] and in early Roman history. These sources indicate that partnerships were frequently used for many commercial activities.

By the late eighteenth century, the courts began to examine the partnership and its activities in the light of contract law. The expansion of trade, both domestic and foreign, during this period and throughout the nineteenth century increased the use of partnerships as a form of business organization. The inevitable litigation associated with such a large number of entities produced a large body of law concerning the relationship. By the latter part of the nineteenth century, the law had reached such a stage of maturity that codification was both possible and desirable. In 1879, Sir Frederick Pollock was asked to prepare a draft bill for the English Parliament that would codify the law of partnership. The bill was duly prepared and presented, but did not come before Parliament until ten years later. The Act was carefully written to reflect the Common Law at the time. As a part of legislation, it retained the rules of equity and Common Law relating to partnerships, except where the rules were inconsistent with the express provision of the Act.[3]

Since the *Partnership Act* was passed in England, it has been adopted with some minor modifications in all of the Common Law provinces of Canada and in many other countries that were part of the British Commonwealth. Partnership legislation permits the formation of partnerships for all commercial enterprises, except for a number of activities such as banking and insurance, where the corporate form (and special legislation) is necessary.[4] By definition, a partnership must consist of at least two persons.[5] While some provinces restrict the number of partners to a specific maximum, no limit is imposed in most.

A final development is the "limited liability partnership" or LLP. It is a special case, a hybrid of a corporation and partnership. The particular rules related to LLPs are discussed in a separate section in this chapter.

NATURE OF A PARTNERSHIP

The essential characteristic of the partnership is that it is a relationship that subsists between persons carrying on business in common with a view to profit. The partnership must be distinguished, however, from other relationships, such as joint or part ownership of property, profit-sharing schemes, the loan of money, and the sharing of gross receipts from a venture. The foregoing associations in themselves do not constitute partnerships. The existence of a partnership turns very much on the particular agreement made between the parties, but, as a general rule, the sharing of the net profits of a business is *prima facie* evidence of the existence of a partnership.[6] In contrast, the remuneration of a servant or agent by a share of the profits of the business would normally not give rise to a partnership,[7] nor would the receipt of a share of profits by the widow, widower, or child of a deceased partner.[8]

2. Ecclesiastes 4:9–10 (about 977 B.C.).
3. *Partnership Act*, 1890, 53 Vict., c. 39.
4. See, for example, *The Partnerships Act*, R.S.O. 1990, c. P.5.
5. ibid., s. 2 and s. 45. Reference in this chapter will be made to specific sections of the Ontario Act, but most other provincial statutes are similar.
6. ibid., s. 3.
7. ibid., s. 3(a).
8. ibid., s. 3(b).

There must usually be something more than simply shared profits before a partnership agreement exists. If the parties have each contributed capital and have actively participated in the management of the business, then these actions would be indicative of the existence of a partnership. Even then, if the "business" simply represents the ownership of a block of leased land, the relationship may not be a partnership. It may, instead, be co-ownership of land; that is, a relationship that closely resembles a partnership but that is treated in a different manner at law. Co-ownership, when examined carefully and compared to a partnership, is a distinct and separate relationship. The principal differences between the two may be outlined as follows:

1. A partnership is a contractual relationship. Co-ownership may arise in several ways — for example, through succession, or through the inheritance of property from a deceased co-owner.
2. A partnership is a relationship that is founded on mutual trust. It is a personal relationship that is not freely alienable. Co-ownership, on the other hand, may be freely alienable without the consent of the other co-owner. For example, Alexander and Bradley may jointly own a block of land. Alexander may sell his interest to Calvin without Bradley's consent, and Bradley and Calvin will then become co-owners. This is not possible in the case of partnership.
3. A partner is generally an agent of all other partners in the conduct of partnership business. A co-owner is normally not an agent of other co-owners.
4. A partner's share in partnership property is never real property. It is always personalty, as it is a share in the assets, which can only be determined (unless there is an agreement to the contrary) by the liquidation of all partnership property. In contrast, co-ownership may be of either personalty or realty. Two persons as co-owners, for example, may purchase a fleet of delivery vehicles or a parcel of land. In each instance they would be co-owners of the property, but they would not be partners in the absence of an agreement to the contrary.
5. A partnership is subject to the *Partnerships Act* in its operation, and dissolution is by the Act. Co-ownership may be dissolved or terminated under legislation that provides for the division or disposition of property held jointly.[9]

A partnership may, in a sense, be established from the point of view of, and for purposes of, third parties by estoppel. If a person holds himself or herself out as being a partner, either by words or conduct, or permits himself or herself to be held out as a partner of the firm, that person may be liable as a partner if the third party, on the strength of the representation, advances credit to the firm. This would apply even if the representation or holding out is not made directly to the person advancing the credit.[10]

> Oxford, an employee of a partnership, holds herself out as a partner, and Baker, a merchant, sells goods on credit to the partnership on the faith of Oxford's representation. In this case, Oxford will be liable to Baker as if she were a partner. The same would hold true if the partnership held out Oxford as a partner and she permitted them to do so, even though she was only an employee. By allowing the firm to hold her out as a partner, she would become liable, as if she were a partner, for any debts if the creditor advanced money in the belief that she was a partner.

In the case of *Lampert Plumbing (Danforth) Ltd. v. Agathos et al.*,[11] Kreizman, the president of Lampert, dealt with Agathos, who represented himself as an owner of a business when in fact he was not an owner, but merely a friend of the owner. Several contracts were negotiated and performed but, eventually, default occurred on a contract. Lampert

9. ibid., s. 3(c).

10. For example, in Ontario under the *Partition Act*, R.S.O. 1990, c. P.4.

11. [1972] 3 O.R. 11; and see the *Partnerships Act*, supra, s. 15(1).

sued Agathos as one of the owners of the business. The court reviewed partnership law as it relates to "holding out" as a partner in the following terms:

> It is clear from s. 15 of the *Partnerships Act* and from the authorities that the person holding himself out as a partner "is liable as a partner" to one who has relied in good faith upon the representations made. Since in my view Agathos did hold himself out as the sole proprietor of the defendant construction company to Kreizman he incurs liability to the plaintiff company as if he were in fact a partner of the construction company or its sole proprietor.

All persons with capacity to enter into contracts have the capacity to become partners. A minor may become a partner, but if he or she should do so, the ordinary rules of contract apply. The contract is voidable at the minor's option. Since it is deemed to be a continuing relationship, the minor must repudiate the agreement either before or shortly after attaining the age of majority. If the minor fails to repudiate, then the minor would be bound by the agreement. The minor would also be liable for the debts of the partnership incurred after he or she reached the age of majority. If repudiation takes place before the minor reaches the age of majority, then the minor will not normally be liable for the debts of the partnership unless he or she committed a fraud. Once repudiation takes place, the minor is not responsible for the liabilities of the partnership. However, the minor is not entitled to a share in the profits until the liabilities have been paid.

CASE IN POINT

Two fishers owned their own fishing boats, and fished for herring. One of the fishers learned of a halibut licence that was available, but his boat was too large for the licence. He informed his friend, who owned a smaller boat, and his friend obtained the licence. The two men shared the cost of the licence on the understanding that the owner of the larger boat could have his money back at any time if he decided not to crew on his friend's boat. They also agreed to equally share in the expenses and profits from fish caught under the licence when they fished together. Some years later, the owner of the larger boat obtained his own licence for halibut, but also claimed to be an equal partner with his friend under his licence.

The court held that no partnership existed, since the owner of the larger boat did not share expenses or profits on those occasions when he did not crew on his friend's boat. The court held that the plaintiff was only entitled to a return of his share of the cost of the licence.

See *Johannes Estate v. Thomas Sheaves* (1996), 23 B.C.L.R. (3d) 283.

Liability of a Partnership for the Acts of a Partner

Joint and several liability
Where partners individually and as a group have liability for a debt of the partnership.

The persons who form a partnership are collectively called the firm, and the business is carried on in the firm name.[12] In a partnership, every partner is the agent of the firm in the ordinary course of partnership business. Partners create **joint and several liability**; in other words, one partner's actions or statements create liability for that partner, *and* for each of the other partners. Every partner may bind the firm in contract with third parties, unless the person with whom the partner was dealing knew that he or she had no authority to do so.[13] The act that the partner performs (for example, if a partner enters into a contract to supply goods or to perform some service) must be related to the ordinary course of partnership business before his or her act binds the other partners. If the act is not something that falls within the ordinary scope of partnership business, then only that partner would be liable.[14]

12. *Partnerships Act*, supra, s. 5.
13. ibid., s. 6.
14. ibid., and see, for example, s. 8, where credit is pledged.

A firm may also be liable for a tort committed by a partner, if the tort is committed in the ordinary course of partnership business.[15] Examples of this situation arise when the partner is responsible for an automobile accident while on firm business, or if the partner is negligent in carrying out a contract and, as a result, injures a third party. Another example is when a partner fraudulently misrepresents a state of affairs to a third party to induce the third party to enter into a contract with the firm. The firm would also be liable if a partner, within the scope of his or her apparent authority, receives money or property either directly from a third party or from another partner, and, while it is in the custody of the firm, misappropriates it or takes it for personal use.[16]

As these examples illustrate, a partnership is generally liable for the acts of the individual partners if committed in the course of partnership business. It is in this respect that a partnership represents a risky form of business organization. For this reason alone, business persons who may wish to engage in business activities using the partnership form should carefully select their partners. In essence, partners expose their total assets to the claims of others who deal with the partnership or who may suffer some injury at the hands of a partner. Consequently, the partnership form of business organization should be supported by a carefully prepared written partnership agreement that clearly delineates the duties and responsibilities of each partner, as well as his or her rights. In addition, business procedures also should be established to reduce the possibility of one partner having the right to commit the partnership to contracts or agreements that could conceivably result in substantial partnership liability.

The admission of a new partner to a firm does not automatically render the new partner liable for the existing debts of the partnership. Under the *Partnerships Act*, a new partner is not liable for any debts incurred before the person becomes a partner.[17] However, a new partner may, if he or she so desires, agree to assume existing debts as a partner, by express agreement with the previous partners. Much the same type of judicial reasoning is applied to the case of a retiring partner. A retiring partner is not relieved of debts incurred while a partner,[18] but if proper notice of retirement is given to all persons who had previous dealings with the firm (and to the public at large) the retiring partner would not be liable for partnership debts incurred after the date of retirement.

CHECKLIST OF ISSUES FOR A PARTNERSHIP AGREEMENT

1. Statement of purpose: to ensure a common focus.
2. Amount of capital contributions: a clear statement of expectation and record of what each partner has invested or will invest in the partnership.
3. Allocation of profit and loss: will it be equal or unequal, in proportion to investment, or otherwise?
4. Responsibilities and authority of partners: who will do what in the daily operation of the enterprise?
5. Signing authority over accounts and assets: one, two or all partners to sign?
6. Growth in number of partners: on what grounds would new partners be added (if any)?
7. Reduction in number of partners: how would a departing partner's share be evaluated, and what mechanism would govern his or her buy-out by other partners?
8. Dispute resolution: how will difficult decisions be made? Is there a provision for arbitration or buy-out?

This list provides basic coverage of chief issues. Every partnership will have other additional issues unique to the needs of its partners and industry.

15. ibid., s. 11.
16. ibid., s. 2.
17. ibid., s. 18(1).
18. ibid., s. 18(2).

Liability of a Partnership for the Acts of its Employees

A partnership may employ others to perform partnership work in much the same way as any employer. In the case of a partnership, however, it is the partnership firm that is the employer, and the partners are responsible for the direction and supervision of the employees in the performance of their duties. The firm may also be bound in contract by the actions of their employees or be liable for torts committed by employees under certain circumstances.

As a general rule, a partnership would be liable for a tort committed by an employee if the tort is committed in the performance of partnership business. For example, if an employee negligently performs work that causes injury to the customer of the partnership, the partnership may be liable for the injury caused by the employee. Similarly, a partnership may be liable on a contract entered into by an employee, if the employee has authority to enter into such contracts. For example, a sales clerk of a partnership may be authorized to enter into a contract to sell certain goods to a customer, but the partnership may fail to deliver the goods as required under the contract. In this case, the partnership would be bound by the contract entered into by the employee, and the partnership may be liable for breach of contract by its failure to deliver the goods.

Rights and Duties of Partners to One Another

The rights and duties of partners are normally set out in a partnership agreement, since most partners wish to arrange their own affairs and those of the partnership in a particular manner. As with most agreements, the parties, within the bounds of the law of contract and the specific laws relating to partnerships, may establish their own rights and duties with respect to each other. They are free to fix their rights and duties in the contract, and they may vary them with the consent of all partners at any time. They need not have a written agreement if they do not wish to have one, but if they do not, their rights and obligations to one another will be defined as in the Act.

Under the Act, all property and rights brought into the partnership by the partners, and any property acquired by the partnership thereafter, become partnership property. They must be held and used for the benefit of the partnership, or in accordance with any agreement of the partners.[19] Any land acquired by, or on behalf of, a partnership, regardless of how the title is held by the partners (or a single partner), is considered to be bought in trust for the benefit of the partnership, unless it is established to be otherwise.[20] Insofar as the partners themselves are concerned, the land is not treated as real property but as personal property, since an individual partner's interest in partnership property is only personalty.[21]

The Act provides a number of rules that, in the absence of any express or implied agreement to the contrary, determine the partners' interests with respect to each other. The rules, as they appear in the Ontario statute,[22] provide as follows:

1. All the partners are entitled to share equally in the capital and profits of the business, and must contribute equally towards the losses, whether of capital or otherwise, sustained by the firm.
2. The firm must indemnify every partner in respect of payments made and personal liabilities incurred by him,
   (a) in the ordinary and proper conduct of the business of the firm; or
   (b) in or about anything necessarily done for the preservation of the business or property of the firm.
3. A partner making, for the purpose of the partnership, any actual payment or advance beyond the amount of capital that he has agreed to subscribe, is entitled to interest at the rate of 5 percent per annum from the date of the payment or advance.

19. ibid., ss. 21(1) and 22.
20. ibid.
21. ibid., s. 23.
22. ibid., s. 24.

4. A partner is not entitled, before the ascertainment of profits, to interest on the capital subscribed by him.
5. Every partner may take part in the management of the partnership business.
6. No partner shall be entitled to remuneration for acting in the partnership business.
7. No person may be introduced as a partner without the consent of all existing partners.
8. Any difference arising as to ordinary matters connected with the partnership business may be decided by a majority of the partners, but no change may be made in the nature of the partnership business without the consent of all existing partners.
9. The partnership books are to be kept at the place of business of the partnership, or the principal place, if there is more than one, and every partner may, when he thinks fit, have access to and inspect and copy any of them.

In addition to these general rules, the Act also provides that in the absence of an express agreement to the contrary, a majority of the partners may not expel any partner from the partnership.[23] The only method that a majority may use if they wish to get rid of a partner would be to terminate the partnership, then form a new partnership without the undesirable partner. Of course, this method of expelling a partner has certain disadvantages, but it is the only procedure available to the partners if they have failed to make express provisions for the elimination of undesirable partners in an agreement.

CASE IN POINT

Three Vancouver business men, Gaglardi, Beedie and Aquilini agreed to work together to purchase an interest in the Vancouver Canucks hockey team. After a few months, Aquilini left the group by agreement. He later tried to re-join the group, but was rejected. Following his rejection, Aquilini decided to negotiate on his own for an interest in the team, and was ultimately successful.

Gaglardi and Beedie then sued Aquilini on the basis that a partnership existed, and they were the rightful owners of the Canucks. They argued that Aquilini had a duty not to compete with them, as their decision to work together had created a partnership amongst the three of them. They also argued that Aquilini had a fiduciary duty to act in their interest as partners.

The court, however, found to the contrary, and held that no partnership existed. By agreement, Aquilini had left the group, and was not permitted to re-join later. Aquilini's relationship with the other parties ended when he departed from the group, and any fiduciary duties he had to the other two individuals also ended at that time. He was free to act on his own if he wished to acquire an interest in the team.

On appeal to the BC Court of Appeal, the appeal was dismissed and the trial court decision was upheld.

Blue Line Hockey Acquisition Co., Inc. v. Orca Bay Hockey Limited Partnership, 2008 BCSC 27 (CanLII).

Because a partnership is a contract of utmost good faith, the partners have a number of obligations that they must perform in the best interests of the partnership as a whole. Every partner must render a true account of any money or information received to the other partners,[24] and deliver up to the partnership any benefit arising from personal use of partnership property.[25]

23. ibid., s. 25.
24. ibid., s. 28.
25. ibid., s. 29(1).

A partnership owns a boat that is occasionally used for partnership business. One of the partners, for personal gain and without the consent of the other partners, uses the boat on weekends to take parties on sightseeing cruises. Any profits earned by the partner using the partnership boat must be delivered up to the partnership, as the earnings were made with partnership property. The same rule would apply when a partner, without the consent of the other partners, uses partnership funds for an investment, then returns the money to the partnership, but retains the profits. The other partners could insist that the profits be turned over to the partnership as well.

In addition to the unauthorized use of partnership property by a partner, the obligation of good faith extends to activities that a partner might engage in that conflict with the business interests of the firm. A partner may not engage, for example, in any other business that is similar to, or competes with, the business of the partnership without the express consent of the other partners. If a partner should engage in a competing business without consent, any profits earned in the competing business could be claimed by the partnership. The partner may also be obliged to provide an accounting for the profits.[26]

A final important matter relating to the duties of a partner to the partnership arises when a partner assigns his or her share in the partnership to another, or changes his or her interest. The assignment does not permit the assignee to step into the position of the partner in the firm. The assignee does not become a partner because of the personal nature of a partnership and only becomes entitled to receive the share of the profits of the partner who assigned the partnership interest. The assignee acquires no right to interfere in the management or operation of the partnership. He or she must be content with receipt of a share of the profits as agreed to by the partners.[27] If the assignment takes place at the dissolution of the partnership, the assignee would then receive the share of the assets to which the partner was entitled on dissolution.[28]

DISSOLUTION OF A PARTNERSHIP

Dissolution

The termination of the partnership relationship.

The parties to a partnership agreement may provide for the term of the agreement and the conditions under which it may be dissolved. A common clause in a partnership agreement is one that provides for a period of notice if a partner wishes to dissolve the partnership. Another common practice is to provide for the disposition of the firm name on dissolution, if some of the partners should desire to carry on the business of the partnership following its termination. Coupled with this provision, the parties may also provide a method of determining the value of the business and the partners' shares if some should wish to acquire the assets of the dissolved business.

Apart from special provisions in a partnership agreement to deal with notice of dissolution, a partnership agreement drawn for a specific term will dissolve automatically at the end of that term.[29] If the agreement was to undertake a specific venture or task, then the partnership would dissolve on the completion of the task or venture.[30] If the agreement is for an unspecified period of time, then the agreement may be terminated by any partner giving notice of dissolution to the remainder of the partners. Once notice is given, the date of the notice is the date of dissolution. However, if no date is mentioned, then the partnership is dissolved as of the date that the notice is received.[31]

A partnership may also be dissolved in a number of other ways. The death or insolvency of a partner will dissolve the partnership, unless the parties have provided

26. ibid., s. 30.
27. ibid., s. 31(1).
28. ibid., s. 31(2).
29. ibid., s. 32(a).
30. ibid., s. 32(b).
31. ibid., s. 32(c).

otherwise.[32] The partners may, at their option, treat the charging of partnership assets for a separate debt as an act that dissolves the partnership.[33] A partnership will be automatically dissolved if it is organized for an unlawful purpose, or if the purpose for which it was organized subsequently becomes unlawful.[34]

In addition to these particular events, there are instances when a partner may believe that a partnership for a fixed term should be terminated before the date fixed for its expiry. If a partner is found to be mentally incompetent or of unsound mind,[35] or if a partner becomes permanently incapable of performing his or her part of the partnership business,[36] the other partner may apply to the courts for an order dissolving the relationship. Relief is also available from the courts in cases where a partner's conduct is such that it is prejudicial to the carrying on of the business.[37] The same applies when the partner willfully or persistently commits a breach of the partnership agreement, or so conducts himself or herself that it is not reasonable for the other partners to carry on the business with such a partner.[38] The courts may also dissolve a partnership if it can be shown that the business can only be carried on at a loss,[39] or where, in the opinion of the court, the circumstances were such that it would be just and equitable to dissolve the relationship.[40] This latter authority of the courts to dissolve a partnership has been included in the *Partnerships Act.* It provides for unusual circumstances that may arise that are not covered by specific provisions in the Act, and that may be shown to work a hardship on the partners if they were not permitted to otherwise dissolve the agreement.

Once notice of dissolution has been given, either in accordance with the partnership agreement or the Act, the assets of the firm must be liquidated, and the share of each partner determined. A partner's share is something that is distinct from the assets of the business and, unless otherwise specified in the agreement, cannot be ascertained until the assets are sold. However, until the share is determined and paid, each partner has an equitable lien on the assets.[41]

Unless the partnership agreement provides otherwise, the assets of the partnership must first be applied to the payment of the debts to persons who are not partners of the firm.[42] Then each partner must be paid rateably what is due for advances to the firm (as distinct from capital contributed).[43] The next step in the procedure is to pay each partner rateably for capital contributed to the firm,[44] and then to divide the residue (if any) amongst the partners in the proportion in which profits are divisible.[45] Any losses are to be paid first out of profits, then out of capital, and, if insufficient funds are available to cover the debts, out of contributions from the partners in the proportions in which they were entitled to share profits.[46] However, the procedure where one partner is insolvent represents an exception to this rule. This results from a case tried some years after the *Partnership Act* was passed in England. The judges in that case stated that if one partner should be insolvent, the remaining solvent partners would be obliged to satisfy the demands of the creditors — not in proportion to the manner in which they share profits, but in proportion to the ratio of their capital accounts at the time of dissolution.[47] The reasoning of the court in reaching this particular conclusion presumably was based upon the assumption that the ratio of the

32. ibid., s. 33(1).
33. ibid., s. 33(2).
34. ibid., s. 34.
35. ibid., s. 35(a).
36. ibid., s. 35(b).
37. ibid., s. 35(c).
38. ibid., s. 35(d).
39. ibid., s. 35(e).
40. ibid., s. 35(f).
41. ibid., s. 39.
42. ibid., s. 44(2)(a).
43. ibid., s. 44(2)(b).
44. ibid., s. 44(2)(c).
45. ibid., s. 44(2)(d).
46. ibid., s. 44(1).
47. ibid., and see *Garner v. Murray*, [1904] 1 Ch. 57.

capital accounts represented a better indicator of the ability of the remaining partners to sustain the loss than the ratio in which profits were shared.

An excerpt of the court's decision follows:

COURT DECISION: Practical Application of the Law

Dissolution of Partnership—One Partner Insolvent—Liability of Other Partners for Deficiencies in Capital

Garner v. Murray, [1904] 1 Ch. 57.

Three persons carried on business in a partnership until it was necessary to dissolve the relationship. One partner had no assets, and the deficiency that resulted in payment of creditors' claims had to be covered by the two remaining partners. The parties had contributed unequal amounts of capital, but had agreed to share profits and losses equally. The question put before the court was: How should the deficiency be paid?

THE COURT: The real question in this case is how, as between two partners, the ultimate deficit, which arises in the partnership from the default of a third partner to contribute his share of the deficiency of the assets to make good the capital, is to be borne by them. We have now in the *Partnership Act*, 1890, a code which defines the mode in which the assets of a firm are to be dealt with in the final settlement of the accounts after dissolution. I do not find anything in [Section 44] that section to make a solvent partner liable to contribute for an insolvent partner who fails to pay his share. Subsection (b) of s. 44 proceeds on the supposition that contributions have been paid or levied. Here the effect of levying is that two partners can pay and one cannot. It is suggested on behalf of the plaintiff that each partner is to bear an equal loss. But when the Act says losses are to be borne equally it means losses sustained by the firm. It cannot mean that the individual loss sustained by each partner is to be of equal amount. There is no rule that the ultimate personal loss of each partner, after he has performed his obligations to the firm, shall be the same as or in any given proportion to that of any other partner.

The application of the court's decision may be best illustrated by way of example:

Suppose that on the dissolution of the ABC partnership, after disposing of all assets and paying all liabilities, the partnership had a loss of $300,000. At that time, the capital accounts of A, B, and C were as follows:

A	B	C
$50,000	$200,000	$300,000

After dividing the $300,000 loss equally ($100,000 each), A's account would be in a deficit position, as A had only a $50,000 capital balance. If A is insolvent and cannot pay the remaining $50,000, the $50,000 deficit would have to be made up by the remaining partners.

According to the law, B and C would be obliged to make up the $50,000 deficit in the ratio of their capital accounts *at the time of dissolution*: i.e., 2/5 for B and 3/5 for C. B would therefore be obliged to pay $20,000 and C $30,000 of A's $50,000 deficit. The capital accounts of the partners after payment of the $300,000 would then look like this:

A	B	C
$0	$80,000	$170,000

Once a partnership has been dissolved, it is necessary to notify all customers of the firm and the public at large. This is particularly important if some of the partners are retiring, and the remaining partners intend to carry on the business under the old firm name. If notice is not given to all customers of the firm, the retiring partners may be held liable by creditors who had no notice of the change in the partnership.[48] The usual practice is to notify all old customers of the firm by letter, and notify the general public by way of a notice published in the official provincial *Gazette*. The notice in the *Gazette* is treated as notice to all new customers who had no dealings with the old firm. Even if the new customers were unaware of the published notice, the retired partner could not be held liable for the debt.[49]

If a partner should die, and the firm thereby dissolve, no notice to the public is necessary. However, the deceased partner's estate would remain liable for the debts of the partnership to the date of that partner's death.

Once dissolution is underway, the business may be conducted by the partners only insofar as it is necessary to close down the operation. This right usually includes the completion of any projects under way at the time of dissolution, but does not include taking on new work. The individual partners may continue to bind the partnership, but only to wind up the affairs of the firm.[50]

After the partnership relationship has terminated, each partner is free to carry on a business similar to the business dissolved. Normally any restriction on the right to do so would be unenforceable unless it is a reasonable restriction limited to a particular term, and within a specified geographic area. Even then, all of the rules of law relating to restrictive covenants would apply.

MANAGEMENT ALERT: BEST PRACTICE

This chapter makes reference to the possible tax advantages of a partnership. If a partnership records a loss, each partner may apply his or her respective share of the loss against (and thereby reduce) his or her taxable income from certain other sources. This feature is not available to the shareholders of a loss-making corporation. Many other tax rules and liability considerations apply, but this ability to "flow through" losses to other income makes the partnership form attractive if such losses are to be expected in the first years of a fledgling business.

LIMITED PARTNERSHIP

Limited partner
A partner who may not actively participate in the management of the firm, but has limited liability.

A limited partnership is one in which a partner, under certain circumstances, may limit his or her liability for partnership debts, and protect his or her personal estate from claims by the creditors of the partnership. The limited partnership bears a resemblance to the European *société en commandite*, which found its way into the Quebec *Civil Code* from French law and into the laws of other provinces by way of the laws of England. All of the Common Law provinces have enacted legislation for the formation of limited partnerships. The provisions of each statute governing the relationship are somewhat similar.

While legislation exists for the formation and operation of limited partnerships (except when certain tax advantages exist), the limited partnership is seldom used for ordinary small businesses. The corporation has been found to be more suited to the needs of persons who might otherwise form a limited partnership. As a result, this type of entity is not commonly found in active small business organizations, other than family-business relationships. However, its use as a special-purpose organization for mining, oil exploration, hotel operations, and cultural activities such as television and film productions is not uncommon, particularly if some of the parties do not wish to engage in an active role in the undertaking. Often these ventures are organized and sold as investments known as "income trusts," where ownership is expressed as an interest in the limited partnership.

General partner
A full partner with unlimited liability for the debts of the partnership.

The legislation pertaining to limited partnerships is not uniform throughout Canada. In general, however, it provides that every limited partnership must have at least one or more **general partners** with unlimited liability and responsibility, both jointly and severally, for

48. ibid., s. 36(1).
49. ibid., s. 36(2).
50. ibid., s. 38.

the debts of the partnership.[51] In addition, the partnership may have one or more limited partners whose liability is limited to the amount of capital contributed to the firm.[52]

Only the general partners may actively transact business for the partnership and have authority to bind it in contract.[53] The name of the limited partner usually must not be a part of the firm name. If a limited partner's name should be placed on letterhead or stationery, in most jurisdictions he or she would be deemed to be a general partner.[54] The limited partner may share in the profits and may examine the partnership books, but must refrain from actively participating in the control of the business. Otherwise the limited partner will be treated as a general partner and lose the protection of limited liability.[55] The limited partner is further restricted with respect to the capital contributed. Once the limited partner has contributed a sum of money to the business, he or she may not withdraw it until the partnership is dissolved.[56]

To provide public notice of the capital contribution of the limited partners, and to identify the general partners in the business, information concerning the limited partnership must be filed in the appropriate public office specified in the provincial legislation. The registration of the notice is very important, as the partnership is not deemed to be formed until the partnership has been registered or the certificate filed.[57] A failure to file, or the making of a false statement in the documents filed, in some jurisdictions renders all limited partners general partners.

While the form of the document filed to register the limited partnership varies from province to province, the information contained generally provides the name under which the partnership operates, the nature of the business, the names of the general and limited partners, the amount of capital contributed by the limited partner, the place of business of the partnership, the date, and the term.[58] Changes in the partnership require a new filing, otherwise the limited partners may either lose their protection or the change is ineffective.

The document is designed to provide creditors and others who may have dealings with the limited partnership with the necessary information to enable them to decide if they should do business with the firm. Alternatively, it also provides information that they might need to institute legal proceedings against the firm if it fails to pay its debts or honour its commitments.

Changes in tax laws have also made certain types of investments attractive if the business is established in the form of a limited partnership. These partnership agreements normally provide for an organization or corporation to carry on the business activity on behalf of the partners, while providing the limited partners the special tax advantages associated with limited partnership ownership and entitlement to the assets and profits.

Notwithstanding the limited use of this type of partnership, some provinces have recently revised their legislation in an effort to clarify and expand the rights and duties of limited partners.[59] How useful these changes will be and whether they stimulate interest in this form of business organization remains to be seen.

LIMITED LIABILITY PARTNERSHIPS

Limited Liability Partnership (LLP)

A partnership where individual partners are liable for the general debts of the partnership and for personal negligence, but not liable for the negligence of other partners.

Some provinces have created legislative provisions for a form of partnership in which all individual partners retain a limited liability status. These **limited liability partnerships (LLP)** are particularly suited to professional practices where one partner cannot hope to know or control the potential professional liability of other partners — for example, lawyers or accountants. When registered as a limited liability partnership, the unlimited

51. For example, the *Partnership Act*, C.C.S.M., c. P30, s. 52.
52. ibid., ss. 52 and 53.
53. ibid., s. 54.
54. ibid., s. 58(1).
55. ibid., s. 63(1).
56. ibid., ss. 60 and 65.
57. ibid., s. 55.
58. See, for example, the *Business Names Registration Act*, C.C.S.M., c. B110, ss. 4(1) and 8(1).
59. See, for example, the *Limited Partnerships Act*, R.S.O. 1990, c. L.16.

liability of each partner is maintained for the general debts of the partnership, and for the partner's personal negligence. Individual partners are not responsible, however, for claims arising from the negligent acts or omissions of the other partners.[60]

> An accounting firm was organized as a limited liability partnership. One partner negligently prepared a set of financial statements for a corporation, overstating the financial health of the corporation. The corporation and its investors consequently sustained great losses when continued operations led to bankruptcy, and sued the partnership for those losses. Only the negligent partner would be liable for the loss. The remaining partners would not be held responsible.

Advanced Business Applications for Partnerships

Joint venture

A business relationship between corporations.

While many partnerships are created for small business purposes, the partnership form serves the needs of large and complex business operations as well. Bear in mind that two or more individuals or corporations can form a partnership, thus large **joint ventures** between major industrial firms is possible as well.

> A petroleum exploration and development firm wishes to establish an oil extractor in a tar-sands project but needs a pipeline operator and refiner. An oil refining company stands ready to build a refinery near a distribution hub but needs an oil supply and a pipeline. A pipeline developer and operator are willing to participate, but need an oil supply and a refinery at each end of the pipeline. A joint venture in the form of a partnership may be the perfect vehicle to bring these parties together.

As suggested earlier in this chapter, the complexities of Canada's *Income Tax Act* also create favourable applications for partnerships. Presently, income earned by a partnership "flows through" the partnership without taxation. It is split up and delivered to each partner in accordance with the share of interest in the partnership that each partner possesses, and then it is taxed according to the tax rate applicable to each partner, be they an individual or corporation. With a little thought, this can serve imaginative investment objectives.

> A limited partnership is created, which acquires ownership of a major existing department store chain. Units of membership in the partnership are then sold to individual personal investors. Accordingly, the profits of the department store chain are regularly distributed to the partners (without intervening taxation) and are taxed as income on the personal tax returns of the investors.

Registration of Partnerships

Limited partnerships and limited liability partnerships are not the only business entities subject to registration requirements. Most provinces require the registration of ordinary partnerships and of sole proprietorships if the sole proprietor is carrying on business under a name other than his or her own. Provincial legislation is not uniform with respect to registration, and some provinces exempt some types of partnerships from registration. For example, professions governed or regulated by provincial bodies are frequently exempt, and, in at least one province, farming and fishing partnerships need not be registered. The purpose of registration is the same as for limited partnerships — that is, to provide creditors and others with information concerning the business and the persons who operate it.

---

60. *The Partnerships Act*, R.S.O. 1990, c. P.5, as amended by S.O. 1998, c. 2, ss. 2 and 6.

Declarations generally require the partners to disclose the name of the partnership, the names and addresses of all partners, the date of commencement of the partnership, and the fact that all partners are of the age of majority (or, if not, the date of birth of the infant partners).[61] The declaration must normally be filed online or in a specified public office within a particular period of time after the partnership commenced operation.[62] Changes in the partnership usually require the filing of a new declaration[63] within a similar time period.

The provinces of Nova Scotia and Ontario provide in their legislation that no partnership or member may maintain any action or other proceeding in a court of law in connection with any contract, unless a declaration has been duly filed.[64] The significance of this particular section looms large in the event that an unregistered partnership wishes to defend or institute legal proceedings. The failure to register would act as a bar to any legal action by the partnership, until such time as registration is effected.

Sole proprietorships normally need not be registered in most provinces. However, a sole proprietor carrying on business under a name other than his or her own is usually required to register in much the same manner as a partnership, since persons doing business with such a business entity would be interested in knowing the identity of the true owner.

In a partnership, all partners required to register under registration legislation usually remain liable to creditors until a notice of dissolution is filed in the proper office. The declaration of dissolution acts as a public notice that a partnership has been dissolved. If afterwards the firm is to continue on, comprising the remaining or new partners, the old partners may still be deemed partners until the declaration of dissolution has been filed.

CLIENTS, SUPPLIERS *or* OPERATIONS *in* QUEBEC

The legal principles governing sole proprietorships and partnerships in Quebec are found in the *Civil Code* (the latter at Articles 2186-2266), but important requirements are also found in the *Legal Publicity Act*, R.S.Q. c. P-45. These rules prescribe quite detailed steps in the registration and administrative maintenance of enterprises.

A QUESTION OF ETHICS

Some jurisdictions forbid members of some professions from practising together in any business form except a partnership. What ethical goal do such policies intend to achieve?

SUMMARY

The sole proprietorship is the simplest form of business organization. The sole proprietor is responsible for all aspects of the business, and makes all management decisions. The sole proprietor is also responsible for all losses that may occur, and is entitled to all profits that the business may make.

The partnership is an ancient form of business organization. It is a relationship of utmost good faith that requires the partners to disclose to one another all information of importance to the firm. Partners are agents of the firm, and one partner acting within the apparent range of his or her authority may bind the partnership in contract

61. *The Business Names Act*, R.S.O. 1990, c. B.17.
62. ibid., s. 2.
63. ibid., s. 4(4).
64. ibid., s. 7, and see the *Partnership and Business Names Registration Act*, R.S.N.S. 1989, c. 335, s. 20.

with third parties. A partnership may also be liable for the torts of partners committed within the ordinary course of partnership business.

Partnership agreements are usually written contracts signed by the parties. A partnership agreement need not be in writing, however, and need not specify the terms of the partnership. However, when no terms are set out, the provisions of the *Partnerships Act* apply to determine the rights and obligations of the partners.

Partners must not use partnership assets or funds to make secret profits at the expense of the partnership. The partners also must not engage in other businesses that compete with the partnership, unless they have the express permission of all of the other partners.

A partnership may be dissolved by the completion of the specific venture for which the partnership was formed or at the end of a specified term, or it may be terminated on notice if no term is specified. It may also terminate on the death or insolvency of a partner, or if the purpose for which it was organized is unlawful. The court under certain circumstances has the right to dissolve a partnership. When a partnership is dissolved, all creditors and the public at large must be notified of the dissolution by the partners if the retiring partners wish to avoid liability for future debts of the partnership.

Subject to certain exceptions, partnerships must be registered. The limited partnership, which may be established in most provinces, depends upon registration to establish the limited liability of the limited partners. The limited liability partnership can be utilized to shield partners from claims arising from the professional negligence of their colleagues.

KEY TERMS

Sole proprietorship, 285
Partnership, 285
Joint and several liability, 288
Dissolution, 292
Limited partner, 295
General partner, 295
Limited Liability Partnership (LLP), 296
Joint venture, 297

REVIEW QUESTIONS

1. Why is registration of a partnership important in certain provinces?
2. If one partner should become personally bankrupt, and the partnership dissolved, how is the liability of each remaining partner for the debts of the partnership determined?
3. Why is a simple sharing of gross profits not conclusive as a determinant of the existence of a partnership relationship?
4. How does a partnership differ from co-ownership?
5. Under what circumstances may a minor be a member of a partnership? What is the extent of the minor's liability?
6. Explain how agency and partnership are related in terms of the operation of a partnership.
7. Under what circumstances would a partnership be liable for a tort committed by a partner?
8. What is the extent of the liability of the partners for the tort of a partner or for contracts entered into by a partner?
9. Under what circumstances may a partnership be dissolved?
10. Is it possible for a partner to sell his or her interest to another person? What is the status of the purchaser of the interest if it should be sold?
11. Explain the rights of creditors of a partnership when the partnership is dissolved.
12. How is a partnership formed?
13. What essential characteristic distinguishes a partnership from other associations of individuals?
14. What is a retiring partner obliged to do in order to avoid liability for future debts incurred by a partnership?

MINI-CASE PROBLEMS

1. A and B purchase a sailboat (which is in poor condition) with the intention of refurbishing it together and selling it when the work is completed. Are A and B partners?
2. X, Y, and Z agree to carry on business in partnership under the firm name XY Plumbing. Y, in the course of partnership business, carelessly injures T. T intends to take legal action against the

partnership for the injury, but is unaware that Z is a partner, since his name is not included in the partnership name. Is Z liable if T sues only X and Y?

3. Albert, Charles and Emily carry on business under the name ACE as a registered partnership. Emily wishes to retire from the partnership and allow Albert and Charles to carry on the business. What advice would you give Emily?
4. Alexander and Betty carry on a partnership business. Carter is an employee of Alexander and Betty. Alexander informs Delbert, a supplier of the business, that Carter is authorized to purchase goods on credit for the partnership. Carter is later discharged by the partnership, but purchases personal goods on credit from Delbert, as Delbert is unaware that Carter is no longer employed by the partnership. Is the partnership liable for payment for the goods purchased by Carter? Explain.

Case Problems for Discussion

CASE 1

Earl, Louise, and Erik carried on business in partnership for many years under the firm name ELE Sales. The partnership was registered under the provincial partnership registration legislation. Earl and Louise were the two partners that dealt with customers and suppliers, and Erik worked for the most part out of sight of everyone in a back office where he did the bookkeeping for the business.

Erik eventually retired, leaving the business without informing either customers or suppliers. Earl and Louise carried on the business for another few years, but without the management skills of Erik, the business ran into financial difficulties. Earl and Louise decided to close the business, and paid most of the creditors. The remaining creditors now wish to take action to recover their debts.

Advise the creditors.

CASE 2

Granite Gold Mine Ltd. and Diamond Drilling Ltd. entered into an agreement whereby Granite Gold Mine Ltd. would provide its mining claims for exploration, and Diamond Drilling would provide drilling services at no cost to Granite Gold Mine Ltd. in an effort to determine the extent of an ore body within the claims area. The two companies agreed to form a third corporation that would operate the mine once the quantity of ore was determined. The third company, to be called Mine Operators Inc. would be jointly owned by Granite Gold Mine Ltd. and Diamond Drilling Ltd. Each would own 50% of the shares of the corporation, each corporation would pay 50% of the start-up cost of the mine.

The two corporations agreed that ownership of the mining claims would remain with Granite Gold Mine Ltd., but all gold mined by Mine Operators Inc. would be owned by Mine Operators Inc. subject to a royalty to Granite Gold Mine Ltd. of 10%.

Exploration by Diamond Drilling Ltd. revealed a large ore body, but before a mine was started, Diamond Drilling Ltd. ran into financial difficulties. Several of its creditors who had supplied services and goods to the mine exploration project decided to take legal action against both Granite Gold Mine Ltd. and Diamond Drilling Ltd. to recover their debts.

Discuss the issue raised in this case, and the arguments each party might raise. Render a decision.

CASE 3

Harold Green and Herbert Green, who were unrelated persons with the same surname, carried on parcel delivery businesses in the same city. Each operated as a sole proprietor, under the name Green Delivery, and carried on business from the same warehouse building. The two men were good friends. They frequently assisted each other by carrying parcels for the other in deliveries in outlying parts of the city. To complicate matters further, the two proprietors would sometimes use the spare truck owned by the other when their own vehicles had breakdowns or required service. Apart from the fact that each had a different telephone number, it was impossible to distinguish between the two firms. Over time, regular customers often referred to the two firms collectively as "Green Brothers Delivery," even though the two men were not related to each other.

Fiona, an antique dealer who frequently used the delivery services of both men, requested Harold Green to deliver an expensive antique chair to her country home. Harold Green, at the time of the request for pickup, advised Fiona that he would send his truck out to pick up the chair. However, Herbert Green picked up the chair at Fiona's place of business and took it to the warehouse for delivery. That day a fire of unknown origin destroyed the warehouse and its contents. The charred remains of the chair were found in the jointly used part of the warehouse after the fire.

The chair had a value of $3,000. Each of the sole proprietors denied liability for the loss.

Advise Fiona as to how she might proceed in this case.

CASE 4

Williams, Oxford, Ogilvie, and Lennox carried on business in partnership for many years as wool merchants. The widespread use of synthetic materials, however, adversely affected the fortunes of the business. Eventually the partners found that the business could no longer be carried on at a profit. Before anything could be done to sell the business, Lennox became insolvent. It was then necessary to wind up the business in accordance with the partnership agreement.

The partnership agreement provided that the parties share losses and profits equally. On dissolution, the capital accounts of the partners were as follows:

Williams	$50,000
Oxford	$30,000
Ogilvie	$20,000
Lennox	—

Creditors' claims at dissolution amounted to $350,000, whereas the total assets of the firm were $250,000.

Explain the nature of the liability of the firm. Calculate the liability of each of the partners as between themselves with respect to creditor claims.

CASE 5

Baker, Felice, and Toby carried on business for many years as clothing retailers under the firm name of Family Clothing Market. The store premises were large and consisted of three separate smaller shops located side by side in the downtown shopping district of a large city. Different clothing lines were sold in each shop, and each shop was managed exclusively by one of the partners. Baker managed the children's clothing shop; Felice, the women's clothing store; and Toby, the men's clothing store. Each shop had its own distinctive name displayed on its shop window and entrance door.

Toby eventually grew tired of the business and decided to retire. He did so on March 1st. The two remaining partners purchased Toby's interest in the partnership and placed an employee (who had previously been the buyer for men's clothing) as manager in charge of the men's clothing store. This employee continued to act as the buyer for men's clothing and to deal with the clothing suppliers, without informing them of the change in the partnership. Nor was any notice of change in the makeup of the partnership filed under the provincial partnership registration legislation.

On April 1st, Baker died. Following his death, the auditors discovered that he had been systematically concealing the true state of the children's clothing operation from the other partners by creating fictitious assets to offset the losses. When the overall loss was calculated, the business was determined to be in serious financial straits.

News of the discovery soon leaked to the suppliers. When all of the creditors' claims were presented, the liabilities of the business exceeded its assets by some $50,000.

Rogers, a supplier, brought an action against the partnership for payment of his account in the amount of $5,000.

Discuss the status and liability of the partners, indicate any defences that might be raised, and render a decision.

CASE 6

Sandra is a successful real-estate agent with HomeBase Realty Inc. (HBRI). She acts for, among many other people, a struggling small contractor, Herbert Homes Limited (HHL). HHL buys land, builds a home on it in the hope of attracting a buyer, and uses Sandra and HBRI as its agent to list the property and make the sale.

HHL was short on cash when a desirable, vacant acre of land came on the market. Sandra was to loan HHL $130,000 (the full price of the lot). Rather than take a mortgage on the land for security, as that could interfere with HHL's ability to borrow bank funds to finance construction, Sandra and HHL agreed that her name would go on the deed as well as that of HHL. They further agreed that when the property was sold with a house on it, she would get her usual 6 percent commission on the sale, together with her original $130,000 loan, plus interest of 12 percent to the sale date.

The property was purchased, Sandra carried on with her other commitments, and HHL built a house, which two months after completion was sold for $400,000. On the sale, Sandra and HHL signed off on the deed to the new owners, and payment was made directly to the lawyer for the vendors. The lawyer for the vendors split up the money: $24,000 to HBRI, $137,800 to Sandra, and the balance to HHL.

As matters turned out, HHL had not built the house up to standard. The roof leaked, costing the purchasers $22,000 in damages. The purchasers sued HHL and Sandra as the vendors. Conclusive evidence was presented that they received a faulty house from the vendors. The Provincial Government New Home Warranty Board was advised of the matter, and its investigation showed that while HHL was a registered builder, as one must be under the law to sell a new home, Sandra was not. She has now been charged with a violation of the law. Such an offence usually results in a $500 fine for a first offence.

Consider the aspects of partnership law as they may apply, and advise Sandra as to how she should proceed. Consider how she is exposed to risk, and what defences or claims she may raise in civil court against HHL or the purchasers, and against the Crown in Provincial Offences Court. If you were in a position to advise the judge in Provincial Offences Court, what would you do?

CASE 7

Sarah, aged 17 years, and Jane, aged 19 years, had been shopping in a large shopping mall. Jane wished to purchase a lottery ticket, but had only $1 in her purse. She turned to Sarah and said, "Do you have $4? They are selling lottery tickets here." Without a word, Sarah took the money from her pocket and gave it to Jane. Jane purchased a ticket that she and Sarah agreed bore a lucky number. The next week, the ticket that Jane had purchased was the winner of $25,000.

When news of the win reached Jane, she immediately visited her friend Sarah and attempted to pay her the $4 that she said she had borrowed. Sarah, who had also heard the good news, refused to accept the $4 and demanded her share of the winnings.

If the above dispute should be brought before a court, describe the arguments that might be raised by each of the parties in support of their respective positions. Indicate how the case might be decided.

CASE 8

In 2007, Tareq, who operated a business under the name Downtown Grocery, employed Marcus as a clerk in his store at a weekly salary of $450. Marcus received regular annual salary increases in the years 2008 to 2012. By 2012, he was earning $700 per week.

Early in 2012, Marcus approached Tareq with a request for a further increase in his wages. Tareq refused on the basis that low business profits limited his ability to pay more than $700 per week. A lengthy discussion followed and the two parties reached the following agreement.

Marcus would receive $700 per week and, in addition, would receive 20 percent of the net profits. He would continue to perform the duties of clerk, but would also assume responsibility for the meat department. He would make all management decisions concerning meat purchases and pricing.

Tareq would continue to handle the general management of the business. He would draw the amount of $1,000 per week and would be entitled to 80 percent of the net profits.

Marcus would be permitted to examine the business account books, and he would be consulted in all major business decisions by Tareq.

A few weeks after the agreement was reached, Tareq discovered that Marcus was purchasing groceries (for his personal use) at a competitor's store. In a rage, he barred Marcus from entering the store and told him that he would send him his severance pay by mail.

Shortly thereafter, Marcus instituted legal proceedings for a declaration that he was a partner in Downtown Grocery.

Discuss the merits of the action taken by Marcus and discuss the arguments that might be raised by the parties. How would you expect the matter to be decided?

The Online Learning Centre offers more ways to check what you have learned so far. Find quizzes, Weblinks, and many other resources at www.mcgrawhill.ca/olc/willes.

CHAPTER 17

Corporation Law

Chapter Objectives

After study of this chapter, students should be able to:

- Understand the nature of a corporation and its formation.
- Recognize the different forms of incorporation.
- Explain the duties and responsibilities of directors.
- Explain how corporations may be sold or dissolved.

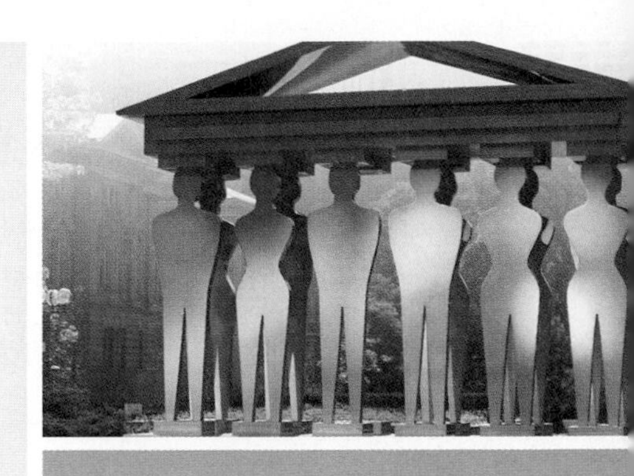

YOUR BUSINESS AT RISK

Conducting business through a corporation is common, and you will probably find yourself doing business with a corporation in future, or investing in one. Even so, it is not the best form for all purposes. It has significantly different operating mechanisms and you must properly manage its responsibilities toward its owners, managers, customers and regulators.

INTRODUCTION

Corporation
A type of legal entity created by the state.

A **corporation** is an entity that has an existence at law that is separate from those who form it. It is also separate from those who from time to time possess shares in it, or who are responsible for its direction and control. A corporation is a creature of the state, and it owes its continued existence to the legislative body responsible for its creation. One of its principal uses is as a vehicle by which large amounts of capital may be accumulated for business purposes. Throughout its history, the corporation has represented a means by which a large number of persons could participate in business transactions requiring more capital than one individual or a small group possessed or cared to risk in a venture.

HISTORICAL DEVELOPMENT OF THE CORPORATION

The corporation is not a new form of business organization. Its history dates back to the creation of the first city-states, when trade became an important activity. Since the earliest times, its creation and operation have generally remained under the control of the sovereign or state. Most of the early corporations were formed for the purpose of carrying out public or quasi-public functions or to provide essential services or goods to the community. It was not until the nineteenth century that modern English trading corporations were generally permitted to incorporate. Even then, it was only during the last half of the century that limited liability was available to the shareholders. The modern form of corporation, incorporated by simply following the procedure set out in a general statute, did not emerge until the middle of the century. Even then, its introduction was decades behind a similar change in the United States and Canada.

In the United States, the War of Independence severed ties between the colonies and England, and the royal prerogative was replaced by special legislation for the incorporation of companies. In 1811, the state of New York passed a statute that in form was a general incorporation act. It was alleged to have been the model for the first incorporation act in the United Provinces of Upper and Lower Canada in 1848–49. Some years later, the *British North America Act*, 1867, established the rights of both the provinces and the federal government to incorporate companies. As a result, legislation at both the federal and provincial levels was introduced to provide for the incorporation and control of corporations. The legislation, unfortunately, was not uniform, and so considerable variation existed from province to province. The newer provinces, as they joined Confederation, passed their own legislation. This again was subject to some variation and, consequently, no uniform legislation exists in Canada today.

THE NATURE OF A CORPORATION

A corporation is neither an individual nor a partnership. It is a separate legal entity in the sense that it has an existence at law, but no material existence. A corporation possesses many of the attributes of a natural person, but it is artificially created and never dies in the natural sense. Its rights and duties are delineated by law, and its existence may be terminated by the state. It has a number of important characteristics, which may be summarized as follows:

Shareholder

A person who holds a share interest in a corporation; a part owner of the corporation.

(a) a corporation is separate and distinct from its **shareholders**, and it acts not through them but through its authorized agents.
(b) a properly authorized agent may bind the corporation in contract with third parties.
(c) the shareholders of a corporation possess limited liability for the debts of the corporation, and the creditors may look only to the assets of the corporation to satisfy their claims.

This latter characteristic particularly distinguishes the corporation from the ordinary partnership or sole proprietorship, where the parties have unlimited liability for debts incurred in the course of business.

The management of a corporation is vested in a small group of shareholders chosen by the general body of shareholders at an annual meeting held for the purpose of electing the management group. This group, known as **directors** of the corporation, in turn selects from their number the principal **officers** of the corporation. These individuals are normally charged with the responsibility to bind the corporation in formal contracts with third parties.

Director

Under corporation law, a person elected by the shareholders of a corporation to manage its affairs.

Officer

A person elected or appointed by the directors of a corporation to fill a particular office (such as president, secretary, treasurer, etc.).

The directors of the corporation are free to carry out the general management functions of the corporation in accordance with the corporation's objects, but their powers are limited by any restriction imposed on them in the articles of incorporation or the charter. To keep the shareholders informed of their activities, the directors are obliged to report to them on a regular basis. The shareholders normally do not participate in the management of the corporation, except when major changes in the corporation are proposed. In these cases, the shareholders usually must approve the proposal before it becomes effective.

Except for private or closely held corporations that do not offer their shares to the public, the shareholders may freely dispose of their interests in the corporation. This is permitted because the relationship is not of a personal nature. The fortunes of the corporation are not dependent upon the personal relationship between shareholders, nor are the creditors of the corporation normally concerned with the shareholders' identity. The shareholders' liability is limited to the amount that they paid or agreed to pay for the shares that they purchased from the corporation.

In general, the corporate form of business organization overcomes many of the disadvantages associated with the partnership. We can compare the differences between these two forms of organization by an examination of each under three major headings: control, limited liability, and transfer of interests.

Control

In a partnership, every partner is an agent of the partnership, as well as a principal. All of the partners have input into how the business may be operated, and on important matters all parties must agree before a change can be made. In a large partnership, these particular rights of each partner often render decision-making awkward and time-consuming, and make general control difficult.

In a corporation, management is delegated by the shareholders to an elected group of directors. The ordinary shareholder does not possess the right to bind the corporation in contract. Only the proper officers designated by the directors may do so. The directors have the authority to make all decisions for the corporation (subject to limitations contained in the Articles of Incorporation or any unanimous shareholders' agreement), although the shareholders may periodically be called upon to approve major decisions at meetings held for that purpose.

Limited Liability

In an ordinary partnership, every general partner has unlimited liability for the debts incurred by the partnership. Since any partner may bind the other partners in contract by his or her actions, the careless act of one partner may seriously affect all. The personal estates of each partner, then, are exposed to the creditors in the event of a loss that exceeds the partnership assets.

The corporate form eliminates this particular risk for ordinary shareholders. Shareholders' losses are limited to their investment in the corporation, and their personal estate may not be reached by creditors of the corporation. The creditors of a corporation must be content with the assets of the corporation in the event of a loss, as creditors are aware at the time (when extending credit to a corporation) that the only assets available to satisfy their claims are those possessed by the corporation.

Transfer of Interests

A partnership is a contractual relationship that is personal in nature and based upon the good faith of the parties in their dealings with each other. The right of each partner to bind the partnership in contract, coupled with the unlimited liability of each of the partners, precludes the unfettered right of a partner to transfer his or her interest in the partnership to another. The retirement of a partner raises a similar problem. Since the retiring partner's interest is not freely transferable, the remaining partners must either acquire the retiring partner's share in the partnership or wind up the business. Neither of these solutions may be entirely satisfactory to either the retiring partner or those who remain. The death of a partner represents another disruptive event that also raises the problem of transfer of interests in the partnership. Unless special provisions are made in the partnership agreement (such as provision for the payment of a deceased partner's share by way of insurance), the retirement or death of a partner will generally have a serious effect on the relationship and the business associated with it.

The corporate form of organization differs substantially in this regard. The rights of the shareholders in a corporation do not include the right to bind the corporation in contract, nor may creditors look to the personal assets of the shareholders in determining whether or not credit should be extended to the corporation. Under these circumstances, the identity of the shareholder becomes unimportant, to some extent, as does the shareholder's personal worth. Once a corporation issues a share to a shareholder and receives payment for it, no further contribution may be demanded from the shareholder. If the shareholder should desire to transfer the share to another, it has no real effect upon the corporation except for a change in the identity of the person holding the share. For these reasons, shares may be freely transferred in a public company, thus overcoming one of the main drawbacks of the partnership.

The corporation has a number of other advantages over the partnership. These are subsequently noted.

Term of Operation of the Business

A partnership's existence is limited by the life of its members. If a partner should die, the partnership is dissolved, but it may be reformed by the remaining parties if the agreement so provides. A partnership may continue as long as its partners wish it to do so, but the death of a partner, or the partner's retirement, is disruptive to the operation.

A corporation, on the other hand, theoretically has an unlimited term of operation. It never dies, even though the persons who own shares in it may do so. The death of a shareholder has no effect on the continued existence of the corporation. A corporation may be dissolved by the state or it may voluntarily be wound up, but, in each case, the act is not dependent upon the life or death of a shareholder. Some corporations have been in existence for many hundreds of years. For example, the Hudson's Bay Company was incorporated in 1670 and remains in existence today. No partnership (at least with original, unincorporated members) is likely to match this record.

Because a corporation's existence is not affected by the fortunes of the shareholders, a corporation is free to accumulate or acquire large amounts of capital — either through the issue of shares, or by the issue of bonds and debentures. These latter forms are special types of security instruments, issued by a corporation, that can be used when large sums of money are required.

Operation of the Business Entity

A partnership is governed by an agreement that establishes the rights and duties of the partners and the manner in which the partnership is to operate. In the absence of an agreement (or if the agreement is silent), the partnership is governed by the provincial *Partnerships Act*[1] and a number of other related statutes.

In contrast, the corporation is governed by the statute under which it is incorporated (as well as a number of other statutes, such as securities legislation) that sets out the conditions and rules that apply to its operation. The rights and duties of the shareholders and directors with respect to the corporation are statutory rather than contractual in nature, although both the corporation and partnership are subject to statute.

Separate Existence of the Corporation

At law, the partnership is in a sense indistinguishable from the partners, yet it does possess many of the attributes of a separate entity. Contracts are made in the firm name, and a partnership in most jurisdictions may be sued in the firm name. However, the fact remains that the parties individually as well as collectively are responsible for its operation. The corporation, in contrast, has a clearly defined separate existence at law.

This particular issue was established in a case that came before the English courts in 1896.[2] The case dealt with the principal shareholder and a corporation that he had incorporated to take over his successful shoe and leather business. The principal shareholder had taken back a debenture from the corporation. When the corporation later became insolvent, the principal shareholder claimed priority in payment of the debenture over the claims of the general creditors. In the litigation that followed, the court held that the corporation was an entity separate and distinct from its shareholders. It permitted the principal shareholder to obtain payment of the debenture from the money available, before claims of the general creditors could be satisfied.

COURT DECISION: Practical Application of the Law

Nature of a Corporation—Limited Liability—Existence Separate from Shareholders

Salomon v. A. Salomon & Co. Ltd., [1897] A.C. 22.

A. Salomon operated a successful leather and footwear business. He incorporated a company and sold the business to the company in return for 20,000 shares and certain debentures. Members of his family each held one share to meet the requirements of the Companies Act, which required seven shareholders. The company later became

1. All provinces have legislation governing this relationship.
2. *Salomon v. A. Salomon & Co. Ltd.*, [1897] A.C. 22.

insolvent, and both the unsecured creditors and Salomon claimed the remaining assets that were allegedly secured by Salomon's debentures. The creditors attempted to establish that the company was really a "front" for the principal shareholder, and that he (Salomon) should be responsible for the debts of the company.

THE COURT: It is to be observed that both Courts treated the company as a legal entity distinct from Salomon and the then members who composed it, and therefore as a validly constituted corporation. This is, indeed, necessarily involved in the judgement which declared that the company was entitled to certain rights as against Salomon. Under these circumstances, I am at a loss to understand what is meant by saying that A. Salomon & Co., Limited, is but an "alias" for A. Salomon. It is not another name for the same person; the company is ex hypothesis a distinct legal persona. As little am I able to adopt the view that the company was the agent of Salomon to carry on his business for him. In a popular sense, a company may in every case be said to carry on the business for and on behalf of its shareholders; but this certainly does not in point of law constitute the relation of principal and agent between them or render the shareholders liable to indemnify the company against the debts which it incurs. Here, it is true, Salomon owned all the shares except six, so that if the business were profitable he would be entitled, substantially, to the whole of the profits. The other shareholders, too, are said to have been "dummies," the nominees of Salomon. But when once it is conceded that they were individual members of the company distinct from Salomon, and sufficiently so to bring into existence in conjunction with him a validly constituted corporation, I am unable to see how the facts to which I have just referred can affect the legal position of the company, or give it rights as against its members which it would not otherwise possess.

What would you consider to be sufficient justification for a court to "pierce the corporate veil" and make a shareholder liable for the acts of the corporation?

Corporate Name

The corporation's name is an asset of the business.[3] The name must not be the same as the name of any other existing corporation. In the case of a corporation incorporated to carry on a business (other than a corporation incorporated under a special statute), the last word in the name must be a word that identifies it as a corporation. A corporation in some jurisdictions may have a numbered name. The last word that may be used varies from jurisdiction to jurisdiction in Canada, but generally must be Limited or Ltd., Incorporated or Inc., or Corporation or Corp. (or the French equivalent). The purpose of the identifying word is to distinguish a corporation from a partnership and/or sole proprietorship. Partnerships are not permitted to use any word reserved for corporation identification or any word that might imply corporate status.

As a general rule, the corporation's name may not be one that denotes any connection with the Crown, nor may the name be obscene, too general, or such that it would cause confusion with other existing names, or the names of well-known incorporated or unincorporated organizations.[4]

A corporation must clearly indicate its name and its place of business on all printed letterhead and business forms. Where required, it must formally execute documents by the use of a seal (which contains the corporation name). The authorized officers of the corporation impress this seal upon all written documents that require the formal "signature" of the corporation.

CLIENTS, SUPPLIERS *or* OPERATIONS *in* QUEBEC

The Quebec *Charter of the French Language*, Articles 63 and 64, requires corporations to utilize the French version of their name within the province. Translations may be offered in less-predominant typeface.

3. *Hunt's Ltd. v. Hunt* (1924), 56 O.L.R. 349.
4. *Tussaud v. Tussaud* (1890), 44 Ch.D. 678.

METHODS OF INCORPORATION

Royal Charter

The original method of incorporation was by a royal charter. The issue of the charter was for the purpose of creating a legal existence for the entity, to permit it to either operate as a monopoly or to own land. The Hudson's Bay Company, for example, was granted rights to all of the land drained by the rivers flowing into Hudson's Bay. The British South Africa Company was granted a similar type of charter with respect to large amounts of land in South Africa.

The royal charter was an exercise of the king's prerogative, and the issue of the charter gave the entity all the rights at law of a natural person (subject to certain exceptions). The charter did not, however, generally give those persons connected with the corporation limited liability if the corporation was a trading company. The royal charter method of incorporation has not been used to incorporate ordinary trading corporations for many years.

Letters Patent

Letters Patent

A government document that creates a corporation as a legal entity.

The Letters Patent system of incorporation is a direct development of the royal charter form. When the Crown ceased the issue of royal charters for trading corporations, a new system was developed whereby legislation was passed that set out the criteria for the issue of **Letters Patent** by a representative of the Crown.[5] The application for incorporation would then be prepared and submitted in accordance with the legislation, and the Crown's representative would issue the incorporating document.

Letters Patent is a government document that grants a special right or privilege. In the case of a corporation, it represents the creation of a legal entity. A corporation created by Letters Patent acquires all the powers of a royal charter corporation, and has, as a result, all of the usual rights at law of a natural person.[6] The federal government, and the provinces of Manitoba, New Brunswick, Ontario, Prince Edward Island, and Quebec used the Letters Patent system for many years. However, after 1970, some of the provinces and the federal government adopted a different system, and, by 1996, only Quebec and Prince Edward Island still retained the Letters Patent system.

Third parties dealing with a Letters Patent corporation may assume that the corporation has the capacity to enter into most ordinary contracts. If the officers of the corporation, with apparent authority to bind the corporation, enter into a contract of a type prohibited by the objects or by-laws of the corporation, the corporation would still be bound. The exception would be if the third party had actual notice of the lack of authority in the offices of the corporation.

Special-Act

Special-Act

A corporation created by an Act of Parliament or a legislature for a specific purpose.

The third type of incorporation represents a quite different form. Parliament may use its powers to create corporations by way of special statute, and these corporations are known as **Special-Act** corporations. Instead of the broad powers possessed by a corporation incorporated under royal charter, the Special-Act corporation has only the powers specifically granted to it by the statute. Special-Act corporations, as the name implies, are corporations incorporated for special purposes. Many are for public or quasi-public purposes, such as for the construction of public utilities, or for activities that require a certain amount of control in the public interest. Banks, telephone companies, railroads, and Crown corporations (such as the Canadian Broadcasting Corporation) are examples of corporations incorporated under special legislation. Special-Act corporations as a general rule do not use the words limited, incorporated, or their abbreviations in the corporate name. For example, chartered banks use no corporate identification at all in the corporation's name.

The legislation sets out the rights and obligations of the corporation. If the corporation should attempt to perform an act that it is not authorized to do under the statute,

5. For example, prior to 1970, under the *Corporations Act of Ontario*, R.S.O. 1960, c. 71, the Provincial Secretary issued the Letters Patent. At the federal level, the Secretary of State issued Letters Patent prior to 1976.

6. *Bonanza Creek Gold Mining Co. v. The King*, [1916] 1 A.C. 566.

the act is *ultra vires* (beyond the powers of) the corporation and a nullity. For example, in a case that involved an English corporation that was limited by Special Act to borrowing £25,000, the corporation exceeded the amount and eventually borrowed £85,000. When a secured creditor attempted to recover money loaned to the corporation, the court held that the plaintiff creditor could not enforce the security because the corporation had no authority to issue it. In rejecting the claim, the court held that all the security issued in excess of the £25,000 authorized by statute was a nullity.[7] The judge in the case described the powers of the Special-Act corporation and its limitations in the following terms:

> ... there is a large difference between (an ecclesiastical) corporation and a corporation such as we are dealing with here, called into existence by statute, the creature of statute for specific purposes, and armed with specific powers which must not be exceeded for the fulfillment of those particular purposes. Looking at the exercise of any alleged power on behalf of a corporation of that kind we must see whether it is either expressly given by the statute, or is necessarily implied in order to the fulfillment of its powers. If we cannot bring it within that — and a necessary implication in such a case differs not at all from expression — then the company cannot exercise the alleged power.

General-Act

General-Act

A form of incorporation whereby a corporation may be created by filing specific information required by the statute.

Memorandum of Association

Incorporation document.

Articles of Incorporation

Incorporation document.

Business corporations under federal jurisdiction are governed by the *Canada Business Corporations Act* while most provincial **General-Act** counterparts are governed by the *Business Corporations Act* in each province. PEI and Quebec still retain the Letters Patent system for business corporations. Even so, incorporation in all jurisdictions has been reduced to a process of disclosures and recitals in a document known and filed as either a **Memorandum of Association** or **Articles of Incorporation**, depending on the province. The federal government uses articles of incorporation for its business corporations. This document is sufficient to establish the fundamental capital structure of the corporation, significant self-imposed restrictions, its location, and the identity of its director(s).

The General-Act corporation, like the Special-Act corporation, is a creature of statute, and its powers are limited to those powers specifically granted under the Act. Unlike the royal charter and letters patent corporations, it does not necessarily have all the powers of a natural person, but only those powers set out in the statute and its memorandum of association. To provide flexibility and greater latitude provincial jurisdictions in most cases have attempted to extend to a General-Act corporation the powers of a natural person. In the remaining provinces, the legislation has been amended to ensure that any limitation on the powers of the corporation does not adversely affect third parties dealing with the corporation.[8]

Relations with Third Parties

To ensure that third parties are not affected by restrictions on the powers of General-Act corporations, the doctrine of *ultra vires* has been abolished insofar as it would apply to corporations incorporated under the Act. As a consequence, third parties dealing with a corporation are not obliged to examine the memorandum of association of the corporation in order to ensure that the powers of the corporation are not restricted with respect to the contemplated transaction.

For corporations generally, and for the business community in particular, the abolition of the doctrine of *ultra vires* and the treatment of corporations as possessing the powers of a natural person are important. In the absence of these provisions the courts would treat both the statute and the memorandum of association as the body of law that delineates the powers of the corporation, and would likely find that any act of the business corporation incorporated by memorandum of association that was beyond its powers set out in the statute or memorandum to be *ultra vires* and a nullity. This would be the case because the statute and the memorandum of association are treated as public documents, and the public is deemed to be aware of any limitations on the power of the corporation to

7. *Baroness Wenlock v. River Dee Co.* (1887), 36 Ch.D. 674 at p. 685.

8. Note that Special-Act corporations are not normally given this power.

contract or to do specific acts. For example, if the memorandum of association prohibits the directors of the corporation from engaging in a particular type of business activity, and the directors proceed to do so, any contract negotiated in that respect would be *ultra vires* the corporation and could not be enforced by the other contracting party. This rule would apply even though the contracting party was unaware of the limitation on the power of the corporation. This rule is known as the **doctrine of constructive notice**. To avoid this problem, most of the General-Act provinces and the federal government have specifically given corporations incorporated in their jurisdictions the powers of a natural person, and all General-Act jurisdictions have either abolished the doctrine of constructive notice, or provided in their legislation that third parties are not deemed to have notice of any unusual limitations on the rights or powers of directors or officers of a corporation.[9]

Doctrine of constructive notice

Presumption at law that everyone has knowledge of the content of all statutes.

The general effect of this clause is to protect third parties from any unusual limitations on the powers of directors or on the corporation itself. It provides that a third party contracting with the corporation is not deemed to have notice of any limitation on the rights or powers of the corporation or its officers. When this clause is coupled with the clause that provides that the corporation has the capacity of a natural person, parties acquire protection similar to that which they would have in contracts with natural persons.

In all cases where a party is dealing with a corporation, the party is entitled to rely on what is known as the **indoor management rule** for the validity of the acts of the officers of the corporation. If the officers of the corporation, for example, are required to obtain the approval of the shareholders before a contract may be effected, the party may accept the evidence submitted by the officers that shareholder approval was obtained. The party may rely on this evidence if it appears to be regular, and need not inquire further into the internal operation of the corporation. It would not apply, of course, if a party is deemed to have notice of a limitation on the powers of the corporation contained in its memorandum of association or in the statute. The application of the rule is directed to the internal operation of the corporation, of which the party normally would have no actual knowledge.

Indoor management rule

A party dealing with a corporation may assume that the officers have the valid and express authority to bind the corporation.

The duties of parties dealing with a corporation were considered in the early case of *Biggerstaff v. Rowatt's Wharf, Ltd.*,[10] where the court summarized the responsibilities in the following terms:

> What must persons look to when they deal with directors? They must see whether according to the constitution of the company the directors could have the powers which they are purporting to exercise. Here the articles enabled the directors to give to the managing director all the powers of the directors except as to drawing, accepting, or indorsing bills of exchange and promissory notes. The persons dealing with him must look to the articles, and see that the managing director might have power to do what he purports to do, and that is enough for a person dealing with him bona fide. It is settled by a long string of authorities that where directors give a security which according to the articles they might have power to give, the person taking it is entitled to assume that they had the power.

THE INCORPORATION PROCESS

In most jurisdictions, the incorporation process begins with the preparation of an application for incorporation that sets out the name of the proposed corporation, the address of the head office and principal place of business, the names of the applicants for incorporation (usually called the incorporators), the object of the corporation (although this is not required in most jurisdictions that have followed the *Canada Business Corporations Act* pattern), the share capital, any restrictions or rights attached to the shares, and any special powers or restrictions that apply to the activities of the corporation. The applicants for incorporation usually must also indicate whether the corporation will offer its shares to the public or whether it will remain "private," in the sense that it will not make a public offering of its securities. This last matter is generally quite important from the point of

9. See, for example, the *Canada Business Corporations Act*, supra, s. 17.

10. [1896] 2 Ch. 93.

view of the incorporating jurisdiction. Corporations that intend to offer their shares to the public usually must follow elaborate procedures under a jurisdiction's securities legislation or corporations Act. This is to ensure that the public is properly advised of all details of the corporation and the purpose for which the shares are offered to the public. A corporation that does not offer its shares to the public, but that may offer them to investors by private negotiation, is usually relieved of many of the special formalities imposed by the legislation to protect shareholders and the public at large. For example, the shareholders of such a corporation may, if they so desire, dispense with the formal audit of the corporation's books each year. The reason is that a small group of shareholders (as few as one person in most jurisdictions) may not require the elaborate examination of the financial records of the corporation that is necessary for a public corporation. A corporation that does not offer its shares to the public may also impose restrictions on the transfer of shares in the corporation. This is to enable the remaining shareholders to exercise some control over persons who become a part of the small group. Corporations that offer their shares to the public must not place restrictions on the transfer of their shares. A number of other differences exist between the two forms of incorporation. The distinction is made to relieve the closely held corporation of many of the onerous obligations imposed on the large public corporation for the protection of its shareholders.

The complete application for incorporation must be submitted to the appropriate office in the incorporating jurisdiction, together with the fee charged for the incorporation. In the case of a jurisdiction that issues letters patent, the company will be incorporated when the letters patent are issued. In General-Act provinces, where a memorandum of association is filed, the filing date becomes the date of incorporation. In those General-Act provinces where articles of incorporation are filed and a certificate of incorporation issued, the corporation comes into existence when the certificate is issued.

Following incorporation, the incorporators, as first directors, proceed in the case of letters patent and certificate corporations with the remaining formalities of establishing by-laws and passing resolutions. These rules for the internal operation of the corporation set out the various duties of the officers and directors. They also provide for banking, borrowing, the issue of shares, and, perhaps, for the purchase of an existing business, if the corporation was incorporated for that purpose.

If the incorporators are not the permanent directors of the corporation, they usually hold a special meeting to resign as first directors. The shareholders then elect permanent directors to hold office until the next annual meeting. The permanent directors will, among themselves, elect the officers of the corporation to hold the offices of president, secretary, treasurer, etc. Once underway, the directors carry on the operation of the corporation until the annual meeting of the shareholders. At the annual meeting, the directors and officers report to the shareholders on their performance since the previous meeting.

SHAREHOLDERS' AGREEMENTS

There are many times when potential incorporators would shy away from creating a corporation because of the vulnerability to which an individual shareholder may be exposed in a "majority-rules" environment.

In a partnership, all partners may work in the business; when a partnership is transformed into a corporation, it is the directors who will decide by majority who does what work, and for what level of pay. Not surprisingly, the creation of **shareholders' agreements** is often important, especially for the protection of minority shareholders, and especially in the initial set-up of the corporate "ground rules" between the investors. Shareholders often create agreements with the new corporation in order to guarantee their right to long-term employment in return for accepting the status of minority shareholder (in contrast, perhaps, to their former position of equal partner). Without such an agreement, a strong personal position as a partner would be transformed into a much weaker minority voice within the group.

Shareholders' agreement

An agreement between shareholders of a private corporation concerning management and/or future reorganization of the corporation such as buy-out of interests.

There are essentially three types of agreements that fall under the title of shareholders' agreements in private corporations. The first is an agreement between the shareholder and the corporation, such as the employment example above. The second is an agreement between the shareholder and other shareholders, such as how they will vote at meetings of shareholders, or the terms on which they will buy out each other's interest in the event of fundamental disagreements. These are not problematic. The third (and more technical) agreement is one between shareholders of a private corporation that in some manner restricts their behaviour or freedom of action when those shareholders sit in their capacity as directors. At law, any agreement between shareholders that restricts their power while acting in their capacity as directors must be a unanimous agreement of all shareholders. For example, shareholders may be concerned about the wisdom of the corporation buying or selling land. Such a decision would normally be one for the directors to make, subject to a normal majority vote. To create a restriction on the directors' power to make such decisions, a shareholders' agreement creating such a prohibition must be unanimous.

Restrictions on directors' powers could also be achieved by placing the restrictions in the articles of incorporation. However, doing so would have the disadvantage of making the restrictions fully public (which may be undesirable), and requiring a more troublesome formal amendment of the articles should the corporation rethink the policy at a later time. Future investors still must, by law, be made aware of the existence of any unanimous shareholders' agreement, despite the fact that an unanimous shareholders' agreement is not evident in the publicly searchable articles of incorporation. This is done through marking the face of share certificates to the effect that shares are subject to an unanimous shareholders' agreement. Unanimous shareholders' agreements act much in the same way that partnership agreements govern a partnership, and as such were recognized by the Supreme Court of Canada in *Duha Printers (Western) Ltd. v. The Queen* (1998), 159 D.L.R. (4th) 457, as part of the "constating documents" (or constitution) of a corporation.

CHECKLIST OF ISSUES ARISING IN SHAREHOLDERS' AGREEMENTS

1. Parties and number/class of shares subscribed and paid.
2. Business objects of the corporation.
3. Restrictions on powers of directors.
4. Special rules on number, nomination, election or vacancies of directors.
5. Required notice of meetings, their place, quorum and voting powers.
6. Special powers of shareholders for special (critical) decisions.
7. Appointment and remuneration of officers.
8. Plans and budgets.
9. Appointment of bankers and auditors.
10. Recordkeeping, access and shareholder rights to information.
11. Policies regarding future financing needs, policy on shareholder loans.
12. Dividend payment / retained earnings policies.
13. Requirements of confidentiality, and non-competition.
14. Terms of employment (and termination) for executives.
15. Dispute resolution mechanism (arbitration, mediation).
16. Dissolution mechanism.

MANAGEMENT ALERT: BEST PRACTICE

In considering the use of a corporate form, business operators have a wide range of competing factors to weigh, and different businesses will reach different conclusions. Limited liability will be a major factor, as is the continuing nature of the corporation. There is also a public perception of the "maturity" of a business to consider. On the other hand, insurance can help mitigate liability, there are different tax treatments to be considered, and director's liability is an issue. Moreover, corporations require much more annual "upkeep" and legal expense than other forms of business. Business operators must discuss their critical business objectives with their lawyer, including growth plans and adversity to risk, to select the best business form.

CORPORATE SECURITIES

Share

The ownership of a fractional equity interest in a corporation.

Corporations may issue a variety of different securities to acquire capital for their operations. One of the most common forms of acquiring capital is to issue **shares** in the organization. A share is simply a fraction of the ownership of the corporation. It represents a part-ownership equal to one part of the total number of shares issued. For example, if a corporation has issued 1,000 shares, each share would represent a one-thousandth part-ownership of the corporation. Unfortunately, corporations may issue many different kinds of shares at different values, and with different rights attached to them. To determine the actual value of the part-ownership that a share represents is sometimes difficult.

Shares may be designated as having either a fixed value or par value, or no par value. In the latter case the value is not fixed, but is the issue price of the share established by the board of directors at the time of the sale of the share. Par value shares have diminished in importance in recent years, and in many of the provinces and federally, par value shares have been abolished.[11] Shares may also be classed as either common or preference shares. All corporations must have some voting common shares, as these are the usual form of shares issued to the shareholders who will elect the directors of the corporation. Corporations sometimes issue preference shares. As the name denotes, these will have special rights attached to them, such as the right to a fixed rate of return in the form of dividends, special voting privileges, or a priority in payment over the common shareholders in the event that the corporation should be wound up.

Floating charge

A debt security issued by a corporation in which assets of the corporation, such as stock-in-trade, are pledged as security. Until such time as default occurs, the corporation is free to dispose of the assets.

Debenture

A debt security issued by a corporation that may or may not have specific assets of the corporation pledged as security for payment.

A corporation may also issue securities that do not represent a share of the ownership of the corporation but a debt. The debt may be either secured by a fixed charge attaching to specific assets of the corporation or a **floating charge**, which does not usually attach to any particular assets of the corporation but simply to the assets in general. Corporate securities that represent a charge against specific assets of the corporation are generally called mortgage bonds. Those that are normally subordinate to mortgage bonds in priority are usually called **debentures**. The degree of security that each type represents depends to a considerable degree on the priority of the rights that it has to the corporation's various assets in the event that the corporation should default on its debts. The rights of these security holders are examined in Chapter 29, "Security for Debt" and in Chapter 30, "Bankruptcy and Insolvency." Securities regulation in terms of the issue and exchange of securities is discussed in greater length in Chapter 18, "Securities Regulation."

THE TAXATION OF CORPORATIONS

While the taxation of corporations is beyond the scope of this text, it is important to note that a corporation is a separate entity at law, and consequently, is subject to taxation on its income from its business activities. The shareholders of the corporation are not subject to this tax, but may in turn be subject to taxation on any dividends that the corporation may declare. In recognition of the fact that both the corporation and the shareholders would be taxed on the part of the corporation's income declared as dividends, tax legislation provides for a dividend tax credit to the shareholders. This credit is designed to reduce the actual tax payable by the shareholders on the dividends received. Separate legislation covers the taxation of corporations due to their complexity, but it is important to recognize that a corporation's earnings are its own for taxation purposes.

DIVISION OF CORPORATE POWERS

Duties and Responsibilities of Directors

Every corporation must have at least one director. In some jurisdictions, notably Quebec and Prince Edward Island, where the Letters Patent method of incorporation is used, the minimum number of directors is three. The directors of a corporation are, in effect, the managers of the business. Unlike a partnership, where all partners may bind the

11. See for example, the *Canada Business Corporations Act*, supra, s. 24(1); *Business Corporations Act*, R.S.S. 1978, c. B-10, s. 24(1); *Business Corporations Act*, R.S.O. 1990, c. B.16, s. 22.

partnership, in a corporation this right is limited to the directors, and in most cases to certain directors who may be officers of the corporation. Thus, in a corporation, the directors, once elected by the shareholders, are responsible for its operation. The various corporation Acts attempt to separate ownership from management insofar as it is practicable, in order to clearly identify the particular rights and duties of shareholders and directors. Most rights with respect to management have been given exclusively to the directors. For example, the right to declare dividends and to conduct the business of the corporation fall exclusively to the directors. Major changes in the nature of the corporation, although initiated by the directors, must generally be referred to the shareholders to confirm. For example, the shareholders must be the ones to ultimately decide if the corporation is to change its objects, wind up, or alter its capital structure. The purpose of the division of powers between shareholders and the directors is essentially one of balancing the need for shareholder protection with the need for freedom to manage on the part of the directors. This balance is reflected in the rights and duties of each group.

The directors are responsible for the day-to-day operations of the corporation. Once the shareholders have elected a board of directors, they have little more to do with the management of the corporation until the next annual general meeting. The ultimate responsibility does rest with the shareholders, however, as they are free to elect new and different directors at the next annual meeting if the directors fail to perform satisfactorily. Under extreme circumstances the shareholders may also take steps to terminate the appointment of directors before the next annual meeting.

A director, once elected, has a duty to conduct the affairs of the corporation in the best interests of the corporation as a whole, rather than in the interests of any particular group of shareholders. This distinction is most important because the directors may be held accountable at law for a breach of their duty to the corporation. Usually the shareholders' interests and the interests of the corporation are the same, but this is not always so. Where they are divergent, the directors must concern themselves with the interests of the corporation alone.

Fiduciary

A relationship of utmost good faith in which a person, in dealing with property, must act in the best interests of the person for whom he or she acts, rather than in his or her own personal interest.

The relationship between a director and the corporation is **fiduciary** in nature. It requires the director to act in good faith at all times in his dealing with, and on behalf of, the corporation. At Common Law, the duty on the part of the director to act in good faith is augmented by a duty to use care and skill in carrying out corporation business. Some jurisdictions have imposed an additional statutory duty on directors to exercise the powers and duties of their office honestly with the care and skill of a reasonably careful and prudent person in similar circumstances.[12]

The fiduciary relationship that a director has with the corporation precludes the director from engaging in any activity that might permit the director to make a profit at the corporation's expense. For example, the director must not use the corporation's name to acquire a personal benefit, nor may he or she use the position in the corporation to make a personal profit that rightfully belongs to the corporation.

COURT DECISION: Practical Application of the Law

Promoters of Corporation—Duty of Disclosure of any Secret Profit Earned

***Gluckstein v. Barnes*, [1900] A.C. 240.**

A number of promoters acquired property, then incorporated a limited company to purchase the property from them. The promoters elected themselves as first directors of the company and then solicited applications for the purchase of shares by way of a prospectus. The prospectus did not disclose the fact that the promoters would earn a secret profit of £20,000 on the sale.

THE COURT: My lords, in this case the simple question is whether four persons, of whom the appellant is one, can be permitted to retain the sums which they have obtained from the company of which they were directors by the fraudulent

12. See, for example, the *Canada Business Corporations Act*, supra, s. 122(1).

pretence that they had paid £20,000 more than in truth they had paid for property which they, as a syndicate, had bought by subscription among themselves, and then sold to themselves as directors of the company. If this is an accurate account of what has been done by these four persons, of course so gross a transaction cannot be permitted to stand. That that is the real nature of it I now proceed to shew. The prospectus by which money was to be obtained from the public disclosed the supposed profit which the vendors were making of £40,000, while in truth their profit was £60,734 6s.1d., and it is this undisclosed profit of £20,000, and the right to retain it, which is now in question. Of the facts there cannot be the least doubt; they are proved by the agreement, now that we know the subject-matter with which that agreement is intended to deal, although the agreement would not disclose what the nature of the transaction was to those who were not acquainted with the ingenious arrangements which were prepared for entrapping the intended victim of these arrangements. My Lords, I decline to discuss the question of disclosure to the company. It is too absurd to suggest that a disclosure to the parties to this transaction is a disclosure to the company of which these directors were the proper guardians and trustees. They were there by the terms of the agreement to do the work of the syndicate, that is to say, to cheat the shareholders; and this, forsooth, is to be treated as a disclosure to the company, when they were really here to hoodwink the shareholders, and so far from protecting them, were to obtain from them the money, the produce of their nefarious plans. I do discuss either the sum sued for, or why Gluckstein alone is sued. The whole sum has been obtained by a very gross fraud, and all who were parties to it are responsible to make good what they have obtained and withheld from the shareholders.

It is important to understand that Common Law remedies are not only the basis for statutory securities regulation today (see Chapter 18), but still exist independently of statute.

A director may, under certain circumstances, engage in a business transaction with the corporation. However, considerable care is necessary, otherwise the director will be in violation of this duty as a director. As a general rule, in any transaction with the corporation, the director must immediately disclose his or her interest in the particular contract or property and refrain from discussing or voting on the matter at the directors' meeting.[13] As well, in some jurisdictions, shareholder approval of contracts in which a director has an interest is required.[14]

It should be noted, however, that as a general rule, if a director has a conflict of interest, the conflict does not prevent the director from exercising his or her rights as a shareholder at any shareholders' meeting called for the purpose of considering any matters related to the conflict of interest. However, the actions of the shareholder at the shareholders' meeting must be such that they are not illegal, fraudulent, or oppressive towards the shareholders who may oppose the decision related to the conflict of interest. For example, in the case of *North-West Transportation v. Beatty*[15] a director and shareholder of the company entered into an agreement with the company to sell it a steamship. The directors approved the purchase, and the purchase was then directed to the shareholders for approval. At the shareholders' meeting, the owner of the steamship voted in favour of the purchase, and because of his shareholding, the purchase was approved. In the case, the court set out the general rule in the following words:

> The general principles applicable to cases of this kind are well established. Unless some provision to the contrary is to be found in the charter or other instrument by which the company is incorporated, the resolution of a majority of the shareholders, duly convened, upon any question with which the company is legally competent to deal, is binding upon the minority, and consequently upon the company, and every shareholder has a perfect right to vote upon any such question, although he may have a personal interest in the subject-matter opposed to, or different from, the general or particular interests of the company.

A director may not normally engage in any business transaction with a third party that might deprive the corporation of an opportunity to make a profit or acquire a particular

13. ibid., s. 120.
14. The *Business Corporations Act*, R.S.O. 1990, c. B.16, s. 132.
15. (1887), 12 App. Cas. 589.

Doctrine of corporate opportunity

The use of corporate information for a personal benefit to the detriment of the corporation.

asset. For example, a director might become aware of an opportunity to acquire a valuable property through his or her position as a director. That property must not be acquired for personal use if the corporation might be interested in obtaining it. To do so would be a violation of the director's duty of loyalty to the corporation. In an instance of this nature, the courts would apply the principle or **doctrine of corporate opportunity**, and find that the director's acquisition of the property was in trust for the corporation. They would treat the corporation as the beneficial owner. If the director had already disposed of the property, the court might require him or her to deliver up any profit made on the transaction to the corporation.

A similar situation may arise when a director trades in shares of the corporation. Directors may lawfully buy and sell shares of the corporation on their own account and retain any profit that they might make on the transactions. However, if directors use information that they acquire by virtue of their position in the corporation, and buy or sell shares using that information to the detriment of others, they may be liable for the losses that those persons suffer as a direct consequence of their actions. In most jurisdictions, the legislation pertaining to corporations requires "insiders" (such as the directors, officers, and persons usually holding over 10 percent of the shares in a public corporation) to report their trading each month to a government regulatory body or official. This information is then made available to the public as a deterrent to directors who might be tempted to use inside information for their own profit.

For example, the issue concerning the particular fiduciary duties of directors and officers of a corporation came before the Supreme Court of Canada for consideration in the case of *Canadian Aero Service Ltd. v. O'Malley et al.*[16]

Two directors of the corporation used information available to them as directors for the purpose of gaining a benefit for a new corporation that they incorporated after resigning their positions. The court considered the extent of their duty to the corporation from which they resigned in the following manner:

> ... the fiduciary relationship goes at least this far: a director or a senior officer like O'Malley or Zarzycki is precluded from obtaining for himself, either secretly or without the approval of the company (which would have to be properly manifested upon full disclosure of the facts), any property or business advantage either belonging to the company or for which it has been negotiating; and especially is this so where the director or officer is a participant in the negotiations on behalf of the company.
>
> An examination of the case law in this Court and in the Courts of other like jurisdictions on the fiduciary duties of directors and senior officers shows the pervasiveness of a strict ethic in this area of the law. In my opinion this ethic disqualifies a director or senior officer from usurping for himself or diverting to another person or company with whom or with which he is associated a maturing business opportunity which his company is actively pursuing; he is also precluded from so acting even after his resignation where the resignation may fairly be said to have been prompted or influenced by a wish to acquire for himself the opportunity sought by the company, or where it was his position with the company rather than a fresh initiative that led him to the opportunity which he later acquired.

16. (1973), 40 D.L.R. (3d) 371.

FRONT-PAGE LAW

Golden Rule's ex-CEO gets jail time, $4-million fine; Hid poor samples while selling shares for millions

Glen Harper, the former chief executive of Golden Rule Resources Ltd., was sent to jail yesterday — albeit for only six hours.

Mr. Harper was handed two concurrent one-year sentences and fined nearly $4-million for insider trading offences, marking the first time in almost a decade regulators have secured jail time for the offence. But by the afternoon he was free on bail, pending an appeal. He was to return to his Calgary home where he was ordered to reside as part of the conditions of his release.

Mr. Harper was found guilty of hiding poor soil sample results at the same time he was selling millions of dollars of Golden Rule shares for his own or his immediate family's personal gain. In addition to the jail time, Mr. Harper received one of the heftiest fines ever imposed on an individual.

From Jan. 3, 1997, to May 6, 1997, Mr. Harper sold shares for $4-million and made purchases through stock options for $1-million, while covering up samples that suggested the company's gold find in Ghana was not as promising as had been suggested.

Who and what are being protected in the prosecution of insider-trading offences? For further reference see Chapter 18.

Source: Excerpted from Garry Marr, *Financial Post*, "Golden Rule's ex-CEO gets jail time, $4-million fine," September 19, 2000.

Personal Liability of Directors

Over the years, there has been a general trend toward holding the directors liable for different events that might take place as a result of their actions. Many of these responsibilities have been imposed on directors as a result of their use of their position or the corporation in such a way that it causes hardship to third parties, employees, or shareholders.

As a general rule, the directors may be held liable for any loss occasioned by the corporation itself, if the directors commit the corporation to an act that is clearly *ultra vires* regarding the corporation's objects clause or contrary to its by-laws. The corporation usually is the body that suffers a loss in such a case, and the shareholders may take action against the directors to recover the loss. The directors would also be liable to the company if they should sell shares at a discount contrary to the statute, or declare a dividend that impairs the capital of the corporation.

More so today than in the past, directors are expected to play a greater and more active role in the operation of the corporation itself. Governments have increasingly imposed obligations on business corporations with respect to activities that have an environmental impact or when economic losses of the corporation result in plant closures and job losses. Penalty or liability provisions in these statutes often are directed at not only the corporation, but the directors as well. For example, a shoe-manufacturing company stored a number of containers of industrial waste of a toxic nature in an outside location without protection from the weather. Some of the containers leaked toxic waste into the soil and groundwater. The company was charged under the provincial environmental-protection statutes, as were the directors of the corporation. The directors were obliged under the statute to prove that they were not negligent, and only one director was successful in doing so. The corporation and the remaining two directors were convicted and personally fined for their breach of the law.[17]

In some jurisdictions, a special liability is imposed on directors with respect to employee wages in the event of bankruptcy of the corporation. The law imposes a duty on the directors to satisfy the amounts owing to employees for unpaid wages if the corporation lacks the funds to do so. The liability for unpaid wages is, in a sense, a contingent liability, because it would only come into play if the assets of the corporation were insufficient to satisfy the employee wage claims.

In addition to these particular liabilities, the directors are exposed to a number of penalties and fines if they should fail to comply with the legislation concerning the filing

17. *Regina v. Bata Industries Ltd., Bata, Marchant and Weston* (1992), 9 O.R. (3d) 329.

of notices, or if they should fail to make returns to the various government agencies that monitor corporation activity.

It is a sad comment on business leaders that, in addition to oversights and failures, repeated corporate scandals have forced shareholders to continually ask whether their directors are involved in fraud. There are a number of mechanisms available to combat director fraud, and the first line of defence lies with the shareholders themselves. Shareholders should not be passive observers at annual general meetings, and should demand full disclosure of all information to which they are entitled. Moreover, a corporation can exempt itself from audit only if all shareholders agree. Thus, if there is any concern of fraud or suspect management practice, shareholders should withhold their consent to audit exemption, and qualified examiners will review the books thereafter. Those same shareholders, if suspicion is borne out, should exercise their right to remove a director or directors, tackling the problem at its source. The shareholders can broaden their offensive with a civil action (in their own name, or possibly as an oppressed shareholding minority, or possibly in the name of the corporation itself) against a director or directors. As a final option, shareholders may seek intervention by the police, which may result in serious fraud charges under the *Criminal Code of Canada*, sections 380-396, or proceeds of crime and money laundering charges under section 462.31. A conviction for either fraud or proceeds of crime can result in a penalty of ten years imprisonment.

Director's Defence of Due Diligence

With director fraud being a feature of the new century, it is also worth considering the limits to director liability. Absent the outright fraud described above, the limits of liability are contingent on two factors: does the action under review lead to absolute liability, or is it one where a defence of due diligence can be raised? In the first instance, it would not even be necessary for a director to know of the offence for his or her liability to result. In the second instance, where the applicable law (environmental, tax, securities regulation, health and safety, etc.) indicates a director must have exercised all reasonable care, the practical question is: *what care amounts to reasonable care?*

Under section 122(1)(b) of the federal *Corporations Act*, directors must "exercise the care, diligence and skill that a reasonably prudent person would exercise in comparable circumstances." Directors therefore must exercise their ability, experience and education in the same way as a reasonably prudent person would do in a similar situation, which would include not being passive or acquiescent to an offence committed by other directors.

Due diligence

The obligation on the directors of a corporation to ensure that effective systems are in place to comply with legislation, and to monitor the systems to ensure compliance.

The actual **due diligence** will vary with the circumstances, but some useful direction can be given. In the shoe-manufacturing/environmental case referred to earlier (*R. v. Bata*), the original trial judge set the following criteria:

> I ask myself the following questions in assessing the defence of due diligence:
>
> Did the Board of Directors establish a pollution prevention "system." Was there supervision or inspection? Was there improvement in business methods? Did he exort those he controlled or influenced?
>
> Did each Director ensure that the Corporate officers have been instructed to set up with a system sufficient within the terms and practices of its industry of ensuring compliance with environmental laws, to ensure that the officers report back periodically to the Board of the operations of the system, and to ensure that the officers are instructed to report any substantial non-compliance to the Board in a timely manner?
>
> The Directors are responsible for reviewing the environmental compliance reports provided by the officers of the corporation but are justified in placing reasonable reliance on reports provided to them by corporate officers, consultants, counsel or other informed parties.
>
> The Directors should substantiate that the officers are promptly addressing environmental concerns brought to their attention by government agencies or other concerned parties including shareholders.

The Directors should be aware of the standards of their industry and other industries which deal with similar environmental pollutants or risks.

The Directors should immediately and personally react when they have notice the system has failed.[18]

Clearly the defence is not simply to say "Well, I tried" — the director must show evidence of knowledge of prevailing standards, and clear action taken to prevent, report, review, and respond to danger, and to react to failure when it occurs.

Outside director

A director who is not an officer or employee of the corporation.

While accepting and exercising due diligence is one effective approach to minimizing director liability, an additional precaution can lie in the use of **outside directors**. Outside directors are paid directors who are not officers or otherwise employed in the day-to-day management of the company. This creates a certain impartiality in the director's deliberations at board meetings. Of course, even impartial directors must adequately inform themselves of the business circumstances of the firm and the questions before the board upon which they are asked to vote.[19]

Director's Duties in Commercial Decisions

Business judgment rule

The reluctance of the court to interfere with decisions of a board of directors.

In the past (prior to the Enron and WorldCom scandals), courts were willing to give great latitude to directors. Now the "**business judgment rule**" — that a court would be reluctant to interfere with decisions of a board of directors — is essentially limited to those cases where the directors can show that they informed themselves of the issue and exercised real judgment. This amounts to a judicial requirement of good faith and due diligence.[20]

The Supreme Court of Canada has spoken with respect to whom a duty is owed in commercial decisions. The 2008 case, *BCE Inc. et al. v. A Group of 1976 Debentureholders*, tested the question whether directors were accountable to others beyond their shareholders for the commercial effects of their decisions. The Court refused to expand that accountability to include a group of debenture holders who were disadvantaged by a commercial decision of the directors, although at time of writing, the reasons for this decision have not been released. In a 2004 case, *Peoples Department Stores v. Wise*, the Court elaborated on the quality of decision making it expected from directors: reasonable in all the circumstances, but not necessarily perfect.

CASE IN POINT

The Supreme Court of Canada: "Directors and officers will not be held to be in breach of the duty of care under s. 122(1)(b) of the *CBCA* if they act prudently, and on a reasonably informed basis. The decisions they make must be reasonable business decisions in light of all the circumstances about which the directors or officers knew or ought to have known. In determining whether directors have acted in a manner that breached the duty of care, it is worth repeating that perfection is not demanded. Courts are ill-suited and should be reluctant to second-guess the application of business expertise to the considerations that are involved in corporate decision making, but they are capable, on the facts of any case, of determining whether an appropriate degree of prudence and diligence was brought to bear in reaching what is claimed to be a reasonable business decision at the time it was made."

Peoples Department Stores Inc. (Trustee of) v. Wise, 2004 SCC 68.

18. ibid.

19. *UPM-Kymmene Corp. v. UPM-Kymmene Miramichi Inc.* (2002), 214 D.L.R. (4th) 496 (Ont. Sup. Ct. J.).

20. ibid., passim.

Corporate Compliance and Director's Obligations — Impact in Canada of U.S. Law

Sarbanes-Oxley Act
A U.S. statute that imposes extensive duties on corporations to ensure accuracy of financial and securities information provided to the public.

In 2002, the United States Congress responded to the earlier financial scandals of Enron and WorldCom and others by creating the ***Sarbanes-Oxley Act***. Named for its sponsors, Senator Paul Sarbanes and Representative Michael Oxley, the Act represents a far-reaching set of accountability and disclosure requirements aimed at rebuilding public confidence in capital markets. While being an American law, the Act affects Canadians as well, as it applies to the operations of publicly-traded U.S. firms (who may have subsidiaries in Canada) as well as any Canadian firm whose securities are traded in U.S. markets, or are already subject to U.S. reporting requirements. The Act has a number of policy targets: creation of a public company accounting oversight board, auditor independence, corporate responsibility for financial reports, enhanced financial disclosure, conflicts of interest by analysts, and corporate and criminal fraud liability.

The oversight board (Section 102) requires registration and supervises compliance by accounting firms that audit public companies, and places significant restrictions on the types of non-audit services that those auditors may offer to their clients. The Act forbids any service which may compromise the independence of the audit, such as record-keeping, advising on financial information systems design or appraisals and valuation opinions. Section 302 requires the Chief Executive Officer (CEO) and Chief Financial Officer (CFO) of publicly-traded firms to review all financial reports, and ensure they are fairly presented and free of misrepresentations, as well making them responsible for internal accounting controls and any deficiencies or changes. Corporate management is also responsible for creating, operating and maintaining an adequate and effective internal control structure. Section 404 requires management reporting on that control procedure and effectiveness, and an accompanying attestation by the firm's independent auditor. Inaccuracy can be punished (with personal liability) ranging up to $1 million or ten years in prison, and deliberate fraud (or attempt, conspiracy or concealment) can attract penalties of $5 million and twenty years in prison. *Sarbanes-Oxley* also imposes requirements for information disclosure to the public, which are further discussed in Chapter 18.

As the United States is leading on this issue, and the influence of its economy and stock markets has significant impact for Canadian business, our jurisdictions are at various stages of creating similar oversight boards for accountants (for example, Ontario's *Public Accounting Act*[21]) and strengthening corporate accountability requirements, including director's liability for misleading disclosures and press releases.

CASE IN POINT

Sarbanes-Oxley applies to the Royal Bank of Canada because RBC shares are publicly traded on U.S. stock markets. Having provided certain non-audit services to RBC, PricewaterhouseCoopers LLP resigned as RBC's auditor in 2003, concerned that continuing as auditor would place it in violation of *Sarbanes-Oxley*. While the non-audit services were probably acceptable, it was judged not to be worth the risk to continue as auditor, given both the strict application of the rules and dire penalties for violation. RBC was fortunate that it had a second auditor under engagement — for almost any other public company, such a resignation would have resulted in a mad scramble to find a replacement auditor, because firms without audited financial statements are suspended from trading on stock markets.

Shareholders' Rights

The shareholders, as owners of the corporation, are entitled at regular intervals to a full disclosure by the directors of the corporation's business activities. In addition to the right to information, the shareholders have the right to elect the directors at general meetings

21. S.O. 2004, c. 8.

of the shareholders and to approve the actions of the directors since the previous annual meeting. Shareholders must also approve all important matters that concern the corporation. These usually take the form of special by-laws. Generally these special by-laws do not become effective until shareholders' approval has been received. In this fashion, the shareholders have ultimate control over all major decisions of the directors affecting the corporation's structure or purpose.

Each year the shareholders review the management of the directors at the annual general meeting of the corporation. At that time they elect directors for the ensuing year. All common shareholders have a right to vote at the meeting, with the number of shares held determining the number of votes that a shareholder might cast.

At the general meeting, the president of the corporation usually explains the activities of the corporation over the past year and may discuss the general plans for the firm's next year. The president or the treasurer may present the financial statements of the corporation and comment on them, as well as answer questions raised by the shareholders. The financial statements are prepared by a public accounting firm that is engaged as an auditor, and the auditor's report is attached to them. The auditor's report is discussed by the shareholders. If necessary, the auditor may be present to explain matters of a financial nature that pertain to the audit. The shareholders also appoint an auditor for the next year at the annual meeting. The auditor's duty is to the shareholders, not to the directors of the corporation. The auditor is responsible for the examination of the books and financial affairs of the corporation on the shareholders' behalf. In order to perform the audit, the auditor has access to all financial accounts and books, but the shareholders do not. The shareholders are only entitled to the financial reports provided by the directors' and auditor's report at year-end. The shareholders do have access, however, to certain other corporate records. A shareholder may, for example, examine the minute-books of shareholders' meetings, the shareholder register, the list of directors, the incorporating documents, and the by-laws and resolutions of the corporation. However, a shareholder may not examine the minute-books of directors' meetings.

Special meetings of shareholders may be convened for a variety of purposes relating to the corporation if a number of shareholders[22] believe that the meetings are necessary. The group of shareholders usually request the directors to call the meeting. But if the directors should fail to do so within a fixed period of time, the shareholders may do so themselves, and the corporation will be obliged to reimburse them for the expenses incurred.[23] At the meeting, the shareholders may deal with the subject matter of the requisition and, where appropriate, take action in the form of a by-law or resolution. An example of a special meeting called by shareholders may be to remove an auditor, or to object to a particular course of action that the directors have taken that might be reversed by a special resolution or by-law of the shareholders.

The fact that shareholders must approve all important decisions of the directors prevents the directors from engaging in certain activities that might be contrary to the interests of the majority of the shareholders. Unfortunately, it does little to protect minority shareholders if the majority of the shares are held by the directors or those who support them. Because "the majority rules" for most decisions within a corporation, a minority shareholder has only limited rights where there is a misuse of power by the majority. At Common Law, it is normally necessary for the complainant to show some injury has occurred as a result of the decision of the majority. This is particularly difficult in the case of a minority shareholder of a corporation, because the corporation, and not the shareholder, is usually wronged by the misuse of power or the breach of duty.[24] When the corporation is controlled by the very majority that has committed the breach of duty or misused its power, the corporation is not likely to take action against that particular group.

To overcome this difficulty, the Common Law courts have recognized a number of exceptions to the "majority rule." They have permitted minority shareholders to take

22. The number is usually presented as a small percentage of the total number of shares outstanding. See, for example, *Canada Business Corporations Act*, supra, s. 143, where the number is set at 5 percent.

23. The *Canada Business Corporations Act*, supra, s. 143.

24. This situation arose in the case of *Foss v. Harbottle*, [1834] 2 Hare 461, 67 E.R. 189.

action on behalf of the corporation against the majority (or the directors) in the following instances:

(a) where the act objected to is *ultra vires* the corporation;
(b) where the act personally affects the rights of the minority shareholders;
(c) where the corporation has failed to comply with the procedural requirements for approval of the act; or
(d) where the act of the majority constitutes a fraud on the minority shareholders.

The last exception to the general rule includes cases where the majority attempt to appropriate the minority interests for themselves, or where they attempt to acquire corporate property at the expense of the corporation.

Unfortunately, a Common Law action taken against the directors by a shareholder who believes that the directors have acted improperly is not an easy matter to bring before the courts. The difficulty lies in the fact that the corporation is the body injured by the alleged improper action of the directors, and the shareholder is only one of perhaps a great many persons indirectly injured or affected. Thus, the shareholder must take action for the corporation on behalf of all persons indirectly injured by the injury to the corporation, using a type of class action known as a derivative action.

The Common Law courts will not proceed, however, with derivative actions of this nature unless the shareholder can first satisfy the court that all internal attempts to have the matter resolved have been exhausted, and that all reasonable demands to rectify the problem were refused by the directors. Then the shareholder must establish that the actions or decisions of the directors were improper, and did in fact cause injury to the corporation. Finally, the shareholder is usually required by the court to provide some security for the costs of the action should the claim fail, as derivative actions tend to be very lengthy and expensive cases.

To overcome the difficulties associated with the Common Law derivative action, corporate legislation patterned after the *Canada Business Corporations Act*[25] usually provides some relief for the shareholder who believes that the corporation has been injured through some act or omission of the directors. For example, under the *Canada Business Corporations Act*,[26] a shareholder may apply to the court for permission to institute the action by satisfying the court that (1) reasonable notice was given to the directors of the shareholder's intention to apply to the court if the directors failed to do so, and the directors refused to act; (2) it would appear to be in the interests of the corporation that the matter be dealt with; and (3) the complainant shareholder is acting in good faith in making the application to the court.

If a shareholder satisfies these conditions, the court may then permit the action to be brought before it. The court may also, if it so desires, order the corporation to pay reasonable legal fees incurred by the complainant as well.[27]

Minority shareholders, particularly those in corporations that do not offer their shares to the public, are also protected in a similar manner by the *Canada Business Corporations Act*[28] and comparable provincial legislation. A minority shareholder who believes, for example, that actions taken by the majority are repressive of the minority shareholders' rights, may apply to the court for relief. If the court is satisfied that the actions of the corporation are oppressive, unfairly prejudicial, or unfairly disregard the interests of any security holder, the court may rectify the matter under the broad powers given to it under the statute. These powers include the right of the court to order the corporation to purchase the securities of the aggrieved shareholder, to restrain the improper conduct, or, in an extreme situation, to order the liquidation or dissolution of the corporation.

The remedial powers of the courts have been examined briefly here in connection with shareholder rights, but it is important to note that complaints may also be brought

25. Certificate-of-incorporation provinces as well as the federal government have this type of provision in their statutes. British Columbia provides for a similar procedure in its legislation.

26. *Canada Business Corporations Act*, supra, s. 239.

27. ibid., s. 240.

28. ibid., s. 241.

by others who might be affected as well. For example, a director or officer of the corporation or one of its affiliates, or any other person the court deems a proper person to make an application, may be considered a complainant under the Act.[29]

CASE IN POINT

Disasters spawn litigation. Trains collide or derail, planes crash, ships sink, lakes and rivers become polluted, chemical factories explode, ordinary people eat, drink, wear or use unhealthy or defective products. People — sometimes hundreds, even thousands — are injured or killed by these events. When the crisis subsides, some of the victims turn to the courts for redress and compensation.

Macpherson J.A. in *Carom v. Bre-X Minerals Ltd.* (2000), CanLII 16886 (Ont. C.A.).

So begins the judgment in the infamous Bre-X gold fraud case. The creation of a "class action," and the specific identification of who is a member of that class (and when) determines who may share in the proceeds of the litigation, if any arise. The class action avoids a multiplicity of individual actions, and creates expediency without a loss of justice. It is common to find class action suits arising in corporate litigation, as the potential number of claimants is often large and geographically diverse.

Dissolution

Theoretically, a corporation has an infinite life span. However, in reality, events may occur that have the effect of limiting the existence of the entity. While these events may take many forms, the most common, at least with respect to the business corporation, is undoubtedly the inability of the corporation to carry on its affairs in a profitable manner. When this situation arises, the corporation, if it is solvent, may wind up its operations and surrender its charter or apply for a certificate of dissolution. If the corporation is in the unfortunate situation of being insolvent, then it may be involuntarily dissolved through corporate winding-up proceedings.

The procedure in both instances is complex, but the result is the same. The corporation ceases to exist when the process is completed. However, when the directors of a solvent corporation make a conscious decision to cease operations and dissolve the entity, the shareholders must approve the proposal. An application may then be made to the incorporating jurisdiction to have the corporation's existence cease. The corporation must follow a specific procedure set out in the statute, depending upon whether it has been operational or not. This usually includes the filing of articles of dissolution (in the case of a certificate of incorporation jurisdiction) along with the appropriate approval of the shareholders, or filing a notice of intention to dissolve. The appropriate procedures are then resumed. They normally include the publication of notice of the intention. When the procedure is completed, a certificate of dissolution is issued, formally putting an end to the existence of the corporation. A somewhat different procedure must be followed for letters patent corporations and memorandum corporations under General-Act legislation.

A QUESTION OF ETHICS

Corporation law shields investors from unlimited liability for business debts, and this is certainly important in making our business world function. On the other hand, an unethical operator might use a corporation to take payments, drain out a dividend, and deliver nothing. How would you balance investor protection and third-party protection? What factors would you want a judge to consider when tempted to "lift the corporate veil" and allow claims against unethical shareholders?

29. ibid., s. 238.

CONSIDERATIONS IN THE PURCHASE AND SALE OF A CORPORATION

In consequence of its shares representing ownership, there are two ways to effectively purchase a "business" that is a corporation: one may purchase its assets directly, or one may purchase its shares. The two different avenues can however create different results for both parties.

In the case of the seller, there is usually a difference in the tax treatment of money received from the corporation's sale of corporate assets versus the sale of an owner's shares, thus from time to time, the vendor will have a tax-driven preference for one form of sale over the other. As for the purchaser, a purchase of assets is just that — equipment, land, inventory, customer lists, intellectual property, and so on. The advantage to the purchaser in an asset purchase is that all things which, from case to case, may be undesirable do not need to be purchased: a parcel of contaminated land, obsolete inventory, certain employment contracts, or lawsuits pending against the company. The purchaser is in a position to pick and choose what he or she desires to buy. On the other hand, in the instance where the purchaser buys the shares of the company, he or she is buying everything, including the undesirables — and other unknown potential liabilities. The shares and their value represent the net of all things good and bad about owning the company, from knowledge of its secret production processes through to the unknown fact that six months from now a lawsuit will be launched against it.

Thus a purchaser may have a general preference for an asset purchase of known things, while a seller may prefer to be rid of everything and lean toward selling shares. A purchaser compelled to buy shares will insist on a lower price to reflect the risk of potential liabilities taken on, while the seller will demand a relatively higher price in an asset purchase, as the buyer is selecting only the gems of the corporation.

CHECKLIST OF CONCERNS IN A SHARE PURCHASE

1. Understanding the corporate structure:
   - Review of all corporate documentation and records.
   - Lists and records of corporate securities outstanding.
   - Searches of public record registries.
2. Understanding the corporation's assets:
   - Real and personal property.
   - Intellectual and intangible property.
   - Associated public searches (where possible) confirming all of above.
3. Understanding the corporation's liabilities:
   - Employees – current contracts and pension obligations.
   - Physical assets – encumbrances (see also asset purchase checklist).
   - Financial – indebtedness and taxation.
   - Regulatory – licences and compliance in good standing.
   - Environmental – possible contingent liabilities.
   - Product – dangerous goods.
   - Litigation – now or pending.
   - Associated public or private searches, or private assurances, confirming all above.
4. Understanding the corporation's business:
   - General operations.
   - Critical success factors.
   - Terms of supplier contracts, customer contracts, labour contracts.
   - Legal obligations arising from the above.

Once each aspect is understood, it can be valued (as an asset or a liability), and a valuation of the shares can therefore be determined.

CHECKLIST OF CONCERNS IN AN ASSET PURCHASE

1. Vendor has title to the assets.
2. Reasonable valuation of assets has been made.
3. Disclosure of any liabilities attached to the assets (debts, liens, mortgages, charges) has been made.
4. Vendor has authority to sell, notice to any third parties has or will be given (tenancies, debts or rights to be assigned).
5. Physical condition and location of the assets is understood.
6. Intangible assets (goodwill, intellectual property) are properly protected and delivered.
7. Government consent is either obtained or unnecessary.

Once each aspect is understood, a reasonable valuation of the assets can be made.

FACTORS RELEVANT TO CHOOSING THE CORPORATE FORM

You may wish to keep in mind the general and relative advantages and disadvantages of the various forms of business organization, as shown below. Individual cases will vary in the relative importance and degree of the individual factors.

Factor	Sole Proprietorship	Agency	Partnership	Corporation
Difficulty in establishment	Easy, little beyond registration	Agreement	Agreement and registration	Significant, can be complex
Cost	Low	Virtually none	Medium	High
Maintenance	Low	Virtually none	Medium	High
Ability to generate capital	None beyond owner's capital and credit	Essentially none	Joint resources of partners in capital and credit	Highest, including share offering
Difficulty in management	Easiest; owner only	Monitoring	Significant	Highest
Liability of owner	Near absolute	Essentially absolute	Shared	Little or none
Transfer of ownership or responsibility	Sale of assets only, with assignment of contracts	Assignment of contract	More complex	Easiest, by sale of shares
Term of operation	Maximum term governed by life of owner	Term of agency contract	Not beyond death or bankruptcy of any partner	Unlimited
Dissolution	Easy	Termination	More complex, can be contentious	Most complex, costly, often contentious

SUMMARY

A corporation is a legal entity that has an existence separate from its owners. The powers that the corporation may acquire vary from something close to the powers possessed by a natural person to very restricted and narrow powers, where the corporation is incorporated by a legislature under a statute for a very specific purpose.

The method of incorporation varies from province to province. It may take a number of different forms, each granting a corporation different rights and powers. In some provinces, incorporation is by the issue of letters patent, which is very similar to the royal charter form of incorporation and gives the corporation virtually all of the powers of a natural person. Other provinces provide for incorporation under a general statute, which limits the powers of the corporation to those powers acquired under the statute and contained in the memorandum of association filed under it. The federal government and several provinces use a General Act for incorporation, but they use a slightly different terminology and have given corporations incorporated under the legislation all the powers of a natural person.

Regardless of the form of incorporation, the corporate body takes on a separate existence from its incorporators and continues in existence until it is wound up, or it surrenders its charter.

A corporation acts through its agents and is liable to its creditors for all corporate debts. The shareholders who own the corporation have limited liability for the debts of the corporation. Their limit is the value of the shares they purchased in the corporation.

In a corporation, the shareholders elect directors annually to manage the affairs of the corporation. The directors, once elected, have a general duty to operate or manage the business in the best interests of the corporation. They owe a duty of loyalty and good faith to the corporation in all of their dealings on its behalf. The shareholders, on the other hand, do not participate in the active day-to-day management of the business, but are entitled to information from the directors on the corporation's performance. If the shareholders disapprove of the directors' management, they may remove the directors at the annual general meeting of shareholders and elect other directors in place of those who previously held the office.

Corporation legislation, in general, attempts to balance the powers of the directors to manage the corporation with the protection of the investment of the shareholders. To some extent, it also considers the rights of the public. It does this through the imposition of special duties and obligations on the directors. It permits the shareholders to take action either within the corporation or through the courts, when the directors fail to comply with the duties and responsibilities imposed upon them.

As well, environmental and employment legislation in some provinces may impose personal liability on directors where the corporation violates the legislation.

KEY TERMS

Corporation, 303
Shareholder, 304
Director, 304
Officer, 304
Letters Patent, 308
Special-Act, 308
General-Act, 309
Memorandum of Association, 309
Articles of Incorporation, 309
Doctrine of constructive notice, 310
Indoor management rule, 310
Shareholders' agreement, 311
Share, 313
Floating charge, 313
Debenture, 313
Fiduciary, 314
Doctrine of corporate opportunity, 316
Due diligence, 318
Outside director, 319
Business judgment rule, 319
Sarbanes-Oxley Act, 320

REVIEW QUESTIONS

1. Describe briefly the relationship between a corporation and its shareholders. How does a shareholder's relationship with the corporation change if the shareholder becomes a director?
2. What drawbacks commonly associated with partnerships are overcome by the use of the corporate form?
3. What is a Special-Act corporation? For what purpose would it be formed? Give two examples.
4. Explain the term "letters patent." In what way or ways would a letters patent corporation differ from a General-Act corporation?
5. Define "corporate *ultra vires*." Explain why most provinces have attempted to eliminate the doctrine as it applies to third parties dealing with a corporation.
6. How does a corporation differ from a partnership?
7. What is the legal nature of a corporation?
8. Explain the "doctrine of corporate opportunity."
9. Define the "indoor management rule" and, by way of example, explain how it is applied.
10. What are the obligations of a director of a corporation when the director has a financial interest in a firm with which the corporation wishes to do business?

11. Indicate how the principle of "majority rule" is applied in the decision-making process of a corporation. What protection is available to a dissenting minority shareholder if a fundamental change in the corporation's object is proposed?
12. Distinguish a "public" corporation from a "private" corporation. Why is this distinction made? What other terms are used for each of these types of corporations?
13. If a corporation wishes to sell its securities to the public, what requirements are imposed upon the promoters, directors, and others associated with the sale and distribution of the securities?

MINI-CASE PROBLEMS

1. A, B, and C are the directors of ABC Corporation. At a directors' meeting, B suggests that the corporation consider the purchase of a block of land owned by the RST Corporation. C is a principal shareholder in the RST Corporation. What are C's obligations to B, A, and the ABC Corporation?
2. X, a director of the DC Corporation, is informed by the corporation's accountant that the decline in the corporation's sales has resulted in a large fourth-quarter loss. Before the corporation's financial results for the year are announced to the public, X sells a large block of his shareholding in the corporation. When the news of the corporation's loss is announced, the price of the shares on the stock exchange falls by $10 per share. What are the rights of Z, who purchased 1,000 shares from X at the higher price?
3. Continuing from the facts in (2), would your answer be different if director X had not sold his shares, but had informed shareholder Y of the expected loss, and shareholder Y (who owned 15% of the corporation's shares) sold his entire shareholding just before the news was released?
4. The directors of B Corporation wished to purchase a vacant property adjacent to their office building. An offer to purchase was made at a price of $120,000. The offer was rejected by the property owner. Director D of the corporation then immediately made a personal offer to purchase the lot for $125,000, and the offer was accepted. What is the position of D? What is the position of B Corporation?

CASE PROBLEMS FOR DISCUSSION

CASE 1

The Board of Directors of Speedy Roofing Ltd. directed the corporation president to draw up and enforce site safety rules for the performance of work at the various kinds of construction sites where the corporation would likely bid for roofing work. In accordance with the directive, the president set out job equipment requirements for three types of projects: small residential home roofing work, apartment and commercial building roofing work, and large industrial and high rise building roofing work. In each case, the site manager was expected to select the proper equipment to perform the work (scaffolding, portable elevators, material handling equipment, etc.) and enforce the safety rules dictated by provincial government work safety regulations. The directors made no further inquiries after the president advised them what rules were in place.

Site managers at small residential roofing jobs frequently used only ladders and minimal scaffolding on the job, and most roofing materials were moved to the roof by manual labour. The reasoning here was that the roofing work was usually completed in only a day or two; and the roof surfaces were usually gentle slopes or relatively flat. Safety equipment was considered unnecessary, and only a rope safety harness was used on the steeper roof surfaces. On a small residential roof repair project on a two-storey home, a worker slipped on some loose shingles and fell over the side of the roof. His safety harness arrested his fall, but in the process he was thrown against the brick wall of the house and suffered a serious head injury. Provincial workplace safety inspectors investigated the accident, and may charge the corporation and management with a violation of the provincial workplace safety legislation.

Advise the directors of the corporation and consider the possible outcome, if the corporation was charged with a violation of the Act.

CASE 2

Ludwig carried on a machine shop business for many years under the name Ludwig's Machine Shop. On advice from his accountant, he transferred the shop and all of its equipment to a corporation that he incorporated with himself as the sole shareholder.

As payment for the shop and equipment, the corporation issued Ludwig 10,000 common shares, and a debenture in the amount of $250,000. The corporation carried on the business as Ludwig's Machine Shop Ltd., with Ludwig as the corporation's president. The large advertising sign on the front of the shop, however, was not changed, and continued to read: Ludwig's Machine Shop. Ludwig continued to operate the business as he had done prior to incorporation, and did not inform his customers or suppliers of the change of ownership of the business. Business invoices nevertheless were issued in the corporation's name and all correspondence was on corporation letterhead.

Several years later, the corporation experienced a loss of business to new competitors, and was soon in financial difficulty. When the corporation found that it could no longer carry on profitably, the corporation ceased operations. Business assets were liquidated for the amount of $175,000. Ludwig, who held the $250,000 debenture, claimed payment in priority over the creditors' claims of $100,000. The creditors claimed priority of payment over Ludwig on the basis that they were unaware of the change of ownership.

Discuss the issues raised in this case, and how the dispute might be resolved.

CASE 3

A corporation owned a parcel of vacant land on which it stored its construction equipment. The land was not large enough for the requirements of the company. When the adjoining landowner expressed a desire to purchase the property from the company, the directors informally considered the offer and agreed to sell the land for $150,000. No directors' meeting was held to formally deal with the matter. However, the secretary-treasurer, on the basis of the informal agreement amongst the directors, contacted the offeror and advised him of the price. The price was acceptable to the purchaser, so the secretary-treasurer then drew up a written purchase agreement that he signed on behalf of the corporation in his capacity as secretary-treasurer. The purchaser also signed the document.

The directors later decided not to carry through with the sale, and the purchaser brought an action against the corporation for specific performance of the contract.

What defences might be raised by the corporation in this case? What legal principles are involved?

Render a decision.

CASE 4

Juana owned 13 shares of the Vermilion Mining Co. Three other shareholders held four shares each. The remainder of the 2,400 shares of capital stock was held by the three directors of the company. The company owned certain mining claims on which some preliminary exploration work had been done, but that required the investment of a large amount of capital in order to establish a mine. Because the company had not been in a position to proceed with the development of the properties, the company faced the prospect in the near future of forfeiture of the mining claims as a result of their forced inaction.

The directors, who were shareholders in another mining company, entered into an agreement to sell the mining claims to that company in exchange for shares in the second company. The share exchange for the mining claims would give Vermilion a 10 percent interest in the other company.

A meeting of shareholders was called to approve the transaction. At that time the directors declared their interest in the other mining company. The directors explained that, in their opinion, the transfer represented fair market value for the claims and they urged approval of the transaction. The directors voted in favour of the sale over the objections of Juana, who was the only dissenting shareholder.

She accused the directors of attempting to confer a benefit on a company in which they had an interest, to the detriment of the company in which they were directors. She eventually brought an action to restrain the directors from completing the sale of the mining claims to the other company.

Discuss the issues raised in this case and render a decision.

CASE 5

Cinema Ltd. owned a theatre that it wished to sell. To make the property more attractive to a prospective purchaser, the directors decided to acquire a second theatre in the same city and offer the two properties as a "package deal."

Some inquiries were made as to the purchase price of a second theatre, and a price of $300,000 was determined for the property. A subsidiary company was incorporated to acquire the second theatre, with the intention that the shares in the subsidiary would be wholly owned by Cinema Ltd. Unfortunately, the lending institutions would only advance Cinema Ltd. $180,000 on its assets. In order to effect the purchase of the second theatre, the three directors of the corporation and a lawyer (who frequently acted for the corporation) each agreed to invest $30,000 to make up the necessary $120,000. The subsidiary corporation issued 300,000 shares valued at $1 each to the parent company and the four investors in return for the $300,000 in cash. It then proceeded with the purchase of the second theatre.

Some time later, a purchaser was found for the two theatres, and a purchase agreement completed. The purchaser, however, insisted on acquiring the second theatre by way of a purchase of the shares in the subsidiary company. The share price was determined at $1.25. This netted Cinema Ltd. a profit on the sale of $45,000, and each of the four investors a profit of $7,500. When details of the sale were revealed to the shareholders, one shareholder demanded that the four individuals pay over their profits to the corporation. When the three directors and the lawyer refused to do so, the shareholders instituted legal proceedings to have the funds paid to the corporation.

Discuss the various legal arguments that might be raised in this case by the parties. Indicate how the case might be decided.

CASE 6

High Rise Apartments Ltd. expressed an interest in the purchase of a large block of land suitable for development as an apartment site. The board of directors asked Sheldon Harris, one of the directors who was also a real-estate broker, to investigate the possibility of the corporation purchasing the land for a reasonable price.

Harris, without revealing that he was a director of High Rise Apartments Ltd., contacted the president of Land Assembly Ltd., the corporation that owned the land. He inquired as to the price the corporation was asking for the property. The president replied that the price was $500,000. Harris then offered to sell the land for Land Assembly Ltd. for his "usual commission" as a real-estate broker. The president of the corporation agreed to have Harris attempt to sell the property on its behalf, so Harris reported back to the board of directors at High Rise Apartments Ltd. that the land was for sale at $525,000.

Unknown to the remaining members of the board of directors, the following events occurred before discussion took place as to whether the corporation should purchase the land at the price of $525,000.

(1) Jeremiah Black and Rodney Jones, both directors of High Rise Apartments Ltd., became interested in the land as a site for a shopping centre. They had incorporated a company for the purpose of buying the land if High Rise Apartments Ltd. should decide not to purchase the property.

(2) Nyssa Green, a director of High Rise Apartments Ltd., was urged by her spouse, who was a minority shareholder in Land Assembly Ltd., to speak and vote against the purchase because he felt that Land Assembly Ltd. was selling the land for less than its true worth.

(3) Sonya Patel, a director of High Rise Apartments Ltd., who was also a principal shareholder in Condominium Construction Company (a corporation interested in the parcel of land as a condominium site), was busy attempting to make an offer to purchase the property. When she heard of the offer from the president of Land Assembly Ltd. to sell the property for $525,000, Patel slipped out of the directors' meeting and telephoned the president of Condominium Construction Company, urging him to call Land Assembly Ltd. with a higher offer.

A meeting of the Board of Directors of High Rise Apartments Ltd. was called for the purpose of considering the purchase of the property at a price of $525,000. Isaac Davis, the Chairman of the Board, called for a vote on the purchase. Only Davis and Harris voted in favour. The remaining members of the board (Patel, Green, Black, and Jones) voted against the motion.

After the purchase was rejected by the Board of Directors at High Rise Apartments Ltd., both the company incorporated by Black and Jones and Condominium Construction Company attempted to purchase the land. When they contacted the president of Land Assembly Ltd., he advised them to contact Harris, who was the real-estate broker engaged to sell the property. Eventually, the property was sold to Condominium Construction Company for $560,000. Harris received a real-estate commission on the sale of $26,000, which was paid to him by Land Assembly Ltd. After the sale was completed, Davis discovered the facts surrounding the sale and the actions of the directors of High Rise Apartments Ltd.

Advise Davis of the legal issues raised in this case and the course of action he might follow. Indicate the arguments that might be raised if the matter came before the court, and render a decision.

CASE 7

Henri Boucher and his son, Gaetan, lived in the same city, where Gaetan ran a small business, an unincorporated restaurant. Henri had operated a small strip plaza comprising a convenience store, a gas station, and a pizzeria/arcade. Under pressure from creditors, Henri had sold the plaza for the amount of his debts, and he began anew with his son's assistance.

They formed a corporation to develop a roadside piece of land into a ten-unit commercial plaza near a residential area.

www.mcgrawhill.ca/olc/willes

Otherwise unemployed, and with Gaetan busy in his restaurant, Henri looked after contracting the majority of the work. This included considerable construction work to build the plaza. As construction progressed there were disturbing signs of discontent among the contractors who were building the plaza. On occasion, they would call the restaurant, asking Gaetan for payment. Gaetan would call his father, who would in turn pay them. Often, however, the calls persisted and Gaetan would find himself paying bills out of his own pocket and keeping a tally of the bills he had paid on behalf of the company.

Eighteen months after incorporation, and 15 months after breaking ground on the project, the plaza had acquired a bad reputation in the town, and suppliers were unwilling to deliver materials. It quickly foundered when the bank called for repayment of the $290,000 that had been borrowed (on demand) by the company. The land had been bought with bank funds for $175,000. Invoices totalling $87,000 had been paid, but the company had no more money in its account. Gaetan's tally showed that he had paid bills out of his own pocket totalling $11,000. Unfortunately, with a half-built plaza and a bad reputation, the only offers for purchase of the site were in the order of $145,000. Henri was despondent and soon left the province, leaving behind Gaetan with the restaurant.

The company lawyer showed Gaetan a letter that had been sent to him a month before, in which Henri had resigned as an officer and director of the company, and had turned in his shares. This left Gaetan as sole officer, director, and shareholder. The lawyer thought Gaetan had known of his father's resignation.

Advise Gaetan with respect to the issues that the company creditors will raise. Discuss Gaetan's rights and/or liabilities and explain any steps he could have taken to protect himself.

CASE 8

The directors of Claridge Supply Company Ltd. approached the Business Bank for a loan. One of the two directors of the corporation attended at the bank and, in the presence of the branch manager, signed the loan application. He then took the documents with him in order to have the corporate seal affixed and to obtain the signature of the other director on the documents. A few days later, the director returned with the signed loan application and documents that purported to be directors' resolutions authorizing the pledging of the corporation's assets as security for the loan. Attached to the resolutions were the secretary's certificates confirming that the resolutions were duly passed by the directors.

Some months later, Claridge Supply Company Ltd. was declared bankrupt, and the trustee in bankruptcy refused to recognize the bank's secured loan on the basis that the corporation's minute books contained no record of any resolutions authorizing the pledging of securities for the loan. The bank then took legal action for a declaration that it was a secured creditor.

Discuss the basis of the bank's claim and the likely outcome of the action.

CASE 9

Acme Forwarding Company leased docking and warehouse facilities at a harbour. The lease gave the company exclusive use of the pier and the buildings, but held the company responsible for the maintenance and repair of the pier.

Tyrone, a director of the Acme Forwarding Company, was responsible for the authorization of use of the pier by ships bringing cargo for off-loading at the company warehouse. The company policy was that only ships handling goods destined for the company warehouse were to use the pier. No authorization would be given unless the ship owners carried adequate insurance to cover any damage that might be done to the pier by careless docking.

Tyrone also held an interest in a shipping firm that wished to off-load a small cargo at the city where the Acme Forwarding Company pier was located. The shipping company, however, wished to place the cargo directly onto two trucks, rather than use the facilities of Acme Forwarding Company. Tyrone, with the intention of accommodating the shipping company, authorized the docking of the ship at the pier.

When the ship attempted to dock with its cargo, it collided with the pier, causing extensive damage to both the ship and the pier. Under the terms of the lease that Acme Forwarding Company had with the property owner, it was obliged to repair the pier. Acme Forwarding Company did so, at a cost of $120,000. Then it looked to the shipping company to recover the cost of the damage caused by the ship.

The shipping company, which had suffered damage to its only ship, was unable to pay for the damage to the pier. Its insurance would cover only a part of the $120,000 cost of the pier repairs. When the directors of Acme Forwarding Company were informed of the shipping company's inability to pay for the damaged pier, they were also informed that Tyrone was a shareholder in the shipping company.

Discuss the rights of the parties in this case and explain how these rights might be enforced.

CASE 10

Model T Motors Ltd. was indebted to Simple Finance for a substantial sum of money. The finance company held a number of mortgages on the corporation's assets, but pressed the corporation for a blanket demand chattel mortgage as additional security. Under pressure from the finance company, one of the principal shareholders, who was also one of the signing officers of the corporation, executed a blanket chattel mortgage to the creditor. The mortgage was not made under the corporation seal, and only one of the two signatures required by the corporation's by-laws was placed on the document.

Some time later, the finance company obtained the corporate seal of Model T Motors for another purpose and affixed it to the chattel mortgage. A few weeks later, when a payment on the loan was overdue, Simple Finance seized the assets of Model T Motors under the blanket chattel mortgage.

Advise Model T Motors of its position in this case. If the matter came before the courts, how would you expect the case to be decided?

The Online Learning Centre offers more ways to check what you have learned so far. Find quizzes, Weblinks, and many other resources at www.mcgrawhill.ca/olc/willes.

www.mcgrawhill.ca/olc/willes

CHAPTER 18

Securities Regulation

Chapter Objectives

After study of this chapter, students should be able to:

- Explain the policy and structure underpinning securities regulation.
- Describe the disclosure obligations imposed upon issuers of securities.
- Identify situations involving illegal securities market activity.
- Explain the use of proxies and the conduct of takeover bids.

YOUR BUSINESS AT RISK

In creating and issuing securities you are giving away rights related to your business — up to and including complete ownership of it. These exceptionally powerful rights can be traded among strangers to your business who may have widely varying intentions in exercising *their* rights over *your* company. Keep in mind that security regulators expect a sophisticated understanding of your corporate legal obligations in this advanced field, and will hold you to that standard.

INTRODUCTION

A corporation or other business entity (such as a partnership or trust) finds its lifespan to be one full of promises. Beyond simply being contracts, many of these promises relate to the division of ownership of the corporation's assets, the distribution of its profits and losses, and its financial obligations to those who have invested money in its future. These promises can be generally called **securities**, and for them to have any meaningful role they must be coupled with an efficient and trustworthy marketplace for money.

Security
A document or other thing that stands as evidence of title to or interest in the capital, assets, property, profits, earnings, or royalties of any person or company, including any document commonly known as a security.

It is an essential element of a developed economy that large-scale users of funds can meet with the many (often smaller) savers of capital to conclude terms of investment or borrowing with both speed and certainty. By the same token, the process must have sufficient integrity to provide the investor with information that allows for a true understanding of the inherent riskiness of a security, and permit an informed pricing decision to be made.

The nature of securities, the method of their distribution and trading, and obligations to disclose information as to their riskiness, are matters of law. So too are identification, registration, and competency requirements for market intermediaries. The body of law that mandates these, and seeks to find the correct balance between market efficiency and market integrity, is known as securities regulation.

HISTORICAL DEVELOPMENT OF SECURITIES REGULATION

The ideal of an efficient and trustworthy securities market has been a long time in coming, and history shows many examples of human foolishness along the way. Kings and commoners alike have long been willing to believe that base metal could be transmuted into gold. In 1636, "tulip-mania" swept the Netherlands, home of one of the world's original stock exchanges, and led investors to trade and pay upwards of what is now $50,000 for a single tulip bulb. Few investments offered today represent such unjustified speculation, but that is only because of effective securities regulation.

Before the Industrial Revolution, when significant assets were largely restricted to land ownership, the investing public was sufficiently well served and protected by the provisions of real-estate law. Even by 1900, very few members of the general public had much concern with anything that would fit into the category of securities. Those who did were primarily the wealthy, who could rely on both their greater sophistication in financial matters and the body of contract and tort law to serve their needs.

Prosperity following World War I led to the "Roaring Twenties," and, for the first time, assets that were held more broadly across the social spectrum. This wash of ready capital, much of it financed through highly leveraged debt, found a home in the stocks and bonds of growing industry. These investments were later wiped out in the Great Depression, to the betterment of a few and the detriment of many.

For the first time, governments were faced with the need for reform and regulation of the financial sector so that they could serve and protect their broadest constituency, the general (and voting) public.

In Canada, government response came at both the federal and provincial levels. Ontario enacted the *Securities Frauds Prevention Act*, 1928,[1] and, by 1930, the Act had been expanded from simply registering brokers and salespersons to creating a Board, which could exercise the powers of the Attorney General in his stead.[2] This Board is the precursor to today's Ontario Securities Commission. This legislative and regulatory response has been followed in all other provinces. By 1932 the original Act, as amended and expanded, was renamed the *Securities Act*, 1930.[3]

At the federal level, the 1930s saw a revised *Bank Act* and the creation of a central bank — the Bank of Canada — to ensure the liquidity and solvency of chartered banks.

Canada has not followed the historical United States practice of using a single national regulator to govern securities markets. However, with the complexities, speed and tight integration of markets today, it is almost certain that Canada will make its own switch to a national regulator. Until then, securities regulation remains a provincial responsibility.[4] The exception is federal regulation of banks, and federally incorporated businesses that are subject to the provisions of the *Canada Business Corporations Act*[5] and, as a consequence, to its sections related to takeover bids, proxies, and insider trading.

Broad similarity exists across all provinces in how the aims of securities regulation are achieved, with identical provisions in many cases. Certain rules do vary, however, particularly in regulations made pursuant to statute, demanding that the Act and regulations of a particular province must be consulted in all specific cases.

CLIENTS, SUPPLIERS *or* OPERATIONS *in* QUEBEC

The Autorité des marchés financiers (Financial Markets Authority) was launched in 2004 to regulate the entire Quebec financial sector. Its mandate covers not only rules and regulation of stock market activity, but also monitors the behaviour of financial institutions, as well as providing deposit and fraud insurance. Additionally, it is intended to act as the single window for consumer complaint processing, and has the goal of streamlining provincial regulation.

1. S.O. 1928, c. 34.
2. The *Securities Frauds Prevention Act*, 1930, S.O. 1930, c. 48.
3. The *Statute Law Amendment Act*, 1932, S.O. 1932, c. 53, s. 36.
4. See, as amended: *Securities Act*, R.S.A. 2000, c. S-4; *Securities Act*, R.S.B.C. 1996, c. 418; *Securities Act*, C.C.S.M., c. S50; the *Securities Act*, S.N.B. 2004, c. S-5.5 and the *Security Frauds Prevention Act*, R.S.N.B. 1973, c. S-6; the *Securities Act*, R.S.N.L. 1990, c. S-13; *Securities Act*, R.S.N.S. 1989, c. 418; the *Securities Act*, R.S.O. 1990, c. S.5; *Securities Act*, R.S.P.E.I. 1988, c. S-3; *Securities Act*, R.S.Q. c. V-1.1; *Securities Act*, 1988, S.S. 1988-89, c. S-42.2; *Securities Act*, R.S.N.W.T. 1988, c. S-5; *Securities Act*, R.S.Y. 2002, c. 201; and the *Securities Act* (Nunavut), R.S.N.W.T. 1988, c. S-5.
5. The *Canada Business Corporations Act*, R.S.C. 1985, c. C-44.

What Is a Security?

A security is an open-ended term, for the purposes of regulation. The open-endedness means that almost anything that could be termed as a security *is* a security. Taking a page directly from its 1928 predecessor, Ontario's *Securities Act*, like securities Acts in other provinces, defines security[6] to *include*:

(a) any document, instrument, or writing commonly known as a security;
(b) any document constituting evidence of title to or interest in the capital, assets, property, profits, earnings, or royalties of any person or company;
(c) any document constituting evidence of an interest in an association of legatees or heirs; and
(d) any document constituting evidence of an option, subscription, or other interest in or to a security, together with 12 other post–1928 additions of items ranging from mutual fund interests through certain annuities and commodities futures to oil and gas leases and royalties, regardless of whether they are related to an existing issuer or a proposed issuer.

In cases where a security would seem to be exempt by definition, such as a contract of insurance, it is only because that example is fully regulated under the terms of another Act (such as the *Insurance Act*). The increase in items defined to be securities is a reflection of the growth in complexity of financial markets over the past century.

The vast majority of securities that will be encountered in public capital markets will be listed securities, made up primarily of government bonds, corporate shares and bonds, debentures, futures, options, derivatives, and units of ownership in mutual funds and similar income trusts. Despite this overwhelmingly large presence of publicly traded securities, the scope of the provincial securities Acts is not limited to public markets. Thus even small-scale enterprises can find that their intended activities fall within their province's securities Act.

CASE IN POINT

Pacific Coast Coin Exchange sold bags of silver coin to investors, most of whom did not take delivery of the bags, but rather left them on account, hoping for a rise in prices before reselling them through PCCE. Many transactions were also conducted on margin, being credit extended by PCCE to the investor, such that only a fraction of the purchase price needed to be paid, with interest and fees being charged by PCCE out of the investors' proceeds of later sale. The Supreme Court of Canada determined that such arrangement was in fact an investment contract, and was "a security" as defined by the Act. As a result, PCCE was obliged to fulfill all applicable compliance requirements of the *Securities Act* before it could conduct such sales.

See: *Pacific Coast Coin Exchange of Canada Limited v. Ontario Securities Commission,* [1978] 2 S.C.R. 112.

Purpose and Administration of Securities Regulation

The purpose of securities regulation was generally stated in the chapter introduction. The principles of efficiency and integrity can be more fully stated as the two chief goals of:

1. providing protection to investors from unfair, improper or fraudulent practices; and,
2. fostering fair and efficient capital markets and confidence in capital markets.

6. The *Securities Act*, R.S.O. 1990, c. S.5, s.1(1).

Of paramount importance is a sense of balance on the part of administrators of the provincial securities commissions. It is possible to create near full confidence in markets by burdening the process of issuing securities with checks and hurdles to create integrity, but, of course, by the time securities come to market, the business opportunities of those issuers may well have passed them by, or other suppliers of capital will have filled the breach. On the other hand, a market that allows, in the name of efficiency, any form of snake oil to be sold will have a steady stream of fraudsters and victims. This inevitably leads to having a reputation as a dangerous place for business, and ultimately no business will be transacted at all.

With the fast-paced change in global capital markets and the novel securities offered in them, it is well and good that legislatures have passed responsibility for securities regulation to administrative boards that have the power, flexibility, and speed to react to these changes. These boards are the provincial securities commissions, or, collectively, Canada's securities administrators. In addition, the efficient and trustworthy operation of markets depends heavily on the participation of its Self Regulatory Organizations (SROs): the Investment Dealers Association, each of Canada's stock exchanges, and the Mutual Fund Dealers Association.

These SROs impose considerable obligations on their member firms (for example, stock brokerages) and, as regards the exchanges, significant conditions on issuers of securities that must be met and maintained for listing. The immediacy and proximity that these regulators and self-regulators have to actual market developments make for a significant contribution to the achievement of the twin goals of the provincial securities commissions.

Industry and investor pressure does exist however, urging abandonment of the multi-jurisdictional regulation of securities in Canada. Such a change to a single set of national regulations — as is the case in the United States — would be a step forward for efficiency and certainty. What harmonization has occurred thus far has resulted from the industry following American and Ontario initiatives, and through the efforts of Canada's Securities Administrators in producing national instruments (rules of application) across all jurisdictions.

MECHANICS OF REGULATION

Understanding the mechanics of securities regulation first requires an understanding of what is, and is not, being regulated. Most importantly, securities regulators (whether provincial administrators or SROs) do not attempt to vet the underlying sensibility or business wisdom of particular investments or securities. Moreover, securities regulators and stock exchanges are in no position to fully prevent financial abuse and outright fraud. The recent cases of Enron, WorldCom, and Bre-X have shown just how impossible such a goal would be. While enforcement measures take place after the harm is done, regulators and stock exchanges do create processes intended to make such abuses difficult, but they cannot insure against the clever manipulator. This, in the final analysis, is left to the investor on the principle of *caveat emptor* — buyer beware.

For a buyer to be in a position to beware, he or she must have confidence in the individuals and service providers who are the public face of the securities markets. To this end, securities regulation first requires registration and licensing of these participants, and prescribes the manner in which trading is to be conducted. Second, the buyer requires timely and accurate material information about the security in question, both at the time of its original issuance, and on an ongoing basis throughout its existence. To satisfy this need, regulation obliges disclosure of material information. Third, the buyer needs special protection in special circumstances, namely, in cases of the solicitation of voting rights (proxies), takeover bids and issuer bids, and insider trading. Each of these needs and responses is detailed below.

Lest it be forgotten that the issuers of securities also expect efficiency in capital markets, there are legislative and regulatory responses to standardize the process of bringing securities to market, and to treat different circumstances and different issuers in a balanced manner.

REGISTRATION

Any company or person acting as an intermediary in the trading of securities must be registered as a dealer or salesperson. There are further provisions of the law to require registration of persons who are officers or directors of registered dealers, and registration is required to engage in underwriting. Underwriting may take the form of purchasing new securities — *en bloc* — from an original issuer as inventory for later resale to the market at large (a "bought deal"), or acting as agent only for an issuer in initial sales to a waiting market.

Beyond simply keeping track of the identities of those engaged in trading, registration also imposes a licensing requirement that must be satisfied through education and training, and examinations of competency, which are administered by the SROs. For example, an investment advisor who intends to take orders for securities and provide advice to investors must complete the Canadian Securities Course on market and investment knowledge, and a Conduct and Practices course on ethical professionalism and account operation, must work under close supervision for six months, and complete further financial study both in the near term and on a continuing basis.

There are limited categories of market participants who are exempt from the requirement of registration as advisors.[7] These exemptions are largely based on the effective governance of their behaviour under other Acts, and subject to the provision that the advice they give is solely incidental to their principal occupation. Accordingly, these exemptions relate to bankers, lawyers, accountants, and financial-media commentators, among others.

DISCLOSURE

Disclosure

The release to the public of information about a corporation that intends to offer its securities to the public.

To foster an environment that allows for an informed public capable of making reasoned investment decisions, regulators insist upon *true, full and plain disclosure of all material facts* relating to securities being issued. The words in italic are law as expressed by the Supreme Court of Canada,[8] echoing the provincial securities Acts.[9] All provincial jurisdictions use words to similar if not identical effect.

There are two important principles to be addressed in ensuring that true, full, and plain disclosure is achieved. The first principle relates to the availability of information pertinent to the security at the time of original issuance, which affects the first investor's desire to purchase the security at all, as well as the price that he or she is willing to pay for it. This process of disclosure is known as "prospectus disclosure," and involves the creation of an omnibus public document stating detailed particulars about the issuer and the security. The second principle relates to material facts or changes in business that affect the security after issuance and purchase during its lifespan or term to maturity. The obligation set upon issuers in this instance is known as "continuous disclosure."

In certain instances, prospectus disclosure is recognized to be unnecessary, or unnecessarily onerous on the issuer. Consequently, there is provision for the filing of an abbreviated prospectus as well as important circumstances that permit a complete exemption, which are more fully discussed below.

Prospectus

A public document required by law before securities are issued, revealing material facts about that security and its issuer, with such true, full, and plain disclosure that a potential investor may make an informed decision as to the riskiness and price of that security.

Prospectus Disclosure

When entirely new securities are to be brought to market — for example, the shares of a formerly privately held corporation — the future investors in the corporation are an entirely new breed. Unlike the former private owners of the company who knew its affairs in great detail, and may well have been its owner/managers, these new investors may reside at great distance from the place of business and in all probability will never visit it or learn of its internal affairs. Thus, unless an exemption can be found, a **prospectus** is

7. For example, The *Securities Act*, R.S.O. 1990, c. S.5, s. 34.

8. *Pacific Coast Coin Exchange of Canada Limited v. Ontario Securities Commission*, [1978] 2 S.C.R. 112.

9. For example, *Securities Act*, R.S.B.C. 1996, c. 418, s. 63(1), or the *Securities Act*, R.S.O. 1990, c. S.5, s. 56.

required to be filed with and accepted by the provincial securities commission before any trading (first issuance and subsequent buying and selling) can take place in that security.[10]

It is important to understand the implication of the word "trading." An accepted prospectus does not confer any rights to stock-exchange trading (although it is one required step closer) and trading in this instance means any trading. Without benefit of an exemption, a distribution of securities to individually approached acquaintances would still require a prospectus, despite no underwriting or stock-exchange involvement.

It is important to note that an exemption from prospectus disclosure relates to the trade in question and not to the security itself. If a particular trade is exempt, and securities are issued pursuant to that exemption, any later trade in those same securities, if not exempt also, will trigger the prospectus-disclosure requirement.

Many requirements need to be met for a prospectus to be acceptable to a regulator, and disclosure can be seen as intrusive into company affairs and secrets. For this reason, many companies that might otherwise issue securities do not, in order that their internal affairs may remain a secret. The price paid for this confidentiality is reduced access to capital markets and potential investors.

Those who choose to go ahead and raise funds through a distribution of securities are known as "issuers." A prospectus for a proposed distribution must, on the basis of true, full, and plain disclosure of material fact, include the price and number of securities to be issued, the net proceeds the issuer expects to raise, and the fees associated with the underwriting of the issue. Having established the amount of the proceeds, the issuer must report how it intends to use the funds, and the business risk factors associated with its enterprise and with the terms of the security itself. The issuer must make public its own financial statements reflecting its financial health both before and after the proposed distribution. Any other relevant reports or opinions must also be provided in the prospectus, such as those of independent auditors or experts in the business being undertaken. As an example of this latter group, a mining venture would be required to provide expert geological and engineering opinions on ore content and the commercial viability of ore extraction. Any and all other factors that are material to an investor's purchase and pricing decision are required, and will vary on a case-by-case basis.

Reporting issuer

A corporation that has issued its shares to the public by way of a prospectus.

The prospectus must be filed with the provincial securities administrator, be accepted, and be made available to all investors prior to trading. Once the securities have been issued, the issuer is then known as a "**reporting issuer**."

CHECKLIST FOR PUBLIC OFFERING VIA PROSPECTUS

1. Assemble complete financial statements.
2. Consider control and tax implications of public offering.
3. Complete appropriate business plan in line with expected share proceeds.
4. Ensure board and management team will meet minimum regulatory requirements for a public company.
5. Begin assembling elements of prospectus from business plan, with additional documentation as required by provincial securities administrator, applying required due diligence.
6. Create advisory team of securities lawyer, auditor, investment dealer.
7. Prepare preliminary prospectus with team, file with regulator and exchange.
8. Apply for listing.
9. Satisfy deficiencies in prospectus noted by regulator and/or exchange.
10. Submit final prospectus and receive approval from regulator.
11. Sale and trading in shares may take place thereafter.

10. For example, *Securities Act*, R.S.B.C. 1996, c. 418, s. 61, or the *Securities Act*, R.S.O. 1990, c. S.5, s. 53.

Prompt Offering Qualification

A short-form prospectus route is available to qualified reporting issuers. Those who have been reporting issuers for three years, are current with all required filings, such as Annual Information Forms, are not in default of their regulatory obligations, and have stock-exchange-listed shares of at least $75 million in market value, may avail themselves of the Prompt Offering Qualification System.[11] This system allows for prospectuses that are essentially limited to disclosure of the purely financial terms of the proposed securities issue. The rationale is simply that the ongoing obligations of continuous disclosure and oversight by stock exchanges should allow such established reporting issuers greater flexibility and speed in conducting their affairs, without unduly exposing the public to risk.

Prospectus and Registration Exemptions

At the other end of the spectrum, Canada's securities regulators recognize that prospectus disclosure would be an unfair hardship on some classes of (usually smaller) issuers and investors. Again in search of a balance between financial transparency and market efficiency, some significant exemptions from prospectus and registration requirements exist. Formerly, a hodgepodge of provincial exemptions existed, some in common and others not, woven into the various provincial securities Acts. Having achieved a consensus for change, and under their powers arising from the provincial Acts, the Canadian Securities Administrators (CSA) have created a significant (but not complete) harmonization of these exemptions across Canada, through National Instrument 45-106 *Prospectus and Registration Exemptions*, which came into force in September of 2005.

The most important exemptions are trades

1. To any one of the 15 classes of "Accredited Investors," which include:
   - Banks
   - Investment dealers
   - Governments
   - Individuals and their spouses owning net financial assets exceeding $1,000,000, or net income before tax of $200,000 (or $300,000 with spouse) in this and the past 2 years
   - Individuals who, either alone or with a spouse, have net assets of at least $5,000,000
   - Persons other than individuals or investment funds (essentially meaning corporations) with net assets of at least $5,000,000

   or

2. By "Private Issuers," whose founding documents contain restrictions on transfer of its securities and whose shares are owned by not more than 50 persons. They may issue securities to:
   - Directors, officers, employees of the issuer
   - Their relatives and close friends and business associates (a rule applicable everywhere but Ontario)
   - Founders and control persons of the issuer, and close relatives of executive officers, directors and founders (the rule applicable in Ontario)
   - Its own existing security holders
   - Accredited investors

   or

3. Minimum Investment Amount of $150,000 — securities may be sold exempt from the prospectus requirement where the cost to the purchaser is at least $150,000.

11. See, for example, the *Securities Act*, R.S.O. 1990, c. S.5, s. 63.

Note: While this is a National Instrument of the Canadian Securities Administrators, the same text was not adopted in each province. In any specific situation, the text as adopted by a particular province must be checked to verify for modifications, rules and limitations specific to that province.

There is a policy theme running through these issuer exemptions from prospectus disclosure. Disclosure is intended to protect the general public from fraud; the expectation in not requiring a prospectus is *not* to suggest that disclosure will not be provided, but rather that persons investing $150,000 will *demand* disclosure on their own, or they will not invest. Likewise, persons with significant income and assets are presumed to know something about investing, and will make similar informational demands. The private issuer exemption suggests that persons who are already "insiders" of a firm are close enough to the necessary information for investment decisions that they can obtain "disclosure" by being part of the business itself. The general obligation of prospectus disclosure should not become a pendulum swinging too far in the name of market integrity that it chokes off all market efficiencies.

Janet has just formed a corporation, and as its sole director and shareholder, has issued one of its five shares to herself. She intends that the corporation will sell the other four shares to other investors. She believes the corporation and the computer software she has designed will create a business worth $1 million. Accordingly, she has valued each of the equal shares at $200,000 each. She finds four persons willing to invest and her corporation accordingly raises $800,000.

"Minimum Investment" exemption applies: Since each new investor will purchase the securities for more than $150,000, each of the four new distributions is exempt from the requirement of prospectus filing.

Mario has just formed a private corporation, and as its sole director and shareholder, has issued 800 shares to himself. He intends that the corporation will sell 200 more shares to other investors. He believes the corporation and the computer software he has designed will create a business worth $1 million. Accordingly, he has valued each of the 1,000 equal shares at $1,000 each. His corporation hires 10 software developers who believe in the design and allows each to purchase 20 shares for a total of $20,000 per employee. The corporation raises $200,000 as a result.

"Private Issuer" exemption applies: Despite the fact that none of the investors reach the Minimum Investment threshold, since Mario's corporation is selling shares to employees, the share distributions are exempt from the requirement of prospectus filing.

The distributions of shares to both Janet and Mario are exempt from the prospectus requirement as they are the initial directors of their respective corporations. If either of their companies attracted the attention of an "accredited investor" any distribution of shares to that investor would also be exempt.

Continuous Disclosure

After becoming a reporting issuer, it is that issuer's responsibility to file with the provincial administrators and make public on a continuous basis all material information that would significantly affect the valuation of its securities. The scope of this disclosure includes both routine and event-specific information. On at least an annual and quarterly basis, financial statements must be disseminated, and an Annual Information Form also must be filed. When a material change in the affairs of the issuer occurs (for better or worse), the issuer is obligated to file a Material Change Report and make a press release. These changes could result from any of a myriad of reasons, such as a court ruling that a patent is invalid, unexpectedly poor operating results, a promising new joint-venture partnership agreement, or the death of a key employee or officer. There is an obligation of timeliness, even immediacy, for these reports, and materiality is a question of fact. It is an objective question of materiality in the eyes of the market and the regulators, though not in the often more subjective eyes of the issuer. As a result, a corporate policy of more disclosure rather than less would be prudent for any reporting issuer.

Liability for misleading disclosures has recently increased for corporations and their directors and officers and more generally, "influential persons." Not only do the U.S. *Sarbanes-Oxley* requirements apply to Canadian firms with U.S. patents or exchange listings, but our own jurisdictions are beginning to impose more significant obligations. These go beyond prospectus and continuous disclosure to expand the definitions of material fact and change, and demand accuracy in press releases and other statements.[12]

CASE IN POINT

At sixteen minutes past noon on February 18, 2011, Cornerstone Capital Resources Inc., a junior mining resource firm, issued a press release. The release stated that the company was not aware of the reason that caused a sudden decline in the price of its shares earlier that day. It further stated there were no undisclosed material changes in the operations of the company or its fundamentals that could account for the change and that it was in contact with regulators for additional clarity.

The sudden price decline represented a significant percentage of the value of this stock. What expectations were imposed upon the management of the firm in these circumstances, and why did they respond in the manner they did?

Electronic Filing and Disclosure

The Canadian Depository for Securities Limited maintains, on behalf of Canada's securities administrators, an electronic filing and disclosure system coupled with public access to its holdings via the Internet. This system, known as the System for Electronic Document Analysis and Retrieval (SEDAR), holds most of the documents that are legally required to be filed with the provincial securities administrators and many documents that may be filed with Canadian stock exchanges by public companies and mutual funds. Filing of prospectuses and continuous disclosure with SEDAR is mandatory for most reporting issuers. While insider-trading reports are not yet available, these may one day join the list, and the site contains a wealth of information and ease for investors and issuers alike. The American equivalent is EDGAR, the Electronic Data Gathering, Analysis, and Retrieval system, and performs the same disclosure role for companies and others required by law to file forms with the U.S. Securities and Exchange Commission (SEC).

CONDUCT OF TRADING

As can be expected of an environment that includes intermediaries, the lack of face-to-face meeting between buyers and sellers could result in abuses by any of the three parties to a trade. Either a buyer or a seller could hide behind a false front to profit from illegally obtained information about a security and its prospects, or an intermediary could have its own undisclosed financial interest in a particular security on offer. Accordingly, the provincial securities Acts and the SROs place responsibilities on intermediaries to "know their client" as well as their client's investment objectives, require disclosure (and avoidance) of conflicts of interest, and set requirements for ethical account operation and an audit trail of account activity.

12. See, for example, Ontario's *Keeping the Promise for a Strong Economy Act (Budget Measures), 2002*, S.O. 2002, c. 22, Part XXVII, amending Ontario's *Securities Act* with new liabilities in 2006. This legislation is also known as "Bill 198" in press parlance, and similar legislation can be expected in future in other provinces.

COURT DECISION: Practical Application of the Law

Improper Trading: Investment Dealers Association Discipline Hearing—H. Pandelidis, Respondent

February 9, 2005 (Alta. Dist. Council).

THE PANEL: The hearing panel and the Respondent agreed as to certain facts and the Respondent admitted to the following...

Count 1. The Respondent used discretion with respect to trades in the accounts of the client, S. J., without the prior written authorization of the client and without approval of the member firm...

Count 2. The Respondent engaged in the improper practice known as "Bucketing," whereby he provided confirmation of a transaction when in fact no trade had been executed, in the account of the client, S. J...

Count 3. The Respondent made offers of compensation for account losses to the client, S.J., on at least five (5) occasions...

Count 4. The Respondent engaged in personal financial dealings with the client, S.J., without disclosing those dealings to the member firm...

Count 5. The Respondent made a representation to his client, S.J., that a particular security would be listed on an exchange, or quoted on a quotation and trade reporting system, and publicly traded on a particular date...

Count 6. The Respondent participated in the distribution of securities when the securities were not approved for distribution in Alberta and, further, he conducted off-book transactions involving the securities without the approval of the member firm... all of which are contraventions of Association By-laws, Regulations, Rulings and Policies, and/or the Alberta Securities Act and/or Standards C and D of the Conduct and Practices Handbook, such conduct unbecoming a registrant and detrimental to the public, contrary to Association By-law 29.1.

[The panel then analyzed the conduct in detail]

Based on the above, the Panel agrees with the recommendations of counsel for the Association and imposes the following penalties:

(a) a fine in the amount of $75,000; (b) a suspension of 5 years from registration in any capacity, with no reinstatement until payment of all monetary penalties and costs have been made; (c) as a condition of re-approval of the Respondent by the Association in any registered capacity that the Respondent re-write and pass the Canadian Securities Course and the examination based on the Conduct and Practices Handbook for Securities Industry Professionals; (d) that upon re-entry to the industry the Respondent be under strict supervision for a 12 month period followed by an additional 12 month period of close supervision; and (e) costs of the Association in the amount of $12,524.

This case serves as a reminder to traders that their actions will be under scrutiny not only by their member firm, but also by their industry association, which will act vigorously in cases of improper conduct.

Fraud on the Market

In receiving and executing client orders, employees of a brokerage house are in the very privileged position of knowing about some or all pending orders, and have direct or indirect access to the marketplace. Other securities-industry professionals may be in a similar position, perhaps with discretionary trading authority over extremely large sums in the form of pension or mutual funds. This knowledge of orders and the ability to "move the market" creates opportunities for abuse and the need for trading regulation.

Structuring trading activity for a trader's own gain or to the preference of one customer over another is prohibited. For example, faced with a client order to buy a large block of a particular security, one almost certain to raise the market price, a trader may be tempted to slip in ahead of it with a personal order or another later-received order from another client. Equally, an unscrupulous dealer might create an illusion of market interest in a security through a series of rapid buy/sell transactions out of its own inventory, or inventory it controls. Once others are induced to join the apparent frenzy at a "pumped" price, the undesirable stock a) looks better on the dealer's books and b) could be "dumped" at a considerable profit. All of this behaviour is prohibited by corporate policies, by the SROs, and by the provincial securities Acts. Such behaviour can become criminal under s. 380 of the *Criminal Code*, which provides for ten years' imprisonment for anyone who, through deceit or falsehood or fraudulent means, affects the public-market price of securities. This criminal sanction would also apply to anyone who attempted to influence security prices through fake press releases or e-mails; the section is not simply confined to actual trading activity.

As a guarantee of a future gain would be a powerful inducement for purchasers of securities, there are specific provisions in all provincial Acts that prohibit intermediaries from making such promises, guarantees, or agreements to refund money or repurchase securities.[13] These prohibitions are backed by the investigation and enforcement powers granted to the provincial securities commissions under their respective Acts, as well as by powers of similar effect exercised by the SROs.

COURT DECISION: Practical Application of the Law

False or Misleading Material Information Released and Uncorrected—Fraud on the Market

Specogna, Re **(1996), 11 C.C.L.S. 276 (British Columbia Securities Commission).**

A director of a corporation that was engaged in the exploration of a potential mining property informed several individuals of favourable drilling results and sent a fax to a broker that contained fabricated drilling results. The director, his mother and his wife sold their shares at the higher share prices that followed the release of the fabricated (and false) drilling information.

THE COMMISSION: Marino Specogna bought and sold shares knowing that false assay results had been disseminated. Client account documentation and brokerage account statements indicate that [in 1993] Marino Specogna had trading authority over at least 13 separate brokerage accounts at eight Exchange member firms. In 626 trades, the Group Accounts purchased 2,794,500 and sold 3,182,000 Doromin shares in 1993 making them net sellers of over 300,000 shares. Although we have found that Specogna Minerals, E. Specogna and L. Specogna have failed to meet certain regulatory requirements, this case centers primarily around the egregious conduct of Marino Specogna. As a director and de facto officer of a reporting issuer he must bear responsibility for that conduct. We have found Marino Specogna responsible for a number of significant violations of securities regulatory standards, many of which were intentional. The Commission considers that it is in the public interest to order: [Exemptions regarding trading and registration] do not apply to Marino Specogna and Specogna Minerals for a period of 20 years from the date of this decision; each of Marino Specogna, E. Specogna and L. Specogna is prohibited from becoming or acting as a director or officer of a reporting issuer until each has successfully completed a course of study satisfactory to the Executive Director concerning the duties and responsibilities of directors and officers; Marino Specogna is prohibited from becoming or acting as a director or officer of a reporting issuer until a period of 20 years has elapsed from the date of this decision and that E. Specogna is prohibited from becoming or acting as a director or officer of a reporting issuer until four years has elapsed from the date of this decision; and the respondents are to pay an administrative penalty: Marino Specogna, $35,000; E. Specogna, $5,000; L. Specogna, $1,000.

In 2003, Marino Specogna published A Convicted Stock Manipulator's Guide to Investing. *(New York: Writers Showcase Press, 2003).*

INSIDER TRADING

Insider trading
The trading in shares of a corporation by a person in a corporation who possesses undisclosed privileged information about the corporation.

When an investor has knowledge of a material fact about a reporting issuer, and that fact has not yet been disclosed to the public, that person is in a powerful position to make money. Material facts would include earnings performance, planned takeover bids, major assets bought and sold, and, perhaps, key staff changes. Whether the fact is good or bad is immaterial, for a trade can be structured to take advantage of the eventual price movement, up or down, in the corporate shares that will occur when the operative

13. For example, the *Securities Act*, R.S.O. 1990, c. S.5, s. 38.

fact is publicly disclosed. The legality of doing so depends entirely on the nature of the relationship between the person with knowledge and the company itself.

If the investor has deduced this fact from afar, by careful observation and analysis, he or she is to be rewarded for his or her investment savvy. However, if the investor is in possession of this fact because he or she is an insider of the reporting issuer or a person in a "special relationship" with the company, this act is generally termed "insider trading" or, more accurately, "trading on undisclosed information," and is illegal in all jurisdictions in Canada.[14]

The "insiders" of a reporting issuer are its directors and senior officers, those directors and senior officers of its parent or subsidiary firms, shareholders with more than 10 percent of outstanding voting rights, and the reporting issuer itself (the company is its own insider).

A "special relationship" exists when a person or company is an affiliate or associate of an insider (business partner, spouse or partner, or relative), a takeover bidder, a professional-services firm acting for the reporting issuer or a takeover bidder, an employee of the reporting issuer, a takeover bidder, or a professional services firm. Finally, a person or company can wind up having a special relationship if they learn of a material fact or change from someone who is already in a special relationship and the existence of that relationship was known or ought to have been known by them.[15]

Anna is a chartered accountant conducting an audit of a reporting issuer. She learns a material fact about the issuer from the company president in the course of performing her audit work. She informs her brother Rudi, who in consequence buys shares in the reporting issuer. Unfortunately, when the material fact is publicly disclosed, it depresses the share price, and Rudi loses his investment.

The president disclosed the fact in the normal course of business, and has no liability. Anna is in a special relationship with the issuer and has disclosed a material fact outside the normal course of business before public disclosure, and has therefore committed the offence of tipping under the provincial securities Act. Rudi knew or ought to have known that Anna was in a special relationship with the issuer. His act of purchasing the securities based on information about the reporting issuer that was not publicly disclosed is insider trading, an offence under the provincial securities Act. His loss or profit is immaterial.

This is not to say that insiders and those in special relationships with reporting issuers cannot trade in the securities of those issuers. In fact, shares in one's company are often an important part of the compensation package of senior executives. The illegality arises in the trading on the basis of undisclosed information. The savvy investor is matching his or her wits with a market that has equal access to that information, but chooses to ignore it or interpret it differently. When insiders take advantage of privileged information to their own benefit, they are taking advantage of individual investors on a tilted playing field, and the law addresses this imbalance with severe sanction. Equally, it is an offence to pass along such privileged information relating to material changes or facts to others, known as "tipping," leaving both the tipper and **tippee** liable to prosecution.

Tippee

A person who receives undisclosed privileged information about a corporation from an insider.

In order to trade legally, insiders and those in a special relationship must wait until the material fact or change has been publicly and generally disclosed. Insiders are further obligated to record and submit the particulars of their monthly trading activity within ten days of that month's end to their provincial securities commission. This information becomes a matter of public record, and, while the insider's motives for trading are not disclosed, the fact that the trade has been made, as well as its quantity and nature (buy or sell), may be of some guidance to other investors and the markets.

14. For example, *Securities Act*, R.S.B.C. 1996, c. 418, s. 86, or the *Securities Act*, R.S.O. 1990, c. S.5, s. 76(5).

15. For example, *Securities Act*, R.S.B.C. 1996, c. 418, s. 3, or the *Securities Act*, R.S.O. 1990, c. S.5, s. 76(5).

The penalty in some jurisdictions for trading on undisclosed information is imprisonment and as much as a $5-million fine, or triple the profit made or loss avoided on the illegal trade, whichever is greater.

While securities regulation is a matter of provincial jurisdiction, the essentially fraudulent character of insider trading has been recognized as a federal crime. In addition to Securities Act prohibitions, it became part of the *Criminal Code of Canada* in 2004 as section 382.1. While the offence applies to any undisclosed information (not just material information), there is a higher burden of proof imposed, that the information was knowingly exploited. Tipping also now has a *Criminal Code* prohibition. It comes with a high standard of proof as well, that the tipper *knew* the information would be acted upon, or that it would be passed on to another who would act upon it. A federal penalty of ten years imprisonment is possible for a conviction under this section of the Code.

CASE IN POINT

In her widely-reported case, style and home diva Martha Stewart was originally charged, along with her broker, by the U.S. Securities and Exchange Commission under U.S. insider trading laws.

Stewart sold stock in a biopharmaceutical company, ImClone Systems, Inc., on Dec. 27, 2001, after receiving an unlawful tip from Peter Bacanovic, at the time a broker with Merrill Lynch, Pierce, Fenner & Smith Incorporated. The Commission further alleged that Stewart and Bacanovic subsequently created an alibi for Stewart's ImClone sales and concealed important facts during SEC and criminal investigations into her trades. In a separate action, the United States Attorney for the Southern District of New York has obtained an indictment charging Stewart and Bacanovic criminally for their false statements concerning Stewart's ImClone trades. Her charges were eventually reduced to obstruction of justice, and upon conviction she served five months imprisonment and five months house arrest.

See: *United States v. Stewart,* S.D.N.Y., 03 Cr. 717 (MGC), 5/5/04.

PROXY VOTING AND PROXY SOLICITATION

It is a fact that few shareholders of listed companies ever attend the annual general meeting of shareholders. Distance, cost, relevance, and the voting impotence of a small shareholder combine to ensure that this situation continues to be the case. Despite this, it is the shareholder's one opportunity to exercise power in reviewing the efforts of the existing Board of Directors and in electing their successors. A Board that is handed this "divide-and-conquer" situation could expect to have almost unlimited tenure. Proxies are intended to address this imbalance.

Proxy

A document evidencing the transfer of a shareholder's voting right to an appointee, either with instructions for voting, or allowing discretion to be exercised by the appointee, at a meeting of shareholders of the corporation.

A **proxy** is a transfer of voting privilege by a shareholder to an agent on the basis of a trust. Consequently, any party acting as an agent who can accumulate enough proxies (and particularly proxies that allow the agent to vote at its own discretion) will have sufficient votes to be a force to be reckoned with at the annual general meeting. In large corporations that are widely held, even a block of votes as small as 5 percent may have sufficient weight to be effective. Naturally, 50 percent plus one vote would represent complete control to vote and install a handpicked Board of Directors.

Proxy rights are therefore important political weapons in corporate politics, and in cases of turbulent governance of a corporation are actively solicited by interested parties. Even under normal conditions, the existing Board of Directors will want to continue in place, and this solicitation still can and does occur.

A solicitation of proxies must be accompanied by an Information Circular,[16] which discloses who is doing the solicitation and his or her ownership or interest in the company. Information requirements depend on whether the solicitation is being made by management or by other parties. Any limitations on revocation of the proxy must be

16. For example, *Securities Act*, R.S.B.C. 1996, c. 418, s. 117, or the *Securities Act*, R.S.O. 1990, c. S.5, s. 86.

stated, as well as the manner in which the proxy is to be employed. The proxy can be used for matters well beyond the election of directors, such as particular acts of management under consideration, the appointment of a particular firm of auditors, property acquisitions, or changes to share capital.

A QUESTION OF ETHICS

In the normal course of business, existing directors of a corporation annually solicit proxies. They do so for a range of reasons, one of which is to ensure their own re-election. Do you consider this to be a conflict of interest? If so, is it a tolerable one? Do you feel that any similar conflict arises when a takeover bid is launched and current management issues a circular on the merits of the bid?

FRONT-PAGE LAW

Ontario Securities Commission goes after Philip Services, seven individuals

TORONTO (CP) — Philip Services Corp. and seven individuals, all former officers or directors of the Hamilton-based company, face disciplinary hearings next month, the Ontario Securities Commission said Wednesday.

The company's co-founders Allen Fracassi and his brother Philip were among the individuals accused of failing to disclose "full, true and plain disclosure of all material facts" in a November 1997 prospectus filed by the company with regulators.

Philip Services, once one of North America's biggest metal recycling companies, recently emerged from bankruptcy protection.

Once a stock-market darling, Philip's shares plunged on North American markets in 1998 after the company revealed massive copper trading losses in its metals division that forced the Hamilton recycler to restate its earnings back several years.

The stock plunge and financial troubles nearly wiped out the company and led to numerous lawsuits by angry investors, who claimed they were misled by the company about its financial troubles while it was issuing new shares in late 1997.

Why does this news article refer to "disciplinary hearings," rather than "criminal charges"? Business can be a risky undertaking, with no guarantee of profits. Are this company and its officers being prosecuted for something that is just a normal business risk that turned out badly?

Source: Excerpted from *Canadian Press*, "Ontario Securities Commission goes after Philip Services, seven individuals," August 30, 2000. (www.nationalpost.com)

TAKEOVER BIDS

Takeover bid
An attempt by a competitor to obtain a controlling interest in a corporation.

When a company has become the target of a **takeover bid**, a market participant (possibly a competitor company) has decided to offer to purchase a significant interest in the target company. This significant interest may, in fact, be a controlling interest, or it may not, but when the offeror's own present holdings together with those that are expected to result from the bid meet or exceed 20 percent of that class of security, the provisions of the provincial securities Acts on takeover bids come into effect.

It is quite possible that such bids could be made in a way that exploits a shareholder's lack of knowledge. A present shareholder is probably ill informed as to the competency and plans of the proposed new management that will likely accompany the new controlling owners, and he or she will have little time to make inquiries. Similarly, the shareholder may be interested in the views of current management, for the bid may even be in the company's best interest, and acceptance may be recommended by the existing board of directors. Time for acceptance of the offer made by the takeover bidder may be so short that an unreasoned and hasty decision could be made. Finally, in an unregulated world, the bidder might accept and pay a premium to buy up the shares of some existing shareholders, consciously and unfairly ignoring the willingness of others who wish to accept the bidder's offer.

The takeover provisions of all provincial securities Acts address each of these concerns. In the case of such a bid, it must be made to all holders of securities of the class being sought, disclose the full financial terms of the proposed purchase, and disclose the offeror's existing interest in the company. The target company's board of directors must issue their own circular within ten days of the bid, providing a reasoned recommendation of either acceptance or rejection of the bid, or no recommendation at all, with reasons for doing so. Investors have a window of 21 days from the bid date to deposit their shares with a trustee, and a further ten days to do so beyond the date of any change made to the terms of the bid.

Ten days after the bid expires, the offeror must take up the securities that have been tendered, and pay for them within another three days. In all cases where the bid was made for less than all of the outstanding issue, and more shares are tendered than the offeror wishes to purchase, what is taken up and paid for must be done so on a *pro rata* basis from all those who tendered into the bid. No preference can be made to buy the shares of one shareholder over another.

Bids of an issuer in its own shares (an issuer bid, which is therefore not an isolated trade and otherwise an exemption from prospectus disclosure) have the same result as a takeover bid, and are dealt with under the securities Acts in a substantially similar way.

MANAGEMENT ALERT: BEST PRACTICE

Becoming a publicly traded company is an enormous step in the evolution of a corporation. Aside from the expense of issue and ongoing compliance, this access to capital is balanced by the heavy onus of disclosure on what were formerly closely held business secrets. For many firms, however, the equation, unfortunately, ends there. The prospect of new capital is so attractive that the fact that its providers will be part-owners (not donors) is conveniently relegated to second place. These new faces may not share the same long-term goals as the original incorporators, are at liberty to make their own takeover bid, or may sell their shares to someone else who is at philosophical odds with the original owners. These latter considerations should be given greater weight in the initial decision to go public, because undoing later by "going private" again will incur an expensive (if not prohibitive) premium.

INVESTIGATION AND ENFORCEMENT

The provincial securities administrators have far-reaching powers of investigation and enforcement, and those of the SROs are no less significant given that suspension of membership, delisting from an exchange, or deregistration of a registrant effectively destroys any hope of that person or firm remaining within the Canadian financial system.

As well as sanctions ranging from two years' imprisonment and five million-dollar fines for contravention of the Acts, the provincial securities Acts also create actionable civil liabilities for misrepresentation in prospectuses, circulars, filings, and like items against issuers, underwriters, and their directors and officers.

Among the powers of the provincial securities administrators are those relating to termination of registrations, cease-trading orders in securities, withdrawal of exemptions, requirements that market participants submit their practices for review, and requirements to hand over documents to either the commission or other persons. These administrators may also make recommendations to prosecute persons or firms. These powers may be exercised by the provincial administrators when, in their opinion, it is in the public interest to do so.

SUMMARY

Securities regulation governs the nature of securities, the method of their distribution and trading, and obligations to disclose information as to their riskiness, in order that investors may make informed purchase decisions. It is an area of provincial responsibility, and seeks to balance the twin needs of capital market efficiency and market integrity. The definition of a security includes any document commonly known as a security, typically evidence of title to or interest in the capital, assets, property, profits, earnings or royalties of any person or company. With a few important exceptions, securities are subject to the obligations of true, full, and plain disclosure within a prospectus, and to continuous disclosure thereafter on a timely basis. Owners of securities are afforded special protections in cases of solicitation of their voting rights, in takeover bids and issuer bids, and in cases of insider trading. Issuers benefit from standardized processes for bringing securities to market, and may be treated on different terms according to their financial standing and track record. The provincial securities administrators have considerable powers to act on specific matters and even more generally in the public interest. The consequences of enforcement action, including imprisonment, fines, and civil liability, can be extremely serious for those found in violation of the legislation.

KEY TERMS

REVIEW QUESTIONS

1. Under what circumstances and on what condition would a takeover bidder not purchase all the shares tendered by a shareholder in response to a bid?
2. What is meant by continuous disclosure?
3. To be registered as an investment advisor, what criteria must an individual meet?
4. What is an SRO and what does it do?
5. Describe the two chief exemptions to the requirement of producing and filing a prospectus for a new issue of corporate shares.
6. What are the twin policy goals of provincial securities legislation?
7. Describe what a proxy is, and why it is important.
8. What is the expected standard of prospectus disclosure?
9. Compared to other bodies of law, why was securities regulation so late in developing?
10. What is a tippee, and what are the consequences of being one?

MINI-CASE PROBLEMS

1. The senior management of Consolidated Moosepastures Mining has discovered errors in their geological reports for their principal mining-claim property. Corrected estimates indicate an ore-body quality revised downward from 9.7 grams of pure unobtainium per ton of rock to 3.5 grams per ton of rock. What should management do, in accordance with the law of securities regulation?
2. Victoria owns 7.5 percent of the shares of Harmony Publishing Corporation, a publicly listed company. On Wednesday, in the locker room of her country club, she overhears one woman telling another that "We've signed Angelica to three, and I've just got to get a dress for Saturday's launch party." Victoria is certain that the speaker is Fran Collins, wife of Harmony's managing editor, whom she had seen at a shareholders' meeting, and that "Angelica" could only refer to Angelica Constance-Smythe, the current diva of romance fiction. Would Victoria be in violation of the law if she bought more shares in Harmony between now and Saturday? Have any laws been broken already?
3. Alex and Bohumir carried on business in partnership for several years, and on the advice of their accountant, decide to incorporate. To grow their business, however, they would require additional capital, and they decide to incorporate as a public company in order to attract investors. What steps must they take in order to offer their shares to the public?
4. Alicia purchased 100 shares of ABC Corporation on the basis of a prospectus issued by the corporation. A few months later, the corporation discovered an error in its reported sales and profits reported in the financial information in the prospectus. Following the corporation's announcement, share prices fell, and Alicia suffered a loss of $1,200. Discuss.
5. Carla, the president of a small corporation, wishes to raise $500,000 in new investment money for the firm. What are her options to do so without the requirement of filing a prospectus? Give examples.

www.mcgrawhill.ca/olc/willes

CASE PROBLEMS FOR DISCUSSION

CASE 1

PR Plastics Inc. was a small publicly traded corporation that manufactured outdoor furniture. Brown and Smyth, a professional accounting firm was for many years their auditors. During most of this time, demand for outdoor furniture remained high, and the corporation experienced significant growth. Because of their knowledge of the corporation's operations, both Brown and Smyth purchased shares in the corporation in the names of their parents and family members. Neither Brown nor Smyth owned any shares in their personal capacity.

During a recent audit, Brown noted a number of irregularities in the firm's invoicing practices and the recording of sales. The firm reported the irregularities in its audit report to the corporation, but prior to the delivery of the report, Brown advised his parents and family members to sell their shares in the corporation.

The corporation in due course issued a news release concerning the accounting irregularities, only with a president's statement that the practices had been corrected on the advice of its auditors. The price of the shares nevertheless dropped by $10.00. An investigation revealed that Brown's parents and family members had sold their shares in the corporation prior to the announcement by the corporation.

Discuss the issues raised in this case.

CASE 2

Mammoth Corporation Ltd. set out a goal to become the dominant firm in the widgets market, and in an effort to reach this goal, decided to make a takeover bid for its principal rival, Widget Corporation Inc. Mammoth Corporation Ltd. expects that the directors of Widget Corporation Inc. may oppose the takeover bid. Discuss the steps that each party must take in order to comply with relevant securities legislation.

CASE 3

Henrick Alfredsson was employed as an investment advisor with a stock brokerage. His wife, Jane Martine, was employed as a wholesale sales agent by a small Canadian private corporation engaged in the manufacture and distribution of automobile-care products, namely wax coatings and polishes. The private corporation engaged Alfredsson to place $200,000 worth of his shares among a small group of investors. To assist him, the corporation gave Alfredsson promotional documents containing its sales projections for the next three years, and advised him that the firm intended to "go public" within another three years. Neither Alfredsson nor Martine had ever seen the manufacturing facility, which they were told was located in the United States, nor did Alfredsson make any further inquiries into the financial health of the company. He duly sold the shares to interested clients who were already customers of his brokerage house. As matters turned out, the Canadian firm did not own a U.S. factory and had only rights of distribution. Moreover, the sales projections were wildly optimistic and, in time, the rights to distribute in Canada were lost altogether. The shares in the Canadian company became worthless as a result, and the investors sought redress from Alfredsson personally, as well as his employer, the brokerage.

Comment on the issues raised in this situation. If this was brought before you as the provincial securities regulator, which issues stand out the most, and what redress or penalty (if any) would you impose?

CASE 4

Julia, a registered investment advisor, was employed at a stock brokerage in a junior capacity as a telemarketer of unlisted shares, making cold sales calls to prospective clients. Once she had made a sale, she passed the client on to a more senior advisor, who finished the paperwork and collected the payments for the shares. Where these shares were sold from the brokerage's own holdings of unlisted shares, she received a commission of 20 percent, and the senior advisor received some further amount, but that percentage was unknown to her. Her firm directed her to verify that prospective clients had an income of at least $20,000 per year, and also that each client possessed a personal net worth of at least $20,000. Julia made this verification in every case. After three years, Julia moved to another brokerage house and applied to the provincial regulator for authorization to sell shares on behalf of the new employer.

Imagine that you are the provincial regulator, and have been made aware of the circumstances of Julia's previous employment. Would you grant or refuse the application? Why?

The Online Learning Centre offers more ways to check what you have learned so far. Find quizzes, Weblinks, and many other resources at www.mcgrawhill.ca/olc/willes.

CHAPTER 19

Employment and Labour Relations

Chapter Objectives

After study of this chapter, students should be able to:

- Understand the nature of the employment relationship.
- Explain how the relationship is established.
- Identify the duties of the employer and employee.
- Understand how employment may be terminated.
- Understand the unionized employment relationship.
- Explain how collective bargaining rights are established and maintained.

YOUR BUSINESS AT RISK

No relationship is more important to a business enterprise than its relationship with its employees. Without these individuals there is no production, management or service. They are entrusted with judgment, trade secrets and finances, yet so often this is the most neglected relationship. The majority of employment-related problems can be avoided or provided for early on — in recruiting, employment contracts, job descriptions and appropriate discipline programs. The consequence of management failure here goes directly to the firm's bottom line.

CONTRACT OF EMPLOYMENT

The origins of the employment relationship have their roots in antiquity. It is impossible to know at what point the first individual voluntarily consented to do the bidding of another in return for a wage. The employment relationship of master and servant developed in early England as slavery declined, and the rise of towns and guilds fostered the climate for the new form of servitude. By the middle of the fourteenth century, there were sufficient numbers of workmen and craftsmen employed in England to warrant legislation to fix their wages.[1]

The first English Act that dealt specifically with the contract of employment was the *Master and Servant Act* of 1867, which was subsequently adopted in Canada by a number of provinces.[2] This statute modified the Common Law contract of employment by imposing limits on the length of time a written contract of employment might bind the parties, and specified many of the terms and conditions that relate to the relationship. Legislation that protected workmen from injury arising out of work accidents in England took the form of a *Workmen's Compensation Act* in 1897. It, too, was introduced in a number of Canadian provinces some time later.

Apart from legislation that had a direct bearing on employment, the Common Law over a long period of time gradually established the relationship as one of contract. The

1. *Statute of Labourers*, 23 Edw. III. (1349). The statute was apparently not a true statute, as Parliament had been prorogued because of the Plague. It was, however, treated as such.
2. For example, the *Master and Servant Act*, R.S.O. 1897, c. 157, and R.S.M. 1891, c. 96.

courts saw the relationship as a form of "bargain" struck between the master and the servant, in which the servant (in return for a wage) agreed to submit to the direction of the master in the performance of his work. The relationship, however, was never considered to be one of pure contract in which the parties, both vested with equal bargaining power, devised an agreement to their mutual satisfaction. The law of master and servant had been around too long to abandon all of the law that had developed over many centuries, and, as a result, the relationship was never treated as being strictly contractual. Masters, for example, had historically been held liable for the torts of their servants committed in the performance of the masters' business, and the breach of a master's contract by a servant at law was the master's default. Since it was first recognized, the relationship has been modified by case law and by statutes that have generally imposed additional duties or limitations on the master. Nevertheless, the basic concept of employment as a contract has remained, for the most part, intact.

The relationship is now also subject to laws relating to occupational safety, minimum employment standards, and hiring practices. All of these impose additional duties and responsibilities on employers, their working conditions, and the employment relationships with their employees. The impact of these laws is dealt with later on in this chapter.

NATURE OF THE RELATIONSHIP

The Common Law contract of employment involves the payment of wages or other remuneration by the employer to the employee in return for the services of the employee. As with other forms of contract, to be enforceable the agreement must contain the essential elements of a contract. The basic characteristic of the relationship, which determines whether a person is an employee or not, is generally considered to be the degree of control that one person exercises over the other. For many years, the courts considered the relationship to be one of employment where the employer had the right to direct what work was to be done and the manner in which it was to be done.[3] This basic test proved to be inconclusive, however, as employment relationships in the twentieth century took on a wide variety of forms. The courts gradually came to realize that control in itself was insufficient to determine the relationship. Eventually, a more complicated test was devised to meet the complex interpersonal relationships that had arisen in modern business. The test was essentially a fourfold one, of which only one of the factors considered was control. The courts added to this three other factors: the ownership of tools, the chance of profit, and the risk of loss.

Fourfold test

A test for employment based upon (1) ownership of tools (2) control (3) chance of profit (4) risk of loss.

The **fourfold test** was described by the court in the following terms:

> In earlier cases a single test, such as the presence or absence of control, was often relied on to determine whether the case was one of master and servant, mostly in order to decide issues of tortious liability on the part of the master or superior. In the more complex conditions of modern industry, more complicated tests have often to be applied. It has been suggested that a fourfold test would in some cases be more appropriate, a complex involving (1) control; (2) ownership of the tools; (3) chance of profit; (4) risk of loss. Control in itself is not always conclusive.[4]

Organization test

A test for employment based upon an examination of the services in relation to the business itself.

More recently, the courts have recognized the limitations of their fourfold test and appear to be groping their way toward an **organization test**, which examines the relationship in relation to the business itself. This latter test is based upon the services of the employee, and whether they represent an integral part of the business or something that is adjunct or accessory to the normal business activities of the employer.[5] In recent cases that have come before the courts, the test's principal application has been to distinguish between employees and independent contractors — not an easy task in today's complex business world.

The independent contractor has usually been distinguished from the employee on the basis that the initiative to do the work and the manner in which it is done are both under

3. *Harris v. Howes and Chemical Distributors, Ltd.*, [1929] 1 W.W.R. 217.
4. *City of Montreal v. Montreal Locomotive Works Ltd. et al.*, [1947] 1 D.L.R. 161, [1946] 3 W.W.R. 748.
5. *Co-operators Ins. Ass'n v. Kearney* (1964), 48 D.L.R. (2d) 1; *Armstrong v. Mac's Milk Ltd.* (1975), 7 O.R. (2d) 478; *Mayer v. J. Conrad Lavigne Ltd.* (1979), 27 O.R. (2d) 129.

the control of the contractor. This has been generally characterized by the right of the contractor to exercise his or her own discretion with respect to any matter not specifically stipulated in the contract.[6] However, this distinction has become blurred in situations where the contractor acts alone.

COURT DECISION: Practical Application of the Law

Master and Servant—Test for Employment Relationship

***Mayer v. J. Conrad Lavigne Ltd.* (1979), 27 O.R. (2d) 129.**

The plaintiff sold television time for the defendant on a straight commission basis. He was required to attend regular sales meetings of the defendant each day and file sales reports. Some direction of the plaintiff's activity was carried out by the sales manager, but this was generally limited to where and to whom to sell. When the defendant failed to pay the salesman vacation pay, he instituted legal proceedings.

THE COURT: The law and cases were much canvassed before us, but as the determination of whether a particular individual is a servant or an independent contractor is completely dependent on the facts, it would serve no useful purpose to review them all only to distinguish them on their facts. The emphasis in the earlier authorities was on the extent of the "control" that the master had over the servant to determine whether there was, indeed, a master-and-servant relationship. The concept of this relationship has, however, been an evolving one, changing with the changes in economic views and conditions. As Lord Wright put it in the leading case of *Montreal v. Montreal Locomotive Works Ltd et al,* [1947] 1 D.L.R. 161 at p. 169: "In the more complex conditions of modern industry more complicated tests have often to be applied." He postulated a fourfold test involving: (1) control, (2) ownership of the tools, (3) chance of profit, and (4) risk of loss. This test has been enlarged by the more recent "organization test" which was approved and applied by Spence, J., in *Co-operators Ins. Ass'n v. Kearney* (1964), 48 D.L.R. (2d) 1. In that case (pp. 22-3), he quoted with approval the following passage from Fleming, *The Law of Torts*, 2nd ed. (1961), at pp. 328-9:

> Under the pressure of novel situations, the courts have become increasingly aware of the strain on the traditional formulation of the control test, and most recent cases display a discernible tendency to replace it by something like an "organization" test. Was the alleged servant part of his employer's organization? Was his work subject to co-ordinational control as to "where" and "when" rather than "how"? [citing Lord Denning in *Stevenson, Jordan & Harrison Ltd. v. MacDonald*, [1952] 1 T.L.R. 101, 111.]
>
> Lord Denning in *Stevenson*, said this: "One feature which seems to run through the instances is that, under a contract of service, a man is employed as part of the business, and his work is done as an integral part of the business; whereas, under a contract for services, his work, although done for the business, is not integrated into it but is only accessory to it."

In my view, the facts as recited satisfy whichever test is used. The appellant had and exercised the control necessary to establish a master-and-servant relationship. The "when" and "where" was within the master's control, and to a certain extent, the "how" when clients were transferred from one salesman's list to another. Because of the training, skill, and experience of the respondent, one would not expect that the appellant would control "how" the respondent sold the air-time, any more than any other skilled or professional servant would be directed how to do his work. Equally, the relationship satisfies the "organization" test. The respondent's work was a necessary and integral part of the appellant's business. It supplied the financial life-blood of the appellant, and his work was subject to the co-ordinational control of management. His work was clearly integrated into the business and not merely accessory to it. In my view, applying the common law tests, the facts establish that the respondent was a servant and employee of the appellant.

How does the organization test differ from the control aspect of the fourfold test?

When the contractor also employs others to do the work, the relationship is generally that of an independent contractor rather than an employee, because the independent contractor exercises the function of an employer as well.[7] It was, however, the difficulties associated with the determination of the true relationship, in cases where no employees

6. *McAllister v. Bell Lumber Co. Ltd.*, [1932] 1 D.L.R. 802.
7. *Re Dominion Shipbuilding & Repair Co. Ltd.; Henshaw's Claim* (1921), 51 O.L.R. 144.

were engaged by the independent contractor, that required the courts to devise an organization test to identify the nature of the contract. The test examines the contractor's role in the context of the employer's business, and the relationship is determined on the basis of whether the work done is a part of the business or something outside of it.

The same test might be applied to distinguish the agency relationship from that of the employer–employee. Generally, the principal has the right to direct the work that the agent is to perform, but not the manner in which it is to be done.[8] Again, this is not always so, particularly if the agent acts only for a single principal, and the principal exercises a substantial degree of control over the agent. Other characteristics of the principal–agent relationship may also apply that may distinguish a case of agency from employment in given circumstances, but the application of the organization test represents a useful tool to identify and distinguish the two relationships.[9]

CHECKLIST OF EMPLOYMENT FACTORS

1. Fourfold Test
   - Degree of control (high)
   - Ownership of tools (low)
   - Chance of profit (low)
   - Risk of loss (low)

   or

2. Overall, integral part of business (high)

   If these tests are not met, then the person may be an independent contractor.

FORM OF THE CONTRACT

A contract of employment need not be in writing to be valid and binding on the parties, but a contract that is to run for a fixed term of more than one year in those provinces with *Statute of Frauds* legislation must be in writing to be enforceable. If the contract may be terminated on proper notice in less than one year, or if the agreement has no fixed term of duration, then it has generally been held to be a contract of indefinite hiring and not subject to the statute.[10]

The impact of the *Statute of Frauds* on employment contracts was examined in the case of *Campbell v. Business Fleets Ltd.*,[11] where the court considered the relationship between an employer and an employee where no written agreement existed. The judge in that case observed:

> In the case at bar, in accordance with the findings of the learned trial judge, the agreement was to continue in force as long as the plaintiff was satisfied with the salary and bonuses, but if not so satisfied the plaintiff could terminate the employment at any time. Therefore, the plaintiff could terminate his employment within the space of one year. Moreover, the contract (if there was no wrongdoing on the part of the plaintiff, and such wrongdoing was not suggested) was to continue for his (the plaintiff's) life, which might or might not be for a period of one day, one month, one year or twenty years. It is to me manifest that this contract must have come to an end at any time the plaintiff was not satisfied with his salary and bonus or on the death of the plaintiff, which might or might not be within the year: *Glenn v. Rudd* (1902), 3 O.L.R. 422.
>
> We are of opinion that on the above authorities the law is that the statute has no reference to cases in which the whole contract may be performed within one year,

8. *Mulholland et al. v. The King*, [1952] Ex. C.R. 233.
9. *Co-operators Ins. Ass'n v. Kearney*, supra.
10. *Campbell v. Business Fleets Ltd.*, [1954] O.R. 87.
11. ibid.

> but there is no definite provision as to its duration, even although it may appear as a fact that the performance has extended beyond that time; that where the contract is such that the whole may possibly be performed within a year and there is no express stipulation to the contrary, the statute does not apply; and that the same principle has been applied to promises in terms of unlimited duration made by or to a corporation when performance of the promise is by the nature thereof limited to the life of the corporation or to the life of the individual ... The Court is of the opinion, therefore, that in the case at bar this verbal contract is without the provisions of s. 4 of The Statute of Frauds ...

When evidence in writing is required to render the contract enforceable, the courts are prepared to accept informal, rather than formal, written evidence. For example, in one case the writing requirement was satisfied by an exchange of letters between the employer and the employee, offering and accepting the employment, but without mention of wages.[12] In another case, the entry of details of the hiring in the corporation minute-book, which had been signed by the corporation officers, was held as sufficient evidence to satisfy the statute.[13]

Employment contracts are often verbal agreements of indefinite hiring, although most employment relationships in which the employee is likely to have access to secret processes or confidential information of the employer are reduced to writing and subject to a restrictive covenant. An employer may insert a restrictive covenant in an employment contract in an effort to protect his or her business secrets, but the ability to enforce such a restriction on the employee is limited to those situations in which the restriction is reasonable and necessary to protect the employer from serious loss. Restrictive covenants usually may not restrict an employee from exercising skills learned on the job, but may limit the employee's use of secret or confidential information if the employee should leave his or her employment.[14] Apart from these few provisions, most employment contracts tend to be informal documents.

Duties of the Employer

The duties of the employer have been the subject of much of the labour legislation since the beginning of the nineteenth century. Laws relating to minimum wages, hours of work, and working conditions have been largely directed against employer abuses in the employment relationship, and override contract terms made contrary to them. For example, each province has passed laws frequently referred to as employment standards or industrial standards, which regulate the terms and conditions of employment and the conditions under which work may be performed. These laws may be divided into two separate classes: those that deal with employee safety and working conditions, and those that deal with the terms of the employment contract.

The former class of laws usually deals with the physical aspect of employment, such as sanitary facilities, and control of dust, fumes, and equipment that might affect employee health and safety in a plant or building. Government inspectors enforce these laws and visit an employer's premises from time to time to make certain that these work hazards are minimized.

Occupational Health and Safety Legislation

Health and safety legislation frequently dictates that the employer must provide employees with safety equipment when hazards are associated with a particular job. The failure on the part of the employer to provide safety equipment normally entitles the employee to refuse to do the work until the equipment is made available. Employers are also obliged to train employees in the safe handling of equipment and substances that pose a safety or health hazard. In addition, the legislation usually imposes stiff penalties on the employer

12. *Goldie v. Cross Fertilizer Co.* (1916), 37 D.L.R. 16.
13. *Connell v. Bay of Quinte Country Club* (1923), 24 O.W.N. 264.
14. *Management Recruiters of Toronto Ltd. v. Bagg*, [1971] 1 O.R. 502.

if he or she should violate the safety requirements. In many provinces, occupational health and safety legislation also imposes fines or penalties on supervisory staff personally if they allow breaches of the legislation to occur.

In most provinces, occupational health and safety legislation requires the employer to ensure that employees are fully protected from occupational and safety hazards on the job, and that all employees are fully trained to safely perform their duties where hazards are associated with the work. In some cases, criminal charges may also result from breaches of the employer's duty.

A sub-contractor was excavating a construction site, and was advised by the general contractor that no services were buried in the area. The sub-contractor's back-hoe struck concrete that encased a hydro duct, and an explosion resulted when an attempt was made to remove the concrete. The sub-contractor was charged and convicted under provincial occupational health and safety legislation for failing to ensure that no services were buried on the site. Reliance on the advice of the general contractor was insufficient to establish a due diligence defence on the part of the sub-contractor.[15]

Employment-Standards Legislation

Employment-standards laws that deal with the employment contract generally impose minimum terms of employment on the parties and allow them to negotiate more favourable terms (from the employee's point of view) if they wish to do so. Most of these provincial statutes establish minimum wage rates, fix maximum hours of work, set conditions under which holiday and vacation pay must be given, and impose minimum conditions for termination of the contract by the parties. While some similarity exists in the legislation, the provinces have generally written their laws to meet their own particular employment needs. Consequently, the laws relating to working conditions, wage rates, and other aspects of employment vary somewhat from province to province.

Human Rights Legislation

Human Rights legislation in most provinces also dictates that employers must not discriminate in their hiring practices on the basis of a person's race, creed, colour, place of origin, nationality, sex, age, or, in some cases, on the basis of physical disabilities or past criminal record. What this usually means to employers is that the selection process they follow must consider factors other than those mentioned in the statute. Where physically handicapped persons are covered by the Human Rights legislation, the handicap may only be considered as a factor if it would affect performance of the job. For example, an employer may reject an applicant for a position that requires a great deal of climbing over and around operating equipment, if the person's handicap would prevent him or her from performing the work in an efficient manner. Apart from these limitations, at Common Law, an employer is normally free to select the person whom the employer believes to be best suited for the position.

Duty to Accommodate

Duty to accommodate

The obligation on an employer to adjust work for an employee with a recognized disability.

Human Rights legislation may also affect the right of employers to terminate employees who develop a disability during employment. In these cases, the employer must make an effort to accommodate the employee's disability in their work assignment. If this can only be done by imposing an undue hardship on the employer, then the right to terminate may be permitted.

15. *The Queen in Right of Ontario v. London Excavators & Trucking Limited* (1998), 40 O.R. (3d) 32.

Under Human Rights legislation, employers may discriminate in their hiring where they can establish *bona fide* occupational requirements for a position that must be met by a job applicant. For *bona fide* occupational requirements to be permitted, the employer must essentially establish that the requirement is reasonable and necessary in order to perform the work, and that the requirement was adopted in good faith by the employer. The employer must also establish that it would not be possible to accommodate employees who lack of the requirement without undue hardship on the employer.

A municipal fire department established as a part of its job requirement that an applicant must be able to climb to the top of a 5 metre ladder without difficulty while carrying a standard fire hose and nozzle, and while wearing prescribed safety equipment and protective clothing. This may well be a *bona fide* occupational requirement for a fire fighter, and if an applicant lacked the ability to successfully perform the test, the applicant may legitimately be rejected.

CLIENTS, SUPPLIERS *or* OPERATIONS *in* QUEBEC

The *Quebec Charter of Human Rights and Freedoms* (R.S.Q., c. C-12) prohibits an employer from discriminating against an employee on the basis of a criminal record. An employer terminated an employee who was unable to report for work because he was incarcerated as a result of a criminal conviction. The employee filed a complaint alleging that he had been terminated on the basis of his criminal conviction contrary to the provisions of the Quebec Human Rights Charter.

When the case reached the Supreme Court of Canada, the court unanimously ruled that the dismissal of the employee for failure to report for work was not discrimination on the basis of a criminal record, as the true reason for termination was his absence from work.[16]

Age Discrimination and Mandatory Retirement

A number of provinces have now amended their human rights code or employment legislation to eliminate the right of an employer to retire employees when they reach age 65. The effect of this type of legislative change permits employees to continue their employment beyond age 65, unless the employer should become entitled to terminate the employee's services for some other cause, or for a *bona fide* occupational requirement. Termination only on the basis of age would constitute discrimination under Human Rights legislation. In some cases, the courts have held that an agreed term in the employment contract that specifies that the contract will terminate when the employee reaches the age of 65 years is enforceable and the employment relationship will terminate at that point in time.[17]

Workplace Discrimination

Human rights considerations also apply to the workplace and the employer's obligation to maintain a discrimination-free work environment. In particular, employers are obliged to control discrimination by employees against other employees in the firm. In one case, an employee was the subject of racial harassment by other employees and complained to the employer. The employer did not investigate the matter, and the harassment continued until the employee lodged a formal complaint under the applicable Human Rights legislation. The human rights tribunal held the employer vicariously liable for the employees' racially discriminating remarks because the employer had failed to investigate the complaint of discrimination.

16. See *Roy v. Maksteel Inc.*, 2003 S.C.C. 68.

17. See, for example, *Johnson v. Global Television Network Inc.*, 2008 BCCA 33 (CanLII).

Employers must also avoid work practices that would constitute discrimination under the Act or code. While employers are not obliged to eliminate *bona fide* or legitimate work requirements of a job, or incur significant costs to satisfy a particular employee, the employer must demonstrate that the job requirement or action is not intentional discrimination. For example, employers are expected to make reasonable efforts to accommodate the religious and other creed-related activities of employees in the scheduling of work, holidays, and vacation time. However, they would not be obliged to close down operations on a normal business day simply because it represented a religious day for certain employees.

An employer had a religious accommodation policy that permitted absence from work for two paid days, and a rescheduling of work to permit one additional day in a three week work cycle period. An employee whose religion required him to observe eleven holy days, requested permission to take the full eleven days without pay. The employer offered several additional work options in an effort to accommodate the employee's request, but no schedule would fit. The employee filed a grievance alleging discrimination on the basis of creed. The case was eventually appealed to the Court of Appeal, where the court concluded that the employer had satisfied its duty to accommodate by its proposals to change its work schedule for the employee, and dismissed the employee's complaint.[18]

Pay Equity Legislation

In all Common Law provinces, private sector employers are not permitted to pay female employees a lower wage than male employees, if both employees are performing the same job. The province of Quebec and the federal government have carried this equal-pay requirement a step further, however, and introduced legislation to provide for equal pay for work of equal value. Some provinces also require all public-sector employers to ensure that certain types of positions that have been traditionally filled by women would be assessed vis-à-vis positions traditionally filled by men in terms of their value, using job-related criteria such as skill, ability, education, working conditions, and effort. Salaries and wages would then be determined on the basis of values assigned to these factors, rather than allowing market forces to determine the wage rates for the respective positions.

Compensation for Work

In addition to the terms of employment imposed by statute, many duties of the employer are implied by Common Law. The most important of these relate to compensation. The employer must pay wages or other remuneration to the employee in return for the employee's services, and generally must indemnify the employee for any expenditures or losses that the employee might incur in the normal course of his or her employment, if made at the employer's direction.[19] For example, if the employer requires the employee to travel to a neighbouring community to carry out some duty on behalf of the employer, the employer would be expected to reimburse the employee for the employee's travel and other expenses associated with the assignment unless customs of the trade or the terms of employment provided otherwise.

Two further duties of the employer are implied in the employment relationship: to provide the employee with sufficient tools to do the work where it is not the custom of the trade for the employee to provide his or her own, and to provide the employee with sufficient information to allow him or her to calculate the remuneration due if the employee is paid through some system other than a salary or hourly rate. For example, the employer who operates an iron mine would be obliged to provide a group of employees with sufficient information to calculate the bonuses due to them for mining over and above a stipulated minimum amount of ore, if the bonus is based upon the tonnage mined.

18. *Ontario (Ministry of Community and Social Services) v. O.P.S.E.U.* (2000), 50 O.R. (3d) 560.

19. *Dugdale et al. v. Lovering* (1875), L.R. 10 C.P. 196.

DUTIES OF THE EMPLOYEE

Apart from specific duties that may be set out in a contract of employment, an employee is subject to a number of implied duties that arise out of the employment relationship. As a general rule, the employee has a basic duty to obey all reasonable orders of the employer that fall within the scope of the employment.[20] In addition, the employee has an obligation to use the property or information of the employer in a careful and reasonable manner.[21] Any confidential information that the employee obtains from the employer must be kept confidential during the course of employment,[22] and afterwards.[23] The employee is also under an obligation to devote the agreed hours of employment to the employer's business, and the employer is entitled to the profits earned by the employee during those intervals of time.[24] The employee's spare time, however, is the employee's own.[25]

If the employee should inform the employer that he or she has a special skill or professional qualification, then it is an implied term of the employment contract that the employee will perform the work in accordance with the standard required of the skill or profession.[26] An employee who professes to be skilled and is negligent in the performance of a skilled task may be liable for damages suffered by the employer as a result of the employee's negligence, provided that there are no intervening factors or special controls exercised over the employee by the employer.[27]

The courts have recently expressed the opinion that senior employees, the executives of a corporation, have a higher duty to their employer than do ordinary employees. According to this opinion, senior executives are, in a sense, in a fiduciary position vis-à-vis their employer. Therefore they owe a clear duty to their employer to devote all of their energy, initiative, and talents in the best interests of the corporation. If they should fail to do so, the employer may treat this failure as grounds for termination. For example, in a case where a senior executive was hired to help improve the profitability of a group of companies, he did so by way of an invention that both improved material handling practices and reduced accounting costs. He then engaged in a protracted disagreement with the employer over the ownership rights to the invention and the right to produce and market it for his own benefit. The employer dismissed the employee. In the legal action that followed, the Court of Appeal, in upholding the employer's right to dismiss the executive without notice, stated that senior employees have an added obligation to make the corporation more profitable. This responsibility consequently requires the employee to place the interests of the corporation before his or her own.[28]

A senior employee and a number of other employees purchased a competitor of their employer. The employees resigned, but proceeded to solicit suppliers and customers of their former employer. When their former employer discovered their activities, it brought an action for damages against the employees for breach of their fiduciary duty to the employer.

The Court of Appeal concluded that the senior employee was in breach of his fiduciary duty by virtue of his senior position in the employer's firm. The court found that the remaining former employees were not in breach of a fiduciary duty, but were jointly and severally liable for the damage caused to the employer because they had participated in the scheme that injured the business of their former employer.[29]

20. *Smith v. General Motor Cab Co. Ltd.*, [1911] A.C. 188.
21. *Lord Ashburton v. Pape*, [1913] 2 Ch. 469.
22. *Bents Brewery Co. Ltd. et al. v. Luke Hogan*, [1945] 2 All E.R. 570.
23. *Robb v. Green*, [1895] 2 Q.B. 315.
24. *William R. Barnes Co. Ltd. v. MacKenzie* (1974), 2 O.R. (2d) 659; *Bennett-Pacaud Co. Ltd. v. Dunlop*, [1933] 2 D.L.R. 237.
25. *Sheppard Publishing Co. v. Harkins* (1905), 9 O.L.R. 504.
26. *Lister v. Romford Ice & Cold Storage Co. Ltd.*, [1957] A.C. 555.
27. *Harvey v. R.G. O'Dell Ltd. et al.*, [1958] 1 All E.R. 657.
28. See *Helbig v. Oxford Warehousing Ltd. et al.* (1985), 51 O.R. (2d) 421.
29. See *Canadian Industrial Distributors Inc. v. Dargue et al.* (1994), 20 O.R. (3d) 574.

Employees have a duty at law to act in the best interests of the employer in the performance of their duties. If they use their position to earn secret profits for themselves, the court may require them to turn over their profits to the employer. The dishonesty associated with such an act would also entitle the employer to dismiss the employee without notice. In the case of *William R. Barnes Co. Ltd. v. MacKenzie*[30] the court described this obligation in the following manner:

> The principle that a dishonest agent is not entitled to a commission from his principal is well recognized as is the right of a principal to any secret profit earned by his dishonest agent. An agent stands in a fiduciary relationship with his principal with his remuneration usually attributable to separate transactions. If he is dishonest in one transaction he forfeits his commission thereon but not on other transactions faithfully performed. In the instant case the relationship is basically that of master and servant rather than principal and agent and the remedy of a master against his defaulting servant is restricted to a right of instant dismissal and to damages which flow from the default. I do not consider wages paid to be such an item of damages and disagree with that part of the judgment in *Protective Plastics Ltd. v. Hawkins* which appears to hold otherwise.
>
> I adopt the view of the trial Judge that wages cannot be recovered if one allows to the plaintiff all secret profits made by the delinquent employee and also holds the employee liable for any loss sustained by the employer as a result of the employee's breach of his employment contract. The argument is that the damages awarded the employer place him in the same position as if the delinquent employee had in fact been performing his duties as a faithful employee and that since all benefits ultimately accrue to the employer, the employee should be compensated for his time and labour in producing such benefits. The employer has already received the fruit of the employee's efforts, honest or otherwise, and cannot repudiate his obligation to pay. I recognize that in an agency situation the principal may take the benefit and refuse to pay the commission, but I am not aware of any binding authority which requires me to extend that principle to a master and servant situation.

Termination of the Contract of Employment

The notice required to terminate a contract of employment has been the subject of legislation in most provinces. Many of these statutes provide a minimum period of notice that varies depending upon the length of service of the employee. This period of notice is generally the minimum requirement, but, in some cases, it may replace the Common Law rule that reasonable notice of termination is required to terminate the contract.

At Common Law, unless the contract stipulates a specific termination date, or a period of notice for the termination of the agreement, both parties are obliged to provide reasonable notice of termination.[31] The adequacy or "reasonableness" of the notice is a matter of fact to be determined from a number of factors, including the nature of the contract, the method of payment, the type of position held by the employee, the length of service, the customs of the business, and even the age of the employee. All of these factors would be considered in the determination of what would constitute a reasonable time period. In some of the older cases, where the employee was unskilled and employed for only a short period of time at an hourly rate, the length of notice was often very short. The trend, however, has been away from short notice since the middle of the twentieth century. A one-week notice period is commonly determined as the minimum for a short service employee, and as much as several years' notice for a long-serving employee or an employee engaged in a senior position in the firm.[32]

30. (1974), 2 O.R. (2d) 659.

31. *Harvard v. Freeholders Oil Co.* (1952), 6 W.W.R. (N.S.) 413.

32. *Campbell v. Business Fleets Ltd.*, [1954] O.R. 87; *Bardal v. The Globe & Mail Ltd.* (1960), 24 D.L.R. (2d) 140.

COURT DECISION: Practical Application of the Law

Wrongful Dismissal—Reasonable Notice—Considerations—Duty to Mitigate Loss

Bardal v. The Globe & Mail Ltd. **(1960), 24 D.L.R. (2d) 140.**

The plaintiff was employed by the defendant newspaper on a contract of indefinite hiring. He was first hired as advertising manager. Sixteen years later, by which time he had become director of advertising, he was terminated without notice. The plaintiff immediately sought other employment, and, some months later, found a new position that paid substantially less in salary and benefits. An action for wrongful dismissal was instituted against his previous employer.

THE COURT: In every case of wrongful dismissal the measure of damages must be considered in the light of the terms of employment and the character of the services to be rendered. In this case there was no stipulated term during which the employment was to last. Both parties undoubtedly considered that the employment was to be of a permanent character. All the evidence goes to show that the office of advertising manager is one of the most important offices in the service of the defendant. In fact, it is by means of the revenue derived under the supervision of the advertising manager that the publication of a newspaper becomes a profitable enterprise. The fact that the plaintiff was appointed to the Board of Directors of the defendant goes to demonstrate the permanent character of his employment and the importance of the office. It is not argued that there was a definite agreement that the plaintiff was employed for life but the case is put on the basis of an indefinite hiring of a permanent character which could be terminated by reasonable notice.

There can be no catalogue laid down as to what is reasonable notice in particular classes to cases. The reasonableness of the notice must be decided with reference to each particular case, having regard to the character of the employment, the length of service of the servant, the age of the servant and the availability of similar employment, having regard to the experience, training and qualifications of the servant. Applying this principle to this case, we have a servant who, through a lifetime of training, was qualified to manage the advertising department of a large metropolitan newspaper. With the exception of a short period of employment as manager of a street car advertising agency, his whole training has been in the advertising department of two large daily newspapers. There are few comparable offices available in Canada and the plaintiff has in mitigation of his damages taken employment with an advertising agency, in which employment he will no doubt find useful his advertising experience, but the employment must necessarily be of a different character.

I have come to the conclusion, as the jury did in the *Sun Printing & Publishing Ass'n* case and as the Court of Appeal agreed, that 1 year's notice would have been reasonable, having regard to all the circumstances of this case. That being true, the next question to decide is what damages have flowed from the failure of the defendant to give a year's notice and how far have those damages been mitigated by the receipt by the plaintiff of a salary from another employer. [The judge then examined the plaintiff's compensation package, and determined the judgment amount.]

Despite the passage of more than 50 years, this case remains good law, and still represents the factors considered by the court in wrongful dismissal cases. Are there other factors you think the court should take into consideration today?

The nature of the work to some extent has become a less important factor in considering the length of notice required for a long service employee, and the courts for the most part have rejected the assumption that more senior employees require a longer notice period to find new employment.[33]

It is also important to note that the character of the industry sometimes establishes the typical notice period, or in some cases, may dictate no notice at all. For example, in the residential construction industry, the completion of the work of a particular craft (such as carpentry or plumbing) may signal the end of the job and the termination of the employee. In the case of *Scapillati v. Rotvin Construction Limited*,[34] the Court of Appeal for Ontario concluded that the seasonal and irregular nature of the on-site residential construction industry often abruptly terminates work, and the custom of the trade has been to terminate without notice. The court noted that it may accept or reject customs of the trade as it sees fit in the determination of reasonable notice, and customs of the trade simply represent a factor to be considered.

33. *Bramble v. Medis Health and Pharmaceutical Services Inc.* (1999), 175 D.L.R. (4th) 385.

34. (1999), 44 O.R. (3d) 737.

Dismissal and Wrongful Dismissal

In the absence of an agreement to the contrary, an employer has the right to dismiss an employee without notice if the employee is incompetent or grossly negligent in the performance of his or her duties. The employer would also be entitled to do so if the employee concurs in a crime against the employer, or if the employee's actions are such that they would constitute a serious breach of the contract of employment. As noted earlier, the failure of a senior executive to devote his or her energies to the exclusive benefit of the employer may also be considered grounds for dismissal in some cases. In each case, however, the onus would be on the employer to establish that the employee's actions were not condoned by the employer, and that termination of the employment relationship was justified. Otherwise, the employee would be entitled to damages against the employer for **wrongful dismissal**.

Wrongful dismissal

The failure of an employer to give reasonable notice of termination of a contract of employment.

Just cause

The onus on the employer to establish grounds for termination of an employee without notice.

More recently, the courts have considered other grounds to justify the dismissal of employees. The disruption of the corporate culture, which encompasses improper employee behaviour towards other employees and customers, may be **just cause** for the dismissal of an employee. For example, a trust company was held to be entitled to dismiss a branch manager because the manager had treated customers in a rude manner and had berated staff to the extent that the branch had a higher-than-normal employee turnover.

The courts may also consider the activities of employees both on and off the job that have a negative impact on the firm as just cause for dismissal. This basis for termination tends to be character-related, and while the courts do not establish a moral standard for the behaviour of employees, if an employee exhibits a serious lack of judgment or integrity in activities off the job that affect the ability of the employee to carry out his or her duties or the employer's dealings with customers or clients of the firm, the activities may be just cause for dismissal.

Some provinces have attempted to clarify the matter of dismissal and termination without notice by setting out in their legislation conditions that permit an employer to terminate an employee without notice. Ontario, for example, provides in its legislation that the notice provisions do not apply to "an employee who has been guilty of willful misconduct or disobedience or willful neglect of duty that has not been condoned by the employer."[35]

This Ontario statutory requirement, however, appears to go well beyond the Common Law in the sense that these obligations must be met in order for the employer to dismiss an employee without notice under the Act. For example, in a recent Ontario case,[36] the court agreed that the employer was justified in terminating a long service employee who was carelessly producing parts in spite of repeated warnings and discipline, but found that the employer had failed to prove that the employee's behaviour was "willful misconduct, disobedience, or willful neglect of duty." The court awarded the employee the statutory notice and severance payments specified in the Act.

Employees may also terminate their employment without notice under certain circumstances. The grounds, however, are limited to, for the most part, those situations where the work has an element of danger attached to it that has caused the employee to believe that it poses a threat to his or her health or life, where the employer has seriously mistreated the employee, or where the employer has failed to perform the employer's part of the employment contract.

If an employee believes that he or she has been wrongfully dismissed, the employee may bring an action against the employer for the failure to give reasonable notice of termination. If the employee should decide to pursue this course, it is important that he or she do everything that a reasonable person might be expected to do to minimize his or her loss. The employee would be expected to seek other employment immediately, and

35. The *Employment Standards Act*, 2000, S.O. 2000, c. 41, s. 55 and Regulation 288/01. It should be noted, however, that this section of the Act does not apply to employees in certain trades, businesses, and professions: see 0. Reg. 288/01.

36. *Oosterbosch v. FAG Aerospace Inc.* 2011 ONSC 1538 (CanLII).

take whatever other steps might be necessary to mitigate financial loss. The employee's actual loss would be the loss that the employee incurred between the time he or she was terminated and the end of a reasonable notice period. For example, if Able was employed by Baker in a responsible position where reasonable notice might be determined as six months, and Baker should wrongfully dismiss Able, Able would be obliged to seek new employment immediately. If Able could not find suitable employment within a six-month period, then his damages would be the lost wages and benefits that he would ordinarily have received from the employer during that period. Had he found employment during the six-month interval, his actual loss would be reduced by the income he received during the period, and that amount would be deducted from the damages to which he would be entitled as a result of the wrongful dismissal.

An employer dismissed an employee, and three months later, the employee found new employment at a higher salary. The employee nevertheless sued the employer for wrongful dismissal. In the wrongful dismissal action, the trial judge fixed reasonable notice and damages at 6 months salary. On appeal, the Court of Appeal held that the employer was only obliged to pay the employee full salary for the three month period before the employee found new employment.[37]

It is important to note that the failure of an employee to immediately seek other employment to mitigate his or her loss can have a serious effect on a damage award for wrongful dismissal. In a 2008 case, the Supreme Court of Canada decided[38] against a long service employee wrongfully dismissed by his employer. The employee was found to have failed to mitigate his loss when he refused the employer's offer of re-employment for a period of time equivalent to reasonable notice. As a result, the employee was held not to be entitled to damages for his wrongful dismissal.

The purpose of damages for wrongful dismissal is to place the injured employee in relatively the same position that the employee would have been in had he or she been given proper notice of termination of the contract. The courts will normally not award punitive damages, nor will they compensate the employee for any adverse effects that the wrongful dismissal might have had on the employee's reputation or stature in the business community.[39] Judges have recently awarded extra compensation, however, where the actions of the employer were such that they caused the employee undue mental distress as a result of the termination.

CASE IN POINT

Keays, a long service employee, was diagnosed with chronic fatigue syndrome. He ceased work and received disability benefits during a 2-year period until his benefits expired. He returned to work on a disability program that permitted absence on doctor's notes confirming the disability-related absence. The employer became concerned about the notes offered by Keays, and asked to have an independent specialist evaluate his illness. The employer then terminated his employment. Keays sued for wrongful dismissal, aggravated damages and punitive damages. Keays was awarded damages equivalent to 15 months notice, but the Supreme Court of Canada refused to allow aggravated or punitive damages for the wrongful dismissal.

Honda Canada Inc. v. Keays 2008 SCC 39 (CanLII).

37. *Eurodata Support Services Inc. v. LeBlanc* (1998), 204 N.B.R. (2d) 179.

38. *Evans v. Teamsters Local Union No. 31*, 2008 SCC 20 (CanLII).

39. See, for example, *Wardell v. Tower Co. (1961) Ltd.* (1984), 49 O.R. (2d) 655.; *Peso Silver Mines Ltd. v. Cropper*, [1966] S.C.R. 673.

Employers who terminate employees by novel or different ways leave themselves open to the possibility of punitive damage awards if the employee successfully maintains a wrongful dismissal action. To be successful in a claim for punitive damages, a dismissed employee must not only be successful in the claim for wrongful dismissal, but must show that the employer's actions were "harsh, vindictive, reprehensible or malicious."[40] For example, in a 1992 case, a branch manager of a bank discovered that he was terminated when he attempted to use his bank credit card at an automated teller machine. The machine would not allow the manager access to his bank account, and flashed a message to him to contact the bank. When he did so, he was advised that the machine was informing him that he had been discharged. Based upon this evidence, the court found that the employee had been wrongfully dismissed, and awarded him 12 months' salary and $40,000 in punitive damages.[41]

Harsh or callous treatment, coupled with sudden termination, in a situation where an employee was led to believe at the time of hiring that the position would be permanent and secure, may result in the employee making a claim for mental distress. To succeed, however, it would appear that the employee would be obliged to establish that the distress was brought on by the failure of the employer to give reasonable notice of termination.[42]

Larger damage awards may also be made by the courts if the employer dismisses an employee in a particularly insensitive and callous manner, or makes false accusations concerning the employee at the time of termination. In a 1997 decision of the Supreme Court of Canada,[43] the court concluded that 24 months would be an appropriate notice period for an employee wrongfully dismissed because the employer had made a number of groundless allegations of cause for dismissal initially, only to drop the claims just before the trial began, an act that caused the court to conclude that the employer had acted in bad faith in the termination of the employee.

A store manager was summarily dismissed after working for four years for the employer, and denied a letter of reference. The employer based the dismissal on a breach of trust and fraud. The manager had apparently had a part-time clerk babysit his child one day on a volunteer basis without payment and without claiming for a lost work shift. The next day, another employee, who was ill, booked off sick, but she felt well enough to babysit for the manager at his home. She filed for sick pay, as the company policy did not require her to remain at home while sick. No attempt was made to hide these actions from other management employees.

The manager sued the company for wrongful dismissal and punitive damages. While the court found that the manager had been wrongfully dismissed, and that a reasonable notice period would be four months, it dismissed the claim for punitive damages on the basis that the employer's actions while "imprudent, were not malicious."[44]

Constructive dismissal

Employer termination of a contract of employment by a substantial, unilateral change in the terms or conditions of employment.

In some cases, the employer need not discharge or terminate an employee directly in order for it to constitute dismissal. Unilateral change in an employee's contract or employment may be considered **constructive dismissal** if the change radically alters the terms of employment or the conditions under which the work of the employee would be performed. Normally, changes instituted by the employer that represent the employee's promotion to a position of greater responsibility and a higher wage are acceptable to the employee, but changes constituting a demotion to a lower-paying or undesirable position may represent constructive dismissal if the employee is unwilling to accept the change.

40. See *Vorvis v. Insurance Corporation of British Columbia*, [1989] 4 W.W.R. 218.

41. *Francis v. Canadian Imperial Bank of Commerce* (1994), 75 O.A.C. 216.

42. See, for example, *Pilon v. Peugeot Canada Ltd.* (1980), 29 O.R. (2d) 711, where the plaintiff was awarded damages to cover the mental distress associated with termination from what he was led to believe was a life-time position with the company. See also *Brown v. Waterloo Regional Board of Police Commissioners* (1983), 43 O.R. (2d) 113.

43. *Wallace v. United Grain Growers* (1997), 152 D.L.R. (4th) 1.

44. See *Kevin Gillman v. Saan Stores Ltd.* (1992), 6 Alta. L.R. (3d) 72.

This may permit the employee to bring an action for wrongful dismissal. For example, if the employer unilaterally and without good reason moves a senior manager from a position of responsibility to that of an ordinary salesperson, with an accompanying substantial reduction in salary, the employee need not accept the new position, but may treat the change as constructive dismissal.

Constructive dismissal may also occur if the employer changes the employee's work environment or facilities so as to render it impossible for the employee to do his or her job. This might take many forms, but, in one case, where the employer rearranged the office and removed the employee's desk, the court held that the change constituted constructive dismissal, because, by removing the desk, the employer indicated to the employee that his services were no longer required. An adjudicator appointed under the *Canada Labour Code* reached a similar conclusion in another instance, where an employee's desk was removed and replaced with a small table after he had several altercations with his supervisors. Even though the employer had grounds for dismissal, the adjudicator held that the harassment and unusual treatment of the employee by the supervisors constituted wrongful dismissal.

While the normal remedy for wrongful dismissal is money damages, employers subject to the *Canada Labour Code*[45] (communications, radio, TV, banks, interprovincial transportation companies, etc.) may seek reinstatement through the adjudication process under the Code. Employees who have been employed for more than 12 months in a non-union position may bring their complaint to an adjudicator appointed to hear the matter under the Code. If the adjudicator finds that the employee was wrongfully dismissed, the adjudicator as a remedy may order the reinstatement of the employee.

A QUESTION OF ETHICS

"Wrongfully dismissed employees should not be given money damages, other than in cases where there needs to be something punitive. In most cases, like in torts, what they lost should be simply restored. They should get their job back, along with wages for any interim period of unemployment." Do you feel there is an ethical element missing from this analysis?

CYBER-LAW ASPECTS OF EMPLOYEE MANAGEMENT

Managerial issues surrounding electronic business and electronic information exchange are increasingly becoming legal issues inside the company itself. The reality is that employees have engaged in harassment through their company e-mail accounts, have created blogs that defame their employer, hack into unauthorized areas of their employer's IT system, and use their employer's IT system for all manner of illegal (or at least unethical) purposes. The first observation should be to note that an employee does not need to act unlawfully to find themselves called to account for their actions. If an employer has set a series of operations policies for use of its IT system, then those policies must be followed like any other firm policy, or progressive discipline and dismissal will result.

The fact that employee IT misuse and abuse is so evident is due to its novelty and newsworthiness, but more importantly, because many believe the issue of free speech and privacy is involved. While these are undoubtedly related, they do not determine the legal outcome. In short, if a defamatory blog or harassing e-mail were read out loud at a street corner, they would result in lawsuits and damages. The fact that the electronic broadcast allows the damage to run further only serves to increase the amount of a damage award. The right of free speech has always been limited to speaking the truth if the words are harmful to the interests of another person.

The incidents where employees have been fired for defaming or otherwise harming the interests of their employer through blogs has yet to reach meaningful levels of litigation in court. When it does, watch for recognition of a right to fair and truthful comment

45. R.S.C. 1985, c. L-2 (Part III).

regarding blogs, but little more. Beyond blogs, where employee misuse of IT systems takes place, watch for recognition that the employer is the owner of the IT system, with the right to exercise control over how it is used by employees.

To that end, an employer should ensure that it can exercise control by having clearly expressed operation policies in place for IT use.

CHECKLIST FOR OPERATIONAL IT POLICIES

1. Authorized use and employer's right to inspect files.
2. Account for exclusive individual non-transferable use.
3. Forbid use of terminal or account of another person.
4. No personal commercial element or application.
5. No attempt to hack privileges beyond those assigned to account.
6. No use of account in illegal activity (e.g., downloading, copyright issues).
7. No publishing or e-mail of obscene, threatening, or harassing material.
8. No interference with other account holders or system operation itself.

EMPLOYER MISREPRESENTATION (WRONGFUL HIRING)

In recent years, a number of employees have taken legal action against their employers because the employer has induced them to join the firm through misrepresentation of the position or job duties. The cause of action in these cases is usually based upon the tort of negligent misrepresentation, and in most instances it arises out of an overstatement of the importance or status of the particular position or job by the employer. The employee accepts the position based upon the description of the position offered by the employer, only to discover on arrival at the firm that the position is very different and often less desirable or with less responsibility and authority than the description stated.

As with negligent misrepresentation generally, the aggrieved employee must establish that the employer clearly misrepresented the position, and that the statements made or the description of the work was fundamental to the employee's decision to accept the new position. The employee must also establish that he or she suffered some loss as a result of the misrepresentation. This usually takes the form of relocation expenses and the loss flowing from the employee's resignation at his or her previous employer's firm.

Promises made to an employee at the time of hiring or subsequent to that date may also affect the period of notice that a court would consider as reasonable. In a recent Ontario case,[46] the Court of Appeal addressed this particular issue in its determination of what it considered appropriate damages for termination. According to the facts of the case, two long-serving employees accepted supervisory positions on the basis that they could return to their former positions on the shop floor "if their new position did not work out." Some ten years later, the employer was obliged to "downsize" his operations, and terminated the two supervisors with severance payments of about nine-months' salary. The employees brought an action for wrongful dismissal on the basis that they were promised a return to their old positions if their new jobs did not work out. The Court of Appeal agreed. The Court awarded the employees damages calculated on the basis of what they would have received had they worked in the shop until their retirement age, discounted for such events as illness or plant closing, and new employment some months after termination. The Court factored these amounts into their decision, and awarded damages in excess of $107,000 to one employee, and $23,000 to the other.

As a consequence of a number of successful actions against employers for wrongful hiring, many employers now take steps to ensure that human-resources managers, executive-recruitment firms, and employment agencies do not exaggerate the job or position descriptions made at the time of hiring.

46. *Leonetti v. Hussmann Canada Inc.* (1998), 39 O.R. (3d) 417.

CASE IN POINT

Cognos, a software company, interviewed an accountant, Queen, telling him that a job opening involved a two year project, although never told him the funding for the project was subject to budgetary approval. The eventual written employment contract included a one-month notice provision for dismissal. After the first year, Queen was dismissed when funding for the project was reduced. Queen claimed negligent misrepresentation in hiring. The Supreme Court of Canada ruled in favour of Queen, that a "special relationship" existed between the parties, and Cognos owed a duty of care to exercise reasonable care and diligence in making representations as to the employment opportunity offered. The misrepresentations made during the interview were negligent, and therefore the duty of care was breached, quite independent of the terms of the employment contract.

See: *Queen v. Cognos Inc.*, [1993] 1 S.C.R. 87, 1993 CanLII 146 (S.C.C.).

FRONT-PAGE LAW

Creative director launches $3M wrongful dismissal suit — sues Gee Jeffery agency

Brett Channer, the former creative director with Gee Jeffery & Partners Advertising Inc., has launched a $3.2-million wrongful dismissal lawsuit against the Toronto ad agency.

In a statement of claim filed June 21 in the Ontario Superior Court of Justice, Mr. Channer alleges he was convinced to turn down an offer from rival agency Wolf Advertising Ltd. last October in return for an increased salary and a new salary for his wife from Gee Jeffery.

The lawsuit, which names the agency and its president, Peter Jeffery, claims $1-million in lost salary, bonus and benefits, $2-million for defamation, $100,000 for damages and $100,000 for emotional distress.

The claim states that, to entice Mr. Channer to stay with the agency, his salary was increased to $185,000 from an undisclosed prior amount and a new salary of $40,000 annually was paid to his wife Becky Channer, who would not be required to work at the agency's offices. According to the claim Mr. Channer, 39, was assured his job was secure despite the impending loss of two major clients at the agency. Three months later, his role was diminished, the suit claims, and on April 17 his employment was terminated.

The filing claims Mr. Channer was denied the chance for a position at Wolf, which would have paid $310,000 annually in addition to company shares estimated to be worth $1,000,000 after three years.

Mr. Jeffery, contacted about the filing, refused comment.

Mr. Channer's claim of defamation stems from a May 1 article in the *National Post* dealing with his termination. The lawsuit alleges a main point of the article, that Mr. Channer intended to sue his former employer, came from Gee Jeffery. The lawsuit also claims Mr. Channer's reputation was further harmed by a company-wide e-mail sent by Mr. Jeffery following the newspaper article.

Does this case — retaining an existing employee — bear similarities to the situation of wrongful hiring? Do you feel it is appropriate to say Mr Channer was "denied" a chance at the Wolf Advertising Ltd. job? What is the effect of his role being "diminished" prior to his April 17th termination?

Source: Excerpted from Paul Brent, *National Post*, "Creative director launches $3m wrongful dismissal suit," July 5, 2000, p. C6.

EMPLOYER LIABILITY TO THIRD PARTIES

A general rule with respect to third-party liability is that an employer may be held liable for any loss or damage suffered by a third party as a result of an employee's failure to perform a contract in accordance with its terms, or for any negligence on the part of the employee acting within the scope of his or her employment that causes injury or loss to the third party. This rule imposes vicarious liability on the employer for the acts of the employee that occur within the scope of the employee's employment.[47] For example, if Arthurs takes her automobile to Gordon's Garage for repairs, and an employee, Smith,

47. See: *McKee et al. v. Dumas et al.; Eddy Forest Products Ltd. et al.* (1975), 8 O.R. (2d) 229, for the difficulty in determining which employer may be vicariously liable when an employee is also under the direction of a temporary employer.

negligently performs the repairs, Arthurs would be entitled to recover from Gordon's Garage for the breach of contract. Similarly, if an employee is sent to a customer's home to repair a defective boiler, and the employee negligently damages the equipment, causing it to explode, the customer would be entitled to look to the employer for the loss on the basis of the employer's vicarious liability for the acts of the employee.

The reason for the imposition of liability on the employer for the acts of the employee has a historical perspective and justification. In the past, employees seldom possessed the financial resources to compensate third parties for any loss suffered as a result of their negligence, and the third party would be unlikely to obtain compensation, even if a judgment was obtained against the employee. However, most employers did have the financial resources to cover a loss that a third party might suffer. In view of the fact that the employee was under the control of the employer, the courts simply carried the liability through the employee to the party primarily responsible for the employee's actions.

Employer liability is limited to those acts of the employee that fall within the ordinary scope of the employee's duties, but does not include acts of negligence that take place outside the employer's normal duties. For example, an employer sends an employee to another city to perform certain services on his behalf for a customer. After the work has been completed, the employee decides to spend the evening in the city, and rents a hotel room for the night. The employee's careless smoking sets fire to the carpet in the room, and results in a loss to the hotel. If the work had been completed in time for the employee to return, and if the employee had been instructed to do so but failed to heed his employer's instructions, the employee would be personally liable for the loss. However, if the employer had required the employee to use the room to display goods to prospective customers and to remain overnight in the room, the employer might be held liable for the loss suffered by the hotel, if it could be established that the employee's occupancy of the room was at the employer's direction and in the course of employment.

Employer Liability for an Employee's Injuries

At Common Law, an employee injured while working for an employer was generally faced with a dilemma. If the injury occurred as a result of the negligence of another employee, it would be necessary to bring an action against the employer for that employee's negligence. The recovery of damages from the employer under such circumstances would be unlikely to enhance the employee's advancement or career with the particular employer. Nor was the prospect of bringing an action against the employer any more promising if the employee was injured by equipment or machinery that the employee was using. If the employee was successful, the employer would be obliged to pay for the injury, but the legal action would likely result in termination of the employee by reasonable notice. On the other hand, if the employer could prove negligence on the part of the employee, or if the employer could establish that the employee had voluntarily assumed the risks that resulted in the injury, the employee would not be successful in the action for damages.

Workers' Compensation Legislation

In spite of a number of attempts by the courts to accommodate the injured employee, the law relating to employee on-the-job injury remained unsatisfactory. It was not until the close of the nineteenth century that legislation in England remedied the situation. In 1897, the *Workmen's Compensation Act* was passed to provide compensation to workmen injured in the course of their employment. In essence, the Act was an insurance scheme similar in concept to ordinary accident insurance. All employees covered by the Act were entitled to compensation without the need to take legal action to prove fault if they were injured in the course of their employment. All employers subject to the legislation were required to contribute to a fund from which the compensation was paid, and the employee was not entitled to take action against the employer if the employee received compensation from the fund. Similar legislation was eventually passed in all provinces and territories in Canada, and in many states of the United States. The legislation has also been further refined since its introduction into England at the end of the nineteenth century. It has virtually eliminated actions against employers for injuries suffered by employees.

MANAGEMENT ALERT: BEST PRACTICE

Too many employers are too casual in documenting the employment relationship. Vague job descriptions, poorly written or non-existent contracts, weak structures for performance review, and a lack of written records of disciplinary action lead to empty file folders that should contain evidence when the relationship finds itself in court. This lack of evidence seriously undermines the employer's position and exposes the business to needless liability. An employer, which is most often the defendant, should expect that it has to refute the claims of the former employee, and must place itself in a position, on an on-going basis, to do so.

LABOUR RELATIONS

Transition to a unionized work environment represents a major alteration in any business model. For many, such an event was never considered in original business planning. Even if so, familiar expectations must be changed. Owners and managers now encounter negotiation in terms of employment, restricted authority, and binding principles of dispute settlement, discipline and seniority. This transition and continuing in the future can easily be mismanaged, with not just direct financial consequences, but extraordinary losses in production due to strikes or lockout.

DEVELOPMENT OF LABOUR LEGISLATION IN CANADA

Little more than a century ago, collective bargaining and labour unions were still looked upon with disfavour, and their activities treated as restraint of trade and contrary to public policy. Since that time, statute law has legitimized many of their actions, and permitted collective bargaining to take place by way of statutory authority and regulation. The process, however, was a slow and, for the most part, evolutionary one, rather than a sudden reversal of public policy. During this period of time, most of the legislation was introduced as a response to the growth and activities of the labour movement, rather than as an attempt to lead labour in a new direction. The laws have simply attempted to deal with the growth of the labour movement and the impact of its activities, as it gradually acquired a prominent position in the conduct of economic activity in Canada.

COLLECTIVE-BARGAINING LEGISLATION

The general approach taken to labour-relations law federally and in each province has been to remove the collective-bargaining relationship from the Common Law and the courts (insofar as possible) and deal with it administratively. The laws normally place employee selection of a union as a bargaining agent, the negotiation of the collective agreement, and the resolution of disputes relating to negotiations (unfair practices and bargaining in bad faith) under the jurisdiction of an administrative tribunal. The rights and duties of the employer, the union, and the employees are also set out in the legislation.

Each of the provinces and the federal government have enacted labour legislation that provides for the control of labour unions and collective bargaining in their respective jurisdictions, and each has established a labour-relations board to administer the law. Labour legislation in each jurisdiction generally assigns a number of specific duties to the respective labour relations boards established under it. The boards normally have the authority to determine the right of a labour union to represent a group of employees, the nature and makeup of the employee group, the wishes of the employees to bargain collectively through a particular union, the certification of a union as a bargaining agent, and the enforcement of the rights and duties of employers, employees, and unions under the legislation. In some jurisdictions, the board is given the power to deal with strikes and lockouts as well.

The general thrust of the legislation is to replace the use of economic power by unions and employers with an orderly process for the selection of a bargaining representative for the employees, and for the negotiation of collective agreements. The use of the strike or lockout is prohibited with respect to the selection and recognition of the bargaining agent, and severely restricted in use as a part of the negotiation process. In most jurisdictions the right to strike or lockout may not be lawfully exercised until the parties have exhausted all other forms of negotiation, and compulsory conciliation or other third-party assistance has failed to produce an agreement.

The Certification Process

Collective bargaining usually begins with the desire on the part of a group of employees to bargain together, rather than on an individual basis, with their employer. To do this, they must first establish an organization to act on their behalf. This may be done by the employees themselves forming their own organization, or, more often, by calling upon an existing labour union for assistance. Most large labour unions have trained organizers whose job consists of organizing employees into new unions or locals affiliated with the larger union organization. These persons will assist the group of employees in the establishment of its own local union. Once the organization is in existence, with its own constitution and officers, its executive and members will then attempt to interest other employees of the firm in joining. When the organization believes that it has the support of a majority of the employees in the employer's plant, shop, or office, it may then approach the employer with a request to be recognized as the bargaining representative of the employees. If the employer agrees to recognize the union as the bargaining representative, it may meet with representatives of the union and negotiate a collective agreement that will contain the terms and conditions of employment that will apply to the group of employees represented by the union. On the other hand, if the employer refuses to recognize the union as the bargaining representative of the employees, the union is obliged to be certified as the bargaining representative by the labour relations board before the employer is required to bargain with it. This process is known as the **certification process**.

Certification process

A process under labour legislation whereby a trade union acquires bargaining rights and is designated as the exclusive bargaining representative of a unit of employees.

The process formally begins with the submission of a written application (by the labour union), to the labour relations board of the correct jurisdiction, for certification as the exclusive bargaining representative of a particular group of employees. On receipt of the union's application, the labour relations board will usually arrange for a hearing. At that time it will normally require a new union to prove that it is a *bona fide* trade union that is neither supported financially (or otherwise) nor dominated in any way by the employer. Once this has been accomplished by the union, the labour relations board will then determine the unit of employees appropriate for collective-bargaining purposes.

Bargaining unit

A group of employees of an employer represented by a trade union recognized or certified as their exclusive bargaining representative.

The group of employees, or the **bargaining unit**, is usually determined by the board in accordance with the legislation or regulations setting out the kinds of employees eligible to bargain collectively. While variation exists from province to province and federally, only "employees" are entitled to bargain collectively. Within this group, some professionals employed in a professional capacity, management employees, and persons employed in a confidential capacity with respect to labour relations, are usually excluded. As well, the legislation does not apply to certain employee groups. Most provinces have special legislation to deal with collective bargaining by persons engaged in essential services and in the employ of government. In some provinces, persons employed in certain activities (such as hunting and trapping) may be excluded from collective bargaining entirely.

Labour relations boards at their hearings will usually receive the representations of the employer, the union, and the employees, if the parties cannot agree upon an appropriate bargaining unit. The determination of the bargaining unit, however, is a board decision. When the decision is made, the board will then proceed with the determination of employee support for the union in the particular unit. It may do this by an examination of union-membership records and of union witnesses at a hearing. However, if any doubt exists, the board will usually hold a representation vote to determine the true wishes of the employees.[48] If a majority of the votes are cast in favour of collective bargaining through the union, then the board will certify the union as the exclusive bargaining representative of all of the employees in the bargaining unit. Certification gives the union the right to negotiate, on behalf of the employees, the terms of their employment and the conditions under which their work will be performed. It also permits the union to demand that the employer meet with its representatives to negotiate the collective agreement that will contain these provisions.

CHECKLIST FOR UNION CERTIFICATION

1. Union proceeds if it believes it has majority support within proposed unit.
2. May be voluntarily recognized by employer at this point.
3. If not voluntarily recognized, then written application made to LR Board.
4. Board determines appropriate bargaining unit.
5. Board examines union membership records for possible immediate certification.
6. If not immediately certified, then representation vote taken among bargaining unit.
7. If simple majority is achieved, union is certified as exclusive bargaining agent for unit.

MANAGEMENT ALERT: BEST PRACTICE

The transition from a non-union workforce to a unionized one through the certification process is one that many employers fear, and one where serious errors in judgment often occur. At the first sign of union activity, employers should seek legal advice as to what management actions (such as changing conditions of work) are, and are not, acceptable. At both extremes, sudden generosity and "retaliation" by the employer can backfire, resulting in charges of "bad faith" and intervention of the provincial labour relations board, often to the employer's detriment.

FRONT-PAGE LAW

Wal-Mart found guilty in Quebec store closure
Labour board rules closing store was reprisal

MONTRÉAL, September 16, 2005—The Québec Labour Relations Board has ruled today in favour of former Wal-Mart workers, following the closing of the Jonquière store.

In April, Wal-Mart management closed the Jonquière store, letting go more than one hundred workers who had succeeded, after a long struggle, in securing union certification. Wal-Mart announced the store closing while negotiation of an initial collective agreement was still underway.

After hearing four representative cases, the Board concluded that Wal-Mart had acted illegally and had dismissed workers for engaging in union activity.

United Food and Commercial Workers local 503 is pleased with the Labour Relations Board's decision. The President of the Québec Council of UFCW, Yvon Bellemare said: "Wal-Mart clearly closed this store because the workers succeeded in unionizing. The Labour Relations Board's decision once again exposes the multinational's anti-union attitude. The momentum is picking up. Wal-Mart employees now realize that if they want a union in their store, Wal-Mart may attempt to but can't stop them." Quebec Federation of Labour President Henri Massé hails the victory: "We are now calling on Wal-Mart to abide by the Board's decision and not to start legal guerrilla tactics by appealing it."

Source: Excerpted from United Food and Commercial Workers Canada Press Release, "Wal-Mart found guilty in Quebec store closure," September 16, 2005. (www.ufcw.ca)

48. Some provinces (such as Ontario) require that a vote be held if the union membership is below a certain percentage of the workforce.

The Negotiation Process

The negotiation process begins when the certified trade union gives written notice to the employer of its desire to meet with representatives of the employer to bargain for a collective agreement. On receipt of the notice, the employer must arrange a meeting with the union representatives and bargain in good faith with a view to making a collective agreement. This does not mean that the employer is obliged to accept the demands of the union, but it does mean that the employer must meet with the union and discuss the matters put forward by it. Nor does it mean that the demands are always one-sided. Employers often introduce their own demands at the bargaining table. For example, employers generally insist upon the insertion in the collective agreement of terms that set out the rights of management to carry on specified activities without interference by the union or employees.

Where the parties reach an agreement on the terms and conditions of employment, and on the rights and duties of the employer, the union, and the employees, the agreement is put in writing and signed by the employer and representatives of the union. When approved by the employees, the agreement then governs the employment relationship during the term specified in the agreement.

Collective agreements must normally be for a term of at least one year. Either the employer or the union may give notice to bargain for a new agreement, or for changes in the old agreement, as the expiry date of the agreement approaches. The minimum term is generally dictated by the governing legislation, and usually cannot be reduced without the consent of the labour relations board. The purpose of the minimum term is to introduce an element of stability to collective bargaining by requiring the parties to live under the agreement they negotiate without stoppage of work for at least a reasonable period of time.

Third-Party Intervention

If the parties cannot reach agreement on the terms and conditions of employment, or the rights and duties of the employer and the union, most jurisdictions provide for third-party intervention in the negotiations to assist the parties. This intervention may take the form of conciliation, mediation, or, in some cases, fact-finding. The purpose of the intervention is to assist the parties by clarifying the issues in dispute and (in the case of mediation) by taking an active part in the process through offers of assistance in resolving the conflict. Only when third-party intervention is exhausted are the parties permitted to strike or lockout to enforce their demands.

The law in most jurisdictions does not permit employers to alter working conditions or the work relationship during the negotiation process. Unions also may not unilaterally alter the relationship. For example, in one case where negotiations had reached an impasse, and before third-party assistance was requested, some of the union members decided to picket at the employer's premises during off-duty hours. The employer then sought an injunction to prohibit the picketing. In handing down its decision the court observed:

> Either party has a definite course to follow if negotiations break down. A request for conciliation services is available to them. I repeat that any act by an employer or a union which at that stage in the negotiations is not in conformity with the "rules" set out in the Act to achieve agreement between the parties, allows drawing an inference of bad faith as it is not pursued by legal and peaceful steps as contemplated by statute. It then becomes an attempt to circumvent the Act and the use of parading, picketing, or in any way affecting the business of the employer or the liberty of the employees is a process or substitute to foster the contract by a procedure not contemplated nor within the purview of the Act.[49]

When the parties are unable to reach an agreement, and the negotiations are for a first agreement between the employer and the union, if the employer's actions or unreasonableness in negotiations have prevented the parties from reaching an agreement, some provinces (e.g., Manitoba, Newfoundland, and British Columbia) have made provision in

49. *Nipissing Hotel Ltd. and Farenda Co. Ltd. v. Hotel & Restaurant Employees & Bartenders International Union C.L.C., A.F. of L., C.I.O., et al.* (1962), 36 D.L.R. (2d) 81.

their legislation for the imposition of the first agreement on the parties. This is usually done by way of a process whereby either an arbitrator or the labour-relations board (depending upon the jurisdiction) will hear the arguments of both sides to the dispute, then impose a collective agreement on the parties. The agreement will normally include the terms agreed upon by the parties plus terms to complete the agreement that address the issues in dispute. The imposed first collective agreement binds the parties for a term of up to two years.

A QUESTION OF ETHICS

The "right to organize and bargain collectively" is sometimes called a second- or third-generation human right. Some nations include this in their equivalent of our *Charter of Rights and Freedoms*. At present, such a right does not exist in Canada. Do you think that such a right, available to all Canadians, needs to be entrenched in our *Charter*? Why?

Strikes and Lockouts

Strike

In a labour relations setting, a cessation of work by a group of employees.

A **strike** is considered to be a concerted refusal to work by the employees of a workplace, although, in some jurisdictions, any slowdown or concerted effort to restrict output may also be considered a strike.[50] A lawful strike under most labour legislation may only take place when a collective agreement is not in effect and after all required third-party assistance has failed to produce a collective agreement. A strike at any other time is usually an unlawful strike, regardless of whether it is called by the union, or whether it is a spontaneous walkout by employees (i.e., a "wildcat" strike).

Lockout

In a labour relations setting, the refusal of employee entry to a workplace by an employer when collective bargaining with the employees fails to produce a collective agreement.

A **lockout** is, in some respects, the reverse of a strike. It is the closing of a place of employment or a suspension of work by an employer. It is lawful when a collective agreement is not in effect and after all required third-party intervention has failed to produce an agreement.

Picketing

The physical presence of persons at or near the premises of another for the purpose of conveying information.

Lawful strikes and lockouts must normally be limited to the premises of the employer that has a labour dispute with its employees. Under a lawful strike, the employees may withhold their services from their employer and, if they so desire, may set up picket lines at the entrances of the employer's premises to inform others of their strike. Lawful **picketing** is for the purpose of conveying information.[51] Any attempt by pickets to prevent persons from entering or leaving the plant may be actionable by law. If property is damaged or persons injured while attempting to enter or leave the employer's premises, the usual action on the part of the employer is to apply for a court order limiting the number of pickets to only a few. In this fashion the lawful purpose of picketing is served, and the likelihood of damage or injury is substantially diminished.

Secondary picketing

Picketing at other than the employer's place of business.

In most cases, striking employees limit their picketing to the employer's place of business. When picketing takes place elsewhere than the employer's place of business, it is known as **secondary picketing**, and, until January 2002, it was considered unlawful except where the employer and a supplier or customer were so closely related that the supplier or customer might be considered involved in the dispute as a part of the employer's overall operation.[52] However, in January 2002, the Supreme Court of Canada was obliged to consider the issue in a Saskatchewan labour dispute.[53] In that case, striking employees of a soft-drink manufacturer picketed retail stores that sold the product, and urged customers of the store to boycott the product. The pickets were clearly engaged in secondary picketing, but the court held that such picketing constitutes freedom of expression, and is protected under the *Charter of Rights and Freedoms*. It should be noted, however, that the court stressed that the freedom was limited to peaceful picketing, and any picketing

50. See, for example, the *Labour Relations Act*, 1995, S.O. 1995, c. 1, s. 49.

51. *Criminal Code*, R.S.C. 1985, c. C-46, s. 423. See also *Smith Bros. Construction Co. Ltd. v. Jones et al.*, [1955] O.R. 362.

52. See, for example, *Canadian Pacific Ltd. v. Weatherbee et al.* (1980), 26 O.R. (2d) 776.

53. *Retail Wholesale and Department Store Union Local 550 v. Pepsi Cola Canada (West) Ltd.*, [2002] 4 W.W.R. 205.

that involved criminal activity, trespass to property, intimidation, defamation, or misrepresentation was unlawful, as was any attempt to interfere with the free access of the public to the premises picketed. The court also held that any effort to pressure the retailer to breach its contract with the employer was also actionable at law in tort.

CASE IN POINT

An employer that operated two stores had a unionized work force, and when the parties could not agree on a collective agreement, the union called a lawful strike. The employer also operated several other stores that were not unionized, and the union members attended at these stores, handing out leaflets, and placing them on the windshields of cars in the parking lots. The union members did nothing else to prevent customers from entering the stores. The employer's response was to seek an order prohibiting the leaflet campaign.

When the case eventually reached the Supreme Court of Canada, the court held that the peaceful distribution of leaflets represented freedom of expression under the *Charter of Rights and Freedoms* and was lawful since the information was accurate, did not interfere with access to the stores by patrons, and did not constitute intimidation or any other unlawful act or tort.

See: *United Food and Commercial Workers, Local 1518 v. Kmart Canada Ltd.* (1999), 176 D.L.R. (4th) 607.

FRONT-PAGE LAW

Politicians urged to end CBC lockout

In 2005, a labour dispute at the Canadian Broadcasting Corporation resulted in a lockout of employees by the Corporation. The lockout was given wide-spread coverage by both the television and print media. This resulted in pressure on politicians in Ottawa to press for legislation to end the lockout by way of back-to-work legislation.

Should government become involved in the settlement of a labour dispute in a non-essential service?

Compulsory Arbitration

The strike or lockout is not available, however, to all employee groups when negotiations break down. Persons employed in essential services, such as hospitals, firefighting, and police work, are usually denied the right to strike. Compulsory arbitration is used to resolve the issues that these groups cannot settle in the bargaining process.

When compulsory arbitration is imposed, an arbitration process replaces the right to strike or lockout and permits work to continue without interruption. Under this system, if the employer and union cannot reach an agreement, they are generally required to have the issues in dispute decided by a representative tribunal called an arbitration board. The tribunal is usually made up of one representative each, chosen by the employer and the union, and an impartial third party, chosen by the representatives (or appointed by the government) who becomes the chairperson of the board. The tribunal will hold a hearing where the arguments of both sides concerning the unresolved issues may be presented. At the conclusion of the hearing, the board will review the presentations and the evidence, then make a decision. The parties will be bound by the decision of the arbitration board. The decision of the board, together with the other agreed-upon terms, will become the collective agreement that will govern the employment relationship for the period of its operation.

While compulsory arbitration is normally applied to employers and employees engaged in activities where the disruption of services by a strike or lockout would be injurious to the public, some jurisdictions have provided for the use of arbitration as an optional means of settling outstanding issues in those industries and services that are not treated as essential. This method of settlement is generally available as a procedure that may be adopted by the parties as a part of their negotiations, or it may be employed as a means of resolving a labour dispute when the parties have been engaged in a lengthy strike or lockout.

CLIENTS, SUPPLIERS *or* OPERATIONS *in* QUEBEC

The Quebec *Collective Agreement Decrees Act*, R.S.Q., c. D-2 permits the government to apply collective agreement terms to entire trades or industries in the province (or regions of the province) where dominant firms in the industry or trade have negotiated similar agreements with their unions. Decrees issued under the Act have applied for the most part to the construction industry, but the Act has also been applied to other sectors of industry as well. The legislation, in effect, imposes the particular terms and conditions of employment on all employees and employers, even if the employees are not in a collective bargaining relationship with their employer.

The Collective Agreement and Its Administration

Collective agreement

An agreement in writing, made between an employer and a union certified or recognized as the bargaining unit of the employees. It contains the terms and conditions under which work is to be performed and sets out the rights and duties of the employer, the employees, and the union.

The **collective agreement** differs from the ordinary Common Law contract of employment in a number of fundamental ways. The collective agreement sets out the rights and duties, not only of the employer and the employees, but of the bargaining agent as well. It is also an agreement that is sometimes negotiated under conditions that would render an ordinary Common Law contract voidable. The imposition of economic sanctions, or the threat of their use, is treated as a legitimate tactic in the negotiation of a collective agreement, but one that would not be tolerated by the courts in the case of a Common Law contract.

Most jurisdictions require the parties to insert special terms in their collective agreement that will govern certain aspects of their relationship. The agreement must usually include a clause whereby the employer recognizes the union as the exclusive bargaining representative of the employees in the defined bargaining unit. This has a twofold purpose. First, it is a written acknowledgment by the employer that the union is the proper body to represent the employees. Second, recognition of the union as the exclusive bargaining representative prevents the employer from negotiating with any other union purporting to act on behalf of the employees while the collective agreement is in existence.

Most jurisdictions also require the parties to provide in their agreement that no strike or lockout may take place during the term of the agreement, should a dispute arise after the agreement is put into effect. Coupled with this requirement is the additional requirement that the parties provide in their agreement some mechanism to settle disputes that arise out of the collective agreement during its term of operation.

Arbitration

A process for the settlement of disputes whereby an impartial third party or board hears the dispute, then makes a decision that is binding on the parties. Most commonly used to determine grievances arising out of a collective agreement, or in contract disputes.

The most common method of dispute resolution is **arbitration**. This is generally compulsory under most collective-bargaining legislation. The law frequently sets out an arbitration process that is deemed to apply if the parties fail to include a suitable procedure in their collective agreement. As a rule, any dispute that arises out of the interpretation, application, or administration of the collective agreement, including any question as to whether a matter is arbitrable, is a matter for arbitration. The procedure would also be used when a violation of the collective agreement is alleged.

Collective agreements usually provide for a series of informal meetings between the union and the employer concerning these disputes (called grievances) as a possible means of avoiding arbitration. The series of meetings, which involve progressively higher levels of management in both the employer and union hierarchy, is referred to as a grievance procedure. It is usually outlined as a series of steps in a clause in the collective agreement. If, after the grievance procedure is exhausted, no settlement is reached, either the employer or the union may carry the matter further and invoke the arbitration process.

The parties under the terms of their collective agreement may provide for the dispute to be heard by either a sole arbitrator or an arbitration board. If the procedure calls for a board, it is usually a three-person board, with one member chosen by the union, and one by the employer. The third member of the board is normally selected by the two persons so nominated and becomes the impartial chairperson. If the parties cannot agree on an independent chairperson, the Minister of Labour usually has the authority to select a chairperson for the arbitration board.

An arbitration board (or sole arbitrator) is expected to hold a hearing where each party is given the opportunity to present their side of the dispute, and to introduce evidence or witnesses to establish the facts upon which they base their case. When all of the evidence and argument has been submitted, the arbitrator (or arbitration board) renders a decision called an award that is binding upon the parties.

Arbitrators and arbitration boards are usually given wide powers under the legislation to determine their own procedure at hearings, to examine witnesses under oath, and to investigate the circumstances surrounding a dispute. However, they are obliged to deal with each dispute in a fair and unbiased manner. If they fail to do so, or if they exceed their jurisdiction, or make a fundamental error of law in their award, their award may be quashed by the courts.

While arbitration is used as a means of interpreting rights and duties of the employer and the union under the collective agreement, it may also be used by the union to enforce employee rights. Employees who are improperly treated by the employer under the terms of the collective agreement, or who believe that they have been unjustly disciplined or discharged, may file grievances that the union may take to arbitration for settlement.

The rights of employees under collective bargaining differ to some extent from the rights of persons engaged in employment under the Common Law. The Common Law right of an employee to make a separate and different contract of employment with the employer is lost insofar as the collective agreement is concerned, but this is balanced by way of new collective-bargaining rights. For example, employees under a collective agreement are subject to different treatment in the case of disciplinary action by the employer. An employer may suspend or discipline an employee (usually for just cause) under a collective agreement in cases where discharge is perhaps unwarranted. This represents an approach to discipline that is not found at Common Law. If the right is exercised, however, the employer's actions may be subject to review by an arbitrator, as they might also be if the employee was discharged by the employer without good reason.

A difference also exists between the remedies available to an arbitrator and the remedies available to the courts in the case of discharge. In most jurisdictions, an arbitrator has the authority to substitute a suspension without compensation when discharge is too severe a penalty or is unwarranted. The arbitrator may also order the reinstatement of an employee wrongfully dismissed, with payment of compensation for time lost. The courts, on the other hand, are normally unwilling to order the reinstatement of an employee wrongfully dismissed. They limit the compensation of the employee to monetary damages.

The Termination of Collective Bargaining

Collective bargaining is an on-going process, but in some cases, if the process fails to serve its purpose for the employees, provision is made in labour legislation to terminate their collective bargaining relationship. This may occur where a union decides to cease acting as the bargaining agent for the employees, or where the majority of the employees no longer wish to be represented by the union in their dealings with their employer.

Termination of bargaining rights usually takes the form of a declaration by the labour relations board after a hearing is held to determine if collective bargaining should end. The process usually begins with a request by a group of employees for decertification of the union as their bargaining agent. A hearing is then held by the labour relations board. At the hearing, the employee group must establish that a majority of the employees support the request, and also establish that the request is not employer-supported or assisted. A vote is usually then held, and if the vote clearly indicates that the employees no longer wish to be represented by the union, the labour relations board will terminate the union's bargaining rights. At that point, the employment relationship will revert to individual employment contracts between each employee and the employer.

It is important to note that the employer must not have instigated or encouraged the employees to apply for decertification of the union. If the employer has done so, the board will reject the application.

An employee who was in line for promotion to management was dissatisfied with the union that represented him and wished to make an application for decertification of the union to the provincial labour relations board. He sought the advice of a member of management. The manager advised the employee that he could not assist him, but put him in touch with a lawyer that was experienced in this type of action. The employee, on the advice of the lawyer, proceeded with the application to the labour relations board.

At the hearing, the labour relations board dismissed the application on the basis that it was made on the advice and assistance of the employer. The employee applied to the court to quash the decision of the labour relations board. The court held that the board's decision was not unreasonable, and dismissed the employee's application to have the board's decision quashed.[54]

THE UNION—MEMBER RELATIONSHIP

As a party to a collective agreement, and as an entity certified by a labour relations board as the exclusive bargaining agent for a group of employees, a trade union is unique. It was initially an illegal organization at Common Law, whose lawful existence was made possible only by legislation. In some provinces, such as Ontario, it is not a suable entity by virtue of statute law.[55] However, in other provinces the courts have held that trade unions have acquired a legal existence through legislation which has clothed them with special rights and powers. As a result, in some provinces their actions may be subject to legal action, except in matters directly related to collective bargaining.[56]

CASE IN POINT

A union called a lawful strike at an employer's supermarket. Picketing of the supermarket took place, and the employer took exception to some of the information displayed on the picket signs which he concluded to be untrue. In response to the picket signs, the employer placed advertisements in the local newspaper stating that false information was being disseminated by the union. In turn, the union filed a grievance against the employer claiming defamation. The Arbitrator held that under provincial statute the union was not a legal entity, and as a consequence, it could not sue for defamation.

See: *Re Fortino's Supermarket Ltd. and United Food and Commercial Workers, Local 175* (2003), 117 L.A.C. (4th) 154.

Apart from its special status as a bargaining agent under labour legislation, a trade union is similar to any club or fraternal organization. As an unincorporated entity, it has no existence separate from its members, and its relationship with its members is contractual in nature.[57] In concept, it consists of a group of individuals who have contracted with one another to abide by certain terms embodied in the organization's constitution (and which form a part of each member's contract) in order to carry out the objects or goals of the organization. The rights of each member, then, are governed by the contract. If a member fails to abide by the contract, the remainder of the members may expel the offending member from the organization.

Most jurisdictions in their collective-bargaining legislation have imposed certain limits on the rights of trade unions to refuse membership or to expel existing members, because of the effect that denial of membership has on an individual's ability to find employment in unionized industries. This is due to the fact that many unions have required employers to insert in their collective agreements a term whereby the employer

54. *Nadon v. United Steelworkers of America* (2004), 244 Sask. R. 255.

55. See the *Rights of Labour Act*, R.S.O. 1990, c. R.33.

56. *Int'l Brotherhood of Teamsters, Chauffeurs, Warehousemen & Helpers, Building Material, Construction & Fuel Truck Drivers, Loc. No. 213 v. Therien* (1960), 22 D.L.R. (2d) 1.

57. *Astgen et al. v. Smith et al.*, [1970] 1 O.R. 129.

agrees to hire only persons who are already union members (a closed-shop clause), or a clause whereby continued employment of an employee is conditional upon union membership (a union-shop clause). In both of these cases, a loss of union membership would either prevent an employer from hiring the person, or oblige the employer to dismiss the employee. To safeguard the rights of individuals, and to protect them from arbitrary action by labour unions, the right to refuse membership or expel an existing member must be based upon legitimate and justifiable grounds. Membership generally may not be denied on the basis of race, creed, colour, sex, nationality, place of origin, or other discriminatory factors, for example. Nor may membership normally be denied or revoked simply because a person belongs to a rival union. Once membership is granted, it may not be withdrawn at the whim of a union officer or executive.

Membership for the union member, like the membership in any club or organization, involves adherence to the rules and obligations set out in the organization's constitution. If the member fails to abide by the terms of the contract made with the rest of the organization, the membership may take steps to end its relationship with the offending member. The expulsion of the member must not, however, be made in an unfair or arbitrary way. The courts generally require that the rules of natural justice be followed by the organization, and the accused be given an opportunity to put his or her case before the membership before any decision is made. This would involve giving the accused member full details of the alleged violation, an opportunity to prepare his or her case, and the conduct of a hearing before the membership that would allow the accused to face his or her accusers and answer their charges by way of evidence and cross-examination.[58] Only then may the decision of the membership be made on the question before them.

It is the obligation of a trade union to treat its members fairly and its duty to expel a member only after giving the member an opportunity to be heard and defend himself or herself before the membership. This may be exemplified by an English case, where a long-time member of a union was expelled and then complained to the courts. The court, in reviewing the actions of the union, summarized the duties in the following manner:

> The jurisdiction of a domestic tribunal, such as the committee of the Showmen's Guild, must be founded on a contract, express or implied. Outside the regular courts of this country, no set of men can sit in judgment on their fellows except so far as Parliament authorizes it or the parties agree to it. The jurisdiction of the committee of the Showmen's Guild is contained in a written set of rules to which all the members subscribe. This set of rules contains the contract between the members and is just as much subject to the jurisdiction of these courts as any other contract.
>
> Although the jurisdiction of a domestic tribunal is founded on contract, express or implied, nevertheless the parties are not free to make any contract they like. There are important limitations imposed by public policy. The tribunal must, for instance, observe the principles of natural justice. They must give the man notice of the charge and a reasonable opportunity of meeting it. Any stipulation to the contrary would be invalid. They cannot stipulate for a power to condemn a man unheard ....
>
> The question in the present case is: To what extent will the courts examine the decisions of domestic tribunals on points of law? This is a new question which is not to be solved by turning to the club cases. In the case of social clubs the rules usually empower the committee to expel a member who, in their opinion, has been guilty of conduct detrimental to the club, and this is a matter of opinion and nothing else. The courts have no wish to sit on appeal from their decisions on such a matter any more than from the decisions of a family conference. They have nothing to do with social rights or social duties. On any expulsion they will see that there is fair play. They will see that the man has notice of the charge and a reasonable opportunity of being heard. They will see that the committee observe the procedure laid down by the rules, but will not otherwise interfere ... It is very different with domestic tribunals which sit in judgment on the members of a trade or profession. They wield powers as great, if not greater, than any exercised by the courts of law. They can deprive a man of his livelihood. They

58. *Evanskow v. Int'l Brotherhood of Boilermakers et al.* (1970), 9 D.L.R. (3d) 715.

can ban him from the trade in which he has spent his life and which is the only trade he knows. They are usually empowered to do this for any breach of their rules, which, be it noted, are rules which they impose and which he has no real opportunity of accepting or rejecting. In theory their powers are based on contract. The man is supposed to have contracted to give them these great powers, but in practice he has no choice in the matter. If he is to engage in the trade, he has to submit to the rules promulgated by the committee. Is such a tribunal to be treated by these courts on the same footing as a social club? I say: "No." A man's right to work is just as important, if not more important, to him than his rights of property. These courts intervene each day to protect rights of property. They must also intervene to protect the right to work.[59]

Duty of Fair Representation

Duty of fair representation

Duty of a union to represent its members in a fair and impartial manner.

A union has a responsibility toward its members to fairly represent them not only in the collective sense, but on an individual basis as well. The *Canada Labour Code* and the legislation in most provinces impose a duty upon unions to act in good faith and in a non-arbitrary, non-discriminatory manner towards individual members in their representation of them and the enforcement of their rights under collective agreements. The duty also extends to services that unions may provide for their members. For example, if a union operates a "hiring hall" (which sends unemployed members to employers who require workers), the union must treat all members fairly in the filling of job openings. It must not give certain members priority over others in the allocation of work.

The most common type of case to arise relates to the right of the union under most collective agreements to process grievances on behalf of employees in the bargaining unit. Because an employee normally does not have the individual right to take a grievance to arbitration, the union has a duty to act in good faith towards an employee with a grievance in any decision to carry a grievance on to the arbitration stage. For example, in a Manitoba case, an employee was accused of theft, but no charges were laid. The employee sought the assistance of his union in the dispute with the employer over the alleged theft. The union reluctantly did so, but made no effort to investigate the charge, nor did it press the employee's grievance. The employee filed a complaint against the union. The labour board held that the union had failed in its duty to fairly represent the employee, and ordered the union to take the employee's grievance to arbitration. The board also ordered the union to pay the employee's legal costs and expenses.[60]

It is important to note, however, that the duty of fair representation does not oblige a union to carry every employee grievance to arbitration. A union is entitled to assess each grievance in terms of its merits and, provided that the assessment is made in good faith and not in an arbitrary or discriminatory manner, it may properly decide against proceeding to arbitration without violating its duty of fair representation.

SUMMARY

The Common Law contract of employment is a special type of agreement that sets out the rights and duties of the employer and employee. It need not be in writing unless subject to the *Statute of Frauds*. The essential characteristic of the employment relationship is the control of the employee by the employer. Originally, the relationship was that of employment if the employer had the right to direct the work to be done, and the manner in which it was to be done. However, modern forms of business organization have required the courts to look beyond the simple control aspect to other determinants, and have included consideration of the ownership of tools, the chance

59. *Lee v. Showmen's Guild of Great Britain*, [1952] 1 All E.R. 1175.
60. *Shachtay v. Creamery Workers Union, Local 1* (1986), 86 C.L.L.C. 16,033 (Man. L.R.B.).

of profit, the risk of loss, and, in some cases, the relationship of the employee's work to the overall operation of the employer. The courts also consider the nature of the work done by the employee in terms of the role of the work in the employer's operation. This latter test is known as the organization test for employment.

Under the contract of employment both the employer and employee have a number of implied duties, the most important being the employer's obligation to pay wages in return for the employee's services, and the employee's obligation to obey all reasonable directions of the employer.

Employment standards legislation and occupational health and safety laws have imposed many obligations on employers. Human Rights legislation has also established clearer duties of employers with respect to employment and subsequent treatment of employees.

If an employee acts in breach of his or her duty, the employer may be entitled to dismiss the employee. However, if the dismissal is unwarranted, the employee may bring an action against the employer for wrongful dismissal. A failure to give reasonable notice of termination, when such is required, would also be grounds for wrongful dismissal and entitle the employee to damages. The employee must act promptly, however, to mitigate loss by seeking employment elsewhere.

An employer may be vicariously liable for the acts of his or her employees, if the acts are done within the scope of the employee's employment. The employer may have a right over and against a skilled or professional employee, however, if the employee was negligent in the exercise of a professional skill in the performance of professional duties.

While labour law in the broad sense refers to all law relating to the employment relationship, it is a term that is often used to refer to the area of employment law concerned with collective bargaining. Collective bargaining is concerned with the negotiation of the terms and conditions of employment by a group of employees with their employer, using a trade union as their bargaining agent. Collective bargaining is conducted under legislation that sets out procedures for the selection of a bargaining agent (the certification process), the negotiation of the collective agreement (the negotiation process), and for the administration of the negotiated agreement (the administration process).

Unlike the Common Law contract of employment, the collective agreement is a special contract of employment that applies to unionized employees. It is not a contract brought before the courts for interpretation, but an agreement that falls within the jurisdiction of a tribunal.

For the most part these tribunals (labour relations boards and arbitration tribunals) deal with issues that arise out of collective bargaining by way of conciliation, mediation and arbitration procedures.

The law relating to the collective agreement is administrative law, rather than Common Law. Labour legislation, of which the law relating to collective bargaining is a part, represents a body of statute law that has grown substantially during the last century to meet the needs of employees in more complex industrial and commercial employment settings.

KEY TERMS

Fourfold test, 350
Organization test, 350
Duty to accommodate, 354
Wrongful dismissal, 360
Just cause, 360
Constructive dismissal, 362
Certification process, 368
Bargaining unit, 368
Strike, 371
Lockout, 371
Picketing, 371
Secondary picketing, 371
Collective agreement, 373
Arbitration, 373
Duty of fair representation, 377

Review Questions

1. How is an employment relationship established? What elements of the relationship distinguish it from agency or partnership?
2. Explain the fourfold test for employment. Why did the courts find it necessary to establish this test?
3. Distinguish a "contract of service" from a "contract for services."
4. If an employee is wrongfully dismissed, explain how a court would determine the money damages that should be paid by the employer for the wrongful act.
5. Outline the general or implied duties of an employee under a contract of employment.
6. Why are employers, under certain circumstances, vicariously liable for the torts of their employees? Identify the circumstances under which vicarious liability would arise.
7. Identify the conditions or circumstances under which an employer would be justified in terminating a contract of employment without notice.
8. What factors must be considered in determining reasonable notice, if an employee or employer should decide to give notice of termination of a contract of indefinite hiring?
9. Why must an employee mitigate his or her loss when wrongfully dismissed?
10. Define a collective agreement.
11. What effect does collective bargaining have on the Common Law employment relationship?
12. Explain how a union acquires bargaining rights.
13. What is a bargaining unit? How is it determined?
14. Outline the steps in the negotiation process and its purpose.
15. Explain briefly the role of a union in collective bargaining.
16. If a collective agreement is negotiated, what methods may be used to resolve disputes that arise out of the collective agreement?
17. Describe the legal obligations a union has towards its members.
18. How are disputes between the parties resolved during the negotiation process if third-party assistance fails?

Mini-Case Problems

1. Marie was employed by Rigney Construction Company as bookkeeper and office manager for over ten years. Without notice or explanation, Rigney informed Marie that her services were no longer required and requested her to leave the premises. At the time, Marie was earning an annual salary of $50,000. What additional information would you wish to know if Marie were to ask your advice as to whether she should take legal action against Rigney Construction for wrongful dismissal?
2. Jones, who was earning an annual salary of $60,000, was summarily dismissed by Smith, his employer. A week later, Jones obtained a new position with another firm at an annual salary of $60,000. If Jones took legal action against Smith for wrongful dismissal, what would his damages be if he were able to prove his claim successfully?
3. Helga was employed for many years as a material handler in an automobile parts warehouse. Her work involved filling orders from auto repair shops for parts weighing from a few grams to as much as 40 kilograms. In a non-work related accident, Helga was permanently injured and unable to walk without a cane. Her employer terminated her employment, giving her two month's wages in lieu of notice. What are Helga's rights in this case?
4. Apex Company is a supplier of goods for the DYNO Company. A lawful strike takes place at the DYNO Company plant, and the picketing employees refuse to allow a truckload of goods sent by Apex Company to enter the DYNO Company plant gates. What are the rights of Apex Company?
5. Suppose that, the next day, the Apex Company finds striking employees of the DYNO Company picketing the Apex Company plant because it is a supplier of DYNO Company. What are the rights of Apex Company?
6. A union was involved in a legal strike, and picketed the employer's plant. Archie, a union member, crossed the picket line and went to work. What might the union do? What are Archie's rights?

Case Problems for Discussion

CASE 1

Creative Advertising Ltd. employed Igor as a copy design editor for many years. The work was very stressful, but Igor enjoyed the work, as it required a great deal of originality and creativity on his part. In particular, he enjoyed the 'presentation to clients' aspect of his job, as it gave him the opportunity to explain his work to clients, and the results he expected from it in terms of product sales.

Unfortunately for Igor, at age 49 he suffered a stroke which left him with a serious speech defect, but otherwise unaffected. The stroke made verbal communication with clients impossible, but he was still able to carry on with his creative work. His employer, however, decided that Igor's services were no longer necessary, and gave Igor notice of termination along with a cheque equivalent to 6 months salary.

Discuss the issues raised in this case, and advise Igor of his options.

CASE 2

CD Transport & Delivery Ltd. engaged the services of Lisa to assist with local delivery of packages to businesses and residents. Lisa was expected to provide her own delivery vehicle, and was paid on a per hour wage basis and a per mile basis for her truck. CD Transport provided magnetic signs that Lisa was expected to attach to the sides of her truck. Lisa was expected to work exclusively for CD Transport and for no other delivery service in the area. Lisa billed for her work hours and services on a monthly basis.

After several years, CD Transport advised Lisa that they would no longer require her services. The notice of termination was to be effective in one month's time.

Lisa objected to the short notice, and claimed that she was an employee of CD Transport.

Outline the position of each of the parties in this dispute, and the arguments each may raise if the case came before the court.

Render a decision.

CASE 3

McKenzie was a qualified and licensed driver of tractor trailers and other heavy types of trucks. He was first employed by FMP Company in 1993, and was steadily employed by the company as a truck driver until 2011. At that time he was voluntarily placed "on loan" to Timber-Hall Trucking under an agreement that provided as follows:

(1) Timber-Hall will provide and maintain trucks and equipment to haul logs from FMP Company logging sites to the FMP Mill.
(2) FMP Company will provide any qualified truck drivers required by Timber-Hall. Drivers will continue to be paid their regular wage rates and benefits by FMP Company.
(3) Timber-Hall will have the right to direct and supervise the work of the drivers, but will not have the right to discharge or discipline drivers provided by FMP Company.

Timber-Hall operated a fleet of 36 trucks used for hauling timber from various logging sites to the FMP mill. Of the drivers, 32 were employees of Timber-Hall and four were "on loan" from FMP Company. All drivers were under supervision of Timber-Hall management at both the loading and unloading points, and the drivers were directed to specific locations by Timber-Hall supervisors.

McKenzie was directed to a particular loading area by Timber-Hall and told to deliver the load to the FMP Company mill some 100 kilometres away. McKenzie picked up the load of timber and set out for the FMP mill. En route, McKenzie encountered icy road conditions as he descended a long hill. Before he could bring the heavy truck under control, it careened from the road and collided with a road-side cabin owned by McGee.

The cabin was demolished, and McGee brought an action against McKenzie, Timber-Hall, and FMP Company for damages. Both Timber-Hall and FMP Company alleged in their defence that they were not the employer of McKenzie.

Discuss the arguments that may be put forward by the defendants and the issues raised by the case.

Render a decision.

CASE 4

Lynne and Leroy were employed as commissioned salespersons by a large computer and electronic company. Each received a basic salary of $700 per week and a commission on gross sales they made on behalf of the company. In addition, the company paid their reasonable travel and living expenses on all authorized business trips involving travel of more than 25 kilometres from the head office building where their own offices were located.

Both Lynne and Leroy were highly productive sales representatives, and earned substantial sales commissions that placed their average earnings close to the $100,000 mark each year. While each was responsible for a specific territory, the two of them took on a proposal of the company concerning a new, remote-community computerized communications system that a government agency had opened to tender. The particular tender required extensive research and preparation, as well as presentation of the proposal for review and approval by several related government agencies before it could be submitted. A number of companies were interested in the project, and Lynne and Leroy put in long hours at their offices to prepare their presentations to the various agencies. Eventually, after much work and effort, they managed to obtain approval for the company project from each agency. A tender on the project was then prepared for submission to senior management for approval. Both

Lynne and Leroy were optimistic that their tender would be successful, in spite of the fact that four other manufacturers were preparing tenders as well.

During the two weeks before Lynne and Leroy submitted the tender for senior management approval, the company initiated a minor reorganization of its marketing department. Then, several days after the submission was in the hands of senior management, both Lynne and Leroy were advised that they would each be promoted to the position of regional manager, a position carrying with it an annual salary of $100,000.

Regional managers were responsible for a number of salespersons who reported to them and were also expected to negotiate some of the larger contracts with customers. However, they received no commissions, since their salaries were designed to cover their efforts as well as their responsibility.

The two employees were initially delighted with the thought of promotion. However, on reflection, they decided that, by accepting it, they would perhaps deny themselves the large commission they would earn if their tender was accepted by the government agency. To clarify their position, they met with the vice-president of marketing, who informed them that the company intended to include their old sales territories with those of other salespersons and that, if they refused to accept the promotions, no sales territory would be open for them. The rationale was that, since, as a result of the reorganization, the sales force would consist of persons with greater service with the company, those salespersons with less service would be terminated or moved to positions of less responsibility and lower salary rates. The vice-president urged both to accept the promotion, since they each had only three years' service with the company, and the company was anxious to retain them. The vice-president also stated that he believed that the two employees would likely have a bright future if they remained with the company.

At this point in the discussion, Leroy demanded to know if they would be entitled to the commission on the government sale if the tender was accepted. The vice-president responded that they would not be entitled to commissions in their new positions. Both Lynne and Leroy then became angry and accused the company of robbing them of what was rightfully theirs. The discussion quickly degenerated into a shouting match, until Lynne ended it by striking the vice-president with her heavy handbag. Both employees were then told to leave the office and "clean out their desks," as they were terminated.

Discuss the possible arguments of the two employees and the employer if the employees were to bring an action for wrongful dismissal.

Render a decision, and in it consider the effect of the success or failure of the company to secure the government contract.

CASE 5

Mall Merchandising Co. operates a department store in a shopping centre located near the outskirts of a large metropolitan city. The store employs a permanent and part-time staff of approximately 120 persons, and is open to the public from 10:00 a.m. to 10:00 p.m. on a six-day-per-week basis. The store is closed on Sundays. Full-time employees work on a shift basis that requires each employee to work on two Saturdays each month, with an equivalent day off during the week that a Saturday shift was scheduled.

Mary K. was first employed by Mall Merchandising Co. in 2000. She was a good employee and had worked in the giftwares department since 2001. Her attendance record was above average, and she seldom lost time due to illness, Until very recently, she willingly worked her Saturday shift.

Some months ago, Mary K. joined a religious denomination that considered Saturdays as their holy day. Saturdays were devoted to religious activities at the denomination's place of worship, and Mary K. refused to work her scheduled Saturday shifts. The store manager was made aware of Mary K.'s refusal to work, and she informed Mary that no employee could be excused from the Saturday work, as it would be unfair to all of the other employees who were required to work. When Mary K. refused to comply with the manager's direction to work on Saturdays, the manager discharged her for insubordination. Mary K. was paid her wages and benefit entitlement up to and including her last day of work.

Advise Mary K. of her rights at law and the course of action (if any) that she might take. If she should decide to take action against her employer, what defences (if any) might the employer raise? How would you expect the case to be resolved?

CASE 6

Jim and Emmanuel were engineers with a mining company that frequently sent employees to remote job sites for long periods of time while a new property was being developed. Because of the time during which employees were required to stay at the remote sites, the company paid its employees extremely well and attracted a large number of unmarried young men to both its professional and labouring positions.

The company had discovered, however, that it had difficulty attracting and keeping more senior and experienced personnel at the job sites. These employees were always in demand for the supervisory tasks and inevitable problems that required the expertise of experienced professionals. After discovering that one of the main factors that discouraged senior employees from accepting such positions was their reluctance to be absent from their families, the company offered to pay all costs of a return flight home once each month for married staff at remote sites.

When Jim and Emmanuel were interviewed for positions with the company they were told of this policy for married staff. Neither man was married and, when both were offered positions a short time later, they understood that they would not be eligible to receive the travel allowance.

One evening Jim and Emmanuel were talking with some of the other employees at the job site about the travel policy. A colleague said that he would like to be able to get home to see his ill father more often, but that he could not afford the cost of a commercial flight on a regular basis and, since he was not married, he was not entitled to the company allowance. Overhearing this, one of the more senior employees commented, "I don't think that policy is very fair. Come to think of it, isn't there some law against that? You guys should go talk to management about it. They're pretty good guys and might make some changes."

Discuss the legal issues raised in this case and the arguments that the respective parties might raise. Explain the factors that a court would consider and what its decision might be.

CASE 7

Calculators Inc. required the services of a bookkeeper for its operations. While it was a very small company, the president believed that the company would grow quickly, and the appropriate employee for the job would be someone who was a qualified professional accountant. He engaged the services of Headhunters Inc., an executive-placement firm, to find a suitable candidate for the position. In describing the position, however, he tended to describe the work and responsibility of the position in terms of what it might be in the near future, if his view of the firm's growth materialized.

Headhunters Inc. staff advertised the position as it was described to them. They received a response to the advertisement from Ms. Take, a chartered accountant, who was currently employed by a large public accounting firm located in a nearby city. The parties met in the nearby city and, following the interview, Ms. Take accepted the position with Calculators Inc. She resigned her position with the accounting firm, paid out the balance of her lease of her apartment, then moved the 90 kilometres to the city where Calculators Inc. was located.

On arrival at her new job, Ms. Take discovered that Calculators Inc. had a workforce of only 40 employees, and her job as head of the accounting and finance department was essentially that of a bookkeeper, since she was the only employee in the department.

Discuss the issues raised in this case and advise Ms. Take of her rights (if any) at law. If you would advise legal action, comment on the possible outcome of the case.

CASE 8

Mario had been well known as a racing-car driver for many years. He had driven a number of different types of racing cars during his racing career, either under the sponsorship of automobile manufacturers or as an "independent."

Following a spectacular race in which Mario had won first prize, a sports-car distributor offered him a position in his organization as director of marketing. The offer included a starting salary of $80,000 per year, participation in a profit-sharing plan open only to senior management, a generous pension plan, and a variety of other benefits including the use of a company-owned car. The position also gave him a place on the board of directors of the company.

As a part of his duties, Mario was expected to enter and drive company racing cars in a number of highly publicized race events held each year.

During his first year with the company, Mario won four of the five races that he had entered and worked hard at all other times to boost sales for the company. As a result of his efforts, sales increased by 20 percent. Mario received a year-end bonus of $10,000 in addition to his salary and share of profits, and was advised by the president of the company that his salary for the next year would be raised to $100,000.

The second year of Mario's employment did not match the previous year. Mario won only two of the five races, and, in spite of spending extra time promoting the employer's sports cars, sales decreased by 3 percent. At a year-end directors' meeting, a bitter argument occurred between Mario and the company president over the poor sales performance of the company. The president blamed the drop in sales on Mario's poor showing on the race circuit, and Mario blamed the unreliability of the new model of the car for his poor performance. The argument ended with the president dismissing Mario.

The next day, Mario received a cheque from the company to cover his salary to that date, along with a formal notice of his termination.

Prior to his dismissal, Mario had arranged for a two-week holiday in Europe. He decided to follow through with these plans, then look for other employment on his return.

Mario searched diligently for a similar position when he returned from his holiday, but could find nothing. Eventually, some six months after his termination, he found a position as a staff writer for a sports car magazine at an annual salary of $48,000.

Under the terms of his employment at the time of his dismissal, Mario was entitled to receive (in addition to his salary of $100,000 per year) pension contributions by the company on his behalf of $6,000 per year, director's fees of $5,000 per year, a profit-sharing plan payment of approximately $6,000, and the use of a company car.

He eventually brought an action for wrongful dismissal against the company and, in addition, alleged damage to his reputation as a professional driver as a result of his summary dismissal by his employer. He claimed $100,000 for damage to his reputation.

Indicate the arguments that might be raised by the parties to this action. Discuss the factors that would be taken into consideration by the court.

Render a decision.

CASE 9

Import Cars Ltd. employs a work force made up of the following employee groups:

6 new car sales staff (New Car Department)
3 used car sales staff (Used Car Department)
3 customer service advisors (Service Department)
8 technicians (Service Department)
4 service employees, unskilled (Service Department)
2 maintenance employees (Building Maintenance Department)
4 collision repair technicians (Auto Body Shop Department)
3 office staff (Office Department)

Each department has a manager, who along with the corporation's president and secretary-treasurer, make up the 'management team.' All of the departments are located in and operate out of a main building except for the auto body shop, which is located in another part of the city about a kilometer distant from the main building.

Representatives from the Auto Technicians Union are currently attempting to interest the 8 technicians in joining a union, and an Auto Employees Union also wishes to organize all of the employees as a local union. Only 3 of the technicians and none of the auto body shop technicians or office staff are interested in joining a union. The new and used car staff, however, has expressed an interest in joining a union.

Outline the steps that the unions must take if they wish to form a union at Import Cars Ltd. What opposition might each union encounter in its membership drive? What steps might the employer lawfully take?

Discuss.

CASE 10

Nigel was employed as a taxi driver by the Rapid Cab and Cartage Company. On May 23, 2011 while driving his taxi, he was involved in a serious collision with a train at a level crossing. The taxi was demolished as a result of the accident, and three passengers riding in the rear seat of the taxi were seriously injured. Nigel, by some miracle, escaped injury.

An investigation of the accident revealed that Nigel had been racing the train to the level crossing and had collided with the side of the engine when the train and the vehicle reached the crossing at the same instant. As a result of the investigation, Nigel was charged with criminal negligence and released on bail pending his trial.

The employees of the company worked under a collective agreement and were represented by a truck drivers' union. The company manager and the union representatives met on May 27 to discuss Nigel's accident. At the request of the union, the company agreed to allow Nigel to continue to drive until his trial.

Nigel had been employed by the company for five years prior to the accident, and had an accident-free driving record until April of that year. During April, Nigel was involved in four minor accidents that were clearly his fault. On May 9 (only two weeks before the accident on May 23) he had crashed his vehicle into the side of the taxi garage, causing extensive damage to both the vehicle and the building. On that occasion, and the two previous occasions, he had been given verbal warnings by the supervisor that he would be dismissed if he continued to drive in a careless manner.

Nigel's case came before the courts on June 10. He was convicted and given a six-month jail term. His driving privileges, however, would be reinstated on his release. Following his conviction, the union, on his behalf, arranged for a six-month leave of absence from the company, subject to the right of the company to review the matter on his return to work.

Before Nigel was released from jail, several large damage claims were made against the company as a result of the accident, and the insurer expressed concern over Nigel's accident-frequency rate. The company management thereupon notified the union of its intention to dismiss Nigel, and advised Nigel that his services would no longer be required on his release.

Upon receipt of the notice of dismissal, Nigel immediately filed a grievance through the union, requesting reinstatement. The grievance was filed in accordance with the time period and procedure outlined in the collective agreement.

The collective agreement contained the following clause:

12.01 The company shall have the right to establish reasonable rules of conduct for all employees, and shall have the right to discipline or discharge employees for just cause, subject to right of grievance as set out in this agreement.

The company had posted the following rule on the office bulletin board in September 2010:

Rule 6 Any failure on the part of a driver to place the comfort and safety of passengers before his (or her) own convenience shall be cause for discipline or dismissal.

As the sole arbitrator in this case, how would you deal with this grievance? Prepare an award and give reasons for your decision.

CASE 11

Haden Manufacturing Ltd. was a large producer of automotive parts and supplies and employed a workforce of over 400 employees. The employees had been represented by the local of a large international union for several years.

An unexpected upturn in consumer demand forced Haden to schedule additional production shifts to meet backlogged orders. This necessitated most employees working at least one overtime shift per week, on either Saturday or Sunday, for which the collective agreement provided that employees be paid time-and-a-half wage rates. When the new work schedule was posted, Susan, who was a parts painter, found herself scheduled to work the overtime shifts on Saturdays. Susan belonged to a religious order that observed Saturday as its day of worship and, as a result, Susan did not wish to work this particular shift.

She went to see the production scheduler to explain her situation and to suggest that he schedule her for the Sunday shift but at the normal, non-overtime rate of pay. Susan felt that her offer was a fair compromise and would be accepted by management. Her suggestion was, in fact, quite acceptable to management, which promised to accommodate her request. The union shop steward was informed of Susan's request and the arrangement which she had made with management. Shortly thereafter the union sent a memo to management stating that it would not permit management to schedule Susan for the Sunday shift without paying her the time-and-a-half rate stipulated by the collective agreement. When management approached Susan about the memo, she stated that she knew nothing about it and that it had not been sent at her request. She also stated that her offer still stood since that arrangement worked out very well for her.

Management took the position that if the union would not waive the overtime pay provision for Susan on Sunday, thereby effectively barring her from that shift, it would have no choice but to require her to work on the Saturday shift. Susan refused to do this and was subsequently dismissed by Haden.

What recourse and rights, if any, does Susan have against the parties in this case? Discuss the arguments each might employ and what remedies might be imposed by a court or arbitrator.

CASE 12

Gear Manufacturing Company carries on business in a part of a factory building in an industrial park located at the outskirts of a large municipality. The remainder of the building is leased by Gear Warehousing Company, a wholly owned subsidiary of Gear Manufacturing Company. Gear Warehousing Company is essentially the storage and marketing subsidiary of Gear Manufacturing Company. It purchases and markets all standard types of gears manufactured by the parent company, even though it is a separate entity.

The employees of Gear Manufacturing Company are represented by the Gear Makers' Union. In the previous year, the union negotiated a collective agreement that expired some months ago. Collective bargaining took place before the expiry of the old agreement, but Gear Manufacturing Company and the union could not agree on the terms of a new collective agreement. They requested conciliation services offered by the Ministry of Labour (which were required before a strike or lockout could take place), but the services failed to produce an agreement. Eventually, the employees went out on strike and set up picket lines at the entrances of the plant of Gear Manufacturing Company. They also set up picket lines at the entrance to Gear Warehousing Company to prevent the shipment of goods from the warehouse.

A few days later, the employees set up a picket line at Transmission Manufacturing Company, an important customer of Gear Warehousing Company, even though the company's only connection with Gear Manufacturing and Gear Warehousing was as a purchaser of Gear products. The pickets prevented Transmission Manufacturing from shipping a large truckload of transmissions to another manufacturer. As a result, the company suffered a loss of $5,000 through its failure to make its delivery on time.

Advise Gear Warehousing Company and Transmission Manufacturing Company of their rights (if any). Suggest a course of action that they might take.

The Online Learning Centre offers more ways to check what you have learned so far. Find quizzes, Weblinks, and many other resources at www.mcgrawhill.ca/olc/willes.

PART 5

THE LAW OF PROPERTY

CHAPTER 20

The Law of Bailment

Chapter Objectives

After study of this chapter, students should be able to:

- Understand the nature of the bailment relationship.
- Identify the various forms of bailment.
- Describe the responsibilities of the parties to a bailment.
- Explain the special status of carriers and storage facilities.

YOUR BUSINESS AT RISK

The business world is filled with grey areas of uncertain responsibility — times when your business owns assets held by someone else (goods given over to a carrier, or to a warehouse), or when you hold goods owned by others — for repair or storage or resale. Before you sell unclaimed items in your hands, or sue to recover goods that may no longer be yours, your business must understand the law of bailment.

NATURE OF BAILMENT

Bailment
The transfer of a chattel by the owner to another for some purpose, with the chattel to be later returned or dealt with in accordance with the owner's instructions.

Bailor
The owner of a chattel who delivers possession of the chattel to another in a bailment.

Bailee
The person who takes possession of a chattel in a bailment.

Bailment is a special arrangement between a person (a **bailor**) who owns or lawfully possesses a chattel, and another person (a **bailee**) who is then given possession of the chattel for a specific purpose. Many business activities involve the transfer of possession of chattels. For example, the leasing of trucks or automobiles by a business constitutes a bailment transaction because the ownership of the vehicles remains with the leasing company, and the persons who have the use of the vehicles have possession only under the lease agreement and not legal title. Similarly, when goods are shipped by a common carrier, such as a highway truck transport company, the carrier of the goods has possession of them only as a bailee. The carrier assumes responsibility for the goods until they are delivered to the designated receiver. Other examples of business transactions that include a bailment would be those transactions that require goods to be left with repair services, such as motor vehicle repair garages, jewellery shops, and appliance repair services.

By definition, bailment consists of the delivery of an article on the express or implied condition that the article will be returned to the bailor or dealt with according to the bailor's wishes as soon as the purpose for which the article was bailed is completed.[1]

A bailment consists of three elements:

(1) the delivery of the goods by the bailor;
(2) possession of the goods by the bailee for a specific purpose;
(3) return of the goods to the bailor at a later time, or the disposition of the goods according to the bailor's wishes.

1. Osborn, P.G., *The Concise Law Dictionary*, 4th ed. (London: Sweet & Maxwell Ltd., 1954), p. 45.

Sub-bailment

Under certain circumstances, a second or sub-bailment may take place. In this case the bailee becomes the sub-bailor, and the person who takes delivery of the goods from the sub-bailor becomes the sub-bailee. Sub-bailment, however, normally must be only by special agreement between the bailor and bailee, or be a custom or practice of the trade relating to the particular type of bailment. The right to make a sub-bailment is not a part of every trade activity. However, the courts have held that bailments involving automobile repairs, the carriage of goods, or the storage of goods are trade activities in which a sub-bailment may customarily be made by the bailee.[2]

In each case, however, the right of sub-bailment may only be made if the bailor is not relying on the special skill of the bailee to perform the work or service. If the bailee makes a sub-bailment under such circumstances, then the bailee would do so at his or her own risk. If a sub-bailment is permissible, either by custom of the trade or by express agreement, the terms of the sub-bailment must be consistent with the original bailment; otherwise it will have the effect of terminating the original bailment. The bailor will then have a right of action against the bailee if the bailee cannot recover the goods from the sub-bailee. In addition, the bailee may be liable to the bailor for any loss or damage to the goods while in the hands of the sub-bailee.

Bailor–Bailee Relationship

Because the essence of a bailment is the delivery of possession of a chattel by one person to another, delivery must take place before the bailor–bailee relationship comes into existence. When the goods are physically placed in the hands of the bailee by the bailor, delivery is apparent. For example, if Smith delivers a book to Jones on the condition that it be returned at a later time, the transfer of possession creates the bailment. The element of delivery becomes less clear, however, if the bailee takes only constructive possession of the goods.

A businesswoman enters a restaurant and places her coat on a coat rack located beside her table. Has she created a bailment by the act of placing her coat on the rack that the proprietor has obviously placed there for that specific purpose?

The basic requirement for a bailment is delivery of possession. If the coat has not been placed in the proprietor's charge, then no bailment may exist. Clearly, if the bailee is unaware of the delivery, and has not consented to it, there may be no bailment. But if the proprietor has either expressly or impliedly requested that the coat be placed upon the rack, then a bailment may have been created by the proprietor's actions. The proprietor under such circumstances may be said to have constructive possession of the goods.[3]

An important characteristic of a bailment is the retention of the title to the goods by the bailor. The bailee receives possession only, and at no time does the title to the goods pass. The rights of the bailee, nevertheless, once delivery has taken place, are much like those of the owner. The bailee has the right to institute legal proceedings against any person who interferes with the property or the bailee's right of possession, even though the bailee does not have the legal title to the goods. The bailee may also recover damages from any person who wrongfully injures the goods, but the money recovered that relates to the damage must be held for the bailor.

The third aspect of bailment is the return of the goods or chattel to the bailor, or the disposition of the goods according to the bailor's directions. The same goods must be returned to the bailor, except fungibles, which are interchangeable commodities such as grain and other natural foodstuffs, fuel oil, gasoline, or similar goods that are frequently stored in large quantities in elevators or tanks. Fungibles of the same grade or quality, and in the same quantity, must be returned in that case. If the bailee refuses to return the bailed goods, the bailor is entitled to bring an action against the bailee for conversion.

2. *Edwards v. Newland*, [1950] 1 All E.R. 1072.

3. *Murphy v. Hart* (1919), 46 D.L.R. 36.

Liability for Loss or Damage

Bailment is a very old area of the law that dates back to the Middle Ages in England. It developed because of the problems that persons had when they entrusted goods to others, or when their goods were at the mercy of a stranger, such as an innkeeper. As the law developed, the courts determined different standards for each of the forms of bailment they were required to consider. As a result, the liability of a bailee for loss or damage to goods while they are in the bailee's possession varies significantly from one type of bailment to another. There are many different general bailment relationships that the courts recognize, and the liability of the bailee differs for each.

Regardless of the standard of care fixed for a bailee, if the bailor can establish that the bailee failed to return the goods, or if the goods when returned were damaged or destroyed (reasonable wear and tear excepted) if the bailee was entitled to use the goods, then the onus shifts to the bailee. He or she must satisfy the court that the standard of care fixed for the particular kind of bailment was maintained, and that the loss or damage was not a result of his or her culpable negligence.

The reason for the placement of the onus on the bailee to show that he or she was not negligent, rather than the normal legal practice of requiring the plaintiff to prove the defendant negligent, is based upon the respective knowledge of the parties. While the goods are in the hands of the bailee, only the bailee is likely to know the circumstances surrounding any damage to the goods. The bailor during this period of time would be unlikely to have any knowledge of how the loss or damage came about, and the courts have, accordingly, recognized this fact. If the bailee is unable to offer any reasonable explanation for the loss, or is unable to establish that negligence did not occur, then responsibility for the loss is likely to fall on him or her. In this sense, the bailor's position is much like that of a person claiming *res ipsa loquitur* in an ordinary tort action: the bailor need only prove the existence of the bailment, and the subsequent loss. The onus then shifts to the bailee to satisfy the court that he or she was not negligent.

Because of the obligation imposed upon bailees to maintain a relatively high standard of care in most bailment relationships, it is not uncommon for bailees to attempt to limit their liability in the event of loss. The usual method used by bailees to limit their liability is to insert a clause that is known as an exculpatory clause in the bailment agreement. An **exculpatory clause** (or exemption clause, as it is sometimes called), if carefully drawn and brought to the attention of the bailor before the bailment is effected, generally has the effect of binding the bailor to the terms of the limited liability (or no liability at all) as set out in the clause.[4] Recent cases, however, have tended to reduce the protection offered by exemption clauses. If the clause is so unreasonable that it amounts to a clear abuse of freedom of contract, the exemption may not be enforced.[5]

Exculpatory clause

A clause in a contract that limits or exempts a party from any liability for damage to the goods.

CLIENTS, SUPPLIERS *or* OPERATIONS *in* QUEBEC

A bailment in Quebec would be treated as a *lease of moveables* under the *Civil Code*.

TYPES OF BAILMENT

Gratuitous Bailment

A gratuitous bailment is a bailment that may be for the benefit of either the bailor or the bailee, or both, and that, as the name implies, is without monetary reward. In the case of gratuitous bailment, the liability for loss or damage to the goods varies with the respective benefits received by the parties to the bailment, unless the parties have fixed the standard

4. *Samuel Smith & Sons Ltd. v. Silverman* (1961), 29 D.L.R. (2d) 98.
5. *Gillespie Bros. & Co. Ltd. v. Roy Bowles Transport Ltd.*, [1973] Q.B. 400; *Davidson v. Three Spruces Realty Ltd.* (1978), 79 D.L.R. (3d) 481.

of care by agreement. If the bailment is entirely for the benefit of the bailor, such as when the bailee agrees to store the bailor's canoe or sailboat without charge during the winter months, then the bailee's liability is minimal. The bailee in such a case is only obliged to take reasonable care of the goods by protecting them from foreseeable risk of harm. The actual standard, unfortunately, appears to vary somewhat, depending upon the nature of the goods delivered. Some years ago, in an English case, the court held that the bailee, in a gratuitous bailment that was entirely for the benefit of the bailor, would only be liable for gross negligence — a degree of carelessness that would constitute neglect in the eyes of a reasonable person.[6] More recent cases, however, have tended to require a somewhat higher standard of care in similar circumstances.[7]

Conversely, if the bailment is entirely for the benefit of the bailee (for example, when a bailor gratuitously loans the bailee his automobile), the bailee would be liable for any damage caused to the goods by the bailee's negligence, reasonable wear and tear being the only exception.

The liability for loss or damage to the goods tends to fall between the two extremes when the bailment is for the benefit of both the bailor and the bailee. For example, Adams stores his sailboat at Burley's cottage, and grants Burley permission to use the boat if he wishes to do so. If the boat should be damaged, the standard that might apply would be that of the ordinary prudent person, and how that person might take care of his own goods.[8]

CASE IN POINT

An automobile mechanic was employed by an automobile dealer under an employment agreement that required the mechanic to supply his own tools. During off-hours the tools were kept in a locked cabinet on the dealer's premises. A number of months later, the employee was injured while on the job, and unable to work. The nature of the injury was such that the mechanic would be unable to return to work for a lengthy period of time, and so the employer terminated his employment. The mechanic requested his tools, but they were not found on the dealer's premises, although a wooden box containing a few of the tools was found in another part of the premises, where the public had access during working hours. The mechanic took legal action against the employer for the loss of the tools, and the court held that the employer, by moving the tools, had taken possession of them as a bailee, and had a duty to take reasonable care of them while they were in his possession. The court held that the employer failed to take reasonable care of the tools and was liable for their loss.

See: *Schatroph v. Preston Chevrolet Oldsmobile Cadillac Ltd.* (2003), B.C.S.C. 265, and *Besser v. Holiday Chevrolet Oldsmobile (1983) Ltd.* (1989), 61 Man. R. (2d) 161.

Licence

The right to use property in common with others.

An important distinction is that between a bailment and a **licence**. While a bailment requires the transfer of possession and a Common Law duty of safekeeping and return, the licence (in this context) lacks either or both of the exclusivity of possession, and the responsibility for protection.

Bailment:

Jane: Can I leave my car in your garage for the weekend?
Tom: Sure.
Jane (handing the keys to Tom): Thanks.

Licence:

Jane: Can I leave my car in your driveway for the weekend?
Tom: Sure.
Jane (putting keys in her purse): Thanks.

---

6. *Martin v. London County Council*, [1947] K.B. 628.
7. *Desjardins v. Theriault* (1970), 3 N.B.R. (2d) 260.
8. *Roy v. Adamson* (1912), 3 D.L.R. 139; *Chaing v. Heppner* (1978), 85 D.L.R. (3d) 487.

Licences are often gratuitous, as in the example on the previous page, or encountered as an ancillary part of another contract. Such cases would include a tenant's use of the parking area around an apartment block, where the landlord has no control over the chattel and little ability to protect it.

Consignment sale

The delivery of a chattel to another person with instructions for its sale.

In the case of a **consignment sale**, an owner delivers a chattel to another person with instructions for its sale. The transaction creates a bailment rather than a licence, with the responsibility not only for protection, but disposal as well. Rather than being a gratuitous bailment, it is for a reward, being a fee deducted from the proceeds of sale. It may be difficult to determine what proportion of the fee deducted is attributable to the sales agency and what is attributable to the storage cost and safekeeping unless the parties make some express provision. They may make this explicit in case the item goes unsold, for example: "Consignment Fee: the greater of 10% of proceeds, or $X.00 per month until sold." For the duration of a consignment the consignor (bailor) retains title, and the consignee (bailee) has possession, until both are delivered to the purchaser. After the purchaser makes payment, the consignee/bailee then has an obligation to account to the consignor/bailor for the proceeds of sale.

Bailment for Reward

Bailment for reward includes a number of different bailment relationships. The bailment may be for storage (or deposit), such as in the case of a warehouse operator, or it may take the form of the delivery of goods to a repair shop for repairs. It may also take the form of a rental of a chattel, the carriage of goods, or the pledge of valuables or securities as collateral for a loan. It would apply, as well, to the safekeeping of goods by an innkeeper. Again, the liability of each of these particular bailees varies, due to the nature of the relationship that exists between each type of bailee and the bailor.

Storage of Goods

The storage of goods for reward may take on many forms, but each represents a bailment if possession and control of the goods pass into the hands of the party offering the storage facility. The bank or trust company that rents a safety deposit box to a customer, the marina that offers boat storage facilities, or the warehouse operator who offers storage space for a person's furniture are bailees for reward. So, too, are the operators of grain elevators, fuel-storage facilities, and parking lots, if the parking lot operator obtains the keys to the vehicle.

Warehouse Storage

The bailee is expected to take reasonable care of the goods while they are in his or her possession, and the standard is normally that which would be expected of a skilled storekeeper.[9] In other words, the bailee would be expected to protect the goods from all foreseeable risks. If the goods have a particular attribute that requires special storage facilities, and the warehouse operator holds himself or herself out as possessing those facilities, then the failure to properly store the goods would render the warehouse operator liable for any loss.

If a company holds itself out as the operator of a cold-storage warehouse, and a bailor delivers to it a quantity of frozen meat that requires the temperature of the goods to be held at some point below freezing, the failure to store the meat at the temperature would render the company (as bailee) liable for any loss if spoilage should occur.

It is important to note that the liability of a bailee for storage is not absolute. The bailee is generally only liable if he or she fails to meet the standard of care fixed by the courts for the nature of the business that the bailee conducts. The bailee firm may be liable for the negligence of its employees if the goods are damaged through their carelessness. However, the courts are unlikely to hold the bailee responsible in cases where the loss or damage could not, or would not, have been foreseen by a careful and vigilant shopkeeper.[10]

9. *Brabant & Co. v. King*, [1895] A.C. 632.

10. *Bamert v. Parks* (1964), 50 D.L.R. (2d) 313.

COURT DECISION: Practical Application of the Law

Bailment—Standard of Care—Liability

***Bexley v. Salmon's Transfer Ltd.*, 2004 BCPC 266.**

Bexley contracted with the defendant to move and store his household goods. The items were stored for a period of about three years. Among the items were three leather sofas and a wall unit, all of them from a high end furniture retailer. They were taken out of storage and delivered to Bexley's new residence and found to be in very poor condition, as if they had been exposed to moisture and then allowed to dry while in storage.

THE COURT: This case involves the law of bailment. The Claimant, as bailor, has delivered his personal goods to the Defendant, as bailee, on trust that they will be returned in their original condition at the end of the bailment. It is said that the legal relationship of bailor and bailee can exist independent of contract, and is created by the voluntary taking into custody of goods which are the property of another. In this case, however, the bailment was not a voluntary acceptance for safekeeping but under a contract with payment to the bailee. In legal terms, it was a bailment for reward.

The law is that where the bailee receives payment for its services, if the bailor pleads that the goods are damaged or lost, the onus of proof rests with the bailee to disprove negligence. The bailee must establish that it exercised due and reasonable care of the goods. As noted by Warren J. in *Tech-North Consulting Group Inc. v. British Columbia*, [1997] B.C.J. No. 1148, "if a thing entrusted to a bailee for reward is lost, then the burden of proof is on the bailee to show that the loss was not the result of his failure to take such care and diligence as a prudent and careful person would take in relation to his own property. But a bailee for reward is not an insurer and in the absence of negligence, a bailee is not liable for a loss which was the result of acts of third parties." For cases where the goods are alleged to be returned in a damaged condition, "the burden of proof is on the bailee to show that the damage was without his fault, or if fault there was it was excused by the exempting clause": *Spurling v. Bradshaw*, [1956] 2 All ER 121 at 125 per Denning J. cited with approval by Southin J. in *BCFP Coast Sawmills Ltd. v. Tritow Systems Ltd.*, [1989] B.C.J. No. 1473.

After considering all the evidence, I conclude that the Defendant has not discharged its onus of proving that the sofa and loveseat were handled with all reasonable care and that the damage is caused by normal wear and tear and not by any act or omission on the Defendant's part. I turn next to the issue of damages. I accept that the damage could not be economically corrected by recovering the sofa and loveseat. On the evidence, the replacement cost of the sofa alone would be in excess of $10,000. I award the Claimant $10,000 for the cost of a replacement sofa. (*ed. — the maximum jurisdiction of the court was $10,000*).

How could the defendant prove, in any case, that it had exercised all reasonable care?

The extent of a bailee's liability for loss arose in a case[11] where a fur-storage company agreed to take a valuable fur coat into storage. The coat was subsequently stolen from the company premises by thieves who broke into the building. The company had taken precautions to secure the building from forced entry. The issue before the court was whether the company, as a bailee, was liable for the loss in spite of the safety precautions taken. The judge hearing the case made the following comment on the standard of care required:

> There can I think be no doubt that the Mitchell Fur Co. Ltd. was a bailee for reward and as such owed a duty to the bailor to take such due and proper care of the bailed goods as a prudent owner, in similar circumstances, might reasonably be expected to take of his or her own goods. In other words, a bailee must, in safeguarding the property of a bailor, do what a reasonably prudent person would do to safeguard his or her own property. And the onus is upon the bailee to show that any loss which occurs did not result from his neglect to use the required degree of care and diligence. However, a bailee is not an insurer. Therefore, if bailed goods are stolen and the bailee can establish that he took reasonable care of the goods, and that he was not guilty of any negligence which was the proximate cause of their loss, the bailee will not normally be held liable.

11. *Longley v. Mitchell Fur Co. Ltd.* (1983), 45 N.B.R. (2d) 78.

In another case,[12] where the standard of liability of the bailee for reward was in issue, the judge reviewed the authorities and established the standard of care in the following manner:

> The question to be determined here is whether any responsibility for the loss of the watch rests with Heppner as a result of his accepting, for reward, the possession of the watch for repair, there being no contract with special provisions and conditions between the parties.
>
> In *Heriteau et al. v. W.D. Morris Realty Ltd.*, [1944] 1 D.L.R. 28, [1943] O.R. 724, it was held that where goods are lost or damaged while in a bailee's possession, the onus is on him to prove that it occurred through no want of ordinary care on his part.
>
> *Leck et al. v. Maestaer* (1807), 1 Camp. 138, 170 E.R. 905, states that a workman for hire is not only bound to guard the thing bailed to him against ordinary hazards, but likewise to exert himself to preserve it from any unexpected danger to which it may be exposed and further, in effect, that where there is need for precaution, the defendant would be answerable for defects of this deficiency.
>
> In essence, what emerges from the authorities is: that the defendant was a bailee for consideration and owed a duty to the plaintiff to exercise that care and diligence which a careful and diligent man would exercise in the custody of his own goods in the same circumstances; that there being loss of the goods bailed the onus is on the defendant (bailee) to show that he exercised a proper degree of care; that failure to observe that duty is negligence; that the unexpected and accidental destruction of the goods while in the possession of the defendant (bailee) for reward, may still render the defendant liable if during his possession of the goods he did not exercise to a reasonable extent the skill or ability which he held out as an expected duty of his calling; and that while no specific period of time may be arranged for the completion of the work to be done, the defendant must perform it within a reasonable time, and if delay in the performance is caused by the defendant's negligence, he is liable.

CASE IN POINT

Hertz left a vehicle for repair with Suburban, who parked the vehicle in an unsecured lot in front of their premises, while a wire fenced and barbed-wire topped compound existed behind the building. Thieves scaled the fenced lot by night, forced a rear entrance, and entered the building. They broke into the area where keys were stored, selected the labelled key, and stole the Hertz vehicle outside. The court found that the bailee Suburban was not an insurer, but had the burden to show on a balance of probabilities that it used reasonable care and diligence. It further found that Suburban took reasonable steps to safeguard the plaintiff's vehicle and storing the keys of customers' cars in a locked room inside the locked premises was not negligent. The loss was not the result of any negligence on the part of the defendant but the result of the efforts of a truly determined third party who surmounted all of the safeguards.

See: *Hertz Canada Limited v. Suburban Motors Ltd.* (2000), B.C.S.C. 667.

Contracts for the storage of goods frequently involve what is known as a warehouse receipt, or evidence of the contract of bailment. The receipt entitles the bearer to obtain the goods from the bailee. Often the original bailor of the goods sells the goods while they are in storage and, as a part of the sale transaction, provides the purchaser with the warehouse receipt. The presentation of the receipt by the new bailor would entitle him or her to delivery of the goods from the bailee. The bill of lading used by carriers of goods performs a similar function when goods are shipped to a purchaser.

Lien

With respect to goods, it is the right to retain the goods until payment is made.

At Common Law, in the absence of a specific right contained in an agreement, the ordinary bailee for storage is not entitled to retain the goods until storage charges are paid. However, all provinces have passed legislation that provides for statutory **lien** that

12. *Chaing v. Heppner et al.*, (1978) 85 DLR (3d) 487.

may attach to the goods in the warehouse operator's possession.[13] The legislation generally provides that the warehouse operator may retain the goods until payment is made, and may sell the goods by public auction if the bailor fails to pay the storage charges. The statutes generally require special care be taken by the bailee with respect to notice and advertisement of the sale to ensure that the bailor has an opportunity to redeem the goods. The statutes also require that the sale of the goods be conducted in a fair manner. The right to a lien, however, is based upon the possession of the goods by the bailee. If the bailee voluntarily releases the goods to the bailor before payment is made, the right to claim a lien is lost.

Parking Lots

The bailment of a motor vehicle for the purpose of parking the vehicle represents one of the most common short-term bailment relationships. However, it is important to distinguish the true bailment of an automobile from the mere use or rental of space for parking. Again, the transfer of possession by the driver of the vehicle to the parking-lot operator is essential to create the bailment. If the operator of the lot accepts the keys to the automobile and parks the vehicle, a bailment is created. The operator has possession of the bailor's property. Similarly, if the operator of the parking lot directs the person to place the vehicle in a certain place on the parking lot, and requests that the keys be deposited with the attendant, the deposit of the keys would also create a bailment. The simple act of parking a vehicle as a "favour" for the patron, however, may not create a bailment that would render the parking-lot operator liable as a bailee if the car should subsequently be damaged or stolen.[14]

If the agreement between the parking-lot operator and the patron is one of rental of a space for parking purposes, and if the patron parks his or her own vehicle and retains the keys, possession does not pass from the patron to the operator of the lot. The retention of the keys by the vehicle owner precludes any control over the vehicle by the parking-lot operator and, consequently, a bailment does not arise. In these cases, the courts generally view the transaction not as a bailment, but as an arrangement whereby the parking-lot operator licenses the use of the parking space by the vehicle driver on a contractual basis.[15]

The enforcement of exculpatory clauses arises frequently in cases concerning the bailment of vehicles. The success of a bailee in avoiding liability by way of an exculpatory clause depends in no small measure on the steps taken to bring the limitation on the bailee's liability to the attention of the bailor either before or at the time that the bailment takes place. The simple printing of a limitation of liability on the back of the parking lot ticket is clearly not enough.[16] The limitation must be forcefully brought to the attention of the bailor, either by direct reference to the limitation or by placing clearly marked signs in conspicuous places where they will not fail to catch the eye of the bailor.[17]

COURT DECISION: Practical Application of the Law

Bailment for Reward—Parking Lot—Signs and Ticket Excluding Liability for Loss

***Samuel Smith & Sons Ltd v. Silverman* (1961), 29 D.L.R. (2d) 98.**

The plaintiff parked his car in a parking lot and left the keys with the lot attendant. He received a ticket in return for payment of the parking fee. The ticket contained a statement that the owner of the parking lot was not responsible for damage to the car or its contents. Several large signs at the entrance to the lot contained the same message. When the plaintiff returned to the lot some time later, he found his car damaged. An action was brought against the owners of the parking lot for damages.

---

13. See, for example, *Warehouser's Lien Act*, R.S.N.L. 1990, c. W-2, s. 3; *Warehousemen's Lien Act*, C.C.S.M., c. W20, s. 2.
14. *Palmer v. Toronto Medical Arts Building Ltd.* (1960), 21 D.L.R. (2d) 181; *Martin v. Town N' Country Delicatessen Ltd.* (1963), 45 W.W.R. 413.
15. *Palmer v. Toronto Medical Arts Building Ltd.*, supra.
16. *Spooner v. Starkman*, [1937] 2 D.L.R. 582.
17. *Samuel Smith & Sons Ltd. v. Silverman* (1961), 29 D.L.R. (2d) 98.

THE COURT: A contract of bailment for reward has been made out. At the time of the delivery of the car to the defendant's servant, Sussman [driver/employee of the car's owner, Smith Ltd.] was given a parking ticket containing the following terms: "WE ARE NOT RESPONSIBLE FOR THEFT OR DAMAGE OF CAR OR CONTENTS HOWEVER CAUSED." These terms are spelled out in bold black type and in letters large enough to dispel any suggestion of an attempt on the part of the defendant to conceal the limiting conditions from the recipient. Had the defendant looked at this ticket he could not possibly have failed to see the terms quoted. When a chattel entrusted to a custodian is lost, injured, or destroyed, the onus of proof is on the custodian to show that the injury did not happen in consequence of his neglect to use such care and diligence as a prudent or careful man would exercise in relation to his own property. It is well settled that a custodian may limit or relieve himself from his common law liability by special conditions in the contract, but such conditions will be strictly construed and they will not be held to exempt the bailee from responsibility for losses due to his negligence unless the words of limitation are clear and adequate for the purpose or there is no other liability to which they can apply.

There was evidence given on behalf of the defendant by the defendant's manager who stated that there were four signs erected on the defendant's lot, two of them at the front near the Victoria St. entrance, and the other two on the rear parking lot. They were at a height of approximately 8 to 10 ft. from the ground, 2 1/2 by 3 ft. in dimension, and contained the following words: "WE ARE NOT RESPONSIBLE FOR THEFT OR DAMAGE OF CAR OR CONTENTS HOWEVER CAUSED." The learned judge accepted this evidence. He found as a fact that there were signs on the lot in the four places indicated, which were lighted at the time in question, and which bore the words set out above.

In *Brown v. Toronto Auto Parks Ltd.*, [1955] 2 D.L.R. 525, this Court had to consider a defence based on limiting conditions contained on signs displayed on the custodian's premises. It was held that the exculpatory signs did not assist the defendant, for while the plaintiff should reasonably have seen them, the words "car and contents at owner's risk" did not suffice clearly to relieve the defendant for liability for negligence. The Court applied the strict rule of construction to which I have referred and supported the judgment for the plaintiff on that ground alone. The words printed on the ticket and the signs in question are not susceptible of this criticism. The clear declaration that the defendant was not responsible for theft or damage of car or contents however caused, is sufficiently broad in its terms to extend to a case where the damage occurred through the negligence either of the defendant or his servants, or the negligence or carelessness of a third party whether lawfully on the premises or not.

Given a fully effective exculpatory sign, under what circumstances do you think the court would disregard it, and find liability against the bailee?

Bailment for Repair or Service

When chattels require repair or service, a bailment takes place if the owner delivers the goods to the repair shop and leaves them with the proprietor. The bailee is expected to protect the goods entrusted to him or her for repair. Even though no charge is made for the bailment separate from the repair charge, the bailment is nevertheless a bailment for reward, and the bailee is expected to take reasonable care of the goods while they are in his or her possession. If the goods are lost or damaged while they are in the bailee's possession, the bailee may be liable if the loss is due to his or her negligence. If the goods are sub-bailed to a sub-bailee in accordance with the customs of the trade, then the bailee may also be liable for loss or damage to the goods by the neglect or willful acts of the sub-bailee.

Smith delivers her automobile to Baker for repairs, and Baker, by way of a sub-bailment, places the car in Carter's possession to have some specialized work done. If Carter negligently damages the car while it is in his possession, Baker may be liable to Smith for the damage. However, if Baker has held himself out to be skilled in the performance of the task, and Smith contracts with Baker for personal performance, the sub-bailment would be improper. Baker would become liable for any damage to the goods, whether caused by Carter's negligence or not. The particular reason for this additional liability for an improper bailment is that Smith, in placing the vehicle in Baker's possession, is accepting a particular set of circumstances or risks relating to the repair of her vehicle, but an unauthorized sub-bailment would be a change in the risk without her consent. To protect the bailor in such cases, the courts have simply imposed liability for any loss or damage to the goods on the bailee.

The bailee who professes to have a particular repair skill is expected to execute the repairs in accordance with the standards set for the skill. The bailee is expected to exercise the duty of care attendant with the skill in the protection or handling of the goods while in his or her possession. Should the bailee be negligent in the repair of the goods, or not possess the particular skill that he or she professed to have, the bailor would be entitled to institute legal proceedings for damages to cover the loss suffered. This claim may cover not only the value of the chattel damaged, but any other loss that would flow from the bailee's breach of the agreement to repair, if the loss was foreseeable at the time the agreement was made.

Bailee's Right to Lien

For this service, the bailee is entitled to compensation that may be either agreed upon at the time the goods are placed in his or her possession, or to a reasonable price for the services when the work is completed. If the bailor refuses to pay for the work done on the goods, at Common Law, the bailee has a right of lien and may retain the goods until payment is made. If payment is not made within a reasonable time, subject to any statutory requirements that set out the rights of the bailee, the bailee may have the goods sold (usually by public auction) to satisfy his or her claim for payment.[18]

A QUESTION OF ETHICS

Many service establishments impose conditions that they may sell unclaimed goods after 30 days, or that they will refuse to deliver up the goods without a claim ticket. In some instances the goods may be worth thousands of dollars, particularly for watch repair and the storage of fur coats. Bearing in mind the "parking lot" cases, what advice would you give to business operators to ensure that these conditions actually provide the protection both parties desire? When these conditions must actually be invoked, what steps should be taken to ensure that the process conforms to a business ethic that the court would find reasonable?

Hire or Rental of a Chattel

The hire of a chattel is a bailment for reward in which the bailor-owner delivers possession of a chattel for use by the bailee-hirer in return for a monetary payment. This type of bailment is usually in the form of a written agreement, with each party's rights and duties clearly set out. However, it need not be in writing to be enforceable, unless by its terms it falls subject to the *Statute of Frauds*.

Under a bailment for the hire of a chattel the bailee is required to pay the rental fee for the use of the chattel that the parties have agreed upon; if no fee was specified at the time the agreement was entered into, then the bailee is required to pay the reasonable or customary price for the use of the goods. If the bailment is for a fixed term, the bailee is usually liable for payment for the full term, unless the bailor agrees to take back the chattel and clearly releases the bailee from any further obligation to pay. Apart from the payment of the rental fee, and except for any specific obligations imposed upon the bailee, the bailee is entitled to possession and use of the goods for the entire rental period.

The bailee at Common Law must not use the goods for any purpose other than that for which they were intended. The bailee must not sub-bail the goods, or allow strangers to use them, unless permission to do so is obtained from the bailor. In the event that the bailee should do any of these things, the bailee would become absolutely liable for any loss or damage to the chattels. Otherwise, the bailee will only be liable if he or she fails to use reasonable care in the operation or use of the goods.[19] The bailee would not be liable for ordinary "wear and tear" that may result from use of the chattel unless the agreement specifically holds the bailee responsible. The maintenance of the equipment in fit condition for the use intended is usually the responsibility of the bailor.

18. Most provinces and territories have legislation covering this type of bailment: see, for example, *Mechanics' Lien Act*, R.S.N.L. 1990, c. M-3, s. 45.

19. *Morris v. C.W. Martin & Sons Ltd.*, [1965] 2 All E.R. 725.

Under an agreement for the hire of a chattel, the prime responsibility of the bailor is to provide the bailee with goods that are reasonably fit for the use intended. The goods must be free from any defects that might cause damage or loss to the bailee when the equipment is put into use. If the bailor knew or ought to have known of a defect when the goods were delivered, the bailor may be liable for the damage caused by the defective equipment.

> Lyndsey hired a truck from Foster for the purpose of delivering crates of eggs to market. If Foster knew or ought to have known that the truck had defective brakes and, as a result of the defect, the truck swerved off the road when the brakes were applied and destroyed Lyndsey's load of eggs, Foster would be liable for Lyndsey's loss. However, if the defect was hidden and would not be revealed by a careful inspection, Foster may not be liable.

If the goods hired have an inherent danger or risk associated with their use, the bailor is normally under an obligation to warn the bailee of the danger, or possible dangers, associated with the use. However, if the bailee is licensed or experienced in the use of the equipment, any loss or damage that may result from the use of the equipment may be assessed in part against the bailee.[20]

MANAGEMENT ALERT: BEST PRACTICE

At first sight, a bailee in possession seems to be in a simple situation: having possession, it must give back the chattel to the owner at the end of the contract. But what if ownership has changed in the interim? The bailee can easily find itself giving the property back to someone who is no longer the owner, only to have the now-rightful owner showing up later to make the same claim. It is vital therefore that bailees (particularly those engaged in warehousing or carriage, but also any more casual bailee) clearly set out what evidentiary basis is acceptable for a valid claim of later ownership. Production of the original receipt given on taking possession must be an essential element. If the receipt is transferable, the bailee must ensure that its production represents a complete release of the bailee's liability.

CHECKLIST FOR BAILMENT

1. Delivery of the goods by the bailor.
2. Possession of the goods by the bailee for a specific purpose.
3. Return of the goods to the bailor at a later time is contemplated, or disposition according to the bailor's wishes.
4. If interchangeable commodities, substitution for original is OK.
5. Action for conversion if no return of goods.
6. Is liability limited by a valid exculpatory clause?

Carriage of Goods

The carriage of goods may include a number of different forms of bailment. The carriage of goods involves the delivery of goods by the bailor to the bailee for the purpose of delivery to some destination by the bailee. As with all bailments, the goods are in the possession of the bailee for a particular purpose, but the title is in someone else.

A carrier of goods is normally a carrier for reward, but this is not always the case. A carrier may be a gratuitous carrier who transports goods without reward, such as a person who agrees to deliver a parcel to the post office for a friend. With a gratuitous carrier, if the service provided is entirely for the benefit of the bailor, the bailee is only expected to use reasonable care in the carriage of the goods.

20. *Hadley v. Droitwich Construction Co. Ltd. et al.*, [1967] 3 All E.R. 911.

There are two classes of carriers for reward: private carriers and common carriers. The standard of care differs for each. A private carrier is a carrier that may occasionally carry goods, but who is normally engaged in some other business activity. A company that is a private carrier is free to accept or reject goods as it sees fit. However, if it should decide to act as a carrier of goods for reward, then it would have a duty to take reasonable care of the goods while they are in its possession.

Common carrier

A transportation business that specializes in the transport of goods.

The **common carrier**, unlike the gratuitous carrier and the private carrier, carries on the business of carriage of goods for reward. It offers to accept any goods for shipment if it has the facilities to do so. For example, a trucking company or railway company that engages in the carriage of goods would be classified as a common carrier. Common carriers are to some extent controlled by statute. The statute generally limits the carrier's ability to escape liability in the event that the goods carried are lost or damaged. The common carrier is essentially an insurer of the goods, and is liable for any damage to the goods, except in certain circumstances.

The principal reason for the very high standard of care required of the common carrier is that the goods are totally within the control of the carrier for the entire period of time that the bailment exists. Unlike other forms of bailment, where the bailor could presumably check on the goods, once the goods are in the hands of the carrier they are no longer open to inspection by the bailor until they reach their destination. Under the legislation pertaining to common carriers, the carrier is usually permitted by contract to limit the amount of compensation payable in the event of loss or damage to the goods. The carrier may also avoid liability if the damage to the goods was caused by an act of God, the improper labelling or packing of the goods by the shipper, or if the nature of the goods was such that they were subject to self-destruction during ordinary handling. The carrier would also be exempt from liability if the damage resulted from the actions of the Queen's enemies in time of war.

COURT DECISION: Practical Application of the Law

Bailment for Reward—Common Carrier—Liability for Loss

***Bear Mountain Logging Ltd. v. K.A.T. Lowbed Service Ltd.*, 1998 CanLII 6730 (B.C.S.C.).**

The plaintiff owned logging equipment which fell off a lowbed while being transported by the defendant. The defendant is a public transport service and moves equipment for hire. In its claim against the defendant, the plaintiff relies on negligence, bailment and breach of contract. The defendant says that it exercised the skill, diligence and judgment expected of it in the circumstances and was therefore not negligent, that the plaintiff did not properly plead bailment.

THE COURT: Damages [to a $450,000 swing yarder] are agreed between the parties at $40,263.11. The hill can be described as a series of switchbacks leading to where the plaintiff was carrying out logging operations, a distance of about six kilometers. Mr. Fontaine [for the plaintiff] telephoned Mr. Lyster [for the defendant] and asked him to pick up the swing yarder the following day, and haul it back down to Menzies Bay. In the telephone conversation, there was some discussion about the steep hill. Mr. Lyster had not been to the site before, but Mr. Fontaine told him about the steep hill. Mr. Lyster expressed some concern about the hill, but Mr. Fontaine told him to "err on the side of safety," and if there was any concern, his yarder operator Earl Haines would "kick it off." By that he meant Mr. Haines could take the machine off the lowbed and walk it down the hill. Mr. Haines walked the yarder onto the flatbed. Mr. Haines asked Mr. Lyster if he was going to chain the yarder onto the lowbed. Mr. Lyster replied that he did not want to chain the yarder down. If the yarder was going to tip over, he wanted it to tip over by itself, and not take his lowbed with it. He knew there was a risk that the yarder could tip over but he was prepared to accept that risk and drive down the hill. As he drove into the corner, Mr. Lyster felt his lowbed start to tilt or swing to the left and then swing back. The lowbed started to tilt again to the left, and this time he felt the entire truck lean over. He stopped the truck expecting the yarder to rock back onto the deck of the lowbed. Instead the lowbed continued to lift and "just kept coming." The yarder fell sideways off the deck of the lowbed onto the ground while the lowbed itself came crashing back onto the road right side up but minus its cargo. While the pleadings are not as clear as they might have been, I must agree with Mr. Bowes [for the plaintiff] when he says that a contract of carriage with a public transport authority is, by its very essence, bailment. The onus of proof is therefore on the defendant to show that the accident and the resulting damage to the plaintiff's machine did not happen as a result of its negligence. If the defendant fails to discharge that onus, the plaintiff is entitled to judgment. I find that the most probable cause of the accident was the speed at which Mr. Lyster turned his lowbed into the curve on the steep hill. I conclude on all of the evidence, and on the basis of common sense, that Mr. Lyster was travelling too fast under the circumstances. On that basis he was negligent. The plaintiff will have judgment against the defendant in the amount of $40,263.11 and costs.

Would an exculpatory clause have helped the defendant in this case? Would it help the defendant enough to avoid liability completely? Why or why not?

Most common carriers are subject to legislation that imposes certain responsibilities on their operation. These statutes usually either set out the liability of the carrier, or set out the carrier's rights and duties in the carriage of goods. In many cases, the rights and duties must be included in the contract of carriage, and these terms are frequently found in small print on the back of the contract. Since separate legislation governs railways, trucking firms, and air carriers, the specific liability tends to vary somewhat for each. The basic liability, however, remains the same.

Under a contract of carriage, the bailor also has certain responsibilities. The bailor is obliged to pay the rates fixed for the shipping of the goods. If the bailor fails to pay, the carrier may claim, or receive under the terms of the contract, the right of lien on the goods until payment is made. If the charges are not paid within a reasonable length of time, the goods normally may be sold to cover the carrier's charges. The bailor is also required to disclose the type of goods shipped, and must also take care not to ship dangerous goods by carrier unless a full disclosure of the nature of the goods is made.

A common occurrence in the carriage of goods is a change of ownership of the goods while in the hands of the carrier. The original bailor is not always the recipient at the destination where goods are to be shipped. Indeed, in most cases, the goods are shipped by the bailor to some other person. The contract with the carrier (sometimes called a bill of lading) names the person to whom the goods are consigned, and the carrier will deliver the goods to the person named as consignee. Goods may be shipped under a second type of contract of carriage, called an order bill of lading. This is essentially a contract, combined with a receipt and document of title, that may be endorsed by the consignee, if the consignee so desires, to some other person. An order bill of lading must be surrendered to obtain the goods from the carrier.

Pledge or Pawn of Personal Property as Security for Debt

Bailment may be associated with debt transactions in the sense that personal property may be delivered to a creditor to be held as security for a loan. The particular personal property may take the form of such securities as bonds, share certificates, or life-insurance policies. These securities may be held by the creditor as collateral to the loan. Because the creditor takes possession of the securities, the transaction represents a bailment, and the creditor as a bailee would be responsible for the property while in his or her possession. When the debt is paid, the same securities must be returned to the bailor. The delivery of securities or similar personal property to the creditor as security for a loan is called a **pledge**. If the bailor–debtor should default on the loan, the bailee–creditor may look to the securities pledged to satisfy the debt. Any surplus from the sale of the securities would belong to the debtor, however, and must be paid over to the debtor by the creditor.

Pledge

The transfer of securities by a debtor to a creditor as security for the payment of a debt.

A **pawn** is similar to a pledge, but is confined to a transaction between a debtor and a pawnbroker. It is concerned with the delivery of goods to the pawnbroker as security for a loan. Pawnbrokers are licensed in Canada. They may accept goods as security under loan agreements that entitle them to sell the goods if default on the debt occurs. While the goods are in the possession of the pawnbroker a bailment exists, and the pawnbroker must take reasonable care of the goods. As with other forms of bailment, the pawnbroker has only possession of the goods, and the bailor-debtor retains the title. The bailment, however, is made on the express condition that the goods may be sold by the creditor

Pawn

The transfer of possession (but not ownership) of chattels by a debtor to a creditor who is licensed to take and hold goods as security for payment of debt.

if default should occur. At that time the creditor may give a good title to the goods to a third party. Any surplus from the sale would belong to the debtor and, conversely, any deficiency would remain as an obligation.

FRONT-PAGE LAW

'Filling a niche' carries a price; Third-generation pawnbroker endeavours to dispel stereotypes that go with the trade

The enduring, if not false, image of a pawnshop is typified by a shady man hunkered down amid a clutter of other people's wedding rings, gold watches and assorted family heirlooms. Whether they're actually pawning an item or just browsing, most people have a difficult time shaking the sense that they're involved in something wrong.

It's this lingering stereotype that Chris Shortt, the president and owner of James McTamney's and Company Inc., is working hard to dispel. "There's the stigma with new customers that they're never going to see their merchandise again," says Mr. Shortt, a third-generation pawnbroker. "We're not selling your stuff," he explains calmly. Pawnbrokers offer quick loans on personal property to a wide variety of customers, he adds.

It works like this: They assess the value of your property and offer you a loan of approximately 5 percent to 20 percent of its actual worth. They then store the property, usually jewellery, watches or VCRs, for a year, during which the customer can return at any time to claim it. To reclaim the item, you have to pay back the loan, plus monthly interest of around 3 percent and a small fee for storage. The large majority of customers do this within the first four months. Only an estimated 10 percent to 12 percent of customers default, the results of which are on sale in the store.

Mr. Shortt has been seeing more and more middle-class family types in his shop, usually double-income folk, caught between paycheques, who need a small loan to make mortgage and bill payments. "It's so expensive to live in this city," he says. "The gap [between rich and poor] is getting bigger. We're just filling a niche that the bank isn't filling. It's quick and easy and there are no forms to fill out. It's just in and out."

You have a good understanding of bailment if you can distinguish this example from the situation where a person enters the shop and sells an item to the pawnbroker. To refine your thinking, consider the pawnbroker's obligations to its bailors if a fire destroyed the shop during the contract period.

Source: Excerpted from Brad Mackay, *National Post*, "'Filling a niche' carries a price," August 3, 2000, p. A23.

INNKEEPERS

The liability of the innkeeper or hotel owner extends back to the Middle Ages in England, to a time when a traveller's goods were at the mercy of the innkeeper. The early English inns (and, for that matter, many of the inns in both Canada and the United States until as late as the nineteenth century) provided a large, single room for sleeping purposes. There were only a few separate bedchambers, so the guests were easy prey for thieves while they slept. The innkeeper was seldom unaware of the pilfering and was often an accomplice in the act. To discourage theft, and to ensure that the innkeeper was not a party to the crime, the Common Law imposed a very high standard of care on the innkeeper with respect to goods brought on the premises by guests. At Common Law, the innkeeper was held to be responsible for any loss, even if it was not the innkeeper's fault. The only exception was when the loss was due to the guest's own negligence. The innkeeper, in effect, was someone who closely resembled an insurer of the goods in the event of loss. This was so in spite of the fact that the goods were not in the innkeeper's possession, and the guests exercised some control over the goods as well.

While the liability of the innkeeper to guests of the hotel was similar to a bailment, the innkeeper's liability at Common Law was essentially based upon "custom of the realm." Later, this was established by statute. This rather unique form of responsibility for the goods of the guest was described by the court to be as follows:

> At Common Law an innkeeper was responsible to his guests if any of their goods were lost or stolen while on his premises. In *Shacklock v. Ethorpe Ltd.*, [1939] 3 All E.R. 372, Lord MacMillan, whose opinion was concurred in by all of the other Law Lords, applies the words of Lord Esher M.R. in *Robins & Co. v. Gray*, [1895] 2 Q.B. 501 at 503, who said:
>
> > The duties, liabilities, and rights of innkeepers with respect to goods brought to inns by guests are founded, not upon bailment, or pledge, or contract, but upon

> the custom of the realm with regard to innkeepers. Their rights and liabilities are dependent upon that, and that alone; they do not come under any other head of law ... the innkeeper's liability is not that of a bailee or pledgee of goods; he is bound to keep them safely. It signifies not, so far as that obligation is concerned, if they are stolen by burglars, or by the servants of the inn, or by another guest; he is liable for not keeping them safely unless they are lost by the fault of the traveller himself. That is a tremendous liability: it is a liability fixed upon the innkeeper by the fact that he has taken the goods in ...[21]

An innkeeper must be distinguished, however, from other persons who offer accommodation to guests, as the special liability applies only to innkeepers. Persons who offer only room accommodation to travellers, or who are selective in offering room and meals (for example, a rooming house), are not usually innkeepers by definition. A restaurant that offers only meals and no sleeping accommodation would not be classed as an "inn." To be treated as an innkeeper, it would appear that a person must offer both meals and room accommodation to the public.[22]

An innkeeper has a public duty to accept any transient person and their belongings as a guest — provided that accommodation exists, and also that the traveller is "fit and orderly" and has the ability to pay. An innkeeper is defined in the Innkeepers Acts of most provinces and territories as a person who offers accommodation and meals to the travelling public. Once the proprietor of the establishment falls within the definition, the liability under the statute also applies. Each province has legislation that sets out the rights and obligations of innkeepers, but, unfortunately, it is not uniform. In general, with respect to the protection of the goods and belongings of travellers, the innkeeper may in most provinces limit his or her liability to a fixed sum, which varies from $40 to $150. This limit applies when the loss or damage is not due to the negligence of, or the willful or deliberate act of, the innkeeper or the innkeeper's employees, or when the goods are not placed in the innkeeper's custody for safe keeping. The legislation in most provinces provides that to obtain the protection of the act, the innkeeper must post the relevant sections in all bedrooms and public rooms in the inn.[23]

Because the innkeeper and the guest to some extent share responsibility for the protection of the guest's goods, the liability of the innkeeper is not absolute. If the innkeeper can establish that the loss of the guest's goods was due entirely to the guest's negligence, the innkeeper may be able to avoid liability.[24] Full liability applies when the goods have been placed in the hands of the innkeeper for safe keeping. In most provinces, full liability will also apply if the innkeeper refuses to accept the goods when requested to do so by a guest.[25]

SUMMARY

A bailment is created by the delivery of possession of a chattel by the bailor (who is usually the owner) to a bailee. Bailment involves the transfer of possession and not title, but a bailee may exercise many of the rights normally exercised by an owner while the goods are in his or her possession. Bailment may be either gratuitous or for reward. Liability is least for a gratuitous bailee who receives no benefit from the bailment. It is highest for special forms of bailment for reward, such as the common carrier of goods, where the bailee is essentially an insurer for any loss or damage. If the agreement between the parties permits a sub-bailment, the bailee may make such a bailment.

21. *Hansen v. "Y" Motor Hotel Ltd.*, [1971] 2 W.W.R. 705.
22. *King v. Barclay and Barclay's Motel* (1960), 24 D.L.R. (2d) 418.
23. Saskatchewan requires only that the sections of the Act be placed in the hall and entrance: *Hotel Keepers Act*, R.S.S. 1978, c. H-11, s. 11. The Manitoba Act does not require the posting of a notice, as the Act does not hold the innkeeper liable for loss except as set out in the statute: see the *Hotel Keepers Act*, C.C.S.M., c. H150.
24. *Loyer v. Plante*, [1960] Que. Q.B. 443; *Laing v. Allied Innkeepers Ltd.*, [1970] 1 O.R. 502; *Hansen v. "Y" Motor Hotel Ltd.*, supra.
25. See, for example, *Innkeepers Act*, R.S.O. 1990, c. I.7, s. 6.

The bailee may also do so in some cases where sub-bailment, in the absence of an agreement to the contrary, may be made by custom of the trade.

Bailment for reward may take the form of bailment for storage, for the carriage of goods, the deposit of goods for repair, the hire of a chattel, and the pledge or pawn of goods to secure a loan. The liability of the bailee in each of the bailment relationships arises if the bailee fails to take reasonable care of the goods while in his or her possession. Innkeepers, under statute, have a special duty to protect the goods of their guests while in the hotel or inn. However, in the case of the common carrier and the innkeeper (who is similar to a bailee), a much higher standard prevails. A bailee may limit his or her liability by an express term in the contract. However, legislation governing such bailees as warehouse operators, carriers of goods, and innkeepers contains specific provisions and limitations that generally govern these special relationships.

Key Terms

Bailment, 386
Bailor, 386
Bailee, 386
Exculpatory clause, 388
Licence, 389
Consignment sale, 390
Lien, 392
Common carrier, 397
Pledge, 398
Pawn, 398

Review Questions

1. In what way or ways is the responsibility of an innkeeper for the safekeeping of the goods of guests similar to that of a bailee for reward?
2. Explain the term "constructive bailment."
3. Explain: (a) fungible, (b) pawn, (c) pledge, and (d) sub-bailment.
4. What rights over a bailed chattel does a bailee possess? Why are these rights necessary?
5. Why should an innkeeper be responsible for the goods of a guest that are brought into the guest's hotel room?
6. Indicate the defences available to a common carrier in the event of loss or damage to goods in the carrier's possession.
7. What standard of care is imposed on a bailor in the hire of a chattel?
8. What essential element distinguishes the rental of space in an automobile parking lot from a bailment of the vehicle? How does this affect the liability of the owner of the parking lot?
9. Define a bailment.
10. Why do the courts impose a greater responsibility for the care of goods on a common carrier than upon a gratuitous carrier?
11. Indicate the effectiveness of an exculpatory clause in a bailment contract for the storage of an automobile. How do the courts view these clauses?
12. To what extent is a bailee for reward entitled to claim a lien for storage costs against the goods?
13. How is the standard of care of a gratuitous bailee determined?

Mini-Case Problems

1. Simple took his power lawn mower to a repair shop to have it repaired. At the shop he signed an "authorization to repair" sheet that directed the repair shop to "repair the engine." The shop did so, but apparently found it necessary to replace most of the internal parts of the engine. The repair bill was $350, almost the price of a new machine. Simple refused to pay the account when he realized the cost of the repairs, and the repair shop refused to release the mower to him.

   Discuss the rights of the parties.
2. Smith entered a clothing shop to purchase a new coat. The clerk was busy, and Smith placed her purse on a table located near the coat racks while she examined several coats for size and fit. Her purse was stolen while she was examining the coats.

   Discuss the question of liability for the stolen purse.
3. Carol checked into the City Hotel, left her suitcase and computer case in her room, and proceeded to the hotel restaurant for lunch. On her return to her room she discovered that her suitcase and computer had been stolen. Advise Carol and the hotel of their rights and responsibilities. How would this case likely be resolved?
4. Daniel was obliged to go abroad on a 6-month work assignment, and arranged with his friend Harry to store his sports car in Harry's garage.

One evening, thieves entered Harry's garage (which was never locked) and made off with the sports car. When Harry discovered the theft, he immediately notified the police. A week later, the police found the vehicle in a seriously damaged condition.

Advise Daniel and Harry.

Case Problems for Discussion

CASE 1

Sleeman had just purchased a new luxury automobile, and decided to use it that evening to drive his wife to a new theatre production in the city centre. As he approached the parking lot adjacent to the theatre, he noticed a long line of vehicles waiting to turn into the lot entrance. The exit lane, however, was clear, and he made the instant decision to enter the lot by the exit lane. Once inside the lot, he stopped his car in order to obtain a parking ticket. An irate lot attendant came over to him, complaining that he was not to enter the lot by the exit lane. Sleeman apologized, and attempted to back out into the street, but at that point, both he and the attendant realized that it would be impossible to do so, given the heavy road traffic.

Exasperated, the attendant took Sleeman's money, obtained a parking ticket from the attendant in the office, and handed it to Sleeman without comment. He then slipped into the driver's seat in order to park the vehicle.

The time of year was late winter, and patches of ice were on the paved lot. The attendant who was now annoyed with Sleeman, accelerated the vehicle down the lot, but was unable to stop the car on the ice. He collided with another vehicle parked on the lot, and seriously damaged Sleeman's vehicle. When the parking lot owner refused to accept responsibility for the damage, Sleeman instituted legal proceedings.

At trial, the parking ticket was introduced as evidence. The ticket contained the following words printed on the back: NOT RESPONSIBLE FOR LOSS OR DAMAGE TO VEHICLES HOWEVER CAUSED.

The lot owner also testified that on the side of the office where vehicles enter, a large metre square sign provided the same message.

Discuss the arguments that each of the parties might raise in this case, and render a decision.

CASE 2

Local Air Ltd. operated a charter airline service that specialized in flying small groups to specified locations. Their fleet consisted of three small eight passenger twin-engine aircraft. While on a charter to a nearby city, one of its aircraft developed a problem with one of its engines, and the pilot, on the authorization of the company, took it to an aircraft repair facility located at the airport.

A second aircraft was dispatched to retrieve the passengers and the pilot when it appeared that the repair would require parts that had to be ordered from the aircraft manufacturer. The parts arrived in due course, but before the engine was repaired, a government air worthiness directive was received by the repair facility that required aircraft owners to make an engine modification in order to receive an airworthiness certificate. In response to the directive, the repair facility made the modifications to both of the engines, then signed off that the aircraft was now airworthy. The repair facility informed Local Air Ltd. of the work done and invoiced Local Air Ltd. for the repair that amounted to $9,825.

When Local Air Ltd. was advised of the repair invoice, they informed the repair facility that in their view the amount was excessive, as the airworthiness repair could have been made at their own shop for half the cost. At this point in the discussion, the repair facility refused to release the aircraft unless the repair account for the aircraft was paid in full.

Advise the parties. What are the issues in terms of the law? Speculate as to the outcome.

CASE 3

Yusef Ali parked his automobile in a parking lot owned by the Dumas Corporation. At the request of the parking-lot attendant, he left his keys at the attendant's office and received a numbered ticket as his receipt for the payment of the parking fee. Before leaving his keys with the attendant, he made certain that the doors of the vehicle were securely locked, as he had left a number of valuable books on the rear seat of the car.

Unknown to Ali, the attendant closed the lot at midnight. Then he delivered the keys to the cars on the lot to the attendant of the parking lot across the street. This lot was also owned by the Dumas Corporation, but it remained open until 2:00 a.m.

Ali returned to the parking lot to retrieve his automobile shortly after midnight, only to discover no attendant in charge and his vehicle missing. By chance, he noticed the attendant on duty at the lot across the street. Ali reported the missing vehicle to him, and found the attendant in possession of his keys.

The police discovered Ali's automobile a few days later in another part of the city. The vehicle had been damaged and stripped of its contents, including Ali's rare books.

Ali brought an action against the Dumas Corporation for his loss. However, the company denied liability on the basis that the ticket (which Ali received at the time of delivery of the keys) read: "Rental of space only. Not responsible for loss or damage to car or contents however caused." Dumas also alleged that the attendant's office had a sign posted near the entrance that bore the same message.

Identify the issues in this case and prepare the arguments that Ali and the Dumas Corporation might use in their claim and defence respectively.

Render a decision.

CASE 4

The Sakuras considered moving to Western Canada from the city of Toronto after Mr. Sakura's retirement. In order to determine an appropriate community in which to reside, they visited a number of west-coast cities by automobile.

On their visit to one community, which appeared to be a delightful place to live, they met the owner of a warehouse business. The warehouse owner suggested that he would be prepared to receive their household goods if they wished to ship them to him. He would hold them in storage until such time as they found a permanent residence.

On their return to Toronto, the Sakuras decided to move immediately. They dispatched their household goods to the warehouse operator that they had met on their visit to the city. Instead of taking up residence immediately, they planned an extensive vacation that would take them across the United States and eventually to that particular community.

While the Sakuras were on vacation, the household goods arrived at the warehouse. The owner issued a warehouse receipt, which he mailed to the Sakuras at the temporary address that they had given him. The warehouse receipt set out the terms and conditions of storage, one item being a condition that read: "All goods stored at owner's risk in case of fire (storage rates do not include insurance)."

On their return from their vacation, the Sakuras found the warehouse receipt in their mail, but did not read the document. They proceeded to obtain a new home. However, before they could retrieve their goods from the warehouse, the building was burned by an arsonist who had apparently gained entry to the building by way of an open rooftop skylight. The household goods that belonged to the Sakuras were totally destroyed by the fire.

When the warehouse operator refused to compensate the Sakuras for their loss, an action was brought claiming damages for the value of the goods.

Discuss the nature of the plaintiff's claim and the defences that the warehouse operator might raise.

Render a decision.

CASE 5

The members of an investment club arranged for a banquet and overnight accommodations for their members at the Municipal Hotel. Most of the members arrived early for the dinner in order to check into their rooms. One member, however, arrived late and, instead of checking in at the desk, went directly to the banquet room where the dinner was about to be served. Before entering the room, she noticed a coat room adjacent to the dining room that contained a number of coats. She hung her fur jacket on a hanger in the coat room. No attendant was in charge of the coat room, although a person wearing a hotel porter's uniform was standing near the doorway to the room.

The club member spent the evening in the banquet room, and at the end of the dinner meeting went to the coat room to retrieve her jacket. The jacket was missing.

As compensation the hotel offered her the sum of $40, the amount that an innkeeper was obliged to pay under the *Innkeeper's Act* of the province. The hotel explained that as a guest in the hotel, this was the extent of its liability to her. The hotel manager pointed out that the club member was aware of the limited liability of the hotel by virtue of the notice to that effect that was posted in all hotel bedrooms.

The club member refused to accept the sum offered as payment. She brought an action against the hotel for $2,800, an amount that she alleged was the appraised value of the fur jacket.

Discuss the issues raised in this case and indicate how the courts might deal with them. Would your answer be any different if the club member's jacket had been stolen from a locked hotel room?

CASE 6

Lacey inherited a large brooch from her grandmother. The brooch contained what appeared to be a number of large precious stones. She was curious as to the value of the piece of jewellery and took it to the B & S Jewellery Shop for examination. The jeweller was busy at the time. He asked Lacey to leave the brooch, saying he would examine it when he had a moment to do so. Lacey filled out a small claim check that required her to set out her name and address. She did so, and was given a portion of it bearing a claim number.

www.mcgrawhill.ca/olc/willes

Lacey placed the claim check in her jewellery box when she returned home. She paid no further attention to the matter, believing that the jeweller would notify her when he had completed his appraisal of the brooch.

Some five years later, while cleaning out her jewellery box, Lacey noticed the claim stub in the bottom of it and remembered that she had left the brooch at the jewellers. She immediately went to the jewellery shop with the ticket to claim her jewellery, but the jeweller was unable to find the brooch. His records indicated that he had written Lacey three weeks after she had brought the brooch to him, to advise her that its value was $6,500. Lacey had not received the letter, however, and, until her return to the jewellery shop, was unaware of its value.

When the jeweller could not produce the brooch, Lacey brought action for damages against the jeweller for the value of the piece.

Indicate the nature of Lacey's action, and discuss the various arguments that both Lacey and the jeweller might raise. Render a decision.

CASE 7

Central Ceramic China Ltd. was an importer of various lines of dishes and tableware that it sold in quantity to hotels and restaurants. Approximately 70 percent of its sales consisted of hotel-grade dishes, 20 percent of fine bone china dishes, and the remaining 10 percent consisted of cutlery and eating utensils.

For many years the firm used the services of Able Transport Co. to deliver its goods to customers who were located in various parts of the country. All goods were shipped in cartons marked: "FRAGILE. CONTENTS BREAKABLE IF ROUGH-HANDLED." The contents were normally packed in a strawlike material to provide some protection in the event of impact or careless handling. This reduced breakage of the shipped china to a minimum acceptable level. Only occasionally would a customer report breakage, and this usually consisted of only one or two pieces in a shipment of perhaps many hundreds.

Central Ceramic recently tested a new type of foam packing material, and decided that its use would permit the contents of a case to withstand a reasonable amount of impact if the case should accidentally be dropped. Management then decided to use the new packing material in cartons that were not marked with a "fragile" label in order to obtain a lower shipping rate. The company informed Able Transport of the removal of the "fragile" notice on the containers and requested a lower shipping rate. Able Transport agreed to handle the goods at a lower rate.

During the month that followed, management of Central Ceramic monitored the breakage rate and noted that it was approximately the same as when the other marked containers were used. The next month the company shipped a very large quantity of china to a distant hotel customer in the new containers. The china was shipped in 36 cases. When it was received by the hotel, almost half of the china was found to be either cracked or broken. An investigation by the carrier revealed that road vibration during the long trip had caused the packing material to shift, allowing the pieces of china to come in contact with each other and break, if the carton received any impact.

Central Ceramic took legal action against Able Transport for damages equal to the loss. Able Transport denied liability.

Discuss the arguments (if any) that the parties might raise in this case.

Render a decision.

CASE 8

Universal Paper Products Ltd. of Vancouver consigned three-and-a-half boxcar loads of goods to RapidMovve Transport Co. to take the goods from Vancouver to Le Havre, France. RapidMovve would normally have made direct arrangements with Canadian Atlantic Railways (CAR) for the Vancouver–Montreal leg, but because there was a half-carload involved, they engaged Railshippers Inc. to put together the Vancouver–Montreal leg. Railshippers could get a better rate from CAR as they specialized in making up full carloads from an assortment of partial carloads. RapidMovve engaged Maritime Containerways to ship from Montreal to France, and a French trucking firm for local delivery in France.

All would have gone well, had it not been for a derailment of the CAR train in Northern Ontario. Unseasonable rains had washed out a section of track and, in the dark, the train was derailed. Most of the product was probably in good shape after the accident, but unfortunately one of the welding crew pressed into service by CAR to clear the blockage of the main line in the emergency ignited the contents of one car as he cut away wreckage. The flames spread to the other cars, destroying $60,000 of paper products.

Three weeks later, Universal Paper became frustrated that RapidMovve could still only provide a garbled, contradictory explanation of what had happened. A newspaper account noted that there had been no salvage of any freight from the wreck. Universal Paper, knowing only that they had delivered goods to RapidMovve, and that none had arrived in France, sued RapidMovve for the value of the goods.

Identify and discuss the legal issues raised in this case and the liability (if any) of the parties.

Render a decision.

CASE 9

Hart operated the Riverside Restaurant and Bakery Shop, which was located on a busy downtown street. The front portion of the premises contained the bakery shop, and the rear part of the building housed the restaurant. Patrons entering the building were required to pass through the store portion to reach the restaurant. In the store area, near the entrance to the restaurant, the owner had installed a number of coat hooks in a recess in the wall of the building. Employees of the shop and restaurant used the alcove to store their overcoats and hats.

Wallinsky, a stranger to the community, entered the shop for the purpose of dining, and proceeded through the shop to the restaurant area. Along the way he noticed the clothing in the alcove and placed his overcoat and hat on one of the unused hooks. He then entered the restaurant and ordered a meal. Some time later, when he was about to leave the restaurant, he discovered that his overcoat and hat were missing.

Hart denied responsibility for the loss. Wallinsky brought an action against him for the value of the hat and coat.

Discuss the arguments that the parties might raise in this case, and identify the legal issue involved.

Render a decision.

Would your decision differ in any way if the coat hooks were located in the restaurant beside Wallinsky's table?

CASE 10

Harriet, a licensed pilot, rented an aircraft from Aircraft Rental Services at a local airport. The purpose of the rental was to fly a friend to a large metropolitan city some 500 kilometres away and return before nightfall. At the time that she arranged for the use of the aircraft, she assured the owner that she would leave the city in ample time to return the aircraft before dark. She paid a deposit for the use of the aircraft and, accompanied by her friend, made an uneventful flight to the distant city. Before returning home, however, she spent some time shopping and lost track of time. Eventually, she realized that she was behind schedule and hurried to the airport.

The weather report for the return trip was not promising, but she nevertheless decided to chance the flight. She took off at 4:45 p.m., some two hours before official nightfall on that particular January night. En route, she discovered that the weather had deteriorated and that visibility was decreased by the combination of sundown and low cloud conditions.

At 7:05 p.m., some 20 minutes after official nightfall, she found that she could proceed no further, as the poor weather and semi-darkness made recognition of her route on the ground virtually impossible. To avoid further difficulties, she made a forced landing in a farmer's field, which resulted in damage to the airplane's undercarriage.

Harriet assumed that the aircraft owner's insurance would cover the cost of the repairs. However, she was surprised to hear that the insurance covered only public liability and not damage to the aircraft itself. The cost of repairs amounted to $11,165. When Harriet refused to pay for the damage, Aircraft Rental Services brought an action for damages against her for the amount of its loss.

In her defence, Harriet alleged no negligence on her part, as the landing was made in accordance with accepted forced-landing procedures and skillfully executed on her part. She argued that in any forced landing, some damage to the undercarriage could be expected, and that the mere fact that damage occurred was not an indication of negligence.

The plaintiff brought out in the evidence that Harriet was licensed to fly under daylight conditions only. She did not have what was called a "night endorsement" on her licence that would permit her to fly after dark. The plaintiff alleged that her act of flying after official nightfall was a violation of Air Regulations under the *Aeronautics Act*.

Discuss the nature of the plaintiff's claim in this case, and the various other arguments that might be raised by the parties. Indicate the issues that must be decided by the court.

Render a decision.

The Online Learning Centre offers more ways to check what you have learned so far. Find quizzes, Weblinks, and many other resources at www.mcgrawhill.ca/olc/willes.

CHAPTER 21

The Sale of Goods

Chapter Objectives

After study of this chapter, students should be able to:

- Explain the nature of the contract of sale and the *Sale of Goods Act*.
- Identify the rights and duties of the seller and the buyer.
- Describe the remedies available to the buyer or seller in the event of breach of the agreement.

YOUR BUSINESS AT RISK

Selling goods should be easy — payment is made and goods are delivered. Unfortunately, the vast majority of trade does not work that way. Commercial sales tend to be on credit terms, production is agreed for the future, and delivery dates rarely correspond to payment dates. Beyond that are the problem cases — strikes, disasters, accidents, delays, misunderstandings and mistakes. Businesses could plan every eventuality into their sales contracts, but each one would fill a book of its own. The law of the sale of goods fills in the gaps that businesses fail to cover.

CODIFICATION OF THE LAW

The law that relates to the sale of goods represents a direct response to the need for clear and precise rules to govern transactions that involve the exchange of money for goods. The existence of laws of this nature are indicative of the stage of development of a society. In a primitive society, where the individual members live at or near the subsistence level, no laws are necessary because the sale of goods seldom occurs. Any surplus production of one product is usually exchanged for other necessary products, usually by barter. It is only when a genuine surplus is produced that a basis for trade is established. Even then, a number of other conditions must be present before laws are necessary to govern the exchanges.

In England, during the Middle Ages, most families and communities were relatively self-sufficient. Any goods that could not be produced within the family were usually acquired by way of exchange or barter with neighbours. Surplus goods of one community were often carried to nearby communities and exchanged for goods not available locally, as the same products were seldom concurrently in surplus supply. The rise of towns, however, set the stage for commerce. Not only did towns provide a ready market for agricultural products, they also produced goods required by the agricultural community and represented a convenient place where exchanges might take place.

Trade was initially by barter at the market "fairs" held in each town, and was largely local. However, as the towns grew in size, foreign merchants began to appear, either with goods to sell or with money to purchase the surplus goods of the community.

Transactions between merchants were first governed by the Law Merchant, and disputes concerning the sale of goods that arose were settled by the merchants themselves.

Dealings between merchants and members of the community, however, were sometimes taken to the Common Law courts. If the particular transaction involved the sale of goods, the courts would often apply the same rules that the merchants used in their own transactions. Over the years, a body of Common Law relating to the sale of goods gradually developed.

The law was far from satisfactory, unfortunately. The methods that a plaintiff was obliged to use to obtain relief were cumbersome. In spite of a desire on the part of the merchants for change, the courts did not re-examine the nature of the transaction until the eighteenth century. At that time, the modern concept of contract emerged, and the sale of goods was treated as a contractual relationship between the buyer and seller.

During the next century, the rules of law relating to the sale of goods developed rapidly. By the late nineteenth century the law had matured to the point where the business community pressed for the law's organization into a simplified and convenient statute. The government responded in 1893 with the codification of the Common Law in the form of a single statute, entitled the *Sale of Goods Act*.[1]

At the request of the government of the day, MacKenzie D. Chalmers, a prominent English county court judge, prepared the draft bill that set out the law relating to the sale of goods as a clear and concise body of rules. Chalmers was familiar with the Common Law relating to the contract of sale and, as a result, the statute became one of the best-drafted laws on the English statute books. Other countries were quick to recognize the advantages of having a codification of this part of the law. It was soon after adopted by the Common Law provinces of Canada and a number of other jurisdictions in the British Empire.

The legal profession in the United States proposed similar legislation there, and a *Uniform Sales Act* was prepared, based upon the English *Sale of Goods Act*. This Act was adopted by many of the states, occasionally with modifications. It eventually found its way into the *U.S. Uniform Commercial Code*, in a somewhat different form, as Article 2. At the present time, the Code has been adopted by all states except Louisiana. This widespread adoption of the English principles and rules relating to the sale of goods reflects the clarity and simplicity of the original law that Mr. Chalmers had so carefully drawn. It remains today in virtually unaltered form on the statute books of many jurisdictions.

NATURE OF A CONTRACT OF SALE

A contract of sale, as the name implies, is a type of contract. Consequently, the rules that relate to the formation, discharge, and impeachment of ordinary contracts also apply to the contract of sale, except where the *Sale of Goods Act* has specifically modified the rules.

A contract of sale is something more than an ordinary contract, however, because the contract not only contains the promises of the parties, but often represents evidence of a transfer of the ownership of the property to the buyer as well. It must, therefore, operate in accordance with the Act to accomplish this purpose.

Under the Act, "a contract of sale of goods is a contract whereby the seller transfers or agrees to transfer the property in goods to the buyer for a money consideration called the price ..."[2] Two different contracts are contemplated by this definition. In the first instance, if the ownership is transferred immediately under the contract, it represents a sale. In the second, if the transfer of ownership is to take place at a future time, or subject to some condition that must be fulfilled before the transfer takes place, the transaction is an agreement to sell. Both the "sale" and the "agreement to sell" are referred to as a contract of sale under the Act where it is unnecessary to distinguish between the two. An agreement to sell may apply to goods that are in existence at the time, or it may apply to a contract where the goods are not yet in existence. An example would be where a farmer enters into a contract with a food processor to sell his entire crop of fruit or vegetables before they are grown.

1. 1893, 56 & 57 Vict., c. 71.

2. *Sale of Goods Act*, R.S.O. 1990, c. S.1, s. 2(1). The *Sale of Goods Acts* of all provinces except Quebec are based upon the British *Sale of Goods Act*, and are virtually identical in content. The numbering and wording may vary, however.

Application of the Act

For Goods

The sale or agreement to sell must be for goods, as distinct from land and anything attached to the land. Buildings, for example, form a part of the land, because they are attached to it, and so, too, would any right to use the land or the buildings. Transactions concerning land are not covered by the *Sale of Goods Act*. The Act, as its name indicates, concerns a sale of goods, but even then some "goods" are excluded. A sale of goods subject to the Act would include tangible things such as moveable personal property, but the term "goods" would not include money or intangible things such as shares in a corporation, bonds, negotiable instruments, or "rights," such as patents or trademarks.

The contract itself must be for the sale of goods, and in this sense it is distinguished from a contract for work and materials. It is sometimes difficult to differentiate between a contract for work and materials and an agreement to sell, if the goods are not yet produced. However, as a general rule, if the contract is for a product of which the cost of the materials represents only a small part of the price, and the largest part of the cost is labour, the contract may be treated as a contract for work and materials. In that case the *Sale of Goods Act* would not apply. For example, if a person engages another person to paint a house, or if a person takes a watch to the repair shop to be cleaned and to have a minor part replaced, the contracts would probably be treated as for work and materials, rather than a sale of goods. In both cases, most of the purchase price would be represented by the "work," rather than the goods themselves.

For Money

A second distinction between a contract of sale and other forms of contract is the requirement that the property in the goods be transferred for a monetary consideration. By this definition, a barter or exchange of goods where no money changed hands would not be a contract of sale within the meaning of the Act; nor would a consignment, where the title to the goods is retained by the owner, and the seller has only possession of the goods pending a sale to a prospective buyer.

Writing

No special form is required for either the sale contract or the agreement to sell. The contract may be in writing, under seal, verbal, or, in some cases, implied from the conduct of the parties. However, if the contract is for the sale of goods valued at more than a particular amount,[3] the agreement must be evidenced by a memorandum in writing, and signed by the party to be charged (or his or her agent) to be enforceable.

The requirement of writing was originally found in the *Statute of Frauds*, and later included in the English *Sale of Goods Act*. When the legislation was adopted by the Canadian provinces and territories, the requirement of writing or now, electronic memorial, was included. Three exceptions are provided in the Act, however, which permit the parties to avoid the requirement of writing. The agreement need not be in writing if the buyer: (1) accepts part of the goods sold; (2) makes a part-payment of the contract price; or (3) gives something "in earnest" to bind the contract.[4]

In each of these cases, the actions of the buyer must relate specifically to the particular contract of sale. The acceptance of part of the goods has been interpreted by the courts to mean any act that would indicate acceptance or adoption of the pre-existing contract, including the ordinary inspection of the goods. The part-payment of the contract price must be just that: a payment of money that relates specifically to the particular contract. The third requirement, the giving of something "in earnest," refers to an old custom of giving something valuable for the purpose of binding the agreement. The object might be an article, or something of value, other than a part-payment of the purchase price. This practice is seldom followed today.

3. The amount in Nova Scotia is $40. Newfoundland, Alberta, and Saskatchewan fixed the amount at $50, and Prince Edward Island placed the amount at $30. British Columbia, Ontario, Manitoba, and New Brunswick have followed the English example and repealed the requirement of writing.

4. *Sale of Goods Act*, R.S.O. 1990, c. S.1, s. 5(1).

COURT DECISION: Practical Application of the Law

Sale of Goods—Application of Act—Requirement of Writing

***N.M. Paterson & Sons Limited v. Lowenburg*, 2005 SKQB 205.**

After calling several times to determine the market price for flax, the defendant farmer agreed on June 25, 2001 to sell to the plaintiff grain merchant 300 metric tonnes of flax at a net price of $264.65 per tonne. The defendant had previously delivered to the plaintiff's elevator 2.043 tonnes of flax which were applied to the contract leaving 297.957 tonnes undelivered. The defendant refused to deliver the balance.

THE COURT: While the defendant at trial denies that such a contract was reached, I do not believe him. His demeanour was evasive and his subsequent behaviour was consistent with the contract having been reached. The plaintiff's agent, Rudy Lepp, was accompanied by his regional supervisor when the contract was made. Both testified. The quantity was substantial and it is not surprising that both have a clear recollection of the event. The defendant stopped calling for prices after the contract was made. During the week of July 2, 2001, the defendant attended at the plaintiff's elevator in Grenfell, Saskatchewan, and acknowledged the terms of the contract. When Mr. Lepp called the defendant on July 19 to advise that he had adequate space to take delivery of the flax he was told that the defendant would not deliver the flax. The market price had risen substantially by that time. The defendant did not deliver the balance of the flax before August, 2001, or at any time, and he is therefore in breach of the contract.

I find that a verbal contract was in force from the time the telephone conversation ended on June 25, 2001. Had the price dropped, the plaintiff would have sustained a loss. As the price rose the plaintiff would have profited by $36.50 per tonne which is the loss of profit upon which the plaintiff's claim is based multiplied by 297.957 and totals $10,875.43. The plaintiff also claims an administrative fee of $10.00 per tonne which, while it may be customary, has not been proven or explained to my satisfaction and I will not allow it. The defendant has advanced the argument that *The Sale of Goods Act*, R.S.S. 1978, c. S-1, requires some writing to support a contract such as this, however, the part performance involving the sale under the contract of 2.043 tonnes is sufficient to meet the requirements of the Act. A document was created to record the sale but as it was never completed by the defendant who denies even receiving it, it has no effect on the contract. The plaintiff will have judgment for the sum of $10,875.43, together with pre-judgment interest from August 1, 2001, and its costs.

What was really at stake between the parties in this case? Why did Paterson succeed? What business practices would you suggest to Paterson to prevent such a "close call" in future?

Transfer of Title

Title

The ownership of the goods.

A final observation with respect to the nature of the contract of sale is that it represents an agreement to transfer property in the goods to the buyer. The "property in the goods" is the right of ownership to the goods, or the **title**. The ownership of the goods normally goes with possession, but this is not always the case. A person may, for example, part with possession of goods, yet retain ownership. It is this attribute that creates most of the difficulties with the sale of goods. The parties in their agreement may determine when the title will pass, and this may differ from the time when possession takes place. Since the risk of loss generally follows the title, in any agreement where the transfer of possession is not accompanied by a simultaneous transfer of ownership, any damage to the goods while the title is not in the person in physical possession of them can obviously raise difficulties.

Goods that are not in a deliverable state (i.e., goods that must be produced, weighed, measured, counted, sorted, or tested before they are identifiable as goods for a particular contract) unless otherwise provided, remain at the seller's risk until such time as they are "ready for delivery." Under the *Sale of Goods Act*, no property in the goods is transferred to the buyer until the goods are in this state.[5] In a contract for goods that are specific or ascertained, the property in the goods may be transferred to the buyer at such time as the parties intend the transfer to take place.[6] In most cases, this intention will be determined by an examination of the contract terms, the conduct of the parties, or the circumstances

5. ibid., s. 17; *Harris v. Clarkson* (1931), 40 O.W.N. 325.

6. *Sale of Goods Act*, R.S.O. 1990, c. S.1, s. 18(1); see also *Goodwin Tanners Ltd. v. Belick and Naiman*, [1953] O.W.N. 641.

under which the contract arose.[7] If the parties specify when the title passes, then the parties themselves have decided who should bear the loss in the event that the goods should be destroyed or damaged before the transaction is completed. If they have not dealt with this matter in their agreement, or if it cannot be ascertained from their conduct (or the circumstances of the case), then the Act provides a series of rules that are deemed to apply to the contract. These rules deal with a number of different common contract situations. The first rule deals with goods that are specific (i.e., identified and agreed upon at the time the contract is made) and in a deliverable state.

Rule 1

If there is an unconditional contract for the sale of specific goods in a deliverable state, the property in the goods passes to the buyer when the contract is made, and it is immaterial whether the time of payment or the time of delivery or both be postponed.[8]

Henderson enters Nielsen's shop and purchases a large crystal bowl that Nielsen has on display in her shop window. Henderson pays for the item and informs Nielsen that he will pick it up the next morning.

During the night, a vandal smashes the shop window and destroys the crystal bowl. The title passed in this case when the contract was made, because the goods were specific and in a deliverable state. Henderson, if he wished, could have taken the bowl with him at the time the contract was made, but he elected not to do so. Since loss follows the title, the destroyed goods belonged to the buyer and not to the seller. It is the buyer who must bear the loss.

Rule 2

The second rule is a variation of Rule 1. It is applicable to a contract where the seller must do something to the goods to put them in a deliverable state. Title in this case does not pass until the seller does whatever is necessary to put the goods in a deliverable state, and notifies the buyer that the goods are now ready for delivery. The rule states: where there is a contract for the sale of specific goods and the seller is bound to do something to the goods for the purpose of putting them in a deliverable state, the property does not pass until the thing is done and the buyer has notice thereof.[9]

Leblanc entered into a contract with Ross to purchase a used car on display at Ross's car lot. The lock on one door was inoperable, so Ross agreed to fix the lock as a term of the contract. Leblanc paid the entire purchase price to Ross. Ross repaired the lock, but before he notified Leblanc that the car was ready for delivery, the car was destroyed by a fire at Ross's garage. Leblanc would be entitled to a return of the purchase price in this case, as the title was still in Ross's name. The title would not pass until Ross notified Leblanc that the car was ready for delivery, and the risk was his until the buyer received the notice.

Rule 3

The third rule is again a variation of Rule 1. It applies where the contract is for the sale of specific goods in a deliverable state, but where the seller must weigh, measure, test, or do something to ascertain the price. Under this rule, the property in the goods does not pass until the act is done and the buyer notified. The rule states: when there is a contract for

---

7. *Sale of Goods Act*, supra, s. 18(2).

8. ibid., s. 19.

9. ibid., s. 19; see also *Underwood Ltd. v. Burgh Castle Brick & Cement Syndicate*, [1921] All E.R. Rep. 515.

the sale of specific goods in a deliverable state, but the seller is bound to weigh, measure, test, or do some other act or thing with reference to the goods for the purpose of ascertaining the price, the property does not pass until such act or thing is done, and the buyer has been notified thereof.[10]

Grange agrees to purchase a quantity of grain that Thompson has stored in a bin in his warehouse. Thompson agrees to weigh the material and inform Grange of the price. If the grain should be destroyed before Thompson notifies Grange of the weight and price, the loss would be the seller's, as the property in the goods would not pass until the buyer has notice. If, however, Thompson weighed the grain and notified Grange of the weight and price, the title would pass immediately. If the goods were subsequently destroyed before Grange took delivery, the loss would be his, even though the goods were still in the seller's possession.

It is important to note with respect to Rule 3 that the seller must have the duty to weigh, measure, or otherwise deal with the goods. In the case where the buyer took the goods and agreed to weigh them on the way home, and then notify the seller, a court held that Rule 3 did not apply to transfer the property interest. The title passed to the buyer when he took the goods.[11]

Rule 4

The fourth rule for the transfer of ownership in goods deals with contracts for the sale of goods "on approval" or with return privileges. This rule is a two-part rule that provides that the title will pass if the buyer, on receipt of the goods, does anything to signify his or her acceptance or approval of the goods, or the adoption of the contract. If the buyer does nothing but retains the goods beyond a reasonable time, then the title will pass at the expiry of that period of time. The buyer must do some act that a buyer would only have the right to do as the owner in order to fall under the first part of this rule. The sale of the goods by the buyer, for example, would constitute an act of acceptance, as it would be an act that only a person who had adopted the contract would normally do. The same rule would hold if the buyer mortgaged the goods. In that case, the title would pass to the buyer the instant that the act of acceptance took place.

Under the second part of the rule, if the buyer simply does nothing after he or she receives the goods, the title will pass when the time fixed for return expires, or, if no time is fixed, after a reasonable time. The purpose of this second part of the rule is to ensure that a buyer cannot retain "approval" of goods beyond a reasonable time. The delivery of goods is frequently a courtesy extended by the seller. To allow the prospective purchaser to retain the goods an unnecessarily long time would only increase the chance of loss or damage to the goods while the risk is still with the seller. The rule states: where goods are delivered to the buyer on approval or "on sale or return" or other similar terms, the property therein passes to the buyer:

(i) when he or she signifies his or her approval or acceptance to the seller or does any other act adopting the transaction; and

(ii) if he or she does not signify his or her approval or acceptance to the seller but retains the goods without giving notice of rejection, then if a time has been fixed for the return of the goods, on the expiration of such time, and if no time has been fixed, on the expiration of a reasonable time; and what is a reasonable time is a question of fact.

10. *Sale of Goods Act*, R.S.O. 1990, c. S.1, s. 19.

11. *Turley v. Bates* (1863), 2 H & C 200, 159 E.R. 83.

Baxter Construction Company purchased a small bulldozer on approval. A few days later, the company pledged it as security for a loan at its bank. The machine was later damaged in a fire. In this case, the buyer, Baxter Construction Company, would be considered to have accepted the goods at the time it pledged the machine as security, and the resulting loss would be the buyer's.

The terms of the contract may alter the liability of the parties, however. In a case where goods were delivered "for cash or return, goods to remain the property of the seller until paid for" it was held that Rule 4 did not apply, as the seller had specifically withheld the passing of the title.[12]

A buyer ordered 140 bags of rice from a seller. The seller delivered 125, with 15 bags to follow. The buyer asked the seller to hold delivery of the remaining 15 bags. After the passing of a reasonable time, the seller asked the buyer if he was appropriating the125 bags, but the buyer did not reply. The seller later sued the buyer for the price of the 125 bags of rice. In this case the court held that the buyer, in failing to reply within a reasonable time, had implied acceptance.[13]

Rule 5

The fifth rule applies to unascertained goods (or goods that are not as yet produced) and that would therefore be the subject matter of an agreement to sell, rather than a sale. Under this rule, as soon as the goods ordered by description are produced and in a deliverable state and are unconditionally appropriated to the contract, either by the seller or by the buyer (with the seller's consent), the property in the goods will pass. Again, this rule is in two parts, which provide:

(i) Where there is a contract for the sale of unascertained or future goods by description, and goods of that description in a deliverable state are unconditionally appropriated to the contract, either by the seller with the assent of the buyer, or by the buyer with the assent of the seller, the property in the goods therein passes to the buyer, and such assent may be expressed or implied, and may be given either before or after the appropriation is made.

(ii) Where, in pursuance of the contract, the seller delivers the goods to the buyer or to a carrier or other bailee (whether named by the buyer or not) for the purpose of transmission to the buyer, and does not reserve the right of disposal, he is deemed to have unconditionally appropriated the goods to the contract.[14]

A pipeline contractor ordered a quantity of special steel pipe from a manufacturer. When the pipe was produced, the contractor sent one of his trucks to the manufacturer's plant with instructions to have the pipe loaded. After the truck was loaded, it was stolen (through no fault of the manufacturer) and destroyed in an accident. The pipe, as a result of the damage suffered in the accident, was useless. Here, the goods were unconditionally appropriated to the contract and the title had passed to the contractor.

Again, the time at which the title passes is deemed to be when the buyer obtains possession of the goods either himself or through his agent, or when the seller loses physical control of the goods.

---

12. *Weiner v. Gill*, [1905] 2 K.B. 172.
13. *Pignitaro v. Gilroy*, [1919] 1 K.B. 459.
14. *Sale of Goods Act*, R.S.O. 1990, c. S.1, s. 19.

A buyer in England ordered certain dyes from a seller in Switzerland, knowing that the seller had them in stock. The seller sent the dyes by mail to the buyer in England, and in so doing was accused of infringement of the English patent. One of the issues in the case was: Where and when did title pass? The court held that, since the buyer had given his implied assent to delivery by mail, as soon as the seller filled the order and placed it in the mail the title passed to the buyer.[15]

This decision is consistent with cases dealing with the use of common carriers to deliver the goods to the buyer, as provided in the second part of the rule. Unless the seller has reserved the right of disposal, goods delivered to the carrier have essentially been disposed of by the seller. Once delivered, the seller no longer has control of the goods, and usually only the buyer may recover the goods from the carrier. Since the seller has effectively transferred control over the goods to the buyer's agent, the rule is sensible in providing for the passing of ownership from the seller to the buyer at the moment when the seller parts with possession.

Withholding title by the seller, or reserving the right of disposal if goods are delivered to the buyer or a carrier, has important implications in the event that the buyer should become insolvent at some point in time during the sale. The general rule is that the trustee in bankruptcy is only entitled to claim as a part of the bankrupt's estate those goods that belong to the bankrupt at the time of the bankruptcy. If the seller has retained the title to the goods, the seller may, in many cases, be in a position to recover the goods or stop their delivery to the bankrupt if they are in the hands of a carrier. Hence the importance, for example, of reserving the title until the goods are paid for in full by the buyer.

MANAGEMENT ALERT: BEST PRACTICE

It is important not to lose sight of the fact that application of the rules for passage of title under the provincial Sale of Goods Acts take effect when the parties fail to provide for timing on their own, in their contract. Since risk of loss is transferred with title to the goods, the first sensible act toward commercial certainty is for the parties to make that timing provision for themselves in their contract. Once this is done, it is prudent to arrange insurance to protect one's interest up to (or from) the point in time of the transfer of title. Bear in mind that this moment is often the riskiest for goods. This is the time when unusual activity is taking place in measurement, transit, handling, and delivery, often by persons with no real knowledge of the contents of the cartons, crates, or containers in their care. In cases of specific and expensive insured goods, it also worth arranging insurance that will run for some time after (or before, in the buyer's case) the intended date of transfer of title. This ensures that there is no accidental gap in coverage resulting from a delayed or early actual transfer.

CLIENTS, SUPPLIERS *or* OPERATIONS *in* QUEBEC

Common Law provinces have taken a step in the direction of Quebec's *Civil Code* through codification of their sale of goods law; so, structurally, there are more similarities than differences. That being acknowledged, the Quebec *Civil Code* covers essentially the same issues as does the Common Law, but with generally greater buyer protection. *Civil Code* articles 1377-1707 on obligations apply to sale contracts, along with special sales rules in articles 1708-1805.

CONTRACTUAL DUTIES OF THE SELLER

The *Sale of Goods Act* permits the parties to include in their contract any particular terms or conditions relating to the sale that they wish, and the seller is obliged to comply with

15. *Badische Analin und Soda Fabrik v. Basle Chemical Works, Bind Schedler*, [1898] A.C. 200.

these terms. Sometimes the contract is one that is not carefully drawn in terms of the particular rights and duties of the parties. In these cases, the Act implies certain obligations. These obligations generally are imposed upon the seller in terms of warranties and conditions with respect to the goods. Under the Act, these terms have particular meanings.

Condition

An essential term of a contract.

Warranty

In the sale of goods, a minor term in a contract. The breach of the term would allow the injured party to damages, but not rescission of the agreement.

A **condition** is a fundamental or essential term of the contract that, if broken, would generally entitle the innocent party, if he or she so elects, to treat the breach as a discharge. The innocent party would then be released from any further performance.

A **warranty** is not an essential term in the contract, but rather, a term that, if broken, would not end the contract, but would entitle the injured party to take action for damages for the breach. A warranty is usually a minor term of the contract, and not one that goes to the root of the agreement.

The *Sale of Goods Act* stipulates the particular terms in the contract of sale that constitute conditions and those that, if broken, would only be warranties. For example, the time for delivery of the goods is treated as a condition, and the promise of payment a mere warranty.

Title

As to the title of the seller, unless the contract indicates otherwise, there is an implied condition that in the case of a sale, the seller has the right to sell the goods. In the case of an agreement to sell, the seller will have the right to sell the goods at the time when the property or the title in the goods is to pass to the buyer.[16] There is also an implied warranty that the goods are free from any charge or encumbrance (such as a chattel mortgage) in favour of a third party, unless the seller has informed the buyer of the charge or encumbrance, either before or at the time the agreement is made.[17] An additional implied warranty relates to the seller's title. It states that the buyer shall have quiet possession of the goods. The term "quiet possession" has nothing to do with solitude; it simply means that no person will later challenge the buyer's title to the goods by claiming a right or interest in them.[18]

Sold by Description

Goods that are sold by description are subject to an implied condition that the goods will correspond with the description.[19] For example, if a buyer purchases goods from a catalogue, where the specifications are given and perhaps a picture of the goods is shown, the goods ordered by the buyer must correspond with the catalogue specifications. Otherwise, the seller will be in breach of the contract, and the buyer will be entitled to reject the goods. If the goods are sold by description as well as by sample, then the goods must correspond to the description as well as the sample.[20]

Sold by Sample

Where goods are sold by sample alone, there is an implied condition that the bulk of the goods will correspond to the sample in quality,[21] and that the buyer will have a reasonable opportunity to examine the goods and compare them with the sample.[22] Even then, there is an implied condition that the goods will be free from any defect (apparent on reasonable examination of the sample) that would render them unmerchantable.[23] For example, a seller sold cloth by sample to a buyer, to be resold by sample to tailors. However, unknown to both the seller and buyer, the cloth dye was such that perspiration would cause the colours to run. When the defect was later discovered, the tailors who

16. *Sale of Goods Act*, R.S.O. 1990, c. S.1, s. 13(a). See also *Cehave N.V. v. Bremer Handelsgesellschaft*, [1975] 3 All E.R. 739; *Wickman Machine Tool Sales Ltd. v. L. Schuler A.G.*, [1972] 2 All E.R. 1173, affirmed [1973] 2 All E.R. 39.
17. *Sale of Goods Act*, R.S.O. 1990, c. S-1, s. 13(c).
18. ibid., s. 13(b).
19. ibid., s. 13; see also *Beale v. Taylor*, [1967] 3 All E.R. 253.
20. *Sale of Goods Act*, R.S.O. 1990, c. S.1, s. 14.
21. ibid., s. 16(2)(a). See also *Buckley v. Lever Bros. Ltd.*, [1953] O.R. 704.
22. *Sale of Goods Act*, R.S.O. 1990, c. S.1, s. 16(2)(b). See also *Godley v. Perry*, [1960] 1 All E.R. 36.
23. *Grant v. Australian Knitting Mills Ltd.*, [1935] All E.R. Rep. 209.

manufactured the overcoats complained to the buyer, who in turn complained to the original seller. The defect was not apparent on ordinary examination of the cloth, but was in both the sample and the bulk of the cloth. When the seller refused to compensate the buyer, the buyer sued the seller for breach of contract. The court held that the examination need only be that which a reasonable person would make. There was no need to conduct elaborate chemical tests. The standard for the examination would be the same as that which a reasonable man buying an overcoat would have made of the material.[24]

In the case of *James Drummond & Sons v. E.H. Van Ingen & Co.*,[25] the judge outlined the law as it relates to sale by sample:

> The sample speaks for itself. But it cannot be treated as saying more than such a sample would tell a merchant of the class to which the buyer belongs, using due care and diligence, and appealing to it in the ordinary way and with the knowledge possessed by merchants of that class at the time. No doubt the sample might be made to say a great deal more. Pulled to pieces and examined by unusual tests which curiosity or suspicion might suggest, it would doubtless reveal every secret of its construction. But that is not the way in which business is done in this country. Some confidence there must be between merchant and manufacturer. In matters exclusively within the province of the manufacturer the merchant relies on the manufacturer's skill, and he does so all the more readily when, as in this case, he has had the benefit of that skill before.
>
> Now I think it is plain upon the evidence that at the date of the transaction in question merchants possessed of ordinary skill would not have thought of the existence of the particular defect which has given rise to this action, and would not have discovered its existence from the sample. It appears to me, therefore, that the sample must be treated as wholly silent in regard to this defect, and I come to the conclusion that if every scrap of information which the sample can fairly be taken to have disclosed were written out at length, and embodied in writing in the order itself, nothing would be found there which could relieve the manufacturer from the obligation implied by the transaction.
>
> I prefer to rest my view on this broad principle. But it seems to me that the obligation of the manufacturer may be put in another way with the same result. When a manufacturer proposes to carry out the ideas of his customer, and furnishes a sample to show what he can do, surely in effect he says, "This is the sort of thing you want, the rest is my business, you may depend upon it that there is no defect in the manufacture which would prevent goods made according to that sample from answering the purpose for which they are required."

CASE IN POINT

An aircraft owner wished to buy an overhauled engine for his aircraft, and purchased one that had been overhauled and certified as airworthy by an aircraft-maintenance engineer. The engine was installed in the aircraft along with an oil pump and oil cooler supplied by the aircraft owner. The cooler and pump were apparently not certified, and there was no evidence that they had been either carefully maintained or properly repaired. On a second test flight, the engine-oil pressure dropped seriously, and the aircraft owner had to make an emergency landing. On examination, extensive internal damage had occurred in the engine. At that time, it was discovered that the engine had one set of unauthorized bearings installed, but this apparently was not the cause of the engine failure.

The aircraft owner took legal action against the engineer and the seller of the engine, alleging negligence and breach of contract.

At trial, the judge dismissed the plaintiff's claim on the basis that the engine had been properly overhauled, and the fault lay with the plaintiff's parts. On appeal, the court allowed the plaintiff's claim for the cost of the improperly sized bearings, but dismissed the balance of his claim, on the basis that the engine was otherwise fit for the use intended.

See: *Shavit v. MacPherson et al.* (1998), 123 Man. R. (2d) 270.

24. *James Drummond & Sons v. E.H. Van Ingen & Co.* (1887), L.R. 12 H.L. 284.
25. ibid.

Caveat Emptor and "Fitness for Use Intended"

Caveat emptor

Latin: "Let the buyer beware."

As to quality and fitness for a particular purpose, the buyer is, to a certain extent, subject to ***caveat emptor*** ("let the buyer beware"). The law assumes that the buyer, when given an opportunity to examine the goods, can determine the quality and the fitness for his or her purpose. The *Sale of Goods Act* does, however, impose some minimum obligations on the seller. Where the seller is in the business of supplying a particular line of goods, and where the buyer makes the purpose for which the goods are required known to the seller, and where the buyer relies on the seller's skill or judgment to supply a suitable product, there is an implied condition that the goods provided shall be reasonably fit for the use intended.[26] This rule would not apply, however, in a case where the buyer requests a product by its patent or trade name, as there would be no implied condition as to its fitness for any particular purpose.[27] This particular proviso means that any time that a purchaser orders goods by "name" rather than leaving the selection to the seller, the buyer will have no recourse against the seller if the goods fail to perform as expected, as the buyer was not relying on the seller's skill to select the proper product.

The importance of distinguishing between reliance on the skill of the seller and merely asking for goods by trade name was examined in the case of *Baldry v. Marshall*,[28] where the judge noted:

> The mere fact that an article sold is described in the contract by its trade name does not necessarily make the sale a sale under a trade name. Whether it is so or not depends upon the circumstances. I may illustrate my meaning by reference to three different cases. First, where a buyer asks a seller for an article which will fulfil some particular purpose, and in answer to that request the seller sells him an article by a well-known trade name, there I think it is clear that the proviso does not apply. Secondly, where the buyer says to the seller, "I have been recommended such and such an article" — mentioning it by its trade name — "will it suit my particular purpose?" naming the purpose, and thereupon the seller sells it without more [sic], there again I think the proviso has no application. But there is a third case where the buyer says to a seller, "I have been recommended so and so" — giving its trade name — "as suitable for the particular purpose for which I want it. Please sell it to me." In that case I think it is equally clear that the proviso would apply and that the implied condition of the thing's fitness for the purpose named would not arise. In my opinion the test of an article having been sold under its trade name within the meaning of the proviso is: Did the buyer specify it under its trade name in such a way as to indicate that he is satisfied, rightly or wrongly, that it will answer his purpose, and that he is not relying on the skill or judgment of the seller, however great that skill or judgment may be?

Merchantable Quality

Merchantable quality

Goods of a quality standard suitable for re-sale.

In general, where goods are bought by description from a seller who deals in such goods, there is an implied condition that the goods shall be of **merchantable quality**. However, if the buyer has examined the goods, the implied condition would not apply to any defect in the goods that would have been revealed by the examination.[29]

---

26. *Sale of Goods Act*, R.S.O. 1990, c. S.1, s. 15(1). See also *Canada Building Materials Ltd. v. W.B. Meadows of Canada Ltd.*, [1968] 1 O.R. 469.
27. *Sale of Goods Act*, supra, s. 15(1). See also *Baldry v. Marshall*, [1925] 1 K.B. 260.
28. *Baldry v. Marshall*, supra.
29. *Sale of Goods Act*, R.S.O. 1990, c. S.1, s. 15(2).

COURT DECISION: Practical Application of the Law

Sale of Goods—Merchantable Quality—Consequential Loss

***Pihach v. Saskatoon Diesel Services Ltd.*, 2004 SKPC 79.**

The plaintiff trucker bought a $10,000 rebuilt diesel engine for his truck from the defendant. Except for the invoice for the sale of the engine there was no written contract. The invoice contained the following:

CPL 1553, ENGINE COMES COMPLETE WITH ALL COVERS, BLOCK HEATER, HARMONIC BALANCER, OIL COOLER, WATER PUMP, INJECTORS, TURBOCHARGER, INJECTION PUMP. ENGINE COMES COMPLETE WITH OIL & FILTER. ***ENGINE INCLUDES A ONE YEAR UNLIMITED MILEAGE WARRANTY ON PARTS & LABOUR.***

Eight days later the engine stopped running while the plaintiff was hauling freight in the U.S.A.

THE COURT: The problem was traced to a leak in the oil cooler. The problem was repaired at a cost of $969.32 (U.S. dollars). The defendant agrees that the plaintiff is entitled to be paid for the repair costs but not for the costs of the hotel expenses of 276.92 (U.S. dollars) the plaintiff had to pay while the engine was being repaired and not for the loss of income that the plaintiff suffered because of the delay caused by the engine repair. Counsel for the defendant argued that no consequential damage or economic loss can be recovered by the plaintiff. He submitted no authority for that proposition. There is nothing on the invoice that would expressly exclude any warranty which *The Sale of Goods Act* provides and there was no evidence that the parties had expressly agreed that the warranty contained in the invoice was the only warranty to govern the sale of the engine. I am unable to agree with the submission of the defendant's counsel that consequential damages could not be awarded. Subsection 52 (2) of *The Sale of Goods Act* provides: "The measure of damages for breach of warranty is the estimated loss directly and naturally resulting in the ordinary course of events from the breach of warranty." The defendant was a seller that dealt in diesel engines and there was no evidence that the defect in the oil cooler could have been detected by an examination by the plaintiff. I find that the engine was not of merchantable quality as that term has been judicially interpreted. The defendant is in breach of warranty provided by subsection 16 (2) of *The Sale of Goods Act* [goods to be of merchantable quality]. In my view, the plaintiff is entitled to damages for his hotel costs and loss of profit. These damages were reasonably foreseeable as a likely consequence of the loss of use of his truck. The defendant is a professional trucker and a loss of income because of a loss of use of his truck resulted directly and naturally in the ordinary course of events from the breach of the warranty. Economic loss due to a loss in profit was awarded in the *Horseshoe Creek Farms Ltd v. Sterling Structures Co. Ltd* case. Judgment for the plaintiff for the following amounts in Canadian dollars: a) Repair cost $1,313.88, b) Motel cost, $378.94, c) Loss of profit $1,000.

Rewrite the warranty so that it gives comfort to purchasers as to merchantable quality but protects the vendor from claims for economic losses.

Delivery

The seller also has a duty to deliver goods as specified in the contract in the right quantity, at the right place, and at the right time. The time of delivery, if stipulated in the contract, is usually treated as a condition. If the seller fails to deliver the goods on time, the buyer may be free to reject them if delivery is late.[30] If no time for delivery is specified, the goods must usually be delivered within a reasonable time.[31]

Delivery of the proper quantity is also important. If the seller should deliver less than the amount fixed in the contract, the buyer may reject the goods, as this generally is a condition of the contract and a right of the buyer under the *Sale of Goods Act*.[32] If the buyer accepts the lesser quantity, then the buyer would be obliged to pay for them at the contract rate.[33] The delivery of a larger quantity than specified in the contract, however, does not obligate the buyer to accept the excess quantity. The buyer may reject the excess, or may reject the entire quantity delivered. However, if the buyer should accept the entire quantity, the buyer usually must pay for the excess quantity at the contract price per unit.[34]

30. ibid., s. 27, and s. 29(1).
31. ibid., s. 28(2).
32. ibid., s. 29(1).
33. ibid., s. 29(1).
34. ibid., s. 29(2).

The importance of exact performance in terms of delivery was described by the court in the following terms:[35]

> ... the right to reject is founded upon the hypothesis that the seller was not ready and willing to perform, or had not performed his part of the contract. The tender of a wrong quantity evidences an unreadiness and unwillingness, but that, in my opinion, must mean an excess or deficiency in quantity which is capable of influencing the mind of the buyer. In my opinion, this excess is not. I agree that directly the excess becomes a matter of possible discussion between reasonable parties, the seller is bound to justify what he has done under the contract; but the doctrine of de minimis cannot, I think, be excluded merely because the statute refers to the tender of a smaller or larger quantity than the contract quantity as entitling a buyer to reject.
>
> I wish to add this. The reason why an excess in tender entitles a buyer to reject is that the seller seeks to impose a burden on the buyer which he is not entitled to impose. That burden is the payment of money not agreed to be paid. It is prima facie no burden on the buyer to have 55 lbs. more than 4,950 tons offered to him, and there is nothing to suggest that these sellers would have ever insisted, or thought of insisting, upon payment of the 4s. over the 40,000£. The sellers' original appropriation appeared to be within the proper quantity. The excess of 55 lbs. appears when the quantity shipped is converted from kilos into tons. If the sellers had expressly or impliedly insisted upon payment of the 4s. upon their view of the contract, the case would have been different; but nothing of that kind can be supposed to have taken place here.

The place for delivery is usually specified in the contract. However, if the parties have failed to do so, the seller is only obliged to have the goods available and ready for delivery at his place of business if he has one, or if not, at his place of residence. If the parties are aware that the goods are stored elsewhere, then the place where the goods are located would be the place for delivery.[36] The place of delivery is often expressly or impliedly fixed when goods are sold. If this should be the case, or if by some custom of the trade the delivery takes place elsewhere than the seller's place, then the seller would be obliged to make delivery there.[37]

When a contract calls for delivery by installments, and the seller fails to make delivery in accordance with the contract, the buyer is often faced with a dilemma. If a lesser quantity is delivered and the contract calls for separate payments for each installment, the buyer may take delivery, if the buyer so desires, and pay for the goods delivered. If the buyer wishes to reject the goods, he or she must take care. The buyer may not treat the failure to deliver the proper amount on a particular installment as a basis for repudiation of the contract, unless the buyer is certain that he could satisfy the courts that the quantity delivered was significantly below the requirement set out in the contract, and that there was a high degree of probability that the deliveries would be equally deficient in the future.[38] This problem is normally limited to installment contracts requiring separate payments, however. If the contract does not provide for a separate payment for each installment, the contract is generally treated as being indivisible, and the buyer is free to repudiate the whole agreement.

Limitation of Liability

In contracts of sale that are not sales to a consumer, the seller may, by an express term in the contract, exclude all implied conditions and warranties that are imposed under the *Sale of Goods Act*. If this is done, however, the seller must comply exactly with the terms of the contract made. For example, a buyer entered into a contract with a seller for the purchase of a new truck for his business. The purchase agreement provided that "all conditions and warranties implied by law are excluded." The truck delivered by the seller was slightly used and did not correspond to the description. The buyer in this

35. *Shipton, Anderson & Co. v. Weil Brothers & Co.*, [1912] 1 K.B. 574.

36. *Sale of Goods Act*, R.S.O. 1990, c. S-1, s. 28(1).

37. ibid., s. 28(1).

38. ibid., s. 30(2).

case was entitled to reject the truck, as he did not receive what he contracted for (a new truck).[39] The particular thrust of most cases on limitation clauses is to limit the extent to which a seller may avoid liability. Many cases of this nature are decided on the basis of fundamental breach, or on the basis of strict interpretation of the seller's duties under the agreement. Additional protection is afforded to consumers in most provinces and territories by limiting or eliminating entirely the seller's right to exclude implied conditions and warranties in contracts for the sale of consumer goods.[40]

The protection of the consumer has been carried one step further in some jurisdictions. Not only is the seller prevented from excluding implied warranties and conditions from the contract in a consumer sale, but any verbal warranties or conditions expressed at the time of the sale not included in a written agreement would also be binding on the seller.[41] In addition, consumer-protection legislation often provides a "cooling-off" period for certain consumer sales contracts made elsewhere than at the seller's place of business. This allows the buyer to avoid the contract by giving notice of his or her intention to the seller within a specified period of time after the contract is made. The most common type of contract of this nature is one in which a door-to-door salesperson sells goods to a consumer in the consumer's home.[42] The purpose of the cooling-off period is to allow buyers to examine the contracts at their leisure after the seller has left. If, after reviewing their actions, the buyers decide that they do not wish to proceed with the contract, they may give the seller notice in writing within the specified period (usually up to ten days) and the contract will be terminated. In each case of this kind, where the legislation applies, the contract is essentially in suspension until the cooling-off period expires. It is only then that it becomes operative.

The general trend in consumer-protection legislation in recent years has been to impose greater responsibility on the seller in the sale of goods. While the rule of *caveat emptor* is still very much alive, the right of the buyer to avoid a contract has been expanded beyond the normal rights of the commercial buyer. The justification for the change is based upon the premise that the buyer and seller are no longer the equals presumed by contract law. Many sales are offered on a "take-it-or-leave-it" basis. In other instances, high-pressure selling techniques or methods have placed the buyer at a particular disadvantage. The widespread use of exemption clauses has also been a factor that prompted legislation to redress the balance in negotiating power and to ensure honesty on the part of sellers in their dealings with buyers.

COURT DECISION: Practical Application of the Law

Sale of Goods—Limitation Clauses—Application and Effect

***Foley v. Piva Contracting Ltd. [Douglas Lake New Holland]*, 2005 BCSC 651.**

The plaintiffs purchased a tractor from the defendant, Piva. The $73,700 tractor required significant warranty service on 18 occasions in 18 months. They ultimately returned it to the dealer in May 2000, rejecting it altogether. The plaintiffs say the tractor was unfit for the use intended, fundamental breach has taken place, and rely on the Sale of Goods Act, and its implied warranties.

THE COURT: Piva's position is that there has been no fundamental breach of contract...and that the document signed excludes any implied warranties, including those contained in the Act. Warranties are referred to in the New Holland Canada Ltd. Warranty and Limitation of Liability — Agricultural Products document. ... in smaller printing, are the words:

> The above and the terms (and the relevant warranty, if any) and on any initialed attachments hereto, shall comprise the entire agreement affecting this purchase and no other agreement, understanding, condition or

39. *Andrews Bros. (Bournemouth) Ltd. v. Singer & Co. Ltd.*, [1934] 1 K.B. 17.
40. *Consumer Protection Act*, 2002, S.O. 2002, c. 30, Schedule A, s. 9(2).
41. *Consumer Protection Act*, 2002, S.O. 2002, c. 30, Schedule A, s. 9(3).
42. *Consumer Protection Act*, 2002, S.O. 2002, c. 30, Schedule A, s. 43(1).

> warranty either expressed or implied by law or otherwise is a part of this transaction, any such agreement, understanding, condition or warranty being hereby expressly excluded.

The New Holland Warranty and Limitation of Liability state that the remedy of repair during the warranty period shall be the purchaser's exclusive remedy. However, that document also says:

> To the extent allowed by law, any implied warranty of merchantability or fitness applicable to this product is limited to the stated duration of this written warranty, neither the company nor the selling dealer shall be liable for loss of the use of the product, loss of time, inconvenience, commercial loss or consequential damages.

This does not say that the implied warranties under the Act do not apply. This only says that they apply during the duration of the written warranty. In the decision *Hunter Engineering Co. v. Syncrude Canada Ltd.* (1989), 57 D.L.R. (4th) 321 (S.C.C.), Wilson J. dealt with an argument that two exclusionary clauses were sufficient to preclude the application of the statutory warranty. In dealing with those two exclusion clauses, she said:

> (1) that an exclusion clause should be strictly construed against the party seeking to invoke it and (2) that clear and unambiguous language is required to oust an implied statutory warranty. I find...the Allis-Chalmers agreement did explicitly and unambiguously oust the statutory warranty by stating: "The Provisions of this paragraph represent the only warranty of the seller and no other warranty or conditions, statutory or otherwise shall be implied."

Given all of these difficulties and problems, I find that this tractor did not perform the way it was expected to, which amounted to a fundamental breach of the contract. The exclusion clause may or may not have been intended to apply to the type of breach which occurs. However, it would have to be stated in clear and unambiguous terms in order to apply. [The exclusion clause] merely limits it to the duration of the written warranty. The document also provides that the selling dealer is not liable for loss of use of the product, loss of time, inconvenience, commercial loss or consequential damages to the extent allowed by law.

The plaintiff was awarded the money he paid for the tractor, but nothing for his claim of commercial loss and consequential damages. Compare the Allis Chalmers (Syncrude) exclusion clause with those of New Holland. Did, in your opinion, the judge do some "stickhandling" to come up with this judgment?

A QUESTION OF ETHICS

Much of the original use of limitation (or exception/exculpatory) clauses came as a response to the *Sale of Goods Act* — a seller's attempt to simply avoid all responsibility for goods after they were sold. The doctrine that fundamental breach could not be excused by a limitation clause was, in turn, the court's response to that attempt. What examples of limitation clauses have you seen in your experiences that seem too high-handed to be acceptable?

CONTRACTUAL DUTIES OF THE BUYER

Apart from the general duty of the buyer to promptly examine goods sent on approval, or to compare goods delivered to a sample, the buyer has a duty to take delivery and pay for the goods as provided in the contract of sale or in accordance with the *Sale of Goods Act*. The delivery of the goods and the payment of the price are concurrent conditions in a sale, unless the parties have provided otherwise.[43] For example, if the contract is silent on payment time and place, then the buyer would be obliged to pay a reasonable price at the time of the delivery of the goods.[44]

Payment is not a condition under the contract unless the parties specifically make it so. Under the Act, payment is treated as a mere warranty.[45] As such, it would not entitle the seller to avoid performance if the buyer failed to pay at the prescribed time. The seller

43. *Sale of Goods Act*, R.S.O. 1990, c. S.1, ss. 26, 27.

44. ibid., s. 9(2).

45. ibid., s. 47.

would, however, have the right to claim against the buyer for breach of the warranty and to recover any damages the seller might have suffered as a result of the buyer's default.[46]

Remedies of the Buyer

Rescission

The seller is subject to a number of conditions and warranties in addition to those that may be set out in the contract itself. The rights of the buyer under the contract are, for the most part, governed by the manner in which the seller fulfills the contract terms, and the manner in which the seller complies with the various implied warranties and conditions. If the seller's breach of the agreement is a breach of a condition or a breach that goes to the very root of the agreement (for example, something that the courts would treat as a fundamental breach), the buyer may be in a position to repudiate the contract and reject the goods. If a buyer is entitled to repudiate the contract, the buyer also has the right to refuse payment of the purchase price; or if the buyer has already paid the price, or a part of it, the buyer may recover it from the seller. The buyer has an alternate remedy in a case where the seller fails to deliver the goods. The buyer may purchase the goods elsewhere, then sue the seller for the difference between the contract price and the price paid in the market.[47]

Damages

In some cases, the seller may be in breach of contract, but only of a minor term, or one that does not go to the root of the contract. For example, in a contract where the seller is obliged to deliver goods according to sample by installments at a fixed price each, the seller may make one delivery that is slightly deficient. In that case the buyer would not be entitled to repudiate the contract as a whole, but only the particular installment.[48] "Microscopic" variation in deliveries, however, would not likely be treated as a breach of contract.[49] If the contract is not severable and the buyer has accepted the goods, or a part of them, or if the contract is for specific goods and the property in the goods has passed to the buyer, the breach of any condition by the seller may only be treated as a breach of warranty. Thus, the buyer would not be entitled to reject the goods or repudiate the agreement unless entitled to do so by an express or implied term in the agreement to that effect.[50] The buyer, if he or she elects to do so, may treat any breach of a condition as a breach of warranty. In that case the contract would continue to be binding on the buyer, but the buyer would be entitled to sue the seller for damages arising out of the breach.[51]

Specific Performance

A third remedy is available to the buyer on rare occasions. If the goods in question have some unique or special attribute or nature, and cannot be readily obtained elsewhere, monetary damages may not be adequate as a remedy if the seller refuses to make delivery. Under such circumstances, the remedy of specific performance may be available to the buyer, at the discretion of the courts.[52] Unless the contract is for the sale of something in the nature of a rare antique or work of art, however, the courts are unlikely to award the remedy, as monetary damages are normally adequate in most sales transactions.

---

46. ibid., s. 47.
47. ibid., s. 49.
48. *Jackson v. Rotax Motor & Cycle Co.*, [1910] 2 K.B. 937.
49. *Shipton, Anderson & Co. v. Weil Bros. & Co.*, [1912] 1 K.B. 574.
50. *Sale of Goods Act*, R.S.O. 1990, c. S.1, s. 12(3). See also *O'Flaherty v. McKinlay*, [1953] 2 D.L.R. 514.
51. *Sale of Goods Act*, supra, s. 12(1).
52. ibid., s. 50.

FRONT-PAGE LAW

N.S. woman sues Coca-Cola for almost $11M

HALIFAX — A Nova Scotia woman is suing Coca-Cola for nearly $11-million because she found shards of glass in a bottle of Fruitopia she drank while she was pregnant in 1997.

In her statement of claim, Sylvia Louise Gillard O'Brien of Shad Bay asks for $7-million in psychological, emotional and physical damage to herself and her three-year-old son. She is also asking the court to award her $1.5-million for special damages listed as "child bonding to mother," and $2.25-million in lost income.

In an affidavit filed with the suit, Ms. O'Brien says she has a "small, struggling" company and is trying to build contacts. The suit says she will consider an out-of-court settlement.

"The impact of the publicity regarding this case would be of great concern to Coca-Cola Bottling Company in terms of sales and profitability to the new Fruitopia product line, not to mention ongoing consumer confidence," Ms. O'Brien writes in her statement of claim.

The suit claims Ms. O'Brien had consumed a third of her Fruit Integration Fruitopia while shopping at the Quinpool Road IGA in September, 1997, when she noticed her lip was bleeding and heard her teeth crunch on something hard. In her lawsuit, filed on Tuesday at Nova Scotia Supreme Court, Ms. O'Brien says she spit shards of glass into a napkin and found more glass in the cover.

She went to the QEII Health Sciences Centre afterward, worried she was going to have her baby prematurely, her suit states. "I was having stomach pains, not contractions, all the the time I was there, and they instructed me to go home to relax. I didn't sleep the whole night," Ms. O'Brien says in her affidavit. "I was very upset and worried."

Her suit does not say when she gave birth, but claims she threw up blood before and after her son's birth by caesarean section. No defence has been filed yet.

This claim apparently refers to a willingness to entertain an out-of-court settlement, and addresses the impact of publicity, whereas most plaintiffs tend to confine themselves to describing the events that led up to their injury. The amounts claimed are higher and different than those traditionally entertained by Canadian courts in similar situations in the past. Do you feel these are in line with a more modern sense of liability, or are they an unwise direction for courts to head?

Source: By Rachel Boomer, "N.S. woman sues Coca-Cola for almost $11 M," courtesy of *The Daily News*.

REMEDIES OF THE SELLER

Lien

The remedies available to the seller in the event of a breach of the contract of sale are to some extent dependent upon the passing of the title to the buyer, as well as the right of the seller to retain the goods. These rights may be exercised either against the buyer personally, or against the goods themselves, depending upon the circumstances and the nature of the remedy. The seller normally may not repudiate the contract in the event of non-payment by the buyer, unless payment has been made a condition in the contract. The seller, however, is not obliged under the *Sale of Goods Act* to deliver the goods, unless payment is made or credit terms are granted by the seller for the purchase in question.[53] In this respect, the seller may claim a lien on the goods. This may be done if the sale is a cash sale, or if the sale is a credit sale and the period of credit has expired, as, for example, where the goods are sold on a "lay-away" plan. The seller may also claim a lien on the goods if the buyer should become insolvent before the goods are delivered.[54] A seller's lien depends, of course, upon possession of the goods. If the seller should voluntarily release the goods to the buyer, the right of lien may be lost.

Action for the Price

If the seller has delivered the goods to the buyer, and the title has passed, the seller may sue the buyer for the price of the goods.[55] An action for the price would also lie where the title has not passed but the seller delivered the goods, and where delivery was refused by

53. ibid., s. 39(1). See also *Lyons (J.L.) v. May & Baker* (1922), 129 L.T. 413.

54. *Sale of Goods Act*, R.S.O. 1990, c. S.1, s. 39(1).

55. ibid., s. 47(1).

the buyer. In this case, the seller has no obligation to press the goods on the buyer, but may simply sue the buyer for the price.[56] The seller must be prepared, of course, to deliver the goods if the seller recovers the price.

Damages

A more common remedy available to the seller is ordinary damages for non-acceptance. This remedy permits the seller to resell the goods to another, and sue the buyer for the loss incurred.[57] The damages that the seller may recover would probably be the monetary amount necessary to place the seller in the same position as he or she would have been in had the transaction been completed. The amount would either be the profit lost on the sale, or perhaps the difference between the disposal price of the goods (if the seller sold them privately) and the contract price.

Retention of Deposit

A feature common to many contracts is a clause that entitles the seller to retain any deposit paid as liquidated damages if the buyer should refuse to perform the contract. A deposit is not necessary in a written agreement to render it binding on the parties. However, its advantages would be to circumvent the requirements of writing under the *Sale of Goods Act* if the contract is unwritten, and if it is for more than the stipulated minimum. The second advantage (from the seller's point of view) is that it represents a fund that the seller might look to in the event of a breach of the agreement by the buyer. If the agreement provides for the payment of a deposit by the buyer and also provides that, in the event of default by the buyer, the seller might retain the deposit as liquidated damages, then if the default should occur, the seller would possess funds sufficient to cover the estimated loss. The amount of the deposit required, however, must be an honest estimate by the parties of the probable loss that the seller would suffer if the buyer should default. If it does not represent an honest estimate, in the sense that the payment is a substantial part of the purchase price rather than a deposit, the seller may not retain the part-payment. However, the seller would be obliged to return the excess over and above the actual loss flowing from the buyer's default.[58]

Stoppage in Transitu

Stoppage in transitu

The right of the seller to stop delivery of goods by the carrier if the buyer is insolvent.

An additional remedy available to the seller, in cases where the seller has shipped the goods by carrier to the buyer, is ***stoppage in transitu***. If the seller has parted with the goods, but discovers that the buyer is insolvent, he or she may contact the carrier and have delivery stopped. "Insolvent" does not mean in this instance the actual bankruptcy of the buyer. It means only that the buyer is no longer meeting his or her debts as they fall due. A particular difficulty associated with this remedy relates to this fact. If the seller should stop delivery, and if the buyer is not insolvent, the buyer may claim compensation from the seller for the loss caused by the wrongful stoppage of the goods. However, if the buyer should be insolvent, and the seller is successful in stopping the carrier before delivery is made, then the title will not pass to anyone who has notice of the stoppage. If the seller fails to contact the carrier in time, and the goods have been delivered to the buyer or the buyer's agent, it is too late, and the title will be in the buyer.[59] The same would hold true if the buyer had sold the goods to a bona fide purchaser for value and without notice of the stoppage.

Recovery of Goods

Amendments to the *Bankruptcy and Insolvency Act*[60] provide a remedy for the seller of goods under certain circumstances. If goods are shipped to the buyer and the buyer becomes bankrupt, the seller may, within 30 days, submit a written demand to the trustee in bankruptcy for a return of the goods, provided that the goods are unsold, still in the

56. ibid., s. 47(2).

57. ibid., s. 48.

58. *Stevenson v. Colonial Homes Ltd.* (1961), 27 D.L.R. (2d) 698; see also *R.V. Ward Ltd. v. Bignall*, [1967] 2 All E.R. 449.

59. *Plischke v. Allison Bros. Ltd.*, [1936] 2 All E.R. 1009.

60. S.C. 1992, c. 27.

bankrupt buyer's possession, identifiable, and in the same condition as when delivered to the buyer. It is important to note that the seller's rights to the goods rank above the claims of any other secured or unsecured creditor to the goods. If the buyer has paid a deposit or partly paid for the goods, the seller may acquire the goods by refunding the amount paid or may repossess that portion of the goods represented by the unpaid portion of the account.

Resale

The act of stopping the goods in transit does not affect the contract between the buyer and the seller. It simply represents a repossession of the goods by the seller. The seller is then entitled to retain the goods pending tender of payment of the price by the buyer. If the buyer does not tender payment, then the seller has the right to resell the goods to a second purchaser, and the second purchaser will obtain a good title to the goods.[61]

CHECKLIST FOR APPLICATION OF THE SALE OF GOODS ACT

1. Is the contract for goods?
2. Are the goods (present or future) tangible goods?
3. Is there monetary consideration for the goods?
4. Is the contract exempt from writing or has the requirement been fulfilled?

If yes to all, and the parties have not made specific provisions of their own, then the Act governs the conduct of the parties or cures the defects of their contract.

ELECTRONIC SALE OF GOODS

Electronic retailing requires new practices and procedures to minimize the chance of disputes and problems, and all aspects of the *Sale of Goods Act* equally apply to electronic contracts. Since there is no face-to-face transaction, or even the perspective of an address or area code, misunderstandings as to the jurisdiction of applicable law often occurs. Most Internet issues are just old problems dressed in new clothes. They can be managed via Web site policies in the same manner that businesses used to preprint their sales terms on the backs of their paper forms. The Management Alert: Best Practice below covers a number of the more important Web practices for electronic sale of goods.

MANAGEMENT ALERT: BEST PRACTICE

1. A provision designating the (provincial) jurisdiction whose law is to govern the contract between the parties. Remember that this will establish a wide range of contractual matters, from what actually constitutes age of majority, through the process of dispute resolution, to which consumer protection laws are applicable.
2. A click-wrap agreement (acceptance or contract terms) whose "manner of presentation" is clear and not misleading as to its significance.
3. Clarity in description of goods or services to minimize questions of a common meeting of the minds between the parties.
4. "CONFIRM" radio-buttons at each of the client's decision steps.
5. Remember, online retail transactions differ little, in principle, from traditional catalogue shopping. Accordingly, adhering to ethical direct-mail principles is a good starting point for electronic retailers.
6. Silence on a particular topic in new (or future) electronic business legislation does not mean a lack of regulation. All existing law, by case or by statute, continues to apply to Web-based businesses.
7. Use caution in offering links from your site, and ensure that your clients know that they are leaving your site when they use them. If a link appears to be simply another page of your site, you risk being drawn into any disputes between your clients and the linked site.
8. Consider the potentially global nature of your clients. Comprehensive explanation should be given regarding currency of payment, sales taxes, shipping costs and delivery times.

61. *Sale of Goods Act*, R.S.O. 1990, c. S.1, s. 46(2).

SUMMARY

The sale of goods represents one of the most common business and consumer transactions. The law of contract in general applies to the sale, but sale of goods legislation sets out the special rules that apply to this type of contract. Two forms of contract of sale are covered by the Act: the sale and the agreement to sell. The former applies to specific goods, and the latter to goods that are not yet manufactured or not yet available for delivery. The property in the goods under a contract of sale passes when the parties stipulate that it will pass. However, if no time is mentioned, then a series of rules in the Act will apply to make this determination. Risk of loss follows the title and, as a result, the time that the title passes is important. Under a contract of sale, a buyer and seller may fix the terms. However, if the parties do not do so, the Act contains a number of implied conditions and warranties that will apply to the contract and to the goods. Implied conditions and warranties, among other things, require the seller to provide goods of merchantable quality at the time for delivery, in the right quantity, and at the right place, and to provide a good title to the goods delivered.

In most provinces and territories, under consumer-protection legislation, consumer-goods contracts may not contain exemption clauses that would eliminate implied warranties and conditions. If a seller acts in breach of a condition (which is a fundamental or major term), under certain circumstances the buyer may treat the contract as at an end and be relieved from any further obligations to perform. A breach of a warranty, which, in contrast to a condition, is only a minor term, simply entitles the buyer to sue for damages. The buyer is not entitled to avoid the contract. The buyer may, at his or her option, treat the breach of a condition as a breach of a warranty, if desired.

If a breach of contract on the part of the seller occurs, the buyer may, in the case of a breach of a condition, obtain either rescission or damages. But for a breach of a warranty, the buyer would only be entitled to damages. In rare instances, a buyer may be granted specific performance of the contract if the subject matter is unique.

The seller has seven possible remedies if the buyer refuses delivery or fails to pay the price. The seller is entitled to take action for the price or claim a lien on the goods until payment is made, if the goods are still in the seller's possession. If the buyer rejects the goods and the seller must resell them, the seller may sue for the loss, or the seller may retain any deposit paid as liquidated damages if the contract so provides. Finally, if the seller ships the goods to the buyer and then discovers the buyer is insolvent, the seller may stop delivery and hold the goods until payment is made, recover the goods in the event of bankruptcy (under certain circumstances), or resell them to a second buyer. If a resale is made, the second buyer will obtain a good title to the goods, notwithstanding the prior sale to the original buyer.

KEY TERMS

Title, 409

Condition, 414

Warranty, 414

Caveat emptor, 416

Merchantable quality, 416

Stoppage in transitu, 423

REVIEW QUESTIONS

1. Under what circumstances would the skill and judgment of the seller give rise to an implied warranty or condition upon which the buyer might rely?
2. Why is the time of passage of title important in the sale of goods?
3. Under what circumstances would a buyer of goods be entitled to rescind the contract? Give an example.
4. Indicate the significance of "notice" in the sale of goods.
5. Outline the contractual duties of a seller under the *Sale of Goods Act*.
6. What implied warranties are part of a sale of goods?
7. Distinguish between a "warranty" and a "condition." Why is this distinction important?
8. Explain the significance of *caveat emptor* in the sale of goods.

9. What are the implications of an unconditional contract for the sale of specific goods in a deliverable state?
10. Outline the remedies available to a seller of goods if the buyer fails to comply with the contract.
11. Explain "*stoppage in transitu.*"
12. Distinguish a "sale" from an "agreement to sell." Why and when is this distinction important?
13. If goods that are the subject matter of a contract for sale are stolen by a thief during the "cooling-off" period, who bears the loss — the buyer or the seller?

MINI-CASE PROBLEMS

1. Andrew wished to buy paint for a dock he had built at his cottage, so he visited Sheldon's Paint Store to obtain something suitable. He asked the clerk for dock paint that would not peel under damp weather conditions. Andrew mentioned that he had heard Brand X was good, but wished to have the clerk's opinion as to its suitability for use on a dock. The clerk said it was "Okay as a paint," and Andrew purchased a quantity.

   Several months after the paint was applied to the dock, it began to peel and fall off the surface.

   If the paint supplier refused to accept responsibility for the suitability of the paint, outline the rights of Andrew.
2. Fresh Fruit Juice Co. agreed to supply 100 250-litre drums of concentrated apple juice to the Institutional Food Produce Corporation, with delivery not later than October 31st.

   On October 30th, the seller delivered 125 200-litre drums of the concentrate to the purchaser's plant. The purchaser rejected the goods because they were in the small drums rather than the 250-litre size.

   Discuss the rights of the parties.
3. Basil Supply Co. shipped on credit twelve cases of goods to Able Wholesale Corporation on June 1st. On June 15th, Able Wholesale Corporation was declared bankrupt and its assets taken over by a trustee in bankruptcy. At that time, the trustee was in possession of eight cases of the goods that were shipped by Basil Supply Co.

   Discuss the rights of the parties.
4. Early on a Monday, Nora agrees to purchase a flat-screen TV from a major chain home-electronics store, to be ordered in from a central warehouse. The purchase order did not specify a delivery date, but the salesperson assured her verbally that "it will be here in a couple of days." Nora left a $100 deposit and her credit card number. On Friday, Nora saw the same model TV offered at another store, with a $500 cheaper sale price, and purchased it. She then called to cancel her order with the first store. Does the *Sale of Goods Act* affect Nora? Will she likely lose her deposit, or should it be refunded? Will she likely end up with two TVs or one? Explain why in each case.
5. Morgan owned logging equipment, which was then working in a forest in a remote area. While in the city, Morgan met and discussed business with Henri, and wound up selling the logging equipment to him. Their agreement stated that payment would be made at the end of the month. A forest fire destroyed the equipment two days before the end of the month.

   Discuss the rights of the parties.

CASE PROBLEMS FOR DISCUSSION

CASE 1

Benjamin owned and operated a dollar store, and decided that adding a full line of greeting cards and party supplies would be desirable. He contacted a wholesaler and placed an order for greeting cards, supplies and display racks. The wholesaler advised him he would receive the manufacturer's "late year package," which would cover Christmas and New Year's theme merchandise. As it was a first order, the wholesaler required payment in full by cash or credit card, and Benjamin provided his credit card number for a payment of $4,000. The delivery date was projected as November 15th. When the goods did not arrive by November 17th, Benjamin called the wholesaler, who apologized and revised the delivery date to December 1st. Benjamin accepted this and waited. December 1st passed without delivery, and increasingly agitated calls resulted in nothing but the wholesaler blaming the manufacturer. Christmas Day came and went, and on December 31st, a truck full of goods arrived. The card boxes were sealed and appeared in good order. The Christmas decorations were in good condition. The racks and equipment were sturdy and of high quality. Benjamin reluctantly accepted delivery, reasoning that he could sell the party merchandise anytime, and could sell the decorations and cards next year. After the truck drove away, Benjamin began opening the boxes. Every piece of the New Year's decorations and cards was printed with the current year.

Advise Benjamin.

CASE 2

Hightower Industries produced 20kg bags of dry dog food, and shipped them on wooden pallets. The firm required an industrial process that could be integrated into its production line that would wrap each loaded pallet in a protective plastic wrap. The firm approached an engineering design firm, Tech-Solutions, which agreed to develop such a machine for a fixed price. In the course of its design work, Tech-Solutions found two routes to such a machine. The more costly approach used stretch wrap and a pallet-spinning platform. The less costly way employed shrink wrap with pallets passing under a heater unit. Both options were offered to Hightower, who chose the less-expensive shrink wrap machine. When the machine was delivered, installation testing revealed that the bags became wet, as the heater condensed water out of trapped air under the shrink wrap, degrading both the bags and the dog food. Hightower rejected the machine. Advise both parties.

CASE 3

Small Parts Manufacturing Co. entered into an agreement with Foremost Forging Co. to have an automated stamping press made for it. The agreement called for the construction of the press and its preparation for pickup by a carrier that Small Parts Manufacturing would designate, not later than March 1st. Payment terms were 50 percent payable at the time of signing the agreement, with the balance of the price payable on March 1st.

On February 25th, the construction of the press was complete. Foremost Forging informed Small Parts Manufacturing that the press was now ready for pickup by the transport company. The parties agreed that the press would be turned over to the carrier as soon as pickup could be arranged.

The press was placed in Foremost Forging's warehouse for the carrier to pick up. However, on February 26th, the press was destroyed when an unknown arsonist set fire to the warehouse building.

Discuss the rights (if any) and the liability (if any) of the parties in this case. Indicate the possible outcome of the case if legal action should be taken.

CASE 4

A wholesaler in Toronto agreed to sell 2,000 5-kilogram cases of walnut pieces to a buyer in Vancouver. The price was to be $1.10 a kilogram with delivery F.O.B. Toronto. The goods were shipped by common carrier in accordance with the buyer's instructions.

The goods were subject to moisture and freezing during transit, and the buyer, on inspection of the goods, found them unfit for his purposes. The goods were then sold by the buyer for 66 cents a kilogram in Vancouver, while the goods were still in the hands of the carrier. In the meantime, the carrier had found a buyer willing to purchase the 10,000 kilograms of walnuts at 88 cents a kilogram. However, the carrier was unable to complete the sale because of the buyer's actions.

The Vancouver merchant later brought an action against the Toronto wholesaler and the carrier for his loss, calculated at 44 cents a kilogram.

Indicate the nature of the plaintiff's claim in this action and the defences that might be raised by the defendants.

Render a decision.

CASE 5

Davy Crockett, a northern Alberta farmer, ordered a set of logs for the construction of a log home from a lumber company that advertised log houses for sale in a back-to-the-land magazine. The magazine advertisement stated that the house was in kit form, and claimed that any qualified builder could construct it in less than ten days. The advertisement recommended hiring a qualified builder. However, it indicated that any person who had experience in house construction could probably do the work, but the result would be his or her own responsibility.

Crockett ordered the log kit early in March for delivery in the second week of April. The logs did not arrive until late May, however, when Crockett was busy planting his crop. He was unable to begin construction during the summer months, due to an injured hand; and during the fall months he was busy with his harvest. When Crockett was ready to build in late October, he unwrapped the logs and discovered that a large number of them were warped and unsuitable for construction. The lumber company normally instructed buyers to construct the house promptly on delivery, or at least within 14 days of receipt, to avoid warping. However, they had failed to do so in Crockett's case because of the late delivery. The company refused to refund Crockett's money or take back the log kit.

Crockett continued to correspond with the company concerning the logs. Eventually, some months later, the company agreed to replace the logs that

www.mcgrawhill.ca/olc/willes

had warped. When Crockett was finally able to begin construction a few weeks later, he discovered that because the replacement logs had experienced a different drying or seasoning time, they would not fit properly with the remainder of the logs in the kit. He then brought an action for rescission against the lumber company.

Discuss the arguments that might be raised in this case.

Render a decision.

CASE 6

A Swiss corporation entered into a contract with the Canadian Dairy Commission to purchase anhydrous milk fat for the production of condensed milk. The contract was executed in Canada by a New York agent of the corporation, who provided that the goods be shipped to Algeria. The Commission was advised that it was to meet the import conditions of the Algerian government, and payment was to be made on presentation of a clean bill of lading and proper certificates of analysis of the goods.

While they met the contract stipulations in Canada, the goods were rejected by the Algerian authorities because the caps on the drums had not been sealed and some of the caps had loosened during shipment, allowing the contents to spoil.

The Swiss corporation then brought an action against the Commission for rescission and reimbursement of the contract price.

Discuss the arguments that the parties might raise, and render a decision.

CASE 7

Stubert operated a produce brokerage, buying agricultural produce and reselling it to any of 30 smaller independent regional distributors. Each distributor served an area no greater than a city, and some competed with one another. The distributors generally sold to independent convenience stores, and they vied for institutional sales such as hospital kitchens.

Stubert visited a farmers' co-operative in an agricultural area, and after some discussion secured a truckload of tomatoes at the wholesale market price for Number 1 Grade Hothouse Tomatoes.

Three weeks later, a commercial-freight company truck arrived in Stubert's part of the province with the tomatoes. Stubert had the driver open the van, and he looked at the frames of cello-packed tomatoes visible from the door. They appeared fine, so he handed over his $4,400 bank draft to the driver in return for the bill of lading.

He endorsed the bill of lading and gave it back to the driver with instructions to him to carry on, as was often the case, to one of his bigger customers, a distributor in the next town. The driver was to turn over the bill of lading against a payment of $6,700.

When the driver returned to Stubert's premises, he had no payment to deliver, but he still had the entire load of tomatoes. The distributor had insisted on unloading the tomatoes before payment. He had found that while the tomatoes near the doors were Number 1 Hothouse, those beyond the doors were at best Number 3 Hothouse, or perhaps even Field Grade. The distributor rejected the shipment, packed it back on the truck, and sent the driver back to Stubert. Stubert demanded a return of his bank draft and ordered the tomatoes to be returned to the co-operative. The driver said his company rule was that a driver is to always leave the load with the last person who pays, and that one never returns money once it is received. Accordingly, he off-loaded "Stubert's" tomatoes despite the protests of Stubert, and drove away.

Advise the parties, including a commentary on the trucking company's policy.

Render a decision.

CASE 8

Maria Henderson contacted Cool Air Contracting Ltd., a local refrigeration contractor with a view to obtaining an air conditioner for use in the beverage room of her hotel. She explained to an employee of the company that she wished to install a device that would not only cool the room, but remove tobacco smoke as well. The employee described a number of room air conditioners (for which the company was the local distributor). He recommended a particular model that he indicated was adequate for a room the size of Henderson's beverage room. Henderson entered into a contract for the model suggested by the contractor and had the equipment installed.

After the equipment was put into operation, Henderson discovered that the equipment did an adequate job of cooling the room, but did not remove the smoke to any significant extent. She complained to the company, but the company was unable to alter the equipment to increase its smoke-removal capacity. When the company refused to exchange the air conditioner for a larger model, Henderson refused to pay for the equipment. The company then brought an action against her for the amount of the purchase price.

Discuss the issues raised in this case and indicate how a judge might decide the matter.

CASE 9

Grant planned to spend his winter vacation at a ski resort in the Rockies. In preparation for the holiday, he purchased a new ski outfit from a local sports-clothing merchant.

The first time he wore his new ski outfit, he noticed that his wrists had become swollen and irritated where the knitted cuffs of the jacket contacted his skin. He wore the jacket the second day, but found that after skiing for a short time, he had to return to the lodge because his wrists had again become badly irritated and had blistered.

Grant required medical treatment for the injury to his wrists. The cause of the injury was determined to be a corrosive chemical that had been used to bleach the knitted cuffs of his jacket. The chemical was one that was normally used to bleach fabric. However, from the evidence, the chemical had not been removed from the material before the cloth was shipped to the manufacturer of the jacket. Neither the manufacturer nor the retailer were aware of the chemical in the cloth, and its existence could not be detected by ordinary inspection.

The injury to Grant's wrists ruined his holiday and prevented his return to work for a week following his vacation.

Discuss the rights (if any) and liability (if any) of Grant, the sports-clothing merchant, the manufacturer of the jacket, and the manufacturer of the cloth.

The Online Learning Centre offers more ways to check what you have learned so far. Find quizzes, Weblinks, and many other resources at www.mcgrawhill.ca/olc/willes.

CHAPTER 22

Interests in Land

Chapter Objectives

After study of this chapter, students should be able to:

- Identify and understand the various estates and interests in land.
- Describe the ways in which interests in land are obtained, transferred and protected.
- Explain the registration system used to identify property interests and their ownership.

YOUR BUSINESS AT RISK

In many cases, the most valuable asset that a business owns is its interest in land. Short of ownership, there are rights of possession that can be just as important, remembering the business maxim of "location, location, location." Failing to understand where those rights begin and end places a critical business asset in constant jeopardy of being lost to competing claimants.

INTRODUCTION

The right to hold property is central to the free-enterprise system. However, even in those few countries that do not permit individuals to own land, the state is obliged to recognize certain basic property rights, if only to permit individuals to occupy some form of shelter or to allow state enterprises or co-operatives to utilize land or buildings free from interference by others. In Canada, business persons may acquire interests in land for virtually any type of business activity. Indeed, some businesses and landlords are largely built around the acquisition, use, and disposition of real property interests.

Interests in land may take on many forms, each designed to serve a particular purpose. For example, a business that requires only a small office space in a downtown location need not necessarily buy land or a building, but may acquire the necessary space by way of a lease of a part of a building. If the business person believes that ownership of a small amount of a space in a large building is desirable, then the acquisition of an office in a commercial condominium is an alternative form of property holding. These property rights are examined in this chapter, but at the outset it is important to note that most property rights are referred to as "estates" in land, which have attached to them specific rights and obligations.

Real property

Land and anything permanently attached to it.

HISTORICAL DEVELOPMENT

Fixture

A chattel that is constructively or permanently attached to land.

Real property is a term used to describe land and everything permanently attached to it. At Common Law, the term real property includes buildings constructed on the land, the minerals or anything else below the surface, and the airspace above. It may, in some instances, include chattels attached to the land in such a way that they have become **fixtures**; as a general rule, however, the term does not include ordinary moveable property. Chattels are personal property. The distinction between real and personal property at law is significant, because the law relating to each form of property evolved in a distinctly different fashion.

The law that relates to land or real property is not unique to North America. It has its origins in the laws of England, dating back to the introduction of the feudal system. Much of the terminology used today in land law and the basic concept of Crown ownership of all land was developed during that time in England.

The feudal system was essentially a system under which land was held as long as the holder of the land complied with a promise to provide the necessary armed men or services in support of the Crown. If the holder of the land failed to comply, then the land would revert (or escheat) to the Crown. Since land was essentially the only source of wealth at the time, the holder of the land was generally always aware of the importance of providing the promised support. Not all grants, however, were made in return for military service: some grants of land were made in return for services such as the supply of agricultural produce, weapons, or administrative functions. Grants of estates in land were also made to the church or religious orders, but these were discouraged by the Crown because they were incompatible with the feudal system. In all cases where the land was granted, the Crown retained ownership and could recover the land from the person in possession. This was seldom done by the Crown, however, and the various estates in land gradually took on a degree of permanency that closely resembled ownership for all but the lowest forms of landholding.

Tenure

A method of holding land granted by the Crown.

The estates carried with them a **tenure** or right to hold the land that was either free or unfree. Estates that were freehold had fixed services attached to them. The type of service, the time, and the place were determined in the grant of the estate. For example, knight service was a form of freehold tenure in which the holder was usually obliged to provide the king with the services of a fully armed knight for 40 days each year, or make an equivalent payment of money.

While freehold estates could take on a number of different forms, the highest estate in land was one that permitted the holder to pass the estate along to heirs-at-law by way of inheritance. This was important at the time, because it meant that the wealth represented by the land could be passed to succeeding generations as long as there were heirs. This form of estate was known as an estate in **fee simple**. The term fee was a derivative of the Latin word *feodum,* meaning fief or estate. Except for persons holding land directly from the Crown, the holder of an estate in fee simple was free to sell the estate if he so desired, or to devise it to another by way of a will or testamentary disposition.

Fee simple

An estate in land that represents the greatest interest in land that a person may possess, and that may be conveyed or passed by will to another, or that on an intestacy would devolve to the person's heirs.

As the feudal system declined in England, the practice of providing the Crown with personal services or particular quantities of produce declined with it. Instead, the services were generally satisfied by a monetary payment, which the Crown used to acquire the necessary services or goods. The development of trade and the gradual increase in importance of the production and sale of goods provided alternative methods of acquiring wealth. As these methods became more widespread, the value of land as a source of wealth and power declined. By the seventeenth century, the personal-service aspects of the feudal system had all but disappeared. In 1660, Parliament passed a statute that eliminated the last vestiges of personal-servitude feudal rights.[1] What remained, however, was the system of landholding based upon Crown ownership of all land, and the holding of land by individuals in the form of estates based upon the estate in fee simple.[2]

ESTATES IN LAND

Fee Simple

In Canada, all land is still owned by the Crown, and estates of land in fee simple are granted by Crown patent to individuals.[3] The patent sets out the conditions subject to which the grant is made. It is not uncommon to find that the Crown in right of the

1. *Statute of Tenures*, 12 Car. II, c. 24.
2. Some of the lesser forms of land tenures remained until the twentieth century; however, see *Law of Property Act*, 1922, 12 & 13, Geo. V, c. 16.
3. In Ontario, for example, all property in the province was granted freehold after 1791. See the *Canada Act*, 1791, 31 Geo. III, c. 31.

province has reserved either the right to all minerals or the rights to certain precious metals in the grant of the land. For example, in the past, the provinces of Ontario and some of the provinces in Eastern Canada frequently reserved all gold, silver, and precious metals. It also reserved all of the white pine trees standing on the property during the period when the British navy and merchant marine used these particular trees as material for the masts of their sailing ships. More recently, the Crown has followed the practice of reserving not only the mineral rights, but all timber standing on the property. The purchasers of the land generally purchase the timber rights separately, but the rights still represent a reservation of the Crown in the patent. In Western Canada, a Crown reservation of mineral rights is a common feature found in the patent of new lands.

Escheat

The reversion of land to the Crown when a person possessed of the fee dies intestate and without heirs.

Expropriation

The forceful taking of land by a government or government agency for public purposes.

Land that is granted by Crown patent seldom reverts to the Crown, because it is freely alienable by way of sale, will, or inheritance. As long as the land may be disposed of in one of these three ways, it does not ***escheat*** (revert) to the Crown. If a person fails to dispose of the property during his or her lifetime and dies without heirs, or dies without a will devising the property to some other person, the land will revert to the Crown.[4] In addition, the Crown may re-acquire the land for public purposes by way of expropriation. **Expropriation** differs from an escheat of the lands to the Crown, because it constitutes a forceful taking of the property, for which the Crown must compensate the person in possession when land is expropriated. The taking usually must be justified as being for some public purpose, but it nevertheless represents the Crown exercising its right of ownership. Apart from expropriation, the land, once granted by Crown patent, remains in the hands of the public, unless it should, by some accident, escheat to the Crown through the failure of an owner without heirs-at-law to provide for its disposition upon the owner's death.

A QUESTION OF ETHICS

From political values rooted in monarchy, Canada was founded on constitutional principles of "peace, order and good government." Citizens of the American colonies forged their nation with "checks and balances" on government, and a declaration of independence that describes "life, liberty and the pursuit of happiness" as unalienable rights. Would you then expect differences in the conduct of expropriations in each of our societies? Which attitudes on either side of the border might be obliged to change in order to accommodate the expectations of modern society?

Deed/transfer

Written or printed instrument effecting legal disposition.

Generally, if a person grants land during his or her lifetime, the grant is by way of a formal document called a **deed**. The grant is embodied in the document in such a way that (1) the execution of the deed by the grantor (under seal in some provinces), and (2) the delivery of the document to the grantee, passes the title to the land to the recipient. The receipt of the deed vests the title in the grantee, and the grantee, as the new freehold tenant, is entitled to exercise all the rights of "owner" with respect to the land. The owner of the land may use the land as he or she sees fit, subject only to the Common Law of nuisance and statutory enactments that restrict the use of property. For example, the owner may farm the land, cut down trees, construct buildings on the land, and use it for any purpose. The owner may also be granted lesser estates in the land.

Life Estate

Life estate

An estate in land in which the right to possession is based upon a person's lifetime.

The highest estate in land that the person in possession of the fee simple might grant (apart from the fee itself) is a **life estate**. A life estate is a freehold estate that may be held by a person other than the owner of the fee simple for a particular lifetime (usually the life tenant's own). This form of grant is frequently made within a family, where the person who possesses the fee simple may wish to pass the property to younger members of the family yet retain the use of the land during his or her lifetime. In this case, the landowner would

4. See, for example, the *Escheats Act*, R.S.O. 1990, c. E.20.

prepare a deed that grants the fee simple to the younger members of the family, but retains a life estate in the land. The effect of the conveyance would be that possession of the land remains with the grantor during his or her lifetime. On death, possession would pass to the grantees. For example, Andrews owns a parcel of land in fee simple and conveys the fee simple to Brown, reserving a life estate to himself. Brown would be the grantee of the fee, subject to the life estate of Andrews. The interest in land that Brown receives in the conveyance would be the remainder or reversion interest. On Andrews's death the life estate would end. Brown would then possess the fee simple and the right to enter on and use the land.

The owner of the land in fee simple may grant many successive life estates in a particular parcel of land if desired.

> Axleson might grant a parcel of land to Baker for life, then to Chapman for life, then to Dawson for life, and the remainder to Emmons. In this case, Baker would be entitled to the land during his lifetime, and then it would pass to Chapman for her lifetime, and then on to Dawson for her lifetime. On Dawson's death, the **remainderman** Emmons would acquire the land in fee simple.

Remainderman

A person who is entitled to real property subject to a prior interest (e.g., a life estate) and who acquires the fee when the prior estate terminates.

A life tenant, while in possession of a life estate, is expected to use the land in a reasonable manner and not to commit waste. The life tenant is under no obligation to maintain any buildings in a good state of repair. However, the tenant cannot tear down buildings, nor is the life tenant permitted to deliberately destroy the property. Normally, a life tenant is not entitled to destroy trees planted to shelter the property from the wind, nor would the tenant be entitled to destroy trees planted for ornamental purposes. The tenant may clear land for cultivation of crops or cut trees to obtain the wood for heating purposes, however. Where land is transferred to a life tenant subject to a mortgage, the life tenant is normally only obliged to pay the interest on the mortgage. The obligation to pay the principal amount rests with the person who holds the remainder or reversion interest. A life tenant is also usually obliged to pay land taxes, but not local improvement taxes charged to the land. Since the remainder interest reaps the benefits of the local improvements, the charge is his or her responsibility. However, the usual practice might be for the life tenant to pay the local improvement levy and recover it from the holder of the reversion interest.

FIGURE 22–1 **Creation of Estates in Land**

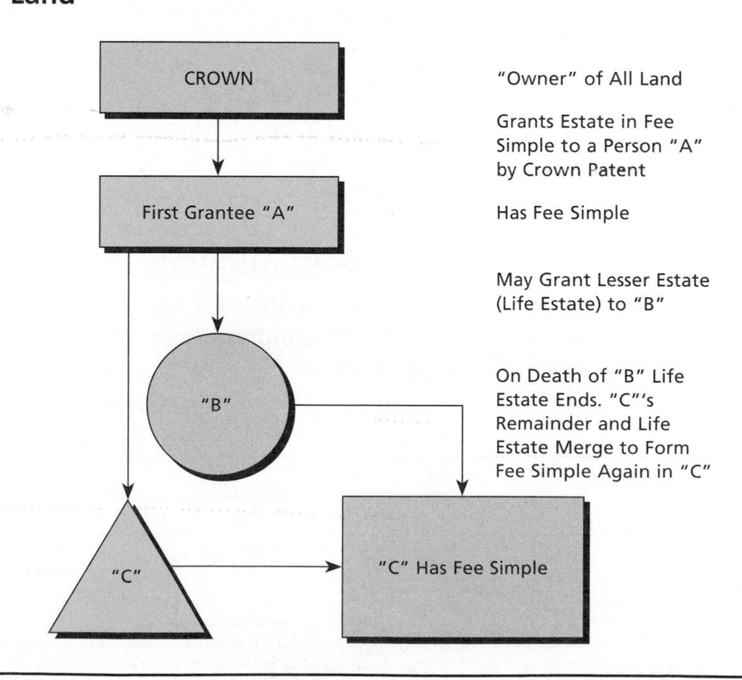

Life estates are not without certain drawbacks or disadvantages. The existence of a life estate frequently renders the property unsaleable. A prospective purchaser of the reversion is unlikely to be willing to wait until the life tenant dies to gain possession. Unless both the life tenant and the holder of the remainder are willing to convey their respective interests in the land, the purchaser would probably not be interested in acquiring the property. The same would hold true for a sale of the life estate. A purchaser of a life estate only acquires the property for the remainder of the life tenant's lifetime. The uncertainty attached to the tenure would not prompt a purchaser to pay a high price for such an estate. More important than the sale of the property, however, is the fact that the life tenant cannot alter the property during the life tenancy, nor may the holder of the remainder, as the holder of the remainder has no right to the property until the life tenant dies. As a result, the life estate is seldom used, except to convey interests in land within families.

CLIENTS, SUPPLIERS *or* OPERATIONS *in* QUEBEC

Large differences exist between the Common Law and Quebec's civil law in the area of interests in land, even if a number of similar configurations of interests can be created in both systems. They approach landholding from completely different philosophical positions. For example, the Common Law recognizes different titles existing at the same time in one property (e.g., legal title — the fee — and beneficial title, a life estate). By contrast, under civil law, there can only ever be one title to property as there is no body of equity law to create beneficial title. Given similar commercial desires however, civil law allows for the dismemberment of components of ownership: the right to use a thing (usus), to profit from its fruits (fructus) and to dispose of it (abusus). Interesting property arrangements can also be created that have no Common Law equivalent, such as selling land while retaining ownership of the buildings that sit upon it (superficies).

Leasehold Estate

Leasehold estate

Grant of the right to possession of a parcel of land for a period of time in return for the payment of rent to the landowner.

A **leasehold estate** differs from a life estate or an estate in fee simple. It represents a grant of the right to possession of a parcel of land for a period of time in return for the payment of rent to the landowner. A leasehold estate, while an ancient form of tenancy, is contractual in nature, and given for a fixed term. The law has always treated leasehold interests in land as being distinctly different from freehold interests, and the body of law applicable to these interests in land remains separate from the law concerning leases. Since a lease is contractual in nature, the parties may insert in the agreement any rights or obligations they wish, provided that they do not violate any law or regulation affecting leasehold interests in the jurisdiction in which the land is located.[5]

A leasehold interest grants the tenant exclusive possession of the property for the term of the lease, provided that the tenant complies with the terms of the lease agreement. Specific legislation in each jurisdiction governs leasehold estates. These are examined in greater detail in Chapter 24.

The Condominium

Condominium

A form of ownership of real property, usually including a building, in which certain units are owned in fee simple and the common elements are owned by the various unit owners as tenants-in-common.

Estates in land may also be "mixed" to create different forms of landholding. One such creation is the **condominium**, which represents exclusive ownership of part of the property and co-ownership of other parts. The condominium concept is used for a variety of different residential, commercial, and industrial uses, where the person or corporation may wish to have exclusive freehold ownership of part of the building or property for residential or business use and shared ownership of those parts of the property used in common with others. The privately owned parts of the building or structure would be the parts used as a residence or business premise. The shared parts would usually include the exterior building structure and walls, hallways, stairwells, elevators, lawn areas, and parking lots.

5. Landlord and tenant legislation in most provinces specifies many of the duties and obligations of the parties. See, for example, *Tenant Protection Act*, S.O. 1997, c. 24.

The Common Law in England and North America has long recognized the right of property ownership in either a horizontal or vertical plane, and of rights to property above and below ground. In England, the ownership of apartments or flats has been the equivalent of condominium ownership, and the rights of these property owners over time have become clearly defined by the law. Under the Common Law, the sale of a part of a building situated above the ground represents a sale of "air rights" or space, but, nevertheless, the sale of a recognized property right. The particular difficulty with the Common Law in this regard, however, is that the establishment of the necessary rights and obligations associated with the "strata title" are extremely complex. For example, to sell the second storey of a three-storey building would require the parties to define with some degree of accuracy the right of access by the owner of the second storey, the right of the owner of the remainder of the building to the support of the third storey, and the right to services, maintenance, and many other considerations.

While the condominium unit in the case of a multi-storey building is similar to an "air right" in property, the owner of the unit usually obtains part-ownership of the common areas of the entire building and the land on which it stands, as a tenant-in-common. Originally, this took the form of co-ownership, and the owners were obliged to jointly maintain the condominium. However, by the twentieth century, the corporation appeared as a more appropriate vehicle to manage and maintain the common elements of the property and to regulate the unit owners' exercise of their rights over those parts of the condominium. Modern condominium legislation in North America, as a result, has generally taken this route, although not universally. In Canada, all of the provinces (except Nova Scotia) provide for a corporation without share capital to come into existence when the condominium is created. In the United States, the corporate form is used in some jurisdictions, and an association or unincorporated body is used in others.

Unfortunately, the documentation required for the creation of a condominium varies from province to province. In some provinces, the condominium comes into existence on the filing or registration of a description and declaration (drawn in accordance with the legislation) at the land registry or land titles office. In other provinces, additional formalities are sometimes required. British Columbia, for example[6] (where a condominium interest is referred to as a strata title), also requires the filing of a prospectus with the Superintendent of Insurance. In all cases, however, the creation of the condominium requires not only the intention to have the property used as a condominium, but the preparation of accurate surveyor's drawings of the overall project and the individual units in sufficient detail to allow the property interests to be identified and dealt with under the land registration system of the particular province. These documents are usually referred to as the description.

The document that embodies the details of the condominium is referred to by various names, depending upon the jurisdiction. In Ontario, for example, it is called a declaration; in British Columbia, a strata plan; and in Alberta, a condominium plan. In each case, the document sets out the interests of the unit owners and provides the general outline for the management and operation of the condominium as a whole. It provides also for the creation of a corporation or society to manage the property for the use and enjoyment of the unit owners. The registration of this documentation in the land registry or land titles office automatically establishes the condominium.

Condominium Development

A condominium frequently begins as a project of a land developer. A parcel of land is acquired, then a building that is specifically designed as a condominium is erected on the land. The property is divided into units (sometimes called apartments) that are generally laid out so that exclusive ownership is confined to the area enclosed by the exterior walls of the unit. Usually, the limits of the unit are described as being to the planes of the centre line of all walls, ceilings, and floors that enclose the space, although this need not always be the case. The developer is free to define it by the planes of the inside surfaces, or any

6. *Real Estate Act*, R.S.B.C. 1996, c. 397, ss. 61–75. See also the *Strata Property Act*, S.B.C. 1998, c. 43.

combination of the two.[7] Sometimes the exclusive-use area may include other parts of the building as well. In many buildings with underground parking or storage facilities, these are sometimes designated as exclusive-use areas and included in the description of the unit. This permits the unit owner to have the exclusive ownership of a parking space or storage locker on the premises, but it has the disadvantage that the management organization loses control over the areas in question. Whatever the extent of the units or exclusive-use areas may be, they must be so designated in the description.

The balance of the property, excluding the units (and exclusive-use areas, if included), is the part of the property known as the common elements, or common-use area. All of the unit owners hold this part of the land jointly as tenants-in-common. Included in this part of the property are, usually, the exterior walls of the building; all hallways, stairwells, and stairs; entrance areas; the building basement; the heating plant; land; and facilities installed for use by all of the unit owners. In some cases, it also includes a unit set aside for use by the building supervisor or manager. All of these areas must be so set out in the description as common elements, however.

In a condominium, the co-ownership rights to the common elements are tied to the ownership of the individual units. If the interest of a person in the unit is transferred to another, the interest in the common elements also will pass, as it is not possible to sever the two by way of a deed. The interest of the individual unit owner in the common element is usually related to the value of the unit owned in relation to the other units, the relative sizes of the units, or in accordance with a formula that the developer might use to establish the interest of each unit. Only two provinces[8] impose initial restrictions on the methods that may be used to calculate this value. As a result, the methods used may vary from province to province and, indeed, from condominium to condominium.

One of the simplest methods of determining the interest of the unit owner in the common elements is to divide the total cost of the condominium by the value of each unit. For example, if the cost of the condominium is $5,000,000 and the value of a particular unit is $250,000, the interest of the unit owner in the common elements would be 5 percent. While this is an oversimplification of the method used, the percentage fixed as the interest of the unit owner would be the amount that he or she would receive from the proceeds of a sale should the property be sold on the termination of its use as a condominium.

The maintenance of the common elements is related to the exclusive-use units, usually by a calculation of the maintenance cost and an apportionment of the cost to each unit on the basis of unit size or value. Included in this cost is usually the municipal property tax, insurance, property maintenance such as cleaning, snow removal, yard maintenance, elevators, and building security, and the operating costs of special facilities for recreation and entertainment. While the cost or charge apportioned to each unit may be the same percentage amount as for the interest of the unit owner in the common elements, this need not necessarily be the case. Sometimes, for example, the cost allocation to unit owners located on the ground floor may recognize the fact that some above-ground floor services are not utilized by them, and their contribution to the operating expenses is correspondingly reduced.

Condominium Management

In all provinces except Nova Scotia, the general management of the condominium is in the hands of a Board of Directors or executive of a condominium corporation. The corporation is a corporation without share capital, but each unit owner has a say in its operation. This is done by way of a right to discuss matters and vote at general meetings held for the purpose of making major decisions affecting the condominium. It is also reflected through the election of members to the Board. The day-to-day management and operation of the condominium, however, are left with the Board.

7. In condominiums registered before 1993, Saskatchewan used the planes of the centre lines. Since then the finished surfaces of the wall and floors mark the boundaries: see the *Condominium Property Act*, R.S.S. 1978, c. C-26, s. 7(2), and the *Condominium Property Act*, S.S. 1993, c. C-26.1, s. 8.

8. Quebec and British Columbia. See, for example, the *Strata Property Act*, S.B.C. 1998, c. 43, s. 99.

Apart from the general duty to manage and to maintain the common elements of the condominium in a good state of repair, the corporation has the following specific obligations:

(1) to protect the premises from damage by way of insurance;
(2) to collect common-element expenses from the unit owners in accordance with the percentage liability of each unit; and
(3) to enforce any rules established for the use and enjoyment of the common areas, and the condominium as a whole.

In addition to the rights granted to the corporation under provincial condominium legislation, the rights and duties of the corporation are governed by by-laws. The by-laws of the corporation delineate its powers and constitute the "rules" for its operation. Some provinces have specified that the buildings be regulated by by-laws and have set out the procedures for their implementation, amendment, and enforcement. Other provinces have simply provided that by-laws may be passed by the corporation, without specifying details as to their nature or implementation, other than in general terms.

The by-laws and the rules govern the use that the unit owners may make of their individual units and of the common elements. For example, by-laws may prohibit alteration of the structure of the unit, restrict the occupancy to one family, prohibit commercial use, prohibit the keeping of animals, prohibit the erection of awnings or shades on the outside of windows, or prohibit the playing of musical instruments on the premises if it constitutes a nuisance or disturbance to neighbouring units. Common-element rules generally govern the behaviour of unit owners and their guests in the common areas, or in the use of facilities such as swimming pools.

Most provinces have vested in the corporation the right of lien[9] against a condominium unit for unpaid expenses, and the right to enforce the lien in the event of non-payment. Provision is normally made in the legislation for the foreclosure or sale of the unit in a manner similar to that for mortgages or charges. The reason for the vesting in the corporation of this particular right is the importance of the contribution of each unit owner of his or her share of the common expenses. A failure on the part of one unit owner to pay shifts the burden of the expense to the remaining unit owners. To allow default to continue would not only burden the remaining unit owners, but would create a serious problem for the owners as a whole if a number of unit owners refused to honour their obligations.

In most other respects the corporations are not unlike other non-profit corporations. The officers of the corporation carry out their duties in accordance with the by-laws and the legislation for the general benefit of the property owners, and for the better use of enjoyment of the common elements. The right to vote for the directors provides a means whereby the unit owners might remove unsatisfactory directors and replace them with others more to the liking of the majority. As with most corporations, it remains in existence until the condominium is terminated, at which time it is dissolved, and any assets that it might hold are then transferred to the unit owners.

Co-operative Housing Corporations

Co-operative housing is a means by which a group of persons may acquire an indirect interest in land through a corporation. Co-operative housing corporations are normally used to establish residential housing units whereby the land and buildings are acquired and owned by the corporation. Members of the co-operative acquire a share in the corporation and the lease of a particular housing unit. In this sense they are shareholders and tenants of the corporation. If a shareholder wishes to "sell" his or her housing unit, the transaction involves the sale of the share in the co-operative and the assignment of the housing unit's lease to the purchaser.

The corporation, as the owner of the building, is responsible for its maintenance and the payment of any mortgage on the property. The shareholder/tenants in turn are responsible for their portion of the expenses of the corporation in much the same fashion

9. All provinces, however such a lien in Saskatchewan is enforced in the same manner as a mortgage: see the *Condominium Property Act*, S.S. 1993, c. C-26.1, s. 63(2)b.

as owners of a condominium unit. If the tenant of a unit defaults in payment of the expenses, the corporation has a variety of remedies that it may pursue, including eviction and foreclosure on the share interest.

Lesser Interests in Land

Easements

Persons other than the owner of land in fee simple may acquire a right or interest in land by an express grant from the owner, by statute, by implication, or by prescriptive right. The interests are usually acquired for the better use and enjoyment of a particular parcel of land called the **dominant tenement**, and represent an interest in a second parcel called the **servient tenement**.

Dominant tenement

A parcel of land to which a right-of-way or easement attaches for its better use.

Servient tenement

A parcel of land subject to a right-of-way or easement.

Interests in the lands of another may be acquired for a wide variety of reasons. For example, a person may wish to travel across the lands of another in order to gain access to a body of water that is not adjacent to the person's own land, or may wish to drain water from his or her own land across the lands of another to a catch basin. A person may also wish to place something such as a telephone line or gas pipeline on or under the lands of another for particular purposes. These rights are known as **easements**.

Easement

A right to use the property of another, usually for a particular purpose.

An easement may be granted by the owner of the fee simple (or servient tenement) to the owner of the dominant tenement by an express grant. This is often done if the owner wishes to obtain a permanent right that will run with the land and be binding on all future owners of the servient tenement. Similar rights may also be acquired by the owners of dominant tenements by way of expropriation rights granted under statute, where the legislation enables the owners to obtain a **right-of-way** across lands for public purposes, such as for a hydro powerline or a pipeline.

Right-of-way

A right to pass over the land of another, usually to gain access to one's property.

Easements may also be implied by law. These easements are sometimes referred to as rights of way of necessity. They usually arise when the grantor of a parcel of land has failed to grant access to the property sold.

If Adams buys a block of land from Baxter that is surrounded by land retained by Baxter, Adams would have no means of access to the land without trespassing on the land of Baxter. In such a case, the courts will imply a right of way for access to the land sold on the basis that a right of access was intended by the parties at the time that the agreement was made.

The same rule would also apply if a grantor sells a parcel of land, then realizes afterwards that the land retained is landlocked and without a right of access. In each situation, however, the land must be truly landlocked. If some other means of access exists, no matter how inconvenient it might be, then a right-of-way of necessity will not be implied.

An easement may also arise as a result of long, open, and uninterrupted use of a right of way over the lands of another. This type of easement is known as a prescriptive right of easement, and it may be acquired in some provinces.[10] A prescriptive right of this nature arises when the person claiming the easement uses the property openly and continuously as if by right, usually for a period of 20 years. The use must be visible and apparent to all who might see it, however. The use by night, without the knowledge of anyone, would not create an easement by prescription, as the use must be open and notorious, and adverse to the owner's rights.

The exercise of the rights in the face of the owner's title is an important component of the acquisition of the prescriptive right of easement. The use must be with the knowledge of the owner of the property, or under circumstances in which the owner would normally be aware of the use. If the owner fails to stop the use by the exercise of ownership rights during the 20-year period, the law assumes that the true owner is prepared to permit the use by the person claiming the prescriptive right. Any exercise of the right of the

10. All except Quebec, Alberta, and Saskatchewan, and lands under the Land Titles System in other provinces.

true owner to exclude the trespasser would have the effect of breaking the time period, provided that the user acknowledges the rights of the true owner and refrains from using the easement for a period of time.

Restrictive Covenants

Restrictive covenant

A means by which an owner of property may continue to exercise some control over its use after the property has been conveyed to another.

A **restrictive covenant** is a means by which an owner of property may continue to exercise some control over its use after the property has been conveyed to another. The covenant that creates the obligation is usually embodied in the conveyance to the party who acquires the land. Normally this covenant takes the form of a promise or agreement not to use the property in a particular way. For example, a person owns two building lots. He constructs a dwelling-house on one lot and decides to sell the remaining lot. He is concerned that the prospective purchaser might use the land for the construction of a multiple-family dwelling. To avoid this, the vendor may include a term in the agreement of sale that the lands purchased may only be used for a single-family dwelling. If the purchaser agrees to the restriction, the vendor may include in the deed a covenant to be executed by the purchaser, which states that the purchaser will not use the land for the construction of anything other than a single-family dwelling. Then, if the purchaser should attempt to construct a multiple-family dwelling on the property, the vendor may take legal action to have the restriction enforced.

Restrictive covenants may be used for a number of other purposes as well. They may be used to prevent the cutting of trees on property sold, control the uses of the land, require the purchaser to obtain approval of the vendor for any building constructed on the property, control the keeping of animals, or a variety of other limitations. Any covenant that attempts to prevent the purchase or use of land by any person of a particular race, creed, colour, nationality, or religion would be void as against public policy, however. But generally, any lawful restriction, if reasonable and for the benefit of the adjacent property owner, may be enforced by the courts if it is described in terms that would permit the issue of an injunction.

Restrictive covenants are used for the better enjoyment and the benefit of the adjacent property owner, usually to maintain the value of the properties, or to maintain the particular character of the area. The widespread use of zoning and planning by municipalities has eliminated the use of restrictive covenants to a considerable degree. However, in many areas they are still used for special purposes where a landowner may wish to control the use of adjacent land in a particular manner.

COURT DECISION: Practical Application of the Law

Restrictive Covenants

***Dube v. Philbrick*, 2003 NBQB 225 (CanLII).**

The restrictive covenants at issue were created twenty years prior by an owner (Central Maine) upon selling the land to a purchaser (Kersey). The covenants run with the land and were passed along to a later owner, Dube. The restrictive covenants only became contentious between Dube and his prospective purchasers, Philbrick. These covenants and easements are typical of those created by developers of housing subdivisions.

THE COURT: The present application is one brought on behalf of Mark Dube. The respondents are John Philbrick and Paula Philbrick. It is an application brought in relation to a requisition on title with respect to restrictive covenants contained in a Deed, which was made on the 1st day of May 1981 from Central Maine Leasing Company to Patrick Kersey. The conveyances by Central Maine Leasing Company to the various persons who acquired the 13 subdivided lots on the north side of the highway contain restrictive covenants. The conveyances are subject to the following conditions, exceptions, reservations and easements:

1. A sixty (60) foot common right of way.
2. A fifteen (15) foot utility easement.

3. That any permanent building constructed on said lot shall not be any less than five hundred (500) square feet with a finished exterior and adequate indoor plumbing facilities.
4. That the Grantee (purchaser) shall connect to the power line prior to making use of said lot.
5. That any temporary dwelling, including trailers, mobile homes, etc. used by the Grantee shall not remain on said lot if the Grantee is not present.
6. That the Grantee cannot subdivide said lot other than a 208' x 300' parcel bound by Highway 118. Any conveyance of said parcel shall:
   (a) Meet with the approval of the Miramichi Planning District Commission;
   (b) Provide that any structure built shall be a permanent structure; and
   (c) Provide that the Grantee shall not have any right to use, for any purpose, the remainder of the area designated as the Central Maine Leasing Subdivision Plan, Lots #1–13 between the Maine Southwest Miramichi River and Highway 118.

The Dube lot was located in the development on the south side of the highway, and thus the covenants did not apply to it. Had the lot been on the north side of the highway, the Philbrick purchasers would have been obliged to comply with the terms of the covenants and easements.

Mineral Rights

The right to the minerals below the surface of land is possessed by the owners of the land in fee simple in most of the older provinces of Canada. In the past it was a practice of the Crown to include in the Crown patent a grant of the right to the minerals (except, perhaps, for gold and silver) along with the surface rights. More recent Crown patents in all provinces usually reserve the mineral rights to the Crown, unless the patentee acquires the mineral rights at the time of issue by way of an express purchase. A person who acquires the mineral rights in the lands of another acquires an interest in land known as a *profit à prendre* that must be in writing, since it is a contract concerning land. In addition, the conveyance of the interest must be in deed form to be enforceable. Unless the owner of the mineral rights owns the surface rights as well, the mining of the minerals carries with it certain obligations to the owner of the surface rights. Since the extraction of minerals normally requires some disturbance of the surface, the owner of the mineral rights must compensate the owner of the surface rights for the interference with the property.

The documents that provide for the removal of the minerals and for the surface use by the persons with rights to the minerals are frequently referred to as "leases." However, they are much more than ordinary leases, even though they relate to the occupancy of a portion of the surface area. It should be noted that the right to remove water from the lands of another is not the same as the right to remove oil, gas, or other minerals. The right to remove water is generally considered to be an easement, rather than a *profit à prendre.*

Riparian Rights

A riparian owner is a person who owns land that is adjacent to a watercourse, or has land through which a natural stream flows either above or below the surface. A riparian owner has certain rights with respect to the use and flow of the water that have been established and recognized at Common Law. These rights include the right to take water from the stream or watercourse for domestic and commercial uses, such as the watering of livestock, generation of power, or for manufacturing purposes. However, the landowner cannot interfere with the flow to downstream users and must return the quantity of water used to the stream (less the amount consumed in the use if it does not appreciably affect the quantity flowing to the downstream landowner). Some provinces control the amount of water that may be diverted from a watercourse for commercial or industrial use.

Riparian owners may erect dams on streams for the purpose of generating hydroelectric power or operating water-powered equipment or machinery. Any such structure must not (except initially), however, restrict the flow of water downstream or force water upon the upstream landowner by the ponding effect of the dam. The dam also must not interfere with fish travelling upstream to spawn. Some provinces (notably New Brunswick, Saskatchewan, British Columbia, and Ontario) control the erection of dams on watercourses and require the construction of "fishways" to ensure unobstructed upstream fish travel.

A riparian owner, while permitted to use water from a natural watercourse, may not do anything to change the quality of the water used. He or she may not raise the temperature of the water or add chemicals or sewage to the water, as the downstream owner is entitled to receive the water in its natural state.

A downstream owner may take legal action against an upstream owner who interferes with the flow or pollutes the water. No proof of damage is necessary, as the landowner need only establish a violation of his or her rights as a riparian owner. An injunction is the usual remedy granted by the court. It is important to note that a downstream landowner may not be successful in enforcing riparian rights if the upstream user or polluter can establish a prescriptive right at Common Law to pollute, due to long and uninterrupted use. Most provinces have now passed legislation to control water pollution, however. This may be something of a mixed blessing to riparian owners, because the legislation frequently permits a certain amount of pollution if the manufacturing operations or use of the water is made in accordance with government standards. Pollution of water that endangers public health is, nevertheless, an offence under the *Criminal Code*.[11]

Possessory Interests in Land

Title to land may be acquired through the possession of land under certain circumstances. In some provinces, the exclusive possession of land for a long period of time — in open, notorious, visible, uninterrupted, and undisputed defiance of the true owner's title — will have the effect of creating a possessory title in the occupier of the land. The possessory title will be good against everyone, including the true owner, if the true owner fails to regain possession of the property by way of legal action within a stipulated period of time. In provinces where a possessory title may be acquired, the time period varies from 10 to 20 years. If the owner fails to take action within that period, the title in the occupier becomes indefeasible.

The period of possession must be continuous and undisputed, but it need not be by the same occupant. For example, one occupant may be in exclusive possession for a part of the time and may convey possession to another occupant for the remainder of the time period. As long as the period of possession is continuous, the time period will run. Any break in the chain of possession, such as where the true owner regains possession for a period of time, will affect the right of the occupant. The time period will begin again only when the possession adverse to the true owner commences for a second time.

Adverse possession

A possessory title to land under the Registry System acquired by continuous, open, and notorious possession of land inconsistent with the title of the true owner for a period of time (usually 10 to 20 years).

Adverse possession requires the occupant in possession to do the acts normally required of an owner of the land. For example, the occupant would be expected to use the land, pay taxes, maintain fences, and generally treat it as the occupant's own. This must be done openly and with the knowledge of the person who has title to the land.

The courts have laid down a number of general requirements that must be met before persons claiming adverse possession may succeed. As one judge described the obligation on the claimants, the claimants must establish:

> (1) actual possession for the statutory period by themselves and those through whom they claim;
> (2) that such possession was with the intention of excluding from possession the owners or persons entitled to possession; and
> (3) discontinuance of possession for the statutory period by the owners and all others, if any, entitled to possession.[12]

A tenant may acquire a possessory title to leased property if the tenant continues to possess the land for the statutory period of time after the lease has expired, provided that

---

11. R.S.C. 1985, c. C-46, s. 180 (common nuisance endangering lives or safety).

12. See Pennell, J., in *Re St. Clair Beach Estates Ltd. v. MacDonald et al.* (1974), 50 D.L.R. (3d) 650.

no act or acknowledgement of the lessor's title is made during the period. During the period of possession following the expiry of the lease, the tenant would be obliged to pay taxes and all other assessments usually imposed on the owner of the land, in addition to maintaining possession of the property to the exclusion of the owner for the statutory period. Any acknowledgement (such as the payment of rent) would terminate the possession time and cause it to begin again from that point. Continuous possession is the essential requirement for the acquisition of a possessory title. Unless continuous, open, and undisturbed possession can be proven, a possessory title may not be acquired.

CASE IN POINT

A pulp and paper company owned a quarter-million acres of Nova Scotia forest. A hunting party built a three-room cabin on less than one acre on a remote lake in 1975. The company was later aware of their presence. In 2000, members of the hunting party sued for a declaration of title based on adverse possession. The trial judge found that the hunter's possession of the land surrounding the cabin was open and notorious (not hidden or covert) peaceful, adverse (against the title-holder), exclusive and continuous for more than 20 years. The appeal court however ruled that the possession had not been exclusive: the cabin was never locked and the hunters never excluded or even interfered with the company's use of the land. The failure of one element of the claim spelled failure for the claim itself.

See: *Spicer v. Bowater Mersey Paper Co.*, 2004 NSCA 39.

Encroachments

Encroachment

A possessory right to the property of another that may be acquired by the passage of time.

An **encroachment** is also a possessory right to the property of another that may be acquired by the passage of time. It most often takes the form of a roof "overhang," where a building has been constructed too close to the property line, or where the building has actually been constructed partly on the lands of a neighbour. If the true owner of the land on which the encroachment is made permits it to exist for a long period of time, the right to demand the removal of the encroachment may be lost. In the case of a building constructed partly on the lands of another, after undisturbed possession for a period of 10 to 20 years (according to the province), the right to object to the encroachment is lost. Encroachments are normally rights in property that may be acquired only in those areas of Canada where land is recorded under the Registry System.

FRONT-PAGE LAW

Yes, our chimneys kiss. We like it that way. Life on a tilt: Faulty foundations on street in Beaches root of the problem

Chris Goddard places a ball on his living room floor and watches as it rolls to the south side of his Beaches home.

This is no poltergeist at work but rather a structural problem that has left his home, and a handful of others along the southeast side of Glen Manor Drive, slightly tilted to one side.

"It was one of the features and characteristics that drew us to it," he said.

The crookedness also had little effect on his neighbours, who were unable to resist the posh two-storey duplexes with a beach for a backyard. Unfortunately, they did not put in proper foundations to counter the effects of a stream that ran below.

The tilt on some homes is more obvious than on others. For example, Derek Ferris's home and his neighbour's home lean toward each other, causing his eavestrough to fit snugly underneath his neighbour's and their chimneys to kiss.

Because both homes are effectively trespassing on each other's property, owners have "encroachment agreements."

He did have some initial concerns when he first saw the place, but those worries were soon dismissed after the house was checked out by a structural engineer and given a clean bill of health.

"Any settlement that will take place will occur in the first few years of life," said John Zimnoch, a sales associate with Re/Max.

Mr. Zimnoch said the crookedness does slightly affect the value of the homes and limits the number of prospective buyers.

"A lot of buyers will come along and see the houses and think there is probably a structural problem," he said. "It takes longer to sell because it has a reduced market."

However, some of the homes have sold for more than $400,000.

This situation is dealt with in a positive way by the parties involved. To have an "encroachment agreement," one first needs to come to an agreement. In the absence of that, what if one chimney wants to kiss and others do not? What about the commercial situation in a business park, where encroachment (of any kind) is not likely to have the residential charm it does here? Can such a cheery result be expected? Consider how an encroachment can affect the value of both properties, and the liability for this diminished value.

Source: Excerpted from Mark Gollom, *National Post*, "Yes, our chimneys kiss. We like it that way,"August 8, 2000, p. A17.

CHECKLIST FOR INTERESTS IN LAND

1. Fee simple
2. Life estate
   - Remainder
3. Leasehold
4. Lesser interests
   - Easements
   - Restrictive covenants
   - Mineral rights
   - Riparian rights
   - Adverse possession
     - Encroachments

(Note: A mortgage also creates certain interests which will be discussed in the next chapter.)

FIXTURES

Fixtures are chattels that are permanently or constructively attached to real property. Real property includes land and all things attached to it in some permanent fashion. For example, a building constructed on a parcel of land becomes a part of the land insofar as ownership is concerned. Some objects, however, are not normally a part of the land, but are sometimes affixed to it. Then the question arises: Did they become a part of the real property or are they still chattels that may be removed? In the early cases, the rule that developed to determine if a chattel had become a fixture and a part of the land was based upon the use and enjoyment of the particular item. Generally, it was thought that any chattel that was attached to the land to improve the land (or building), even if only slightly attached, became a part of the realty, but anything attached for the better use of the chattel did not.[13]

In the case of *Stack v. T. Eaton Co. et al.* (1902), 4 O.L.R. 335, the judge set out in the following manner the basic tests at Common Law that apply to fixtures:

(1) That articles not otherwise attached to the land than by their own weight are not to be considered as part of the land, unless the circumstances are such as shew that they were intended to be part of the land.

(2) That articles affixed to the land even slightly are to be considered part of the land unless the circumstances are such as to shew that they were intended to continue as chattels.

13. *Haggert v. Town of Brampton et al.* (1897), 28 S.C.R. 174; *Stack v. T. Eaton Co. et al.* (1902), 4 O.L.R. 335; *Re Davis*, [1954] O.W.N. 187.

(3) That the circumstances necessary to be shewn to alter the *prima facie* character of the articles are circumstances which shew the degree of annexation and object of such annexation, which are patent to all to see.
(4) That the intention of the person affixing the article to the soil is material only so far as it can be presumed from the degree and object of the annexation.
(5) That, even in the case of tenants' fixtures put in for the purposes of trade, they form part of the freehold, with the right, however, to the tenant, as between him and his landlord, to bring them back to the state of chattels again by severing them from the soil, and that they pass by a conveyance of the land as part of it, subject to this right of the tenant.

These tests over the years have resulted in an unusually confusing series of cases that have failed to provide a clear rule as to what may constitute a fixture. For example, carpet in a hotel was held to be a fixture,[14] but a mobile home on a concrete foundation and attached to a septic tank and drainage field was held not to be a part of the land.[15] As a result of these and similar cases, what constitutes a fixture turns very much on the particular facts of the case. The degree of annexation to the land is generally important, as is the ability of the person in possession to remove the chattel without causing serious damage to either the chattel or the building. The particular use of the chattel is also important, since some obviously have little value except as a part of the land. Made-to-measure storm windows, designed for a particular house, for example, would normally be constructively attached to the property as fixtures. These windows would be used specifically for the better use of the particular building, and in themselves, as chattels, would have very little value. The same would hold true for fences and trees planted on the land.

Where chattels are brought on the property and affixed by a tenant, the chattels are treated differently by the courts. Fixtures that are firmly affixed to the land or building become a part of the property and may not be removed, but items that are classed as trade fixtures may generally be removed by the tenant. Trade fixtures include chattels such as display cabinets, shelving, signs, mirrors, equipment, and machinery. These normally may be removed at the termination of the lease, provided that the tenant does so promptly and repairs any damage caused by their removal. Prompt removal is important, however. Trade fixtures left on the premises for a long period of time may eventually become a part of the realty, and the tenant may not later claim them.

MANAGEMENT ALERT: BEST PRACTICE

Tenant-supplied trade fixtures (showcases, permanent custom shelving, and the like) are fertile ground for disputes in commercial leases, particularly if the tenant's business runs into financial difficulties. The landlord (presumably with rent in arrears) has an interest in seeing them treated as true fixtures and part of the land, while the supplier of the trade fixtures may have a secured interest in them (to get them back, if they are yet to be paid in full). As well, the tenant will advocate that they are general assets of the business, available for liquidation to be generally applied to all creditors. The terms of the lease and the terms of the sale contract for the fixtures may be completely irreconcilable on their face.

TITLE TO LAND

Estates in land may be held by either an individual or a number of persons. If a number of persons hold title to property, the interests of each need not be equal, depending upon the nature of the conveyance.

14. *La Salle Recreations Ltd. v. Canadian Camdex Investments Ltd.* (1969), 4 D.L.R. (3d) 549.
15. *Lichty et al. v. Voigt et al.* (1977), 80 D.L.R. (3d) 757, 17 O.R. (2d) 552.

Joint Tenancy

Joint tenancy

The joint holding of equal interests in land with the right of the surviving tenant to the interest of a deceased joint tenant.

When land is conveyed to persons in **joint tenancy**, the interests of the grantees are always equal. Joint-tenancy interests in land are identical in time, interest, and possession with respect to all joint tenants. A joint tenant acquires an undivided interest in the entire property conveyed. A joint tenancy must also arise out of the same instrument, such as a deed or will, and possession must arise at the same time. For example, two parties may be granted land as joint tenants in a deed or devised land under a will as joint tenants, but they cannot become joint tenants through inheritance in the sense that a person inherits the share of a joint tenant on his or her death. Joint-tenancy interests vest in the surviving joint tenants on the death of a joint tenant. Consequently, a joint-tenancy interest may not be devised by will to another to create a new joint tenancy. A joint tenancy may be terminated by the sale of the interest of a joint tenant to another party.

Tenancy-In-Common

Tenancy-in-common

The joint holding of interests in land that need not be equal.

A second type of tenancy is a **tenancy-in-common**, which differs from a joint tenancy in that the right of survivorship does not attach to the interests of the tenants, nor do the interests necessarily need to be equal. For example, two individuals might receive a grant of land as tenants-in-common. The grant may be of equal interests, in which each would acquire an undivided one-half interest in the land. The grantor, however, could convey unequal interests in the property to each person, in which case the interests would be unequal but in the whole of the land. In the example above, one might receive an undivided three-quarters interest, and the other might receive an undivided one-quarter interest in the whole. It should also be noted in the case of tenancy-in-common that a part of the tenancy may be inherited or may be devised by will. Since the right of survivorship does not exist, when a tenant-in-common dies, the interest of the tenant passes by way of the tenant's will or by way of intestacy to the devisee or heirs-at-law. The interest does not vest in the surviving tenant-in-common.

If the tenants wish to divide the property and they cannot agree upon the division, the division may be made by the courts or under the *Partition Act* in those provinces with partition legislation.[16] The tenancy may also be dissolved by the acquisition by one tenant of the interest of the other tenant (or tenants), as the union of the interest in one person will convert the tenancy-in-common into a single fee simple interest.

REGISTRATION OF PROPERTY INTERESTS

In the past, it was necessary for the individual landowner to closely guard all documents related to the title of land in order to establish ownership rights to the property. The list of title documents began with the Crown patent and extended down through a chain of deeds from one owner to another to the deed granting the land to the present owner. If the list of deeds contained no flaws or breaks in the chain of owners, the present owner was said to have a "good title" to the land. If for some reason the title documents were destroyed through fire or were stolen, the landowner faced a dilemma. The documented legal right to the land in the form of a chain of title was gone, so a prospective purchaser was obliged to rely on the landowner's word (and perhaps the word of the neighbours) that he or she in fact had title to the lands that were being sold. The difficulties attached to this system of establishing land ownership eventually gave way to a system of land registration in which all of the land in a country or district was identified. A public record office was established to act as the recorder and custodian of all documents pertaining to the individual parcels of land. Then a prospective purchaser could simply go to the public record office to determine if the vendor had a good title to the property that was offered for sale.

Surprisingly, the widespread use of the public record-office approach for the custody of documents pertaining to land and the recording of the instruments in land registers first took place in North America, rather than in England. In 1862, a Land Registry Office

16. For example, Ontario and British Columbia.

was established in London, but for many years the registration of title documents was voluntary. Registry offices were established in other areas of England as well, but it was not until comparatively recent times that compulsory registration was extended to all areas of England. In this respect, Canada and the United States were many years ahead of the country in which their system of land tenure developed.

In Canada, the registration system did not develop immediately, but evolved gradually, with the public office first acting as custodian of the title documents, then providing for the registration of documents. Over the years, this was refined further to the point where, in some provinces of Canada, the province itself now certifies the title of the owner of the land.

The public-registration system and the certification of titles is designed to reduce to an absolute minimum the chance of fraud in land transactions and to eliminate the need for safeguarding title documents by the individual. All provinces have a system for the registration of interests in land and have public record offices where a person may examine the title to property in the area.

All interests in property require registration to protect them. Otherwise, any person who does not have actual notice of the interest of another in the land may acquire an interest in the land in priority over the interest of the holder of an unregistered instrument or deed. All unregistered interests in land would therefore be void as against a person who registers a deed to the property and who had no actual notice of the outstanding interest.

Two distinct systems of registration of land interests exist in Canada, the Registry System and the Land Titles System. Newfoundland and Prince Edward Island use only the Registry System. All other jurisdictions either use the Land Titles System exclusively, or are in successive stages of conversion from the Registry System to the Land Titles System, as is the case in parts of New Brunswick, Nova Scotia, and Ontario. As will be seen, the Land Titles System is ideally suited to computerized mapping, a property database and online registration of interests.

Registry System

Registry System

A provincial government operated system for the registration of interests in land.

The **Registry System** is the older of the two systems of land registration. Under the Registry System, a register is maintained for each particular township lot or parcel of land on a registered plan of subdivision. All interests in land that affect the particular parcel or lot are recorded in the register that pertains to the lot, and may be examined by the general public. Any person may present for registration an instrument that purports to be an interest in the land, and it will be registered against the land described in the document. For this reason, the prospective purchaser or investor must take care to ascertain that the person who professes to be the owner of the land has, in fact, a good title.

Under the Registry System, to determine the right of the person to the property it is necessary to make a search of the title at the Registry Office to ascertain that a good "chain of title" exists. This means that the present owner's title must be traced back in time through the registered deeds of each registrant to make certain that each person who transferred the title to the property was in fact the owner of the land in fee simple at the time of transfer.

In the Registry System, it is usually necessary to establish a good chain of title for a 40-year period before the title of the present owner may be said to be "clear." Each document registered against the land must be carefully examined to make certain that it has been properly drawn and executed, and that no outstanding interests in the land exist that are in conflict with the present registered owner's title. Under the Registry System, the onus is on the prospective purchaser or investor to determine that the registered owner's title is good. Consequently, the services of a lawyer are usually necessary to make this determination. If a person fails to examine the title and later discovers that the person who gave the conveyance did not have title to the property, or that the property was subject to a mortgage or lien at the time of the purchase, the purchaser has only the interest (if any) of the vendor in the land. The only recourse of the purchaser under the circumstances would be against the vendor (if the person could be found) for damages.

Land Titles System

Land Titles System

A provincial government operated system for the registration of interests in land where the government confirms and warrants the particular interests in land.

The **Land Titles System** differs from the Registry System in a number of important aspects. Under the Land Titles System, the title of the present registered owner is confirmed and warranted by the province to be as it is represented in the land register. It is not necessary for a person to make a search of the title to the property to establish a good chain of title. This task has already been performed by the Land Registrar, and the title of the last registered owner as shown in the register for the particular parcel of land is certified as being correct. To avoid confusion, instruments pertaining to land under the Land Titles System are given different names. A deed, for example, is called a transfer, and a mortgage is referred to as a charge. The legal nature and the differences between these latter two instruments are dealt with in Chapter 23, "The Law of Mortgages." As well, a number of other differences exist between the two systems. One of the more notable differences is that in land-titles jurisdictions an interest in land may not normally be acquired by adverse possession.

The advantage of the Land Titles System over the older Registry System is the certainty of title under the newer system. If, for some reason, the title is not as depicted in the Land Titles Register, the party who suffered a loss as a result of the error is entitled to compensation from the province for the loss.

In an effort to streamline the land-registration system further, Land Titles System regions are establishing under legislation computerized systems consisting of two computerized databases: (1) a Title Index database, and (2) a Property Mapping database.

The Title Index database is designed to provide a computerized version of the existing records, organized on a property ownership basis with immediate update of ownership registration. The index system permits computer searching of the title to properties and provides printouts of information on properties if the search person has a property identifier number, an address, the owner's name, or an instrument-registration number.

The Property Mapping database provides a computerized property-map file organized by property-identifier number. Using these numbers, a search person may obtain a computer-generated map of a property showing survey and property lines to assist in searching the title to a property. This project improves the system by simplifying the methods of registration and certifying the title to lands on plans of subdivision, and generally provides greater ease and certainty in the examination of the title to land. The project of modernization also includes the extensive use of computerized information, storage, and indexing, and completely new forms for deeds, mortgages, and all other documents used to convey or deal with interests in land. The goal is to gradually convert all Registry areas into the Land Titles System.

Every parcel of land will soon be recorded with a property-identification number and map, and as the conversion nears completion, the automated system is replacing the manual system of property registration.

The final phase of the registration process has been the establishment of the electronic-registration system. Under this system, the Parcel Identification Number (PIN) allows the title to properties to be examined by computer from remote locations. The system, where fully operational, also allows the registration of documents electronically and the transfer of funds between lawyers related to the transaction.

SUMMARY

Real property includes land and everything attached to it in a permanent manner. The Crown owns all land, but has conveyed estates in land by way of Crown patents. The highest estate in land is an estate in fee simple. When someone states that he or she owns land, the estate to which the person refers is ownership in fee simple. A life estate is another estate in land that is limited to a particular lifetime, and afterwards reverts to the grantor or the person in possession of the remainder or reversion.

A condominium is a unique estate in land. A person who acquires a condominium acquires exclusive ownership of a part of it (a unit) and part of it in co-ownership with all other unit owners.

An individual may possess land either alone, or jointly. This may be either by a joint tenancy or tenancy-in-common. In both cases, the interests are in the entire property. However, in the case of tenancy-in-common, the interests may be unequal.

Interests in land, other than estates, exist as well. A person may acquire an easement or right-of-way over the land of another, or may exercise control over land granted to another by way of a restrictive covenant in the conveyance. Restrictive covenants are generally used to protect adjacent property by controlling the use that the grantee may make of the property.

Land or interests in land may be acquired in some parts of Canada by way of adverse possession. This usually requires the open and undisputed adverse possession of the land or right for a lengthy period of time, but once the right has been established, the lawful owner can do nothing to eliminate it.

All instruments concerning land must be registered in order that the public may have notice of the interest in land and the identity of the rightful owner. Each province maintains public registry offices where the interests in land are recorded, either under the Registry System or the Land Titles System. Persons must satisfy themselves as to the title to lands under the Registry System. In contrast, under the Land Titles System, the province certifies the title to be correct as shown in the land register for the particular parcel. Under both systems, an unregistered conveyance or interest in land is void as against a person who has registered his or her interest without actual notice of the unregistered instrument.

KEY TERMS

Real property, 430
Fixture, 430
Tenure, 431
Fee simple, 431
Escheat, 432
Expropriation, 432
Deed/transfer, 432
Life estate, 432
Remainderman, 433
Leasehold estate, 434
Condominium, 434
Dominant tenement, 438
Servient tenement, 438
Easement, 438
Right-of-way, 438
Restrictive covenant, 439
Adverse possession, 441
Encroachment, 442
Joint tenancy, 445
Tenancy-in-common, 445
Registry System, 446
Land Titles System, 447

REVIEW QUESTIONS

1. How are condominiums normally established?
2. Explain the term "freehold estate." How does this term apply to land?
3. What lesser estates may be carved out of an estate in fee simple?
4. Indicate how a condominium organization deals with the problem of a unit owner who fails to contribute his or her share of the cost of maintaining the common elements of the condominium.
5. Define the terms "fee simple," "escheat," "life estate," "tenants-in-common," and "prescriptive right."
6. How does a life estate differ from a leasehold estate?
7. Describe briefly how landholding developed in Canada, and identify the system upon which it is based.
8. What method is generally used to determine the unit owner's interest in the common elements?
9. Once land is granted by the Crown, how is it recovered?
10. In what way (or ways) would an easement arise?
11. Under what circumstances would a restrictive covenant be inserted in a grant of land? Give three common examples of this type of covenant.
12. What characteristic distinguishes a fixture from other chattels brought onto real property?

13. Explain the term "adverse possession" and describe how it might arise.
14. What is the purpose of a land registry system?
15. Explain how the Land Titles System differs from the Registry System.
16. What special advantages attach to the Land Titles System?
17. Why is a "good chain of title" important under the Registry System?
18. Distinguish joint tenancy from tenancy-in-common.

Mini-Case Problems

1. Southside Land Development Corp. offered to sell Trend Contracting Ltd. a small block of vacant land in a large city for $50,000. Southside Land Development Corp. presented a deed describing the property and showing Southside Land Development Corp. as the owner in fee simple. What information should Trend Contracting Ltd. obtain before delivering the $50,000 to Southside Land Development Corp.?
2. A conveyed a parcel of land to his daughter B and son C. The deed recited, in effect: to B for life, and then to C in fee simple. C would like to sell the land to D. Advise C.
3. Able acquired a condominium unit that included a parking space on a surface lot facing a sidewalk and street. Able leased the space to his friend Samantha, who parked her chip wagon in the space. She sold french fries and soft drinks to the public from the location. The other residents of the condominium objected. Advise Able of his rights (if any).
4. Lot 'A' was a lot that had attached to it a 5 metre wide right-of-way over Lot 'B' to enable the owner of Lot 'A' to reach a lake front beach where the owner could launch a canoe. The owner of Lot 'A' decides to leave his canoe on the right-of-way at the beach rather than carry it back to Lot 'A' after each use. What are the rights of the parties in this situation?
5. A lumber company for over 50 years had owned a 3,000 hectare tract of forested land. In a far corner of the tract an old fur trapper had built a cabin, where he had lived for 40 years. Discuss.

Case Problems for Discussion

CASE 1

Iron Mining Corporation held the mineral rights to a large area of land in a municipality where the surface rights were owned for the most part by cottage property owners. In fact, Iron Mining Corporation had the mineral rights to all of the land around a lake where 30 surface rights land owners had built large, expensive cottages. Several years after the cottages had been built, a geologist at Iron Mining Corporation examined some recent mineral surveys of the area, and decided that the area warranted further investigation. He accordingly sent out a drilling crew to drill some core samples "to see what was under the land by the lake." The drilling crew arrived at the lake, and made an effort to find a cottage owner present, but because it was early April, no cottage owners were at their cottages. The drilling crew then proceeded to drill a large number of drill holes in the yards of the cottages. While little evidence remained of their drilling work, the cottage yards were torn up by the heavy machinery and trucks. When the cottage owners arrived in May to open their cottages, they discovered the extensive damage to their former manicured lawns, ornamental trees and gardens.

Advise the parties.

CASE 2

Dillon and Alexei owned a 20 hectare parcel of farm land as tenants-in-common. For many years they operated a vegetable growing operation, each farming and harvesting a 10 hectare part of the parcel. Each made his own decisions as to crops planted and harvested. They shared equally the cost of property taxes, the only expense related to their ownership of the land. Each kept the profits earned from their respective farming operations, but Alexei usually earned double the profits of Dillon because he specialized in growing unusual and high value vegetables. Just before one harvest season, Alexei was killed in a traffic accident. To save the crop, his widow and only heir-at-law, helped Dillon harvest both of the 10 hectare crops. When the crops were sold, Alexei's widow asked Dillon for the proceeds from the sale of the crop from Alexei's 10 hectare parcel. Dillon offered her half of the net proceeds from the total 20 hectares after the deduction of his production expenses.

Discuss the rights of the parties. How would a court likely decide the issue?

www.mcgrawhill.ca/olc/willes

CASE 3

Maya and Louis were unit owners in the Happy Times Condominium complex, along with 95 other unit owners. Both Maya and Louis considered themselves to be creative cooks and bakers. Their cooking odours or the smoke from their burnt baked goods constantly filled the hallways and common elements, because both insisted on leaving the doors to their units open during the baking or cooking process to allow the air to circulate through their units.

The remaining unit owners objected to their cooking activities and demanded that the Happy Times Condominium Corporation prohibit Maya and Louis from engaging in their hobbies on the premises. The solicitor for the corporation, however, advised that cooking or baking, as long as it was not for commercial purposes, could not be prohibited in the complex without affecting everyone else as well.

When the unit owners discovered that the cooking could not be stopped, a number of the unit owners suggested that the condominium be terminated by selling the complex to a buyer who was interested in operating it as a rental apartment building. Maya and Louis objected vigorously to the suggestion of terminating the condominium. However, in the end, all unit owners except Maya and Louis agreed that steps should be taken to do so.

Advise Maya and Louis of their rights. Describe the procedure that must be followed to terminate the condominium.

CASE 4

Winston County Condominium Corp. No. 221 (WCCC #221) was formed just over a year ago, with all the usual condominium documentation. Contained in its declaration was a reference that common expenses included municipal water charges, unless the same were separately metered for each unit. There were 91 units in the building, one of which was a ground-floor restaurant unit. The restaurant represented 10 percent of the floor space, and therefore 10 percent of common expenses. Each of the other 90 dwelling units would bear 1 percent of common expenses.

After examination of the accounts for the first year of operation, WCCC #221 found that the restaurant accounted for 47 percent of water charges, and the amount budgeted by the corporation of $50,000 would fall short of actual costs by $18,500. The directors passed a motion for a special levy on the restaurant unit, and a motion that a meter be installed on the pipes to and from the restaurant unit. The action was ratified by the unit holders 90 to 0, with one abstention in protest.

The owner of the restaurant unit came before the courts for relief, stating that water rates had figured into her calculations on whether to purchase the unit, and that the same calculations must have figured into WCCC #221's decision to sell the unit to her. WCCC #221 had the power to write what it wished into its declaration, and she now holds it to what it has written. The corporation had set the price for her commercial unit, knowing it would contain a restaurant. The restaurant owner acknowledged she was prepared to suffer her fate, should WCCC #221 on a vote decide to install meters to all units.

Elaborate on the issues in the arguments and render a decision on behalf of the court.

CASE 5

Baxter owned a large block of forested land that surrounded a small lake. The lake was fed by a small stream that crossed the property, and another that drained the lake into a larger body of water several kilometres away.

With the intention of eventually constructing a resort on the lot, Wilson purchased from Baxter a parcel of land fronting on the lake. On the payment of the purchase price he received a deed to the land from Baxter. Without examining it, he placed it in his safety deposit box.

Wilson used the property as a personal campsite for several years while he searched for financial backers for his resort. Because the lake was several hundred metres from the road, each time he visited the lake he would leave his automobile parked at the roadside and carry his camping equipment through the woods to his property.

Five years after he purchased the land from Baxter, he was in a position to build the resort on his lot. No road access was available to the land, but Wilson assumed that the pathway that led to his property was his access route. He engaged a logger to cut trees to widen the path in order that a truck carrying his building materials could reach his lot. No sooner had Wilson's logger cut the first tree than Baxter appeared and ordered him to stop cutting. When Wilson arrived, Baxter ordered him to leave the property.

Wilson protested that he was entitled to clear the trees from the access route to his land, but Baxter replied that he had sold him only the lot and not a roadway. According to Baxter, Wilson had water access by way of the stream if he wished to enter or leave his property. The surrounding land belonged to Baxter.

Wilson had travelled the stream with his canoe on a number of occasions. While it was possible to gain access to his lot in that fashion, it would not be possible to transport the heavy building materials into the property as the stream was too shallow to allow the use of larger watercraft. Rather than continue his argument with Baxter, Wilson decided to examine his deed to determine if Baxter was correct in his position on the access route. Wilson returned home and read the description of the property contained in the conveyance. It described the lot only and made no mention of a roadway to the property. According to the deed, Baxter owned all of the land surrounding Wilson's property. His only access appeared to be by way of the small stream to the lake.

Examine the rights of the two parties in this case. If either party should decide to take legal action to enforce his rights, explain the nature of the action and indicate the probable outcome.

CASE 6

Suburban Land Development Ltd. owned a block of land that it wished to develop as a residential housing subdivision. The land was heavily treed and bordered the shore of a lake. To preserve the woodland setting of the area as each house lot was sold, the corporation inserted in the deed the following clauses:

The grantee agrees

(1) to construct a house on the premises with a floor space of not less than 237 square metres and a construction value of not less than $200,000.

(2) that construction shall not begin until the grantor approves in writing the architectural drawings or plans for any proposed dwelling.

(3) that no trees shall be cut on the property without the express consent in writing of the grantor.

(4) that no pigs, chickens, or other domestic animals may be kept on the property.

(5) that the above covenants shall run with the land for a period of 20 years, and shall be binding on the heirs, executors, and assigns of the grantee.

Casey purchased a large, heavily wooded lot in the subdivision. She received a deed to the lot in fee simple that was registered in the Land Registry Office and contained the above covenants. Without constructing a house or dealing in any way with the property, she sold it to MacGregor, who received a deed to the lot that did not contain the covenants. He registered that deed in the Land Registry Office without examination of the title to the property.

Some time later, MacGregor attempted to cut some of the trees on the property to clear a place where he intended to build a small building to house his racing pigeons.

When Suburban Land Development Ltd. became aware of MacGregor's activity, it immediately informed him that he must stop cutting the trees until approval was given, and also that the company would not permit the construction of any building on the property except a dwelling-house.

MacGregor decided to ignore the prohibition, on the basis that he was unaware of the restrictions and had not agreed to them.

Advise MacGregor and Suburban Land Development Ltd. What might be the possible outcome, if legal action was taken by Suburban Land Development Ltd. to enforce the restrictions?

CASE 7

The Golf and Country Club owns a large block of land at the edge of a municipality that the club uses as an 18-hole golf course. A small stream runs through the property and eventually drains into a lake some distance away. The stream also passes through the municipality that is located upstream from the Golf and Country Club property.

The municipality installed new storm sewers in an area of the city and constructed them in such a way that, in a heavy rain, overflow from the sewers would drain into the stream.

Shortly after the construction of the new sewers, several days of heavy rains resulted in a large quantity of water from the storm sewers being discharged into the stream. This in turn produced flooding of the stream and serious erosion of the banks of the stream where it passed through the golf course. Damage was estimated at $350,000.

The Golf and Country Club instituted legal proceedings against the municipality for the damage.

Discuss the arguments that might be raised by each of the parties, and render a decision.

CASE 8

Samuels, an elderly widower, conveyed his farm land to his son Peter for life, then to his grandson Paul in fee simple. In the deed, he reserved to himself the right to continue to live on the farm and to receive 50 percent of the income from the farm during the rest of his lifetime.

Peter operated the farm with the assistance of both Samuels and Paul for a number of years, during which time he paid his father 50 percent of the farm income. Eventually, Peter decided to seek employment in industry and took a job in a nearby manufacturing plant. At that point he decided to cease farming. A short time later, the barn was accidentally destroyed by fire and the farm machinery destroyed. Samuels, Peter, and Paul each claimed to be entitled to the proceeds of the fire insurance on the barn and the farm machinery.

Discuss the rights (if any) of each of the parties to the insurance funds in this dispute.

CASE 9

Smith owned a large farm in eastern Ontario. Part of the farm, which consisted of a woodlot, fronted on an unimproved township road. In 1978, Crockett, a middle-aged bachelor, constructed a small log cabin in the woodlot for use as a fishing and hunting camp, with Smith's permission. For several years Crockett occupied the cabin on weekends, while fishing in the area in the summer months, and for a few weeks in the fall of each year during the hunting season.

During the summer of 1981, Crockett took a month's vacation and spent the time at the cabin. He planted a small vegetable garden and constructed a fence around both cabin and garden to keep animals away from his flowers and vegetables. During the hunting season of the same year he cut down a number of small trees and extended the fenced-in area to a parcel of land 23 metres by 30 metres, and he built a gate in the fence where it faced the roadway.

Smith noticed the fence and gate shortly after it was constructed and asked Crockett why it was necessary. Crockett replied that the animals in the area were damaging the flowers that he had planted around the cabin, and he felt that the fence would probably keep them out.

The next year, Crockett decided to accept early retirement from the firm where he was employed. He spent the period from May 1st to November 30th at the cabin. Crockett planted a vegetable garden, fished, and helped Smith with the planting of his crops and his fall harvest. At the end of November, he left his belongings in the cabin and spent the winter in a warmer climate.

He returned to the cabin the next April, only to be met by the local tax assessor, who asked him if the cabin was his. He replied in the affirmative, and, some time later, received a municipal tax bill issued in his name. He paid the municipal taxes for that year (1983).

Crockett continued to live in the cabin, spending only the coldest of the winter months away from the premises. He paid the taxes on the land and building each year. In 1994, he moved the fences to include an area 30 metres by 45 metres in order to enclose a larger vegetable garden. Smith did not object to the new location of the fence, but warned Crockett not to cut down two large hickory nut trees in the enclosed area. Crockett agreed to leave the trees standing.

In the summer of 2002, during a thunderstorm, lightning struck and damaged one of the large hickory trees. Without consulting Smith, Crockett cut down the damaged tree.

Several months later, Smith noticed that the tree was missing. In a rage, he ordered Crockett from the property. Crockett refused to leave, claiming he was the owner of the parcel of land.

Discuss the rights of the parties. Evaluate the arguments and evidence that each might raise if the matter should be brought before the courts to determine the rights of the parties in the land.

Render a decision.

The Online Learning Centre offers more ways to check what you have learned so far. Find quizzes, Weblinks, and many other resources at www.mcgrawhill.ca/olc/willes.

www.mcgrawhill.ca/olc/willes

CHAPTER 23

The Law of Mortgages

Chapter Objectives

After study of this chapter, students should be able to:

- Describe the nature of a mortgage.
- Identify the rights and duties of the parties under a mortgage.
- Explain the effect of assignment of a mortgage.
- Outline the default, foreclosure and sale provisions of mortgages.
- Identify the business uses of mortgages.

YOUR BUSINESS AT RISK

While land and buildings are often the most valuable assets of a business, mortgages are just as often the largest liabilities. If mortgages are not well understood, or entered into without regard to their technical provisions, a default may trigger immediate repayment obligations. If the mortgage cannot be refinanced, the usual result is insolvency, and cascading difficulty in paying other trade debts may lead to bankruptcy proceedings.

INTRODUCTION

A mortgage is a very old method of securing payment of indebtedness. In its simplest form, it involves the transfer of the debtor's title or interest in property to the creditor on the condition that the title or interest will be reconveyed to the debtor when the debt is paid. The **mortgage** is the formal agreement made between the parties, under which the debtor is referred to as the mortgagor, and the creditor, the mortgagee. The written agreement effects the transfer of the mortgagor's interest in the property or title to the mortgagee, and it contains the terms of the debt and the conditions under which the property interest will be returned to the mortgagor.

Mortgage

An agreement made between a debtor and a creditor in which the title to property of the debtor is transferred to the creditor as security for payment of the debt.

From a conceptual point of view, a modern mortgage has a number of distinct characteristics:

(1) It transfers the title of the mortgagor's property to the mortgagee.
(2) The mortgagor retains possession of the property until the debt is paid or default occurs.
(3) The document that transfers the title also contains a proviso that entitles the mortgagor to a reconveyance of the title to the property when the debt is paid.
(4) The document imposes certain obligations on the mortgagor to protect and maintain the property while the debt remains unpaid.
(5) The document also contains terms that permit the mortgagee to take steps to terminate the mortgagor's rights and interest in the property if the mortgagor defaults in payment.

Charge

A secured claim (similar to a mortgage) registered against real property under the Land Titles System.

Under the Land Titles System, a land or real property "mortgage" is called a **charge**. While similar to the mortgage in purpose, it does not transfer the title of the property to the creditor, but merely, as the name states, charges the land with payment of the debt. The obligations imposed on the debtor or chargor are much the same as those imposed on the mortgagor. If default occurs, the creditor in most provinces may take steps to acquire the title and possession of the land or have the land sold to satisfy the debt.

Under the Land Titles System, a mortgage in the ordinary sense of the word does not exist, as a charge does not involve the transfer of the title to the property to the mortgagee. Consequently, the remedies available to the mortgagee and the procedure that must be followed to realize on the debt in some provinces may differ from the procedure in provinces under the Registry System. At the present time, the charge under the Land Titles System may be found in parts of Ontario and Manitoba, and in Saskatchewan, Alberta, British Columbia, the Yukon, and the Northwest Territories. The Registry System and the mortgage are used in parts of Ontario,[1] Manitoba, and the provinces lying to the east of the province of Quebec. Unfortunately, with both systems, substantial provincial and territorial variation exists, and the legislation of each jurisdiction must be consulted for the law applicable to the particular securities in question.

CLIENTS, SUPPLIERS *or* OPERATIONS *in* QUEBEC

The mortgage is not found in Quebec. Instead, an instrument somewhat similar to the charge, a hypothec, is used to secure debt. This particular instrument, prepared by a notary, gives the creditor a security interest in the lands of the debtor and permits the debt to be satisfied by way of a sale of the land if the debtor defaults in payment.

Land is not the only form of property that may be used as security for debt. The mortgage may also be used to establish a security interest in personal property. Chattels such as automobiles, boats, aircraft, furniture, and equipment of all kinds are frequently used as security for indebtedness, in which case the security instrument is called a chattel mortgage. Because of the many varied and specialized uses of this form of mortgage, it is treated separately along with a number of other forms of security for debt in Chapter 29.

HISTORICAL DEVELOPMENT

The fact that land has a value and represents a source of wealth is the principal reason why the mortgage has acquired widespread acceptance as a security instrument. When land is used for agricultural or silvicultural purposes, it provides income in the form of produce or timber. If buildings are situated on the land, the buildings may be leased to provide income in the form of rents. Even in a raw state, land has value based upon the potential uses that may be made of it.

Land was particularly important as a source of wealth in the past. Until the Industrial Revolution, it represented one of the most important forms that wealth might take. As such, land was naturally considered by creditors and debtors alike as something that might be conveniently pledged as security for the payment of a debt. As a result, some legal mechanism for the pledge of land as security was considered soon after individual rights to property emerged in Western Europe and England. The modern form of mortgage nevertheless differs substantially from its ancient predecessors, although the concept has changed very little. The first mortgages were conceived as a means of securing debt by way of a transfer of an interest in land, and this is essentially what a mortgage does today. The older forms of mortgages, however, were subject to different formalities in execution and effect.

Since the mortgage involved a change in title, the legal nature of the transaction had to comply with procedural requirements of the law. At Common Law, the agreement made between the mortgagor and mortgagee was strictly enforced. For the usual type

1. In Ontario, a mortgage under the Registry System does not transfer legal title to the mortgagee, but rather creates a charge on the land in the same manner as a Land Titles charge. See *Land Registration Reform Act*, R.S.O. 1990, c. L.4.

of mortgage, the title passed to the mortgagee, subject to a condition subsequent that allowed the mortgagor to acquire the title back upon payment of the debt on the due date. From the point of view of the Common Law courts, a failure to pay on the due date would extinguish the rights of the mortgagor; after that the right to recover the property was forever lost. The debt remained, however, and (unless the agreement provided otherwise) the mortgagor was still liable to the mortgagee for payment of the money owing.[2]

FIGURE 23–1 **Mortgage Transaction and Discharge of Mortgage**

MORTGAGE TRANSACTION

"A" Transfers Title of Land to "B" but Retains Possession of Land

"A" (mortgagor)
Owner of Land in Fee Simple

"B" (mortgagee)

"B" Loans Money to "A" on Land as Security

DISCHARGE OF MORTGAGE

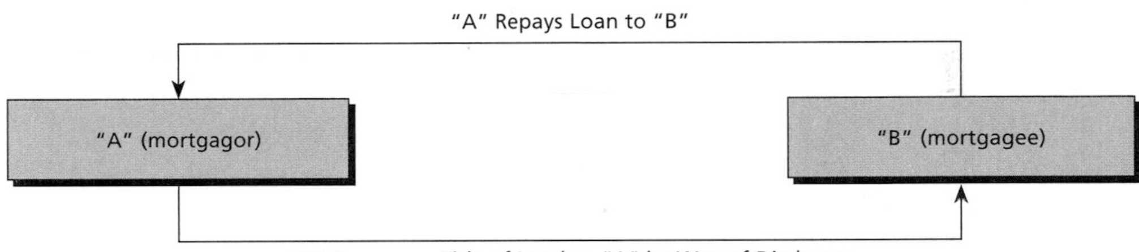

In the case of a charge, the creditor has only a secured interest in the debtor's land, and on repayment of the loan, the cessation of charge, given by the creditor as receipt of payment, extinguishes the claim against the debtor's land.

The strict approach to the agreement that the Common Law courts followed, and the inevitable hardships that it placed on unfortunate mortgagors, led many to seek relief from forfeiture in the Court of Chancery, where equitable relief might be had. This was done because, during the sixteenth century, the attitude of the Court of Chancery had changed with respect to mortgage actions. At the beginning of that century, the general rule was that the mortgagor must pay the mortgage debt by the date fixed for payment; otherwise, a decree of foreclosure would be issued.[3] Later, the Court of Chancery was prepared to provide relief for the mortgagor who failed to make payment on the due date, if the mortgagor could satisfy the court that the failure to make payment on time was due to some unforeseen circumstance or delay. Eventually, the court was prepared to provide relief regardless of the mortgagor's reason for the delay, so long as the mortgagor offered payment to the mortgagee within a reasonable time after the due date.[4] By this time — toward the end of the century — the court recognized the fact that the true nature of the contract was that of debt, rather than that of a transfer of land. From that time on, the court treated the transaction as a form of contract in which the land was simply held as security and viewed as something that could be used to satisfy the debt if payment was not made by the debtor.[5]

2. See, for example, comments by Lord Haldane in *Kreglinger v. New Patagonia Meat & Cold Storage Co. Ltd.*, [1914] A.C. 25 at p. 35.
3. Holdsworth W.S., *A History of English Law,* 3rd ed., vol. 3 (London: Methuen & Co. Ltd., 1923), p. 129.
4. See Holdsworth, vol. 5 (1st ed.), pp. 330–331.
5. ibid., p. 331.

The establishment of an equitable right to redeem in the mortgagor by the court also affected the mortgagee's rights. Having taken the position that a mortgagor had the right to redeem the property within a reasonable time after the contractual right was lost, the court unwittingly placed the mortgagee in a position of uncertainty with respect to the property. This in turn prompted mortgagees to seek relief from the same court. In response to petitions from mortgagees for recognition of their titles to mortgaged lands, the court engaged in a practice of granting decrees of foreclosure, which confirmed the mortgagee's absolute title and precluded the mortgagor from later demanding the right to redeem.[6]

Over time, a change also took place with respect to possession of the land by the mortgagee. The inconvenience of the mortgagor losing possession during the term of the mortgage was gradually replaced by an agreement that allowed the mortgagor to remain in possession of the property, provided that the mortgage terms were complied with and the mortgagor was not in default in payment. This change permitted the mortgagor to retain what was often the source of income necessary to pay the debt. It also reflected the true nature of the transaction: the use of the property as security for a debt.

The concept of a mortgage was further refined by the courts as a security instrument (rather than a transfer of title), by requiring the mortgagee to look first to the property mortgaged for payment of the debt. This relieved the mortgagor of the double disaster of not only losing the land, but remaining liable for the debt as well, and introduced the final element of fairness in the transaction. Thereafter, the mortgage as a form of security for debt changed little in its basic form and effect until the late nineteenth century.

In Canada and the United States, the mortgage arrived with the first settlers in the colonies. It developed along similar lines to those in England until the late nineteenth century, with the introduction of the Land Titles System of land registration. The adoption of this system by many of the states of the United States and some of the provinces of Canada further changed the form of mortgage to that of a charge on the land, rather than a transfer of title to the property. The late nineteenth and early twentieth century also saw the introduction of legislation relating to mortgages and charges, which clarified and modified the rights of the parties. As a result, the modern law of mortgages has become a blend of historic legal concepts, Common Law rules, equitable principles, and statute law.

THE NATURE OF MORTGAGES

Legal mortgage

A first mortgage of real property whereby the owner of land in fee simple transfers the title of the property pledged as security to the creditor on the condition that the title will be reconveyed when the debt is paid.

Equitable mortgage

A mortgage subsequent to the first or legal mortgage. A mortgage of the mortgagor's equity.

Equity of redemption

The equitable right of a mortgagor to acquire the title to the mortgaged property by payment of the debt secured by the mortgage.

The most common form that a mortgage may take is that of a first mortgage or **legal mortgage**. Only the first mortgage of a parcel of land may be the legal mortgage, since it represents a transfer of the title of the property to the mortgagee. Where the mortgagor holds an estate in fee simple, the mortgagee acquires the fee, leaving the mortgagor with possession of the property and the right to redeem the title in accordance with the terms of the instrument.

Assuming that a willing mortgagee may be found, a mortgagor may also pledge the same property as security for debt by a second or **equitable mortgage.** A second mortgage differs from a first or legal mortgage in the sense that the mortgagor does not have a title to transfer to the mortgagee as security. The mortgagor has only an **equity of redemption**, or the right to redeem the first mortgage. It is this right that the mortgagor mortgages, hence the name equitable mortgage. In essence, it is the equitable right of the mortgagor that the mortgage attaches as security, a right made possible because the Court of Chancery over time treated the right of redemption as an estate in land[7] and, hence, open to mortgage. See Figure 23–2.

6. The right is described in *Cummins v. Fletcher* (1880), 14 Ch.D. 699.

7. See *Kreglinger v. New Patagonia Meat & Cold Storage Co. Ltd.*, supra, footnote 1. See comments by Lord Parker at pp. 47–48. See also *Casborne v. Scarfe* (1737), 1 Atk. 603 at p. 605, 26 E.R. 377 (Lord Hardwicke).

FIGURE 23–2 Difference Between First and Second Mortgages

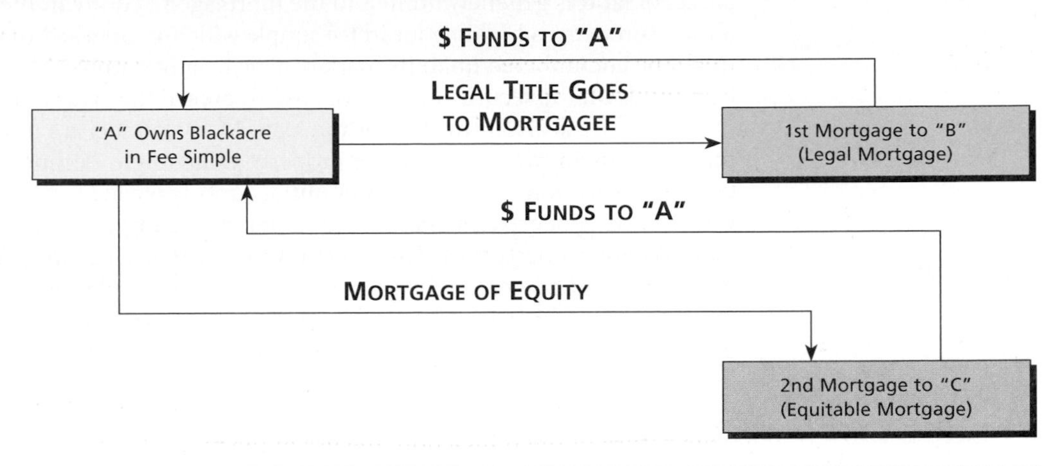

A second mortgagee, therefore, obviously assumes substantially greater risks with respect to the property than does the first mortgagee. The second mortgagee must be always vigilant and prepared for default on the part of the mortgagor. Default on the first mortgage would entitle the first mortgagee to take steps to foreclose the interests of not only the mortgagor, but the second mortgagee as well. The second mortgagee must, as a result, be in a position to make payment of the first mortgage if the mortgagor should fail to do so. This greater risk associated with second mortgages explains the reluctance on the part of some lenders to advance funds under this type of mortgage, and the reason why lenders, when they do agree to make such a loan, charge a higher rate of interest.

Mortgages subsequent to a second mortgage are also possible, as the mortgagor does not entirely extinguish his or her equity of redemption by way of the second mortgage. A mortgagor may, for example, arrange a third mortgage on the same property, which would be essentially a mortgage of the right to redeem the prior mortgages. Again, the risk associated with a third mortgage would be correspondingly higher than that associated with the second mortgage, as the third mortgagee must make certain that the mortgagor maintains the first and second mortgages in good standing at all times. The third mortgagee must also be prepared to put the prior mortgages in good standing or pay them out if the mortgagor should default. See Figure 23–3.

FIGURE 23–3 Mortgage Priorities Under Registry System*

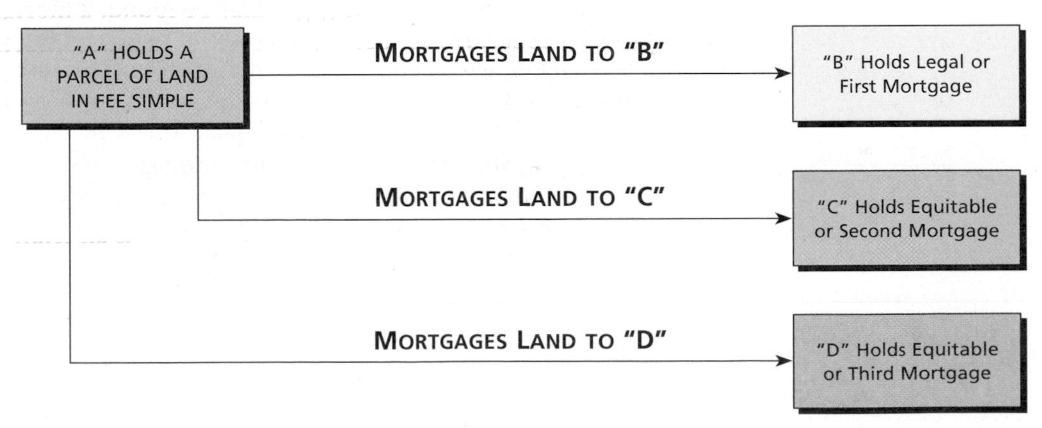

*Assuming registration in the above order.

The number of equitable mortgages that might conceivably be placed on a single parcel of land is generally limited to the mortgagor's equity in the property. For example, if a person owns a building lot in fee simple with an appraised market value of $50,000, it might be encumbered up to its market value by any number of mortgages. Lenders rarely lend funds on the security of land that would exceed the value of the property, unless some other security is also provided for the loan. Many lenders are unwilling to lend up to the market value in the event that the property value might decline. It is not uncommon for lending institutions to limit the amount of a mortgage loan to 75 percent of the appraised value of the property, unless some other guarantee of payment is also associated with the loan. Where property is used as security for more than one mortgage, the mortgagor must, of course, maintain each mortgage in good standing at all times. Otherwise, the default on one mortgage would trigger foreclosure proceedings on that particular mortgage and produce a parallel reaction among all other mortgagees holding mortgages subsequent to it.

Priorities Between Mortgages

If a mortgagor encumbers a parcel of land with more than one mortgage, the priority of the mortgagees becomes important. Since, in theory, only the mortgage first in time constitutes a legal mortgage (and all subsequent mortgages are merely mortgages of the mortgagor's equity), it is essential that some system be used to establish the order of the instruments and the rights of the parties.

In all provinces, the order of priority is established for the most part by the time of registration of the mortgage documents in the appropriate land registry office.[8] Assuming that a *bona fide* mortgagee has no actual notice of a prior unregistered mortgage, the act of registration of a valid mortgage in the registry office would entitle a mortgagee to the first or legal mortgage with respect to the property. It should be noted, however, that a mortgagee must undertake a search of the title to the property at the land registry office in order to determine the actual status of the mortgage, since the registration of a document, such as a mortgage, is deemed to be notice to the public of its existence.[9] Any prior mortgage of the same property, if registered, would take priority. Consequently, a mortgagee as a rule does not accept only the word of the mortgagor that the land is unencumbered, but makes a search of the title at the land registry office as well. This is done in advance of the registration to determine the status of the mortgage before funds are advanced under it.

Rights and Duties of the Parties

Under mortgage law, a mortgagor is entitled to remain in possession of the mortgaged property during the term of the mortgage, provided that the mortgagor complies with the terms and conditions set out in the mortgage. (If the mortgagor defaults in any material way, the mortgage usually provides that the mortgagee is entitled to take action for possession or payment). A mortgagor also has the right to demand a reconveyance of the title to the property, or a discharge of the mortgage when the mortgage debt has been paid in full. In contrast to these rights, a mortgagor has a number of duties arising out of the mortgage itself.

A mortgage, apart from being an instrument that conveys an interest in land, is contractual in nature. It contains a number of promises or covenants that the mortgagor agrees to comply with during the life of the mortgage. While the parties are free to insert any reasonable covenants they may desire in the mortgage, the instrument normally contains four important covenants on the part of the mortgagor. These are:

(1) a covenant to pay the mortgage in accordance with its terms;
(2) a covenant to pay taxes or other municipal assessments;
(3) a covenant to insure the premises, if the property is other than vacant land; and
(4) a covenant not to commit waste, and to repair the property if any damage should occur.

8. See, for example, the *Registry Act*, R.S.O. 1990, c. R.20, s. 70.
9. ibid., s. 74.

Payment

Both the mortgagor and the mortgagee are vitally interested in the payment of the debt. The mortgagee wants to receive back the principal and interest earned on the loan, and the mortgagor wishes to obtain a reconveyance of the title to the property.

Mortgage-payment terms are negotiated by the parties and may take many forms. The most common arrangement calls for the repayment of the principal together with interest in equal monthly payments over the term of the mortgage. The particular advantage of this form of repayment over provisions that provide for quarterly, half-yearly, annual, or "interest-only" payments during the life of the agreement is that any default on the part of the mortgagor will be quickly noticed by the mortgagee, and steps may then be taken to remedy the situation. Mortgages that call for periodic payments generally contain an acceleration clause that is triggered by default of payment of a single installment, thereby rendering the whole of the outstanding balance due and payable immediately. This permits the mortgagee to demand payment of the full amount. Failing payment by the mortgagor, the default would entitle the mortgagee to take steps to realize the debt from the property by way of foreclosure or sale proceedings.

A mortgagee seldom demands payment in full simply because a mortgagor fails to make a monthly payment on the due date. However, a failure to make a number of payments would probably result in action taken by the mortgagee on the basis of the mortgagor's breach of the covenant to pay.

Payment of Taxes

Under most tax legislation, taxes on real property levied by municipalities (or in some cases by the province) provide that the claim of the municipality for unpaid taxes becomes a claim against the property in priority over all other encumbrances. To ensure that the municipality does not obtain a prior claim against the property, mortgages usually contain a covenant on the part of the mortgagor to pay all such taxes levied when due. Again, if the mortgagor fails to pay the taxes, he or she is treated as being in default under the mortgage for breach of the covenant concerning taxes, and the mortgagee may then take action under the mortgage.

In the recent past and, to some extent, today, mortgagees (especially lending institutions) inserted a clause in their mortgages requiring the mortgagor to make payments to the mortgagee over the course of the year (usually on a monthly basis) of sufficient funds to pay the municipal taxes on the property. The mortgagee would then use the funds to pay the taxes when levied by the municipality. In this fashion, the mortgagee was in a position to maintain the mortgage free of any claims for taxes. As an alternative, mortgagees who do not collect tax payments and pay the taxes directly may still require the mortgagor to submit the paid municipal tax bill each year for examination in order to substantiate payment and to ensure that no claims for taxes may take priority over their mortgages.

Insurance

The obligation on the part of the mortgagor to insure only applies if a building or structure is erected on the mortgaged land. The purpose of the covenant to insure is obvious: much of the value of the property rests in the existence of the building or structure, and its loss, through damage by fire or other causes, in most cases would substantially lessen the security for the mortgagor's indebtedness.

Careful mortgagees generally arrange their own insurance on the mortgaged premises in order to satisfy themselves that the insurance is adequate and that it covers their interest in the event that loss or damage should later occur. Other mortgagees simply rely on the mortgagor to maintain the required insurance. The liability for the payment of the premium remains, however, with the mortgagor even though the insurance covers both the mortgagor and mortgagee. In the event of a loss, the proceeds of the insurance would be payable first to satisfy the principal and interest owing to the mortgagee, and the balance, if any, would then be payable to the mortgagor. The land, of course, would remain intact. If the insurance proceeds paid the mortgage in full, the mortgagor would be entitled to a discharge of the mortgage and a return of the title to the property.

Waste

Closely associated with insurance, but of a different nature, is the mortgagor's covenant not to commit waste. Waste is a legal term used to describe any act that would reduce the value of the property. It would include, in the case of a mortgage, an act by a mortgagor to reduce the security available to the mortgagee. For example, a mortgagor who demolishes a building on the mortgaged property would be committing waste, because the building would represent part of the value of the property. Similarly, a mortgagor who strips and sells the top soil from the land would commit waste if the act was done without the consent of the mortgagee.

The covenant by the mortgagor not to commit waste is essentially a term that preserves the property intact during the time that the mortgage is in existence. It is inserted in the agreement to protect the mortgagee by requiring the mortgagor to maintain the property value. A violation of the covenant by the mortgagor would entitle the mortgagee to take action under the terms of the mortgage. Waste does not include, however, ordinary deterioration of property as a result of lack of repair. To cover this, mortgages may also include a term that obliges the mortgagor to maintain the property in a good state of repair while the mortgage is in effect.

SPECIAL CLAUSES

Not every term in a mortgage imposes an obligation on the mortgagor. Frequently, the parties will agree that the mortgage should contain the privilege of prepayment. The parties will insert in the mortgage a provision whereby the mortgagor, while not in default, may pay the whole or any part of the principal amount owing at any time without notice or without the payment of a bonus to the mortgagee. This is often a valuable privilege for the mortgagor, as the mortgagee is not required to accept payment of the mortgage money owing except in accordance with the terms of the mortgage.[10]

A mortgagor may also have the privilege of obtaining discharges of parts of the mortgaged land on the part-payment of stipulated amounts of principal, if such a right is inserted in the mortgage. The particular advantage of this clause only arises if the mortgagor intends to sell parts of the mortgage land, either to pay the mortgage or as a part of a development scheme for the property.

MANAGEMENT ALERT: BEST PRACTICE

Be careful of special clauses that can cut both ways. An adjustable rate mortgage may allow a mortgage interest rate to ratchet down along with (perhaps expected) decreases in the market's prime lending rate; however, the same provision can drive the mortgage rate (unexpectedly) upward. Do not let unwarranted optimism regarding market rates to allow you to make unsustainable financial decisions. Watch out as well for any terms that allow the lender to make arbitrary or structural changes to the mortgage. A fixed "introductory rate" on mortgages for one year, then to be replaced by "the lender's prevailing rate," was one of the causes of the U.S. credit crisis of 2008, when the later rate was beyond the repayment ability of many mortgagees.

DISCHARGE OF MORTGAGE

If the mortgagor complies with the terms and conditions set out in the mortgage and makes payment of the principal and interest owing as required and when due, the mortgagor is entitled to a discharge of the mortgage. Mortgage legislation, in provinces where the Registry System is in effect, generally provides that a mortgagee may release all right, title, and interest in the property subject to the mortgage by providing the mortgagor with a discharge of mortgage. This particular instrument, when properly executed and

10. See the *Interest Act*, R.S.C. 1985, c. I-15, s. 10. A mortgagee in a long-term mortgage is obliged to accept payment in full, however, if a non-corporate mortgagor tenders full payment and an additional three months' interest at any time after the mortgage has been in effect for five years.

delivered to the mortgagor, along with the registered duplicate original copy of the mortgage, constitutes a receipt for payment. When registered, the discharge of mortgage acts as a statutory reconveyance of the title to the mortgagor. The document, in effect, releases all claims that the mortgagee may have in the land under the mortgage and acknowledges payment of the debt. In provinces where the Land Titles System is used, the discharge of mortgage is replaced by a cessation of charge, which has a somewhat different effect. This document acts as an acknowledgement of payment of the debt and removes the charge from the title to the property when it is registered in the land titles office where the land is situated. On receipt of the cessation of charge, the office amends the title to the parcel of land to reflect the change and to show the title free of the particular charge.

ASSIGNMENT OF MORTGAGE

A mortgagee may assign a mortgage at any time after the mortgage is executed by the mortgagor. A mortgage is unlike an ordinary debt, however, in that it represents an interest in land. Consequently, it must be made in a form that complies with the legislation pertaining to mortgages in the particular jurisdiction where the land is situated. This is because the assignment must not only assign the debt, but also transfer the assignor's interest in the property. Consent to the assignment by the mortgagor is not required, but actual notice of the assignment to the mortgagor is essential if the assignee wishes to protect the right to demand payment of the balance owing on the mortgage from the mortgagor.

As with the assignment of any contract debt, if notice of the assignment is not given to the mortgagor, the mortgagor may quite properly continue to make payments to the original mortgagee (the assignor). All other rules relating to the assignment of debts normally apply to a mortgage assignment. The mortgage is assigned as it stands between the mortgagor and the mortgagee at the time of the assignment, for example. Any defence that the mortgagor might raise to resist a demand for payment by the assignor would also be effective as against the assignee. The assignee must, therefore, be certain to promptly give the mortgagor notice of the assignment following the assignment, and determine the status of the mortgage between the mortgagor and the mortgagee immediately. In most cases, an assignee will determine this from the mortgagor prior to the assignment.

SALE OF MORTGAGED PROPERTY

A mortgagor is free to sell or otherwise dispose of the equity of redemption in mortgaged land at any time during the term of the mortgage, unless the mortgage provides otherwise.[11] The disposition of the property does not, however, relieve the mortgagor of the covenants made in the mortgage. In the event that the purchaser should default on the mortgage at some later time, the mortgagee may, if the mortgagee so desires, look to the original mortgagor for payment in accordance with the original covenant to pay. It should be noted, however, that this particular rule of law only applies to the original mortgagor, and not to a purchaser who subsequently sells the property subject to the mortgage before default occurs.

Smith gives Right Mortgage Company a mortgage on a parcel of land that she held in fee simple. Smith later sells her equity in the land to Brown. If Brown defaults in payment of the mortgage, Right Mortgage Company may either claim payment from Smith under her covenant to pay the mortgage, or it may take steps to have the debt paid by way of foreclosure or sale of the property. If the mortgagee pursues the latter course, Brown would lose his interest in the land. However, if Brown does not default, but sells his equity in the land to Doe, and Doe allows the mortgage to go into default, Right Mortgage Company does not have a claim against Brown, but only against Smith on the covenant, or Doe, who is in possession of the land.

11. Occasionally, a mortgagee may insert a term in the mortgage that requires the mortgage to be paid in full if the mortgagor should desire to sell the property.

Mortgagees frequently require the purchaser of the mortgagor's equity to sign an agreement to assume the mortgage and all its covenants in order to increase the security for payment. A purchaser who covenants to pay the mortgage in accordance with its terms becomes liable on the covenants as a result. The purchaser is placed in much the same position as the original mortgagor with respect to payment.

COURT DECISION: Practical Application of the Law

Assignment of Mortgage—Release from Covenant to Pay

***Omista Credit Union Ltd v. Bank of Montreal*, 2004 NBQB 107.**

Eugene & Yolande Leaman mortgaged a property to the Bank in 1996. In 1999 they separated and the Bank agreed to release Eugene from his personal covenant as he conveyed his interest in the property to Yolande. She did her banking business with the Credit Union, and decided to refinance the property through it. It requested the Bank to execute an assignment of the mortgage. The Bank did not inform the Credit Union that Eugene was released from his personal covenant to pay under the mortgage. Yolande later went bankrupt. The Credit Union exercised its power of sale under mortgage, and a deficiency arose. Eugene refused to pay, as his personal obligation had been released. The Credit Union sued the Bank to recover the deficiency claiming it breached its obligation in contract and in tort to inform them that Eugene had been released from his personal obligation to pay under the mortgage.

THE COURT: The Credit Union argues that it thought it was obtaining not just the mortgagees' interest in the property, but also the personal covenants of the two original mortgagors, including Eugene Leaman, to make payment. Although there was no formal document in the Bank's file that showed the release of Eugene's personal covenant…it was information known to the Bank and I find it was information required to be provided to the Credit Union as a condition of its agreeing to take an assignment of the mortgage. The Bank says the Credit Union had an obligation to exercise due diligence and responsibility for any failure to acquire knowledge of the release of Eugene's personal covenant rested with the Credit Union. Good business practice might support this position, but I cannot find that any obligation of inquiry would relieve the Bank's obligation to disclose changes in the mortgage security of which it not only had knowledge but in which it was a primary party. The evidence is that the Credit Union did not require a credit application at the time it took the assignment and that it appears to have relied on a credit application completed by Yolande and Eugene a year previously, before they had separated, settled the distribution of their property and debts and negotiated Eugene's release from the mortgage. Again this may not have been the best business practice, but I fail to see how it relieves the Bank of its obligation to fulfill the conditions under which the assignment was taken by the Credit Union. Which brings me to the final issue.

If the Credit Union knew at the time it took the assignment of the mortgage that Eugene had been released from his personal covenant to pay, it may be inferred that it did not require notice from the Bank that this had occurred. In this situation, the Credit Union would have taken the security with full knowledge of its worth and cannot be seen to complain of any failure of the Bank to provide documentation or notice in this regard. Even though a breach of condition might be alleged, the result would be no depreciation of the recognized security and no increase of risk to the Credit Union. The result is that in such a circumstance the Credit Union cannot be seen to have suffered damages for which the Bank would be responsible. The evidence of Eugene, which I accept, is that both he and Yolande met with Sharon Duguay, who was a loans officer of some considerable experience with the Credit Union, for the purpose of discussing how he was to "get his name off" several mortgages held by the Credit Union on properties owned jointly with his wife which were to be transferred to her under their separation agreement. It is difficult to believe that at least the intention to have Eugene relieved of his covenant to pay under the Bank of Montreal mortgage was not known to the Credit Union at the time it was being requested to give effect to the separation agreement with respect to the personal obligations to pay under the mortgages on other properties held by the Credit Union. When the Credit Union was approached to take an assignment of the Bank's mortgage, it was approached by Yolande only. Eugene had nothing to do with the request. The letter of direction to deliver an Assignment Agreement to the Credit Union was prepared by the Credit Union but signed by Yolande only. The Statement of Mortgage Loan Account for Payout Purposes was sent to Yolande only. The payout letter from the Credit Union refers to an assignment/transfer of mortgage referring to Yolande only. The Credit Union was in possession of the tax bill on the property for the year 2000 which showed Yolande to be the sole owner. It seems apparent to me that the Credit Union was well aware that the person responsible under the mortgage was Yolande alone. Given the nature of this correspondence; given the prior discussions with the Leamans concerning their borrowings at the time of their separation; and given the fact that the Credit Union had knowledge of the separation agreement, I am led to the conclusion that the Credit Union knew that Eugene was no longer responsible to pay the mortgage debt. I find that on a balance of probabilities the Credit Union accepted the security without the personal covenant of Eugene as security for the loan it extended to Yolande. On this finding, as stated previously, the Credit Union cannot be seen to have suffered damages for which the Bank would be responsible...and the claim is dismissed.

The case clearly illustrates the merits of a release, and the importance of making complete inquiries as a business practice. Recalling chapters on tort and contract, what would the Credit Union claim as the specific tort and contractual breach?

A QUESTION OF ETHICS

Where is the fairness in keeping an original mortgagor (borrower) "on the hook" under the covenant to pay a mortgage, after they have sold their interest in the property to another person who has assumed the mortgage? How would you attack or defend this practice?

DEFAULT, FORECLOSURE, AND SALE

Acceleration clause

A clause in a debt instrument (e.g., a mortgage) that requires the payment of the balance of the debt on the happening of a specific event, such as default on an installment payment.

If the mortgagor defaults in the payment of principal or interest under the terms of a mortgage, or causes a breach of a covenant (such as the commission of waste), the mortgagee may, at his or her option, call for the immediate payment of the full amount owing by way of the **acceleration clause** in the mortgage. The demand for payment in full does not necessarily mean, however, that the mortgagor has no choice but to find the full amount required to pay the mortgage. In most provinces, the legislation pertaining to mortgages permits the mortgagor to pay the arrears, or correct the breach of covenant along with the payment of all related expenses, and thereby put the mortgage in good standing once again.[12]

If financial difficulties prevent the mortgagor from correcting the default, the mortgagee is usually obliged to take action in order to recover the debt secured by the property. In this respect, the mortgagee has a number of options open. The most common courses of action that the mortgagee may follow are:

(1) sale of the property under the power of sale contained in the mortgage;
(2) foreclosure;
(3) judicial sale; and
(4) possession of the property.

Sale Under Power of Sale

Mortgages drawn in accordance with most provincial statutory forms generally contain a clause that permits the mortgagee to sell or lease the property subject to the mortgage if the mortgagor's default in payment persists for a period of time.[13] To exercise the power of sale under the mortgage, the mortgagee must give written notice to the mortgagor and to any subsequent encumbrancers, then allow a specified period of time to elapse before the property may be sold. The purpose of the notice and the delay is to provide the mortgagor and any subsequent encumbrancers with an opportunity to put the mortgage in good standing once again, or to pay the full sum owing to the mortgagee.

12. See, for example, the *Mortgages Act*, R.S.O. 1990, c. M.40, ss. 22 and 23. See also *Township of Scarborough v. Greater Toronto Investment Corp. Ltd.*, [1956] O.R. 823.

13. In Ontario, for example, the usual clause provides that the mortgagee may begin the exercise of the power of sale when default exceeds 15 days. See also the *Mortgages Act*, R.S.O. 1990, c. M.40, s. 32.

If the mortgagor is unable to place the mortgage in good standing, the mortgagee is free to proceed with the sale and to use the proceeds first in the satisfaction of the mortgage debt, and the remainder in the payment of subsequent encumbrances and any execution creditors. Any surplus after payment to the creditors belongs to the mortgagor. Should the mortgagor be in possession prior to the sale, the mortgagee would be obliged to bring an action for possession in order to render the property saleable. If the mortgagee is successful in this action, the courts will render a judgment for possession, which in turn would enable the mortgagee to obtain a writ of possession. The writ of possession is a direction to the sheriff of the county where the land is situated, authorizing the sheriff to obtain possession of the property for the mortgagee.

Since the mortgagor is entitled to the surplus in a sale under a power of sale (and also liable for any deficiency), the mortgagee is under an obligation to conduct the sale in good faith and to take steps to ensure that a reasonable price is obtained for the property. The effort normally required would include advertising the property for sale, and perhaps obtaining an appraisal of the property to determine its approximate value for sale purposes. The mortgagee is not obliged, however, to go to great lengths to obtain the best price possible.[14]

CASE IN POINT

A mortgage lender sold a mortgaged property (in default) under power of sale. It had appraisals of $300,000 and $370,000 and received letters of interest in the amounts of $380,000 and $385,000. It thereafter accepted an offer of $410,000, against a mortgage debt of $418,000, and commenced action against the mortgagor for the shortfall. Within three months, the purchaser under power of sale had resold the property for $500,000. The defendant mortgagor counterclaimed against the lender that this was proof that it had been improvident in accepting the $410,000 offer. The court disagreed, stating that a mortgagee selling under power of sale is under a duty to take reasonable precautions to obtain the true value of the mortgaged property at the date on which a decision is made to sell it. This does not mean that the mortgagee must obtain true value, only to take reasonable precautions. Perfection is not required, and it does not matter if the moment may be unpropitious and that, by waiting, a higher price could be obtained.

See: *Canada Trustco Mortgage Co. v. Casuccio*, 2005 CanLII 25887 (Ont. S.C.).

Foreclosure

Except in Quebec, an alternative to sale under the power of sale contained in the mortgage is an action for foreclosure. While much procedural variation exists between provinces, this type of action, if successful, results in the issue by the courts of a final order of foreclosure, which extinguishes all rights of the mortgagor (and any subsequent encumbrancer) in the property. An action for foreclosure, however, gives the mortgagor and any subsequent encumbrancers the right to redeem. As well, the courts will provide a period of time (usually six months) to enable any party who makes such a request the opportunity to do so. If the party fails to redeem the property within the time provided, or if no request for an opportunity to redeem is made, the mortgagee may then proceed with the action and eventually obtain a judgment and final order of foreclosure. In spite of its name, the final order of foreclosure is not necessarily final in every sense of the word. The mortgagor may, under certain circumstances, apply to have it set aside after its issue, if the mortgagor acquires the necessary funds to redeem and the mortgagee has not disposed of the property.

14. *Farrar v. Farrars, Ltd.* (1889), 40 Ch.D. 395.

Judicial Sale

As an alternate method of obtaining payment of the mortgage, the mortgagee may apply to the court for possession, payment, and sale of the property. Under this procedure, the property would be sold and the proceeds distributed — first in the payment of the mortgage, then in payment of the claims of subsequent encumbrancers, and, finally, the payment of any surplus to the mortgagor. The sale under this procedure differs substantially from a sale under the power of sale. However, the most common procedure is to have the property sold by tender or public auction, usually subject to a reserve bid. The method of sale does vary, since in most provinces the courts generally have wide latitude in this respect.

A sale may also be requested by the mortgagor when foreclosure action is instituted by the mortgagee. In this case, the mortgagor is usually expected to pay a sum of money to the courts to defray a part of the expenses involved in the sale. This is normally required at the time that the request for a sale is made. Once the request is received, however, the foreclosure action becomes a sale action, and the action then proceeds in the same manner as if a sale had been requested originally by the mortgagee. The advantage to the mortgagor of a sale is that the proceeds of the sale will also be applied to the payment of the claims of subsequent encumbrancers and not just to the claim of the first mortgagee. In this manner, the mortgagor is released from all or part of the claims of the subsequent encumbrancers, rather than only the claim of the first mortgagee.

Possession

When default occurs, the mortgagee has a right to possession of the mortgaged premises. If the premises are leased units, the mortgagee may displace the lessor and collect any rents that are payable by the tenants.

The rents collected must be applied, however, to the payment of the mortgage, and the mortgagor is entitled to an accounting of any monies collected. If the mortgagor has vacated the mortgaged property, the mortgagee may move into possession. But usually this is not the case, and the mortgagee must normally apply to the courts for an order for possession. This is generally done when the mortgagee institutes legal proceedings on default by the mortgagor. The usual relief requested by the mortgagee is either foreclosure, payment, and possession, or sale, payment, and possession.

CASE IN POINT

In one of the highest profile mortgage-fraud cases, Susan Lawrence found herself confronted by a mortgage company intending to evict her from her home. A person posing as "Susan Lawrence" had presented a lawyer with a fraudulent Offer to Purchase for Susan's home, executed by another fraudster, a buyer named "Thomas Wright." "Thomas Wright" had earlier obtained approval from a legitimate lender, Maple Trust, for a mortgage on the property. In return for the mortgage money, the lawyer prepared a deed to "Thomas Wright," and a mortgage in favour of Maple Trust. The fraudulent "Susan Lawrence," the mortgage proceeds, and "Thomas Wright" subsequently disappeared, leaving Maple Trust to descend upon the oblivious (and real) Susan Lawrence. The Ontario Court of appeal ruled that "Wright" never became a registered owner as he took his title by fraud, and thus could not give a valid charge upon the land. The mortgage to Maple Trust was void against the real Susan Lawrence and the land. Ontario has enacted Bill 152 (Government Services and Consumer Protection) to create statutory protection in these situations, to streamline resolutions, and to make such frauds more difficult. Time will tell.

Lawrence v. Maple Trust Company, 2007 ONCA 74.

CHECKLIST FOR MORTGAGES

1. Types
   - Legal (first) mortgage.
   - Equitable (second, and later) mortgage(s).
2. Duties of Mortgagor
   - Payment.
   - Taxes.
   - Insurance.
   - Not to waste.
3. Rights of Mortgagor / Duties of Mortgagee
   - Quiet possession.
   - Discharge on payment.
4. Mortgagee's Remedies
   - Acceleration.
   - Sale under Power of Sale.
   - Foreclosure.
   - Sale under Court Order.

BUSINESS APPLICATIONS OF MORTGAGE SECURITY

The purchase of real property represents a substantial investment by most purchasers, and they often do not have sufficient funds available to pay the purchase price in full. Many financial institutions and other investors are prepared to provide these funds by way of a mortgage, because a mortgage permits them to secure their investment in the property. Consequently, real-property purchase transactions frequently involve three parties: a vendor, a purchaser, and a mortgagee.

A simple personal transaction, such as the purchase of a dwelling-house by an individual, might be conducted in the following manner:

> The purchaser enters into an agreement with the vendor to purchase the property for a fixed amount, say $160,000, subject to the condition that the purchaser obtain suitable financing. The purchaser would then contact a mortgage lender, such as a bank, trust company, mortgage corporation, or private investor, to arrange for a loan to be secured on the land by way of a mortgage. If the lender agrees to provide financing (usually up to 75 percent of the property value), the purchase transaction would then proceed, with the purchaser paying the vendor $40,000 (25 percent) and the lender advancing $120,000 (75 percent) for a total of $160,000. A mortgage to the lender securing the $120,000 would be registered against the property at the time that the purchaser's deed was registered. The purchaser would then obtain the property (subject to the mortgage) and would pay the mortgage amount to the lender over the period of time specified in the mortgage.

A more complex commercial example using mortgage financing is represented by the land-development mortgage. Land developers usually buy large tracts of land suitable for development as housing or for commercial buildings (an office or industrial condominium, for example). The developer may finance the development by way of a mortgage that contains a clause allowing the developer to obtain the release of parts of the land covered by the mortgage as the property is developed.

Land Development Corporation may purchase a 20-hectare parcel of land zoned for residential housing. If the purchase price is $1,500,000, the developer may finance the transaction by making a $500,000 cash payment and obtaining a mortgage for $1,000,000 from a financial institution. The mortgage could contain a term that allows the developer to obtain partial discharges of the mortgage at the rate of $25,000 per building lot.

Once the developer acquires the raw land, the developer would lay out the property into perhaps 150 building lots and obtain the necessary approvals for a housing subdivision. The developer would then build roads, install services, and generally prepare the parcels of land for sale as housing sites. As each lot was sold to a purchaser, the developer would pay the mortgagee the required $25,000 to obtain a partial discharge of the lot from the mortgage. This would allow the developer to convey a clear title in fee simple to the lot to the purchaser, and at the same time reduce the mortgage debt. In this case, when 40 lots were sold, the mortgage would be fully paid. The remaining lots in the development would be released from the mortgage when the mortgagee received payment for the last lot of the 40, and it would then give the developer a full discharge of the mortgage.

MANAGEMENT ALERT: BEST PRACTICE

If a business obtains a mortgage to build or renovate a structure, the mortgagee (lender) will provide the funds under the mortgage not as a block, but as a series of advances as construction progresses. Each advance will be subject to a "holdback," often of 10 percent for 45 days, to shield the mortgage lender (by law) from claims of unpaid accounts of the building contractor or subcontractor doing the work. This may put the business (mortgagor) in a cash-flow squeeze, because it must pay these accounts, but may not have sufficient funds due to the holdback — a "Catch 22" situation. It is up to the mortgagor to ensure that the construction and accounts-payable timetables do not conflict with the timetable and amounts for the mortgage advances and the release of the holdback funds.

Mortgages may also be used to finance the construction of buildings, under a type of mortgage known as a building mortgage. With this mortgage, only a small portion is advanced initially to the property owner (usually an amount not exceeding the value of the building lot). As the building construction proceeds, the mortgagee would advance further sums to enable the property owner to pay for the construction work done. These advances are usually made at specific points in the construction process, such as when the foundation is completed, the exterior walls and roof are finished, the plumbing and electrical work are completed, and the interior is finished. The final payment would be made when the landscaping was completed and the building was ready for occupancy. Mortgage financing of a very large commercial building complex is usually handled in a somewhat different manner, but the general concept underlying the use of the mortgage as a means of financing the building project remains the same.

Mortgages may also be used as a means of providing security to support other debt instruments, such as promissory notes. In these situations, the debtor, under a promissory note, may offer as collateral security a mortgage on real property. The mortgage given in this instance is related to the negotiable instrument and is called a collateral mortgage. If the debtor (mortgagor) defaults on payment of the promissory note, then the creditor (mortgagee) may look to the collateral security that is the mortgage on the land, as default on the note would constitute default on the mortgage. Payment of the debt could then be realized by taking action under the mortgage by way of sale or foreclosure.

CASE IN POINT

A development corporation (A. Ltd.) was advanced money from a government-mortgage corporation for the construction of a multi-unit residential-housing project. The loan was secured by a mortgage, guaranteed by the principal shareholders of the corporation.

The corporation later sold the project to another corporation (B. Ltd.), which ran into financial difficulty. In an effort to assist the new owner, B. Ltd., the government-mortgage corporation lowered the interest rate on the mortgage to 6 percent without the agreement of A. Ltd. or its principal shareholders. The government-mortgage corporation eventually foreclosed on the mortgage, and sued A. Ltd. and its principal shareholders on their guarantees.

The court held that A. Ltd. and its principal shareholders were still liable on their guarantees, as novation did not take place, and while the interest rate was changed, A. Ltd. and its principal shareholders were not prejudiced by the lower interest rate.

See: *Alberta Mortgage and Housing Corporation v. Strathmore Investments Ltd.* (1992), 6 Alta. L.R. (3d) 139.

SUMMARY

A mortgage is an instrument that utilizes land as security for debt. In most provinces, the mortgagee acquires an interest in the land by way of the mortgage, either in the form of a transfer of the title to the property as security (such as in Eastern Canada, and parts of Ontario and Manitoba under the Registry System) or in the form of a charge on the land in those provinces under the Land Titles System. Quebec law provides for a somewhat similar instrument (called a hypothec) that is in the nature of a lien on the land.

A mortgage, in addition to conveying an interest in the land, contains the details of the debt and the provisions for its repayment. In the instrument, the mortgagor covenants to protect the property by payment of taxes and insurance, and promises not to diminish the mortgagee's security by waste or non-repair. If the mortgagor defaults in payment, or fails to comply with the covenants in the mortgage, the mortgagee may institute legal proceedings to have the mortgagor's interest in the property foreclosed or sold. Foreclosure is not available in some provinces. However, in all provinces, if the mortgagor fails to pay the debt after default, the property may be sold, and the proceeds used to satisfy the indebtedness.

If more than one mortgage is registered against a parcel of land, the mortgagees are entitled to payment in full in the order of their priority. This is usually determined by the time of registration of each instrument under the land registry system in the province. If a surplus is available after the claims of all mortgagees and other creditors with claims against the land are satisfied, the sum goes to the mortgagor.

If no default occurs during the term of a mortgage, and the indebtedness is paid by the mortgagor, the mortgagee must provide the mortgagor with a discharge of the mortgage. This acts as a statutory reconveyance of the title and receipt for payment for a mortgage under the Registry System, or a release of the claim against the property in the case of a cessation of charge under the Land Titles System. In both instances, the discharge given to the mortgagor extinguishes the debt and releases the property as security from the mortgagee's claim against it.

Because mortgage law developed in a different fashion in each province, the nature of mortgage instruments and the rights of the parties vary substantially from province to province. The basic concept of land as security for debt is common, however, to all provincial systems.

Mortgages represent a very useful tool in the financing of many business transactions related to land, particularly in land development and building construction.

Key Terms

Mortgage, 453
Charge, 453
Legal mortgage, 456
Equitable mortgage, 456
Equity of redemption, 456
Acceleration clause, 463

Review Questions

1. What is the normal procedure used to re-vest the legal title of a property in the mortgagor when the mortgage debt is paid? Does this differ in the case of a charge?
2. In what way does a mortgagee's interest differ from that of a person who holds land in fee simple?
3. In what way (or ways) does a mortgage differ from a charge?
4. Explain the relationship that exists between a mortgagee and a person who acquires the mortgaged property from the mortgagor. Does the original relationship of mortgagor–mortgagee continue as well?
5. How does a "first" mortgage differ from a "second" mortgage?
6. What factors must be considered by a person who wishes to extend a loan of money to another on the security of a second mortgage?
7. What is the nature of the covenants that a mortgagor agrees to in a mortgage?
8. Why do mortgages usually contain an acceleration clause? What is the effect of the clause if default occurs?
9. Outline the nature of a mortgagor's interest in the mortgaged land.
10. Indicate what the rights of a mortgagee would be if a mortgagor defaulted on the payment of the mortgage.
11. Outline the rights of an assignee of a mortgage from a mortgagee. What steps must be taken to ensure that the mortgagor makes payment to the assignee after the assignment takes place?
12. What are the rights of a mortgagor if, on default of payment, the mortgagee commences foreclosure proceedings?
13. How does a sale under a power of sale differ from a sale action?
14. What rights, if any, are available to a mortgagor after foreclosure takes place?
15. Define the term "mortgage" as an interest in land.
16. If the original mortgagor sold the mortgaged lands to a purchaser, and the purchaser failed to make payments on the mortgage, explain the possible courses of action that the mortgagee might take.

Mini-Case Problems

1. B mortgages "Blackacre" to A for $100,000. Some time later, B defaults on payment, and A begins foreclosure proceedings. If Blackacre has a value of $500,000, what should B do?
2. If B did nothing in the above case until after foreclosure proceedings were completed, then wished to pay the amount owing, could B still do so to re-acquire the property? Would your answer be any different if A had sold the land to C?
3. Rosa gives a mortgage on Green Acres to Shelley, and it is duly registered. Rosa later gives a mortgage to Tina, and it is registered. Rosa, a year later, defaults on the mortgage to Shelley. What is the position of Tina and Rosa?
4. William owns a house and lot, and gives a mortgage on the property to Wallace. William later sells the house and lot to Black, with Black assuming the mortgage.

   A year later, Black defaults on the mortgage. Advise Wallace.

Case Problems for Discussion

CASE 1

Smith, Jones and Davis carried on business in partnership as SJD Building Contractors. They arranged for the purchase of 3 adjacent lots in a subdivision at a price of $80,000 per lot, financing the transaction with a mortgage on the three lot parcel for $180,000, and the partnership supplying the balance of the funds. The mortgage was in the name of the three partners carrying on business as SJD Building Contractors. Smith signed the mortgage on behalf of the partnership.

A month later, Davis decided to retire from the partnership, but wished to build a home for himself and his family on one of the lots. The partnership approached the mortgagee and requested a release of the lot from the mortgage on payment to the mortgagee of $60,000, being approximately one third of the amount of the mortgage. The mortgagee agreed, and Davis personally paid the $60,000 for a release of the lot.

The housing market, however, took a downturn, and Smith and Jones found themselves shortly thereafter in financial difficulty. Unable to make mortgage payments, the mortgage fell into arrears. The value of the two lots, unfortunately, had fallen in value and were worth approximately $50,000 each. The mortgage at this point, with arrears, was $118,000.

Advise each of the parties (including Davis) of their rights and obligations (if any).

CASE 2

Headrick owned a house and lot in a very desirable residential neighbourhood of a large city. In order to purchase a new luxury motor vehicle, new boat, and pay for a luxury vacation, he arranged for a mortgage on the house and lot for $330,000. The property at the time had an appraised value of $375,000. Once in funds, Headrick left on his vacation. While on vacation, he decided to extend his travels, even though his vacation time had ended. A month later, he returned home, only to find that he had been terminated by his employer. At this point in time he had spent all of his money on the automobile, boat and vacation, and was unable to make payments on the mortgage. Over the next number of months, Headrick searched for employment without success. He ignored letters from the mortgagee demanding payment, and eventually, the mortgagee decided to sell the property under its power of sale.

The mortgagee contacted an appraiser and requested an appraisal of the property "at a fire sale price" as it was anxious to get rid of the property. The appraiser gave an appraisal at $300,000, well below the value of the property if it was offered in a normal real estate listing. The mortgagee proceeded with the sale under the power of sale, and quickly sold the property at the appraised value of $300,000. The mortgagee then demanded the difference between the sale price and the amount of the mortgage, which was still at $330,000.

Headrick had protested the listing of the property in the power of sale advertisements at such a low price, and had advised the mortgagee that his friend Esson was prepared to purchase the property for $345,000, but his objections had been ignored by the mortgagee, who simply wanted to get rid of the property. He is now angry and upset that the mortgagee is demanding payment of the additional $30,000 from him.

Advise Headrick.

CASE 3

Agricola sold his farm to Ambrose for $100,000. Ambrose arranged for a purchase-money mortgage from the Agricultural Loan Company for $60,000, and Agricola agreed to take back a mortgage in the amount of $20,000 in order that Ambrose could acquire the property. On the date fixed for closing, Agricola gave Ambrose the deed to the property and received a cheque from Ambrose in the amount of $20,000. He also received a cheque from the loan company for $60,000 when the company registered its mortgage immediately after the registration of the deed to Ambrose. Agricola then registered his mortgage for $20,000. At the time, Agricola transferred the fire-insurance policy (which covered the buildings) to Ambrose. The policy transfer named Ambrose as the new owner, subject to the interest of Agricola as mortgagee. Through an oversight, the Agricultural Loan Company was not named as an insured on the policy.

Some time later, Ambrose defaulted on the mortgage to Agricola, and it was necessary for Agricola to foreclose on the mortgage. Agricola continued to make the mortgage payments each month to Agricultural Loan Company and allowed Ambrose to remain on the property to work the farm on a crop-sharing basis.

Not long after Agricola had foreclosed on his mortgage and taken back the property, a serious fire destroyed a large barn on the premises. The barn had a value of $50,000. The insurer noted that the fire-insurance policy listed Ambrose as the owner of the property, and Agricola as the mortgagee. However, before the insurance company made payment, all three parties — Agricola, Ambrose, and the Agricultural Loan Company — claimed the insurance proceeds.

Discuss the nature of the rights that each party might raise. Discuss the possible outcome.

CASE 4

An elderly woman who could only read with difficulty operated a rooming house for students. Her home was large and in close proximity to the college. As such, it was a valuable piece of residential property.

Her nephew, Herman, had on numerous occasions urged her to retire and suggested that she sell the property. For many years she refused to consider the idea. However, as a result of Herman's insistence, she eventually agreed that she would list the property with a local real-estate agent to see what the market might be. A few days later, Herman appeared at her home with some papers that he said were the forms that he had obtained from the real-estate agent for the listing of her property. In reality the forms were mortgage forms. Herman represented the forms as "only a formality, to let the real-estate agent have authority to show the house to prospective buyers." His aunt signed the forms, believing them to be copies of the real-estate listing agreement.

Herman later registered the mortgage, which was drawn for $50,000, and assigned it to a finance company for $45,000. He intended to use a part of the money to make the payments on the mortgage himself; he planned to use the balance for a trip to Las Vegas, where he expected to make a fortune by employing a new system for placing bets at the gambling tables in a casino.

Herman's scheme failed, however, and he returned to Canada penniless. He soon spent the funds, which he had originally set aside to make a few payments on the mortgage, and again found himself without funds. The mortgage, as a result, went into default, and the finance company instituted foreclosure proceedings against the property. At that point his aunt suddenly became aware of the mortgage.

Indicate the action that the aunt might take in this case.

Discuss the position of the finance company. What might be the outcome of its foreclosure proceedings?

CASE 5

Zalinski was the owner in fee simple of a house and lot. She mortgaged the property to the Home Bank for $50,000. Then she sold the house and lot to Steele for $75,000, of which $50,000 represented the mortgage Steele assumed and the remaining $25,000 was payment to Zalinski for her equity.

Shortly after that, Steele borrowed $10,000 from Gray, giving Gray a $10,000 second mortgage on the property as security for the loan. Steele then sold the property to Allen for $80,000, of which $50,000 was the first mortgage to the Home Bank, $10,000 the mortgage to Gray, and $20,000 cash payment for Steele's equity.

The house caught fire and burned to the ground a few days after Allen acquired the property. The house was insured for $45,000. The policy named Allen as the insured, the Home Bank as first mortgagee, and Gray as second mortgagee. After the fire, Allen abandoned the property and left the country.

Advise the Home Bank and Gray as to their legal rights. Speculate as to how the parties might proceed toward protecting their respective interests.

CASE 6

Penfield owned property that he mortgaged to The Bank of Regina for $100,000 on a term of three years, amortized over 25 years, at a rate of 8 percent per annum. His payments were $894.49 monthly.

At the end of one year he sold the property to Carson, who assumed the Bank of Regina mortgage. At the end of two more years, when the mortgage came up for renewal, Carson renewed the mortgage for another three years, at the going rate of 9 percent per annum, amortized over 25 years. The monthly payments became $962.53.

Two years later, Carson defaulted on the mortgage, and the bank sued not only Carson (who was penniless) but Penfield as well. In defence, Penfield claimed novation.

What is the source of the bank's claim against Penfield, and how does Penfield construct novation as a defence? Who is likely to be successful?

CASE 7

Hambly was the owner of a block of land in fee simple. He arranged a mortgage on the property with Blake for $50,000, and the mortgage was duly registered in the appropriate Land Registry Office. Hambly used the funds for the renovation of an existing building on the premises, but discovered that he had insufficient funds to complete the changes he wished to make.

A few months later, he borrowed the sum of $10,000 from his friend Clark and gave a second mortgage on the property as security. Clark did not register the mortgage; instead, she placed it in her safety deposit box, with the intention of registering it at some later date.

When the renovations to the building were completed, Hambly decided to install a swimming pool on the grounds. He borrowed $5,000 from Simple Finance Co. to pay the pool contractor for the installation. Simple Finance, as security for its loan to Hambly, took a mortgage on the property. The mortgage was registered the same day that the funds were given to Hambly.

Shortly thereafter, Hambly arranged a party to celebrate the completion of the swimming pool and invited his many friends to attend. At the party, Clark mentioned to Anderson, another friend of Hambly, that she held a mortgage on Hambly's property, and that she would like to dispose of it in order to have the funds available for another more attractive investment.

Anderson expressed an interest in the purchase of the mortgage and, after some discussion, agreed to give Clark $9,000 for it. The next day, Anderson paid Clark the $9,000 for the mortgage (on which the full $10,000 principal was owing) and received an assignment of the mortgage. When Anderson realized that the mortgage itself had not been registered, he had the documents registered immediately. Unfortunately, he failed to notice the mortgage to Simple Finance in his examination of the title to the property.

Hambly was killed in an automobile accident a few days before he was scheduled to make his first payments on the mortgages. His only asset was his interest in the property. Blake instituted foreclosure proceedings when the first payment required under the mortgage became overdue. An appraisal of the property indicated that it had a market value of approximately $60,000.

Discuss the position of the parties in this case. Indicate their rights in the foreclosure action.

Comment on the possible outcome of the case.

CASE 8

Parker Construction Co., a building contractor, constructed a single-family dwelling on a building lot in a suburban area. To obtain the necessary funds to build the house, Parker Construction Co. entered into a mortgage with Green Mortgage Co. for the principal amount of $85,000, repayable in monthly installments of $900 each.

Some time later, Parker Construction Co. sold the property subject to the mortgage to Baker at a purchase price of $155,000. Baker paid Parker $70,000 cash and assumed the mortgage. He continued to make payments on the mortgage while he possessed the house, but two years after he had purchased the property, he was transferred to another city by his employer. He sold the property at that time to Brown, who assumed the mortgage and paid Baker $160,000 for the property. The mortgage had a principal balance outstanding of $81,500 at the date the property was sold.

Unfortunately, Brown found herself overextended financially soon after she had purchased the property. She was forced to let the monthly mortgage payments fall into arrears in order to pay more pressing debts. In spite of repeated requests for payment by the mortgage company, Brown refused to do so. Eventually, Green Mortgage Co. was obliged to take action. Instead of foreclosure, however, it brought an action against Parker Construction Co. for payment.

Discuss the possible reasons why the mortgage company decided to take action against Parker Construction Co. rather than institute foreclosure proceedings. On what basis could it do so?

Discuss the rights and obligations of the parties in light of this action by the mortgagee.

The Online Learning Centre offers more ways to check what you have learned so far. Find quizzes, Weblinks, and many other resources at www.mcgrawhill.ca/olc/willes.

CHAPTER 24

Leasehold Interests

Chapter Objectives

After study of this chapter, students should be able to:

- Understand the legal nature of a tenancy and the rights and duties of the landlord and tenant.
- Explain the effect of a breach of a lease agreement.
- Understand shopping-centre leases.

YOUR BUSINESS AT RISK

There are times when owning the perfect business location is impossible because the owner will not sell, or it is financially unworkable for cash-flow or taxation reasons, or in cases where ownership would mean taking on unwanted responsibilities. In these situations, possession without ownership may be the preferred (or only) option, and the mechanism that will achieve this goal is the leasehold interest. All other options would be less than optimal.

LEASEHOLD INTEREST

A leasehold interest in land arises when a person who owns an estate in land grants, by way of either an express or implied contract, possession of the land to another for a fixed period of time. The owner of the estate in land holds the "lordship" over the particular parcel of land and is known as the landlord or **lessor**. The person granted possession of the property is called the tenant or **lessee**; the contract between the landlord and tenant is called a **lease** or **tenancy**. While the contract may be verbal, written, or, sometimes, implied, the terms of the tenancy may vary, from a simple verbal agreement giving the tenant possession in return for a periodic rent, to a complex and lengthy lease of shopping-centre premises that grants the landlord a percentage of the profits of the tenant's business and substantial control over a number of aspects of the tenant's activities.

Lessor
A landlord.

Lessee
A tenant.

Lease
An agreement that constitutes a grant of possession of property for a fixed term in return for the payment of rent.

Tenancy
A relationship between parties governed by a lease.

The contract is, in a sense, more than an ordinary contract, because it amounts to a conveyance of a part of the landlord's interest in the land to the tenant for the term of the lease. The lease creates a privity of estate between the landlord and tenant, since each has an interest in the same land. In addition, the contract itself creates a privity of contract between the parties. If the landlord and tenant retain their interest for the entire period of the tenancy, privity of estate and privity of contract would remain between them. However, if either party should assign his or her interest in the land, the privity of estate ceases to exist when the new party acquires the interest in the land. The contract between the landlord and tenant may continue to bind the parties to the covenants in the contract if it is expressly set out, even though the tenant has parted with the estate or interest in the land.

Privity of estate in itself creates certain rights in the landlord and in the tenant. For example, if a lease is assigned to a new tenant, by virtue of the privity of estate that exists between the landlord and tenant the new tenant would be obliged to pay the rent and also perform all covenants that run with the land.

Reversion

The return of the right to possession to the landlord at the end of the lease.

The creation of a tenancy gives rise to two concurrent interests in land: the leasehold and the **reversion**. The tenant acquires the exclusive possession of the land under the tenancy, and the landlord retains the reversion or the title to the property until the lease terminates. At the expiration of the term of the lease, the two interests (possession and title) merge, and the landlord's original estate becomes whole once again. In the interval, however, the tenant has exclusive possession of the property. Unless an agreement is made to the contrary, the tenant may exclude everyone from the land, including the landlord.

A leasehold interest is an interest in land. Consequently, anything that is attached to the land becomes a part of the leasehold interest during the currency of the lease. For example, if a person leases a parcel of land that has a building upon it, the building becomes the possession of the tenant for the time. As well, any rents or benefits that the building produces would belong to the tenant, unless the parties had agreed that the rents would be used or applied in a different manner.

Commercial leases, in particular, tend to be negotiated leases, where the parties include in the lease those terms specific to their relationship and needs. For example, a commercial tenant may wish to make extensive changes to the interior or exterior of the building to suit its business operations, and the commercial lease will be written to permit such changes to the building or premises. Similarly, a landlord may enter into an agreement with a tenant to construct a building on a parcel of land that is designed specifically for the tenant's business. The tenant under these circumstances will usually agree, in return, to lease the property for a lengthy period of time. A variation of this type of lease arrangement is the sale and lease-back, where a firm will acquire land and construct a building on the land suitable for its business activities. In order to free up the capital investment in the new building for use in the business, the firm may sell the property to an investor and in turn lease the property back from the investor under a long-term lease.

Historical Development

A lease is a very old method of acquiring an interest in land. Long before the modern concept of contract was developed, the leasehold interest was recognized, and the rights and duties attached to it defined. Many changes have naturally taken place over the years that have altered the relationship of landlord and tenant, but the basic concept has remained the same.

Leases existed in Europe and England long before the feudal system of land tenure was adopted, and the leasehold interest was never (in England at least) a feudal tenure. It was always an interest in land that was based upon an agreement between the parties, and it was seldom considered to be something permanent in nature.

When the further subdivision of estates in England became impossible in the late thirteenth century, the lessor lords turned to leases as a means of creating interests in land that would, in turn, provide them with benefits. The leasehold interest, while it lacked the permanency of a tenured estate in land, did have certain advantages. Since it was for a fixed term, the land would eventually return to the landlord. Thus, the wealth associated with the land remained with the lord. In the meantime, the land produced revenue in the form of rents or a share of the produce. It also gave the landlord some protection against inflation (particularly after the mid-fourteenth century), because the rents could be renegotiated at the end of the term of the lease. The advantage to the tenant under a lease, apart from the right to possession of the land, was the fact that the land did not carry with it the burdens or restrictions of feudal duties that accompanied some grants of feudal estates.

A significant change in leases took place in 1677 with the passing of the *Statute of Frauds*. The statute, discussed at Chapter 10 of this text, affected leases as well as other interests in land. It required all leases to be in writing and made under seal unless the term was for a period of less than three years.[1] Under the Act, a lease for more than three years that had not been made in writing was void.[2] The requirements imposed by this statute have continued, in one form or another, to the present time in most Common Law provinces of Canada.

In Canada, the law relating to the landlord and tenant relationship followed the English law, but, due to the availability of low-cost land, was less concerned with agricultural leases. The Canadian legislation, as a consequence, has tended to be more streamlined, with the emphasis on leases of land for business or residential purposes, rather than agriculture. As in England, much of the law relating to landlord and tenant has been incorporated into provincial statutes that set out the nature of the relationship and the rights and duties of the parties.[3]

CREATION OF A TENANCY

A lease is a contract made between a landlord and a tenant that gives the tenant exclusive possession of the property for a specified period or term. A lease is distinct from a licence in that a lessee is entitled to exclusive possession of the property, whereas a licence grants the licensee the right to use the property in common with others, but does not create an interest in land. For example, a property owner may permit certain persons to use the property from time to time for the purpose of hunting, but the permission to do so would not create an interest in land. It would only give the licensee the right to lawfully enter on the property for the particular purpose set out in the licence. A lease, on the other hand, would give the tenant exclusive possession of the property and an interest in land. The extensive use of licences for various purposes has, unfortunately, made the distinction between a lease and a licence unclear with regard to the question of when rights granted under a licence become an interest in land. The courts generally look at the intention of the parties in order to distinguish between the two, since possession alone is no longer a deciding factor.[4] Nevertheless, if the agreement gives the occupier of the land exclusive possession of the property and the right to exclude all others, including the owner, the courts generally conclude that the relationship established by the agreement is one of landlord and tenant.[5]

Contractual Foundation

A lease is contractual in nature, and may be either an express agreement (verbal or written) or be implied from the conduct of the parties. The terms of the lease may set out the specific rights and duties of the parties. In that case (provided that the rights and duties are lawful), they will be binding on the lessor and lessee and delineate the tenancy. If specific terms are not set out, then the rights of the parties and their duties may be determined by statute or the Common Law. Since the nature of the relationship is contractual, the law of contract, as modified by the Common Law and statutes relating to leasehold interests, will apply. A lease then must normally meet the requirements for a valid contract to be enforceable. It must contain an offer and acceptance, consideration (in the form of rent and premises), and legality of object. The parties must also have the capacity to contract and the intention to create a legal relationship. See Figure 24–1 for an example of the lease relationship.

1. The *Statute of Frauds*. See also, for example, R.S.O. 1990, c. S.19, s. 3.
2. ibid., s. 1(2).
3. See, for example, the *Landlord and Tenant Act*, C.C.S.M., c. L70.
4. *Errington v. Errington*, [1952] 1 All E.R. 149; *Lippman v. Yick*, [1953] 3 D.L.R. 527.
5. *Re British American Oil Co. Ltd. and De Pass* (1959), 21 D.L.R. (2d) 110.

FIGURE 24–1 Lessor–Lessee Relationship

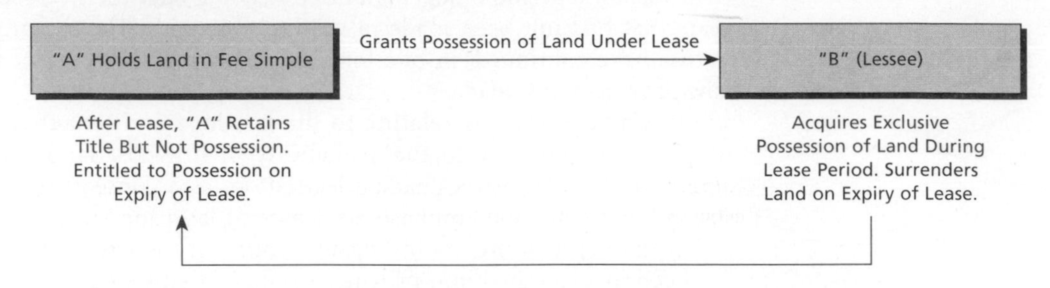

Where the *Statute of Frauds* (or similar legislation pertaining to leases) applies to lease agreements, it requires any lease for a term of more than three years to be made in writing; otherwise, it is void.[6] Leases for a term of less than three years need not be in writing, provided that the tenant goes into possession.

A minor may enter into a lease as a tenant, but unless the accommodation is necessary, the lease would be voidable at the minor's option. The minor must repudiate the lease promptly on attaining majority, however, as the contract of lease is an agreement of a continuing nature, and to continue to acknowledge the relationship may render the lease binding.[7]

A lease by a drunken or mentally impaired person is generally held to be binding, unless the person in that state at the time of the execution of the lease is in a position to prove that he or she was drunk or mentally impaired and that the other party knew of this condition.[8] This rule is simply the application of the ordinary contract rule with respect to agreements made by persons incapable of appreciating the nature of their acts. However, it is generally necessary to show as well that the person of sound mind took unfair advantage of the person with the disability in cases where the leased premises would be treated as a necessary. As with the ordinary law of contract, leases must be repudiated promptly by drunken or mentally impaired persons on their return to sanity or sobriety.[9]

Legislative Development

Legislation in a number of provinces has distinguished between residential and other tenancies in the determination of the rights and duties of the parties to a lease. To some extent the legislation has changed the nature of the lease itself. The general thrust of the new legislation with respect to residential tenancies has been to provide greater security of tenure for the tenant and to put additional obligations on the landlord to maintain safe premises for the tenant. The particular rights and obligations generally apply to all residential tenancies, and the parties usually may not contract out of the statutory requirements. Commercial and other tenancies are subject to the ordinary Common Law rules for landlord and tenant and to the general provisions of the legislation pertaining to the tenancy relationship. Many of the rules relating to the relationship are the same, but the special provisions concerning residential tenancies have substantially altered the rights of the parties.

Characteristics of a Lease

Term certain
The fixed term of a lease.

A characteristic of a lease is that it is a grant of exclusive possession for a **term certain**. The lease itself must stipulate when the lease will begin, and when it will end. If it is for a term that begins at some future time, it is said to be an agreement for a lease rather than a lease and, with the exception of Ontario, it must be in writing to be enforceable. In Ontario, the requirement of writing would only appear to apply if the agreement to lease relates to

6. For example, the *Statute of Frauds*, R.S.O. 1990, c. S.19, ss. 1–3.
7. *Sturgeon v. Starr* (1911), 17 W.L.R. 402.
8. See, for example, *Hunt v. Texaco Exploration Co.*, [1955] 3 D.L.R. 555.
9. *Seeley v. Charlton* (1881), 21 N.B.R. 119.

a period of more than three years.[10] In addition to the requirement of writing, long-term leases must be registered in most provinces if the tenant wishes to retain priority over subsequent purchasers or mortgagees.[11]

Periodic tenancy

A lease that automatically renews at the end of each rent payment period until notice to terminate is given.

If the lease does not specify a definite term, the tenancy may be a **periodic tenancy**, in which the lease period may be yearly, monthly, or weekly. A periodic tenancy automatically renews at the end of each period, and continues until either the landlord or the tenant gives the other notice to quit. The type of periodic tenancy is usually determined from the agreement of the parties, but, in the absence of evidence to the contrary, the tenancy is usually related to the rent payment interval. For example, unless otherwise indicated, if the rent is paid monthly, the tenancy is generally considered to be a monthly tenancy. A periodic tenancy may also arise on the expiration of a lease for a term certain. If the tenant continues to occupy the premises and the landlord accepts the rent, the lease will become a periodic tenancy that will continue until either party gives notice to terminate.

A tenancy for a term certain may also give rise to another form of tenancy that is not a tenancy in the true sense, but an occupancy of the property only. If a tenant at the end of the term remains in possession after notice to quit has been given and the notice period expired, then the occupancy of the premises is known as a tenancy at sufferance. Under this form of occupancy, no rent is payable, since a tenancy does not exist. However, if the landlord so desires, the landlord may demand compensation for the overholding tenant's possession of the property.

A form of tenancy may also be established by a landowner who permits another to enter on the premises and occupy them where no express lease is made, and where the purpose of the occupancy is usually related to another transaction. This tenancy is known as a tenancy-at-will, and no rent is payable. The occupier remains on the premises at the pleasure of the landlord, and the landlord may at any time order the occupier from the property.

Amanda agrees to a five-year lease for shop space owned by Roger. The shop must first be renovated, which is expected to take two months. Roger allows Amanda to store inventory and use office space at one of his other properties (for free) while the renovations are completed. Roger does so to cement the deal, as Amanda would have otherwise entered into a long-term lease for someone else's property which was ready for immediate occupancy. The interim arrangement is a tenancy-at-will.

Sublet

The lease of leased premises by a tenant-in-chief to another tenant for a shorter term than the original tenancy.

A special form of tenancy arises when a tenant enters into a lease with another for a term that is less than the tenancy that the tenant holds. A lease of a leasehold interest of this nature is called a subtenancy (or under-tenancy); the tenant in the subtenancy is called a subtenant. The lease creating the subtenancy is called a sublease. It may contain terms and obligations that differ substantially from those of the lease under which the tenant-in-chief is bound. A subtenancy, nevertheless, must be consistent with the term of the original tenancy in the sense that it is for a lesser term. To have a subtenancy, the tenant-in-chief must possess a reversion that would entitle him or her to regain possession before the expiry of the original lease with the landlord.

A tenant may lease premises for a term of five years, then immediately enter into a lease with a subtenant to sublet the premises for a period of three years. The tenant-in-chief would be liable to the landlord under the original lease, and the subtenant would be liable to the tenant under the sublease. Each would be obliged to perform their particular obligations to their respective "landlords." On the termination of the sublease, the tenant-in-chief would regain possession of the premises and, in turn, would deliver up possession to the landlord on the expiry of the original lease.

10. The *Statute of Frauds*, supra, s. 3; *Manchester v. Dixie Cup Co. (Canada) Ltd.*, [1952] 1 D.L.R. 19.

11. In Ontario, a lease for more than seven years must be registered to preserve priority under the Registry System. *Registry Act*, R.S.O. 1990, c. R.20, s. 70(2). Under the Land Titles System, a lease must be registered to preserve priority over subsequent purchasers of the fee. See the *Land Titles Act*, R.S.O. 1990, c. L.5, s. 111(5).

Rights and Duties of the Landlord and Tenant

The landlord and tenant usually specify their rights and duties in the lease agreement, which they agree will be binding upon them for the duration of the lease. Most written leases set out these rights and duties in the form of promises or covenants that apply to the tenancy. However, if the lease is merely verbal, the Common Law and the statutes pertaining to landlord and tenant in most provinces will incorporate in the lease a number of implied terms to form the basis of the tenancy. As previously indicated, residential tenancies in a number of provinces are now subject to special rights and obligations that have been imposed on landlords and tenants, and that distinguish residential tenancies from the ordinary landlord–tenant relationship. Some of the more important differences are noted under the following topics, which deal with the general rights and duties of the parties to a lease.

Rent

The covenant that comes to mind first when a lease is suggested is the covenant to pay rent. Rent is usually paid in the form of legal tender or a cheque of a money amount, but it is not restricted to money. Rent may take the form of goods in the case of an agricultural lease, where the landlord is to receive a share of the crop grown on the land,[12] or it may be in the form of services, such as when a person living in an apartment building agrees to provide cleaning or janitorial services in return for an apartment in the building.[13] In rare circumstances, such as a tenancy-at-will, no rent may be payable at all.

Commercial and industrial leases usually have their rental amounts determined upon the basis of the square footage leased, with the rent calculated at a dollar amount per square foot.

A small accounting or business-service firm may lease 2,000 square feet of office space in a building at a rental rate of $5/square foot. This would represent a yearly rental of $10,000 for the premises. Residential tenancies are usually based upon a "flat-rate" rental fee, for example, an apartment at a rate of $800 per month.

Commercial leases are often drawn for lengthy periods of time, and it is difficult under these circumstances to establish a rental amount for the full term. A common method of determining the rental amount on a long-term lease is to provide for periodic adjustments (for example, at five-year intervals). The parties will, in effect, negotiate a new rental amount, based upon the rental market at the time the new rental rate becomes negotiable. If the parties are unable to agree upon a new rental rate for the next period, the lease will usually provide that the dispute may be submitted to arbitration, where the arbitrator will fix the rent payable, based upon the current market for such a property.

The method expressed for the payment of rent may indicate the nature of the tenancy, if the parties have not expressly agreed on the term. For example, the rent may be expressed as a lump sum for the entire lease period, or it may be expressed as an annual amount. In the former example, the payment method would indicate a lease for the term; in the latter (if no agreement to the contrary), an annual lease is indicated. If the parties agree to a periodic tenancy with the rent payable monthly, a monthly tenancy will generally be inferred. If the rent is payable weekly, it usually indicates a weekly tenancy. Tenancies for longer terms than a month often specify a total rent amount or lump-sum payment, then break the amount down into monthly rent payments. However, in this type of lease agreement the requirement for monthly payments would not change the term of the lease to a monthly tenancy.

12. *Kozak v. Misiura*, [1928] 1 W.W.R. 1.

13. *Robertson v. Millan* (1951), 3 W.W.R. (N.S.) 248.

Time for Payment

At Common Law, the time for payment of rent may be either express or implied from the agreement of the parties, or it may be determined by custom. If the rent is to be paid in advance, the parties must generally agree to the advance payment either expressly or by their conduct.[14] In the case of residential tenancies, the landlord may demand rent in advance, but in some provinces the practice of demanding the equivalent of several months' rent as a "security" or "damage" deposit has been prohibited. Instead, the landlord may retain an amount equal to the last month's rent under the lease as a deposit, but must pay interest on the sum for the period of time that the funds are in the landlord's possession.[15]

As a general rule, if no place for the payment of rent is specified, the tenant must seek out the landlord on the day on which the rent is due and make payment as required under the lease.[16] If, however, the landlord is in the habit of collecting the rent at the leased premises, the tenant is not in default so long as he or she is ready and willing to make payment on demand.[17]

The covenant to pay rent affects both the landlord and the tenant. During the term of the tenancy, unless the lease provides otherwise, the rent is fixed and may not be raised by the landlord.

Damage or Destruction

The tenant at Common Law is generally liable for the payment of rent for the entire term in cases where the land is leased, even if some of the buildings on the land may be destroyed by an act of God, or through no fault of the tenant. Where residential tenancies are concerned, some provinces have included in their legislation that the doctrine of frustration applies if the property is seriously damaged or destroyed by fire. In these jurisdictions, the lease may be terminated.[18] In other provinces, the doctrine of frustration would not apply. However, the courts have suggested that in apartment buildings and other multi-storied buildings, where the landlord is responsible for parts of the building, if there is destruction of a part of the building, the rent would cease until repairs were completed, or, if the landlord did not repair, then the lease would be terminated.[19] Some attempts have been made to clarify the law in this area, but only residential tenancies in some provinces would appear to be fully protected, and then only if the premises are completely destroyed.[20] No great pressure for legislative reform has taken place in this regard, probably because most commercial leases specifically provide for this, as do most formal leases for residential tenancies. The only leases that might not cover this eventuality would be short-term leases, and monthly or weekly periodic tenancies. In each of these situations, only a small amount of money would be in issue if the premises should be destroyed, and the question of the tenant's continued liability for rent would not likely be a matter that would come before the courts.

Quiet Possession

In return for the payment of rent, the tenant is entitled to possession of the premises, undisturbed by any person claiming a right to the property through or under the landlord. This entitlement is in the form of an express or implied covenant by the landlord that the tenant will have quiet possession of the leased premises. In the case of a leasehold, the landlord covenants that he or she has a right to the property that is such that he or she is entitled to make the lease. The landlord also promises not to enter on the premises or

---

14. *Brunner v. Pollock*, [1941] 4 D.L.R. 107.
15. See, for example, Ontario: the *Tenant Protection Act*, S.O. 1997, c. 24, s. 118.
16. *Chemarno v. Pollock*, [1949] 1 D.L.R. 606.
17. *Browne v. White*, [1947] 2 D.L.R. 309.
18. See, for example, Ontario: the *Tenant Protection Act*, S.O. 1997, c. 24, s. 10.
19. *Dunkelman v. Lister*, [1927] 4 D.L.R. 612.
20. See, for example, Ontario Law Reform Commission, Report on Landlord and Tenant Law, 1976, pp. 209–211.

interfere with the tenant's possession, except as authorized by law. The covenant extends as well to any activities of the landlord that would be actionable in nuisance.

If the landlord:

a) visits unnecessarily or creates a nuisance at law, or,
b) allows other tenants (e.g., in an apartment building) to persist in nuisance, or,
c) a third party shows a valid claim to ownership,

then, the landlord is in breach of the covenant of quiet possession.

Repairs

The obligation to repair is usually set out in the lease, because at Common Law neither the landlord nor the tenant would be liable to make repairs, unless the lease specifically required one or the other to do so. If neither party is obliged to make repairs, the landlord has an obligation to warn the tenant at the time the tenancy is made of any dangers that exist as a result of the non-repair of the premises. However, if the tenant causes any subsequent damage to the property, the tenant must repair it and must not deliberately commit waste (such as the demolition of buildings, or the cutting of shade or ornamental trees).[21]

If the premises are leased as furnished, the property must be fit for habitation at the beginning of the tenancy, but, at Common Law, the landlord is under no obligation to maintain the property in that state. Landlord and tenant legislation in many provinces has altered the Common Law rule, however, and landlords are normally required to maintain the safety of the premises. This is particularly true where provinces have imposed an obligation to repair on the landlord with respect to residential tenancies.[22] The landlord's obligation, even under recent legislation, does not extend to damage caused to the premises by the tenant's deliberate or negligent acts, but only to ordinary wear and tear or structural defects. Even then, the landlord would only be obliged to repair those defects brought to the landlord's attention by the tenant.

Sublet and Assignment of Leasehold Interests

Most commercial leases provide for the tenant's right to assign or sublet leased premises. Unless the lease contains an express prohibition, a tenant may assign or sublet if he or she wishes. At Common Law, the tenant is entitled to assign a lease, as the assignment does not affect the tenant's liability under the lease agreement. The tenant is still liable under the express covenants in the lease. The normal practice for leases is to include a right to assign the lease with the consent of the landlord, and to provide further that the landlord may not unreasonably withhold the consent. A number of provinces have included this change in their landlord and tenant legislation. Where special legislation with respect to residential tenancies is in force, the right to assign or sublet with the landlord's consent is usually expressly provided.[23]

Taxes and Insurance

Most leases also provide for the payment of municipal taxes and insurance by the tenant, but unless the lease so provides, there is no obligation on the tenant to be concerned with either of these expenses. In the absence of an express covenant to pay taxes, the landlord is usually obliged to cover the cost, but if the tenant pays the taxes, depending upon the province, the tenant may deduct the expenses from the rent payable. Municipal charges assessed for property improvements — such as sewer and water lines, sidewalks,

21. *McPherson v. Giles* (1919), 45 O.L.R. 441.
22. See, for example, Ontario: the *Tenant Protection Act*, supra, s. 24.
23. The *Commercial Tenancies Act*, R.S.O. 1990, c. L.7, s. 23(1).

or road paving — however, are improvements to the lands. Generally, these charges are a responsibility of the landlord, regardless of any obligation on the tenant in the lease to pay ordinary municipal taxes. Municipal business taxes, which may be levied against a business occupying leased premises, are the responsibility of the tenant and represent a tax separate from the property tax itself.

Insurance may be an obligation on the tenant by an express term in the lease, but apart from an express requirement, there is no obligation on the tenant to insure the premises. Most tenants, if careful and prudent, would at least insure their own chattels and provide for liability insurance in the event of an injury to a guest on the premises. Landlords similarly insure their buildings to protect themselves from loss or damage through the negligence of the tenant in possession.

MANAGEMENT ALERT: BEST PRACTICE

Common lease descriptors are shown below. In each case, the cost of heat, water and electricity would be borne by the commercial tenant, unless otherwise provided for in the lease. While it is convenient to use these terms in casual discussion among real estate agents, landlords, and tenants, parties must ensure that the meaning of these terms is clearly supported by the terms of the lease, and not left to "local practice" for interpretation.

- Gross (or full service) lease: → Monthly rent payable by tenant, landlord pays all expenses.
- Net lease (single net): → Monthly rent plus property taxes.
- Net/Net lease (double net): → Monthly rent, plus property taxes, plus building insurance.
- Net/Net/Net lease (triple net): → Monthly rent, plus property taxes, plus building insurance, plus repair and maintenance costs.
- Absolute Triple Net lease: → Lease of underlying land only; all expenses, including reconstruction of existing buildings, is the responsibility of the tenant.

Fixtures

Trade fixtures

Chattels attached to leased premises by a commercial tenant that may be removed at the end of the tenancy by the tenant.

A tenant may bring chattels on leased premises during the currency of a lease, and unless the chattels become a part of the realty, the tenant may remove them on departure. If the chattels have become attached to the realty in the form of improvements to the building, such as walls, plumbing fixtures, or similar permanently attached chattels, they may not be removed on the expiry of the lease. Some fixtures, called **trade fixtures**, may be removed by the tenant, provided that any minor damage to the premises that occurs during removal is repaired. The nature of a fixture is dealt with in some depth in Chapter 22, "Interests in Land." It is sufficient to note with respect to the tenant's fixtures that goods that are normally considered trade fixtures may be removed, if the tenant does so either on, or immediately after, vacating the premises.

Rights of a Landlord for Breach of the Lease

The rights of a landlord in the event of a breach of the lease depend to some extent on the nature of the breach committed by the tenant. The most common breach by a tenant is the breach of the covenant to pay rent. If the tenant fails to pay rent, the landlord has three remedies available. However, as a general rule, the breach of a term or covenant in the commercial lease by the landlord will not entitle the tenant to withhold payment of rent. Residential-tenancies legislation does permit tenants to withhold rent under certain circumstances, but unless a term in a commercial lease expressly permits the tenant to withhold rent, the tenant must continue to make rental payments and select an appropriate available remedy for redress.

The first remedy is the right to institute legal proceedings to collect the rent owing. This is known as an action on the covenant. In a commercial tenancy the landlord may also distrain against the goods of tenants until the rent is paid, or if the rent is not paid, have the goods of the tenant sold to cover the rent owing.

Distress

Distrain/distress

The right of a landlord to seize and sell the chattels of a tenant if arrears of rent are not paid.

The second remedy, the right of **distress (to distrain)**, is very similar to a claim for lien in the sense that the landlord may seize the chattels of the tenant (subject to certain exceptions) and hold them as security for payment of the rent owing. If it is necessary to sell the goods to cover the arrears of rent, and the proceeds are insufficient to pay the arrears, the landlord may then take action on the covenant against the tenant for any difference in the amount.[24] In some jurisdictions, the right of distress against the goods of the tenant is no longer applicable to residential tenancies.

Re-entry

Re-entry

The act of a landlord to move into possession of leased property.

A third remedy that may be exercised by the landlord in the event of non-payment of rent is the right of **re-entry**. Under landlord and tenant legislation, in a number of provinces, the landlord's right to re-enter arises when rent is in arrears for a period of time. On the expiry of the time, the landlord may repossess the premises. The exercise of the right of re-entry has the effect of terminating the tenancy, since the tenant no longer has possession. It should be noted, however, that the right of re-entry in a sense is an alternative to the right of distress. A landlord may not distrain against the goods and re-enter at the same time. The act of re-entry terminates the tenancy, and with it the right of distress. Consequently, the landlord must distrain first, then later re-enter, or choose between the two remedies.

Notice to Correct

Eviction

The court ordered removal of a tenant from leased premises.

In commercial tenancies, if the tenant's breach is of a covenant or term other than the covenant to pay rent, the landlord may give notice to the tenant to correct the breach (if possible) within a reasonable time. If the tenant fails to do so, the landlord may take action to regain the premises. The legislation in most provinces provides that the courts may relieve against forfeiture. If the matter comes before the courts, the courts may order the tenant to correct the breach or pay damages to the landlord for the breach of the covenant. A court may also issue an injunction to restrain any further breach by the tenant. A landlord may also have the tenant evicted by court order if the court believes that such an order should be issued. In the case of residential tenancies (in some provinces) the right of re-entry for breach of a covenant is restricted, and the landlord may only regain possession by way of an order of the court. Often, in the case of a residential tenancy, the landlord must not only notify the tenant of a breach (other than non-payment of rent), but also give the tenant a short period of time (20 to 30 days) to correct the matter. If the tenant fails to do so, the landlord may then apply to a provincial rental housing tribunal for an order to have the tenant evicted. The grounds upon which the landlord may evict the tenant at the present time are limited to non-payment of rent, undue damage, overcrowding the premises, disturbing or interfering with the enjoyment of the premises of the landlord or other tenants, and interfering with the safety or rights of other tenants.[25]

COURT DECISION: Practical Application of the Law

Commercial Leases—"Net/Net" Leases

***1163133 Ontario Ltd. v. Lazer Mania Inc.*, 2005 CanLII 18845 (Ont. S.C.).**

Lazer signed a commercial offer to lease business premises, which was accepted by the landlord (the numbered company), forming a binding lease agreement with a 60 month term. Lazer occupied the premises and after 26 months found its heating and air conditioning (HVAC) unit required repair. When the landlord refused to repair it, Lazer abandoned the property; the landlord sued for lost rent over the remaining term of the lease, repair of the HVAC unit, and clean-up of the premises in preparation of a new tenant.

24. *Naylor v. Woods*, [1950] 1 D.L.R. 649.

25. See, for example, the *Tenant Protection Act*, supra, ss. 61–67.

THE COURT: The Offer to Lease in question was of a commercial/industrial nature and provided the following clauses:

5. SERVICES AND BUSINESS TAXES *"The Tenant shall pay its own hydro, gas, water, heating costs, air-conditioning costs and for all other services and utilities as may be provided to the premises. The Tenant shall arrange with the local authority for connection of gas, electricity and water in the name of the Tenant. The Tenant shall pay its own business taxes."*

6. ADDITIONAL RENT AND CHARGES *(a) "The Tenant shall additionally pay monthly a proportionate share of all costs and expenses incurred by the Landlord in maintaining, operating, cleaning, insuring and repairing the property and, without limiting the generality of the aforesaid, such costs and expenses shall include the costs of: (iii) Heating, ventilating and air-conditioning and providing hot and cold water and other utilities and services to and operating the common areas of the property, and maintaining and repairing the machinery and equipment providing such utilities and services..."*

Mr. Bennie Graham was the person who originally negotiated the Offer to Lease...[H]e gave evidence that the negotiation was on the basis that the Offer to Lease was on a net/net basis. I accept that evidence as being accurate. He is an experienced landlord and in his view net/net on this lease meant that the tenant would be responsible for the replacement of the HVAC unit. Lazer never requested that 133 replace the HVAC unit, but merely repair the units which, in my view, is clearly the responsibility of Lazer. There were two HVAC units, both totally dedicated to the demised premises. Neither unit was used for a common area referred to in paragraph 6 (iii) of the Offer to Lease. The Offer to Lease is a net/net lease and the crucial paragraph for interpretation is 5. Had the defendant Lazer wanted to avoid liability to replace the HVAC units, they should have amended the Offer to Lease or signed a lease wherein their position would be protected, ie: not to be responsible to replace the HVAC units. I find that the Offer to Lease was a binding agreement between 133 and Lazer. In the publication called The Commercial Lease: A Practical Guide, 3rd edit., prepared by Harvey M. Haber, Q.C. under article 2 Intent and Interpretation, he refers to a net lease as something that the landlord is attempting to make clear to the Tenant is, that it is not responsible for any expenses under the lease. I accept that as custom in the trade. I also accept that there is a current trend amongst tenants to attempt to reduce their exposure to capital expenditures... it is incumbent upon Lazer in this case to limit or reduce its exposure by including in the Offer to Lease or a lease that they will not be responsible for HVAC replacement, otherwise that responsibility is theirs. I also find that it would be inappropriate in the case before me for 133 to replace an HVAC unit and charge the pro-rata share to Lazer since this does not apply to a common area, but to the whole of the demised premises. I am therefore of the view that the responsibility to replace the HVAC units is that of Lazer. I am satisfied that the damages for repairs to the demised premises are $48,316.11. I am satisfied that the loss of rental income (*while completely vacant*) is 13 months times $7,251.90 a month for a total of $94,274.70. From September 1st, 2003 to the end of the Offer to Lease, (*a new tenant having then been found*) 2,000 square feet of the 6,150 square feet occupied by Lazer remains vacant. 2,000 square feet equals 32.52 percent of the total space formally occupied by Lazer and that amounts to $2,358.31 per month times 21 months for a total of $49,524.67.

Judgment for the landlord amounted to over $192,000. Be absolutely sure of the implications of terms in any commercial lease (e.g.: "it's a net/net lease") before signing it. A "triple net" lease means the lessee is to pay rent, and all of maintenance, insurance and property taxes. None of the important automatic protections offered by law to residential tenants apply to commercial tenancies.

RIGHTS OF A TENANT FOR BREACH OF THE LEASE

A tenant is entitled to enforce all covenants made for the tenant's benefit in the lease. If the landlord fails to comply with the covenants, the tenant may take advantage of three possible remedies. He or she may bring an action for damages against the landlord if the landlord's actions constitute a breach of the lease. For example, if a landlord wrongfully evicts a tenant from farm property, the tenant may be entitled to damages as compensation for summer fallow work done, crops planted, and the estimated future profits for the term of the lease.[26] If the interference does not constitute eviction from the premises, the tenant may obtain relief from the courts in the form of an injunction to restrain the landlord from interfering with his or her possession and enjoyment of the property. This

26. *Haack v. Martin*, [1927] 3 D.L.R. 19.

remedy might, for example, be sought if the landlord conducts an operation on premises adjacent to the tenant's land that creates a nuisance.

A third remedy is also available to a tenant when the landlord's breach of the lease is such that the interference with the tenant's possession amounts to eviction: the tenant may seek to terminate the lease. For example, in the case of a residential tenancy, the landlord's refusal to repair the building and maintain it in a safe condition would constitute a breach that entitles the tenant to apply to the courts to have the lease terminated.[27]

TERMINATION

Surrender

The agreement by the parties to terminate a lease.

A lease may be terminated in a number of different ways. A commercial lease for a fixed term will terminate when the term ends, or when the landlord and tenant agree to terminate the lease before the date on which it is to expire. Where the parties agree to terminate the lease, the agreement is called a **surrender**. If the lease is made in writing and under seal, the surrender normally must take the same form.[28] A lease may also terminate if the parties agree to replace the existing lease with a new lease, or if the tenant, at the landlord's request, voluntarily gives up possession to a new tenant, and the new tenant takes possession of the premises.[29]

In the case of a periodic tenancy, a lease may be terminated by the giving of proper notice to quit. It should be noted, however, that recent legislation in a number of provinces has limited the right of a landlord to obtain possession on the expiry of a lease, or to give effective notice to quit in the case of residential tenancies. Generally, such legislation limits the landlord's rights to enforce the termination (apart from non-payment for rent) to those cases where the tenant has damaged the premises, or where the landlord requires possession for his or her own use, or to change the nature of the property. In some provinces, the landlord is obliged to obtain possession through the courts even under these circumstances, if the residential tenant refuses to deliver up possession.

For commercial tenancies, the point in time when the notice to quit is given is important and must be carefully adhered to, if the notice is to be effective in terminating the tenancy at the time required by the person giving the notice. In the absence of a term in a lease that specifies the notice period, a party who wishes to terminate a periodic tenancy must give notice equal to a full tenancy period. In other words, notice must be given before the end of one tenancy period to be effective at the end of the next tenancy period. For example, if a monthly periodic tenancy runs from the first day of the month to the last day of the month, notice to terminate must be given not later than the last day of one month to be effective on the last day of the next month. To give notice on the first day of the month would be too late to terminate at the end of that month, and would not be effective until the end of the following month. The reasoning here is that the tenancy would have renewed on the first day, and notice on that day would not give the other party the required full notice period.

Where a periodic tenancy is yearly in nature, a full year is not required as notice. Instead, most provincial statutes provide for a lesser period of time. Quebec and the Maritime provinces of New Brunswick, Nova Scotia, and Prince Edward Island require three months' notice, Alberta requires only 60 days' notice, and the remainder of the provinces specify six months' notice. Residential tenancies in most provinces have different notice requirements and procedures for termination.

27. *Tenant Protection Act*, S.O. 1997, c. 24, s. 34(1).

28. By virtue of the requirements of the *Statute of Frauds* (or equivalent legislation) in most provinces. See, for example, Ontario: R.S.O. 1990, c. S.19, c. 2.

29. *Wallis v. Hands*, [1893] 2 Ch. 75.

CASE IN POINT

A tenant leased eight floors (100,000 square feet) of an office tower in 1995 under a long-term lease. In 1998, the tenant leased an additional floor of the building. Both leases required the consent of the landlord if the tenant wished to assign the lease. The 1998 lease, however, contained a clause that provided that any default of the 1998 lease would also constitute a default of the 1995 lease, and entitle the landlord to terminate both leases.

In mid-2000, the parent corporation of the tenant sought bankruptcy protection in the United States, and the tenant assigned the 1995 lease to two associated corporations with the full knowledge of the landlord, and with the landlord's tacit consent. The 1998 lease was not mentioned at the time, however, but also was assigned to the associated corporations by the tenant.

In late 2000, the landlord claimed that the 1998 lease was in default, because it was assigned without consent, and on this basis both leases were in default. The landlord then applied to the courts for a declaration that the leases were terminated.

The court held that while the 1998 lease was terminated because it was assigned without consent, the 1995 lease was not in default, and could not be terminated, because the landlord had, in effect, consented to the assignment. The court held that the provision in the 1998 lease tying default in it to the 1995 lease would not be applicable under the circumstances.

See: *Greenwin Construction Co. Ltd. v. Stone & Webster Canada Ltd.* (2001), 55 O.R. (3d) 345.

The breach of a covenant may give the landlord the right to treat the lease as being at an end. For example, if the tenant is in arrears of payment of rent, the landlord may, by complying with the statute, move into possession of the property and terminate the tenancy. Similarly, if the tenant abandons the property during the currency of the lease, the lease is not terminated. However, any act of the landlord that would indicate that the landlord has accepted the abandonment as a surrender on the part of the tenant would constitute termination. For example, if the landlord moved into the premises abandoned by the tenant, or if the landlord leased the premises to another tenant without notice to the original tenant that the premises were re-let on the tenant's account,[30] the lease would be treated as at an end.

CHECKLIST FOR LEASEHOLD INTERESTS

1. Tenancy
   - For a term certain.
   - Periodic.
2. Reciprocal rights
   - Lessee — right of exclusive possession, general repair.
   - Lessor — payment of rent, right of reversion.
3. Essential elements for negotiation
   - Amount of rent.
   - Payment of utilities.
   - Payment of taxes and insurance.
   - Sublet and assignment.
   - Fixtures.

Lessor's Remedies: Action for rent, distrain, injunction, re-entry, eviction.
Lessee's Remedies: Action for damages, injunction, termination.

---

30. *Green v. Tress*, [1927] 2 D.L.R. 180.

CLIENTS, SUPPLIERS *or* OPERATIONS *in* QUEBEC

While Common Law impairs ownership of land subject to a lease (by transferring a leasehold interest to the tenant), again, civil law does not. Ownership of leased land under the Quebec *Civil Code* remains quite clear (unchanged), the tenant is only receiving a right of claim to enjoyment (Articles 1851-2000, CCQ). At first this distinction may seem only to be of interest to lawyers, but our culturally ingrained notions of rights attached to ownership and use of land suggests further consideration. The specific leasehold rights vary between the systems, as do the attitudes of landlords and tenants, which spell trouble for business persons who ignore the differences.

SHOPPING-CENTRE LEASES

In contrast to the typical commercial lease of an office, plant, or warehouse, the shopping-centre lease is usually more complex, since it frequently provides for greater landlord involvement in the tenant's business activities. The complexity of the lease varies, however, depending upon the type of shopping-centre premises that are the subject matter of the lease. For small neighbourhood shopping plazas, the lease may not differ substantially from the ordinary commercial lease of retail premises. On the other hand, for premises in large shopping centres that contain major department stores, supermarkets, and chain retail outlets, the leases generally require the tenant to contribute toward the cost of maintaining the parking and common areas, to play an active part in the centre's merchants' association, and to participate in all promotional activities of the centre.

Landlords who build large shopping centres actively seek a desirable mix of retailer tenants in order to provide the widest variety of goods and services possible for the type of consumer that the centre wishes to attract. Because these tenants are often large retail chains, the landlord and tenant generally have equal bargaining power, which is reflected in the leases that the parties negotiate. Certain clauses in the lease must nevertheless remain uniform for all tenants, but apart from these, many of the terms of the lease are negotiated on an individual basis.

The shopping-centre lease varies from the ordinary commercial lease in that it must cover the use of premises outside the retailing area. Most shopping-centre leases will cover parking-area maintenance; the use of storage, shipping, and receiving areas; maintenance of the common areas of the centre; participation in the advertising and promotional activities of the centre; and a contribution to the cost of these activities. Shopping-centre leases normally provide for landlord participation in the profits of the tenant as well. This latter term in the lease usually requires the tenant to pay a minimum rent, plus an additional percentage rent if the tenant's sales exceed a certain dollar amount in a specified time period (usually monthly).

Shopping-centre leases generally require all tenants to remain open for business during the hours that the landlord designates as the centre's hours of operation. This type of lease frequently contains a clause that prohibits a tenant from operating a similar retail outlet using the same business name within a specified distance of the shopping centre. The lease may also contain a use clause that sets out the use and type of products that the tenant may sell in the leased premises, in order to avoid tenant disputes and maintain as wide a variety of goods as possible for the consumer.

In addition to these specific clauses, most of the more common commercial-lease provisions may also be present. These would include the term of the lease and the options to renew; a description of the area leased; payment of taxes, insurance, and utilities; repairs to premises; assignment of the lease and the right to sublet; notice requirement; and such provisos as tenant guarantees and landlord responsibilities.

The negotiation of shopping-centre leases usually requires the assistance of legal counsel, due to the complexity and long-term nature of the lease.

MANAGEMENT ALERT: BEST PRACTICE

The frame of reference for leasing among most business owners comes from their experiences in their own residential leases. It cannot be stressed enough that the legal world of commercial leases is much more draconian than that of residential tenancies. Gone are the protections of the provincial residential-tenancies acts. Commercial lessors usually have far greater bargaining power, and this is reflected in their leases, often given on a "take-it-or-leave-it" basis. Finally, courts are far less sympathetic to the commercial tenant, reasoning that this area of law is just as open to freely negotiated commercial rights and responsibilities as any other, and deserves just as much contractual legal certainty.

SUMMARY

The landlord and tenant relationship is a contractual relationship that gives rise to two concurrent interests in land: exclusive possession of the property by the tenant, and a reversion in the landlord. A leasehold is an interest in land for a term, and at the end of the term, possession reverts to the landlord. During the term of the lease the tenant is entitled to exclusive possession, however, and may then exercise many of the rights that are normally possessed by a landowner.

Because it is contractual in nature, a lease may contain express terms that delineate the relationship and the rights and duties of the parties; where the lease is silent, however, some terms may be implied by law. The most common terms or covenants are the covenants to pay rent, repair, pay taxes, have quiet enjoyment, and assign or sublet as desired. In most provinces, leases for more than three years must be in writing and be under seal to be enforceable. In addition, the lease or a notice of lease must normally be registered to protect the tenant's interest as against subsequent mortgagees or purchasers without notice.

A failure to perform the covenants in a lease by either party may give rise to an action for damages or injunction, or perhaps permit the injured party to terminate the relationship. The particular rights of the parties, and, in particular, those of the landlord, have been altered in some provinces by legislation to protect tenants under leases of residential property. These changes have granted tenants greater security of tenure and have imposed a number of statutory duties on landlords to maintain residential premises in a good state of repair. In most cases, these new laws permit the landlord to repossess residential property only through court order if the tenant refuses to deliver up possession.

Apart from termination arising out of a breach of the lease, a lease may be terminated automatically at the end of its term by surrender or abandonment, or by notice to quit in the case of a periodic tenancy. Again, residential tenancies in some provinces may only be terminated for limited reasons and with the permission of the courts.

KEY TERMS

Lessor, 473
Lessee, 473
Lease, 473
Tenancy, 473
Reversion, 474
Term certain, 476
Periodic tenancy, 477
Sublet, 477
Trade fixtures, 481
Distrain/distress, 482
Re-entry, 482
Eviction, 482
Surrender, 484

Review Questions

1. What remedies are available to a landlord when a tenant fails to comply with the terms of the lease?
2. In what way (or ways) does a tenancy differ from a licence to use property?
3. In a commercial lease, in most provinces, landlords may distrain against the chattels of the tenant for non-payment of rent. What does this mean, and how is it accomplished?
4. What is a reversion in a lease?
5. Explain how the term of a tenancy may be determined where the tenancy agreement is not specifically set out in writing.
6. Define or explain the term "tenancy-at-will" and indicate how it differs from a "tenancy at sufferance."
7. How does a subtenancy differ from an assignment of a lease by a tenant? What rights are created in the sub-tenant by the granting of a subtenancy by the tenant?
8. Distinguish "privity of estate" from "privity of contract." Explain how the rights of the assignee and the landlord are affected when an assignment of lease is made by a landlord.
9. Outline the covenants that a tenant makes in an ordinary lease. Explain the effect of a tenant's non-compliance with these terms.
10. What are the rights of a landlord if a tenant abandons the leased property?
11. What is the legal nature of a leasehold interest, and how does it arise?
12. How do residential tenancies differ from commercial tenancies in most provinces? Why was this change in the law necessary?
13. Explain the legal significance of "surrender of a lease." How is this effected?
14. Explain the legal nature of a covenant of quiet enjoyment as it pertains to a leasehold. Give an example of a case where breach of the covenant would arise.

Mini-Case Problems

1. X entered into a verbal monthly tenancy with Y, on June 1, to rent Y's shop. X paid Y the first month's rent and also moved into possession on the same day. Some months later, on November 1, X gave Y notice of his intention to vacate the premises on November 30 and paid Y the November rent. Y demanded an additional month's rent in lieu of notice. Is Y entitled to the additional rent?
2. The Regal Co. leased a large commercial building for its business. The lease called for monthly payments of $5,000 per month on a two-year lease. At the end of the first year, the Regal Co. fell into arrears on its monthly rent payments. Three months' rent is now due and owing. What action might the landlord take against the Regal Co.?
3. A tenant rented a small store from the owner of a commercial building under a five-year lease. The tenant vacated the premises at the end of the first year, and refused to make any further lease payments. The owner of the building did nothing to find a new tenant for over two years, then finally rented the property again. What are the rights (and liabilities) of the parties?
4. A landlord rented office space in a building to a commercial tenant under a five-year lease. Two years later, the tenant vacated the office space, and stopped rental payments. The landlord immediately rented the space to a new tenant at a higher rent. The landlord then took legal action against the former tenant for the rent owing on the balance of the lease.

   Advise the parties.

Case Problems for Discussion

CASE 1

Basic Warehousing Inc. leased a 300 square metre warehouse building to Fabrics & Lace Inc., a cloth and fabric wholesaler, under a five year lease. The lease provided for an annual rental of $15,000, payable monthly, with the tenant responsible for all costs related to the maintenance of the building except taxes. During the course of negotiations, the president of Fabrics & Lace Inc. mentioned that it would be necessary to have insurance of goods on the premises, and the president of Basic Warehousing Inc. commented that in his opinion the tenant would probably be able to obtain insurance coverage for the goods. Fabrics & Lace Inc. moved its goods into the warehouse, and shortly thereafter, applied for insurance coverage. An inspector from the insurer examined the building and concluded that it could not accept the risk of insuring the type of goods stored by the tenant. When insurance coverage was refused, Fabrics & Lace Inc. immediately moved its goods out of the warehouse, and ceased making its monthly rental payments.

Advise the parties. What would be the nature of the claims and arguments if the case came before the court?

Render a decision.

CASE 2

Down Country Mall Ltd. owns a small shopping mall that consists of eight small, leased shops offering various products and services to the public, and one large shop that it had leased to Mini Department Store Ltd. The lease to Mini Department Store Ltd. was for a ten-year term, with no provision for termination or notice. Mini Department Store Ltd. initially found business at the location to be good, but after a few years, it noticed that the suburban neighbourhood was becoming more affluent, and less interested in its array of goods. Sales were gradually declining during this period, and by the end of the sixth year, the store experienced a loss. Revival of sales appeared to be unlikely, and Mini Department Store Ltd. vacated the premises without giving notice of its departure to Down Country Mall Ltd. Down Country Mall Ltd. immediately set out to find a new tenant, but was largely unsuccessful. Eventually, after much searching, it found a new tenant, but at a rental that was 30% less than the rent amount paid by Mini Department Store Ltd. The lease with the new tenant was for a five-year term. Down Country Mall Ltd. has decided to take action against Mini Department Store Ltd. for breach of its lease, and seeks damages for its loss. It also claims for the rent owing for the four remaining years of the lease. Discuss the position and arguments of each party.

Render a decision.

CASE 3

Bingham leased a small shop from Wright under a tenancy agreement that provided for a three-year term at a monthly rental of $800 per month. The lease did not contain an option to renew, but following the expiry of the lease on October 31, 2010, Bingham continued to pay the monthly rental of $800 to Wright.

The term of the tenancy was never discussed between the parties, nor was the lease arrangement discussed until June 2011, when Wright gave Bingham a written notice that read: "I have sold the property in which you presently occupy space, and the purchaser will require vacant possession on December 31, 2011. This letter gives you written notice to vacate in six months' time."

In response, Bingham wrote Wright and advised her that he was in possession under a lease that did not expire until October 31, 2013, and that the notice given did not apply to his present tenancy.

Wright gave Bingham a second written notice on October 25, 2011, now demanding vacant possession of the premises by November 30, 2011.

When Bingham refused to vacate the shop, Wright brought an action for a writ of possession.

Outline the arguments that might be raised by the parties in this case.

Render a decision.

CASE 4

Quinn leased a small retail shop from Chaplin for the purpose of establishing a fruit and vegetable market. The lease was drawn for a three-year term, commencing May 1, 2011, and provided for a total rental of $36,000, payable at $1,000 per month. The first and last months' rent were due on May 1. Quinn paid the two months' rent and moved into possession.

A month later, and a few days after the rent for the month was due, Chaplin discovered that Quinn had sold the business to Rizzoto. The sale was contrary to the terms of the lease, which permitted assignment of the lease only on consent. Chaplin immediately went to the shop and, when Rizzoto arrived, told him that he was not willing to have anyone but Quinn operate a shop on the premises. Chaplin advised Rizzoto that Quinn was in breach of the lease by assigning it without his consent and suggested that Rizzoto seek out Quinn to get his money back.

Chaplin then contacted a licensed bailiff and gave him authority to collect the rent owing. The sheriff went to the store and made an inventory of the stock and equipment, which he valued at $5,000 and $6,000 respectively. He then changed the locks on the door and posted a notice on the premises that informed the public that the landlord had taken possession for non-payment of rent. The next day he notified Quinn and Rizzoto that he had distrained the chattels in the shop on behalf of the landlord. He advised the two parties that they had five days to redeem the chattels by payment of the arrears of rent, otherwise the chattels would be sold. Quinn and Rizzoto made no attempt to pay the rent.

The bailiff later attempted to sell the business, but was unsuccessful. Eventually his services were terminated by Chaplin.

Chaplin did not attempt to rent the premises and retained the stock and equipment. In December 2011, he brought an action against Quinn for damages for breach of the lease.

Discuss the particular issues that are raised by this case and indicate the arguments that the parties might present with respect to each. Render a decision.

www.mcgrawhill.ca/olc/willes

CASE 5

Sheila verbally agreed to lease Beverly's furnished apartment from her for the months of May through to August, while Beverly would be in another city attending a university summer program. Under the agreement, Sheila would pay the rental payments to the landlord at the beginning of each month and "take good care of the apartment" in Beverly's absence. Sheila was expected to vacate the apartment on August 30.

Sheila moved into the apartment on May 1 and lived in it until July 30, when she found a new apartment located closer to her place of employment. On July 30, she permitted several of her friends to move into Beverly's apartment on the condition that they pay the August rent and vacate before the end of the month, when Beverly was expected to return to the city.

The new tenants did not pay the rent as they had agreed to do. On August 25 they held a party at the apartment that resulted in $1,000 damage to the premises and $800 damage to Beverly's furniture. The new tenants vacated the apartment the next day, leaving no forwarding address.

When Beverly returned to the city on August 31, she discovered her apartment in ruins. A few days later, she received a notice from her landlord demanding the overdue rent for the month of August.

Advise Beverly of her rights and outline a course of action for her to follow.

CASE 6

The Acme Co. leased the second floor of a three-storey office building to the High Finance Company for a three-year term. The lease contained a clause in which the tenant acknowledged that the building was in a good state of repair. The lease made no mention, however, of responsibility to repair subsequent damage to the rental premises.

Before moving into the premises, the High Finance Company made extensive changes to the leased premises by adding several partition walls, special electrical wiring for its computer operation, and an air-conditioning system. Three months later, a fault in the electrical wiring caused a serious fire that destroyed the interior of the High Finance Company's rental premises, and caused serious damage to the third floor of the building and water damage to the first-floor tenant's equipment and merchandising business.

The High Finance Company agreed to pay for the damage to the part of the premises that it had leased, but refused to pay for the general damage to the building on the basis that the landlord had agreed to the changes in the electrical wiring that had resulted in the fire. The High Finance Company also refused to pay rent for its part of the building until the premises were again fit for occupation.

Discuss the rights of The Acme Co., the third-floor tenant, and the first-floor tenant. On what basis, if any, might the High Finance Company refuse to pay its rent?

CASE 7

New Tomorrows Inc. is a registered, non-profit charitable corporation that runs a group home allowing ex-convicts, homeless persons, and others who are "down on their luck" a chance to "get on their feet" in an understanding environment. Applicants sign a rehabilitation agreement and receive the use of a bedroom and common kitchen space. Failure to pay any month's modest fee on time was to be taken, under the agreement, to be notice by the resident of his or her intention to leave the home in five days' time. This point was reiterated each month, as those who had given their notice were quickly replaced by other people.

After residing there for three months, Henry was late in the payment of his fees on the first day of the new month. Five days later, he and his belongings were "helped" to the street and, as was so often the case, he was replaced by someone else. Henry went to the provincial ministry responsible for tenancies to complain.

Discuss the legal issues in this case, and the arguments and principles that the parties may rely upon.

What rights, if any, does Henry have?

CASE 8

In 1981, Relax Retreat Inc. leased a 20-hectare parcel of land for use as a recreational "park" from Land Holdings Co. on a 30-year lease. During the term of the lease, the corporation constructed a number of small one- and two-room cabins on the property, and eventually installed an inground fibreglass swimming pool. The cabins were of wood construction and set on concrete blocks, in order that they could be occasionally moved as the site was developed. In 1984, a concrete-block building was constructed on a reinforced concrete pad to house shower and washing facilities and an office for the park manager.

When the lease expired in 2011, Relax Retreat Inc. informed Land Holdings Co. that the cabins, main office/shower building and pool, and all of the facilities would be removed. Land Holdings Co. objected to the removal of any structure from the property.

Advise the parties of their respective rights and, assuming that the case came before the court, render a decision.

CASE 9

The Washa-Matic Company carried on business as the owner of coin-operated washing machines. The company entered into an agreement with 108 Suite Apartments to provide washing machines for use by tenants of the apartment building.

The agreement was entitled "Lease Agreement" and provided that the landlord "demise and lease the laundry room on the ground floor of the building to the tenant for a monthly rental equal to $1 per machine installed." The agreement was drawn for a five-year term and provided for free access to the room by all tenants. The agreement also allowed employees of Washa-Matic the right of access to the premises "at all reasonable times" to repair or service the machines.

Some time after the machines were installed, the owner of the building sold the premises to a new owner, and the new owner asked Washa-Matic to remove the washing machines. Washa-Matic refused and argued that it was the lessee of the laundry room under the lease agreement.

The new owner removed the washing machines owned by Washa-Matic and installed new equipment. Washa-Matic Company then brought an action against 108 Suite Apartments for damages for breach of the lease and for lost profits.

Discuss the legal issues raised in this case.

Render a decision.

CASE 10

The Chens leased an apartment suite from Broughton Road Apartments for a one-year term, commencing July 1. The Chens were particularly attracted by the location of the apartment building, since it was a long, low building of Tudor design, surrounded by rather spacious grounds. The grounds were important to them because they required a place where their two-year-old child might have a safe place to play.

A few weeks after the tenants moved into their apartment, the owner of the building decided to remove the roof of the building and replace it with new roofing boards and shingles. The noise of the construction work, which was carried on from approximately 8:00 a.m. to 4:00 p.m., interfered with the normal sleeping hours of Mr. Chen, who worked a second shift as a night security guard at a nearby industrial plant. It also interfered with their child's customary afternoon nap.

In addition to the noise of the construction, the Chens discovered that another tenant in the building owned a large pet snake, which was permitted to roam at will over the lawns of the property.

The Chens protested to the landlord that the noise of the roof repairs interfered with Mr. Chen's sleep, and the presence of the snake made the use of the grounds impossible, since they both feared the reptile, even though it was of a harmless species.

When the landlord refused to limit the construction and failed to control the snake, the Chens moved from the apartment. They had been in possession for less than two weeks. The landlord then brought an action to recover apartment rent and other expenses that were alleged to be owing as a result of the breach of the lease by the tenants.

Discuss the arguments that the defendants might raise in this action and determine the issues that the court must deal with before a judgment might be given.

Render a decision.

The Online Learning Centre offers more ways to check what you have learned so far. Find quizzes, Weblinks, and many other resources at www.mcgrawhill.ca/olc/willes.

CHAPTER 25

Commercial and Residential Real-Estate Transactions

Chapter Objectives

After study of this chapter, students should be able to:

- Describe the steps encountered in a real-estate transaction.
- Identify the tools available to shift or mitigate risk in real-estate transactions.
- Describe the role of an agent, surveyor, appraiser, and the legal professional in the conduct of a real-estate transaction.
- Recognize aspects of real estate law that are unique to commercial transactions.

YOUR BUSINESS AT RISK

Without certainty in land-ownership rights, major business assets (and personal assets securing business debt) are exposed to risk. Complex rights of ownership have led to complex processes to secure those rights. You will engage professionals to navigate these processes, but it is important for you to be aware what services and performance you can expect in return for the professional fees you pay.

INTRODUCTION

The purchase of real property — whether it involves vacant land, business premises, a house and lot, or a cottage property — has, in recent years, become a complex transaction for which the advice and services of a member of the legal profession is not only desirable but virtually a necessity. In one sense, the transaction has now turned full circle. In the past, the transfer of property was characterized by much formality and ritual, and required the services of a person not only literate, but knowledgeable of the process of transfer. The process has remained so in England, but, in North America, by the nineteenth century and early twentieth century, the transaction acquired a large degree of informality, particularly in rural areas. During this period of time, in many localities, deeds were drawn up by persons who had only a limited knowledge of the formalities required to transfer title. Often the transaction consisted of little more than the preparation and delivery of a deed to the property, with no thought given to the examination of the vendor's title to the property or the accuracy of the boundaries of the land transferred.

The informality of real-estate transactions during this period may be easily explained. In rural North America, land was relatively inexpensive, and, when a sale of property did occur, the parties were generally familiar with the land and its past owners. All dealings with the property were frequently common knowledge to the community at large, and the boundaries of the lands were usually settled and known. It was also a simpler age, when government planning and interference with the use of property was virtually non-existent. By the middle of the twentieth century, however, the introduction of planning legislation and land-use control gradually increased the complexity of real-estate transactions to the point where, today, expert assistance is usually essential to properly carrying out the necessary searches, clearances, and formalities required to effect the transfer of a good title to a purchaser.

The sale of freehold land in England was originally characterized by a formal ceremony, without a written conveyance. The ceremony known as livery of seisin could take one of two forms: livery in law or livery in deed. Livery in law simply involved the owner of the freehold, in the presence of witnesses, pointing to the land and stating to the intended recipient that it was his intention to give the recipient the particular property for himself and his heirs. The transfer was complete when the party entered into possession of the land.[1] Livery in deed followed a slightly different pattern, which required an act (or deed) on the part of the owner of the land. The event was usually ceremonial and symbolic, with the owner, in the presence of witnesses, delivering to the intended recipient some element of property associated with the land itself (such as the keys to a building on the land), as well as providing vacant possession of the property for the new owner.[2]

Gradually, documentary evidence of the transaction also became a part of the ceremony, first in the form of a written record of the ceremony itself, and then as a written statement of the transaction in which the grantor described the property and the grantee to whom the property was delivered, accompanied by the expression of the intention to transfer and deliver up the land. The written document gradually assumed a greater significance in the ceremony. Eventually, the document became an important part of the transfer itself.

The execution of the document under seal has a long history in connection with the transfer of land. After the Norman Conquest, the practice of affixing a seal as a means of execution became a common method used by grantors to evidence their intention to be bound by the covenants in the conveyance. Literate persons would sign their names as well.

For many years, the written document was used as a symbolic element in the ceremony by a practice that required the grantor in the presence of the witnesses to sign, seal, and deliver the deed to the grantee at the property site. Gradually, this too changed, as the written document assumed greater importance, and the ceremonial aspects became a mere formality. Eventually, the written document became the essential element of the transfer of the title to property. Even today, in some provinces that call for registration under the Registry System, the deed must be "signed, sealed, and delivered" to transfer the title to the property. Today, widespread acceptance of electronic-registration regimes and the employment of secure electronic signatures is fast replacing the need for a seal and witnesses.[3]

Modern Real-Estate Transactions

The purchase and sale of real property takes place with much greater frequency at present than it did in the past. Until very recently, property (and in particular agricultural lands) often remained in the hands of families for many generations. With the mobility that developed in the twentieth century as a result of technological advances, however, this is not the case today. Current employment and business practices often require persons to move regularly. As a result, residential property may change hands every few years. This change in lifestyle and in business practices has created a need for the services of a large number of persons with expertise in the negotiation and completion of land transactions. Some of these experts, such as land surveyors and members of the legal profession, have long been associated with this type of activity. But more recently, others, such as real-estate agents, property appraisers, and (in some cases) public accountants, now play an important role in the purchase and sale of property.

The Role of the Real-Estate Agent

The complexity of the modern real-estate transaction virtually obliges the vendor and purchaser to obtain the services of professionals in order to transfer the property interest.

1. P.G. Osborn, *The Concise Law Dictionary,* 4th ed. (London: Sweet & Maxwell Ltd., 1954), p. 206.
2. The significance of the ceremonial aspects associated with the transfer of property are described in W.S. Holdsworth, *A History of English Law,* 3rd ed., vol. 3 (London: Methuen & Co. Ltd., 1923), pp. 222–225.
3. See *Electronic Transactions Act,* S.B.C. 2001, c. 10.

For example, some services, such as those of the real-estate agent, are often essential to initiate the transaction. Most property owners rarely have the time to search for a prospective purchaser if they should desire to sell their property. Consequently, the successful sale is often dependent upon the efforts of persons whose expertise lies in the area of seeking out interested buyers and bringing such parties into a contractual relationship with the property owner. This, in essence, is the role of the real-estate agent.

A real-estate agent (or broker, as he or she is sometimes called) is a particular type of agent who (except in special circumstances) normally acts for the vendor of the property. A real-estate agent may act for a purchaser, however, if the agent is engaged for the specific purpose of acquiring a property for a purchaser. The relationship between the property owner and the real-estate agent is usually established by a contract between the two, called a listing agreement. Under the terms of this agreement, the agent agrees to seek out prospective buyers for the land and, often, for the business enterprise on it, and bring the two parties together in contract. In return for this service, the property owner usually agrees to pay the agent a commission (based upon a percentage of the sale price, usually in the 5–10 percent range, or a fixed sum) for the agent's services, if a sale is negotiated. Sales of commercial and industrial properties usually command a higher commission rate than residential property, and usually fall in the 8–10 percent range.

Agent Services

The services that a real-estate agent is expected to provide under a typical listing agreement are as follows:

(1) inspect and value the property or, where an expert opinion as to the value is required, arrange to have an evaluation made by a professional appraiser;
(2) actively seek out prospective purchasers of the property by way of advertisement (paid for at the agent's expense), personal contact, and, in some cases, by way of notification to other agents of the availability of the property for sale;
(3) arrange to have prospective purchasers inspect the property and advise the purchasers of the vendor's terms of sale, and deliver appropriate financial statements in the case of a sale of a small business;
(4) prepare a written offer to purchase for execution by the prospective purchaser and deliver it to the owner for consideration;
(5) in extremely rare cases, execute the purchase agreement on behalf of the owner under a grant of express authority (a power of attorney); and
(6) hold all deposits paid by the prospective purchaser in trust pending the completion or termination of the transaction, and either pay the funds over to the vendor on completion of the sale or return the money to the prospective purchaser, if the offer is rejected by the vendor.

It is important to note that the real-estate agent, under a listing agreement, is expected to act at all times in the best interests of the vendor who engaged the agent to sell the property. The rules of law pertaining to agency (subject to a few minor exceptions) generally apply to the relationship. The agent is not a "middle man" who is used simply to bring the parties together, but is the agent of the vendor.[4] For example, a real-estate agent must never act for both parties in a transaction without the express consent of both (a rare situation) and, while engaged by the owner of the property as his or her agent, must never attempt to obtain a benefit for the purchaser that would be to the vendor's detriment. If the agent should attempt to obtain a benefit for the purchaser that is detrimental to the agent's principal, the agent would not only be liable for the amount of the loss, but would also not be entitled to claim a commission from the vendor on the sale. This would arise from the fact that the agent would be in breach of duty to the client.[5]

Assuming that the real-estate agent brings together a willing vendor and purchaser, the agent's duties are generally completed when the written offer, which the agent draws and

4. See, for example, *D'Atri v. Chilcott* (1975), 55 D.L.R. (3d) 30; *Len Pugh Real Estate Ltd. v. Ronvic Construction Co. Ltd.* (1973), 41 D.L.R. (3d) 48.

5. *Len Pugh Real Estate Ltd. v. Ronvic Construction Co. Ltd.*, supra.

the purchaser signs, is accepted by the vendor. At that time, the parties are bound in contract for the purchase and sale of the property that is the subject matter of the agreement.

CASE IN POINT

The vendor and purchaser of a parcel of land entered into an agreement for the sale of the property that contained a clause that stipulated that time was of the essence. The agreement also contained a clause that stated that the purchaser must pay a cash deposit within 5 days of acceptance of the offer. The purchaser inadvertently failed to pay the deposit within the 5 days stipulated in the agreement, and the vendor refused to accept the tender of the deposit when it was later offered. He informed the purchaser that he considered the agreement at an end. The purchaser sued for completion of the deal, but the action was dismissed by the court. The court held that the vendor was entitled to treat the agreement as at an end because the agreement stipulated that time was of the essence with respect to the payment of the deposit.

See: *1473587 Ontario Inc. v. Jackson* (2005), 74 O.R. (3d) 539; 75 O.R. (3d) 484.

The contract only establishes the contractual relationship, however. It does not effect a transfer of the title to the purchaser, as the parties must both carry out a number of duties under the agreement before the title changes hands. In addition, the offer to purchase usually provides the purchaser with a period of time to make a number of searches and determinations, in order to be satisfied that the title to the property is in order.

Appraisal

Appraisal
An estimate of the value of a property prepared by a person trained in the evaluation of the worth of property.

Property value is usually estimated by the real-estate agent at the time that the vendor enters into a listing agreement. The valuation of an ordinary residential property is usually well within the expertise of the experienced real-estate agent. However, in the case of a commercial or industrial property, the services of a professional property appraiser are usually obtained, in order to fix a value for sale purposes. Appraisers are frequently engaged by purchasers as well, especially for commercial or industrial purchases. The appraisal, from the purchaser's point of view, is important for two reasons:

(1) The appraisal provides the purchaser with an expert opinion as to the value of the property and provides some guidance in the negotiation of a price.
(2) The appraisal provides the purchaser with a general estimate of the amount of financing that might be obtained from a lending institution to assist in the financing of the purchase, since most lending institutions will fix the mortgage principal amount that they will lend as a percentage of the appraised value of the property.

Unlike the real-estate agent, a property appraiser charges a fixed fee for appraisal services. This service usually consists of an examination of the property and the formulation of an opinion as to its market value, based upon a number of factors. The most important of these are the condition of the building, the value of the land, the zoning of the property, the present rate of return (in the form of rents, business traffic, etc.), the potential for development, and the known market value of similar properties. The ultimate value placed on the given property depends a great deal upon the expertise of the appraiser, but the report is particularly useful for the purchaser in the determination of the purchase price to include in the offer.

Offer to Purchase

The offer to purchase is only an offer until such time as the vendor accepts it. At that point, the agreement becomes a binding contract. Nevertheless, the usual offer to purchase generally contains a number of provisions that, if unsatisfied by the vendor, render the agreement null and void at the option of the purchaser. For example, a standard form of an offer to purchase usually provides that the purchaser will have a period of time following the acceptance of the offer by the vendor in which to examine the title to the property at his or her own expense. If any valid objection to the vendor's title is raised within that time that the vendor is unable or unwilling to correct, the agreement will then become null and void, and any deposit paid by the purchaser must be refunded. The right to rescind the agreement may also be available to the purchaser if the property itself is not as described. This may arise, for example, if the purchaser discovers that the vendor does not have title to all the lands that he or she has agreed to sell. This latter discrepancy is frequently determined by way of a survey of the property and a comparison of the surveyor's description of it with the description of the land contained in the vendor's deed.

CHECKLIST OF COMMON TERMS IN AN OFFER TO PURCHASE

1. Description of property.
2. Price and deposit.
3. Conditions which must be performed by the vendor:
   - To provide clear or satisfactory title (including liens and taxes).
   - To provide a survey.
   - To deliver up included chattels (if any).
4. Time allowed for purchaser's examinations of title.
5. Conditions and time limits imposed upon the purchaser (or for its benefit):
   - To obtain satisfactory purchase financing.
   - To be satisfied as to quality of water and septic system (if rural).
   - To be satisfied as to results of inspections or environmental audit.
   - Suitability of zoning.
6. Work to be completed on property.
7. Closing date.

All of the conditions and terms of the accepted offer to purchase must be complied with in a real-estate transaction. The failure on the part of either party to fully perform the agreement would entitle the injured party (usually at the purchaser's option) to terminate the agreement, or bring an action for either specific performance or for damages. To satisfy themselves that the terms of the agreement are complied with, and to carry out their own duties and responsibilities, most parties to real-estate transactions engage the services of legal experts. The experts should be engaged before the offer to purchase is signed. If this is not done, however, it is wise to do so immediately after the contract is made, to ensure that rights are protected during the course of the transaction.

COURT DECISION: Practical Application of the Law

Land Transactions—Conditions Precedent and Warranties

***Springhill Gardens Developments Inc. v. Kent*, 2004 CanLII 15057 (Ont. S.C.).**

Kent agreed to purchase a luxury home which was nearing completion of construction. The agreement required the plaintiff Springhill to deliver a survey within 14 days and complete a list of repairs and changes, and allow a purchaser's walk-through to establish a list of further repairs or items for completion. The survey was delivered late as was the walk-through, and the repairs were completed to industry standard, but not to the level of perfection expected by the purchaser. The purchaser refused to complete the transaction, and the developer was forced to later sell the house for $52,000 less.

THE COURT: The legal test to determine whether a term is a condition precedent or a warranty is well established. A term of an agreement is a condition precedent if its breach will deprive the purchaser substantially of the whole benefit of the agreement as expressed in the contract. The issue will turn upon a finding of fact: what was the intention of the parties as expressed in the contract? The language of the agreement does not incorporate a subjective standard. It does not say the repairs shall be to the complete satisfaction of the purchaser. It is not even clear that the list is to be made by the purchaser alone: odds and ends of construction are to be listed on a walk-through by the vendor and the purchaser. The use of a non-standard builder's agreement and the lack of extensive reference to the [*Provincial home warranty scheme*] does not establish that failure to complete all odds and ends of construction and all repairs prior to closing would deprive the purchaser of substantially the whole benefit of the contract. In the absence of any special language or circumstance, it is reasonable to conclude that the intention of the parties as expressed in the language of the agreement was that the repairs would be completed to reasonable objective standards. It would not be fair to make a party to an agreement subject to the subjective sensibilities of the other party if the language or other circumstance does not put the vendor on notice that only the other party has the discretion to determine whether a special standard has been met. A contract is an agreement between two parties and it is the intention of both parties as disclosed by the contract that dictates the conduct of each party. In these circumstances the position taken by the purchaser both prior to closing and at this trial was not reasonable. While the vendor did not provide the survey or the walk-through on time, it did so by mid October promptly after the purchaser complained. The delay did not jeopardize the purchaser's ability to close on time given that she took the position that the contract had been repudiated and she did not ask for an extension of the closing. Furthermore, I have found that the vendor had satisfied its obligations. The purchaser did not feel that the vendor was respectful of their requirements and standards. However, I am satisfied that the terms added by the purchaser to the agreement of purchase and sale were not conditions precedent. At most they were warranties that would have survived closing and entitled them to damages. Even if the vendor had breached the terms, it would not entitle the purchaser to repudiate the contract and refuse to close. I am further satisfied that the standard for the repairs and odds and ends of construction was a reasonable objective industry standard and not the subjective satisfaction of the purchaser. I am also satisfied that the vendor provided the survey and made all reasonable repairs. Because the vendor insisted upon reasonable industry standards and was not prepared to make the repairs to Kent's satisfaction, the defendants took the position that they were not required to close and did not take the opportunity to inspect the home prior to closing to see if the repairs were satisfactory. I am satisfied that the vendor had satisfied its obligations and was ready to close the transaction. The purchaser did not request an extension for the closing and cannot rely upon the vendor's failure to provide the survey and the walk-through within the time specified for its failure to tender and close. For these reasons, judgement shall go for the plaintiff against the defendant Kent in the amount of $52,404.32 for breach of contract.

If you have particular expectations or standards, then they must be clearly expressed in the agreement. Otherwise there is no basis to expect anything beyond customary or industry standards. There is certainly no ability to avoid completion of the contract on this basis.

Expert Property Inspection

In the past twenty years, it has become much more frequent for purchasers to demand that the vendor permit an expert inspection of the property or structure. The results of this inspection must be wholly satisfactory to the purchaser before the purchaser's offer can become unconditional and binding. In residential cases these inspections can range from a critical review to estimate the extent and cost of repair of obvious deficiencies, through a search for latent defects, to simply being an "escape hatch" for the purchaser to exit the deal. The same may be true in commercial business transactions, but additionally, many inspections are conducted to ensure that the proposed business can be conducted in the premises from an engineering standpoint. For example, a purchaser may wish to have expert assessment of an abandoned warehouse for conversion to loft apartments. In another instance, inspection of floors and foundations may be carried out to determine if heavy mechanical production equipment can be properly installed and operated.

Environmental Audit

Environmental audit

A procedure whereby real property is examined to determine if environmental hazards exist on the property and to assess the clean-up costs, if any are found.

Commercial and industrial properties occasionally may be contaminated with hazardous or toxic substances that may have leaked from containers or were allowed to drain into the soil and groundwater, where they created an environmental hazard. Since the owner of a property is primarily liable for the cleanup costs, careful prospective purchasers usually have the property examined for contamination before the purchase is made or make their offer conditional upon a clean **environmental audit**. For example, a block of land that was formerly used as a service station may have soil contaminated with gasoline if old, buried fuel tanks have rusted and allowed the fuel to leak into the surrounding soil. This would constitute an environmental hazard that could be identified by an environmental audit. This would give the purchaser the opportunity to consider the cleanup cost in deciding whether to purchase or avoid the property altogether. Environmental laws and their implications to property owners are more fully examined in Chapter 34, "Environmental Law."

FRONT-PAGE LAW

A title for your castle: New homeowners not keen to spend the money on a property survey are turning to title insurance instead

Sometimes, the worst part about buying a new home is having to meet your new neighbours. That was certainly Candice Pilon's experience. Ms. Pilon had been living in her Hamilton, Ont., row house for a few months before she met the owner next door. That's when she learned that the people she had bought her house from had removed the firewall in the attic and built a bathroom across the property line.

"A week later I got a letter from [my neighbour] saying that if it wasn't fixed, he was going to sue me," she recalls. Fortunately, Ms. Pilon bought a title insurance policy when she purchased her house two years ago. And it saved her a lot of money. After she was threatened with a lawsuit, she submitted a claim to her insurer, First Canadian Title Co. Ltd. The insurer forked over $10,000 to rectify the problem and put her up in a hotel during the construction period.

When you purchase a title insurance policy you're shifting a lot of the risk associated with buying a home to the insurer. So, you're covered if you discover your house's former owner didn't pay his property taxes or he had built an illegal addition that needs to be torn down. Title insurance is fast — and cheap.

Now all of this sounds pretty compelling. But, unfortunately, title insurance isn't always a perfect solution. Sure, title insurance can really expedite a closing (which is probably why it is favoured by so many real-estate agents). But that's because it glosses over known or subterranean problems. And that could come back to haunt you when you decide to put your property up for sale.

Apart from residential applications, title insurance is often considered for business properties. The disruption in Ms. Pilon's life is small compared to a factory or office closure, even if it is for a short period of time. Lost business profits (and customers who go elsewhere) may not be recoverable. While title insurance is valuable, particularly where old and unseen industrial environmental contamination is possible, there is no substitute for a thorough title search, survey, and environmental audit. It is better to avoid a problem in the first place than to rely on compensation later.

Source: Excerpted from "A title for your castle," by James Pasternak, a Toronto-based writer whose work appears nationally in a variety of publications (jpasternak@Canada.com).

Survey

Survey

A drawing of the boundaries of a property prepared by a professional surveyor.

While an appraisal of the property is frequently obtained by a prospective purchaser before the purchase agreement is entered into, a **survey** is not usually obtained until after the parties have signed the contract. The reason for this procedural difference is due to the nature of the transaction. Most purchase agreements provide that the vendor must have a good title to the lands described in the agreement. A survey is used to determine if the vendor's land as described in the deed actually coincides with the boundaries and area of land that the vendor claims to possess.

Surveys are prepared by persons trained and licensed to perform the service pursuant to provincial legislation. Statutes govern not only the procedure for the conduct and preparation of surveys, but the licensing of persons who may perform the work. One statute usually governs the technical aspects of survey work; another governs the training and licensing of the profession.

The importance of obtaining a survey (from the purchaser's point of view) is that it will uncover any encroachments on the vendor's property by adjoining property owners, and any easements or rights of way that might affect the land. It will also establish the accuracy of the boundaries of the property as described in the deed. This will not only be determined by the surveyor from the vendor's title, but by "ground reference," because the surveyor generally marks the boundaries of the property by driving steel bars or pins into the ground at each of the corners of the parcel of land. In this fashion, the legal description of the property is translated into ground area for the purchaser to examine.

If the survey reveals a boundary discrepancy, the vendor is usually obliged to have the discrepancy corrected, in order that the purchaser will receive the amount of land described in the purchase agreement. If the discrepancy is significant and the vendor cannot correct the title, the purchaser may agree to take the property, but with some abatement of the purchase price to compensate for the deficiency. However, if the difference should be substantial, either in quantity or in its effect on legal use, the purchaser would be in a position to avoid the transaction. In this regard, the survey is used by the purchaser's solicitor to assess the vendor's title to the property and to determine the vendor's right to the property described in the agreement.

The Role of the Legal Profession

Lawyers who engage in the practice of law concerning land transactions have historically been solicitors. Both the vendor and the purchaser usually require the services of a solicitor to perform the many searches and duties associated with the transfer of the title in a sale of land. The services vary substantially depending upon whether they are for the vendor or the purchaser, and whether the transaction is one that involves the sale of residential or commercial property.

In the sale of a property, the purchaser's solicitor (depending upon the province in which the land is situated) will usually be called upon to make the following searches:

(1) For the title to the property at the land registry office in the county or district where the land is located. If the land is under the Registry System, then the solicitor is obliged to confirm the validity of the title held by a chain of previous owners, for a lengthy period of time.[6] This search would involve an examination of all deeds, mortgages, discharges of mortgage, and other documents registered against the title to the property to determine that the vendor's title, when traced back, represents a right in the vendor to convey a good title to the purchaser that would be free of any conflicting claims to the vendor's interest in the land.

Under the Land Titles System, the solicitor's search is much easier, as the province certifies the title of the vendor as being as it stands on the register. Any charges, liens, or other instruments that might affect the title of the vendor would be recited, and the solicitor might then ascertain their status and call on the vendor's solicitor to take steps to have the encumbrances or interests cleared from the title before it is transferred to the purchaser.

On the completion of the search (under both systems) the purchaser's solicitor notifies the vendor's solicitor of any problems concerning the title by way of a letter of requisition, which requires the vendor to clear the rights or encumbrances from the title before the closing date. The vendor's solicitor, on receipt of the letter, must make every effort to satisfy requisitions that represent valid objections to the vendor's title. If the objection cannot be satisfied, the purchaser, under the terms of the contract, is usually free to reject the property, and the purchase agreement becomes null and void.

(2) Of documents at the local municipal office, to determine
   (a) if the municipal taxes on the property have been paid and that no arrears of taxes exist;
   (b) that the zoning of the property is appropriate for the existing or intended use of the property;
   (c) that the property conforms to the municipal building and safety standards, and that no work orders are outstanding against the property;
   (d) in the case of land under development, that any obligations that the builder–developer may have under any subdivision or development agreement have been fully satisfied, or funds deposited with the municipality to cover the completion of the work;
   (e) where services (such as electric power) are provided by the municipality, that the payment for these services is not in arrears. For this and other municipal searches, the solicitor will obtain a certificate, or some other written confirmation, as to the status of the property.

6. In those parts of Ontario still under the Registry System the chain of title must cover a 40-year period. See the *Registry Act*, R.S.O. 1990, c. R.20, s. 112.

(3) For liens for unpaid corporation taxes if the vendor is a corporation, or if one of the vendor's predecessors in title was a corporation. The solicitor will contact the appropriate government office to obtain this information. If a lien for the unpaid tax is claimed by the province, the purchaser's solicitor will require the payment of the taxes before the transaction is closed in order to clear the lien from the property. This is not a universal requirement, however, as in some provinces the lien does not attach to the land unless a notice of the lien is registered against the title to the property in the appropriate land registry office.
(4) Where a province has a new housing warranty scheme in effect and the property entails a new dwelling, to determine if the builder is registered under the scheme, and that the particular dwelling unit is enrolled under the program. Not all provinces provide this form of consumer protection, but if the transaction concerns the sale of a new housing unit, and such a scheme is in operation, the solicitor for the purchaser would be anxious to ensure that the property was properly covered by the building warranty under the program.
(5) For any claims or encumbrances against chattels or fixtures sold with the property. Depending upon the province, these might be maintained in the local land registry office, the county or district court office, or in a central registry for the entire province. A certificate is usually obtained by the solicitor that would indicate any creditor claims against the goods in question.
(6) For any judgments or writs of execution that might attach to the lands of the vendor. A certificate would be obtained by the solicitor from either a provincial database or directly from the county or district sheriff if no writs of execution were in the sheriff's hands. Because a writ of execution can attach to the property under the Registry System as soon as it is filed, it is essential that this particular search be made as close to the time of transfer of the property as possible. As a result, the purchaser's solicitor normally makes a second or final search for executions at the moment of registration of the conveyance of the property to the purchaser. This is now usually a part of electronic document registration. Under the Land Titles System, the execution does not affect the land until it is placed in the hands of the Land Titles office, and a final search is also part of electronic registration.

In addition to the many searches that the purchaser's solicitor makes (some of which are outlined above), the solicitor attends to a number of other duties associated with the transaction itself. Insurance coverage for the property is obtained for the purchaser; the status of any mortgage or charge to be assumed by the purchaser is determined; draft copies of the conveyance and other title documents are checked; a mortgage back to the vendor (if a part of the transaction) is prepared; and the funds are obtained for the closing of the transaction.

While the purchaser's solicitor is busy with the many searches and preparations for the closing, the vendor's solicitor usually prepares[7] draft copies of the deed or transfer and a statement of adjustments (setting out the financial details of the transaction). These documents are then sent to the purchaser's solicitor for approval. The vendor's solicitor is obliged to answer any requisitions that the purchaser's solicitor may raise concerning the title to the property. The vendor's solicitor must also attempt to correct them in order that the vendor may pass a good title to the purchaser. If the transaction involves the assumption of an existing mortgage by the purchaser, the vendor's solicitor must obtain a mortgage statement from the mortgagee at the time that it is assumed by the purchaser. The vendor's solicitor must also collect all title documents, surveys, tax bills, keys, proof of ownership of chattels or business inventory to be transferred, and other documents pertaining to the sale for delivery to the purchaser on closing. The solicitor must also prepare a bill of sale for any chattels to be sold as a part of the transaction. This document, along with the deed or transfer, must then be properly executed by the vendor in the presence of a witness. These documents are collected together in preparation for delivery on the date fixed for the closing of the transaction. If the property to be sold is occupied

7. This is not a universal practice, however. It is not uncommon to have draft documents prepared by the purchaser's solicitor.

by tenants, the vendor's solicitor will attend to preparing assignment of the leases to the purchaser and to obtaining from the tenants an acknowledgement as to the status of the lease agreements. Copies of these documents will also be examined by the purchaser's solicitor, and the lease agreements verified. Notices to the tenants of the change of ownership of the property would also be prepared by the vendor's solicitor for delivery to the tenants on or after the closing of the sale transaction.

A QUESTION OF ETHICS

Land ownership, unlike most other business dealings, is highly emotional for many people. Add to this that factual knowledge about a particular property can be very one-sided (or polarized) and real estate can bring out the best and worst in people. Most ethical traps in real estate arise from non-disclosure and/or unrealistic expectations about property and prices. Common problem situations include non-disclosure of defects not readily apparent, expectations on repairs prior to completion, impending changes in a neighbourhood or its zoning, or sudden increases or decreases in prices of similar properties. Each of these situations (and many others) tempts vendors and purchasers toward misrepresentation or drives them to enforce or avoid completion of transactions that are no longer what they first appeared to be.

The Closing of the Transaction

Closing
The completion of a real-estate transaction where the documents relating to the property interest are exchanged for the payment amount.

The **closing** of the real-estate transaction today lacks much of the ceremony associated with a transfer of property in the past. The transaction is now usually completed electronically by the solicitors for the vendor and the purchaser. After exchanging an electronic copy of the transfer and being satisfied as to its content, the solicitors electronically submit the transfer to the local land registry office for registration. A search of executions is usually an integral part of the submission. Just prior to registration, but to be held in trust until registration, the vendor's solicitor usually delivers to the purchaser's solicitor a bill of sale for chattels (if any), along with supplementary documents and the keys. In return, the vendor's solicitor receives the purchase monies also to be held pending successful registration. The purchaser's solicitor attends to the registration of any mortgage the purchaser has arranged to finance the deal.

MANAGEMENT ALERT: BEST PRACTICE

For business purchases of land or buildings, it is important to discuss the nature of the intended business with a lawyer before approaching a real-estate agent. The offer must be made conditional on obtaining municipal-zoning permission to use the property as the purchaser intends. Many municipalities have a range of commercial zoning, where one business use (e.g., retail) may be permitted, and another (e.g., industrial) is forbidden. Therefore, a property offered by a real-estate agent and described as "zoned commercial," is not sufficient. If the offer is signed without the specific condition, it will still be binding, and the property will be useless to the purchaser. Even if the property is presently being used for the purchaser's desired purpose, it may be subject to a one-time municipal exception requiring that use to end when the current owner sells the property.

The registration of the documents completes the transaction, and the title at that point is registered in the purchaser's name. Vacant possession of the property is normally a part of the transaction, and the delivery of the keys to the premises is symbolic only of the delivery of possession to the purchaser. In most cases, the vendor and purchaser arrange between themselves the time when the vendor will vacate the premises, and when the new purchaser will move into possession. This is usually on the same day that the transaction closes.

Following the closing of the transaction and registration of title documents, the solicitors attend to the many final tasks associated with the change of ownership. These tasks usually take the form of providing notice of the change of ownership to municipal tax offices, assessment offices, mortgagees, public utilities, insurers, and any tenants. When these duties have been completed, the solicitors make final reports to their respective clients.

Closing Commercial Transactions

When a purchase and sale agreement involves the sale of a commercial building or a business, the transaction is considerably more complicated, for the purchaser had contracted for not only the purchase of land and a building, but stock-in-trade, trade fixtures, goodwill, and other assets of the business. Such a sale frequently involves the transfer or acquisition of special licences, as well as compliance with legislation governing the sale of business assets "in bulk." Because most businesses have trade and other types of creditors, a sale of a business in bulk (which means a sale of stock and/or the assets of a business such as equipment, trucks, furnishings, etc.) requires the vendor's solicitor to notify the seller's creditors of the sale, and make arrangements for the payment of their accounts out of the proceeds of the sale. This procedure may be either simple or lengthy, depending upon the number of creditors and the nature of their claims, but the bulk-sales legislation must be complied with; otherwise, the sale may be overturned by the creditors. For this reason, the sale of a business or business assets tends to be more complex, and the solicitors' work under the circumstances includes additional duties and responsibilities on behalf of their clients in order to effect the transfer of ownership of the business in accordance with this special legislation. The legislation concerning the sale of business assets is the subject matter of part of the next chapter.

CLIENTS, SUPPLIERS *or* OPERATIONS *in* QUEBEC

For all practical purposes, businesses should engage a notary to conduct land transactions in Quebec. Purchases and sales are customarily handled by the profession, and it is a legal requirement that documents pertaining to hypothecs (mortgages) and condominiums must be drafted by a Quebec notary.

Summary

The transfer of the title to real property was effected originally by way of ceremony and the expression of intent. This took the form of livery of seisin. Over time, the embodiment of the grantor's intentions in a written document gradually replaced the ceremony. Today, electronic transfers and registration systems have largely replaced paper-based registration. Registration of a transfer completes the transaction in terms of public notice, and, in the case of a transfer under the Land Titles System, serves as provincial certification of the purchaser as the owner of the land.

Real-estate agents assist the owner of land by providing a service that brings together the owner and prospective purchasers in an effort to establish a purchase and sale agreement. When property is listed with a real-estate agent, the agent must act in the best interests of the principal at all times. The general rules of agency normally apply to the relationship between the principal and the agent.

In any sale of real property, it is essential that the lands described in a deed or transfer correspond with the land actually occupied by the title holder. To determine this relationship, a survey is used to identify the land to ascertain if any discrepancy exists between the two. Surveys are prepared by persons trained and licensed under provincial legislation, and must be prepared in accordance with statutory requirements as well.

Land ownership is presently subject to a great many controls and regulations that, if ignored, can adversely affect the owner. For this reason, the parties to real-estate transactions usually engage the services of members of the legal profession to assist in the completion of the transfer of title to the purchaser. In addition to the preparation of the formal documents associated with the transaction, the lawyers (called solicitors) involved perform the many services necessary to carry out the terms of the contract.

On the purchaser's part, this usually involves a number of searches in addition to an examination of the title to the property, in order that the title passed to the purchaser will be good and marketable in the future.

KEY TERMS

Appraisal, 495
Environmental audit, 497
Survey, 498
Closing, 501

REVIEW QUESTIONS

1. Why must a purchaser make certain that no writs of execution are attached to the property to be purchased?
2. Explain the significance of the vendor providing vacant possession at the time of transfer of ownership. Is this still important today?
3. What role does a real-estate broker or agent play in a modern real-estate transaction?
4. How is the agency relationship established?
5. Why is a survey important in a land transaction? What does it establish?
6. Explain the purpose and use of an exclusive listing agreement.
7. In what way (or ways) does an appraiser assist in a real-property transfer?
8. Outline the role of a land-registration system in modern real-estate transactions.
9. How does the land-registration system aid a prospective purchaser of real property?
10. What constitutes a good and marketable title to real property? How is this determined under the Registry System?
11. In what way or ways does the Land Titles System simplify the determination of the vendor's title to lands offered for sale?
12. What are the duties of a real-estate agent? Do they differ in any way from that of an agent in an ordinary principal–agent relationship?
13. Describe the role of a lawyer or solicitor in a land transaction.
14. What other searches in addition to a title search are necessary in a land transaction in order to protect the purchaser? How are these searches usually made?
15. Explain the significance of the process whereby a deed of land is "signed, sealed, and delivered."
16. Why were the ceremonial aspects of the transfer of land important in early real-estate transactions in England?
17. Indicate the importance of the registration of a transfer of land in the proper land registry office as soon as possible after the documents of ownership change hands.

MINI-CASE PROBLEMS

1. A agreed to sell B a building lot for $25,000. Without searching the title, B gave A the $25,000 and received a deed to the property in fee simple. B then discovered that the property was registered in A's wife's name and not in A's name.

   What are B's rights?
2. The ABC Co. intends to offer to purchase a large parcel of vacant land from D, a farmer. The company plans to erect a manufacturing plant on the land after the purchase is completed.

   What conditions or provisos should the company include in the offer to purchase, and what searches should it make to ensure that it may use the land for its intended purpose?
3. Rudolf owned a house and lot. He engaged the services of Victoria, a real-estate agent, to sell the property for him, and signed a 3-month listing agreement with a commission of 4% payable if she sold the property. A month before the listing agreement expired, Victoria took George, an interested buyer to see the house. While Victoria was adjusting a 'for sale' sign on the lawn, Rudolf told George that if he was interested in buying the property, he could sell it to him in a month's time after the listing agreement had expired. Rudolf indicated that he would sell it 'less commission,' which would save George about $10,000. A month later, George returned, and Rudolf and George arranged for the sale of the property at the listed price, less $10,000.

   What are Victoria's rights in this case?
4. Jason purchased a building at an intersection of two highways that had once been a service station and gasoline storage facility. Haggard, the seller, assured Jason that the property was 'clean' and there were no environmental problems. A year after the purchase was completed, the owner of the land next to Jason's property complained to Jason that his well water was now contaminated with gasoline and oil.

   Advise Jason.

www.mcgrawhill.ca/olc/willes

Case Problems for Discussion

CASE 1

Albert, a real-estate agent, entered into an exclusive listing agreement with Amelia whereby he agreed to find a buyer for a block of development land owned by Amelia. The property had a listed value of $750,000, an amount that Albert had indicated was a reasonable price for the property. Amelia, who was inexperienced in business matters, accepted Albert's appraisal as being a fair market value for the land. In actual fact the fair market value was approximately $1,000,000. Albert made no effort to put the parcel of land on the market, but instead, incorporated a company for the purpose of purchasing the property. When the new corporation was operational, Albert prepared an offer on behalf of the corporation, and presented it to Amelia. At the time of presentation of the offer, he told her that he knew the president of the corporation. Since the price was exactly $750,000, Amelia readily signed the agreement to complete the contract. The sale was completed in due course, and Amelia gave her deed to the property in return for the payment of the purchase price. At that time, she also paid Albert the 5% commission that the listing agreement had specified as the agent's commission. A few months after the sale of the land had been completed Amelia discovered that Albert was the president of the corporation that had purchased her property.

Advise Amelia. If Amelia should take legal action against Albert, what would be the nature of her claim?

Render a decision.

CASE 2

A municipality owned a large tract of land that had been a municipal landfill site many years before. Custom Construction Inc., a land development corporation, moved into the municipality from another city, and was anxious to establish itself in the community as a quality developer. When it discovered that the municipality had a tract of land for sale, it approached the municipality to determine if the land could be purchased for residential development. The municipality was anxious to dispose of the property, and agreed to sell to Custom Construction Inc. A price was agreed upon, and the purchase was completed after the lawyer for Custom Construction Inc. had examined the title to the property, and found that the municipality had a good title to the tract of land. During the preparation of the land for house building, Custom Construction Inc. employees noticed that a methane gas smell was rising from the excavations for house basements, and also that a coloured liquid was seeping into a small stream that crossed the property. When Custom Construction Inc. queried the methane gas smell, and seepage into the stream at the municipal office, it was advised at that time that the land had previously been the municipal landfill site. The revelation that the property had been a landfill site in the past effectively precluded Custom Construction Inc. from proceeding with its housing development.

Discuss the issues raised in this case, and advise Custom Construction Inc. Speculate as to the outcome of this case if Custom Construction Inc. should decide to take matters to court.

CASE 3

In December of 2011, Fanshawe agreed to purchase a house and lot from Miovsky for a purchase price of $258,000. The property was in an area under the Registry System, and Fanshawe determined from the registry office that Miovsky had what appeared to be a good title to the land. Unfortunately, he had failed to notice a mortgage that had been registered against the property to secure the indebtedness of Williams, Miovsky's predecessor in title. The mortgage had been assumed by Miovsky as a part of his purchase from Williams, but that fact had not been revealed to Fanshawe at the time that the offer was drawn.

The offer to purchase that Fanshawe had accepted contained the following clause that read in part: "The purchaser shall have 10 days to examine the title at his own expense .... Save as to any valid objections made to the vendor's title within that time, the purchaser shall be conclusively deemed to have accepted the title of the vendor."

Without knowledge of the existing mortgage, Fanshawe proceeded to pay over to Miovsky the $258,000 and received a deed to the property. He registered the deed on January 6, 2012. Some time later, Fanshawe was contacted by the mortgagee and advised that the sum of $85,000 remained due and owing on the mortgage. As Fanshawe was the new owner of the property, the mortgagee would look to him for payment.

To compound Fanshawe's problems, a finance company that had obtained a judgment for $12,000 against Miovsky in October of 2011 had filed a writ of execution with the sheriff of the county where the property was located. Fanshawe had not searched in the sheriff's office at the time of closing the transaction. He was surprised to discover that the finance company now claimed that the execution had attached to the land that he had purchased.

Advise Fanshawe of his legal position in this case. What are the rights of the mortgagee and the execution creditor? What action could they take against Fanshawe or the property?

CASE 4

Samuel Reynolds was the registered owner of a 200-hectare farm. In July 1991, he retired from farming and gave his son Jacob a deed to the farm property. Jacob did not register the deed, but instead placed it in a safety deposit box that both he and his father rented at a local bank.

Samuel Reynolds died in 2011. His will, dated, May 3, 1988, devised the farm to his daughter, Ruth, who was living on the farm at the time, and who maintained the house for her father. The will also named Ruth as the sole executrix of his estate. When the contents of the will were revealed, Jacob announced to Ruth that their father had given him a deed to the farm some years before, and that he was the owner of the property.

No further discussion of the farm took place between the brother and sister. However, after the debts of the estate were settled, Ruth had a deed to the farm prepared and executed in her capacity as the executrix of her father's estate. Then Ruth delivered the deed to the registry office. The deed conveyed the farm property to her in her personal capacity.

Some months later, Jacob entered into an agreement of purchase and sale for the farm with Smith. Smith made a search at the registry office and discovered that the title to the property was not registered in the name of Jacob, but in the name of his sister, Ruth. Smith refused to proceed with the transaction.

Discuss the rights of the parties in this case. Indicate, with reasons, the identity of the lawful owner of the property. If Jacob should show his deed to Smith and insist that he has title to the property, what argument might Smith raise to counter Jacob's claim?

CASE 5

Abner, who was quite elderly, owned a large number of vacant building lots in different parts of a city. He was approached by Drossos to purchase a particular lot that Abner agreed to sell for $25,000. After making certain that Abner had a good title to the property, Drossos paid over the purchase price to Abner and received a deed to the land in fee simple. Drossos didn't register the deed, but placed it in his safety deposit box instead, as it was his intention to sell the lot himself at a later date.

A year later, Abner was approached by Carol's Construction Corporation with an offer to purchase the lot he had sold to Drossos. The price was $35,000, and Abner, who had forgotten that he had previously sold the lot to Drossos, agreed to sell it to Carol's Construction Corporation.

The solicitor for Carol's Construction Corporation made the appropriate searches in the land registry office and, from the records there, determined that Abner was the owner of the land in fee simple. The money was then paid to Abner, and Abner delivered a deed to the property, which the lawyer registered at the land registry office.

Some time later, Drossos entered into an agreement with Davis to sell her the lot for $38,000. When Davis searched the title to the property, she discovered that the lot was registered in the name of Carol's Construction Corporation. Davis then demanded that Drossos obtain a conveyance of the land from Carol's Construction Corporation to enable him to provide her with a good title in fee simple.

If Davis brought an action for specific performance against Drossos, discuss the arguments that the parties might raise. Explain the respective positions of Abner and Carol's Construction Corporation in relation to the claim. Speculate as to the ultimate outcome of the situation.

CASE 6

For 50 years, Metal Plate Inc. manufactured metal signs in a manufacturing plant located in the industrial area of a city. In 2000, a large container of used cleaning solvent that was stored behind the plant leaked into the soil. At the time, the solvent was not classified as hazardous waste material under the provincial *Environmental Protection Act*, and nothing was done to retrieve the solvent, except to dispose of the leaking container. Several years later, the solvent was classified as a registerable hazardous material.

In 2008, traces of the solvent were found in the surface-water drainage sumps in a neighbouring property, and the source was identified on the Metal Plate land. The company notified the Ministry of the Environment, and, as a compliance method, the company drilled several collector wells and installed a pumping system to monitor and collect the contaminant as it moved through the soil. The system was approved by the Ministry, although the Ministry could require the removal of the contaminated soil at any time, if it should be necessary to do so.

In 2010, the owner of a business in the area offered to purchase the plant and land for $1,500,000, and Metal Plate Inc. agreed to sell the property. The purchaser was given a full opportunity to inspect the property before the closing of the transaction, but did not do so at the time. The purchaser intended to rezone the property as a residential property and build a large condominium there, but did not disclose his plans at the time of purchase.

Some months after the new owner had purchased the property, soil tests were made to assess the suitability of the property for the construction of the residential condominium complex. The test engineers reported that the soil structure would permit the construction of the proposed building, but noted in their report the contaminated soil and the monitoring system. They estimated the cost of removal of the contaminated soil at $100,000.

The new owner did nothing about the reported problem and proceeded with his proposed plans to construct the condominium. His efforts were in vain, however, as the municipality refused to rezone the land to residential use because of the location in an industrial area. Several years later, the new owner abandoned his plans. He demanded a return of his money from Metal Plate Inc., on the basis that the lands were contaminated and useless for his purposes.

Metal Plate Inc. refused to return the purchaser's money, and the purchaser instituted legal proceedings against Metal Plate Inc.

Discuss the issues raised in this case and render a decision.

CASE 7

Ibrahim was the registered owner of several adjoining parcels of vacant land that he had purchased some 12 years earlier. During that period of time, the property had appreciated substantially in value.

Recently, Ibrahim was approached by a real-estate agent who suggested that the property might be of interest to a number of developers who had just begun construction in the immediate area. After some discussion, Ibrahim entered into a listing agreement with the agent, and the agent agreed to seek out prospective purchasers for the property. Ibrahim established $200,000 as the selling price he would accept for the land.

For several months, the agent attempted to find a buyer for the property, but without success. When the developers in the area were not interested in the property, the agent returned to Ibrahim and suggested that a corporation in which he had an interest might be willing to purchase the land. To this suggestion Ibrahim replied that it did not matter to him who the purchaser was, so long as the purchaser was prepared to pay his price for the land.

A week later, the agent returned with an offer to purchase from the corporation in which he had an interest. The offer price was $200,000, and was described by the agent as a "clean deal — all cash." The offer was prepared on a standard real-estate offer-to-purchase form and contained a clause that read: "Any severance or impost fee plus any expenses for water and sewer connections to be included in the purchase price." Ibrahim queried the clause, and the agent explained that it meant that the cost of obtaining permission to use the three parcels of land as separate building lots, and the hook-up costs of water and sewer lines to them, would be deducted from the purchase price. He added that this "usually did not cost much."

At the agent's urging, Ibrahim signed the offer. Some weeks later, Ibrahim discovered to his sorrow that the severance fees and the water and sewer connections would cost close to 10 percent of the sale price. The municipality required the payment of 5 percent of the value of the property as part of the severance fee, and the water and sewer connections accounted for the remainder. When Ibrahim refused to proceed with the transaction, the purchaser instituted legal proceedings for specific performance, and Ibrahim, on the advice of his solicitor, settled the action. As a result, he received only $180,000 for the property, from which the real-estate agent demanded a selling commission of 5 percent based upon the $200,000 selling price.

Ibrahim refused to pay the agent and demanded that the agent compensate him for the $20,000 loss that he had suffered. Eventually, the agent brought an action against Ibrahim for the commission that he claimed was due and owing. Ibrahim, in turn, filed a counterclaim for payment of the $20,000 loss that he had suffered.

Discuss the arguments that might be raised by the parties in this case. Render a decision.

CASE 8

Justin Abernathy was the registered owner of a building lot in a residential area of a large city. He entered into an agreement of purchase and sale with Karina Giltay, who wished to build a house on the land.

Following the execution of the agreement, Giltay engaged a land surveyor to prepare a survey of the property. When the survey was completed, it disclosed that the building lot had a frontage of only 18.15 metres. The agreement of sale stated that the lot had "frontage of approximately 18.46 metres and a depth of 38.46 metres, more or less." Abernathy's deed described the lot as being 18.46 metres by 38.46 metres.

The discrepancy between the two measurements was apparently due to the fact that the owner of the adjacent lot had erected a fence that encroached on Abernathy's property. The fence had been erected some 15 years before and was taken by the surveyor as the property line.

Determine the rights of the parties in this case. Indicate how the problem might be decided if Giltay should refuse to proceed with the agreement.

The Online Learning Centre offers more ways to check what you have learned so far. Find quizzes, Weblinks, and many other resources at www.mcgrawhill.ca/olc/willes.

CHAPTER 26

Intellectual Property, Patents, Trademarks, Copyright and Franchising

Chapter Objectives

After study of this chapter, students should be able to:

- Describe the unique nature of intellectual property.
- Distinguish between patents, trademarks and copyrights.
- Explain the legal protection offered to each of these types of property.
- Describe the chief elements encountered in license and franchise agreements.

YOUR BUSINESS AT RISK

Intellectual property — whether a trade name, secret process, design or a manuscript — represents a very important business asset. This property must be properly protected to ensure your ownership rights are enforceable. To commercialize your intellectual property (IP), you may wish to transfer a right to use the property without selling it outright — after all, you may wish to concurrently use the IP yourself. In these cases, a licence agreement will be required in order to transfer a right to use the IP while retaining ownership and control over it.

INTRODUCTION

Patent

The exclusive right granted to the inventor of something new and different to produce the invention for a period of 20 years in return for the disclosure of the invention to the public.

Trademark

A mark to distinguish the goods or services of one person from the goods or services of others.

Copyright

The right of ownership of an original literary or artistic work and the control over the right to copy it.

Patents, trademarks, and copyright are essentially claims to the ownership of certain types of industrial and intellectual property. A **patent** is a right to a new invention; a **trademark** is a mark used to identify a person's product or service; and a **copyright** is a claim of ownership and the right to copy a literary or artistic work. Bridging trademarks and copyright is a fourth form of protection, known as an industrial design, that is simply the right to produce in quantity some artistic work, such as a piece of furniture or article of unique design. Legislation has been passed by the Federal Government to address these particular rights.

From a public-policy point of view, the purpose and intent of each statute is somewhat different. However, each statute attempts to balance the right of public access to, and the use of ideas and information, with the need to foster new ideas and new literary and artistic works. As a result, the legislation incorporates special benefits or rights to promote the particular activities, along with appropriate safeguards to protect the public interest.

Patent legislation is designed to encourage new inventions and the improvement of old ones by granting the inventor monopoly rights (subject to certain reservations) for a period of time. Copyright laws are also designed to encourage literary and artistic endeavour by vesting in the author or creator of the artistic work the ownership and exclusive right to reproduce the work over a lengthy period of time. Registered-design legislation has a similar thrust. Trademark legislation, on the other hand, has a slightly different purpose and intent. It is designed to protect the marks or names that persons use to distinguish their goods or services from those of others, and to prevent unauthorized persons from using them.

Originally, the rights associated with intellectual and industrial property were not subject to legislation. Inventors, authors, and artists had very little or no protection for their creative efforts. The general need for protection did not arise until the invention of printing and the Industrial Revolution in England, however. These two changes created an economic environment in which the creators of industrial and intellectual property were in a position to profit from their creativity. The changes also created an opportunity for others to reap the rewards of an inventor's or author's endeavours without providing the creators with compensation for their loss. The Common Law proved unequal to the task of establishing and enforcing ownership rights to inventions and creative works of a literary or artistic nature. It was eventually necessary for the English Parliament to deal with each specific property right by way of legislation.

CLIENTS, SUPPLIERS *or* OPERATIONS *in* QUEBEC

As a matter of federal jurisdiction, intellectual property legislation and the conventions to which Canada is a party, apply uniformly across Canada.

TRADE SECRETS AND NON-DISCLOSURE AGREEMENTS

Trade secret

A commercial secret regarding a product or process.

Non-disclosure agreement

A contract to maintain a trade secret.

While there are many cases in which intellectual property deserves formal legal protection, one of the best forms of protection is simply to keep the knowledge as a secret. In some cases it is easy to keep a "**trade secret**," where examination of the product does not give away the secret knowledge. Typical of these would be secret sauces or soft drink recipes, or processing techniques that create desirable textures, finishes or special attributes. As long as the employees who know the recipe or technique are held accountable to keep the secret, then secrecy may be the best protection for that intellectual property. Agreements to keep trade or commercial knowledge secret are known as confidentiality agreements or **non-disclosure agreements** (**NDA**). Remember however, that if the employee breaches an agreement not to disclose a secret, the recoverable damages from the employee may not represent much compensation.

Beyond just employees, non-disclosure agreements are also used to protect intellectual property in cases where a company is dealing with strangers. Potential partners, suppliers and customers may all be interested in doing business with a firm, but require some knowledge about its products and processes before that is possible. Since business negotiations may properly result in disclosure of corporate secrets, both parties may wish to bind the other with an NDA. As such, whether or not a deal is struck, neither party may make use of the knowledge that they gain in course of negotiations. In some cases it can be appropriate to provide for automatic expiration of the agreement, chiefly in relationships where rapidly changing technology will quickly make many secrets obsolete.

CHECKLIST OF COMMON NON-DISCLOSURE CLAUSES

1. Parties who are bound by the agreement.
2. Acknowledgement of the duty to maintain confidentiality.
3. Description of the intellectual property not to be further disclosed.
4. Covenant not to use knowledge, or description of permitted uses.
5. Term of agreement or date of automatic expiration (if appropriate).
6. Exception for any prior knowledge of a party.
7. Exception for knowledge later learned by other legitimate means.
8. Exception for knowledge now or later falling into the public domain.
9. Exception for any disclosure compelled by law.
10. Acknowledgement of a right to seek an injunction or other remedy.

PATENTS

Historical Development of Patent Law

Early patents in England were essentially monopoly rights granted by the Crown under Letters Patent. These rights were granted to individuals or guilds and gave them exclusive rights to deal in the particular commodity, or the right to control a particular craft or skill. Most of these grants were made ostensibly for the purpose of fostering the trade or the skill. However, in many cases, they were simply privileges bestowed upon a subject by the Crown for the general enrichment of the individual.

The issue of monopoly rights to encourage the production of new products was justified on the basis of public policy generally, since England lacked the special skills and equipment necessary to produce the many kinds of goods available on the Continent, especially cloth and metal wares. To encourage English and foreign entrepreneurs to bring in artisans with the necessary skills to produce similar goods in England, monopoly rights were frequently granted. The justification was that the new skills would be learned by native craftsmen, and the country as a whole would benefit. Most of the early "patents" stipulated that the holders of the patent must provide particular quality products in sufficient supply to satisfy the market. The patentees, as a consequence, were obliged to establish production facilities and train their workers in order to comply with the stipulations in the patents. The incentive to do so, in many cases, was a proviso in the patent to the effect that a failure to comply with the conditions set out in the document would result in a revocation of the grant.

The province of Lower Canada introduced legislation in 1823 relating to patents by the statutory recognition of the right of inventors to the exclusive making, use, and selling of their new inventions, provided that they were British subjects and residents of the province.[1] The province of Upper Canada established similar legislation three years later.[2] In 1869, two years after Confederation, the first federal law was introduced that took a different approach to patent rights. This was due, in part, to events that occurred in the United States and, in part, to the fact that patents in England at the time were still issued under the Great Seal of the Crown.

As a result of the severance of its ties with England by the Revolutionary War, the United States introduced its own patent law in 1790.[3] The statute blended the right of the individual to the fruits of his or her labours with the right of the state to permit free trade for the benefit of the public. The product of this blend was essentially a bargain struck between the state and the inventor, whereby the inventor, by revealing the secrets of the invention, would obtain monopoly rights to its use and manufacture for a fixed period of time. When the time period expired, the invention became public property.

Canada adopted the procedure outlined in the United States' law, but retained the Common Law for the interpretation and expression of patent rights. In this sense, it incorporated English case law as authoritative for the exercise of the patent rights. The Canadian Act[4] was subsequently amended on a number of occasions, the most important ones being the introduction of a compulsory licence requirement in 1903,[5] and the complete revision of the Act in 1923 to set out the rights and duties of the patentee and to prevent abuse of the system.[6] These amendments were based upon the 1919 English *Patents and Designs Act*, with the result that the Canadian law became a unique blend of both English and United States' laws.[7] Since that time, the statute has been subject to numerous amendments that were designed to change the rights of the patentees or protect the public interest. Amendments that became effective in 1989 moved the Canadian patent law closer to the laws of the European Patent Convention countries and Japan.

1. *An Act to Encourage the Progress of Useful Arts In This Province*, 4 Geo. IV, c. 25 (Lower Canada).
2. *An Act to Encourage the Progress of Useful Arts Within the Province*, 7 Geo. IV, c. 5 (Upper Canada).
3. The *Patent Act*, 1790, 1 Stat. 109, ch. 7 (U.S.).
4. The *Patent Act* of 1869, 32-33 Vict., c. 11 (Can.).
5. *An Act to Amend the Patent Act*, 1903, 3 Edw. VII, c. 46 (Can.).
6. The *Patent Act*, 13-14 Geo. V, c. 23 (Can.).
7. *Patent Act*, R.S.C. 1985, c. P-4, as amended.

The *Patent Act*

The present patent legislation is available to inventors of any "new and useful art, process, machine, manufacture, or composition of matter, or any new or useful improvement of the same."[8] To *invent*, however, means to produce something new and different — something that did not exist before.[9] What is created must be more than what a skilled worker could produce, in the sense that it must be something more than mere mechanical skill that is the subject matter of the patent.[10] It must also be new in terms of time. As a general rule, an invention must be kept secret from the public before the filing of the application for a patent takes place. Any invention that has been described or disclosed by the inventor more than a year before the date of application for the patent is not patentable, as it is deemed to be in the public domain. In general, it is not possible to patent something that is only a vague idea or an abstract theory, nor is it possible to patent a very slight improvement in an existing invention.[11] It should also be noted that any invention that has an unlawful purpose is not subject to patent protection. For example, a new "five-in-one burglary tool" would not be granted a patent, no matter how handy or useful it might be for a burglar.

The legal nature of a patent and the justification for its issue was described by the court in the case of *Barter v. Smith*[12] in the following terms:

> It is universally admitted in practice, and it is certainly undeniable in principle, that the granting of letters-patent to inventors is not the creation of an unjust or undesirable monopoly, nor the concession of a privilege by mere gratuitous favor; but a contract between the State and the discoverer.
>
> In England, where letters-patent for inventions are still, in a way, treated as the granting of a privilege, more in words however than in fact, they, from the beginning, have been clearly distinguished from the gratuitous concession of exclusive favors, and therefore were specially exempted from the operation of the statute of monopolies.
>
> Invention being recognized as a property, and a contract having intervened between society and the proprietor for a settlement of rights between them, it follows that unless very serious reasons, deduced from the liberal interpretation of the terms of the contract, interpose, the patentee's rights ought to be held as things which are not to be trifled with, as things sacred in fact, confided to the guardianship and to the honor of the State and of the courts.
>
> As it is the duty of society not to destroy, on insufficient grounds, a contract thus entered upon, so it is the interest of the public to encourage and protect inventors in the enjoyment of their rights legitimately, and sometimes painfully and dearly, acquired. The patentee is not to be looked upon as having interests in direct opposition to the public interest, an enemy of all in fact.

Patent Procedure

An application for a patent may be made by the inventor or the inventor's agent[13] at the Patent Office. Under present legislation,[14] the first inventor to file for a patent is entitled to the patent. Foreign inventors, as well as Canadians, may apply for patent protection for their inventions, as it is always possible that someone in Canada or elsewhere in the world may become aware of an inventor's invention, or may develop essentially the same device or process. Consequently, a prompt application for a patent is important.

The usual practice in a patent application is for the inventor to engage the services of a patent attorney (or patent agent, as they are called in Canada) to assist in the preparation of the documentation and in the processing of the patent. Patent agents are members of

8. ibid., s. 2.
9. ibid.
10. *Can. Raybestos Co. Ltd. v. Brake Service Corp. Ltd.*, [1927] 3 D.L.R. 1069.
11. *Lightning Fastener Co. v. Colonial Fastener Co.*, [1936] 2 D.L.R. 194, reversed on appeal [1937] 1 D.L.R. 21.
12. (1877), 2 Ex. C.R. 455.
13. Patent agents (or attorneys) are registered with the Patent Office under the *Patent Act*, R.S.C. 1985, c. P-4.
14. *Patent Act*, supra.

the legal profession who specialize in patent work. In addition to training in the area of law, most agents usually have a specialized professional background. Patent agents frequently possess professional engineering degrees, or advanced training in another field of science, and are skilled in the assessment of inventions in terms of their being new and useful.

Search

The patent agent will generally make a search at the Patent Office for any similar patents before proceeding with an application on behalf of the inventor. This search is a useful first step in the patent process, because any patents already issued that cover a part of the invention (or possibly all of it) would indicate that the invention may not be patentable at all, or subject to patent for only those parts that are new. The same would hold true if the search revealed that the same invention had been patented some time ago and the invention was now in the public domain. Patent agents frequently conduct a similar search at the U.S. Patent Office to determine if a patent has been issued there for all or a part of the invention.

Application

If the search reveals that the invention is in fact something new and open to patent, the next step is that the inventor makes an application for patent protection for the invention. This is done by the preparation of an application or petition for the patent, which the inventor must submit, along with detailed specifications of the invention. A part of this must be a claims statement that indicates what is new and useful about the invention. A drawing of the invention is usually required if it is something of the nature of a machine, product, etc., that has a shape or parts that must be assembled. To complete the application, the inventor must submit the patent-filing fee and a short abstract of the disclosure written in simple language, capable of being understood by the ordinary technician. Each of these documents is important from the applicant's point of view and must be carefully prepared.

Claim

The most important document is the specifications and claims statement, which describes the invention in detail and sets out what is new and useful about the discovery. It must contain a description of all important parts of the invention in sufficient detail to enable a skilled worker to construct the patented product from the information given when the patent protection expires. If the inventor intentionally leaves out important parts of the invention in order to prevent others from producing it, the patent may be void. Hence, accuracy is important to obtain patent protection.

The claims statement is equally as important, as it sets out the various uses of the invention and what is new in the product or process that would entitle the inventor to a patent. The claim must also be accurate, as too broad a claim could cause difficulties for the inventor later if the invention should fail to live up to its claims.

The abstract that accompanies the application is simply a brief synopsis of the detailed submission to enable a person searching later to determine the general nature of the invention and its intended uses. It is written in non-technical language and seldom exceeds a few hundred words in length.

Examination and Issue

Once the material filed is in order, the staff of the Patent Office proceed with a detailed examination of the material and the Patent Office records to determine if the invention infringes on any other patent. If the patent staff determine that the invention is indeed new and different, then a patent is issued to the inventor. In the past, the search and issue sometimes took long periods of time. For example, a patent for the manufacture of a type of plastic dinnerware was applied for in 1935, but the patent was not issued until 1956, some 21 years later. Today, however, the period of protection is from the date of application, and the process is expected to take only a short period of time. Delay in issue, however, is seldom a serious matter, as most manufacturers of the product for which the patent has been applied may institute a special process for rapid examination and issue of the patent if some other manufacturer should produce the product without the inventor's consent. The prompt issue of the patent under these circumstances would place the other manufacturer in the position of infringing on the patent and liable to the inventor for damages once the patent is issued. As a result, few manufacturers would likely trouble themselves to tool up for the manufacture of goods knowing that a patent has been applied for. The marking of goods "patent pending" or "patent applied for" has no other purpose than to notify others that the application has been made for the patent. It has no special significance at law, as the rights of the inventor only arise on the issue of the patent.

The issue of a patent under the present Act provides the inventor with exclusive rights to the invention and its manufacture and distribution for a period of 20 years from the date of the application.[15] Inventors granted a patent are required to pay both an issuing fee and an additional annual fee to maintain the patent. To enable others to know that a product is protected by a patent, the article or product may often be marked with the date of issue of the patent, but it is not a requirement under the Act.[16]

Patents for pharmaceutical products are subject to special provisions under the *Patent Act* that protect the public from excessive prices being set for the products by patent holders. A Patent Medicines Prices Review Board may review drug prices to ensure that drugs are available at the lowest cost to the public yet still recognize the research and development costs of the drug manufacturers. The purpose of the Board is to attempt to balance these interests and provide an incentive for manufacturers to undertake the high cost of research to develop new drug products.

Sample Patent (aircraft flight simulator)

Patent: CA 2196605

METHOD OF MANUFACTURING A MOTION SIMULATOR, AND A MOTION SIMULATOR

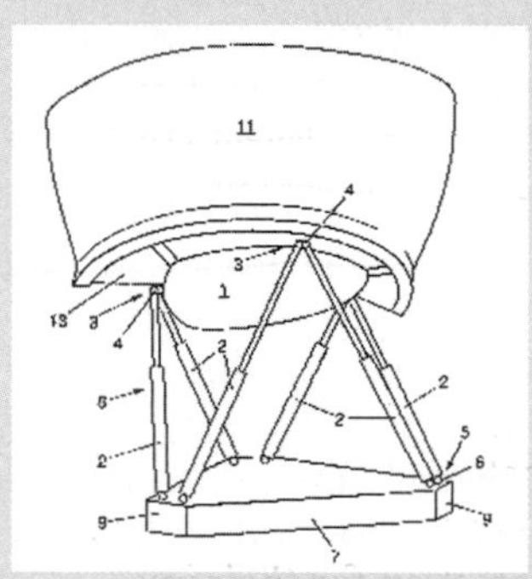

ABSTRACT: A method of manufacturing a motion simulator, which motion simulator has a deck and a number of deck-supporting legs that are pivotally connected with the deck in first pivot points, the legs being actively and continuously length-adjustable, such that the deck is capable of describing a motion envelope comprising all desired deck positions, wherein a leg envelope is determined for each leg within which the possible positions of the relevant leg are located, whereafter the common surrounding space of the legs, and in particular the interspace included between the leg envelopes, is determined, whereafter a shell is then designed that at least carries the deck, which shell defines an inner space extending at least partly within the interspace between the legs defined by the leg envelopes, in such a manner that in each position of the simulator, the legs are clear of the shell.

CLAIMS:
1. A multi-purpose motion simulator comprising:
a) a deck;
b) a base plate;
c) a number of deck-supporting legs, each of the deck-supporting legs having a first end pivotally connected with the deck in a first pivot point, and a second end being pivotally connected to the base plate in a second pivot point;
d) control means and drive means for actively and continuously adjusting the lengths of the deck-supporting legs, such that the deck can describe a motion envelope comprising all desired deck positions, each of the decksupporting legs defining, during use, a leg envelope within which the possible positions of the relevant deck-supporting

15. *Patent Act*, R.S.C. 1985, c. P-4, s. 44.
16. Patent marking is used at times to clearly warn others not to copy the goods.

leg are located, said leg envelopes defining a common surrounding space of the legs, and an interspace enclosed between the leg envelopes wherein the control means and drive means allow movement of the deck in six degrees of freedom; and
e) a shell defining an inner space for accommodating operating and regulating means and enclosing the deck wherein the inner space extends partly within the interspace between the deck-supporting legs defined by the leg envelopes, such that for each position of the simulator, the deck-supporting legs are clear of the shell, wherein the shell extends between the deck-supporting legs so that a center of gravity of the deck and shell of the simulator is near a plane defined by at least three of the first pivot points, between at least three first pivot points and wherein part of the deck and shell of the simulator is movable above and over the legs during use.
[*Twenty-one further claims follow*]

Inventors (Country):	**BEUKERS, ADRIAAN** (Netherlands) **VAN BATEN, TOM JACOBUS** (Netherlands) **ADVANI, SUNJOO KAN** (Netherlands)
Owners:	**TECHNISCHE UNIVERSITEIT DELFT** (Netherlands)
Applicants	**TECHNISCHE UNIVERSITEIT DELFT** (Netherlands)
Agent:	**BORDEN LADNER GERVAIS LLP**
Issued:	**Oct. 25, 2005**
PCT Filing Date:	**Aug. 1, 1994**
PCT Publication Date:	**Feb. 15, 1996**
Examination requested:	**July 27, 2001**
International Class (IPC):	**G09B 9/02**
Patent Cooperation Treaty (PCT):	**Yes**

Source: Canadian Intellectual Property Office, Patent Database. The reproduction is not represented as an official version of the materials reproduced, or as having been made, in affiliation with or with the endorsement of CIPO.

Foreign Patent Protection

During the nineteenth century, patent protection was the subject of discussion at a number of meetings held by industrialized countries in an effort to devise a system whereby inventors might obtain patent protection for their inventions in countries other than their place of residence. Eventually, at the meeting in Paris, France, in 1883, agreement was reached whereby an inventor, who had applied for patent protection in his or her own country, could make an appropriate application in any other country that was a party to the agreement within 12 months after the original application. The application in the foreign country would have the same filing date as that of the first filing. This permitted any inventor residing in a country that belonged to the Union Convention of Paris, 1883, to obtain a uniform filing date in all countries where patent protection was applied for. For example, if an inventor applied for a patent in Canada on February 1 and for a patent on the same invention in the United Kingdom on June 1, the effect of the Convention would be that the inventor's application in the United Kingdom would be back-dated to the date of the original filing in Canada (i.e., February 1). Amendments to the *Patent Act* and a more recent (1970) international treaty now require a foreign applicant to claim treaty priority within 12 months of filing, and provide the Commissioner of Patents with information concerning the foreign application.[17]

Compulsory Licences

One of the obligations of a patentee is that he or she must work the invention to satisfy public demand for the new product or process. This is essentially a public duty that the

17. *Patent Act*, S.C. 1983, c. 15, s. 28.1, and the Patent Co-operation Treaty of 1970, of which Canada is a signatory.

patentee must perform in return for the grant of monopoly rights to his or her discovery. Most inventors are usually only too happy to perform this duty, either by the production of the product themselves or by licensing others to manufacture the product in return for a royalty payment. If the patentee so desires, he or she might also assign the patent rights to another, in which case the obligation to work the patent would shift to the assignee.

Compulsory licence

A directive by the Commissioner of Patents requiring the patentee to license others to produce a product for the general benefit of the public.

The Act provides that, for certain inventions, a **compulsory licence** may be in order for the general benefit of the public. Compulsory licensing may arise, for example, when the work of the patent is dependent upon the right to produce a part covered by an earlier patent. If such a licence is required to work the later patent (usually an improvement in some part of the original patent), the patentee of the improvement may apply to the Commissioner of Patents for the issue of a compulsory licence.

If a patentee fails to work a patent to meet public demand for the invention, or if the price for the product is unreasonably high, any interested party may apply to manufacture the invention under licence at any time after the patent has been in effect for three years. If the patentee cannot refute the claim that he or she cannot supply the demand for the invention, a licence may be issued to the applicant on whatever terms would appear reasonable in the circumstances. This usually means the issue of a licence to manufacture on a royalty basis. However, depending upon the circumstances, a failure or refusal to work a patent in the face of demonstrated public demand for the invention could also result in a revocation of the patent.

Infringement

Infringement

The unlawful interference with the legal rights of another.

The issue of a patent is essentially a grant of a monopoly to the patentee for a fixed period. During this time, the patentee, subject to certain public interest limitations, has the exclusive right to deal with the invention. The production of a product or use of the process covered by the patent by any person not authorized to use it would constitute **infringement**, and would entitle the patentee to take legal action against the unauthorized producer, user, or seller. Infringement is very broad in its application. It includes not only unauthorized production of the invention but the importation of the product, or any other working of the patent without the consent or payment of royalties to the patentee. The remedies available when a patent has been found to be infringed include not only money damages, but an injunction and an accounting for all profits made from the sales of products that infringed upon the patent.

To succeed against the unauthorized producer or importer of the invention, the patentee (assignee or licensee) must prove that infringement has taken place, since damages do not automatically flow from the production of an invention that is subject to a patent. For example, a defence against a claim for infringement might be that the patent is invalid, or that the patent had expired before the goods were produced. A patentee may also be faced with the defence of estoppel if the validity of the patent was successfully attacked by another person prior to the patentee's claim of infringement, or if the patentee had allowed the infringement to take place with tacit approval for some time before claiming infringement. Infringement cases tend to be very complex, and infringement itself is very much a question of fact; consequently, the defences can be many and varied.

FRONT-PAGE LAW

U of T files suit over drug patent

The University of Toronto has filed a lawsuit against Draxis Health Inc., a Mississauga-based pharmaceutical company, claiming $100-million in damages.

In its statement of claim, the university alleges that Draxis has ignored a patent held by two University of Toronto scientists on use of the drug deprenyl.

"We claim they are practising our patent," said George Adams, chief executive of the University of Toronto Innovations Foundation, responsible for protecting the intellectual property of university employees.

Yesterday afternoon, Draxis put out a statement saying that it considers the University of Toronto's claim to be without merit and that it will defend itself vigorously.

Mr. Adams said the dispute, which surfaced only recently, has been simmering for some time. He said the university's patent was filed back in the early '90s and concerned a discovery that deprenyl not only protects but actually rescues damaged brain cells in both animals and people.

This excerpt refers to "the university's patent" and "the patent held by two ... scientists." What sort of conflict over the patent could arise between the university and the scientists? How might an employer prevent such a conflict from arising?

Source: Excerpted from John Greenwood, *Financial Post*, "U of T files suit over drug patent," July 29, 2000, p. D3.

TRADEMARKS

Historical Development of the Law

A trademark is a mark that may be used by a producer or merchant to distinguish his or her goods or services from those of others. It may take the form of either a trademark or a trade name, but the purpose of the mark or name is the same: to identify the goods or services of the owner of the mark.

The use of trademarks to distinguish or identify the wares of a producer or seller would appear to be a practice with a long history. Early craftsmen, such as the brickmakers of early Babylon and the water-pipe manufacturers of ancient Rome, marked their wares with their distinctive symbols or signatures. Later, members of some of the early guilds established their own special marks for the goods they produced.

In England, one of the first recorded cases dealing with trademark infringement concerned an action in which a manufacturer of cloth claimed damages for deceit from another who had passed off his goods by marking the cloth with the plaintiff's mark.[18] The court found in favour of the plaintiff, but some doubt existed as to whether the plaintiff was in fact the cloth manufacturer or a purchaser of the cloth who was deceived by the improper use of the mark. At Common Law, if the plaintiff was the purchaser, the action was an ordinary action of deceit, but if the plaintiff was the manufacturer, the case would be Common Law recognition of the right of action for infringement. The matter remained confused for over two centuries following this initial case. However, the need for legislation to establish ownership of trademarks was finally recognized by 1875, when the *Trade Marks Registration Act*[19] was introduced to remedy the situation.

The Act provided that, by registration, the owner of a trademark would establish the *prima facie* right to use the mark exclusively. After five years, the right became absolute, so long as the owner used the mark in business. The Act set out the various requirements for registration and a general outline of the type of marks that were registrable.

Trademark legislation in Canada followed a similar pattern of development, with the first statute introduced in 1868.[20] The Act provided for the registration of marks under a procedure that gave the registered user the right to exclusive use of the mark. An unusual feature of the Act was the sharing of the fine for infringement on a 50–50 basis between the Crown and the party injured by the infringement. The Act, along with legislation pertaining to patents and copyright at the time, was placed under the administration of the Department of Agriculture.[21]

Trademark legislation passed through a number of amendments and changes during the latter part of the nineteenth century, with a complete revision of the act in 1879.[22] The revision repealed and replaced the prior legislation except for those acts dealing with the marking of timber,[23] and the law dealing with the fraudulent marking of merchandise.[24] While not following the English legislation in detail, the law in Canada underwent a

18. *Southern v. How* (1618), Popham 143, 79 E.R. 1243.
19. 1875, 38 & 39 Vict., c. 91.
20. The *Trade Mark and Design Act* of 1868, 31 Vict., c. 55 (Can.) (1868).
21. 31 Vict., c. 53 (1868) (Can.).
22. The *Trade Mark and Design Act* of 1879, 42 Vict., c. 22 (Can.).
23. *An Act Respecting the Marking of Timber*, 33 Vict., c. 36 (Can.).
24. The *Trade Mark Offences Act*, 1872, 35 Vict., c. 32 (Can.).

number of changes to expand the nature of the protection available to users of marks by including certification marks (marks used to identify goods or services produced or performed under controlled conditions, or of a certain quality), the importation of trade-marked goods, the licensing of users of trademarks by the "owner," and a revision of the penalties for unfair practices and infringement. The last major overhaul of the Act was in 1953, but, more recently, the legislation has been subject to minor revision.[25]

Trade Marks Act

The present *Trade Marks Act*[26] is a federal statute that governs the use of all trademarks and trade names in use in Canada. The Act defines a trademark as any mark "used by a person for the purpose of distinguishing or so as to distinguish wares or services manufactured, sold, leased, hired or performed by him from those manufactured, sold, leased, hired or performed by others."[27] The Act also provides for the registration of trademarks and maintains a register of marks at the Trade Marks Office. Protection under the Act is provided by a registration process, and any mark that is not descriptive, in use by another prior user, confusing with existing marks, and contrary to the public interest, may be registered.

At the present time, there are a number of other different types of marks that may also be registered under the Act:

(1) **Service marks:** marks that are used by service industries, such as banks, airlines, and trucking companies, where the principal business is that of providing a service to the public. The mark may also be applied, however, to any product that the user might sell, such as (in the case of an airline) flight bags or toy models of an aircraft.

(2) **Certification marks:** marks used to distinguish goods or services of a certain quality that, in the case of goods, are produced under certain working conditions or, in the case of services, performed by a certain class of persons, or goods or services produced in a particular area. Most certification marks are used for franchise operations where the owner of the mark does not produce the goods or perform the services directly, but merely sets and enforces the standard for the goods or services to which the mark is applied. Certification marks are essentially "quality marks."

(3) **Distinctive guise**: a trademark that takes the form of a particular shape, to distinguish it from the products of others. Distinguishing guises are generally in the form of the package, or the shape of the product itself, and may be protected by the Act. Utilitarian features of the guise, however, may not be protected — for example, a moulded or built-in handgrip on a bottle or box.

(4) **Trade name:** generally, a name coined or chosen to describe a business. It must not be a name that might be confused with any other name. As a rule, a person is not prohibited from using his or her own name simply because it is the same as that of a well-known establishment.

Trademark

- Kleenex® brand facial tissue is a trademark of the Kimberly-Clark Corporation.

Service mark

- Walmart℠ is a service mark of Wal-Mart Stores, Inc.

25. R.S.C. 1985, c. T-13, as amended.
26. ibid.
27. ibid., s. 2.

Certification mark

- The Canadian Standards Association issues its mark (a "C" enclosing the letters "SA") to manufacturers of goods compliant with its standards.

Distinctive guise

- The contour bottle is a trademark of The Coca-Cola Company.

Trade name

- Many "Smith's Grocery" and "Smith's Hardware" stores operate across Canada with no relation to one another, other than a common trade name.

A classic case on the right to a trade name or trademark was *Singer Mfg. Co. v. Loog*,[28] where the judge described the right to mark goods in the following language:

> ... no man is entitled to represent his goods as being the goods of another man; and no man is permitted to use any mark, sign or symbol, device or other means, whereby, without making a direct false representation himself to a purchaser who purchases from him, he enables such purchaser to tell a lie or to make a false representation to somebody else who is the ultimate customer. That being, as it appears to me, a comprehensive statement of what the law is upon the question of trade-mark or trade designation, I am of opinion that there is no such thing as a monopoly or a property in the nature of a copyright, or in the nature of patent, in the use of any name. Whatever name is used to designate goods, anybody may use that name to designate goods; always subject to this, that he must not, as I said, make directly, or through the medium of another person, a false representation that his goods are the goods of another person. That I take to be the law.

Registration Requirements

A trademark must be distinctive and used in order to be registrable under the Act. However, special provision is made for proposed marks that may be cleared as suitable trademarks, then later registered, once they are put in use. The Act does not permit all marks to be registered that are distinctive and used, as certain marks are prohibited by the statute. Prohibited marks are usually associated with royalty, governments, or internationally known agencies, and may not be used without their consent. A mark must also be such that it cannot be associated with any famous or well-known living person (or a person who has died within the previous 30 years), and it cannot be an offensive symbol. The "distinctive" requirement is often the most difficult to meet, as the mark must be such that it cannot be confused with the mark of another.

Marks that are searched in the register and found to be acceptable are advertised in the *Trade Marks Journal* to advise the public of the intended registration of the mark. If no objections arise as a result of the public notice, the mark may then be registered. If its distinctiveness is not challenged within the next five years, it becomes incontestable unless it can be shown that the applicant knew of other users prior to the application for registration. A mark that should not have been registered in the first place would, of course, be open to challenge as well. An example of the latter objection would be a mark that was later discovered to be an offensive symbol. Registration is valid for a period of 15 years, and the registration may be renewed, as long as the mark is in use.

28. (1880), 18 Ch.D. 395.

CASE IN POINT

Molson Breweries applied for trademark registration of the word "Export" for one of its beers. A competitor, John Labatt Limited, objected to the registration on the basis that the word "Export" was descriptive and generic, not distinctive. The word was commonly used in the brewing industry to describe a type, standard or grade of beer. Labatt's position was that Molson was using the term "Export" in conjunction with the Molson name and was descriptive of its beer, and that Molson should not be allowed to have exclusive right to use of the word. Other brewers had "Export" products as well, and Molson had not developed a stand-alone trademark. Molson entered evidence to suggest that the term had acquired a distinct status as "Molson," but the court felt the standard was not met, and the registration of the mark "Export" was refused. [*Willes — Perhaps "Molson's Ex" might have fared better?*]

Molson Breweries, a Partnership v. Labatt (John) Ltd. et al. (2000), 252 N.R. 91.

Enforcement

A person who has registered a trademark is entitled to protect the mark by taking legal action to prevent the use of the mark by another. The usual remedy is an injunction. However, if unauthorized goods or services were sold under the registered mark, an accounting for the lost profits due to the use of the mark also may be had. The forgery of a trademark, or the passing off of goods as being the goods of another, is also a criminal offence.[29] Criminal penalties may be imposed if criminal proceedings are taken against the unlawful user of the trademark.

If the trademark is no longer used, or has lost its distinctiveness because the product has become so successful that the name has become generic, the user may no longer claim exclusive rights to the use of the name. The trademark "Linoleum," for example, became the generic word through public use of the term to mean all floor coverings of that type, and the word lost its distinctiveness with respect to its user's product. The name "Aspirin" suffered the same fate in the United States, but to date it is a registered trade name in Canada. Users of well-known trademarks, as a result, are careful to guard their trademarks and names to prevent the word from being used by the news media as a generic term for all products of a similar type.

COURT DECISION: Practical Application of the Law

Trademark Validity—Infringement—Distinctiveness

***A & W Food Services of Canada Inc. v. McDonald's Restaurants of Canada Ltd.*, 2005 FC 406.**

Since 1987, A & W has sold a grilled chicken sandwich called a "Chicken Grill." It registered a trade-mark for that name and product in 1988. Since 2001, McDonald's has sold a grilled chicken sandwich which it calls "Chicken McGrill." A & W alleges that McDonald's has infringed its trademark, used a confusingly similar mark and preyed upon its goodwill in the marketplace. In turn, McDonald's claims that A & W's trademark is invalid because it lacks the essential element of distinctiveness.

THE COURT: Trade-mark owners have the right to use their marks exclusively, throughout Canada, for the products or services to which those marks relate, unless shown to be invalid (*Trade-marks Act*, s. 19). In this case, A & W clearly has the exclusive right to use its registered trade-mark "Chicken Grill" for a grilled chicken sandwich, unless McDonald's can prove that A & W's mark lacks the essential element of distinctiveness. A & W argues that McDonald's use of "Chicken *Mc*Grill" for a grilled chicken sandwich infringes its trade-mark because McDonald's mark is essentially the same; indeed, it wholly encompasses A & W's mark. I do not see in the Canadian cases any room for A & W's argument

29. *Criminal Code*, R.S.C. 1985, c. C-46, ss. 406–411.

that McDonald's use of "Chicken *Mc*Grill" infringes the trade-mark for "Chicken Grill." It seems clear that s. 19 of the Act deals only with identical marks. A & W argues that if McDonald's use of "Chicken *Mc*Grill" does not amount to trade-mark infringement, then McDonald's would be given a license to put the "Mc" prefix in front of any company's trade-mark and adopt it as its own. By that reasoning, McDonald's could sell a "McWhopper" or "McPepsi." This strikes me as a strong argument in principle, but not one that finds favour in any case law that was cited to me. According to current Canadian law, McDonald's use of "Chicken *Mc*Grill" does not infringe A & W's trade-mark for "Chicken Grill." The main issue, therefore, is whether there is confusion. Having examined the expert reports and testimony thoroughly, I cannot conclude that the survey evidence shows confusion. All that we know is that when people hear the name "Chicken *Mc*Grill," most of them will think of McDonald's. Further, I am struck by the fact that almost as many people thought that the "Chicken Grill" came from KFC or Burger King as from McDonald's. This is apparent from the raw figures in Dr. Corbin's study. Dr. Reich also testified that the "Mc" prefix has a clear, well-known and specific meaning: it means "from McDonald's." Finally, in Dr. Reich's view, the "Mc" prefix clearly distinguishes the "Chicken *Mc*Grill" from the "Chicken Grill" making it unlikely that consumers would confuse them. Overall, considering the evidence of both experts, I find that the "Mc" prefix is meaningful and, in the circumstances of this case, significantly reduces the likelihood of confusion about the source of each party's product. Both parties are in the same business, but target different segments of the population. A & W targets mainly adult consumers, while McDonald's focusses more on children. In my view, this reality makes it very unlikely that a person who purchases a "Chicken Grill" would think that the product comes from McDonald's even though the name is somewhat similar to McDonald's "Chicken *Mc*Grill." Nor would a consumer be likely to think that a "Chicken *Mc*Grill" comes from A & W. The competitive nature of this industry militates against any potential for confusion, whether forward or reverse. It follows from my analysis of the issue of confusion that there is little evidence here showing that consumers make a connection between the marks of the two parties. Further, A & W has not presented any evidence of damage to its reputation. The onus [*in proving the invalidity of the A & W mark*] lies on McDonald's to show that A & W's trade-mark lacks the essential characteristic of distinctiveness. A registered mark is presumed to be valid until proved otherwise. In my view, McDonald's has succeeded in showing that the "Chicken Grill" mark is not widely known or readily associated with A & W. However, this is not sufficient for purposes of s. 18(1)(*b*) of the Act. McDonald's has not shown that A & W's trade-mark is so devoid of distinctiveness that it fails to distinguish A & W's product from the wares of other restaurants. This court's judgment is that the plaintiff's action is dismissed with costs, and the defendant's counterclaim is dismissed with costs.

"Grill" is sufficiently distinctive that it shielded A&W from the McDonald's counterclaim, but "Grill" lacked some other feature that prevented A & W from using it as a sword. What is missing? As A & W, what argument would you raise against a McTeen Burger? *A bit more generically, what would Dairy Queen say about competing against a* McBrazier?

Foreign Trade Marks

As with patents, Canada is a member of an international convention concerning trademarks that permits a user of a trademark in a foreign country to apply for registration there. If an application is also made in Canada within six months thereafter, the application in Canada will be dated as of the date of application in the foreign country. Special filing requirements are imposed upon the foreign trademark applicant, however.

CASE IN POINT

A corporation applied for trademark registration of the name "Micropost" for its point-of-sale computer terminals, which performed a variety of cash-related functions. The "post" part of the name was an acronym for Point-Of-Sale Terminal. Canada Post Corporation opposed the registration on the basis that the word would be confusing, however, and would suggest that the terminals were in some way connected with Canada Post. The Registrar of Trademarks dismissed Canada Post's objection, and the issue was taken to the courts by Canada Post. The matter eventually went to the Federal Court of Appeal.

The Federal Court of Appeal dismissed Canada Post's opposition to the registration of the trademark on the basis that the Canada Post monopoly on the word for postal services did not go beyond that, and did not extend to the use of the word "post" in connection with equipment or other services.

See: *Canada Post Corp. v. Micropost Corp.* (2000), 253 N.R. 314.

FRANCHISES

Franchise

A business relationship that licenses the use of trade names, trademarks, and operating procedures to operate a similar business.

A **franchise** is a special business relationship that is founded on a contract between a franchisor that operates a particular type of business and a franchisee, whereby the franchisee agrees to operate a similar business using the trade names, trademarks, copyright material, and products or services of the franchisor under licence and in accordance with the terms of the agreement.

Franchises are most often found in the retail area, and are adaptable to many types of business operations. Some of the most common are fast-food restaurants, grocery stores, automotive parts and accessory stores, and other retail establishments where the business name and quality or selection of the products sold may be important or attractive to the customer.

Some Well-known Franchises in Canada[30]
(Additional corporately-owned outlets in parentheses)

• Canadian Tire	475 franchises since 1934 (0)
• Tim Hortons	3000 franchises since 1964 (18)
• McDonald's	1136 franchises since 1968 (297)
• Mister Transmission	85 franchises since 1969 (0)
• East Side Mario's	93 franchises since 1979 (5)
• First Choice Haircutters	190 franchises since 1982 (173)

A franchise is usually based upon a lengthy and sometimes very complex contract under the terms of which the franchisor agrees to provide the franchisee with a licence to use the business name and/or brand names of the franchisor's business and products as well as the franchisor's operating systems and procedures. Much of the agreement concerns the operation of the business, and even though the franchisee is an independent business person, the franchisee must agree to operate the business in strict accordance with the agreement.

CHECKLIST OF COMMON FRANCHISE CLAUSES

1. The franchisor grants the franchisee an exclusive licence to use the trade name and trademarks of the franchisor's products or services within a defined geographical area.
2. The franchisor agrees not to grant a similar franchise to any other person or corporation within the defined area of the franchise without the consent in writing of the franchisee.
3. The franchisor agrees to supply the franchisee with complete operating-system manuals and training for the franchise operation, and in some cases may agree to oversee the construction or preparation of the building for the operation of the business. The franchisor may also provide management services on a fee basis.
4. The franchisor agrees to supply the products used in the operation of the business, or to identify the suppliers of products for the business, and the franchisee agrees to purchase only from the franchisor or as directed by the franchisor.
5. The franchisee agrees to keep confidential all operational information (including operating manuals) and further agrees that the confidentiality agreement will continue to bind the franchisee after the franchise has been terminated.
6. The franchisee agrees to permit the franchisor to inspect the books and records of the franchisee, and to make operational inspections to ensure that the franchise is operated properly.

---

30. Source: Canadian Franchise Association, www.cfa.ca, August 2011.

In addition to these basic clauses, franchise agreements may contain many other clauses related to the operation of the franchise and the rights of the parties. These clauses may require the franchisee to contribute to advertising promotions, general product or trade name advertising, to lease or sub-lease premises from the franchisor, or to provide particular services to the public.

Payment for the franchise usually consists of an upfront fee payment to the franchisor that may range from a relatively modest amount to hundreds of thousands of dollars. In addition to this payment, the franchisee also agrees to pay a royalty to the franchisor for the use of the trade names or trademarks. This royalty fee is usually based upon monthly sales revenue, and is paid on a monthly basis to the franchisor.

Because franchise agreements have from time to time imposed onerous terms on franchisees, several provinces (e.g., Ontario[31] and Alberta[32]) have passed legislation that requires franchisors to be more fair and open in their dealings with franchisees.

MANAGEMENT ALERT: BEST PRACTICE

Potential franchisees should exercise guarded business judgment in taking up a franchise contract. Most are heavily one-sided agreements favouring the franchisor, despite provincial legislation aimed at greater fairness. Even if the business is profitable, it may be the franchisor that ends up with the largest share, from revenues off the top and from payments among the expenses. A legal right to use the franchise does not guarantee the size of the bottom line, and it all turns on the fine print.

COPYRIGHT

Copyright is a term that means what it says: the "right to copy." The law pertaining to copyright is concerned with the control of the right to copy. It recognizes the right of the original creator of any writing or artistic work to control the reproduction of the work. Included in the type of work that copyright covers is all writing in the form of books, articles, and poems, as well as written work of every description, including musical compositions (both music and lyrics), dance choreography, live performances by performers and musicians, sound recordings and broadcasters' signals on both radio and television, and dramatic works. The right also covers all forms of artistic work in the nature of sculpture, paintings, maps, engravings, sketches, drawings, photographs, and motion pictures, including video recordings. Because reproduction in the case of music and dramatic works involves, in many cases, the recording of the music or work on phonograph record, tape, or film, the right extends to the right to record the work by electronic or mechanical means. The same holds true for the reproduction of any literary or other work photocopied or stored in a computer-retrieval system. Copyright protection also extends to computer programs that may not be reproduced (except for the making of a single backup copy by the owner of the copy of the program).[33] Subject to certain exceptions, the law protects the original author's right to control all reproduction of his or her work.

Historical Development of the Law

At Common Law, the right of the author of a literary or artistic work was not entirely clear. This was due in part to the fact that reproduction of a literary or artistic work prior to the invention of printing was a laborious undertaking, and so issues of copyright were not likely to be of much concern before the courts. After the invention of the printing press, written work acquired a special commercial value, and the question of ownership of written work and the right to reproduce it became important (at least from the author's point of view). The establishment of a printing press and the printing of the first book in

31. *Arthur Wishart Act* (Franchise Disclosure), S.O. 2000, c. 3.

32. *Franchises Act*, R.S.A. 2000, c. F-23.

33. *Copyright Act*, R.S.C. 1985, c. C-42, as amended.

England in 1477 by William Caxton created the need for a law that would determine the right of an author to control the reproduction of his or her work.

At Common Law, the right of an author to control unpublished work was generally settled: the author was entitled to do as he wished with the material, because it was his, and his alone.[34] The right to control the copy of the work once published, however, was another matter. On this point, the law was unclear, so much so that many legal authorities believed that the right did not exist once the work was published, as publication was, in a sense, placing the work in the public domain.

In 1709, the first statute was passed to establish the rights of authors to their published works.[35] A series of statutes followed; they extended copyright to musical and dramatic works, and, later, to photographs and other products of the advancing technology of the nineteenth and early twentieth centuries.

After the *British North America Act*, 1867, Canadian legislation pertaining to copyright was passed from time to time in the years that followed, but it was not until 1921 that a comprehensive law was introduced.

The 1921 Act,[36] which came into force in 1924, covered virtually all literary and artistic endeavour, and provided for a Registrar of Copyrights and a Copyright Office. The Act provided for the registration of copyrights and set out an elaborate list of material that was subject to copyright. It also included penalties for the infringement of copyright.

In 1988, new legislative changes were introduced that updated the law pertaining to copyright. The new changes addressed some of the technological advances that have had an impact on the reproduction of copyrighted work, and included new types of work that required protection under the Act.

In 1997, changes in the legislation protected copyright in live performances by performers and broadcasters' communication signals, as well as sound recordings, in order to bring Canadian copyright law in line with the 1961 Rome Convention to protect performers, broadcasting organizations, and the producers of phonograms. The Act also provides an organizational structure for the collection of fees to compensate performers, musicians, and recording companies for the recording of their works by individuals for personal use by way of a levy on blank recording tapes. In an effort to balance the rights of copyright owners and the access to information by the public, educational institutions, libraries, museums, and archives (as well as by persons with perceptual disabilities) are given limited rights to copy material for research purposes, for protection of originals from excessive wear, for individual study, and for other purposes outlined in the Act. The right to copy by these institutions and persons is clearly delineated and must be carefully adhered to, otherwise the reproduction or copying would violate the Act.

Amendments to the Act also modernize the copyright-registration system and provide remedies for exclusive distributors of books against importers of parallel books.

The *Copyright Act*

The present legislation[37] provides that the sole right to publish or reproduce an original work of a literary or artistic nature is in the original author of the work. The protection of the right extends for the life of the author and for 50 years after the author's death. Work not published during the author's lifetime is subject to a copyright for a period of 50 years. In the case of recorded works, copyright runs for a period of 50 years from the date the recording is cut or first made. Registration of the copyright is not essential in order to claim copyright. However, registration is public notice of the copyright. It becomes proof of ownership of the work if the author should be required to bring an action for damages against a person who copies the work without permission.

---

34. *Jefferys v. Boosey* (1854), 4 H.L. Cas. 815, 10 E.R. 681.

35. *An Act for the Encouragement of Learning by Vesting Copies of Printed Books in the Authors and Purchasers of Such Copies*, 8 Anne, c. 19.

36. The *Copyright Act*, 1921, 11 & 12 Geo. V., c. 24.

37. *Copyright Act*, R.S.C. 1985, c. C-42, as amended. At the time of writing (2011) significant amendments to Canadian copyright law are expected to soon be forthcoming.

It is important to note that the author is the first owner of the work and entitled to claim copyright in it, unless the author was employed by another for the purpose of producing the work, painting, photograph, etc., provided that the parties did not agree to the contrary. Only the arrangement of the words or the expression of the idea is subject to copyright, however; the idea or the subject matter of the work is not subject to protection. For example, two authors might each write an article for a magazine dealing with energy conservation, each article containing the same ideas or suggestions. Each would be entitled to claim copyright in the arrangement of the words, but the ideas contained in the articles, although identical, would not be subject to a claim of copyright by either of the writers.

Protection of copyright usually takes the form of registration of the work and the marking of material by the symbol ©. This is followed by words to indicate the date of first publication and the name of the author. Canada has been a member of the Universal Copyright Convention since 1962,[38] and the marking of published material in this manner is notice to all persons in those countries that are a part of the convention that copyright is claimed in the marked work. The enforcement or protection of the copyright is the responsibility of the owner, however.

An author or artist is entitled to assign a copyright, either in whole or in part, to another person. However, to be valid, the assignment must be in writing. Assignments of copyright are normally registered as well, to give public notice of the assignment. The Act also provides for the issue of licences to print published works in Canada where a demand exists and the author has failed to supply the market. When a licence is issued, the publisher is expected to pay a royalty to the author as compensation. If a work is printed or copied without either the permission of the author or a licence, the reproduction of the work may constitute infringement. It may also expose the unauthorized publisher to penalties under the Act and/or an action for infringement by the holder of the copyright.

Infringement

Infringement consists of unauthorized copying of the protected work except for "fair dealing" with the work by others for the purpose of private study, research, criticism, review, or newspaper summary. Certain other exceptions exist as well, such as when short excerpts from a copyright work are read in public, or where the work is performed for educational or charitable purposes by unpaid performers.[39] In the case of infringement, the copyright owner is usually entitled to an injunction and an accounting, as well as damages.

Infringement action may also arise when the moral rights of the author have been affected by persons dealing with the author's work.[40] In particular, the author has a "moral right" to have his or her name associated with the work. Under the Act this is described as a "right to the integrity of the work." The author also has the moral right to have a pseudonym associated with the work, or to remain anonymous. The right to integrity also encompasses a right of action if distortion or mutilation of works would prejudice the honour or reputation of the author.

Moral rights may not be assigned by the author, but may be waived and pass on the author's death to his or her beneficiaries under a will or to heirs-at-law. Infringement of a moral right of the author would entitle him or her to take legal action against the violator for damages.

Plagiarism
The taking of the copyrighted work of one person by another.

If a person takes the work of another person and claims it to be his or her own, the act of doing so not only violates the *Copyright Act*, but constitutes **plagiarism**, as it is essentially the theft of the copyrighted property of the author of the work. Plagiarism is a tort actionable at law, and may constitute a criminal offence as well.

Electronic Copyright Infringement

The ease of cutting and pasting or downloading from a Web site can create the mistaken impression that these activities are exempt from copyright. Just as with any other copying

38. Canada is also a member of the Berne Copyright Convention, which provides an author with protection in member countries.
39. See, for example, *Copyright Act*, R.S.C. 1985, c. C-42, Part III.
40. ibid.

technique, however, they may be done only with permission of the copyright holder or his or her agent, and often only on the express terms of a royalty agreement that provides for payment in return for the copy. Unauthorized duplication constitutes copyright infringement, depriving the true owner of his or her property or royalty fee.

Some Web content is exempt from infringement and may be freely duplicated. These are postings made by the creator of that content, or someone who has legally obtained the rights to the content from its creator, who then asserts no copyright, with the intention that the work will be freely disseminated. In other cases, while copyright may be claimed, the content owner may limit that to the moral right to be identified as the author of the work, and/or grant (or even encourage) its free distribution. This latter case occurs often with the distribution of new, "limited-feature" software, in the hope of gaining widespread public acceptance and leading to sales of a more fully functional version. Even then, a free copy made by an individual and then resold would be an infringement of the creator's copyright.

One of the most problematic aspects of the Internet is the fact that a virtually anonymous individual can infringe on a copyright by posting, for example, a hit song or other written work, giving the means for millions of other innocent (or not-so-innocent) parties to make copies of it. The original owner of the work suffers a massive loss, which is difficult to recover from one wrongdoer, and he or she is left with the impossible task of chasing countless eventual recipients.

A satisfactory solution is a long way off. An early belief was that the "host" Web site, or Internet Service Provider (ISP) should be held liable. This reaction is impractical, however, and would be the equivalent of holding a courier company liable because criminals forwarded stolen goods inside the parcels. In the United States, a compromise has been reached, providing for liability when an ISP continues to host infringing content after being served notice of the fact by the copyright holder. We can expect that this lead will be followed in Canada. ISPs that do more than passively host content, and actively participate in what ought to be known as infringing activity, also risk being held liable.

In the past decade, abortive attempts have been made to revamp aspects of Canada's copyright law to better cope with the challenges and opportunities afforded by the digital age. Current efforts are likely to be aimed at resolving the following priority issues:

- Posting copyright material online without authority would be an offense.
- A fair dealing exemption for making backup copies and non-commercial use.
- Prohibitions against circumventing copy protection measures.
- Responsibilities of Internet Service Providers when advised of infringing content.
- Data retention and disclosure obligations of Internet Service Providers.

Quite apart from artistic works, is the question of treatment of simple Web site content. This too is showing itself, quite properly, to be within the protection of the court, as the following case indicates.

A real estate association posted listings copied from the Web site of another firm. The association sought an injunction to stop the defendant from continuing the act. It argued a breach of contract case based upon the Terms of Use indicated on its Web site. The Quebec Supreme Court upheld the Terms of Use of which the defendant should have been aware, and issued the injunction.[41]

41 *The Canadian Real Estate Association/L'Association Canadienne d'immeuble v. Sutton (Québec) Real Estate Sevices Inc.* (2003), CanLII 22519 (Que. S.C.).

CASE IN POINT

Google, the online search engine giant, is developing its "Library Project." It is scanning hundreds of books whose copyright has expired to provide full text, free online. Its ambitions are reported to include providing brief extracts of books under copyright, with links to booksellers. Four lawsuits by U.S. publishers have resulted thus far, based on Google's claim that, in order to properly index the book and provide a small segment, it must first digitize the entire book. Each alleges that harm will result if Google makes such copies of copyright material. The case poses the interesting question of what harm would result if Google's files are hacked and these full manuscripts are illegally distributed.

Follow developments in: *McGraw-Hill, Pearson, Penguin, Simon & Schuster, and John Wiley v. Google Inc.*, United States District Court for the Southern District of New York, Docket 05 CV 8881 and future rulings at www.nysd.uscourts.gov.

Performing-Rights Societies and Collective Societies

Performing-rights societies obtain by way of assignment the performing rights to musical and dramatic works, and in turn grant performing licences for a fee to organizations that may wish to perform the works (such as a particular play or musical). The fees collected are paid in part to the copyright owner, and the balance is retained by the society to cover its operating expenses. The fee schedules of performing-rights societies are filed with the Copyright Board and are subject to public review through advertisement in the *Canada Gazette*. The fees are finalized after the public has had an opportunity to review the schedule, and remain in effect for a prescribed period of time (usually one or more years).

Collective societies are societies that operate licensing schemes for the performance of works in places other than theatres. These societies acquire the rights to a repertoire of a performer's performances, sound recordings, and the communication signals of broadcasters. The collective will license the performance of the works in return for a royalty or fee. Collective societies must also provide the Copyright Board with their tariff schedule for approval.

A QUESTION OF ETHICS

Copyright law and enforcement can never keep pace with copying techniques. What can a copyright holder do in the Internet age? Sue everyone, and get a dollar from each for the trouble? Success in one prosecution may mean the biggest offenders go to another country, and everyone still gets a copy. What is the solution — an Internet tax, a hard-drive tax, higher music prices, or an international agreement? Is the hope that consumers will behave ethically simply naive?

INDUSTRIAL DESIGNS

Industrial-design legislation[42] applies to certain artistic works produced by an industrial process.The Act defines a registerable design as one that has in a finished article features of shape, pattern or ornament, configuration, or any combination of these "that appeal to and are judged solely by the eye."[43] An industrial design is normally a design that would be the subject matter of copyright if it was not for the fact that it is reproduced by an industrial process. For example, new furniture designs would require registration under the *Industrial Design Act* in order to be protected. Not all products must be registered, as the Act does not apply to some "artistic" products produced by industrial processes. The Act also does not apply to features that are purely utilitarian, such as a handle on a container. However, the design of the handle may be protected if it is unique in appearance.

42. *Industrial Design Act*, R.S.C. 1985, c. I-9.

43. *Industrial Design Act*, R.S.C. 1985, c. I-9, s. 2.

Registration gives the owner of the design exclusive rights to produce the design for a period of ten years. The design must be original, however, and not something that is likely to be taken for the design of another. In this respect, the requirements for registration are similar to those for a patent, but the investigative process is not nearly so exhaustive. The design also must be registered within 12 months of its first publication in Canada in order to acquire protection.

Once the design is registered, the owner of the design is obliged to notify the public of the rights claimed in the design by marking the goods (or by printing a label with "Rd"). The date and the design owner's name should also appear.

The ownership of the design, as with patents or copyright, may be assigned, or rights to manufacture may be granted under licence. Any unauthorized manufacture, however, would entitle the owner of the design to take legal action for infringement.

Licence Agreements

The licensing of intellectual property is a common means to commercialization. If a manufacturer has protected its product by way of a patent or trademark, or has a copyright or design right on a work, other firms with the facilities to produce and market the product may be licensed to do so. The licence is usually granted in return for payment of a royalty calculated on the basis of production or sales.

CHECKLIST OF COMMON LICENSING CLAUSES

1. The names of the parties.
2. The ownership of the patent, design, trademark, or other rights subject to the licence, and an acknowledgement of the ownership by the other party to the agreement.
3. The royalty rate and its method of calculation.
4. The quantities to be produced and the quality standards.
5. The duration of the licence agreement.
6. The method of dispute resolution should a dispute arise.
7. The disposition of stock and special equipment used in the production of the product on termination of the agreement, or the ownership of plates, moulds, or masters for copyright works when termination arises.
8. Technical or other assistance provided to the licensee.
9. The right to assign or sub-license by the licensee.
10. The territorial boundaries where the licence covers sale as well as manufacture (to protect both the domestic market and other licensees).
11. The protection of confidential information and manufacturing know-how not protected by patent.
12. The right to improvements in the product made by both the licensor and the licensee.

Often licensing is undertaken because the licensor has limited skill or interest in the undertaking. For example, the Walt Disney Company devotes its energies and expertise to producing animation and movies. Merchandise of all kinds are a profitable spin-off of its movies, but Disney has no interest or ability in operating a thousand factories turning out everything from lunchboxes to toys and beach towels. Licensing is the perfect solution.

The licensing arrangement has become an attractive business model because it usually requires a minimum investment in time and expertise on the part of the licensor. Apart from the negotiation and legal work associated with the preparation of the licensing agreement, the licensor's obligations are generally limited to monitoring the agreement and providing technical assistance to the licensee. The capital investment in plant and equipment, recruitment of personnel, and compliance with foreign laws are the responsibility of the licensee. Nevertheless, licence agreements are not without some disadvantages. Royalty arrangements may not produce the same levels of profits for the licensor that might otherwise be obtained (such as through a joint venture) because the licensee may not operate an efficient manufacturing facility or sales force. Under these circumstances, the licensor would be unable to directly control or correct the problems affecting the overall profitability of the venture.

SUMMARY

The protection of rights to intellectual and industrial property is covered by a number of federal statutes. Each Act recognizes and protects a special property right and confirms ownership rights in it. A patent, which is the exclusive right of ownership granted to a first inventor of a new and useful product, lasts for 20 years from time of filing. During this time the holder of the patent (subject to certain exceptions related to the public interest) is granted monopoly rights in the invention. Patents, in some cases, are subject to compulsory licensing requirements, and may also be revoked if they are not "worked" to meet public demand. Once a patent is issued, unauthorized production of patented works constitutes infringement and would entitle the patentee or those claiming rights through the patentee to bring an action for damages against the unauthorized producer.

Trademarks are rights to the exclusive use of marks that distinguish the wares or services of one person from the wares or services of another. Under the *Trade Marks Act*, distinctive marks used to identify a person's product or service may be protected by registration. Registration permits the user of the name or mark to prevent others from using the mark without express permission. The unauthorized marking of goods by a person, for the purpose of passing them off as being those of the authorized owner of the trademark, constitutes the criminal offence of "passing off." It would also leave that person open to an action for damages and an injunction by the owner of the mark.

Copyright is the "right to copy" original literary or artistic work. Copyright legislation recognizes the author or composer of the work as the owner of the copyright and the person entitled to benefit from any publication of the material. The statute provides the exclusive right in the owner for the owner's lifetime plus 50 years; however, in the case of some types of copyright material, the right is limited to only 50 years. A copyright may be assigned or licences may be granted for copyright work. However, any unauthorized publication or performance of the work (subject to certain exceptions) constitutes infringement and would entitle the owner of the copyright to receive the profits on the unauthorized publication, damages, and an injunction, depending upon the circumstances.

An industrial design is somewhat similar to copyright in the sense that it is an artistic work that is reproduced by an industrial process. Registration of the design protects the owner from unauthorized reproduction of the same design by others. An industrial design is protected for ten years from the date of registration.

A licence agreement is often desirable to commercialize intellectual property. The licence contract conveys a right to use the property in return for a royalty payment, without selling the property outright. It thus allows the licensor to retain ownership and control while having one or more licensees rapidly produce and distribute products and services on its behalf.

KEY TERMS

Patent, 507
Trademark, 507
Copyright, 507
Trade secret, 508
Non-disclosure agreement, 508
Compulsory licence, 514
Infringement, 514
Franchise, 520
Plagiarism, 523

Review Questions

1. Explain the meaning of "patent pending."
2. Outline the steps that a Canadian inventor must follow in order to obtain patent protection for an invention in a foreign country.
3. How is the public interest protected under patent legislation?
4. What steps must an inventor follow in order to acquire patent protection for an invention?
5. For what public purpose did the Crown originally grant monopolies for certain products?
6. If an inventor had reason to believe that someone was producing a product that infringed on his or her patent, what would the inventor's rights be? What remedies are available for infringement?
7. Under modern patent legislation, what is the purpose of granting a patent for a new product?
8. Describe briefly the purpose of trademark legislation. Why has it been necessary?
9. How does a trademark differ from a trade name?
10. Distinguish between a service mark and a certification mark.
11. Explain the term "distinguishing guise."
12. What must a person who has a proposed mark do in order to establish rights to the mark?
13. What constitutes infringement of a trademark? What steps must the owner of a trademark take in order to prevent further infringement?
14. Outline the purpose of copyright legislation. What type of work is it intended to protect?
15. How is notice of copyright usually given?
16. What defences may be available to a person who is accused of infringing on a copyright work?
17. Where infringement is established, what remedies are available to the owner of the copyright?
18. What is an industrial design? How does it differ from copyright?
19. Explain the protection that an industrial design offers the owner of the design. How is this enforced?

Mini-Case Problems

1. A engaged the services of B, a professional photographer, to take a series of photographs of his power boat. B did so and was paid $200 for his services. Some time later, A discovered that B had sold one of the pictures to a boating magazine for the cover of one of its issues. B received $500 from the magazine for the picture.

   Is A entitled to the $500?
2. X produced a cola beverage that he sold for many years under the trade name Krazy Kola. If Y decided to produce and sell a cola beverage in the same area under the name Crazy Cola, would Y's actions constitute a violation of X's trade name?
3. Kitchen Products Inc. produced a unique design utensil that would peel, core and slice apples. It applied for a patent, but before the patent could be issued, a competitor copied the design and flooded the market with its copy of the product.

   Advise Kitchen Products Inc., assuming that its application for a patent was valid, and a patent would eventually be issued.
4. Furniture Design Ltd. produced a modified French Provincial chair. It was concerned that competitors might copy it once it was put on display at an upcoming furniture trade show.

   How could Furniture Design Ltd. protect its product?

Case Problems for Discussion

CASE 1

Karen wrote a novel that she self-published. She offered her novel for sale using an advertisement in a literary magazine. The advertised price of the novel was $9.95, providing her with an after-cost profit of $2.00 per copy. The novel sold well, and had received favourable reviews by book critics. A year later, she noticed that sales of the novel were falling rapidly, and she decided to investigate. To her surprise, she found a small website that was offering her novel for sale in a downloadable form for $5.00 a copy. The website proudly announced that it had sold over 10,000 copies of the novel to date.

Advise Karen. What would be the nature of her claim, and how would the court likely decide the case?

CASE 2

Dimitri carried on business as an independent consulting engineer. In addition to providing the usual engineering work for firms, he also designed a number of production processes and unique production equipment used in the production processes. He obtained patent protection on both the production process and the equipment. He arranged with a manufacturer to produce the 'package,' which he would supply and license users to use in their production of goods.

Several years later, the manufacturer of Dimitri's equipment carefully examined his design, and developed a new and more efficient type of machinery that would perform the same work as that of Dimitri's equipment. The manufacturer applied for a patent on the equipment, and when Dimitri discovered that the manufacturer had designed a new product, he contacted the manufacturer and informed him that he would not permit the manufacturer to use the new design with his production process.

Discuss the arguments of the parties. How is the dispute likely to be resolved?

CASE 3

Grigori Denton, an electronics engineer, worked on the development of a miniature hearing aid in his spare time. After much experimentation, he was successful in developing what he wanted. Denton was a member of a local service club that frequently assisted persons with hearing problems. For a special meeting of the club, he was invited to give the members a short lecture on hearing aids and a demonstration of how his device operated. At the meeting, he described how the device was constructed and demonstrated its effectiveness even though it was still in the experimental stage. The meeting was later reported in the local newspaper, along with some general information on Denton's presentation at the meeting. Another member of the club, who was also an electronics engineer, wrote a brief note on Denton's presentation and submitted it to a scientific journal that subsequently printed the note in its "New Developments" section.

Several years later, Denton finally perfected his hearing device and applied for a patent. He then set up facilities for its production, marking each unit produced with the words "patent pending." The product sold well in all parts of the country except British Columbia. When Denton investigated the market in that area, he discovered that a west-coast manufacturer was producing hearing aid units that incorporated the particular design that he had developed. His competitor had been selling the similar models for almost a year before Denton had gone into production. Unknown to Denton, the manufacturer had apparently developed his own hearing aid model from information that he had read in the scientific journal report of Denton's presentation to his service club.

Denton had the Patent Office expedite his patent application. On its issue, he instituted legal proceedings against the west-coast manufacturer for infringement.

Discuss the arguments that might be raised by the parties in this case. Render a decision.

CASE 4

Pia Myers, a part-time news writer for a local newspaper, attended an air show at a local airport. While watching two aircraft performing synchronized aerobatics, she noticed that the wings of the two aircraft were exceptionally close to each other. She photographed the aircraft at the instant that the two aircraft collided and took a second photograph of the pilots as they parachuted to the ground. Myers wrote a brief description of the accident and submitted the two pictures and the written material to the local newspaper for publication. The pictures and the report were published in the next edition of the newspaper, in which she received credit for the pictures and the story in a byline. She was paid her regular rate for the written material, and $50 for each picture. Myers later submitted the same pictures and story to an aviation magazine, and the material was subsequently published. Myers was paid $300 for the pictures and story by the magazine.

When the newspaper discovered the magazine article it instituted legal proceedings against the magazine and Myers, for copyright infringement, claiming that the copyright belonged to it.

Discuss the arguments that might be raised by the parties. Indicate how the case might be decided.

CASE 5

Holtzkopf, the President of Holtzkopf Furniture Design Ltd., a furniture manufacturer, engaged the services of Adrienne, a professional photographer, to attend a furniture exhibit and take a number of photographs of a particular chair that Larsen, a competitor, had on display. Adrienne did so and delivered the photographs to Holtzkopf, along with her invoice for $500. Holtzkopf paid the invoice and began the production of a chair that was very similar in appearance to the competitor's chair, but that had a different structural design beneath the fabric outer cover.

Holtzkopf advertised his chair in a trade magazine. He used several of Adrienne's photographs (which were black-and-white prints) because, although the chairs were similar in appearance, the exact design and colour of the fabric could not be ascertained from the photographs to identify the chair as Larsen's.

Larsen noticed Holtzkopf's advertisement and brought an action against him for violation of the industrial design that he held on the chair. Holtzkopf's defence was that he did not copy the design, he only used Adrienne's photograph of the chair, which was too small to permit him to make an exact reproduction of the design. He also argued that the structural design of his chair was completely different as well.

When Larsen discovered that Adrienne had photographed his chair, he included her as a co-defendant with Holtzkopf, claiming that she was a part of a conspiracy to infringe on his industrial design.

Adrienne had also noticed the advertisement in the trade magazine and determined that Holtzkopf had used her photographs of the chair in the advertisement. She immediately brought an action against Holtzkopf for infringement of her copyright in the pictures that she had taken of Larsen's chair.

Discuss the issues raised by the facts and the arguments that the parties might raise. Indicate how the court might decide the matter.

CASE 6

For several years, the residents of Smallville had been served by one prominent pizza franchise, Lotza Pizza, which had operated in Smallville with the telephone number 456-1010. This telephone number corresponded to that of the parent company, which was 123-1010 and was a registered trademark of the company. The telephone number figured prominently in the company's advertising and jingles.

A rival franchise of a competing pizza company then moved into Smallville and established a similar operation. That company, Better Pizza, also had a trademarked telephone number, which was 222-0234. Just like Lotza Pizza, the telephone number played a large role in the company's promotions and was one of the major factors of customer recognition. In Smallville, Better Pizza had obtained the number 457-0234. Smallville had only two exchanges, 456 and 457.

Quon, a resident of Smallville, who had the telephone number 456-0234 for almost 15 years, began to receive a large number of inadvertent calls intended for Better Pizza. Quon soon tired of receiving these calls and approached the local Lotza Pizza franchise. He told the manager about his telephone number and the recurring problem. Shortly thereafter, the Smallville Lotza Pizza franchise acquired Quons' telephone number 456-0234 and used it in its business.

The local franchise of Better Pizza, upon discovering the use by Lotza Pizza of the number 456-0234, brought legal action against both the Smallville Lotza Pizza franchise and its parent company.

Discuss the nature of the action and the rights and liabilities, if any, of the various parties involved. What arguments and/or defences may be used, and what would be the likely outcome?

CASE 7

Natasha Holdsworth produced a beautiful drawing of a French Provincial loveseat at the request of Classical Furniture Manufacturing Company. Classical Furniture paid Holdsworth $500 for the drawing and used it as the design for its loveseat in a current furniture collection. Six months after it acquired the drawing, and several months after it produced its first production models of the loveseat, the company applied for registration of the design.

Shortly after Classical Furniture registered its design for the loveseat, it discovered that Antique Furniture Co. had a similar loveseat on display in its collection at a furniture exhibit. Classical Furniture immediately accused Antique Furniture of copying its design.

As a defence, Antique Furniture Co. argued that the design was not original. It also came out in the course of discussion that it had acquired its own design by purchasing it from a designer by the name of Holdsworth.

If Classical Furniture should institute legal proceedings against Antique Furniture, what arguments might the parties raise on their own behalf. What is the position of Holdsworth, and what are her rights (if any) or liability (if any)?

Speculate as to the outcome of the action.

CASE 8

The Cod Oil Drug Company produced a concentrated vitamin product that it sold in capsule form to its customers. To identify its products, it produced its capsules with three broad red bands — the first bearing the letter "C," the second, an "O," and the third, a "D." Each letter was printed in white against the red background to identify the company and its product that was derived from cod liver oil. The centre red band, bearing the letter "O," also acted as a seal that held the two parts of the capsule together. The Cod Oil Drug Company applied for a patent on the method of sealing the two parts of the capsule together, and for registration of the three bands with the letters imprinted as a trademark for its product.

In the course of its application for a trademark, Cod Oil Drug Company was faced with an objection to its use of the trademark by Careful Drug Company, a competitor that produced its product in capsule form and bearing two blue bands, one on each part of the capsule, the first bearing a white "C" and the second, a white "D" against the blue bands. A thin blue band was used to join the capsule together, but it bore no letter. The design had been used by Careful Drug for many years before Cod Oil Drug developed the marking of its capsules.

On what basis would Careful Drug argue that the trademark should not be issued? What might Cod Oil Drug argue in response? How successful would the patent application likely be?

The Online Learning Centre offers more ways to check what you have learned so far. Find quizzes, Weblinks, and many other resources at www.mcgrawhill.ca/olc/willes.

PART 6

SPECIAL LEGAL RIGHTS AND RELATIONSHIPS

CHAPTER 27

Consumer-Protection Legislation

Chapter Objectives

After study of this chapter, students should be able to:

- Describe consumer safety, information, and product quality/performance protection.
- Explain consumer protection related to business practices.
- Discuss credit-granting and credit-reporting consumer protection.

YOUR BUSINESS AT RISK

In an effort to balance the commercial playing field, our governments have armed consumers with significant legal rights. Business owners must recognize and accommodate these rights. If the firm fails in its responsibilities, or takes advantage of consumers, it will be called to account with significant remedies and penalties available to both consumers and government. Moreover, breaches of consumer-protection legislation are usually well-publicized, creating a severe penalty: damage to the public image of the firm, which is sometimes irreparable.

INTRODUCTION

Business organizations, for the most part, attempt to establish sound and ongoing relationships with their suppliers and customers, because continuing relationships with customers and suppliers represent the most efficient and profitable way for a firm to operate in today's competitive market. To achieve this goal, most businesses follow policies of fair dealing and honesty in their contractual relationships with customers and the advertising of their goods or services. However, not all business organizations adhere to the high ethical standards of fairness and honesty, and some legislative control is necessary in order to protect the public from unscrupulous operators.

Unfortunately, consumer-protection legislation is not uniform throughout Canada, as it falls partly within provincial jurisdiction and partly within the federal sphere of legislative powers. The result has been a complex "mix" of federal and provincial statutes, each designed to redress some real or perceived unfairness in the marketplace.

HISTORICAL DEVELOPMENT

Laws protecting the consumer are not a new phenomenon. The state has always attempted to protect its constituents from both real and imaginary harm at the hands of unscrupulous merchants. Many of the early laws were concerned with the control of suppliers of food and clothing, rather than durable goods. They reflected the particular concerns of the populace in the marketplace at that time. Bakers in Paris, France, for example, were subject to inspection as early as 1260. Any bread they produced for sale that was of insufficient weight was subject to confiscation and distribution to the city's poor. How uniform weights were determined undoubtedly raised some difficulties for the bakers. However, by 1439, under the reign of Charles VII, the problem was solved in part by an ordinance requiring the municipal magistrates to designate a place for the weights to be kept, and grain and flour to be weighed. By 1710, the consumer-information movement was underway, with the bakers required to mark each loaf of bread with its weight. Again, any loaf that failed to correspond to the actual weight was confiscated,[1] and the baker subject to a fine. Similar consumer-protection laws were placed on the statute books in England, and later in the colonies. They were the forerunners of present-day consumer-protection legislation, and serve to remind us that consumer protection is not something unique to our modern consumer society.

Governments have seldom been reluctant to pass legislation governing the activities of parties to commercial transactions where the safety or the welfare of the citizenry was concerned, but consumer protection has been subject to varying degrees of emphasis by lawmakers throughout English history. Merchants were subject to much control during the guild period, and, after that, during the fifteenth to eighteenth centuries, to a somewhat lesser degree of control. The rise of *laissez-faire* and the concept of a contract as a bargain struck between individuals resulted in a shift to *caveat emptor* as a consumer-rights philosophy, however. In general, the Common Law courts only attempted to inject an element of fairness into contracts between buyers and sellers when bargaining power was relatively unequal. What they did not do, however, was protect the careless buyer, as the courts saw (and to some extent still see) little reason why the law should do so.

MODERN DEVELOPMENT

Modern consumer-protection legislation is essentially a response to changes in technology and marketing practice. The major changes in technology, manufacturing, and distribution that had their beginnings in the late nineteenth century brought with them fundamental changes in the sale of goods to consumers. To an increasing extent, throughout the first half of the twentieth century, goods became more complex. New scientific advances spawned a vast array of durable goods for household use, sometimes so complex that they were not easily understood and were not self-serviceable. As mechanical products became more technical, so too did the chances of breakdown and costly repairs. Concurrent with the development of new household goods was the widespread use of limited warranties and exclusionary clauses to eliminate the warranty protection offered by the *Sale of Goods Act*.[2]

Political response was not uniform throughout Canada: rather, it reflected the major complaints of consumers in particular jurisdictions. In most provinces, the initial changes took the form of laws that prevented sellers from excluding the implied warranties of the *Sale of Goods Act* in contracts for the sale of consumer goods.[3] Other legislation, particularly in Western Canada, required manufacturers to provide parts and service for equipment in the province and, in some provinces, to warrant that the equipment would last for a reasonable period of time in use.[4]

Concern for the safety of users of consumer products also resulted in legislation at both the federal and provincial levels, in an attempt to protect consumers from products that had an element of hazard associated with their use. In addition, the 1960s and 1970s saw amendments to the *Combines Investigation Act* (now called the *Competition Act*), which was designed to control misleading advertising, double-ticketing of consumer products, bait-and-switch selling, and a number of other questionable selling techniques.

During the same period, legislation was also introduced to deal with a number of other business practices that had developed with respect to credit reporting, credit selling, and selling door to door. All governments dealt with these problems, but, again, not in a uniform fashion. As a result, considerable variation in consumer-protection legislation exists in Canada today. The different laws, for the most part, have much the same general

1. An interesting account of the development of the ordinances under which bakers in Paris were obliged to carry on business may be found in P. Montague, *Larousse Gastronomique* (New York: Crown Publishers Inc., 1961), pp. 78–80.
2. E.g., R.S.O. 1990, c. S.1.
3. See, for example, the *Consumer Protection Act*, R.S.O. 1980, c. 87, now S.O. 2002, c. 30, Schedule A, s. 9(3).
4. See, for example, the *Consumer Protection Act*, S.S. 1996, c. C-30.1, s. 48(g), (h).

thrust in those jurisdictions where they have been introduced. They may be classified in terms of laws relating to product safety, product quality and performance, credit granting and credit reporting, and laws directed at business practices in general. Depending upon the nature of the protection required, the laws have generally taken five different approaches: (1) disclosure of information to the consumer; (2) expanded consumer rights at law; (3) minimum standards for safety, quality, and performance; (4) control of sellers and others by way of registration or licensing of the activities and individuals; and (5) the outright prohibition of certain unethical practices. In many cases, the legislation may employ two or more of these approaches to protect the consumer. For example, consumer-credit reporting organizations must be licensed or registered and, in addition, are subject to certain disclosure rules for consumer credit information. Since only licensed or registered organizations may carry on consumer-credit reporting activities, a failure to comply with the legislation could have as a consequence the loss of the licence to carry on the activity. The various methods of control are examined in greater detail with respect to each of the different types of consumer-protection legislation.

Consumer Safety

Common Law remedies are available to consumers injured by defective goods when the seller or manufacturer owes a duty not to injure, but the rights arise only after injury occurs. Governments long ago were quick to realize that consumer protection from hazardous products or services, to be effective, must not only compensate for injury, but must contain an incentive for the manufacturer or seller to take care. Consequently, governments everywhere have generally controlled products injurious to the health of consumers, or imposed a duty on the manufacturer or seller of the products to warn the consumer of the hazards associated with the products' use or consumption.

COURT DECISION: Practical Application of the Law

Consumer Protection—Duty to Warn Consumer of Risks Associated with Use—Extent of Duty
***Buchan v. Ortho Pharmaceutical (Canada) Ltd.* (1986), 54 O.R. (2d) 92.**

The plaintiff was prescribed a birth-control drug manufactured by the defendant company. As a result of taking the drug, the plaintiff alleged that she suffered a stroke and was left partially paralyzed. Medical evidence indicated a link between the drug and the injury suffered by the plaintiff.

THE COURT: ... the plaintiff's case is that Ortho failed to warn of the danger of stroke inherent in the use of the oral contraceptive and that that failure caused or materially contributed to her injuries. There is no question of any defect or impropriety in the manufacture of the oral contraceptive, nor of its efficacy when taken as prescribed. While Ortho acknowledges that manufacturers of prescription drugs are subject to a common law duty to warn prescribing physicians of the material risks involved in using their drugs of which they know or should know, it denies any duty to warn consumers directly. In the alternative, Ortho argues that its compliance with the statutory standard of disclosure established under the Food and Drugs Act satisfies any duty to consumers, and it is under no obligation to provide consumers with supplementary information or to issue additional warnings. With respect to physicians, Ortho contends that the plaintiff's physician was aware of the then current medical information on the relationship between oral contraceptives and stroke when he prescribed the pill for the plaintiff, and any further warning by Ortho would have been redundant. In any event, Ortho says, in the circumstances any lack of warning on its part to prescribing physicians was not the proximate cause of the plaintiff's injuries. Furthermore, Ortho argues, a reasonable person in the plaintiff's position would have accepted her doctor's advice and taken the pill even if properly warned. Therefore, the argument concludes, the trial judge erred in finding the necessary causal link between Ortho's alleged breach of the duty to warn and the plaintiff's use of the pill. As a matter of common law, it is well settled that a manufacturer of a product has a duty to warn consumers of dangers inherent in the use of its product of which it knows or has reason to know. Once a duty to warn is recognized, it is manifest that the warning must be adequate. A manufacturer of prescription drugs occupies the position of an expert in the field; this requires that it be under a continuing duty to keep abreast of scientific developments pertaining to its product through research, adverse reaction reports, scientific literature and other available methods. When additional dangerous or potentially dangerous side-effects from the drug's use are discovered, the manufacturer must make all reasonable efforts to communicate the information to prescribing physicians. Unless doctors have current, accurate and complete information about a drug's risks, their ability to exercise the fully informed medical judgment necessary for the proper performance of their vital role in prescribing drugs for patients may be reduced or impaired. A reading of Ortho U.S.'s warnings to physicians makes it manifest that Ortho was aware or should have been aware of the association between oral contraceptive use and stroke. Ortho U.S. provided American physicians with data from and the conclusions of the studies in Britain and the United States, and warned of the risk of cerebral damage posed by the pill. Yet, in Canada, Ortho chose not to provide physicians with any similar warning. Why the medical profession in this country, and, through it, consumers in this country, should be given a less explicit and meaningful warning by the Canadian manufacturer of the same drug is a question that has not been answered. Be that as it may, I think it evident that Ortho failed to give the medical profession warnings commensurate with its knowledge of the dangers inherent in the use of Ortho-Novum; more specifically it breached its duty to warn of the risk of stroke associated with the use of Ortho-Novum ....

What do you think of the other defences that Ortho raised regarding redundant advice and acceptance of a doctor's advice?

In Canada, the provinces and the federal government have established legislation to control hazardous activities and products. The most notable legislation, however, has been passed at the federal level, in the form of a number of statutes relating to consumer goods. These include the *Food and Drugs Act*[5] and the *Hazardous Products Act*.[6] Some overlap exists between the two statutes with respect to false or deceptive labelling of products, but the intent of the legislation in each case is to protect the consumer from injury. Both are regulatory in part, and quasi-criminal in nature. The *Food and Drugs Act*, for example, does not confer a civil right of action as a result of a breach of the statute,[7] but, instead, imposes strict liability and penalties under the Act where a breach occurs. A manufacturer would be strictly liable, therefore, in the case of false or deceptive labelling of a product.[8]

Both statutes are designed to enhance the public's safety. The *Food and Drugs Act* primarily controls harmful products that could cause injury or illness if improperly used or ingested by consumers. Under the Act, many drugs are controlled, in an effort to limit their possession and application to proper medical purposes. The legislation also safeguards the purity of food products and regulates matters such as packaging and the advertisement of food and drug products.

The *Hazardous Products Act* takes a slightly different approach to consumer safety. As the name implies, it is concerned with hazardous products. It either prohibits the manufacture and sale of products of an extremely dangerous character or regulates the sale of those that have the potential to cause injury. Hazardous products sold to the public are usually subject to regulation with respect to packaging, and must bear hazard warnings, depending upon their particular nature. In addition to written warnings, most of these products must depict the type of danger inherent in the product by way of warning symbols. Products that are stored under pressure, corrosive substances such as acids, and products that are highly flammable or explosive are required to have these warning symbols printed on their containers.

CASE IN POINT

A purchaser of a fibreglass repair kit suffered serious injury to his face and eyes when a tube of chemical hardener burst in his hands and splashed onto his face. The tube contained a warning in small print, and made no mention of the need to wear protective eye cover or the danger associated with opening the tube.

The purchaser sued the manufacturer of the kit for damages for the injury. At trial, it was found that the testing of the tubes for defects was inadequate, as the testing procedure did not include testing for rupture of the tube. The court also found that the warnings on the tube as to use were inadequate. The manufacturer was found liable at trial, and the finding was confirmed on appeal.

See: *LeBlanc v. Marson Canada Inc.* (1995), 146 N.S.R. (2d) 392.

5. R.S.C. 1985, c. F-27, as amended.
6. R.S.C. 1985, c. H-3, as amended.
7. See, for example, *Heimler v. Calvert Caterers Ltd.* (1974), 49 D.L.R. (3d) 36.
8. *R. v. Westminster Foods Ltd.*, [1971] 5 W.W.R. 300.

Some products are subject to special legislation at the federal level. The *Motor Vehicle Safety Act*[9] provides for the establishment of safety standards for motor vehicles and vehicle parts, and for notice to consumers when unsafe parts or other defects are discovered through use or testing. Similar legislation applies to aircraft in Canada,[10] with elaborate testing procedures that must be undertaken and satisfied before the aircraft may be certified as safe to fly. The statute also governs the use and maintenance of all powered and non-powered aircraft, in an effort to protect the public from injury. The Act not only deals with the safety of the product, but governs the qualifications and licensing of all persons associated with the flying or maintenance of aircraft, since safety is related not only to the maintenance of the product itself, but to the skills of those engaged in its use.

FRONT-PAGE LAW

Candy linked to choking deaths still in stores. Loophole in Food and Drug Act allows 'killer candy' to be sold

A jelly candy linked to the choking deaths of a dozen children around the world, including a young girl in Toronto last spring, remains on store shelves across the city due to a loophole in federal safety standards.

The imported Asian candies, sold in packages under various names, such as Jelly Cups, resemble rounded-off coffee creamers and are filled with flavoured gelatin and a chunk of fruit.

The Canadian Food Inspection Agency has singled the product out as a choking hazard for young children, seniors and anyone with difficulties swallowing.

In North America, the candies have led to the deaths of a two-year-old boy in Seattle and a four-year-old Toronto-area girl last April. It was that girl's death that sparked the recent action by the federal regulatory body. A Toronto doctor complained to Health Canada, which, holding no authority over food, passed the case over to the CFIA.

Mr. Marcynuk explained that the *Food and Drug Act* restricts them from regulating anything based on its size or shape, which, ironically, is exactly why the product is dangerous.

Recently dubbed a "killer candy" by a Vancouver newspaper, the treats are widely available in Asian grocery stores throughout the city and are often prominently displayed near the cash register alongside other candies.

Most packaging does contain some warning, usually focusing on children under three, though some are not that reassuring. One package contained this grim, yet cheery, statement: "This product is insured for $30-million liability. Please eating [sic] it without worry."

Mr. Marcynuk wishes his agency could recommend a risk assessment, which is the first step toward a voluntary recall, but until the legislation is changed he feels his hands are tied.

Undoubtedly, an aide in the Prime Minister's Office circled this news item for attention. Insurance is no comfort at all to a parent who has lost a child. As Prime Minister, who would you call to create action? Your own Consumer Affairs Minister, or those of the provinces? As this is an imported item, should a protective responsibility lie with the Canada Border Services Agency? Perhaps demanding a recall by the importer, or better labelling of the product is in order. What do you think?

Source: Brad Mackay, *National Post*, "Candy linked to choking deaths still in stores," September 13, 2000, p. A21.

CONSUMER INFORMATION

Consumer information is closely related to both consumer safety and consumer protection from deceptive or unfair practices. For this reason, much of the legislation designed to protect consumers is concerned with either the disclosure of information about the product or service, or the prohibition of false or misleading statements by sellers. Some laws, however, are designed to protect consumers by providing standards by which the consumer may make direct comparisons of products and prices. The *Weights and Measures Act*[11] is one such statute. It is designed to establish throughout Canada a uniform system of weights and measures that may be applied to all goods sold. This Act fixes the units of measure that may be lawfully used to determine the quantity of goods and to calculate the price. The statute also provides for the testing and checking of all measuring devices used for such purposes.

9. S.C. 1993, c. 16, as amended.
10. The *Aeronautics Act*, R.S.C. 1985, c. A-2.
11. R.S.C. 1985, c. W-6.

A complementary statute at the federal level is the *Consumer Packaging and Labelling Act*[12] that, from a consumer-protection point of view, has as its purpose the protection of the public from the labelling and packaging of products in a false or misleading manner.[13] This Act provides penalties for violation, but does not provide a civil cause of action for consumers misled by the false labelling. The right to damages for any injury suffered as a result of the misleading label, however (depending upon the circumstances), may be available at Common Law, or under one of the provincial statutes that provide for such a right of action.

CONSUMER-PRODUCT QUALITY AND PERFORMANCE PROTECTION

The first action to protect consumers in the product-performance area took the form of consumer-protection legislation that prohibited sellers from exempting sales of consumer goods from the implied conditions and warranties available under the *Sale of Goods Act*. While these moves helped to balance the rights of buyers with those of sellers at the point of sale, after-sale service and provision for repairs were not affected. The change also did nothing to provide persons who obtained consumer goods by way of gift with enforceable rights under the original sale agreement. Manufacturers of inferior goods continued to enjoy relative protection from consumer complaints through the rules relating to privity of contract. For the most part, only the sellers were directly affected in actions for breach of contract. To overcome some of the difficulties faced by consumers, any redress in the balance of rights between buyers and sellers had to clearly come through new legislation directed at specific abuses in the marketplace.

A comparatively recent trend in consumer-protection legislation has been towards the expansion of buyers' rights and sellers' obligations with respect to consumer goods that fail to deliver reasonable performance, or that prove to be less durable or satisfactory than manufacturers' claims indicate. New Brunswick[14] and Saskatchewan[15] have both passed legislation of this nature, and other provinces have passed comprehensive business-practices legislation that provide enhanced protection for consumers.[16]

As a method of addressing consumer complaints about automobile warranties, some provinces have established dispute-resolution mechanisms patterned after the U.S. state automobile "lemon laws." These are motor-vehicle arbitration plans whereby the automobile manufacturers voluntarily agreed to resolve warranty complaints related to the operational or reliability qualities of their vehicles through binding arbitration. Since 1994 access to arbitration has been available across Canada through the Canadian Motor Vehicle Arbitration Plan (CAMVAP). Under this process, if a new automobile has reliability problems that the manufacturer is unable to repair, or if the vehicle possesses numerous defects, the purchaser of the vehicle must first give the manufacturer the opportunity to repair, and if this proves unsuccessful, the dispute may be taken before an arbitrator. The arbitrator hears both sides of the dispute, and then renders a decision. The arbitrator has the authority to direct the manufacturer to repair the defects, take back the vehicle and repay all or a part of the purchase price to the buyer, or dismiss the complaint if it is frivolous or unwarranted. The program applies only to new vehicles during the warranty period, and for a fixed time thereafter. In the case of a vehicle that has so many problems or defects that a manufacturer buy-back is appropriate, the arbitrator is required to determine a usage charge as a deduction from the price if the vehicle has been in the buyer's possession for more than a year. This deduction recognizes the fact that the buyer has had the use of the vehicle during that period of time.

The Saskatchewan legislation, the *Consumer Protection Act*, substantially alters the contractual relationship between the consumer and the seller by expanding the class of persons entitled to protection under the Act, and by imposing heavy burdens on sellers and manufacturers of consumer goods who fail to provide products capable of meeting

12. R.S.C. 1985, c. C-38.

13. *R. v. Steinbergs Ltd.* (1977), 17 O.R. (2d) 559.

14. The *Consumer Protection Warranty and Liability Act*, S.N.B. 1978, c. C-18.1.

15. The *Consumer Protection Act*, S.S. 1996, c. C-30.1.

16. See, for example, *Business Practices Act*, C.C.S.M., c. B120.

advertised performance claims. The legislation applies to all sales of consumer goods and, in addition, many goods not normally considered to be products of a consumer nature.[17] It also covers used goods sold by second-hand dealers,[18] but permits dealers in used goods to exempt themselves from many of the obligations imposed on sellers under the Act if they expressly exclude the warranties at the time of the sale.[19]

The Saskatchewan Act defines a consumer in a very broad way in order that not only the immediate purchaser of a consumer product, but also persons who subsequently acquire the goods, may enforce statutory warranty rights under the Act. For example, persons who obtain consumer goods by way of gift or inheritance would be entitled to enforce a breach of a statutory warranty, even though no consideration was given to acquire the goods and no direct contractual link may exist between them and the seller. In order to achieve this end, the Act provides that a manufacturer may not claim a lack of privity of contract as a defence against a claim by an owner of goods where a breach of a warranty under the Act is alleged.[20]

Statutory warranties

Express warranties on goods provided by statute.

The legislation sets out a number of **statutory warranties** that apply to all sales of consumer goods.[21] The majority of these resemble sections of the Sale of Goods legislation in most other provinces. They include a warranty that the retailer has the right to sell the goods and that the goods are free from any liens or encumbrances. Included, as well, are the usual sale-of-goods warranties as to fitness, etc. The legislation goes beyond the usual types of warranties, however, and includes a requirement that the goods be durable for a reasonable period of time. It also requires a warranty that spare parts and repair facilities will be available for a reasonable time after the date of the sale, if the product is one that normally may be expected to require repair.

In addition to the statutory warranties, the Act imposes an obligation on the seller to comply with any or all other warranties or promises for performance made either through advertising, writing, or statements made at the time of sale.[22] These statements or representations are treated as express warranties and are actionable in the event of breach.

A farmer purchased a harvester on the basis of the salesperson's representation that it would harvest "at least 60 hectares a day." The manufacturer's brochure stated the machine was designed to harvest 40 hectares a day. In this case, if the harvester failed to perform as claimed, the salesperson's representation would be an express warranty upon which the farmer could rely in a later action for breach of the agreement.

The Act provides a number of different remedies, depending upon the nature of the breach, and includes exemplary damages as a remedy if the seller or manufacturer willfully acts contrary to the statute.[23] In an effort to reduce litigation arising as a result of the Act, provision is made for mediation of disputes by officials of the Saskatchewan Department of Consumer Affairs,[24] and for binding arbitration when the parties agree to have the matter decided by an arbitrator.[25]

Consumers who wish to exercise rights under the Act are obliged to do so within the basic provincial limitation period. No action may be brought that alleges a violation of the Act unless it is commenced within two years after the alleged violation took place.[26] However, since the rights set out in the Act are in addition to any other rights that the person may have at law,[27] the time limit may not affect the ordinary Common Law remedies available to the consumer.

---

17. For example, certain goods used in agriculture and fishing.
18. The *Consumer Protection Act*, S.S. 1996, c. C-30.1, ss. 42–43.
19. ibid.
20. ibid., s. 14.
21. ibid., s. 48.
22. ibid., s. 47.
23. ibid., s. 65.
24. ibid., s. 27.
25. ibid., s. 34.
26. *Limitations Act*, S.S. 2004, c. L-16.1, s. 5.
27. *Consumer Protection Act*, S.S. 1996, c. C-30.1, s. 34.

CASE IN POINT

A consumer purchased a new truck that some time after purchase burst into flames due to a manufacturing defect in the running lights module. Both the manufacturer and dealer denied liability, and the consumer sued both for damages. At trial the evidence indicated that the manufacturer was aware of the default, but did nothing to inform consumers of the risk of fire.

At trial, the judge found both the manufacturer and dealer liable for breach of the provincial statutory warranties, and awarded the consumer both general and exemplary damages. On appeal to the Supreme Court of Canada, the court upheld the trial judge's award.

Prebushewski v. Dodge City Auto (1984) Ltd. [2005] 1 S.C.R. 649.

While consumer groups have advocated similar legislation in all provinces, some concern has been expressed over the introduction of laws if they should vary from province to province. Uniform legislation has been urged upon the provinces, if only to provide common consumer rights throughout the country. Whether other provinces heed this admonition remains to be seen.

CLIENTS, SUPPLIERS *or* OPERATIONS *in* QUEBEC

Consumer protection-legislation in Quebec is more robust than anywhere else in Canada. Starting from the fact that it governs not only goods but services as well, and allows for specific performance in addition to damages, it requires many more small aspects of compliance that are in favour of the consumer. For example, an advertisement offering goods must indicate the name of the merchant, and its full street address, not just a post office box. Such a provision may at first seem trivial, but the collection of requirements amounts to full disclosure; a merchant who does not comply raises a red flag, and in doing so exposes itself to immediate enforcement action.

MANAGEMENT ALERT: BEST PRACTICE

It is common for computer-produced or bulk-purchased paper business forms to contain preprinted warranties or limitations. Business owners may also attempt to limit their own liability by including other specific provisions. When conducting consumer transactions it is important, however, for owners to realize that their efforts are trumped by legislation. Consumers cannot sign away these statutory rights to their detriment, and any warranty offered in excess of the statute will be enforced against the seller. While including such limitations in contracts or on receipts is not necessarily pointless, a prudent business owner should seek legal advice as to any particular statutory warranties that are applicable to his or her line of business.

CONSUMER PROTECTION RELATED TO BUSINESS PRACTICES

Itinerant Sellers

Itinerant sellers

Sellers of goods who sell goods at the buyer's residence.

Door-to-door sellers have always presented a special problem for consumers because of the conditions under which the selling takes place. The door-to-door seller conducts business in the prospective buyer's home, and, as a result, the sale is not initiated by the buyer, but rather by the seller. One of the particular difficulties with door-to-door selling is the fact that the buyer cannot leave the premises if the product is not what he or she needs or wants. As a consequence, the buyer often feels uncomfortable or vulnerable. Under these circumstances, high-pressure or persuasive selling techniques may result in the buyer signing a purchase contract on impulse, or, under pressure, simply to get rid of the seller.

While many products sold by door-to-door sellers are of high quality and are sold by reputable firms, the selling practices of the less reputable eventually resulted in consumer demands to have this form of selling brought under legislative control. Most provinces, as a part of their consumer-protection legislation, now require door-to-door sellers to be licensed or registered in order to conduct their selling practices and to ensure compliance with the statute and regulations. While variation exists from province to province, door-to-door sales are now usually subject to a "**cooling-off period**" after the purchase agreement is signed. During this period, the contract remains open to repudiation by the buyer without liability. It is only after the cooling-off period has expired that a firm contract exists between the buyer and the seller.[28]

Cooling-off period

A time interval after a contract is signed that allows the buyer to repudiate the contract without liability.

In addition to the imposition of a cooling-off period, the contract states that the sale of goods by door-to-door sellers, if it exceeds a specified sum, must be in writing. It must also describe the goods sufficiently to identify them, provide an itemized price, and give a full statement of the terms of payment. If a warranty is provided, it must be set out in the agreement, and if the sale is a credit sale, a full disclosure of the credit arrangement, including details of any security taken on the goods, must be provided.[29]

The general thrust of consumer-protection legislation of this nature is to nullify or eliminate the use of questionable selling techniques, and to provide sufficient information to the consumer to allow the consumer to review the agreement during the cooling-off period. By providing the consumer with the necessary information, and an opportunity to contemplate the transaction without the presence of the seller, the law encourages the buyer to make a rational buying decision.

Unfair Business Practices

Unfair practices

Business practices that mislead or treat the buyer unfairly.

Honest sellers, as well as consumers, suffer when questionable practices are used by unethical merchants to induce consumers to purchase goods. As a result, consumer-protection legislation is frequently designed to not only protect the consumer, but to maintain fair competition between merchants in the marketplace. Consumer-protection legislation concerning unfair business practices may take the form of general legislation, or it may be directed at specific areas of business activity or sectors of business. Motor-vehicle repairs are an example of a sector of business where legislation has been directed toward protecting consumers. Both Quebec and Ontario have specifically targeted automobile repairs for control,[30] and require repair shops to provide written estimates (on request, in the case of Ontario). Repair charges cannot exceed the estimate by more than 10 percent and a detailed invoice must be provided. The work must also be guaranteed for a period of time or mileage, and any breakdowns due to faulty work may be charged back to the shop.

While specific legislation is common, the general trend has been for provinces to pass broad legislation concerning all sectors of consumer-related business. The province of Ontario, for example, has a *Consumer Protection Act*[31] that sets out a list of activities deemed to be unfair practices. These activities include false, misleading, or deceptive representations to consumers as to quality, performance, special attributes, or approval that are designed to induce consumers to enter into purchases of consumer goods or services.[32] This Act also covers the negotiation of unconscionable transactions that take advantage of vulnerable consumers, or that result in one-sided agreements in favour of the seller. The kinds of transactions that are considered unconscionable include: (1) those that take advantage of physical infirmity, illiteracy, inability to understand the language, or the ignorance of the consumer; (2) those that have a price that grossly exceeds the value of similar goods on the market; and (3) those contracts in which the consumer has no reasonable probability of making payment of the obligation in full.[33] In addition, transactions that are excessively one-sided in favour of someone other than the consumer, and those in which the conditions are so adverse to the consumer as to be inequitable, are treated in the same fashion. The same part of the Act treats misleading statements

28. See, for example, *Consumer Protection Act*, 2000, S.O. 2002, c. 30, Schedule A, s. 43(1).
29. ibid., s. 77.
30. In their *Consumer Protection Acts*.
31. *Consumer Protection Act*, 2000, S.O. 2002, c. 30, Schedule A.
32. ibid., s. 14.
33. ibid., s. 15.

of opinion upon which the consumer is likely to rely, and the use of undue pressure to induce a consumer to enter into a transaction, as unfair practices.[34]

The list of unfair practices is not limited to those set out in the Act, for the legislation provides that additional unfair practices may be proscribed by regulation. The general thrust of the law, however, is to eliminate the specified unfair practices. The legislation provides that any person who engages in any of the enumerated practices commits a breach of the Act.[35] While fines are provided as a penalty, the most effective incentive to comply may be found in the sections of the Act that permit a consumer to rescind an agreement entered into as a result of the unfair practice, or to obtain damages where rescission is not possible.[36] The Act also provides that the courts may award exemplary or punitive damages in cases where the seller has induced the consumer to enter into an unconscionable or inequitable transaction.[37]

The minister and the director responsible for the administration of the Act have wide powers of investigation,[38] and may issue cease-and-desist orders to prevent repeat violations. An added penalty, where violations persist, is the right to cancel the registration of the seller if the seller is engaged in a business that requires registration or a licence to carry on the activity. Safeguards are included in the Act to prevent the arbitrary exercise of powers under the Act, and limitation periods are included. These require action on the part of the consumer within a period of time after an unfair practice occurs, if the consumer wishes to obtain the relief provided by the legislation.[39]

This approach has been incorporated in part in the legislation of a number of other provinces, but, again, the provinces have not made a concerted effort to establish uniform laws in the area of unfair business practices. At the federal level, however, certain practices have been dealt with under the anti-competition legislation that has nationwide application.

Some provinces have also attempted to consolidate consumer protection legislation into a more manageable package for enforcement. The province of Ontario, for example, in 2005 brought into effect new consumer protection legislation that incorporated six existing laws: The *Consumer Protection Act, Consumer Protection Bureau Act, Motor Vehicle Repair Act, Prepaid Services Act,* the *Law Brokers Act,* and the *Business Practices Act.* The protection provided under the new Act also provides expanded protection for consumers, including an extension of the cooling-off period to 10 days for contracts over $50 negotiated in the consumer's home, cooling-off periods for memberships in clubs such as fitness, dance or martial arts clubs, and a limit of 1 year on prepaid membership contracts. The Act also broadens the seller's duty to provide full information about all service charges on credit transactions and how they would be calculated. Other transactions are also affected. For example, a failure to deliver goods within 30 days of the promised delivery date would entitle the consumer to cancel the contract.

COURT DECISION: Practical Application of the Law

Unfair Practices—Misleading Statements

***R. v. St. James International Academy Ltd.*, 2005 OCJ 369.**

The defendants were Pellegrini and the private school named St. James International Academy. Pellegrini advised parents that his school offered high school credit education.

THE COURT: The Business Practices Act endeavours to take the guesswork out of deciding what kind of conduct, through either action or omission, could be an unfair practice. In other words, the Act provides an extensive listing in

34. ibid.
35. ibid., s. 17.
36. ibid., s. 18.
37. ibid., s. 18(11). This section provides for this form of penalty for unfair practices of the type found in Part III of the Act.
38. ibid., Part XI.
39. ibid., s. 18. The Act provides that steps to rescind the agreement must be taken by the consumer within one year of the date the transaction was entered into.

s. 2 of prescribed actions, which includes the deliberate failure to reveal a material fact to a consumer and the making of improper or unconscionable representations that would ultimately affect a consumer's decision, and designates or deems them to be unfair practices. In particular, the parents believed that their children would be attending a school that granted high school credits for an Ontario Secondary School Diploma. They were not aware that St. James could not grant high school credits at the time they initially enrolled their children. Nor were they aware that St. James had not obtained the authorization from or had even sought the approval from the Ministry of Education to grant high school credits. Many of the parents were angry that an entire academic year or semester had been wasted because their children did not or would not receive any credits towards a high school diploma for any of the courses they were taking or had completed while attending St. James. Pellegrini had represented to the parents that St. James was eligible or able to grant high school credits. He also argued that he had fulfilled the minimum Ministry requirement to be "able" to grant high school credits by filing the Notice of Intention to Operate a Private School and checking off on the notice document that the school intended to offer high school credit courses. Therefore, Pellegrini submits that he did not mislead the parents about St. James's ability to issue credits. However, receiving authorization to grant credits is not simply contingent on filing a government form. Having the authorization to issue credits is a material fact. Thus, either St. James had the authority to grant credits or it did not at the time the students were initially registered. Being able, or in other words, having the potential to issue credits is not the same as having the actual authority. Hence, Pellegrini's use of words that St. James was able to issue credits is ambiguous and can be easily construed the wrong way by the public that St. James actually had the authorization from the Ministry to issue high school credits. Therefore, telling parents that St. James is able or eligible to grant credits is not only ambiguous, but also misleading and deceptive, as it does not mean that St. James had Ministry approval to actually grant credits. Therefore, Pellegrini knowingly made representations about St. James's ability to issue credits that he knew to be misleading or untrue and intended to deceive the parents or complainants. Pellegrini in response to this prosecution…insists that the parents knew that their children were already having learning problems at their previous schools and were therefore not academically at the high-school-credit level when they enrolled, so they should not be surprised that their children were not able to work at the level needed to obtain credits. Alas, this is exactly what the Business Practices Act is about, to protect consumers from unconscionable and false misrepresentations about the type or quality of service offered. In other words, it does not matter that the parents knew their children were poor students or had learning disabilities, it is what they were actually told or promised by Pellegrini that is at issue. Thus, if someone tells the parents that they can help their children (and work miracles) and that their children would receive high school credits, then those misstatements or representations once shown to be false and intentionally given to mislead would be acts or omissions designated under this governing statute to be an unfair practice.

The intention of such legislation is to ensure that substance of a transaction triumphs over its form. While this is "consumer" protection, don't other areas of the marketplace deserve such protection? How do you think this might be happening?

Restrictive Trade Practices

Restrictive trade practices

Business practices that restrict trade to the detriment of the consumer or competitors.

The *Competition Act*[40] specifically prohibits false and misleading advertising with respect to both price and performance. Provisions in the Act also prohibit deceptive practices, such as bait-and-switch selling techniques, referral selling, and the charging of the higher price where two price stickers are attached to goods. Resale price maintenance and monopoly practices detrimental to the public interest are also prohibited. A more complete description of these consumer-protection measures, and others relating to restrictive trade practices, is presented in Chapter 32, "Restrictive Trade Practices."

Regulation by Licence or Registration

In most provinces, many business activities have been subject to licensing and special rules in an effort to control unfair practices that are contrary to the public interest. The sellers of securities, real-estate and business brokers, mortgage brokers, motor-vehicle dealers, and persons dealing in hazardous products, to name a few, are often subject to laws regulating their activities and practices. By imposing a licensing requirement on the particular activity, compliance with the law becomes necessary in order to maintain the licence, and violations of the statute are accordingly minimized. As a result, most provinces provide for licensing or registration as a means of control of the particular activities in the public interest.

40. R.S.C. 1985, c. C-34, as amended.

Collection Agencies

A particular business organization that has been singled out by most provinces for the purpose of consumer protection and control is the collection agency. Collection agencies play a useful role in the collection of debts, often from delinquent consumers, but many of their collection methods in the past aroused the ire of debtors. As a result of complaints to provincial governments, all provinces now regulate collection agencies by way of licences or registration, and their activities are subject to a considerable degree of control.[41]

In general, collection agencies are not permitted to harass or threaten the debtor in any way, nor are they permitted to use demands for payment that bear a resemblance to a summons or other official legal or court form. The legislation also prohibits the agency from attempting to collect the debt from persons not liable for the debt, such as the debtor's family, or by harassing persons other than the debtor in an effort to pressure the debtor into payment. As well, the agency is usually not permitted to communicate with the debtor's employer, except to verify employment, unless the debtor has consented to the contact. These are but a few of the limitations on collection techniques of the agencies, but they serve to indicate the attempts by the provincial legislatures to balance the legitimate rights of creditors to obtain payment with protection of the debtor from undue pressure to make payment. Again, the laws relating to this form of consumer protection lack uniformity, but in each case the method of control of the activity remains similar. Agencies that persistently violate the Act may find that the province has revoked their licence to operate.

CREDIT-GRANTING CONSUMER PROTECTION

The granting of credit by a lender or seller depends to a large extent upon the credit rating of the consumer who wishes to borrow or the buyer who wishes to purchase goods on credit.

For many years, lenders of money and sellers of goods on credit were not obliged to disclose to the borrower or buyer more than a minimum amount of information concerning the credit extended. Even then, it was usually only in the form of the promissory note, mortgage, or other documents that secured or evidenced the actual loan or credit sale. In many cases, the borrower was unaware of the true cost of credit, because the lender often made special charges for arranging or servicing the loan and added these to the amount that the debtor was obliged to pay. The lack of information, together with the inability to fully understand the documents signed to secure the loan, left the debtor bewildered and often at the mercy of the lender or seller.

The widespread use of consumer credit following World War II, along with consumer complaints, prompted governmental review of lending practices and credit selling. This resulted in legislation that requires the lender or seller to disclose the true cost of credit to the borrower or credit buyer at the outset of the transaction. During the next three decades, virtually all provincial legislatures established the requirement that the borrower be provided with a written statement that discloses the total dollar cost of credit, including any charges, bonuses, or amounts that the borrower must pay in addition to the interest, as well as the interest amount. The cost of the credit must also be displayed as an annual percentage rate.

On a Monday morning, Pay-daze Loans Company offers to exchange a worker's next expected Friday afternoon paycheque for 90% of the expected amount given in immediate cash. There is also a $20 fee payable for the transaction. On an $800 paycheque, the worker would receive $720, and Pay-daze would receive $80, or $100 with the fee included. On the discounted amount alone, this five day "loan" represents a simple annual rate of 730% interest, or 912.5% with the fee also treated as interest. Litigation on this type of transaction is presently beginning its journey through the courts. What is your opinion of its merits?

41. *Collections Agencies Act*, R.S.O. 1990, c. C.14.

The general thrust of the legislation is toward consumer credit in the form of consumer loans and credit purchases. Consequently, long-term financing, such as for housing or other substantial purchases that utilize a land mortgage as security, are treated as exempt transactions in a number of provinces.[42] The penalties imposed for a failure to comply with the disclosure requirements vary, unfortunately, from province to province. However, the legislation generally prevents the lender from collecting the full amount of interest set out in the loan document. Some provinces prohibit the lender from claiming other than the principal amount of the loan; one province limits the lender to the disclosed cost of borrowing if the lender should attempt to recover on the loan;[43] and another limits the lender to the legal rate.[44]

Credit-Reporting Consumer Protection

Disclosure of a different nature that is nevertheless related to consumer credit is credit reporting. For many years, credit-reporting agencies have provided an important service to lenders and credit sellers by supplying credit reports on borrowers or credit buyers. The widespread use of credit, coupled with the relatively impersonal nature of credit sales, created a need for quick and accurate information about prospective debtors to enable the lender or seller to decide promptly if credit should be extended. Agencies providing this type of service keep files on persons using credit. They generally include in the file all information that might have an effect on a person's ability to pay. The information is usually stored in a computer, and, through nationwide hook-ups, credit-reporting organizations are usually in a position to provide credit information on borrowers anywhere in the country on relatively short notice.

The potential for error becomes greater, however, as the amount of information on an individual increases. Concern over the uses made of the information, and its accuracy, have resulted in new laws designed to control the type and use of the collected information, and to enable the consumer to examine the information for accuracy. Once again, variation exists from province to province, but generally the laws are designed to license the consumer credit-reporting agencies and limit access to the information to those persons who have the consent of the debtor, or to persons with a legitimate right to obtain the information. The nature of the information stored or revealed is also usually subject to the proviso that it be the best reasonably obtainable, and that it be relevant. If a consumer credit-reporting agency has collected information on a person, it must permit the person to examine the file and challenge or counter any inaccurate information by way of insertion in the file of other information of an explanatory nature. In most provinces, the agency must also provide the person with the names of all persons who received credit reports during a particular interval of time, although the specifics of this obligation vary from province to province.

Persons who want credit reports usually obtain the prospective debtor's permission to do so, but this is not always necessary in all provinces. In many cases, the creditor need only inform the prospective debtor of his or her intention, and the name and address of the agency that the creditor intends to use.[45]

If credit is refused, or if credit charges are adjusted to reflect a poor credit rating, and the action is based upon a report received from a consumer credit-reporting agency, the creditor must generally so advise the person and supply the name and address of the credit-reporting agency. The purpose of this latter provision in the legislation is to enable the person refused credit the opportunity to determine if the report was inaccurate in any way, and to take steps to correct it. The law is enforced by way of penalties, but serious repeated violations may be dealt with by the revocation of the agency's licence to operate.

---

42. All provinces except Alberta, British Columbia, and Manitoba.

43. *Consumer Protection Act*, R.S.N.S. 1989, c. 92, s. 18.

44. The *Consumer Protection Act*, C.C.S.M., c. C200.

45. See, for example, the *Consumer Reporting Act*, R.S.O. 1990, c. C.33.

A QUESTION OF ETHICS

Computer capacity and ability to collect and store information on consumers grows exponentially, such that traditional credit-reporting agencies may be easily eclipsed by the amount of data held by a wide range of commercial enterprises, from banks to grocery stores. Likewise, loyalty-point programs can generate hauntingly accurate profiles from purchase patterns. These factors have led governments to bolster data-protection laws to further protect consumers. Frequently, however, consumers must consent to information exchange to participate in the benefits offered by those holding the data. This represents a major shift from the more limited purpose of credit granting to the wider purpose of consumer profiling. In your opinion, is this consent really freely given, and what limits do you think should be imposed on the use of such profiles?

CHECKLIST FOR CONSUMER PROTECTION

Identifying particular causes of action

1. Are aspects of the matter within federal jurisdiction (for example, restrictive trade practices, weights and measures or product safety)? If yes, also consult federal legislation.
2. Regarding those aspects within provincial jurisdiction, review –
   a. Form of business (proprietorship, partnership or corporation) and registration.
   b. Legislation governing the industry or profession specifically at issue.
   c. Consumer protection legislation.
   d. (Unfair) business practices legislation.
   e. Statutory warranties legislation.
   f. *Sale of Goods Act.*
   g. Any municipal bylaws regulating conduct of the particular commercial activity.

SUMMARY

Laws to protect consumers from deceptive or unfair business practices are not new. At Common Law, for example, a contract entered into as a result of misrepresentation, whether innocent or fraudulent, is voidable at the option of the injured party. In addition, the law of torts provides a remedy when a person is injured as a result of a breach of duty of care by the seller or manufacturer. Canadian legislation takes a different approach. It is designed to protect the consumer by regulations discouraging deception or unfair practices. The laws attempt to do this by way of penalties that may be imposed upon dishonest merchants who engage in such practices. The legislation also attempts to broaden the group of persons entitled to relief by protecting not only purchasers but the users or recipients of consumer goods as well.

The laws generally fall into a number of different classifications: those designed to protect consumers from hazardous or dangerous products, those designed to provide accurate and useful information and to prohibit deception, and those designed to control activities associated with the actual sale of goods, such as credit and credit-information services. Control is generally exercised by licensing persons engaged in the particular activities where such control is considered necessary, or by way of penalties for violation of the legislation where licensing is impractical or unworkable as a means of control.

Unfortunately, there is much overlap in the various statutes at the two levels of government, and a lack of uniformity in the approaches taken to consumer protection among the provinces themselves. As a result, in Canada, no uniform consumer-protection legislation exists, and consumer rights and protection vary from province to province.

KEY TERMS

Statutory warranties, 538
Itinerant sellers, 539
Cooling-off period, 540
Unfair practices, 540
Restrictive trade practices, 542

REVIEW QUESTIONS

1. Why was it necessary for provincial legislatures to introduce comprehensive consumer-protection laws during the post-World War II period?
2. What form did the consumer-protection laws take?
3. What practices of some collection agencies led to legislation controlling the collection of debts generally?
4. Describe the impact of much of the consumer-protection legislation on exemption clauses in the sale of goods.
5. Assess the statement: "Consumer-protection legislation has increased the cost of selling and, in turn, the price the buyer must pay for the goods purchased. It does nothing to protect the negligent or careless buyer."
6. Why was it necessary for the province of Saskatchewan to introduce its *Consumer Protection Act*?
7. Has consumer-protection legislation carried consumer protection too far in terms of the onus it places on the seller? Does this not simply increase the cost of goods to the buyer?
8. Explain the need for legislative control over the selling practices of door-to-door sellers.
9. What is the purpose of the "cooling-off period" that the consumer-protection legislation frequently imposes on contractual relations between buyers and door-to-door sellers?
10. Describe some of the practices of credit-reporting agencies that resulted in legislative control over their activities.
11. What controls were generally imposed on sellers of durable goods?
12. The general thrust of consumer-protection legislation has been to provide accurate information or disclosure of essential terms to the buyer. Has consumer-protection legislation generally met this goal?
13. How has consumer-protection legislation addressed exaggerated advertising claims?

MINI-CASE PROBLEMS

1. An automobile owner purchased a container of cleaning solvent that was designed for removing rust and dirt from corroded metal parts. The directions indicated that it should be dissolved with 10 parts water to 1 part solvent, and stated that it should not be used at full strength. The label bore the symbol for corrosive material and the words: "For Industrial Use Only." The automobile owner diluted the solvent and applied it as directed but, in doing so, accidentally splashed the chemical into his eyes, causing him to lose sight in one eye. He brought an action for damages against the manufacturer for his injury.

   Discuss and render a decision.

2. Clarissa purchased a new Super 8 Sedan from Super 8 Motors Ltd., a licensed dealer of the manufacturer of the automobile. The vehicle developed a number of problems after delivery, each requiring extensive repairs by the dealer under warranty. After the twelfth breakdown of the vehicle, Clarissa approaches you for advice as to what she should do.

   Advise Clarissa.

3. A magazine salesperson called at Angela's home and offered to sell her a 'package' of 6 magazines at a special 1 year subscription price of $360. Angela gave the salesman her cheque for $30, being the first of the 12 payments under the agreement. The next day, Angela discovered that the price she would be paying was equivalent to the regular newsstand price for each issue.

   Advise Angela.

4. Max purchased a used motor vehicle from Used Cars Inc. on a time payment arrangement where he paid the seller $300 per month. The vehicle broke down shortly after the purchase, and when Used Cars Inc. refused to make repairs, Max stopped making payments. A few months later, Max was turned down on a credit purchase because he was a 'bad credit risk.' Apparently, Used Cars Inc. had reported his default to the Credit Bureau.

   Advise Max.

CASE PROBLEMS FOR DISCUSSION

CASE 1

Selma purchased a home entertainment and satellite dish system from a local TV and electronics store. A part of her purchase included a large 50 inch TV. Selma paid the store in full with her credit card. The store delivered the system, and an employee of the store assembled it. Picture quality, however, was poor at the time, and the employee suggested that the poor quality reception was due to atmospheric conditions that interfered with the satellite signal. Over the next few days the picture quality did not improve, and Selma complained to the store owner. A technician came out to Selma's home, and adjusted the satellite dish, but the change only marginally improved reception. A series of telephone complaints by Selma followed, and each time a technician visited the house to adjust the equipment, with only limited success. Eventually, a technician advised Selma that the location of her home made it difficult to place the satellite dish to get good reception. He recommended a cable hook-up for better reception, but Selma did not wish to have the cable wiring hook-up.

Advise Selma.

CASE 2

Black's Furniture Ltd. was a high-end home furnishings establishment in a large city. Hilary, a mid-level executive in a large corporation, visited the showroom, and saw an expensive bedroom suite that she felt would fit well with her other furniture in her town-house. Hilary enquired to determine if Black's Furniture Ltd. offered a time-payment plan for purchases, and was advised that the company did so for 'good customers.' She was also advised that purchases on the time-payment plan required a 10% first payment, with the balance payable over the next 12 months in equal payments plus interest at 12% per annum. Hilary signed a purchase agreement for the $16,000 bedroom suite and provided a cheque for the 10% deposit. The agreement included authorization for a credit report.

Later in the day, while at work, Hilary was informed by the receptionist for her department that the furniture store had called wanting a mailing address to return her cheque, saying that they would not sell her the bedroom suite due to a 'bad credit report.'

Hilary was embarrassed by the news, and immediately called the store. She was then advised by a store employee that the credit report had revealed that she had defaulted on a car loan, and was delinquent in payments on a personal loan. No other information was provided, but Hilary recalled that on the car loan, she had returned the vehicle to the car dealer due to defects, and the loan had been cancelled. She also recalled that while she had missed a few payments on the personal loan, it was eventually paid in full.

Discuss the issues raised in this case, and the rights and responsibilities of all parties.

CASE 3

Harvey purchased 1 kilogram of ground meat from Alice's Meat Market. The package was labelled "lean ground beef" and had been located in a freezer under a sign that advertised "Special sale: $2.99 for 1 kilogram."

Harvey's friend, who was a meat inspector at a local packing house, dropped by for a visit while he was preparing to barbecue patties made from the ground meat. His friend examined the meat and advised Harvey that in his opinion the meat was not lean ground beef, but ordinary "hamburg" that contained close to 40 percent fat.

Harvey checked with Alice's Meat Market and was told that a clerk had mislabelled the meat as lean ground beef, and that the meat was actually hamburg. The special sale, however, was for hamburg at $2.99.

Discuss the issues raised in this case, and discuss the legal position of Harvey and of Alice's Meat Market.

CASE 4

Carter carried on a part-time business of lending money to his friends to enable them to purchase consumer goods. He would also lend money to strangers who had been directed to him by his friends. The loans were generally for a short term and were written up in a casual way. Usually the document set out the name of the party and referred only to the principal amount borrowed and the lump-sum interest amount payable on the due date.

On March 1st, Jerry Messier approached Carter in order to borrow $800 for the purchase of a stereo system. Carter loaned him the money and had him sign a document that read as follows:

March 1, 2011

I promise to pay S. Carter on the first day of each month the sum of $200 until the total amount of $1,000 has been paid.

$ 800 principal
$ 200 interest

$1,000
Value received
"J. Messier"

A few weeks later, Messier advised Carter that he had no intention of paying him the money, as the paper he signed was worthless and the debt unenforceable.

Advise Carter and Messier. What issues (if any) might arise if Carter should decide to institute legal proceedings against Messier?

CASE 5

John Smith lived at 221 Pine Avenue in a large city. He had no debts and had never previously purchased goods on credit. He did, however, wish to purchase a particular power boat, so he entered into negotiations with the owner of a marina to obtain the boat on credit. He consented to the marina owner making a credit check before the transaction was completed, and was dismayed when the marina owner refused to proceed with the transaction because he was a "poor credit risk."

The credit-reporting agency apparently had provided a credit report on a John Smith who some months before had resided at 212 Pine Street in the same city, and who had defaulted on a number of substantial consumer debts. John Smith knew nothing of the other John Smith, nor had he resided at 212 Pine Street.

What avenues are open to John Smith in this case to rectify the situation?

CASE 6

Mary Dwight purchased a vacuum cleaner from a salesman who represented himself as a sales agent for Speedy Vacuum Cleaners. The salesman gave a demonstration of the vacuum in Mary's living room, and the machine appeared to do an excellent job of cleaning dust and dirt from her carpets. At the conclusion of the demonstration, the salesman produced a form contract that called for a deposit of $50, and monthly payments of $50 each until the full purchase price of $400 was paid. Mary paid the $50 deposit and signed the contract. Later that day, the salesman delivered the new vacuum to her residence. On his departure, he stated that he was certain that Mary would find the vacuum satisfactory, as the particular model was "the finest model that the company had produced."

The machine did operate in a satisfactory manner for some seven months. Then one day while Mary was using the vacuum to clean her automobile, she noticed a wisp of black smoke seeping from a seam in the casing. She immediately unplugged the machine and threw it in the swimming pool.

A few minutes later she retrieved the machine from the pool and returned it to the Speedy Vacuum Cleaner store. The repair man examined the machine and explained to her that the smoke had been caused by the melting of a small electrical part in the machine. He offered to replace the part free of charge even though the six-month written warranty had expired, but refused to provide free replacement for several other electrical parts that had been damaged by the machine's immersion in the swimming pool. The cost of repairs amounted to $80, and Mary paid the account. At the end of the month, however, she refused to make the final $50 payment under the purchase agreement, because she felt that the company should cover at least a part of the cost of the repairs to the machine.

Eventually, the company brought an action against her for the $50 owing under the purchase agreement.

Discuss the defences (if any) that Mary might raise in this case. Indicate the possible outcome.

CASE 7

Wily Willie sold kitchen gadgets door to door. One of his products was a tomato slicer that he stated would slice tomatoes "paper thin." His sales display included a picture of a tomato cut into slices of a uniform one-millimetre thickness. The caption on the picture stated: "Look at what our slicer does to a firm ripe tomato!" The photograph was of a very firm variety of tomato, noted for its uniformity. The instruction sheet that accompanied the gadget stated that the user should "select only firm tomatoes that have not fully ripened." Users were cautioned against using fully ripe or over-ripe tomatoes.

Charlie purchased one of the tomato slicers at a price of $19.95 and attempted to slice a tomato for his lunch. He ignored the instruction sheet and simply selected a tomato from his refrigerator. The gadget mashed the tomato instead of slicing it. Charlie tried to slice a second tomato and, when the machine mashed the second tomato as well, he became angry and smashed the slicer. He then sought out Willie, who was at the next house, attempting to sell his products to Charlie's neighbour. Charlie threw the smashed slicer at the salesman's feet and demanded his money back. When Willie refused, Charlie turned to the neighbour and said: "Don't buy anything from this crook! The junk he sells doesn't work!"

Discuss the legal issues raised in this case and advise Charlie and Willie of their rights.

CASE 8

Adrienne had been annoyed with the paint peeling from the iron railing on the stairs of her front porch. She had purchased some inexpensive paints in the past, and each time, after two or three months, rust had bubbled up from beneath the paint.

Exasperated, she returned to the hardware store. On this occasion, the store had a glossy cardboard end-of-aisle display of a premium-priced paint made by Protecto Paints Ltd. Printed on the display were the words "stops rust," and on the labels of the cans were the words "prevents rust."

Adrienne bought the paint and set out to apply it to the railing. The directions called for the removal of all prior paint and primer. For the most part, she was successful in removing the prior paint, but not the primer beneath.

After two years, the rust returned, flaking the paint. Adrienne informed the consumer ministry, who brought suit against Protecto Paints.

An internationally recognized expert on paint gave evidence that no paint known to the industry can stop rust indefinitely. The ability to stop rust ends when the seal is broken, and some paints keep a seal better than others. The expert advised the court that the Protecto formulation was the finest known to the industry, using the finest possible ingredients.

Render a decision on behalf of the court.

CASE 9

Luther Green was employed by an aircraft maintenance and repair company to repair and modify aircraft airframes and interiors. Green possessed the necessary Department of Transport licences to perform the type of work for which he was engaged. Since much of the work involved metal repair and refinishing, a certain amount of the work consisted of grinding and polishing, using power grinders and finishers.

While engaged in the grinding of a metal seat bracket in a large jet aircraft, Green decided to change grinding wheels on his power grinder in order to speed up the shaping of the part. He replaced the fine grit wheel on his grinder with a coarse grit wheel that bore the following warning on the package: "DO NOT USE AT MACHINE SPEEDS IN EXCESS OF 6,000 RPM."

Oliver Brown, a fellow employee, picked up the grinder after Green had completed the grinding work on the seat bracket, and began the grinding of a part of the wing assembly. He set the machine speed first at 5,000 rpm, but later increased the speed to 9,000 rpm, a common grinding speed. No sooner had the speed increased then the grinding wheel disintegrated, causing injury to Brown and a nearby worker.

What consumer-protection issues are raised by this incident? What rights (if any) would Brown have at law?

CASE 10

Jean Hamilton admired a used car that Honest Harry had on display at his car lot. Hamilton took the car for a test drive and found the vehicle to be ideal for her purposes. When she inquired about the previous owner, the salesman told her that it was his understanding that the last owner had been an elderly schoolteacher, who usually used the automobile only on weekends. The odometer on the automobile indicated that the vehicle had been driven only 80,000 kilometres.

Hamilton purchased the automobile, but discovered a few months later that the vehicle had been used as a taxi before it was purchased by the schoolteacher. The automobile, in effect, had been driven 100,000 kilometres further than the odometer indicated, as it registered only five digits before returning to zero — the true distance that the vehicle had been driven was 180,000 kilometres.

The automobile had given Hamilton no trouble during the time she had owned it, and she had driven the vehicle over 5,000 kilometres. She was annoyed, however, that the vehicle had had so much use, even though it still had a "like-new" appearance.

Hamilton brought an action for rescission of the contract when Honest Harry refused to take back the automobile and return the purchase price.

Discuss the argument that each party might raise in this case. Render a decision.

The Online Learning Centre offers more ways to check what you have learned so far. Find quizzes, Weblinks, and many other resources at www.mcgrawhill.ca/olc/willes.

CHAPTER 28

Law of Negotiable Instruments

Chapter Objectives

After study of this chapter, students should be able to:

- Differentiate among the various kinds of negotiable instruments.
- Explain the *Bills of Exchange Act*.
- Discuss claims for payments and the defences to certain claims.
- Explain consumer protection related to negotiable instruments.

YOUR BUSINESS AT RISK

Negotiable instruments are paper promises creating rights and duties, and can represent enormous sums of money. Since our markets and transactions demand both efficiency and integrity, a full set of rules has been developed to govern them, and because certainty is an equally important market concern, these rules are not flexible or open to much interpretation. As a result, failure to understand rules of negotiable instruments can cause significant and irretrievable losses, sometimes for just a technicality. It is up to you to know the rules, not for the system to make special exceptions for ignorance.

INTRODUCTION

Most business transactions are based upon the exchange of money for either goods or services — for example, goods are sold at a price that represents a money amount. Similarly, employees work for a wage or salary at an agreed-upon money rate, payable at specific intervals of time. Except for the simplest transactions, these contracts are generally settled by some means other than the actual payment of money by one party to the other.

Negotiable instruments play a significant role in commercial transactions, and in the creation of commercial credit and loan transactions. These latter two kinds of transactions affect most businesses, are the least visible to most observers, and often involve immense sums of money. Because these transactions depend upon an understanding of the rights and defences associated with negotiable instruments, they are a critical component in the study of business law.

Negotiable instrument

An instrument in writing that, when transferred in good faith and for value without notice of defects, passes a good title to the instrument to the transferee.

The most common form of payment used is the **negotiable instrument**, which for payment purposes is frequently in the form of a cheque, a generally accepted substitute for the actual payment of money. A cheque is only one type of negotiable instrument, however. Business persons may use other types, depending upon the form of settlement specified in the contract or agreement. Negotiable instruments in their simplest form are written promises or orders to pay sums of money to the holders of the instruments. They are represented in today's business world by cheques, promissory notes, and bills of exchange. The widespread use of these documents is due to the convenience and reduced risk that has been attached to them since early times. The nature of these "money substitutes" and the rights that attach to them are governed for the most part by the *Bills of Exchange Act*,[1] a federal statute.

Historical Development of the Law

A negotiable instrument is a written document that passes a good title to the rights contained in the document from a transferor to a transferee if the transferee takes the instrument in good faith and for value, without notice of any defect in the transferor's title.[2] The law relating to this unique instrument has its roots in early mercantile customs that date back to, and perhaps beyond, the time of the Roman Empire. The law developed from the practices of merchants in their dealings with each other, and from the decisions made to settle their disputes, first by their own guild members, and, later, by the courts.

Negotiable instruments were first used in international transactions between merchants. During the Middle Ages, a European merchant who sold his goods at the various "fairs" held throughout Europe might be reluctant to carry on his person the gold he received for his goods. In that case, he would arrange with a merchant in the fair town who had a business connection with a merchant in his own town to provide a note or order authorizing the payment to him (on presentation) of the amount set out in the document. In return for the note, the foreign merchant would pay over his gold to the local merchant and carry the note with him until he returned home. There, he would present the note for payment to the merchant named in the document and receive from him the required amount of gold. Over time these notes became known as **bills of exchange**.

Bill of exchange

An instrument in writing, signed by the drawer and addressed to the drawee, ordering the drawee to pay a sum certain in money to the payee named therein (or bearer) at some fixed or determinable future time, or on demand.

By the late nineteenth century, the law relating to negotiable instruments was relatively well settled, but found only in a myriad of court decisions, many being at slight variance with one another on the same point of law. To simplify and render certain the law, the English Parliament passed the *Bills of Exchange Act* in 1882.[3] The Parliament of Canada passed a statute in 1890 that was essentially the English statute of 1882, with a few minor modifications.[4]

The *Bills of Exchange Act*

The federal *Bills of Exchange Act* sets out the general rules of law in Canada that relate to bills of exchange, cheques, and promissory notes. Very few changes have been made to the Act since its introduction in 1890. Its provisions are very similar to the laws relating to these instruments in both the United Kingdom and the United States.

Today, the important features of bills of exchange are the particular features that made them so attractive to merchants centuries ago. In particular, a bill of exchange reduces the risk involved in transporting money from one place to another. Merchants no longer carry gold coin from place to place. However, its modern counterpart, legal tender of the Bank of Canada, would be required if some of the more convenient forms of negotiable instruments were not available for use in its place. In a sense, it is a convenient substitute for money.

A second advantage of a negotiable instrument is that it may be used to create credit. A great deal of modern commercial activity is based upon credit buying. Without the ease attached to the creation of credit by way of a bill of exchange, credit buying would not be as widespread as it is today.

> A Company wishes to purchase goods from B Company but will not be in a position to pay for the goods for several months. If it is agreeable to B Company, A Company may give B Company a promissory note payable in 60 days. A Company will receive the goods but must be prepared to honour the note later when B Company presents the note for payment.

1. R.S.C. 1985, c. B-4.
2. P.G. Osborn, *The Concise Law Dictionary*, 4th ed. (London: Sweet & Maxwell Ltd., 1954), p. 230.
3. 45 & 46 Vict., c. 61.
4. *Bills of Exchange Act*, 1890 (53 Vict.), c. 33, now R.S.C. 1985, c. B-4, as amended.

A third advantage of a bill of exchange is its negotiability, a particular attribute that permits it to be more readily transferred than most contractual obligations. In addition, in some circumstances a transferee of a bill of exchange may acquire a greater right to payment than the transferor of the bill. In this respect, the transferee of a bill encounters less risk in taking an assignment of the instrument than the assignee of an ordinary contract. Recent consumer-protection amendments, however, have altered this particular attribute of a bill of exchange if it is issued in connection with a consumer purchase. The change was designed to prevent consumer abuse through the use of bills of exchange by unscrupulous businesses. However, apart from this limitation, it remains as a method of reducing risk in an ordinary business transaction.

The *Bills of Exchange Act* deals at length with three general types of negotiable instruments: the **promissory note**, the **cheque**, and the bill of exchange. A cheque is a special type of bill of exchange. As a result, much of what might be said in general about a bill of exchange would apply to a cheque as well. A promissory note, on the other hand, differs in form and use from both the cheque and the bill of exchange.

Promissory note

A promise in writing, signed by the maker, to pay a sum certain in money to the person named therein, or bearer, at some fixed or determinable future time, or on demand.

Cheque

A bill of exchange that is drawn on a banking institution, and payable on demand.

Endorsement

The signing of one's name on the back of a negotiable instrument for the purpose of negotiating it to another.

Holder

The person in possession of a negotiable instrument.

Each of these instruments has particular features that lend themselves to specific commercial uses. Because they developed as a separate branch of the Common Law, much of the legal terminology associated with the instruments differ from that used in the law of contract. For example, a contract at Common Law is assigned by an assignor to an assignee, usually by a separate contract. A negotiable instrument, on the other hand, is negotiated by an endorser to an endorsee on the document itself. The endorser and endorsee are roughly the equivalent of the assignor and the assignee of a contract. The endorser is a person who holds a negotiable instrument and transfers it to another by signing his or her name on the back, and delivering it to the endorsee. The **endorsement**, together with delivery, gives the endorsee the right to the instrument.

> Jones is indebted to Brown and gives Brown a cheque for the amount of the debt. Brown is named as the payee on the cheque (i.e., he is named as the person entitled to payment). If Brown is indebted to Smith, he may endorse the cheque by signing his name on the back and delivering it to Smith. On delivery, Smith becomes the endorsee and the **holder** of the cheque. If Brown endorsed the cheque by signing only his name on the back, the cheque then becomes a bearer cheque, and Smith becomes the bearer, since Smith is in physical possession of the instrument.

The same terminology would apply if the cheque had been made payable to "bearer" instead of to Brown, since the person in possession of a cheque made payable to bearer is also called by that name.

The person in possession of a negotiable instrument is sometimes called a holder, but, to be a holder, the party must be either a bearer, a payee, or an endorsee. In addition to an ordinary holder, a person who paid something in return for the instrument is referred to as a holder for value, to distinguish such a holder from one who received the instrument as a gift. Since every party whose signature appears on a bill or note is presumed to have acquired the instrument for value, unless it can be established to the contrary, a holder of the instrument is usually considered to be a holder for value.[5]

Holder in due course

A person who acquires a negotiable instrument before its due date that is complete and regular on its face, and who gave value for the instrument, without any knowledge of default or defect in the title of prior holders.

A third type of holder is one who obtains special rights under a negotiable instrument. If a holder takes an instrument that is complete and regular on its face, before it is overdue, without any knowledge that it has been previously dishonoured, and, if the holder took the bill in good faith and for value and, at the time, had no notice of any defect in the title of the person who negotiated it, the holder would be a **holder in due course**.[6]

5. *Bills of Exchange Act*, R.S.C. 1985, c. B-4, s. 53(1).

6. ibid., s. 55(1).

The particular advantage of being a holder in due course of a negotiable instrument is the greater certainty of payment. Many of the defences that may be raised against an ordinary holder claiming payment are not available against a holder in due course. The particular advantages of being this type of holder are examined in greater detail in the part of this chapter that deals with defences available to the parties to a negotiable instrument.

CLIENTS, SUPPLIERS *or* OPERATIONS *in* QUEBEC

As a federal law, the *Bills of Exchange Act* applies uniformly across Canada.

BILLS OF EXCHANGE

The modern bill of exchange bears a close resemblance to earlier negotiable instruments, and while its use has declined to some extent in favour of cheques, it remains an important form of trade credit between commercial businesses. Its function combines formal security for goods (since one can sue for payment upon the promise it contains) with the creation of a credit arrangement, deferring payment for the goods secured by it (the future determinable date). For example, assume Wholesaleco in Vancouver imports clothing from abroad and ships it across Canada, selling it to Retailco in New Brunswick. It offers credit terms of three months, to allow the retailer to sell the goods in order to generate funds for payment. Wholesaleco wants an actionable, convenient promise for the eventual payment, however, and selects a bill of exchange.

Wholesaleco draws the bill to its own specifications, stating the amount (including the appropriate interest and cost of the goods), and then delivers it along with the goods and invoice. The buyer, Retailco will then "accept" the bill by signing it in the upper-left-hand corner on the diagonal line, most often as a condition of having the goods released from the truck. The bill is then returned to Wholesaleco, which may "cash" it at its own bank when the bill becomes payable (like a cheque), or send it to Retailco's bank for collection.

Bills of exchange are used by businesses because they are exchangeable. Wholesaleco may endorse it over, to pay its own accounts, to any other party willing to accept the creditworthiness of Retailco. In some cases, the creditworthiness of Retailco may be very attractive to others. For example, Wholesaleco may be a smaller import business in Vancouver, whereas Retailco is a major retail corporation, such as the Hudson's Bay Company. The added confidence of this payment assurance may be quite attractive to Wholesaleco's own bankers or suppliers of services.

Under the *Bills of Exchange Act*, "a bill of exchange must be an unconditional order in writing, addressed by one person to another, signed by the person giving it, and requiring the person to whom it is addressed to pay either on demand, or at a fixed or determinable future time, a sum certain in money to, or to the order of a specified person or to a bearer."[7] The Act is very specific that the document alleged to be a bill of exchange must meet these requirements. If the document fails to comply, or if it includes some other thing that a person must do in addition to the payment of money (except as provided in the Act) then the document will not be a bill of exchange.[8] Because a bill of exchange is not used by ordinary business or individuals nearly as often as cheques or promissory notes, a sample is shown in Figure 28–1.

7. ibid., s. 16(1).
8. ibid., s. 16(2).

FIGURE 28–1 Example of a Bill of Exchange

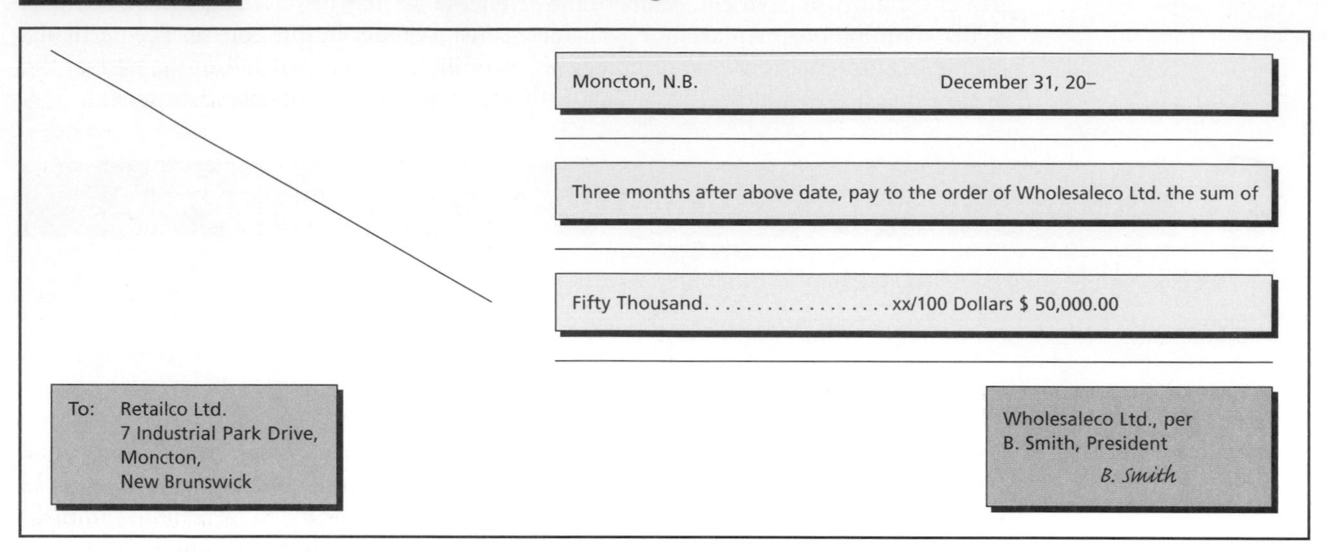
Moncton, N.B. December 31, 20–

Three months after above date, pay to the order of Wholesaleco Ltd. the sum of

Fifty Thousand. . . . . .xx/100 Dollars $ 50,000.00

To: Retailco Ltd.
7 Industrial Park Drive,
Moncton,
New Brunswick

Wholesaleco Ltd., per
B. Smith, President
B. Smith

The bill of exchange is a document in writing. It may be partly printed, partly handwritten, and partly typewritten. Each of these methods of writing is permissible, but it is essential that all of the important terms be evidenced in writing.

The bill is an unconditional order, as indicated by the words, "pay to the order of," and it is addressed by one person to another ("To: Retailco Ltd., 7 Industrial Park Drive, Moncton, New Brunswick). It is signed by the person giving it ("Wholesaleco Ltd., per B. Smith"), and it requires the person to whom it is addressed to pay (in this case at a fixed or determinable future time — "Three months after above date"). The bill is drawn for a sum certain in money ($50,000) and it is payable to the order of a specific person (in this case, Wholesaleco Ltd.).

The party who prepares the bill is called the drawer (Wholesaleco Ltd., per B. Smith); the drawee (Retailco Ltd.) is the party to whom it is addressed. The payee named is the party to receive the money, and may be the drawer (as in this case), a bank, or some other party.

This last point sometimes causes confusion, as we tend to think in terms of cheques, and the example above causes query, because the drawer and the payee are the same, giving the appearance of someone writing a cheque payable to themselves. Remember that the drawer Wholesaleco is merely setting out the terms of the transaction (being payment to Wholesaleco for Wholesaleco's goods), and it is up to Retailco to "accept" and be bound to make the payment as ordered.

Accordingly, once the bill is drawn, it is sent to the drawee for acceptance. The drawee "accepts" the bill by writing its acceptance of the bill across its face or in a corner of the bill set aside for acceptance (as in this case). In this example Retailco would write "Accepted" along with the date and where payable then sign by its officer in the corner of the bill. The drawee must deliver (i.e., return) the signed bill before acceptance is completed.[9]

Before the date for payment, a bill may be negotiated to other persons, known as holders. A holder usually negotiates a bill by endorsement. This is the act of placing one's signature on the back of the note, then delivering it to the new holder. The act of endorsement in effect represents an implied promise to compensate the holder or subsequent endorsers in the event that the bill is dishonoured.

It should be noted that the time for payment of bills of exchange is subject to certain "rules" that may add three days to the time specified for payment, depending upon the type of bill drawn. In this example, the bill is payable at a particular time that may be calculated from the information given, but the bill may also be made payable on demand or at sight. If the bill is payable "on demand" or "on presentation" or if it does not set

9. ibid., s. 38.

out a time for payment, it is considered to be a demand bill. It will be payable without acceptance by the drawee, unless it is payable other than at the drawee's place of residence or place of business.[10] A cheque is an example of a demand bill that is always drawn on a bank. In Canada, three days' grace is added to the payment date, except in the case of a demand bill, which, like a cheque, is payable immediately on presentation. A sight bill, which is similar to a time bill, stated that it is payable "at sight" or at a specific number of days after sight. Sight means "acceptance," and since three days' grace would be added in the case of a sight bill, the payment date in effect becomes three days after the date it is presented for acceptance. The bill may specify that the three days' grace will not apply, in which case it becomes payable on presentation.

It should be noted that a bill of exchange will not be invalid if it is not dated, has no place fixed for payment, no mention of consideration,[11] or if there is a discrepancy between words and figures.[12] However, if the bill is not dated, it must state when it is due, or it must contain the information necessary to calculate the due date. Under certain circumstances, the date may be added later if the bill is undated.[13]

The payee named in the illustration is also the drawer, but any person except the drawee may be named, or it may simply be made payable to the bearer. If the bill should be made payable to a fictitious person (such as Santa Claus or the Easter Bunny), the bill is still valid and becomes a bill payable to the bearer.[14]

Liability of the Parties to a Bill of Exchange

Acceptance of a bill of exchange by the drawee renders the drawee liable to pay the bill at the time and place fixed for payment, or, in the case of a demand bill, within a reasonable time after its issue.[15] The bill must also be presented for payment by the holder or an authorized representative at a reasonable hour on a business day[16] at the place specified in the bill. However, if no place is specified, then it may be presented at the drawee's address. When payment is made, the drawee is entitled to a return of the bill from the holder in order that it might be cancelled or destroyed.

If payment is refused, then the holder must act quickly if he or she wishes to hold the drawer and any other endorsers liable on the bill. If the bill is dishonoured by non-payment, the holder can sue the drawer, acceptor, and endorsers.[17] However, in order to hold the parties liable, the holder must give them an opportunity to pay the bill by giving each of them (except the acceptor) notice of the dishonour.[18] To be valid, the notice must be given not later than the juridical or business day next following the dishonour of the bill.[19] The notice may be either in writing or by personal communication (such as telephone, or other verbal means) but it must identify the bill and indicate that it has been dishonoured by non-payment.[20] The drawer or any endorser who does not receive notice of the dishonour is discharged from any liability on the bill,[21] unless the holder is excused from giving immediate notice as a result of circumstances beyond his or her control.[22] As soon as the cause for the delay ends, however, notice must promptly be given.[23] As a result, it is in the interests of all parties to make certain that the person from whom they acquired the bill receives notice. Consequently, all endorsers will generally give notice to prior endorsers to preserve their own rights.

10. ibid., s. 74(2).
11. ibid., s. 26.
12. ibid., s. 27.
13. ibid., s. 29.
14. ibid., s. 20(5).
15. ibid., s. 85(1)(b).
16. ibid., s. 77, for the time for presentation for acceptance. The time for payment would probably be somewhat similar.
17. ibid., s. 94(2).
18. ibid., s. 97.
19. ibid., s. 96.
20. ibid., s. 97(1)(d).
21. ibid., s. 95(1).
22. ibid., s. 104(1).
23. ibid., s. 104(2).

Endorsers who receive notice have the same length of time as the holder to give notice of the dishonour to those liable to them,[24] but it is common practice as well to give notice to all parties liable. Notice of dishonour may also be dispensed with to certain parties under certain circumstances,[25] such as to the drawer if the drawer has countermanded payment,[26] or to an endorser if the endorser is the person to whom the bill is presented for payment.[27] If the bill is a foreign bill of exchange, a special procedure must be followed in the event of non-payment. A formal procedure called protest is used for notice, and the protest must generally be made on the same day that the dishonour occurs.[28]

CHEQUES

A cheque is a form of bill of exchange that is payable on demand, where the drawee is always a bank. The bank, however, is a special type of drawee in the case of a cheque, because it does not become liable to a holder in the same way that an ordinary drawee does when a bill is presented for payment. A bank need only honour the cheque when it is presented in proper form and sufficient funds of the drawer are on hand at the bank to cover the cheque. If the drawer has insufficient funds on deposit to permit the bank to make payment, it may refuse to honour the cheque, and the holder will be obliged to look to the drawer (or other endorsers) for payment. The only circumstances under which the bank might be liable would be if the cheque was properly drawn and the drawer had sufficient funds on deposit to cover payment. Even then, it would be liable to the drawer rather than to the holder. An example of a cheque is shown in Figure 28–2.

FIGURE 28–2 Example of a Cheque

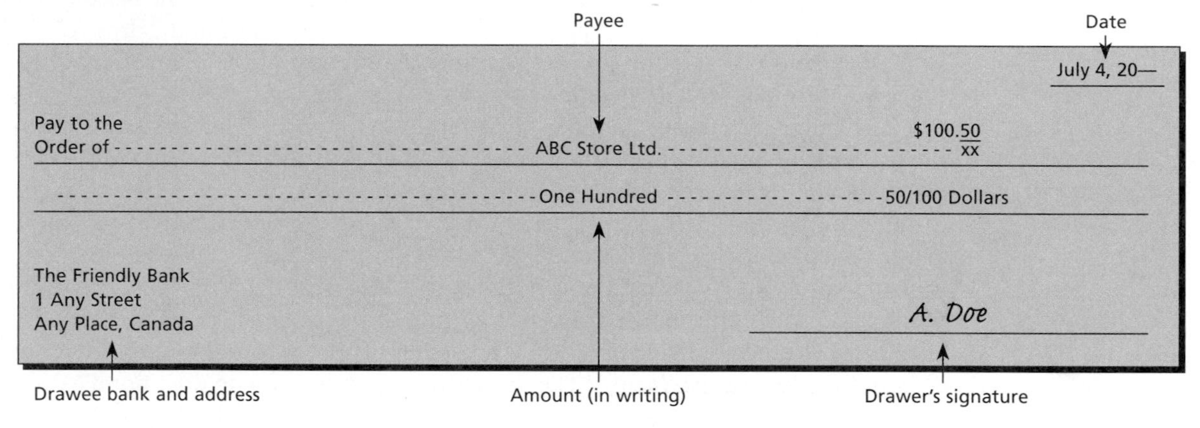

Note: Drawer's account number (not shown in the example) is printed on the cheque for electronic processing.

Certification

Certification
Of a cheque, an understanding by a bank to pay the amount of a cheque on presentation.

Cheques are sometimes presented to the bank for **certification**. The certification of cheques is a U.S. practice that is not covered by the *Bills of Exchange Act*, but one that has developed as a usage in Canada. As a result, the effect of certification on the rights of the parties to a cheque has for the most part been determined by the courts. This procedure alters the position of the bank with respect to the holder. The bank, on the presentation of the cheque, will withdraw the amount of the cheque from the drawer's account and place the funds in an account of the bank set aside for the purpose of payment. Once the funds

24. ibid., s. 100.
25. ibid., ss. 106 and 107.
26. ibid., s. 106(e).
27. ibid., s. 107.
28. ibid., s. 111 et seq.

have been removed, the bank is in a somewhat similar position to a person accepting a bill of exchange. Two forms of certification may take place: certification at the request of the drawer, and certification at the request of the payee or holder. If a bank is prepared to certify a cheque at the request of the drawer, the bank does so before delivery of the instrument, and in this sense it differs from ordinary acceptance of a bill of exchange. If the drawer should decide to return the cheque for cancellation before delivery, the drawer is entitled to do so, at which time the drawer may have the funds returned to his or her account. Countermanding the certified cheque before delivery would also have the effect of terminating the bank's obligation to honour the cheque.[29] However, if the cheque should be delivered to the payee, the right to countermand or stop payment on the cheque would appear to be lost, as once delivery is made to a holder, the bank would be obliged to honour the cheque when presented for payment.[30]

Certification at the request of the holder renders the bank liable for payment when the certified cheque is presented for payment at a later date. This is so because if the cheque is presented to the bank by the holder, the holder at that time would be entitled to payment. If the holder requests certification of the cheque instead, and the bank complies, the bank would become liable, and the drawer would be discharged from all liability. Since the drawer would in effect have made payment of the funds, the drawer would no longer be in a position to stop payment.[31] The bank would be obliged to pay the cheque when presented for payment at a later time, however.

As a result of recent cases, the courts have basically concluded that certification obliges the bank to pay the holder of a certified cheque. For example, Paisley, J., of the Ontario Court (General Division), in his judgment in *Centrac Inc. v. Canadian Imperial Bank of Commerce*,[32] summarized the similarity between cheques certified by the drawer and holder with the words:

> There appears to be no valid reason given in the submissions I have heard from coming to any other conclusion than, in a mercantile context, a properly certified cheque is a cash substitute and is irrevocable. There should be no need for the holder of a certified cheque to rush to a bank to deposit it, which would be the prudent course if a different conclusion were to apply where the cheque was certified by the drawer rather than the holder.

COURT DECISION: Practical Application of the Law

Cheques—Certification—Responsibilities of Bank

***Re Maubach and Bank of Nova Scotia* (1987), 60 O.R. (2d) 189.**

The drawer prepared a cheque and had the cheque certified by the bank where he maintained his account. The bank certified the cheque. The drawer subsequently lost the cheque before it was delivered to the payee. The drawer requested the bank to issue a new cheque to replace the missing cheque, but the bank was not willing to do so unless the drawer was prepared to give the bank a bond of indemnity to cover the lost cheque in the event that it should be found and presented for payment. The drawer refused to do so.

THE COURT: The binding effect of cheque certification is well established. The precise legal impact of certification is nevertheless quite controversial. The better view is to treat the certification of the cheque as an acceptance. ... I return again to the words of Sir Henry Strong at pp. 285-6 in the *Gaden* case, supra, after speaking of the effect of certification.

> The only effect of certifying is to give the cheque additional currency by showing on the face that it is drawn in good faith on funds sufficient to meet its payment, *and by adding to the credit of the drawer that of the bank on which it is drawn.* (Emphasis added by me.)

29. *Gaden v. Newfoundland Savings Bank*, [1899] A.C. 281 (P.C.).
30. *Centrac Inc. v. Canadian Imperial Bank of Commerce* (1994), 20 O.R. (3d) 105.
31. *Commercial Automation Ltd. v. Banque Provinciale du Canada* (1962), 39 D.L.R. (2d) 316.
32. (1994), 20 O.R. (3d) 105.

After referring to the law in the United States, Lord Wright, at pp. 186–7 in *Bank of Baroda v. Punjab National Bank*, [1944] A.C. 176, speaks thus:

> Certification makes the banker the debtor of the holder, and discharges the drawer altogether if the certification is not made by his procurement. Certification adds a new party, the bank, as primary debtor, and necessarily involves readjusting the legal position of the original parties, drawer and payee. A similar rule has been adopted, it seems, by the courts of Canada, on the basis of the custom in Canada judicially recognized by this Board in *Gaden v. Newfoundland Savings Bank.*

The conclusion I have reached is that although certification by the drawee bank is not "acceptance" of the cheque by the drawee bank within the meaning of s. 35 of the Bills of Exchange Act, certification is equivalent to an acceptance with the result that the drawee bank is liable on a certified cheque to the payee and any holder thereof. This follows from the authorities I have quoted. By certification, the respondent upon presentation undertakes to pay the cheque: *Campbell v. Raynor* (1926), 59 O.L.R. 466 at p. 470; *Commercial Automation Ltd.*, supra, at p. 321; *Boyd at al. v. Nasmith* (1888), 17 O.R. 40 at p. 46. I remind myself that in this application, I am not adjudicating on the rights of some holder of the cheque and I also remind myself that the respondent which seeks the bond of indemnity is at risk. Examples were given to me of situations wherein the respondent might suffer a loss but no good purpose would be served by canvassing them in detail. If the cheque is certified, the respondent cannot simply honour the applicants' countermand.

If that is the situation, should the bank be entitled to the indemnity it was seeking from the drawer upon replacement of the cheque? Explain why.

Effect of Presentation

An uncertified cheque, like a bill of exchange, is not legal tender and, if offered as payment to a creditor, represents only conditional payment of the debt. If the cheque should be dishonoured, the debt remains, and the creditor then may take action for payment on either the debt itself or the dishonoured cheque. A creditor or seller need not accept payment by way of a cheque. However, if the creditor or seller should decide to do so, and the cheque is honoured, the debt will be extinguished, and the drawer will have evidence of payment in the form of the creditor's endorsement on the back of the cheque. For this reason, a debtor giving a cheque in full or part-payment of a debt will often identify the purpose of the payment on the back of the cheque. The creditor's signature will then indicate that payment of the amount of the cheque had been received with respect to that specific account.

A cheque, being payable on demand, must be presented for payment within a reasonable time after its issue. This may vary depending upon the nature of the instrument, the customs of the trade, and the banks in the particular instance.[33] However, the fact remains that cheques should be promptly cashed on receipt; otherwise, circumstances could affect payment.[34]

Endorsement

Unless the cheque is made payable to the bearer, it is negotiated by endorsement in the same manner as any other bill of exchange. The endorsement may be in blank, in which case the endorser would sign only his or her name on the back. The cheque then could be passed from one person to another without further endorsement in the same fashion as a bearer instrument. If, however, the endorser wishes to restrict the payment to one person only, the endorsement might, for example, take the form "Pay to J. Brown only," followed by the endorser's signature. This type of endorsement is called a restrictive endorsement, and would prevent any further endorsement. Only J. Brown would be permitted to present the cheque to the bank for payment. A person may also use a restrictive type of endorsement to prevent the theft and cashing of a cheque. By writing the words "for deposit only to the account of ...," followed by the person's name, the cheque may not be cashed — it may only be deposited to that person's bank account. A third general type of endorsement is called a special endorsement. This type requires the person named in the endorsement

33. *Bills of Exchange Act*, supra. Most banks consider a "reasonable time" to be less than six months.

34. For example, the customer might die, in which case the duty of the bank to pay the cheque would have terminated. See *Bills of Exchange Act*, supra, s. 167(b).

to endorse the cheque before it may be negotiated to anyone. A special endorsement would read "pay to the order of J. Brown," followed by the endorser's signature.

Other forms of endorsement exist as well. A person may endorse a cheque for the purpose of identifying the person or the signature of the person negotiating a cheque. In this case, the party would not incur liability on the cheque, since the endorsement would be for identification purposes only. A typical endorsement of this nature might read, "J. Brown is hereby identified," followed by the signature of the person making the identification.

An endorsement may be qualified, which would limit the liability of the endorser (provided that the other party should be willing to accept such an endorsement) if dishonour should later occur. This type of endorsement is usually called an endorsement without recourse. It might read, "without recourse," followed by the signature, or it might limit the time for recourse — "without recourse unless presented within 10 days." The unwillingness of subsequent endorsers to accept this type of endorsement has limited its use, since subsequent endorsers must look to prior endorsers for payment in the event of non-payment by the drawer of the cheque.

By signing the back of a cheque or bill of exchange, an endorser, under the *Bills of Exchange Act*, impliedly contracts that he or she will compensate the holder or any other subsequent endorser in the event that the cheque is not honoured when presented for payment, provided that the necessary proceedings are followed by the holder on dishonour.[35] Endorsement also precludes the endorser from denying to a holder in due course the regularity in all respects of the drawer's signature, and that of all previous endorsements.[36] In addition, an endorser is precluded from denying to immediate or subsequent endorsers that the bill was a valid bill at the time of the endorsement, and that the endorser had a good title to it.[37]

Promissory Notes

A promissory note differs from a bill of exchange or a cheque in that it is, as its name implies, a promise to pay, rather than an order. It differs also in its form and acceptance, as well. By definition, "a promissory note is an unconditional promise in writing, signed by the maker of the note, to pay to, or to the order of, a specific person or bearer on demand, or at a fixed or determinable future time, a sum certain in money."[38] A promissory note in a business context is a negotiable instrument that is used extensively for the creation of credit on either a single-time loan basis, such as the purchase of a truck or equipment, or as evidence of a revolving-credit commercial loan negotiated with a bank.

From this definition, several differences between a promissory note and a bill of exchange are readily apparent. A note does not contain an order to pay, nor does it have a drawee who must accept the instrument. Instead, a note is a promise to pay, which is signed by the party who makes the promise. A simple promissory note might appear as in Figure 28–3.

FIGURE 28–3 Example of a Promissory Note

January 3, 20—

On demand after date, I promise to pay to The Friendly Bank or order,

-- Five hundred------------------------- xx/xx DOLLARS

At The Friendly Bank here, with interest at the rate of 10 percent per annum as well after as before maturity, until paid.

Value received J. Smith

35. ibid., s. 132.
36. ibid., s. 132.
37. ibid., s. 132.
38. ibid., s. 176(1).

To be negotiable, the note must meet the essentials of negotiability as set down in the *Bills of Exchange Act* and outlined in the definition. It is important that the time for payment be clearly determined if the instrument is other than a demand note, and also that the promise to pay is unconditional, and for a sum certain in money. It cannot, for example, be payable "if I should win a particular lottery," nor can it be payable in merchandise or goods, as neither of these stipulations would meet the definition of time or payment.

A promissory note, then, is signed by the maker, and contains a promise to pay a sum certain in money on certain terms. The note is incomplete until it has been signed and delivered to the payee or bearer.[39] But once this act has been accomplished, the maker of the note (with legal capacity) becomes liable to pay the note according to its terms to the holder.

Effect of Presentation

Like a bill of exchange or cheque, a promissory note that is payable on demand must be presented for payment within a reasonable time; otherwise, any endorser of the note may be discharged. But, if the note with the consent of the endorser is used as collateral or continuing security, then it need not be presented for payment as long as it is held for that purpose.[40]

The place of payment of a promissory note is normally set out in the body of the note, and, if so, presentation for payment must take place there if the holder of the note wishes to hold any endorser liable. If no place is specified, then usually the maker's known place of business or residence would constitute the place for payment. The time for payment is also important if the holder wishes to hold endorsers liable. As with bills of exchange, three days' grace would be added in the calculation of the time for payment for all promissory notes except those payable on demand.

Endorsement

Endorsers of promissory notes are in much the same position as endorsers of bills of exchange. The maker of a promissory note, by signing it, engages that he or she will pay the note according to its original terms, and is precluded from denying to a holder in due course the existence of the payee and the payee's capacity to endorse at the time.[41] If a promissory note is dishonoured when properly presented for payment at the date on which payment is due, the holder is obliged to immediately give notice of the dishonour to all endorsers if the holder wishes to hold the endorsers liable on the note. The *Bills of Exchange Act* provides for notice to the endorsers of a promissory note along the same procedural lines as set out for bills of exchange. The important difference is that the maker is deemed to correspond with the acceptor of a bill, and the first endorser of the note is deemed to correspond to the drawer of an accepted bill of exchange.[42]

Installment Payments

Promissory notes, unlike bills of exchange or cheques, frequently provide for installment payments. An installment note is often used as a means of payment for relatively expensive consumer goods, such as large household appliances, automobiles, and boats. The seller often may take a security interest in the goods as collateral security to the promissory note, or may simply provide in the note that title does not pass until payment is made in full. The advantage of using the promissory note for this purpose is that the note initially facilitates the sale by the seller to the buyer. The buyer need not pay the full price at the time of purchase, but may spread the cost of the purchase over a period of time. The advantage to the seller of the promissory note is that it is a negotiable instrument, and the seller may negotiate the note to a bank or other financial institution and receive a payment of money immediately. This method permits the seller to avoid having large amounts of his

39. ibid., s. 178.
40. ibid., s. 181.
41. ibid., s. 185.
42. ibid., s. 186.

or her own money tied up in credit transactions. Promissory notes of this nature normally provide for the payment of interest by the maker. Consequently, the financial institution that receives the note collects the interest as its compensation for its investment.

A promissory note that provides for installment payments usually provides that each installment payment is a separate note for payment purposes. However, if default should occur, the whole of the balance immediately becomes due and payable. The reason for this special clause is that, in its absence, the holder would only be entitled to institute legal proceedings to recover overdue payments as they occurred. This clause, which is known as an acceleration clause, permits the holder to sue for the entire balance of the note if default should occur on any one installment payment.

FRONT-PAGE LAW

BCI gets $580 million by selling notes issued through sale of Hansol

MONTREAL (CP) — Bell Canada International Inc. (BCI) said Tuesday it has raised $580 million in cash as a followup to the sale of its 21-per-cent interest in Korean wireless phone provider, Hansol M.com.

BCI said it received the funds by selling promissory notes that were issued to it by Korea Telecom as part of the sale of its interest in Hansol. BCI sold its stake in Hansol for a total of $1.7 billion in a deal announced in June. BCI is the international mobile phone subsidiary of Montreal-based telecom giant BCI Inc.

This excerpt demonstrates the versatility of promissory notes. While they are used at the consumer level, they are frequently found also in the largest of transactions. Being promissory notes, they can accommodate installment payments of both principal and interest at a series of future dates. They may be resold by BCI in whole or part to other investors, although they remain promises of Korea Telecom to pay in accordance with their terms.

Source: Excerpted from *Canadian Press*, "BCI gets $580 million by selling notes issued through sale of Hansol," August 22, 2000.

CHECKLIST FOR NEGOTIABILITY

A Bill of Exchange is an		A Promissory Note is an
unconditional order		unconditional promise
	in writing	
signed by drawer		signed by maker
addressed to drawee requiring drawee		
	to pay	
	on demand or at a fixed or determinable time	
	to the order of a specific person or bearer	
	a sum certain	
	in money	

Defences to Claims for Payment of Bills of Exchange

The holder of a negotiable instrument, whether it be a cheque, bill of exchange, or promissory note, is entitled to present the document for payment. If an instrument has two or more endorsers, each endorsement is deemed to have been made in the order in which it appears on the instrument, unless an endorser can prove that the contrary is the case.[43] In the event

43. ibid., s. 64.

of default, prior endorsers must indemnify subsequent endorsers. Liability to some extent follows the order of signing or endorsement, with the last person to receive the bill or note normally entitled to payment if the bill or note is properly presented for payment.

Not every holder may successfully receive payment when the instrument is presented, for instruments may be made from time to time under circumstances or contain defects that would entitle a party to the instrument to resist payment. Because a negotiable instrument is similar in many respects to a contract, the courts have applied many of the rules of contract to negotiable instruments. In spite of this, however, the rights of the holder differ substantially from those of an ordinary promisee in a contractual setting.

One of the advantages that a holder of a bill of exchange may have over an assignee of contractual rights is a right that is particular to bills of exchange in general. For reasons that are largely historical, and that arose out of early merchants' need for certainty in payment, a holder of a bill of exchange may, under certain circumstances, obtain a better right to payment of a bill than an ordinary assignee of contractual rights would acquire.

An ordinary assignee of contractual rights takes the rights of the assignor, subject to any defects that may exist in the assignor's title. If the contract was obtained as a result of some fraud or undue influence, or if the promisor had a right of set-off against the assignor, the assignee's right to payment from the promisor might be thwarted by such a defence. With a negotiable instrument, this is not always the case. If a negotiable instrument such as a bill of exchange, cheque, or promissory note is negotiated to a party for value and without notice of any defect in the instrument or the title of the prior holder, the holder who took the instrument under these circumstances may enforce the instrument against all prior parties in spite of any fraud, duress, undue influence, or set-off that may have existed between the original parties. The only case where such a holder (called a holder in due course) would be unsuccessful would be where the prior parties could establish that the instrument was essentially a nullity due to some defect such as forgery or the minority of the maker.

These special rights and how they developed were described by the judge in the case of *Federal Discount Corp. Ltd. v. St. Pierre and St. Pierre*[44] in the following manner:

> The rights which accrue to a holder in due course of a bill of exchange are unique and distinguishable from the rights of an assignee of a contract which does not fall within the description of a bill of exchange. The assignee of a contract, unlike the holder in due course of a bill of exchange, takes subject to all the equities between the original parties, which have arisen prior to the date of notice of the assignment to the party sought to be charged.
>
> The special privileges enjoyed by a holder in due course of a bill of exchange are quite foreign to the common law and have their origin in the law merchant.
>
> There is little difficulty in appreciating how trade between merchants required that he who put into circulation his engagement to pay a specified sum at a designated time and place knowing that it was the custom of merchants to regard such paper much as we do our paper currency, should be held to the letter of his obligation and be prevented from setting up defences which might derogate from the apparently absolute nature of his obligation.
>
> At first the customs prevailing amongst merchants as to bills of exchange extended only to merchant strangers trafficking with English merchants; later they were extended to inland bills between merchants trafficking with one another within England; then to all persons trafficking and finally to all persons trafficking or not.
>
> Thus in time the particular conditions which were recognized as prevailing amongst merchants became engrafted onto the law generally applicable and came to be looked on as arising from the document itself rather than from the character of the parties dealing with the document. It is significant, however, that the transition did not affect the legal position as to one another of immediate parties and that as between any two immediate parties, maker and payee, or endorser and endorsee, none of the

44. (1962), 32 D.L.R. (2d) 86.

> extraordinary conditions otherwise attaching to the bill, serve to affect adversely the rights and obligations existing between them as contracting parties. The document itself becomes irreproachable and affords special protection to its holder only, when at some stage of its passage from payee or acceptor to holder, there has been a bona fide transaction of trade with respect to it wherein the transferee took for value and without any notice of circumstances which might give rise to a defence on the part of the maker. Unless the ultimate holder or some earlier holder has acquired the instrument in the course of such a transaction the earlier tainting circumstances survive and the holder seeking to enforce payment of it must, on the merits, meet any defence which would have been available to the maker. Thus it appears that the peculiar immunity which the holding in due course arises not from the original nature of the document itself but from the quality which had been imparted to it by at least some one transfer of it.

The defences that may be raised on a negotiable instrument vary with the relationship that exists between the parties. Defences to an attempt to enforce a negotiable instrument may be divided into three separate classes, each good against a particular type of holder, if the defence can be proven.

Real Defences

Of the three classes of defences, the most effective are called real defences. Real defences are defences that go to the root of the instrument, and are good against all holders, including a holder in due course. These defences include the following:

Forgery

If the signature of a maker, drawer, or endorser is forged on a negotiable instrument, the holder may not enforce payment against any party through the forged signature unless the party claiming that it is forged is precluded from raising it as a defence either by conduct or negligence.[45] For example, White prepares a bearer cheque and forges Black's name as drawer, then negotiates it to Brown in return for goods sold to her by Brown. If Brown takes the cheque without knowledge of White's forgery of Black's signature, and presents it for payment, Black may raise the forgery of his signature as a defence to payment even though the holder, Brown, was innocent of the forgery. Brown's only right in this case would be to look to White for compensation.

Incapacity of a Minor

A minor cannot incur liability on a negotiable instrument; hence, it is a real defence against any holder, including a holder in due course.[46] If the party is insane, the same defence may apply in some circumstances, as the capacity to incur liability in the case of a negotiable instrument is co-extensive with the party's capacity to contract.[47]

Lack of Delivery of an Incomplete Instrument

If a drawer or maker signs an incomplete negotiable instrument, but does not deliver it, the lack of delivery of an incomplete instrument may be a real defence if another party should complete the instrument, and either negotiate it or present it for payment. Both elements must be present, however, as the lack of delivery alone does not constitute a real defence. For example, Smith signs a promissory note but does not fill in the amount. Later the note is stolen by Brown. If Brown fills in the amount and any other blanks, then negotiates it to Green, Smith may raise as a real defence the lack of delivery of an incomplete instrument. This defence would be good against all parties, even a holder in due course.[48]

---

45. *Bills of Exchange Act*, R.S.C. 1985, c. B-4, s. 48(1). Note that an agent without express or apparent authority to sign a negotiable instrument would render the instrument void insofar as the principal is concerned.
46. ibid., s. 46(1).
47. ibid.
48. J.D. Falconbridge, *The Law of Negotiable Instruments in Canada* (Toronto: Ryerson Press, 1923), p. 149.

Material Alteration of the Instrument

Under certain circumstances, a person may be able to raise as a real defence the material alteration of the instrument. This defence is limited to the changes made, however, and does not affect the enforcement of the instrument according to its original tenor. For example, if Martin draws a cheque payable to Baker for $100 and Baker alters the amount to $1,100 and negotiates the cheque to Doe, Doe may only be entitled to enforce the cheque for its original amount. The material alteration may be raised as a real defence by Martin, unless Martin was negligent by drawing the cheque in such a way that Baker could easily alter it.[49]

Fraud as to the Nature of the Instrument

Fraud as a real defence to payment is limited to those cases where *non est factum* may be raised as a defence. Fraud is normally not a real defence because a person signing a negotiable instrument owes a duty of care to all others who may receive the instrument. However, if the fraud is such that the person signing the instrument is unable to ascertain the true nature of the instrument as a result of infirmity, advanced age, or illiteracy, and is induced to sign the instrument honestly believing it to be something else, then fraud might be raised as a real defence.[50]

Cancellation of the Instrument

Cancellation of an instrument, if the cancellation is apparent on the face of the instrument, would be a defence against a claim for payment by a holder.[51] If, however, payment should be made before the due date, and the cancellation is not noted on the instrument, the careless handling of the instrument may allow it to fall into the hands of another who may negotiate it to a holder for value. Under these circumstances, the holder may be able to require payment by the maker or drawer of the instrument a second time, and the defence of cancellation would not hold.

Defect of Title Defences

Real defences are good against all holders, including a holder in due course. However, there are a number of other defences that are related to the title of a person that may be good against every holder except a holder in due course. A title may be defective if it is obtained by fraud, duress, or undue influence, or if the instrument is negotiated to another by way of a breach of trust or a promise not to negotiate the instrument after maturity. It may also arise where the consideration for the instrument is illegal, or where there is a total failure of consideration.[52] While fraud may be a real defence in cases where it is serious enough to constitute *non est factum*, it may be a defect of title defence as well, if the fraud is insufficient to constitute a real defence. For example, if a person is induced to sign a promissory note on the strength of false representations made by the payee, the defence of fraud may be raised by the maker as against the payee. However, it would not apply if the payee negotiated the note to a holder in due course. Duress and undue influence, as in the case of ordinary contract law, would be a good defence against a payee, or any other party to the instrument except a holder in due course.

A defect of title defence may also be available if a person charged with the responsibility for filling in the blanks on a negotiable instrument fills them in improperly, or releases an instrument to a holder when instructed not to do so. Similarly, if a maker or drawer prepares and signs a bill or note, and it is stolen in completed form, the absence of delivery would constitute a defect of title defence good against a holder, but not against a holder in due course.

---

49. See *Bills of Exchange Act*, R.S.C. 1985, c. B-4, ss. 144(1), 145 for examples of material alteration, but see *Will v. Bank of Montreal*, [1931] 3 D.L.R. 526, for negligence on the part of the drawer.

50. See, for example, *Foster v. MacKinnon* (1869), L.R. 4 C.P. 704.

51. *Bills of Exchange Act*, supra, s. 142.

52. ibid., s. 55(2).

Personal Defences

A third type of defence is a defence that is effective only as against an immediate party, and not as against a remote party. The principal personal defence is set-off, which entitles a party to raise as a defence the indebtedness of the party claiming payment.

> White owes Black $1,000, and gives Black a note for that amount due in 30 days' time. In the interval, Black becomes indebted to White for $500. If, on the due date, Black claims payment of the $1,000 note, White may set-off Black's indebtedness of $500 and pay only the remaining balance to Black.

A number of other personal defences also exist. The absence of consideration may be a defence that a drawer or maker may raise against a party who obtains the negotiable instrument. However, this may only be raised as a defence if the person holding the instrument has not given consideration for it, and no prior holder did either. If any prior holder gave consideration for the negotiable instrument, then the maker or drawer may not raise absence of consideration as a defence. The reason for this is that a holder, even though he or she did not give consideration, may enforce the instrument on the basis that the holder acquired all the rights of the prior holder.

Release or payment before maturity are also considered to be personal defences, when a release had been given or payment had been made before maturity. However, in each of these cases, the defences would only apply against the party who gave the release or who received the payment. The defence would not apply as against subsequent holders who had no notice of the release or payment.

TABLE 28-1 Defences to Claims for Payment of a Bill of Exchange

DEFENCE	DEFENCE EFFECTIVE AGAINST		
	HOLDER	HOLDER IN DUE COURSE	ENDORSER
Real Defences	X	X	X
Forgery			
Incapacity of Minor			
Lack of delivery of an incomplete instrument			
Material alteration			
Fraud as to the nature*			
Cancellation of instrument			
Defect of Title Defences	X		X
Fraud			
Duress			
Undue influence			
Total failure of consideration			
Personal Defences	X*		X (if immediate party)*
Set-off			
Absence of consideration			
Payment before maturity			

*Subject to exception with respect to certain types of defences.

CASE IN POINT

A company bookkeeper had signing authority for cheques drawn on the company bank account. Over several years, she prepared and signed a number of cheques made payable to fictitious persons with names similar to a subcontractor of the company, whose name could be confused with that of her spouse. She then deposited the cheques without endorsement in a bank account she held jointly with her spouse. The bank accepted the cheques, contrary to its policy of requiring endorsement on all cheques.

When the company eventually discovered her conversion of the money, she was dismissed, and the company sued the bank for the return of the money.

The issue before the court was whether the bank could avoid repayment of the money on the basis that it was a holder in due course. The court held that while the cheques were drawn payable to a fictitious person, and therefore would normally be treated as payable to the bearer, in this instance it would only apply if the bearer was entitled to the cheque as either the payee or a legitimate endorser when the cheques were deposited. Since this was not the case, the bank was not a holder in due course, and was liable to repay the money.

See *Boma Manufacturing Ltd. v. Canadian Imperial Bank of Commerce* (1996), 140 D.L.R. (4th) 463.

Consumer Protection and Negotiable Instruments

In 1970, the *Bills of Exchange Act* was amended to provide for two new types of negotiable instruments — consumer bills and consumer notes. These instruments are ordinary bills of exchange or notes that arise out of a consumer purchase. According to the Act, a consumer purchase is defined as one that is a purchase of goods or services other than a cash purchase from a person in the business of selling or providing consumer goods and services. It does not include the purchase of goods by merchants for resale, nor does it include the purchase of goods for business or professional use.[53]

The need for special legislation to govern negotiable instruments that arise from consumer purchases was recognized by the federal government in the 1960s. During that period, credit buying expanded rapidly, and companies that financed retail purchases frequently arranged with sellers to provide financing to consumers who purchased their goods. In those situations, the seller would agree to sell the goods on credit and, as a part of the sale, have the purchaser sign a promissory note to finance the purchase. The note would either be made payable directly to the finance company, or the seller would later sell the note to the finance company. In the latter case, the finance company would claim to be the holder in due course of the promissory note. In the event that the goods were defective or misrepresented by the seller, the purchaser could not withhold payment of the note to pressure the seller to correct the situation. Because the finance company was a remote party, and could enforce payment regardless of any breach of the contract of sale by the seller, the purchaser's only remedy was to take action against the seller (if the seller could be found).

Often, if the finance company took action against the purchaser, the courts would attempt to tie the seller and the finance company together in an effort to assist the hapless purchaser,[54] but in many cases, it could not do so. In the end, only legislation appeared to be the solution. In response to consumer demand, remedial legislation was introduced at the provincial and federal levels. The change at the federal level was reflected in the *Bills of Exchange Act* by identifying the particular negotiable instruments as consumer bills and consumer notes.

A consumer bill is a bill of exchange, including a cheque, that is issued in respect to a consumer purchase in which the purchaser or anyone signing to accommodate the purchaser is liable as a party. However, it does not include a cheque that is dated the day

53. ibid., s. 188.

54. See, for example, *Federal Discount Corp. Ltd. v. St. Pierre and St. Pierre* (1962), 32 D.L.R. (2d) 86.

of issue (or prior thereto), or a cheque that is post-dated not more than 30 days.[55] A consumer note is a promissory note that is issued in respect of a consumer purchase on which the purchaser or anyone signing to accommodate the purchaser is liable as a party.[56] Both of these instruments are deemed to arise out of a consumer purchase if the funds secured by the note are obtained from a lender who is not dealing at arm's length with the seller.[57] In other words, if the seller directed the purchaser to a particular lending institution, or arranged the loan from the lending institution to enable the purchaser to make the purchase, the note or bill signed by the purchaser would still be treated as arising out of a consumer purchase.

Under the Act, every bill or note arising out of a consumer purchase must be marked with the words "consumer purchase" before or at the time that the bill or note is signed. A consumer bill or note that is not so marked is void except in the hands of a holder in due course who had no notice of the fact that the bill arose out of a consumer purchase.[58] The Act provides penalties for violation, in addition to rendering the note void as against the purchaser. The principal thrust, however, is to eliminate collusion between sellers and lenders to avoid consumer-protection legislation. The *Bills of Exchange Act* provides that the holder of a negotiable instrument arising out of a consumer purchase is subject to any defences by the consumer that might be raised against the seller of the goods if the goods prove to be defective or unsatisfactory.[59] This amendment to the Act also facilitates the operation of the provincial consumer-protection laws by requiring the seller (or lender) in a consumer purchase to so identify the negotiable instrument.

MANAGEMENT ALERT: BEST PRACTICE

When a business gives a promissory note or cheque (for example, to a supplier in return for inventory for later sale), it is important not to underestimate the power of holders in due course. If the supplier negotiates the note or cheque to its bank or other holder in due course, that holder does not care if the inventory was not delivered or is defective. Unfulfilled promises of the supplier do not change the note's promise to pay. The rights (and limits to them) of the holder in due course are derived from the face of the note, not the underlying contract. As useful and important as it is, payment made via a negotiable instrument invites unforeseen players — who may have powerful rights of their own — into a commercial relationship.

SUMMARY

Negotiable instruments in the form of bills of exchange, cheques, and promissory notes are governed by the *Bills of Exchange Act*. Each of these instruments developed to meet the particular needs of merchants. Their operation was first governed by the law merchant; later, this body of law was absorbed into the Common Law. As a result, the law distinguished the negotiable instrument from an ordinary contract, particularly with respect to transfer. To be negotiable, an instrument must possess the essentials for negotiability. The instrument must be an unconditional order or promise in writing, signed by the maker or drawer, requiring the maker or the person to whom it is addressed to pay to a specific person or bearer, on demand, or at some fixed or determinable future time, a sum certain in money.

55. *Bills of Exchange Act*, supra, s. 189(1).
56. ibid., s. 189(2).
57. ibid., s. 189(2).
58. ibid., s. 190.
59. ibid., s. 191.

www.mcgrawhill.ca/olc/willes

If an instrument meets the requirements for negotiability, it may be negotiated by the holder to another person by way of delivery (if a bearer instrument) or by endorsement and delivery. The endorsement of a negotiable instrument renders the person making the endorsement liable to the holder in the event that it is dishonoured by the maker, drawer, or acceptor. A holder acquires greater rights under a negotiable instrument than an ordinary assignee of a contractual right. This is particularly true if the person who holds the instrument is a holder in due course. A holder in due course generally is entitled to claim payment even though a defect of title may exist between prior holders. The only defences good against a holder in due course are defences that may be called real defences (forgery, incapacity of a minor, and others that render the instrument a nullity). Under the Act, special instruments called consumer bills and notes must be so marked to distinguish them from other negotiable instruments. These particular instruments arise out of consumer purchases and, by marking them as a "consumer purchase," may limit the right of a holder to claim the rights of a holder in due course in the event that payment is resisted by the maker or drawer for a breach of the contract of sale from which the instrument arose.

Key Terms

Negotiable instrument, 550
Bill of exchange, 551
Promissory note, 552
Cheque, 552
Endorsement, 552
Holder, 552
Holder in due course, 552
Certification, 556

Review Questions

1. Why is acceptance of a bill of exchange important?
2. How does a cheque differ from the usual type of bill of exchange?
3. Define a "bill of exchange." Indicate how it is determined to be "negotiable."
4. What is the purpose of a bill of exchange in a modern commercial transaction?
5. Distinguish a "sight bill" from a "demand bill."
6. Define a "holder in due course." Explain how a holder in due course differs from an ordinary holder of a bill of exchange.
7. Outline the procedure to be followed when a bill of exchange is dishonoured by non-payment.
8. When a holder in due course of a promissory note attempts to enforce payment, what types of defences might be raised by the maker named in the note?
9. Define a "promissory note." Distinguish it from a "bill of exchange."
10. Explain how an endorsement in blank differs from a restrictive endorsement. Explain the circumstances under which each might be used.
11. Promissory notes that call for installment payments often contain acceleration clauses. Why is this so, and what is the purpose of such a clause?
12. Indicate the different treatment at law that is given to a cheque certified at the request of the holder as opposed to a cheque certified at the request of the drawer.
13. What is a "defect of title" defence? What type of holder of a promissory note or bill of exchange would this type of defence be effective against?
14. Outline the various personal defences available. Indicate the type of holder that they might be raised against.

Mini-Case Problems

1. X gave Y a post-dated cheque for $3,000 as payment for a well that Y drilled on X's farm. The cheque fell due in 30 days' time. Y negotiated the cheque to Z for $2,800, a few days after it was given as payment for the well drilling. Before the end of the month, the well ran dry, and X stopped payment on the cheque. Advise Z of his rights as the person in possession of the cheque.
2. B purchased an automobile from C for $500. B gave C a cheque for $500 that C, a minor aged 16, endorsed to D as payment for D's motorcycle that C had purchased. D presented the cheque for payment, only to discover that B had insufficient funds in the bank for payment, and the cheque was dishonoured.

   Advise D of the law in this instance.
3. McMullan opened an account at a bank and received a box of personalized cheques bearing his name and address. Later, he closed the account. Several months after he closed the account, he wrote a cheque on the account and gave it to a friend as payment of a loan he had received from the friend. Discuss the issues raised in this case.

4. Duarant purchased a new automobile, giving a cash down payment and a promissory note for the balance. The automobile dealer assigned the note to a local finance company. The vehicle proved to be totally unreliable, and when the dealer was unwilling to take back the vehicle, Duarant stopped making payments on the promissory note. The finance company sued him for the balance owing.

   Discuss the issues that will be raised at trial.

CASE PROBLEMS FOR DISCUSSION

CASE 1

Subdivision Construction Ltd. was a developer of residential construction projects. Much of its construction work was contracted out to specialty firms. Joseph Walters Co-Carpeting Ltd. was one of the subcontractors and for many years supplied and installed all of the flooring and carpeting in the houses built by Subdivision Construction Ltd. Joseph Walters Co-Carpeting Ltd. billed on a per house basis, and in most months billed for perhaps a dozen houses. Harvey, the accounts payable clerk at Subdivision Construction Ltd., prepared the cheques to pay the accounts, and had them signed by an authorized signing officer for the corporation. Because of the length of the corporate name of the flooring contractor, everyone in the office referred to the contractor as 'Co-Carpeting Ltd.,' and Harvey used that abbreviated name on its payment cheques.

Many months later, Harvey realized that he could defraud his employer. He registered a business name as 'Co-Carpeting Sales,' and opened a bank account at his own bank in that name. From time to time thereafter he would prepare a cheque payable to Co-Carpeting Sales and include it in the group of cheques for Co-Carpeting Ltd. The cheque always fell within the mid-amount of the billing by Co-Carpeting Ltd., and the signing officer signed the pile of cheques without noticing the different payee name. Harvey would deposit the cheque in his Co-Carpeting Sales account, and when the paid cheque was returned to Subdivision Construction Ltd. by its bank, Harvey would remove the cheque and destroy it. Several years later, an audit discovered Harvey's scheme, but by this time Harvey had spent all of the money and was virtually penniless. Subdivision Construction Ltd. then decided to take legal action against Harvey's bank for return of the money, which was calculated to be $200,000.

Discuss the arguments of the parties, and render a decision.

CASE 2

Custom Canoe Ltd. offered canoes for sale by way of a newspaper advertisement. Carlo, age 17, responded to the advertisement, and visited the Custom Canoe showroom. There, he saw a canoe that fitted his needs, and he wrote a cheque for the full amount of the canoe. He took immediate delivery, put the canoe on the roof of his car, and drove home. The next day, he set out on a whitewater canoe trip that would take up his full two weeks vacation. In the meantime, Custom Canoe Ltd. deposited Carlo's cheque in its bank, and its bank duly presented it for payment at Carlo's bank. Because Carlo had insufficient funds in his account, the bank returned the cheque 'NSF.' After Carlo had set out on his vacation, he realized that he had insufficient funds in his account to cover the cheque, but decided to leave the matter until his return home. While on vacation, the canoe did not perform as well as expected, and on several occasions was damaged by rocks in the river. Carlo pondered what he should do about his purchase that was now seriously damaged.

Advise the parties.

CASE 3

Casey purchased a used four-wheel-drive truck from Sam's Off-Road Vehicles for $26,000. The vehicle was licensed as a commercial vehicle, but Casey intended to use it primarily as transportation to and from his employment at a local manufacturing plant. Apart from this type of driving, he expected to use it occasionally in his part-time work as a fishing guide.

He signed a promissory note to Sam's Off-Road Vehicles for $25,000 that called for payments of principal and interest of $1,000 per month over a 30-month term. Sam's Off-Road Vehicles immediately sold the note to Easy Payment Finance Co. for $24,000. A few days later, Casey was notified by letter to make all payments on the note to Easy Payment Finance Co.

Before the first payment was due, Casey discovered that the truck was in need of extensive repairs and returned it to Sam's Off-Road Vehicles. The company refused to take back the truck and return Casey's $1,000 down payment. Casey then refused to make payments on the promissory note.

Some months later, Easy Payment Finance Co. brought an action against Casey for the amount owing on the note.

On what basis would Easy Payment Finance Co. claim payment? What defences might be available to Casey?

Render a decision.

CASE 4

Hanley Supply Co. sold Roberts Retail Inc. a quantity of goods for $10,000 on 30 days' credit. As agreed by the parties, Hanley Supply Co. drew a bill of exchange on Roberts Retail Inc., naming itself as payee. The bill was payable in 30 days' time. Roberts Retail Inc. accepted the bill and returned it to Hanley Supply Co. Hanley Supply Co. then endorsed the bill to Smith Manufacturing to cover its indebtedness for goods purchased from the company. Smith Manufacturing, a small firm, endorsed the bill in blank to Brown, the company's office manager as a retirement gift, rather than wait until the bill became due to obtain the funds. Brown, on receipt of the bill, delivered it without endorsing it to her friend, Jones, whom she owed a sum of money. Jones, in turn, endorsed the bill and sold it to Doe for $9,000. On the due date, Doe presented the bill for payment, and it was dishonoured.

Advise Doe of his rights. Explain the liability (if any) of each of the parties.

CASE 5

Carla bought a number of products from Scoville Limited in a single mail order, and enclosed a cheque for $130 with the order, drawn on the Bank of Hamilton. In the interval between mailing the order and the arrival of the products, Carla noticed a few of them (totalling $65) were available locally at a much lower price.

On the day the products arrived, she visited her bank and was pleased to see that her cheque had not yet been cashed. She placed a stop-payment order on her cheque. In filling out the request slip, she placed the words "goods unsatisfactory" in the box allotted for the reason for the request. She decided she would send back the half of the order that she had now bought more cheaply elsewhere, and assumed that Scoville would send her a new invoice for $65.

The Bank of Hamilton failed to immediately enter the request into its computer system, and as a result, on the arrival of the cheque a day later, it paid Carla's cheque out of her account in the normal manner. Carla discovered this error in the course of using an automated cash machine a few days later, and asked the bank to correct the error. The bank put $130 back into Carla's account and told her that they would collect back the $130 that they had paid Scoville's bank, The Bank of Manitoba. The Bank of Hamilton returned the cheque in the clearing system, now marked "Payment Stopped," and demanded $130 from the Bank of Manitoba.

The Bank of Manitoba refused the stopped cheque and would not make payment back, stating that by accepted banking convention, too much time had elapsed between acceptance by The Bank of Hamilton and the return of the item. While this had been going on, Scoville Limited had received the goods returned by Carla and had mailed her a refund cheque for $65, for as far as they knew, they had been paid in full.

Carla was pleased. Clearly a computer error had sent her a $65 cheque rather than a $65 invoice, and she ignored the whole matter.

Assume another week passes. Discuss the events that follow, and the positions of the parties, with respect to the law of negotiable instruments as it is written. In advising the banks, what would you suggest that they add to their standard form account-operation agreements?

CASE 6

Transport Corp. entered into an agreement to purchase three trucks from a competitor, Delivery Inc. Payment was made by way of a cash payment and a promissory note for $120,000, bearing interest at 6 percent, and due in ten months from the date of issue. The note provided that the principal was repayable in 12 monthly payments of $10,000 each, with the interest payable at the end of the 12 months. The note did not contain an acceleration clause in the event of default.

Transport Corp. made the first monthly payment of $10,000. At that time, Delivery Inc. suggested that the note be replaced by a new promissory note that contained an acceleration clause, as this proviso had been discussed during negotiations, but omitted in error when the transaction was finalized. Transport Corp. agreed to do so, and a week later they forwarded by mail a new promissory note for $110,000 on the same repayment terms and containing the requested acceleration clause. A covering letter stated that the new note was given in accordance with an agreement to have it replace the existing promissory note.

Before the next monthly payment became due, a dispute arose over the condition of some of the purchased vehicles, and Transport Corp. refused to make the monthly payment when it became due and payable. During the course of the discussion that followed, Delivery Inc. threatened to implement the acceleration clause and sue for the balance of the debt owing. Transport Corp. replied that the second promissory note had been signed by the office receptionist in error, so the note was not enforceable, as it was not signed in the corporation's name, nor was it signed by an officer of the corporation. It also informed Delivery Inc. that the receptionist was a 17-year-old minor.

Discuss the issues raised in this case. Advise Delivery Inc. as to its rights (if any) and its position on the debt owed by Transport Corp.

CASE 7

Sugar Confectionery Ltd. borrowed a sum of money from its banker, the Big Business Bank, to purchase certain production equipment. The promissory note that the bank prepared and that James Anawak, the president of the corporation, signed, reads as follows:

> April 1, 2011
>
> I hereby promise to pay on demand to the Big Business Bank, Metro Branch, the sum of Fifty Thousand Dollars ($50,000) together with interest thereon at 10 percent per annum calculated from April 1st, 2011, until the date of payment.
>
> Value Received
> James Anawak
> President

The bank placed the $50,000 in the Sugar Confectionary Ltd. bank account, and the corporation drew a cheque on the amount to pay the equipment supplier.

Some months later, the shareholders of Sugar Confectionary Ltd. removed James Anawak as president of the company and elected Jane Bellamy in his place. Shortly thereafter, the Big Business Bank endorsed the note to Big Finance Company in return for the sum of $48,000.

On September 1st, Big Finance Company contacted Sugar Confectionery Ltd. and demanded payment. Sugar Confectionery Ltd. refused to pay and stated that it was not indebted to Big Finance Company or, for that matter, to any other creditor.

Discuss this situation, evaluate the claim of Sugar Confectionery Ltd., and outline the nature of the arguments the parties might raise if the matter should come before a court.

Render a decision.

CASE 8

Ascot was in the process of negotiating the purchase of an oil painting from The Macey Art Gallery. As a result of a number of telephone calls to the gallery owner, he eventually convinced the owner to sell the painting to him for $1,000. He prepared a cheque in the amount of the purchase price and signed it. However, because he was uncertain as to the exact spelling of the gallery's name, he left that part of the cheque blank. He placed the signed cheque in his office desk drawer, with the intention of making a telephone call to the gallery later in the day for the information necessary to complete it.

Ascot determined the gallery's name while at lunch, but when he returned to the office, he discovered that the cheque had been stolen.

Hines, a fellow employee of Ascot, had taken the cheque, filled in the cheque payable "to bearer," and used it to purchase items at a store where Ascot frequently shopped. The store owner accepted Ascot's cheque without question, as he was familiar with his signature, and later presented it to Ascot's bank for payment.

Within minutes after the bank had paid the cheque, Ascot telephoned to have the bank stop payment.

Advise the parties of their respective rights (if any) and liability (if any).

The Online Learning Centre offers more ways to check what you have learned so far. Find quizzes, Weblinks, and many other resources at www.mcgrawhill.ca/olc/willes.

CHAPTER 29

Security for Debt

Chapter Objectives

After study of this chapter, students should be able to:

- Distinguish between the various forms of security for debt.
- Describe the statutory protection available to creditors.
- Explain the chief rules that establish priority between competing interests of creditors.

YOUR BUSINESS AT RISK

A credit relationship may be an essential term in a sale or service contract, or it may be a contract in itself, fundamental to the capital structure of your firm. As is the case in many advanced areas of the law, technical rules often prevail, and the equity (fairness) argument is difficult to make if you have not made the effort to learn the rules. Given the power of compound interest, firms can quickly find that they have either committed to too much debt, or too little security, and they are caught by repayment obligations they agreed to without full knowledge of their scope or limit. By the same token, prudent firms that know their rights as debtor or creditor are in a position to use this knowledge to their competitive advantage.

INTRODUCTION

The extension of credit or the loan of money by a lender on the strength of the borrower's unsupported promise to repay carries with it considerable risk and a blind faith in the integrity of the borrower. Because debtors have not always been as good as their word, creditors have looked beyond the assurances of the debtors to their lands and goods in order to ensure payment.

Secured creditor

A creditor that may look to particular assets of the debtor to ensure payment of the debt.

The importance to lenders of becoming a **secured creditor** (as opposed to an unsecured creditor) cannot be stressed enough, in every instance where the time, effort and business sensibility can possibly permit (or demand) it. Eventually you will encounter a borrower — a credit customer — who cannot pay, and only if you are a secured creditor will you stand a chance of recovering all or a significant portion of what is owed to you, because you may look to the security to repay the debt. If you are an unsecured creditor your claim for payment ranks behind all secured creditors. Only after the claims of secured creditors have been satisfied will you share in the remainder (if any) of the borrower's assets.

One of the earliest methods of securing payment of a debt was the mortgage transaction, in which an interest in the debtor's lands would be transferred to the creditor on the condition that it be returned when the debt was paid. In this arrangement, the creditor could retain the property in satisfaction of the debt if the debtor failed to pay. Many debtors did not possess land, however. In those instances, creditors were obliged to consider the debtor's **chattels** (such as stock-in-trade, furniture, animals, tools, and jewellery) as

Chattels

Moveable property.

security for the debt. This security was effected by either the physical transfer of possession of the goods to the creditor, or by the grant of an interest in the property by way of a transfer of title. Both of these methods are in use today.

The possession of the debtor's property by the creditor forms the basis for the pledge or pawn bailment situation. However, as a debt relationship, possession is for the most part limited to smaller, expensive chattels, such as jewellery and other valuables. The second type of security relationship, the transfer of an interest in the goods, is perhaps the most versatile from a business point of view in that it may use virtually any type of identifiable chattel to secure debt. Of these two methods, the transfer of physical possession of the goods to the creditor is probably the oldest, and one of the first to fall subject to statutory control in England.[1]

The expansion of commerce during the nineteenth century created a greater demand for credit and for new means to protect the interests of creditors. During this period, the courts came to recognize the different interests of the debtor and creditor in chattels, and of the need to utilize goods as security for debt. This change in attitude was reflected in the development of a variety of legal instruments designed to create security interests in chattels as well as real property.

In England, the chattel mortgage was the first security instrument (apart from the pledge or pawn) to receive statutory recognition and general acceptance. Closely following the chattel mortgage was the hire-purchase agreement. Unlike the chattel mortgage, which was similar in purpose and effect to the land mortgage, the hire-purchase agreement represented a means of facilitating the sale of goods on credit, whereby the seller retained title to the goods until the goods were paid for, but the buyer acquired possession. While the chattel mortgage could also be used for this purpose, the hire-purchase agreement provided a much simpler procedure for the seller to follow in the event of default by the buyer.

During this same time, economic activity of a rapidly growing business community in Canada produced a demand for additional means of securing credit, and a parallel desire by creditors for means whereby the payment of the debts might be secured. The various security instruments used in England were quickly adopted by Canadian debtors and creditors, occasionally in modified form. The hire-purchase agreement, in particular, found favour in a slightly modified form as the conditional-sale agreement. In addition, new security instruments were developed to facilitate the expansion of business and the sale of goods on credit. Chartered banks in Canada were permitted, under the *Bank Act*,[2] to acquire a security interest in the present and future goods of a debtor in priority over subsequent creditors by following a simple filing procedure set out in the legislation. Laws were also passed to give workmen, suppliers, and building contractors security interests in land by way of **mechanics' lien** legislation. Shortly after the turn of the last century, creditor protection in the form of a *Bulk Sales Act* was introduced to protect the interests of unsecured trade creditors, where a sale "in bulk" was made of business assets.

Mechanics' lien

A lien exercisable by a worker, contractor, or material supplier against property upon which the work or materials were expended.

The rapid growth of commerce during the last century and the tremendous expansion of consumer credit, coupled with the mobility of the population, particularly after 1950, eventually created a need for more modern legislation to deal with security interests.

Forms of Security for Debt

Apart from the mortgage of real property (which is examined in Chapter 23, "The Law of Mortgages"), there are a number of methods that may be used by the parties to credit transactions in order to provide security for the debt. The most common forms of security with respect to chattels are the chattel mortgage, the conditional-sale agreement, and the bill of sale. These may be used in appropriate forms for commercial- and consumer-credit transactions. In addition, a creditor of a commercial firm may take an assignment of book debts to secure the indebtedness of the merchant, while a chartered bank, under the *Bank*

1. *An Act Against Brokers* (1604), 1 Jac. I, c. 21.
2. 1890, 53 Vict., c. 31, s. 74 (Can.).

Act, may acquire a security interest in the inventory of wholesalers, retailers, manufacturers, and other producers of goods. Corporations may pledge their real property and chattel assets as security by way of bonds and debentures (including a floating charge). Each of these instruments creates or provides a security interest in the property that a creditor might enforce to satisfy the debt in the event that the debtor should default in payment.

Other special forms of security are available to certain types of creditors, or to creditors in certain circumstances. These include the right of lien available to workers, suppliers, and contractors in the construction industry, and the statutory protection of unsecured trade creditors of a business where the merchant makes a sale in bulk. All of these forms of security are subject to legislation in each province or territory. Although variation exists with respect to each under the different statutes, the nature of the particular security is usually similar in purpose and effect.

Chattel Mortgage

Chattel mortgage

A mortgage in which the title to a chattel owned by the debtor is transferred to the creditor as security for the payment of a debt.

Before the development and judicial recognition of the validity of the **chattel mortgage**, the most common method of using chattels as security for debt was for the creditor to take physical possession of the goods until payment was made. This early form of security involved the transfer of possession, but not title, to the creditor, with an arrangement whereby the creditor might keep the goods or dispose of them if the debtor should default. This form of security for debt has been subject to statutory regulation since as early as 1604 in England, and remains so today. The obvious drawback of this method is that the debtor is deprived of the use of the goods, and the creditor is obliged to care for and protect the goods while they are in his or her possession. To overcome the disadvantages of this form of security, the chattel mortgage was devised. This permitted the debtor to retain the goods, but granted the creditor title until the debt was paid.

Operation

In concept, the chattel mortgage is essentially the same as the real-property mortgage after which it was designed. Under a chattel mortgage, the debtor transfers the entire interest in the property to the creditor, subject to the right to possession while not in default, and the right to redeem. In Canada, chattel mortgages are subject to legislation in all of the Common Law provinces and territories. Where Personal Property Security Acts are in force, the rights of the parties with respect to chattel security interests are governed by these statutes. Though provincial variation exists, the nature of the instrument is similar in each province and territory, and the mortgage form itself is largely standardized.

Under a chattel mortgage, the title to the property is transferred to the chattel mortgagee, and the mortgagor retains possession. The mortgage sets out the covenants of the mortgagor, the most important being the covenant to pay the debt and the covenant to insure the goods for the protection of the mortgagee. The mortgage also sets out the rights of the parties in the event of default.

If the mortgagor fails to pay the debt as provided in the chattel mortgage, the mortgagee normally has the right to take possession of the mortgaged goods, and either sell the goods by public or private sale, or proceed with foreclosure.[3] In most jurisdictions, the mortgagee, after taking possession of the goods, must give the mortgagor an opportunity to redeem the goods before the sale or foreclosure may take place. If the goods are sold, any surplus after payment of the debt and the costs of the sale belongs to the mortgagor. However, if the proceeds of the sale are insufficient to cover the debt, the mortgagor remains liable for any deficiency.

The mortgagee's right to foreclosure in the case of a chattel mortgage is much like that of a mortgagee under a land mortgage, but the procedure is less formal.[4] The mortgagor is entitled to an opportunity to redeem under the foreclosure procedure. However, if the mortgagor fails to redeem, the mortgagee then obtains foreclosure, which vests the ownership of the property absolutely in the mortgagee. Foreclosure is seldom used by

3 *Rennick v. Bender*, [1924] 1 D.L.R. 739.

4. *Carlisle v. Tait* (1882), 7 O.A.R. 10. See also *Warner v. Doran* (1931), 2 M.P.R. 574.

mortgagees, not only because the procedure is more involved than sale proceedings, but also because the mortgagee is generally interested in receiving payment of the debt rather than acquiring the chattel.

Unlike the real-property mortgagor, who is usually free to sell or dispose of the mortgaged property without the consent of the mortgagee, the chattel mortgagor may not sell the mortgaged chattels without consent. If a mortgagor should sell the goods without the mortgagee's consent, and the chattel mortgage had been properly registered to provide notice to the buyer, the buyer of the goods takes them subject to the mortgage. For this reason, a person who purchases goods from other than a merchant should make certain that the goods are free from encumbrances, and that the seller is the lawful owner before the purchase is effected. This may be done by way of a search in the appropriate public office where registers of chattel mortgages and other security interests in chattels are maintained.

A chattel mortgage may be assigned by a chattel mortgagee if the mortgagee so desires, and this may be done without the consent of the mortgagor. The formalities associated with assignments must be complied with, however, and the mortgagor must be properly notified of the assignment to the assignee. The assignment of a chattel mortgage in this regard is much like the assignment of a real-property mortgage.

Registration

The use of a chattel mortgage as a means of securing debt required the provinces to put in place a number of procedures to protect the creditor and other interested parties. Since a chattel mortgagor retains possession of goods and has the equity of redemption, it was necessary to provide third parties with notice of the mortgagee's interest in the goods.

To make this information available to the public at large, all jurisdictions have established either a central registry for the province or regional registries (usually at a land registry office), where a chattel mortgagee might register the mortgage. The registration must be made within a short time after the mortgage is executed, and has the effect of giving public notice of the creditor's security interest. The failure to register renders the chattel mortgage void as against any person who purchases the goods for value and without notice of the mortgage, or as against any subsequent encumbrancer who has no actual notice of the prior chattel mortgage. It does not affect the validity of the mortgage as between the mortgagor and the mortgagee, however.

Conditional-Sale Agreement

Conditional-sale agreement

An agreement for the sale of a chattel in which the seller grants possession of the goods, but withholds title until payment for the goods is made in full.

The **conditional-sale agreement** differs from a chattel mortgage in that it is a security interest that arises out of a sale rather than a conventional debt transaction. The conditional-sale agreement had its beginnings in the English hire-purchase agreement. The hire-purchase agreement was devised in the early nineteenth century[5] to permit a prospective purchaser of goods to acquire possession of the goods immediately under a lease agreement, with an option to purchase the goods at a later date. If the option was exercised, the payments made under the hire agreement were applied to the purchase price, and the title to the goods then passed to the buyer. This type of agreement was developed in the mid-nineteenth century, but it was the adoption of the hire-purchase agreement by the Singer Manufacturing Company for the purpose of selling sewing machines to the general public that publicized its use as a method of securing debt. It was adopted not long after by the sellers of wagons, carriages, furniture, and most other durable goods.

Operation

The conditional-sale agreement differs from the hire-purchase agreement in one important aspect. Under the hire-purchase agreement, the "hirer" has the option to purchase the goods. It is only when the option is exercised that the title passes to the hirer, who at that point becomes the buyer. The conditional-sale agreement contains no such option. Instead, the installment payments are applied to the purchase price from the outset, but

5. See, for example, *Hickenbotham v. Groves* (1826), 2 C. & P. 492, 172 E.R. 223, where reference is made to hotel furniture purchased under a hire-purchase agreement.

the title does not pass to the buyer until the final payment under the agreement had been made. Under the early conditional-sale agreements, if the buyer defaulted on an installment payment, the seller was free to treat the payments as rent for the use of the goods and to recover the goods from the purchaser, since the title had never passed. The passage of legislation governing the use and application of conditional-sale agreements altered the rights of the parties and imposed certain duties on sellers in the event of default by the buyer. As has been the case with most matters within the jurisdiction of the provinces, each province established legislation for its own perceived needs. Consequently, the law relating to conditional-sale agreements is not uniform throughout the country.

Registration

In all provinces, conditional-sale agreements fall under Personal Property Security Acts, and must be in writing and signed. The agreement need not be in any special form, but must set out a description of the property and the terms and conditions relating to the sale. Since a conditional-sale agreement is by definition a sale in which the goods are paid for over time, consumer-protection legislation, which, for example, requires the true interest rate and the cost of credit to be revealed, must be complied with in the preparation of the sale agreement. As with chattel mortgages, the conditional buyer has possession of the goods, but not the title. To protect the creditor against subsequent creditors' claims or against a sale of the property to an unsuspecting purchaser, evidence of the conditional-sale agreement must normally be registered in a public record office (usually the land registry office) in order that the public be made aware of the seller's title to the property. The legislation, with certain exceptions, provides that a failure to register the agreement renders the transaction void as against a *bona fide* purchaser for value of the goods without notice of the seller's interest, or as against a subsequent encumbrancer without notice of the prior claim. The failure to register the agreement does not render it void as between the buyer and the seller, however.

The registration of a conditional-sale agreement between a manufacturer (or wholesaler) and a retailer is not effective against a purchaser of the goods from the retailer who sells the goods in the ordinary course of business. The retailer would give good title to the goods to the purchaser, because the retailer had purchased the goods from the manufacturer *for the purpose of resale.*

> A snowmobile manufacturer sold a snowmobile to a retailer of snowmobiles under a conditional-sale agreement that was subsequently registered. A purchaser of the snowmobile acquires good title to the machine from the retailer, even though the conditional-sale agreement existed between the manufacturer and the retailer.

An exception to the rule exists in the case of *used* goods, however.

> A snowmobile retailer has a new snowmobile for sale. The retailer's title need not be questioned (as per the example above) because the goods are new, and are presumed to have been bought by the retailer for resale.
>
> The new snowmobile is bought by an individual subject to a conditional-sale agreement. Since the individual is not buying for resale, but buying for enjoyment, there can never be a future presumption that this individual has good title. In fact the individual does not receive good title until the last conditional-sale payment is made.
>
> Assume that the individual does enjoy it and soon hopes to sell it used to a second snowmobile retailer, perhaps as part of a trade-in deal. The individual has not made the last payment and therefore has received no actual title to give. The individual is not saved by presumption, for the machine had not been bought by him or her for

resale in the first place. If the second snowmobile retailer buys the used machine, it does not receive good title, nor does anyone else who buys it used from that second retailer. Title will have remained all along with the first retailer who sold it to the individual under the conditional-sale agreement.

It should be noted, however, that the seller under the Sale of Goods Act of the province would be subject to the implied warranty that the goods were free from encumbrance. The buyer would therefore be in a position to demand that the seller pay off the secured creditors' claim, and recover the goods.

A QUESTION OF ETHICS

Why does the distinction exist between a retail sale and consumer resale regarding a presumption of title? Why is so much more protection offered by the presumption of title in one case, and not in the other?

Assignment

A conditional seller may assign to a third party the title to goods covered by a conditional-sale agreement. The rules applicable to the assignment of ordinary contracts also apply to conditional-sale agreements. A common practice of merchants who sell goods by way of conditional-sale agreements is to arrange with a financial institution (usually a finance company) to purchase the agreements once signed by the buyers, and to collect the money owing. The merchant with this type of arrangement assigns the agreements to the financial institution that would then register the agreements, give notice to the conditional buyers of the assignment, and collect the installments as they fall due. As with all contracts, the assignee takes the agreement as it stands between the conditional buyer and the seller. Any defence that the buyer might have against the seller could also be raised against the assignee.

Buyer's Relief

Apart from the protection of the seller's interest in the goods, conditional-sale agreement legislation is generally designed to provide the buyer with relief against sellers who place onerous terms in the agreement itself. This usually takes the form of providing the buyer with time to redeem after default, since sellers frequently provide in their conditional-sale agreements that they are entitled to repossess and sell the goods immediately on default.

Except in those provinces with consumer-protection legislation that prohibits repossession when a substantial portion of the purchase price has been paid,[6] or when the goods are exempt from repossession,[7] the seller is generally free to repossess the goods when default occurs. Again, this is subject to qualification, as some provinces require that a "notice of intention to repossess" be first given, or a judge's order be obtained before the goods may be taken. However, in no case may force be used to acquire possession of the goods. Once acquired, however, the seller may then proceed with the sale of the chattels to recover the amount owing on the debt. To provide relief from a seller's right of immediate sale, and to introduce an element of fairness into the relationship, a buyer in default is given a period of time to find the funds necessary to complete the payment. Most provincial statutes require the seller who repossesses goods on default to hold the goods for a period of time before selling them. The time interval varies depending upon the province and method of repossession, and ranges from 14 to 30 days, with 20 days being the most common time period for the buyer to redeem the goods.

6. New Brunswick, Nova Scotia, and Ontario, for example, prohibit repossession after two-thirds of the purchase price has been paid unless a court order is obtained. Manitoba fixes the amount at 75 percent of the selling price.
7. Saskatchewan, for example, restricts the right to repossess certain goods, such as agricultural implements and certain household goods.

A seller who repossesses goods with the intention of resale must comply strictly with the resale procedure requirements of the statute. For example, depending upon the particular province, the seller must provide written notice to the buyer of his or her intention to sell the goods. The notice must contain a detailed description of the goods, the amount owing and required in order for the buyer to redeem, and the time period in which the buyer has to make payment. In order to conduct a valid sale, the seller must carefully comply with the notice requirements, then wait the full statutory period before proceeding with the sale.

If the buyer fails to pay the balance owing within the prescribed time period, the seller may then proceed with the sale, and the proceeds obtained would be applied to the outstanding indebtedness. In some provinces, the seller may look to the buyer for any deficiency if the proceeds of the sale fail to pay the balance owing in full. However, this right is only available when the seller has expressly established the right in the conditional-sale agreement. Any surplus that the conditional seller may receive as a result of the sale must be turned over to the buyer. All of the Common Law provinces and territories provide an alternate remedy to the conditional seller in the event of default by permitting the seller to take legal action on the contract to recover the amount owing.

CLIENTS, SUPPLIERS *or* OPERATIONS *in* QUEBEC

The same mortgage-like security (the hypothec) used in financing real estate in Quebec is used for moveable property as well. With few exceptions, any moveable rights relating to a business enterprise must be registered, the details of which may be filed and searched electronically. Where private individuals are acquiring property, moveable real rights do not need to be registered unless they pertain to vehicles, watercraft or aircraft. As with personal property security regimes elsewhere in Canada, a search of the register will reveal encumbrances on the items to be acquired or financed.

Bill of Sale

A bill of sale is a contract in which the title to goods passes to the buyer. However, it is important from the buyer's point of view that certain formalities be followed when goods are purchased, and possession of the goods remains with the seller. Under these circumstances, if the buyer wishes to protect his or her interest in the goods, the bill of sale must be registered in accordance with the requirements of the provincial legislation relating to this type of transaction.

The territories and all provinces except Quebec require a buyer of goods who takes title but not immediate possession to register the bill of sale in accordance with the legislation (usually the *Personal Property Security Act*) in a designated public registry office within a specified time after the bill of sale is executed. The time for registration varies from province to province, and ranges from 5 days to 30 days. Most statutes require that the bill of sale be accompanied by affidavit evidence of the buyer to the effect that the sale was a *bona fide* sale, and not for the purpose of defeating the claims of the seller's creditors. In addition, most provinces require affidavit proof that the seller signed the bill of sale, and some provinces impose special requirements on the parties when the bill of sale is for an automobile.

The purpose of the registration of the bill of sale is twofold. First, it protects the interest or title of the buyer in the goods in the event that the seller should attempt to sell the goods a second time to another, unsuspecting buyer. Second, the registration provides public notice that the title has passed to the buyer, and creditors of the seller will be made aware that the ownership of the goods in the hands of the seller lies elsewhere. The affidavit of *bona fides* attached to the bill of sale is designed to discourage sellers from transferring the ownership of goods in their possession to a third party by way of a bill of sale to defeat their creditors. The fact that the document is registered in a public record office represents notice to the public at large of the transaction that has taken place.

A failure to register would permit an innocent third party who purchases the goods from the seller to obtain a good title to the goods, leaving the unregistered owner with only a right of action against the seller.

Assignment of Book Debts

The assignment of book debts is a method whereby a creditor of a merchant might take an assignment of the accounts receivable of the merchant and collect what is owed to the merchant from customers. The assignment is similar in many respects to the ordinary assignment of a contract in the sense that the debtors must be notified to make payment to the assignee, if the assignee should decide to have the debts paid directly to him or her. Because merchants are not enthusiastic at the thought that their customers will be notified by the creditor to make payments of their accounts directly to the creditor, the creditor and the merchant frequently agree that the general assignment of book debts will not be acted upon by the creditor while the merchant is not in default under the debt-payment arrangement between them.

The assignment of book debts can be sheltered under provincial legislation, in order that the merchant's creditor might preserve his or her claim to the book debts as security.[8] Once an assignment is registered it is effective against persons who may have a later cause to claim against the accounts.[9] Most other provinces have similar registration provisions available. Registration of the security interest is usually required in a province-wide central registry under personal property security legislation.[10] The mere fact that the assignment has been registered, however, does not give the creditor under a general assignment priority over a subsequent creditor who obtained an assignment of a specific debt and gave notice to the debtor of the merchant to make payment of the debt to him or her.[11] A particular advantage of the properly executed and registered assignment of book debts, however, is that the assignee will acquire a secured claim to the book debts over the trustee in bankruptcy should the merchant become insolvent.

Personal Property Security Legislation

PPSA

Personal Property Security Act.

By 1990, Ontario, Manitoba, Saskatchewan, Alberta, and the Yukon Territory[12] had modernized their legislation concerning the use of chattels and other personal property as security for debt by the introduction of personal property security legislation. By 2001, all provinces had passed somewhat similar legislation: Personal Property Security Acts.

The system is based upon the United States' *Uniform Commercial Code, Article 9.* In concept, this legislation represents a complete overhaul of the older systems of registration of security interests and the elimination of a number of conflicting rules with respect to creditor's rights. The general thrust of the new legislation is to simplify transactions associated with securing debts, and to provide a simple system for the registration of all personal property security interests. It recognizes the fact that all of the older security devices had a common purpose: to provide the creditor with a security interest in personal property. To simplify the process of establishing the security interest, the legislation abolished the registration requirements of the older security devices and replaced them with a single registration procedure for the security interest. This interest would represent, for example, the rights of the mortgagee under a chattel mortgage, or the rights of a conditional seller under a conditional-sale agreement. The property to which this security interest attaches is called the collateral, which may be almost any type of personal property.

Perfection

Perfect

The act of registration of a security interest under personal property security legislation.

Under the legislation, the proper registration of the security agreement (usually called a financing statement) perfects the security interest in the creditor. The registration establishes the creditor's priority right to the security interest in the personal property.

8. This is provided for under personal property security legislation in some provinces. See, for example, Ontario, the *Personal Property Security Act*, R.S.O. 1990, c. P.10, s. 2(b).
9. *Personal Property Security Act*, supra, ss. 19, 20, 23.
10. See, for example, Ontario, the *Personal Property Security Act*, supra.
11. *Snyder's Ltd. v. Furniture Finance Corp. Ltd.* (1930), 66 O.L.R. 79.
12. The Yukon Territory introduced personal property security legislation in 1982, which became fully effective on May 31, 1985. For some years, Alberta maintained a central registry for motor vehicles and certain chattel registrations, until 1988 when it introduced a PPSA, now the *Personal Property Security Act*, R.S.A. 2000, c. P-7.

In each province the statute establishes a procedure whereby security interests are registered at either a dedicated office or land registry or titles office, and the information is then transmitted to a central, computer-based storage system. This enables any person to make a search to determine if a security interest is claimed in personal property located anywhere in the province. A failure to register, as required under the Act, would allow a subsequent *bona fide* purchaser to obtain a good title to the goods, or a subsequent creditor to obtain a security interest in the goods in priority over the unregistered security interest.

"After-Acquired" Property and Purchase Money Security Interests

After-acquired property

A term in a security agreement that permits the security interest to attach to goods acquired later by the debtor.

PMSI

Purchase money security interest.

A particular concern arises among priorities when **"after-acquired" property** exists. Consider a situation where a firm borrows money or buys equipment on credit. This investment may be secured by a registered (and thus perfected) security interest, over all its present and future (i.e., after-acquired) general assets. If the firm later wishes to purchase other assets from another supplier, how should priorities be allocated between the two seller–creditors? The first will take the position that it has priority within the terms of its after-acquired clause, and, if that were true, the second supplier would refuse the transaction, for it would have no security at all. Clearly this would greatly impair the ability of the firm to source assets on credit. The solution is found in each province's Personal Property Security Act, which contains rules related to purchase money security interests, also known as **PMSI**. Under these rules, in the face of a previously perfected security interest, a later supplier has priority over an earlier interest to the extent that it is supplying credit for the purchase of this after-acquired property. The later supplier must also, however, perfect its interest through registration within set time limits. The earlier creditor's interest is therefore not impaired, and the later creditor is not given priority beyond the assets it has been willing to supply. Note that the after-acquired clause is not rendered meaningless. It still covers anything the debtor purchases outright with its own funds, but does not cover those assets that the debtor firm purchases and finances from another seller. It is critical to be aware that it is not enough to simply create a security interest by executing a document with a set of terms on its face. For a secured creditor to gain the protection and rights offered by the PPSA legislation it is essential that the creditor perfect the security interest with timely registration in the system.

Default

If a debtor defaults under a security agreement, depending upon the nature and provisions of the security agreement, the creditor may seize the collateral and dispose of it by public or private sale in accordance with the procedure set out in the statute. In the case of consumer goods, the creditor must normally proceed with the sale within 90 days of repossession when the debtor has paid at least 60 percent of the amount owing. If the sale provides a surplus after all expenses have been paid, the balance must be paid over to the debtor. The right to redeem, however, unless otherwise provided in writing, is available to the debtor until the collateral is sold by the creditor, or until the creditor signifies in writing his or her intention to retain the goods in full satisfaction of the debt.

Secured Loans under the *Bank Act*, s. 427

In addition to ordinary secured loans that a bank may make under the *Bank Act*,[13] the legislation gives a bank the right to lend money to wholesalers, retailers, shippers, and dealers in "products of agriculture, products of aquaculture, products of the forest, products of the quarry and mine, products of the sea, lakes, and rivers, of goods, wares and merchandise, manufactured or otherwise" on the security of such goods or products, and to lend money to manufacturers on their goods and inventories. The Act also makes special provision for such loans to farmers, fishers, and forestry producers on their equipment and goods as well as their crops and products.[14]

The *Bank Act* provides that the borrower must sign a particular bank form and deliver it to the bank in order to vest in the bank a first and preferential lien on the goods or equipment similar to that which the bank might have acquired if it had obtained a

13. S.C. 1991, c. 46.
14. ibid., s. 427(1).

warehouse receipt or bill of lading for the particular goods.[15] The bank usually extends the security interest to include after-acquired goods of a similar nature. To perfect its security interest, the bank need only register with the Bank of Canada a notice of intention to take the goods as security for the loan at any time within the three years prior to the date on which the security is given.[16] The registration is designed to give the bank priority over subsequent creditors and persons who acquire the goods (except for *bona fide* purchasers of goods) or equipment, and over the trustee in bankruptcy in the event that the debtor should become insolvent. However, the priority is not absolute. The failure to register the notice of intention as required under the Act would render the transaction void as against subsequent purchasers and mortgagees in good faith and for value. Moreover, the provision that extends the security to after-acquired goods does not have first priority where the debtor has previously given an interest in after-acquired goods to an earlier creditor.

The obvious advantage of the procedure under the *Bank Act* is the ease by which the security interest is established, and the broad reach of the claim to after-acquired goods. Its use, however, is limited only to those lenders set out in the legislation. Although this type of loan is available to a large group in terms of the commercial or business community, it represents only a small number, when compared to the consumer–borrower group that is not eligible for loans under this particular section of the Act.

Bank Credit Cards

Credit cards issued by chartered banks and other financial institutions represent a type of payment instrument frequently used in consumer purchases of chattels and services. While not security instruments in themselves, credit cards provide for security of payment to the merchant who sells goods to the cardholder. A credit-card transaction is supported by two separate agreements:

(1) a contract between the bank and the merchant, whereby the merchant agrees to accept the credit cards issued by the bank (when offered by the cardholder) as payment for purchases,

(2) a contract between the bank and the applicant of a credit card, which provides that the applicant, when issued a credit card, will pay the bank for all debts incurred by the cardholder through the use of the card.

The contract between the bank and the merchant assures the merchant of prompt payment of all purchases made by the cardholder using the bank card, as the bank in effect guarantees payment of the bank card amounts. In return for this security of payment the merchant pays the bank a small percentage charge based upon the amount of the bank card sales.

In the contract or agreement between the bank and the cardholder, the cardholder agrees to compensate the bank for all purchases made using the bank card. The bank usually issues monthly statements listing card purchases, and if the cardholder pays the statements promptly (usually within a stipulated time period of 10 to 21 days) no interest is charged on the amount of the statement. Amounts unpaid are subject to interest charges and cardholders are expected to make certain minimum payments on account of their outstanding indebtedness on a monthly basis. Most credit cards have an established upper limit on the amount of credit the bank will extend or allow to be charged against the credit card.

While security of payment is provided to merchants who agree to honour the bank's credit cards, the credit extended by the bank to the cardholder is normally unsecured, as the cardholder does not pledge any particular security as collateral for what is essentially a loan extended to the cardholder. For this reason, credit limits for cardholders are often fixed at relatively low amounts (e.g., $5,000). The credit limit may vary substantially, however, depending upon the creditworthiness of the cardholder or the type of card issued by the financial institution.

It should be noted also that if a primary credit cardholder requests a supplementary card issued to another person, both cardholders usually become jointly liable for all amounts charged to the cards.

15. ibid., s. 427(2).

16. ibid., s. 427(3).

COURT DECISION: Practical Application of the Law

Unsecured Debt—Joint and Several Liability

***Bank of Montreal v. Demakos* (1997), 31 O.R. (3d) 757.**

The defendant obtained a credit card with a credit limit of $2,200. His son held a supplementary card, made extensive use of it, and the card limit was raised to $20,000 over a relatively short period of time. When the debt reached $20,000 the son paid the account with a fraudulent cheque drawn on a non-existent bank. The card issuer cancelled the cards and sued both cardholders for the debt. The primary cardholder argued he was not responsible for the fraud, and not responsible for any indebtedness beyond the initial, agreed-upon $2,200 credit limit.

THE COURT: In my opinion, a careful reading of the agreement discloses no ambiguity regarding the liability of Mr. Demakos as the primary cardholder. The following are the agreement provisions which must be considered:

3. Each Cardholder agrees not to use a MasterCard card in a manner which would permit the aggregate of all advances of money, interest and other charges to exceed, at any time, the credit limit established from time to time by the Issuer with respect to the MasterCard account, regardless of the number of MasterCard cards on the same MasterCard account. No Cardholder shall use a MasterCard card in an Instabank transaction to make any withdrawal or payment from a chequing or savings account exceeding the balance of the account, except to the extent such Cardholder has made overdraft arrangements with the Issuer.

4. Each Cardholder is responsible for the repayment to the issuer of the amount of any advance of money obtained through the use of any MasterCard card and for the payment of interest and charges pursuant to the terms of this Agreement.

17. If (an)other MasterCard card(s) on the MasterCard account is/are requested and issued to any other person(s), each such other person shall also be a Cardholder for the purposes of this Agreement and there shall be one MasterCard account for all such MasterCard cards. Each such Cardholder shall be jointly and severally liable with the other(s) for any indebtedness incurred through use of any such MasterCard card. Withdrawal of rights and privileges of any such Cardholder or cancellation of any such MasterCard card shall apply to each such MasterCard card and each such Cardholder and the loss of the benefit of the terms by any such Cardholder by application of Section 10 hereof shall entail loss of the benefit of the term by the other(s).

Paragraph 3 imposes an express duty on each cardholder — i.e., the primary cardholder and any subsidiary cardholders — not to run the expenditures under the cards in excess of the credit limit of $2,200 established for the primary card account. Paragraph 4 goes on to create liability on the part of each cardholder for "repayment" of any advance of money obtained through the use of any cards which have been issued. The last clause of this paragraph emphasizes joint responsibility for "the amount by which all outstanding advances, charges and interest exceed the authorized credit limit". Finally, para. 17 reinforces the joint responsibility of all cardholders by saying that "Each such Cardholder shall be *jointly and severally liable* with the other(s) for *any indebtedness* incurred through use of any such MasterCard card" (emphasis added). I see no escape for any cardholder under this most explicit language of the agreement even where the expenditures have been rung up through the fraudulent activities or scheme of one of the cardholders. The reason for this broad liability is, it seems to me, self-evident; if the internal law of the agreement were otherwise, the Bank would be forced to chase individual cardholders forever, and often futilely, in an effort to recover expenditures rung up by imprudent or even dishonest cardholders. To protect itself, the Bank has created sweeping liability for all cardholders under a particular credit card account; if a primary cardholder wishes to authorize the issuance of subsidiary cards to family members or friends, he and they must expect that they will be commonly liable for any expenditures which are rung up. Someone once said that "life isn't necessarily fair." The same could be said about credit card liability. However, so long as the stringency of the agreement language is clear-cut and is free of unconscionablity, it must be enforced. This language is not unconscionable and it is unambiguous in its reach over all cardholders. Accordingly, it must be given its full effect.

Judgment was entered against the defendant (father) for the full amount. Situations involving joint and several liability, regardless of the type of security interest, must be treated with the gravity attached to a guarantee of payment.

Bonds, Debentures, and Floating Charges

Corporations may use a variety of different methods to acquire capital using the assets of the corporation as security for the debt. It may, for example, mortgage its fixed assets, or pledge its chattels as security under a chattel mortgage. In addition to these various methods, a corporation has open to it the opportunity to raise funds by way of the issue of bonds and debentures.

Bond

A debt security issued by a corporation in which assets of the corporation are usually pledged as security for payment.

Debenture

A debt security issued by a corporation that may or may not have specific assets of the corporation pledged as security for payment.

Floating charge

A debt security issued by a corporation in which assets of the corporation, such as stock-in-trade, are pledged as security.

Bonds and **debentures** are instruments issued by corporations that represent a pledge of the assets of the corporation or its earning power as security for debt. The two terms are used interchangeably to refer to debt obligations of the corporation, since no precise legal definition exists for either of these terms. Nevertheless, the term "bond" is generally used to refer to a debt that is secured by way of a mortgage, as a charge on the assets of the corporation. Such bonds are sometimes referred to as mortgage bonds.

The term "debenture" is frequently used with reference to debt that is unsecured or secured to the extent that it takes priority over unsecured creditors, but subsequent to secured debt such as first-mortgage bonds. The practice, however, of issuing bonds that secure debt by way of a mortgage on the fixed assets and chattels, and a floating charge on all other assets, leaves subordinate security holders with very little security in the event that the corporation becomes insolvent. As a result, holders of subordinate obligations are sometimes in much the same position as ordinary, unsecured creditors.

When the security is given to a single creditor, such as a bank or other financial institution, a single instrument is prepared and executed. However, if the amount of capital desired through the issue of the bonds or debentures is substantial, and the terms of the debt obligation lengthy, the common practice is to prepare a trust indenture that embodies all of the terms and conditions of the indebtedness. Then shorter, less detailed debentures are issued that incorporate the terms of the trust indenture by reference. These are then sold to the public as a means of acquiring capital for the corporation, and the purchasers of the securities become creditors of the corporation.

The practice of including a **floating charge** in the debt instrument provides the bond or debenture holders with added security. A floating charge is a charge that does not affect the assets of the corporation or its operation so long as the terms and conditions of the security instrument are complied with. But in the event that some breach of the terms of the debt instrument occurs (such as a failure to make a payment on the debt) the charge ceases to float and crystallizes. At that point it becomes a fixed charge and attaches to the particular security covered by it. For example, finished goods of a manufacturer may be covered by a floating charge. While the charge remains as an equitable charge, the corporation is free to sell or dispose of the goods, as it has the title and the right to do so. Default will crystallize the charge, however, and it will immediately attach to the remaining goods in possession of the corporation. The goods will then become a part of the security that the holders of the debt obligation may look to for the purpose of satisfying their claims.

The nature of the floating charge and its operation as security both before and after default was succinctly described in the case of *Evans v. Rival Granite Quarries Ltd.*[17] where the court said:

> A floating security is not a future security; it is a present security, which presently affects all the assets of the company expressed to be included in it. On the other hand, it is not a specific security; the holder cannot affirm that the assets are specifically mortgaged to him. The assets are mortgaged in such a way that the mortgagor can deal with them without the concurrence of the mortgagee. A floating security is not a specific mortgage of the assets, plus a licence to the mortgagor to dispose of them in the course of his business, but is a floating mortgage applying to every item comprised in the security, but not specifically affecting any item until some event occurs or some act on the part of the mortgagor is done which causes it to crystallize into a fixed security. Mr. Shearman argued that it was competent to the mortgagee to intervene at any moment and to say that he withdrew the licence as regards any particular item. That is not in my opinion the nature of the security; it is a mortgage presently affecting all the items expressed to be included in it, but not specifically affecting any item till the happening of the event which causes the security to crystallize as regards all the items. This crystallization may be brought about in various ways. A receiver may be appointed, or the company may go into liquidation and a liquidator be appointed, or any event may happen which is defined as bringing to an end the licence to the company to carry on business. There is no case in which

17. [1910] 2 K.B. 979.

> it has been affirmed that a mortgagee of this description may at any moment forbid the company to sell a particular piece of property or may take it himself and keep it, and leave the licence to carry on the business subsisting as regards everything else. This would be inconsistent with the real bargain between the parties, which is that the mortgagee gives authority to the company to use all its property until the licence to carry on business comes to an end.

Securities legislation in each province requires compliance and filings to be made by corporations that issue securities or pledge their assets to the public through debt obligations.[18] However, some debt obligations (most often, certain issues to a single creditor) are exempt from securities regulation. In such cases, the creditor should register the obligation under the appropriate provincial personal property securities legislation. Failure to do so may render it void against subsequent purchasers or encumbrancers for value without notice of the prior debt obligation.

FRONT-PAGE LAW

US$275M in funding may be on the way for Laidlaw
Analysts: 'It's going to be a challenge, but I think it's possible'

Laidlaw Inc.'s debenture holders appear likely to approve US$275-million in new borrowing by the near-bankrupt firm although it is not a certainty, say debt analysts.

The Burlington, Ont., transport and health care company took the rare step on Tuesday of hosting a two-hour conference call for noteholders in which Stephen Cooper, vice-chairman and chief restructuring officer, begged noteholders to approve the financing by Oct. 10.

"Bondholders I've talked to have said they thought it was a good presentation and basically accepted the premise, but no one said, 'I've definitely decided I've got to do this,'" said one analyst, asking not to be identified.

Another analyst, who also asked not to be named, noted the bonds are trading at a two-thirds discount and it is in the bondholders' interests to co-operate because a bankruptcy would diminish the value of Laidlaw's assets.

Still, he added, the company is facing a credibility issue with some debt holders who are accusing it of violating its 1992 indenture by improperly giving banks priority in the debt structure. Earlier this week, a class-action suit was begun against Laidlaw and its underwriters, Goldman Sachs & Co. and Bear Stearns & Co., related to these allegations.

Laidlaw, which has suspended principal and interest payments while it restructures US$3.5-billion bank and public debt, said the new money, in the form of two revolving debt facilities, will enable it to avoid seeking court protection.

There are four indentures to be voted on. Three require two-thirds majority while the fourth needs a simple majority.

Raising capital through share issues brings new owners into the management structure of a business, in return for a share of uncertain dividends. Capital raised through debt issue brings in funds without that voice, in return for a contractually fixed interest payment. A preference of one for the other is simply a management choice, until default occurs. On default, the true power of the debt holder is felt. Not only has a breach of contract occurred, which may lead to a lawsuit, but the debt holder can assume a powerful management role. It does not have to consider the best interests of the business debtor — it only looks after itself. One should choose wisely.

Source: Excerpted from Peter Fitzpatrick, *National Post*, "US$275M in funding may be on the way for Laidlaw," September 28, 2000, p. C7.

Statutory Protection of Creditor Security

In addition to the secured rights that a creditor might obtain by taking a security interest in the property of the debtor, most provinces have provided the creditors of certain debtors with special statutory rights in cases where the actions of the debtor could seriously affect the rights of the creditors to payment. The first of these statutes is the *Bulk Sales Act*; the second is the *Mechanics' Lien Act* or *Construction Lien Act*. Both are designed to protect the rights of creditors in transactions in which the creditor may not be a party, and both provide the creditor with rights that were not originally available at Common Law.

Bulk Sales Act

Statutes pertaining to sales "in bulk" refer to the sale of all or large quantities of stock by a merchant in a transaction that is not in the ordinary course of business, or the sale of

18. For example, the *Securities Act*, R.S.B.C. 1996, c. 418, or the *Securities Act*, R.S.O. 1990, c. S.5.

assets and equipment of the business itself to a purchaser. The purpose of the legislation is to protect the creditors of a merchant by requiring the merchant to comply with the procedure set down in the Act for all sales not in the ordinary course of business.

Canadian bulk-sales legislation had its origins in the laws of the United States, rather than the laws of England. The first bulk-sales statute was enacted in Ontario in 1917,[19] and most other provinces followed the Ontario example. At the present time, however, the Western provinces have repealed their bulk-sales legislation, leaving only the provinces of Ontario and Eastern Canada with bulk-sales legislation in place. Provincial variation exists, in spite of attempts to draft uniform legislation.

Under bulk-sales legislation, a prospective buyer of the goods or assets of a business is obliged to follow the procedure set out in the Act; otherwise, the sale may be declared void, and the buyer perhaps unable to recover the purchase price from the seller. The procedure set down for bulk sales generally permits the prospective buyer to make only a small deposit or down payment on the goods or equipment. The buyer must withhold the balance until the seller in bulk either provides an affidavit to the effect that all creditors have been paid, or that arrangements have been made to pay the creditors in full from the proceeds of the sale. In the latter case, unless the creditors waive their rights in writing, the prospective buyer is obliged to have the proceeds paid to a trustee, who will arrange for the payment of the creditors.[20]

In some provinces, a statement of indebtedness must also be provided to the prospective buyer, showing the amounts owing to each creditor of the seller. If the creditors' claims do not exceed a stipulated amount (in Ontario, unsecured and secured creditor claims must each be less than $2,500)[21] the buyer may then pay over the balance. In some provinces, it is also possible to obtain an order from the court to complete a sale in bulk if the court can be convinced that the transaction is in the best interests of the creditors.[22]

If the buyer fails to comply with the Act, an unpaid creditor may, within a stipulated time (six months in most provinces),[23] attack the sale in bulk and have it declared void. The effect of such a declaration would be to make the goods purchased by the buyer available to satisfy the creditor's claims, and leave the buyer only with recourse against the seller who may, under the circumstances, be difficult to find.

CASE IN POINT

A tax-advisory service sued its creditor (which was providing unsecured financing) for breach of contract. The creditor counterclaimed for $205,295 plus interest.

Before the case came to trial, the tax-advisory service sold its assets to a competitor for $800,000. The service did not comply with the *Bulk Sales Act* on the sale, but the company and its president promised the purchasing corporation that it would indemnify the purchaser from any liability arising from its non-compliance with the Act. The proceeds of the sale were used to pay several secured creditors of the tax-advisory service. The unsecured creditor was not paid, because of the pending legal proceedings.

Some time later, the case with the unsecured creditor was heard, and the creditor was granted $327,625 on its counterclaim. It also commenced proceedings under the *Bulk Sales Act* to have the sale declared void, and the proceeds made available to the remaining unsecured creditors.

The court held that the purchaser of the assets was liable to the creditor for the debt of the tax-advisory service at the time of the sale, because the parties had not complied with the provisions of the *Bulk Sales Act.*

See: *National Trust Company v. H. & R. Block Canada Inc.* (2002), 56 O.R. (3d) 188.

19. The *Bulk Sales Act*, 1917 (Ont.), 7 Geo V, c. 33.
20. See, for example, Ontario, the *Bulk Sales Act*, R.S.O. 1990, c. B.14, s. 9.
21. ibid., s. 8.
22. ibid., s. 3.
23. ibid., s. 19.

Mechanics' or Construction Liens

A mechanics' lien (which also may be referred to as a builder's or construction lien) is a statutory right of a worker or contractor to claim a security interest in property to ensure payment for labour or materials applied to land or a chattel. A mechanics' lien is a creature of statute. Hence, it is not a right available to a party except as provided under the legislation responsible for its creation and application. The lien takes two forms: a lien against real property, and a lien against chattels. Each is distinct and separate in the manner in which it is claimed and enforced. Consequently, some provinces have established separate legislation to govern the two distinct types of liens. Some provinces have incorporated both in the same statute, however, but distinguished the procedure applicable to each. Where separate legislation has been enacted, the law related to chattels is usually described as a mechanics' lien act, and the real-property lien legislation as a construction (or builder's) lien act.

A mechanics' lien, like a sale in bulk, is subject to legislation that found its way into the Common Law provinces of Canada by way of the United States. The right of lien under the circumstances provided in mechanics' lien statutes is not of Common Law origin, but rather has its roots in Roman law, and, later, in the civil codes of European countries. As a result, the province of Quebec was the first Canadian province to possess legislation similar to the Mechanics' Lien Acts that was eventually enacted by the Common Law provinces. Manitoba and Ontario were the first Common Law provinces to adopt mechanics' lien legislation,[24] and the other provinces, over time, followed suit. At present, all provinces and territories in Canada have legislation that provides workers with a right of lien against land. Although the legislation is not uniform, the general thrust and application of the law are similar everywhere.

A right of lien to protect the labour and materials of workers and contractors was found to be necessary in the latter part of the nineteenth century, when the vulnerability of persons engaged in the construction industry became evident. The construction of buildings, regardless of size, usually requires large quantities of materials and the special skills of many craftspeople and professionals. At the time, these parties were tied together by a series of contracts under which each was required to invest labour or materials in a property in which they had no security interest to protect their investment. If the owner sold the property and failed to pay the contractor, the contractor's only recourse was legal action for breach of contract against the owner. Subcontractors were equally vulnerable. If the principal contractor became insolvent during the course of construction, the subcontractors and workers had no recourse against the owner of the property for payment, because their contract was with the principal contractor. In order to protect all parties who expended labour and materials on property, the Mechanics' Lien Acts were passed to give each worker or contractor a right to claim a lien against the property as security for payment, regardless of the relationship between the party and the property owner.

Right to Lien

Mechanics' lien legislation as it pertains to chattels varies to some extent from province to province, but the general thrust of the various statutes is to allow a person who repairs a chattel to lawfully retain possession of the goods until payment is made. The legislation usually provides that where the owner does not pay for the repairs within a specific time (usually, a number of months) the repair person on notice may advertise the goods for sale by public auction. If the public-auction sale does not provide sufficient money to cover the repair account, the owner may still be liable for the shortfall in the amount payable for the repairs.

With respect to building construction and similar land-based projects, the construction or builder's lien statutes of all provinces broadly define the term "owner" to include not only the person who holds property in fee simple, but persons with lesser legal or equitable interests in property. They may also include a mortgagee or a tenant who enters

24. The *Mechanics' Lien Act* of 1873, 36 Vict., c. 27 (Ont.); the *Mechanics' Lien Act*, 36 Vict., c. 31 (Man.) (1873).

into a construction contract. The broad definition of the term "owner" is compatible with the intent of the Act: to prevent the person entitled to an interest in the real property from obtaining the benefit of the labour and materials expended upon the land without providing compensation for the benefit received.

The class of persons entitled to claim a lien is equally as broad in most jurisdictions. It includes wage earners, subcontractors, material suppliers, suppliers of rental equipment used for construction purposes (in some provinces), the principal or prime contractor, and, under certain circumstances, the architect.[25] In addition to providing the right of lien to persons who expend labour and materials to enhance the value of real property, the legislation also provides a simplified procedure for the enforcement of lien rights. Low cost and general compliance with formalities are emphasized in the enforcement procedure, but certain aspects, such as the time limits set out in the Act, are rigidly enforced.

The right to claim a lien arises when the first work is done on the property by the claimant, or when the material supplier delivers the first supplies to the building site. Thereafter, the worker, subcontractor, or material supplier may claim a lien at any time during the performance of the contract, and until a stipulated time after the work has been substantially performed. The time limit following the date on which the last work was done (or material supplied) varies from province to province, and ranges from 30 days to 60 days, depending upon the nature of the work and the province. For example, in British Columbia, the time for registration of a lien is 45 days from the date on which the last work was done.[26] In Ontario, the time limit is also 45 days.[27] In order to preserve the right to a lien, a lien claim must be registered in the land registry office in the jurisdiction in which the land is situated, and notice of the lien claim given to the owner of the property.[28] In some provinces, however, a lien claimant's rights may be protected if another claimant has instituted lien proceedings and filed a certificate of action within the time limits appropriate to protect the unregistered claimant. The registration of a lien gives the lien claimant priority over subsequent encumbrancers and subsequent mortgages of the lands.

Following the registration of a lien claim, a lien action must be commenced within a relatively short period of time based upon the date when the last work was done, materials supplied, or lien filed (depending upon the province). An informal legal procedure then follows to determine the rights of the lien claimants and the liability of the contractor and "owners" of the property. Lien claimants are treated equally in a lien action and, with the exception of wage earners, who are entitled to all or a part of their wage claims in priority over other lien claimants, are entitled to a share *pro rata* in the funds or property available.

Lien Hold-backs

In order to avoid disputes that may arise between the contractor and subcontractors, or others, the owner may avoid liability for payment of lien claims by complying with the **hold-back** provisions of the Act. These sections require the owner to withhold a certain percentage of the monies payable to the contractor (10 percent in Ontario)[29] for a period of time following the completion of the contract (45 days in Ontario).[30] The hold-back replaces the land for lien purposes. If claims for lien should be filed within the period that the "owner" is obliged to hold back the funds, the owner may pay the lien claims (or pay the money into court if the claims exceed the hold-back) and obtain a discharge of the lien or a vacating order to clear the liens from the title of the property. The failure of an owner to hold back the required funds would oblige the owner to pay the amount necessary to clear the liens (up to the amount of the required hold-back) in order to free the property from lien claims. If the "owner" should be insolvent or unable to pay the hold-back, the lien claimants may proceed with the lien action and have the property sold by the court to satisfy their claims.

Hold-back

The retention of a part of the contract price by the owner as required under construction lien legislation to ensure payment of subcontractors and suppliers of materials.

25. *Re Computime Canada Ltd.* (1971), 18 D.L.R. (3d) 127.
26. *Builders Lien Act*, S.B.C. 1997, c. 45, s. 20.
27. *Construction Lien Act*, R.S.O. 1990, c. C.30, s. 31.
28. ibid., s. 34.
29. ibid., s. 22(1).
30. ibid.

Some provinces, notably British Columbia, Manitoba, New Brunswick, Ontario, and Saskatchewan, provide additional protection to subcontractors and wage earners by declaring in their legislation that all sums received by the contractor from the "owner" constitute trust funds for the benefit of the subcontractors, workers, and material suppliers. These funds must be used first for the payment of the suppliers of labour and materials before the contractor is entitled to the surplus. A failure to distribute the funds in this fashion would constitute a breach of trust on the part of the contractor.

The trust provision of the various lien Acts has been the subject of much litigation, in order to establish the rights of creditors and other parties to the funds. Nevertheless, the provision remains as an added security for the payment of those who depend on the contractor for payment for their goods or services.

CHECKLIST OF STEPS IN A SAMPLE CONSTRUCTION LIEN ACTION

Given: 45-day period for each of preservation and perfection of lien.
NOTE: Period and timings vary from province to province.

August 31: Date of last work on $100,000 contract project. Contractor paid $90,000. Property owner holds 10% of project value in trust ($10,000 hold-back) for benefit of unpaid subcontractors or suppliers.

September 1: 45-day window to expiry of any unpaid subcontractor/supplier lien begins (to end of October 15). Unpaid subcontractors may register (preserve) lien claims against property at registry or land titles office where land is situated.

October 15: Unpaid subcontractor's right to preserve lien expires when registry or land titles office closes (end of 45th day following August 31).

IF NO LIEN PRESERVED:

October 16:
Contractor paid remaining $10,000 by property owner. Unpaid subcontractors or suppliers have no right of action against property owner, having only a right of action for payment against contractor. END.

IF LIEN(S) PRESERVED:

October 16:
45-day window to expiry (October 16 – November 29) of preserved but unperfected lien begins.

1. Claim settled

ATTENTION: *Property owner must not pay settlements to claimants prior to October 16. The 10% trust is held for all potential lien claimants, and claimants may preserve up to and including October 15. If a lien is preserved before October 15, and land title must be cleared for sale/mortgage purposes, then property owner pays the $10,000 into court office.*

October 16 (or later):
Property owner pays all preserved unpaid subcontractor(s) or supplier(s) from $10,000 trust money. If $10,000 is insufficient for claim(s), this amount is paid to court office. In either case, lien(s) are discharged. After discharge of lien(s), property owner pays remaining balance of $10,000 (if any) to contractor. END.

2. Claim not settled

November 29:
Preserved lien(s) expires if a court action is not commenced (lien perfected) by end of this date (45 days after the end of the last day to preserve, October 15). If one preserved lien is perfected in time, all other preserved liens shelter under the one court action, seeking a court ordered sale of property to satisfy claim(s). Property owner pays $10,000 into court and obtains court order clearing title.

OR

November 30:
If no preserved lien was perfected, property owner obtains a court order to clear title. Property owner may then pay the $10,000 hold-back to the contractor (deducting cost of obtaining order). END.

MANAGEMENT ALERT: BEST PRACTICE

Many business owners use the corporate form to insulate their personal assets as owner/shareholder from the claims of creditors of the business. This is a prudent move, yet many owners then proceed to destroy this protective wall. The first casualty is often the personal guarantee. Many lenders make this a condition of the loan because they desire as much security as possible. If the lender wants to earn interest by making the loan, it might not insist on a personal guarantee if the borrower resists the demand. Lenders who do not lend soon go out of business. Owners also expose themselves to risk by failing to ensure that loan documents show that the business is doing the borrowing, and that the owner's signature is in his or her capacity as a corporate officer, and not in a personal capacity. This happens surprisingly often, giving the creditor an easy attack on personal assets should the business default. It is then up to the debtor to prove the contrary in court, to rebut what can be clearly read on the face of the loan document. Owners soon learn that the best way to stay out of court is to keep the door to it firmly closed.

SUMMARY OF PRIORITIES BETWEEN SECURITY INTERESTS

A number of generalizations can be offered about priorities between security interests which will settle 95% of questions. The remaining 5% however, generate litigation as a result of the particular or peculiar facts of a given situation.

- A secured creditor will trump an unsecured creditor.
- A registered security interest trumps an unregistered security interest.
- An earlier registered security interest trumps a later registered security interest.
- A *bona fide* purchaser of an asset, for value, without notice of a security interest in that asset, receives good title to the asset. **Warning**: notice does not need to be actual awareness, and a registered security interest constitutes valid notice to the world.
- A registered lien does not take priority over earlier registered mortgages, but does take priority over later mortgage advances as well as any later claims to the property.
- A lender who today registers a purchase money security interest (PMSI) that enables a borrower to buy a particular asset will trump most, if not all, of the borrower's creditors who have an earlier blanket security which was drawn with the intention of attaching to all assets later acquired by the borrower.
- A security interest under s. 427 of the *Bank Act* trumps any other security interest in goods except either a purchase money security interest, or after-acquired goods that have already been pledged under an earlier security interest.[31]

SUMMARY

The desire of creditors to reduce risk in debt transactions is reflected in the development of the large and varied assortment of debt instruments available to the business community today. With the exception of the land mortgage and the pledge of goods as security for debt, most of these instruments have been developed within the past few centuries. The chattel mortgage was one of the first security instruments to be developed and recognized as a means of securing debt against chattels by way of a transfer of the title to the creditor. The conditional-sale agreement was used in a somewhat similar fashion, but the seller simply retained the title and gave the conditional

31. *Royal Bank of Canada v. Radius Credit Union Ltd.*, 2010 SCC 48 (CanLII).

buyer possession. When the price was paid, title passed to the buyer. Both of these instruments were used to facilitate purchases of goods.

Corporations are permitted to use a number of different methods of raising capital on the security of their assets. These take the form of bonds and debentures, which may represent charges on specific assets by way of mortgage, or they may be simply unsecured debentures or debentures secured by way of a floating charge.

The need for business to finance the purchase of goods resulted in the development of other security interests. The assignment of book debts permitted creditors to obtain security for debt on loans to merchants. As well, under the federal *Bank Act*, banks were entitled to make loans to wholesalers, manufacturers, lumber businesses, farmers, and fishers on a security interest in inventories, equipment, crops, and machinery. In addition to these forms of creditor security, statutory protection for unsecured creditors also developed, in the form of legislation designed to establish creditors' rights to payment when a sale of the assets of a business was made in bulk, or when subcontractors, workers, and material suppliers increased the value of property by their labour and materials. In the former case, unsecured creditors in a bulk sale were given the right to have a sale declared void if they were not paid. In the latter case, mechanics' lien legislation established a statutory right of lien against the property to secure payment of the claim of subcontractors and workers.

Apart from legislation designed to establish special rights for unsecured creditors and persons in the construction industry, all security interests are designed to protect creditors' claims to the debtor's assets in the event of default. Because of the many confusing procedures and security instruments associated with these interests, attempts have been made by some provinces to streamline and simplify the procedures. These statutes, which are commonly referred to as Personal Property Security Acts, are already in place in most provinces.

Key Terms

Secured creditor, 572
Chattels, 572
Mechanics' lien, 573
Chattel mortgage, 574
Conditional-sale agreement, 575
PPSA, 579
Perfect, 579
After-acquired property, 580
PMSI, 580
Bond, 583
Debenture, 583
Floating charge, 583
Hold-back, 587

Review Questions

1. Why was legislation for mechanics' lien (sometimes referred to as construction lien) necessary?
2. What must an "owner" of property under construction do to protect against mechanics' lien claims attaching to the property?
3. Describe the procedure that a chattel mortgagee may follow to recover the debt, if default should occur.
4. How does a conditional-sale agreement differ from a chattel mortgage?
5. Explain how a conditional seller may realize on the security if default in payment occurs.
6. Describe the procedure that a bank must follow to secure a loan under s. 427 of the *Bank Act*.
7. What is the purpose and effect of an assignment of book debts? Identify the circumstances where registration must take place.
8. Why must a bill of sale be registered in certain circumstances? Identify the circumstances where registration must take place.
9. Outline the special types of security instruments that may be issued by corporations as security for debt.
10. Define "bond," "floating charge," and "debenture."
11. What types of assets may be used as security by chartered banks for loans made under s. 427 of the *Bank Act*?
12. Describe the effect of personal property security legislation on chattel mortgages and conditional-sales agreements in those provinces where such legislation has been introduced.
13. What is a bank credit card? In what way does it secure a debt?

14. Explain the purpose of the procedure set down in the *Bulk Sales Act*. What is the effect of a sale that does not comply with the Act?
15. How does a chattel mortgage differ from a pledge or pawn of a chattel?
16. Outline the general procedure that a subcontractor would follow to secure payment under mechanics' lien legislation.
17. What procedure must a chattel mortgagee follow to preserve the security of the mortgage against subsequent encumbrancers or purchasers?

MINI-CASE PROBLEMS

1. X purchased an automobile under a conditional-sale agreement from Y. Y registered the agreement in accordance with the provincial security registration requirements. X sold the automobile to Z without revealing the fact that it was subject to a conditional-sale agreement. X defaulted on the payments to Y. Advise Y of his rights at law.
2. A engaged B to build a garage for her on her property. B constructed the garage with materials purchased on credit from C Co. A paid B. B did not pay C Co. Advise C Co. of its rights.
3. Alymer had a credit card with a $25,000 credit limit. He arranged for a supplementary card on his account for Annya, his 18-year-old daughter. Annya used the card to purchase goods worth $3,000, but Aylmer disapproved of her purchases. He demanded that she return them, but Annya refused to do so. An identity thief obtained details from Annya's card and ran up another $25,000 in charges. Who is responsible for payment of the card account, and in what proportions?
4. A corporation borrowed $200,000 from a commercial lender by way of a mortgage on its plant and building. The corporation later borrowed $1,000,000 from a bank, and as security for the loan gave the bank a debenture containing a floating charge. The corporation later defaulted on the payment of both the loan and the debenture. The total corporate assets are $1,000,000.

   Discuss the rights of the security holders and shareholders.

CASE PROBLEMS FOR DISCUSSION

CASE 1

Yvonne purchased a 2 hectare building lot in a rural township that permitted property owners to place house trailers or mobile homes on their land as dwelling houses. Instead of building a house on her lot, she decided to purchase a mobile home. Because the lot was in a rural area, and not serviced by municipal water and sewer, it was necessary for her to drill a well, and install a septic system. To fund these two projects, she arranged for a mortgage on her lot in the amount of $25,000 with the local bank. The mortgage was duly registered on the title.

Shortly thereafter, Yvonne arranged for the purchase of a mobile home from Mobile Home Sales Inc., a local business that was prepared to sell her a mobile home on the basis of the $10,000 down payment and a chattel mortgage on the mobile home in the amount of $30,000. The chattel mortgage was duly registered in accordance with provincial personal property security legislation. Yvonne moved the mobile home to her lot, removed the wheels, and placed the structure on concrete block supports. She also had the water supply and septic system attached to the mobile home. A few months later, Yvonne lost her job, and found she was unemployed. The mortgage on her lot and the chattel mortgage on the mobile home fell into arrears. The bank instituted foreclosure proceedings on the property, claiming the mobile home as a part of the real property. Mobile Home Sales Inc. wished to seize the mobile home, claiming it was a chattel.

Discuss the competing claims. What would you expect the court's decision to be?

CASE 2

In order to start a cartage business, Morgan purchased a 3-year-old truck from Used Trucks Ltd., a local truck dealer, for $15,000. Morgan paid $5,000 down, and financed the balance of the purchase price by way of a conditional-sale agreement for the remaining $10,000. The agreement was duly registered in accordance with the provincial personal property security legislation. Business was good, but several months later the vehicle experienced a break-down, and Morgan had the vehicle transported to Ali's Auto Service to have the vehicle repaired. After a careful examination of the problem with the truck, Ali's Auto Service reported that the truck would require a new transmission as well as extensive engine repair. Morgan instructed Ali's Auto Service to proceed with the repairs.

A week later, Ali's Auto Service reported to Morgan that it had replaced the transmission, and repaired the engine. The cost of the repair amounted to $6,300. Morgan was not in a position to pay for the repairs, and Ali's Auto Service informed Morgan that it would not release the truck to him until he paid their account. Without the truck, Morgan was unable to carry on his business, and unable to make his payments to Used Trucks Ltd., or to Ali's Auto Service for the repairs. When Used Trucks Ltd. and Ali's Auto Service realized that Morgan was unable to pay them, each decided to exercise their security rights.

Discuss the nature of the claims, and how the claims may be resolved.

CASE 3

Smith & Co. carried on business as a wholesaler and operated a fleet of delivery trucks to supply its customers. The company found the cost of maintaining the delivery equipment excessive and decided to sell its fleet of trucks. The company arranged to have all delivery work handled by a local cartage company and advertised for a buyer for its delivery equipment.

Eventually, Smith & Co. found a buyer willing to purchase the trucks for $180,000, and the sale was immediately completed.

Some months later, a creditor of Smith & Co. discovered that it no longer possessed a fleet of trucks. The creditor complained that the trucks should not have been sold without his permission, even though he was only an unsecured creditor to whom Smith & Co. owed $6,000 on a trade account.

What are the rights of the parties in this case? What steps should Smith & Co. have taken to avoid creditor complaints?

CASE 4

Baxter owned a block of land that fronted on a large lake. On May 1, he entered into a contract with Cottage Construction Company to have a custom-designed cottage constructed on the site. Cottage Construction Company fixed the contract price at $50,000, $10,000 payable on the signing of the agreement, and the balance on the completion of the contract. Baxter signed the contract and urged the building contractor to begin construction immediately. On May 1, he gave the contractor a cheque in the amount of $10,000.

Cottage Construction Company entered into the following subcontracts for the construction work:

(a) A $3,000 contract with Abe Excavation to excavate and prepare the foundation, the work to be completed on June 1.

(b) A $15,000 contract with Larch Lumber Company for materials, the last to be delivered by July 1.

(c) A $10,000 contract with Ace Framing Contractors to provide labour only to erect and close in the cottage by July 20.

(d) A $5,000 contract with Roofing Specialty Company to install and shingle the building roof by July 20.

(e) A $5,000 contract with Volta Electrical for wiring and electric heating equipment, the work to be completed by August 1.

Cottage Construction Company agreed in its contract to have the cottage completed and ready for occupancy by August 6. Work progressed on schedule, and each subcontractor completed its work on the agreed finish date. By August 1, the cottage was almost ready for occupancy — only the door trim and eavestroughing remained unfinished.

On August 1, the proprietor of Cottage Construction Company approached Baxter and asked him if he might receive the balance of the contract price, as he wished to use the funds to pay his subcontractors. Baxter gave him a cheque for the remaining $40,000, confident that the contract would be completed.

On August 2, a dispute arose between Cottage Construction Company and Ace Framing Contractors over the terms of the contract between them, and Cottage Construction refused to make payment. Ace Framing Contractors registered a construction lien against the cottage lot later the same day. All of the other subcontractors immediately became aware of the lien claim, and registered liens on August 3.

The next day, Cottage Construction Company was found to be insolvent, without having paid the subcontractors.

Discuss the legal rights (if any) of the various subcontractors, Baxter, and Cottage Construction Company. Indicate the probable outcome of the case.

CASE 5

The Mammoth Housing Corporation required capital in order to finance certain land acquisitions for its proposed housing projects. The corporation made a $2,000,000 bond issue to acquire the funds necessary for working capital and to cover a 20-percent down payment on the purchase of a large block of land that it purchased for $5,000,000. The balance of the land transaction was in the form of a first mortgage back to the vendor for $4,000,000.

Once the land was acquired, Mammoth Housing Corporation entered into a building contract with High Rise Construction Company to construct a large apartment building on the site. The contract was for the sum of $15,000,000, which Mammoth Housing Corporation expected to finance by a construction loan of $18,000,000 from Apartment Finance Limited. The money was to be advanced as construction of the building progressed. The building mortgage was registered as a second mortgage, on the understanding that the part of the funds remaining after the building was constructed (plus the corporation's working capital) would be used to discharge the first mortgage.

After the contractor had completed $1,000,000 worth of work on the building, and after Mammoth Housing Corporation received a $1,000,000 advance on the building mortgage, Mammoth Housing Corporation decided to stop construction due to a sudden decline in demand for apartment units in the city.

Assuming that the bonds issued contain a floating charge, and assuming that the contractor files a construction lien against the property for $1,000,000, discuss the rights of the various creditors if Mammoth Housing Corporation decides to abandon the project and allow its bonds and mortgage obligations to go into default, even though it has cash in the amount of $1,000,000 and assets (excluding land) in the amount of $500,000.

CASE 6

Betty Blaine, owner of Ad Consult Ltd., acquired five new computers and two laser printers from Computing Supplies Inc., a major wholesale supplier of word-processing equipment to business and institutions. Each piece of equipment was acquired pursuant to a lease that Betty signed on behalf of her company. The cost of each computer was $2,700, and the printers were $3,600 each. Betty's Company was required to pay $75 monthly for each computer lease and $100 monthly for each printer, for three years.

In addition to the monthly payments Betty's company was to make, Betty was required by Computing Supplies Inc. to sign a personal guarantee for payment on each lease in the event that her company should default. The wording of the guarantee stated that Betty would be liable to make payments under the lease even if the lease turned out to be void or voidable against her company or its creditors.

Shortly after the acquisition of the computing equipment, Betty borrowed a sum of money from the bank for some improvements to her offices. As security for the bank, she executed a general security agreement over all assets of her business.

After about a year of making regular payments on the computer leases and to the bank, Betty's business began to slow down considerably. She struggled to make the payments for a few more months, but eventually found herself unable to continue. First, Betty failed to make payments on the leases. The following month she defaulted on her bank loan. The bank immediately seized the computing equipment, and the other assets of the business, pursuant to the terms of its general security agreement.

In the legal argument that followed, it became apparent that the bank had registered its security agreement under provincial personal property security legislation, but Computing Supplies Inc. had not registered any of its leases.

Identify the legal issue or issues that have arisen, and the arguments that each party will rely upon, including the legal principles upon which they are based. Render a decision.

CASE 7

Hazel purchased a sewing machine from the Easy-So Company under a conditional-sale agreement that required her to make 36 equal monthly payments of $15 each in order to fully pay for the machine. Easy-So Company assigned the conditional-sale agreement to Easy Finance immediately after the agreement was signed by Hazel. Easy Finance registered the agreement in accordance with the provincial legislation pertaining to security instruments of this type, and notified Hazel of the assignment.

Hazel used the machine for several months, during which time she found that the machine required constant adjustment by the seller. Eventually, Hazel came to realize that the sewing machine was unsuitable for her purpose. She arranged with Easy-So to take back the machine as a trade-in on a different type of sewing machine that the dealer also sold. Hazel paid the cost difference of $100 and took the new machine home.

Without advising Easy Finance of the change in the transaction, Easy-So Company sold the trade-in model to Henrietta for $350 cash.

Some time later, Hazel defaulted in her payments to Easy Finance, and the finance company repossessed her sewing machine. When the finance company indicated that it intended to sell the machine to satisfy the debt, Hazel demanded the return of the machine on the basis that it was not the sewing machine described in the conditional-sale agreement. When Easy Finance confirmed the error, it traced the machine covered by the conditional-sale agreement to Henrietta, then seized the proper sewing machine.

Both Hazel and Henrietta brought a legal action against Easy Finance for a return of their respective sewing machines.

Advise all parties of their legal position in this case and indicate the possible outcome. What is the legal position of the Easy-So Company?

CASE 8

Casey, a resident of the United States, visited Canada on his sailboat. While in Canada, he sold the boat to his friend Donald, who resided in Toronto, Ontario. The friend purchased the sailboat for $10,000. Some time later, Donald purchased a power boat from a dealer and used the sailboat as a trade-in to cover part of the purchase price. The dealer made a search for security interests under the provincial *Personal Property Security Act* and found no claims against the sailboat. The boat dealer sold the sailboat some time later to Morgan, under a conditional-sale agreement, and registered the security interest. Morgan later sold the sailboat to Kidd for $8,000 and moved to the province of Alberta.

Kidd did not make a search for creditor claims at the time of the purchase. He had paid over the money unaware of the boat dealer's registered security interest in the property.

The conditional-sale agreement went into default when Morgan neglected to make a payment to the boat dealer. However, before the dealer could find the boat, Customs and Excise claimed that the sailboat had been illegally brought into Canada, and the property in the goods, as a result, had vested in the Crown.

Discuss the rights of the parties, including the Crown, in this case.

The Online Learning Centre offers more ways to check what you have learned so far. Find quizzes, Weblinks, and many other resources at www.mcgrawhill.ca/olc/willes.

CHAPTER 30

Bankruptcy and Insolvency

Chapter Objectives

After study of this chapter, students should be able to:

- Explain the reasons that underpin the need for bankruptcy legislation.
- Describe the actions of a debtor that constitute acts of bankruptcy.
- Describe the steps in bankruptcy proceedings.
- Describe the outcomes in bankruptcy for both debtors and creditors.

YOUR BUSINESS AT RISK

Situations of insolvency and bankruptcy are the most stressful of business circumstances. Even so, critical decisions must be made quickly, decisions that can lead to the preservation, recovery or termination of the business. These decisions can determine the extent to which assets will be exposed to bankruptcy proceedings and the personal liability of directors and parties related to the business. In approaching this topic, the first warning is to realize that the legal definition of bankruptcy is far broader than the accounting definition of liabilities in excess of assets.

INTRODUCTION

Bankruptcy and insolvency are matters of concern to both debtors and creditors. Persons who carry on business must accept the risk that their business venture may prove unsuccessful, either through their own poor management decisions or through the actions of their competitors. When a business fails, not only do the operators of the business frequently suffer economic loss, but their creditors may do so as well if insufficient assets remain to pay the debts. Bankruptcy legislation is designed to provide a procedure whereby the assets of the unfortunate debtor are divided in a fair and orderly way amongst the creditors. It promotes commerce as well, in the sense that it is a means by which a failing business may be ended, and the entrepreneur permitted to start again. Because the assets on bankruptcy seldom cover the full amount of creditors' claims, most creditors will attempt to minimize their risk in extending credit (when possible, by taking security) while a business is in operation.

The debt transaction carries with it a risk that the loan or debt will not be repaid, either as a result of some misfortune that may fall upon the debtor, or because of the debtor's deliberate refusal to make payment. The secured transaction is essentially an attempt by a creditor to ensure repayment, regardless of the circumstances that might affect the debtor's ability to repay the debt in the future. However, secured transactions, in terms of time, are relatively recent phenomena. The more common forms of security, such as chattel mortgages, conditional-sale agreements, and mechanics' liens, did not receive judicial or statutory (in the case of mechanics' liens) recognition until the nineteenth century. While these forms of security assist creditors in many instances, a great many trade-debt transactions are unsecured because the very nature of business often involves an element of trust. Consequently, the decision to extend credit is frequently based upon the reputation of the debtor. Nevertheless, debtors with the best of intentions sometimes encounter financial difficulties due to unforeseen illness or injury, loss of employment, or a decline in business, and find themselves unable to pay their debts. To compound the problems of creditors, persons who deliberately set out to defraud creditors still seem to be ever-present.

Historical Background

Roman law recognized the distinction between the honest but unfortunate debtor and the debtor who obtained credit by fraud or deception. The law established separate remedies for the creditor in each case. The debtor who could establish that his inability to repay the debt was due to matters beyond his control could avoid execution by delivering up his assets to his creditors. The dishonest debtor, on the other hand, was not granted this privilege.

The term "bankruptcy" did not appear until the Middle Ages and was of Italian origin. With the rise of trade and the general increase in commercial activity, many small businesses developed. Invariably, some tradesmen and artisans encountered financial difficulties. The solution adopted by creditors during that period was to go to the place of business of the debtor and break up his workbench. The term bankruptcy was derived from the Italian *bankarupta*, which, literally translated, means "broken bench."

While the *bankarupta* process of the early Italian creditors was no doubt an effective method of demonstrating their displeasure with the debtor's default, it did little to satisfy their financial loss. The Roman law, however, still applied, and the debtor's assets (other than his workbench) presumably remained open to seizure by the creditors to satisfy their debts.

In England, the expansion of trade, particularly during the nineteenth century, carried with it an expansion of credit and the inevitable problem of default by debtors (either by accident or design) to the sorrow of their creditors. The Common Law at the time did not address itself to the problems of trade creditors. It provided only a rather complex procedure that a creditor was obliged to follow in order to attach the property of a debtor. The only effective remedy was through legislative enactment of procedures to deal with a bankrupt debtor's property.

Canada introduced its own insolvency legislation in 1869.[1] Some years later, it repealed the Act,[2] leaving the problems of insolvency and debt with the provinces until 1919, when it once again occupied the field.[3]

Insolvency

The inability of a person or corporation to pay their debts as they fall due.

Bankruptcy

A condition that arises when a person commits an act of bankruptcy under the *Bankruptcy and Insolvency Act*.

The *Bankruptcy Act* of 1919 provided for the liquidation of the debtor's assets and the release of the honest debtor, with no protection for him or her in cases of fraud or willful wastage of assets. Over the next 30 years, the Act proceeded through a series of amendments, with greater official supervision of the liquidation and disposition process added by each change. By 1949, however, an overhaul of the legislation was necessary in order to clarify and simplify the procedures and the application of the Act. The revisions became the *Bankruptcy Act* in 1949.[4] In 1992, the Act was extensively revised again to reflect and recognize the changes that have taken place in business. The new *Bankruptcy and Insolvency Act*[5] constitutes the principal bankruptcy and insolvency legislation in Canada at the present time, subject to amendments made in 2007 which came into force in 2009.

Insolvency versus Bankruptcy

Insolvency and **bankruptcy** are easily confused because they are most often related, and both have financial definitions and legal definitions. In financial terms, insolvency occurs

1. *Insolvent Act* of 1869, 32-33 Vict., c. 16 (Can.).
2. *An Act To Repeal the Acts Respecting Insolvency Now in Force in Canada*, 43 Vict., c. 1 (Can.).
3. *Bankruptcy Act*, 9-10 Geo. V, c. 36 (Can.) (1919).
4. *Bankruptcy Act*, 1949, 13 Geo. VI, c. 7 (Can.).
5. S.C. 1992, c. 27.

when cash flow is insufficient to pay one's current liabilities, and bankruptcy occurs when total liabilities exceed total assets of a person or corporation. At law, insolvency is the inability of an individual or corporation to pay debts as they fall due. It frequently represents a financial condition that precedes bankruptcy, but is only one of ten possible conditions that can help trigger bankruptcy. Bankruptcy at law arises when the debtor has debts in excess of $1,000, and a creditor has filed a petition in bankruptcy against the debtor, with the debtor having committed one of the ten acts of bankruptcy set out in the Act within the six months prior to the petition. The debtor can also voluntarily become bankrupt without the petition of a creditor.

BANKRUPTCY LEGISLATION IN CANADA

Purpose and Intent

The bankruptcy legislation is multi-purpose. It is designed first of all to provide honest but unfortunate debtors with a release from their debts if the debtors deliver up all of their assets to their creditors in accordance with the Act. The second general thrust of the legislation is to eliminate certain preferences and provide a predictable and fair distribution of the assets of the debtor amongst the creditors in accordance with the priorities set out in the statute. A third purpose of the law is to uncover and punish debtors who attempt to defraud creditors by various means. The legislation, being federal law, has the added benefit of providing a uniform system for dealing with bankruptcy throughout Canada.

The general thrust of the legislation is to promote the survival of the debtor's business, rather than simply provide a procedure for its dissolution and the distribution of the debtor's assets to its creditors. The shift in emphasis from creditors to debtors represents a marked change from past legislation. The Act contains additional measures to preserve business firms and the employment they support in the economy. The change, nevertheless, significantly reduces creditors' rights and their security interests in the course of bankruptcy proceedings. The Act does, however, enhance the rights of certain unsecured creditors, particularly unsecured trade creditors, by giving them a greater role in the process. In the past, these creditors frequently suffered the most when their customers became insolvent. These groups, along with the debtors, benefit most under the legislation.

The importance of the release of the honest but unfortunate debtor and the underlying reasoning for the release was described by the court in the following terms:

> I think it is a fair statement to say that originally bankruptcy and insolvency legislation was designed to enable the assets of an honest debtor to be equitably distributed among his creditors, and to enable him to have a livelihood, in business or otherwise, which he might have difficulty in doing unless he was freed of the burden of his debts, and, with respect to a debtor who was not engaged in business but who had been imprisoned for non-payment of his debts, that he might be released from imprisonment. Bankruptcy laws are now of wider application and provide for the right of persons who are not engaged in business to make assignments of their property to trustees for distribution among their creditors and for discharge from further liability with respect to their debts, subject, of course, to certain rules and conditions. But, as has been said, The Bankruptcy Act should not be regarded as a clearing-house for the liquidation of debts solely, irrespective of the circumstances in which they were incurred. As to the position of the debtor, if it is a case of a person being so weighed down by his debts as to be incapable of properly earning a living or of performing the ordinary duties of citizenship, including the support of a wife and family, that is one thing; but where the only object to be served is the comfort and convenience of the debtor and his being freed from the necessity for using his earning-capacity or any property he may acquire or be able to acquire in the future for the discharge or partial discharge of his debts, it is quite another.[6]

6. *Re Buell*, [1955] O.W.N. 421.

CASE IN POINT

Lai and another investor in a corporation had serious differences of opinion over Lai's operation of the corporation which resulted in losses for the corporation and the investor. Lai also had a number of personal creditors. The investor incorporated a company for the sole purpose of purchasing the personal debts of Lai, and once the corporation had acquired the debts, the corporation (of which the investor was the only shareholder) petitioned the court to place Lai in bankruptcy.

At the hearing, it was established that the investor (through his corporation) had purchased Lai's debts for the sole purpose of punishing Lai for the loss incurred by the other corporation in which both had been investors. The court dismissed the petition on the basis that it was brought for an improper purpose, and not for the intended purpose of the bankruptcy legislation.

See: *In the Matter of the Bankruptcy of Edward Lai* (2005), 75 O.R. (3d) 451.

CLIENTS, SUPPLIERS *or* OPERATIONS *in* QUEBEC

As a federal law, the *Bankruptcy and Insolvency Act* applies uniformly across Canada.

Application

The *Bankruptcy and Insolvency Act* is administered by a Superintendent of Bankruptcy who appoints and exercises supervision over all trustees who administer bankrupt estates under the Act. In 1966,[7] the Superintendent was given additional powers to investigate suspected violations of the Act, particularly in the case of complaints concerning fraud, and this provision was carried over to the new legislation.

The Act designates a particular court, usually the highest trial court in each province or territory, as the court to deal with bankruptcy matters in that jurisdiction.[8]

Bankruptcy falls within the jurisdiction of the federal government under the Canadian constitution and, as a consequence, the law applies to all parts of the country. The *Bankruptcy and Insolvency Act* is not, however, the only statute that pertains to bankruptcy and insolvency, nor does it apply to all persons and corporations. Proceedings under the *Winding Up and Restructuring Act*[9] are available to creditors of a corporation, and the *Companies' Creditors Arrangement Act*[10] is available to corporations with bondholders, if the corporation is in financial difficulties.

The *Companies' Creditors Arrangement Act* (CCAA) applies only to corporations that have outstanding issues of bonds or debentures and find themselves in financial difficulty. If a corporation cannot meet its current obligations to its creditors, it may apply to the court for time to submit a plan for its reorganization and restructuring of its financial obligations. If the court grants the order, the order usually will stay any action by creditors until the corporation's plan for reorganization has been brought before the creditors and the creditors given the opportunity to deal with the plan. If the required number of creditors in each class approve the plan, then the plan becomes binding on all creditors, and all must accept the rescheduled debt-payment arrangements.

7. *An Act to Amend the Bankruptcy Act*, 14-15 Eliz. II, c. 32 (Can.) (1966).
8. For example, in Newfoundland, the court is the Trial Division of the Supreme Court; in Alberta, it is the Court of Queen's Bench; and in Ontario, it is the Ontario Court (General Division).
9. R.S.C. 1985, c. W-11, as amended.
10. R.S.C. 1985, c. C-36.

Air Canada entered into bankruptcy protection under the *Companies' Creditors Arrangement Act* (CCAA) on April 1, 2003. The airline then began rounds of negotiation with its creditors on restructuring its debt, and eventually obtained access to new funding from GE Capital, Deutsche Bank, and other investors. The original shareholders lost their investment. Over 88 million shares in a new company, ACE Aviation Holdings Inc., were issued to former creditors and new investors and began trading on October 4, 2004, as the airline and its assets emerged from CCAA proceedings.

The *Companies' Creditors Arrangement Act* is a very broadly worded statute that allows the courts a great deal of latitude in dealing with the debt problems of corporations in financial difficulty. The Act differs substantially from the *Bankruptcy and Insolvency Act*, and is normally used only by very large corporations as the process is usually much more expensive than under the *Bankruptcy and Insolvency Act*. It should be noted, however, that corporations entitled to use the *Companies' Creditors Arrangement Act* are also free to use the *Bankruptcy and Insolvency Act* if they are in financial difficulty, as are their creditors. More debtor corporations may now use the *Bankruptcy and Insolvency Act* not only because it will probably be less expensive to do so, but because the Act places greater emphasis on the preservation of businesses that may be viable if their debts are restructured. Proceedings, once underway under the *Companies' Creditors Arrangement Act*, would probably forestall action under bankruptcy legislation pending the completion of the process.

The *Bankruptcy and Insolvency Act* does not apply to certain persons or to corporations of special kinds. For example, the Act at the present time does not apply to farmers or fishers, or any wage earner or commission salesperson whose income is less than $2,500 per annum, although these persons may make a voluntary assignment in bankruptcy. Nor does the Act apply to any chartered bank, trust or loan company, insurance company, or railway, as special provision is made in the legislation governing each of these if they should become insolvent. Farmers also have special legislation available to them in the case of financial difficulties in the form of the *Farm Debt Mediation Act*.[11] The *Bankruptcy and Insolvency Act* does apply, however, to most persons and corporations, and takes precedence over the *Winding Up and Restructuring Act* in its application.

Contrasts in U.S. Legislation

The *U.S. Bankruptcy Code* is the governing law on bankruptcy in the United States, and thanks to American media, the term "Chapter 11" (of the Code) is known globally. Chapter 11, *Reorganization*, applies to U.S. firms (and qualifying individuals) with a similar intention as that served in Canada by the *Companies' Creditors Arrangement Act*. The firm in Chapter 11 protection attempts to work out changes to the term, interest rate and payments with its creditors, but the principal owing remains the same — leading to a "restructuring" of its debts. While an appointed trustee supervises the assets, the company continues operations in a hope of working itself out of debt under the resulting terms. A major point of contrast between Canadian and U.S. law is the ability in the U.S. to re-write labour contracts with the firm's workforce, a course of action which is not routinely followed in Canada. If the firm's debt cannot be restructured, or fails in its action plan thereafter, Chapter 7 applies to liquidations (corporate and individual), which operates in a manner similar to Canada's *Bankruptcy and Insolvency Act*.

Bankruptcy in the United States has never shared the social stigma found in Canada, to the point that a bankruptcy record in the U.S. is seen by some as indicative of one's entrepreneurial inclination. While this may be the case (but not evidence of entrepreneurial skill, one assumes), it led to widespread abuse of the system, and the law earned a reputation as a means to shed responsibility for reckless spending. In its first major

11. S.C. 1997, c. 21.

reform in thirty years, the United States moved in October 2005 to restrict access to Chapters 11 and 7. Only low-income earners are absolved from repayment, higher earners face repayment terms, fewer assets are exempt, and creditors are given a stronger voice in bankruptcy court.

Acts of Bankruptcy

The failure on the part of a debtor to pay a creditor does not automatically render the debtor bankrupt, nor does it establish that the debtor is insolvent. A debtor may have good reason not to pay a particular creditor, or circumstances may prevent the debtor from doing so. For example, a debtor may, through oversight, fail to have sufficient funds available when a debt falls due. Nevertheless, the debtor may possess assets worth many times the value of the indebtedness. Under such circumstances, the debtor's problem would be one of liquidity, rather than bankruptcy.

The particular activities that the Act[12] defines as acts of bankruptcy are as follows:

> 42(1) A debtor commits an act of bankruptcy in each of the following cases:
> (a) if in Canada or elsewhere he makes an assignment of his property to a trustee for the benefit of his creditors generally, whether it is an assignment authorized by this Act or not;
> (b) if in Canada or elsewhere he makes a fraudulent conveyance, gift, delivery, or transfer of his property or of any part thereof;
> (c) if in Canada or elsewhere he makes any conveyance or transfer of his property or any part thereof, or creates any charge thereon, that would under this Act be void as a fraudulent preference;
> (d) if with intent to defeat or delay his creditors he does any of the following things: namely, departs out of Canada, or being out of Canada, remains out of Canada, or departs from his dwelling-house or otherwise absents himself;
> (e) if he permits any execution or other process issued against him under which any of his property is seized, levied upon or taken in execution to remain unsatisfied until within five days from the time fixed by the sheriff for the sale thereof or for fifteen days after such seizure, levy or taking in execution, or if the property has been sold by the sheriff, or if the execution or other process has been held by him for fifteen days after written demand for payment without seizure, levy or taking in execution or satisfaction by payment, or if it is returned endorsed to the effect that the sheriff can find no property whereon to levy or to seize or take, but where interpleader proceedings have been instituted with respect to the property seized, the time elapsing between the date at which such proceedings were instituted and the date at which such proceedings are finally disposed of, settled or abandoned shall not be taken into account in calculating any such period of fifteen days;
> (f) if he exhibits to any meeting of his creditors any statement of his assets and liabilities that shows that he is insolvent, or presents or causes to be presented to any such meeting a written admission of his inability to pay his debts;
> (g) if he assigns, removes, secretes or disposes of or attempts or is about to assign, remove, secrete or dispose of any of his property with intent to defraud, defeat or delay his creditors or any of them;
> (h) if he gives notice to any of his creditors that he has suspended or that he is about to suspend payment of his debts;
> (i) if he defaults in any proposal made under this Act;
> (j) if he ceases to meet his liabilities generally as they become due.

When a debtor with debts in excess of $1,000 has performed any of these acts within the preceding six months, then any creditor may file a petition for the bankruptcy of that debtor.

12. *Bankruptcy and Insolvency Act*, R.S.C. 1985, c. B-3, s. 42(1).

Bankruptcy Proceedings

The *Bankruptcy and Insolvency Act* distinguishes between commercial and consumer bankruptcies, and provides a somewhat more streamlined process for the latter. Commercial debtors, on the other hand, are given the opportunity if they so desire to restructure their financial affairs. There are essentially three routes that debtors in financial difficulties may follow to resolve their financial problems. A debtor may (1) make a proposal to his or her creditors; (2) make a voluntary assignment in bankruptcy; or (3) permit the creditors to petition for a receiving order. Of the three methods, only the first two may be undertaken by a debtor as a voluntary act; the third represents an involuntary procedure from a debtor's point of view.

Proposal

Proposal

A plan prepared by a commercial debtor to restructure the affairs of the business to enable the business to continue.

In the case of a commercial business, if the business finds itself in financial difficulty, or if secured creditors have notified the debtor of their intention to realize on their security, the debtor may file with the Official Receiver a notice of intention to make a **proposal** to its creditors.[13] The filing then provides the debtor with a 30-day period during which a plan for the restructuring of debts may be prepared for presentation to the creditors. If the debtor is unable to prepare a proposal within the 30-day period, 45-day extensions may be obtained from the court (up to a maximum of five months) to allow time for the preparation of the proposal. Secured creditors may oppose extensions of the time if the extension will jeopardize their security. The court, however, is free to impose conditions on the creditors in limiting the debtor's time to prepare a proposal.

Once the proposal is prepared, it is filed and presented to the creditors for approval. A meeting of creditors must be held within 21 days after the proposal is filed, and both secured and unsecured creditors are entitled to vote on the proposal. Each group of creditors votes as a separate group or class, but the votes of the unsecured creditors are the most important. The emphasis on the wishes of the unsecured creditors is recognition of the fact that these creditors as a group usually have the most to lose if the business is closed. Secured creditors may look to specific assets to recover their money, but the unsecured creditors in many cases have few unencumbered assets to cover their debt, and would receive little or nothing if the business terminated. If the unsecured creditors as a class vote two-thirds or more in favour of the proposal, then the proposal will be binding on the creditors and the debtor.

The approval by the court has the effect of binding the parties to the terms of the proposal. Compliance with the proposal by the debtor will preclude any creditor from taking independent proceedings against the debtor. The successful performance of the agreement by the debtor would have the same force and effect as if the debtor had paid the debts in full. If the proposal is rejected by the unsecured creditors, the debtor is deemed to have made an assignment in bankruptcy, and proceedings for the administration of the debtor's estate begin.

Voluntary Assignment

Assignment

The transfer of a debtor's property to an Official Receiver under a voluntary assignment in bankruptcy.

An insolvent person may make a voluntary assignment[14] in bankruptcy as an alternative to a proposal. A voluntary assignment differs from creditor-instituted proceedings only at the outset. Under a voluntary assignment, the debtor files with the Official Receiver an **assignment** of his or her property for the general benefit of the creditors, with the assignee's name left blank.[15] The Official Receiver then selects a trustee to accept the debtor's property and to proceed with the bankruptcy. Once this is done, the *Bankruptcy and Insolvency Act* procedure comes into play, and the administration of the bankrupt's estate begins.

13. *Bankruptcy and Insolvency Act*, R.S.C. 1985, c. B-3, s. 50.

14. ibid., s. 49.

15. ibid., s. 42(1)(a). An assignment for the general benefit of creditors constitutes an act of bankruptcy.

Petition

The request by a creditor to the court to institute bankruptcy proceedings against a debtor.

Trustee

A person licensed to receive and dispose of the assets of a bankrupt, then distribute the proceeds to the creditors.

Petition

The third method of instituting proceedings under the Act is through the action of a creditor. When a debtor, as defined in the Act, has debts owing to one or more creditors in excess of $1,000 and has committed an act of bankruptcy, a creditor may at any time, within the six months after the act of bankruptcy occurred, file a **petition** for a receiving order with the Registrar in Bankruptcy of the provincial or territorial court designated under the *Bankruptcy and Insolvency Act* to hear such matters.[16] If the debtor does not object to the petition (or consents), a receiving order is issued by the registrar that determines the debtor to be bankrupt and that permits the appointment of a Licensed **Trustee** to administer the estate of the bankrupt. If the debtor objects, the matter is heard by a judge, and the debtor may then present evidence to satisfy the court that he or she is not bankrupt. If the debtor is successful, then the judge will dismiss the petition; if not, the receiving order is issued. The various ways of initiating proceedings are illustrated in Figure 30–1.

FIGURE 30–1 Initiation of Bankruptcy Proceedings

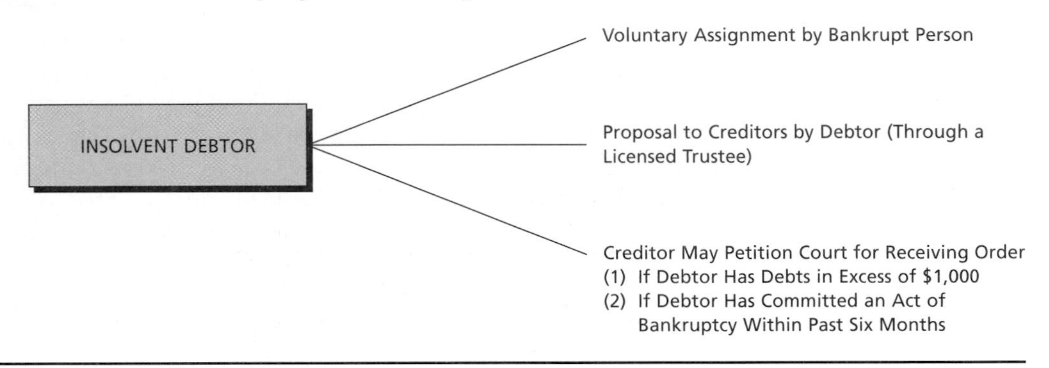

Provision is made under the Act for the appointment of an interim receiver if it should be necessary to preserve the assets or the business of the debtor pending the hearing of the petition by the court. The interim receiver usually becomes the trustee who administers the debtor's estate if the receiving order is later issued by the court.

Following the issue of the receiving order, the appointed trustee has a duty to call together the creditors of the bankrupt. At that time the trustee's appointment is affirmed or a new trustee is appointed. The trustee then reports the assets of the debtor and determines the amount of the creditors' claims. The debtor must be present at the first meeting of creditors. At that time the creditors are free to examine the debtor as to the state of his or her affairs and the reasons for the insolvency. At the first meeting the creditors also appoint inspectors (not exceeding five), who assume responsibility for the supervision of the trustee on behalf of the rest of the creditors. The inspectors usually meet with the trustee following the first meeting of creditors and instruct the trustee on all matters concerning the liquidation of the bankrupt debtor's estate. The trustee, even though an officer of the court and subject to its direction, must act in accordance with the inspectors' instructions, as they must authorize all important decisions concerning the realization of the assets. This is so, provided that their decisions are consistent with the *Bankruptcy and Insolvency Act.*

Unpaid suppliers are entitled to reclaim goods supplied to the bankrupt business if the goods are recognizable in inventory, and were delivered within 30 days preceding the bankruptcy. Farmers, fishers, and aquaculturalists who have supplied a bankrupt business with their goods during a 15-day period prior to the bankruptcy may claim a special-priority security interest on the debtor's inventory for the value of the goods so delivered.[17]

16. ibid., s. 43(5).
17. ibid., s. 81.2(1).

The trustee usually collects all assets of the bankrupt and converts them to cash. Assets that are subject to security interests, such as land mortgages, chattel mortgages, or conditional-sale agreements (to name a few), must be made available to the secured creditor. If the goods are sold, any surplus goes to the trustee, to be included in the estate for distribution to the creditors. If the proceeds from the disposition of the particular security are insufficient to satisfy a secured creditor's claim, the secured creditor is entitled to claim the unpaid balance as an unsecured creditor. The assets not subject to secured creditors' claims, and any surplus remaining from the disposition of assets subject to security interests when liquidated, are distributed by the trustee in accordance with the priorities set out in the *Bankruptcy and Insolvency Act.* The legislation provides that certain preferred creditors be paid before the unsecured general creditors, in the following order:[18]

136(1) Subject to the rights of secured creditors, the proceeds realized from the property of a bankrupt shall be applied in priority of payment as follows:

(a) in the case of a deceased bankrupt, the reasonable funeral and testamentary expenses incurred by the legal personal representative of the deceased bankrupt;

(b) the cost of administration, in the following order,

(i) the expenses and fees of any person acting under a direction to preserve or protect the assets.

(ii) the expenses and fees of the trustee.

(iii) legal costs;

(c) the levy payable under section 147; (costs paid to the court).

(d) wages, salaries, commissions or compensation of any clerk, servant, travelling salesman, labourer or workman for services rendered during six months immediately preceding the bankruptcy to the extent of two thousand dollars in each case, together with, in the case of a travelling salesman, disbursements properly incurred by that salesman in and about the bankrupt's business, to the extent of an additional one thousand dollars in each case, during the same period, and for the purposes of this paragraph commissions payable when goods are shipped, delivered or paid for, if shipped, delivered or paid for within the six-month period, shall be deemed to have been earned therein;[19]

(e) municipal taxes assessed or levied against the bankrupt within two years next preceding his bankruptcy and that do not constitute a preferential lien or charge against the real property of the bankrupt, but not exceeding the value of the interest of the bankrupt in the property in respect of which the taxes were imposed as declared by the trustee;

(f) the landlord for arrears of rent for a period of three months next preceding the bankruptcy and accelerated rent for a period not exceeding three months following the bankruptcy if entitled thereto under the lease, but the total amount so payable shall not exceed the realization from the property on the premises under lease, and any payment made on account of accelerated rent shall be credited against the amount payable by the trustee for occupation rent;

(g) the fees and costs referred to in subsection 70(2) but only to the extent of the realization from the property exigible thereunder;

(h) in the case of a bankrupt who became bankrupt before the prescribed date, all indebtedness of the bankrupt under any Act respecting workers' compensation, under any Act respecting unemployment insurance or under any provision of the *Income Tax Act* creating an obligation to pay to Her Majesty amounts that have been deducted or withheld, rateably;

(i) claims resulting from injuries to employees of the bankrupt in respect to which the provisions of any Act respecting workers' compensation do not apply, but only to the extent of moneys received from persons guaranteeing the bankrupt against damages resulting from such injuries; and

18. ibid., s. 136(1).

19. As of July 2008, the federal government's Wage Earner Protection Program guarantees direct payment of up to $3,000 in outstanding wages, in return for which, qualified employees subrogate their $2,000 priority claim to the government.

(j) in the case of a bankrupt who became bankrupt before the prescribed date, claims of the Crown not mentioned in paragraphs (a) to (i), in right of Canada or of any province, rateably notwithstanding any statutory preference to the contrary.

The unsecured creditors, being at the bottom of the list, share *pro rata* in any balance remaining. This amount is usually calculated in terms of "cents on the dollar." For example, if after the payment of secured and preferred creditors in a bankruptcy the sum of $4,000 remains, and unsecured creditors' claims amount to $10,000, the creditor's individual claims would be paid at the rate of 40 cents for each dollar of debt owing to the creditor.

CASE IN POINT

A fleet of cars was leased from a Quebec dealer by a debtor corporation, for use in a commercial enterprise. The dealer failed to register public notice of the leases as required under Quebec law. The debtor eventually became bankrupt and the cars were taken by the trustee in bankruptcy and sold to satisfy the debtor's creditors. As the leases had not been properly registered, the trustee took the position they were invalid, and the dealer fell into the category of an unsecured creditor, only to take a share in the worth of the cars. The Supreme Court of Canada disagreed, ruling that the lease did nothing to alter the dealer's true ownership of the cars, and that they were not the property of the debtor available for the other creditors. The entire proceeds from the sale of the cars were ordered to be turned over to the dealer.

See: *Lefebvre (Trustee of) [DaimlerChrysler Services Canada v. Lebel et al.]*, [2004] 3 S.C.R. 326.

MANAGEMENT ALERT: BEST PRACTICE

Bankruptcy illustrates the importance of being a secured creditor, whenever possible, over being an unsecured creditor. When legislation permits registration of a security interest, it should be done. The time and money in legal expenses saved by not registering a security interest will be small compared to both the legal expense of being represented later and the loss of priority in payment, if the debtor goes bankrupt.

CHECKLIST FOR PETITION INTO BANKRUPTCY

1. Petitioner is a creditor of the debtor?
2. Debtor is not an exempt person or corporation?
3. Debtor has debts in excess of $1,000?
4. Debtor has committed an act of bankruptcy within past six months?

If yes to all, creditor may file petition for bankruptcy of debtor.

FRONT-PAGE LAW

FTC sues to stop failed dot-com selling data

MOUNTAIN VIEW, Calif. — In the first case of its kind, U.S. regulators are suing a bankrupt e-commerce firm to try to prevent the company from selling its customer database.

The Federal Trade Commission said in a complaint filed yesterday that Toysmart.com Inc. is violating its own privacy policy and has misrepresented itself.

Toysmart.com's privacy policy, still posted on the company's Web site, states that "Personal information ... such as name, address, billing information and shopping practices, is never shared with a third party."

"Even failing dot-coms must abide by their promise to protect the privacy rights of their customers," Robert Pitofsky, FTC chairman, said in a statement.

Toysmart.com stopped taking orders in May and filed for bankruptcy protection last month after burning through US$45-million in private financing in less than three years. "Historically, bankrupt companies have been able to sell their customer lists," said Mark Stein, president of Ozer Internet Services, a Needham, Mass.-based consulting firm that oversees the liquidation of bankrupt Internet companies. "However, the FTC's suit may have a chilling effect on such sales in the future," he said.

In the Information Age, data are increasingly important business assets. For some, data may be the firm's only significant asset. Has the information age created a difference between a comprehensive database and simple customer lists? Where do you stand? Should the database be sold to help satisfy debts, or should the privacy promise be upheld?

Source: Excerpted from Simon Avery, *National Post*, "FTC sues to stop failed dot-com selling data," July 11, 2000, p. C10.

Discharge

Discharge

The release from obligations imposed by the *Bankruptcy and Insolvency Act*.

Until a bankrupt is discharged by the court, the bankrupt is not released from his or her debts, although there are some automatic discharges if unopposed by creditors. In effect, any earnings or other income received by the debtor before the **discharge** may be applied to the payment of the creditors if the court so orders. Apart from this, the bankrupt must not engage in any business without disclosing that he or she is an undischarged bankrupt. As well, the bankrupt must not purchase goods on credit except for necessaries (and then only for amounts under $500 unless the bankrupt discloses the undischarged bankrupt status).

Bankruptcy also places certain limitations on the activities of the bankrupt until a discharge is obtained. The debtor may not become a director of any limited liability corporation, nor may the debtor accept an appointment to the Senate (a matter unlikely to be of concern to most bankrupts in any event).

The trustee will generally arrange for the discharge of the bankrupt shortly after bankruptcy proceedings are under way. In most cases, if the bankrupt was an honest but unfortunate debtor who had done nothing to defraud the creditors and who had complied with the Act and the debtor's duties and obligations under it during the course of the proceedings, a discharge will normally be granted on application by the trustee. This usually occurs from three to six months after proceedings were instituted, but normally not later than 12 months. A debtor may, however, make an application for discharge on his or her own behalf and at his or her own expense if so desired. However, the debtor is not likely to succeed unless he or she can satisfy the court that the creditors have received at least 50 cents on the dollar as payment of their debts, and that no fraud existed. Whether the bankruptcy was the debtor's first, and whether the debtor carried out the debtor's duties under the Act, are also factors considered by the court in reaching a decision. Even then, the court has wide powers to impose conditions on the bankrupt. The conditions to some extent are governed by the circumstances that led to the bankruptcy and the debtor's conduct thereafter. A debtor who has never been declared bankrupt before is entitled to an automatic discharge nine months after bankruptcy proceedings were instituted unless creditors, the superintendent, or the trustee objects to the discharge.

A discharge releases the bankrupt from all debts and obligations except those arising from the debtor's wrongdoing and those associated with the debtor's marital obligations. All fines and penalties imposed by law and any obligation that arose out of the fraud of the debtor or a breach of trust would remain, as would any personal obligation arising out of a maintenance or alimony agreement or order.[20] Any debts incurred for necessaries would not be released, if the court so ordered.

"Necessaries" have been described by the court in the following terms:

> Unfortunately there is no definition of the term "necessaries of life" in the Bankruptcy Act. Rarely has the judicial mind placed so many and varied interpretations on a particular phrase in its attempts to determine the intention of the legislators. The legion of cases in no way restricts the term to mean the basic necessaries of life, such

20. *Bankruptcy and Insolvency Act*, R.S.C. 1985, c. B-3, s. 136(1).

as food, shelter, clothing and the like, but would stretch its coverage to such things as "gas for the family Cadillac". Perhaps this is a reflection of the regional disparity that exists in this country of ours. In any event the general judicial consensus appears to be that the term envisions all goods appropriate to a person according to his or her particular lifestyle.[21]

A corporation, unlike an individual, is not entitled to a discharge unless all of the creditors' claims are paid in full.[22] The various steps in the bankruptcy process are shown in Figure 30–2.

FIGURE 30–2 Bankruptcy Procedure

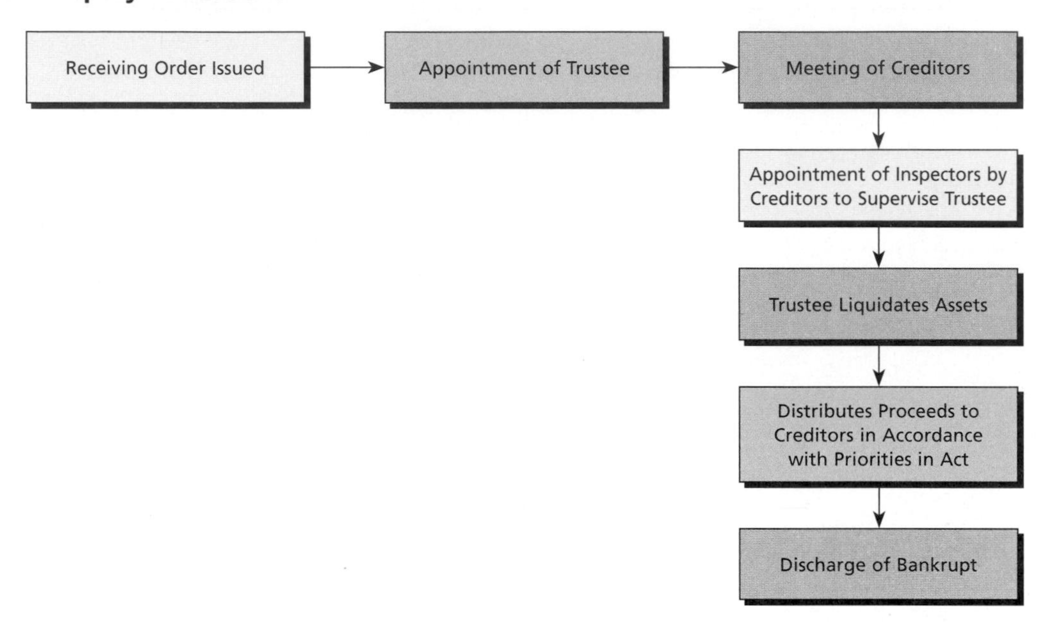

COURT DECISION: Practical Application of the Law

Repeat Bankrupt—Denial of Discharge

***Re: Pufahl (Bankrupt)*, 2005 ABQB 23.**

The debtor made an assignment in bankruptcy in the USA in 1982. He came to Canada, incurred debt and was declared bankrupt in 1990. He was discharged on February 22, 2004. Having managed to incur debt prior to his discharge, he declared bankruptcy again the next day, February 23, 2004.

THE REGISTRAR IN BANKRUPTCY: ...Mr. Pufahl is a third time bankrupt. It is also clear from a review of the file that Mr. Pufahl obtained his discharge from his 1990 bankruptcy in order to file this bankruptcy. As Justice Anderson stated in re *Hardy* (1979) 30 C.B.R. (N.S.) 95 at paragraph 3:

In my view, a third bankruptcy is one too many. The well-recognized principle underlying bankruptcy law is that a debtor may, in proper circumstances, be relieved of his obligations and enabled to re-establish himself financially. I do not consider that he should be entitled to do so on a reoccurring basis. The process of the Act and of the court should not be considered to bestow a licence to incur debts and be purged of them at periodic intervals.

21. *Amherst Central Charge Limited v. Hicks* (1979), 29 C.B.R. 313.

22. *Bankruptcy and Insolvency Act*, supra, s. 169(4).

In considering an application for discharge under section 173, Justice Hallett in re *Crowley* (1984) 54 C.B.R. 303 discusses various principles the court ought to consider.

> *In recent years there has been a trend by this court to impose conditions of payment on the bankrupt as the price of his discharge. This reflects the feeling of the public as stated through the decisions of this court that abuses of the bankruptcy process are perceived. While the vast majority of the public are wrestling with their finances to make ends meet, there is a small percentage, albeit a large number of persons, who are availing themselves to the provisions of the Bankruptcy Act and, in particular, the discharge provision, to walk away from the debts they have accumulated.*

An order for discharge may be refused where the bankrupt's conduct is particularly reprehensible. While Mr. Pufahl's conduct is not reprehensible this is his third bankruptcy. Furthermore, during his second bankruptcy Mr. Pufahl failed to complete his duties and obligations under the Bankruptcy Act in a timely fashion. He failed to co-operate with his trustee. [T]he bankrupt stated that he failed to file three years of tax returns for 1987 to and including 1989. He further stated that he had made arrangements with his trustee to pay $120.00 per month into the estate. Voluntary payments were not made ... [T]here is no information suggesting why the defaults continued from 1995 to 2003 [when he] was earning approximately $3,000.00 net per month. In this bankruptcy his statement of finances indicate a Canada Revenue Agency indebtedness of $2,700.00; credit card and bank debt of $38,050.00; an unsecured loan from a family member for $5,000.00. While the bankrupt may have lost his job in January 2004 as a result of the BSE crisis, he was re-employed in about a month. His real cause of bankruptcy was an over extension of credit. Mr. Pufahl obtained extensive credit prior to 2004 and has likely been in breach of the Bankruptcy Act, section 199(b) which states as follows:

> *An undischarged bankrupt who... (b) obtains credit to a total of five hundred dollars or more from any person or persons without informing such persons that the undischarged bankrupt is a undischarged bankrupt...is guilty of offence punishable on summary conviction and is liable to a fine not exceeding five thousand dollars or to imprisonment for a term not exceeding one year, or to both.*

In order to maintain the integrity of the bankruptcy system, Mr. Pufahl will be required to pay a substantial amount to his trustee for distribution to his creditors. He has the ability to pay $400.00 per month to the estate and still leave available sufficient funds to maintain himself and his wife. The monthly payments will commence February 2005 and continue until such time the estate has received a total of $22,000.00. In the event the bankrupt defaults in any monthly payment, the trustee has leave to issue a garnishee against any employer pursuant to the Bankruptcy & Insolvency Act, section 68. There will be a further condition that the bankrupt continue to file monthly statements of income and expenses with the trustee until such time as he obtains his discharge.

> *Canada has always placed greater expectations upon bankrupts than has been the case in the United States. Only recently has the U.S. made changes to its system to discourage serial bankruptcies. See the section on* Contrasts in U.S. Legislation.

Consumer Bankruptcy Summary Proceedings

The 1949 *Bankruptcy Act* provided a simplified procedure for the administration of small estates where the assets of the bankrupt debtor amounted to $500 or less. This procedure was frequently used by consumer or non-trader bankrupt debtors as a means of making a fresh start, as the procedure provided for a prompt discharge. Under the current Act, non-trader or consumer insolvencies, where the individual has assets of less than $5,000, may be eligible for a summary administration procedure. This procedure may be used if the consumer debtor has total debts excluding those secured by the person's principal residence of less than $250,000. In July 2008, two significant elements were proclaimed in force. First, RRSP assets (other than contributions made within a year of the bankruptcy) will be exempt from creditors, and second, student loans will be automatically discharged if the bankrupt has been out of school for seven years or more.

Under the consumer-proposal provisions of the Act, the consumer begins the process by obtaining the assistance of an administrator (a person designated to act in this capacity by the superintendent or a trustee) to assist in the preparation of a proposal to creditors. The proposal must provide for its performance to be completed in not more than five years, and also provide for the priority payment of certain debts and of the fees and expenses of the administrator and debt-counselling services. If the proposal is accepted by the creditors, and no objection is raised, it is deemed to be approved by the court. The administrator then proceeds to receive all moneys payable under the proposal, pays the expenses, and distributes funds in accordance with the proposal. When the proposal has been fully performed by the debtor, the administrator provides the debtor with a certificate to that effect.

In an effort to reduce the possibility of debtors falling into financial difficulties in future, consumer debtors are expected to attend financial-counselling sessions; this requirement is usually included in the proposal.

Bankruptcy Offences

The *Bankruptcy and Insolvency Act* is not only designed to provide an orderly procedure for the distribution of a debtor's assets to his or her creditors, but to identify and punish debtors who attempt to take advantage of their creditors by fraud or other improper means. The legislation therefore addresses the problem by establishing a series of offences punishable under the Act by way of fine or imprisonment, as well as by the withholding of discharge from the bankrupt.

Under the Act, the superintendent has wide powers to investigate fraudulent practices and allegations of violation of the Act by bankrupt debtors. For example, if a debtor, once aware of the serious state of her finances, transfers or conveys assets to members of her family in an effort to hide the assets from her creditors, she would in effect have committed a bankruptcy offence by making a fraudulent conveyance of her assets. The principal offences under the legislation include:[23]

> 198(1). Any bankrupt who
>
> (a) fails, without reasonable cause to do any of the things required of him ...
> (b) makes any fraudulent disposition of his property before or after bankruptcy;
> (c) refuses or neglects to answer fully and truthfully all proper questions put to him at any examination held pursuant to this Act;
> (d) makes a false entry or knowingly makes a material omission in a statement or accounting;
> (e) after or within twelve months next preceding his bankruptcy conceals, destroys, mutilates, falsifies, makes an omission in or disposes of or is privy to the concealment, destruction, mutilation, falsification, omission from or disposition of a book or document affecting or relating to his property or affairs unless he proves that he had no intent to conceal the state of his affairs;
> (f) after or within twelve months next preceding his bankruptcy obtains any credit or any property by false representations made by him or made by some other person to his knowledge;
> (g) after or within twelve months next preceding his bankruptcy fraudulently conceals or removes any property of a value of fifty dollars or more or any debt due to or from him; or
> (h) after or within twelve months next preceding his bankruptcy pawns, pledges or disposes of any property which he has obtained on credit and has not paid for, unless in the case of a trader such pawning, pledging or disposing is in the ordinary way of trade and unless in any case he proves that he had no intent to defraud;
>
> is guilty of an offence punishable on summary conviction and is liable to a fine not exceeding five thousand dollars or to imprisonment for a term not exceeding one year or to both, or is guilty of an indictable offence and is liable to a fine not exceeding ten thousand dollars or to imprisonment for a term not exceeding three years or to both.

23. ibid., s. 198(1).

As a general rule, investigations are conducted by an official receiver under the direction of the Superintendent of Bankruptcy. These investigations are usually to determine the cause of the bankruptcy; in a case where fraud or a criminal act is suspected, however, the investigation may extend to those matters as well.[24] If the court suspects that the debtor might attempt to leave Canada with assets to avoid paying his or her debts, or to take other similar action to avoid creditors, the court has the power to order the debtor's arrest.[25]

A QUESTION OF ETHICS

As the manager of a corporate creditor, you must act in the best interests of your own shareholders in deciding to trigger bankruptcy proceedings against another corporate debtor. Doing so means jeopardizing the jobs and livelihood of tens, hundreds, or even thousands of the debtor's employees, to recover part of what might be (in the circumstances) a relatively small debt. You have every right to do this. Would it trouble you?

SUMMARY

Bankruptcy legislation is a federal statute and, in its present form, is an attempt to provide an honest but unfortunate debtor with an opportunity to reorganize his or her debts or to start afresh and free of debts. The procedure requires the debtor to deliver up his or her property to a receiver or court-appointed trustee for the purpose of distribution to the creditors in accordance with certain specified priorities. The Act also deals with fraudulent and improper actions by debtors who attempt to defraud their creditors. Penalties are provided in the statute for persons found guilty of these bankruptcy offences.

Before persons may be subject to bankruptcy proceedings, they must first commit an act of bankruptcy. The commission of any one of the acts by a debtor would entitle a creditor (or creditors) owed at least $1,000 by the debtor to petition for a receiving order at any time within six months after the commission of the act of bankruptcy. If the debtor fails to convince the court that he or she is not bankrupt, the receiving order is issued, and a trustee establishes control of the debtor's assets for liquidation and distribution to the creditors. Inspectors, appointed by the creditors, supervise the trustee until the estate has been fully distributed. The debtor, if he or she has acted without fault, is entitled to a discharge of all debts except such debts as fines, court maintenance orders for support or alimony, and funds acquired by fraudulent means or by breach of trust.

A voluntary procedure and two summary procedures are available to debtors, in addition to the creditor-initiated process. Recent changes in the legislation permit debtors to make proposals for the restructuring of the debtor's business, rather than the immediate division of its assets amongst the creditors.

KEY TERMS

Insolvency, 596
Bankruptcy, 596
Proposal, 601
Assignment, 601
Petition, 602
Trustee, 602
Discharge, 605

24. ibid., ss. 161 and 162.
25. ibid., s. 168(1).

REVIEW QUESTIONS

1. Describe the role of the Superintendent of Bankruptcy in bankruptcy proceedings.
2. Under what circumstances could a person have assets in excess of liabilities, yet be bankrupt?
3. In what way (or ways) does bankruptcy affect the rights of secured creditors? How would they recover their debts if the security they held was insufficient to cover the amount owing?
4. Describe the acts of bankruptcy that would entitle a creditor to institute bankruptcy proceedings.
5. Outline the requirements a creditor must satisfy in order to institute bankruptcy proceedings against a debtor.
6. If a debtor makes a proposal to his or her creditors, then fails to comply with the proposal at a later date, what steps may be taken by the creditors?
7. What is a "preferred" creditor, and how does this status affect the right to payment?
8. Why are "inspectors" appointed by creditors at their first meeting in bankruptcy proceedings?
9. What should a person who finds it impossible to carry on business any longer without incurring further losses do?
10. Under what circumstances would a debtor be permitted to make a voluntary assignment for the benefit of his or her creditors?
11. Outline the order of priority of payment to preferred creditors in a bankruptcy.
12. Explain the duties of an undischarged bankrupt.
13. What is the effect of a discharge on a bankrupt debtor's obligation to pay his or her creditors?
14. Why is an insolvent person not necessarily a bankrupt?

MINI-CASE PROBLEMS

1. Alexander was insolvent and attempted to hide from his creditors his only asset, a valuable painting worth $10,000. Assuming that he is two months in arrears on his apartment rent of $1,000 per month, owes his housekeeper back wages of $1,000 for one month's work, and owes unsecured creditors $12,000, what would the unsecured creditors receive in cents on the dollar if the trustees' fees and the costs of the bankruptcy were $2,000?
2. Baker owed $3,000 to Able, $42,000 to Alice and $1,000 to Henry. All of the debts were in default, but Henry believed that Baker had assets that would cover the debts.

   What steps could Henry take to obtain payment of his debt?
3. Shelley was in financial difficulty, and made a voluntary assignment in bankruptcy. At the time she owned a $100,000 property subject to a mortgage in the amount of $70,000, and had personal unsecured debts of $40,000.

   How will the assets be distributed if the property was sold for $93,000, and trustee fees were $3,000?

CASE PROBLEMS FOR DISCUSSION

CASE 1

For many years, Commercial Canners Ltd. supplied Gibson Food Wholesale Ltd. with canned food products. These were usually shipped directly to the Gibson Food Wholesale Ltd. main warehouse, but occasionally the goods were directed by Gibson Food Wholesale Ltd. to a storage warehouse owned by Pierson Storage Inc., where Gibson Food Wholesale Ltd. leased storage space.

Gibson Food Wholesale Ltd., unfortunately, ran into financial difficulty, and was eventually forced into bankruptcy by Pierson Storage Inc. who had been unpaid on its storage charges for many months. A number of other suppliers of Gibson Food Wholesale Ltd. were unpaid creditors, including Commercial Canners Ltd. When Commercial Canners Ltd. was notified of the bankruptcy, it examined its records and discovered that in the 2 weeks prior it had shipped 100 cases of canned goods to Gibson Food Wholesale Ltd., and 80 cases of canned goods to the Pierson Storage Inc. warehouse. Commercial Canners Ltd. valued the canned goods at $20 per case (average) for a total value of $3,600. Commercial Canners Ltd. contacted the trustee in bankruptcy, and discovered that 50 cases of its canned goods were still in the Gibson Food Wholesale Ltd. warehouse, and 80 cases were in the Pierson warehouse. The trustee also advised Commercial Canners Ltd. that Pierson Storage was claiming a lien on all Gibson Food Wholesale Ltd. goods in its warehouse.

Advise Commercial Canners Ltd. How will the claim likely be resolved?

www.mcgrawhill.ca/olc/willes

CASE 2

Martin Electrical Limited carried on business as an electrical contractor for many years. Martin was the sole shareholder of the corporation. Three years ago, the corporation purchased a warehouse building for $150,000. The building was financed in part by a mortgage loan on the building from Martin's spouse in the amount of $125,000. A year later, Martin purchased a new truck for $35,000, financing it by a conditional-sale agreement to the truck dealer, who subsequently assigned it to Auto Finance Ltd. Over the past year, business gradually began to slow down for the corporation, and in an effort to help the corporation meet its debts, Martin personally loaned the corporation $30,000, in return for a promissory note for the amount loaned.

When business did not improve, Martin realized that the business could not continue, and notified his creditors that operations would cease. The principal creditor, Big Business Bank, who was owed $25,000 secured by a chattel mortgage on inventory and a unsecured line of credit of $30,000, joined unpaid suppliers (who were owed $45,000) in a bankruptcy petition. Accounts receivable in the amount of $15,000, inventory of $30,000 in stock, and cash in the bank of $1,000 were the only other assets of the corporation. Martin, as an employee of the corporation was owed unpaid wages of $3,000. The trustee in bankruptcy fees would probably be $5,000.

If the property and assets were sold for the amounts as indicated, determine how the claims of the creditors would be assessed by the trustee.

What objections might be made to some of the creditors' claims?

CASE 3

For many years the Acme Company carried on business as a manufacturer of consumer products. In 2005, it embarked on an ambitious program of expansion that involved the acquisition of a new plant and equipment. Financing was carried out by way of real-property mortgages, chattel mortgages, and conditional-sale agreements, with very few internally generated funds being used for the expansion.

The general decline in demand for its product line as a result of the poor economic climate placed the company in a serious financial situation by 2011. As a result of a failure to pay a trade account to one creditor, bankruptcy proceedings were instituted. Acme did not object to the proceedings and did not make a proposal to its creditors.

The trustee disposed of the assets of the company and drew up a list of creditors entitled to share in the proceeds. His preliminary calculations were as follows:

Sale of assets	
Sale of land and buildings	$350,000
Sale of production equipment	35,000
Sale of trucks & automobiles	25,000
Sale of inventory of finished goods	30,000
Accounts receivable	45,000
Cash	3,000
	$488,000

Expenses and creditor claims (all secured claims properly registered)	
First mortgage on land and buildings	$290,000
Second mortgage on land and buildings	45,000
Third mortgage on land and buildings	40,000
First chattel mortgage on vehicles	22,000
Second chattel mortgage on vehicles	40,000
Bank claim under s. 427 of *Bank Act*	25,000
Unsecured trade creditors	60,000
Unpaid wages (ten employees at $300 each)	3,000
Unpaid commissions to salespeople (1 at 1,500)	1,500
Bankruptcy expenses, fees, and levy	39,000
Unpaid municipal taxes	9,000
Equipment's conditional-sale agreement	10,000
	$584,500

Calculate the distribution of the funds to the various creditors and calculate the cents-per-dollar amount that the unsecured trade creditors would receive.

CASE 4

Simple purchased a small business from a well-established proprietor for $200,000. To finance the transaction, he borrowed $150,000 by way of a mortgage on the premises, and he prevailed upon the proprietor to accept a chattel mortgage on the equipment for the balance. Both mortgages were duly registered. He then arranged with the trade suppliers to sell him his inventory on credit.

The business was a high-volume, low mark-up type of business. A large amount of money passed through Simple's hands each day, even though the portion that represented his profit was small. During the first few months of operation, he purchased a new, expensive automobile, refurnished his apartment, and took a quick four-day holiday to Las Vegas, where he lost several thousand dollars at the gaming tables.

When his suppliers began pressing him for payment of their accounts, he managed to pacify them by staggering payments in such a way that each received the payment of some accounts, but their total indebtedness remained about the same. He accomplished this in part by seeking out other suppliers and persuading them to supply him with goods on credit.

A few months later, it became apparent to Simple's creditors that he was in financial difficulty, and several creditors threatened to institute bankruptcy proceedings. To forestall any action on their part, Simple paid their accounts in full. The threats of the creditors brought his desperate financial position forcefully to his attention, however. He promptly transferred $10,000 to his wife, and he placed a further $10,000 in a bank account that he opened in another city.

A few days later, Simple purchased two one-way airline tickets for a flight to Brazil that was scheduled for the next week. Before the departure date, a creditor, to whom Simple owed a trade account in excess of $5,000, became aware of his plans and instituted bankruptcy proceedings against him.

Discuss the actions of Simple and indicate how the provisions of the *Bankruptcy and Insolvency Act* would apply. What steps may be taken to protect the creditors in this case?

CASE 5

Able carried on business as a service-station operator. In addition to repairing automobiles, he maintained a franchise for the sale of a line of new automobiles. He also sold gasoline and the usual lines of goods for the servicing of vehicles. Business was poor, however, and Able made a voluntary assignment in bankruptcy, in which he listed as assets:

Land and building	$50,000
New automobile (1)	24,000
Gasoline and oil	3,000
Parts, supplies, and equipment	3,000
Accounts receivable	2,000
Bank	100
Personal assets (furniture, etc.)	1,900
	$84,000

His creditors' claims were as follows:

First registered mortgage	$ 40,000
Second registered mortgage	7,000
Registered conditional-sale agreements on cars	22,000
Due and owing to fuel supplier	5,000
Due and owing to other trade creditors	18,000
Municipal taxes owing	1,000
Personal debts (unsecured)	10,000
	$103,000

When the trustee went to Able's place of business, he discovered that: (1) the new cars had been taken by the manufacturer; (2) the fuel tanks had been emptied by the fuel supplier; and (3) Baker, an employee of Able's, was on the premises and in the process of removing an expensive set of tools that he maintained had been given to him by Able in lieu of wages for his previous week's work.

Discuss the steps that the trustee might take as a result of the discoveries.

The Online Learning Centre offers more ways to check what you have learned so far. Find quizzes, Weblinks, and many other resources at www.mcgrawhill.ca/olc/willes.

CHAPTER 31

Insurance Law

Chapter Objectives

After study of this chapter, students should be able to:

- Recognize the various forms of insurance.
- Understand the nature of the contract of insurance and the concept of indemnity for loss.
- Identify the parties associated with insurance contracts and their rights and duties.

YOUR BUSINESS AT RISK

The crippling risks of fire, liability for negligent acts and business interruption are self-evident, as is the need to protect against them. The contract of insurance is an equally obvious answer, but not enough business persons realize that it is a contract of utmost good faith. As such, all the protection that insurance represents may be lost if the insured has not revealed all material facts to the insurer. Even if the contract remains enforceable, there still may be significant exclusions and limitations to coverage. This requires the business operator to make sure he or she has professional assistance in the first place to obtain adequate coverage across all major areas of risk, and to understand the triggers and limits to compensation.

HISTORICAL DEVELOPMENT

The reduction of risk, and, in particular, the risk of loss from unforeseen dangers, has been a quest of humanity since the beginning of time. Originally, risk was reduced by members of a family banding together to protect one another and their possessions. As society developed, however, the community tended to act as an expanded family when misfortune struck one of its members. The loss of a limb or the destruction of a dwelling often triggered a community response to aid the unfortunate individual.

Protection from the financial loss that frequently accompanies a misfortune was recognized by the early guilds in England and Continental Europe during the Middle Ages. The benefits of protection from loss due to injury were not, however, the primary reason why craftsmen and merchants joined guilds, but they did represent an attractive advantage of membership. Originally, the members of the group made a common pledge to compensate one another in the event of loss. Later, the individuals in the group contributed sums of money to form a pool or fund from which a loss incurred by a member would be paid. As the fund was gradually reduced by the payment of compensation for losses, additional levies were made on the group members to replenish the fund. Eventually, nearly all guilds used this method as a basis for their insurance. It should be noted, however, that insurance schemes were not restricted to the guilds alone. Some early merchants operated indemnity funds as well, and charged fees in return for a promise to indemnify a fund member in the event of loss.

In the late medieval period, some forms of insurance developed that did bear a certain resemblance to the modern concept of insurance. The practice of spreading the risk of loss for maritime adventures among a number of individuals was apparently carried on in the early Middle Ages. By the twelfth century, merchants in the Lombard area of what is now Italy insured against some of the perils of navigation. Italian merchants who visited England to trade brought the custom with them, and, by the fourteenth century, a form of maritime insurance was available for English adventurers to cover loss at sea. Disputes between the parties were normally settled by the merchants themselves, in accordance with the customs that had developed to deal with these early forms of insurance.

Throughout the early stages of development of the concept of insurance, the law did little to encourage the practice of reducing risk. The early merchants settled their own insurance disputes using their own law or custom, but non-members of the guilds had to be content with a Common Law and with a court system that was unsuited to the enforcement of insurance agreements. It was not until the middle of the eighteenth century that the widespread use of insurance prompted the courts to examine the nature of insurance and to make the Common Law responsive to the legal problems associated with insurance and the enforcement of claims. By this time, the courts began to view the insurance relationship as one of contract, and treated it as such in their application of the law. The relatively favourable climate that existed after that time permitted the creation of many types of insurance based upon this concept. A general body of law soon developed, based in part on the law of contract.

Forms of Insurance

There are many different kinds of insurance available to the business executive today. Nearly every conceivable form of risk may be insured, the only exception being certain activities that might encourage carelessness if insured, and deliberate acts that may cause injury or loss.

Modern insurance is based upon statistical calculation of the likelihood of a particular loss occurring. As a result of accurate recordkeeping over a long period of time, insurers can determine the frequency of occurrence of different types of losses. By these records they may establish the amount of money they require from each insured in order to maintain a fund sufficiently large at all times to cover losses as they occur. Because some of these funds are invested by the insurer, the income earned is included in the fund as well to cover the insurer's expenses and profits, and to reduce the amount that the insured must pay for the insurance coverage.

Life insurance differs, to some extent, from other forms of insurance, in that the insurer will eventually be obliged to pay the face value of all policies of insurance in force at the time of death of the insured person. Statistical data on the probable lifespan of individuals, called actuarial tables, are used to determine the likelihood of loss due to the premature death of policyholders, and to determine the premium required to cover this unexpected event. The tables are also used to calculate the expected payout of the value of the policy, if the policyholder dies at the end of a normal lifespan.

Some life-insurance policies may be used for investment purposes, as well as for protection of the beneficiaries in the case of the unexpected death of the insured. For life insurance of this type, the premiums include not only an amount to cover the cost of coverage for an unexpected loss of life, but an amount to provide the insured a sum of money at the end of a specified period of time.

Fire Insurance

This type of insurance is designed to indemnify a person with an interest in property for any loss that might occur as a result of fire. Normally, any person with an interest in the property may protect that interest by fire coverage. The owner of the property, and any secured creditors or tenants (to the extent of their interest), may obtain this form of protection. Fire coverage is not limited to buildings only, as chattels contained in a building may also be insured. Fire-policy protection is normally extended to damage caused as a result of the fire, as in the case of smoke and water damage. Insurance policies usually only indemnify the insured against loss from "hostile" fires, i.e., a fire that is not in its proper place. In contrast, a fire in a fireplace, for example, would be classed as a "friendly" fire, rather than a "hostile" one, as it is a fire deliberately set in its proper place. Insurers distinguish between the two types of fires, because a friendly fire is usually not insured. If a friendly fire becomes a hostile fire, the resultant loss is usually covered by fire insurance, unless the actions of the insured with respect to the fire were such that the fire may be classed as arson, or as a deliberate attempt to destroy the insured premises by that fire.

Life Insurance

As the name indicates, life insurance is insurance on the life of a person, be it one's own life, or that of another person in which one has an insurable interest. Life insurance, in its simplest form, is payable on the death of a particular person. Unlike other forms of insurance, it is based upon an event that will eventually occur. The only uncertainty attached to life insurance is the timing of the death that will render the policy payable. It differs from other forms of insurance in that the person upon whose life the insurance is placed does not receive the proceeds of the insurance, although they may be made payable to the deceased's estate if no specific beneficiary is named in the policy.

Life-insurance policies usually include an application for the insurance, in which the insured sets out all the information required by the insurer to determine if the risk should be accepted and, if so, the premium payable for assuming the risk. The application is usually incorporated in the policy and becomes a part of the contract. Fraudulent statements by the applicant generally permit the insurer to avoid payment under the policy when the fraud is discovered.

Under provincial legislation, a life-insurance policy may take on a variety of different forms, from simple term insurance to special-purpose policies. The legislation generally does not determine the specific kinds of policies that a life insurer may issue but, rather, the terms that must be contained in the policy respecting such matters as lapse, renewal, time for payment of proceeds, and proof of death of the insured. The legislation also covers other aspects of life insurance, such as life insurers themselves and the operation of their businesses. As with most insurers, life insurers are required to follow strict rules regarding the investment of their funds, in order to make certain that the company remains solvent, and that it is in a position to pay all claims under the policies.

Sickness and Accident Insurance

Insurance for sickness and accident represents a type of insurance that protects against or reduces the loss that a policy holder might incur through sickness or accident. The amounts payable usually vary, but upper limits on sickness benefits are normally set at an amount less than a person's normal income, payable on presentation of proof of the illness. Accident benefits that cover loss of limb or eyesight, or other permanent injuries, are generally fixed in the policy at specific dollar amounts. As with other forms of insurance (other than life), this type of insurance is designed to provide indemnity for the loss incurred.

Liability and Negligence Insurance

Liability and negligence insurance is designed to indemnify persons for liability for losses due to negligence in the performance of their work, profession, or actions, or the use of their premises. The policy, by nature, covers specific losses, such as those that may arise from negligence in the operation of a motor vehicle, a business establishment, or even a residence. These policies are designed to compensate for losses due to the torts of the individual, rather than for direct losses that an individual might suffer through no fault of his or her own. Of the many forms of negligence or liability insurance, automobile insurance has become so important, and its use so widespread, that it is treated separately under insurance legislation in most provinces. A standard policy form has also been devised as a result of interprovincial cooperation by those provinces that do not maintain their own compulsory, government-administered automobile-insurance schemes.

Apart from automobile insurance, liability insurance is normally used to protect against claims of loss arising out of the use of premises (i.e., occupier's liability), manufacturer's product liability, professional negligence, and third-party liability for the acts of servants or agents.

More recently, many firms have turned to insurance as a means of protection from claims under new environmental laws. Policies may be obtained to cover environmental accidents such as product spills causing ground or water pollution, pollution damages caused by customer use of products manufactured by the insured, or the insured's negligence in the design of products for others that in turn causes environmental damage. Of these types of policies the most important is probably the policy that covers the cost of cleanup in cases where the government orders a business to remove contaminants from soil or water.

As well, most professional persons carry liability insurance to cover professional errors and omissions. Physicians and surgeons generally obtain coverage for claims that may arise out of improper treatment of patients' illnesses, or the failure to perform medical procedures in accordance with accepted standards of care. Similarly, professional accountants and lawyers carry liability insurance to cover errors or omissions they may make in the performance of their work on behalf of clients, or on the advice they may offer, if it should prove to be negligently given. Engineers, architects, and other professionals may also obtain coverage for errors they might make in the conduct of their professional duties.

Title Insurance

Title insurance, which provides protection to purchasers of property and mortgagees from any legal challenges that might arise to their interests in real property, has long been in use in the United States and some other jurisdictions. However, its use in Canada has been limited until very recently. Insurance of this type does not replace the services of a lawyer to search the title of the property and to make all other searches to ensure that a purchaser or mortgagee has a good title or interest in property, but it does protect against errors or oversights of interests or the failure to register title documents or mortgages. Current title insurance usually does not cover claims related to environmental hazards, aboriginal land claims, or some types of governmental action unless specifically applied for by the applicant.

Additional Business-Specific Forms of Insurance

In addition to these general forms of insurance coverage, insurance is also available for many specialized purposes, related to the unique risks of business operations. The principle at work in each of these areas is an intention to indemnify the insured in the event of a loss or in a claim for compensation.

In the previous sections on fire, liability, and negligence insurance, the indemnity is focused on replacing the physical asset (for example, a factory) or a person outside the business (for example, a customer, in a case of product liability). Still, there are intangible risks inside the company, which must be protected.

Closely related to fire insurance is business-interruption insurance. While fire insurance will replace the physical premises of a factory or shop, consideration should be given to the fact that the business will be out of business for the months or even years that it takes to rebuild the premises. During this time, fixed expenses, such as interest on business debts, will continue, and may force the bankruptcy of the company. Business-interruption insurance, which covers more than fire insurance, including damage from flood or forced evacuation, is the appropriate response. Premiums will reflect the cash flow to be insured and the period during which any interruption could be foreseen to exist.

As well, the business depends upon key personnel for its success. The death of a particularly skilled employee, partner, or shareholder who is also instrumental in the operation and success of the business may lead the firm toward insuring itself against the loss of that person's life (and, hence, his or her services). A particularly difficult and often unforeseen problem arises in situations involving the death of partners and shareholders, even if they are not considered key to business operations. This is most acute in small, private corporations with little ability to raise large blocks of capital financing quickly. Here, the estate of the deceased will likely have the right to liquidate his or her business interest, and the firm may not have sufficient liquidity to pay this demand without selling off or dramatically mortgaging its assets. The solution is for the firm to take out life insurance on the life of each person whose estate is in a position to make this demand. Upon that person's death, the business will then have the liquidity to meet the demand, without financial crisis.

Unscrupulous employees pose a risk to the health of a business, and insurance is available against this risk. This type of insurance is known as bonding. Bonding provides indemnity for loss suffered by the company, and may give a measure of confidence to customers who deposit valuables (for storage, repair, or transport) with the firm. The premium payable for an employee bond will reflect the potential fraud or theft that could be executed, as well as the internal management controls in place to prevent it. Often, particular management controls will be dictated under the terms of the bond (e.g., double signatures, joint custody of assets) and the bond will be void if these procedures are not in place at the time of loss.

CASE IN POINT

In 2000, a software system design company entered into an agreement with a community college to design and install software for an academic management information system, and to maintain the system until a date in 2005. The system was completed in May of 2002, but problems continued with the system.

In 2002, the company applied for and received an Information Technology Errors and Omissions policy from an insurer to run for 12 months, ending in February 2003. In January 2003 serious problems had developed with the system, but this was not reported to the insurer, and a new policy was issued to run until March 2004. In April 2003, the college abandoned the contract and took legal action against the company for losses sustained due to the failure of the system.

The insurer denied coverage on the basis that the company had not disclosed the potential claim by the college at the time that the policy was issued.

The court agreed with the insurer, and stated that the company had an obligation to inform the insurer of the potential claim of the college.

Agresso Corporation v. Temple Insurance Corporation, 2007 B.C.C.A. 559.

FRONT-PAGE LAW

Stolen construction machinery shipped overseas, consultant warns

TORONTO — Heavy-construction equipment, from small bulldozers to $1-million excavators, is being stolen in record numbers across Canada, often ending up in east Asia or South America, an expert on the phenomenon said yesterday.

Among the customers for the machines are Colombian drug cartels who use filched earth movers to clear bush for illicit coca plantations, said security consultant George Kleinsteiber.

A $750,000 bulldozer stolen from Orangeville, north of Toronto, showed up recently at China's Three Gorges Dam construction project, said Mr. Kleinsteiber.

The thefts can have a devastating impact on small construction companies without insurance, resulting in bankruptcy, house foreclosures and even suicide attempts in some cases, he said.

"The biggest things you can think of, they're out there stealing: the excavators for houses, bulldozers, the real big ones. We've had stuff stolen out of the north that would only fit on a train or on a large float truck."

About 5,000 pieces of heavy equipment, costing anywhere from $30,000 to more than $1-million, were stolen across Canada last year, said Mr. Kleinsteiber.

Hopefully, the Canadian construction firm in Orangeville was not the one that tracked down its bulldozer in China. Who did, and what were their technical legal grounds for doing so and demanding its surrender?

Source: Excerpted from Tom Blackwell, *Southam News*, "Stolen construction machinery shipped overseas, consultant warns," September 21, 2000.

THE NATURE OF THE INSURANCE CONTRACT

Utmost Good Faith

Disclosure

The obligation for an insured to disclose all material information to the insurer.

The contract of insurance, as the name implies, is a contractual relationship to which the general rules of contract, and a number of special rules, apply. It is treated by the courts as a contract of utmost good faith. This means that the applicant for insurance must disclose all information requested to enable the insurer to decide if it should assume the risk. The insurer–insured relationship has also been the subject of much control through legislation. Each province has legislation governing the contract of insurance in its various forms and, with the exception of the province of Quebec, the legislation has tended to become uniform for most types of insurance. A number of provinces have special legislation that provides for provincially controlled automobile insurance, or for "no-fault" insurance for automobile-accident cases. For the remainder, the general legislation and the Common Law rules apply. Changes in standard form contracts are effected by riders or endorsements that represent changes or additions to the standard terms and coverage in the agreement. While these two terms are frequently used to refer to changes made to a standard form contract, a **rider** is an additional clause attached to the contract that adds to, or may alter, standard form coverage. A rider is normally included in the agreement at the time the contract is written. An endorsement, on the other hand, is a change the parties agree to make to an existing contract and that, to save rewriting the contract, is simply attached to it.

Rider

A clause altering or adding coverage to a policy.

The contract of insurance is a special type of contract called a policy that is made between an insurer and an insured, whereby the insurer promises to indemnify the insured for any loss that may flow from the occurrence of any event described in the agreement. In return for this promise, the insured pays, or agrees to pay, a sum of money called a premium.

The contract of insurance bears a resemblance to a simple wagering agreement, in that the insurer must pay out a sum of money on the occurrence of a particular event. The resemblance, however, is only superficial, because there are substantial differences between the two types of agreements — the most important difference being the interests of the parties. In the case of a simple wager, the basis of the agreement is generally the occurrence of an event that will not directly result in a loss to either party (except for the amount of the wager that each party has pledged). In contrast to this, under an insurance policy, the insured receives nothing until he or she suffers some loss. Even then, the insured will only receive a sum that will theoretically place the insured in the same position that he or she was in before the loss occurred. The exception here is life insurance, where the insured must die to collect. However, even here, payment is not made unless the insured suffers the loss.

COURT DECISION: Practical Application of the Law

Lack of Good Faith—Punitive Damages

***Whiten v. Pilot Insurance Co.*, [2002] 1 S.C.R. 595.**

The appellant's home burned while covered by the respondent's policy of insurance. Despite evidence to the contrary, the respondent maintained that the cause was arson and refused to pay the claim.

THE COURT: This case raises once again the spectre of uncontrolled and uncontrollable awards of punitive damages in civil actions. The jury was clearly outraged by the high-handed tactics employed by the respondent, Pilot Insurance Company, following its unjustified refusal to pay the appellant's claim under a fire insurance policy (ultimately quantified at approximately $345,000). Pilot forced an eight-week trial on an allegation of arson that the jury obviously considered trumped up. It forced her to put at risk her only remaining asset (the insurance claim) plus approximately $320,000 in legal costs that she did not have. The denial of the claim was designed to force her to make an unfair settlement for less than she was entitled to. The conduct was planned and deliberate and continued for over two years, while the financial situation of the appellant grew increasingly desperate. Evidently concluding that the arson defence from the outset was unsustainable and made in bad faith, the jury added an award of punitive damages of $1 million, in effect providing the appellant with a "windfall" that added something less than treble damages to her actual out-of-pocket loss.

The respondent argues that the award of punitive damages is itself outrageous. The appellant, Daphne Whiten, bought her home in Haliburton County, Ontario, in 1985. Just after midnight on January 18, 1994, when she and her husband Keith were getting ready to go to bed, they discovered a fire in the addition to their house. It was minus 18 degrees Celsius. Mr. Whiten gave his slippers to his daughter to go for help and suffered serious frostbite to his feet for which he was hospitalized. He was thereafter confined to a wheelchair for a period of time. The fire totally destroyed the Whitens' home and its contents, including their few valuable antiques and many items of sentimental value and their three cats. The appellant was able to rent a small winterized cottage nearby for $650 per month. Pilot made a single $5,000 payment for living expenses and covered the rent for a couple of months or so, then cut off the rent without telling the family, and thereafter pursued a hostile and confrontational policy which the jury must have concluded was calculated to force the appellant (whose family was in very poor financial shape) to settle her claim at substantially less than its fair value. The allegation that the family had torched its own home was contradicted by the local fire chief, the respondent's own expert investigator, and its initial expert, all of whom said there was no evidence whatsoever of arson. The respondent's position, based on wishful thinking, was wholly discredited at trial. Pilot's appellate counsel conceded here and in the Ontario Court of Appeal that there was no air of reality to the allegation of arson.

A majority of the Ontario Court of Appeal allowed the appeal in part and reduced the punitive damage award to $100,000. In my view, on the exceptional facts of this case, there was no basis on which to interfere with the jury award. The award, though very high, was rational in the specific circumstances disclosed in the evidence and within the limits that a jury is allowed to operate. The appellant was faced with harsh and unreasoning opposition from an insurer whose policy she had purchased for peace of mind and protection in just such an emergency. The jury obviously concluded that people who sell peace of mind should not try to exploit a family in crisis. Pilot, as stated, required the appellant to spend $320,000 in legal costs to collect the $345,000 that was owed to her. The combined total of $665,000 at risk puts the punitive damage awards in perspective. An award of $1 million in punitive damages is certainly at the upper end of a sustainable award on these facts but not beyond it. I would allow the appeal and restore the jury award of $1 million in punitive damages.

The sword of good faith cuts in both directions — insured persons must be able to rely on the fact that their insurer will pay their claims when such a loss occurs. This case stands as a warning to insurers and a reminder of their duties.

Insurable Interest

Insurable interest

An interest that would result in a loss on the occurrence of the event.

The loss that the insured suffers must relate to what is known as an **insurable interest**. This interest must be present in every insurance contract. It may be defined as anything in which the insured has a financial interest that on the occurrence of some event might result in a loss to him or her. An insurable interest may arise from ownership or part-ownership of a chattel or real property, or a security interest in either of them, or it may be one's own life, the life of one's spouse or child, or the life of a debtor or anyone in whom a person may have a pecuniary interest (for example, a partner or a key employee). It may also arise out of a person's profession, or activity to protect income or assets. Most insurers, however, will not insure persons against the willful acts that they commit against themselves or against their insured interests. For example, an insured person may not obtain fire coverage on a home, then deliberately burn the premises to collect the insurance proceeds. Nor would an insurer normally be obliged to pay out life insurance on the life of an insured who committed suicide. However, it should be noted that under insurance legislation in some jurisdictions, the beneficiaries may be entitled to the insurance proceeds in the case of a suicide where the policy so provides or has been in effect for some time.[1] In general, an insurable interest is anything that stands to benefit the insured person by its continued existence in its present form, and that, if changed, would represent a loss. With the exception of life insurance, the insurable interest must exist both at the time the contract of insurance is made and when the event occurs that results in a loss.

If Arthurs places a policy of insurance on a house she owns, then later sells the house to Bond for cash, and the house is subsequently destroyed by fire, Arthurs would not be permitted to collect the insured value of the house. By selling the house, she

1. Ontario, for example, provides that payment shall be made if the policy contains an undertaking to pay if the insured should commit suicide.

divested herself of the interest she had in the property, and she no longer had an insurable interest in the property at the time of the loss. Nor would Bond be entitled to recover under the policy, because he was not a party to the insurance contract.[2]

In the case of life insurance, the person who takes out a policy of insurance on the life of another need only establish an insurable interest in the life of that person at the time the policy of insurance was issued. For example, if a creditor arranged for the issue of a policy of life insurance on the life of a person indebted to him, the creditor could show an insurable interest at the time of issue of the policy. The creditor, however, need not establish an insurable interest at the time of the debtor's death to receive the proceeds of the policy.

Disclosure

In addition to the requirement that the insured possess an insurable interest, the contract of insurance, being a contract of utmost good faith, requires full disclosure on the part of the applicant for the insurance of all material facts that might affect the decision of the insurer to accept the risk and to determine the appropriate premium. With respect to disclosure, the courts have reasoned that the insurer knows nothing, and the applicant everything; hence, the obligation on the part of the applicant to disclose all material facts.

The right of the insurer to be apprised of all material facts is important. The insurer is undertaking a risk that is frequently determined from the information supplied by the applicant. Consequently, honesty on the part of the applicant is essential. If the applicant fails to disclose material facts, then the insurer may later refuse to compensate the insured if a loss occurs.

If the true owner of a motor vehicle arranges with a friend to have the vehicle registered in his name for the purpose of obtaining insurance, the insurance protection may not extend to the true owner if the true owner was driving the vehicle at the time of an accident that involved him, and for which he was responsible.[3]

The question of what represents an innocent misrepresentation of a material fact or non-disclosure was discussed by the courts in the case of *Mutual Life Ins. Co. of N.Y. v. Ontario Metal Products Co.*,[4] where the judge commented:

> The main difference of judicial opinion centres round the question what is the test of materiality? It is the insurers who propound the questions stated in the application form, and the materiality or otherwise of a misrepresentation or concealment must be considered in relation to their acceptance of the risk. All of the questions may be presumed to be of importance to the insurer who causes them to be put, and any inaccuracy, however unimportant in the answers, would, in this view, avoid the policy. Suppose, for example, that the insured had consulted a doctor for a headache or a cold on a single occasion and had concealed or forgotten the fact, could such a concealment be regarded as material to the contract? Faced with a difficulty of this kind, the appellants' counsel frankly conceded that materiality must always be a question of degree, and therefore to be determined by the Court, and suggested that the test was whether, if the fact concealed had been disclosed, the insurers would have acted differently, either by declining the risk at the proposed premium or at least by delaying consideration of its acceptance. If the former proposition were established in the sense that a reasonable insurer would have so acted, materiality would, their Lordships think, be established, but not in the latter if the difference of action would have been delay and delay alone. In their view it is a question of fact in each case whether if the matters concealed or misrepresented had been truly disclosed, they

2. *Rowe v. Fidelity Phenix Fire Insurance Co. of New York*, [1944] O.W.N. 387, 600.
3. *Minister of Transport et al. v. London & Midland General Ins. Co.*, [1971] 3 O.R. 147.
4. [1925] 1 D.L.R. 583, [1925] A.C. 344, [1925] 1 W.W.R. 362.

would, on a fair consideration of the evidence, have influenced a reasonable insurer to decline the risk or to have stipulated for a higher premium.

At Common Law, the non-disclosure or misrepresentation of a material fact would entitle the insurer to later avoid liability when the non-disclosure or misrepresentation was discovered. This has been altered, to some extent, by statute in various provinces, but for the most part the rule still holds. The exception that the legislation makes relates generally to innocent misrepresentation or innocent non-disclosure. However, where the non-disclosure or the misrepresentation amounts to fraud, then the Common Law rule still holds.

The legislative modification of the Common Law position has, as its justification, the unfairness of an insurer refusing payment of a loss where the insured without intention to deceive failed to disclose a fact, or stated an untruth as something that he or she honestly believed to be true. In these cases, the Common Law requirements for a contract of utmost good faith have been modified to require the insurer to carry out the policy terms if the policy has been in effect for a considerable period of time before the loss occurs (usually several years). For example, Ontario legislation provides that innocent non-disclosure by an applicant for life or health and accident insurance may not be a basis for the insurer to avoid payment of a claim made after the policy issued on the basis of the application has been in force for a period of more than two years.[5]

A contract of insurance differs from an ordinary contract in a number of other ways as well. It tends to be an ongoing relationship that usually requires the insured to advise the insurer of any substantial changes in the risk covered by the policy. Fire-insurance policies usually require the insured to notify the insurer if the insured premises will be left unoccupied for more than a specified period of time. Insured business people are expected to notify the insurer if the risks associated with the conduct of their business change substantially. For example, if a manufacturer of children's toys decides to change his product line to include the manufacture of fireworks or some other dangerous product, he would be obliged to notify the insurer that a new, higher-risk activity was to take place on the premises.[6]

CASE IN POINT

Marche bought and insured a house and converted it to apartments. For a time the house was vacant before a tenant was found. The house subsequently burned, but Halifax denied the claim because the insured had failed to inform them of the earlier vacancy. It argued that the earlier vacancy was a material change in risk which invalidated the coverage under Statutory Condition #4, which is a condition of all policies, created by the *Insurance Act* for the benefit of insurers. While the insured may have been in breach by not advising of vacancy, the court found they should be relieved from its consequences, as s. 171 of the *Insurance Act* states that a policy condition is not binding if a court holds it to be "unjust or unreasonable." The Court of Appeal reversed the decision, saying that s. 171 did not apply to statutory conditions, but applied only to contractual conditions. The Supreme Court of Canada ruled that provisions such as s. 171 must be interpreted broadly and apply to terms inserted by legislation as well as those created by the contract itself. The trial decision was restored.

See: *Marche v. Halifax Insurance Co.*, 2005 S.C.C. 6.

CLIENTS, SUPPLIERS *or* OPERATIONS *in* QUEBEC

Much of insurance litigation in Common Law jurisdictions revolves around finding outcomes that are "fair," recognizing the role of equity in contracts that are of utmost good faith. Quebec law is certainly fair, but avoids the ambiguities attached to equity, which does not form a part of civil law systems. Consequently, Supreme Court of Canada judgments involving equity are difficult to apply in Quebec, and much more reliance must be placed on the detailed provisions of Quebec's *Insurance Act*, R.S.Q., c. A-32, and the *Civil Code* provisions of general application.

5. *Insurance Act*, R.S.O. 1990, c. I.8, s. 309.
6. *Poapst v. Madill et al.*, [1973] 2 O.R. 80, 33 D.L.R. (3d) 36.

A QUESTION OF ETHICS

Randall is 77 years of age, in failing health. He is not required to submit to mandatory provincial driver testing of seniors until age 80. He has recently had a number of parking lot "fender-benders." In each case he has paid the other drivers' damages out of his own pocket. Some were less than his deductible, others were in excess. None were reported to his insurer. All were under the amount provincially required for reporting to police. Can this situation be reconciled with the contractual requirement of utmost good faith owed to Randall's insurer?

THE CONCEPT OF INDEMNITY FOR LOSS

Indemnity

The concept that distinguishes a contract of insurance from a wager.

The particular feature that distinguishes the contract of insurance from a wager is the fact that it is a contract of **indemnity**. With the exception of life insurance and, to some extent, accident insurance, all contracts of insurance prevent the insured from making a profit from a loss. A number of special insurance concepts ensure that the insured will only be placed in the position that he or she was in before the event occurred that caused the loss. For some forms of loss, which concern third parties, no special protection is needed for the insurer. For example, if Smith should injure Jones by her negligence, Smith's insurer will compensate Jones for his loss, or pay any judgment that Jones might obtain against Smith for her carelessness. Only the injured party will be compensated, and then only for the actual loss suffered.

With respect to chattels or property owned by the insured, three special rights of the insurer apply in the event of loss in order to prevent the insured from receiving more than the actual loss sustained. If the property is not completely destroyed, the insurer has the option to repair the chattel, or pay the insured the full value of the property at the time of loss. If the insurer pays the insured the value of the chattel, then the insurer is entitled to the property. This particular right is known as **salvage**, and it gives the insurer the right under the policy to demand a transfer of the title to the damaged goods.

Salvage

The right of the insurer to the property where the insured is compensated for the loss.

McKay owns a truck insured by the Car Insurance Company. The truck is involved in an accident and is badly damaged. If the Car Insurance Company compensates McKay for the value of the truck, then McKay must deliver up the damaged truck to the insurer in return for the payment. The insurance company may then dispose of the wreck to reduce the loss that it has suffered through the payment of McKay's claim.

The same principle would apply in the case of goods stolen from the insured. If the insurer pays the insured the value of the stolen goods, and if the goods are subsequently recovered, the goods will belong to the insurer and not the insured. By the terms of the policy of indemnity, the goods become the goods of the insurer on the payment of the claim. In a sense, the contract bears some resemblance to a purchase of the goods by the insurer.

Subrogation

The substitution of parties whereby the party substituted acquires the rights at law of the other party, usually by way of contractual arrangement.

A second form of protection for the insurer is the doctrine of **subrogation**. Subrogation concerns the right of the insurer to recover from another person that which the insured recovers from the insurer. The doctrine of subrogation arises when the insured is injured or suffers some loss due to the actionable negligence or deliberate act of another party.

If an insured vehicle is damaged by the negligence of another driver, the owner would have a right of action against the other driver for the damage caused by the other driver's negligence. If the insurer compensates the owner for the damage to his or her vehicle, then, by the doctrine of subrogation, the insurer is entitled to take over the owner's right of action against the negligent driver.

Contracts of insurance may contain a subrogation clause that specifically provides that the insured cedes the right to proceed against the party causing the injury to the insurer, or it may require the insured to proceed against the wrongdoer on behalf of the insurer, if the insurer pays the insured for the loss that the insured suffered.

The doctrine of subrogation represents an important insurance concept. Without the right of subrogation, the insured would be entitled to payment twice: once from the insurer under the contract of insurance, and a second time in the form of damages that the insured might obtain by taking legal action against the negligent party for the injury suffered. The right of subrogation precludes double payment to the insured, and places the liability for the loss upon the person responsible for it. Subrogation has an additional beneficial side effect: the right of the insurer to recover losses from the negligent party substantially reduces the premiums that the insured must pay for insurance coverage.

Contribution

The right of insurers (more than one) to share a loss where the policies so provide.

A third factor that limits the insured's compensation to the actual amount of the loss is the right of **contribution** between insurers. Persons sometimes have more than one policy of insurance covering the same loss. However, if the policies contain a clause that entitles the insurer to contribution, then each insurer will only be required to pay a portion of the loss.

If an insured has insurance coverage with three different insurers against a specific loss and suffers a loss of $1,000, the insured will not be permitted to collect $1,000 from each of the insurers. He or she will only be entitled to collect a total of $1,000 from the three (i.e., $333.33 each). Each insurer would only be required to pay its share of the loss suffered by the insured.

Co-insurance clause

A clause that may be inserted in an insurance policy that renders the insured an insurer for a part of the loss if the insured fails to maintain insurance coverage of not less than a specified minimum amount or percentage of the value of a property.

In some cases, if the policy so provides, the insured may become an insurer for a part of the loss if the insured fails to adequately insure the risks. With some risks, the likelihood of a total loss may sometimes be small. To prevent the insured from placing only a small amount of insurance to cover the risk, the insurer may, in the policy of insurance, require the insured to become a co-insurer in the event of a partial loss. Generally, a minimum amount of insurance will be specified in the policy. If the insured fails to maintain at least that amount, then the insured becomes a co-insurer for the amount of the deficiency. For example, if the policy contains an 80 percent **co-insurance clause**, then the insured must maintain insurance for at least that amount of the value of the property (or, if the insurance is burglary insurance, not less than a stated sum). The formula applied in the event of a partial loss is:

$$\frac{\text{actual amount of insurance carried}}{\text{minimum coverage required}} \times \text{loss} = \text{insurer's contribution}$$

Thus, if the property is worth $100,000, and the insurance coverage is $60,000, a loss of $10,000 would be calculated as follows if the policy contains an 80 percent co-insurance clause (80 percent of $100,000 = $80,000 minimum coverage required).

$$\text{insurer's contribution} = \$10{,}000 \times \frac{\$60{,}000}{\$80{,}000} = \$7{,}500$$

In this example, the insurer would only be obliged to pay $7,500 of the $10,000 loss. Because the insured failed to maintain a minimum of 80 percent coverage, the insured would be required to absorb the remainder of the loss as a co-insurer. If the loss had exceeded $80,000, however, then the full amount of the insurance would be payable by the insurer. Co-insurance only applies when the insured suffers a partial loss of less than the required amount of insurance coverage. Figure 31–1 illustrates the insurance relationship.

FIGURE 31–1 **Compensation for Loss — Rights of Insurer**

1. **Contribution Among Insurers (If More Than One)**
2. **Right of Salvage (Title to Insured Goods)**
3. **Right of Action Against Party Causing the Loss (Subrogation)**

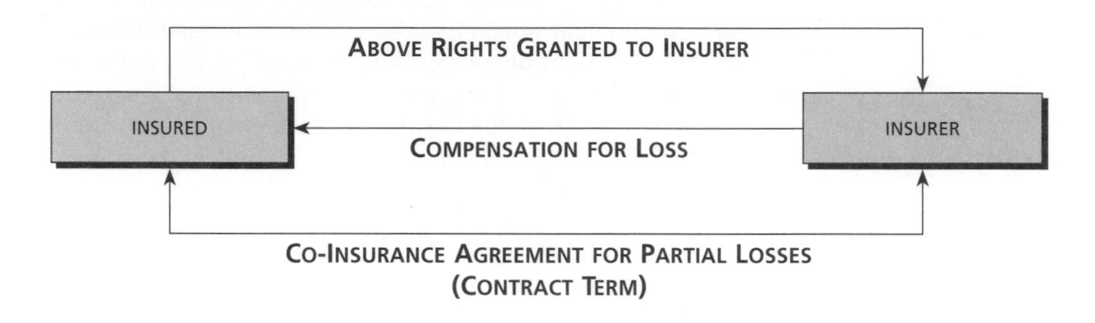

THE PARTIES ASSOCIATED WITH INSURANCE CONTRACTS

Apart from the insurer and the insured, a number of other parties may be involved in either the negotiation of the contract of insurance or the processing of claims under it. Most insurance is negotiated through agents or employees of the insurer, and these persons have varying degrees of authority to bind the insurer in contract. Agents are liable to the insurer for their actions. However, in cases where the insured has relied on the statements of the agent that the policy written by the agent covers the risks that the insured wished to have insured, and this later proves not to be the case, the insured may have a cause of action against the agent if a loss should occur.[7]

This occurred in the case of *Fine's Flowers Ltd. et al. v. General Accident Assurance Co. of Canada et al.*,[8] where the insured requested insurance coverage for specific risks from an agent of the insurers. The agent failed to obtain the required coverage in the insurance policy. When a loss occurred, the insured discovered that the policy did not cover the loss. The insured brought an action against the agent for the agent's failure to include the requested coverage in the policy. In finding the agent liable for the loss, the court stated:

> The agent's duty, counsel submits, is "to exercise a reasonable degree of skill and care to obtain policies in the terms bargained for and to service those policies as circumstances might require."
>
> I take no issue with counsel's statement of the scope of the insurance agent's duty except to add that the agent also has a duty to advise his principal if he is unable to obtain the policies bargained for so that his principal may take such further steps to protect himself as he deems desirable. The operative words, however, in counsel's definition of the scope of the agent's duty, are "policies in the terms bargained for."
>
> In many instances, an insurance agent will be asked to obtain a specific type of coverage and his duty in those circumstances will be to use a reasonable degree of skill and care in doing so or, if he is unable to do so, "to inform the principal promptly in order to prevent him from suffering loss through relying upon the successful completion of the transaction by the agent": *Ivamy, General Principles of Insurance Law*, 2nd ed. (1970), at p. 464.
>
> But there are other cases, and in my view this is one of them, in which the client gives no such specific instructions but rather relies upon his agent to see that he is protected and, if the agent agrees to do business with him on those terms, then he

7. *Fines Flowers Ltd. et al. v. General Accident Assurance Co. of Canada et al.* (1974), 49 D.L.R. (3d) 641.
8. *Fines Flowers Ltd. et al. v. General Accident Assurance Co. of Canada et al.* (1977), 81 D.L.R. (3d) 139.

cannot afterwards, when an uninsured loss arises, shrug off the responsibility he has assumed. If this requires him to inform himself about his client's business in order to assess the foreseeable risks and insure his client against them, then this he must do. It goes without saying that an agent who does not have the requisite skills to understand the nature of his client's business and assess the risks that should be insured against should not be offering this kind of service. As Mr. Justice Haines said in *Lahey v. Hartford Fire Ins. Co.*, [1968] 1 O.R. 727 at p. 729, 67 D.L.R. (2d) 506 at p. 508; varied [1969] 2 O.R. 883, 7 D.L.R. (3d) 315:

> The solution lies in the intelligent insurance agent who inspects the risks when he insures them, knows what his insurer is providing, discovers the areas that may give rise to dispute and either arranges for the coverage or makes certain the purchaser is aware of the exclusion.

I do not think this is too high a standard to impose upon an agent who knows that his client is relying upon him to see that he is protected against all foreseeable, insurance risks.

Broker

A person who may act for either the insurer or the insured in the placement of insurance.

Adjuster

A person employed by the insurer to determine the extent of a loss incurred by an insured.

Brokers may also place insurance with insurers. They may act either for the insured or the insurer. Persons with complex insurance needs may use a broker to determine the various kinds of insurance that they require. The broker will determine the risks, then arrange for the appropriate coverage by seeking out insurers who will insure the risks for the client.

Insurance **adjusters** are persons employed by an insurer to investigate the report of loss by an insured and to determine the extent of the loss incurred. Insurance adjusters report their findings to the insurer, and, on the basis of the investigation, the insurer frequently settles insurance claims. When, as a result of the adjuster's investigation, the issue of liability is unclear, the insurer may carry the matter on to the courts for a decision before making payment for the loss.

MANAGEMENT ALERT: BEST PRACTICE

As in any contract, business owners must read and understand the provisions of a contract of insurance. This is particularly important when "standard form" coverage is offered. Regardless of size, all businesses are unique, and standard coverage sufficient for one may be insufficient for another. Time and money are best spent providing a broker with a written (and complete) description of the business activities and requesting appropriate coverage — to eliminate a later dispute as to the coverage demanded. Watch for narrowly defined terms in coverage — a fire insurance policy may generously replace destroyed premises and equipment, but provide no coverage for lost business income while the premises are rebuilt.

CHECKLIST FOR INSURANCE

Insurer's obligation to pay

1. The policy covers the event?
2. The coverage equals or exceeds the value of the loss?
3. The insured has an insurable interest in the event?
4. Have all material facts been disclosed by the insured?
5. If there is any failure by the insured, is it cured by equity?
6. The insured is not a co-insurer?
7. No other policy covers the same event?

If the answer to each is true, then the insurer will be obligated to pay the insured an amount equal to the extent of the loss.

SUMMARY

With the exception of life insurance, the contract of insurance is a special type of contract designed to indemnify an insured if the insured should suffer a loss insured against in the policy. A contract of insurance differs from a wagering agreement in that it is only designed to indemnify the insured for the actual loss sustained. It differs also in that the insured must have an insurable interest in the property or activity before the loss becomes payable. The contract of insurance is a contract of utmost good faith. The full disclosure of all material facts must be made to the insurer if the insured wishes to hold the insurer bound by the policy. Life insurance differs from other forms of insurance in that it is not payable to the person on whose life it is placed.

Because insurance (except life and accident insurance) is designed only to indemnify the insured for losses suffered, the insurer is entitled to the rights of salvage, subrogation, and contribution to limit the loss that it suffers as an insurer. Where an insured underinsures, some policies also make the insured a co-insurer for partial losses.

Insurance coverage is usually obtained by businesses through agents of the insurer, or through independent brokers. Brokers are expected to be knowledgeable of the needs of the insured and acquire appropriate coverage when directed to do so.

KEY TERMS

REVIEW QUESTIONS

1. Explain the doctrine or concept of salvage. Give an example of how it might apply.
2. Explain an insurable interest as it applies to a contract of insurance.
3. Why is a contract of insurance a contract of utmost good faith?
4. At what point in time does a life-insurance policy become effective?
5. Describe the right of contribution and, by way of example, show how insurance companies use it to determine their liability.
6. What right of the insurer prevents an insured party from making a profit by a loss?
7. Is it possible for a creditor to insure the life of a person indebted to him or her? Explain.
8. Explain the concept of insurance and indicate how it differs from wagering.
9. In what way does the right of subrogation ultimately benefit the insured?
10. What limitations or exceptions permit an insurer to avoid payment of loss claims caused by deliberate acts of the insured?
11. What mathematical principles are used to determine premium rates for life-insurance policies?
12. A creditor insured the life of a debtor to cover the amount of the debt owed. Two years later the debtor died, having paid back over half the debt. Is the creditor entitled to the full amount of the policy?

MINI-CASE PROBLEMS

1. A homeowner insured her home for $50,000, knowing that its true value was $100,000. An accidental fire in her kitchen caused $10,000 damage. Her fire insurance policy contained an 80 percent co-insurance clause.

   Calculate the liability of the insurer.
2. Assume a situation identical to that described in the above question, except that damage caused by the fire exceeded $50,000.

   How would you calculate the liability of the insurer in this case?
3. A husband and wife took out life insurance policies naming each other as beneficiaries. The wife had a medical problem at the time, but did not reveal the true nature of the problem in her insurance application. Six months later, the wife died as a result of the particular medical problem. The insurer denied payment under the policy.

   Discuss.

www.mcgrawhill.ca/olc/willes

4. Samuel insured his automobile for theft. He later claimed that his automobile had been stolen and damaged. The insurer was suspicious, and believed that Samuel had deliberately damaged the vehicle to claim the insurance proceeds. The insurance company refused to pay.

   Advise Samuel and the insurer. If Samuel should sue for breach of the insurance contract, what would be the arguments of the parties?

   Render a decision.

CASE PROBLEMS FOR DISCUSSION

CASE 1

Acme Furnishings Ltd. operated a small manufacturing plant that was housed in a warehouse building that was owned by the corporation, and valued at $500,000. A recession in the house furnishings market forced the corporation to borrow $300,000 from its bank, using the building as security for the loan. Acme Furnishings Ltd. gave the bank a mortgage on the property, and as a part of the transaction, the bank required the corporation to insure the building for fire for its full value and include the bank as an insured. A year later, Acme Furnishings Ltd. was in serious financial difficulty, and the bank foreclosed on the mortgage, but did not evict Acme Furnishings Ltd. from the premises. A few months after the bank's foreclosure was finalized, a fire broke out in the plant, and the building was destroyed. The bank and Acme Furnishings Ltd. both filed claims for payment under the fire policy. Advise the insurer.

Discuss the nature of the claims, and how the claims may be resolved.

CASE 2

Ashley carried on a gift shop business in a neighbourhood with many boutique shops. Her shop was part of a mini mall that housed several shoe stores, a dress shop, and a competing gift shop business operated by Helga. Helga resented Ashley's presence in the mall, and did her utmost to persuade patrons to not patronize Ashley's shop. One winter day, after a verbal dispute with Ashley, Helga took her delivery van, and crashed it through the front window of Ashley's shop, destroying much of Ashley's inventory. Helga's excuse was that she lost control of the van on the icy parking lot, and could not prevent the "accident." Ashley claimed on her insurance for the damage to her shop and for business interruption while inventory was replaced and the shop repaired. Her insurer paid Ashley for the $35,000 claimed loss in full. Ashley then took legal action against Helga in tort for damaging her business and the destruction of her inventory.

If Ashley succeeds in her claim against Helga, and recovers damages in the amount of $40,000, discuss the rights and responsibilities of the parties.

CASE 3

Benton carried on a successful restaurant business in a large city. The restaurant had an excellent reputation. This was for the most part due to the skill of Benton's gourmet chef, Simmons.

Benton realized that his business would be adversely affected if he should lose Simmons through accident or injury. For that reason he arranged for a life-insurance policy in the amount of $100,000 on Simmons's life, and named himself as the beneficiary. The annual premium in the amount of $500 was paid by Benton.

Some months later, at the end of a busy day, Benton and Simmons became involved in a violent argument. Simmons left the restaurant saying, "Don't expect me to work tomorrow. I quit. I'll be in to pick up my personal belongings at 8:00 a.m."

On his way home from the restaurant, Simmons was involved in a serious automobile accident and was killed. Benton claimed the $100,000 under the life-insurance policy.

Should the insurer pay the claim? What defences might it raise to resist the demand for payment?

CASE 4

Hector and Keech carried on a fishing business together as a partnership. Each partner's life was insured for $200,000 under an insurance policy that named the other partner as beneficiary.

One afternoon, while the two partners were in their boat, fishing close to the shore, a sudden storm came up. Hector started the engine to run the boat back to harbour. Keech wished to remain and insisted that they ride out the storm. The two men exchanged words, and then both struggled for the controls of the boat. In the process, Hector was pushed overboard and into the water. Before Keech could turn the boat around, Hector had disappeared beneath the surface of the choppy water.

Keech was charged with criminal negligence causing the death of Hector and was convicted. Because of the circumstances surrounding the death, he was given only a light prison sentence.

On his release, Keech claimed the $200,000 under the insurance policy on Hector's life in which he was named beneficiary.

Should Keech be entitled to the insurance proceeds?

CASE 5

Rosa Rugrosa Flowers Ltd. carried on business as a florist. It operated a large greenhouse in which it grew most of its flowers during the cold winter months. To protect its business, the company contacted its insurance agent for insurance coverage against loss or damage by fire, and for theft of stock. Because the building was largely glass and steel, the agent placed its value at $25,000, when in fact its actual value was approximately twice that amount. The contents, also covered by the policy, were similarly undervalued.

The insurance agent placed the insurance coverage, and a policy was issued that contained an 80 percent co-insurance clause in the event of fire damage.

A few months later, on a cold winter night, vandals broke into the greenhouse, smashed some of the panes of glass by throwing potted plants against the sides of the building, and pulled the furnace flue pipe from the chimney. Smoke from the furnace filled the greenhouse, damaging most of the flowers and other plants. The vandals took with them equipment valued at $1,000.

The company claimed the following amount from its insurer:

Damage to building by vandals	$ 3,000
Smoke damage to plants, etc.	10,000
Equipment stolen	1,000
	$14,000

The insurer agreed to compensate the company for the stolen equipment in the amount of $1,000 and to cover the damage to the building. However, pointing to the 80 percent co-insurance clause in the fire-insurance policy, the insurer argued that it was only responsible for a part of the damage and the loss of the plants, as the damage was caused by fire.

A few days later, the vandals were apprehended by the police and admitted causing the damage. Unfortunately, they had sold the equipment and had spent the money before they were caught. The vandals had no personal assets.

Discuss the rights and obligations of the parties in this situation. Explain how the loss should be borne.

CASE 6

Major Manufacturing Company produced a variety of children's toys in a small plant located in a multiple-unit industrial complex. Most of the products manufactured were either of a plastic composition or painted metal. Consequently, relatively large quantities of flammable solvents and paint products were stored on the premises.

As a tenant of the building, Major Manufacturing Company carried fire and liability insurance on its operations in the amount of $500,000, as well as business-interruption insurance designed to compensate the company for any losses arising from the interruption of the operation due to fire damage. The fire-policy agreement restricted the storage of flammable products to a single room of the plant area, and it prohibited smoking in that area. Containers of flammable products in the remainder of the plant were to be kept to a minimum, and no container was to be opened in the storage area.

Employees in the plant followed the insurer's directions by taking the large storage drums out of the storage room. Once outside the room, they would open them and fill smaller containers from them for distribution to the various painting areas, then return the drum to the storage area.

One day, while an employee was filling smaller containers from an open drum, he was visited by an executive of a company that purchased toys from Major Manufacturing Company. The executive was making a tour of the manufacturing facilities with the marketing manager of Major Manufacturing Company. At the moment when the marketing manager opened the fire door to show the visitor the storage area, the visitor, without thinking and without noticing the large NO SMOKING sign, took out a cigarette and lit it. The fumes in the area immediately ignited, and the resulting fire destroyed the entire complex. The visitor, the employee, and the marketing manager were all seriously burned in the accident, but there was no loss of life.

Discuss the ramifications of this incident. Speculate as to how the loss may be determined, assuming that the building owner, all tenants, and the visitor's company were insured for liability, fire, and business-interruption losses.

CASE 7

Jennifer and Suzanne were the sole equal shareholders in the operation of a company that owned a department store. The owner's equity in the store amounted to $4 million.

At age 52, both Jennifer and Suzanne knew that no one in either of their families had any interest in taking over their shares in the business in the event that either Jennifer or Suzanne died. At the same time, neither wanted to see the other saddled with a new partner should the family of the deceased sell the inherited half-share to someone undesirable. Accordingly, they made a buyout agreement, and resolved to buy insurance sufficient that, on the death of either of them, the survivor would have enough cash to buy up the deceased's shares in the company, and that survivor would then own the company outright.

Carlyle, an agent for Solid Life Insurance Co., had been Jennifer's agent for the better part of 25 years. Carlyle wrote two policies; one on each life, with the other named as the beneficiary in the amount of $2 million. A medical exam was required, and in the course of Suzanne's examination, she was asked by the doctor if she had smoked in the last twelve months. She said that she had not.

A year later, Suzanne was killed in an auto accident. After investigation, Solid Life Insurance Co. refused to pay because it had discovered that Suzanne had, in fact, been a smoker at the time the policy was issued. There was a policy available for smokers at the time of original issuance, but it carried a higher premium.

Jennifer sued Carlyle (with Solid Life Insurance Co. as a co-defendant) in a suit for negligence.

Identify the issues involved and render a decision.

CASE 8

Speedy Goliath had a poor driving record and found that insurers were reluctant to insure his automobile. Part of the reason for his high accident rate was the fact that he enjoyed using his automobile in car rally contests. By ignoring the driving rules in the races, he frequently became involved in minor accidents. When he purchased a new sports car, he decided that he might obtain a lower insurance rate if the ownership of the vehicle was placed in his friend's name. His friend consented to the arrangement, and insurance coverage on the vehicle was arranged.

A short time later, while Speedy was driving his friend to work, he carelessly backed up and collided with a parked car.

The owner of the parked car demanded damages in the amount of $4,000 for Speedy's carelessness, but the insurer refused to make payment.

The owner of the damaged vehicle brought an action against Speedy as the driver of the automobile, and his friend as its owner. He obtained a judgment against them for $4,000.

Discuss the position of Speedy, his friend, and the insurer in this case.

CASE 9

Timber Limits Inc. owned an extensive tract of forested land from which it selectively cut particular species of trees on an ongoing basis for custom wood buyers. The company employed a full-time forester to manage and supervise cutting on the tract, and it had constructed two cabins on the property to be used by the forester from time to time during his travels throughout the property. The cabins were insured as "seasonal dwellings" under the company's fire-insurance policy covering the buildings, and the endorsement on the policy provided that the buildings would not be covered for fire if either building was abandoned or remained vacant for more than 30 days prior to a loss.

One of the cabins eventually became uninhabitable due to a leaking roof, and the company sent in a work crew to remove the furnishings, make the necessary repairs, and generally refurbish the cabin. Before the work could be completed, the work crew was directed elsewhere, and the cabin remained empty for several months. Before work commenced on the cabin, a member of the work crew was dispatched to examine the cabin and determine the materials needed to refurbish the interior.

The worker examined the cabin and prepared a list of materials for completion of the repair, then left the premises. The next day the cabin was destroyed by a fire of undetermined origin, and the company applied for compensation for the loss under its policy.

The fire-insurance company refused to pay for the loss, and Timber Limits Inc. instituted legal proceedings against the insurer.

Discuss the basis of the arguments of the parties and render a decision.

The Online Learning Centre offers more ways to check what you have learned so far. Find quizzes, Weblinks, and many other resources at www.mcgrawhill.ca/olc/willes.

CHAPTER 32

Restrictive Trade Practices

Chapter Objectives

After study of this chapter, students should be able to:

- Identify restrictive trade practices.
- Explain the legislation as it relates to advertising and the promotion of products.
- Describe offences and "reviewable activities."

YOUR BUSINESS AT RISK

Competitive markets are in the interest of businesses and consumers alike. Some business practices strike at the heart of this policy objective, and have attracted legislative and regulatory responses. As the rules are not necessarily self-evident, and the penalties can be very severe (in addition to negative publicity), it is important for businesses to understand how to remain in compliance with competition regulation.

INTRODUCTION

The law relating to restrictive trade practices is based upon the premise that the forces of competition and the free market should regulate industry, rather than governments or dominant members of the business community. As a consequence, both the Common Law and restrictive trade-practices legislation have as their main thrust the preservation or protection of competition. Only those activities that tend to restrict or interfere with competition are controlled by the law, and industry is left to regulate itself through the market forces that are created by the free-enterprise system.

Restrictive trade practices were originally governed by the Common Law, and all restraints of trade that were considered unreasonable or contrary to the public interest were actionable at law. The Common Law, unfortunately, was not adequate to ensure that the forces of competition remained free from manipulation by those in industry who possessed substantial economic power, so protection of competition (in the form of legislation) was necessary.

Most of the control of anti-competition activity is now found in the *Competition Act*.[1] The law, in general, prohibits combinations or conspiracies that prevent or lessen competition unduly, and reviews mergers or monopoly actions that may operate to the detriment of the public. It establishes a number of unfair trade practices, such as resale price maintenance (where manufacturers attempt to control or set retail prices), price discrimination (selling at different prices to different buyers), discriminatory promotional allowances, false advertising, and bid-rigging as criminal offences. The Act applies to both federal and provincial Crown corporations, as well as those in the private sector.

1. R.S.C. 1985, c. C-34, as amended.

The current law represents an attempt by government to eliminate those forces that interfere with free competition and thereby minimize the need for direct government regulation of activities in the marketplace. The general thrust of anti-competition legislation represents a means of control of industry and trade by way of prohibition of only those activities that interfere unduly with free enterprise. The *Competition Act* describes the intent of the legislation in the following terms:

> The purpose of this Act is to maintain and encourage competition in Canada in order to promote the efficiency and adaptability of the Canadian economy, in order to expand opportunities for Canadian participation in world markets, while at the same time recognizing the role of foreign competition in Canada, in order to ensure that small and medium-sized enterprises have an equitable opportunity to participate in the Canadian economy in order to provide consumers with competitive prices and product choices.[2]

This philosophical approach was not always taken toward trade and commerce. The most significant change took place at the beginning of the industrial period in England, when the country adopted the philosophy of *laissez-faire.* Adam Smith's views of competition and its benefits were embraced by the courts and the populace. Restraint of trade at Common Law became *prima facie* void, unless circumstances could justify some "reasonable restraint." In general, the law as a matter of public policy prohibited any unjustified interference with a person's right to trade, and all conspiracies to willfully injure business became actionable at law.

While it was not unlawful at Common Law for an owner of a business to purchase a competitor's business, and lessen competition accordingly, the development of large trusts and businesses that had acquired monopoly powers did alarm the governments of the United States and Canada in the late nineteenth century. In both of these countries, the adverse effects on the public of large-scale business acquisitions by investment trusts, and the vertical integration of business activities that gave particular firms virtual monopolistic power over the supply of goods and services, prompted legislative action. The response by the governments took the form of the *Sherman Act*[3] in the United States, and a statute in Canada that prohibited any combination or conspiracy that had the effect of limiting competition unduly in a trade or manufacture.[4]

The Canadian legislation was essentially criminal law. It made the conspiracy or combination an offence punishable on conviction. The statute became the foundation of restrictive trade-practices legislation, and still remains as the core of the present Act. Over time, various other practices that interfered with competition were added. Price-fixing and price discrimination were prohibited, and predatory pricing designed to destroy competition was also declared contrary to public policy.

By 1960, it was necessary to consolidate the law pertaining to restrictive trade practices. The statute, entitled the *Combines Investigation Act*,[5] incorporated the provisions of the *Criminal Code* in the new legislation. It added a number of additional trade practices to the prohibited list, and defined misleading price advertising and discriminatory promotional allowances as restrictive trade practices. No other major changes were made in the law until 1976, when the Act was subjected to a thorough review and a number of major revisions were made. The changes represented an attempt by Parliament to preserve and encourage free competition and was the first of a two-part overhaul of the law relating to restrictive trade practices. The changes were not finalized until 1986, when the second phase of the law was passed by Parliament. In 1999, as the *Competition Act*, it was again amended, and deceptive telemarketing was added as an offence.

CLIENTS, SUPPLIERS *or* OPERATIONS *in* QUEBEC

As a federal law, the *Competition Act* applies uniformly across Canada.

2. ibid., as amended by S.C. 1992, c. 2, s. 44 (new section 1.1 of the Act).
3. 26 Stat. 209, as amended 15 U.S.C. §§ 1 and 2.
4. *An Act for the Prevention and Suppression of Combinations Formed in Restraint of Trade*, S.C. 1889, c. 41.
5. *An Act to Amend the Combines Investigation Act and the Criminal Code*, S.C. 1960, c. 45.

NATURE OF THE LEGISLATION

The *Competition Act* represents a blend of both criminal and administrative approaches to the regulation of restraint of trade, with certain trade practices prohibited and subject to criminal law proceedings and penalties, and others subject to review and control. Included in the legislation are civil remedies that may be pursued by persons or businesses injured as a result of violations of the Act or orders issued pursuant to it.

Prohibited trade practices (some 15 in number) are designated as criminal offences. The enforcement of the Act with respect to these remains subject to the criminal law standard of proof, which requires the Crown to prove beyond any reasonable doubt that the offence was committed by the accused. The onus in these instances was described in *R. v. British Columbia Sugar Refining Co. Ltd. and B.C. Sugar Refinery Ltd.*[6] in the following terms:

> As this is a criminal prosecution there are certain principles that I must apply to its consideration. They are: (1) The onus is on the crown throughout to prove its case and every essential part of it by relevant and admissible evidence beyond a reasonable doubt; (2) This onus never shifts; (3) There is no onus on the accused to prove their innocence; (4) To the extent that the guilt of the accused depends on circumstantial evidence, that evidence must be consistent with the guilt of the accused and inconsistent with any other rational conclusion; (5) In the construction of a penal statute, such as the Combines Act, if there are two or more reasonable interpretations possible, the interpretation most favourable to the accused must be adopted.

Investigation

The Commissioner of Competition (formerly Director of Investigation and Research), as the title implies, is primarily responsible for the investigation of any complaint that a violation of the *Competition Act* has taken place. While criminal law standards of proof apply to prohibited trade practices, the Act provides for broad investigative powers that the Commissioner may utilize in the gathering of evidence. These powers include very wide powers of search and seizure, and the right to compel parties to provide information. A complaint from a private individual to the Commissioner often results in an investigation, but the Act provides that the Commissioner must investigate any allegation of a violation of the Act that is brought to his or her attention in the form of an application for inquiry requested by six residents of Canada.[7]

The Act permits the Commissioner or the Commissioner's agents to enter on the premises of any person that the Commissioner believes may have evidence related to the inquiry. However, usually the Commissioner or agent must first obtain a search warrant from the Federal Court or a Provincial Supreme or County Court to authorize the search and seizure of evidence.[8] The Commissioner cannot, however, use the search and seizure powers simply to engage in a "fishing expedition" for possible evidence of violation. The Commissioner must do so only in accordance with an inquiry pursuant to a complaint of an alleged violation. The Commissioner's powers extend beyond the mere right to search. The Act empowers the Commissioner to apply to the court for an order to interrogate corporate officers or require them to furnish affidavit evidence relating to the inquiry.[9]

If at any time during the inquiry the Commissioner decides that further investigation is unwarranted, he or she may discontinue the inquiry.[10] If, however, the Commissioner finds evidence of a violation of the Act, he or she may either deliver the evidence to the Attorney-General of Canada for consideration of possible criminal charges, or the Commissioner may bring the matter before the Competition Tribunal.

6. (1960), 32 W.W.R. (N.S.) 577.
7. The *Competition Act*, R.S.C. 1985, c. C-34, ss. 9 and 10, as amended by R.S.C. 1985 (2nd Supp.), c. 19, s. 22, and by S.C. 1999, c. 2, ss. 6(2) and 7.
8. ibid., s. 15.
9. ibid., s. 11.
10. ibid., s. 22.

Competition Tribunal

The Competition Tribunal is the second component of the *Competition Act* enforcement process. The tribunal was established under the *Competition Tribunal Act*[11] in 1986. The tribunal is a rather unique court of record that consists of both lay members and judges of the Trial Division of the Federal Court. The tribunal is presided over by a chairperson who supervises the tribunal and assigns the work to its members. All matters brought before the tribunal are heard by a panel of between three and five members, presided over by one of the judges.

Proceedings before the tribunal are normally brought by the Commissioner concerning trade practices that are designated under the Act as reviewable practices. In addition, the tribunal is empowered to deal with matters concerning foreign laws and judgments, foreign suppliers, and specialized agreements and mergers. The tribunal has the authority to issue appropriate orders after hearing all of the evidence and the submissions of the Commissioner and the parties involved. An order of the tribunal is similar to a judgment of the court, and the statute provides that a failure to comply with an order of the tribunal constitutes contempt of the order or a criminal offence. An appeal from an order of the tribunal lies with the Federal Court of Appeal.

The Competition Tribunal has the right to investigate and review certain business activities and make rectification orders to restore competition. Reviewable marketing activities include market restriction, exclusive dealing, "tied" selling, consignment selling, and the refusal to supply goods.[12] The common remedy here might be an order to cease the restrictive activity. The tribunal also has the authority to investigate and deal with abuse of dominant position[13] and mergers.[14] In these cases, the tribunal may review the practices of persons in a dominant or monopoly position and make whatever order it deems necessary to restore competition. In each of these situations, the Commissioner must first make an inquiry and then, if the circumstances warrant, recommend that a hearing be held into the practice. Again, a full opportunity to be heard must be given to any person affected. In addition to the right to be heard, the Act also entitles such persons to cross-examine other witnesses. However, unlike an inquiry into an ordinary restrictive trade practice, the tribunal does not make recommendations. In the case of a reviewable practice, if the results of the hearing dictate some action on the part of the tribunal, it may make an order prohibiting the practice engaged in by the party, or it may establish procedures that the party must follow to restore competition. Under the Act, a failure to obey the order would constitute contempt or a criminal offence.

RESTRICTIVE TRADE PRACTICES

The *Competition Act* applies to both goods and services. Only those services or goods that fall under the control of a public regulatory body would appear to be exempt from the legislation. Any seller or supplier whose services or goods are sold at prices reviewed or determined by a government body or commission, even if the seller is in a monopoly position, would probably not be subject to prosecution under the *Competition Act* for any marketing activity carried on under the direct control of the regulatory body. The Act, of course, would still apply to activities of the organization that fall outside the direct control of the regulatory body, and to any action designed to prevent the regulatory body from protecting the public interest.[15]

Restrictive trade practices subject to the Act may be divided into three separate categories:

(1) practices related to the nature of the business organization itself;
(2) practices that arise out of dealings between a firm and its competitors; and
(3) practices that arise out of dealings between a firm and its customers.

11. S.C. 1986, c. C-26.
12. *Competition Act*, R.S.C. 1985, c. C-34, Part VIII.
13. ibid.
14. ibid.
15. See, for example, *R. v. Canadian Breweries Ltd.*, [1960] O.R. 601.

MERGERS AND FIRMS IN A DOMINANT POSITION

Dominant position

A firm that controls a major segment of a market for a product or service.

The first category subject to the Act (see previous page) is related to the nature of the firm, if the firm should become dominant in a particular field of business or industry. This may arise in one of two ways: a firm may gradually eliminate all competition by aggressive business activity, or it may merge with other competitors to assume a **dominant position**. Neither of these methods of growth or dominance is in itself improper. However, under the *Competition Act*, any merger or monopoly activity that is likely to lessen competition to the detriment of the public would be subject to review and intervention by the Competition Tribunal.[16] The rationale behind these provisions is that mergers or monopolies that substantially control the market have the potential for abuse, in that the price-reducing effects of free competition no longer apply to a company product or service. While it is difficult to pinpoint when a merger becomes contrary to the public interest, or may have the effect of lessening competition to the detriment of the public, any merger that gives a single organization in excess of half the market for a particular product might very well come under scrutiny by the Commissioner. If the merger is found to be one that would result in a substantial lessening of competition, the tribunal has the power to intervene and modify (or prohibit) the activity.[17]

The issue of market concentration arose in a case where a single corporate interest acquired control of the two leading newspapers in the B.C. Lower Mainland. When the matter came before the Competition Tribunal, the corporation was directed to divest itself of one of the two newspapers. On appeal before the Federal Court of Appeal, the corporation's appeal was dismissed, and the ruling was left to stand.[18]

In the past, when criminal prosecution was the only choice, the courts were reluctant to convict in the case of mergers and monopolies, because of the many factors that must be considered in the determination of what constitutes a lessening of competition "unduly." Apart from one case, in which a monopoly firm was so blatant in its conduct of restrictive trade practices that competition was clearly lessened "unduly" and the public interest adversely affected,[19] the Crown has had little success in the enforcement of the merger and monopoly sections of the Act. Some of the reasons put forward by the courts in dismissing the Crown's cases have been the control or regulation of prices by a public body,[20] the potential for competition from large firms in other areas of the country,[21] and the fact that substitutes for the product were available.[22]

The *Competition Act* moves away, however, from the criminal approach to the protection of competition. It provides the Competition Tribunal with the power to review the practices of business firms in a dominant position (such as a monopoly) on a non-criminal basis. After an examination of the practice, the tribunal may make an order that will restore competition if the practice is determined to be an abuse of the dominant position.[23] In the case of a merger, if after a review it should determine that the merger would result in a substantial lessening of competition, the tribunal may prohibit or modify the proposed change.[24]

The importance of preserving competition by intervention was noted in an early case,[25] where the court expressed the need for control in the following manner:

> The right of competition is the right of every one, and Parliament has now shewn that its intention is to prevent oppressive and unreasonable restrictions upon the exercise

16. The *Competition Act*, R.S.C. 1985, c. C-34, ss. 79 and 92.
17. ibid., s. 79.
18. *Canada (Director of Investigation and Research) v. Southam Inc.*, [1995] 3 F.C. 557.
19. *R. v. Eddy Match Co. Ltd.* (1952), 13 C.R. 217, affirmed (1954), 18 C.R. 357.
20. *R. v. Can. Breweries Ltd.*, [1960] O.R. 601.
21. *R. v. British Columbia Sugar Refining Co. Ltd. et al.* (1960), 32 W.W.R. (N.S.) 577.
22. *R. v. K. C. Irving Ltd.* (1974), 7 N.B.R. (2d) 360, affirmed [1978] 1 S.C.R. 408.
23. ibid.
24. The *Competition Act*, R.S.C. 1985, c. C-34, s. 92(1).
25. *R. v. Elliott* (1905), 9 O.L.R. 648.

of this right; that whatever may hitherto have been its full extent, it is no longer to be exercised by some to the injury of others. In other words, competition is not to be prevented or lessened unduly, that is to say, in an undue manner or degree, wrongly, improperly, excessively, inordinately, which it may well be in one or more of these senses of the word, if by the combination of a few the right of the many is practically interfered with by restricting it to the members of the combination.

A QUESTION OF ETHICS

"A company may consistently produce better products than its competitors, products that are preferred by the consumer to the exclusion of all others. It can grow to a size that allows the firm to enjoy economies of scale that it can pass on in the form of lower prices. It receives its final reward in the form of a government investigation." Comment on the validity of this statement.

CONSPIRACIES AND COMBINATIONS IN RESTRAINT OF TRADE

Conspiracy

Agreement between firms to unduly lessen competition.

The general thrust of the present legislation is to prohibit conspiracies and combinations that unduly lessen competition. The relative seriousness of offences relating to these activities may be underscored by reference to the penalties imposed for contravention of this part of the Act: a breach of any of the sections related to combinations and conspiracies carries with it a fine of up to $10 million, or imprisonment for up to five years.[26] The "conspiracy and combination" (s. 45) of the Act[27] provides that:

> everyone who conspires, combines, agrees or arranges with another person:
> (a) to limit unduly the facilities for transporting, producing, manufacturing, supplying, storing or dealing in any product,
> (b) to prevent, limit or lessen, unduly, the manufacture or production of a product, or to enhance unreasonably the price thereof,
> (c) to prevent, or lessen, unduly, competition in the production, manufacture, purchase, barter, sale, storage, rental, transportation or supply of a product, or in the price of insurance upon persons or property, or
> (d) to otherwise restrain or injure competition unduly, is guilty of an indictable offence and is liable to imprisonment for five years or a fine of ten million dollars or to both.

The obligation on the Crown to prove a violation of the Act is alleviated to some extent by an exemption in the legislation. This provision states that it is not necessary to prove that the combination, conspiracy, or agreement if carried into effect would be likely to completely or virtually eliminate competition in the market to which it relates, or that it was the object of the parties to eliminate, completely or virtually, competition in that market.[28] However, the Act would not apply if the combination or agreement between the parties relates only to one of the following activities:[29]

(a) the exchange of statistics;
(b) the defining of product standards;
(c) the exchange of credit information;
(d) the definition of terminology used in a trade, industry, or profession;
(e) co-operation in research and development;
(f) the restriction of advertising or promotion, rather than a discriminatory restriction directed against a member of the mass media;
(g) the sizes or shapes of the containers in which an article is packaged;
(h) the adoption of the metric system of weights and measures; or
(i) measures to protect the environment.

26. The *Competition Act*, R.S.C. 1985, c. C-34, s. 45.
27. ibid., s. 45.
28. ibid., s. 45(2).
29. ibid., s. 45(3).

If the arrangement or agreement to carry out any of these activities restricts (or is likely to restrict) any person from entering the business, trade, or profession, or has the effect of lessening (or is likely to lessen) competition with respect to prices, markets or customers, channels or methods of distribution, or the quantity or quality of production, then the parties would still be subject to conviction under the Act.[30]

Exemptions

The Act normally applies only to conspiracies, combinations, or agreements in restraint of trade on a domestic basis. If the activity relates wholly to the export of products from Canada, the restraint of trade restrictions would not apply,[31] unless the agreement or arrangement (1) has resulted, or is likely to result in a reduction or limitation of the real value of exports of a product; (2) has restricted or injured, or is likely to restrict any person from entering into or expanding the export business; or (3) has lessened or is likely to lessen competition unduly in the supply of service facilitating the export of products from Canada.[32] The purpose of the exception is to allow Canadian business firms maximum latitude in their activities with respect to the export of goods from Canada, and to limit their actions only where the activity would harm other Canadian firms or have a negative impact on the domestic market.

In the case of services, a further exception is made. The courts are not to convict an accused if the conspiracy, combination, agreement, or arrangement relates only to the standards of competence and integrity reasonably necessary for the protection of the public in either the practice of the trade or profession, or in the collection and dissemination of information relating to such services.[33] As a result, firms engaged in certain activities, such as skilled trades (e.g., master electricians) or professions (e.g., accountants, lawyers) would not violate the Act if they "conspired" to set professional standards for their services or to provide the public with information about their services.

Banks are also covered by the *Competition Act*. Any conspiracy or arrangement between banks to establish rates of interest for deposits on loans, the service charges to customers, the amount or kind of loan to a customer, or the classes of persons to whom loans or other services would be provided or withheld is a violation of the Act. It constitutes an indictable offence subject to a penalty of up to five years in prison or a fine of up to $10 million, or both.[34] Certain exceptions are made with respect to some bank activities to reflect the realities of banking and the making of loans to persons outside of Canada.[35]

The Act exempts affiliated corporations from the conspiracy provisions. As a consequence, if a wholly owned subsidiary of a corporation enters into an agreement with that corporation that would otherwise be a conspiracy, it would not be subject to charges under this part of the Act.[36] However, if the parent corporation is a foreign corporation and requires the Canadian subsidiary to enter into an agreement with another firm outside Canada that would constitute a violation of the conspiracy provisions of the Act if the agreement had been made in Canada, the directors or officers of the Canadian corporation may be liable, even if unaware of the agreement.[37]

Bid-Rigging

Bid-rigging

A practice whereby contractors, in response to a call for bids or tenders, agree amongst themselves as to the price or who should bid or submit a tender. A restrictive trade practice, unless the person calling for the bids is advised of the arrangement.

The practice of **bid-rigging**, which is any agreement or arrangement among two or more persons where all but one undertakes not to submit a bid in response to a call for bids or tenders (and where the person calling for bids is unaware of the arrangement), is prohibited under the Act.[38] The practice was made an offence under the Act in 1976 in an

30. ibid., s. 45(4).
31. ibid., s. 45(5).
32. ibid., s. 45(6).
33. ibid., s. 45(7).
34. ibid., s. 49(1).
35. ibid., s. 49(2).
36. ibid., s. 45(8).
37. ibid., s. 46(1).
38. ibid., s. 47.

effort to encourage greater competition by the elimination of secret arrangements. The offence differs to some extent from other restrictive trade practices in that it would not be necessary for the Crown to prove that the bid-rigging represents an undue restraint of trade. An important point to note with respect to this activity, however, is the fact that a bidding arrangement is only an offence if the fact is not revealed to the person calling for the bids, either before or at the time the bid is made. The purpose for this exemption is to allow parties to undertake projects jointly, provided that the nature of the arrangement is revealed beforehand to the other party.

A municipality wishes to reconstruct 8 kilometres of its roads. The four road contractors in the area secretly agree that only one contractor would bid (and would bid high) on the project, and later the successful bidder would divide up the work amongst them by sub-contract, 2 kilometres each. This agreement would constitute bid-rigging.

Professional Sports

The *Competition Act* prohibits conspiracies relating to professional sports where the conspiracy is intended to limit unreasonably the opportunities for any person to participate as a player or competitor in a professional sport, or to impose unreasonable terms on persons who so participate. It also applies to any attempt to limit unreasonably the opportunity for any person to negotiate with and (if an agreement is reached) to play for the team or club of his or her choice in a professional league.[39] This provision in the legislation applies only to professional sport. It requires the courts to take into consideration the international aspects of the activity and the unique relationship that exists between teams or clubs that compete in the same league.[40] Nevertheless, the law has necessitated a change in a number of activities associated with professional sport, the most notable being the practice of tying a player to a club by way of a special reserve clause.

OFFENCES RELATING TO DISTRIBUTION AND SALE OF PRODUCTS

Offences relating to distribution are generally designed to prevent sellers from granting special concessions to large buyers and, conversely, to prevent large buyers from insisting upon special concessions from sellers. Special concessions, usually in the form of lower prices or special allowances, would grant one buyer a particular competitive advantage over other buyers. They carry with them the potential for a restriction on competition. The Act, consequently, has identified and prohibited a number of distribution activities that affect competition.

For example, a seller must not make a practice of discriminating between competing purchasers with respect to the price of goods sold.[41] This activity only constitutes an offence if the seller makes a practice of price discrimination between competing firms, when the goods sold are of the same quality, in the same quantity, and are sold at approximately the same time.[42] Isolated sales to meet competition, or sales between affiliated firms, would probably not constitute offences under the Act.[43]

In a similar fashion, a seller is prohibited from granting buyers special rebates, promotional allowances, or grants for the advertising or promotion of goods unless the allowance or amount is made on a proportional basis.[44] Once again, the purchasers must be in competition with one another, and the seller must not discriminate. Under the Act,

39. ibid., s. 48.
40. ibid., s. 48(2) and (3).
41. ibid., s. 50(1)(a).
42. ibid., s. 50(2).
43. ibid., ss. 50(2) and 45(8).
44. ibid., s. 51.

an allowance would be treated as proportional if it is based upon the value of sales to each competing purchaser, or, if it is in the form of services, in accordance with the kinds of services that purchasers at each level of distribution would ordinarily be able to perform.[45]

A seller must not engage in a policy of selling products in any area of Canada at prices lower than those elsewhere if the sales would have the effect of substantially lessening or eliminating competition in that area,[46] or if the policy of low prices is established for the purpose of lessening or eliminating competition.[47] In both of these cases, the Act is not attempting to prohibit lower prices, but rather to make it an offence if a seller uses lower prices to eliminate or lessen competition either on a regional or broader basis. The practice of selling goods at a low price normally would not offend the Act, but if the price is unreasonably low for the purpose of lessening or destroying competition, then the practice would probably be in contravention of the Act.

Price discrimination

A practice of selling goods in an unfair or improper manner such as to lessen competition.

The underlying thought behind each of these prohibitions is that a seller must treat all competing buyers of his or her products in a fair and impartial manner, and that the selling of products must be done without some unlawful motive, such as the elimination of competition. A seller is not obliged to treat non-competing buyers in the same fashion, however, and a seller may establish separate prices and discounts for each type of non-competing buyer.

Price maintenance

An attempt by the seller to control the resale price of a product by the retailer.

A quite different sales activity is also covered by a section of the Act that prohibits the seller from controlling the prices at which the seller's goods may be sold by others. A seller is prohibited from attempting, either directly or indirectly, by any threat or promise or any other inducement, to influence the price upwards of his or her products, or from discouraging price reductions by the purchasers of the products for resale.[48] The offence is not limited to cases where a seller attempts to fix the price at which the product may be sold, but also applies to any attempt to enhance the price or influence the price upwards. The practice by sellers of providing a "suggested retail price" for advertising, or for price lists or other material, would probably violate the Act unless the seller clearly indicates that the buyer is under no obligation to resell the goods at the suggested price, and that the goods may be resold at a lower price.[49]

A manufacturer of computer printers produced a new model, and in an effort to prevent retailers from selling it at discount prices, required dealers to sign an agreement stating that they would only advertise the printer for sale at the price specified by the manufacturer. The agreement was held to be a violation of the *Competition Act*.

Loss leader selling

A practice of selling goods not for profit but to advertise or to attract customers to a place of business.

A seller may not refuse to supply goods to a buyer in an attempt to prevent the buyer from reselling the goods to others who maintain a policy of selling the goods at lower prices. However, a seller would have the right to refuse to supply goods if the buyers make a practice of engaging in **loss leader selling**, and not for the purpose of profit.[50] The same would be the case if the goods required certain services, and the person was not making a practice of providing the level of service that a purchaser would normally expect.[51]

The Act also prohibits a number of schemes used by sellers to promote sales that tend to discourage competition. There is also a prohibition of "pyramid selling," a practice involving the payment of fees or commissions not based upon the sale of a product, but upon the recruitment or sales of others.[52] Multi-level marketing programs, as distinct from pyramid selling, are permitted under the Act if they meet prescribed conditions.[53]

45. ibid., s. 51(3).
46. ibid., s. 50(1)(b).
47. ibid., s. 50(1)(c).
48. ibid., s. 61.
49. ibid., s. 61(3).
50. ibid., s. 61(10)(a).
51. ibid., s. 61(10)(d).
52. ibid., s. 55.1.
53. ibid., s. 55(1).

REVIEWABLE ACTIVITIES

In addition to prohibited activities relating to the sale of goods and services, the Competition Tribunal may review a number of different selling methods. These include abuse of dominant position, a refusal to supply goods, consignment selling, exclusive dealing, "tied" selling, and market restriction. This "tied" selling refers to an offer (usually of a wholesaler to a retailer) of a very attractive price or quantity of goods, but only if the purchaser also agrees to also buy other less desirable stock as well. The tribunal may also review foreign directives to Canadian subsidiaries and foreign arrangements in restraint of trade that affect Canadian business.[54] A review that confirms that the activity has taken place, and that the activity is carried on for a purpose specified in the Act, may be ordered stopped, or a remedy set out in the legislation for that particular activity may be applied. For example, the tribunal may order a major supplier to cease exclusive dealing arrangements if the arrangement is likely to: impede entry into or expansion of a firm in the market; impede the introduction of a product into the market; impede an expansion of sales of a product in the market; or have any other exclusionary effect in the market,[55] with the result that competition is or is likely to be lessened substantially. The tribunal is also permitted, in the case of exclusive dealing, to include in the order any other requirement necessary to overcome the effects of the exclusive dealing, or to include any other requirement that might be necessary to restore or stimulate competition.[56]

MANAGEMENT ALERT: BEST PRACTICE

Activity in setting "industry standards" that is not against public policy in one field of endeavour may have the effect of eliminating competition in another. Business operators should understand that competition policy is more an art than a science. It is important, therefore, in everything from information exchange to merger negotiation, to consider not only the letter of the law but the spirit of it as well. Business operators must objectively understand the competitive realities of their own market and not underestimate the market potency of individual firms or a combination of them, as it may be seen through the eyes of the regulator.

OFFENCES RELATING TO PROMOTION AND ADVERTISING OF PRODUCTS

Misleading or false advertising and a number of other promotional activities are subject to the *Competition Act.* The Act makes any representation to the public that is false or misleading in any material respect,[57] or any materially misleading representation to the public concerning the price at which a product or products have been, are, or will be sold, an offence under the Act. The Act, with respect to false or misleading advertising, is broadly written to include cases where the information may be technically correct but where the impression given would mislead the public in some material way.

CASE IN POINT

Competition Bureau: Abtronic Muscle Stimulators Removed from Market.

Canadian consumers will no longer be subjected to false claims of weight loss and muscle toning, inducing them to purchase the Abtronic and Abtronic Pro, two electronic muscle stimulation devices, from Thane Direct Canada Inc. (Thane). As part of a consent agreement registered with the Competition Tribunal, Thane has agreed to refund consumers the full value of the devices. Thane sold these two devices via television infomercials and their

54. ibid., s. 46.
55. ibid., s. 77.
56. ibid.
57. ibid., s. 52.

Web site for approximately $120 each to hundreds of thousands of Canadians, giving the false impression that without performing any physical exercise, a person could lose weight, obtain an athletic physique with well-defined abdominal muscles, replace the workout benefits of a fully equipped gymnasium and increase their strength. After being advised that the Competition Bureau had commenced an inquiry into this matter, Thane requested the issue be resolved by Consent Agreement. According to this agreement, Thane has agreed to stop selling and marketing the devices. Furthermore, the company will not market any similar device that offers weight loss or muscle toning without exercise, unless the Competition Bureau agrees that the claims are based on adequate and proper tests. The company has also agreed to pay a $75,000 administrative penalty.

See: Industry Canada Press Release, "Abtronic Muscle Stimulators Removed from Market by Competition Bureau," December 12, 2002 (www.ic.gc.ca).

Initially, the legislation attempted to impose a standard of absolute liability on persons who violated the misleading advertising provisions of the Act. But, in a 1991 case, the Supreme Court of Canada held that such a standard violated s.7 of the *Charter of Rights and Freedoms* and concluded that a strict liability standard was appropriate.[58] This lesser standard permitted an accused to avoid liability if it could be shown that the violation was due to an error and reasonable precautions had been taken to avoid its occurrence. To avoid this defence, amendments to the Act were made in 1999 to clarify the standard of liability and to provide that it was not necessary to prove that any person was deceived or misled by the misleading advertising of the product. The Act now provides both a criminal and civil approach to the offence, and the Commissioner may choose which approach to take for an alleged violation of the Act.

COURT DECISION: Practical Application of the Law

Reviewable Conduct—Sale Pricing

***Commissioner of Competition v. Sears Canada Inc.*, 2005 Comp. Trib. 2.**

The Commissioner alleged that, during three sales events, Sears Canada employed deceptive marketing practices in connection with price representations with certain lines of tires it offered for sale. Only a few tires had been offered and sold at a higher "regular" price (a volume test) and that these periods of regular pricing had been short (a time test).

THE TRIBUNAL: The Commissioner asserts that this constituted reviewable conduct contrary to the Competition Act. The advertisements contained "save" and "percentage off" statements. For example, Sears advertised "Save 45%. Our lowest prices of the year on Response RST Touring '2000' tires," and advertised comparisons between Sears' regular prices and its sale prices. The Commissioner asserts that the prices referred to by Sears as being its regular prices were inflated because: i) Sears did not sell a substantial volume of these tires at the regular price featured in the advertisements within a reasonable period of time before making the representations; and, ii) Sears did not offer these tires in good faith at the regular price featured in the advertisements for a substantial period of time recently before making the representations. The remedies sought by the Commissioner include an order prohibiting such reviewable conduct for a period of 10 years, the publication of corrective notices, and the payment of an administrative monetary penalty in the amount of $500,000.00. Sears is an "off-price" (also called a "high-low") retailer, which means that Sears relies on discounting and promotions to build in-store traffic and generate sales. An off-price or high-low retailer typically charges a higher "regular" price for its merchandise and then, from time to time, offers merchandise "on-sale" at event-driven discount sales. Sears offered the Tires for sale at the following ...price points:

a) Sears' "regular" price was the price of a single unit of any Tire offered by Sears, when that particular tire was not promoted as being "on sale." This was the price used as the reference price in advertisements when the Tires were promoted as being "on sale" by Sears.

58. *R. v. Wholesale Travel Group Inc.*, [1991] 3 S.C.R. 154.

b) Sears' "2For" price was the price at which Sears would sell two or more of a given tire to consumers when that tire was not being offered at a "sale" price. Sears' "2For" price for a given tire was always lower than its regular price for a single unit. Sears did not use its "2For" price as a reference price in any of the sales representations at issue and did not advertise its "2For" price when promoting retail sales. The "2For" price came into effect when a customer bought more than one tire and the customer was only informed of the discount on a purchase of multiple tires by the sales associate at the store.

[*ed. — Three hundred paragraphs of analysis then followed, on pricing, sales volumes and expert opinion.*] The question to be determined, therefore, is whether the impression created by the price comparisons and/or the save stories would constitute a material influence in the mind of a consumer. First, the magnitude of the exaggerated savings. Returning to the Michelin RoadHandler advertisement, for the smallest tire size advertised, an ordinary citizen considering the purchase of four tires would reasonably believe, in my view, their savings to be $248.00 or ($153.99 – $91.99) × 4. In fact, the 2For price for each tire was $94.99. Accordingly, the actual savings would be $12.00 or ($94.99 – $91.99) × 4. In this example, the savings were substantially exaggerated. Because Sears' 2For price was always substantially lower than its regular price, it follows that the savings were similarly substantially overstated in every Ordinary Selling Price (OSP) representation made concerning the Tires. Thus, on the whole of the evidence, Sears has failed to establish that its OSP representations were not false or misleading in a material respect. Harm is not a necessary element of reviewable conduct. As the Court noted in *Kellys on Seymour*, the "criteria is, did in fact the person think that what he was buying was, to the ordinary purchaser, in the ordinary market, worth the price it is purported to be worth, and from which it is reduced." Whether or not a consumer in fact got a bargain or paid less than what the consumer would ordinarily have paid is not the criteria. Sears admitted that it did not meet the requirements of the volume test and I have found that the tires were not offered at Sears' regular price in good faith and that Sears failed to meet requirements of the time test for four of the five tire lines. I have also found that Sears failed to establish that the representations at issue were not false or misleading in a material respect. It follows that the allegations of reviewable conduct have been made out and the Tribunal finds Sears to have engaged in reviewable conduct.

Sears was ordered to pay an administrative penalty of $100,000, plus costs fixed in the amount of $387,000, for a total of almost one-half million dollars.

Sales above the advertised price constitute an offence under the *Competition Act*, and a seller who advertises a product at a particular price in a geographic area would be expected to sell the goods to all persons in that general area at the advertised price. The Act recognizes, however, that errors do occur in the advertisement of goods. It provides that where a false or misleading advertisement is made with respect to the price at which goods are offered for sale, prompt action by the advertiser to correct the error by placing another advertisement advising the public of the error would exempt the advertiser from prosecution under the Act.

Double ticketing

A practice of attaching several different price tickets to goods. Under the *Competition Act*, only the lowest price may be charged for the goods.

A practice somewhat related to misleading price advertising is the **double ticketing** of goods for sale. This sometimes occurs in self-serve establishments. To discourage the practice, the Act provides that the seller must sell the goods at the lowest of the marked prices; otherwise, the sale would constitute an offence.

The Act also discourages the rather dubious selling technique of "bait-and-switch," whereby the seller advertises goods at a bargain price for the purpose of attracting customers to the establishment when there is not an adequate supply of the low-priced goods to sell. The practice is an offence under the Act unless the seller can establish that he or she took steps to obtain an adequate supply of the product, but was unable to obtain such a quantity by reason of events beyond the seller's control. Another defence would arise where the seller did not anticipate the heavy demand for the advertised product. Here, to avoid a violation of the Act, the seller would be obliged to prove that he or she obtained what was believed to be an adequate supply and, when the supply was exhausted, undertook to supply the goods (or similar goods) at the same bargain price within a reasonable time to all persons who requested the product.

Civil Actions Under the *Competition Act*

Apart from the right to maintain a Common Law civil action for restraint of trade activities not covered by the legislation, the *Competition Act* provides that a civil action may be maintained by a party injured as a result of a breach of the *Competition Act* or the violation of a Competition Tribunal order. The party affected by the breach may claim damages suffered as a result of the breach of the Act, but the amount that may be recovered is limited to the actual loss.[59] In this sense, Canadian legislation differs from that of the United States, where triple damages may be recovered in restrictive trade-practice cases.

The burden of proof that is imposed upon the private plaintiff in the civil action would not be the criminal burden of "beyond any reasonable doubt," but the lesser civil law burden based upon a balance of probability. The civil plaintiff, however, would be entitled to use the record of any criminal proceedings against the defendant as evidence in the civil action, provided that the action is commenced within two years of the final disposition of the criminal case.[60]

While some doubt was initially raised as to the validity of the part of the *Competition Act*[61] that creates a civil cause of action, Federal Court and Ontario Court of Appeal judgments have held the provisions of the Act to be a valid exercise of federal powers under the constitution.[62] More recently, challenges were also made as to the constitutionality of the Competition Tribunal and its powers, but the Supreme Court of Canada has held that the Competition Tribunal is a constitutionally valid body with the power to enforce its orders by contempt proceedings.[63]

FRONT-PAGE LAW

Courier UPS pursues Canada Post under NAFTA, alleging unfair competition

HAMILTON (CP) — An American courier company is pressing ahead with charges of unfair competition against Canada Post and the federal government. United Parcel Service, in a suit filed under the North American Free Trade Agreement, is seeking $230 million in damages. In a meeting with members of the Hamilton Spectator's editorial board, UPS Canada spokesmen said Tuesday they expect a tribunal to start work on the complaint later this year.

UPS launched the action after complaining that Canada Post was abusing its monopoly as the country's carrier of letter mail to undercut private couriers.

Susan Webb, public relations manager of UPS Canada, alleged the Crown corporation uses its retail outlets, collection boxes and sorting facilities to give unfair market advantages to its Xpresspost and Priority Post services — direct competitors to 1,400 private couriers — and to Purolator Inc., a courier service in which Canada Post holds a 95 per cent stake.

UPS says that's a violation of NAFTA chapters that require countries not to discriminate against investments made by other member countries and not to allow state-controlled monopolies to compete unfairly.

Beyond NAFTA, state-controlled monopolies are under the microscope in a world that sees ever-increasing privatization of government services. In most nations, state monopolies exist for the carriage of letters. Where the state monopoly further competes with private enterprise, questions arise about whether government funding of the monopoly side in fact subsidizes the "competitive" side. Here, UPS believes that postal revenue, facilities, or other government funding tilts the field against it, allowing a lower price for Xpresspost and Priority Post services. Does partial or full privatization distort the market? What measures would help to ensure it does not?

Source: Excerpted from Steve Arnold, *Canadian Press*, "Courier UPS pursues Canada Post under NAFTA, alleging unfair competition," September 19, 2000.

59. *Competition Act*, supra, s. 36(1).
60. ibid., s. 36(4).
61. *Seiko Time Canada Ltd. v. Consumers Distributing Co. Ltd.* (1980), 29 O.R. (2d) 221; *Racois Construction Inc. v. Quebec Red-i-Mix Inc.* (1979), 105 D.L.R. (3d) 15; *Vapor Canada Limited et al. v. MacDonald et al.* (1977), 66 D.L.R. (3d) 1.
62. See, for example, *City National Leasing v. General Motors of Canada Limited* (1986), 28 D.L.R. (4th) 158n; *Attorney-General for Canada v. Quebec Ready Mix Inc. et al.* (1985), 25 D.L.R. (4th) 373.
63. See, for example, *Canada (Competition Tribunal) v. Chrysler Canada Ltd.*, [1992] 2 S.C.R. 394; *R. v. Nova Scotia Pharmaceutical Society*, [1992] 2 S.C.R. 606; see also *R. v. Wholesale Travel Group Inc.*, [1991] 3 S.C.R. 154.

CHECKLIST FOR COMPETITION ACT

The Act covers:

<table>
<tr><th>A. Competition Offences</th><th>B. Deceptive Marketing Practices</th><th>C. Restrictive Trade Practices</th></tr>
<tr><td>Limiting dealing</td><td>Misrepresentations to the public</td><td>Refusal to deal</td></tr>
<tr><td>Manipulating price</td><td>Control of regular versus sale pricing</td><td>Consignment selling</td></tr>
<tr><td>Injuring competition</td><td>Use of testimonials</td><td>Exclusive dealing</td></tr>
<tr><td>Discriminatory pricing</td><td>Bargain pricing and “bait and switch”</td><td>Market restriction</td></tr>
<tr><td>Misleading representations</td><td>Sales in excess of advertised price</td><td>Tied selling</td></tr>
<tr><td>Deceptive telemarketing</td><td rowspan="4">D. Notifiable Transactions
Acquisition of firms with either assets or annual sales in excess of 400 million dollars.</td><td>Abuse of dominant position</td></tr>
<tr><td>Double ticketing</td><td>Delivered pricing</td></tr>
<tr><td>Pyramid selling</td><td>Specialization agreements</td></tr>
<tr><td>Price maintenance</td><td>Mergers substantially lessening competition</td></tr>
</table>

SUMMARY

The purpose of restrictive trade-practices legislation is to maintain free competition. The law is designed to permit the forces of competition to regulate trade and industry, rather than government or dominant members of an industry.

The general thrust of the law is to review mergers and monopolies and prohibit those that are contrary to the public interest, and to ban any combination or conspiracy that might unduly lessen competition. The law also prohibits certain activities (on the seller's part) designed to drive the prices of goods and services upward, as well as to prevent price discrimination, along with certain other practices that might restrict competition.

The *Competition Act* is, in part, criminal in nature. However, parts of the Act are regulatory to cover a number of reviewable activities, and provide for the right of civil action for persons injured as a result of a breach of the Act.

The legislation has worked reasonably well in controlling selling practices that are contrary to the public interest, but until recently has failed to deal adequately with mergers and monopolies. The particular problem with the merger–monopoly parts of the legislation was related to the criminal nature of the law, and the burden of proof that it imposes upon the Crown. Changes in this area of the law now permit the Competition Tribunal to review mergers and monopoly actions with power to protect the public interest by way of modification or prohibition of the activities.

KEY TERMS

Dominant position, 634
Conspiracy, 635
Bid-rigging, 636
Price discrimination, 638
Price maintenance, 638
Loss leader selling, 638
Double ticketing, 641

REVIEW QUESTIONS

1. What activities are considered prohibited trade practices?
2. Outline the implications of the *Competition Act* for an advertiser of goods. What types of advertising would likely be affected by the Act?
3. What effect has the criminal nature of the law had upon the ability of the Crown to control restrictive trade practices?
4. Mergers of corporations or businesses are not unlawful *per se*. Under what circumstances would a merger likely be subject to review under the *Competition Act*?
5. Why did the Canadian government find it necessary to introduce restrictive trade-practices legislation?
6. What activities are not "prohibited" but "reviewable" practices?
7. Under what circumstances would an investigation under the *Competition Act* be instituted?
8. Must a manufacturer of goods sell its products to all retailers? If not, why not? Give an example of a case where a manufacturer might lawfully refuse to do so.
9. Why was it originally necessary to make the restrictive trade-practices laws criminal in nature?
10. Explain the following terms: bait-and-switch, loss leader, bid-rigging, exclusive dealing, predatory pricing, tied selling.
11. What is the significance of a price advertised by a manufacturer as a "maximum retail price"? How does this differ from a "suggested retail price"?

MINI-CASE PROBLEMS

1. A retailer advertised television sets for sale with banner headlines that stated, "Special Sale! Brand X Model XX 20" colour TV only $199 with trade-in!" At his store he had three model XX sets for sale at $199, but required an older working model of the same type of TV as a trade-in. At his shop he would also urge customers to buy a different brand of TV at $599, since brand X, in his opinion, was poor quality and not really worth buying. A and B went to the retailer's shop to buy a brand X TV set, but neither had a brand X TV set for trade-in purposes.

   Discuss the issues raised in this case.
2. Workers' Clothing Co. conducted a promotional contest in which the prize offered was described as a new motorcycle with a retail value of $12,995. Investigation revealed that the motorcycle had a suggested retail price of $12,495 and was a new, but previous year's model.

   Discuss the implications of this information in light of the *Competition Act*.
3. Alter Company Ltd. and Bounder Company Ltd. manufacture the same product, each under their own brand name. Alter Company Ltd.'s product represents 46% of the domestic market, and Bounder Company Ltd.'s product represents 48% of the domestic market. Alter Company Ltd. and Bounder Company Ltd. are considering a merger, with the merged company continuing to sell the two brands as separate products.

   What issues are raised by these facts?
4. MFG Inc. sold specialized medical equipment through wholesalers to hospitals. Wholesalers signed exclusive dealership agreements proposed by MFG Inc. that required all sales to hospitals to be at prices set by MFG Inc. Middleman Ltd., a wholesaler, sold several machines at less than the prices set by MFG Inc. On discovery, MFG Inc. cancelled their exclusive selling agreement with Middleman Ltd. and refused to supply equipment to it.

   Advise Middleman Ltd.

CASE PROBLEMS FOR DISCUSSION

CASE 1

Generics Ltd., a manufacturer of pharmaceutical drugs, marketed a generic relaxant drug that was dispensed by pharmacists under prescription. For many years the drug was sold under its brand name, and its use gradually expanded until it represented over half of tranquillizer market sales. A U.S. pharmaceutical manufacturer investigated the size of the Canadian market for tranquillizers, and decided to introduce its own generic product through its Canadian subsidiary, under its own brand name. In an effort to protect its market share, Generics Ltd. decided to provide its product to all Canadian hospitals on a free basis. This was done in expectation that patients who were given the drug in hospital would then request it by brand name from their physicians, or that physicians would simply continue to prescribe the same product that had been used in hospital. When the competitor discovered that Generics Ltd. was providing drugs on a free basis to hospitals, it lodged a complaint under the *Competition Act*.

Discuss the nature of the complaint, the effect of the complaint, and the possible outcome.

CASE 2

A soft drink manufacturer entered into an exclusive dealership agreement with a large university whereby the manufacturer would supply its product line to the university cafeteria, coffee shop and snack bar. The manufacturer would also supply soft drink dispensing machines in all campus residence and classroom buildings. The agreement was to run for a period of five years, and provided that the university would not permit any soft drinks to be sold on campus except those of the manufacturer. In return, the university was paid a large up-front cash payment, and entitled to a percentage of the sales of all of the soft drinks sold on campus. A competitor of the soft drink manufacturer consults you for your opinion on this agreement between the manufacturer and the university.

What issues are raised in this case?

What advice might you give?

CASE 3

Retailers from time to time advertised widgets at extremely low prices as an advertising gimmick to attract customers to their stores, much to the annoyance of the marketing manager of World Widget Co. Ltd. To clarify its position on "loss leaders," the company decided to issue a price list and a memorandum to discourage the use of its product in this fashion. The new price list read as follows:

Standard Model	Distributor Net Price	Regular Dealer Price	Minimum Profitable Resale Price	Fair Retail Value
(each)	$21.07	$24.47	$29.95	$34.95

All widget distributors were advised by the company that the sale of widgets at prices lower than fair retail value would be investigated, and any sale at a price lower than the minimum profitable resale price might be considered a loss leader sale. The company indicated that it would "assess such a sale as it related to the marketing of World Widget Co. Ltd. products."

The memorandum further stated that it was the opinion of the company that a person loss leads widgets when he or she sells the product at a gross margin less than the average cost of doing business plus a reasonable profit.

In the months that followed the issue of the memorandum, the World Widget marketing manager noted that two retailers continued to sell widgets at very low prices. One of the retailers, a large retail chain, regularly advertised and sold widgets at a unit price that would amount to $21.00 each. Since the retailer was a purchaser at the distributor price of $21.07, the product was sold at slightly less than actual cost.

The marketing manager stopped shipments to the retailer on the completion of his investigation. He advised the customer that no further shipments would be made until World Widget had some assurance that the retailer would not loss lead widgets. The retailer eventually agreed to notify all branch managers that widgets should be sold at the regular price, and loss leader selling would be discontinued. A copy of the memorandum was sent to World Widget, and on its receipt, shipments of widgets to the retailer resumed.

The other retailer, who purchased widgets at $24.47 each, sold widgets at a price equivalent to $24.90. World Widget considered this to be loss leader selling and refused to make further shipments until the retailer agreed to stop selling widgets as loss leaders. The retailer eventually agreed to stop selling widgets in this manner, and further agreed to sell the product at a price not less than the "minimum profitable resale price." Shipments of widgets were resumed when this agreement was reached.

Assess the actions of the marketing manager in this case.

CASE 4

In an attempt to diversify its product line, World Widget Co. Ltd. purchased all right, title, and interest in an automobile "jet-ignition unit with transistors" from a Miami, Florida, inventor. The unit consisted of a small metal container, two transistors, a small spring, and a blob of tar. The designer of the unit claimed that the device would give "better automobile gas mileage, easier starting, and better performance." In support of his claim, he provided 625 testimonial letters from users of the product who reported reduction in gas consumption ranging from 10 percent to 30 percent.

World Widget engineers were skeptical of the performance claims made for the jet-ignition unit, but their concern was brushed aside by the general manager when he discovered that the selling price represented a 100 percent markup over cost.

During 2010, the "jet-ignition unit with transistors" had been advertised in Canada by the previous manufacturer through several metropolitan television stations, and World Widget arranged for continued television advertising for 2011. The advertising indicated that the product would increase automobile gas mileage by up to 30 percent and would improve engine performance. It was sold under a money-back guarantee if the product was returned to the manufacturer within 30 days of purchase.

In January 2011, a motorist purchased a unit. When it was installed in his car, the engine would not start. He reported his experience to a government agency, which then tested the motorist's unit in a number of its own vehicles. The tests indicated that the unit had no noticeable effect on engine performance or fuel consumption. The agency informed World Widget of its findings.

In the same month, another motorist wrote the company the following letter: "I purchased one of your jet-ignition units recently, and I am very pleased with the change in performance of my automobile. Starting is much easier, and gasoline mileage has improved 10 percent. Two of my friends purchased ignition units and have obtained similar results. I heartily recommend your product."

The company received no complaints concerning its product from purchasers, and no user requested a return of the purchase price under the money-back guarantee.

Should the company continue to market the product?

CASE 5

Retail Furs Limited carried on business as a furrier in a large metropolitan city. The general manager attributed much of the company's sales volume to the use of extensive advertising and frequent "sales." As a general practice, the store held four sales a year: a summer (off-season) sale, a fall sale, a New Year's sale, and a spring sale. In addition, the general manager would occasionally hold a special sale if sales volume was below expectations for the year. Each sale normally lasted a month, although extensions were occasionally made to clear models that proved to be poor sellers.

During the past year, the company held five sales, and each featured a standard type of mink jacket at 50 percent off the regular price. On the last day of the final sale for the year, a customer entered the store and wished to purchase one of the advertised jackets. The regular price was stated as $6,000, and the sale price $3,000. The customer, however, argued that the regular price was really $3,000 for at least 5 of the previous 12 months.

The general manager denied that $3,000 was the regular price and refused to sell the jacket for $1,500.

An investigation into the ordinary selling prices of similar jackets in the area found that prices ranged from $2,995 to $7,500, depending upon the quality of the fur pelts, the cut, and the style.

Discuss the issues raised in this case.

CASE 6

Best Appliance Co. produces a complete line of small and large domestic kitchen appliances, about 30 different products in all, ranging from small electric toasters to automatic dishwashers and laundry appliances. Most of the small appliances are sold through wholesalers, while the larger appliances are sold directly to "authorized dealers." The authorized dealers normally carry only the larger appliance line consisting of about 12 different products, but the occasional dealer would carry the full line of both large and small appliances.

In recent years, the company experienced problems with retailers selling its smaller appliances as "loss leaders." It considered either selling only through authorized dealers who would agree to carry the full line of products, or to dealers on a consignment basis (whereby the company would control the retail price). A third alternative would be to notify all wholesalers who carried the small-appliance lines that they must refrain from selling to retailers who used the company's products as loss leaders. Then, if the wholesaler failed to monitor retail sales and keep retailers from loss leader selling, the company would no longer supply them with stock.

The company officers, who wish to have the legal implications of these proposals examined, ask for advice.

Prepare a response to their request and include suggestions as to how they might deal with their problem.

CASE 7

Gargantuan Gravel Corporation carries on business as a producer and supplier of gravel and crushed rock for road-construction and building-construction purposes. It owns and operates pits and quarries located in a 160-kilometre radius of its head-office operation (which is located in a large metropolitan city). It supplies its products to approximately 55 percent of the users in that area.

Crushed Rock Corporation, the second largest producer in the area, owns most of the remaining quarries. It holds approximately 42 percent of the market. The remaining 3 percent is controlled by 26 operators who own small gravel pits. These operators tend to sell to customers located in the immediate area of their pits, or supply only their own construction projects.

For many years, Gargantuan Gravel followed a practice of setting its prices for gravel and crushed rock on January 2nd of each year. Crushed Rock Corporation usually established its prices a few weeks later, and the prices were the same or slightly lower than those set by Gargantuan. The 26 smaller producers normally followed price structures of the two larger firms. The asset values of both of the larger firms are in the multi-million dollar range. Both firms have extensive aggregate holdings and operate large fleets of trucks. Sales are proportionally large for both firms, and fall in the multi-million dollar range for each.

The directors of the two large firms recently entered into negotiations whereby Gargantuan Gravel would acquire all of the shares of Crushed Rock Corporation. However, some of the customers of Gargantuan Gravel are concerned about the takeover of Crushed Rock Corporation, and the effect that it might have on them financially.

Advise the customers of their rights and how their concerns might be dealt with if they elect to take action.

CASE 8

Hewson Jewellers Limited were agents for, among other famous names, one of the finest of the Swiss watchmaker–jewellers. As Hewson was undergoing rough financial times due to an economic downturn, it decided for the first time in its history to have a general 50-percent-off sale. A newspaper advertisement was prepared, and to entice the public it mentioned by name a number of the product lines "to be slashed by half."

The advertisement ran for one day in a national daily newspaper before it was discovered by the Canadian business representative and wholesaler of the Swiss firm. The representative–wholesaler, CanJewel Ltd., telephoned Hewson, and threatened to stop shipment of further stock, and to call Hewson's debt for stock previously shipped, unless the ad was stopped and the name of the Swiss line dropped from the copy. CanJewel reminded Hewson that "image was everything," and the Swiss line must never appear to the public to be "on sale." CanJewel extracted a signed agreement from Hewson that all future advertising copy would be reviewed by CanJewel on behalf of the Swiss manufacturer. Hewson complied, and the second advertisement made no mention of the Swiss line.

When government authorities became aware of the matter, a charge was laid under the *Competition Act*. State what the charge would likely be, upon whom it would be laid, and the possible defences.

Render a decision on behalf of the court.

CASE 9

Prior to 1927, widgets were manufactured in Canada by three companies: World Widgets Ltd., Canadian Widgets Ltd., and E.M.C. Widgets Ltd. In 1927, the three firms merged to form the World Widget Co. Ltd., a new firm incorporated that year. As a result of the merger, World Widget Co. Ltd. became the only producer of widgets in Canada.

In January 1928, the company issued new price lists for its products to jobbers and distributors. The new lists raised the case price from $18.00 (the price before the merger) to $21.91.

In March 1928, Columbia Widgets Ltd. was incorporated in New Brunswick, and started widget production in October of that year. The price of its product (per case) to jobbers was $19.75.

In October of the same year, World Widget Co. Ltd. introduced its product under a new brand name at a very low price to jobbers in those areas where Columbia widgets were sold. World Widget also granted confidential prices or special rebates to large jobbers and distributors that agreed to handle only the World Widget product.

In 1932, Columbia Widgets Ltd. closed its doors and sold its assets to D. Widget, an American firm controlled by World Widget. A few weeks later, the Commonwealth Widget Company Ltd. was established by D. Widget on the site occupied previously by Columbia Widgets. Prices established by the new company were in line with those published by World Widget.

In 1931, Canada Widget Ltd., a competitor, was formed by a group of Quebec businessmen, and a plant to manufacture widgets was established in Quebec. World Widget gradually acquired control of this new firm by the acquisition of shares through its subsidiary, Widget Holdings Ltd. Prices of widgets sold by Canada Widget Ltd. were brought in line with World Widget prices following the acquisition of Canada Widget Ltd. by Widget Holdings Ltd.

In 1936, Federal Widget Ltd. was organized, and a plant established in Quebec. Following the establishment of this company, World Widget offered its regular product under another brand at a lower price in all areas where Federal widgets were sold. World Widget also established special prices and rebates on a confidential basis with jobbers prepared to carry World Widgets exclusively.

In the years that followed, World Widget maintained the selling price of its product in those areas where no competition existed. It entered into agreements with distributors whereby the distributor would receive special rebates if they would certify from time to time in writing that all widgets sold by them were sold above specified minimum prices.

In addition to these practices, World Widget established an elaborate network of "contacts" within the Federal Widget firm, and with individuals in the firm's channel of distribution. Information sent to World Widget gave World Widget a complete picture of the operation of Federal Widget. World Widget was aware of the quantities of goods received or shipped by the firm, the destination of shipments, the names of the recipients, and the selling prices of the goods shipped. Documentation of the internal operation of Federal Widget was so complete that World Widget was in possession of the Federal Christmas bonus list, and aware of the Christmas presents to be given by Federal to its employees and customers before the presentations were made.

Federal Widget found competition with World Widget difficult, and eventually came under the control of Widget Holdings Ltd. through the acquisition of shares of Federal Widgets by Widget Holdings Ltd.

In British Columbia, the Western Widget Company, a new competitor, was established. Western Widget was interested in the widget market west of Ontario and instituted an active campaign to enter this market.

World Widget, once aware of its new competition, flooded the western market with its product, using special brands and special discounts to distributors to encourage the exclusive handling and sale of World widgets.

Western Widget tried in vain to offset the activities of World Widget with its own discounts, but failed. Eventually, Western Widget was purchased by World Widget for $210,000.

World Widget continued aggressive marketing activity until complaints triggered an investigation of its past practices. The investigation confirmed the previous actions of the company and revealed the following information:

(1) On each "take-over" (purchase of assets or shares), World Widget Co. Ltd. or Widget Holdings Ltd. would obtain an agreement from the vendor company and its officers whereby the company and its officers would agree not to engage in the widget business in Canada either directly or indirectly, for a specified period of time (usually 20 years).

(2) World Widget Co. Ltd. restricted the distribution of its special brand-name widgets to areas where competition existed. These products were priced lower than regular widgets bearing the World Widget name, and were offered to distributors at prices that gave them a larger margin. World Widget also entered into agreements with distributors that required the distributor to offer the special brand-name products to retailers at reduced prices.

Assuming that a widget is an essential consumer and industrial product of standard design, difficult to differentiate, and with a very limited useful product life (purchased frequently), discuss the legality of the activities carried on by the firm.

Identify any difficulties that might arise with respect to interpretation and enforcement of the law.

CASE 10

In an effort to increase sales, the marketing manager of the World Widget Co. Ltd. introduced a co-operative advertising campaign, in which it agreed to pay 50 percent of the widget-advertising expense of any retailer selling World Widget products, provided that the retailer agreed to advertise widgets at the "suggested retail price." No assistance would be given to the retailer if the retailer did not include the price of the product in its advertisement. The 50-percent payment would be based upon the dollar cost of the advertising and would be open to all who sold the company product, regardless of sales volume.

The memorandum to retailers advised all retailers taking part in the campaign that they were free to sell the product at any price they wished to establish. The memorandum also stated that retailers were free to advertise at their own expense, if they wished to advertise widgets at a price different from the suggested retail price. The proclaimed purpose of the advertising campaign, according to the marketing manager, was to promote the product on a national basis, and not to advertise the product as a special sale item.

Assess the activities of the company in the light of current restrictive trade-practices legislation.

The Online Learning Centre offers more ways to check what you have learned so far. Find quizzes, Weblinks, and many other resources at www.mcgrawhill.ca/olc/willes.

CHAPTER 33

International Business Law

Chapter Objectives

After study of this chapter, students should be able to:

- Describe Canada's regulation of its imports and exports.
- Describe the principles that underpin the WTO and NAFTA.
- Distinguish between the major vehicles used to conduct international business.
- Describe the elements of an international contract of sale.
- Explain the process and advantages of arbitration of international business disputes.

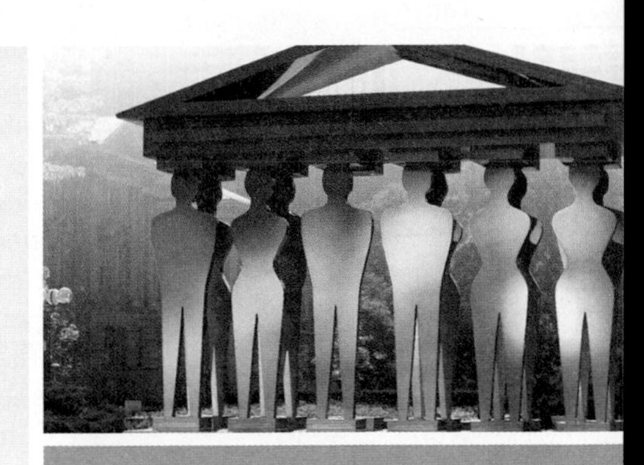

YOUR BUSINESS AT RISK

Business transactions abroad introduce three new elements: a foreign party, another legal jurisdiction, and international rules that apply to the relationships between countries. Since each of these factors affects the business transaction, the element of risk increases significantly compared to domestic business situations. The first solid risk management tool is to understand the policy and application of these rules affecting international business transactions.

INTRODUCTION

Canada is a trading nation that depends to a large extent upon foreign trade for its economic well-being. Initially, as a colony, Canada was a supplier of furs and fish. Later, as a fledgling dominion, it was a source of a wide variety of raw materials. After the turn of the twentieth century, Canada gradually developed a manufacturing base, and eventually moved into the export of manufactured goods. Today, raw materials, agricultural products, and lumber continue to represent a substantial part of Canadian exports, but manufactured goods and services, particularly those in high-tech fields, represent a vital and growing segment of Canada's foreign trade.

Canadian businesses that engage in the export and import of goods operate in a different legal environment from those firms that carry on business in the domestic market only. The laws that affect a Canadian firm trading on an international basis fall roughly into two categories: (1) Canadian laws, and those negotiated between nations to control or facilitate international trade; and (2) private laws that govern the transactions between the parties when one contracting party is not a domestic firm.

The Importance of Trade in the Canadian Economy

As a yardstick of the importance of trade, Statistics Canada indicates that in 2010, Canadians generated a Gross Domestic Product of $1.6 trillion. This figure is comprised of everything we consumed, investments, our government spending and the difference between our exports and imports. By comparison, in 2010, we exported goods and services valued at $476 billion (equivalent to 29% of GDP) and imported $507 billion (31% of GDP).

Of this, our biggest trading partner is the United States, which purchased $299 billion of our exports (74%) and provided us with $203 billion of our imports (50%). Whatever we would prefer our relationship with the United States to be, the fact remains that our jobs and economic security are tied to the United States until such time as we develop other customers and partners around the world.

THE IMPORTATION OF GOODS INTO CANADA

Tariff

The duty (payment) charged by the Federal Government on goods imported into Canada.

The import (and export) of goods is subject to a number of federal statutes. The most important of these Acts are the *Customs Act*[1] and the *Customs Tariff Act*.[2] The *Customs Act* is an administrative statute that sets out the various powers and duties of customs officers, the procedures for the importation of goods, and the rules for the collection of customs duties and the payment of refunds. The Act also provides appeal procedures that may be taken by importers who disagree with alleged customs violations or duty rate decisions. Included in the Act are penalties that may be imposed upon persons who violate the customs rules or who attempt to avoid payment of duty properly imposed.

The *Customs Tariff Act* sets out the various duty rates applicable to goods brought into Canada. This statute sets out not only the generally applicable rates, but preferential rates that may apply to goods imported from certain countries. It also contains a list of goods that may not be imported into Canada (i.e., prohibited commodities).

World Customs Organization

An organization that provides internationally-recognized identification numbers for goods traded internationally.

Canada's *Customs Tariff* (1,724 pages) is based on the **World Customs Organization**'s "Harmonized System." This system assigns internationally-recognized identification numbers (HS numbers) to essentially all traded commodities, so that nations can have a common basis for the identification of goods. With this, Canada publishes the rate of duty it will levy on each category of goods. Some examples are provided below:

Ch 16 — Preparations of Meat, Fish or Crustaceans	MFN Rate of Duty*
16.05 — Crustaceans, Molluscs — prepared or preserved	
16.05.10 — Crab	5%
16.05.20 — Shrimp	Free
...	
16.05.90 — Other	
16.05.90.10 — Squid, Octopus and Cuttlefish	Free
16.05.90.20 — Oysters	
16.05.90.20.10 ..... in cans or glass jars	2%
16.05.90.20.90 ..... other	2%
16.05.90.30 — Clams	
16.05.90.30.10 ..... Atlantic	6.5%
16.05.90.30.90 ..... other	6.5%

*See "Most Favoured Nation" discussed in this chapter in the section entitled World Trade Organization. Lower rates may apply to goods sourced from free-trade partners or nations participating in other multilateral or bilateral trade arrangements.

COURT DECISION: Practical Application of the Law

Classification of Goods—Interpretation

***Canada (Attorney General) v. Suzuki Canada Inc.*, 2004 FCA 131.**

The Canadian International Trade Tribunal classified Suzuki ATVs under HS number 87.11 (motorcycles) which enter Canada duty-free. The government appealed the tribunal decision to the court, arguing ATVs should enter Canada under classification number 87.03 (motorcars), attracting 6.1% duty.

1. R.S.C. 1985 (2nd Supp.), c. 1, as amended.
2. S.C. 1997, c. 36, as amended.

THE COURT: An ATV is a four-wheeled, motorized, off-road vehicle with a tube chassis, and is used to transport a person and goods over rough terrain. Like virtually all other four-wheeled vehicles, the steering system on an ATV turns the inside wheels at a slightly sharper angle than the outside wheels so that the wheels track a straight line. This is known as the "Ackerman principle." Handlebar movement and operator weight shift also contribute to steering an ATV. The competing headings for the ATVs, together with their Explanatory Notes, are as follows:

87.03 ***Motor cars and other vehicles principally designed for the transport of persons (other than those of heading No. 87.02), including station wagons and racing cars.***

This heading covers motor vehicles of various types (including amphibious motor vehicles) designed for the transport of persons ... The heading also includes "Four-wheeled motor vehicles with tube chassis, having a motor-car type steering system (e.g. a steering system based on the Ackerman principle)."

87.11 ***Motorcycles (including mopeds) and cycles fitted with an auxiliary motor, with or without side-cars;***

The World Customs Organization has also issued a classification opinion, indicating that ATVs are to be classified within heading 87.03. It reads: ***8703.21*** *Four-wheeled (two wheel-driven) All Terrain Vehicle ("A.T.V.") with tube chassis, equipped with a motorcycle type saddle, handlebars for steering and off-the-road balloon tyres. Steering is achieved by turning the two front wheels and is based on a motor-car steering system (Ackerman principle).*

The respondents argued that ATVs do not have the characteristics of a motor-car type steering system and therefore cannot be classified under that heading. In making its determination, the Tribunal stated that it did not see why having a steering system based on the Ackerman principle would determine the ATVs' exclusion from heading 87.11 and their inclusion in heading 87.03, even though it also found as a fact that ATVs do have a steering system based on the Ackerman principle. The Tribunal found that ATVs were more specifically described by heading 87.11, on the basis of their historic evolution from, and certain similarities to, motorcycles, such as handlebar steering and being "ridden" rather than driven. In essence, the Tribunal selected its own meaning of the term "motor car type steering system" and, in so doing, redefined the wording established by the World Customs Organization, whose mandate includes customs harmonization. Furthermore, the phrase "a steering system based on the Ackerman principle," which the Tribunal rejected as constituting a "motor-car type steering system," was specifically used on three occasions by the World Customs Organization. In my view, this repetition emphasizes that the reference to the Ackerman principle was chosen deliberately and was clearly intended to illustrate what constitutes a "motor car steering system." On a proper interpretation of the Explanatory Notes and based on the undisputed evidence that the steering system utilizes the Ackerman principle, the Tribunal acted unreasonably in concluding that the ATVs could be classified other than under heading 87.03. Accordingly, I would allow the appeal with costs, and set aside the decision of the Tribunal dated May 2, 2003.

Sadly, Suzuki Canada must pay 6.1% duty, rather than import its merchandise duty-free. The case shows the importance of knowing your classifications before importing goods. This 6.1% might represent all the profit available to an importer and it may be impossible to raise prices to compensate for this sudden cost if price commitments have already been made to customers.

Dumping

The selling abroad of goods at prices lower than the prices of the goods sold domestically in the country of origin.

Canadian industries and businesses are protected from the **dumping** of foreign goods into Canada by the *Special Import Measures Act.*[3] Under this statute, a special duty is levied on goods that are imported at a lower price than the price that the same commodities are sold for in the normal course of business in the country of origin. The statute is also intended to prevent foreign sellers from selling goods produced under government subsidy to the Canadian market, if the sales would cause injury to Canadian producers of similar goods, or if the subsidized goods would prevent or retard the development of the production of the goods in Canada. The "dumping" of goods must be established before the special duty rates would apply to the goods.

In an effort to protect Canadian industries from competition from countries with extremely low production and labour costs, the importation of certain types of goods is subject to control under the *Export and Import Permits Act.*[4] In contrast to the *Special Import Measures Act*, which is designed to deal with dumping situations, the *Export and Import Permits Act* attempts to control the flow of goods into Canada from those countries where goods may be produced at prices so low that Canadian firms would be

3. R.S.C. 1985, c. S-15, as amended.

4. R.S.C. 1985, c. E-19, as amended.

unable to meet such competition on a fair basis. In an effort to provide Canadian firms with some protection, the statute essentially imposes limits on the quantities of certain goods brought into Canada by requiring importers to obtain import permits for specific goods that have been identified under the Act. The Act also requires exporters to obtain permits before they may export certain controlled goods from Canada. Goods subject to export-permit requirements are generally goods of military importance, or goods classed as "strategic" commodities. The sale of these goods to countries that the government may list from time to time under Orders-in-Council is usually prohibited, or, where export is permitted, may only be exported with a permit.

Canadian businesses engaged in the importation of goods from foreign countries are obliged to carry on their business within the framework of these four statutes. Because a special knowledge of the legislation is often necessary to import goods with a minimum of time and effort, many firms will use the services of business firms that specialize in dealing with customs officers and persons who administer the legislation affecting the importation of goods. Those firms are generally known as **customs brokers**, and they play an important role in the importation of goods into Canada.

Customs broker

A firm with specialized knowledge of customs duties that will assist firms in the importation of goods.

FRONT-PAGE LAW

U.S. appliance firms guilty of dumping in Canada: ruling

OTTAWA — Prices for some major appliances could more than double as a result of a ruling that the U.S. companies behind Whirlpool, Frigidaire, Kelvinator and Amana have unfairly dumped exports on the $1-billion-a-year Canadian appliance market.

The Canadian International Trade Tribunal has confirmed a ruling from the Canadian Customs and Revenue Agency that refrigerators, dishwashers and dryers were sold at below-market prices.

Camco Inc. of Hamilton, which makes appliances under the GE, Hotpoint, Kenmore, Beaumark and other brand names, complained last November that it was losing sales to unfair competition from White Consolidated Industries Inc. and Whirlpool Corp., two major U.S. manufacturers.

A four-month investigation, following a probe by customs officials, confirmed the goods had been sold for less than the price in the home market or at a loss.

"It will have a substantial positive impact for Camco. We're confident we will regain a substantial share of the profit we've lost over the past two to three years," James Fleck, Camco president, said yesterday. "It's a level playing field and an opportunity to regain the market share we've lost."

Depending on the model, prices will have to increase anywhere from a few dollars to almost 150 percent. Camco, the only major Canadian manufacturer of the appliances, blamed dumping for a drop in 1999 net income to $5.5-million, compared with $9-million the previous year. It employs more than 2,000 workers at plants in Montreal and Hamilton.

The tribunal provided few details on its ruling, but the earlier government investigation found dumping was cutting prices on some models by as much as 60 percent.

This excerpt illustrates the dynamics of a dumping case, where a balance between low-priced appliances for Canadians, the survival of a Canadian industry and jobs, and a level playing field for competition must be addressed. While dumped goods appear to be priced attractively, the savings are just a short-run temptation. Once the domestic producer is driven out of business, Canada is left with unemployed workers and the price of the imports will increase, reflecting the lack of competition.

Source: Excerpted from Ian Jack, *Financial Post*, "U.S. appliance firms guilty of dumping in Canada: ruling," August 9, 2000, p. C3.

THE EXPORT OF GOODS FROM CANADA

Canadian firms that engage in the export of goods to foreign purchasers face many trade and tariff barriers, not unlike those that have been erected by Canada to control the flow of goods into our country and protect domestic industry. Apart from the *Export and Import Permits Act*, there are few Canadian laws that restrict Canadian exporters from selling their goods abroad. Indeed, the export of goods is encouraged by the Canadian government, because it creates jobs and wealth at home. Some Canadian laws that would normally control or prohibit particular business practices for domestic firms would not apply to foreign or export operations. For example, the *Competition Act* specifically prohibits combination in restraint of trade domestically, but permits such combinations formed to engage in export market activities, provided that the combination does not adversely affect the domestic market.[5]

Many of the challenges faced by Canadian exporters are related to the trade barriers that foreign countries have erected to protect their own manufacturing and production sectors. The movement of goods into foreign countries often requires the services of foreign firms not unlike Canadian customs brokers. Efforts have been made, however, to reduce international trade barriers and provide a common structure or framework for the consideration of tariff rates and the control of such activities as dumping. These international agreements have been an important factor in the growth of international trade over the past few decades.

CLIENTS, SUPPLIERS *or* OPERATIONS *in* QUEBEC

As a matter of federal jurisdiction, border measures, the laws relating to international trade, and the conventions to which Canada is a party, apply uniformly across Canada.

INTERNATIONAL TRADE REGULATION

Most countries control the import and export of goods to some degree. They have laws in place that either regulate trading in particular goods or impose duties or taxes on goods moving across their borders. However, because of the use of tariff barriers and controls by virtually all nations, a number of important international agreements have been established, whereby the signing nations have agreed to limit their controls and duties on goods in accordance with the terms of the particular treaties.

World Trade Organization

World Trade Organization (WTO)

A multi-nation organization that provides a forum for the negotiation of trade rules, and provides a mechanism for the resolution of international trade disputes.

Most Favoured Nation (MFN)

The obligation of a member of the WTO to impose the same lowest rate of duty granted on goods from one member state to the same goods from all other member states.

National Treatment (NT)

The prohibition on imposing special taxes or duties on goods after import that exceed those of domestic production.

Canada has a strong record in the pursuit of freer trade among nations, and it has demonstrated this through its full membership in the **World Trade Organization (WTO)** and its predecessor, the General Agreement on Tariffs and Trade (GATT). GATT was established on January 1, 1948, and represented a multinational agreement, which grew from an initial founding membership of 23 nations (including Canada) to over 100 nations by the 1990s. As of January 1, 1995, the GATT was replaced by the 128-member WTO. This successor organization is a more potent international forum in areas of policy-making and dispute settlement, but the original principles of GATT are alive and well, and form the basis of the WTO, which has grown to 153 members. The original GATT rules had as their initial thrust the reduction of trade restrictions between countries. GATT sought to accomplish this by the establishment of a number of rules that would govern the various duties and import charges that a country could fairly impose on imports. Chief among these are **Most Favoured Nation (MFN)** status, and **National Treatment (NT)**. MFN status means that the lowest rate of duty imposed by a member nation on a product coming from another member country (its most-favoured trade partner) must be extended to imports of like products arriving from all other member nations. In short, all imports of like products will attract the same, and lowest, rate of duty. NT takes effect after imported goods cross the border. This rule forbids member nations from creating special taxes or treatment for imported goods that exceed measures that apply to their own domestic production. Again, it means that imports must be treated on the same footing as domestic goods once they are in circulation within the country. The combined objective of the two rules is to treat all imports equally on international terms (to avoid special preferences among trade partners), and then to treat imports and domestic products alike, once both are competing on the internal market. A range of exceptions exists to accommodate historical relationships with former colonies, the needs of the developing world, national

5. *Competition Act*, R.S.C. 1985, c. C-34, s. 45(5), as amended.

emergencies, and the creation of free-trade arrangements, but the two principles of MFN and NT are the cornerstones of the GATT and WTO regimes. GATT also provided a procedure for the determination of the fair value of goods subject to duty and charges and set out rules to prevent the dumping of goods in export markets.

The GATT agreement provided a general framework for the reduction of barriers to the free movement of goods between nations. At the same time, it attempted to accommodate the particular economic and political goals of the individual countries. Canada signed GATT and, as a member nation, had an obligation to follow the rules that the agreement set down in the determination of Canada's own customs duties and tariffs. As well, Canada was, and is, expected to abide by the GATT rules and those promulgated by the WTO with respect to any restrictions that Canada may impose on imports and exports of goods. The new WTO goes beyond the scope of GATT rule-making to include a multilateral dispute-settlement mechanism, the establishment of administrative bodies to oversee the implementation of agreements, and a forum for the negotiations themselves. Canadian business firms engaged in international trade also must, of course, comply with the Canadian laws that may impose restrictions on the importation or the export of goods, and they must be aware of the various duties and controls imposed as a part of their overall business planning and decision-making.

Bilateral Agreements

Ever more rapid globalization of trade has produced a range of multinational and bilateral trading relationships among nations. The most basic of these is the bilateral trade agreement, binding between two nations, and confined only to their dealings with each other. Bilateral agreements represent regulatory rules that may affect international trade in specific types of goods. While some bilateral agreements may establish a general framework for trade regulations between two countries, bilateral agreements often deal with specific types of goods or sectors of trade. Bilateral agreements are frequently used to regulate the quantity or flow of specific goods. Consequently, these agreements frequently call for export licences or permits in order that the governments may monitor compliance with the agreement. Where such permits or licences are required, dealing in the specific goods for which a licence or permit is required is essentially contingent on the government. Entry into a particular market or the importation of the controlled goods may or may not be possible, depending upon the necessary approval. Information and advice on licences and permits is generally available from the federal ministries concerned with international trade and commerce.

NAFTA

North American Free Trade Agreement.

While the Canada–United States Free Trade Agreement[6] of 1989 has been rolled into an expanded North American Free Trade Agreement (**NAFTA**),[7] it serves as a further example of a bilateral agreement that provided a broad agreement concerning trade between the two countries.

> The Canada–U.S. FTA covered specific goods, services, business travel and investment, financial services, dispute-settlement provisions, the protection of industries adversely affected by the agreement, procedures to deal with dumping, and countervailing duties.

The agreement was quite lengthy, but the general thrust of the various provisions was to promote trade and create an expanded market for an extensive range of goods and services of both countries. The basic principle underlying the agreement was that each country would treat the other country's goods, services, investors, and investments in the same manner as their own with respect to the goods and services covered by the agreement. All substantive aspects of this agreement have been preserved in NAFTA.

6. *Canada–United States Free Trade Implementation Act*, S.C. 1988, c. 65.

7. *North American Free Trade Implementation Act*, S.C. 1993, c. 44.

North American Free Trade Agreement

NAFTA, which came into force in 1994, was born from the successful free-trade experience of Canada and the United States, and the three-way interest of these two nations and Mexico in expanding the agreement. In addition to a phased reduction and elimination of tariffs, NAFTA provides for a Trilateral Trade Commission to resolve specific disputes, together with a provision for the enforcement of such resolutions. It is a wide-ranging agreement that includes provisions on intellectual-property rights, investments, market access, and standards, and it requires each government to open up its own procurement process to enterprises of the member nations.

CASE IN POINT

A waste management corporation intended to build a facility in Mexico, with the support of Mexican state officials, who assured the corporation that all necessary permits could and would be obtained in due course. When the facility was complete, the municipal (rather than state) government refused to permit the facility to open on the basis that it had not issued an operating permit. As the case spiralled toward litigation, the state governor declared the entire region around the facility to be an environmentally-protected area. In an appeal pursuant to the NAFTA dispute mechanism, the panel found that the permits and environmental decree were without due process and amounted to expropriation of the facility. It further found that the Government of Mexico itself was liable for the acts of lower levels of government for the injury to the corporation.

See: *Metalclad v. United Mexican States,* ICSID ARB(AF)/97/1, August 30, 2000.

Strict rules-of-origin requirements must be met to qualify for the preferential NAFTA treatment; in general terms, the goods must be substantially of North American origin. Those parts not of North American origin must be sufficiently transformed (worked upon) in North America so as to be of a new and different tariff classification after manufacturing, in order to qualify for duty-free treatment.

The Republic of Chile has experienced very rapid economic growth, and has been accepted in principle by the United States, Canada, and Mexico as the next member of NAFTA. Negotiations continue between the nations of North, Central, and South America for a hemispheric arrangement, a Free Trade Agreement of the Americas. One day, the agreement may stretch from Alaska to Argentina. Canada has, however, enjoyed a bilateral free trade agreement with Chile since 1997.

Free-trade area

Two or more member territories for which tariffs on trade between them are abolished.

Customs union

A free trade area whose members apply a uniform schedule of tariffs on imports from non-member territories.

Common market

A customs union which further allows barrier-free movement of services, workers and finance among member territories.

In considering these types of trading relationships, it is important to recognize the degree of limitation on national sovereignty imposed on nations. In its simplest form — the bilateral (or even multilateral) trading agreement — the participants define a relationship between themselves, but place no limitation upon their liberty to deal with third parties. This is typical of the **free-trade area** agreements described above, where the right to deal in any manner with nations outside the agreement remains unfettered: a free-trade relationship exists between the members, but there is no common external tariff applied to non-member states. For example, under NAFTA, Canada, the United States, and Mexico have determined a particular arrangement among themselves, but they are free to deal with other nations in any manner they wish. In a more advanced variety of a trading relationship, the parties also define mutual economic boundaries — perhaps a common external tariff — with other nations that are not a party to the agreement. This type of relationship is a **customs union**: the early economic stages of the European Union represent such an example. The signatories not only agree to treat each other in a particular manner, but go further and agree that they will treat all other nations in an even manner as well. This entails a considerable abridgment of national economic liberty in pursuit of a higher-order goal. Beyond the customs union is the notion of a **common market**, which allows for freedom of further factor mobility (labour and capital) without restriction. The European Union has progressed through this stage, whereas NAFTA, while it permits some factor mobility, does not alter existing restrictions on population migration for residency. A final stage of integration is being pursued in Europe but not as yet — nor is it even likely — in North America. This is complete integration of currency and monetary policy, requiring member states to relinquish all policy levers in favour of unified economic decision making. This devolution of national liberty goes far beyond the trade arena and can be described as **monetary union**.

Monetary union

An area, most successfully a common market, issuing its own currency for use among its member territories.

Canada has engaged in negotiations with the European Union to establish a free trade area between Canada and the EU area. To be clear, this does not mean that Canada desires to join the European Union; it means only that Canada and the EU members wish to see tariff-free trade between them. So while Canada almost certainly will never be a party to the EU agreement (it being a monetary union) we may have our own agreements with the EU. Such a single bilateral free-trade agreement between Canada and the EU would eliminate the need for Canada to negotiate a free-trade agreement with each EU member state.

Other agreements to which Canada is not a party can also shape the international environment for business, for example, the Organization of Petroleum Exporting Countries (OPEC). This oil-trading cartel had a significant trading impact on Canadian firms in the oil import business during the 1970s. However, more recently, the organization has been unable to maintain an effective price/production agreement, due to the internal problems of its own member states.

Extraterritoriality

Outside of the realm of bilateral and multilateral agreements, attention must be paid to the extraterritorial nature of some nations' laws, in particular those of the United States. The application of U.S. laws outside that country is a right asserted by the U.S. Congress, requiring compliance by foreigners unless they are prepared to incur sanction from the U.S. legal system. One such example is the 1996 Helms–Burton law, enacted by the U.S. Congress. This law prohibits any enterprise — worldwide — from dealing with assets in Cuba formerly owned by American interests prior to their nationalization by the Communist Cuban government. The sanctions against those found to be in contravention of the Helms–Burton Act range from barring travel into the United States to recovery through U.S. court judgment of profits thought to have been generated in foreign hands from these expropriated properties.

Considerable pressure from America's trading partners has been brought to bear in opposition to Helms–Burton, so much so as to lead the U.S. President to suspend the right to sue for these profits. The final disposition of Helms–Burton remains in some doubt, as it is being also challenged under both NAFTA and WTO rules. Because Canada has many business interests in Cuba, the direct response of the Canadian government lies in the application of our *Foreign Extraterritorial Measures Act*[8] (FEMA), which applies not only in the case of Helms–Burton, but to any situation of similar nature. While, generally speaking, Canada will recognize foreign judgments where treaties to do so exist, Canada will not recognize judgments of foreign courts that arise from legislation of extraterritorial application that is itself not recognized by Canada. Further, the Act gives Canadians the right to counter-sue in Canada against Canadian-based foreign interests should such a foreign judgment be issued against Canadian interests abroad. In any event, it serves to make the point that familiarity with all laws of foreign nations and their agreements with nations trading in similar products is essential for Canadian firms trading on the international scene, if only to assess their own trading position in the market.

Government Trade Assistance

Apart from legislation regulating the import and export of goods, a number of federal-government agencies or bodies have been established to assist Canadian firms that may

8. R.S.C. 1985, c. F-29, as amended by S.C. 1996, c. 28.

wish to establish export markets. These agencies or government departments may provide assistance in a number of ways:

(1) by providing financial assistance to Canadian firms that may wish to explore the possibility of export selling. This may take the form of organizing trade missions abroad, the cost-sharing of feasibility studies, or marketing research by firms interested in a particular export market or country.
(2) by providing security for the payment for goods sold under certain export transactions.
(3) by providing loan guarantees to enable Canadian exporters to fund sales or operations internationally.

The Canadian International Development Agency (CIDA) and the Department of External Affairs, under its Program for Export Market Development (PEMD), offer trade assistance to firms through the organization of trade missions, and travel abroad to international trade fairs where Canadian firms may either display their wares or contact foreign buyers or sellers. Financial assistance in the form of cost-sharing for market studies and similar activities may also be provided by these agencies. Government involvement for the latter usually consists of some form of dollar-for-dollar sharing of the costs.

At the present time, the Export Development Corporation, a Crown corporation, provides Canadian firms with insurance against many of the risks associated with foreign business transactions. These may range from protection from loss on export transactions to compensation if the foreign government seizes the Canadian firm's foreign assets or prevents the transfer of money or property from the country in question. Loss from war or revolution are also covered by the Export Development Corporation. In general, the Crown corporation's mandate is to encourage foreign trade by offering a wide range of protective services to reduce many of the risks associated with international business transactions.

A QUESTION OF ETHICS

Moving business production overseas can reduce corporate costs, but at what ethical price? Child labour and appalling conditions of work exist in many countries. Only a tiny percentage of the world's workers enjoy the legal protection Canada's workforce takes for granted. Would you turn a blind eye and accept the savings, or pay Canadian rates, possibly causing economic disruption abroad among families, towns, and regions? Are you prepared to suffer the bad publicity that may come if your products are tainted by exploitation?

INTERNATIONAL TRADING RELATIONSHIPS

In its simplest form, international trade consists of a single transaction whereby a Canadian retailer may import goods for resale to the Canadian public. More often, however, international trade takes the form of established business relationships with foreign buyers or sellers, or the establishment of business organizations in other countries to carry on trading operations alone or in concert with others. While we tend to think of foreign trade in terms of Canadian firms selling goods abroad, it is also important to note that international trade is essentially two-way in its nature, and includes foreign sellers establishing business relationships with Canadian firms as well. Canadian firms operating foreign-automobile dealerships represent examples of the latter type of business relationship.

Risks in International Operations

The risk attached to international operations lies on a spectrum from fairly low (export operations) through extremely high (physical investments in unstable foreign markets). Risk also varies inversely with the degree of management control that can be exercised. For example, a lone exporter exercises near-absolute control over its business, while those who take on partners abroad (licensing and joint ventures) must share their management control. Other matters cannot be controlled at all in a direct sense. When a Canadian business operates in foreign markets and serves foreign customers, those customers are usually paying in their local currency. This creates a risk to the Canadian supplier through fluctuating exchange rates, and various hedge instruments are available to soften these effects. It is possible (and advisable) to reduce each of these risks through choosing wisely among possible foreign partners and associates and creating good legal agreements that clearly set out the responsibilities of both parties. Beyond the legal documents, allowance must be made for misunderstandings of culture in all cases. This is particularly true in joint ventures, where management responsibilities may be shared on a day-to-day basis.

In many instances, the foreign business activity abroad cannot be returned to Canada at all if events turn for the worse (for example, in the case of a mining operation), but even under ideal conditions, business assets abroad are exposed to risk. These risks arise from local political and economic instability abroad, and the exposed position of being a foreigner in a foreign market. Using government-backed insurance schemes (such as EDC) is advisable, as well as private insurance, to reduce this risk to acceptable levels.

Foreign-Distribution Agreements

Apart from the single-transaction type of purchase of goods from a foreign seller, most international trading arrangements tend to be undertaken by manufacturers or wholesalers of goods where an ongoing relationship with the foreign firm (or firms) is established. These relationships may take on many forms, but the most common are the foreign-distribution agreement, the foreign branch plant or sales office, the joint venture to sell or manufacture abroad, and the licensing of a foreign firm to use patents or technological information to produce goods in the foreign country. Many variants of these four basic relationships also exist, as do many purely service-oriented activities, such as the management of foreign businesses and the provision of advice on the manufacture or preparation of goods for sale in the Canadian market.

Foreign-distribution agreements are contracts between Canadian exporters and firms that undertake the marketing of the exporters' products on an international basis or in a particular country. This form of distribution is frequently used by smaller Canadian firms that lack the necessary funding to support an international sales staff, or firms that believe that their interests in those markets are best served by distributors native to the particular countries. Foreign-distribution agreements may also be negotiated by Canadian exporters with Canadian or foreign firms that specialize in the marketing of products in many foreign countries, or they may be negotiated with wholly owned subsidiary organizations. In the latter case, the subsidiary distribution firm would essentially be the international marketing organization of the parent manufacturing firm, but charged with the responsibility for foreign sales.

Since foreign-distribution agreements are contractual, considerable care is required in their negotiation. Such agreements will normally very clearly set out the product or products subject to the agreement, and the area or territory in which the foreign distributor has the right or exclusive rights (if permitted under the foreign country's restrictive trade-practices laws) to sell the products. These agreements will usually include an obligation on the distributor to provide a sales staff of a particular size, and the efforts that the distributor will make to develop a market in the territory for the products. An obligation to keep information and special "know-how" confidential, both during and after termination of the agreement, is usually included as well. If the product requires servicing, the agreement may also set out the distributor's obligation to provide parts inventories of a particular size and service facilities or service depots where warranty work and the general service needs of the product may be satisfied. Most agreements of this type would also set out the terms and conditions under which the goods might be sold in the territory.

Foreign-distribution agreements are seldom one-sided. They will also include the Canadian exporter's obligations. These usually include the obligation to supply the goods (with, perhaps, certain quality standards specified) and replacement parts, and to supply technical advice, advertising material, or catalogues. The right of the distributor to use the manufacturer's or exporter's trade names for the duration of the agreement is generally a term of the contract.

Most agreements are negotiated for a fixed term, with provision for renewal or termination on notice. A common provision permits termination if the sales volume fails to meet or be maintained at a specified minimum level. It also provides for the disposition of the distributors' inventories if termination takes place. A *force majeure* clause that would permit termination of the agreement in the event of major strikes, riots, or social disorder is often included in the agreement.

Foreign-distribution agreements are generally complex agreements that will also set out the precise relationship between the Canadian firm and the foreign distributor. Due to their complexity and international nature, they will usually set out the governing law that will apply in the event of a dispute between the parties. Most international agreements of this type will usually provide that the disputes be resolved by way of arbitration rather than the courts, and thus an arbitration procedure will usually be set out in the agreement.

Foreign Branch Plants or Sales Offices

Foreign branch plants or sales organizations represent two distinct alternatives to foreign-distribution agreements, but both frequently require a commitment of Canadian resources and personnel to the international venture. Both also require a more intimate knowledge of the laws of the particular countries in which the plants or sales offices will be established, but permit the Canadian firm to exercise a greater measure of control over the product and its marketing abroad. A knowledge of the laws of the particular jurisdiction before the venture is begun is essential because, once committed to the project, the branch plant or sales office must operate subject to all of the laws of the particular country, both national and local. These laws often control capital flow, as well as impose restrictions on technology that might conceivably have an adverse future effect on the Canadian firm. As a result, foreign branch plants often take the form of assembly facilities rather than full-scale manufacturing operations, because the capital commitment is, usually, substantially less. Foreign laws that must be considered are those controlling the flow of capital, technology, and material both into and out of the country, employment laws that may have an impact on the staffing and the health and safety of employees, and laws relating to business transactions, including consumer protection and the pricing of goods. The complexity of these laws tend to vary in terms of both the economic sophistication and the political orientation of the country in question.

International Joint Ventures and Licensing

Problems associated with the wholly owned foreign branch plant or subsidiary may be reduced, or to some extent avoided, by way of a joint venture with a foreign national firm. Joint ventures of this type may take the form of unincorporated joint ventures (where the relationship would be based upon an agreement) or incorporated joint ventures (where a corporation would be created in the foreign jurisdiction and each of the parties to the venture would acquire a share interest in the corporation). The corporation so formed would carry on the manufacturing operations or the business activity. Share interests in the corporation and, indeed, the corporation itself, would be subject to the laws of the foreign country. In many cases where foreign ownership is of national concern, shareholding requirements imposed by the law may fix the share interests of the parties to certain percentages that would give the national party or parties in the joint venture effective control over the corporation. In these instances, supporting agreements are normally required to ensure that the joint venture corporation has its full energies directed to the production and sale of products in such a way that the objectives of the Canadian shareholders' interests would be achieved and protected. This may sometimes be accomplished in part by the Canadian party to the joint venture licensing the joint venture corporation to produce products that the Canadian firm has protected by patent or design rights on an international basis. Licensing may be an effective international strategy in its own right, and is discussed at length in Chapter 26, "Intellectual Property, Patents, Trademarks, Copyright and Franchising."

International Contracts of Sale

Returning now to the export sale contract, we can examine the component parts of the transaction in greater detail.

The law of contract plays an important role in international trade because a contract is the heart of the export sale. Indeed, a basic export sale generally consists of four documents, each serving a distinct purpose in the overall transaction. These are: (1) the contract of sale, (2) the bill of lading, (3) the contract of insurance, and (4) the commercial invoice. While these four contracts are the core documents, other contracts, such as bank financing agreements and guarantee agreements with the Export Development Corporation, frequently form a part of the contract package associated with the transaction.

Contract of Sale

International or export sale agreements frequently differ from domestic contracts for the sale of goods in that they must address a number of elements of the sale that have international importance. For example, trade terms and terminology must be clearly understood to have the same meaning to both parties. To avoid misunderstanding, contracts will often make reference to published interpretations of international trade terms, such as those available from the International Chamber of Commerce. Export contracts will also usually refer to the governing law, as well as the time when title to the goods will pass. Apart from special terms, international sale agreements tend to be more detailed in that they will clearly set out the quantities of the goods and their quality, the unit prices (as well as total price), delivery dates, mode of shipping, type of packaging, the time and method of payment and currency to be used, financing arrangements, insurance, provision of any required licences or permits applicable to the sale, a *force majeure* clause, and, usually, an arbitration clause to resolve disputes.

Export sales are generally the result of a series of negotiations that often take the form of inquiries, quotations, orders, and acknowledgements, and may include a variety of other forms of correspondence. Because each of these documents may have a different legal significance in each of the countries, export sellers often clearly define what may or may not constitute an offer, and the conditions under which it may be revoked or expire. From the outset then, the international sale is conducted in a manner different from that of the domestic sale, where the *Sale of Goods Act* would apply.

Bill of Lading

Bill of lading

A contract entered into between a bailor and a common carrier of goods (bailee) that sets out the terms of the bailment and represents a title document to the goods carried.

The **bill of lading** is an essential part of an international sale. It is a contract between the seller and the carrier of the goods that sets out the carrier's responsibilities to protect and deliver the goods to the purchaser. The bill will generally set out the name of the seller (shipper) and the consignee (usually the buyer or the buyer's agent), a description of the goods, the aircraft (or vessel's name if by ship), export licence numbers or permit numbers, and any other information that the particular entry state may require on the bill. Apart from the use of the bill of lading as a contract between the shipper and carrier, the bill of lading also represents a title document. Once the goods are placed in the hands of the carrier, the carrier, as a bailee, has a duty to deliver up the goods only to the consignee named in the bill. In this sense, the bill of lading becomes a title document, as the shipper will send a copy of the bill of lading to the consignee (assuming financing or payment has been settled between the parties), and the consignee on receipt of the bill of lading may present it to the carrier to receive the goods. Because the contract of sale usually provides that title will pass upon delivery of the bill of lading, the risk of loss generally follows with the bill. Since the bill of lading essentially represents the title to the goods, a buyer on receipt of the bill of lading may use the bill to acquire financing by using the bill as security for the loan.

The bill of lading may also be coupled with a sight draft if the seller wishes to retain title or perhaps maintain control of the goods until payment is assured. This procedure allows the seller to obtain a negotiable instrument (the sight draft) from the buyer in return for the title documents to the goods. Sellers who may wish to do this will usually use a negotiable bill of lading and send it to the buyer's bank along with the sight draft and other documents required under the terms of the sale agreement. The bank will acknowledge receipt of the documents and will not release the bill of lading and documentation until the amount of the sight draft has been paid to the seller.

Insurance

The third type of contract generally associated with an international sale is the contract of insurance. Because of the hazards associated with the shipment of goods, most agreements will provide for insurance against the loss or damage to the goods while in transit. The cost of insurance will normally vary according to the risks that the seller or buyer may wish to protect against. Insurers that specialize in insuring international trade agreements offer cargo insurance that covers either a specific shipment or insures on a blanket or open basis, covering all cargo that may be shipped by the particular seller. Because the contract of sale will usually specify that either the buyer or seller will arrange for the insurance, the party not required to provide the coverage may often acquire contingency coverage in the event that the other party neglects or fails to obtain cargo insurance. Contingency insurance is usually obtained by the seller, where the buyer has responsibility for obtaining insurance under the contract of sale. Sellers may also obtain political-risk insurance in some instances where goods are shipped to buyers on a consignment or deferred payment basis, if the country in question is politically or economically unstable.

Commercial Invoice

In addition to the contracts related to the export sale, a commercial invoice is usually necessary, and often required by the buyer's customs office. The invoice form and content may vary from country to country, and in some cases must be prepared in the language of the foreign country. The commercial invoice frequently represents both an invoice for the goods sold and a customs document that sets out details of the goods, to enable customs officials to set the tariff classification and rate applicable to the goods. As noted previously, other documentation may also be required by the government of Canada, the buyer's government, or the buyer. These documents may include export permits or licences, or certificates relating to purity or analysis with respect to certain types of goods (such as some prepared food products or chemicals).

Sellers also may require the delivery of certain documents as well. If an export sale provides for payment before shipment, acceptance of a time draft or sight draft, or provision of a letter of credit, these matters must be attended to and provided by the buyer in accordance with the terms of the contract of sale.

Choice of Law

In Canadian domestic sales contracts there is little need to make a conscious choice between bodies of law that might govern the operation and interpretation of a contract. The parties to any contract may well reside in the same province, making any such concern irrelevant. By the same token, if the parties to a contract are not located in the same province, any failure to make a choice of laws apparent on the face of the contract is not likely to be catastrophic. The commonality in principle between laws of different provinces of Canada in broad measure ensures that there will be no truly repugnant result from such a failure. In contrast, however, failure to specify a choice of governing law in an international contract virtually ensures disaster if any problem arises that requires third-party intervention by a judge or an arbitrator. What principles will serve to guide a person attempting to interpret the contract or assign liability under its terms?

In most cases, each contracting party desires to have its own national laws serve as the governing law of the contract. This arises from the comfort level generated by the familiar, and, therefore, in most cases, the choice of law becomes a bargaining point itself. There is no obligation for the parties to select one of their own national bodies of law to govern their contract, however, and they are free to select and have applied those of any other nation.

When an explicit choice has been made, courts or other forums hearing any dispute will generally respect and enforce that choice. When no explicit choice has been made, and no clear inference can be drawn as to intention of the parties, a court will make a choice for the parties based on either requirements of its own statutory code, or by determining which nation's laws are most closely related to the dispute at hand. In this latter instance, the court would consider in varying measure the citizenship, residency, or place of incorporation of the parties, the location at which the contract was negotiated, the place where performance was to have been made, the current location of any subject matter in dispute, or the place where damages have been suffered as a result of breach. The significance of each of these factors will vary from case to case.

Consider, however, that obtaining a judgment in any court is only worth as much as the defendant's assets that are present inside that territory that may be executed upon (seized). For example, an importer may have imported defective goods from a foreign exporter. If the importer seeks a judgment from his or her home court, what is he or she to do with it? The exporter has no assets in the importer's territory, and if there was not a means of international seizure (by mutual recognition of court judgments), the plaintiff importer will have engaged in a pointless exercise of its legal rights.

ARBITRATION OF INTERNATIONAL TRADE DISPUTES

Agreements pertaining to international trade present a number of problems for the contracting parties because of the complexity of the transaction in terms of enforceability. In many cases, the parties each operate under different political systems, and, not infrequently, a government or state organization may be one of the contracting parties or a direct player in the negotiations. These differences, in an international trade context, often dictate some form of dispute-resolution mechanism other than the courts of one country or the other. Commercial arbitration is frequently the method that the parties incorporate in their agreements to resolve any disputes that may arise.

Commercial arbitration, by definition, is a method of resolving disputes arising out of an agreement made by two or more parties. This method employs one or more third parties who impartially decide the dispute by rendering an interpretation of the agreement or a decision that becomes binding on the parties to the agreement. The authority of the third-party decision-makers arises out of the agreement, although the procedural methods the third parties may use and the enforcement of their decision may either be incorporated in the agreement by reference to statute or code, or by state adoption of a particular international model arbitration law.

Commercial arbitration generally involves the resolution of commercial disputes by persons experienced in the particular branch of the trade or business where the dispute arose. Business persons have used this method of dispute resolution since the early years of commercial trading to quickly and effectively resolve their differences. It is also a means whereby the disputes would be resolved largely in private. Today, it represents a common form of dispute resolution utilized by business persons in Canada and other countries that have legislation pertaining to the arbitration of business disputes.[9] In recent years, efforts have also been made to establish international arbitration laws to facilitate the use of arbitration on an international basis and to enforce international arbitration awards.[10]

Arbitration Process

Arbitration is a method that the parties at any time may mutually agree upon to resolve a dispute that arises out of a contract. However, to make arbitration a required method of resolution of future disputes that may arise out of the agreement, the arbitration process must be included in the agreement. For international trade agreements it is also necessary to set out the process in some detail, in order that the arbitration itself will represent an appropriate and effective dispute-resolution mechanism. This is so because a number of issues must be addressed in the preparation of the agreement clause to avoid problems with the implementation of the process and to confirm matters of a procedural and

9. See, for example, for commercial disputes in Ontario, the *Arbitration Act*, S.O. 1991, c. 17.

10. See *United Nations Commission on International Trade Law — Model Arbitration Law.*

legal nature. For example, many international trade transactions involve not only private organizations but state or state agencies as well. A Canadian firm that engages in a transaction with a foreign state agency must ensure that its rights under the transaction may be enforced if the foreign country should attempt to exercise its sovereign power to revoke the particular trading rights or confiscate/expropriate the property of the Canadian firm or its assets in the foreign country. To protect itself in this example, the Canadian firm might insist at the time of negotiation that the agreement shall be subject to arbitration in a country other than the foreign state, under internationally recognized arbitration rules. The firm might also provide in the agreement itself that unilateral actions by the foreign state, such as expropriation, new controls on repatriation of capital, more onerous customs duties, and other changes, be subject to arbitration.

Because the effectiveness of an arbitration clause is dependent upon its terms, most clauses will include reference to the composition of the arbitration board, the place or country where the arbitration would be held, the applicable law, the language to be used in the proceedings, the procedure to be followed by the arbitration board, and its powers or jurisdiction. The enforcement of the arbitration award may also be addressed, depending upon the particular laws relating to arbitration in the jurisdictions involved.

As a general rule, arbitration boards usually consist of three persons, with each of the parties to the agreement appointing one member of the board. The two appointees then select the third member, who may be characterized as a truly impartial member. The third person so selected is usually designated as the chairperson of the tribunal. If the parties are unable to agree on the third member, the agreement should provide a mechanism for the selection or appointment of a third member, or reference should be had to a statute or code that does provide for the selection. Unless specified in the agreement, the language of the arbitration is usually the language of the chairperson selected (or the single arbitrator, as the case may be), even if the language is not the native language of either of the parties to the arbitration.

Most arbitration agreements will state the place where the arbitration will take place and the governing law. This is an important term in the arbitration clause, as the governing law provides the procedural rules applicable to the arbitration. In the absence of specific reference to the governing law, the laws of the place where the arbitration is held will normally apply. The contracts, however, may specify one of the internationally recognized arbitration laws, such as the United Nations Commission On International Trade Law (UNCITRAL) rules for arbitrations.

If one party invokes arbitration and the other fails to assist in creating the arbitration panel, or doesn't respond at all, arbitration would be pointless if it were so easily paralyzed. For that reason, arbitration laws and private arbitration institutes provide for default proceedings, whether for lack of participation in creation of a panel, for failing to provide documents, or for failing to attend to give evidence.

For a sample trigger process for appointment of an arbitrator, see UNCITRAL Rules, Articles 6–8. A typical example of ex-parte (unilateral) powers is found in Article 9 of the Rules of the Hong Kong Arbitration Centre:

> The Arbitrator shall be entitled to proceed with the arbitration notwithstanding the failure or refusal of any party to comply with these Rules or with the Arbitrator's written orders or written directions. Such power shall extend to the Arbitrator proceeding ex-parte providing the Arbitrator has given due written notice of his intention to so proceed.

ENFORCEMENT OF ARBITRATION AWARDS

The arbitration process as a means of international trade dispute resolution is essentially a creature of the contract negotiated by the parties. In a sense, the parties may, within bounds, determine in their agreement the powers of the arbitrator or board of arbitration, and the manner in which the award may be enforced. However, for the most part, the enforcement of the award is something that most developed countries have dealt with by statute. Enforcement of an award, then, will generally fall outside the agreement and may vary from state to state, depending upon their legislation.

In early times, when international trade was conducted at "fairs," the merchants resolved their disagreements by peer adjudication of the disputes, a process not unlike modern commercial arbitration. At that time, the enforcement of the decision was largely by the merchant guild or organization itself, with the threat of expulsion from the group as the principal sanction. Over time, however, merchants sought other means of enforcement with rather limited success, and it has fallen to the governments to establish enforcement mechanisms by statute. Canada adopted the *United Nations Commercial Arbitration Code* in 1986.[11] The Code defines an arbitration agreement, provides for the appointment of arbitrators in cases where one party fails to act, sets out the jurisdiction of the arbitration tribunal, the place and procedure, the recognition of the award, recourse against it, and its enforcement. Under the Code, the award may be enforced by applying to the designated court and then proceeding on it through the court process against the defaulting party.

Canada also adopted the United Nations convention on the recognition and enforcement of foreign arbitral awards in 1986.[12] Enforcement of an arbitration award is effected by application to the designated court. In Canada, the designated courts are the Federal Court and any superior, district, or county court of a province.[13]

In 1992, Canada joined the Vienna Sales Convention, an international agreement that establishes a single set of rules that automatically apply to international trade contracts for the sale of a wide variety of goods. The convention also deals with the forum and law that will be applied in the event of a dispute. Business firms are not obliged to follow the Convention, but may opt out of the rules if they so desire, and specify in their sale contracts the particular nation's laws that they wish to have govern their agreement and its enforcement.

In many jurisdictions, the most logical enforcement mechanism has been the court system, which not infrequently has also been charged with the duty of ensuring that the arbitration process itself has been conducted fairly and in accordance with the applicable law. For example, the enforcement process may consist of a filing of the arbitration award with the office of a designated court of the country, and on this basis obtaining a judgment of the court. The same court may also be called upon to review the arbitration process itself if some misconduct or unfairness is alleged on the part of an arbitrator or member of the arbitration board. Judicial review may also take place in some jurisdictions where the arbitrators have made an error in law or exceeded their authority.

The enforcement of arbitration awards in international trade transactions continues to remain complex in many jurisdictions. However, the adoption of model laws by many countries in recent years has generally reduced the enforcement process and made the process itself more uniform.

MANAGEMENT ALERT: BEST PRACTICE

When Canada becomes a party to an international agreement, it is duty-bound to adopt domestic law giving effect to these obligations. The scope of these agreements keeps getting wider, and affects more and more Canadian businesses as time goes on. Canadian industry associations, speaking with a common voice, and a few Canadian firms acting alone, are large enough to shape Canadian policy and take a role in the process. The remaining firms must accept these new laws as they arrive. By monitoring international negotiations, the long time lag between the start of negotiations and the time when an agreement enters into force can be used by managers to forecast and adapt the business to the changing legal reality.

11. *Commercial Arbitration Act*, R.S.C. 1985, c. 17 (2nd Supp.) as amended by S.C. 1986, c. 22.

12. *United Nations Foreign Arbitral Awards Convention Act*, R.S.C. 1985, c. 16 (2nd Supp.) as amended by S.C. 1986, c. 21.

13. ibid., s. 6.

SUMMARY

Canada has become increasingly involved in international trade. Business firms engaged in this trade must be familiar with not only the Canadian laws that affect their business, but the laws of those countries with which they trade. This is particularly important if the Canadian firm has established a manufacturing facility or sales office in a foreign country.

The import or export type of transaction usually involves custom tariff legislation or laws requiring special permits to import or export certain goods. The purchase or sale itself is by contract, but the contract normally must address a number of aspects of the sale that are often unimportant in a domestic sale. The international contract usually includes not only the complete details of the transaction, but a clause whereby the parties agree to resolve any dispute arising out of the transaction by binding arbitration. Other documentation is also required, including a bill of lading, insurance, and a commercial invoice. Where required, special permits for customs clearance and certificates as to purity or analysis may also form a part of the transaction.

When a firm decides to do business in another country, it may do so by establishing a manufacturing facility or sales office, either on its own or as a joint venture with a local partner. A knowledge of local laws of the foreign country is essential in either case, but the advantage of a local partner might be its familiarity with its national laws. An alternative approach might be to license a foreign manufacturer to produce or sell the goods.

Where a dispute arises between the parties, arbitration is the usual method of resolving the matter. Arbitration clauses in the contract usually set out the procedural details or refer to arbitration in accordance with a particular internationally recognized procedure or set of rules.

KEY TERMS

Tariff, 650
World Customs Organization, 650
Dumping, 651
Customs broker, 652
World Trade Organization (WTO), 653
Most Favoured Nation (MFN), 653
National Treatment (NT), 653
NAFTA, 654
Free-trade area, 655
Customs union, 655
Common market, 655
Monetary union, 656
Bill of lading, 660

REVIEW QUESTIONS

1. How are arbitration awards enforced?
2. Explain the different ways in which a Canadian firm may establish an international trading relationship.
3. How does the *Special Import Measures Act* affect foreign sellers?
4. Why was it necessary for Canada to pass an *Export and Import Permits Act*?
5. Identify the usual documents required for an international contract of sale. What is the purpose of each of these documents?
6. Outline the general thrust and purpose of the World Trade Organization.
7. Distinguish a bilateral trade agreement from a multinational trade agreement.
8. What assistance does the Canadian government provide to Canadian firms that may wish to enter the international market?
9. Explain the difference between the *Customs Act* and the *Customs Tariff Act*.
10. What are the advantages and disadvantages of a foreign trading relationship in the form of a joint venture?
11. What is the role of a customs broker in a transaction whereby goods are imported by a Canadian firm?
12. Explain why commercial arbitration is used as a means of dispute resolution in international agreements.
13. Why do most international trade contracts provide that arbitration will take place in a country other than the country of either of the contracting parties?
14. What is the effect of Canadian laws on the importation of goods into Canada?

www.mcgrawhill.ca/olc/willes

Mini-Case Problems

1. A Canadian manufacturer that has been successful in supplying the domestic market with its products is considering the sale of its products in Europe. Since this would be a 'first step abroad' it is considering a foreign distribution approach. What are the advantages and disadvantages of this approach?
   What are the legal implications?

2. The Canadian manufacturer of a product wishes to expand the manufacture and sale of its product to China under an international joint venture agreement.
   What points or issues should be covered in the agreement to protect the Canadian firm?

Case Problems for Discussion

CASE 1

CRT Manufacturing Ltd. produced a line of specialized machine tools used in the production of parts for the automotive industry. The corporation's product line was well-known in North America for quality and durability, but virtually unknown in Europe. With the trend in the automobile industry towards the production of vehicles for a world market, CRT Manufacturing Ltd. decided to expand its operations to Europe. Management at CRT Manufacturing Ltd. has narrowed its approach to either a foreign-distribution agreement or a branch plant. A branch plant would require a significant capital commitment, as production equipment for the specialized machine tools would be expensive, as would be the training of employees in production methods. Since a certain amount of expertise would also be required on the part of the sales staff, the question of training would be important in order to sell the products in the new market.

Advise CRT Manufacturing's management on the range of legal and contract issues that would be associated with the two proposed courses of action.

CASE 2

Using the facts in Case 1, if CRT Manufacturing Ltd. should further decide to investigate a joint venture approach to its European business expansion, what foreign law considerations arise?

Given its unfamiliarity with the European market, which of the three approaches might be best for CRT, considering the legal issues that it may face?

CASE 3

Kyoto Mfg. Inc. was a producer of sensitive measuring equipment for the steel industry. Kyoto was a Japanese company and had its principal place of business in Japan. Kyoto had no assets in Canada.

A Canadian firm, Concepts Mfg. Limited, located in Vancouver, entered into a licensing agreement with Kyoto for the right to manufacture and market several of Kyoto's products in Canada. Concepts was to pay a royalty to Kyoto based upon a formula set out in the contract, and Kyoto was to provide technical support and know-how. Concepts was further entitled under the licence to receive information and support promptly whenever Kyoto made improvements to the products as the result of technological advances.

Two of the clauses contained in the agreement stated, in part, the following:

4.1 The validity and interpretation of this Agreement and of each clause or part thereof shall be governed by the laws of Japan ...

5.1 Any and all disputes arising from this Agreement shall be amicably and promptly settled upon consultation between the parties hereto; however, in case of failure of settlement, the disputes shall be settled by arbitration in Tokyo, Japan, in accordance with the rules of the Japan Commercial Arbitration Association and the award shall be final and binding upon both parties. In no case shall any award against Concepts Mfg. Limited exceed all royalty fees due by Concepts Mfg. Limited at the date of commencement of the arbitration hearing.

The parties co-operated under the agreement for several years, then a dispute arose as to the entitlement of Kyoto to certain royalties. Concepts alleged that Kyoto had failed to provide adequate technical support and advice concerning certain improvements it had made to its technology. Kyoto denied this allegation and insisted on receiving its royalty payments. Concepts responded by alleging fundamental breach of the contract by Kyoto.

Following an unsuccessful attempt to resolve the matter, Kyoto submitted the dispute to the Japan Commercial Arbitration Association for arbitration. Both parties were sent notices concerning the names of the arbitrators appointed to the arbitration tribunal and the date, place, and time of the hearing. On the day set for the hearing, only Kyoto attended and made submissions. There was no correspondence from Concepts before or during the proceedings.

The arbitration tribunal awarded Kyoto the sum of $150,000 in royalties to be paid by Concepts, together with interest and a portion of the costs of the arbitration process. Although Concepts was sent a copy of the award, it neither acknowledged its receipt nor made any payments in accordance with it.

After six months of unanswered communication, Kyoto brought an action in the British Columbia courts for a declaration that the Japanese arbitration award was valid and enforceable in Concepts' jurisdiction. Kyoto argued that it had complied with all procedures necessary to have the award enforced in Canada. Moreover, the parties had a written contract to submit disputes to arbitration, that the subject matter of the dispute was not outside that contemplated to be settled by arbitration, and, further, that Concepts had never taken any steps to dispute either the jurisdiction of the arbitration or the merits of Kyoto's claim.

Concepts argued that the award could not be enforced against it since the subject matter of the dispute, namely, fundamental breach of contract, is not a matter that can be settled by arbitration in its province. Rather, this is a question that could only be determined by a court. It further argued that to enforce the Japanese award would be contrary to public policy since it would allow a foreign company to receive benefits under the contract while, at the same time, preventing the Canadian company from seeking any remedy for its damages caused by the fundamental breach by the foreign company.

If you were the judge hearing this case, how would you decide and why? On what legal principles would you base your decision?

The Online Learning Centre offers more ways to check what you have learned so far. Find quizzes, Weblinks, and many other resources at www.mcgrawhill.ca/olc/willes.

www.mcgrawhill.ca/olc/willes

CHAPTER 34

Environmental Law

Chapter Objectives

After study of this chapter, students should be able to:

- Describe the Common Law provisions in environmental protection.
- Describe the principal elements of environmental protection legislation.
- Identify the chief intersections between business activity and environmental concern.

YOUR BUSINESS AT RISK

We may think globally, but we still act locally, for ill as well as good. When businesses treat the environment carelessly in the disposal of effluent or waste, or directly harm other people, considerable liability can result. This liability is not only large in dollar terms, but it can extend to individual directors and officers as well as the company itself. The environment must also be considered early in the business planning processes, as many projects can be sidelined before they begin as a result of undesirable environmental impact.

THE COMMON LAW

At Common Law, injury to the environment has generally been considered by the courts on a relatively personal level, in the sense that actions of one individual that interfere with the property or rights of another are actionable at law. If a property owner pollutes a watercourse and causes injury to the downstream user, the downstream user (or riparian owner) may take action against the upstream owner for the injury caused. Similarly, if someone interferes with the lands of his or her neighbour by contaminating the neighbour's soil or groundwater, the contamination may be an actionable tort of nuisance. Creating contaminated smoke where the particles fall on neighbouring properties and cause injury would also be actionable.[1] Even making excessive noise that interferes with a neighbour's enjoyment of his or her property may be treated as a tort.[2]

In one case, a foundry operated for many decades, producing smoke from its operations without complaint from its industrial neighbours. An automobile transport company then acquired vacant lands next to the foundry for the purpose of storing new automobiles pending shipment. Particles from the smoke caused damage to the finish of the stored automobiles, and the transport company took legal action against the foundry for the damage caused by its smoke emissions. The foundry argued that it had acquired

1. *Russell Transport Ltd. v. Ontario Malleable Iron Co. Ltd.*, [1952] 4 D.L.R. 719.
2. *340909 Ontario Ltd. v. Huron Steel Products (Windsor) Ltd. and Huron Steel Products* (1990), 73 O.R. (2d) 641.

the right to emit the smoke on the basis of the passage of time. In finding against the foundry, the court addressed the various defences raised in the following way:[3]

(1) A defendant cannot claim that the plaintiffs came to the nuisance.
(2) A defendant cannot claim that even though the nuisance caused injury to the plaintiff, it is a benefit to the public at large.
(3) A defendant cannot claim as a defence that the place where the nuisance operates is a suitable one for carrying on the operation in question, and that no other place that is suitable would result in less of a problem.
(4) The defendant may not claim that all possible care and skill were used to prevent the operation from being a nuisance, because nuisance is not a part of the law of negligence.
(5) The defendant cannot argue that its actions would not amount to a nuisance because other firms acting independently of it were doing the same thing.
(6) A defendant cannot say as a defence that it is merely making a reasonable use of its property, as no use of property is reasonable if it causes substantial discomfort to others or causes damage to their property.

In this case, the polluter was held responsible for the damages caused by the smoke particles, and given a brief period of time to correct the pollution problem. While the case effectively ended the environmental damage caused by the foundry, the cumulative damage to the environment was not, and could not be addressed by the court in its judgment. This was so because the Common Law relief was limited to those individuals who could show damage and establish their right to compensation in court. In effect, the Common Law and the relief it offered could only address injury to property or persons on an individual basis. Protection of the public from damage to the environment in general, insofar as the courts were concerned, was a matter for the government to address by legislation.

Early Reforms

A further drawback of the Common Law in cases of damage to the environment was the matter of standing before the courts. In this situation, the applicable law was the law related to nuisance, as manifested by interference with a person's enjoyment of their property. Environmental groups concerned about pollution of air or water could seldom establish that they suffered injury or damage, as in most cases the injury (if it could be established) was to the property of the Crown, and not directly to the individuals in question. In this regard, the Common Law was limited for the most part to an individual, rather than public, action as a means of controlling or eliminating pollution to the environment.

In a case heard in 1917, the limitations of the Common Law vis-à-vis broader public policy issues were raised by the court. In that case,[4] a mining company used an open-roasting method for the smelting of its ore. The smoke and fumes from its operations damaged the crops of neighbouring farmers, and a farmer who suffered damage applied to the court for an injunction to stop the damage. The mining company was clearly at fault, but the court faced a dilemma. To issue an injunction to stop the damage to the farmer's crops would essentially eliminate the employment of the majority of the residents of the city that had developed around the mine. The court recognized the broader economic issue raised by the case, and awarded money damages to the farmer, but refused to grant an injunction to stop the smoke damage. The provincial government responded with legislation shortly thereafter that dealt with the need for ongoing compensation for the smoke damage caused by the smelter. The legislation did nothing, however, to address the problem of ongoing environmental pollution caused by the smelting operation itself. This approach was typical of the response by governments at the time. While recognizing that environmental damage was broader than the individuals involved, the tendency was for the legislators of the day to treat environmental damage as a localized matter, rather

3. *Russell Transport Ltd. v. Ontario Malleable Iron Co. Ltd.*, supra.
4. *Black v. Canada Copper Co.* (1917), 12 O.W.N. 243.

than a broader public policy problem. Attitudes changed following World War II, as the magnitude of the problem of damage to the environment began to unfold.

Problems of a Complex Society

The necessities of war produced a host of new products and chemical compounds that had peacetime uses and applications. In addition, many new developments in the years that followed were later discovered to have harmful effects on the environment, either through their manufacturing processes or when the products were discarded. Under Common Law it was not easy to provide relief to those affected by these products or processes because it was often difficult to pinpoint the source of the pollution. In many cases, the pollution may have originated in a number of sources. For example, a downstream user of water might be affected by contaminants in the water supply, but not be in a position to identify the particular polluter if the same water supply was used by numerous upstream commercial or industrial users. Long-forgotten waste-disposal sites also might be leaking into the stream, or municipal storm sewer runoff allowed to enter the water supply may compound the problem. The runoff from farmlands adjacent to the stream may contain harmful chemicals that had been used by farmers for weed control or fertilizer. In some cases, the discharge of chemicals not harmful in themselves might combine to form pollutants with an unidentifiable source. In these situations, the Common Law could not adequately address the problem, and, perhaps more importantly, could not provide an appropriate remedy.

A further difficulty of the Common Law was the limitation of the remedy to address the problem of cleanup of the polluter's own lands. The landowner injured by the pollution would receive compensation for the damage suffered, but the court would not be in a position to order a cleanup of the polluter's own property, and the source of the pollution would remain.

Governments recognized the limitations of the Common Law as a means of control and abatement of environmental damage, and began a proactive role in environmental protection. The Common Law, nevertheless, has continued to be a useful and effective means of dealing with individual and localized instances of injury to property. Its use, however, has largely been overshadowed by legislative regulation and control measures.

CASE IN POINT

A mushroom-growing operation produced compost odours that neighbouring residents complained were causing them health problems and mental distress as well as interference with the enjoyment of their properties. The residents instituted legal proceedings against the farm operators for negligence and nuisance. The operators argued that they were not liable for nuisance, as provincial legislation protected farmers from nuisance claims provided that they followed "normal farm practices."

The court found that the mushroom operation was commenced after the residential development of the area in close proximity to the farm, and the farm was not operating in accordance with similar operators who conduct this type of farming in close proximity to residential neighbours. The farm operator was held liable in nuisance, but not negligence.

See: *Pyke et al. v. Tri Gro Enterprises Ltd.* (2001), 55 O.R. (3d) 257.

ENVIRONMENTAL LEGISLATION

The legislative approach overcomes most of the difficulties related to identification of source, and control and abatement (or prohibition) of pollution, and provides protection for the environment in general. Environmental-protection laws recognize that a great many human endeavours produce some form of pollution of the air, water, or land. They also recognize that many necessary business activities can only be carried out through the production of waste, and, in some cases, hazardous waste. What environmental protection legislation attempts to do is minimize the pollution through control and monitoring procedures and, where necessary, prohibition of former production or waste-disposal practices.

The legal aspects of protection of the environment fall within the jurisdiction of the federal, provincial and territorial governments, as well being the subject of extensive municipal regulation. This shared responsibility illustrates how wide ranging environmental regulation must be. A single issue such as the preservation of fish habitat may involve overlapping regulation; for example, federal fisheries law and provincial natural resources regulation, as well as provincial and municipal regulation of farm and lawn herbicide that winds up in water run-off through rivers to oceans.

Stakeholder approach
Consideration of the needs of the broadest community of persons interested in an issue.

Precautionary principle
Where serious damage may occur, lack of full scientific certainty shall not be used to postpone cost-effective measures to prevent environmental degradation.

Sustainable development
Outcomes should meet the needs of the present generation without compromising the ability of future generations to meet their own needs.

The law in Canada today also draws much of its inspiration from concepts of environmental awareness that have arisen around the world over the past fifty years. The adoption of a **stakeholder approach** to environmental issues has steered the law toward considering the rights and needs of the broadest community of interested persons, rather than narrow classes of persons who can show proximate cause in an existing harmful action. This is partially in recognition of the need to move toward prevention of harm rather than its remediation, including use of the "**precautionary principle**," and the goal of **sustainable development** in both policy and in practice. Concepts such as these have taken society a great distance from former times where individual rights exercised on private lands were nearly inviolate, now to considerations of the rights of generations yet to come.

Few people would have thought that environmental issues would have accelerated at the pace they have in recent years. Whether the Al Gore documentary "An Inconvenient Truth" is either a catalyst for action or simply a symbol of the times is a matter of opinion, but more responsive and even proactive behaviour is everywhere. As society changes, our law changes with it, although often at a slower and more measured pace. However, as early as 2001, and perhaps ready for more widespread application now, is the following case regarding municipal action in environmental matters. It prepares the ground for much more action in the future.

CASE IN POINT

A Quebec town enacted a by-law prohibiting the aesthetic use of pesticides. Commercial lawn care firms challenged the by-law as outside municipal authority. On final appeal to the Supreme Court of Canada, the Court endorsed municipal powers in promoting the "general welfare" of citizens, that simultaneous provincial and municipal regulation can co-exist as long as they do not contradict each other, and that subsidiarity (regulation by a government closest to those persons effected) was preferable. Moreover, the Court observed that "everyone is aware that individually and collectively, we are responsible for preserving the natural environment," and that "environmental protection [has] emerged as a fundamental value in Canadian society." The Court acknowledged the importance of the international law "precautionary principle": environmental measures must anticipate, prevent, and attack the causes of environmental degradation. Where there are threats of serious or irreversible damage, lack of full scientific certainty should not be used as a reason for postponing measures to prevent environmental degradation.

114957 Canada Ltée (Spraytech, Société d'arrosage) v. Hudson (Town), 2001 SCC 40.

These principles — prevention, precaution, sustainable development and "polluter pays" — are the guiding principles recognized in the federal *Canadian Environmental Protection Act, 1999*.[5] The Act provides for a wide range of regulations, capping or mandating reductions in toxic substances, risk assessment, regulating emissions, hazardous waste, emergencies, enforcement, public participation and co-operation between jurisdictions.

5. S.C. 1999, c. 33.

Businesses must carefully consider how and where their operations intersect with the environment, whether it is in obtaining, using or storing dangerous raw materials, right through the entire product life cycle, to disposal and recycling. Businesses must also be aware be that in enforcement of the provisions of the Act, officers have all the rights of a peace officer to enter premises and examine and seize evidence. The Act provides for a wide range of resolutions, from warnings and fines to orders for remedial measures, all consistent with the "polluter pays" principle.

Recalling the tragedy of the 2010 oil spill from a drilling rig in the Gulf of Mexico, what aspects of that situation can you see reflected in the principles of the *Canadian Environmental Protection Act, 1999*? Alternatively, what cautions would you offer if an application was made to drill for oil in the northern waters of Canada's Beaufort Sea?

"Polluter Pays" Regulation

Polluter pays principle

The obligation on a polluter to pay for environmental damage that results from a violation of environmental legislation.

Apart from the "grey areas" of jurisdiction, for many business activities, provincial legislation is the applicable law, and it is this legislation that must be carefully adhered to in the conduct of business activity. Each province has addressed environmental protection in its own way, but the legislation has a common thrust and purpose: to limit or prohibit those business (or individual) activities that either harm or degrade the environment. The laws, therefore, deal with the discharge of harmful substances into the air, water and ground, and, in some cases, also address the cleanup of past pollution of ground and water. For example, under the *Fisheries' Act*,[6] private landowners who damage fish habitat, even inadvertently, may be subject to severe penalties. Nevertheless, environmental-protection legislation recognizes that economic activity in many cases cannot be carried out without causing some environmental damage. As a consequence, rather than prohibiting the business activity entirely, the legislation takes a regulatory approach. These laws for the most part are concerned with the discharge of environmentally harmful substances into the air or water. They tend to be specific about the quantity of a pollutant that may be discharged in a certain period of time. The amounts may be expressed in parts per million of the specific substance in a specific volume of water or air. Some laws also require that the business carrying out the activity that causes the pollution monitor and record the discharge to ensure that the pollution does not exceed the allowable limits. In some jurisdictions, devices that cause pollution (such as equipment for burning materials) are subject to licensing requirements, and, if the operators fail to contain the levels of pollutants produced to within the limits set out in the legislation, the licence to operate the equipment may be revoked. As an example of pollution regulation on a more individual level, automobile engines must be equipped with air-pollution-control devices that limit pollutants in engine exhaust to specific levels, and the vehicle owners may not alter or remove the equipment as long as the vehicle is licensed.

The general approach taken in the enforcement of this regulatory type of legislation is inspection and monitoring for compliance. In order to ensure compliance, enforcement officers are generally given wide powers of inspection, and the authority to examine and seize records where a violation of the Act is suspected. Offenders are punished by fines where damage to the environment is established that contravenes the Act or where the allowable pollution limits have been exceeded without excuse. In some cases, if pollution is serious, or if immediate action is required to prevent environmental damage, environmental enforcement officers have the authority to order the polluter to cease operations until the pollution problem can be corrected.

6. R.S.C. 1985, c. F-14, as amended.

COURT DECISION: Practical Application of the Law

Environmental Law—Private Prosecution—Municipal Liability

***R. v. Kingston (Corp. of the City)*, 2004 CanLII 39042 (Ont. C.A.).**

The City of Kingston operated a municipal dump on the shore of the Cataraqui River, from the early 1950s to the early 1970s. After the dump was closed, the City did little to address the environmental problems created by the dump, despite public demands for action and studies that showed that the site was of serious concern. After testing liquids emanating from the site, Janet Fletcher, an environmentalist, laid charges against the City by means of a private citizen's information. Separately, the Ontario Ministry of the Environment did so as well. Convictions and acquittals resulted on different counts, all of which were later set aside. Fletcher and the Ministry appealed.

THE COURT: On four separate dates, Ms. Fletcher had samples taken of leachate entering the Cataraqui River from the landfill site. The Fletcher samples were analysed for "acute lethality" to rainbow trout fingerlings. Rainbow trout is the standard test species for this type of analysis. All of the trout fingerlings that were exposed to the Fletcher samples died within twenty-four hours. Ms. Schroeder testified that the effluent collected in the Fletcher samples was acutely lethal to fish. After being advised of the analysis results from the testing of the Fletcher samples, the Ministry took its own samples of leachate from the landfill site on four separate dates. Mr. Lee testified that based on the results from "acute lethality" testing involving rainbow trout of the Ministry leachate samples, there was no doubt in his mind that the leachate was poisonous to aquatic life.

The trial judge had no difficulty in finding that the City created and owned the landfill site, was responsible for the site's ongoing operation and maintenance, and had deposited or permitted the deposit of a substance in the Cataraqui River, which was water frequented by fish. As the trial judge noted, the issue that was "hotly contested" was whether the substance in question — the leachate — was deleterious. The trial judge refused to convict on the first count in the Ministry information because of confusion over the date of the chemical analysis of the Ministry sample.

The court rejected the [City of Kingston] due diligence defence. Relying on *R. v. Sault Ste. Marie (City)*, [1978] 2 S.C.R. 1299, the trial judge stated that the defence of due diligence involves the characterization of efforts taken to prevent the act or event, including the history of the defendants' efforts for a reasonable period before the charge dates. He found that the City [was] aware that the leachate was flowing into the Cataraqui River and that they chose to ignore the problem. With regard to the prosecution brought by the Ministry against the City, the court ordered a fine of $30,000, to be paid within ninety days. In addition, the City was ordered to, within twelve months, provide the Ministry with a plan for the capping of the site in accordance with current standards of practice.

This was a very difficult trial. Although the trial judge's reasons are not exhaustive, his reasons nevertheless demonstrate a full understanding of the complex issues of scientific evidence that were before him. I therefore conclude that the record does not disclose a lack of appreciation of relevant evidence. Accordingly, I ...would restore the convictions and acquittal at first instance in the Ministry's action.

While Fletcher's action ultimately did not result in a conviction, she precipitated all of the events which did lead to a conviction in the Ministry action. The vigilance of a private citizen is often required to ensure that those with direct responsibility to preserve the environment do, in fact, discharge their duties.

Environmental Assessment

Environmental assessment

A procedure undertaken to determine the effect on the physical environment of a particular undertaking or activity.

Environmental-protection laws may also require governments, organizations, or businesses to engage in **environmental assessments** of certain kinds of activities if the activity has the potential to cause environmental damage.

The federal Act, the *Canadian Environmental Assessment Act*,[7] regulates the requirement and conduct of these reviews, which are intended to predict the impact on the environment of a particular initiative before it is executed. These assessments are to inform a decision on whether such a project should go ahead from an environmental standpoint, and should not be confused with site assessments for contamination or audits for compliance with environmental regulations.

7. S.C. 1992, c.37.

The projects contemplated by the Act are physical works for which a federal government authority must exercise its powers for the project to be carried out. The list is therefore a broad one, encompassing (in principle) anything the government regulates on an individual basis.

A mining project requiring dynamite blasting will need a federal permit under the *Explosives Act*. As this is an exercise of power in granting the permit, the related project will require environmental assessment. Some exemptions exist to ensure that projects with a minimal environmental footprint are not unduly hindered.

The assessment process can be very expensive, very long, and can require considerable technical skill, although provisions do also exist for simpler screening versus comprehensive assessment. Not surprisingly, major works such as hydroelectric projects, water diversion, mines, pipelines and waste-disposal sites attract more involved and detailed study. The process usually provides for public input as well, before approval is granted for the project to proceed.

Storing and Handling of Hazardous Products

The storage and transportation of hazardous products or other materials that would cause environmental damage is generally subject to legislation that directs care in storage and handling, and in most cases requires notification to the appropriate government body (usually the Ministry of the Environment or its designated agency) in the event that hazardous products or contaminants are spilled or released causing air, ground, or water contamination. In most cases, the legislation requires the person or business that caused the pollution to pay the cost of the cleanup, either by assuming responsibility of the cost directly, or by compensating the government authority that performed the cleanup for the costs that it incurred.

The legislation is often non-specific in terms of how parties must ensure the protection of the environment. The method of storage of products that may contaminate ground or water is not always specified, but a very high standard of care is imposed on the user. Products that are improperly stored or allowed to leak into the ground or water may result in charges under most environmental laws dealing with hazardous materials, as the laws tend to be couched in terms of a prohibition of certain types of pollution. In some provinces, the legislation holds the directors and officers personally responsible for allowing the pollution to occur, unless they can show that they used due diligence in their efforts to prevent the pollution from taking place. The *Environmental Protection Act*[8] of the province of Ontario, for example, places a heavy responsibility on officers and directors:

> Every director or officer of a corporation that engages in an activity that may result in a discharge of a contaminant into the natural environment contrary to this Act or the regulations has a duty to take all reasonable care to prevent the corporation from causing or permitting such unlawful discharge.

A number of federal-government laws related to environmental protection also hold directors and officers of corporations personally liable. The *Canadian Environmental Protection Act*[9] provides that:

> If a corporation commits an offence under this Act, any officer, director or agent of the corporation who directed, authorized, assented to, acquiesced in or participated in the commission of the offence is a party to and guilty of the offence, and is liable to the punishment provided for the offence, whether or not the corporation has been prosecuted or convicted.

8. R.S.O. 1990, c. E.19, s. 194(1).
9. *Canadian Environmental Protection Act*, 1999, S.C. 1999, c. 33, s. 280.

The *Transportation of Dangerous Goods Act*[10] and the *Hazardous Products Act*,[11] both federal statutes related to environmental matters, contain director and officer liability provisions similar to the above-noted section of the *Canadian Environmental Protection Act*. As with the Ontario *Environmental Protection Act*, the violations tend to be strict liability offences, where intent is not a factor that permits a corporation or its directors to avoid liability. The only defence for a director would appear to be due diligence.

To be effective as a defence in this kind of situation, due diligence means much more than the directors or officers of the company issuing directives to management to carefully store hazardous products or potential contaminants. It requires followup efforts to ensure that employees are properly trained in the safe use, handling, and storage of potentially hazardous products. It probably also means that careful personal monitoring or inspection of the premises should occur from time to time to ensure that the directives are enforced, and that no potentially risky conditions exist. In effect, due diligence probably requires the directors to satisfy the court that control and responsibility were not simply delegated to management on the assumption that compliance would take place, but that the policies were monitored by the directors on an ongoing basis.[12]

A QUESTION OF ETHICS

"Fines are not appropriate for breaches of environmental law. The only acceptable standard is to require cleanup. Anything short of this is a business licence for the right to hurt people." Comment on this ethical stand. As this statement could be levelled at any law that provides for a fine, is there something special about the environment that provokes this reaction?

CLIENTS, SUPPLIERS *or* OPERATIONS *in* QUEBEC

As with all other provinces, Quebec has enacted legislation in support of federal action in environmental matters, the focal point being the province's *Environment Quality Act*. The Act limits certain emissions and contaminations, and prohibits those which would adversely impact the heath and safety of people. Specific regulations are provided for the operation of a range of industrial activities. In creating a permit regime, the Quebec Ministry of the Environment requires the submission and approval of plans where proposed activity includes emissions or contamination, such as mines and pulp and paper mills. This approach is generally reflected in legislation and regulation in Common Law provinces as well.

RESPONSIBILITY FOR EXISTING CONTAMINATION

Environmental damage has generally been considered to be the responsibility of the party that caused the damage. This is not always the case, however, particularly if the contamination involves land or water. If contamination is found to exist on land, the legislation in most provinces permits the government agency or ministry to order the current owner to clean up the premises. The discovery of contamination on a land site may, in some circumstances, result in an order to clean the site, and the cleanup costs may exceed the value of the property. Consequently, careful legal practitioners will strongly recommend to a client interested in a land purchase that an environmental audit be made of the property before the purchase is finalized. Most commercial transactions of this nature now include a "clean" environmental audit as a condition precedent to the purchase of the land. Even when an environmental audit concludes that a property is clean, some risk remains, as no

10. S.C. 1992, c. 34.

11. R.S.C. 1985, c. H-3, as amended.

12. *Regina v. Bata Industries Ltd., Bata, Marchant and Weston* (1992), 9 O.R. (3d) 329.

standards have been determined for many contaminants, and the government may later require a higher standard of cleanliness. Nevertheless, the audit is a useful tool in reducing the risks associated with commercial-property purchases. An audit may reveal long-forgotten buried fuel-storage tanks, waste-disposal sites, and, sometimes, soil contaminated with hazardous products produced in the distant past by previous owners of the site.

Contamination of property also poses a risk (in some jurisdictions) to lenders who look to land and buildings as security for debt, because a mortgagee may be required to move into possession of the property to realize on its security, and, in doing so, it may fall within the definition of owner, and become responsible for cleanup costs.[13] To avoid this danger, banks and other financial institutions may require **environmental audits** before making a secured loan on property. A "clean" environmental audit would allow the mortgagee to seize the property of the debtor business on default without undue concern about hidden environmental risks associated with the land.

Environmental audit

A site-specific inspection and analysis to determine the presence of existing environmental hazards or contamination.

In some cases, environmental legislation may create situations where the risk is so great that no business, lender, or lower-level government would be willing to acquire or deal with properties that have become seriously contaminated.

> An old foundry operation went into receivership. The property remained vacant for many years as it could not be sold, and the mortgagee was unwilling to move into possession because of suspected land contamination from the foundry operations. Municipal-tax arrears entitled the municipality to dispose of the land by tax sale, but the municipality was unwilling to do so, as it did not wish to assume any responsibility for cleanup. As a result, the property remained unsaleable and unusable.

Environmental legislation does not effectively address this scenario, nor does it provide for government cleanup at public expense in this type of situation — other than through a direct government initiative to resolve an environmental problem.

MANAGEMENT ALERT: BEST PRACTICE

Environmental law is one area that has managed to "pierce the corporate veil" and hold directors personally liable for damages caused by their corporations. While this situation is not unique at law, the extent of the possible liability is grossly unpredictable, compared to director's liability in other areas of the law. The best protection lies in taking the steps of reasonable care demanded by the provincial Acts (due diligence), conducting environmental audits to root out unseen problems, and to have the firm carry director's-liability insurance to minimize the residual risk.

FRONT-PAGE LAW

Arsenic, PCBs forcing major cleanups at hydro stations; grounds of Toronto station among those contaminated

More than 600 electrical stations across Ontario — including one in the Toronto area — are standing on ground contaminated with arsenic, PCBs and old insulation oil.

The transformer and distribution stations belong to Hydro One, the provincially owned transmission company that used to be one arm of Ontario Hydro before it was broken up last year.

Hydro One has identified 26 stations, including Fairbanks Station on Roselawn Avenue, as being polluted with enough arsenic to require cleanup. The utility has spent between $20-million and $25-million on site investigations and cleanups so far, and says it has not finished. In some cases, Hydro One employees have spent weeks digging out soil, trucking it away to waste disposal sites, and replacing it with clean soil.

The rest of the 600 tainted sites will just sit there, however — arsenic, PCBs and all, with no cleanup action. Hydro One says there is no need to do anything because

13. *Ontario (Attorney-General) v. Tyre King Tyre Recycling Ltd.* (1992), 9 O.R. (3d) 318.

the arsenic is not moving onto any neighbouring property. The Ontario Environment Ministry agrees. As long as the underground poison does not move off Hydro One's property to neighbouring soil, wells or rivers, it says the toxins can remain.

"As long as the underground poison does not move" is apt to be little comfort to a neighbour who seemingly must suffer before cleanup will be undertaken. Should the fear of damage be enough suffering on its own to allow a neighbouring landowner to demand action, either by legislation or in the courts?

Source: Excerpted from Tom Spears, *Ottawa Citizen*, with files from the *National Post*, "Arsenic, PCBs forcing major cleanups at hydro stations," September 13, 2000.

International Obligations

Kyoto Protocol

A commitment by Canada and other countries to reduce greenhouse gas emissions.

Canada has made a number of international commitments with respect to environmental damage and management over the past three decades. Most visible of these, and most contentious is the 1997 **Kyoto Protocol.** It is an international agreement, which Canada signed in 1998 and ratified in 2002, to reduce greenhouse gas emissions by six percent below 1990 levels in the five-year commitment period of 2008 to 2012. By 2007 however, the federal government announced it felt the targets were unreachable, and essentially turned away from the international commitments it had made. This signalled that it intends to advance its own policy, although no comprehensive legislation with direct control of emissions has yet become law.

SUMMARY

Protection of the environment was initially left with the individual to enforce through tort laws, but this was only satisfactory when the contamination was localized and directly affected the person bringing the action. Even then, the remedies were limited to damages and an injunction, and did not address pollution problems that caused more fundamental damage to the environment. The problem required legislative initiative and more effective solutions than the Common Law could offer.

Most environmental legislation is designed to control or eliminate pollution and environmental hazards either by regulation of the quantity of pollutants produced or prohibition of their production. The legislation generally shifts the responsibility for pollution of the environment to the person causing environmental damage by requiring the polluter to cover the cost of the cleanup. It encourages compliance by holding directors and officers of corporations personally responsible for any pollution violations by their corporation.

Polluted or contaminated property represents a serious risk for purchasers and mortgagees unless they take steps to ensure that the lands are free from contaminating substances. Environmental audits are usually used to determine the "cleanliness" of lands before purchase.

KEY TERMS

www.mcgrawhill.ca/olc/willes

Review Questions

1. Why should mortgagees of industrial property be concerned when securing their mortgages?
2. Where environmental damage is prohibited under legislation, what defence may be available to the directors and officers of the corporation?
3. Why does the purchase of lands previously used for industrial purposes pose a risk to the purchaser?
4. Outline the various ways that legislation addresses environmental pollution.
5. To what extent does the legislation recognize the fact that environmental damage cannot be eliminated from certain industrial processes?
6. Identify the remedies available to the court to control damage to a person's property by his or her neighbour's actions.
7. Why was it necessary for governments to introduce legislation to control environmental damage?
8. What steps may be taken by prospective purchasers of property to reduce the risk of facing an environmental cleanup order?
9. "At Common Law, damage to property or the environment is actionable, but restricted in terms of the type of case that may be brought before the court."
   Explain.
10. Outline the method used by governments to ensure that large industrial projects such as hydroelectric dams or large land-development undertakings result in a minimum of environmental damage.

Mini-Case Problems

1. ABC Company stored steel drums of contaminated waste products behind its plant. Some of the drums leaked contaminants into the soil, and these contaminants eventually found their way into a neighbour's drinking-water supply. What are the rights of the neighbour?
2. C was a director of the ABC Company in the above example. As a director, C rarely visited the plant, so he was unaware of the storage of the waste product behind the plant. However, a year previous, at a directors' meeting, he raised the issue of establishing a company directive to management that would require managers to ensure the safe storage of contaminants at all company plants. Advise C on whether a government examination of the plant site should take place.
3. Ashley purchased a house in a rural area. A farmer operated a cattle farm behind her property, and for many years had stored cattle manure at the top of a small hill near her property. Some months after Ashley had purchased the property, she had her well-water tested. The test indicated that the water was contaminated with e-coli bacteria. What advice would you give Ashley?
4. Two motor vehicles were in an accident on a highway. Both vehicles leaked oil, anti-freeze and gasoline on the highway. Police at the scene directed a local municipal fire department vehicle to clean the mess from the highway to avoid the danger of a fire. The fire department tanker used its water supply to flush the roadway. This drained into the ditch, then into a stream, then into a wildfowl nesting area.
   Who should have responsibility for the environmental clean-up and damage in this case?

Case Problems for Discussion

CASE 1

Wilbur operated a farm on the outskirts of a small rural village. His farm was located near the water reservoir that supplied drinking water to the small community. Wilbur's farm was divided by a township gravel road that he used to travel to and from his crop fields. A small stream that fed the village water reservoir also crossed his farm property and the road that separated Wilbur's fields. The bridge over the stream was in poor repair, and in spite of requests by Wilbur to have the bridge repaired, the municipality did nothing.

One day, while Wilbur was transporting his tank sprayer to his fields, the bridge collapsed, and Wilbur, his tractor, and tank sprayer fell into the stream. Wilbur was killed in the accident, and the tank sprayer ruptured, sending 3,000 litres of a potent carcinogenic weed killer into the stream. 50 litres of diesel fuel from the fuel tank of the tractor also leaked into the stream.

The village was immediately alerted concerning the contamination of their drinking water supply, and the clean-up was done under the direction of provincial authorities. Bottled water was supplied to the village, as was an expensive cleansing of the water reservoir. Eventually, the contamination was eliminated.

Discuss the issue of responsibility for the clean-up costs, as well as liability under the provincial environmental-protection legislation.

CASE 2

Haber Wood Products Ltd. had a permit from the provincial government to extract water from a nearby river for manufacturing purposes. It was also permitted to discharge the water, which now contained harmless chemicals, into the river. The discharge was continuously monitored for other contaminants to ensure that the discharge only contained the permitted harmless chemicals. MD Manufacturing Inc. also had a permit to draw water from the river, and under its permit it was entitled to discharge certain non-poisonous chemicals into the river. The discharge water was also monitored.

While the chemical discharge from each plant was harmless, and the combination was non-toxic to humans, higher concentrations of the resulting mixture were toxic to fish. Downstream from the two plants, Olsen operated a fish hatchery and fish farm. One day he discovered his entire stock of fish dead or near death. He immediately contacted the government ministry to determine the cause of his loss.

Discuss, the issues raised in this case. How would liability (if any) for Olsen's loss be determined?

CASE 3

The Savellis purchased a rural house and lot located near the intersection of two highways. Adjacent to the intersection, Carlton Fuels Ltd. had operated a gas bar and service station for many years. Over time, spilled gasoline and other petroleum products may have seeped into the soil. The fuel tanks, however, had been removed some years before the Savellis purchased their home, and the service-station property had been sold to a plumbing-supply company that used the buildings and grounds to store plastic pipe and copper plumbing fittings.

Some time after the purchase of their home, the Savellis began to notice a strange taste and odour in their drinking water. Their water supply was obtained from a well on their premises. Tests of the water supply indicated that the water was contaminated by gasoline.

Advise the Savellis of their rights at law and the possible course of action open to them.

CASE 4

McDonald operated a small farm where he raised pigs for market. The farm was located at the outskirts of a small city, and was surrounded on three sides by rural housing developments. McDonald stored pig manure in the corner of one field until late in the autumn of each year when he used it to fertilize his crop fields. The manure pile was located behind Black's lot, and attracted an enormous number of flies and insects during the summer months. In addition, the pungent odour of fresh pig manure prevented Black from using his backyard for any kind of social or recreational purpose.

Black complained to McDonald about the storage of manure next to his lot, but McDonald refused to change the location, as it was the most convenient storage place from his point of view. Black then took his complaint to the office of the Ministry of the Environment in the city, but was told that manure was not considered a hazardous waste in a farm setting, and was apparently being handled and stored in accordance with standard agricultural practices.

Advise Black of his rights (if any) and the possible responses of McDonald to any action on Black's part.

CASE 5

Gerry and Janet Weisberg had been married for 15 years and had operated a family business from their home. They lived on 50 acres of land just outside a small town, and their business consisted of breeding, training, and boarding German shepherd dogs. Formally, Gerry was the sole owner of the property and of the business. Janet, however, had been active throughout their marriage in the management and daily activities of the business.

Since the Weisbergs frequently transported their dogs to shows, the veterinarian, or to handlers, they maintained three vans for this purpose. They also had a gasoline pump with an underground storage tank installed on their farm to fuel the vehicles as a convenience for their business.

The Weisbergs were having trouble maintaining the lawn above the tank. For several years the grass above the tank died despite repeated efforts to reseed and water the area each spring. They sought the advice of a landscaper who suggested that the problem might be gasoline spillage from the pump nozzle when the fuel tanks of the vehicles were being filled. This would cause the soil to be soaked with gasoline at the surface and would burn the grass roots.

Gerry eventually called a pump-equipment service company who sent a representative out to the farm. On inspection of the equipment he discovered a crack in the underground tank which appeared to have been leaking leaded gasoline into the surrounding soil for quite some time. The company representative notified environmental authorities who came to the Weisberg's farm to test the surrounding soil. They found levels of soil contamination far in excess of approved standards, and they ordered the Weisbergs to remove the damaged tank and clean up the site.

Partly as a result of the strain and financial burden that this incident placed on them, Gerry and Janet separated several months later. In the arguments that followed concerning the division of their property, the farmland and the business played a prominent role.

Under the applicable family law legislation, all the property of the husband and wife, whether jointly owned or not, was to be pooled for the purpose of valuation, and the value then divided equally between the spouses. Assets were generally valued at fair market value as at the date of separation. The division of assets often required that some be liquidated in order to ensure that equal shares could be given to each spouse.

The farm property was Gerry's primary asset. A year before the soil contamination had been discovered, Gerry had had the farm appraised at $175,000. After the separation, Gerry engaged the services of an environmental expert to prepare a report estimating the cost of cleaning up the site. The expert reported that a cost of at least $200,000 would have to be incurred in order to meet approved standards.

Janet then engaged her own environmental specialist, who reported that the cleanup could be done at much less cost by using older technology. Her expert quoted a figure around $50,000.

There was no question that the farm would have to be sold in order to equally distribute the value of the couple's property. Gerry's shares in the business were his only other substantial asset and had been valued two years earlier at approximately $25,000.

Discuss the legal issues that arise in this case. Identify the arguments that the various parties may rely on and discuss their significance in the context of both environmental and business law. How would you resolve this matter?

CASE 6

Hazardous Waste Trucking Company carried on business as a transporter of industrial-waste products to licensed waste-disposal sites. Liquid Waste Disposal Company carried on a similar type of business, but handled only liquid waste. On a clear winter day, a transport truck owned and operated by Hazardous Waste Trucking Company collided on an icy patch of highway with a truck operated by Liquid Waste Disposal Company. The drivers of both vehicles had been operating their respective vehicles in accordance with the provincial Highway Traffic Act, and the patch of ice on the highway was totally unexpected. Both drivers had lost control of their vehicles on the ice, and their trailers that carried the waste products collided, causing their contents to spill on the highway.

The contents of each trailer did not constitute a toxic waste in itself, but the mixing of the two products produced a toxic compound hazardous to fish and animals. A local fire department answered the accident call and flushed the substance from the highway, instead of simply containing the waste mixture. As a result, a small stream was contaminated by the runoff.

Environmental inspectors ordered a cleanup, but both companies refused to do so, blaming each other and the fire department. The government arranged for the cleanup at a cost of $135,000.

Discuss the issues raised in this case on the basis that the "owners" of a contaminant under the environmental-protection legislation are responsible for any environmental damage that it may cause and the cost of any cleanup required.

The Online Learning Centre offers more ways to check what you have learned so far. Find quizzes, Weblinks, and many other resources at www.mcgrawhill.ca/olc/willes.

Canadian Charter of Rights and Freedoms

CANADIAN CHARTER OF RIGHTS AND FREEDOMS

Schedule B
Constitution Act, 1982

Enacted as Schedule B to the Canada Act 1982 (U.K.) 1982, c. 11, which came into force on April 17, 1982

PART I
Canadian Charter of Rights and Freedoms

Whereas Canada is founded upon principles that recognize the supremacy of God and the rule of law:

Guarantee of Rights and Freedoms

Rights and freedoms in Canada

1. The *Canadian Charter of Rights and Freedoms* guarantees the rights and freedoms set out in it subject only to such reasonable limits prescribed by law as can be demonstrably justified in a free and democratic society.

Fundamental Freedoms

Fundamental freedoms

2. Everyone has the following fundamental freedoms:
a) freedom of conscience and religion;
b) freedom of thought, belief, opinion and expression, including freedom of the press and other media of communication;
c) freedom of peaceful assembly; and
d) freedom of association.

Democratic Rights

Democratic rights of citizens

3. Every citizen of Canada has the right to vote in an election of members of the House of Commons or of a legislative assembly and to be qualified for membership therein.

Maximum duration of legislative bodies

4. (1) No House of Commons and no legislative assembly shall continue for longer than five years from the date fixed for the return of the writs of a general election of its members.

Continuation in special circumstances

(2) In time of real or apprehended war, invasion or insurrection, a House of Commons may be continued by Parliament and a legislative assembly may be continued by the legislature beyond five years if such continuation is not opposed by the votes of more than one-third of the members of the House of Commons or the legislative assembly, as the case may be.

Annual sitting of legislative bodies

5. There shall be a sitting of Parliament and of each legislature at least once every twelve months.

Mobility Rights

Mobility of citizens

6. (1) Every citizen of Canada has the right to enter, remain in and leave Canada.
(2) Every citizen of Canada and every person who has the status of a permanent resident of Canada has the right

Rights to move and gain livelihood

a) to move to and take up residence in any province; and
b) to pursue the gaining of a livelihood in any province.

Limitation

(3) The rights specified in subsection (2) are subject to
a) any laws or practices of general application in force in a province other than those that discriminate among persons primarily on the basis of province of present or previous residence; and
b) any laws providing for reasonable residency requirements as a qualification for the receipt of publicly provided social services.

Affirmative action programs

(4) Subsections (2) and (3) do not preclude any law, program or activity that has as its object the amelioration in a province of conditions of individuals in that province who are socially or economically disadvantaged if the rate of employment in that province is below the rate of employment in Canada.

Legal Rights

Life, liberty and security of person

7. Everyone has the right to life, liberty and security of the person and the right not to be deprived thereof except in accordance with the principles of fundamental justice.

Search or seizure

8. Everyone has the right to be secure against unreasonable search or seizure.

Detention or imprisonment

9. Everyone has the right not to be arbitrarily detained or imprisoned.

Arrest or detention

10. Everyone has the right on arrest or detention
a) to be informed promptly of the reasons therefor;
b) to retain and instruct counsel without delay and to be informed of that right; and
c) to have the validity of the detention determined by way of *habeas corpus* and to be released if the detention is not lawful.

Proceedings in criminal and penal matters

11. Any person charged with an offence has the right
a) to be informed without unreasonable delay of the specific offence;
b) to be tried within a reasonable time;
c) not to be compelled to be a witness in proceedings against that person in respect of the offence;
d) to be presumed innocent until proven guilty according to law in a fair and public hearing by an independent and impartial tribunal;
e) not to be denied reasonable bail without just cause;
f) except in the case of an offence under military law tried before a military tribunal, to the benefit of trial by jury where the maximum punishment for the offence is imprisonment for five years or a more severe punishment;
g) not to be found guilty on account of any act or omission unless, at the time of the act or omission, it constituted an offence under Canadian or international law or was criminal according to the general principles of law recognized by the community of nations;
h) if finally acquitted of the offence, not to be tried for it again and, if finally found guilty and punished for the offence, not to be tried or punished for it again; and
i) if found guilty of the offence and if the punishment for the offence has been varied between the time of commission and the time of sentencing, to the benefit of the lesser punishment.

Treatment or punishment

12. Everyone has the right not to be subjected to any cruel and unusual treatment or punishment.

Self-crimination

13. A witness who testifies in any proceedings has the right not to have any incriminating evidence so given used to incriminate that witness in any other proceedings, except in a prosecution for perjury or for the giving of contradictory evidence.

Interpreter

14. A party or witness in any proceedings who does not understand or speak the language in which the proceedings are conducted or who is deaf has the right to the assistance of an interpreter.

Equality Rights

Equality before and under law and equal protection and benefit of law

15. (1) Every individual is equal before and under the law and has the right to the equal protection and equal benefit of the law without discrimination and, in particular, without discrimination based on race, national or ethnic origin, colour, religion, sex, age or mental or physical disability.

Affirmative action programs

(2) Subsection (1) does not preclude any law, program or activity that has as its object the amelioration of conditions of disadvantaged individuals or groups including those that are disadvantaged because of race, national or ethnic origin, colour, religion, sex, age or mental or physical disability.

Official Languages of Canada

Official languages of Canada

16. (1) English and French are the official languages of Canada and have equality of status and equal rights and privileges as to their use in all institutions of the Parliament and government of Canada.

Official languages of New Brunswick

(2) English and French are the official languages of New Brunswick and have equality of status and equal rights and privileges as to their use in all institutions of the legislature and government of New Brunswick.

Advancement of status and use

(3) Nothing in this Charter limits the authority of Parliament or a legislature to advance the equality of status or use of English and French.

English and French linguistic communities in New Brunswick

16.1. (1) The English linguistic community and the French linguistic community in New Brunswick have equality of status and equal rights and privileges, including the right to distinct educational institutions and such distinct cultural institutions as are necessary for the preservation and promotion of those communities.

Role of the legislature and government of New Brunswick

(2) The role of the legislature and government of New Brunswick to preserve and promote the status, rights and privileges referred to in subsection (1) is affirmed.

Proceedings of Parliament

17. (1) Everyone has the right to use English or French in any debates and other proceedings of Parliament.

Proceedings of New Brunswick legislature

(2) Everyone has the right to use English or French in any debates and other proceedings of the legislature of New Brunswick.

Parliamentary statutes and records

18. (1) The statutes, records and journals of Parliament shall be printed and published in English and French and both language versions are equally authoritative.

New Brunswick statutes and records

(2) The statutes, records and journals of the legislature of New Brunswick shall be printed and published in English and French and both language versions are equally authoritative.

Proceedings in courts established by Parliament

19. (1) Either English or French may be used by any person in, or in any pleading in or process issuing from, any court established by Parliament.

Proceedings in New Brunswick courts

(2) Either English or French may be used by any person in, or in any pleading in or process issuing from, any court of New Brunswick.

Communications by public with federal institutions

20. (1) Any member of the public in Canada has the right to communicate with, and to receive available services from, any head or central office of an institution of the Parliament or government of Canada in English or French, and has the same right with respect to any other office of any such institution where

a) there is a significant demand for communications with and services from that office in such language; or

b) due to the nature of the office, it is reasonable that communications with and services from that office be available in both English and French.

Communications by public with New Brunswick institutions

(2) Any member of the public in New Brunswick has the right to communicate with, and to receive available services from, any office of an institution of the legislature or government of New Brunswick in English or French.

Continuation of existing constitutional provisions

21. Nothing in sections 16 to 20 abrogates or derogates from any right, privilege or obligation with respect to the English and French languages, or either of them, that exists or is continued by virtue of any other provision of the Constitution of Canada.

Rights and privileges preserved

22. Nothing in sections 16 to 20 abrogates or derogates from any legal or customary right or privilege acquired or enjoyed either before or after the coming into force of this Charter with respect to any language that is not English or French.

Minority Language Educational Rights

Language of instruction

23. (1) Citizens of Canada

a) whose first language learned and still understood is that of the English or French linguistic minority population of the province in which they reside, or

b) who have received their primary school instruction in Canada in English or French and reside in a province where the language in which they received that instruction is the language of the English or French linguistic minority population of the province,

have the right to have their children receive primary and secondary school instruction in that language in that province.

Continuity of language instruction

(2) Citizens of Canada of whom any child has received or is receiving primary or secondary school instruction in English or French in Canada, have the right to have all their children receive primary and secondary school instruction in the same language.

Application where numbers warrant

(3) The right of citizens of Canada under subsections (1) and (2) to have their children receive primary and secondary school instruction in the language of the English or French linguistic minority population of a province

a) applies wherever in the province the number of children of citizens who have such a right is sufficient to warrant the provision to them out of public funds of minority language instruction; and

b) includes, where the number of those children so warrants, the right to have them receive that instruction in minority language educational facilities provided out of public funds.

Enforcement

Enforcement of guaranteed rights and freedoms

24. (1) Anyone whose rights or freedoms, as guaranteed by this Charter, have been infringed or denied may apply to a court of competent jurisdiction to obtain such remedy as the court considers appropriate and just in the circumstances.

Exclusion of evidence bringing administration of justice into disrepute

(2) Where, in proceedings under subsection (1), a court concludes that evidence was obtained in a manner that infringed or denied any rights or freedoms guaranteed by this Charter, the evidence shall be excluded if it is established that, having regard to all the circumstances, the admission of it in the proceedings would bring the administration of justice into disrepute.

General

Aboriginal rights and freedoms not affected by Charter

25. The guarantee in this Charter of certain rights and freedoms shall not be construed so as to abrogate or derogate from any aboriginal, treaty or other rights or freedoms that pertain to the aboriginal peoples of Canada including

a) any rights or freedoms that have been recognized by the Royal Proclamation of October 7, 1763; and

b) any rights or freedoms that now exist by way of land claims agreements or may be so acquired.

Other rights and freedoms not affected by Charter

26. The guarantee in this Charter of certain rights and freedoms shall not be construed as denying the existence of any other rights or freedoms that exist in Canada.

Multicultural heritage

27. This Charter shall be interpreted in a manner consistent with the preservation and enhancement of the multicultural heritage of Canadians.

Rights guaranteed equally to both sexes

28. Notwithstanding anything in this Charter, the rights and freedoms referred to in it are guaranteed equally to male and female persons.

Rights respecting certain schools preserved

29. Nothing in this Charter abrogates or derogates from any rights or privileges guaranteed by or under the Constitution of Canada in respect of denominational, separate or dissentient schools.

Application to territories and territorial authorities

30. A reference in this Charter to a Province or to the legislative assembly or legislature of a province shall be deemed to include a reference to the Yukon Territory and the Northwest Territories, or to the appropriate legislative authority thereof, as the case may be.

Legislative powers not extended

31. Nothing in this Charter extends the legislative powers of any body or authority.

Application of Charter

Application of Charter

32. (1) This Charter applies

a) to the Parliament and government of Canada in respect of all matters within the authority of Parliament including all matters relating to the Yukon Territory and Northwest Territories; and

b) to the legislature and government of each province in respect of all matters within the authority of the legislature of each province.

Exception

(2) Notwithstanding subsection (1), section 15 shall not have effect until three years after this section comes into force.

Exception where express declaration

33. (1) Parliament or the legislature of a province may expressly declare in an Act of Parliament or of the legislature, as the case may be, that the Act or a provision thereof shall operate notwithstanding a provision included in section 2 or sections 7 to 15 of this Charter.

Operation of exception

(2) An Act or a provision of an Act in respect of which a declaration made under this section is in effect shall have such operation as it would have but for the provision of this Charter referred to in the declaration.

Five year limitation

(3) A declaration made under subsection (1) shall cease to have effect five years after it comes into force or on such earlier date as may be specified in the declaration.

Re-enactment

(4) Parliament or the legislature of a province may re-enact a declaration made under subsection (1).

Five year limitation

(5) Subsection (3) applies in respect of a re-enactment made under subsection (4).

Citation

Citation

34. This Part may be cited as the *Canadian Charter of Rights and Freedoms.*

The British North America Act, 1867

The British North America Act, 1867
[Consolidated with amendments]

An Act for the Union of Canada, Nova Scotia, and New Brunswick, and the Government thereof; and for Purposes connected therewith.
(29th March, 1867)

Statutes of Great Britain (1867), 30 & 31 Victoria, chapter 3

VI. DISTRIBUTION OF LEGISLATIVE POWERS

Powers of the Parliament

91. It shall be lawful for the Queen, by and with the Advice and Consent of the Senate and House of Commons, to make laws for the Peace, Order, and good Government of Canada, in relation to all Matters not coming within the Classes of Subjects by this Act assigned exclusively to the Legislatures of the Provinces; and for greater Certainty, but not so as to restrict the Generality of the foregoing Terms of this Section, it is hereby declared that (notwithstanding anything in this Act) the exclusive Legislative Authority of the Parliament of Canada extends to all Matters coming within the Classes of Subjects next hereinafter enumerated; that is to say, —

1. Repealed.
1A. The Public Debt and Property.
2. The Regulation of Trade and Commerce.
2A. Unemployment insurance.
3. The raising of Money by any Mode or System of Taxation.
4. The borrowing of Money on the Public Credit.
5. Postal Service.
6. The Census and Statistics.
7. Militia, Military and Naval Service, and Defence.
8. The fixing of and providing for the Salaries and Allowances of Civil and other Officers of the Government of Canada.
9. Beacons, Buoys, Lighthouses, and Sable Island.
10. Navigation and Shipping.
11. Quarantine and the Establishment and Maintenance of Marine Hospitals.
12. Sea Coast and Inland Fisheries.
13. Ferries between a Province and any British or Foreign Country or between Two Provinces.
14. Currency and Coinage.
15. Banking, Incorporation of Banks, and the Issue of Paper Money.
16. Savings Banks.
17. Weights and Measures.
18. Bills of Exchange and Promissory Notes.
19. Interest.
20. Legal Tender.
21. Bankruptcy and Insolvency.

22. Patents of Invention and Discovery.
23. Copyrights.
24. Indians, and Lands reserved for the Indians.
25. Naturalization and Aliens.
26. Marriage and Divorce.
27. The Criminal Law, except the Constitution of Courts of Criminal Jurisdiction, but including the Procedure in Criminal Matters.
28. The Establishment, Maintenance, and Management of Penitentiaries.
29. Such Classes of Subjects as are expressly excepted in the Enumeration of the Classes of Subjects by this Act assigned exclusively to the Legislatures of the Provinces.

And any Matter coming within any of the Classes of Subjects enumerated in this section shall not be deemed to come within the Class of Matters of a local or private Nature comprised in the Enumeration of the Classes of Subjects by this Act assigned exclusively to the Legislatures of the Provinces.

Exclusive Powers of Provincial Legislatures

92. In each Province the Legislature may exclusively make Laws in relation to Matters coming within the Classes of Subject next hereinafter enumerated; that is to say, —

1. Repealed.
2. Direct Taxation within the Province in order to the raising of a Revenue for Provincial Purposes.
3. The borrowing of Money on the sole Credit of the Province.
4. The Establishment and Tenure of Provincial Offices and the Appointment and Payment of Provincial Officers.
5. The Management and Sale of the Public Lands belonging to the Province and of the Timber and Wood thereon.
6. The Establishment, Maintenance, and Management of Public and Reformatory Prisons in and for the Province.
7. The Establishment, Maintenance, and Management of Hospitals, Asylums, Charities, and Eleemosynary Institutions in and for the Province, other than Marine Hospitals.
8. Municipal Institutions in the Province.
9. Shop, Saloon, Tavern, Auctioneer, and other Licences in order to the raising of a Revenue for Provincial, Local, or Municipal Purposes.
10. Local Works and Undertakings other than such as are of the following Classes: —
    (*a*) Lines of Steam or other Ships, Railways, Canals, and other Works and Undertakings connecting the Province with any other or others of the Provinces, or extending beyond the Limits of the Province;
    (*b*) Lines of Steam Ships between the Province and any British or Foreign Country;
    (*c*) Such Works as, although wholly situate within the Province, are before or after the Execution declared by the Parliament of Canada to be for the general Advantage of Canada or for the Advantage of Two or more of the Provinces.
11. The Incorporation of Companies with Provincial Objects.
12. The Solemnization of Marriage in the Province.
13. Property and Civil Rights in the Province.
14. The Administration of Justice in the Province, including the Constitution, Maintenance, and Organization of Provincial Courts, both of Civil and of Criminal Jurisdiction, and including Procedure in Civil Matters in those Courts.
15. The Imposition of Punishment by Fine, Penalty, or Imprisonment for enforcing any Law of the Province made in relation to any Matter coming within any of the Classes of Subjects enumerated in this Section.
16. Generally all Matters of a merely local or private Nature in the Province.

Non-Renewable Natural Resources, Forestry Resources and Electrical Energy

92A. (1) In each province, the legislature may exclusively make laws in relation to
(*a*) exploration for non-renewable natural resources in the province;
(*b*) development, conservation and management of non-renewable resources, natural resources and forestry resources in the province, including laws in relation to the rate of primary production therefrom; and
(*c*) development, conservation and management of sites and facilities in the province for the generation and production of electrical energy.

(2) In each province, the legislature may make laws in relation to the export from the province to another part of Canada of the primary production from non-renewable natural resources and forestry resources in the province and the production from facilities in the province for the generation of electrical energy, but such laws may not authorize or provide for discrimination in prices or in supplies exported to another part of Canada.

(3) Nothing in subsection (2) derogates from the authority of Parliament to enact laws in relation to the matters referred to in that subsection and, where such a law of Parliament and a law of a province conflict, the law of Parliament prevails to the extent of the conflict.

(4) In each province, the legislature may make laws in relation to the raising of money by any mode or system of taxation in respect of
(*a*) non-renewable natural resources and forestry resources in the province and the primary production therefrom, and
(*b*) sites and facilities in the province for the generation of electrical energy and the production therefrom,

whether or not such production is exported in whole or in part from the province, but such laws may not authorize or provide for taxation that differentiates between production exported to another part of Canada and production not exported from the province.

(5) The expression "primary production" has the meaning assigned by the Sixth Schedule.

(6) Nothing in subsections (1) to (5) derogates from any power or rights that a legislature or government of a province had immediately before the coming into force of this section.

The following list contains brief definitions of many of the legal terms used in the text. For a full and complete definition of each term, reference should be made to the appropriate chapter of the text, or to a legal dictionary.

acceleration clause: a clause in a debt instrument (e.g., a mortgage) that requires the payment of the balance of the debt on the happening of a specific event, such as default on an installment payment. p. 463.

acceptance: a statement or act given in response to and in accordance with an offer. p. 120.

Act of God: an unanticipated event that prevents the performance of a contract or causes damage to property. pp. 91, 231.

***ad hoc* tribunal:** a tribunal established to deal with a particular dispute between parties. p. 42.

adjuster: a person employed by the insurer to determine the extent of a loss incurred by an insured. p. 625.

administrative law: a body of rules governing the application of statutes to activities regulated by administrative tribunals or boards. p. 13.

administrative tribunals: agencies created by legislation to regulate activities or do specific things. pp. 13, 50.

adverse possession: a possessory title to land under the Registry System acquired by continuous, open, and notorious possession of land inconsistent with the title of the true owner for a period of time (usually 10 to 20 years). p. 441.

after-acquired property: a term in a security agreement that permits the security interest to attach to goods acquired later by the debtor. p. 580.

agency by conduct: an agency relationship inferred from the actions of a principal. p. 267.

agency by estoppel: A representation by words or conduct that a person is an agent cannot be later denied if a third party relies on the representation. p. 271.

agency by express agreement: an agency relationship established by an express oral or written agreement. p. 267.

agency by operation of law: agency that may arise in certain circumstances out of necessity where it is not possible to obtain the authority of the principal to act. p. 274.

agent: a person appointed to act for another, usually in contract matters. p. 266.

anticipatory breach: an advance determination that a party will not perform his or her part of a contract when the time for performance arrives. p. 245.

apparent authority: the ability of an agent to bind a principal where the principal has not notified third parties of the restricted or terminated authority of the agent. p. 273.

appraisal: an estimate of the value of a property prepared by a person trained in the evaluation of the worth of property. p. 495.

arbitration: a process for the settlement of disputes whereby an impartial third party or board hears the dispute, then makes a decision that is binding on the parties. Most commonly used to determine grievances arising out of a collective agreement, or in contract disputes. pp. 42, 373.

Articles of Incorporation: incorporation document. p. 309.

assault: a threat of violence or injury to a person. p. 62.

assignment: the transfer of a debtor's property to an Official Receiver under a voluntary assignment in bankruptcy. p. 601.

***assizes*:** sittings of the court held in different places throughout the province. p. 30.

attorney: a lawyer. p. 45.

bailee: the person who takes possession of a chattel in a bailment. p. 386.

bailment: the transfer of a chattel by the owner to another for some purpose, with the chattel to be later returned or dealt with in accordance with the owner's instructions. p. 386.

bailor: the owner of a chattel who delivers possession of the chattel to another in a bailment. p. 386.

bankruptcy: a condition that arises when a person commits an act of bankruptcy under the *Bankruptcy and Insolvency Act.* p. 596.

bargaining unit: a group of employees of an employer represented by a trade union recognized or certified as their exclusive bargaining representative. p. 368.

barrister: a lawyer who acts for clients in litigation or criminal court proceedings. p. 45.

battery: the unlawful touching or striking of another person. p. 62.

bid-rigging: a practice whereby contractors, in response to a call for bids or tenders, agree amongst themselves as to the price or who should bid or submit a tender. A restrictive trade practice, unless the person calling for the bids is advised of the arrangement. p. 636.

bill: a proposed law presented to a legislative body. p. 11.

bill of exchange: an instrument in writing, signed by the drawer and addressed to the drawee, ordering the drawee to pay a sum certain in money to the payee named therein (or bearer) at some fixed or determinable future time, or on demand. p. 551.

bill of lading: a contract entered into between a bailor and a common carrier of goods (bailee) that sets out the terms of the bailment and represents a title document to the goods carried. p. 660.

bond: a debt security issued by a corporation in which assets of the corporation are usually pledged as security for payment. p. 583.

breach of contract: the failure to perform a contract in accordance with its terms. p. 230.

broker: a person who may act for either the insurer or the insured in the placement of insurance. p. 625.

business judgment rule: the reluctance of the court to interfere with decisions of a board of directors. p. 319.

Canon Law: the law developed by the church courts to deal with matters that fell within their jurisdiction. p. 10.

causation or "proximate cause": a cause of injury directly related to an act of a defendant. p. 81.

***caveat emptor*:** Latin: "Let the buyer beware." p. 416.

certification: of a cheque, an understanding by a bank to pay the amount of a cheque on presentation. p. 556.

certification process: a process under labour legislation whereby a trade union acquires bargaining rights and is designated as the exclusive bargaining representative of a unit of employees. p. 368.

charge: a secured claim (similar to a mortgage) registered against real property under the Land Titles System. p. 453.

chattel mortgage: a mortgage in which the title to a chattel owned by the debtor is transferred to the creditor as security for the payment of a debt. p. 574.

chattels: moveable property. p. 572.

cheque: a bill of exchange that is drawn on a banking institution, and payable on demand. p. 552.

choses in action: a paper document that represents a right or interest that has value (e.g., a share certificate). p. 219.

***Civil Code*:** a body of written law that sets out the private rights of the citizens of a state. p. 12.

class action: an action where a single person represents the interests of a group, who will share in any award. p. 39.

click-wrap agreement: an Internet click box of "I agree," which constitutes valid acceptance of enumerated contractual responsibilities. p. 126.

closing: the completion of a real-estate transaction where the documents relating to the property interest are exchanged for the payment amount. p. 501.

co-insurance clause: a clause that may be inserted in an insurance policy that renders the insured an insurer for a part of the loss if the insured fails to maintain insurance coverage of not less than a specified minimum amount or percentage of the value of a property. p. 623.

collateral agreement: an agreement that has its own consideration, but supports another agreement. p. 184.

collective agreement: an agreement in writing, made between an employer and a union certified or recognized as the bargaining unit of the employees. It contains the terms and conditions under which work is to be performed and sets out the rights and duties of the employer, the employees, and the union. p. 373.

common carrier: a transportation business that specializes in the transport of goods. p. 397.

Common Law: the law as found in the recorded judgments of the courts. p. 8.

common market: a customs union which further allows barrier-free movement of services, workers and finance among member territories. p. 655.

compulsory licence: a directive by the Commissioner of Patents requiring the patentee to license others to produce a product for the general benefit of the public. p. 514.

condition: an essential term of a contract. pp. 245, 414.

condition precedent: a condition that must be satisfied before a contract may come into effect. pp. 183, 236.

condition subsequent: a condition that alters the rights or duties of the parties to a contract, or that may have the effect of terminating the contract if it should occur. p. 231.

conditional-sale agreement: an agreement for the sale of a chattel in which the seller grants possession of the goods, but withholds title until payment for the goods is made in full. p. 575.

condominium: a form of ownership of real property, usually including a building, in which certain units are owned in fee simple and the common elements are owned by the various unit owners as tenants-in-common. p. 434.

***consensus ad idem*:** agreement as to the subject or object of the contract. p. 116.

consideration: something that has value in the eyes of the law, and which a promisor receives in return for a promise. p. 136.

consignment sale: the delivery of a chattel to another person with instructions for its sale. p. 390.

conspiracy: agreement between firms to unduly lessen competition. p. 635.

constitution: the basis upon which a state is organized, and the powers of its government defined. p. 14.

constructive dismissal: employer termination of a contract of employment by a substantial, unilateral change in the terms or conditions of employment. p. 362.

contempt of court: refusal to obey a judge's order. p. 93.

contingency fee: a lawyer's fee payable on the condition of winning the case. p. 39.

contribution: the right of insurers (more than one) to share a loss where the policies so provide. p. 623.

conversion: the refusal to deliver up a chattel to its rightful owner by a bailee. p. 71.

cooling-off period: a time interval after a contract is signed that allows the buyer to repudiate the contract without liability. p. 540.

copyright: the right of ownership of an original literary or artistic work and the control over the right to copy it. p. 507.

corporation: a type of legal entity created by the state. p. 303.

customs broker: a firm with specialized knowledge of customs duties that will assist firms in the importation of goods. p. 652.

customs union: a free trade area whose members apply a uniform schedule of tariffs on imports from non-member territories. p. 655.

debenture: a debt security issued by a corporation that may or may not have specific assets of the corporation pledged as security for payment. pp. 313, 583.

deceit: a tort that arises when a party suffers damage by acting upon a false representation made by a party with the intention of deceiving the other. pp. 74, 202.

deed/transfer: written or printed instrument effecting legal disposition. p. 432.

defamation: false statements that injure a person's reputation. p. 66

director: under corporation law, a person elected by the shareholders of a corporation to manage its affairs. p. 304.

discharge: the release from obligations imposed by the *Bankruptcy and Insolvency Act.* p. 605.

disclosure [chapter 18]: the release to the public of information about a corporation that intends to offer its securities to the public. p. 336.

disclosure [chapter 31]: the obligation for an insured to disclose all material information to the insurer. p. 618.

dissolution: the termination of the partnership relationship. p. 292.

distrain/distress: the right of a landlord to seize and sell the chattels of a tenant if arrears of rent are not paid. p. 482.

doctrine of constructive notice: presumption at law that everyone has knowledge of the content of all statutes. p. 310.

doctrine of corporate opportunity: the use of corporate information for a personal benefit to the detriment of the corporation. p. 316.

doctrine of laches: an equitable doctrine of the court which provides that no relief will be granted when a person delays bringing an action for an unreasonably long period of time. p. 91.

dominant position: a firm that controls a major segment of a market for a product or service. p. 634.

dominant tenement: a parcel of land to which a right-of-way or easement attaches for its better use. p. 438.

double ticketing: a practice of attaching several different price tickets to goods. Under the *Competition Act,* only the lowest price may be charged for the goods. p. 641.

due diligence: the obligation on the directors of a corporation to ensure that effective systems are in place to comply with legislation, and to monitor the systems to ensure compliance. p. 318.

dumping: the selling abroad of goods at prices lower than the prices of the goods sold domestically in the country of origin. p. 651.

duress: the threat of injury or imprisonment for the purpose of requiring another to enter into a contract or carry out some act. p. 207.

duty of care: the duty not to injure another person. p. 80.

duty of fair representation: duty of a union to represent its members in a fair and impartial manner. p. 377.

duty to accommodate: the obligation on an employer to adjust work for an employee with a recognized disability. p. 354.

easement: a right to use the property of another, usually for a particular purpose. p. 438.

employer vicarious liability: the liability of an employer for acts of his or her employees in the course of business. p. 64.

encroachment: a possessory right to the property of another that may be acquired by the passage of time. p. 442.

endorsement: the signing of one's name on the back of a negotiable instrument for the purpose of negotiating it to another. p. 552.

environmental assessment: a procedure undertaken to determine the effect on the physical environment of a particular undertaking or activity. p. 673.

environmental audit [chapter 25]: a procedure whereby real property is examined to determine if environmental hazards exist on the property and to assess the cleanup costs, if any are found. p. 497.

environmental audit [chapter 34]: a site-specific inspection and analysis to determine the presence of existing environmental hazards or contamination. p. 676.

equitable assignment: an assignment that could be enforced if all parties could be brought before the court. p. 219.

equitable mortgage: a mortgage subsequent to the first or legal mortgage. A mortgage of the mortgagor's equity. p. 456.

equity: rules originally based on decisions of the king rather than on the law, and intended to be fair. p. 10.

equity of redemption: the equitable right of a mortgagor to acquire the title to the mortgaged property by payment of the debt secured by the mortgage. p. 456.

***escheat*:** the reversion of land to the Crown when a person possessed of the fee dies intestate and without heirs. p. 432.

estoppel: a rule whereby a person may not deny the truth of a statement of fact made by him or her when another person has relied and acted upon the statement. p. 145.

eviction: the court ordered removal of a tenant from leased premises. p. 482.

examination for discovery: a pretrial oral or written examination under oath, of a person or documents. p. 36.

exculpatory clause: a clause in a contract that limits or exempts a party from any liability for damage to the goods. p. 388.

exemplary/punitive damages: damages awarded to "set an example" or discourage repetition of the act. p. 93.

express term: discharge by the occurrence of an event specified in the contract. p. 231.

expropriation: the forceful taking of land by a government or government agency for public purposes. p. 432.

fee simple: an estate in land that represents the greatest interest in land that a person may possess, and that may be conveyed or passed by will to another, or that on an intestacy would devolve to the person's heirs. p. 431.

fiduciary: a relationship of utmost good faith in which a person, in dealing with property, must act in the best interests of the person for whom he or she acts, rather than in his or her own personal interest. p. 314.

fiduciary duty: a duty to place a client's interest above the professional's own interests. p. 101.

fixture: a chattel that is constructively or permanently attached to land. p. 430.

floating charge: a debt security issued by a corporation in which assets of the corporation, such as stock-in-trade, are pledged as security. Until such time as default occurs, the corporation is free to dispose of the assets. pp. 313, 583.

***force majeure*:** a major, unforeseen, or unanticipated event that occurs and prevents the performance of a contract. p. 231.

forcible confinement: confinement against a person's will. p. 65.

fourfold test: a test for employment based upon (1) ownership of tools (2) control (3) chance of profit (4) risk of loss. p. 350.

franchise: a business relationship that licenses the use of trade names, trademarks, and operating procedures to operate a similar business. p. 520.

fraudulent misrepresentation: a false statement of fact made by a person who knows, or should know, that it is false, and made with the intention of deceiving another. p. 202.

free-trade area: two or more member territories for which tariffs on trade between them are abolished. p. 655.

frustrated contract: a contract under which performance by a party is rendered impossible due to an unexpected or unforeseen change in circumstances affecting the agreement. p. 232.

fundamental breach: a breach of the contract that goes to the root of the agreement. p. 247.

General-Act: a form of incorporation whereby a corporation may be created by filing specific information required by the statute. p. 309.

general damages: restitution for losses naturally expected from a breach of contract. p. 251.

general partner: a full partner with unlimited liability for the debts of the partnership. p. 295.

guarantee: a collateral promise (in writing) to answer for the debt of another (the principal debtor) if the debtor should default in payment. p. 178.

hold-back: the retention of a part of the contract price by the owner as required under construction lien legislation to ensure payment of subcontractors and suppliers of materials. p. 587.

holder: the person in possession of a negotiable instrument. p. 552.

holder in due course: a person who acquires a negotiable instrument before its due date that is complete and regular on its face, and who gave value for the instrument, without any knowledge of default or defect in the title of prior holders. p. 552.

implied term [chapter 10]: the insertion by the court of a standard or customary term omitted by the parties when the contract was prepared. p. 183.

implied term [chapter 13]: discharge by the occurrence of an event that by custom of the trade would normally result in exemption from liability. p. 231.

indemnity: the concept that distinguishes a contract of insurance from a wager. p. 622.

indoor management rule: a party dealing with a corporation may assume that the officers have the valid and express authority to bind the corporation. p. 310.

infant: a person who has not reached the age of majority. p. 151.

informed consent: the full and understandable explanation of the risks associated with a course of action, and the clear understanding by the client or patient. p. 103.

infringement: the unlawful interference with the legal rights of another. p. 514.

injunction: an equitable remedy of the court that orders the person or persons named therein to refrain from doing certain acts. pp. 88, 256.

injurious falsehood: false statements about a firm, its products or business practices intended to dissuade others from doing business with the firm. p. 72.

innocent misrepresentation: a false statement of a material fact made by a party that honestly believed the fact to be true. p. 200.

insider trading: the trading in shares of a corporation by a person in a corporation who possesses undisclosed privileged information about the corporation. p. 342.

insolvency: the inability of a person or corporation to pay their debts as they fall due. p. 596.

insurable interest: an interest that would result in a loss on the occurrence of the event. p. 619.

intention to be bound: the assumption at law that strangers intend to be bound by their promises. p. 115.

itinerant sellers: sellers of goods who sell goods at the buyer's residence. p. 539.

joint and several liability: where partners individually and as a group have liability for a debt of the partnership. p. 288.

joint tenancy: the joint holding of equal interests in land with the right of the surviving tenant to the interest of a deceased joint tenant. p. 445.

joint venture: a business relationship between corporations. p. 297.

judgment: a decision of the court. p. 35.

judicial review: the judicial process whereby the decision of an administrative tribunal is examined by the court to determine if the decision-making process and decision was unfair or flawed. p. 56.

just cause: the onus on the employer to establish grounds for termination of an employee without notice. p. 360.

Kyoto Protocol: a commitment by Canada and other countries to reduce greenhouse gas emissions. p. 677.

Land Titles System: a provincial government operated system for the registration of interests in land where the government confirms and warrants the particular interests in land. p. 447.

lapse: the termination of an unaccepted offer by the passage of time, a counteroffer, or the death of a party. p. 128.

the law: the body of rules of conduct that are obligatory in the sense that sanctions are normally imposed if a rule is violated. p. 5.

Law Merchant: the customs or rules established by merchants to resolve disputes that arose between them, and that were later applied by Common Law judges in cases that came before their courts. p.10.

lease: an agreement that constitutes a grant of possession of property for a fixed term in return for the payment of rent. p. 473.

leasehold estate: grant of the right to possession of a parcel of land for a period of time in return for the payment of rent to the landowner. p. 434.

legal mortgage: a first mortgage of real property whereby the owner of land in fee simple transfers the title of the property pledged as security to the creditor on the condition that the title will be reconveyed when the debt is paid. p. 456.

lessee: a tenant. p. 473.

lessor: a landlord. p. 473.

Letters Patent: a government document that creates a corporation as a legal entity. p. 308.

libel: defamation in some permanent form, such as in writing, a cartoon, etc. p. 66.

licence: the right to use property in common with others. p. 389.

lien: with respect to goods, it is the right to retain the goods until payment is made. p. 392.

life estate: an estate in land in which the right to possession is based upon a person's lifetime. p. 432.

Limited Liability Partnership (LLP): a partnership where individual partners are liable for the general debts of the partnership and for personal negligence, but not liable for the negligence of other partners. p. 296.

limited partner: a partner who may not actively participate in the management of the firm, but has limited liability. p. 295.

liquidated damages: a *bona fide* estimate of the monetary damages that would flow from the breach of a contract. p. 254.

lockout: in a labour relations setting, the refusal of employee entry to a workplace by an employer when collective bargaining with the employees fails to produce a collective agreement. p. 371.

loss leader selling: a practice of selling goods not for profit but to advertise or to attract customers to a place of business. p. 638.

material alteration: the major alteration of an agreement that has the effect of discharging the contract and replacing it with another. p. 238.

mechanics' lien: a lien exercisable by a worker, contractor, or material supplier against property upon which the work or materials were expended. p. 573.

Memorandum of Association: incorporation document. p. 309.

merchantable quality: goods of a quality standard suitable for re-sale. p. 416.

mistake: a state of affairs in which a party (or both parties) has formed an erroneous opinion as to the identity or existence of the subject matter, or of some other important term. p. 194.

mistake of fact: mistake as to the existence of the subject matter of a contract or the identity of a party. p. 195.

mitigation: the obligation of an injured party to reduce the loss flowing from a breach of contract. p. 253.

monetary union: an area, most successfully a common market, issuing its own currency for use among its member territories. p. 656.

mortgage: an agreement made between a debtor and a creditor in which the title to property of the debtor is transferred to the creditor as security for payment of the debt. p. 453.

Most Favoured Nation (MFN): the obligation of a member of the WTO to impose the same lowest rate of duty granted on goods from one member state to the same goods from all other member states. p. 653.

motion: the decision to read a bill a first time. p. 11.

mutual mistake: a mistake where both parties have made mistaken assumptions as to the subject matter of the agreement. p. 197.

NAFTA: North American Free Trade Agreement. p. 654.

National Treatment (NT): the prohibition on imposing special taxes or duties on goods after import that exceed those of domestic production. p. 653.

natural justice: procedural fairness in decision-making. p. 57.

negligent misrepresentation: negligent misstatements made by a professional to a client. p. 104.

negotiable instrument: an instrument in writing that, when transferred in good faith and for value without notice of defects, passes a good title to the instrument to the transferee. pp. 223, 550.

***non est factum*:** a defence that may allow illiterate or infirm persons to avoid liability on a written agreement if they can establish that they were not aware of the true nature of the document, and were not careless in its execution. p. 196.

non-culpable dismissal: dismissal of an employee where the inability to perform is not self-induced but due to frustrating factors. p. 233.

non-disclosure agreement: a contract to maintain a trade secret. p. 508.

novation [chapter 12]: the substitution of parties to an agreement, or the replacement of one agreement by another agreement. p. 218.

novation [chapter 13]: a mutual agreement to amend the terms or parties to an existing agreement. p. 238.

nuisance: interference with the enjoyment of real property or, in some cases, material interference with a person's physical comfort. p. 87.

offer: a tentative promise subject to a condition. p. 119.

officer: a person elected or appointed by the directors of a corporation to fill a particular office (such as president, secretary, treasurer, etc.). p. 304.

option: a separate promise to keep an offer open for a period of time. p. 129.

order of replevin: court action that permits a person to recover goods unlawfully taken by another. p. 93.

organization test: a test for employment based upon an examination of the services in relation to the business itself. p. 350.

outside director: a director who is not an officer or employee of the corporation. p. 319.

parol evidence rule: a rule that prevents a party from introducing evidence that would add to or contradict terms of a contract. p. 183.

part performance: a doctrine that permits the courts to enforce an unwritten contract concerning land where certain conditions have been met. p. 181.

partnership: a legal relationship between two or more persons for the purpose of carrying on a business with a view to profit. p. 285.

patent: the exclusive right granted to the inventor of something new and different to produce the invention for a period of 20 years in return for the disclosure of the invention to the public. p. 507.

pawn: the transfer of possession (but not ownership) of chattels by a debtor to a creditor who is licensed to take and hold goods as security for payment of debt. p. 398.

perfect: the act of registration of a security interest under personal property security legislation. p. 579.

periodic tenancy: a lease that automatically renews at the end of each rent payment period until notice to terminate is given. p. 477.

personal information: information about an identifiable individual. p. 187.

petition: the request by a creditor to the court to institute bankruptcy proceedings against a debtor. p. 602.

picketing: the physical presence of persons at or near the premises of another for the purpose of conveying information. p. 371.

plagiarism: the taking of the copyrighted work of one person by another. p. 523.

pleadings: written statements prepared by the plaintiff and defendant that set out the facts and claims of the parties in a legal action, and are exchanged prior to the hearing of the case by the court. p. 35.

pledge: the transfer of securities by a debtor to a creditor as security for the payment of a debt. p. 398.

PMSI: purchase money security interest. p. 580.

polluter pays principle: the obligation on a polluter to pay for environmental damage that results from a violation of environmental legislation. p. 672.

power of attorney: a legal document usually signed under seal in which a person appoints another to act as his or her attorney to carry out the contractual or legal acts specified in the document. p. 176.

PPSA: Personal Property Security Act. p. 579.

precautionary principle: where serious damage may occur, lack of full scientific certainty shall not be used to postpone cost-effective measures to prevent environmental degradation. p. 671.

price discrimination: a practice of selling goods in an unfair or improper manner such as to lessen competition. p. 638.

price maintenance: an attempt by the seller to control the resale price of a product by the retailer. p. 638.

***prima facie*:** "on first appearance." p. 162.

principal: a person on whose behalf an agent acts. p. 266.

private law: the law relating to the relationship between individuals. p. 22.

privity: a person cannot incur liability under a contract to which he or she is not a party. p. 214.

procedural law: the law or procedures that a plaintiff must follow to enforce a substantive law right. p. 22.

proclaimed: when a law becomes effective. p. 11.

professional: a person with special skills not possessed by most individuals. p. 99.

promissory note: a promise in writing, signed by the maker, to pay a sum certain in money to the person named therein, or bearer, at some fixed or determinable future time, or on demand. p. 552.

proposal: a plan prepared by a commercial debtor to restructure the affairs of the business to enable the business to continue. p. 601.

prospectus: a public document required by law before securities are issued, revealing material facts about that security and its issuer, with such true, full, and plain disclosure that a potential investor may make an informed decision as to the riskiness and price of that security. p. 336.

proxy: a document evidencing the transfer of a shareholder's voting right to an appointee, either with instructions for voting, or allowing discretion to be exercised by the appointee, at a meeting of shareholders of the corporation. p. 344.

public law: the law relating to the relationship between the individual and the government. p. 22.

public policy: the unwillingness of the courts to enforce rights that are contrary to the general interests of the public. p. 162.

punitive damages: damages awarded by a court to punish a wrongdoer. Not normally awarded for ordinary breach of contract. p. 254.

***quantum meruit*:** "as much as he has earned." A quasi-contractual remedy that permits a person to recover a reasonable price for services and/or materials requested, where no price is established when the request is made. pp. 143, 256.

ratification: the adoption of a contract or act of another by a party who was not originally bound by the contract or act. p. 154.

real property: land and anything permanently attached to it. p. 430.

reasonable person: a standard of care used to measure acts of negligence. p. 81.

rectification: the correction of a mistake in an agreement that would have rendered the agreement impossible to perform. p. 199.

re-entry: the act of a landlord to move into possession of leased property. p. 482.

Registry System: a provincial government operated system for the registration of interests in land. p. 446.

regulations: procedural rules made under a statute. p. 13.

release: a promise not to sue or press a claim, or a discharge of a person from any further responsibility to act. p. 91.

remainderman: a person who is entitled to real property subject to a prior interest (e.g., a life estate) and who acquires the fee when the prior estate terminates. p. 433.

reporting issuer: a corporation that has issued its shares to the public by way of a prospectus. p. 337.

repudiation [chapter 9]: the refusal to perform an agreement or promise. p. 154.

repudiation [chapter 14]: a refusal to perform a contract. p. 244.

***res ipsa loquitur*:** "the thing speaks for itself." p. 84.

rescission: the revocation of a contract or agreement. p. 200.

***restitutio in integrum*:** to restore or return a party to an original position. p. 251.

restraint of trade: agreement between firms to fix prices, injure competition, or prevent others from entering a market. p. 73.

restrictive covenant [chapter 9]: a contractual clause limiting future behaviour. p. 163.

restrictive covenant [chapter 22]: a means by which an owner of property may continue to exercise some control over its use after the property has been conveyed to another. p. 439.

restrictive trade practices: business practices that restrict trade to the detriment of the consumer or competitors. p. 542.

reversion: the return of the right to possession to the landlord at the end of the lease. p. 474.

revised statutes: updated or amended statutes. p. 11.

revocation: the termination of an offer by notice communicated to the offeree before acceptance. p. 128.

rider: a clause altering or adding coverage to a policy. p. 618.

right-of-way: a right to pass over the land of another, usually to gain access to one's property. p. 438.

royal assent: needed in order for a bill to become law. p. 11.

salvage: the right of the insurer to the property where the insured is compensated for the loss. p. 622.

***Sarbanes-Oxley Act*:** a U.S. statute that imposes extensive duties on corporations to ensure accuracy of financial and securities information provided to the public. p. 320.

seal: a formal mode of expressing the intention to be bound by a written promise or agreement. This expression usually takes the form of signing or affixing a wax or gummed paper wafer beside the signature, or making an engraved impression on the document itself. p. 138.

secondary picketing: picketing at other than the employer's place of business. p. 371.

secured creditor: a creditor that may look to particular assets of the debtor to ensure payment of the debt. p. 572.

security: a document or other thing that stands as evidence of title to or interest in the capital, assets, property, profits, earnings, or royalties of any person or company, including any document commonly known as a security. p. 332.

servient tenement: a parcel of land subject to a right-of-way or easement. p. 438.

set-off: when two parties owe debts to each other, the payment of one may be deducted from the other, and only the balance paid to extinguish the indebtedness. p. 221.

share: the ownership of a fractional equity interest in a corporation. p. 313.

shareholder: a person who holds a share interest in a corporation; a part owner of the corporation. p. 304.

shareholders' agreement: an agreement between shareholders of a private corporation concerning management and/or future reorganization of the corporation such as buy-out of interests. p. 311.

slander: defamatory statements or gestures. p. 66.

slander of goods: a statement alleging that the goods of a competitor are defective, shoddy or injurious to the health of a consumer. p. 71.

slander of title: an untrue statement about the right of another to the ownership of goods. p. 72.

sole proprietorship: a business where the sole owner is responsible for the management and the debts of the business. p. 285.

solicitor: a lawyer whose practice consists of the preparation of legal documents, wills, etc., and other forms of non-litigious legal work. p. 45.

solicitor–client privilege: the duty of a lawyer to keep confidential information provided by a client. p. 108.

Special-Act: a corporation created by an Act of Parliament or a legislature for a specific purpose. p. 308.

special damages: specific damages that would flow from a breach of contract. p. 251.

specific performance: an equitable remedy of the court that may be granted for breach of contract where money damages would be inadequate, and that requires the defendant to carry out the agreement according to its terms. p. 255.

stakeholder approach: Consideration of the needs of the broadest community of persons interested in an issue. p. 671.

statute law: a law passed by a properly constituted legislative body. p. 8.

statutory assignment: an assignment of rights that an assignee may enforce if certain conditions are met by the assignment. p. 220.

statutory warranties: express warranties on goods provided by statute. p. 538.

***stoppage in transitu*:** the right of the seller to stop delivery of goods by the carrier if the buyer is insolvent. p. 423.

strict liability: responsibility for loss regardless of the circumstances. p. 83.

strike: in a labour relations setting, a cessation of work by a group of employees. p. 371.

sublet: the lease of leased premises by a tenant-in-chief to another tenant for a shorter term than the original tenancy. p. 477.

subrogation: the substitution of parties whereby the party substituted acquires the rights at law of the other party, usually by way of contractual arrangement. p. 622.

subsequent agreement: an agreement made after a written agreement that alters or cancels the written agreement. p. 185.

substantive law: all laws that set out the rights and duties of individuals. p. 22.

substantive right: an individual right enforceable at law. p. 22.

surrender: the agreement by the parties to terminate a lease. p. 484.

survey: a drawing of the boundaries of a property prepared by a professional surveyor. p. 498.

sustainable development: Outcomes should meet the needs of the present generation without compromising the ability of future generations to meet their own needs. p. 671.

takeover bid: an attempt by a competitor to obtain a controlling interest in a corporation. p. 345.

tariff: the duty (payment) charged by the Federal Government on goods imported into Canada. p. 650.

tenancy: a relationship between parties governed by a lease. p. 473.

tenancy-in-common: the joint holding of interests in land that need not be equal. p. 445.

tender: the act of performing a contract or the offer of payment of money due under a contract. p. 227.

tenure: a method of holding land granted by the Crown. p. 431.

term certain: the fixed term of a lease. p. 476.

tippee: a person who receives undisclosed privileged information about a corporation from an insider. p. 343.

title: the ownership of the goods. p. 409.

trade fixtures: chattels attached to leased premises by a commercial tenant that may be removed at the end of the tenancy by the tenant. p. 481.

trade secret: a commercial secret regarding a product or process. p. 508.

trademark: a mark to distinguish the goods or services of one person from the goods or services of others. p. 507.

trespass: a tort consisting of the injury of a person, the entry on the lands of another without permission, or the seizure or damage of goods of another without consent. p. 69.

trial court: the court in which a legal action is first brought before a judge for a decision. p. 29.

trust: an agreement or arrangement whereby a party (called a trustee) holds property for the benefit of another (called a beneficiary or *cestui que* trust). p. 215.

trustee: a person licensed to receive and dispose of the assets of a bankrupt, then distribute the proceeds to the creditors. p. 602.

***ultra vires*:** an act that is beyond the legal authority or power of a legislature or corporate body. p. 157.

undue influence: a state of affairs whereby a person is so influenced by another that the person's judgment is not his or her own. p. 205.

unfair business practices: business practices designed to take advantage of consumer inexperience or ignorance. p. 74.

unfair practices: business practices that mislead or treat the buyer unfairly. p. 540.

unilateral mistake: a mistake by one party to the agreement. p. 197.

vicarious liability: the liability at law of one person for the acts of another. p. 83.

vicarious performance: a performance of a contract by a third party, where the contracting party remains liable for the performance. p. 219.

***volenti non fit injuria*:** voluntary assumption of the risk of injury. p. 89.

waiver: an express or implied renunciation of a right or claim. pp. 91, 237.

warranty: in the sale of goods, a minor term in a contract. The breach of the term would allow the injured party damages, but not rescission of the agreement. pp. 249, 414.

World Customs Organization: an organization that provides internationally-recognized identification numbers for goods traded internationally. p. 650.

World Trade Organization (WTO): a multi-nation organization that provides a forum for the negotiation of trade rules, and provides a mechanism for the resolution of international trade disputes. p. 653.

wrongful dismissal: the failure of an employer to give reasonable notice of termination of a contract of employment. p. 360.

General Reference

Holdsworth, Sir William S., *A History of English Law*, London: Methuen; Sweet & Maxwell Limited, several editions, multiple volumes.

James, S. Phillip, *Introduction to English Law*, 13th ed., London: Butterworths, 1996.

Maitland, Frederick W. and Pollock, Sir Frederick, *The History of English Law*, 2nd ed. (re-issued), London: Cambridge University Press, 1968.

Woodley, et al., *Osborn's Concise Law Dictionary*, 11th ed., London: Sweet & Maxwell Limited, 2009.

Agency

Bowstead, William, *Bowstead and Reynolds on Agency*, 19th ed. (P. Watts), London: Sweet & Maxwell Limited, 2010.

Harvey, Cameron, *Agency Law Primer*, 4th ed., Toronto: The Carswell Company, 2009.

Bailment

Palmer, N.E., *Bailment*, 2nd ed., Sydney, Australia: Law Book Co., 1991.

Bankruptcy

Houlden, Lloyd W., Morawetz, C.H., and Sarra, J., *The 2011 Annotated Bankruptcy and Insolvency Act*, Toronto: The Carswell Company, 2010.

Condominiums

Loeb, Audrey M., *Condominium Law and Administration*, 2nd ed., Toronto: The Carswell Company, 2000.

Constitutional Law

Hogg, Peter W., *Constitutional Law of Canada 2010*, Toronto: The Carswell Company, 2010.

Contracts

Anson, Sir William R., *Anson's Law of Contract*, 28th ed. (A.G. Guest), Oxford: Clarendon Press, 2002.

Waddams, Stephen M., *The Law of Contracts*, 6th ed., Toronto: Canada Law Book Inc., 2010.

Corporations

Buckley, E.H., and Conelly, M.Q., *Corporations: Principles and Policies*, 3rd ed., Toronto: Emond Montgomery Publications Limited, 1995.

Welling, Bruce L., *Canadian Corporate Law*, 3rd ed., Toronto: Butterworths, 2006.

E-Commerce & Data Protection Law

Handa, S., Marseille, C., and Sheehan M., *E-Commerce Legislation and Materials in Canada 2008 Edition*, Toronto: Butterworths, 2007.

McNairn, C., *A Guide to the Personal Information Protection and Electronic Documents Act, 2010 Edition*, Markham: Lexis Nexis Butterworths, 2010.

Employment and Collective Bargaining

Ball, Stacey R., *Canadian Employment Law*, Aurora: Canada Law Book Inc., 1996 (Update Series).

Carter, D., England, G., Etherington, B., and Trudeau, G., *Labour Law in Canada*, 5th ed., Toronto: Lexis Nexis Butterworths, 2001.

Parry, Robert M., *Employment Standards Handbook*, 3rd ed., Toronto: Canada Law Book Inc., 2002.

Environmental Law

Benidickson, J., *Environmental Law*, 3rd ed., Concord, Ont.: Irwin Law, 2008.

Thompson, Geoffrey, *Environmental Law and Business in Canada*, Toronto: Canada Law Book Inc., 1993.

Franchise Law

Levitt, E., *Canadian Franchise Legislation* (2002 Update Series), Toronto: Lexis Nexis Butterworths, 2002.

Insurance

Brown, Craig, *Introduction to Insurance Law*, Toronto: Butterworths, 2003.

Brown, Craig and Menezes, Julio, *Insurance Law in Canada*, Toronto: The Carswell Company, 1999.

International Business Law

Willes, J.H. and Willes, J.A, *International Business Law: Environments and Transactions*, Chicago: McGraw Hill, 2005.

Landlord and Tenant

Balfour, Richard J., *Landlord and Tenant Law*, Toronto: Emond Montgomery Publications Limited, 1991.

Mortgages

Falconbridge, John D., *Falconbridge on Mortgages*, 5th ed., Agincourt: Canada Law Book Co., 2002.

Roach, Joseph E., *The Canadian Law of Mortgages of Land*, Toronto: Butterworths Ltd., 1993.

Negotiable Instruments

Baxter, Ian F., *The Law of Banking*, 4th ed., Toronto: The Carswell Company, 1992.

Falconbridge, John D., *Crawford and Falconbridge on Banking and Bills of Exchange*, 8th ed. (Bradley Crawford), Toronto: Canada Law Book Company, 1986.

Partnership

Lindley, N., Anson Banks, R.C.I., *Lindley & Banks on Partnership*, 19th ed., London: Sweet & Maxwell Limited, 2010.

Patents, Trademarks, Copyright

Handa, S., *Copyright Law in Canada*, Toronto: Butterworths, 2002.

Hughes, R., and Peacock, S., *Hughes on Copyright and Industrial Design* (Update Series), Toronto: Lexis Nexis Butterworths, 2002.

Vaver, David, *Intellectual Property Law*, 2nd ed., Concord, Ont.: Irwin Law, 2011.

Real Property

Haber, Harvey M., *The Commercial Lease, A Practical Guide*, 4th ed., Toronto: Canada Law Book Inc., 2004.

Megarry, Sir Robert E., and Wade, W.H.R., *The Law of Real Property*, 7th ed., London: Sweet & Maxwell, 2007.

Reiter, B., Risk, R., and McLellan, B., *Real Estate Law*, 5th ed., Toronto: Emond Montgomery Publications Limited, 1998.

Restrictive Trade Practices

Nozick, Robert S., *The 2008 Annotated Competition Act*, Toronto: The Carswell Company, 2008.

Sale of Goods

Fridman, Gerald H.L., *Sale of Goods in Canada*, 5th ed., Toronto: The Carswell Company, 2004.

Security for Debt

Bennett, Frank, *Bennett on Creditors' and Debtors' Rights and Remedies*, 5th ed., Toronto: The Carswell Company, 2006.

Dunlop, C.R.B., *Creditor–Debtor Law in Canada*, 2nd ed., Toronto: The Carswell Company, 1995.

Macklem, Douglas N. and Bristow, David I., *Construction and Mechanics' Liens in Canada*, 7th ed., Toronto: The Carswell Company, 2005.

MacLaren, Richard H., *Secured Transactions in Personal Property in Canada* (Update Series), Toronto: The Carswell Company, 1989.

Torts

Fleming, John G., *The Law of Torts*, 9th ed., Sydney, Australia: The Law Book Company, 1998.

Linden, Allen M. and Feldthusen, B., *Canadian Tort Law*, 9th ed., Toronto: Lexis Nexis Butterworths, 2011.

Waddams, Stephen M., *Products Liability*, 5th ed., Toronto: The Carswell Company, 2011.

Credits